ADVANCED WASTE TREATMENT

Fifth Edition

A Field Study Training Program

prepared by

Office of Water Programs
College of Engineering and Computer Science
California State University, Sacramento

in cooperation with

California Water Environment Association

❧ ❧ ❧

Kenneth D. Kerri, Project Director
Bill B. Dendy, Co-Director
John Brady, Consultant and Co-Director
William Crooks, Consultant

❧ ❧ ❧

2006

Cover: Rock Creek Advanced Wastewater Treatment Facility, Rock Creek, Oregon.
Photo courtesy of Bruce Johnson, CH2M Hill Consulting Engineers.

In recognition of the need to preserve natural resources, this manual is printed using recycled
paper. The text paper and the cover are composed of 10% post-consumer waste. The Office
of Water Programs strives to increase its commitment to sustainable printing practices.

Funding for this operator training manual was provided by the Office of Water Programs,
California State University, Sacramento. Mention of trade names or commercial products
does not constitute endorsement or recommendation for use by the Office of Water
Programs or California State University, Sacramento.

ISBN
978-1-59371-035-4

www.owp.csus.edu

OFFICE OF WATER PROGRAMS

The Office of Water Programs is a nonprofit organization operating under the California State University Sacramento Foundation, to provide distance learning courses for persons interested in the operation and maintenance of drinking water and wastewater facilities. These training programs were developed by people who explain, through the use of our manuals, how they operate and maintain their facilities. The university, fully accredited by the Western Association of Schools and Colleges, administers and monitors these training programs, under the direction of Dr. Ramzi J. Mahmood.

Our training group develops and implements programs and publishes manuals for operators of water treatment plants, water distribution systems, wastewater collection systems, and municipal and industrial wastewater treatment and reclamation facilities. We also offer programs and materials for pretreatment facility inspectors, environmental compliance inspectors, and utility managers. All training is offered as distance learning, using correspondence, video, or computer-based formats with opportunities for continuing education and contact hours for operators, supervisors, managers, and administrators.

Materials and opportunities available from our office include manuals in print, CD, or video formats, and enrollments for courses providing CEU (Continuing Education Unit) contact hours. Here is a sample:

- Industrial Waste Treatment, 2 volumes (print, course enrollment)
- Operation of Wastewater Treatment Plants, 2 volumes (print, CD, course enrollment)
- Advanced Waste Treatment (print, course enrollment)
- Treatment of Metal Wastestreams (print, course enrollment)
- Pretreatment Facility Inspection (print, video, course enrollment)
- Small Wastewater System Operation and Maintenance, 2 volumes (print, course enrollment)
- Operation & Maintenance of Wastewater Collection Systems, 2 volumes (print, course enrollment)
- Collection System Operation & Maintenance Training Videos (video, course enrollment)
- Utility Management (print, course enrollment)
- Manage for Success (print, course enrollment)

These and other materials may be ordered from:

Office of Water Programs
California State University, Sacramento
6000 J Street
Sacramento, CA 95819-6025
(916) 278-6142 – phone
(916) 278-5959 – FAX

or

visit us on the web at www.owp.csus.edu

ADDITIONAL VOLUMES OF INTEREST

Operation of Wastewater Treatment Plants, Volume I

The Treatment Plant Operator
Why Treat Wastes?
Wastewater Treatment Facilities
Racks, Screens, Comminutors, and Grit Removal
Sedimentation and Flotation
Trickling Filters
Rotating Biological Contactors
Activated Sludge
Wastewater Stabilization Ponds
Disinfection and Chlorination

Operation of Wastewater Treatment Plants, Volume II

Activated Sludge
Sludge Digestion and Solids Handling
Effluent Disposal
Plant Safety
Maintenance
Laboratory Procedures and Chemistry
Applications of Computers for Plant O & M
Analysis and Presentation of Data
Records and Report Writing
Treatment Plant Administration

Industrial Waste Treatment, Volume I

The Industrial Plant Operator
Industrial Wastewaters
Regulatory Requirements
Preventing and Minimizing Wastes at the Source
Industrial Waste Monitoring
Flow Measurement
Preliminary Treatment (Equalization, Screening, and
    pH Adjustment)
Physical–Chemical Treatment Processes (Coagulation,
    Flocculation, and Sedimentation)
Filtration
Physical Treatment Processes (Air Stripping and
    Carbon Adsorption)
Treatment of Metal Wastestreams
Instrumentation
Safety
Maintenance

Industrial Waste Treatment, Volume II

The Industrial Plant Operator
Fixed Growth Processes
Activated Sludge Process Control
Sequencing Batch Reactors
Enhanced Biological Treatment
Anaerobic Treatment
Residual Solids Management
Maintenance

Treatment of Metal Wastestreams

Need for Treatment
Sources of Wastewater
Material Safety Data Sheets (MSDSs)
Employee Right-To-Know Laws
Methods of Treatment
Advanced Technologies
Sludge Treatment and Disposal
Operation, Maintenance, and Troubleshooting
Polymers
Oxidation-Reduction Potential (ORP)

Pretreatment Facility Inspection

The Pretreatment Facility Inspector
Pretreatment Program Administration
Development and Application of Regulations
Inspection of a Typical Industry
Safety in Pretreatment Inspection and Sampling Work
Sampling Procedures for Wastewater
Wastewater Flow Monitoring
Industrial Wastewaters
Pretreatment Technology (Source Control)
Industrial Inspection Procedures
Emergency Response

PREFACE

The fifth edition of *ADVANCED WASTE TREATMENT,* developed by the Office of Water Programs at California State University, Sacramento, contains the most current information for operators of advanced wastewater treatment facilities. *ADVANCED WASTE TREATMENT* covers the topics of odor control, activated sludge, solids handling and disposal, solids removal from secondary effluents, phosphorus removal, nitrogen removal, nutrient control, wastewater reclamation and reuse, and instrumentation.

Operators of advanced wastewater treatment plants (tertiary facilities) continue to convey to us that they need a training program that covers biological treatment processes as well as physical–chemical treatment processes. This training manual was developed to meet those training needs and to serve as an operations, maintenance, and troubleshooting reference.

Like its companion manuals, *OPERATION OF WASTEWATER TREATMENT PLANTS,* Volumes I and II, *ADVANCED WASTE TREATMENT* is regularly updated as technology, regulations, and processes advance, to provide operators with the most current information on the technology of wastewater treatment. Whenever the Office of Water Programs revises one of our operator training manuals, the material is updated in accordance with the comments and suggestions received from the operators enrolling in the courses that use the manual. While you are reading the material in this manual, please make notes of questions and areas where you would improve the material. By sending your comments and suggestions to us, operators who use the manual in the future will benefit from your knowledge and experience. Thanks.

Kenneth D. Kerri
Office of Water Programs
California State University, Sacramento
6000 J Street
Sacramento, CA 95819-6025
(916) 278-6142 – phone
wateroffice@csus.edu – e-mail

2006

USES OF THIS MANUAL

Originally this manual was developed to serve as a home-study course for operators in remote areas or persons unable to attend formal classes either due to shift work, personal reasons, or the unavailability of suitable classes. This home-study training program used the concepts of self-paced instruction where you are your own instructor and work at your own speed. In order to certify that a person had successfully completed this program, objective tests and special answer sheets for each chapter are provided when a person enrolls in this course.

Once operators started using this manual for home study, they realized that it could serve effectively as a textbook in the classroom. Many colleges and universities have used the manual as a text in formal classes often taught by operators. In areas where colleges were not available or were unable to offer classes in the operation of industrial wastewater treatment plants, operators and utility agencies joined together to offer their own courses using the manual.

Occasionally a utility agency has enrolled from three to over 300 of its operators in this training program. A manual is purchased for each operator. A senior operator or a group of operators are designated as instructors. These operators help answer questions when the persons in the training program have questions or need assistance. The instructors grade the objective tests, record scores, and notify California State University, Sacramento, of the scores when a person successfully completes this program. This approach eliminates any waiting while papers are being graded and returned by the university.

This manual was prepared to help operators run their treatment plants. Please feel free to use it in the manner that best fits your training needs and the needs of other operators. We will be happy to work with you to assist you in developing your training program. Please feel free to contact:

Project Director
Office of Water Programs
California State University, Sacramento
6000 J Street
Sacramento, CA 95819-6025
(916) 278-6142 – phone
(916) 278-5959 – FAX
wateroffice@csus.edu – e-mail

INSTRUCTIONS TO PARTICIPANTS
IN HOME-STUDY COURSE

Procedures for reading the lessons and answering the questions are contained in this section.

To progress steadily through this program, you should establish a regular study schedule. For example, many operators set aside two hours during two evenings a week for study.

The study material is contained in 9 chapters. Some chapters are longer and more difficult than others. For this reason, many of the chapters are divided into two or more lessons. The time required to complete a lesson will depend on your background and experience. Some people might require an hour to complete a lesson and some might require three hours; but that is perfectly all right. The important thing is that you understand the material in the lesson.

Each lesson is arranged for you to read a short section, write the answers to the questions at the end of the section, check your answers against suggested answers; and then *YOU* decide if you understand the material sufficiently to continue or whether you should read the section again. You will find that this procedure is slower than reading a typical textbook, but you will remember much more when you have finished the lesson.

Some discussion and review questions are provided following each lesson in the chapters. These questions review the important points you have covered in the lesson. Write the answers to the discussion and review questions in your notebook.

In the appendix at the end of this manual, you will find some comprehensive review questions and suggested answers. These questions and answers are provided as a way for you to review how well you remember the material. You may wish to review the entire manual before you attempt to answer the questions. Some of the questions are essay-type questions, which are used by some states for higher-level certification examinations. After you have answered all the questions, check your answers with those provided and determine the areas in which you might need additional review before your next certification or civil service examination. Please do not send your answers to California State University, Sacramento.

You are your own teacher in this program. You could merely look up the suggested answers at the end of the chapters or comprehensive review questions or copy them from someone else, but you would not understand the material. Consequently, you would not be able to apply the material to the operation of your plant or recall it during an examination for certification or a civil service position.

You will get out of this program what you put into it.

SUMMARY OF PROCEDURE

OPERATOR (YOU)

1. Read what you are expected to learn in each chapter; the major topics are listed at the beginning of the chapter.

2. Read sections in the lesson.

3. Write your answers to questions at the end of each section in your notebook. You should write the answers to the questions just as you would if these were questions on a test.

4. Check your answers with the suggested answers.

5. Decide whether to reread the section or to continue with the next section.

6. Write your answers to the discussion and review questions at the end of each lesson in your notebook.

ORDER OF WORKING LESSONS

To complete this program you will have to work all of the lessons. You may proceed in numerical order, or you may wish to work some lessons sooner.

Safety is a very important topic. Everyone working in a treatment plant must always be safety conscious. You must take extreme care with your personal hygiene to prevent the spread of disease to yourself and your family. Operators daily encounter situations and equipment that can cause a serious disabling injury or illness if the operator is not aware of the potential danger and does not exercise adequate precautions. Safe procedures are always stressed.

ADVANCED WASTE TREATMENT
COURSE OUTLINE

CHAPTER 1

ODOR CONTROL

by

Tom Ikesaki

Revised by

Russ Armstrong

TABLE OF CONTENTS
Chapter 1.   ODOR CONTROL

OBJECTIVES
Chapter 1.   ODOR CONTROL

Following completion of Chapter 1, you should be able to:

1. Determine the source and cause of odors.

2. Respond to odor complaints.

3. Solve odor problems.

WORDS

Chapter 1.   ODOR CONTROL

ABSORPTION (ab-SORP-shun)   ABSORPTION

The taking in or soaking up of one substance into the body of another by molecular or chemical action (as tree roots absorb dissolved nutrients in the soil).

ADSORPTION (add-SORP-shun)   ADSORPTION

The gathering of a gas, liquid, or dissolved substance on the surface or interface zone of another material.

AEROBES   AEROBES

Bacteria that must have dissolved oxygen (DO) to survive. Aerobes are aerobic bacteria.

AMBIENT (AM-bee-ent) TEMPERATURE   AMBIENT TEMPERATURE

Temperature of the surroundings.

ANAEROBES   ANAEROBES

Bacteria that do not need dissolved oxygen (DO) to survive.

BENZENE   BENZENE

An aromatic hydrocarbon (C_6H_6) that is a colorless, volatile, flammable liquid. Benzene is obtained chiefly from coal tar and is used as a solvent for resins and fats and in the manufacture of dyes. Benzene has been found to cause cancer in humans.

BIOMASS (BUY-o-mass)   BIOMASS

A mass or clump of organic material consisting of living organisms feeding on wastes, dead organisms, and other debris. Also see ZOOGLEAL MASS and ZOOGLEAL MAT (FILM).

BLOCKOUT   BLOCKOUT

The physical prevention of the operation of equipment.

ELECTROLYTE (ee-LECK-tro-lite)   ELECTROLYTE

A substance that dissociates (separates) into two or more ions when it is dissolved in water.

ELECTROLYTIC (ee-LECK-tro-LIT-ick) PROCESS   ELECTROLYTIC PROCESS

A process that causes the decomposition of a chemical compound by the use of electricity.

FACULTATIVE (FACK-ul-tay-tive) BACTERIA   FACULTATIVE BACTERIA

Facultative bacteria can use either dissolved oxygen or oxygen obtained from food materials such as sulfate or nitrate ions. In other words, facultative bacteria can live under aerobic, anoxic, or anaerobic conditions.

INDOLE (IN-dole)   INDOLE

An organic compound (C_8H_7N) containing nitrogen that has an ammonia odor.

MATERIAL SAFETY DATA SHEET (MSDS)

A document that provides pertinent information and a profile of a particular hazardous substance or mixture. An MSDS is normally developed by the manufacturer or formulator of the hazardous substance or mixture. The MSDS is required to be made available to employees and operators or inspectors whenever there is the likelihood of the hazardous substance or mixture being introduced into the workplace. Some manufacturers are preparing MSDSs for products that are not considered to be hazardous to show that the product or substance is not hazardous.

MERCAPTANS (mer-CAP-tans)

Compounds containing sulfur that have an extremely offensive skunk-like odor; also sometimes described as smelling like garlic or onions.

MICRON (MY-kron)

μm, Micrometer or Micron. A unit of length. One millionth of a meter or one thousandth of a millimeter. One micron equals 0.00004 of an inch.

MOLE

The name for a number (6.02×10^{23}) of atoms or molecules. See MOLECULAR WEIGHT.

OBLIGATE AEROBES

Bacteria that must have atmospheric or dissolved molecular oxygen to live and reproduce.

ODOR PANEL

A group of people used to measure odors.

OLFACTOMETER (ALL-fak-TOM-utter)

A device used to measure odors in the field by diluting odors with odor-free air.

OXIDATION

Oxidation is the addition of oxygen, removal of hydrogen, or the removal of electrons from an element or compound; in the environment and in wastewater treatment processes, organic matter is oxidized to more stable substances. The opposite of REDUCTION.

OXIDATION-REDUCTION POTENTIAL (ORP)

The electrical potential required to transfer electrons from one compound or element (the oxidant) to another compound or element (the reductant); used as a qualitative measure of the state of oxidation in water and wastewater treatment systems. ORP is measured in millivolts, with negative values indicating a tendency to reduce compounds or elements and positive values indicating a tendency to oxidize compounds or elements.

OXIDIZED ORGANICS

Organic materials that have been broken down in a biological process. Examples of these materials are carbohydrates and proteins that are broken down to simple sugars.

OZONATION (O-zoe-NAY-shun)

The application of ozone to water, wastewater, or air, generally for the purposes of disinfection or odor control.

PHENOLIC (fee-NO-lick) COMPOUNDS

Organic compounds that are derivatives of benzene. Also called phenols (FEE-nolls).

REDUCTION (re-DUCK-shun)

Reduction is the addition of hydrogen, removal of oxygen, or the addition of electrons to an element or compound. Under anaerobic conditions (no dissolved oxygen present), sulfur compounds are reduced to odor-producing hydrogen sulfide (H_2S) and other compounds. In the treatment of metal finishing wastewaters, hexavalent chromium (Cr^{6+}) is reduced to the trivalent form (Cr^{3+}). The opposite of OXIDATION.

SEPTIC (SEP-tick) or SEPTICITY SEPTIC or SEPTICITY

A condition produced by bacteria when all oxygen supplies are depleted. If severe, the bottom deposits produce hydrogen sulfide, the deposits and water turn black, give off foul odors, and the water has a greatly increased oxygen and chlorine demand.

SKATOLE (SKAY-tole) SKATOLE

An organic compound (C_9H_9N) that contains nitrogen and has a fecal odor.

STRIPPED ODORS STRIPPED ODORS

Odors that are released from a liquid by bubbling air through the liquid or by allowing the liquid to be sprayed or tumbled over media.

THRESHOLD ODOR THRESHOLD ODOR

The minimum odor of a gas or water sample that can just be detected after successive dilutions with odorless gas or water. Also called ODOR THRESHOLD.

ZOOGLEAL (ZOE-uh-glee-ul) MASS ZOOGLEAL MASS

Jelly-like masses of bacteria found in both the trickling filter and activated sludge processes. These masses may be formed for or function as the protection against predators and for storage of food supplies. Also see BIOMASS.

CHAPTER 1.   ODOR CONTROL

1.0   NEED FOR ODOR CONTROL

Odor control in wastewater collection systems and at wastewater treatment plants is very important. With the increased demand for housing, collection systems are being extended farther and farther away from the treatment plant. Longer collection systems create longer flow times to reach the treatment plant. Increased travel times cause the wastewater to become septic and thus cause odor and corrosion problems in collection systems and treatment plants. To complicate matters, the larger buffer areas around wastewater treatment plants have all but disappeared. Land values and increased population have made it impossible to continue to have large buffer areas around most plants. As homes and businesses become neighbors to existing plants, what was a minor odor problem now becomes a major problem. No longer can even the smallest trace of odor exist without complaints from neighbors. Thus, preventing the emission of odors has become a prime operating consideration.

1.1   ODOR GENERATION

In order to control odors more effectively, an understanding of odor generation is needed. Understanding the problem and the causes will lead to a more effective solution.

1.10   Biological Generation of Odors

The principal source of odor generation is the production of inorganic (no or one carbon atom in formula, H_2S) and organic (more than one carbon atom in formula, C_8H_7N) gases by microorganisms in the collection system and treatment processes. Odors also may be produced when odor-containing or odor-generating materials are discharged into the collection system by industries and businesses.

The main concerns of operators are the inorganic gases hydrogen sulfide (H_2S) and ammonia (NH_3). These two gases give off the most offensive odors. As little as 0.5 ppb (parts of gas per billion parts of air) of hydrogen sulfide can be detected by the human nose and cause odor complaints. Hydrogen sulfide has an extremely offensive smell and has the odor produced by rotten eggs. Ammonia has a very sharp, pungent smell and also is very offensive. Other inorganic gases found in wastewater treatment plants are: carbon dioxide (CO_2), methane (CH_4), nitrogen (N_2), oxygen (O_2), and hydrogen (H_2). Normally found in nature, these gases are the products of normal respiration and biological activity of plants and animals and are not odorous.

Organic gases usually are formed in the collection system and in the treatment plant by the anaerobic decomposition of nitrogen and sulfur compounds. Organic gases also can derive their odors from industrial sources. Examples of organic gases found around treatment plants are *MERCAPTANS*,[1] *INDOLE*,[2] and *SKATOLE*.[3] These odorous compounds contain nitrogen or sulfur-bearing organic compounds.

In the normal biological oxidation of organic matter, the microorganisms remove hydrogen atoms from the organic compounds. In the process, the microorganisms use the bound sources of oxygen to gain energy. The hydrogen atoms are then transferred through a series of reactions that are sometimes called "hydrogen transfer" or "dehydrogenation."

1.11   Hydrogen Transfer Schemes

The following reactions illustrate the role of the hydrogen atom in the formation of both odorless and odorous compounds or end products.

$$\begin{matrix} \text{Hydrogen} \\ \text{Acceptor} \end{matrix} + \begin{matrix} \text{Hydrogen} \\ \text{Atoms} \\ \text{Added} \end{matrix} \rightarrow \begin{matrix} \text{End} \\ \text{Product} \end{matrix}$$

---

[1] *Mercaptans* (mer-CAP-tans).   Compounds containing sulfur that have an extremely offensive skunk-like odor; also sometimes described as smelling like garlic or onions.

[2] *Indole* (IN-dole).   An organic compound (C_8H_7N) containing nitrogen that has an ammonia odor.

[3] *Skatole* (SKAY-tole).   An organic compound (C_9H_9N) that contains nitrogen and has a fecal odor.

AEROBIC REACTION

$$O_2 + 4 H^+ \rightarrow 2 H_2O$$

Molecular
Oxygen

Water
(Odorless)

ANAEROBIC REACTIONS

$$2 NO_3^- + 12 H^+ \rightarrow N_2 + 6 H_2O$$

Nitrate

Nitrogen Gas
(Odorless)

$$CO_2 + 8 H^+ \rightarrow CH_4 + 2 H_2O$$

Carbon
Dioxide

Methane Gas
(Relatively Odorless)

$$SO_4^{2-} + 10 H^+ \rightarrow H_2S + 4 H_2O$$

Sulfate

Hydrogen Sulfide Gas
(Odorous)

$$Organized + n H^+ \rightarrow Lower\ Organics$$

Organics[4]

(Odorous)

The order in which microorganisms break down compounds containing oxygen in nature is: molecular oxygen (free dissolved oxygen), nitrate, sulfate, oxidized organics, and carbon dioxide.

There are some organisms that can only use molecular oxygen and cannot use the other forms. These microorganisms are called "strictly aerobic microorganisms" or "obligate aerobes." "Facultative microorganisms" can use molecular oxygen and combined (or bound) sources of oxygen such as nitrate. Still others, "strictly anaerobic microorganisms" or "anaerobes," can only use bound sources such as nitrate.

1.12   Hydrogen Sulfide Generation

The main cause of most odors in wastewater systems is hydrogen sulfide. Hydrogen sulfide can be detected by the human nose at a concentration as low as 0.00047 ppm (parts per million parts). This gas has a very characteristic rotten egg odor. The conditions that lead to hydrogen sulfide production also are conditions that produce other odors and other problems. These problems include dangers from explosive gas mixtures and haz-

ards to the respiratory system to persons working in confined or close spaces, and corrosion or deterioration of concrete sewer structures (such as pipelines and manholes). For these reasons, special attention is given to hydrogen sulfide generation.

The most common source of sulfide in wastewater is biological activity in the collection sewer or treatment plant. Sulfide compounds can develop in the natural breakdown of sulfur-bearing organic compounds. The source for the breakdown is protein that is discharged in wastes in the forms of undigested amino acids as part of feces and in urine as part of the unstable urea protein. This natural breakdown accounts for only a minor portion of all the sulfide compounds in the system. The major part of odor-producing sulfide results from the breakdown of inorganic sulfur compounds.

The principal sulfur compound found in wastewaters is sulfate. Sulfate compounds find their way into wastewaters from the public water supply and from industrial sources. The presence or absence of oxygen establishes whether or not hydrogen sulfide will exist. When dissolved oxygen is present, the sulfate ions will remain as sulfate. If more than 1.0 mg/L of oxygen is present, any sulfide that has been produced by anaerobic bacteria will be aerobically oxidized to thiosulfate, sulfate, and elemental sulfur. If the wastewater does not contain dissolved oxygen, the biological breakdown will reduce sulfate to sulfide, using the oxygen in the sulfate compound for energy to break down organic matter. Although sulfate and all essential elements may be present, sulfide is not produced in all wastewater systems. The pH of the wastewater is an important condition. Hydrogen sulfide is extremely pH dependent. Sulfide can exist in wastewater in three forms depending on the pH: S^{2-} ion, HS^- ion, or H_2S gas. When sulfide is in an ionic form, it is in solution so that it cannot escape as a gas. Odors are formed and released when sulfide is in the gaseous form (H_2S). The pH of the solution has to be below pH 7.5 in order for hydrogen sulfide gas to escape. At a pH below 5, all sulfide is present in the gaseous H_2S form and most of it can be released from wastewater and may cause odors, corrosion, explosive conditions, and respiratory problems.

SULFIDE FORMS (also see Figure 1.1)

below pH of 5	neutral pH of 7	pH of 9 to 11	pH of 14
hydrogen sulfide (H_2S) gas, 100%	H_2S, 50% HS$^-$, 50%	HS$^-$ ion, 100%	S^{2-} ion, 90%

The temperature of the system is an important factor that must be considered because the rate of bacterial metabolism is related to temperature. Areas where the temperature is normally above 65°F (18°C) have more problems than those with lower temperatures. Hydrogen sulfide generation is the greatest at temperatures around 85°F (30°C) and above.

---

[4] *Oxidized Organics.*   Organic materials that have been broken down in a biological process. Examples of these materials are carbohydrates and proteins that are broken down to simple sugars.

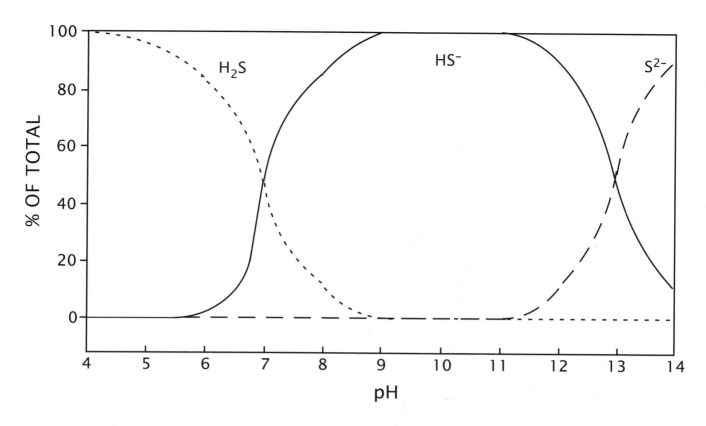

Fig. 1.1   Effect of pH on hydrogen sulfide-sulfide equilibrium

Figure 1.2 shows the sulfur cycle. The arrows indicate all the ways sulfide can be produced in collection systems and treatment plants.

QUESTIONS

Write your answers in a notebook and then compare your answers with those on page 39.

1.0A   Why is wastewater tending to become more septic and thus causing odor and corrosion problems?

1.1A   How are odors produced?

1.1B   What are the main inorganic gases of concern to operators?

1.1C   What is the order in which microorganisms break down compounds containing oxygen in nature?

1.1D   What is the major source of inorganic, odor-producing sulfate compounds found in collection systems and treatment plants?

1.1E   Hydrogen sulfide causes problems at what pH range?

1.2   ODOR IDENTIFICATION AND MEASUREMENT

In order to control odors effectively, the operator should know where odors originate and the cause. Odor detection in the past has been very unscientific because it relied on the human sense of smell. While our noses are more sensitive than most instruments or detection devices, each person has a different tolerance level for various odors. Occasionally, what smells good to one individual smells bad to another.

Odors can be detected, measured, and identified by several methods. Gas detection devices can be used to detect the presence of specific gases that cause odors. Today odors can be measured by the use of an *OLFACTOMETER*,[5] an *ODOR PANEL*,[6] or possibly by analytic testing. The olfactometer can measure odors in the field by diluting the odors with odor-free air. The number of dilutions required to reduce an odor to its minimum detectable *THRESHOLD ODOR*[7] concentration (MDTOC) provides a quantitative measure of the concentration or strength of an odor. The odor concentration is reported as the number of dilutions to "MDTOC." The results are reproducible within reasonable levels but there is the possibility of individual interpretation.

---

[5] *Olfactometer* (ALL-fak-TOM-utter).   A device used to measure odors in the field by diluting odors with odor-free air.

[6] *Odor Panel.*   A group of people used to measure odors.

[7] *Threshold Odor.*   The minimum odor of a gas or water sample that can just be detected after successive dilutions with odorless gas or water. Also called ODOR THRESHOLD.

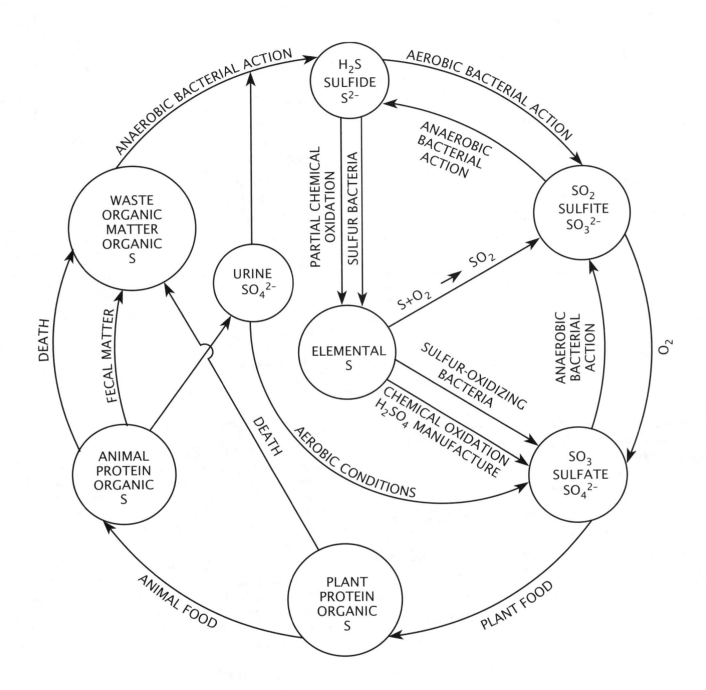

Fig. 1.2   Sulfur cycle

The most common method used to evaluate odor nuisances is the odor panel. This method of odor measurement involves a group of people (usually eight or more) in the evaluation of odors. The panel members are given diluted odorous gas samples and are asked to indicate whether they can or cannot detect an odor at various dilutions. The dilution(s) at which the odor is detected may be used to assess its strength or "threshold odor concentration."

Analytic testing to identify relative components of a gas sample is also an option in the odor identification process. Gas chromatography (GC) and mass spectrometry (MS) have been used for this purpose. However, this quantitative analysis may be difficult to interpret because the threshold concentrations of some odors are below the GC/MS detection level.

Certain types of odors have significant effects on people and animals. These odors have a major health and economic impact on those affected. For these reasons, the identification of odors is important. Once the odor has been identified, the solutions can be studied.

The following are some facts that can help in odor classification:

1. Almost all individuals have a sense of smell.

2. Individuals respond differently to the same odor.

3. Some odors are objectionable and others are pleasant.

4. Odors travel great distances with the direction of the wind.

5. Small concentrations of odors can be offensive.

6. Similar compounds do not have the same odor.

7. The human nose rapidly becomes fatigued (insensitive) to odors.

The best that most operators can do when recording odors is to classify the odor in some reasonable fashion. Sometimes a person not working in a plant will have to identify odors because an operator's nose can become insensitive. Usually offensive odors can be classified into the following groups:

Ammonia

Decayed cabbage

Decayed flesh

Fecal

Fishy

Garlic

Medicinal

Rotten egg

Skunk

With this information the operator can begin attacking the source of the problem. Skunk odors are frequently organic gases that contain sulfur compounds. These odors are usually from crotyl mercaptans. Decayed cabbage odors come from organic compounds with sulfur compounds attached. Usually the organic compound associated with this smell is dimethyl sulfide. Fishy smells are produced by organic compounds that have nitrogen compounds attached; dimethyl amine is a typical compound producing such a smell. Ammonia odors also come from organic compounds with nitrogen attached; indole is such a compound. Fecal odors are derived from skatole, which is an organic compound with nitrogen attached. Rotten egg odors are from the hydrogen sulfide molecule. The smell of decayed flesh comes from diamines, which are another ammonia-type compound.

All of these compounds are chemically similar, but they all smell different. Once the odors are identified, the source may be controlled and possibly eliminated. Solutions to odor problems are different if there are mixtures of odors because different compounds are involved.

Table 1.1 summarizes the odors we can detect from various substances and the threshold odor concentration (the level at which our nose first detects an odor).

TABLE 1.1 ODOR CHARACTERISTICS[a]

Substance	Remarks	Typical[b] Threshold Odor, ppm
Alkyl Mercaptan	Very disagreeable, garlic-like	0.00005
Ammonia	Sharp, pungent	0.037
Benzyl Mercaptan	Unpleasant	0.00019
Chlorine	Pungent, irritating	0.010
Chlorophenol	Medicinal	0.00018
Crotyl Mercaptan	Skunk	0.000029
Dimethyl Sulfide	Decayed vegetables	0.0001
Diphenyl Sulfide	Unpleasant	0.000048
Ethyl Mercaptan	Odor of decayed cabbage	0.00019
Ethyl Sulfide	Nauseating	0.00025
Hydrogen Sulfide	Rotten egg	0.00047
Methyl Mercaptan	Decayed cabbage	0.0011
Methyl Sulfide	Decayed vegetables	0.0011
Pyridine	Disagreeable, irritating	0.0037
Skatole	Fecal, nauseating	0.0012
Sulfur Dioxide	Pungent, irritating	0.009
Thiocresol	Rancid, skunk-like	0.0001
Thiophenol	Putrid, garlic-like	0.000062

[a] *MOP 11*, Chapter 27, "Odor Control," Water Pollution Control Federation, Washington, DC, 1976.
[b] Various references will list slightly different threshold odor concentrations.

1.3 ODOR COMPLAINTS

Periodically, all wastewater treatment plants will cause some odors. These will be detected by the public and must be handled by the operator. All complaints should be answered promptly and courteously. The public pays for your services and indirectly is your boss.

When responding to odor complaints, maintain a positive attitude. Beginning a conversation with a negative attitude will quickly upset the public. Even if you cannot detect the odor when you answer a complaint, that does not mean that the odor was not there or is not there now. Your nose may not be as sensitive as the nose of the person filing the complaint. Also, your nose may be accustomed to the smell and may no longer be able to detect the offensive odor.

The greatest complication develops if you do not properly handle the problem. If the public unites against the plant and becomes very odor conscious, even the slightest odor can cause an uproar. Remember that the person filing the complaint called because of a problem. You must be a diplomatic listener. Invite those who have complained to visit the plant and offer them a tour. While you are showing them around the plant, they may indicate to you where the odor that is bothering them is the strongest. This information may help you identify and control the cause of the odor problem.

Whenever an odor complaint is investigated, a record should be made of the visit and the important facts should be recorded (Figure 1.3). Investigations in the neighborhood near the location of an odor complaint can be very helpful. Odors can be coming from a nearby sewer, storm drain, trash pile, home plumbing problem, or dead animal. If an odor complaint is repeated and the source cannot be located, consider sending personnel to the site during the time of day when odors are a problem to determine the source.

QUESTIONS

Write your answers in a notebook and then compare your answers with those on page 39.

1.2A How can odors be measured?

1.2B List as many groups or types of odors or smells as you can recall.

1.3A When responding to odor complaints, maintain a _____ attitude.

1.3B When investigating an odor complaint, why might you be unable to detect an odor that is disturbing to the person complaining?

1.4 SOLUTIONS TO ODOR PROBLEMS

In order to solve any odor problem, a good systematic problem analysis is essential. To identify the source or cause of an odor problem and to select a solution, follow the procedures outlined below:

1. Make an on-site inspection and investigation of the problem areas.

2. Attempt to identify the source or cause of the problem.

3. Review plant housekeeping.

4. Review plant operations.

5. Review plant performance.

6. Evaluate plant performance.

7. Review engineering or design features of the plant.

8. List and review all solutions to the problem.

9. Put into practice the best possible solution.

Solving the problem may create complications including new odors and operating problems. Many times the odor is produced by a number of different gases in combination. When this happens, a single solution may not be the answer. Chemicals that counter the highest odor may cause other odors that are even more offensive. Solutions to odor problems may substantially increase operating and chemical costs. Often the solution will not be a textbook answer, but a combination of solutions developed by trial and error and technical aid from others.

To solve odor problems, try to identify the source and correct the problem when and where the odors are being produced. Some possible solutions you can use to control odors include the following:

- Minimizing hydraulic detention times in pipes and wet wells.

- Maintaining dissolved oxygen in the wastewater.

- Ensuring sufficient flow velocities to prevent solids deposition in pipelines and channels.

- Routinely cleaning structures to remove slime, grease, and sludge accumulations.

- Treating liquid and solid recycle streams (for example, chlorination, aeration).

- Changing or enforcing sewer-use ordinances.

- Routinely and frequently disposing of screenings and grit.

- Immediately removing floating scum/solids.

- Promptly and thoroughly cleaning process units as they are removed from service.

If the production of odors cannot be stopped, then the sources and chemical composition of the odors to be treated must be firmly defined before selecting a treatment method. Local regulatory requirements for air quality must also be considered when selecting a treatment method.

ODOR COMPLAINT FORM

COMPLAINT INFORMATION

Location: _____

Person filing complaint: _____ Phone: _____

Address: _____

Date of complaint: _____ Time of complaint: _____

Nature of complaint: _____

_____

INVESTIGATION INFORMATION

Name of investigator: _____

Date of investigation: _____ Time of investigation: _____

Strength of Odor	**Description of Odor**	**Wind Direction**	**Wind Velocity**
❏ Undetectable	❏ Ammonia	❏ North	❏ Calm
❏ Slight	❏ Decayed Cabbage	❏ South	❏ Mild Breeze
❏ Definite	❏ Fecal	❏ East	❏ Gusty
❏ Strong	❏ Fishy	❏ West	❏ Strong
❏ Intense	❏ Garlic	❏ None	❏ Very Strong
	❏ Medicinal		
	❏ Rotten Egg		
	❏ Skunk		
	❏ Other _____		

Comment on whether odor occurs during any specific time of the day, day of the week, or weather conditions.

Comments: _____

_____

_____

ACTION

Reviewed by: _____ Date: _____

Corrective action: _____

_____

_____

Fig. 1.3   Odor complaint form

The two major methods of odor control are chemical addition to the wastewater and, where odorous vapors can be contained and collected, odorous air treatment. Variations of both methods are discussed in the following sections.

1.40   Chemical Treatment of Odors in Wastewater

1.400   Chlorination

Chlorination is one of the oldest and most effective methods used for odor control. Chlorine is used in the disinfection process and is readily available at the wastewater treatment plant. Because of this availability, chlorine is frequently used to control odors. Chlorine is a very reactive chemical and, therefore, oxidizes many compounds in wastewater. The reaction between chlorine and hydrogen sulfide and ammonia has been studied by many researchers.

The reaction between chlorine and hydrogen sulfide is:

$$H_2S + 4\,Cl_2 + 4\,H_2O \rightarrow H_2SO_4 + 8\,HCl$$

The reaction of ammonia with chlorine is:

$$NH_3 + Cl_2 \rightarrow NH_2Cl + HCl \quad \text{(monochloramine, } NH_2Cl)$$

$$NH_2Cl + Cl_2 \rightarrow NHCl_2 + HCl \quad \text{(dichloramine, } NHCl_2)$$

$$NHCl_2 + Cl_2 \rightarrow NCl_3 + HCl \quad \text{(trichloramine, } NCl_3)$$

The most important roles that chlorine plays in controlling odors are to (1) inhibit the growth of slime layers in sewers, (2) destroy bacteria that convert sulfate to sulfide, and (3) destroy hydrogen sulfide at the point of application. This control requires less chemical than trying to oxidize the odor once formed. This means that the chlorine should be added in the collection system ahead of the plant.

Odors are not always removed by the use of chlorine. The reaction of chlorine with certain chemicals can cause a more odorous gas. One example is the reaction of chlorine with *PHENOL*[8] to form chlorophenol, a medicinal-smelling substance.

Experience has shown that a dosage as high as 12 to 1 of chlorine to dissolved sulfide (12 mg/L chlorine per each 1 mg/L sulfide) may be needed to control the generation of hydrogen sulfide in sewers. Do not determine the chlorine dose on the basis of the concentration of H_2S in the sewer atmosphere. Use the procedure in *"STANDARD METHODS"* and determine the chlorine consumed during a five-minute contact period. The correct dosage will yield a measurable chlorine residual at the end of five minutes, which is sufficient time for all the chlorine-sulfide reactions to go to completion.

Sodium hypochlorite has been used like chlorine to control odors. The chemical reactions with other substances are very similar.

1.401   Hydrogen Peroxide

For a number of years, hydrogen peroxide (H_2O_2) has been used as an oxidant to control odors. Hydrogen peroxide reacts in three possible ways to control odors.

1. Oxidant action: Oxidizes the compound to a non-odorous state. An example of this is the conversion of hydrogen sulfide to sulfate compounds.

$$H_2S + H_2O_2 \rightarrow H_2O + \text{sulfate compounds}$$

In actual practice, a dose of 2:1 to 4:1 of H_2O_2 to S^{2-} is needed for control.

2. Oxygen producing: Acts to prevent the formation of odor compounds. This is accomplished by keeping the system aerobic.

3. Bactericidal to the sulfate-reducing bacteria: Kills the bacteria that produce odors. Without biological activity, odors will not be generated. This high dose of H_2O_2 is probably not economically feasible.

Advantages of hydrogen peroxide use include its effectiveness as an oxidant, its ability to inhibit the regeneration of sulfate-reducing microorganisms, and the lack of toxic by-products. Disadvantages of hydrogen peroxide include its inability to treat ammonia or odorous organics, the contact time required for effective odor control (15 minutes to 2 hours), and its high cost.

QUESTIONS

Write your answers in a notebook and then compare your answers with those on page 39.

1.4A   Outline the systematic steps you would follow to solve an odor problem.

1.4B   What are the most important roles that chlorine plays in controlling odors?

1.4C   What are the three possible ways hydrogen peroxide reacts to control odors?

1.402   Oxygen and Aeration

Oxygen has been used for odor control with a great deal of success. The most common practice with oxygen is to use air to aerate the wastewater and try to keep the wastewater aerobic. The transfer of oxygen to the wastewater will increase its *OXIDATION-REDUCTION POTENTIAL (ORP)*[9] and thus reduce the formation of odorous gases. With more oxygen in the wastewater, the ORP is increased, and the sulfate ion is not used as an oxygen source; therefore the odor is reduced.

---

[8] *Phenolic* (fee-NO-lick) *Compounds.*   Organic compounds that are derivatives of benzene. Also called phenols (FEE-nolls).

[9] *Oxidation-Reduction Potential (ORP).*   The electrical potential required to transfer electrons from one compound or element (the oxidant) to another compound or element (the reductant); used as a qualitative measure of the state of oxidation in water and wastewater treatment systems. ORP is measured in millivolts, with negative values indicating a tendency to reduce compounds or elements and positive values indicating a tendency to oxidize compounds or elements.

Upstream aeration will cause hydrogen sulfide to be stripped out (carried out by the air) of the liquid, if it is present, and thus reduce the release of odors at the plant when water flows over weirs or other locations of high turbulence. *STRIPPED ODORS*[10] may be collected from above the surface where aeration takes place and be treated. If these odors are not properly handled, localized corrosion and odor problems can result.

Use of high-purity oxygen to maintain aerobic conditions in force mains can be a very effective way to control odors. However, three conditions must be met in order to successfully use this method:

1. The wastewater and oxygen must be thoroughly mixed.

2. The force main must have a continuous uphill grade from the point of application.

3. There must be adequate pressure within the main (typically greater than 15 psi gauge pressure) to force the oxygen into solution.

1.403   Ozone

Ozone is a powerful oxidizing agent that effectively removes odors. Ozone has limited use because an effective concentration may be too costly to use at large treatment plants. Ozone works well when used to remove odors from air collected over sources of odors (Section 1.47, "Ozonation," page 35).

An advantage of ozone is the fact that there have been no known deaths resulting from the use of ozone. Ozone can cause irritation of the nose and throat at a concentration of 0.1 ppm, but some people can smell ozone at concentrations around 0.01 to 0.02 ppm. Another advantage of ozone is that you can manufacture what you need at the plant site and do not have to handle bulky containers. Ozone is not available in containers because it is relatively unstable and cannot be stored.

1.404   Chromate

Chromate ions can effectively inhibit the sulfate reduction to sulfide. However, this method introduces heavy metals into the sludge and wastewater, and this may cause an even more offensive odor. Heavy metal ions, such as chromate, cause serious toxic conditions that limit their usefulness.

1.405   Metallic Ions

Certain metallic ions (mainly zinc) have been used to form precipitates with sulfide compounds. These precipitates are insoluble and have a toxic effect on biological processes such as sludge digestion. Therefore, this process has its limitations.

1.406   Nitrate Compounds

The first chemicals used in the anaerobic breakdown of wastes are nitrate ions. If enough nitrate ions are present, the sulfate

ions will not be broken down. The cost of this type of treatment to halt hydrogen sulfide production is very high and, at present, is not practical.

1.407   pH Control (Continuous)

Increasing the pH of the wastewater is an effective odor control method for hydrogen sulfide. By increasing the pH above 9, biological slimes and sludge growth are inhibited. This, in turn, halts sulfide production. Also, any sulfide present will be in the form of the HS^- ion or S^{2-} ion (above pH 11), rather than as H_2S gas, which is formed and released at low pH values.

1.408   pH Control (Shock Treatment)

Short-term, high pH (greater than 12.5) slug dosing with sodium hydroxide (NaOH) is effective in controlling sulfide generation for periods of up to a month or more depending on temperature and sewer conditions. At pH 13 the HS^- ion and S^{2-} ion are approximately of equal concentration and no H_2S is present. Care must be exercised in selecting the length of dosing so that downstream treatment plant biological systems will not be seriously impaired.

QUESTIONS

Write your answers in a notebook and then compare your answers with those on page 39.

1.4D   How is oxygen used to control odors?

1.4E   What is a limitation of using metallic ions to precipitate sulfide?

1.4F   How can pH adjustment control odors from hydrogen sulfide?

1.41   Biological Odor Control Systems

Biological odor control systems rely on the biological oxidation of odorous air. Odorous air has been fed to activated sludge systems, passed through trickling filters, and used as source air for dedicated biological towers.

Biological odor removal systems are not particularly effective in the removal of organic odors and relatively few of these systems are in use today. However, given adequate contact time and balanced environmental conditions, they have proven effective in removing a variety of inorganic compounds from process foul air streams.

Advantages of biological odor removal systems include their simple operation, lack of chemical usage, and their ability to treat high volumes of gas economically. Disadvantages include the space required for system installation, limited gas transfer capability, media fouling, the need for balanced environmental conditions, and their sometimes questionable reliability.

---

[10] *Stripped Odors.*   Odors that are released from a liquid by bubbling air through the liquid or by allowing the liquid to be sprayed or tumbled over media.

1.410   Odor Removal Towers (ORT) (Figure 1.4)

An odor removal tower is essentially a deep bed trickling filter that is lightly loaded (0.5 lb BOD/day/cu yd). The hydraulic loading should be maintained between 2 to 3 GPM/sq ft to produce a nitrifying biological *ZOOGLEAL MASS*[11] on the filter media. Foul air and off gases from the treatment plant process systems are captured and piped into the bottom of the odor removal tower. As this odorous air passes up through the filter media, the odors may be oxidized to an acceptable odor level and discharged to the atmosphere at the top of the tower. The odor removal tower (ORT) has two flow streams. One flow stream is liquid, to maintain the biomass on the media, and the other is air, to carry the odors.

1.411   Odor Removal Tower Parts

LIQUID STREAM

1. Filter Media.   Usually, plastic media is used to provide surface area for the biological slimes or biomass to attach themselves. The filter media bed may range from 20 to 30 feet (6 to 9 m) in depth.

2. ORT Sump.   A recess at the bottom of the filter where the applied liquid (primary or secondary effluent) is collected to be pumped back over the filter media to sustain the biomass.

3. Sump Overflow.   An outlet weir that prevents the ORT sump from filling too high and returns the overflow back to the plant.

4. ORT Circulation Pump.   A pump that recycles applied liquid from the treatment plant to the top of the filter. This water is applied through spray nozzles to the filter bed. The filter feed may be secondary effluent, primary effluent, or a blend of both. This blend is essential to maintain the proper BOD loading on the filter to support the biomass in a nitrifying stage.

5. Spray Nozzles.   Nozzles placed over the top of the filter media to ensure an even distribution over the filter of the recirculated effluent. These spray nozzles perform the same function as the rotating distributor arm on a trickling filter.

AIR STREAM

1. Supply Fans (Blowers).   Fans that transport foul air and off gases from the treatment plant process units through ducts and pipes to diffusers at the bottom of the odor removal tower.

2. Mist Eliminator.   A device located at the top of the odor removal tower above the filter bed. This device separates as much moisture as possible from the gas stream before it enters the exhaust fan of the odor removal tower to be discharged to the atmosphere.

3. Tower Exhaust Fan.   This fan pulls air from the filter bed and tower column and discharges scrubbed air to the atmosphere.

1.412   Odor Removal Tower Loading Rates

Recommended loading rates to maintain a nitrifying biomass are summarized in Table 1.2. Odor removal towers operated in accordance with Table 1.2 should use secondary effluent and a single-pass operation. There are two advantages to this method of operation. One, plugging of the spray heads is minimized by using secondary rather than primary effluent. Two, a single-pass operation prevents buildup of sulfuric acid (H_2SO_4), which could eventually corrode the structure and equipment. The pH of the spray water must be maintained above 6.0 so it will not cause corrosion damage. Caustic soda or an appropriate compatible chemical can be added to the spray (feed) water if necessary to increase the pH.

NOTE: Chlorinated secondary effluents may contain enough residual chlorine to be toxic to the biomass or they may not contain enough BOD to support the biomass. Therefore, carefully monitor the use of chlorinated secondary effluents to prevent a marked decline in odor removal efficiency.

TABLE 1.2   ODOR REMOVAL TOWER LOADING RATES

Organic Loading	= 0.5 lb BOD/day/cu yd media (0.3 kg BOD/day/m³ media)
Hydraulic Loading	= 2.3 to 3.0 GPM/sq ft media surface (1.6 to 2.0 L/sec/sq m media surface)
Foul Air Application	= 125 CFM/sq ft media surface (0.63 m³ air/sec/sq m media surface)
Average Air Velocity	= 150 ft/min (46 m/min)
Maximum Feed Water Recirculation	= 7:1
Minimum Air Retention Time	= 10 sec

1.413   Start-Up

1. Check to determine if all items such as hatches, grates, and duct work are secure.

2. Check sump for debris and dirt. Wash down the sump and inspect the overflow line.

3. Check filter spray nozzles for operation and position of flushing valves. Flushing valves should be closed.

4. Check drain lines from mist eliminators. They should be free of obstructions and the drain valves should be opened.

---

[11] *Zoogleal* (ZOE-uh-glee-ul) *Mass.*   Jelly-like masses of bacteria found in both the trickling filter and activated sludge processes. These masses may be formed for or function as the protection against predators and for storage of food supplies. Also see BIOMASS.

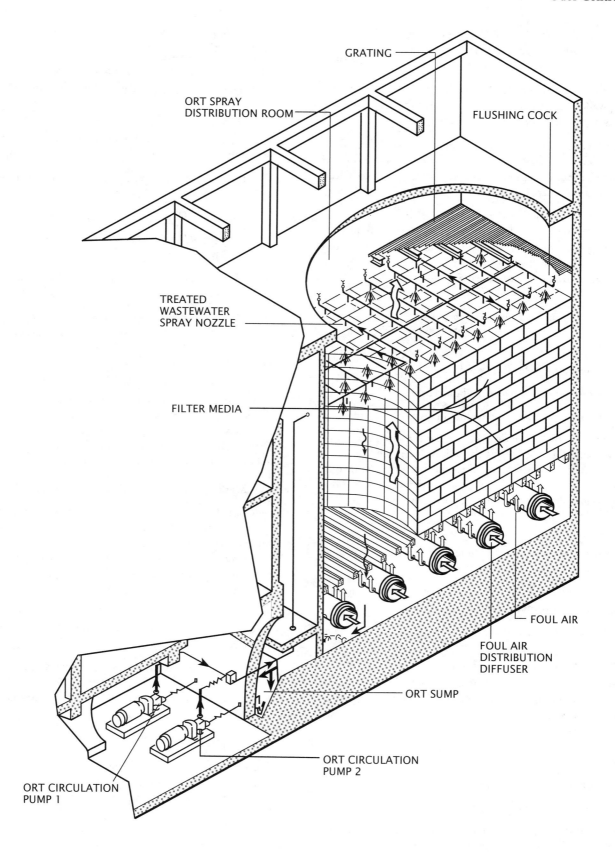

Fig. 1.4   Odor removal tower (ORT)
(Permission of Sacramento Area Consultants)

5. Inspect condition of supply and discharge fans. These fans should rotate freely and have the proper lubricants and belt tension.

6. Fill sump with selected liquid to the proper level. Regulate supply flow to the sump to make up for the small amount being returned to the plant through the overflow box.

7. Start recirculation pump that supplies water to the filter media spray nozzles. Examine spray nozzle operation and water distribution patterns on the top of the filter bed. Clean any nozzles that are plugged or restricted.

8. Start tower discharge and supply fans. Note that the tower will not remove odors from the air stream until a biomass is established on the filter media. The fans do not have to run until the biomass is established; however, some air must be ventilated through the filter bed during start-up.

1.414   Odor Removal Tower Monitoring

DAILY

1. Check operation of fans and pumps.

2. Check for proper sump level and makeup water feed supply.

3. Measure pH of sump feed recirculation water to the filter.

Do not allow the pH to drop below 6.0. If the pH is too low, either increase the makeup water addition rate (increase sump overflow rate) or raise pH by the addition of a compatible chemical (caustic soda). Water with a low pH can cause corrosion damage and may inhibit biological activity.

WEEKLY

1. Calculate the BOD loading of the filter (lbs BOD/day/cu yd media).

2. Check spray nozzle distribution pattern. The supply fans may have to be turned off during this check.

3. Lubricate fans and pumps.

QUARTERLY

1. Inspect sump for silt and debris. Supply fans and recirculation pumps may have to be turned off during this inspection. Use approved confined space entry procedures if you must enter the sump.

ANNUALLY

1. Take odor removal tower out of service.

2. Check and clean spray nozzles and distribution lines.

3. Check air ducts, plenums, fan housings, diffusers, and mist eliminators for corrosion. Clean and paint as required.

4. Clean sump. Use approved confined space entry procedures if you must enter the sump.

5. Flush filter media. Inspect media for fit (secureness) and deterioration. Inspect grates, valves, and other appurtenances.

QUESTIONS

Write your answers in a notebook and then compare your answers with those on page 40.

1.4G   How are off gases and foul air treated in a biological odor removal tower?

1.4H   How is the filter feed spread over the media?

1.4I   Why should the pH of the spray water not be allowed to drop below 6.0?

1.4J   How can the pH of the spray water be increased if the pH becomes too low?

1.4K   Why must caution be used when applying chlorinated secondary effluent to a biological odor removal tower?

1.42   Treatment of Odors in Air

The practice of treating air containing odors from a treatment process may be more economical than treating the wastewater. This type of odor control is becoming more and more popular. The methods of controlling odors in air include masking and counteraction (counter masking), combustion, absorption, activated carbon adsorption, and *OZONATION.*[12]

1.43   Masking, Modification, and Counteraction

Odor masking has been used with limited success for many years. Odor masking is accomplished by mixing the odorous compound with a control agent. The masking agent or chemical has a stronger and supposedly more pleasant odor quality which, when mixed with the odorous compound, results in a more pleasant odor than the odorous compound.

"Counteraction" is the control of odors by adding nonodor-producing reactive chemicals to the odor by spraying the air over the odor-producing area.

Caution must be exercised when considering the application of masking chemicals because they are usually chlorinated *BENZENE*[13] compounds. These compounds may be undesirable from an environmental standpoint.

1.44   Combustion

Industry has been removing odorous gases by combustion for years. The problem with using combustion to remove odors from a wastewater treatment plant is that the concentrations of odors in the gases are extremely low and the combustibility of the gases is so low that fuel costs are high unless the odors are combusted with treatment plant solids.

---

[12] *Ozonation* (O-zoe-NAY-shun).   The application of ozone to water, wastewater, or air, generally for the purposes of disinfection or odor control.

[13] *Benzene.*   An aromatic hydrocarbon (C_6H_6) that is a colorless, volatile, flammable liquid. Benzene is obtained chiefly from coal tar and is used as a solvent for resins and fats and in the manufacture of dyes. Benzene has been found to cause cancer in humans.

The key to the process is temperature. It has been reported that the best temperature is greater than 1,500°F (820°C). This temperature can be reduced by using catalysts. If insufficient temperatures are used and incomplete combustion occurs, odors that are not completely oxidized can be more obnoxious than the original odor. If sufficient temperatures are maintained, the combustion process can oxidize odor compounds not otherwise treatable by chemical or biological methods.

1.45 Absorption

"Absorption" is the process in which the odorous components are removed from a gas by being taken in or soaked up by a chemical solution.

Odors may be absorbed through a process called "scrubbing." This process is one of the most economical methods used today. The odorous compound is absorbed into a solution by the physical process of solubilization or by changing the chemical structure of the odorant. Odors may be scrubbed with chemicals such as potassium permanganate, sodium hypochlorite, caustic soda, and chlorine dioxide.

The air stream must be brought into contact with the absorbing compound. Usually, air is moved through ducts to a scrubbing unit. A scrubbing unit is a device that provides for the contact of the air with the scrubbing compound. This device may be a simple spray chamber, packed tower, or any similar unit. Spray chambers usually are not as effective as other methods due to the short contact time. Contact time can be increased through the use of packed media towers or tray towers.

Regardless of the chemical used, the arrangement of the units is very similar. Some of the odorous gases removed by a hypochlorite process are hydrogen sulfide, mercaptans, sulfur dioxide, ammonia, and organic gases. Absorption using this method (hypochlorite) is very rapid. Unfortunately, all odorous gases are not removed by this method or by the same chemicals.

A brine solution absorption system is shown in Figure 1.5. Salt is dissolved in water in the brine tank. This brine solution flows into the recycle tank. The solution is pumped from the recycle tank directly over the media or to electrolytic cells. As the brine solution passes through electrolytic cells, oxidizing agents are formed and this solution is sprayed over the media. As the solution comes in contact with odorous gases, they are oxidized to form less objectionable gases. Odorous air may flow horizontally or vertically through the media in a column. This system has been used successfully to reduce odors significantly from organically overloaded plastic-media trickling filters.

1.450 Electrolytic Chemical Scrubber Units for Foul Air Treatment

This absorption process removes odors from gases produced at various treatment processes and locations in a treatment plant. Odors are removed by *ELECTROLYTIC PROCESSES*[14]

that generate oxidizers, which destroy or convert odors into harmless, nonodorous gases. The system is very similar to the old radio vacuum tubes that use an anode-cathode assembly.

The anode-cathode assembly is the key in the scrubber system, and the heart of the assembly is the anode. Each assembly is activated when the rectifier transfers electrons from the anode (leaving it positively charged) and forces them on the cathode making it negatively charged. Raising the voltage causes an excess of electrons on the cathode to seek a means of reaching the electron-deficient anode. Electrons on the cathode will use any available means to cross the gap between the two surfaces.

Should pure water fill the void between the anode and cathode, only a few electrons would escape to the anode. If, however, soluble salts are present, the conductivity of the water is immensely increased. As the electrons flow across the gap, they chemically convert the salt to sodium hypochlorite and dissociate some water into oxygen and hydrogen. As the concentration of soluble salts increases, electron flow resistance diminishes. Subsequently, less voltage is required to force the transfer of electrons from the cathode to the anode, and the chemical changes occurring to the dissolved salts are increased.

Chemical reactions occur at both the anode and the cathode. Reactions at the cathode result primarily in the decomposition of water into hydrogen and the hydroxyl ion, whereas the following oxidative reactions occur simultaneously at the anode:

1. Oxidation of chloride to hypochlorite.

2. Formation of other highly oxidative species, namely ozone, singlet oxygen, and peroxides.

3. Electrolysis of water to produce normal gaseous oxygen.

All of these reactions are completely dependent upon the amounts of soluble salts in the circulation stream and on the output of the rectifier.

1.4500 MAJOR COMPONENTS

The major components of an electrolytic chemical scrubber unit are summarized in the following paragraphs.

Brine Distribution System. Consists of a separate tank and metering equipment for converting sodium chloride crystals (common salt) into a brine solution. A measurable quantity of the prepared brine is overflowed into the scrubber basin to form the desired concentration of *ELECTROLYTE*.[15]

Cell Recycle System. Composed of an electrically driven, noncorrosive, magnetic pump and associated piping for transferring the solution from the scrubber basin through the cells and then to the scrubbing tower where the solution returns to the basin by gravity.

Electrolytic Unit. Consists of anode-cathode assemblies that are activated through the application of DC power. For

---

[14] *Electrolytic* (ee-LECK-tro-LIT-ick) *Process.* A process that causes the decomposition of a chemical compound by the use of electricity.

[15] *Electrolyte* (ee-LECK-tro-lite). A substance that dissociates (separates) into two or more ions when it is dissolved in water.

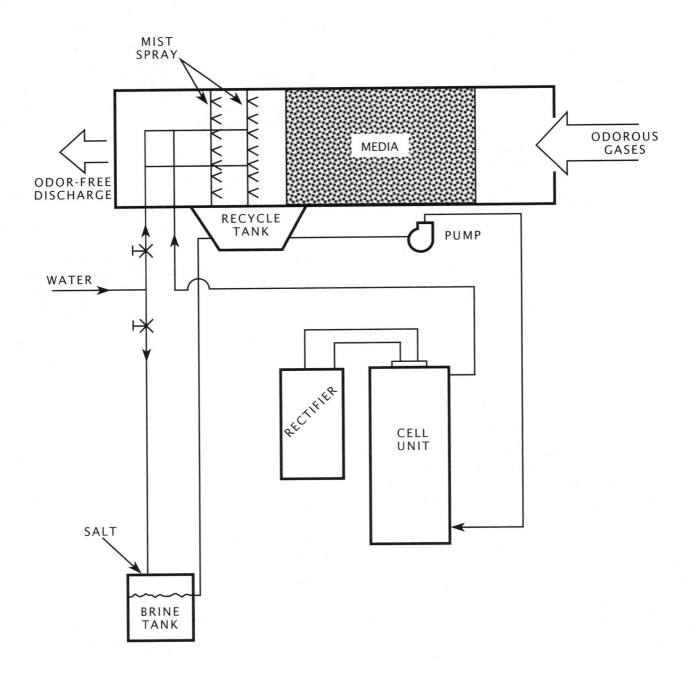

Fig. 1.5   Brine solution absorption system
(Permission of PEPCON)

simplicity and readability, the word CELL will be used interchangeably with the phrase ANODE-CATHODE ASSEMBLY throughout the remainder of this section.

Fresh Water System. Composed of an incoming line for supplying fresh water to the brine tank and the scrubbing tower.

Rectifier. Comprised of the electrical unit used in converting AC power to DC power.

Scrubbing System. Comprised of noncorrosive recycle pump, vertical tower, a set of spray nozzles, a packing bed, and a collection basin.

1.4501 STARTING PROCEDURE

The scrubber system is designed so that sodium chloride (salt) is converted to hypochlorite in the electrolytic cells at the rate required to produce the oxidant for destruction or conversion of odorous compounds in the absorption scrubber. The hypochlorite is regenerated after the oxidation of H_2S and other odorous materials. Salt makeup is only required to replace physical losses of chlorine compounds (mainly salt) from the system.

Use either a high-quality rock salt, one that does not contain an appreciable amount of impurities, or the type of salt recommended for water softeners. Do not use finely granulated table salt as it has a tendency to compact and dissolve slowly.

1. Place about 200 pounds (91 kg) of salt in the brine tank and replenish periodically in order to keep at least 12 inches (30.5 cm) of salt in the tank.

2. Fill the remaining portion of the brine tank with water.

3. Fill the scrubber basin with sufficient water to completely cover the pump intake nozzles and add about 20 pounds (9 kg) of the same type of salt as was used in filling the brine tank. An initial source of electrolyte is now available.

4. For achieving an optimum conversion of brine into hypochlorite, the solution in the scrubber basin should contain 30 grams per liter of sodium chloride. Laboratory analyses are required to accurately determine the concentration.

5. If the quantities and types of oxidizable impurities in the air stream are not known, refer to Table 1.3 for operating guidelines to determine the optimum level of operation for a 12,000 CFM (5.66 m^3/sec) scrubber.

6. Referring to Table 1.3, the initial operation of the unit should begin at an estimated level of 5 ppm of hydrogen sulfide within the air stream. This suggests setting the brine tank flowmeter at 50 milliliters per minute.

7. Open the valve in the cell pump intake line. Place the cell pump discharge valve in the ¾ open position.

8. Push the "Pump Start" button on the face of the rectifier. The cell pump will begin circulating the solution in the scrubber basin through the cells. Operate the pump for more than a minute before pushing the rectifier control switch to ON. Flow through the cells can be regulated by adjusting the pump discharge valve. The magnetic pump may become overloaded if the valve is closed completely.

CAUTION: Since the pump works on a magnetic principle and there is little head pressure between the basin and the pump, it may be necessary to momentarily close the discharge valve when starting. Thus the pump creates internal pressure and orients itself magnetically, resulting in almost immediate circulation as the valve is slowly reopened to the ¾ position.

TABLE 1.3   OPERATING GUIDELINES FOR A 12,000 CFM CHEMICAL SCRUBBER

| H₂S, ppm | H₂S, lbs per day[a] | NaCl, lbs per day[a] | Fresh Water to Brine Tank | | Fresh water through tower spray nozzles in gallons | Total overflow from scrubber basin in gallons per day[b] |
			Gallons per day[b]	Milliliters per minute		
1	1.28	7.7	3.7	9.7	9.1	12.8
2	3.56	15.4	7.4	19.4	18.2	25.6
5	6.40	38.5	18.5	48.7	45.5	64
10	12.8	77	37	97.2	91	128
20	25.6	154	74	197.3	182	256

[a] lbs/day × 0.454 = kg/day

[b] gal/day × 3.785 = L/day

9. Check that the rectifier current control switch is at zero. Solution must be flowing through the cells before power is applied. Flow of solution is easily determined by checking the vinyl discharge hoses on the cells. Turn the current control switch slowly until the desired operating level is achieved. At the assumed 5 ppm of H_2S within the air stream, set the rectifier at about 500 amps. Adjustments, if necessary, can then be made up or down in 50-amp increments.

   CAUTION: Follow the manufacturer's instructions regarding rectifier operation. Do not exceed either the amperage or voltage limits.

10. Open both the intake and discharge valves on the scrubber recycle pump.

11. Start the scrubber recycle pump.

12. If an undesirable odor is detected near the treatment site, increase the output of the rectifier by about 50 amps. After about two hours, determine whether or not the increase in power was sufficient to eliminate the undesirable odor. If not, increase the output by another 50 amps and continue this procedure until the undesirable odor is eliminated. Similarly, if a hypochlorite odor (smell of household bleach) is detectable near the treatment site and the gas stream is otherwise odor free, *DECREASE* the rectifier setting in 50-amp increments until the hypochlorite odor disappears.

13. Under the assumed operating level of 5 ppm of H_2S in the gas stream, add approximately 46 gallons (174 liters) of fresh water through the spray nozzles at the top of the tower each day. In order to achieve the desired daily level of overflow (64 gallons or 242 liters) from the scrubber basin without hindering the operator's schedule, the fresh water spray valve on the spray line to the upper tower section should be completely opened once each 8-hour shift. During each opening, the operator will want to overflow about 16 gallons (61 liters) of water from the scrubber basin.

    This quantity of water is easily determined by collecting the overflow in a container of known quantity. For example, if 2 gallons (7.6 liters) are collected per minute, then it can be assumed 16 gallons (61 liters) will overflow the scrubber basin in 8 minutes.

14. Generally, two days of continuous operation are necessary to remove any fluctuation that may occur.

1.4502 SHUTDOWN PROCEDURE

1. Turn the current control knob on the rectifier to zero.

2. Push the "Pump Stop" button on the rectifier panel. Both the pump and the rectifier are simultaneously turned off when this button is pushed.

3. Turn the main rectifier switch to the OFF position.

4. Stop the scrubber recycle pump.

5. Shut the flowmeter valve if the unit is to be off for more than two hours.

1.4503 OPERATIONAL CHECKS AND MAINTENANCE

This unit requires minimum operator attention; however, several routine checks should be accomplished each day, especially during the initial operating period. The following items can be performed while the system is operating:

1. Keep the vertical hole in the exposed end of each anode filled with silicone oil during the first six weeks of operation. More than likely, the addition of oil will be required once each week during the six-week interval. Use the oil supplied by the manufacturer for servicing the anodes.

2. Once each week for the first six weeks, check for loose nuts on the bus bar and bus bar pieces, and for loose banks around the tops of the cells. Tighten any connections that are loose.

3. Once each day check for the presence of gas bubbles in the vinyl (plastic) discharge tube at the top of each cell. Gas bubbles in the discharge stream reveal the cell is functioning properly and conversely a lack of bubbles suggests the cell is inoperative.

4. Once each day the operator should feel the outside surface of the copper cathode to determine if the assembly is operating at a temperature comparable to the other assemblies and to the temperature during the last daily inspection.

5. If a cell feels warm (hot enough to cause an immediate withdrawal of your hand), the condition more than likely reveals a plugged or shorted assembly.

6. Check for a noticeable increase (2 or more volts) in the output of the rectifier, which may signify an increase of electrical resistance within the cells. Resistance of this type can result from the formation of a scale-like deposit on the inner walls of the cathode.

   If an increase in DC voltage occurs and continues for a period of a day, you can assume a deposit has formed on the cathodes. Cathode scale can be easily removed by flushing the system with dilute nitric acid.

Most of the material in this section is reproduced with the permission of PEPCON.

QUESTIONS

Write your answers in a notebook and then compare your answers with those on page 40.

1.4L   How can odors in air be treated?

1.4M   When operating a chemical scrubber unit using a brine solution, how would you determine if the rectifier output is set properly or is set too high or too low?

1.451 Chemical Mist Odor Control System (Figure 1.6)

This hypochlorite absorption process removes hydrogen sulfide (H_2S) from atmospheres generated by the treatment process. The hydrogen sulfide is removed with a mixture of soft water, air, and a solution of sodium hypochlorite and sodium hydroxide in a single, nonrecirculation-type system. Hydrogen sulfide is oxidized to sulfuric acid (H_2SO_4) or to sulfur (S) and water (H_2O), depending upon conditions at the time of reaction.

The chemical mist odor control system consists of a reaction vessel (vertical tower), atomizer nozzles, chemical feed system, diluted chemical liquid distribution system, air system, water softener, electrical control panel, and controls for the odor control system.

Advantages of a chemical mist system include its capacity for accommodating high flow rates, low pressure drop, high transfer efficiency, and no chemical regeneration. Disadvantages include high energy and maintenance requirements, mist carryover in the discharge stream, and slow response to changing concentrations of odor compounds.

1.4510 PROCESS DESCRIPTION

The odor-carrying air enters the reaction vessel (Figure 1.7) at one end and mixes with a chemical solution mist from the atomizing nozzles. The angle at which the air stream enters the reactor results in an even mixture of the stream and chemical mist throughout the vessel.

The air/chemical mixture travels toward the vessel's discharge while the hydrogen sulfide is absorbed and oxidized by the chemical mist. Residual odors are neutralized as the mixture nears the discharge of the reaction vessel. The H_2S concentration (ppm) at the discharge point is automatically monitored and provides a control signal to regulate chemical addition.

The air must pass through a mist eliminator near the outlet of the vessel. Droplets are entrained (trapped) on the mist eliminator and fall to the bottom of the vessel for discharge through the drainage system. Unlike the packed bed scrubber (Section 1.452), the scrubbing chemical is not recirculated. Drainage is returned to the facility for treatment.

The atomizing nozzles use a high-velocity, compressed air stream to atomize a mixture of water and sodium hypochlorite into 5- to 20-*MICRON*[16] drops. Sodium hydroxide can also be added to the mixture if odor strength is high and a low pH exists. Sodium hydroxide injection is controlled to maintain a pH of 8.0 to 9.0 in the drainage system. Metering pumps inject the chemicals into the water and are controlled automatically. Soft water (less than 1 grain/gal or 17.1 mg/L hardness) is supplied to the nozzles to reduce calcium buildup.

1.4511 CHEMICAL REACTIONS

Two reactions may occur with hydrogen sulfide (H_2S) and sodium (NaOCl), depending upon the conditions at the time of reaction. Under alkaline conditions (pH greater than 7.0) and the presence of excess sodium hypochlorite, H_2S is oxidized to sulfuric acid (H_2SO_4). The sulfuric acid is then neutralized by NaOH to sodium sulfate as follows:

$$4\,NaOCl + H_2S \rightarrow 4\,NaCl + H_2SO_4$$

$$H_2SO_4 + 2\,NaOH \rightarrow Na_2SO_4 + 2\,H_2O$$

(8.76 lbs NaOCl + 2.35 lbs NaOH per lb of H_2S reacted)

If the reaction conditions are acidic (pH less than 7.0) or neutral and only a limited amount of NaOCl is available, H_2S is oxidized to sulfur and water as follows:

$$NaOCl + H_2S \rightarrow NaCl + H_2O + S$$

(2.19 lbs NaOCl per lb of H_2S reacted)

1.4512 TYPICAL OPERATING GUIDELINES

The following guidelines are typical design guidelines for operation of a chemical mist odor control system.

Foul Air Capacity, CFM[a] 23,500

H_2S Inlet Concentration, ppm 10

Required H_2S Outlet Concentration, ppm 0.1

Inlet Odor Units 500

Required Outlet Odor Units, maximum 50

Gas Residence Time, seconds, minimum 12

Exit Velocity, fpm,[b] minimum 4.000

---

[a] CFM × 1.6990 = m³/hr
[b] fpm × 0.3048 = m/min

1.4513 OPERATIONAL AND MAINTENANCE CHECKS

To ensure optimal odor removal efficiency of the system, the following tasks should be performed at the intervals indicated. (Note that operational and maintenance considerations will vary depending upon the specific process, design, and location.)

DAILY

1. Check outlet H_2S concentration

2. Check compressed air pressure

3. Check water flow rate and pressure

4. Check chemical feed pumps (to be sure they are pumping at the proper rate)

---

[16] *Micron* (MY-kron). μm, Micrometer or Micron. A unit of length. One millionth of a meter or one thousandth of a millimeter. One micron equals 0.00004 of an inch.

*Fig. 1.6   Chemical mist odor control tower
and foul air supply fan*

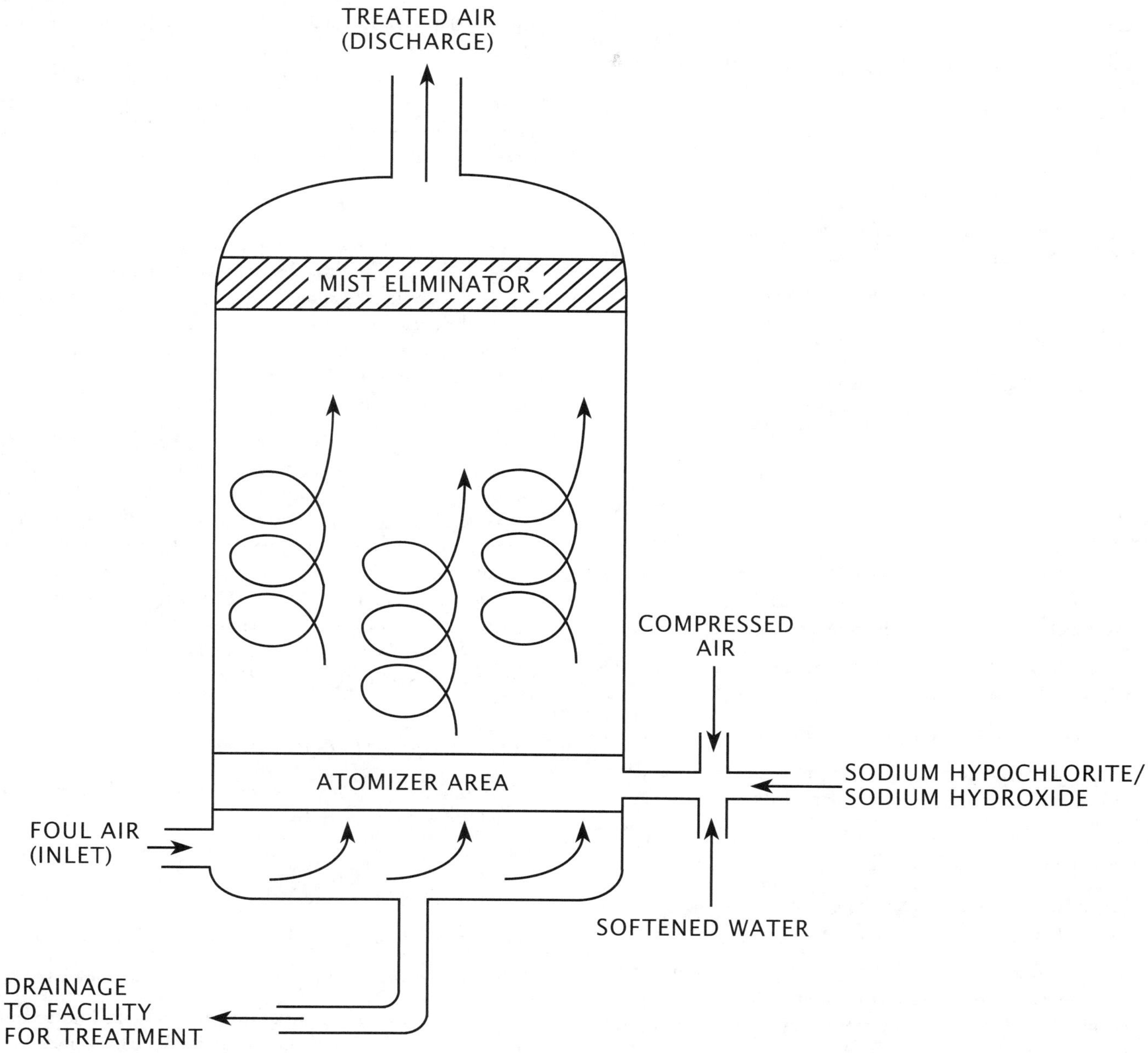

Fig. 1.7   Typical chemical mist odor control system

5. Check chemical tank levels

6. Check alarm indicator lights

7. Check drain pH

WEEKLY

1. Check salt supply in the water softener system

2. Check the water filter for plugging

MONTHLY

1. Clean atomizer nozzles

2. Check compressor air filter

3. Check and maintain compressor oil level (change as recommended by the manufacturer)

4. Test water supply from the softener for hardness

QUARTERLY

1. Change H_2S analyzer sensor if required

2. Replace H_2S analyzer permeation tube if required

1.4514 SAFETY

The following safety precautions should be taken when working with or installing chemical mist odor control systems:

1. Personal Protective Equipment (PPE), as specified on the chemical *MATERIAL SAFETY DATA SHEETS (MSDS),*[17] should be used when handling process chemicals.

2. Proper lockout/*BLOCKOUT*[18] procedures should be used when working on any equipment.

3. Pressures above operating guidelines should be prevented.

4. The system should be routinely inspected for leaks. Any leak(s) detected should be repaired immediately.

5. An emergency eye wash/deluge shower should be provided in an easily accessible location near the installation.

6. Secondary containment should be provided for the chemical storage tanks.

7. The chemical system should be designed to prevent operators from manually transferring, diluting, or rehandling the chemicals after delivery. (Systems based on dilution of powdered chemicals are undesirable.)

1.452 Packed Bed Scrubber Odor Control System

The packed bed is the most common type of wet scrubber system that uses chemical absorption to control odors. The system consists of a contact chamber with inert packing material, scrubbing liquid, a recirculating system, soft water makeup system, air blower, ductwork, and controls. A packed bed wet scrubber system differs from the chemical mist scrubber in its use of packing material in the contact chamber and recirculation of the scrubbing chemical.

Advantages of a packed bed scrubber include a high mass transfer efficiency, the ability to effectively handle changes in odorous compound concentrations, and economical treatment of high gas flows. Disadvantages include the recycling of odorous compounds, high chemical regenerant usage, chemical carryover in the discharge stream, and potentially high maintenance requirements.

1.4520 PROCESS DESCRIPTION

Odor-carrying air enters the scrubber and passes through the packed bed. Air flow may be vertical countercurrent (Figure 1.8) or horizontal crossflow (Figure 1.9). The inert material used in the packed bed provides increased surface area for contact between the odorous gases and the liquid scrubbing chemical. The bed also promotes turbulent mixing of liquid and gas and thus increases the gas-liquid mass transfer rate.

The oxidant used in packed bed scrubbers may be chlorine, chlorine dioxide, sodium hydroxide, sodium hypochlorite, or hydrogen peroxide. The scrubbing liquid collects at the bottom of the scrubber and is recirculated to the flow distribution system located above (and optionally, below) the packed bed. Makeup chemical is added to the recirculated chemical stream as needed to maintain the desired pH or solution strength. Softened water makeup is used to maintain the liquid level with a percentage blowdown to limit the total dissolved solids (TDS). Treated air passes through a mist eliminator and is discharged to the atmosphere.

1.4521 CHEMICAL DOSAGES

The following data represent typical oxidant dosages for the packed bed wet scrubber:

Chlorine: 2 to 4 *MOLES*[19] of Cl_2 per mole of H_2S

Sodium Hydroxide: 2 to 4 moles of NaOH per mole of H_2S

Sodium Hypochlorite: 2 to 4 moles of NaOCl per mole of H_2S

Hydrogen Peroxide: 3 to 8 moles of H_2O_2 per mole of H_2S

These dosage rates depend on the pH and take into account the typical mix of organic odorants and other compounds along with hydrogen sulfide.

NOTE: Hydrogen peroxide has a slower reaction time and requires a higher dosage than chlorine. A pH of 9 should also be maintained for the process to be most effective.

---

[17] *Material Safety Data Sheet (MSDS).*   A document that provides pertinent information and a profile of a particular hazardous substance or mixture. An MSDS is normally developed by the manufacturer or formulator of the hazardous substance or mixture. The MSDS is required to be made available to employees and operators or inspectors whenever there is the likelihood of the hazardous substance or mixture being introduced into the workplace. Some manufacturers are preparing MSDSs for products that are not considered to be hazardous to show that the product or substance is not hazardous.

[18] *Blockout.*   The physical prevention of the operation of equipment.

[19] *Mole.*   The name for a number (6.02×10^{23}) of atoms or molecules. See MOLECULAR WEIGHT.

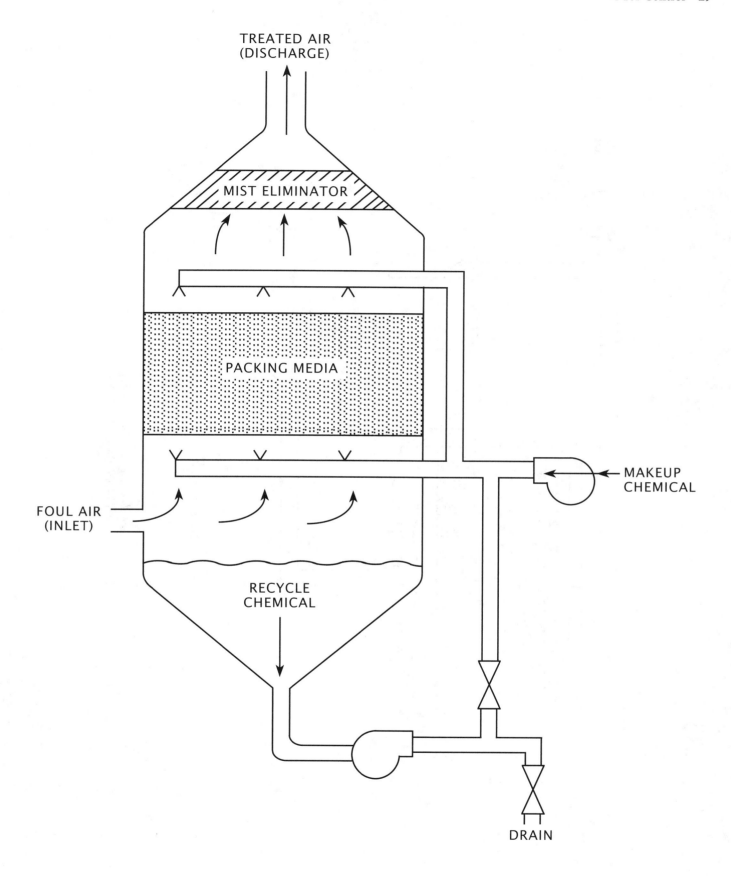

*Fig. 1.8   Typical vertical packed bed scrubber odor control system
(countercurrent air flow)*

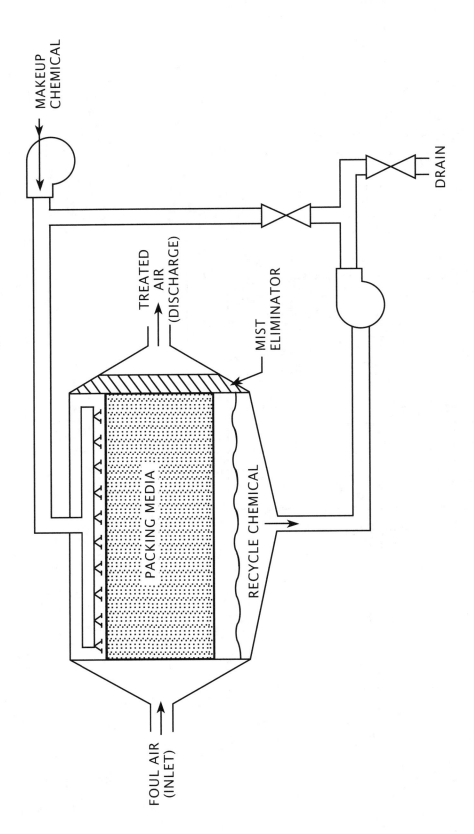

Fig. 1.9   Typical horizontal packed bed scrubber (cross air flow)

1.4522   TYPICAL OPERATING GUIDELINES

The following guidelines are typical design guidelines for operating a packed bed wet scrubber odor control system.

Liquid Application Rate	15 GPM[a]/1,000 CFM[b] of air flow
Gas Residence Time	1.5 to 2.0 seconds in packing (assuming empty), depending on safety factor
Packing Depth	10 feet[c] optimum
Scrubbing Liquid pH	9.5 to 10.5
Pressure Drop Across Packing Bed	0.5 inch[d] w.c.[20]/foot of packing
Outlet Velocity	2,500 fpm[e]
Gas Flow Velocity	1.2 to 3.0 m/sec

[a] GPM × 3.785 = L/min
[b] CFM × 1.6990 = m^3/hr
[c] ft × 0.3048 = m
[d] in × 2.54 = cm
[e] fpm × 0.3048 = m/min

1.4523   OPERATIONAL AND MAINTENANCE CHECKS

To ensure optimal odor removal efficiency with a packed bed scrubber, the following tasks should be performed at the frequencies indicated. (Note that operational and maintenance considerations will vary depending upon the specific process, design, and location.)

DAILY

1. Check outlet H_2S concentration.

2. Check chemical feed and recirculation pumps (to be sure they are pumping at the proper rate).

3. Check chemical tank levels.

4. Check alarm indicator lights.

5. Check the pH of recycle scrubbing liquid.

WEEKLY

1. Check head loss across the bed. (Surface fouling may result from carbonate deposits and acid washing may be required.)

2. Check salt supply in the water softener system.

3. Check the water filter for plugging.

MONTHLY

1. Test water supply from the softener for hardness.

QUARTERLY

1. Change H_2S analyzer sensor, if required.

2. Replace H_2S analyzer permeation tube, if required.

1.4524   SAFETY

The following safety items should be considered when working with or installing packed bed scrubber odor control systems:

1. If H_2O_2 (hydrogen peroxide) is used as the oxidant, passivated aluminum tanks and 316 stainless-steel piping and pumps are required. (A passivated tank is one that has undergone a metal plating process involving immersion in an acid solution to cause the formation of a protective film on the metal surface.)

2. Other safety considerations for a packed bed scrubber are essentially the same as those for a chemical mist system using similar chemicals; see Section 1.4514, "Safety."

1.46   Activated Carbon Adsorption

"Adsorption" is the process in which the odorous components are removed from a gas through adherence to a solid surface (Figure 1.10). The attractive force holding the gaseous molecule at the surface may be either physical (physical adsorption) or chemical (chemisorption). Any adsorption process should include a solid with an extremely large capability to adsorb gases. Activated carbon is such a solid.

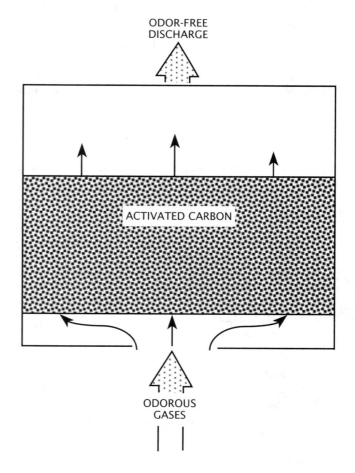

Fig. 1.10   Activated carbon adsorption process

---

[20] *w.c.*   Water column or pressure drop equal to pressure from 0.5 inch column of water.

Activated carbon is a highly porous material. Adsorption takes place upon the walls of the pores within activated carbon. Due to the nonpolar nature of its surface, activated carbon has the ability to adsorb organic and some inorganic materials in preference to water vapor. The materials and amounts adsorbed are dependent upon the physical and chemical makeup of the compound. Adsorption is affected by the molecular weight and boiling point of a compound. Higher molecular weight compounds are usually more strongly adsorbed than lower ones. Activated carbon will adsorb most organics that have molecular weights over 45 and boiling points over 41°F (5°C). Also, activated carbon can be manufactured from several different materials that have considerable variation in their adsorption characteristics.

Activated carbon is an effective means for controlling odors from all areas of the system, such as primary sedimentation and trickling filtration. Odorous air is collected and directed through activated carbon beds where the odor-causing organic gases and some inorganic gases (hydrogen sulfide) are removed from the air and adsorbed on the carbon. Activated carbon has been found to remove hydrocarbons before removing compounds such as hydrogen sulfide. The use of sand or soil for adsorption of odorous gases from pumping stations has seen limited use to date and the role of bacteria in these systems is not well understood at this time.

Some advantages of activated carbon adsorption include its fairly consistent reliability, simple operation, the ability to increase adsorbent capability by the use of additive compounds, and accommodation of high gas flow rates by use of multiple units. Disadvantages of activated carbon adsorption include potentially high regeneration costs, short use time due to high concentrations of odorous compounds, reactivity of caustic-impregnated material, fouling of the adsorbent by particulate matter, and disposal of the spent carbon if it is classified as a hazardous waste.

1.460   Process Description

The equipment for an activated carbon adsorption process consists of a foul air collection system, ducting, blowers, and activated carbon beds. Foul air containing odors from various sources in the treatment plant is collected and the odors are removed from the air as the odorous air passes through the activated carbon beds. The activated carbon beds consist of several feet of granular activated carbon (Figure 1.10).

Activated carbon odor control units may be fixed (Figure 1.11) or portable (Figure 1.12). Because carbon is most effective on higher molecular weight polar molecules (such as organic sulfur compounds), it is well suited for use as a "polishing" unit in support of other odor control systems that are typically ineffective in removing these odorants.

Figure 1.13 is a diagram of the components of a portable activated carbon odor control unit, referred to as a Mobile Ventilation Unit (MVU). This portable unit consists of an activated carbon filter, a 12,000 CFM exhaust fan, a flexible intake duct, a 500-CFM (850 m^3/hr) equipment ventilation system, and a control panel. All of the components are mounted on a semi-trailer that is moved by a fifth wheel tractor. The unit's mobility makes it useful as a backup to fixed odor control systems or as the primary ventilation and odor control unit for covered tank dewatering operations.

The MVU has a bottom inlet and side inlet. During normal operation, the bottom inlet is used. Use of the side inlet requires that the bottom inlet be blocked (which may be done by using the blind flange from the side inlet).

Start-up, shutdown, and operational considerations (with the exception of in-place carbon regeneration) are essentially the same for fixed or mobile activated carbon odor control units.

1.461   Start-Up

1. Inform operating personnel of intent to start blowers.

2. Make sure that the air blower motor electric switches are off and tagged out.

3. Check to be sure that the blower rotates freely.

4. Make sure all covers are properly in place and are secure.

5. Unlock electric switches at the main power control center for the blower motor. Turn on electrical power for the blower.

6. Observe blower to make sure it is operating properly.

7. Check carbon bed for air flow.

8. After initial start-up, measure air flow above carbon beds with a probe and record the velocity.

1.462   Shutdown

1. Turn electrical power off at blower. If for short duration, this is good enough but it should be tagged. For long duration, turn power off at main power switch and tag and lock out.

2. Inform all operating personnel of status of activated carbon units.

1.463   Operational Checks

DAILY

1. Inspect operation of air blower and motor.

2. Check air flow through the carbon bed.

WEEKLY

1. Measure air flow on the discharge side of the carbon bed with a velocity meter (like the ones used to measure air flow in air conditioning ducts). Compare against initial readings.

   If the air flow is low, the activated carbon is becoming plugged and will require replacement if the head loss cannot be reduced through mechanical cleaning methods.

   Backwashing carbon beds with water is not recommended in vapor phase applications. Application of water can significantly reduce the effectiveness of the carbon because the water will occupy adsorption sites on the carbon. The current

Fig. 1.11   Fixed activated carbon odor control unit

Fig. 1.12   Portable odor control unit
(Mobile Ventilation Unit, MVU)

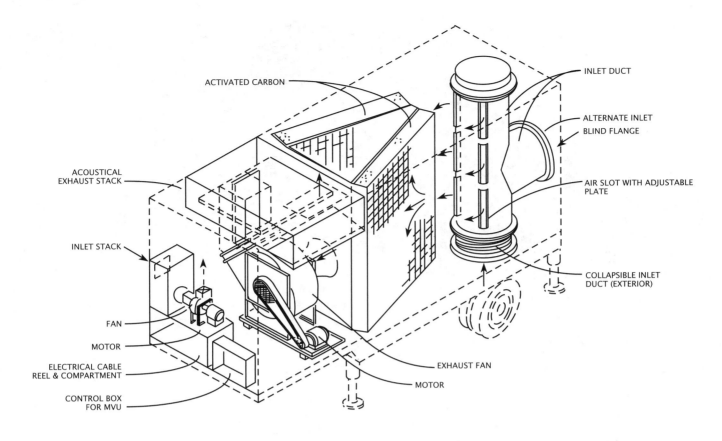

Fig. 1.13 Components of a mobile ventilation unit
(Source: Sacramento Area Consultants, Regional O & M Manual)

trend in operating practice is to dehumidify (for example, 70 to 75 percent humidity) odorous air before treating it by activated carbon adsorption processes.

QUARTERLY

1. Measure H_2S levels in the air at various depths in the activated carbon.

   Draw off samples and use an H_2S gas detector. If H_2S levels are found almost all the way through the activated carbon bed, this is an indication that the bed may fail soon. Usually activated carbon beds will regenerate themselves during periods when the odor levels are low.

   If an activated carbon bed fails, the carbon must be replaced or regenerated.

1.464 Carbon Disposal/Regeneration

Spent carbon can be disposed of or regenerated.

DISPOSAL

The hazard classification of the spent carbon must be established when considering disposal. The spent carbon may be a hazardous waste, depending upon the type and quantity of contaminants adsorbed during its use. Costly and extensive laboratory analysis is required to confirm the hazard classification of spent carbon; however, analysis may be required only once if future application and contaminant levels remain fairly stable. If the spent carbon is determined to be a hazardous waste, disposal will be very costly when compared to disposal in a normal landfill. The hazard class must then be a major consideration in the disposal/regeneration decision process.

REGENERATION

Regeneration of activated carbon temporarily eliminates any concerns about hazardous waste disposal and, as a secondary benefit, valuable landfill areas are conserved.

Spent carbon can be regenerated in place by the use of hot air or chemicals or it can be removed and thermally regenerated. The cost of thermal regeneration is somewhat less than the cost of purchasing an equal volume of new carbon. Approximately 5 to 10 percent of the carbon is lost in the regeneration process and must be replaced by virgin carbon. Regenerated carbon has less capacity than virgin carbon because of the reduction in available pore space. (The carbon is usually replaced after three regeneration cycles.) This reduced capacity must be considered when evaluating the cost effectiveness of thermal regeneration versus disposal. Decreased capacity will result in less run time and therefore more frequent handling of the carbon will be required.

Caustic-impregnated carbon (which has an enhanced capacity to remove hydrogen sulfide and methyl mercaptan odors due to chemical reactions) can be partially restored by in-place sodium hydroxide regeneration. The regeneration process also partially reduces the adsorbed organics. The efficiency of chemical regeneration and the number of regeneration cycles depend on the constituents of the odorous gas. Some high-molecular-weight organic odorants are not removed by the caustic soda regeneration process and may build up on the carbon. This can lead to organic odor breakthrough on later cycles. The carbon must be replaced if this occurs.

Chemical regeneration is accomplished by thoroughly flushing the carbon bed and containment vessel with water, draining the vessel, and adding new sodium hydroxide. The process is difficult to accomplish and has resulted in bed fires and explosions. As a result, the use of impregnated carbon is not recommended except in very specific applications.

Hot air regeneration may be used when volatile organic compounds (VOCs) are nonflammable or have a high ignition temperature (so as not to pose a risk of carbon fires). Hot air (350°F) is applied through the carbon unit countercurrent to the adsorption flow. Inert gas may be used rather than air if the off gas contains compounds that pose a fire risk at high temperatures. As the temperature rises, desorbed organics transfer to the regeneration gas stream. The gas stream is subsequently treated by a thermal oxidizer, where the VOCs are destroyed. The bed is allowed to cool to *AMBIENT*[21] temperature by stopping the flow of hot regeneration gas and continuing to pass ambient air through the bed. The regeneration and cool down times depend on the amount of carbon and the load on the carbon.

Contact your local or state hazardous waste authority for assistance in hazard classification. Your vendor should be able to provide assistance in replacement or regeneration of the activated carbon for your specific process.

NOTE: After activated carbon has been in service for a long period of time and then is taken out of service, the carbon may develop a gray appearance. This gray appearance is usually caused by salts, which form crystals when the activated carbon dries out. These crystals will cause a gray color, but the activated carbon will still be black.

1.465   Safety

The following safety precautions should be taken when working with or installing activated carbon odor control units:

1. Implement engineering controls to eliminate dust inhalation hazards during carbon handling operations. Use approved respiratory protection if dust exposure reaches the action level (following implementation of engineering controls).

2. Use Personal Protective Equipment (PPE) as specified on the product Material Safety Data Sheet (MSDS).

3. Make available in the work area an emergency eye wash/deluge shower and washing facilities.

4. Prevent contact between carbon and strong oxidizing agents (for example, chlorine, ozone).

5. Adequately ground carbon treatment systems. (In certain systems, high voltage static electrical charges may accumulate to levels of shock or ignition hazard.)

6. Use approved confined space procedures for vessel entry. (Under certain process conditions, activated carbon has shown an affinity for (attraction to) atmospheric oxygen or may interact with process streams to generate potentially toxic or hazardous levels of hydrogen sulfide, methane, ethanol, carbon dioxide, and other gases.)

1.47   Ozonation

Ozonation is an oxidation process (Figure 1.14). Ozone is a powerful oxidant and can effectively oxidize odor-causing compounds to less objectionable forms. Air is collected from the sources and directed into a mixing chamber where this odorous air is mixed with ozone. The odorous air is oxidized, and the odors are eliminated. Ozone, being relatively unstable, must be manufactured on site. Controlling the dosage of ozone to prevent overdosing is critical and expensive.

Successful treatment of odors with ozone depends on the type of odor, intensity or strength of odor, flow rate of odorous air to be treated, and the size of the ozone contact chamber. The longer the contact time, the more effective the treatment. Therefore, control of the ozonation process is achieved by regulating the speed or number of suction fans. Fifteen seconds is considered the minimum mixing and contact time for effective treatment of odors. Ozonation systems have experienced serious mechanical problems and few remain in use.

1.48   Good Housekeeping

The best solution to odor problems is to prevent odors from ever developing. One step to help achieve this is good housekeeping. Regularly clean all baffles, weirs, troughs, diversion boxes, channels, and all exposed clarifier mechanisms where scum and solids could accumulate and decompose. Cleaning is accomplished by the use of deck brushes with long handles and hosing with a high-velocity jet of water. These efforts will help keep odor problems from becoming a serious public issue.

---

[21] *Ambient* (AM-bee-ent) *Temperature.*   Temperature of the surroundings.

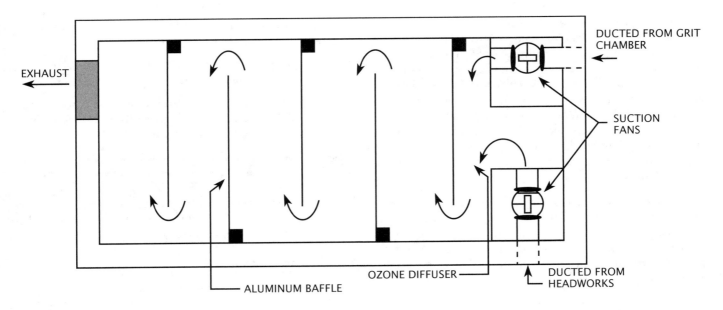

EXHAUST

DUCTED FROM GRIT CHAMBER

SUCTION FANS

OZONE DIFFUSER

DUCTED FROM HEADWORKS

ALUMINUM BAFFLE

Fig. 1.14   Ozone contact chamber

1.5   TROUBLESHOOTING ODOR PROBLEMS

Plant Area	Odor Problem	Possible Solution[a]
Influent	H_2S odor	• Chemical addition.
	Organic odor	• Air stream treatment.[b]
Headworks	H_2S odor	• Chemical addition.
	Organic odor	• Air stream treatment.[b] • Correct faulty plant operation.
Primary Sedimentation	H_2S odor	• Chemical addition at headworks. • Air stream treatment. • Remove sludge faster.[b] • Chemical addition. • Cover tanks, vent odors, and treat.
Biological System		
Activated sludge	Organic odor	• Eliminate short-circuiting in aeration chamber by increasing aeration rates.[b] • Prevent sludge deposits from developing.
Biological filter	H_2S odor	• Air stream treatment. • Chemical addition (H_2O_2 in the liquid prior to the filter). • Chlorinate filter influent for short periods, preferably at low flows. • Provide even air distribution to the filter to avoid anaerobic areas.[b] • Clean filter vents and underdrain system. • Correct any design deficiencies.
	Decayed organic odor (ammonia, fishy, rotten cabbage)	• Correct any design problems. • Chemical addition (H_2O_2, Cl_2, or NO_3^-). • Air stream treatment.

1.5   TROUBLESHOOTING ODOR PROBLEMS *(continued)*

Plant Area	Odor Problem	Possible Solution[a]
Biological System *(continued)*		
Secondary sedimentation	Organic odor	• Correct design problem.
	Inorganic odor	• Increase sludge removal rate. • Reduce turbulence that is causing stripping of odor. • Increase air in aeration systems. • Improve operations.
Anaerobic digestion	Decaying organic odor	• Light waste gas burner if not lit.

	Inorganic odor	• Fixed cover digesters:
		    a. Add water to water seal.
		    b. Seal cracks in roof.
		• Floating cover digesters:
		    Lower cover until sufficient annular space seal is obtained.
		• Gas equipment:
		    Find and correct leaks around valves, flame assembly, sample wells, and access hatch.
		• Look for sources of leaks and correct around pressure relief valves, vents, drain lines, and piping.
		• Check digester operation.

Solids Handling		
Sludge drying beds	Decaying organic odor	• Chemical addition (countermasking, masking). • Apply dry lime if partially digested (green sludge) was withdrawn to drying bed. • Check operation of digester. • Review operations such as withdrawal rate from digester.
Vacuum filters or filter presses	Organic odor	• Check operations. • Check operations at digester area. • Remove solids from area. • Improve housekeeping. • Air stream treatment.
Centrifuges	Organic odor	• Check operations. • Check operations at digesters. • Improve housekeeping. • Treatment of odors in air removed from centrifuge area.
Sludge retention basins	Organic odor	• Review operations.
	Inorganic odor	• Develop an aerobic layer on the surface.
Ponds or lagoons	Ammonia odor	• Check operations. • Review operational data. • Chemical addition.
Disinfection	Chlorine smell	• Check mixing. • Increase/optimize water flow through injector. • Lower dosage rate. • Check diffuser. • Look for chlorine leaks.
	Ammonia odor	• Remove sludge deposits in chamber. • Source control. • Improve housekeeping.

[a] Each solution listed may be applied to all odor problems in the plant area where the problem is occurring.

[b] Investigate industrial waste sources as a possible cause of the problem if this item appears to be the correct solution.

1.6   REVIEW OF PLANS AND SPECIFICATIONS

Odor control facilities require a careful review of the plans and specifications similar to the review given other treatment processes.

1. When clarifiers and other large tanks and areas are enclosed to control odors, be sure provisions are made to remove tank covers for inspection and maintenance purposes. Movable cranes or hoists are helpful.

2. Hydrogen sulfide gas combines with moisture to form sulfuric acid, which is very corrosive. All concrete must be protected from corrosion. All pipes, vents, screens, grates, support systems, and other materials exposed to odorous air must be made of corrosion-resistant materials.

3. Be sure provisions are made for ventilation of enclosed spaces with fresh air before anyone enters for any reason. Assume they are confined spaces and use extreme caution at all times. Equipment and instruments must be available to detect oxygen concentration, explosive conditions, and hydrogen sulfide in enclosed atmospheres before entry and while anyone is working in the enclosed area.

QUESTIONS

Write your answers in a notebook and then compare your answers with those on page 40.

1.4N   What are the advantages of a chemical mist odor control system?

1.4O   How does a packed bed wet scrubber system differ from the chemical mist scrubber?

1.4P   What is a solid that is used in an adsorption process to remove odors from air?

1.4Q   What is a secondary benefit derived from regenerating carbon?

1.4R   What are two methods for regenerating carbon in place?

1.7   ADDITIONAL READING

1. *MOP 11*, Chapter 13,* "Odor Control."

2. *CONTROL OF ODORS AND EMISSIONS FROM WASTEWATER TREATMENT PLANTS* (MOP 25). Obtain from Water Environment Federation (WEF), Publications Order Department, 601 Wythe Street, Alexandria, VA 22314-1994. Order No. MOP425. ISBN 1-57278-189-0. Price to members, $76.75; nonmembers, $91.75; price includes cost of shipping and handling.

3. *A CRITICAL REVIEW OF ODOR CONTROL EQUIPMENT FOR TOXIC AIR EMISSIONS REDUCTION.* Obtain from Water Environment Federation (WEF), Publications Order Department, 601 Wythe Street, Alexandria, VA 22314-1994. Order No. D53010. ISBN 1-57278-004-5. Price to members, $36.75; nonmembers, $46.75; price includes cost of shipping and handling.

4. *ODOR AND CORROSION PREDICTION AND CONTROL.* Obtain from Water Environment Federation (WEF), Publications Order Department, 601 Wythe Street, Alexandria, VA 22314-1994. Order No. P01100. ISBN 1-57278-169-6. Price to members, $68.70; nonmembers, $90.45; price includes cost of shipping and handling.

5. *TOXIC AIR EMISSIONS FROM WASTEWATER TREATMENT FACILITIES.* Obtain from Water Environment Federation (WEF), Publications Order Department, 601 Wythe Street, Alexandria, VA 22314-1994. Order No. P01025. ISBN 1-57278-012-6. Price to members, $32.25; nonmembers, $40.75; price includes cost of shipping and handling.

6. *SAMPLING FOR MEASUREMENT OF ODOURS.* Obtain from IWA Publishing, www.iwapublishing.com. ISBN 1843390337. Price to members, $82.50; nonmembers, $110.00.

7. *ODOURS IN WASTEWATER TREATMENT.* Obtain from IWA Publishing, www.iwapublishing.com. ISBN 1900222469. Price to members, $133.50; nonmembers, $178.00.

* Depends on edition.

Please answer the discussion and review questions next.

DISCUSSION AND REVIEW QUESTIONS

Chapter 1.   ODOR CONTROL

The purpose of these questions is to indicate to you how well you understand the material in the chapter. Write the answers to these questions in your notebook.

1. How should an odor complaint be handled?

2. What can happen if an odor complaint is handled improperly?

3. What procedures would you follow to identify the source or cause of an odor problem and to select a solution?

4. How can odors in air be treated?

5. What would you do if persons living near your treatment plant complained of a chlorine smell?

SUGGESTED ANSWERS
Chapter 1.   ODOR CONTROL

Answers to questions on page 11.

1.0A   Wastewater is tending to become more septic, and thus causing odor and corrosion problems, because collection systems are being extended farther and farther away from treatment plants.

1.1A   The principal source of odor generation is the production of inorganic and organic gases by microorganisms. Odors also may be produced when odor-containing or odor-generating materials are discharged into the collection system by industries and businesses.

1.1B   The main inorganic gases of concern to operators are hydrogen sulfide (H_2S) and ammonia (NH_3).

1.1C   The order in which microorganisms break down compounds containing oxygen in nature is: molecular oxygen (free dissolved oxygen), nitrate, sulfate, oxidized organics, and carbon dioxide.

1.1D   The major source of inorganic, odor-producing sulfate compounds found in collection systems and treatment plants is sulfate compounds from the public water supply and from industrial sources.

1.1E   Hydrogen sulfide causes problems at lower (acidic) pH ranges. At a pH below 5, all sulfide is present in the gaseous H_2S form; most of it can be released from wastewater and may cause odors, corrosion, explosive conditions, and respiratory problems.

Answers to questions on page 14.

1.2A   Odors can be measured by the use of an olfactometer, an odor panel, or possibly by analytic testing.

1.2B   Usually odors can be classified into the following groups:

1. Ammonia
2. Decayed cabbage
3. Decayed flesh
4. Fecal
5. Fishy
6. Garlic
7. Medicinal
8. Rotten egg
9. Skunk

1.3A   When responding to odor complaints, maintain a positive attitude.

1.3B   You might not be able to detect an odor that is disturbing to a person complaining because:

1. Your nose may not be as sensitive as the nose of the person complaining.
2. Your nose may be accustomed to the smell and may no longer be able to detect the offensive odor.

Answers to questions on page 16.

1.4A   Systematic steps to follow to solve an odor problem include:

1. Make an on-site inspection and investigation of the problem areas.
2. Attempt to identify the source or cause of the problem.
3. Review plant housekeeping.
4. Review plant operations.
5. Review plant performance.
6. Evaluate plant performance.
7. Review engineering or design features of the plant.
8. List and review all solutions to the problem.
9. Put into practice the best possible solution.

1.4B   The most important roles that chlorine plays in controlling odors are to (1) inhibit the growth of slime layers in sewers, (2) destroy bacteria that convert sulfate to sulfide, and (3) destroy hydrogen sulfide at the point of application.

1.4C   Three possible ways hydrogen peroxide reacts to control odors are (1) oxidant action, (2) oxygen producing, and (3) bactericidal to the sulfate-reducing bacteria.

Answers to questions on page 17.

1.4D   Oxygen is used to control odors by aerating wastewater and attempting to keep it aerobic. Also, aeration can strip odors out of wastewater. High-purity oxygen may also be used to maintain aerobic conditions in force mains.

1.4E   A limitation of using metallic ions to precipitate sulfide is the toxic effect of the precipitates on biological processes such as sludge digestion.

1.4F   Odors can be controlled by increasing the pH. At pH levels above 9.0, biological slimes and sludge growth are inhibited. Also, any sulfide present will be in the form of HS^- ion or S^{2-} ion, rather than as H_2S gas, which is formed and released at low pH values.

Answers to questions on page 20.

1.4G   Off gases and foul air are treated in a biological odor removal tower by passing this air up through the filter media where the odors are oxidized to an acceptable odor level and discharged to the atmosphere at the top of the tower.

1.4H   The filter feed is spread over the media by the use of spray nozzles.

1.4I   The pH of the spray water must be maintained above 6.0 so it will not cause corrosion damage.

1.4J   The pH of the spray water can be increased by the addition of caustic soda or an appropriate compatible chemical.

1.4K   Chlorinated secondary effluents may contain enough residual chlorine to be toxic to the biomass or they may not contain enough BOD to support the biomass. Odor removal efficiency may decline sharply.

Answers to questions on page 24.

1.4L   Odors in air can be treated by masking and counteraction, combustion, absorption, adsorption, and ozonation.

1.4M   If the rectifier output is set too high, a hypochlorite odor (smell of household bleach) is detectable. If the output is set too low, an undesirable odor is detectable. No odors are detectable if the rectifier is set properly.

Answers to questions on page 38.

1.4N   Advantages of a chemical mist system are: the capacity for accommodating high flow rates; low pressure drop; high transfer efficiency; and no chemical regeneration.

1.4O   A packed bed wet scrubber system differs from the chemical mist scrubber in its use of packing material in the contact chamber and recirculation of the scrubbing chemical.

1.4P   Activated carbon is a solid that is used to remove odors from air by the adsorption process.

1.4Q   A secondary benefit derived from regenerating carbon is the conservation of landfill area.

1.4R   Two methods for regenerating carbon in place are the use of hot air or chemicals (sodium hydroxide).

CHAPTER 2

ACTIVATED SLUDGE

Pure Oxygen Plants and Operational Control Options

by

Ross Gudgel

and

Larry Peterson

Further information related to this topic may be found in:

OPERATION OF WASTEWATER TREATMENT PLANTS

Volume I, Chapter 8
Package Plants and Oxidation Ditches

Volume II, Chapter 11
Operation of Conventional Activated Sludge Plants

TABLE OF CONTENTS

Chapter 2.   ACTIVATED SLUDGE

Pure Oxygen Plants and Operational Control Options

OBJECTIVES

Chapter 2.   ACTIVATED SLUDGE

Pure Oxygen Plants and Operational Control Options

Following completion of Chapter 2, you should be able to:

1. Safely operate and maintain a pure oxygen activated sludge plant.

2. Review the plans and specifications for a pure oxygen system.

3. Describe the various methods of determining return sludge and waste sludge rates and select the best method for your plant.

4. Operate an activated sludge process that must treat both municipal and industrial wastes.

5. Operate an activated sludge process that must treat strictly an industrial waste.

6. Operate an activated sludge process to produce a nitrified effluent.

WORDS

Chapter 2.   ACTIVATED SLUDGE

Pure Oxygen Plants and Operational Control Options

ABSORPTION (ab-SORP-shun)   ABSORPTION

The taking in or soaking up of one substance into the body of another by molecular or chemical action (as tree roots absorb dissolved nutrients in the soil).

ACTIVATED SLUDGE   ACTIVATED SLUDGE

Sludge particles produced in raw or settled wastewater (primary effluent) by the growth of organisms (including zoogleal bacteria) in aeration tanks in the presence of dissolved oxygen. The term "activated" comes from the fact that the particles are teeming with bacteria, fungi, and protozoa. Activated sludge is different from primary sludge in that the sludge particles contain many living organisms that can feed on the incoming wastewater.

ACTIVATED SLUDGE PROCESS   ACTIVATED SLUDGE PROCESS

A biological wastewater treatment process that speeds up the decomposition of wastes in the wastewater being treated. Activated sludge is added to wastewater and the mixture (mixed liquor) is aerated and agitated. After some time in the aeration tank, the activated sludge is allowed to settle out by sedimentation and is disposed of (wasted) or reused (returned to the aeration tank) as needed. The remaining wastewater then undergoes more treatment.

ADSORPTION (add-SORP-shun)   ADSORPTION

The gathering of a gas, liquid, or dissolved substance on the surface or interface zone of another material.

AERATION (air-A-shun) LIQUOR   AERATION LIQUOR

Mixed liquor. The contents of the aeration tank, including living organisms and material carried into the tank by either untreated wastewater or primary effluent.

AERATION (air-A-shun) TANK   AERATION TANK

The tank where raw or settled wastewater is mixed with return sludge and aerated. The same as aeration bay, aerator, or reactor.

AEROBES   AEROBES

Bacteria that must have dissolved oxygen (DO) to survive. Aerobes are aerobic bacteria.

AEROBIC (air-O-bick) DIGESTION   AEROBIC DIGESTION

The breakdown of wastes by microorganisms in the presence of dissolved oxygen. This digestion process may be used to treat only waste activated sludge, or trickling filter sludge and primary (raw) sludge, or waste sludge from activated sludge treatment plants designed without primary settling. The sludge to be treated is placed in a large aerated tank where aerobic microorganisms decompose the organic matter in the sludge. This is an extension of the activated sludge process.

AGGLOMERATION (uh-glom-er-A-shun)   AGGLOMERATION

The growing or coming together of small scattered particles into larger flocs or particles, which settle rapidly. Also see FLOC.

AIR LIFT PUMP   AIR LIFT PUMP

A special type of pump consisting of a vertical riser pipe submerged in the wastewater or sludge to be pumped. Compressed air is injected into a tail piece at the bottom of the pipe. Fine air bubbles mix with the wastewater or sludge to form a mixture lighter than the surrounding water, which causes the mixture to rise in the discharge pipe to the outlet.

AMBIENT (AM-bee-ent) TEMPERATURE

AMBIENT TEMPERATURE

Temperature of the surroundings.

ANAEROBES

ANAEROBES

Bacteria that do not need dissolved oxygen (DO) to survive.

ANOXIC (an-OX-ick)

ANOXIC

A condition in which the aquatic (water) environment does not contain dissolved oxygen (DO), which is called an oxygen deficient condition. Generally refers to an environment in which chemically bound oxygen, such as in nitrate, is present. The term is similar to ANAEROBIC.

BACTERIAL (back-TEER-e-ul) CULTURE

BACTERIAL CULTURE

In the case of activated sludge, the bacterial culture refers to the group of bacteria classified as AEROBES and FACULTATIVE BACTERIA, which covers a wide range of organisms. Most treatment processes in the United States grow facultative bacteria that use the carbonaceous (carbon compounds) BOD. Facultative bacteria can live when oxygen resources are low. When nitrification is required, the nitrifying organisms are obligate aerobes (require oxygen) and must have at least 0.5 mg/L of dissolved oxygen throughout the whole system to function properly.

BATCH PROCESS

BATCH PROCESS

A treatment process in which a tank or reactor is filled, the water (or wastewater or other solution) is treated or a chemical solution is prepared, and the tank is emptied. The tank may then be filled and the process repeated. Batch processes are also used to cleanse, stabilize, or condition chemical solutions for use in industrial manufacturing and treatment processes.

BENCH-SCALE ANALYSIS (TEST)

BENCH-SCALE ANALYSIS (TEST)

A method of studying different ways or chemical doses for treating water or wastewater and solids on a small scale in a laboratory. Also see JAR TEST.

BIOASSAY (BUY-o-AS-say)

BIOASSAY

(1) A way of showing or measuring the effect of biological treatment on a particular substance or waste.

(2) A method of determining the relative toxicity of a test sample of industrial wastes or other wastes by using live test organisms, such as fish.

BIOMASS (BUY-o-mass)

BIOMASS

A mass or clump of organic material consisting of living organisms feeding on wastes, dead organisms, and other debris. Also see ZOOGLEAL MASS and ZOOGLEAL MAT (FILM).

BIOSOLIDS

BIOSOLIDS

A primarily organic solid product produced by wastewater treatment processes that can be beneficially recycled. The word biosolids is replacing the word sludge when referring to treated waste.

BREAKPOINT CHLORINATION

BREAKPOINT CHLORINATION

Addition of chlorine to water or wastewater until the chlorine demand has been satisfied. At this point, further additions of chlorine will result in a free chlorine residual that is directly proportional to the amount of chlorine added beyond the breakpoint.

BULKING

BULKING

Clouds of billowing sludge that occur throughout secondary clarifiers and sludge thickeners when the sludge does not settle properly. In the activated sludge process, bulking is usually caused by filamentous bacteria or bound water.

CATHODIC (ka-THOD-ick) PROTECTION

CATHODIC PROTECTION

An electrical system for prevention of rust, corrosion, and pitting of metal surfaces that are in contact with water, wastewater, or soil. A low-voltage current is made to flow through a liquid (water) or a soil in contact with the metal in such a manner that the external electromotive force renders the metal structure cathodic. This concentrates corrosion on auxiliary anodic parts, which are deliberately allowed to corrode instead of letting the structure corrode.

CAUTION

CAUTION

This word warns against potential hazards or cautions against unsafe practices. Also see DANGER, NOTICE, and WARNING.

CHEMICAL OXYGEN DEMAND (COD)

CHEMICAL OXYGEN DEMAND (COD)

A measure of the oxygen-consuming capacity of organic matter present in wastewater. COD is expressed as the amount of oxygen consumed from a chemical oxidant in mg/L during a specific test. Results are not necessarily related to the biochemical oxygen demand (BOD) because the chemical oxidant may react with substances that bacteria do not stabilize.

CILIATES (SILLY-ates)

CILIATES

A class of protozoans distinguished by short hairs on all or part of their bodies.

COAGULATION (ko-agg-yoo-LAY-shun)

COAGULATION

The clumping together of very fine particles into larger particles (floc) caused by the use of chemicals (coagulants). The chemicals neutralize the electrical charges of the fine particles, allowing them to come closer and form larger clumps.

COMPOSITE (PROPORTIONAL) SAMPLE

COMPOSITE (PROPORTIONAL) SAMPLE

A composite sample is a collection of individual samples obtained at regular intervals, usually every one or two hours during a 24-hour time span. Each individual sample is combined with the others in proportion to the rate of flow when the sample was collected. Equal volume individual samples also may be collected at intervals after a specific volume of flow passes the sampling point or after equal time intervals and still be referred to as a composite sample. The resulting mixture (composite sample) forms a representative sample and is analyzed to determine the average conditions during the sampling period.

CONING

CONING

Development of a cone-shaped flow of liquid, like a whirlpool, through sludge. This can occur in a sludge hopper during sludge withdrawal when the sludge becomes too thick. Part of the sludge remains in place while liquid rather than sludge flows out of the hopper. Also called coring.

CONTACT STABILIZATION

CONTACT STABILIZATION

Contact stabilization is a modification of the conventional activated sludge process. In contact stabilization, two aeration tanks are used. One tank is for separate reaeration of the return sludge for at least four hours before it is permitted to flow into the other aeration tank to be mixed with the primary effluent requiring treatment. The process may also occur in one long tank.

CRYOGENIC (KRY-o-JEN-nick)

CRYOGENIC

Very low temperature. Associated with liquified gases (liquid oxygen).

DANGER

DANGER

The word *DANGER* is used where an immediate hazard presents a threat of death or serious injury to employees. Also see CAUTION, NOTICE, and WARNING.

DENITRIFICATION (dee-NYE-truh-fuh-KAY-shun)

DENITRIFICATION

(1) The anoxic biological reduction of nitrate nitrogen to nitrogen gas.

(2) The removal of some nitrogen from a system.

(3) An anoxic process that occurs when nitrite or nitrate ions are reduced to nitrogen gas and nitrogen bubbles are formed as a result of this process. The bubbles attach to the biological floc and float the floc to the surface of the secondary clarifiers. This condition is often the cause of rising sludge observed in secondary clarifiers or gravity thickeners. Also see NITRIFICATION.

DIFFUSED-AIR AERATION

DIFFUSED-AIR AERATION

A diffused-air activated sludge plant takes air, compresses it, and then discharges the air below the water surface of the aerator through some type of air diffusion device.

DIFFUSER

DIFFUSER

A device (porous plate, tube, bag) used to break the air stream from the blower system into fine bubbles in an aeration tank or reactor.

DISSOLVED OXYGEN

DISSOLVED OXYGEN

Molecular oxygen dissolved in water or wastewater, usually abbreviated DO.

ENDOGENOUS (en-DODGE-en-us) RESPIRATION

ENDOGENOUS RESPIRATION

A situation in which living organisms oxidize some of their own cellular mass instead of new organic matter they adsorb or absorb from their environment.

F/M RATIO

F/M RATIO

See FOOD/MICROORGANISM RATIO.

FACULTATIVE (FACK-ul-tay-tive) BACTERIA

FACULTATIVE BACTERIA

Facultative bacteria can use either dissolved oxygen or oxygen obtained from food materials such as sulfate or nitrate ions. In other words, facultative bacteria can live under aerobic, anoxic, or anaerobic conditions.

FILAMENTOUS (fill-uh-MEN-tuss) ORGANISMS

FILAMENTOUS ORGANISMS

Organisms that grow in a thread or filamentous form. Common types are *Thiothrix* and *Actinomycetes*. A common cause of sludge bulking in the activated sludge process.

FLOC

FLOC

Clumps of bacteria and particles, or coagulants and impurities, that have come together and formed a cluster. Found in flocculation tanks, sedimentation basins, aeration tanks, secondary clarifiers, and chemical precipitation processes.

FLOCCULATION (flock-yoo-LAY-shun)

FLOCCULATION

The gathering together of fine particles after coagulation to form larger particles by a process of gentle mixing. This clumping together makes it easier to separate the solids from the water by settling, skimming, draining, or filtering.

FOOD/MICROORGANISM (F/M) RATIO

FOOD/MICROORGANISM (F/M) RATIO

Food to microorganism ratio. A measure of food provided to bacteria in an aeration tank.

$$\frac{\text{Food}}{\text{Microorganisms}} = \frac{\text{BOD, lbs/day}}{\text{MLVSS, lbs}}$$

$$= \frac{\text{Flow, MGD} \times \text{BOD, mg}/L \times 8.34 \text{ lbs/gal}}{\text{Volume, MG} \times \text{MLVSS, mg}/L \times 8.34 \text{ lbs/gal}}$$

or by calculator math system

$$= \text{Flow, MGD} \times \text{BOD, mg}/L \div \text{Volume, MG} \div \text{MLVSS, mg}/L$$

or metric

$$= \frac{\text{BOD, kg/day}}{\text{MLVSS, kg}}$$

$$= \frac{\text{Flow, M}L/\text{day} \times \text{BOD, mg}/L \times 1 \text{ kg/M mg}}{\text{Volume, M}L \times \text{MLVSS, mg}/L \times 1 \text{ kg/M mg}}$$

HEADER

HEADER

A large pipe to which the ends of a series of smaller pipes are connected. Also called a MANIFOLD.

HETEROTROPHIC (HET-er-o-TROF-ick)

HETEROTROPHIC

Describes organisms that use organic matter for energy and growth. Animals, fungi, and most bacteria are heterotrophs.

INTERFACE

INTERFACE

The common boundary layer between two substances, such as water and a solid (metal); or between two fluids, such as water and a gas (air); or between a liquid (water) and another liquid (oil).

MCRT

MCRT

Mean Cell Residence Time. An expression of the average time (days) that a microorganism will spend in the activated sludge process.

$$\text{MCRT, days} = \frac{\text{Total Suspended Solids in Activated Sludge Process, lbs}}{\text{Total Suspended Solids Removed From Process, lbs/day}}$$

or

$$\text{MCRT, days} = \frac{\text{Total Suspended Solids in Activated Sludge Process, kg}}{\text{Total Suspended Solids Removed From Process, kg/day}}$$

NOTE: Operators at different plants calculate the Total Suspended Solids (TSS) in the Activated Sludge Process, lbs (kg), by three different methods:

1. TSS in the Aeration Basin or Reactor Zone, lbs (kg)

2. TSS in the Aeration Basin and Secondary Clarifier, lbs (kg)

3. TSS in the Aeration Basin and Secondary Clarifier Sludge Blanket, lbs (kg)

These three different methods make it difficult to compare MCRTs in days among different plants unless everyone uses the same method.

MLSS

MLSS

Mixed Liquor Suspended Solids. The amount (mg/L) of suspended solids in the mixed liquor of an aeration tank.

MLVSS

MLVSS

Mixed Liquor Volatile Suspended Solids. The amount (mg/L) of organic or volatile suspended solids in the mixed liquor of an aeration tank. This volatile portion is used as a measure or indication of the microorganisms present.

MANIFOLD

MANIFOLD

A large pipe to which the ends of a series of smaller pipes are connected. Also called a HEADER.

MEAN CELL RESIDENCE TIME (MCRT)

MEAN CELL RESIDENCE TIME (MCRT)

See MCRT.

MECHANICAL AERATION

MECHANICAL AERATION

The use of machinery to mix air and water so that oxygen can be absorbed into the water. Some examples are: paddle wheels, mixers, or rotating brushes to agitate the surface of an aeration tank; pumps to create fountains; and pumps to discharge water down a series of steps forming falls or cascades.

METABOLISM

METABOLISM

All of the processes or chemical changes in an organism or a single cell by which food is built up (anabolism) into living protoplasm and by which protoplasm is broken down (catabolism) into simpler compounds with the exchange of energy.

MICROORGANISMS (MY-crow-OR-gan-is-ums)

MICROORGANISMS

Living organisms that can be seen individually only with the aid of a microscope.

MIXED LIQUOR

MIXED LIQUOR

When the activated sludge in an aeration tank is mixed with primary effluent or the raw wastewater and return sludge, this mixture is then referred to as mixed liquor as long as it is in the aeration tank. Mixed liquor also may refer to the contents of mixed aerobic or anaerobic digesters.

MIXED LIQUOR SUSPENDED SOLIDS (MLSS)

MIXED LIQUOR SUSPENDED SOLIDS (MLSS)

The amount (mg/L) of suspended solids in the mixed liquor of an aeration tank.

MIXED LIQUOR VOLATILE SUSPENDED SOLIDS (MLVSS)

MIXED LIQUOR VOLATILE SUSPENDED SOLIDS (MLVSS)

The amount (mg/L) of organic or volatile suspended solids in the mixed liquor of an aeration tank. This volatile portion is used as a measure or indication of the microorganisms present.

MOVING AVERAGE

MOVING AVERAGE

To calculate the moving average for the last 7 days, add up the values for the last 7 days and divide by 7. Each day add the most recent day's value to the sum of values and subtract the oldest value. By using the 7-day moving average, each day of the week is always represented in the calculations.

NITRIFICATION (NYE-truh-fuh-KAY-shun)

NITRIFICATION

An aerobic process in which bacteria change the ammonia and organic nitrogen in water or wastewater into oxidized nitrogen (usually nitrate).

NOTICE

NOTICE

This word calls attention to information that is especially significant in understanding and operating equipment or processes safely. Also see CAUTION, DANGER, and WARNING.

OXIDATION

OXIDATION

Oxidation is the addition of oxygen, removal of hydrogen, or the removal of electrons from an element or compound; in the environment and in wastewater treatment processes, organic matter is oxidized to more stable substances. The opposite of REDUCTION.

POLYELECTROLYTE (POLY-ee-LECK-tro-lite)

POLYELECTROLYTE

A high-molecular-weight (relatively heavy) substance, having points of positive or negative electrical charges, that is formed by either natural or synthetic (manmade) processes. Natural polyelectrolytes may be of biological origin or obtained from starch products or cellulose derivatives. Synthetic polyelectrolytes consist of simple substances that have been made into complex, high-molecular-weight substances. Used with other chemical coagulants to aid in binding small suspended particles to larger chemical flocs for their removal from water. Often called a POLYMER.

POLYMER (POLY-mer)

POLYMER

A long-chain molecule formed by the union of many monomers (molecules of lower molecular weight). Polymers are used with other chemical coagulants to aid in binding small suspended particles to larger chemical flocs for their removal from water. Also see POLYELECTROLYTE.

PROTOZOA (pro-toe-ZOE-ah)

PROTOZOA

A group of motile, microscopic organisms (usually single-celled and aerobic) that sometimes cluster into colonies and generally consume bacteria as an energy source.

PURGE

PURGE

To remove a gas or vapor from a vessel, reactor, or confined space, usually by displacement or dilution.

RAS (pronounce as separate letters, or RAZZ)

RAS

Return Activated Sludge. Settled activated sludge that is collected in the secondary clarifier and returned to the aeration basin to mix with incoming raw or primary settled wastewater.

REDUCTION (re-DUCK-shun)

REDUCTION

Reduction is the addition of hydrogen, removal of oxygen, or the addition of electrons to an element or compound. Under anaerobic conditions (no dissolved oxygen present), sulfur compounds are reduced to odor-producing hydrogen sulfide (H_2S) and other compounds. In the treatment of metal finishing wastewaters, hexavalent chromium (Cr^{6+}) is reduced to the trivalent form (Cr^{3+}). The opposite of OXIDATION.

RETURN ACTIVATED SLUDGE (RAS)

RETURN ACTIVATED SLUDGE (RAS)

Settled activated sludge that is collected in the secondary clarifier and returned to the aeration basin to mix with incoming raw or primary settled wastewater.

RISING SLUDGE

RISING SLUDGE

Rising sludge occurs in the secondary clarifiers of activated sludge plants when the sludge settles to the bottom of the clarifier, is compacted, and then starts to rise to the surface, usually as a result of denitrification, or anaerobic biological activity that produces carbon dioxide and/or methane.

ROTIFERS (ROTE-uh-fers)

ROTIFERS

Microscopic animals characterized by short hairs on their front ends.

SECCHI (SECK-key) DISK

SECCHI DISK

A flat, white disk lowered into the water by a rope until it is just barely visible. At this point, the depth of the disk from the water surface is the recorded Secchi disk transparency.

SEIZING or SEIZE UP

SEIZING or SEIZE UP

Seizing occurs when an engine overheats and a part expands to the point where the engine will not run. Also called freezing.

SEPTIC (SEP-tick) or SEPTICITY

SEPTIC or SEPTICITY

A condition produced by bacteria when all oxygen supplies are depleted. If severe, the bottom deposits produce hydrogen sulfide, the deposits and water turn black, give off foul odors, and the water has a greatly increased oxygen and chlorine demand.

SHOCK LOAD

SHOCK LOAD

The arrival at a treatment process of water or wastewater containing unusually high concentrations of contaminants in sufficient quantity or strength to cause operating problems. Organic or hydraulic overloads also can cause a shock load.

(1) For activated sludge, possible problems include odors and bulking sludge, which will result in a high loss of solids from the secondary clarifiers into the plant effluent and a biological process upset that may require several days to a week to recover.

(2) For trickling filters, possible problems include odors and sloughing off of the growth or slime on the trickling filter media.

(3) For drinking water treatment, possible problems include filter blinding and product water with taste and odor, color, or turbidity problems.

SLUDGE AGE

SLUDGE AGE

A measure of the length of time a particle of suspended solids has been retained in the activated sludge process.

$$\text{Sludge Age, days} = \frac{\text{Suspended Solids Under Aeration, lbs or kg}}{\text{Suspended Solids Added, lbs/day or kg/day}}$$

SLUDGE DENSITY INDEX (SDI)

SLUDGE DENSITY INDEX (SDI)

This calculation is used in a way similar to the Sludge Volume Index (SVI) to indicate the settleability of a sludge in a secondary clarifier or effluent. The weight in grams of one milliliter of sludge after settling for 30 minutes. SDI = 100/SVI. Also see SLUDGE VOLUME INDEX.

SLUDGE VOLUME INDEX (SVI)

SLUDGE VOLUME INDEX (SVI)

A calculation that indicates the tendency of activated sludge solids (aerated solids) to thicken or to become concentrated during the sedimentation/thickening process. SVI is calculated in the following manner: (1) allow a mixed liquor sample from the aeration basin to settle for 30 minutes; (2) determine the suspended solids concentration for a sample of the same mixed liquor; (3) calculate SVI by dividing the measured (or observed) wet volume (mL/L) of the settled sludge by the dry weight concentration of MLSS in grams/L.

$$\text{SVI, mL/gm} = \frac{\text{Settled Sludge Volume/Sample Volume, mL/L}}{\text{Suspended Solids Concentration, mg/L}} \times \frac{1,000 \text{ mg}}{\text{gram}}$$

STABILIZED WASTE STABILIZED WASTE

A waste that has been treated or decomposed to the extent that, if discharged or released, its rate and state of decomposition would be such that the waste would not cause a nuisance or odors in the receiving water.

STEP-FEED AERATION STEP-FEED AERATION

Step-feed aeration is a modification of the conventional activated sludge process. In step-feed aeration, primary effluent enters the aeration tank at several points along the length of the tank, rather than at the beginning or head of the tank and flowing through the entire tank in a plug flow mode.

STRIPPED GASES STRIPPED GASES

Gases that are released from a liquid by bubbling air through the liquid or by allowing the liquid to be sprayed or tumbled over media.

SUPERNATANT (soo-per-NAY-tent) SUPERNATANT

Liquid removed from settled sludge. Supernatant commonly refers to the liquid between the sludge on the bottom and the scum on the surface.

SURFACTANT (sir-FAC-tent) SURFACTANT

Abbreviation for surface-active agent. The active agent in detergents that possesses a high cleaning ability.

TOC (pronounce as separate letters) TOC

Total Organic Carbon. TOC measures the amount of organic carbon in water.

TURBIDITY (ter-BID-it-tee) METER TURBIDITY METER

An instrument for measuring and comparing the turbidity of liquids by passing light through them and determining how much light is reflected by the particles in the liquid. The normal measuring range is 0 to 100 and is expressed as nephelometric turbidity units (NTUs).

TURBULENT MIXERS TURBULENT MIXERS

Devices that mix air bubbles and water and cause turbulence to dissolve oxygen in the water.

VISCOSITY (vis-KOSS-uh-tee) VISCOSITY

A property of water, or any other fluid, that resists efforts to change its shape or flow. Syrup is more viscous (has a higher viscosity) than water. The viscosity of water increases significantly as temperatures decrease. Motor oil is rated by how thick (viscous) it is; 20 weight oil is considered relatively thin while 50 weight oil is relatively thick or viscous.

VOLUTE (vol-LOOT) VOLUTE

The spiral-shaped casing that surrounds a pump, blower, or turbine impeller and collects the liquid or gas discharged by the impeller.

WAS WAS

See Waste Activated Sludge.

WARNING WARNING

The word *WARNING* is used to indicate a hazard level between *CAUTION* and *DANGER*. Also see CAUTION, DANGER, and NOTICE.

WASTE ACTIVATED SLUDGE (WAS) WASTE ACTIVATED SLUDGE (WAS)

The excess quantity (mg/L) of microorganisms that must be removed from the process to keep the biological system in balance.

ZOOGLEAL (ZOE-uh-glee-ul) MASS ZOOGLEAL MASS

Jelly-like masses of bacteria found in both the trickling filter and activated sludge processes. These masses may be formed for or function as the protection against predators and for storage of food supplies. Also see BIOMASS.

CHAPTER 2.   ACTIVATED SLUDGE

Pure Oxygen Plants and Operational Control Options

(Lesson 1 of 4 Lessons)

NOTE: Review Volume I, Chapter 8, and Volume II, Chapter 11, of *OPERATION OF WASTEWATER TREATMENT PLANTS* for additional basic information on the activated sludge process.

2.0   THE ACTIVATED SLUDGE PROCESS

Research and operational experience are gradually revealing how the activated sludge process treats wastes and how to control the process. One objective of this chapter is to provide operators with a better understanding of the factors that can upset an activated sludge process and how to control the process to produce a high-quality effluent. Pure oxygen systems dissolve oxygen into wastewater with a high efficiency for use by microorganisms treating the wastes. This allows the use of smaller aeration (reactor) tanks than air activated sludge systems. Operators need greater skill and knowledge to operate pure oxygen systems than conventional systems.

Operation of either pure oxygen or conventional aeration activated sludge processes is very complex. The quality of your plant's effluent depends on the characteristics of the plant's influent flows and wastes, as well as how the actual process is controlled. Two very important factors are:

1. RETURN ACTIVATED SLUDGE (RAS) RATE

2. WASTE ACTIVATED SLUDGE (WAS) RATE

Several methods have been developed to help operators select the proper rates. This chapter reviews some of these methods and their advantages and limitations. You must remember that each of these factors affects the others and the impact on all process variables must be considered before changing one variable.

Some NPDES permits require the removal of ammonia from plant effluents. Biological nitrification (converting ammonium (NH_4^+) to nitrate (NO_3^-)) is the most effective way to remove ammonia unless total nitrogen removal is necessary. If total nitrogen removal is required, biological nitrification is the first step of the biological nitrification-denitrification approach to nitrogen removal. The biological nitrification process is an extension of the activated sludge process and is operated on the basis of the same concepts.

Industrial wastes are becoming more common in many municipal wastewaters. Whether you operate an activated sludge plant in a small town or a large city, you must know how to treat the industrial wastes that may be present with your municipal wastewaters. Many industries pretreat their own wastewaters before discharge to municipal collection systems while other industries treat all of their wastewaters rather than discharge into municipal collection systems. Whether you are treating strictly industrial wastewaters or a mixture of industrial and municipal wastewaters, this chapter will provide you with the information you need to know to safely and effectively treat these different types of wastewaters using the activated sludge process.

QUESTIONS

Write your answers in a notebook and then compare your answers with those on page 146.

2.0A   Why are pure oxygen systems used instead of conventional aeration methods?

2.0B   What treatment process can be used to remove ammonium (NH_4^+) from wastewater, but not total nitrogen?

2.1   PURE OXYGEN

2.10   Description of Pure Oxygen Systems

The pure oxygen system (Figure 2.1) is a modification of the activated sludge process (Figure 2.2). The main difference is the method of supplying dissolved oxygen to the activated sludge process. In other activated sludge processes, air is compressed and released under water to produce an air-water *INTERFACE*[1] that transfers oxygen into the water (dissolved oxygen). If compressed air is not used, surface aerators agitate the water surface to drive air into the water (dissolved oxygen) to obtain the oxygen transfer. In the pure oxygen system, the only real differences are that pure oxygen rather than air is released below the surface or driven into the water by means of surface aerators and the aerators are covered.

In the pure oxygen system, oxygen is first separated from the air to produce relatively high-purity oxygen (90 to 98 percent oxygen). Pure oxygen is applied to the wastewater as a source of

---

[1] *Interface.* The common boundary layer between two substances, such as water and a solid (metal); or between two fluids, such as water and a gas (air); or between a liquid (water) and another liquid (oil).

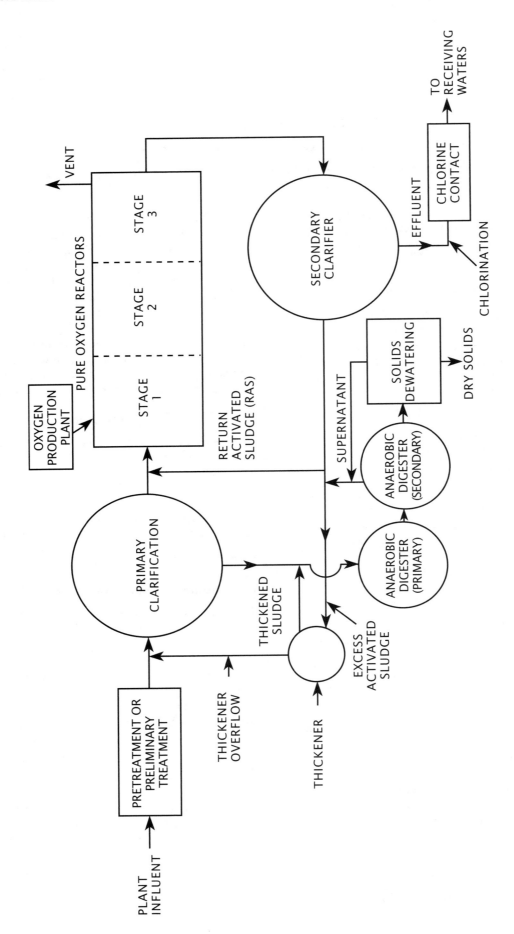

Fig. 2.1   Plan layout of a typical pure oxygen activated sludge plant

TREATMENT PROCESS

FUNCTION

PRELIMINARY TREATMENT

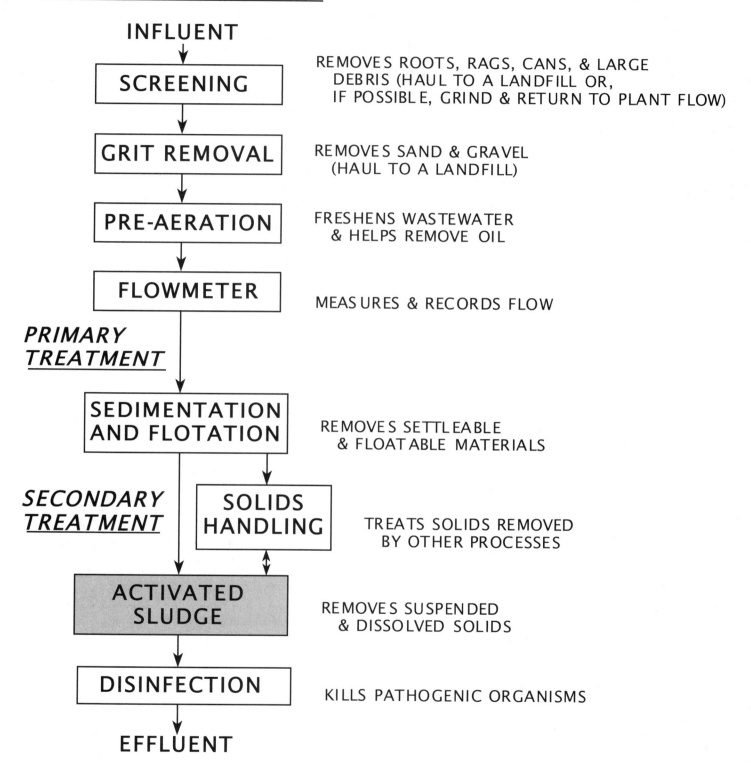

INFLUENT

SCREENING — REMOVES ROOTS, RAGS, CANS, & LARGE DEBRIS (HAUL TO A LANDFILL OR, IF POSSIBLE, GRIND & RETURN TO PLANT FLOW)

GRIT REMOVAL — REMOVES SAND & GRAVEL (HAUL TO A LANDFILL)

PRE-AERATION — FRESHENS WASTEWATER & HELPS REMOVE OIL

FLOWMETER — MEASURES & RECORDS FLOW

PRIMARY TREATMENT

SEDIMENTATION AND FLOTATION — REMOVES SETTLEABLE & FLOATABLE MATERIALS

SECONDARY TREATMENT

SOLIDS HANDLING — TREATS SOLIDS REMOVED BY OTHER PROCESSES

ACTIVATED SLUDGE — REMOVES SUSPENDED & DISSOLVED SOLIDS

DISINFECTION — KILLS PATHOGENIC ORGANISMS

EFFLUENT

Fig. 2.2   Flow diagram of a typical plant

oxygen for the microorganisms treating the wastes. As with forced-air activated sludge systems, the pure oxygen must be driven into the water. This is accomplished by a diffuser mechanism or by mechanical agitation consisting of *TURBULENT MIXERS*[2] and surface aerators. The agitators also supply the energy to mix the reactor (aeration tank) contents to distribute the waste food (measured as BOD or COD) to the activated sludge microorganisms in the mixed liquor suspended solids (MLSS) and to prevent buildup of MLSS deposits in the reactor.

The pure oxygen reactors are staged (divided into two to five sections by baffles as shown in the three-stage system in Figure 2.3) and are completely covered to provide a gas-tight enclosure. The wastewater, return sludge, and oxygen are fed into the first stage. The mixed liquor and atmosphere above it flow in the same direction from the first stage to the last. Staging the pure oxygen reactors increases the efficiency of oxygen use by capturing the head space gases from prior stages and recycling them to subsequent stages.

When pure oxygen is driven into the water, it behaves like air and the bubbles rise to the water surface. During this rise to the surface, only a small portion of the oxygen is absorbed into the mixed liquor. Covering the reactor and sealing it from the outside atmosphere allows the oxygen that is not dissolved into the water (mixed liquor) to be used again. This contained gas over the water is still relatively high in oxygen concentration, the main contamination consisting of carbon dioxide gas given off by the respiration (breathing) of the activated sludge microorganisms and the nitrogen stripped from the incoming wastewater. The number of stages and the methods of contacting the liquid and oxygen vary from one system to another. In one type of design, the oxygen-rich gas that accumulates in the space between the mixed liquor water surface and the roof of the reactor is removed, compressed, and recycled back to the diffuser of the reactor. In another type of design, surface aerators are used to drive the oxygen-rich gas into the water. In deep reactors (40 feet or 12 m), the surface aerator device may also be equipped with an extended shaft and submerged impeller to keep the tank contents well mixed. A third type of design incorporates both submerged diffusers and surface aeration. Some reactors vent the excess oxygen along with the carbon dioxide. Uncovered reactors skim the surface sludge for wasting.

In all types of design, a valve opens automatically and admits more pure oxygen to the first-stage reactor whenever sufficient oxygen is removed from the gas space of the first-stage reactor to drop the pressure below the required 1 to 4 inches (2 to 10 cm) of water column pressure. This constantly replenishes the oxygen supply and the pressure is sufficient to force gas movement through the succeeding stages. This pressure prevents air from leaking into the reactors, diluting the oxygen concentration, and possibly creating an explosive mixture. Oxygen leaking from a reactor can create an explosive condition on the roof or around the reactor. Potentially explosive conditions inside the reactor from a mixture of hydrocarbons and oxygen are avoided by an automatically activated analysis and purge system.

In each of the succeeding stages, the gas above the mixed liquid in that stage is reinjected into the mixed liquor of the same stage (by compressor or surface aerator). As the oxygen-rich gas passes from one stage to the next, the oxygen is used by the activated sludge microorganisms and the atmosphere becomes more and more diluted by the carbon dioxide produced by the organisms and nitrogen *STRIPPED*[3] from solution. The last stage in the reactor is equipped with a roof vent controlled by a valve mechanism that is called an oxygen vent valve. This valve vents gas from the last stage to the atmosphere and is normally set to vent gas when the oxygen concentration drops below 50 percent. As gas is vented from the last stage, more pure oxygen is released into the first stage to maintain the desired 1 to 4 inches (2 to 10 cm) of water column pressure.

Two methods are commonly used to produce pure oxygen. One is the Pressure Swing Adsorption (PSA) Oxygen Generating System and the other is the *CRYOGENIC*[4] Air Separation Method.

QUESTIONS

Write your answers in a notebook and then compare your answers with those on page 146.

2.1A Why are the pure oxygen reactors staged?

2.1B How is the pure oxygen diluted as it passes from one stage to the next stage?

2.1C What two methods are commonly used to produce pure oxygen?

2.11 PSA (Pressure Swing Adsorption) Oxygen Generating System (Figure 2.4)

The PSA Oxygen Generating Systems are usually installed in smaller plants. They take air from the atmosphere and compress it to 30 to 60 psi (2 to 4 kg/sq cm) and cool the compressed air in a water-cooled heat exchanger called an after cooler. The after cooler condenses and removes the moisture from the air stream. Next the air passes through an adsorbent vessel filled with a molecular sieve. Under pressure, this molecular sieve has the ability to adsorb nitrogen and other impurities from the atmospheric air, thus allowing the remaining pure oxygen to be used in the reactor. While one adsorber vessel is separating air into oxygen and nitrogen, the other two vessels are in various stages of desorption (or cleanup). The cleanup cycle consists of depressurizing and *PURGING*[5] with some product oxygen. The last step involves pressurizing with compressed air before going back on stream. During this process, product oxygen is flowing continuously to the activated sludge plant.

---

[2] *Turbulent Mixers.*   Devices that mix air bubbles and water and cause turbulence to dissolve oxygen in the water.

[3] *Stripped Gases.*   Gases that are released from a liquid by bubbling air through the liquid or by allowing the liquid to be sprayed or tumbled over media.

[4] *Cryogenic* (KRY-o-JEN-nick).   Very low temperature. Associated with liquified gases (liquid oxygen).

[5] *Purge.*   To remove a gas or vapor from a vessel, reactor, or confined space, usually by displacement or dilution.

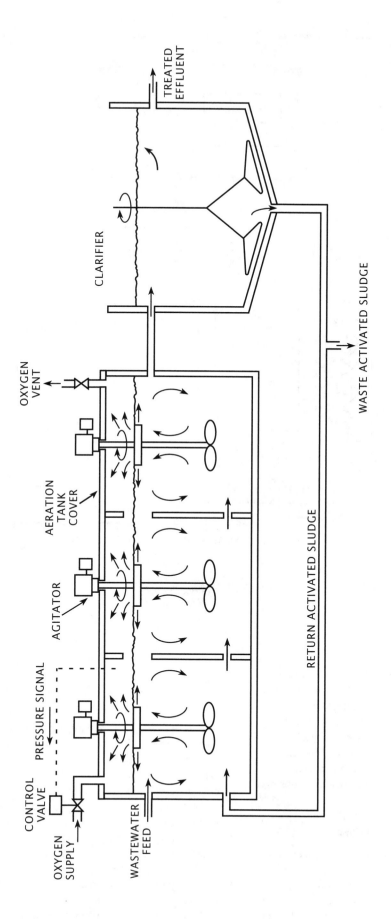

Fig. 2.3 Schematic diagram of pure oxygen system with surface aerators (3 stages shown)

(Permission of Union Carbide Corporation)

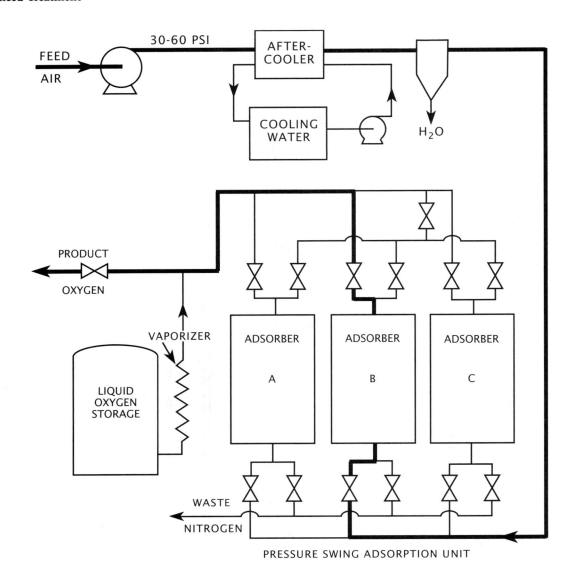

Fig. 2.4 Flow diagram of a three-bed PSA (Pressure Swing Adsorption) oxygen generating system
(Permission of Union Carbide Corporation)

The PSA unit can be turned down to 25 percent of its rated oxygen throughout without a significant loss of efficiency. Compressor and valve maintenance can be scheduled so as not to have more than one or two days of downtime per year by the use of multiple compressors. A backup tank of liquid oxygen provides oxygen (after evaporation) to handle peak loads or downtime oxygen demand. The switching valves are selected on the basis of their ability to withstand very severe conditions over long periods of time.

2.12 Cryogenic Air Separation Method (Figure 2.5)

Oxygen produced by the cryogenic air separation method uses low temperatures (cryogenic) ($-297°F$, $-183°C$) to condense oxygen into a liquid to separate it from air. Air is filtered, compressed, cooled to remove moisture, and then routed to the cold box or "cryo" plant tower. These towers are heavily insulated to

conserve energy by minimizing heat leaks or losses. In Figure 2.5 all the items contained in the dashlined box are located in the "cryo" plant tower.

The reversing heat exchanger primarily removes carbon dioxide and water. This heat exchanger has two directional gas flows; one of air going into the tower and the other of nitrogen being exhausted to the atmosphere. As the flowing air removes carbon dioxide and water, ice will form in the heat exchanger and restrict the air flow through the heat exchanger. After several minutes of operation, a valve is activated that interchanges the gas stream flows by reversing the direction of flow. The nitrogen exhaust is routed through the inlet air passage and the partially blocked passages. A small portion of the water and carbon dioxide leaves as tiny ice particles. The inlet air then travels through the previous nitrogen exhaust side until once again the ice builds up in the passage. Again the valve is activated and switches the

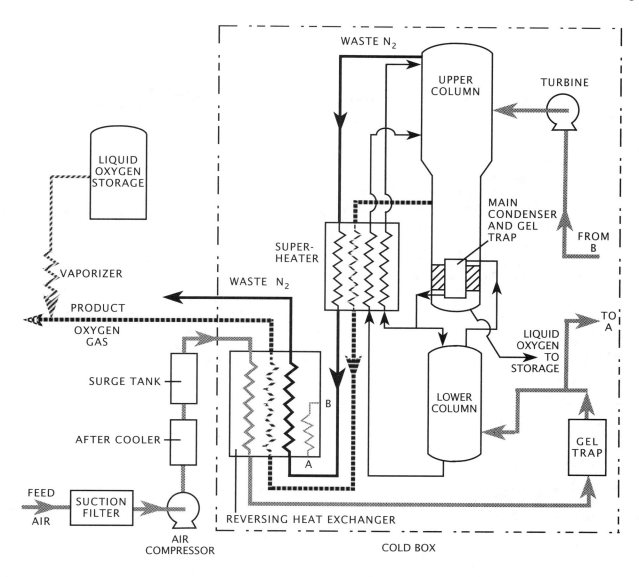

Fig. 2.5   Flow diagram of a cryogenic oxygen generating system
(Permission of Union Carbide Corporation)

routes of the two gas flows. This cycle usually varies from 5 to 20 minutes depending on the system. The exiting pure oxygen is also heat exchanged against the incoming air. In this case, however, the passages are never switched. This allows the oxygen to remain at high purity (about 98 percent pure oxygen).

A silica gel trap adsorbs any remaining moisture that may have gotten past the reversing heat exchanger. Trace hydrocarbons are also picked up by the gel trap. The clean, cold air is liquified and separated into oxygen and nitrogen by fractional distillation in a two-column arrangement. The lower high-pressure column produces pure liquid nitrogen to use as reflux (flow back) in the low-pressure upper column. Nitrogen, the most volatile component of air, is taken from the top of the upper column. Pure oxygen, the less volatile component, is taken from near the bottom of the

upper column. A 98 percent purity oxygen stream is sent to the activated sludge process after heat exchange against the incoming air to recover its refrigeration (cool incoming air). Refrigeration to run the process comes from expanding a portion of the cooled and cleaned incoming air through an expansion turbine before it enters the upper column.

Approximately three percent of the capacity of the oxygen plant is available as liquid oxygen. This liquid can be used to keep the stored liquid oxygen backup tank full and ready to supply oxygen during peak loads or plant start-up.

To start an *AMBIENT TEMPERATURE*[6] cryogenic plant producing oxygen without liquid oxygen requires about three to five days. If liquid oxygen is available, a few hours is all that is

---

[6] *Ambient* (AM-bee-ent) *Temperature.*   Temperature of the surroundings.

required to place the plant in production. The oxygen production rate of a plant is determined by the oxygen demand in the activated sludge plant. As less production is required, the oxygen plant air compressors are partially unloaded.

Cryogenic plants are usually shut down once a year for approximately five to seven days for maintenance. During this period the gel traps are warmed to drive off moisture and hydrocarbons. By the use of multiple compressors and operational maintenance thaws, this downtime can be reduced to two or three days per year. During downtime, oxygen vaporized from the backup liquid oxygen storage tank is used. Sometimes in larger plants more than one oxygen-producing facility is supplied to minimize the use of backup liquid oxygen.

QUESTIONS

Write your answers in a notebook and then compare your answers with those on page 146.

2.1D   What does cryogenic mean?

2.1E   How often and for how long are cryogenic plants shut down for maintenance?

2.13   Process and System Control

Pure oxygen systems may be used to supply oxygen to any of the activated sludge process modes—conventional, step-feed, complete mix, or contact stabilization.

2.14   System Start-Up

Pure oxygen system start-up is basically much the same as starting conventional air systems. Individual components and starting procedures are usually outlined in the O & M manual or the manufacturer's literature. Take special care with the reactor because flow and organic loadings must be determined prior to start-up. Overloading or underloading may cause problems. Careful review of design data usually provides sufficient information to initially start the system. After start-up, the system is "fine-tuned" to prevailing conditions in the wastewater.

2.15   Control Guidelines

1. *REACTOR VENT GAS*—A mixture of unused oxygen (about 5 to 10 percent of the oxygen supplied), inert gases, and carbon dioxide is continually discharged from the last stage of the reactor. The vent purity, or percentage of oxygen contained in the mixture of gases, is an indicator of oxygen use efficiency. A low oxygen purity reading (25 percent or below) indicates that sufficient oxygen is not present and adequate BOD removal may not be accomplished. A high purity reading (50 percent or higher) indicates that too much oxygen is being wasted with the by-product gases. A manually controlled vent valve is adjusted to control vent purity. If purity is low, the valve could be opened further and closed down if purity is high. In normal operation (after start-up), "fine-tuning" of this setting usually is not changed unless there is a drastic change in either the quantity or strength of the wastewater entering the plant.

2. *REACTOR GAS SPACE PRESSURE*—Gas space pressure is set by controlling the vent rate in the last stage. This will automatically establish the pressure level throughout the reactor. Gas pressure will vary to some extent within the reactor, dropping as more oxygen is vented and rising as venting is decreased or consumption is reduced. Pressure is usually preset at two inches (5 cm) water column in the first stage and the system will automatically feed oxygen at the required rates to maintain this condition. However, during high loading periods, the operator can increase oxygen transfer and production by increasing the pressure set point from 2 to 4 inches (5 to 10 cm) of water column, providing the vent valve setting is not changed. Relief valves on the first and last stage of each reactor prevent overpressurization. Similarly, during periods of unloading, a vacuum release provided by these same valves prevents a negative pressure.

3. *DISSOLVED OXYGEN*—A dissolved oxygen probe is sometimes located in the diversion box prior to the secondary clarifier or in the last stage of the reactor. It indicates the amount of DO in the mixed liquor. Typical oxygen systems usually operate with a DO range of 4 to 10 mg/L of dissolved oxygen. If the organic load increases over an extended period, which would tend to drop the dissolved oxygen level below 4 mg/L, the operator should adjust the vent valve to a more open position, which will increase oxygen production. Above a DO of 10 mg/L, the amount of oxygen being produced could be decreased if this is anticipated to be a long-term condition.

By measuring and monitoring these control guidelines, the operator can establish the most efficient treatment method on the basis of plant performance and experience. Operation of the secondary clarifiers, return rates, wasting rates, and other control guidelines, are much the same for the pure oxygen system as they are for the conventional air activated sludge system.

2.16   Process Safety

Potentially explosive or flammable conditions can be present when pure oxygen gas is mixed with any hydrocarbon such as gasoline, fuel oil, and lubricating oils. In addition to normal safety devices found on motors, compressors, electrical components, and control mechanisms, the pure oxygen system uses safety devices to ensure process safety when working with oxygen gas. These safety devices are:

1. *LOWER EXPLOSIVE LIMIT (LEL) COMBUSTIBLE GAS DETECTOR*—Indicates potential explosive conditions within the reactor, and analyzes samples collected from the first and last stage of each train in the reactor. Readings are made based on all components being analyzed as propane. If a hydrocarbon spill gets through the primary treatment system without being diverted and causes a reading of more than 25 percent of the LEL, an alarm will sound. The product valve from the oxygen system will shut down and air will automatically be directed to the reactor gas space to purge the system. The purge will continue until normal readings are obtained. If the spill is so large that the LEL continues to rise to the 50 percent level, in addition to an alarm sounding,

an electrical restart of the purge blower will occur. The mixers will shut down to stop hydrocarbon stripping and they cannot be restarted until combustible gas readings are below 10 percent LEL.

No electrical work is ever installed below the roof nor are there any metal-to-metal contact potentials present. The mixers pass through the roof through a water seal. Operation during freezing weather requires that the seal be filled with a mixture of ethylene glycol and water. This eliminates the potential for sparks and sources of ignition. By eliminating sources of ignition and any chance of ignitable mixtures, the chances of an explosion become virtually zero. To date the safety record at activated sludge plants using pure oxygen has been excellent. Also, having a deck (roof) over the reactor provides a safe and easily accessible place for maintenance work and further minimizes the chances of having accidents. People should not be allowed on reactor covers except for essential work.

One way to help prevent explosive conditions from developing is to install a lower explosive limit (LEL) combustible gas detector in the plant headworks. This detector will trigger an alarm whenever hydrocarbons reach the headworks so action can be taken to prevent hydrocarbons from reaching reactors containing pure oxygen. Wastewater containing hydrocarbons can be diverted to emergency holding ponds if available.

2. *LIQUID OXYGEN (STORAGE TANK) LOW-TEMPERATURE ALARM*—Provides an alarm and shutdown of the liquid storage system in the event heated water recirculation within the vaporizer reaches a low temperature level. A temperature monitor measures temperature levels of the oxygen gas and if the vapor falls below −10 degrees Fahrenheit (−23°C), an alarm will sound on the instrument panel. An indication of −20 degrees Fahrenheit (−29°C) triggers an alarm and the liquid system will shut down until the temperature returns to normal conditions, but must be manually reset.

3. *EMERGENCY TRIP SWITCH*—In the event that any other unsafe condition should arise within the compressor system, liquid oxygen system, or electrical panels, an emergency trip switch may be manually pulled. When pulled, the entire oxygen system shuts down and must be reset manually and each major piece of equipment restarted. This safety switch is not commonly used. It is only used as a last resort if safety systems fail or a major problem exists within the system that threatens the well-being and the safety of personnel. This switch is usually located away from a source of danger and at an obvious and easily accessible location. As with any treatment plant, the operator must follow safety precautions established by the manufacturers and design engineers.

CAUTION[7] and *WARNING*[8] signs should be posted in areas where danger is present.

2.17 Operator Safety

Special safety rules must be applied when operating and maintaining pure oxygen systems because of the unique properties of high-purity oxygen. Cold liquid oxygen (LOX) can cause skin burns. Safety goggles or a face shield should be worn if liquid oxygen releases or splashing may occur or if cold gas may be released from equipment. Clean, insulated gloves that can be easily removed and long shirt sleeves for arm protection are recommended. Cuffless trousers should be worn outside boots or overshoes to shed spilled liquid oxygen. Sampling is the only time operators need to handle liquid oxygen.

Pure oxygen supports and accelerates combustion more readily than air. Therefore, all types of hydrocarbons and other flammable materials must be kept from mixing with oxygen. Many materials that are considered absolutely safe under normal conditions may become highly dangerous when saturated with liquid oxygen. Under certain conditions mixtures of powdered organic material and liquid oxygen can explode. The following precautions are intended to eliminate the possibility of combustion and explosions.

1. Special non-hydrocarbon lubricants as specified by the manufacturer should be used for equipment in contact with oxygen.

2. Tools and equipment must be specially designed to be compatible for use in oxygen service.

3. Flammable materials must be kept far away from oxygen systems and storage tanks.

4. Grease and oil must be removed from tools and equipment by the use of an approved solvent for the material to be cleaned.

5. Smoking and open flames are prohibited near oxygen systems and storage tanks.

6. Use only the gasket and pipe fitting compounds recommended specifically for oxygen plant equipment.

7. When installing field fabricated piping, be sure a suitable safety valve is installed in each section of piping between the shutoff valves.

Liquid oxygen is delivered by specially designed trucks and transferred by specially trained technicians. Therefore, the chances of liquid oxygen spills are remote. If a liquid oxygen spill occurs, the liquid could saturate a combustible material and this material could ignite or explode. Ignition can be caused by hot objects, open flames, glowing cigarettes, embers, sparks, or impact such as might be caused by striking with a hammer, dropping a tool, or scuffing with your heel. Typical combustible

---

[7] *Caution.* This word warns against potential hazards or cautions against unsafe practices. Also see DANGER, NOTICE, and WARNING.
[8] *Warning.* The word *WARNING* is used to indicate a hazard level between *CAUTION* and *DANGER*. Also see CAUTION, DANGER, and NOTICE.

materials that are dangerous when saturated with spilled liquid oxygen include asphalt in blacktop pavements, humus in soil, oil or grease on concrete floors or pavements, articles of clothing, or any other substance that will burn in air. Any equipment involving liquid oxygen should be constructed on a concrete pad to avoid the potential of soaking a blacktop surface with liquid oxygen.

Every possible effort must be made to prevent the spillage of liquid oxygen. If a spill does occur, the following procedures must be followed:

1. No one may set foot in any area still showing frost marks from an oxygen spill.

2. The affected area must be roped off as soon as possible. When rope, barricades, and signs are not immediately available, someone must stay at the area to warn other persons of the hazard.

3. No tank car or truck movements are allowed over an area still showing frost marks from an oxygen spill.

These procedures apply to any oxygen spillage on any surface, including cement, gravel, blacktop, or dirt either inside buildings or outdoors. *NO ONE IS ALLOWED TO STEP ON ANY AREAS WHERE FROST MARKS EXIST FROM A SPILL.*

When a considerable amount of liquid oxygen is exposed to the atmosphere, a fog will develop because of water vapor being condensed by the cold oxygen vapor. If this occurs, all potential ignition sources in the vicinity should be removed from service. This includes electric motors, engines, electric switch gear (if not rated for an explosive atmosphere), and any equipment or operation that may create a spark. Barricade the area and prevent access by personnel and vehicles.

If liquid oxygen is spilled and falls into low-lying areas such as pits, pipe trenches, and sewers, the resulting cold oxygen gas, being heavier than air, will tend to stay in the low area. This situation will require forced ventilation of the area to disperse the concentrated oxygen gas. (*NOTE:* When liquid oxygen at atmospheric pressure warms to room temperature, the liquid vaporizes and expands 862 times in volume.)

There is no special procedure for cleaning up a spill of liquid oxygen. A gaseous spill will dissipate into the atmosphere very quickly. A liquid spill will produce a cloud or steaming appearance while vaporizing. The spill will take longer to vaporize and dissipate in low areas. Barricade the area and stay out of the area. An oxygen tester can be used to measure oxygen levels in the atmosphere. Do not enter an area with an oxygen level greater than 23 percent. Ventilation can help clear a liquid oxygen spill area. Wait until it is safe to enter the area.

A portable atmospheric monitor with a 0 to 25 percent oxygen range should be used to check the atmosphere in any area where it is necessary to make repairs. Any concentration of oxygen, either above or below that of air (20.9 percent oxygen), should be regarded as dangerous and requires further investigation before work proceeds.

2.18   Pure Oxygen System Maintenance

Maintenance of a pure oxygen production system is specialized for the specific equipment. However, this equipment is similar to the equipment found in other types of activated sludge plants, including air compressors, valves, and instruments. Manufacturers commonly aid the operator during start-up with training sessions and field work. A maintenance contract with the supplier can be used to provide the technical services needed. As with any large-scale production system, equipment preventive maintenance ensures proper operation and greater efficiency.

The primary difference between maintenance of equipment in a pure oxygen activated sludge plant and other types of activated sludge plants lies in the thorough cleaning requirements for piping and equipment handling pure oxygen. Five basic steps are necessary to ensure that equipment is *"cleaned for oxygen service."*

1. Selection of a suitable cleaning agent (depends on the material to be cleaned).

2. Removal of contaminants by using an effective, approved cleaning method(s).

3. Removal of all cleaning agent residuals and thorough drying of equipment.

4. Inspection of equipment for cleanliness by approved method(s).

5. Protection of clean parts and equipment from contamination at all times (during storage, handling, and installation).

QUESTIONS

Write your answers in a notebook and then compare your answers with those on page 146.

2.1F   What special measurements are used to control pure oxygen systems?

2.1G   How can hydrocarbons be detected before they reach the reactor?

2.19   Acknowledgment

This section was reviewed by R. W. Hirsch. The authors thank Mr. Hirsch for his many helpful comments and suggestions.

Please answer the discussion and review questions next.

DISCUSSION AND REVIEW QUESTIONS
Chapter 2.   ACTIVATED SLUDGE
Pure Oxygen Plants and Operational Control Options
(Lesson 1 of 4 Lessons)

At the end of each lesson in this chapter you will find some discussion and review questions. The purpose of these questions is to indicate to you how well you understand the material in the lesson. Write the answers to these questions in your notebook.

1. Why does the pure oxygen process normally use sealed reactors?

2. How is pure oxygen separated from impurities and other gases in the PSA system?

3. What safety hazards might an operator encounter when working around a pure oxygen system?

4. What safety systems are found around pure oxygen systems to protect operators and equipment?

CHAPTER 2.   ACTIVATED SLUDGE

Pure Oxygen Plants and Operational Control Options

(Lesson 2 of 4 Lessons)

NOTE: The next two sections, Section 2.2, "Return Activated Sludge," and Section 2.3, "Waste Activated Sludge," are provided to familiarize you with different ways to control both the pure oxygen and air activated sludge processes. You, the operator, will have to determine which ways will work best for your plant. Once a particular procedure is selected, every operator on every shift must follow the same procedure. If the procedure does not produce satisfactory results, then new procedures must be developed and tested for everyone to follow.

Abbreviations used in this section include:

1. MLSS, Mixed Liquor Suspended Solids, mg/*L*

2. MLVSS, Mixed Liquor Volatile Suspended Solids, mg/*L*

3. RAS, Return Activated Sludge, mg/*L* (may also be % flow)

4. F/M, Food to Microorganism Ratio, lbs BOD or COD added per day per lb of MLVSS or kg/day per kg MLVSS

5. Q, Flow, MGD or m³/sec

2.2   RETURN ACTIVATED SLUDGE

2.20   Purpose of Returning Activated Sludge

To operate the activated sludge process efficiently, a properly settling mixed liquor must be achieved and maintained. The mixed liquor suspended solids (MLSS) are settled in a clarifier and then returned to the aeration tank as the Return Activated Sludge (RAS) (Figure 2.6). The RAS makes it possible for the microorganisms to be in the treatment system longer than the flowing wastewater. For conventional activated sludge operations, the RAS flow is generally about 20 to 40 percent of the incoming wastewater flow. Changes in the activated sludge quality will require different RAS flow rates due to settling characteristics of the sludge. Table 2.1 shows typical ranges of RAS flow rates for some activated sludge process variations.

2.21   Return Activated Sludge Control

Two basic approaches that can be used to control the RAS flow rate are based on the following:

1. Controlling the RAS flow rate independently from the influent flow

2. Controlling the RAS flow rate as a constant percentage of the influent flow

TABLE 2.1   A GUIDE TO TYPICAL RAS FLOW RATE PERCENTAGES[a]

Type of Activated Sludge Process	RAS Flow Rate as Percent of Incoming Design Average Wastewater Flow to Aeration Tank	
	Minimum	Maximum
Conventional or Standard Rate	15	100
Carbonaceous Stage of Separate Stage Nitrification	15	100
Step-Feed Aeration	15	100
Complete Mix	15	100
Contact Stabilization	50	150
Extended Aeration	50	150
Nitrification Stage of Separate Stage Nitrification	50	200

[a] *RECOMMENDED STANDARDS FOR WASTEWATER FACILITIES (10 STATE STANDARDS),* Great Lakes-Upper Mississippi River Board of State Sanitary Engineers, 2004 Edition. Available from Health Education Services, PO Box 7126, Albany, NY 12224. Price, $17.00 (includes shipping).

2.210   Constant RAS Flow Rate Control

Setting the RAS at a constant flow rate that is independent of the aeration tank influent wastewater flow rate results in a continuously changing MLSS concentration. The MLSS will be at a minimum during peak influent flows and at a maximum during low influent flows. This occurs because the MLSS are flowing into the clarifier at a higher rate during peak flow when they are being removed at a constant rate. Similarly, at minimum influent flow rates, the MLSS are returned to the aeration tank at a higher rate than they are flowing into the clarifier. The aeration tank and the secondary clarifier must be looked at as a system in which the MLSS are stored in the aeration tank during minimum wastewater flow and then transferred to the clarifier as the wastewater flows initially increase. In essence, the clarifier has a constantly changing depth of sludge blanket as the MLSS move from the aeration tank to the clarifier and vice versa. The advantage of using this approach is simplicity because it minimizes the amount of effort for control. This approach is especially effective for small plants with limited flexibility.

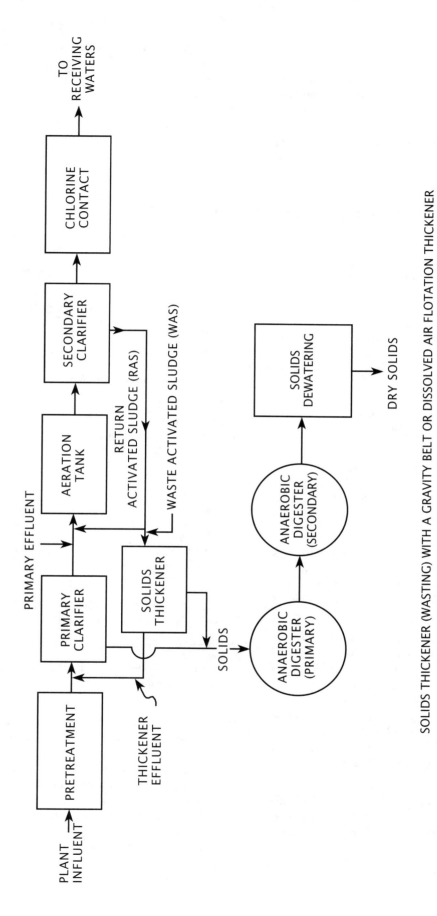

SOLIDS THICKENER (WASTING) WITH A GRAVITY BELT OR DISSOLVED AIR FLOTATION THICKENER

Fig. 2.6   Return and waste activated sludge flow diagram

2.211   Constant Percentage RAS Flow Rate Control

The second approach to controlling RAS flow rates to the aeration tank requires a programmed method for maintaining a RAS flow rate that is a constant percentage of the influent wastewater flow rate. The program may consist of an automatic flow measurement device, a programmed system, or frequent manual adjustments. The programmed method is theoretically designed to keep the MLSS more constant through high- and low-flow periods.

2.212   Comparison of RAS Control Approaches

The advantages of the constant RAS flow approach are the following:

1. Simplicity.

2. Maximum solids loading on the clarifier occurs at the start of peak-flow periods.

3. Requires less operational time.

The advantages of the constant percentage RAS flow approach are the following:

1. Variations in the MLSS concentration are reduced and the F/M ratio varies less.

2. The MLSS will remain in the clarifier for shorter time periods, which may reduce the possibility of denitrification in the clarifier.

A limitation of using the constant flow approach is that the F/M ratio is constantly changing. The range of F/M fluctuation due to short-term variations in the MLSS (because of hydraulic loading) is generally small enough so that no significant problems arise.

The most significant limitation of the constant percentage flow approach is that the clarifier is subjected to maximum hydraulic loading when the reactor contains the maximum amount of sludge. This may result in solids washout with the secondary effluent.

In general, it appears that most activated sludge operations perform well and require less attention when the constant RAS flow rate approach is used. In many plants, it is much simpler for the operator to let the MLSS fluctuate, as long as adequate treatment can be maintained. Larger, more complex plants may have to vary the RAS to keep the MLSS close to the target value. Activated sludge plants with flows of 10 MGD (37,850 m³/day) or less often experience large hydraulic surges and performance of these plants will benefit the most from the use of the constant RAS flow rate approach.

Procedures for monitoring and maintaining RAS flow rates are presented in Table 2.2. This table can be used to develop detailed standard operating procedures for your own treatment plant.

QUESTIONS

Write your answers in a notebook and then compare your answers with those on page 146.

2.2A   What words do the letters in the following abbreviations represent?

　　1. MLSS　　　3. RAS

　　2. MLVSS　　4. F/M

2.2B   What are the two basic approaches that can be used to control the RAS flow rate?

2.22   Methods of RAS Flow Rate Control

For either RAS flow rate control approach discussed above, there are a number of techniques that may be used to set the rate of sludge return flow. The most commonly used techniques are listed below:

1. Monitoring the depth of the sludge blanket

2. Settleability approach

3. SVI (Sludge Volume Index) approach

2.220   Sludge Blanket Depth

Monitoring the depth of the sludge blanket in the clarifier is a direct method for determining the RAS flow rate. The sludge blanket depth and uniformity may be checked by any of the following methods:

1. A series of air lift pumps mounted within the clarifier at various depths.

2. Gravity flow tubes located at various depths.

3. Electronic sludge level detector (a light source and photoelectric cell attached to a graduated handle or drop cord. The photoelectric cell actuates a buzzer when in contact with the sludge).

4. Sight glass finder (a graduated pipe with a light source and sight glass attached to the lower end).

5. Plexiglass core sampler.

6. Some type of portable pumping unit with a graduated suction pipe or hose (siphon).

The blanket depth should be kept from one to three feet (0.3 to 1 m) as measured from the clarifier bottom at the sidewall. The operator must check the blanket depth on a routine basis, making adjustments in the RAS to control the blanket depth at a level that provides the highest effluent quality for the particular plant. If it is observed that the depth of the sludge blanket is increasing, however, an increase in the RAS flow can only solve the problem on a short-term basis. Increases in sludge blanket depth may result from having too much activated sludge in the treatment system, or because of a poorly settling

TABLE 2.2 STANDARD OPERATING PROCEDURES FOR RAS CONTROL

Process	RAS Control Method	Mode of Operation	What to Check	Frequency of Adjustment	When to Check	Condition	Probable Cause	Response
Complete Mix or Plug Flow	Constant Flow	Manual	Sludge Blanket	Daily	Every 8 Hr	High	Low RAS Rate	Increase Return
						Satisfactory	—	Continue Monitoring
						Low	High RAS Rate	Decrease Return
	Constant % of Influent Flow	Manual	% of Influent Flow	2 Hr	Every 2 Hr	High	Variations in Daily Influent Flow	Adjust to Desired % of Influent Flow
						Satisfactory		
						Low		
			Sludge Blanket	Daily	Every 8 Hr	High	% of Flow Too Low	Increase % of Flow
						Satisfactory	—	Continue Monitoring
						Low	% of Flow Too High	Decrease % of Flow
	Constant % of Influent Flow	Automatic	Sludge Blanket	Daily	Every 8 Hr	High	% of Flow Too Low	Increase % of Flow
						Satisfactory	—	Continue Monitoring
						Low	% of Flow Too High	Decrease % of Flow
	Control by Sludge Blanket Level	Automatic	Sludge Blanket	Daily	Every 8 Hr	High or Low	Controller Malfunction	Fix Controller or Manually Adjust Accordingly
						Satisfactory	—	Continue Monitoring
Reaeration or Contact Stabilization	Constant % of Flow	Automatic	Ratio of MLSS/RAS$_{SS}$ (Centrifuge Test)	Every 2 Hr	Every 2 Hr	High Ratio	Return Too High	Decrease Return
						Satisfactory	—	Continue Monitoring
						Low Ratio	Return Too Low	Increase Return

sludge. Long-term corrections must be made that will improve the settling characteristics of the sludge or remove the excess solids from the treatment system. If the sludge is settling poorly, increasing the RAS flow may even cause more problems by further increasing the flow through the clarifier. If the sludge is settling poorly due to *BULKING,*[9] the environmental conditions for the microorganisms must be improved. If there is too much activated sludge in the treatment system, the excess sludge must be wasted.

Measurements of the sludge blanket depth in the clarifier should be made at the same time each day so the depth will be measured under the same conditions of flow and solids loading. The best time to make these measurements is during the period of maximum daily flow because the clarifier is operating under the highest solids loading rate. The sludge blanket should be measured daily, and adjustments to the RAS rate can be made as necessary. Adjustments in the RAS flow rate should only be needed occasionally if the activated sludge process is operating properly.

An additional advantage of monitoring the sludge blanket depth is that problems, such as improperly operating sludge collection equipment, will be observed due to irregularities in the blanket depth. A plugged pickup on a clarifier sludge collection system would cause sludge depth to increase in the area of the pickup and decrease in the areas where the properly operating pickups are located. These irregularities in sludge blanket depth are easily monitored by measuring profiles of blanket depth across the clarifier.

---

[9] *Bulking.* Clouds of billowing sludge that occur throughout secondary clarifiers and sludge thickeners when the sludge does not settle properly. In the activated sludge process, bulking is usually caused by filamentous bacteria or bound water.

QUESTIONS

Write your answers in a notebook and then compare your answers with those on page 146.

2.2C   What techniques are used to determine the rate of RAS flow?

2.2D   The sludge blanket depth should be kept to what depth?

2.2E   When should the sludge blanket depth be measured and why?

2.221   Settleability Approach

Another method of calculating the RAS flow rate is based on the results of the 30-minute settling test. Settleability is defined as the percentage of volume occupied by the sludge after settling for 30 minutes. The settleability test is run on a sample collected from the aeration basin effluent.

EXAMPLE 1

Determine the return activated sludge (RAS) flow as a percentage of the influent flow and in MGD and GPM when the influent flow is 7.5 MGD (28,390 m³/day) and the sludge settling volume (SV) in 30 minutes is 275 mL/L.

Known	Unknown
Infl Flow, MGD = 7.5 MGD	1. RAS Flow as a Percent of Infl Flow, %
Sl Set Vol (SV), mL/L = 275 mL/L	2. RAS Flow, MGD and GPM

1. Calculate RAS flow as a percent of influent flow.

$$\text{RAS Flow, \%} = \frac{\text{SV, mL/L} \times 100\%}{1,000 \text{ mL/L} - \text{SV, mL/L}}$$

$$= \frac{275 \text{ mL/L} \times 100\%}{1,000 \text{ mL/L} - 275 \text{ mL/L}}$$

$$= \frac{275 \text{ mL/L} \times 100\%}{725 \text{ mL/L}}$$

$$= 38\% \text{ of influent flow rate}$$

2. Calculate RAS flow, MGD and convert to GPM.

$$\text{RAS Flow, MGD} = \text{RAS Flow, decimal} \times \text{Infl Flow, MGD}$$

$$= 0.38 \times 7.5 \text{ MGD}$$

$$= 2.85 \text{ MGD} \times 694 \text{ GPM/MGD*}$$

$$= 1,978 \text{ GPM}$$

---

* This is a factor for converting MGD to GPM.

The settleability method assumes that measurements made with a laboratory settling cylinder will accurately reflect the settling in a clarifier. This assumption will seldom (if ever) be true because of the effects of the cylinder walls and the quiescent (still or lack of turbulence) nature of the liquid in the cylinder. Some operators have found that gently stirring (1 to 2 RPMs) the sludge during the settling test reduces these problems.

Another way to calculate the RAS flow as a percentage of the influent flow is by using the chart in Figure 2.7 below. First, locate the SV value (from the 30-minute sludge settling test, 275 mL/L) on the bottom scale. Draw a vertical line up to the curve and draw a horizontal line from that point to the left vertical axis. The value (38%) is the RAS flow as a percentage of the influent flow. To find the RAS flow in MGD, multiply the R/Q value (0.38) by Q (7.5 MGD).

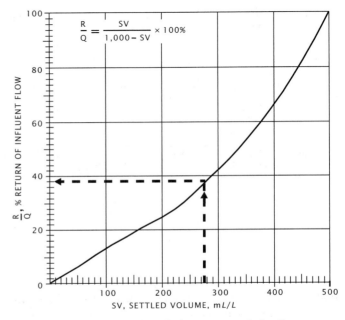

Fig. 2.7   Chart for calculating return sludge flow

2.222   SVI[10] Approach

To determine the RAS flow rate using the Sludge Volume Index (SVI), it is necessary to include the settleability approach. Therefore, this method is subject to the same limitations as the settleability method.

As you can see by the following example, the SVI method uses the Sludge Volume Index to estimate the suspended solids concentration of the return activated sludge (RAS). In calculation #1, RAS Suspended Solids = 8,333 mg/L. This value

---

[10] *Sludge Volume Index (SVI).* A calculation that indicates the tendency of activated sludge solids (aerated solids) to thicken or to become concentrated during the sedimentation/thickening process. SVI is calculated in the following manner: (1) allow a mixed liquor sample from the aeration basin to settle for 30 minutes; (2) determine the suspended solids concentration for a sample of the same mixed liquor; (3) calculate SVI by dividing the measured (or observed) wet volume (mL/L) of the settled sludge by the dry weight concentration of MLSS in grams/L.

$$\text{SVI, mL/gm} = \frac{\text{Settled Sludge Volume/Sample Volume, mL/L}}{\text{Suspended Solids Concentration, mg/L}} \times \frac{1,000 \text{ mg}}{\text{gram}}$$

(8,333 mg/L) is then used to find the RAS flow rate (calculation #2; 2.4 MGD). Next, calculate the RAS flow as a percentage of influent flow (calculation #3 in the example; 32%).

EXAMPLE 2

Determine the return activated sludge (RAS) flow in MGD and as a percentage of influent flow when the influent flow is 7.5 MGD (28,390 m³/day), the mixed liquor suspended solids (MLSS) are 2,000 mg/L, and the SVI is 120.

Known		Unknown
Infl Flow, MGD	= 7.5 MGD	1. RAS Flow, MGD
MLSS, mg/L	= 2,000 mg/L	2. RAS Flow as a Percent
SVI	= 120	of Influent Flow

1. Calculate the RAS suspended solids based on the SVI.

$$\text{RAS Susp Sol, mg}/L = \frac{1,000,000}{SVI}$$

$$= \frac{1,000,000}{120}$$

$$= 8,333 \text{ mg}/L$$

2. Calculate RAS Flow, MGD, based on SVI.

$$\text{RAS Flow, MGD} = \frac{\text{Infl Flow, MGD} \times \text{MLSS, mg}/L}{\text{RAS Susp Sol, mg}/L - \text{MLSS, mg}/L}$$

$$= \frac{7.5 \text{ MGD} \times 2,000 \text{ mg}/L}{8,333 \text{ mg}/L - 2,000 \text{ mg}/L}$$

$$= \frac{15,000 \text{ MGD} \times \text{mg}/L}{6,333 \text{ mg}/L}$$

$$= 2.4 \text{ MGD}$$

3. Calculate the RAS flow as a percent of influent flow.

$$\text{RAS Flow, \%} = \frac{\text{RAS Flow, MGD} \times 100\%}{\text{Infl Flow, MGD}}$$

$$= \frac{2.4 \text{ MGD} \times 100\%}{7.5 \text{ MGD}}$$

$$= 32\% \text{ of Influent Flow}$$

The real value in the SVI is not in calculating the RAS flow, but in its use as a process stability indicator. Changes in the SVI at a constant MLSS are more important than the SVI value. Never be concerned too much about comparing the SVI of different treatment plants because the SVI value that indicates good operation in one plant may not necessarily apply to good operation in other plants. In general, typical pure oxygen activated sludge plants have SVIs from 50 to 100 and air activated sludge plants have SVIs from 100 to 300.

2.23 Return Rates With Separate Sludge Reaeration

In the sludge reaeration (contact stabilization) variation of the activated sludge process, the return sludge rate is very significant. This is true because the rate of return directly affects the ratio of sludge concentration between the contact portion of the process

and the stabilization or reaeration portion. Generally, a higher rate of return will shift solids from the stabilization portion of the process to the contact portion of the process. Adequate theories for making rational adjustments of the contact/stabilization ratio are available; however, you must depend on rules of thumb or on operating experience to determine which levels are appropriate. These rules of thumb include the following:

1. The suspended solids concentration in the reaeration portion will eventually equal the RAS suspended solids concentration. Therefore, the RAS flow rate should be controlled on the basis of maintaining the desired suspended solids concentration in the reaeration portion of the process.

2. The contact portion suspended solids concentration may be determined by the following formula:

$$\text{Contact MLSS, mg}/L = \frac{\text{RAS Flow, MGD} \times \text{RAS Susp Sol, mg}/L}{\text{Infl Flow, MGD} + \text{RAS Flow, MGD}}$$

3. RAS flow may be determined by the following formula:

$$\text{RAS Flow, MGD} = \frac{\text{Infl Flow, MGD} \times \text{MLSS, mg}/L}{\text{RAS Susp Sol, mg}/L - \text{MLSS, mg}/L}$$

4. If the SVI remains constant or begins to drop, it indicates that the solids inventory in the process may be too high and wasting should be increased. If the SVI increases in conjunction with a rising sludge blanket in the clarifier, sludge bulking may occur. Refer to Table 2.3 for some suggested solutions under these conditions.

2.24 Acknowledgment

This section was adapted from *PROCESS CONTROL MANUAL FOR AEROBIC BIOLOGICAL WASTEWATER TREATMENT FACILITIES,* US Environmental Protection Agency. Obtain from National Technical Information Service (NTIS), 5285 Port Royal Road, Springfield, VA 22161. Order No. PB-279474. EPA No. 430-9-77-006. Price, $86.50, plus $5.00 shipping and handling per order.

QUESTIONS

Write your answers in a notebook and then compare your answers with those on pages 146 and 147.

2.2F   How long is the sludge allowed to settle in the settleability test?

2.2G   Calculate the return activated sludge (RAS) flow rate in MGD when the influent flow is 4 MGD and the sludge settling volume (SV) in 30 minutes is 250 mL/L.

TABLE 2.3   EFFECTS OF CHANGES ON RETURN SLUDGE AERATION TIME (RSAT)

Process Control Guidelines	Change Made		Effect on SVI		Effect on Nitrification	
			Increase	Decrease	Increase	Decrease
Step Change	Increase RSAT			x		x
	Decrease RSAT		x		x	
Return Sludge Flow	Increase		x			x
	Decrease			x	x	
Process Air Rate (DO)	Increase			x	x	
	Decrease		x			x
Plant Loading, lbs COD or BOD per lb total plant MLVSS	Increase	Increase Plant Q	x			x
		Decrease MCRT				x
	Decrease	Decrease Plant Q		x	x	
		Increase MCRT			x	

NOTES:
1. Changes made in the aeration system are reflected in the secondary clarifiers by how well the sludge settles.
2. When making a process change, you must consider what changing one control guideline does to the others.
   EXAMPLE: a. An increase in the return sludge flow will decrease the Return Sludge Aeration Time (RSAT).
   b. An increase in the process air rate is like increasing the RSAT.
   c. An increase in plant Q can result in a return sludge increase.
   d. An increase in the return sludge flow can cancel a step change that was made to increase the RSAT.

End of Lesson 2 of 4 Lessons on ACTIVATED SLUDGE

Please answer the discussion and review questions next.

DISCUSSION AND REVIEW QUESTIONS

Chapter 2.   ACTIVATED SLUDGE

Pure Oxygen Plants and Operational Control Options

(Lesson 2 of 4 Lessons)

Write the answers to these questions in your notebook. The question numbering continues from Lesson 1.

5. Different RAS flow rates will be required as the result of what two activated sludge conditions?

6. What is the difference between the constant method of RAS control and the constant percentage method of RAS control?

7. What are the advantages of the constant RAS flow approach and the constant percentage RAS flow approach?

8. Why is the following statement true and how can this problem be corrected on a long-term basis? If you observe that the depth of the sludge blanket is increasing, an increase in the RAS flow can only solve the problem on a short-term basis.

CHAPTER 2.   ACTIVATED SLUDGE

Pure Oxygen Plants and Operational Control Options

(Lesson 3 of 4 Lessons)

Abbreviations used in this section include:

1. RAS, Return Activated Sludge

2. WAS, Waste Activated Sludge

3. MCRT, Mean Cell Residence Time

4. MLVSS, Mixed Liquor Volatile Suspended Solids

5. F/M, Food to Microorganism Ratio

2.3   WASTE ACTIVATED SLUDGE

2.30   Purpose of Wasting Activated Sludge

One of the most important controls of the activated sludge process is the amount of activated sludge that is wasted. The amount of waste activated sludge (WAS) removed from the process affects all of the following items:

1. Effluent quality

2. Growth rate of the microorganisms

3. Oxygen consumption

4. Mixed liquor settleability

5. Nutrient quantities needed

6. Occurrence of foaming/frothing

7. Possibility of nitrifying

The objective of wasting activated sludge is to maintain a balance between the microorganisms under aeration and the amount of incoming food as measured by the COD or BOD test. When microorganisms remove BOD from wastewater, the amount of activated sludge increases (microorganisms grow and multiply). The rate at which these microorganisms grow is called the growth rate and is defined as the increase in the amount of activated sludge that takes place in one day. The objective of sludge wasting is to remove just the amount of microorganisms that grow in excess of the microorganism death rate. When this is done, the amount of activated sludge formed by the microorganism growth is just balanced by that which is removed from the process. This allows the total amount of activated sludge in the process to remain somewhat constant. This condition is called "steady state" and is a desirable condition for operation. However, "steady state" can only be approximated because of the variations in the nature and quantity of the food supply (BOD) and of the microorganism population. The objective of process control is to approach a particular "steady

state" by controlling any one or a combination of the following control guidelines:

1. Sludge Age

2. F/M or Food to Microorganism Ratio

3. MCRT, Mean Cell Residence Time

4. Volatile Solids Inventory

5. MLVSS Concentration

The best mode of process control will produce a high-quality effluent that meets NPDES permit requirements with consistent treatment results at a minimal cost.

Wasting of the activated sludge (Figure 2.6, page 69) is usually achieved by removing a portion of the RAS flow. The waste activated sludge is either pumped to thickening or dewatering facilities and then to a digester or incinerator, or to the primary clarifiers where it is pumped to a digester with the raw sludge. Procedures for making WAS adjustments are presented in Table 2.4, which you can use to develop standard operating procedures for your treatment plant.

An alternative method for wasting sludge is taking it from the mixed liquor in the aeration tank. There are much higher concentrations of suspended matter in the RAS than there are in the mixed liquor. Therefore, when wasting is practiced from the aeration tank, larger sludge handling facilities are required. Wasting from the RAS takes advantage of the gravity settling and thickening of the sludge that occurs in the secondary clarifier. However, wasting from the aeration tank has the advantage of not wasting excessive amounts of sludge since a large quantity of mixed liquor is involved. The extra security of wasting from the

TABLE 2.4 STANDARD OPERATING PROCEDURES FOR WAS CONTROL

Method of Control	Process Operation	What to Check	When to Check	Calculation	Frequency of Adjustment	Condition	Probable Cause	Response[a]
F/M	High Rate Conventional Rate Extended Aeration	MLVSS and Influent COD	Daily	F/M Based on $\dfrac{\text{7-Day Avg COD}}{\text{7-Day Avg MLVSS}}$	Daily	Actual F/M: High Satisfactory Low	Excessive Wasting — Insufficient Wasting	Reduce WAS — Increase WAS
MLVSS	High Rate Conventional Rate Extended Aeration	MLVSS and Influent COD or BOD	Daily	Volatile Solids Inventory	Daily	Actual MLVSS: High Satisfactory Low	Insufficient Wasting — Excessive Wasting	Increase WAS — Reduce WAS
MCRT	High Rate Conventional Rate Extended Aeration	MLSS, WAS_{SS}, Q_{WAS}, $EFFL_{SS}$	Daily	7-Day Avg[b] Solids Inventory 7-Day Average[b] of Solids in WAS 7-Day Average[b] of Solids in Effluent	Daily	Actual MCRT: High Satisfactory Low	Insufficient Wasting — Excessive Wasting	Increase WAS — Reduce WAS
Sludge Age	High Rate Conventional Rate Extended Aeration	Influent SS and MLSS	Daily	7-Day Avg of SS Inventory and SS in Influent	Daily	Actual SA: High Satisfactory Low	Insufficient Wasting — Excessive Wasting	Increase WAS — Reduce WAS

[a] Response—Calculations should be made to determine the WAS rate. However, when increasing or decreasing daily WAS rates, any changes should not exceed 10 to 15 percent of the previous day's WAS rate. This is necessary to allow the process to stabilize.

[b] When calculating the MCRT, determine the desired MCRT (5 days) and use the moving average for the number of days (7 days) to calculate values used in the formula to determine the desired MCRT.

aeration tank should not be underestimated. Unfortunately, many plants do not have the flexibility to waste from the aeration tank nor are there sufficient sludge handling facilities to handle the more dilute sludge.

2.31 Methods of Sludge Wasting

Wasting of the activated sludge can be accomplished on an intermittent or continuous basis. The intermittent wasting of sludge means that wasting is conducted on a batch basis from day to day.

2.310 Sludge Age Control

Sludge age is a measure of the length of time a particle of suspended solids has been undergoing aeration in the activated sludge process. As you can see in this formula for calculating sludge age, it is based on a ratio between the solids in the aeration tank and the solids in the incoming wastewater.

$$\frac{\text{Sludge Age,}}{\text{Days}} = \frac{\text{Suspended Solids Under Aeration, lbs or kg}}{\text{Suspended Solids Added, lbs/day or kg/day}}$$

Using sludge age as a control technique, the operator wastes just enough sludge to maintain the sludge age that produces the best quality effluent. In most activated sludge plants, sludge age ranges from 3 to 8 days. Difficulties are commonly experienced using the sludge age control technique when the BOD or COD to solids ratio in the wastewater changes. This is because sludge age is based on the assumption that the BOD (or COD)/solids ratio is fairly constant. By realizing that the BOD or COD to solids ratio does change and adjusting the sludge age when the ratio changes, the sludge age is a useful process control technique. Calculate the sludge age as shown in the following example.

EXAMPLE 3

Determine the sludge age for an activated sludge plant with an influent flow of 7.5 MGD (28,390 m³/day). The primary effluent suspended solids concentration is 100 mg/*L*. Two aeration tanks have a volume of 0.6 MG (2,270 m³) each and a mixed liquor suspended solids (MLSS) concentration of 2,200 mg/*L*.

Known		Unknown
Infl Flow, MGD	= 7.5 MGD	Sludge Age, days
Prim Effl SS, mg/*L*	= 100 mg/*L*	
Tank Vol, MG	= 0.6 MG/tank	
MLSS, mg/*L*	= 2,200 mg/*L*	
No of Tanks	= 2 tanks	

$$\text{Sludge Age, days} = \frac{\text{Solids Under Aeration, lbs}}{\text{Solids Added, lbs/day}}$$

1. Calculate the solids under aeration, lbs.

$$\text{Solids Under Aeration, lbs} = \frac{\text{No}}{\text{Tanks}} \times \frac{\text{Tank Vol,}}{\text{MG/tank}} \times \text{MLSS, mg/}L \times 8.34 \text{ lbs/gal}$$

$$= 2 \text{ tanks} \times 0.6 \text{ MG/tank} \times 2,200 \text{ mg/}L \times 8.34 \text{ lbs/gal}$$

$$= 22,000 \text{ lbs}$$

2. Calculate the solids added, lbs/day.

$$\text{Solids Added, lbs/day} = \text{Infl Flow, MGD} \times \text{Prim Effl SS, mg/}L \times 8.34 \text{ lbs/gal}$$

$$= 7.5 \text{ MGD} \times 100 \text{ mg/}L \times 8.34 \text{ lbs/gal}$$

$$= 6,255 \text{ lbs/day}$$

3. Determine sludge age, days.

$$\text{Sludge Age, days} = \frac{\text{Solids Under Aeration, lbs}}{\text{Solids Added, lbs/day}}$$

$$= \frac{22,000 \text{ lbs}}{6,255 \text{ lbs/day}}$$

$$= 3.5 \text{ days}$$

Calculate the waste activated sludge (WAS) flow rate using the sludge age control technique as shown in the following example.

EXAMPLE 4

Determine the waste activated sludge (WAS) flow rate in MGD and GPM for an activated sludge plant that adds 6,255 lbs (2,837 kg) of solids per day (from Example 3). The solids under aeration are 33,075 pounds (15,000 kg), the return activated sludge (RAS) suspended solids concentration is 6,300 mg/*L*, and the desired sludge age is 5 days. Current sludge waste rate is 4,455 lbs (2,020 kg) per day.

Known		Unknown
Solids Added, lbs/day	= 6,255 lbs/day	WAS Flow, MGD and GPM
Solids Aerated, lbs	= 33,075 lbs	
RAS Susp Sol, mg/*L*	= 6,300 mg/*L*	
Desired Sludge Age, days	= 5 days	
Current WAS Rate, lbs/day	= 4,455 lbs/day	

1. Calculate the desired pounds of solids under aeration (MLSS) for the desired sludge age of 5 days.

$$\begin{aligned}\text{Desired Solids Under Aeration, lbs} &= \text{Solids Added, lbs/day} \times \text{Sludge Age, days} \\ &= 6,255 \text{ lbs/day} \times 5 \text{ days} \\ &= 31,275 \text{ lbs}\end{aligned}$$

2. Calculate the additional WAS flow, MGD and GPM, to maintain the desired sludge age.

$$\begin{aligned}\text{Additional WAS Flow, MGD and GPM} &= \frac{\text{Solids Aerated, lbs} - \text{Desired Solids, lbs}}{\text{RAS Susp Sol, mg/}L \times 8.34 \text{ lbs/gal}} \\[6pt] &= \frac{(33,075 \text{ lbs} - 31,275 \text{ lbs})/\text{day}}{6,300 \text{ mg/}L^* \times 8.34 \text{ lbs/gal}} \\[6pt] &= \frac{1,800 \text{ lbs removed per day}^{**}}{52,542} \\[6pt] &= 0.034 \text{ MGD}^{***} \times 694 \text{ GPM/MGD} \\[6pt] &= 24 \text{ GPM}\end{aligned}$$

---

*   Remember that mg/*L* is the same as lbs/M lbs.
** Remove an additional 1,800 lbs during a 24-hour period.
*** If the actual solids under aeration are less than the desired solids, reduce your current wasting rate.

3. Add the current WAS flow to the additional WAS flow, MGD.

$$\begin{aligned}\text{Total WAS Flow, MGD and GPM} &= \frac{\text{Current WAS}}{\text{Flow, MGD}} + \frac{\text{Additional WAS}}{\text{Flow, MGD}} \\[6pt] &= \frac{\text{Solids Wasted, lbs/day}}{\text{RAS Susp Sol, mg/}L \times 8.34 \text{ lbs/gal}} + \text{Flow, MGD} \\[6pt] &= \frac{4,455 \text{ lbs/day}}{6,300 \text{ mg/}L \times 8.34 \text{ lbs/gal}} + 0.034 \text{ MGD} \\[6pt] &= 0.085 \text{ MGD} + 0.034 \text{ MGD} \\[6pt] &= 0.119 \text{ MGD} \times 694 \text{ GPM/MGD} \\[6pt] &= 83 \text{ GPM}\end{aligned}$$

QUESTIONS

Write your answers in a notebook and then compare your answers with those on page 147.

2.3A   What is the objective of wasting activated sludge?

2.3B   How is wasting of the excess activated sludge usually achieved?

2.3C   Calculate the change in the waste activated sludge (WAS) flow rate in MGD and GPM for an activated sludge plant that adds 4,750 lbs of solids per day. The solids under aeration are 41,100 pounds and the return activated sludge (RAS) suspended solids concentration is 5,800 mg/L. The desired sludge age is 8 days.

2.311   F/M (Food/Microorganism) Control

F/M control is used to ensure that the activated sludge process is being loaded at a rate that the microorganisms in the mixed liquor volatile suspended solids (MLVSS) are able to use most of the food supply in the wastewater being treated. If too much or too little food is applied for the amount of microorganisms, operating problems may occur and the effluent quality may drop.

There are four facts that should be remembered regarding the F/M method of control:

1. The food concentration is estimated with the COD (or BOD) tests. The oxygen demand tests provide crude but reliable approximations of the actual amount of food removed by the microorganisms.

2. The amount of food (COD or BOD) applied is important to calculate the F/M ratio.

3. The quantity of microorganisms can be represented by the quantity of MLVSS. Ideally, the living or active microorganisms would simply be counted, but this is not feasible, and studies have shown that the MLVSS is a good approximation of the microorganism concentrations in the MLSS.

4. Operation by or calculations of the F/M ratio should not be based on daily tests because flows and organic concentrations can vary widely on a day-to-day basis. One way to handle these variations is to calculate the F/M ratio based on a seven-day *MOVING AVERAGE*[11] of food (COD, BOD, or TOC), flow, and microorganisms (MLSS).

The range of organic loadings of activated sludge plants is described by the F/M ratio. Different ranges of organic loadings are necessary for conventional, extended aeration, and high-rate types of activated sludge systems. These ranges have been shown to produce activated sludge that settles well.

Table 2.5 presents the ranges of F/M values that have been used successfully with the three loading conditions. The F/M values shown are expressed in terms of BOD, COD, and Total Organic Carbon (TOC). The TOC is an additional means of estimating organic loading. The values indicated are guidelines for process control, and they should not be thought of as minimums or maximums.

TABLE 2.5   TYPICAL RANGES FOR F/M LOADINGS

	Conventional AS Range F/M	Extended Aeration F/M	High-Rate Range F/M
BOD	0.1 to 0.5	0.05 to 0.1	0.5 to 2.5
COD[a]	0.06 to 0.3	0.03 to 0.06	0.3 to 1.5
TOC[b]	0.25 to 1.5	0.1 to 0.25	1.5 to 6.0

[a] Assumes BOD/COD for settled wastewaters = 0.6.
[b] Assumes BOD/TOC for settled wastewaters = 2.5.

The F/M ratio is calculated from the amount of COD or BOD applied each day and from the solids inventory in the aeration tank.

EXAMPLE 5

Determine the food to microorganism (F/M) ratio for an activated sludge plant with a primary effluent COD of 100 mg/L applied to the aeration tank, an influent flow of 7.5 MGD (28,390 m³/day), and 33,075 lbs (15,000 kg) of solids under aeration. Seventy percent of the MLSS are volatile matter. All known values are seven-day moving averages.

Known		Unknown
Infl Flow, MGD	= 7.5 MGD	F/M, lbs COD/day/ lb MLVSS
COD, mg/L	= 100 mg/L	
Solids Under Aeration, lbs	= 33,075 lbs	
MLSS VM, %	= 70%	

---

[11] *Moving Average.*   To calculate the moving average for the last 7 days, add up the values for the last 7 days and divide by 7. Each day add the most recent day's value to the sum of values and subtract the oldest value. By using the 7-day moving average, each day of the week is always represented in the calculations.

1. Calculate the food to microorganism ratio.

$$\text{FM, } \frac{\text{lbs COD/day}}{\text{lb MLVSS solids}} = \frac{\text{Flow, MGD} \times \text{COD, mg/}L \times 8.34 \text{ lbs/gal}}{\text{Solids Under Aeration, lbs} \times \text{VM Portion}}$$

$$= \frac{7.5 \text{ MGD} \times 100 \text{ mg/}L \times 8.34 \text{ lbs/gal}}{33,075 \text{ lbs} \times 0.70}$$

$$= \frac{6,255 \text{ lbs COD/day}}{23,150 \text{ lbs MLVSS}}$$

$$= 0.27 \text{ lb COD/day/lb MLVSS}$$

The determination of WAS flow rates using the F/M control technique is calculated in the same manner as for the sludge age technique. However, the solids inventory for the aeration tank can be more logically determined based on the COD or BOD concentration of the wastewater to be treated when using the F/M control technique for process control. This procedure is shown in the following calculations.

EXAMPLE 6

Determine the desired waste activated sludge (WAS) flow rate using the F/M control technique. The influent flow is 7.5 MGD (28,390 m³/day), total aeration tank volume is 1.2 MG (4,542 m³), COD to aeration tank is 100 mg/L, the mixed liquor suspended solids (MLSS) are 3,300 mg/L and 70 percent volatile matter, the RAS suspended solids are 6,300 mg/L, and the desired food to microorganism (F/M) ratio is 0.29. Current WAS flow rate is 0.085 MGD.

Known		Unknown
Infl Flow, MGD	= 7.5 MGD	WAS Flow, MGD and GPM
Tank Vol, MG	= 1.2 MG	
COD, mg/L	= 100 mg/L	
MLSS, mg/L	= 3,300 mg/L	
MLSS VM, %	= 70%	
RAS Susp Sol, mg/L	= 6,300 mg/L	
Desired F/M, $\frac{\text{lbs COD/day}}{\text{lb MLVSS}}$	= 0.29 $\frac{\text{lb COD/day}}{\text{lb MLVSS}}$	
Current WAS, MGD	= 0.085 MGD	

1. Determine COD applied in pounds per day.

$$\text{COD, lbs/day} = \text{Flow, MGD} \times \text{COD, mg/}L \times 8.34 \text{ lbs/gal}$$

$$= 7.5 \text{ MGD} \times 100 \text{ mg/}L \times 8.34 \text{ lbs/gal}$$

$$= 6,255 \text{ lbs COD/day}$$

2. Determine the desired pounds of MLVSS.

$$\text{Desired MLVSS, lbs} = \frac{\text{COD Applied, lbs/day}}{\text{F/M, lbs COD/day/lb MLVSS}}$$

$$= \frac{6,255 \text{ lbs COD/day}}{0.29 \text{ lb COD/day/lb MLVSS}}$$

$$= 21,569 \text{ lbs MLVSS}$$

3. Determine the desired pounds of MLSS.

$$\text{Desired MLSS, lbs} = \frac{\text{Desired MLVSS, lbs}}{\text{MLSS VM Portion}}$$

$$= \frac{21,569 \text{ lbs}}{0.70}$$

$$= 30,813 \text{ lbs}$$

4. Determine actual MLSS pounds under aeration.

$$\text{Actual MLSS, lbs} = \text{Tank Vol, MG} \times \text{MLSS, mg/}L \times 8.34 \text{ lbs/gal}$$

$$= 1.2 \text{ MG} \times 3,300 \text{ mg/}L \times 8.34 \text{ lbs/gal}$$

$$= 33,026 \text{ lbs}$$

5. Calculate the additional WAS flow, MGD and GPM, to maintain the desired food to microorganism (F/M) ratio.

$$\begin{array}{l}\text{Additional WAS} \\ \text{Flow, MGD} \\ \text{and GPM}\end{array} = \frac{\text{Solids Aerated, lbs} - \text{Desired Solids, lbs}}{\text{RAS Susp Sol, mg/}L \times 8.34 \text{ lbs/gal}}$$

$$= \frac{33,026 \text{ lbs} - 30,813 \text{ lbs}}{6,300 \text{ mg/}L^* \times 8.34 \text{ lbs/gal}}$$

$$= \frac{2,213 \text{ lbs removed per day}^{**}}{52,542}$$

$$= 0.042 \text{ MGD}^{***} \times 694 \text{ GPM/MGD}$$

$$= 29 \text{ GPM}$$

---

* Remember that mg/L is the same as lbs/M lbs.

** Remove 2,213 lbs during a 24-hour period, starting slowly.

*** If the actual solids under aeration are less than the desired solids, reduce or stop your current wasting rate.

6. Calculate the total WAS flow in MGD and GPM.

$$\begin{array}{l}\text{Total WAS Flow,} \\ \text{MGD and GPM}\end{array} = \begin{array}{l}\text{Current WAS} \\ \text{Flow, MGD}\end{array} + \begin{array}{l}\text{Additional WAS} \\ \text{Flow, MGD}\end{array}$$

$$= 0.085 \text{ MGD} + 0.042 \text{ MGD}$$

$$= 0.127 \text{ MGD} \times 694 \text{ GPM/MGD}$$

$$= 88 \text{ GPM (target total rate—} \\ \text{approach over several days)}$$

The F/M control technique for sludge wasting is best used in conjunction with the MCRT control technique. Control to a desired MCRT is achieved by wasting an amount of the aeration tank solids inventory, which, in turn, fixes or provides an F/M ratio.

2.312 MCRT (Mean Cell Residence Time) Control

By using the MCRT, the operator can control the organic loading (F/M). In addition, the operator can calculate the amount of activated sludge that should be wasted in a logical manner.

Basically, the MCRT expresses the average time that a microorganism will spend in the activated sludge process. The MCRT value should be selected to provide the best effluent quality. This value should correspond to the F/M loading for which the process is designed. For example, a process designed to operate at conventional F/M loading rates may not produce a high-quality effluent if it is operating at a low MCRT because the F/M ratio may be too high for its design. Therefore, you must find the best MCRT for your process by relating it to the F/M ratio as well as to the effluent COD, BOD, and suspended solids concentrations.

The MCRT also determines the type of microorganisms that predominate in the activated sludge because it has a direct influence on the degree of nitrification that may occur in the process. A plant operated at a longer MCRT of 15 to 20 days will generally produce a nitrified effluent. A plant operating with an MCRT of 5 to 10 days may not produce a nitrified effluent unless wastewater temperatures are unusually high (above 77°F or 25°C). Table 2.6 presents the typical range of MCRT values that will enable nitrification at various wastewater temperatures. MCRTs below the values listed in Table 2.6 are also possible under more optimum conditions; similarly, under less favorable conditions, a higher MCRT may be required. The determination of a correct MCRT is only the first of many factors to be considered in operating an activated sludge plant to achieve nitrification. Nevertheless, the values shown have been used successfully to produce nitrified effluents at numerous plants where the removal of ammonia from the effluent is required but total nitrogen removal is not required.

TABLE 2.6 MCRT NEEDED TO PRODUCE A NITRIFIED EFFLUENT AS RELATED TO THE TEMPERATURE

Temperature, °C	MCRT, Days
10	30
15	20
20	15
25	10
30	7

As stated earlier, MCRT expresses the average time that a microorganism spends in the activated sludge process. The MCRT and the WAS flow rate for maintaining a desired MCRT are shown in the following example.

EXAMPLE 7

Determine the waste activated sludge (WAS) flow rate using the MCRT technique. The influent flow is 7.5 MGD (28,390 m³/day), total aeration tank volume is 1.2 MG (4,542 m³), mixed liquor suspended solids (MLSS) are 3,300 mg/L, RAS suspended solids are 6,300 mg/L, effluent suspended solids are 15 mg/L, and the desired mean cell residence time (MCRT) is 8 days.

Known		Unknown
Infl Flow, MGD	= 7.5 MGD	WAS Flow, MGD and GPM
Tank Vol, MG	= 1.2 MG	
MLSS, mg/L	= 3,300 mg/L	
RAS SS, mg/L	= 6,300 mg/L	
Effl SS, mg/L	= 15 mg/L	
Desired MCRT, days	= 8 days	

$$\text{MCRT,* days} = \frac{\text{Suspended Solids in Aerator, lbs}}{\text{Susp Sol Wasted, lbs/day} + \text{Susp Sol in Effl, lbs/day}}$$

1. Determine suspended solids in aerator in pounds.

$$\text{SS in Aerator, lbs} = \text{Aerator, MG} \times \text{MLSS, mg/L} \times 8.34 \text{ lbs/gal}$$
$$= 1.2 \text{ MG} \times 3,300 \text{ mg/L} \times 8.34 \text{ lbs/gal}$$
$$= 33,026 \text{ lbs}$$

2. Determine suspended solids lost in effluent in pounds per day.

$$\text{SS Lost in Effl, lbs/day} = \text{Infl Flow, MGD} \times \text{Effl SS, mg/L} \times 8.34 \text{ lbs/gal}$$
$$= 7.5 \text{ MGD} \times 15 \text{ mg/L} \times 8.34 \text{ lbs/gal}$$
$$= 938 \text{ lbs/day}$$

3. Determine the desired suspended solids wasted in pounds per day.

$$\text{MCRT, days} = \frac{\text{SS in Aerator, lbs}}{\text{SS Wasted, lbs/day} + \text{SS in Effl, lbs/day}}$$
$$\text{SS Wasted, lbs/day} = \frac{\text{SS in Aerator, lbs}}{\text{MCRT, days}} - \text{SS in Effl, lbs/day}$$
$$= \left(\frac{33,026 \text{ lbs}}{8 \text{ days}}\right) - 938 \text{ lbs/day}$$
$$= 4,128 \text{ lbs/day} - 938 \text{ lbs/day}$$
$$= 3,190 \text{ lbs/day}$$

---

* NOTE: Some operators use volatile suspended solids instead of suspended solids. The percent volatile of the MLSS, WAS, and effluent are usually very close and their exclusion makes little difference in the MCRT calculation.

MCRT may be calculated three different ways; however, be consistent with the method you select.

1. SS in Aerator, lbs (kg)
2. SS in Aerators and Secondary Clarifiers, lbs (kg)
3. SS in Aerators and Secondary Sludge Blankets, lbs (kg)

4. Determine the waste activated sludge (WAS) flow rate, MGD and GPM.

$$\text{SS Wasted, lbs/day} = \text{WAS Flow, MGD} \times \text{RAS SS, mg/}L \times 8.34 \text{ lbs/gal}$$

$$\text{WAS Flow, MGD and GPM} = \frac{\text{SS Wasted, lbs/day}}{\text{RAS SS, mg/}L \times 8.34 \text{ lbs/gal}}$$

$$= \frac{3,190 \text{ lbs/day}}{6,300 \text{ mg/}L \times 8.34 \text{ lbs/gal}}$$

$$= 0.06 \text{ MGD} \times 694 \text{ GPM/MGD}$$

$$= 42 \text{ GPM}$$

This means that for the next 8 days, approximately 60,000 gallons per day should be wasted from the RAS system. However, the WAS flow rate should be calculated and adjusted daily to maintain the desired MCRT.

QUESTIONS

Write your answers in a notebook and then compare your answers with those on pages 147 and 148.

2.3D What four facts should be remembered regarding the F/M method of control?

2.3E Calculate the desired pounds of MLSS if the desired F/M ratio is 0.30 lb COD/day/lb MLVSS, 7,000 lbs of COD per day are added, and the volatile matter is 70 percent of the MLSS.

2.3F What does the Mean Cell Residence Time (MCRT) represent?

2.313 Volatile Solids Inventory

Wasting of activated sludge requires operators to be aware of the conditions in the activated sludge process and to carefully evaluate any changes, regardless of whether wasting rates are determined using sludge age, food/microorganism, or mean cell residence time. Critical guidelines include aeration tank or system mixed liquor suspended solids or mixed liquor volatile suspended solids, return activated sludge (RAS) suspended solids or return volatile suspended solids, and sludge settleability in the final clarifier.

If wasting is done from RAS, the operator must measure the volatile suspended matter in the RAS to obtain average concentrations. Also, the operator must determine whether the volatile suspended matter is constant, increasing, or decreasing in the aeration tank or the aeration system. First, let us assume the volatile solids matter in the aeration tank is staying relatively *constant*. If the volatile content in the RAS suspended solids concentration is *decreasing*, the WAS flow rate must be increased proportionally to waste the proper amount of volatile suspended solids to maintain the constant volatile suspended solids in the aeration tank. Similarly, if there is an increase in the RAS volatile content, the WAS flow rate must be decreased proportionally to avoid losing or wasting too many microorganisms and still maintain the constant volatile suspended solids in the aeration tank.

Second, let us assume the volatile solids matter is *changing* in the aeration tank. The volatile solids could be decreasing due to inert matter in stormwater inflow to the system or the volatile solids could be increasing due to an increase resulting from the start of the canning season. Consider the situation where the volatile solids in the aeration tank are *decreasing* and causing a *decrease* in the volatile content of the RAS. The decrease in volatile RAS will require a decrease in RAS wasting to maintain the microorganisms needed in the aeration tank. If an *increase* in volatile content in the aeration tank causes an *increase* in RAS volatile solids, then an *increase* in WAS wasting will be required to maintain the desired level of microorganisms needed to treat the wastes.

However, a *decrease* in aeration tank volatile solids due to stormwater inflow or an *increase* in aeration tank volatile solids due to the start of the canning season *may* require a *change* in the target sludge age, food/microorganism (F/M), or mean cell residence time (MCRT). Operators need to change the target values when there are changes in temperature or changes in the type of volatile solids in the stormwater runoff or cannery waste discharges. Ultimately, the success of the activated sludge process depends on operators who understand the process and know how to properly regulate WAS flow rates.

When continuous wasting is practiced, the operator should check the RAS volatile suspended solids at least once every shift and make the appropriate WAS flow adjustment.

Summary

Assume aeration tank volatile solids matter constant.

MLVSS CONTENT	RAS VOLATILE CONTENT		WAS FLOW RATE
Constant	Decreases	→	Increase
Constant	Increases	→	Decrease

Assume aeration tank volatile solids matter changes.

MLVSS CONTENT	RAS VOLATILE CONTENT		WAS FLOW RATE
Decrease	Decrease	→	Decrease
Increase	Increase	→	Increase

EXAMPLE 8

An activated sludge plant is currently wasting 0.05 MGD (35 GPM or 2.18 L/sec). The return activated sludge (RAS) volatile suspended solids (VSS) on day 1 are 6,000 mg/L and on day 2 (the next day) the RAS VSS are 7,500 mg/L. Determine the adjusted waste activated sludge (WAS) rate based on the increase in return activated sludge (RAS) volatile suspended solids (VSS) from 6,000 to 7,500 mg/L.

Known	**Unknown**
WAS Flow, MGD = 0.05 MGD	Adjusted WAS Flow, MGD and GPM
RAS VSS, mg/L (day 1) = 6,000 mg/L	
RAS VSS, mg/L (day 2) = 7,500 mg/L	

1. Calculate the adjusted waste activated sludge (WAS) flow in MGD and GPM.

$$\text{Adj WAS Flow, MGD and GPM} = \frac{\text{RAS VSS for day 1, mg/}L \times \text{WAS Flow, MGD}}{\text{RAS VSS for day 2, mg/}L}$$

$$= \frac{6,000 \text{ mg/}L \times 0.05 \text{ MGD}}{7,500 \text{ mg/}L}$$

$$= 0.04 \text{ MGD} \times 694 \text{ GPM/MGD}$$

$$= 28 \text{ GPM}$$

When intermittent wasting is practiced, the operator must check the RAS volatile suspended solids to calculate the necessary WAS flow. In addition, this calculation must be readjusted for the reduced time of wasting.

EXAMPLE 9

The waste activated sludge (WAS) pumping period is 4 hours per day and the calculated WAS flow is 0.04 MGD or 28 GPM (1.75 L/sec). Calculate the WAS flow, MGD and GPM.

Known	**Unknown**
WAS Flow, MGD = 0.04 MGD	WAS Flow, MGD and GPM for 4 hours/day wasting period
Wasting Time, hr/day = 4 hr/day	

1. Determine the WAS flow for a 4 hours/day wasting period.

$$\text{New WAS Flow, MGD and GPM} = \frac{\text{WAS Flow, MGD} \times 24 \text{ hr/day}}{4 \text{ hours of wasting/day}}$$

$$= 0.04 \text{ MGD} \times \frac{24 \text{ hr/day}}{4 \text{ hr/day}}$$

$$= 0.24 \text{ MGD} \times 694 \text{ GPM/MGD}$$

$$= 167 \text{ GPM}$$

The operator would repeat the WAS flow calculation for each wasting period to take into account the RAS volatile suspended solids variations.

Intermittent wasting of sludge has the advantage that less variation in the suspended matter concentration will occur during the wasting period and the amount of sludge wasted will be more accurately known. The disadvantages of intermittent wasting are that the sludge handling facilities in the treatment plant may be loaded at a higher hydraulic loading rate and that the activated sludge process will be out of balance for a period of time until the microorganisms regrow to replace those wasted over the shorter period of time. Intermittent wasting usually is not practiced in plants treating more than 10 MGD (37,800 m³/day).

In using either of these methods for wasting, the operator does not have complete control of the amount of activated sludge wasted due to the solids lost in the effluent. This "wasting" of activated sludge in the effluent must be accounted for with any method of process control or the system will always be slightly out of balance. The loss of activated sludge in the effluent generally accounts for less than five percent of the total solids that need to be wasted; however, it is necessary to be aware of this loss and to be able to take it into account by the methods shown in Section 2.312, "MCRT (Mean Cell Residence Time) Control." The need for taking into account the solids lost in the effluent is especially important if one encounters situations where large concentrations of suspended solids are washed out in the secondary effluent, as in the case of sludge bulking.

Proper control of the WAS will produce a high-quality effluent with minimum operational difficulties.

2.314   MLVSS (Mixed Liquor Volatile Suspended Solids) Control

This technique for process control is used by many operators because it is simple to understand and involves a minimum amount of laboratory control. The MLVSS control technique usually produces good-quality effluent as long as the incoming wastewater characteristics are fairly constant with minimal variations in influent flow rates.

With this technique, the operator tries to maintain a constant MLVSS concentration in the aeration tank to treat the incoming wastewater organic load. To put it in simple terms, if it is found that an MLVSS concentration of 2,000 mg/L produces a good-quality effluent, the operator must waste sludge from the process to maintain that concentration. More sludge is wasted until the desired level is reached again.

The laboratory control tests and operational data involved in using this technique include the following:

1. MLVSS concentration

2. RAS Volatile Suspended Solids (VSS) concentration

3. Influent wastewater flow rate

4. Volume of the aeration tank

Whether a new plant is being started or the operation of an existing plant is being checked, this control technique is used to indicate when activated sludge should be wasted. In most cases, it is not the most reliable technique because it ignores process variables such as F/M and microorganism growth rate, which are necessary for maintaining optimum system balance. When operational problems occur, the operator is unable to make rational process adjustments due to the lack of process control data.

The control technique is implemented by choosing an MLVSS concentration that produces the highest quality effluent while maintaining a stable and economical operation. WAS flow rates are determined as shown in Example 10.

EXAMPLE 10

A 1.2 MG (4,542 m^3) aeration tank has a mixed liquor suspended solids (MLSS) concentration of 3,300 mg/L. The return activated sludge (RAS) suspended solids concentration is 6,300 mg/L. The volatile portion of both suspended solids is 70 percent. Experience has shown that the desired mixed liquor volatile suspended solids (MLVSS) in the aeration tank is approximately 21,250 pounds (9,639 kg). Determine the desired waste activated sludge (WAS) flow rate if the current WAS flow rate is 0.15 MGD.

Known		Unknown
Tank Vol, MG	= 1.2 MG	Desired WAS Flow, MGD and GPM
MLSS, mg/L	= 3,300 mg/L	
RAS SS, mg/L	= 6,300 mg/L	
Volatile Portion	= 0.70	
Desired MLVSS, lbs	= 21,250 lbs	
Current WAS Flow, MGD	= 0.15 MGD	

1. Determine the pounds of mixed liquor volatile suspended solids (MLVSS) under aeration.

$$\text{Actual MLVSS, lbs} = \text{Tank Vol, MG} \times \text{MLSS, mg/}L \times \text{Volatile} \times 8.34 \text{ lbs/gal}$$

$$= 1.2 \text{ MG} \times 3,300 \text{ mg/}L \times 0.70 \times 8.34 \text{ lbs/gal}$$

$$= 23,120 \text{ lbs}$$

2. Determine the pounds of volatile solids to be wasted.

$$\text{Amt Wasted, lbs} = \text{Actual MLVSS, lbs} - \text{Desired MLVSS, lbs}$$

$$= 23,120 \text{ lbs} - 21,250 \text{ lbs}$$

$$= 1,870 \text{ lbs to be wasted per day}$$

3. Determine the additional waste activated sludge (WAS) flow rate in MGD and GPM.

$$\text{Amt Wasted, lbs/day} = \text{WAS Flow, MGD} \times \text{RAS SS, mg/}L \times \text{Vol} \times 8.34 \text{ lbs/gal}$$

or

$$\text{WAS Flow, MGD} = \frac{\text{Amount Wasted, lbs/day}}{\text{RAS SS, mg/}L \times \text{Volatile} \times 8.34 \text{ lbs/gal}}$$

$$= \frac{1,870 \text{ lbs/day}}{6,300 \text{ mg/}L \times 0.70 \times 8.34 \text{ lbs/gal}}$$

$$\text{WAS Flow, GPM} = 0.05 \text{ MGD} \times 694 \text{ GPM/MGD}$$

$$= 35 \text{ GPM}$$

4. Determine the desired WAS flow rate in MGD and GPM.

$$\text{Desired WAS Flow, MGD and GPM} = \text{Current WAS Flow, MGD} + \text{Additional WAS Flow, MGD}$$

$$= 0.15 \text{ MGD} + 0.05 \text{ MGD}$$

$$= 0.20 \text{ MGD} \times 694 \text{ GPM/MGD}$$

$$= 140 \text{ GPM}$$

2.315 Respiration Rate as a Control Guideline

Most of us have used many tests and observations to help control the activated sludge process, including Food to Microorganism (F/M) Ratio, Mixed Liquor Suspended Solids (MLSS), surface appearance, odor, color, settleability, dissolved oxygen (DO), Mean Cell Residence Time (MCRT), BOD, and TOC.

The reason for all the tests and observations is to confirm that the organic matter in the wastewater has been properly stabilized and that the biological solids (MLSS) will settle properly in the secondary clarifier. Many things affect these objectives, such as strength of the influent wastestream, DO, hydraulic loading, health and diversity of the biological population, mixing, temperature, and presence of toxics.

Activated sludge quality and secondary effluent quality vary with sludge characteristics. Overstabilized organic matter generally produces fast-settling sludge, which may cause high effluent turbidity resulting from pin floc. Understabilized organic matter generally results in a sludge that is slow to settle and hard to keep in the clarifier, and that may even turn septic in the return sludge system.

Some tests used to monitor the activated sludge process take hours and days to complete. The Respiration Rate for activated sludge (RR$_{AS}$) test gives some insight into the health of microorganisms in the activated sludge and activity of the MLSS in the activated sludge process. The RR$_{AS}$ test takes only a few minutes to perform, and the data can be analyzed and used for process control within an hour. This is a process control method that correlates well with settleability, F/M, and MCRT and one that should be conducted at least once per week and preferably daily.

The RR$_{AS}$ test can be used to alert you to changes required in waste sludge flow rates, return sludge flow rates, aeration tank detention times, and the quantity of oxygen needed to maintain a healthy activated sludge process. Like the F/M ratio, no one specific RR_{AS} value is best for every activated sludge process. Once you identify a range of RR_{AS} values that work well in your particular operation, they should be used as an index to monitor the activated sludge process.

To determine an RR_{AS} value (expressed as milligrams oxygen consumed per hour per gram MLSS, or mg O_2/hr/gm MLSS) requires the measurement of two values: (1) the Oxygen Uptake Rate (OUR), which is the amount of dissolved oxygen used by the aeration system's microorganism population in a given amount of time (expressed as milligrams per liter oxygen consumed per minute, or mg O_2/L/min), and (2) the concentration (mg/L) of MLSS in the aeration system. OUR values may also be used to gauge the strength and quality of the aeration system influent (food), the acceptability of a new industrial wastewater to your system, the oxygen demands in your aeration system, and the quality of the secondary effluent. Thus, RR_{AS} and OUR values can be used as additional tools to further aid your understanding of activated sludge process performance in your plant.

2.3150 RESPIRATION RATE TEST

RR_{AS} values are typically determined from tests conducted on a sample of aeration tank effluent (mixed liquor). The purpose of testing a mixed liquor effluent sample is to examine the activity of the microorganisms with respect to all food sources available to the microorganisms during the preceding aeration period. A very large use of oxygen per minute generally indicates that the microorganisms are aggressively using large quantities of oxygen to stabilize the food supply. In extreme cases, however, high oxygen uptake rates may lead to septic sludge conditions in the secondary clarifier or return sludge system. Low RR_{AS} values may indicate unhealthy or inactive microorganism conditions, or a sludge that settles rapidly in the clarifier yet is associated with a turbid secondary effluent caused by pin floc.

1. SAMPLE REQUIREMENTS

  a. Sample Type

    (1) Properly collected grab samples should be obtained from various locations within the process depending upon the test method you select. Sample locations include: (1) mixed liquor effluent from the aeration system, (2) fresh return activated sludge collected just as the sludge enters the aeration system, (3) influent (food) as it is introduced to the aeration system, and (4) unchlorinated secondary effluent.

  b. Sample Volume

    (1) One liter of sample from each location listed above should be adequate for routine monitoring tests.

  c. Sample Holding Time

    (1) The test should be started immediately after sample collection.

2. REQUIRED APPARATUS AND EQUIPMENT

  a. Apparatus

    (1) 300-mL BOD bottles.

    (2) 500-mL glass graduated cylinder.

    (3) Diffused aeration stone (aquarium type).

    (4) BOD bottle adapter (for transferring contents of BOD bottles back and forth) or a 500-mL wide-mouth Erlenmeyer glass flask with stopper.

    (5) 1,000-mL glass beaker for use in maintaining sample temperature during the test.

  b. Equipment

    (1) DO probe with built-in temperature and agitator features.

    (2) Electric/magnetic stirrer if DO probe does not have an agitator feature.

    (3) Magnetic stirring bars (small).

    (4) Small air compressor (aquarium type).

    (5) Stopwatch.

3. PROCEDURE

  a. Calibrate and "zero" your DO probe.

  b. Collect a fresh 1-liter sample of aeration system effluent (mixed liquor). Immediately take the temperature of the sample. This is the temperature of the aeration system. Record the sample temperature on the Respiration-Oxygen Uptake Rate Bench Sheet (Figure 2.8).

    NOTE: It is essential that the temperature of the mixed liquor sample not change between the time the sample is collected and the end of the test. This usually means you will need to place the sample-filled bottle in a temperature maintenance device as soon as the sample is collected and during the test. Operators frequently use a 1,000-mL glass beaker containing either ice or warm water for this purpose. If the initial temperature of the sample is cool, use ice in the beaker to keep it cool. If the initial temperature is warm, use warm water in the beaker.

  c. Place an aeration stone or other aeration device in the mixed liquor sample container and aerate the sample. If an aeration device is not available, shake the sample container vigorously 5 to 15 times to saturate the sample with dissolved oxygen. Take care to fully aerate the sample initially so that the DO is established between 6.0 and 10.0 mg/L, depending on sample temperature. Upon completion of the test procedure, at least 1.0 mg/L of DO should remain in the sample. The test will not be valid otherwise.

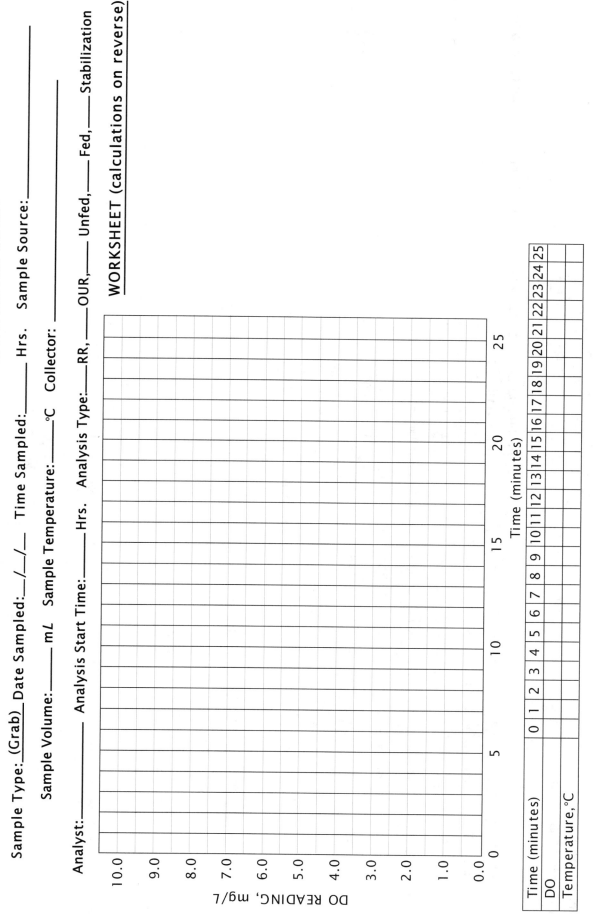

Fig. 2.8 *Respiration-oxygen uptake rate bench sheet*

1. RESPIRATION RATE:

A. ΔDO FOR: RR ——, OUR ——, UNFED ——, FED ——, STABILIZATION ——

Minute —— DO, —— mg/L – minute —— DO, —— mg/L = —— ΔDO[1]

Minute —— DO, —— mg/L – minute —— DO, —— mg/L = —— ΔDO

Minute —— DO, —— mg/L – minute —— DO, —— mg/L = —— ΔDO

Minute —— DO, —— mg/L – minute —— DO, —— mg/L = —— ΔDO

Minute —— DO, —— mg/L – minute —— DO, —— mg/L = —— ΔDO

Minute —— DO, —— mg/L – minute —— DO, —— mg/L = —— ΔDO

Minute —— DO, —— mg/L – minute —— DO, —— mg/L = —— ΔDO

B. OUR mg O_2/L/minute:

$$\left(\frac{——\ \Delta DO\ @\ min.\ ——\ -\ ——\ \Delta DO\ @\ min.\ ——}{——\ minutes} \right) = ——\ OUR,\ mg\ O_2/L/minute$$

C. OUR, mg O_2/L/hour:

—— OUR, mg O_2/L/minute x 60 minutes/hour = —— OUR, mg O_2/L/hour

D. Respiration Rate, mg O_2/hour/gm MLSS:

$$\left(\frac{——\ OUR,\ mg\ O_2/L/hour\ x\ 1{,}000\ mg}{——\ mg/L\ MLSS} \right) = ——\ RR,\ mg\ O_2/L/hr/gm\ MLSS$$

E. Load Ratio:

$$\frac{——\ Fed\ OUR,\ mg\ O_2/L/minute}{——\ Unfed\ OUR,\ mg\ O_2/L/minute} = ——\ Load\ Ratio$$

2. RETURN SLUDGE VOLUME REQUIRED FOR FED OR UNFED OUR PROCEDURE:

A. Method #1:

$$\left(\frac{\%\ Return\ Sludge\ Solids}{(1.0\ +\ ——\ \%\ Return\ Sludge\ Solids)} \right) x\ 300\ mL = ——\ mL\ Return\ Sludge\ OUR\ Sample$$

B. Method #2:

$$\left(\frac{MGD,\ Return\ Sludge\ Flow}{——\ MGD,\ Return\ Sludge\ Flow\ +\ ——\ MGD,\ Influent\ Flow} \right) x\ 300\ mL = ——\ mL\ Return\ Sludge\ OUR\ Sample$$

C. Method #3:

$$\left(\frac{\%\ Aeration\ System\ Solids}{\%\ Return\ Sludge\ Solids} \right) x\ 300\ mL = ——\ mL\ Return\ Sludge\ OUR\ Sample$$

[1] Multiply all delta DOs by 2 if DO readings were taken every ¹/₂ minute because of rapid oxygen uptake.

Fig. 2.8   Respiration-oxygen uptake rate bench sheet (continued)

d. After confirming that the mixed liquor sample DO is between 6.0 and 10.0 mg/L, pour a well-mixed portion of the sample into a 300-mL BOD bottle until the bottle is full. Add a magnetic stirring bar if your DO probe does not have an agitator.

e. Place the BOD bottle containing the mixed liquor sample in the 1,000-mL temperature maintenance device (to keep the sample cool, use ice; to keep the sample warm, use warm water). If your DO probe does not have an agitator, place the 1,000-mL beaker containing the sample-filled 300-mL bottle on the electric/magnetic stirrer. Place the DO probe in the BOD bottle making sure not to trap air bubbles between the bottle and the probe shaft.

f. Start the DO probe agitator (or electric/magnetic stirrer). The stirrer should be adjusted to provide just enough gentle stirring to keep all the mixed liquor solids in suspension. Immediately determine the sample temperature followed by the DO. The DO and temperature determined at this point in the test are the "zero minute" DO and temperature and should be recorded on the Bench Sheet. Also plot the DO value on the Bench Sheet graph. Leave the DO probe agitator (or electric/magnetic stirrer) running. Activate the stopwatch for an elapsed time countdown of one minute.

g. After one minute has elapsed (continue to let the stopwatch run), read and record the temperature and DO of the sample again. After the second one-minute time period has elapsed, read and record the temperature and DO of the sample again while the stopwatch continues to run.

h. Continue performing the steps outlined in Item g above until a fairly consistent rate of DO change per minute (known as delta (Δ) DO) over a given period of time is observed, or until the sample DO reaches 1.0 mg/L.

The example below shows a fairly consistent ΔDO rate of change:

minute 1 = 0.2 mg/L

minute 2 = 0.3 mg/L

minute 3 = 0.4 mg/L

minute 4 = 0.4 mg/L

minute 5 = 0.4 mg/L

minute 6 = 0.4 mg/L

minute 7 = 0.4 mg/L

minute 8 = 0.3 mg/L

minute 9 = 0.25 mg/L

The most consistent rate of DO change per minute (0.4 mg/L) occurs from minute 3 through minute 7.

i. ΔDO is calculated as follows:

minute 1 DO minus minute 2 DO = ΔDO

minute 2 DO minus minute 3 DO = ΔDO

minute 3 DO minus minute 4 DO = ΔDO, etc.

NOTE: If DO readings are taken every ½ minute (which will be the case in tests where the rate of oxygen uptake is rapid), multiply the ΔDO by 2 to arrive at the DO per minute.

j. The DO value obtained at the time period when the rate of fairly consistent DO change began (minute 3 in the ΔDO example), together with the DO value obtained through the time period just before consistency in DO values changed (minute 7 in the ΔDO example), and the total time of the test in which the DO change was consistent (4 minutes) are used to calculate the slope (Bench Sheet graph) or oxygen uptake rate (OUR) expressed as mg O_2/L/min:

$$\text{OUR, mg } O_2/L/\text{min} = \frac{\text{DO at minute 3} - \text{DO at minute 7}}{4 \text{ minutes}}$$

k. The OUR, mg O_2/L/min, must then be converted to mg O_2/L/hour:

OUR, mg O_2/L/hr = OUR, mg O_2/L/min × 60 min/hr

l. Upon completion of the DO determination portion of the test, pour off a sufficient volume of the mixed liquor sample and perform the Total Suspended Solids (TSS) test to determine the mg/L MLSS.

m. Once the mg/L MLSS value has been determined, the RR$_{AS}$ value, expressed as mg O_2/hr/gm MLSS may be calculated:

$$\text{Respiration Rate, mg } O_2/\text{hr/gm MLSS} = \frac{(\text{OUR, mg } O_2/L/\text{hr} \times 1,000 \text{ mg/gm})}{\text{mg/}L \text{ MLSS}}$$

NOTE: To relate the OUR value to the volatile content of the mixed liquor, use mg/L MLVSS.

2.3151 INTERPRETATION OF RESPIRATION RATE VALUES

The RR$_{AS}$ value, by itself, is useful to indicate the relative biological activity of the aeration system microorganisms. Typical RR$_{AS}$ values for activated sludge systems (mg O_2/hr/gm MLSS) are:

• Less than 4 mg O_2/hr/gm MLSS = Biological population not stable and healthy. Possible toxic load applied to the aeration system.

• 4 to 9 mg O_2/hr/gm MLSS = Typical of extended aeration processes; *ENDOGENOUS RESPIRATION*[12] activity. Also found where organic matter is slow to biodegrade.

---

[12] *Endogenous* (en-DODGE-en-us) *Respiration.* A situation in which living organisms oxidize some of their own cellular mass instead of new organic matter they adsorb or absorb from their environment.

- 10 to 20 mg O_2/hr/gm MLSS = Normal for most conventional activated sludge processes.

- More than 20 mg O_2/hr/gm MLSS = Rapidly biodegradable organic matter; typical of high-rate activated sludge processes. Could also indicate understabilized organic matter.

Low RR_{AS} values, generally 4 to 9 mg O_2/hr/gm MLSS, typically indicate overstabilized organic matter. This could be caused by:

1. High MLVSS values.

2. Low F/M ratio (for example, F/M of 0.1—not enough food to support balanced microorganism activity).

3. High MCRT (more than 15 days) as could occur in extended aeration type processes.

4. Long aeration system detention times. A low RR_{AS} value could also result from a toxic shock load reaching the aeration tank.

Higher RR_{AS} values generally indicate aggressive microorganism activity coupled with rich organic matter. On the other hand, higher values may indicate insufficient organic matter stabilization in the aeration system. Insufficient organic matter stabilization could occur for several reasons:

1. Low MLVSS concentration (for example, not enough viable organisms to oxidize the food source within the aeration system).

2. High F/M ratio (for example, too much food, which overwhelms the microorganism population).

3. Short MCRT (for example, microorganism population indicative of a "young sludge").

4. Short aeration system detention time (for example, the food and microorganisms contact time is insufficient to stabilize the incoming food; hydraulic overload).

High respiration rates may also indicate that continued stabilization is taking place in the secondary clarifier resulting in low DO values in the return sludge or development of *ANOXIC*[13] conditions in the clarifier.

The best way to use RR_{AS} values is to measure them daily and prepare a trend chart of the RR_{AS} values and other process guidelines. By preparing a trend chart, you can readily see fluctuations in RR_{AS} values, establish the best range of RR_{AS} values for your particular process, and implement operational control strategies to improve process performance.

2.3152   RESPIRATION RATE TREND INTERPRETATIONS

If mixed liquor settleability tests indicate a normal or slow settling sludge, an increase in the RR_{AS} value trend may indicate

that the organic matter in the wastewater is less stabilized than before. More time or more organisms are needed to properly stabilize the incoming food source. Three control strategies may correct this problem: (1) decrease the return sludge flow rate to minimize hydraulic loading through the aeration system (permits more time for stabilization), (2) decrease the waste sludge flow rate to increase the MCRT, and (3) if your system is so configured, increase the aeration system detention time (step feed is a good process modification for this latter control strategy approach).

If the RR_{AS} value trend plot indicates a decreasing RR_{AS} value (organic matter is more stabilized than before), the return sludge flow rate can be increased, the waste sludge flow rate can be increased, or the aeration system detention time can be decreased. These control strategies will increase hydraulic loading through the aeration system and reduce the MCRT by removing excess microorganisms.

If you are operating an activated sludge system with a significant sludge blanket (using the secondary clarifier as storage for microorganisms) and you have a rapidly settling sludge, then process control based on RR_{AS} is as follows: As the trend of the RR_{AS} value increases (indicating lower levels of organic matter stabilization), increase the return sludge flow rate to move organisms out of the clarifier and into the aeration system to feed on the incoming food. Next, decrease the waste sludge flow rate to increase the MLVSS concentration.

If the RR_{AS} value trend is decreasing (indicating higher levels of organic matter stabilization), decrease the return sludge flow rate, thus storing more microorganisms in the clarifier, and increase the waste sludge flow rate in order to decrease the MLVSS concentration.

NOTE: Make one process control change at a time and allow time for the results to be seen in the overall process. As a "rule of thumb": A change in the return sludge flow rate noticeably affects the process in one-half to one aeration tank detention time (several hours); a change in the waste sludge flow rate noticeably affects the process in one-half to one MCRT (several days).

QUESTIONS

Write your answers in a notebook and then compare your answers with those on page 148.

2.3G   What problems are caused by overstabilized and understabilized organic matter?

2.3H   The respiration rate (RR_{AS}) test can be used to alert operators to changes required in what operating guidelines?

2.3I   What can low respiration rate (RR_{AS}) values indicate?

---

[13] *Anoxic* (an-OX-ick).   A condition in which the aquatic (water) environment does not contain dissolved oxygen (DO), which is called an oxygen deficient condition. Generally refers to an environment in which chemically bound oxygen, such as in nitrate, is present. The term is similar to ANAEROBIC.

2.316   Oxygen Uptake Rate (OUR) Tests

You can obtain a fairly good estimate of effluent quality, as well as food quality and strength, by calculating the oxygen uptake rate (OUR) value (expressed as mg O_2/L/min). Two different test procedures are commonly used:

1. The unfed OUR procedure, in which unchlorinated secondary effluent is mixed with return sludge.

2. The fed OUR procedure, in which aeration system influent (food) is mixed with return sludge.

Test results can be obtained in 20 minutes or so and may be correlated with plant influent and secondary effluent BOD_5. Thus, you can have a "real time" process control tool that can help you achieve a higher degree of operational control.

2.3160   GENERAL PROCEDURES

1. Determine the volume of return sludge sample that will be needed to perform the unfed or fed OUR procedure. This volume can be calculated in several ways, but the general objective is to arrive at a volume of return sludge in the same proportion as found in the aeration system. Listed below are three different formulas that may be used to determine the appropriate volume of return sludge to use:

   a. If you know the percent solids in the return sludge, the formula is:

   Return Sludge Sample Volume, mL =

   $$\left(\frac{\% \text{ Ret Sludge Solids}}{(1.0 + \% \text{ Ret Sludge Solids})}\right) \times \begin{array}{c} 300 \text{ m}L, \text{ BOD} \\ \text{bottle volume} \end{array}$$

   b. If you know the return sludge flow and the aeration system influent flow rates, the formula is:

   Return Sludge Sample Volume, mL =

   $$\left(\frac{\text{Ret Sludge Flow, MGD}}{(\text{Ret Sludge Flow, MGD} + \text{Infl Flow, MGD})}\right) \times \begin{array}{c} 300 \text{ m}L, \text{ BOD} \\ \text{bottle volume} \end{array}$$

   c. If you usually use a centrifuge to determine sludge solids concentrations, conduct a 15-minute spin on the aeration system mixed liquor and the return activated sludge:

   Return Sludge Sample Volume, mL =

   $$\left(\frac{\% \text{ Aeration System Solids}}{\% \text{ Ret Sludge Solids}}\right) \times \begin{array}{c} 300 \text{ m}L, \text{ BOD} \\ \text{bottle volume} \end{array}$$

   Whichever calculation method you prefer to use, use it all of the time. Consistency is the key to reliable data.

2. Calibrate and "zero" your DO probe.

3. Collect fresh, individual 1-liter samples of: (1) unchlorinated secondary effluent for the unfed procedure, (2) aeration system influent for the fed procedure, and (3) return activated sludge as it enters the aeration system. Immediately measure the temperature of the return activated sludge sample. For purposes of this test, this temperature will be the temperature of the aeration system. Record the temperature information on the Bench Sheet.

NOTE: It is essential that the temperature of the mixed liquor sample not change between the time the sample is collected and the end of the test. This usually means you will need to place the sample-filled bottle in a temperature maintenance device as soon as the sample is collected and during the test. Operators frequently use a 1,000-mL glass beaker containing either ice or warm water for this purpose. If the initial temperature of the sample is cool, use ice in the beaker to keep it cool. If the initial temperature is warm, use warm water in the beaker.

4. For the return activated sludge sample, place an aeration stone or other aeration device in the sample container and aerate the sample. If an aeration device is not available, shake the sample container vigorously 5 to 15 times to saturate the sample with dissolved oxygen. Care must be taken to fully aerate the sample initially so that the DO is established between 6.0 and 10.0 mg/L, depending on sample temperature. Upon completion of the test procedure, at least 1.0 mg/L of DO should remain in the sample. The test will not be valid otherwise.

2.3161   UNFED OXYGEN UPTAKE RATE (OUR) PROCEDURE

1. Pour into a BOD bottle a sample of well-mixed activated sludge. You calculated the sample volume in Section 2.3160, Step 1, above. Add a magnetic stirring bar if your DO probe does not have an agitator. Then fill the bottle to the top with unchlorinated secondary effluent. Do not overfill. Place the BOD bottle adapter into the full bottle and connect an empty BOD bottle upside down on the adapter. Aerate by inverting (turning the bottles so the empty one is on the bottom) and shaking the contents from the full BOD bottle into the empty bottle. When the liquid transfer is complete, again rotate the BOD bottles to their original positions and shake the contents from the full BOD bottle back into the original bottle. Carefully remove the BOD bottle adapter.

   If you do not have a BOD bottle adapter, use the following method to aerate the sample. Using a graduated cylinder, carefully measure the calculated volume of return activated sludge and the volume of unchlorinated secondary effluent that is required to bring the total combined sample volume to 300 mL. Pour these volumes (a total of 300 mL) into a single, 500-mL capacity, wide-mouth Erlenmeyer glass flask. Stopper the flask and shake the flask vigorously to thoroughly mix and aerate the combined sample volumes. Next, pour the well-mixed and fully aerated sample into a 300-mL BOD bottle.

2. Place the BOD bottle containing the combined return activated sludge and unchlorinated secondary effluent in the 1,000-mL temperature maintenance device (to keep the sample cool, use ice; to keep the sample warm, use warm water). If your DO probe does not have an agitator, place the 1,000-mL beaker containing the sample-filled 300-mL

bottle on the electric/magnetic stirrer. Place the DO probe in the BOD bottle making sure not to trap air bubbles between the bottle and the probe shaft.

3. Start the DO probe agitator (or electric/magnetic stirrer). The stirrer should be adjusted to provide just enough gentle stirring to keep all the solids in suspension. Immediately determine the sample temperature followed by the DO. The DO and temperature determined at this point in the test are the "zero minute" DO and temperature and should be recorded on the Bench Sheet. Also plot the DO value on the Bench Sheet graph. Leave the DO probe agitator (or electric/ magnetic stirrer) running. Activate the stopwatch for an elapsed time countdown of 1 minute.

4. After 1 minute has elapsed (continue to let the stopwatch run), read and record the temperature and DO of the sample again. After the second 1-minute time period has elapsed, read and record the temperature and DO of the sample again while the stopwatch continues to run.

5. Continue performing the steps outlined in Item 4 above until a fairly consistent rate of DO change per minute (ΔDO) over a given period of time is observed, or until the sample DO reaches 1.0 mg/L.

The example below shows a fairly consistent ΔDO rate of change:

minute 1 = 0.2 mg/L

minute 2 = 0.3 mg/L

minute 3 = 0.4 mg/L

minute 4 = 0.4 mg/L

minute 5 = 0.4 mg/L

minute 6 = 0.4 mg/L

minute 7 = 0.4 mg/L

minute 8 = 0.3 mg/L

minute 9 = 0.25 mg/L

The most consistent rate of DO change per minute (0.4 mg/L) occurs from minute 3 through minute 7.

6. ΔDO is calculated as follows:

minute 1 DO minus minute 2 DO = ΔDO

minute 2 DO minus minute 3 DO = ΔDO

minute 3 DO minus minute 4 DO = ΔDO, etc.

NOTE: If DO readings are taken every ½ minute (which will be the case in tests where the rate of oxygen uptake is rapid), multiply the ΔDO by 2 to arrive at the DO per minute.

7. The DO value obtained at the time period when the rate of fairly consistent DO change began (minute 3 in the ΔDO example), together with the DO value obtained through the time period just before the consistency in DO values changed (minute 7 in the ΔDO example), and the total time

of the test in which the DO change was consistent (4 minutes) are used to calculate the slope (Bench Sheet graph) or oxygen uptake rate (OUR) expressed as mg O_2/L/min:

$$\text{OUR, mg } O_2/L/\text{min} = \frac{\text{DO at minute 3} - \text{DO at minute 7}}{4 \text{ minutes}}$$

8. Typical OUR values (mg O_2/L/min) for unfed samples are commonly interpreted as follows:

- OUR value is less than 0.2 = Old sludge; overstabilized organic matter; possible pin floc; toxic load may have killed the microorganisms. Increase the waste sludge flow rate.

- OUR value is between 0.3 and 0.7 = A stable effluent. Good settling characteristics in the secondary clarifier.

- OUR value is greater than 1.0 = Young sludge; poor settleability; high effluent BOD; poor organic matter stabilization; high F/M and low MCRT. Increase the return activated sludge flow rate.

2.3162 FED OXYGEN UPTAKE RATE (OUR) PROCEDURE

1. Pour into a BOD bottle a well-mixed sample of aerated return activated sludge. You calculated the sample volume in Section 2.3160, Step 1. Add a magnetic stirring bar if your DO probe does not have an agitator. Then fill the bottle to the top with aeration system influent. Do not overfill. Place the BOD bottle adapter into the full bottle and connect an empty BOD bottle upside down on the adapter. Aerate by inverting (turning the bottles so the empty one is on the bottom) and shaking the contents from the full BOD bottle into the empty bottle. When the liquid transfer is complete, invert the BOD bottles again and shake the contents from the full BOD bottle back into the original bottle. Carefully remove the BOD bottle adapter.

If you do not have a BOD bottle adapter, use the following method to aerate the sample. Using a graduated cylinder, carefully measure the calculated volume of return activated sludge and the volume of aeration system influent that is required to bring the total combined sample volume to 300 mL. Pour these volumes (a total of 300 mL) into a single, 500-mL capacity, wide-mouth Erlenmeyer glass flask. Stopper the flask and shake the flask vigorously to thoroughly mix and aerate the combined sample volumes. Next, pour the well-mixed and fully aerated sample into a 300-mL BOD bottle.

2. Place the BOD bottle containing the combined return activated sludge aeration system influent in the 1,000-mL temperature maintenance device (to keep the sample cool, use ice; to keep the sample warm, use warm water). If your DO probe does not have an agitator, place the 1,000-mL beaker containing the sample-filled 300-mL bottle on the electric/magnetic stirrer. Place the DO probe in the BOD bottle making sure not to trap air bubbles between the bottle and the probe shaft.

3. Start the DO probe agitator (or electric/magnetic stirrer). The stirrer should be adjusted to provide just enough gentle stirring to keep all the solids in suspension. Immediately

determine the sample temperature followed by the DO. The DO and temperature determined at this point in the test are the "zero minute" DO and temperature and should be recorded on the Bench Sheet. Also plot the DO value on the Bench Sheet graph. Leave the DO probe agitator (or electric/magnetic stirrer) running. Activate the stopwatch for an elapsed time countdown of 1 minute.

4. After 1 minute has elapsed (continue to let the stopwatch run), read and record the temperature and DO of the sample again. After the second 1-minute time period has elapsed, read and record the temperature and DO of the sample again while the stopwatch continues the run.

5. Continue performing the steps outlined in Item 4 above until a fairly consistent rate of DO change per minute (ΔDO) over a given period of time is observed, or until the sample DO reaches 1.0 mg/L.

The example below shows a fairly consistent ΔDO rate of change:

minute 1 = 0.2 mg/L

minute 2 = 0.3 mg/L

minute 3 = 0.4 mg/L

minute 4 = 0.4 mg/L

minute 5 = 0.4 mg/L

minute 6 = 0.4 mg/L

minute 7 = 0.4 mg/L

minute 8 = 0.3 mg/L

minute 9 = 0.25 mg/L

The most consistent rate of DO change per minute (0.4 mg/L) occurs from minute 3 through minute 7.

6. ΔDO is calculated as follows:

minute 1 DO minus minute 2 DO = ΔDO

minute 2 DO minus minute 3 DO = ΔDO

minute 3 DO minus minute 4 DO = ΔDO, etc.

NOTE: If DO readings are taken every ½ minute (which will be the case in tests where the rate of oxygen uptake is rapid), multiply the ΔDO by 2 to arrive at the DO per minute.

7. The DO value obtained at the time period when the rate of fairly consistent DO change began (minute 3 in the ΔDO example), together with the DO value obtained through the time period just before the consistency in DO values changed (minute 7 in the ΔDO example), and the total time of the test in which the DO change was consistent (4 minutes) are used to calculate the slope (Bench Sheet graph) or oxygen uptake rate (OUR) expressed as mg O$_2$/L/min:

$$\text{OUR, mg O}_2/L/\text{min} = \frac{\text{DO at minute 3} - \text{DO at minute 7}}{4 \text{ minutes}}$$

8. Typical OUR values (mg O$_2$/L/min) for fed samples are as follows:

- Normal OUR values range between 2.0 and 2.5.

- Lower OUR values indicate a possible toxic shock load to the aeration system.

2.317 Load Ratio

The Load Ratio (LR) can be used for further process evaluation. The LR is calculated using the fed and unfed OUR values as follows:

$$\text{Load Ratio (unitless)} = \frac{\text{Fed OUR, mg O}_2/L/\text{min}}{\text{Unfed OUR, mg O}_2/L/\text{min}}$$

Load ratio values are commonly interpreted as follows:

- If the LR value is less than 1.0, this typically indicates a toxic substance has entered the aeration system influent and is inhibiting microorganism activity.

- An LR value between 1.0 and 2.0 typically indicates that:

  - The aeration system influent is weak in organic strength, or

  - The return activated sludge is unhealthy, or

  - The aeration system influent contains organic matter that takes longer to stabilize, or

  - The aeration system influent contains a toxic substance. Microscopic examination of the microorganisms in the aeration system would help you verify the types of organisms present and their condition.

- An LR value between 2.0 and 5.0 has been found to indicate normal activated sludge system operating conditions.

- If the LR is greater than 5.0, this typically indicates poor organic matter stabilization (anticipate poor sludge settleability), or the aeration system influent is very strong in terms of organic strength.

2.318 Organic Matter Stabilization Test

The organic matter stabilization test is a procedure you may use to determine how long a particular strength and quality of organic matter in wastewater will take to stabilize in the aeration system. The unfed OUR value and the OUR value from the organic matter stabilization test are used to determine the time it will take for organic matter to become properly stabilized.

The organic matter stabilization test is particularly useful when a new industrial waste discharger is considering discharging to your treatment plant. The test is also valuable for determining the best aeration system detention time for complete organic matter stabilization if your plant is subjected to seasonal organic load variations. It is also useful for managing the step-feed aeration system process. The organic matter stabilization test takes several hours to complete and the information obtained is not applicable for daily process control purposes. Usually, the test is only run when there is a specific need for the information it provides.

PROCEDURE

You will need to construct a bench-top aeration tank system that works like the aeration tank system in your plant. You can use a 10-gallon capacity aquarium tank (or other 10-gallon capacity container) equipped with an aeration system. Mark the five-gallon liquid level on the side of the container or tank.

1. Perform the unfed OUR test on your aeration system as described in Section 2.3161.

2. Determine the volume of return activated sludge sample that will be needed to conduct the organic matter stabilization test. This volume is calculated in the manner described in Section 2.3160, Step 1, except that a "5-gallon container volume" value is used (multiplied) in place of the "300-mL BOD bottle volume."

3. Obtain a sample of return sludge equal to the volume you calculated above and pour this sample volume into the tank or container. Start the aeration system.

4. Next, obtain a sample of the wastewater you want to use in the stabilization test such as that from an industrial process. Pour this sample volume into the tank or container until the liquid level in the tank or container reaches the five-gallon liquid level mark. Allow the contents in the tank or container to mix well and become diffused with between 6.0 and 10.0 mg/L DO.

5. Obtain a 300-mL sample of the well-mixed contents of the tank or container and determine the OUR of the sample. Plot and record the oxygen consumption (ΔDO) on the Bench Sheet. Upon completion of the test, pour the 300-mL sample volume back into the tank or container.

6. Allow 15 minutes to elapse, collect another 300-mL sample from the tank or container and determine the OUR of the sample. Plot and record the oxygen consumption (ΔDO) on a new Bench Sheet. Upon completion of the test, pour the 300-mL sample volume back into the tank or container.

7. Repeat Steps 5 and 6 every 15 minutes until the OUR, mg O_2/L/min equals the unfed OUR, mg O_2/L/min value obtained in Step 1 above.

8. The time it takes for the stabilization test OUR to equal the unfed OUR is known as the Stabilization Time. This Stabilization Time is the period it will take for the organic matter in the test sample to become properly stabilized in your aeration system.

2.319 Acknowledgment

Portions of Sections 2.315 through 2.318 were provided by Douglas Lee Miller, Regional Manager of Operations, Operations Management International (OMI), Inc.

QUESTIONS

Write your answers in a notebook and then compare your answers with those on page 148.

2.3J What information can be obtained from the oxygen uptake rate (OUR)?

2.3K What does a high load ratio (LR greater than 5) indicate?

2.3L What is the purpose of the organic matter stabilization test?

2.32 Microscopic Examination

Some operators use a microscope to examine the microorganisms in the mixed liquor for an indication of the condition of the activated sludge treatment process. The majority of the BOD is removed by common zoogleal microorganisms. The microorganisms that are important indicators in the activated sludge process are the *PROTOZOA*[14] and *ROTIFERS*.[15] The protozoa eat the bacteria and help produce a clear effluent. The presence of rotifers is an indication of a stable effluent. If excessive *FILAMENTOUS ORGANISMS*[16] are observed, you usually can expect an activated sludge that settles poorly.

If most of the microorganisms in the mixed liquor suspended solids are protozoa (ciliates) and rotifers, you can expect an activated sludge with good settling characteristics. By using the proper return activated sludge (RAS), waste activated sludge (WAS), and aeration rates, you can produce an effluent with a BOD of less than 10 mg/L.

Apparently, some filamentous organisms are good, but too many are bad. Filamentous organisms can form a network or backbone upon which clumps of activated sludge can gather; this produces a floc with excellent settling characteristics. If the

---

[14] *Protozoa* (pro-toe-ZOE-ah).   A group of motile, microscopic organisms (usually single-celled and aerobic) that sometimes cluster into colonies and generally consume bacteria as an energy source.

[15] *Rotifers* (ROTE-uh-fers).   Microscopic animals characterized by short hairs on their front ends.

[16] *Filamentous* (fill-uh-MEN-tuss) *Organisms*.   Organisms that grow in a thread or filamentous form. Common types are *Thiothrix* and *Actinomycetes*. A common cause of sludge bulking in the activated sludge process.

filaments become excessive, however, a bridging mechanism forms and prevents the numerous small clumps of sludge from gathering or packing together. If the floc are prevented from clumping together, sufficient particle mass will not be produced to achieve good settling rates.

Three groups of protozoa are important to the operator of an activated sludge process.

1. Amoeboids

2. Flagellates

3. Ciliates

AMOEBOIDS (Figure 2.9)

Look for amoeboids in the mixed liquor suspended solids floc during start-up periods or when the process is recovering from an upset condition.

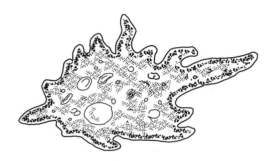

Fig. 2.9 Amoeboids

FLAGELLATES (Figure 2.10)

Flagellates are usually found with a light, dispersed, straggler floc, a low population of microorganisms, and a high organic (BOD) load. With a high organic (BOD) load, other microorganisms will start to thrive, a more dense sludge floc will develop, and the flagellate population will decrease.

Fig. 2.10 Flagellates

CILIATES

Ciliates are usually found in large numbers when the activated sludge is in a fair to good settling condition. Ciliates are classified into two basic groups, free-swimming ciliates (Figure 2.11) and stalked ciliates (Figure 2.12).

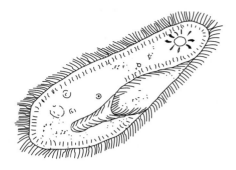

Fig. 2.11 Free-swimming ciliates

Fig. 2.12 Stalked ciliates

Free-swimming ciliates are usually present when there is a large number of bacteria in the activated sludge. These organisms feed on bacteria and help produce a clear effluent. They are associated with a good degree of treatment.

Stalked ciliates are usually present when the free-swimming ciliates are unable to compete for the available food. A large number of stalked ciliates and rotifers (Figure 2.13) will indicate a stable and efficiently operating activated sludge process.

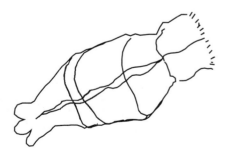

Fig. 2.13 Rotifers

The types of microorganisms and the numbers of microorganisms can be used as a guide in making activated sludge process control adjustments. Figure 2.14 can help you determine whether the mean cell residence time (MCRT) should be increased or decreased. A decline in microorganisms, especially ciliates, is frequently a warning of a poorly settling sludge. If a

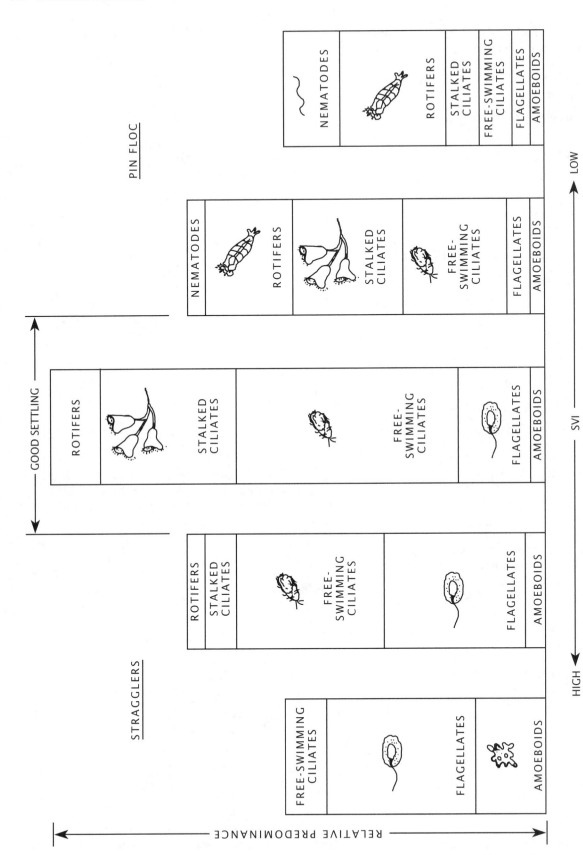

Fig. 2.14   Relative number of microorganisms versus sludge quality

(Bar graph taken from *EPA PROCESS CONTROL MANUAL FOR AEROBIC BIOLOGICAL WASTEWATER TREATMENT FACILITIES*, 1977) relatively large number of amoeboids and flagellates are observed, try increasing the MCRT. If the numbers of microorganisms are relatively low with rotifers predominating and you have a pin floc, try decreasing the MCRT.

These observations can allow an operator to detect a change in organic loading or in the level of a toxic chemical before the activated sludge process becomes upset. The changes in type and number of microorganisms should be compared with observations of the settling characteristics of the mixed liquor suspended solids in the 30-minute settleability test and with the calculated food to microorganism ratio.

In summary, large numbers of ciliates and rotifers are an indication of a stable activated sludge process that will produce a high-quality effluent. For additional information on microscopic examination of activated sludge, see Section 11.10, "Microbiology for Activated Sludge," by Paul V. Bohlier in *OPERATION OF WASTEWATER TREATMENT PLANTS,* Volume II.

Another good source of information about the microorganisms found in activated sludge processes is *WASTEWATER MICROBIOLOGY* by Gabriel Bitton. This book can be purchased from John Wiley & Sons, Inc., Customer Care Center (Consumer Accounts), 10475 Crosspoint Boulevard, Indianapolis, IN 46256. ISBN 0-471-65071-4. Price, $99.95, plus $5.00 shipping and handling.

2.33 The Al West Method

Section 2.2, "Return Activated Sludge," and Section 2.3, "Waste Activated Sludge," have outlined various methods operators use to "control" their activated sludge process. Al West has correctly observed that the activated sludge "process is *NOT* controlled by attempting to achieve *PRECONCEIVED* levels of *INDIVIDUAL* variables such as mixed liquor sludge concentration, mean cell residence time, and food to microorganism ratios. *CONTROL* tests such as final clarifier sludge blanket depth determinations, mixed liquor and return sludge concentrations (by centrifuge), and sludge settleability are used to define sludge quality and process status and to determine process adjustments." Mr. West worked continuously to develop better ways for operators to control the activated sludge process.

2.34 Summary of RAS and WAS Rates

How should you operate your activated sludge process? Only you can answer this question. In Chapters 8 and 11 of Volumes I and II of *OPERATION OF WASTEWATER TREATMENT PLANTS,* we outlined what we consider are simple and direct procedures for operating package plants, oxidation ditches, and conventional activated sludge plants. In Chapter 2 of this manual, various alternative methods were outlined on how to control the activated sludge process.

What counts is the effluent quality from your activated sludge plant. The effluent quality is influenced by influent characteristics and conditions in the aeration tank and secondary clarifier. Observe these characteristics and conditions; you must be alert for any change and make appropriate adjustments to control the activated sludge process. Every operator on every shift must follow the same procedures. Try not to adjust your RAS and WAS rates by more than 10 or 15 percent from one day to the next day. Select a method you understand, record and analyze data, stick with your method, and you can make it work to produce a good effluent.

2.35 Acknowledgment

Major portions of Section 2.3 were adapted from *PROCESS CONTROL MANUAL FOR AEROBIC BIOLOGICAL WASTEWATER TREATMENT FACILITIES,* US Environmental Protection Agency. Obtain from National Technical Information Service (NTIS), 5285 Port Royal Road, Springfield, VA 22161. Order No. PB-279474. EPA No. 430-9-77-006. Price, $86.50, plus $5.00 shipping and handling per order.

QUESTIONS

Write your answers in a notebook and then compare your answers with those on page 148.

2.3M Which microorganisms are important indicators in the activated sludge process?

2.3N In the Al West method, what important activated sludge control tests are used to define sludge quality and process status?

END OF LESSON 3 OF 4 LESSONS

on

ACTIVATED SLUDGE

Please answer the discussion and review questions next.

DISCUSSION AND REVIEW QUESTIONS
Chapter 2.   ACTIVATED SLUDGE
Pure Oxygen Plants and Operational Control Options
(Lesson 3 of 4 Lessons)

Write the answers to these questions in your notebook. The question numbering continues from Lesson 2.

9. What items are affected by the amount of waste activated sludge (WAS) removed from the process?

10. How will you know when you have established the best mode of process control for your plant?

11. What is the basis for the F/M method of controlling the activated sludge process?

12. How would you adjust the MCRT to produce a nitrified effluent?

13. What are the advantages and limitations of intermittent wasting of activated sludge?

14. How would you select the RAS and WAS rates for your activated sludge plant?

CHAPTER 2. ACTIVATED SLUDGE

Pure Oxygen Plants and Operational Control Options

(Lesson 4 of 4 Lessons)

2.4 TREATMENT OF BOTH MUNICIPAL AND INDUSTRIAL WASTES

2.40 Monitoring Industrial Waste Discharges
(Also see *PRETREATMENT FACILITY INSPECTION* in this series of training manuals.)

Industrial manufacturing processes of most types generate some waste materials. These waste materials take the form of liquid, gaseous, or solid residuals. In most cases, the deliberate and indiscriminate disposal of these waste materials to the collection system will have a deteriorating effect upon the activated sludge process and possible detrimental effects on the treatment plant effluent receiving waters.

As discussed in Chapter 11, Section 11.040, *OPERATION OF WASTEWATER TREATMENT PLANTS*, Volume II, you should become acquainted with the various industrial facilities in your area to determine what, if any, adverse impact they may create on your activated sludge process.

2.400 *Establishing a Monitoring System (Pretreatment Inspection)*

Regulations at both state and federal levels usually require that industrial waste monitoring be established as an important part of an industrial waste control and treatment system.

An industrial waste monitoring program is valuable for the following reasons:

1. To assure regulatory agencies of industrial compliance with discharge requirements and implementation schedules set forth in the discharge permit.

2. To maintain sufficient control of treatment plant operations to prevent NPDES permit violations.

3. To gather necessary data for the future design and operation of the treatment plant.

In establishing a monitoring program, one of the first tasks should be an examination of the wastewater characteristics of each industry that discharges to the collection system. Awareness of the specific types of harmful waste materials that may enter the collection system will help you prepare a monitoring program to protect the treatment plant and receiving waters.

Some industries pretreat wastewaters before discharge to the collection system to recover valuable materials, to reduce sewer-use charges, and to keep undesirable constituents out of the sewers and treatment plant. However, if undesirable constituents are known to be present in the municipal wastewater stream, pretreatment of the industrial wastewater portion must be enforced to reduce the constituents to acceptable levels. Proper monitoring at the site of an industrial wastewater pretreatment plant is essential. Additionally, if it is likely that an accidental spill or unlawful discharge may escape pretreatment and enter the collection system, a sophisticated monitoring system should be installed at your treatment plant. The reasons for wastewater monitoring at the treatment plant itself include establishing a last point of measurement of certain problem constituents before entering the unit processes so that the operator can start corrective operational measures where possible.

2.401 *Automatic Monitoring Units*

Automatic monitoring of several wastewater characteristics has been a dependable method of alerting the operator to abnormal influent wastewater conditions. Automatic devices are commonly used to monitor flow, pH, oxidation potential (conductivity), suspended solids (SS), and dissolved oxygen (DO). Other useful automatic devices include wastewater samplers and gas-phase hydrocarbon analyzers.

Normally, data are recorded on a strip chart recorder or transmitted directly to a computer data management system. In addition, other equipment, such as pumps or valves, may be activated by these monitoring devices. The monitoring systems should also be combined with an alarm system that will give the operator warning when a high concentration of an undesirable water quality indicator reaches the monitoring station.

Numerous water quality indicators are used for operational controls, yet the number of water quality indicators that can be automatically measured without difficulty is limited. Therefore, it is essential that the operator also rely on the appearance and odor of the influent wastewater as part of the monitoring program at the treatment plant.

2.41 Common Industrial Wastes

Reduction or elimination of harmful waste constituents from industrial discharges may be controlled by enforcing your municipal sewer ordinance. Several objectionable industrial waste constituents and the possible effects of each waste are discussed below.

1. Flammable Oils.   Examples of flammable oils are crude gasoline, benzene, naphtha, fuel oil, and mineral oil. These substances are not soluble (do not dissolve in water); they tend to collect in pools, thus creating potentially explosive conditions. When methane gas is mixed with flammable oils, a very powerful explosion may result.

2. Toxic Gases.   Toxic gases such as hydrogen sulfide (H_2S), methane (CH_4), and hydrogen cyanide (HCN) are often present or may be formed in industrial discharges. Wastewater with high quantities of sulfate can cause problems in anaerobic decomposition, due to the formation of H_2S. Also, cyanide combines with acid wastes to form the extremely toxic gas, HCN.

3. Oils and Greases.   A municipal plant generally does not have facilities for the removal of significant quantities of oils and grease. Pretreatment of wastewater may be desirable to reduce the total concentration of oils and grease (hexane extractables). In general, emulsified oils and greases of vegetable and animal origin are biodegradable and can be successfully treated by a properly designed municipal treatment facility. However, oils and greases of mineral origin may cause problems and these are the constituents generally requiring pretreatment.

4. Settleable Solids.   Settleable solids cause obstructions in the sewer system by settling and accumulating. At places where wastewater accumulates, anaerobic decomposition may take place, producing undesirable products such as hydrogen sulfide and methane. High settleable solids concentrations may also overload the capacity of the treatment plant.

5. Acids or Alkalies.   Acids or alkalies are both corrosive and may also interfere with biological treatment. Even neutral sulfate salts may cause corrosion, since the sulfate can be biologically reduced to sulfide and then oxidized to sulfuric acid.

6. Heavy Metals.   Heavy metals may be toxic to biological treatment systems or to aquatic life in the receiving water and may adversely affect downstream potable (drinking) water supplies.

7. Cyanides.   Cyanides are toxic to bacteria and may cause hazardous gases in the sewer.

8. Organic Toxicants.   Pesticides and other extremely toxic substances in wastewater are objectionable except in very small concentrations. Even if the biological treatment systems are not affected by higher concentrations, toxicants may still damage receiving water quality.

2.42   Effects of Industrial Wastes on the Treatment Plant Unit Processes

When undesirable industrial constituents enter the municipal wastestream, certain adverse effects on common unit treatment processes can be expected. For example, acids and corro-

sive materials would damage the conveyance system of pipes and pumps. Dangerous gases and explosive materials create hazards to plant personnel. Other constituents, such as heavy metals or toxic organics, may actually inhibit or kill the microorganisms at the treatment plant.

Some of the more common adverse effects that industrial waste constituents have on unit processes are listed below.

1. Sewer System

   a. Corrosion caused by acids
   b. Clogging due to fats and waxes
   c. Hydraulic overload by discharge of cooling waters
   d. Potential explosion danger with gasolines and other fuels

2. Grit Channels

   a. Overloading with high grit concentrations
   b. Increased organic content of grit
   c. Intermittent flow reduces removal efficiency

3. Screens and Comminutors

   a. Overload with excess solids
   b. Excessive wear on comminutor cutting surfaces by hard materials

4. Clarifiers

   a. Fluctuating hydraulic loadings reduce removal efficiency
   b. Scum problems from excessive quantities of oils
   c. Impaired effluent quality caused by finely divided suspended solids
   d. Excessive sludge quantities with high suspended solids concentrations

5. Sludge Digesters

   a. Negative effects on sludge digestion caused by inorganic solids
   b. Overload caused by excessive solids
   c. Increased scum layers caused by excessive organic solids
   d. pH problems from an industrial wastewater with a high sugar content
   e. Toxicants inhibit or kill microorganisms

6. Activated Sludge

   a. Deterioration in quality with transient loading
   b. Excessive carbohydrate concentrations can cause bulking or poor sludge settling
   c. Toxicants inhibit or kill microorganisms
   d. Foaming problems

2.43   Operational Strategy

2.430   Need for a Strategy

Adverse effects to the activated sludge process can be significantly reduced if the operator has a plan of operation or an operational strategy ready to implement when adverse industrial constituents enter the treatment plant. This section will discuss the observations and corrective actions (operational strategy) taken at a typical activated sludge plant when (1) a toxic waste (cyanide), and (2) a high BOD waste (milk) enter the treatment plant mixed with the domestic wastewater.

This typical plant monitors the influent pH and the aeration tank DO. These are the only two water quality indicators that are monitored continuously. The operators at this plant rely heavily on observations of the appearance and odor of the influent wastewater, aeration tank mixed liquor, and the secondary effluent as the first indicators of shock or overload conditions at the plant. These observations allow the proper corrective actions to be implemented before significant damage to the activated sludge process occurs.

The typical treatment plant is designed to operate in the contact stabilization modification of the activated sludge process. This mode of operation is very desirable when industrial waste constituents that are harmful to the organisms in activated sludge may occur unexpectedly in the treatment plant influent. The advantage of operating in this mode is discussed in Chapter 11, Section 11.81 of *OPERATION OF WASTEWATER TREATMENT PLANTS,* Volume II.

The average operating data for our typical plant are as follows:

Peak influent flow:	17.0 MGD
Raw wastewater COD:	400 to 450 mg/*L*
Primary effluent feed to the aeration tanks:	Three-point step feed
Aeration contact basin F/M:	0.5 to 0.8 lb COD/day/lb MLVSS
DO through the system:	0.05 to 5.0 mg/*L*
Sludge aeration time:	7 to 12 hours to keep most of the biomass under aeration
Ammonia oxidation:	30 to 50 percent
Secondary effluent nitrate:	0.3 to 1.5 mg/*L*
Secondary effluent SS:	5 to 10 mg/*L* (varies with the waste)
Air-to-flow ratio:	1.8 to 2.5 cu ft/gal
RAS flow rate control:	Flow paced and allow "zero" sludge blanket in secondary clarifier

2.431 Recognition of a Toxic Waste Load

In our sample plant, the first indication of a toxic waste load within the treatment plant is recognized by observing the aeration basin DO recording device. As the toxic load moves into and through the aeration basin, the DO residual will increase significantly. A DO increase without an air input increase indicates that the toxic waste load is killing the microorganisms in the aeration tank, thus reducing the oxygen uptake (respiration) by the microorganisms.

A second indication of a toxic waste reaching the plant may be observed in the secondary clarifier effluent. The effluent will

begin to exhibit floc carryover (an indication of cell death). The degree of carryover will depend on the substance and quantity of the toxic waste.

When a toxic substance is known to have entered the treatment plant, the operator should make every effort to obtain a sample of the wastewater and have the sample analyzed as soon as possible to determine the toxic constituents. A record of these upset conditions and the constituents involved is very important so that if uncontrollable problems develop at the treatment plant, the records can be reviewed in an attempt to determine the input source.

2.432 Operational Strategy for Toxic Wastes

The operator's primary mission in the case of toxic wastes is to save the activated sludge system. When the operator in our sample plant recognizes a toxic waste condition, the RAS flow is reduced significantly. If this action is taken promptly, it isolates in the secondary clarifiers most of the bacteria affected by the toxic waste. The operator then significantly increases the WAS flow to purge the activated sludge process of the toxic waste and the sick or dead microorganisms. Additionally, every attempt is made to process the toxic waste flow through the plant as fast as possible to reduce contamination of other unit processes such as anaerobic or aerobic digesters. The toxic waste processing time through our typical plant may vary from 30 minutes to 2 hours.

2.433 Recognition of a High Organic Waste Load

The first indication of a high organic waste load within the treatment plant is recognized by observing the aeration basin DO recording device. As the high organic load moves into and through the aeration basin, the DO residual will decrease significantly. A DO decrease without an air input decrease indicates that the high organic waste load is too great for the available microorganisms to properly assimilate and metabolize the waste (food to microorganism ratio is out of balance because of a greater BOD (food)).

A second indication of high organic waste reaching the plant may be observed in the secondary clarifier effluent. The effluent will become more turbid (less clear) indicating that the waste flow has not been adequately treated.

2.434 Operational Strategy for High Organic Waste Loads

The operator's primary mission in the case of high organic loads is to improve the microorganism treatment efficiency. The RAS flow must be significantly increased to provide more microorganisms to the aeration contact basin to adequately treat the high organic waste. The rate of RAS increase must be accomplished gradually so that both design hydraulic and solids loading rates for the secondary clarifiers are not exceeded. In addition, every attempt should be made to increase the air or oxygen input to maintain proper DO levels in the aeration basins.

If the high organic waste load is significant, the nutrient content of the municipal waste may be inadequate for proper biological activity. Nitrogen and phosphorus should be added to provide the additional nutrients needed by the microorganisms. The quantity of nitrogen and phosphorus required for a waste can be estimated from the quantity of sludge produced per day. The pounds of nitrogen required per day will be about 10 percent of the volatile solids (dry weight) produced each day. The phosphorus requirement will be one-fifth of the nitrogen requirement. The amounts of nitrogen and phosphorus added daily are equal to the differences between the quantity required and the quantity in the wastes.

$$\text{Additional Nitrogen, lbs/day} = \text{Nitrogen Required, lbs/day} - \text{Nitrogen in Wastes, lbs/day}$$

High organic waste loads may be caused by industrial waste discharges containing excessive amounts of fats, oils, and greases. The operator should prepare for increased scum accumulations and removal of the scum under these conditions.

QUESTIONS

Write your answers in a notebook and then compare your answers with those on page 148.

2.4A Why is an industrial waste monitoring program valuable for the operator of an activated sludge plant?

2.4B List five wastewater characteristics commonly monitored with automatic monitoring devices.

2.4C What would you do if a high organic waste load entered your activated sludge plant?

2.5 INDUSTRIAL WASTE TREATMENT

2.50 Need to Treat Industrial Wastes

As the operator of an industrial wastewater treatment plant, it is your responsibility to ensure that required sewer-use standards or effluent quality standards are achieved in order that the production system may remain on line. Because of tighter restrictions on the quality of industrial wastewater that can be discharged to municipal sewer systems or receiving streams, your industrial facility may be required to pretreat the industrial process wastewaters. This pretreatment can be costly, but failure to properly pretreat these wastewaters may result in excessive sewer use fees and treatment charges, in severe fines for violations, and possible production shutdowns. Many industries are finding it more economical to build, operate, and maintain their own treatment facilities than to pay for pretreatment and use of municipal treatment plants.

This section will provide you with information necessary to increase your process awareness and alert you to precautions required in operating your activated sludge industrial wastewater treatment plant. Although there are a wide variety of applications of the activated sludge process to industrial wastewater treatment in operation today, this section will concentrate on some typical plants that treat wastes from fruit and vegetable processing, paper product manufacturing, petroleum refining, and dairy products processing.

2.51 Characterization of Influent Wastes

The character of any industrial wastewater depends on the particular production process and the way it is operated. The first step to successful operation of your treatment plant is to characterize the wastewaters through various analyses. These characteristics describe the concentrations or amounts of wastes in the wastewaters and include BOD, total suspended solids, settleable solids, COD, pH, DO, temperature, total solids, dissolved solids, chloride, nitrogen, and phosphorus. These pollutants are often used to determine the fees paid by industry for use of municipal collection systems and treatment plants. Other pollutants that are measured include toxic substances such as arsenic, zinc, and copper. Toxic substances usually must be pretreated by industry. The procedures for measuring the concentrations or amounts of these wastes are outlined in Chapter 16, Volume II, *OPERATION OF WASTEWATER TREATMENT PLANTS*.

Another important characteristic of the wastewater being treated is the flow. Flow rate should be described in terms of the daily average and also the fluctuations. Because many industrial facilities do not generate consistent waste flows and constituents over a 24-hour day or 7-day week, you must determine when the variations can be expected and operate your activated sludge process in a manner that anticipates these variables and does not allow these fluctuations to cause a deterioration of the quality of the plant effluent. Effective sewer-use ordinances require that industry notify the operator of a municipal treatment plant whenever a significant accidental discharge (spill or dump) or process failure might upset a treatment plant. By warning the operator of the volume and nature of the discharge, provisions can be made to handle the waste when it reaches the plant. Process supervisors should notify the operators of their industrial wastewater treatment plants when a potentially harmful spill or dump occurs.

2.52 Common Industrial Wastewater Variables

2.520 Flow

Flow measurements are a basic requirement at every treatment plant. Recorded flow data are essential because these records allow you to establish correct operating guidelines (loadings) and determine if your treatment plant is properly sized to process the hydraulic loads.

In many industrial production facilities, the wastewater flows are generally higher during the day-shift hours of Monday through Friday. Significant flow reductions may be anticipated in the evening hours with possible "zero" flow conditions during weekends, holidays, and annual production maintenance shutdown periods.

In cases where flows to your treatment plant approach or exceed the hydraulic design capacity, every effort should be taken to implement a production facility pollution prevention program. The main goals of such a program should be to reduce water consumption and minimize waste generation through proper management of processing and production operations.

2.521 pH and Temperature

The pH of production facility wastewaters may vary from 2.5 to 13.0 depending upon the product being processed and the type of operations conducted within the facility. Natural waters generally have pH values between 6.5 and 8.0.

If the wastewater pH varies greatly from neutral (7.0), the wastewater should be adjusted (neutralized). Neutralization may be accomplished in a tank of sufficient detention time (15 to 20 minutes). A compatible acid (to lower pH, such as sulfuric) or base (to increase pH, such as caustic) addition to the wastestream may be controlled with pH probes and controllers for rough and fine pH adjustment.

Extreme variations in the temperature of influent wastestreams can upset microorganisms and prevent their degrading organic material.

2.522 BOD and Suspended Solids

The BOD test measures the rate of oxygen uptake from the wastewater by microorganisms in biochemical reactions. These microorganisms are converting the waste materials to carbon dioxide, water, and inorganic nitrogen compounds. The oxygen demand is related to the rate of increase in microorganism activity resulting from the presence of food (organic wastes) and nutrients. Microorganism activity may be hindered in some industrial wastes due to the presence of toxic wastes. Industrial wastewaters may contain levels of BOD from 500 mg/L to over 10,000 mg/L.

Suspended solids information may be used to determine the quantity of solids that will require removal in the activated sludge treatment system. Typical suspended solids concentrations for industrial wastewaters vary from 125 mg/L to 3,000 mg/L.

For any waste, the concentrations of pollutants can be readily reduced by simply using more water, but even the increased volume will contain the same number of total pounds of pollutants. The organic load, consisting of both BOD and suspended solids, can only be effectively reduced by reductions in, or removal of, pounds of pollutants generated at production facility sources.

2.523 COD

Chemical oxygen demand (COD) is an alternative to biochemical oxygen demand (BOD) for measuring the pollutional strength of wastewaters. When considering the use of COD for measuring the strength of wastewater, you must bear in mind that the BOD and COD tests involve separate and distinct reactions. Chemical oxidation measures the presence of carbon and hydrogen, but not amino nitrogen in organic materials. Furthermore, the COD test does not differentiate between biologically stable and unstable compounds. For example, cellulose is measured by chemical oxidation, but is not measured biochemically under aerobic conditions.

The primary disadvantage of the COD test is that chloride may interfere with the chemical reactions. Thus, wastewaters containing HIGH salt concentrations, such as brine, cannot be readily analyzed without modification.

2.524 Nutrients

Aside from carbon in the organic matter (which is measured largely as BOD), the nutrients required for reproduction of microorganisms are nitrogen (N) and phosphorus (P). In unusual cases, other elements may also be critical, such as iron, calcium, magnesium, potassium, cobalt, and molybdenum. Since many production facility wastewaters are deficient in nutrients for biological treatment, nutrients can be added to optimize your biological wastewater treatment system efficiency.

The amount of BOD, nitrogen, and phosphorus required for a treatment process depends both on the age of the microorganisms and the growth rate during the reduction of BOD. A BOD/nitrogen/phosphorus ratio of 100:5:1 is usually adequate. However, high-rate systems with no available nitrogen or phosphorus in the wastewater may require a ratio of 100:10:2. Ratios lower than 100:5:1 may be adequate for aerated ponds and systems with a very long sludge age. All nutrients need to be in a soluble form to be used by the microorganisms.

2.525 Toxicity

The most common causes of wastewater toxicity are excessive amounts of free ammonia, residual chlorine, detergents, paints, solvents, and biocides. Other wastes that can upset microorganisms include heavy metals, chlorinated hydrocarbons, petroleum products, acids, and bases (caustics). These toxic materials can enter the wastewater stream through indiscriminate dumping, improper handling of toxic materials, leaking vessels and pipes, or accidental spills.

QUESTIONS

Write your answers in a notebook and then compare your answers with those on page 148.

2.50A List three types of industries whose wastes could be treated by an activated sludge plant.

2.51A What factors influence the character of industrial wastewaters?

2.52A How would you attempt to solve a hydraulic overloading problem at an industrial wastewater treatment plant?

2.52B How can the pH of an industrial wastewater be adjusted (1) upward, and (2) downward?

2.52C Chemical oxidation (COD) measures the presence of (1) _____ and (2) _____, but not (3) _____ in organic materials.

2.52D List the three major nutrients required by microorganisms and four other elements that might be critical.

2.52E What are the most common causes or kinds of toxicity in industrial wastewaters?

2.53 Flow and Pretreatment Considerations

Because industrial wastewater characteristics may change significantly over a given period of time, a program for sampling, testing, and measuring the flow is essential.

Variations in production activity will change wastewater characteristics. Samples should be collected during each operating shift and during different stages of the finished product and raw product runs. Flows should be monitored continuously, even during cleanup and on weekends.

Daily or seasonal shutdown and start-up of a processing/production facility usually cause wastewater characteristics to vary greatly. This variation often causes problems in a treatment system. Biological treatment systems perform best on a uniform supply of a given source of food (BOD). If the food supply changes greatly, the biological process may not be able to adjust to the change. The impact of frequent shutdowns and start-ups on a treatment system should be carefully evaluated. Flow regulation procedures using holding tanks may be necessary to smooth out the flows (see Section 2.531, "Flow Control").

2.530 Flow Segregation

Considerable treatment cost savings may be achieved by processing only the wastewaters that contain pollutant materials in quantities exceeding the limits set forth in your local, state, or federal discharge permit. Consider classifying the processing/production facilities' wastewaters into three categories (low strength BOD, medium strength BOD, and high strength BOD). You may find that it is possible to reuse the low strength BOD waters for cleanup purposes or dispose of them by discharge directly to a sanitary sewer system, by spray on fields, or by irrigation of pasture land.

Treatment of the medium strength BOD wastewaters (usually resulting from cleanup) may be possible by using a screen system to remove large solids and then processing the wastewater through an air flotation unit, plate and frame filter press, or similar device. Since only 30 to 40 percent BOD and suspended solids reductions may be obtained from this treatment method, the dewatering process effluent (low in BOD/SS) may then be routed to the activated sludge system where it will aid in diluting the high strength BOD wastewaters.

2.531 Flow Control

Wide flow variations often cause problems in a treatment system. Depending on the daily operating mode of the processing/production facility, variations in instantaneous flow can be from very small to very great (a maximum of 10 times the minimum). Large variations in flow may be smoothed out with a surge tank or storage tank of about 10 to 20 percent of the total daily flow volume. Settling of solids will be a significant problem in a tank of this size, so the tank must either be mixed or some means must be provided for solids removal.

Control measures for water use in the process/production facility should be implemented. One of the most important methods of water use reduction is a complete and comprehensive training program for everyone involved in process/production. Additionally, roof gutters, downspouts, and facility storm drains may discharge to the treatment plant. These should be relocated or rerouted to appropriate drainage systems to eliminate charges for treating this water.

2.532 Screening

Discrete waste solids (such as trimmings, rejects, corn meal, and pulp) are effectively separated from the wastewater flow by various types of screens. Screening has many objectives including recovery of useful solid by-products; providing first-stage primary treatment; or pretreating wastestreams for discharge to a municipal wastewater treatment system.

Screens should be located as close as possible to the process/production activity producing the waste. The longer the solids are in contact with water and the rougher the flow is handled (more turbulent), the more material will pass through the screens and the more the solids will become dissolved.

2.533 Grit, Soil, Grease, and Oil Removal

Fruit and vegetable, meat and fish, paper, and petroleum processing/production introduce large amounts of grit, soil, grease, and oil to the wastestream. This material will accelerate equipment wear, settle in pipelines, accumulate in the treatment system, and create odors if not removed.

An aerated tank or lagoon can be provided to remove these wastes from the wastestream. This aeration system is usually designed to pre-aerate the wastewater while it aids the release of free and emulsified grease for surface collection. Additionally, separation and settling of grit, soil, and oil sludge are accomplished.

QUESTIONS

Write your answers in a notebook and then compare your answers with those on page 149.

2.53A How often should samples from industrial wastewaters be collected?

2.53B How can large variations in flow be reduced?

2.53C What kinds of waste solids can be effectively removed by various types of screens?

2.53D Why should screens be located as close as possible to the process/production activity producing the waste?

2.534 Central Pretreatment Facilities

Microorganisms that treat wastes in the activated sludge process will not work unless they are provided a suitable environment and are properly acclimated (adjusted to the wastes). A suitable environment may be provided by treating undesirable or toxic constituents in industrial wastes at the source where they are produced or at a central pretreatment facility (Figure 2.15).

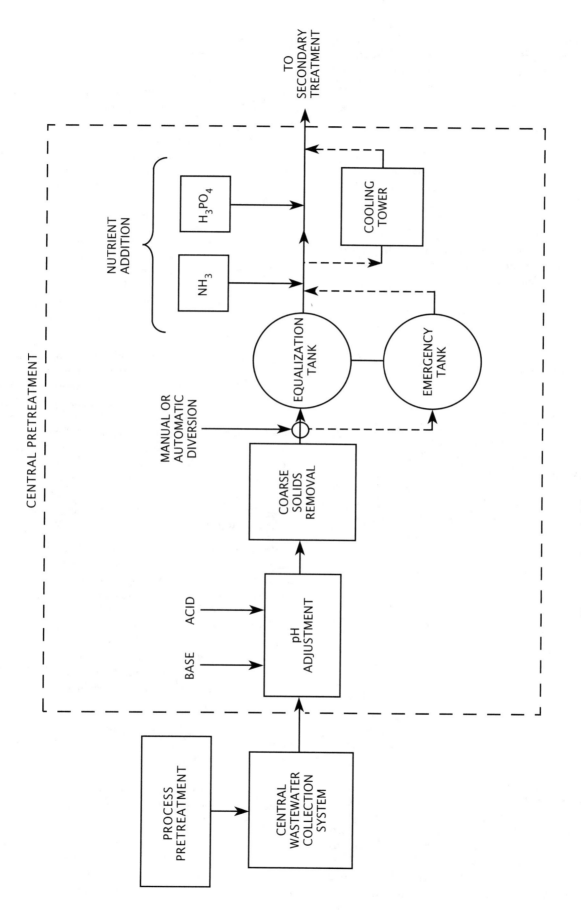

Fig. 2.15   Control of the influent hydraulic and organic loading to the secondary biological treatment process

Frequently, the most economical method of treating toxic or undesirable wastes is at the source. If possible, do not allow these wastes to enter the plant wastewater or, if they do, treat the waste in as concentrated form as possible, before it becomes diluted with other wastewaters. Source pretreatment is appropriate for extreme pH levels, inert suspended solids, oil, grease, or toxic materials (such as heavy metals).

Figure 2.15 shows a typical industrial pretreatment facility. The processes and the order of treatment will depend on both the type of industry and waste constituents. The first process is pH adjustment. Many manufacturing processes produce either high or low pH wastestreams on either a continuous or batch basis. If source pretreatment does not reduce these pH variations to within acceptable ranges, central pH control facilities must adjust the wastewater pH to near neutral levels. Neutralization chemicals may be added and mixed in pipelines or in neutralization tanks. Figure 2.15 shows the pH adjustment process before the equalization tank. Some plants adjust the pH after the equalization tank. Location of the neutralization process depends on the type of industry and type of wastes being treated. See Chapter 7, "Preliminary Treatment," Section 7.6, "Need for pH Adjustment of Industrial Wastewaters," *INDUSTRIAL WASTE TREATMENT*, Volume I, for more details.

Industry frequently uses special screens and microscreens to remove floatable and settleable solids instead of using primary clarifiers. These coarse or large solids are usually removed before the flow equalization tank. Removing these solids now will reduce unnecessary clogging and wear on downstream pipes, pumps, aerators, and clarifier mechanisms. Also, this process will help avoid odor problems that could develop from the settling out of solids in the equalization and emergency basins. For more information, see Section 7.5, "Screening," in Chapter 7, "Preliminary Treatment," in the *INDUSTRIAL WASTE TREATMENT* manual, Volume I.

Biological treatment processes work best if they receive fairly constant hydraulic and organic loadings. Equalization and emergency tanks are used by industry to store peak loads for release and treatment during periods of low loadings. Fluctuating flows (hydraulic loads) are usually smoothed out by allowing all wastewater to flow into the equalization tank and by pumping the wastewater to be treated out of the equalization tank at a constant flow rate. Variations in organic loadings are smoothed out by mixing the wastewater flowing into the equalization tank with the water already in the tank. By the use of effective mixing (such as with propellers, stirrers, or aeration), the organic content of the wastewater in the equalization tank can be kept fairly constant, thus providing a constant organic load to the activated sludge process.

Emergency basins or storage tanks are usually kept empty or almost empty so that there is plenty of room to store any process spills or wastewater diverted when treatment processes are upset. Wastewater may be diverted automatically on the basis of signals from continuous monitoring equipment (total organic carbon, pH, or conductivity analyzers). Also, diversion can be done manually based on results from continuous analyzers, grab samplers, or from a verbal warning of upset conditions from process operating personnel.

Generally, continuous monitoring and automatic diversion is the ideal approach for detection and isolation of process upsets. Unfortunately, continuous monitoring produces some severe operational problems in the form of sampling system and analyzer difficulties. Solids can plug the sampling lines, and the analyzers may require considerable maintenance. To avoid these problems, some plants rely on the analysis of grab samples and notification by process operators when spills occur. Whether you use continuous monitoring or grab samples to detect spills and undesirable influent conditions, you must protect your biological treatment processes from unsuitable conditions.

Frequently, industrial wastewaters lack the proper amounts of nutrients for healthy microorganism growth and reproduction in the activated sludge process. The major nutrients of concern are nitrogen and phosphorus. Other nutrients such as calcium, magnesium, sodium, potassium, iron, chloride, and sulfur are needed by the microorganisms, but sufficient quantities are usually found in most process waters.

Nitrogen and phosphorus are needed for biological growth and reproduction in quantities approximated by a $BOD_5:N:P$ ratio of 100:5:1 or a COD:N:P ratio of 150:5:1. If the nitrogen or phosphorus levels in the wastewater being treated are less than the values indicated by these ratios, more nitrogen can be added in the form of ammonia (usually as 30 percent aqueous) and phosphorus as phosphoric acid (usually as 75 percent aqueous). When these nutrients must be added, they are metered into the wastewater from the equalization tank or recycle sludge streams before the aeration tank. The additions are based on the difference between the desired amount and the actual amount in the wastewater being treated.

When the wastewater being treated has been heated by a production process, the water may have to be cooled before biological treatment, especially during the warmer summer months. Biological activity and treatment effectiveness usually drop rapidly at water temperatures above 99°F (37°C). Cooling towers are sometimes installed where heated discharges can cause problems. Usually, a portion of the influent flow, or in extreme cases, the entire influent flow, is directed over the cooling towers during the warmest summer months. Proper operation of these towers allows sufficient evaporative cooling for control of the aeration basin temperature.

Sometimes sufficient water cooling will occur from aeration in the equalization and aeration basins. Mechanical aerators can be especially effective. Diffused air aeration systems, however, can increase the temperature by 5 to 7°F (3 to 4°C) because of the hot air from the blowers. If excessive cooling is caused by surface aerators in the cold winter months, shut down the aerators (if possible) or add heat by using heat exchangers or steam.

Troubleshooting problems in central pretreatment facilities is similar to troubleshooting other treatment processes. If a spill occurs with a high initial oxygen demand, mechanical aerators may not be able to supply enough oxygen to maintain a minimum dissolved oxygen (0.5 mg/L) in the equalizing basin. When this happens, the basin may become septic and give off undesirable odors. If the aerators cannot deliver more oxygen, additional oxygen may be provided in the form of hydrogen peroxide.

Another possible problem with mechanical floating aerators is excessive cooling of the wastewater during cold winter months. If the water temperature becomes too low, the activity of the activated sludge organisms and the performance of the secondary clarifier may be reduced greatly. A partial solution to this problem is to shut off some of the aerators completely or to shut off all of the aerators periodically. You must keep the contents of the equalizing basin well mixed to avoid surges of waste organic loads or solids settling on the bottom of the basin. If a proper schedule is developed for turning the aerators on and off, you can provide sufficient mixing and minimize the heat lost from the surface of the basin.

If the aerators fail, carefully monitor the flows from the equalizing basin to the aeration tank using continuous analyzers or frequent grab samples. If the variation in organic loading, hydraulic loading, or temperatures becomes much greater than normal, divert all or some of the flow to the emergency basin. When the aerators are returned to operation, the flow to the aeration basin should be increased gradually to avoid shocking the biological system.

When the nutrient flow stops, try to get the flow back on line as soon as possible. If the nutrient flow can be restarted in 24 hours or less, feed nutrients at twice the normal rate for the same period of time that the nutrient flow was off. If the nutrient flow will be off for more than 24 hours, try adding nutrients from bags such as agricultural or garden fertilizers.

2.535 Start-Up or Restart of an Industrial Activated Sludge Process

Once the wastewater has been properly pretreated and is ready for the activated sludge process, the activated sludge biological culture must be ready to treat the wastes. Whether you are starting a new activated sludge process or trying to get an existing process back on line after the culture has been wiped out by a toxic waste, the procedures to develop a new activated sludge culture are similar.

New activated sludge cultures are started by obtaining "seed activated sludge" or activated sludge microorganisms from either a nearby municipal or industrial wastewater treatment plant. The amount of seed activated sludge needed depends on the hydraulic and organic loadings on your treatment plant. The greater the load, the more tank trucks of activated sludge seed will be needed.

Once the activated sludge seed has been added to the aeration basin, the level of the water in the aeration basin is usually maintained just below the overflow level. The aeration system is in operation. You want to increase the population of activated sludge microorganisms. This is accomplished by feeding the "bugs" a solution they can eat quickly and use efficiently for building more microorganism cells. The most common chemical feed is sodium acetate. These microorganisms also require nutrients for fast and healthy growth. The key elements are nitrogen and phosphorus, which must be provided in the form of chemicals added to the wastewater. Usually, ammonium sulfate and potassium dibasic phosphate (available in a dry form in bag quantities) are used because they are relatively pure and can provide an adequate buffer for pH control in the aeration basin. These chemicals are easily dissolved in small amounts of water.

Start-up procedures will vary with each industrial waste treatment plant. Usually, the chemicals are added to achieve a ratio of COD:N:P of 150:5:1 or BOD:N:P of 100:5:1. Chemicals are added on a batch basis directly to the aeration basin (or indirectly through pump wells) until the microorganisms multiply to approach an MLSS level of 2,000 mg/L. Usually, enough sodium acetate (food) is added in a batch to allow the bugs to feed for one to three days. Between batch feedings, periodic analyses of the aeration basin contents will indicate the rate at which the bugs are consuming the organic materials (oxygen uptake rate readings and filtered COD or TOC analyses) and the rate of growth of the bugs (MLSS analyses). These measurements will indicate the timing for future batch feedings. Experience indicates that you usually have to feed chemicals for about one week to produce a flocculating sludge.

Once an MLSS level of over 2,000 mg/L has been reached, the industrial wastewater may be introduced into the aeration basin. This step must be very gradual. Pump approximately 10 percent of the total industrial waste flow from the equalizing basin. Chemical feeding should be continuous if at all possible (even after wastes are introduced). This procedure allows the activated sludge microorganisms to slowly adapt or become acclimated to the industrial wastes so the wastes will not shock the delicate growth patterns of the activated sludge microorganisms.

Periodically, take samples from the aeration basin in order to monitor the degree of biological growth and removal of organic wastes (oxygen uptake rate, MLSS, and filtered COD). After two or three days, if the monitoring results are favorable (that is, you have high COD removal, MLSS is increasing, and oxygen uptake rates are steady), the industrial waste flow input can be increased (20 to 25 percent of the total flow) and the chemical feed (food) can be reduced by a similar amount. Also, at this time, the permanent nutrient feed system (if any) should be started to ensure an adequate supply of nitrogen and phosphorus.

This procedure is followed until all of the industrial wastewater flow (with no chemical feed) is being fed to the activated sludge process. The time for the acclimation process to be complete will vary with the industrial wastes being treated, but usually two to three weeks are required. Once the activated sludge process starts to generate excess activated sludge above the desired MLSS level, sludge wasting should be started to control the MLSS.

Portions of this section were taken from *BACKGROUND DOCUMENT, DU PONT ACTIVATED SLUDGE TREATMENT.* Permission to use the material in this section is sincerely appreciated.

QUESTIONS

Write your answers in a notebook and then compare your answers with those on page 149.

2.53E   In general, where is the most economical location to treat toxic wastes?

2.53F   Why do some industries require a waste pretreatment facility?

2.53G   Why does industry use screens to remove coarse solids?

2.53H   How can variations or fluctuations in influent organic loadings be smoothed out before the aeration basin?

2.53I   What chemicals are used to provide nutrients in the form of nitrogen and phosphorus?

2.53J   Where would you obtain a "seed activated sludge" to get an existing activated sludge process back on line after being wiped out by a toxic waste?

2.53K   Why must industrial wastes initially be added gradually to the aeration basin?

2.54   Operational Considerations (Activated Sludge)

The following discussion will center on specific operational problems experienced at various industrial wastewater treatment plants. Corrective action methods are also presented.

2.540   Neutralization

pH control systems may be installed to treat various process/production wastewaters. pH adjustment of this nature is usually done as a pretreatment process. However, lime, alum, and iron salts may be used in a pre-secondary chemical clarification process. In this case, it is necessary to provide additional pH adjustment facilities to neutralize the normally high pH effluent (8.5 to 11.0) from the chemical clarification process. Abnormally high wastewater pH will result in loss of activated sludge system efficiency and cause settling problems in the secondary clarifier. A consistently alkaline waste (high pH) can be neutralized by using carbon dioxide (CO_2). Boiler stack gas and a compressor delivery system could be a source of carbon dioxide.

If a high pH wastewater is a problem because you use lime, alum, or iron salts as a settling aid, consider the use of POLYMERS [17] instead.

If the pH is too low (acid), the pH can be increased by the addition of lime ($Ca(OH)_2$) or caustic soda ($NaOH$). The activated sludge process usually operates satisfactorily in a pH range from 6.5 to 8.0.

2.541   Nutrients

Nutrients, especially nitrogen and phosphorus, can be critical in the performance of the activated sludge system. The exact point at which nutrients become critical depends on your type of treatment process and how you operate it.

Remember that all applied nutrients must be in a soluble form to be used by the microorganisms. The addition of nutrients should be accomplished at a point where the incoming wastewater is highly mixed, preferably in the aeration basin.

The quantity of nutrients added to the treatment system should be based on the desired BOD reduction and the extent of the nutrient deficiencies. The amount of nitrogen and phosphorus required to treat a waste can be estimated from the quantity of sludge produced per day. The pounds of nitrogen required per day is equal to 10 percent of the volatile solids (on a dry weight basis) produced each day. Phosphorus requirements are one-fifth of the nitrogen requirements. The amount of nutrients that have to be added each day is determined by the difference between the quantity required and the quantity available in the wastes.

Typically, supplemental nitrogen is provided by using aqueous ammonia or anhydrous ammonia. Supplemental phosphorus is provided by using dissolved triple superphosphate, phosphate fertilizer, or phosphoric acid (a waste acid from aluminum bright dipping facilities).

When the supply of nitrogen and phosphorus from the wastewater is below that required by the microorganisms in your biological treatment process for extended periods of time, filamentous organisms may begin to predominate and cause sludge bulking in the secondary clarifier.

2.542   Daily System Observations

Daily observations of the bacterial population must be made to observe developing system changes and stress conditions so that proper action may be taken before overall treatment efficiency is affected. Under normal conditions, the bacterial population is composed primarily of small bacteria with large numbers of stalked ciliates and many free-swimming ciliates and rotifers (Section 2.32). The presence of these higher forms indicates that the process is operating properly. The following conditions have been found to cause the disappearance of the higher forms.

1. DO levels below 3.0 to 4.0 mg/L: With a high mixed liquor solids, it is possible that the oxygen transfer efficiency is impaired at lower DO concentrations. Additionally, filamentous growth may predominate.

2. High organic loadings: When the process is in the rapid or accelerated growth phase, it has been observed that the higher forms do not compete as well as the simpler bacterial forms.

3. Toxic substances or nutrient deficiencies: These conditions will affect the growth and maintenance of the higher forms.

4. Lack of pH control: If the pH control system is not operating properly to adjust the pH to near neutral, ciliates and rotifers will not adapt to pH fluctuations and will disappear.

---

[17] Polymer (POLY-mer).   A long-chain molecule formed by the union of many monomers (molecules of lower molecular weight). Polymers are used with other chemical coagulants to aid in binding small suspended particles to larger chemical flocs for their removal from water. Also see POLYELECTROLYTE.

2.543   Return Activated Sludge (RAS)

In most cases, the RAS values for industrial waste treatment systems are the same as for municipal systems as discussed in Section 2.2. However, some systems experience a sludge with low compacting characteristics (high SVI). This is usually, yet not always, the result of conditions shown in Section 2.23, Table 2.3, on page 74.

2.544   Waste Activated Sludge (WAS)

An activated sludge plant, operating at a high rate (or low sludge age), will produce 0.5 to 1.0 pound of microorganisms for each pound of BOD removed. In addition to the production of up to 1.0 pound of microorganisms per pound of BOD removed, added sludge results from the nonbiodegradable suspended solids, both volatile and nonvolatile, in the influent to secondary treatment. Consequently, it is common for the total secondary sludge production to be 0.8 to 1.0 pound per pound of BOD removed. Sludge from secondary treatment systems is still biologically active and will putrefy. This can cause an intolerable odor. If the sludge contains no domestic wastes, it may be possible to spread and dry the sludge quickly on a disposal site or agricultural land, and then plow it into the soil.

Secondary sludge is difficult to dewater as compared with primary sludge. Raw, undigested secondary sludge has a total solids content of only one-half to one percent in air systems and one to three percent in pure oxygen systems. In addition, the cellular matter in the sludge is only 15 percent solids. Unless the cell membranes are ruptured, microorganisms cannot be dewatered to greater than 10 percent solids. Cells can be ruptured by heating or slow freezing although natural freezing can be used in some climates.

An alternative method of handling WAS from a treatment system that operates seasonally is to divert WAS to a lagoon. At the end of the season, the solids from the lagoon are periodically returned to the aeration system for "complete" oxidation during the off-season period.

2.545   Clarification

In industrial waste treatment systems, sludge settling often is significantly affected by influent characteristics variations. As a result of these variations, the settling characteristics of the sludge are extremely critical. Therefore, clarification rates of less than 400 GPD/sq ft (16 m^3 per day/sq m) may be necessary to obtain proper operation.

If you observe a gradual decrease in percent solids removal over a period of time, this reduction may be the result of grit and silt accumulation in the aeration basin. The grit and silt are not carried out in the effluent, remain in the basin, and reduce aeration volume and detention time. A higher volatile content of the sludge would be indicative of this condition. A lower volatile content would occur only if the aeration basin's contents are well mixed. In this case, there should be no settling or accumulation of grit and silt.

If grit and silt are accumulating, make every attempt to increase the efficiency of your grit removal process. If you do not have grit removal facilities, make every effort to increase the mixing in the aeration system to prevent the settling of solids in "blind" or "dead" areas (typically corners). If none of the above control methods are available to you, consider cleaning your aeration basins on a yearly basis if necessary. If you operate a municipal treatment facility and find grit and silt are coming from an industrial source, make the industry remove the grit and silt by pretreatment methods. Industry should not be allowed to dump this material into a sewer.

QUESTIONS

Write your answers in a notebook and then compare your answers with those on page 149.

2.54A   Where should nutrients be added to a wastewater?

2.54B   List some typical sources of supplemental nitrogen and phosphorus.

2.54C   What factors or conditions have been found to cause the disappearance of higher forms of microorganisms from the activated sludge process?

2.54D   What items could cause a gradual decrease in percent solids removal from an activated sludge process over a period of time?

NOTICE

Sections 2.55 through 2.58 were prepared by the operators of actual industrial wastewater treatment plants. These operators have described procedures they use to treat industrial wastes by the activated sludge process. You must remember that there are many types of industries and that the waste flows and constituents will be different at every treatment plant. The intent of the following sections is to provide you with ideas that might work for your industrial waste treatment activated sludge plant.

2.55   Pulp and Paper Mill Wastes
by James J. McKeown

Industrial wastewater treatment plants are designed in much the same manner as municipal plants; however, there are some important differences in treating pulp and paper mill wastes. Before any wastestream can be treated, both the average and range of fluctuations of flows and waste constituents must be measured, recorded, and analyzed.

2.550   Need for Recordkeeping

The format for recordkeeping will usually be prepared by the environmental control supervisor. There are three basic reasons for keeping accurate records. The first reason is to keep a history that helps troubleshoot problems that arise. Each entry onto the log sheet is made at the frequency needed to help locate trouble. The operator also needs to be alert for trends that occur in between the log entries. Most log sheets call for pertinent remarks and the operator should make use of this column to note unusual conditions or to simply enter "all is well." Your record will be read by the person relieving you in order to pick up the operation. You know the things most helpful to you when starting a shift and it is simply a matter of doing likewise for your relief.

The second reason for keeping accurate records has to do with accounting for the operation of the plant on a weekly or monthly basis. If the operator should fail to make an entry and must estimate a value, the value should be footnoted accordingly. Failure to enter even routine data may prevent a calculation from being made. Serious mechanical problems or chronic symptoms of potential problems should also be noted so the work orders can be issued to correct the problem.

The last reason for keeping accurate records is that records are legal documents that will protect the company from unjust claims. In other words, the company may have to prove that its operations were normal during a certain time period. The only way this can be accomplished is by use of the actual operating data. Therefore, it is very important to keep accurate records. Do not forget that all recorder charts are also a part of the official record. You may be asked to make notations on these charts as part of the accounting procedure and it is usually a good idea to initial the note. Another good idea is to make a note on your log sheet each time you mark the recorder chart. If information is recorded in a computer data management system, you still must record important observations.

2.551 Wastes Discharged to the Plant Collection System

In most industrial systems, separate departments or small networks of sewers are monitored individually. The control of wastewater in these networks is often the responsibility of production personnel. Experienced production personnel know that certain conditions in their area will result in abnormal discharges to the sewer. With proper communications between operating and waste treatment plant personnel, an alarm can be sounded either before or right after a spill actually happens. A quick phone call will allow emergency measures to be implemented in time to prevent operating problems. Everyone must encourage the necessary communications to give waste treatment personnel all the advance notice they need to properly operate the plant.

Several key sewers may be monitored continuously. In such a system, the wastewater treatment plant operator can usually read the recorders to see if all the sewers contain normal quantities of wastewater. The use of centralized instrument panels both in the mill production area as well as the treatment plant will allow confirmation of a particular condition.

Operators should understand the reason why each water quality indicator is being monitored and, if necessary, how the sensor works. In highly automated systems, instrumentation personnel may calibrate and maintain the sensors. However, even in this case, the operator should have a basic understanding of the instrument or electrode.

The most common control sensors used in pulp and paper mills are conductivity, flow, pH, and turbidity. Special situations may be monitored to determine temperature, color, sodium, COD, TOC, or TOD (Total Oxygen Demand). Oxygen consumption (organism respiration) may also be monitored to detect overloads or spills. Continuous BIOASSAYS [18] may be required at certain installations and DO is used to control aeration in activated sludge plants. Treatment plant personnel should be aware of the meaning of changes in the values of each water quality indicator at each sampling point.

The operator should be familiar with certain mill waste flow and constituent recorder patterns (variations) that are likely to occur in each sewer, so that appropriate control actions can be initiated. For example, a modest increase in conductivity accompanied by a low pH indicates that acid is probably being discharged into the mill sewer. However, if both conductivity and pH are quite high, then an organic discharge from the pulp digester is the probable cause. Further, if flow measurement is available, the operator should be able to determine the potential severity of the discharge. The operator should know what combinations of readings to look for and what actions are necessary in each case to control the situation. Many mills use continuous recorders with signal alarms to alert the operators that a particular sewer is exceeding its normal limit. However, whether communication is automatic or by person-to-person contact, most situations have a history of occurrence and can be properly handled using predetermined control measures.

If a spill is detected by the sensor system, the operator should be able to verify the spill. A phone call to the mill operating area may be sufficient to learn how long the event is likely to continue. With this information, a decision can then be made on the procedures to use to control the spill. Industrial control systems often use emergency storage tanks or spill ponds that can be filled quickly to protect the treatment plant.

Knowledge about the nature and duration of an emergency upset is a necessity for proper management of reserve storage capacity. These diverted wastes can later be bled into the treatment plant at controlled rates.

2.552 Variables Associated With the Treatment of Paper Mill Wastes

NUTRIENT CONTROL

Paper mill wastewater usually does not have enough nitrogen and phosphorus to support bacterial growth. Thus, facilities are often installed to add these nutrients to the wastewater before

---

[18] Bioassay (BUY-o-AS-say).   (1) A way of showing or measuring the effect of biological treatment on a particular substance or waste. (2) A method of determining the relative toxicity of a test sample of industrial wastes or other wastes by using live test organisms, such as fish.

biological treatment. The proper amounts of these nutrients must be added because too little or too much may create problems. The operators should be familiar with their nutrient feeding systems and, in cold climates, the temperatures at which certain nutrient chemicals will freeze. Many wastewaters only require nitrogen addition because there is enough phosphorus in the wastewater as a result of boiler water corrosion control and cleaning operations.

Experience with BOD removal at the lowest possible nutrient addition rates will allow the operator to properly control the addition rate. Soluble ammonia nitrogen and phosphorus generally need not exceed 0.5 and 0.25 mg/L, respectively, in the final effluent. Many systems are providing adequate treatment at effluent concentrations of 20 to 50 percent of these values. If final effluent nutrient values are considerably lower than these values, then additional nutrients may be necessary.

FOAM CONTROL

Another additive commonly found in pulp mill wastewater is a defoamer. Some pulp mill effluents foam because the wood soaps are not fully captured in either the tall oil or recovery systems. The chemicals that cause foams are oxidized during biological treatment. However, if foaming is not controlled in the biological plant while treatment is taking place, a variety of problems may result, including frothing of mixed liquor solids. Frothing of the mixed liquor solids causes a loss of microorganisms from the system. Foam can also engulf and short out motors and other electrical equipment. Also, foam will cause certain level recorders to misinterpret the true liquid level and this can be troublesome if flow measurement is affected. Foam may also cause an air pollution problem, especially if it dries and becomes wind-borne.

Foam is generally controlled by water sprays or dosing with an antifoam chemical. The object of either approach is to cause an uneven distribution of surface tension in the foam. The physical impact of the spray also breaks up the foam and creates drainage channels through the foam.

Defoaming chemicals are very expensive and their applications should be carefully controlled. The delivery system can be by overhead spray or by using the aerator to distribute the chemical into its own spray. Both systems have been used with success in the paper industry. Most paper mills use metering pumps to feed defoamer. The chemical is added continuously at a predetermined rate. If foam volume increases, the chemical dose is increased for a short period of time and then reduced to the baseline again. Excessive use of defoamers may also add an undesirable oxygen demand on the system. Automatic systems respond to conductivity or photoelectric sensors. When the foam reaches the sensor, the defoamer chemical is applied in steps until the foam recedes below the sensor. Chemical suppliers will gladly recommend procedures for applying their products in specific situations.

pH CONTROL

The plant may be equipped with a neutralization system where acid or caustic is added prior to the mixed liquor tank.

Most facilities add these chemicals prior to the primary clarifiers. However, some plants bypass the clarifiers with large volumes of bleach plant wastes and, in this case, pH adjustment is accomplished in a blend tank where bypassed wastes join primary effluent.

Hydrated Lime, $Ca(OH)_2$, and caustic, NaOH, are two of the common alkaline solutions used to increase pH while sulfuric acid, H_2SO_4, is the acid solution commonly used to lower pH. Control of pH variations is probably more important than the actual pH value. However, most mills have a high and low pH target value that must not be exceeded and pH should be adjusted to fall within this range.

Most pH control systems are automatic and are based upon pH and flow sensors placed downstream from the blending point. Acid and caustic makeup requires that the operators follow prescribed safety rules because of the danger involved in handling these concentrated chemicals.

PRODUCT-RELATED SITUATIONS

Treatment of pulp and paper mill wastewaters is similar in many ways to the treatment of municipal wastes. The machinery is identical in many cases. However, the operator who worked in a municipal plant will quickly become aware of some notable differences that depend on the type of pulp and paper being produced.

The paper industry has several types of pulp mills and manufactures a large number of grades of paper. To a large extent, the waste treatment system's operations will reflect the particular mix of pulp and paper being made each shift. Thus, a knowledge of which grades are being produced can be an important factor in the operation of the wastewater treatment plant. We cannot cover all of these product-related differences in this section. However, the topics of fiber types, flow, settleable solids, color, and odor will be discussed in general terms.

1. FIBER—The paper industry manufactures products made primarily of cellulose. The cellulose must be treated in the mill in a variety of ways that include cleaning, blending, refining, screening, bleaching, trimming, and drying. In many of these operations, the cellulose is pumped from one unit process to another. In order to efficiently process the cellulose into paper, the fibers must be separated (diluted) for some unit processes and matted (molded) for others. The result is that the cellulose is mixed with large quantities of water at various points in the process.

   This water is later removed from the cellulose and much of it is reused. The recycle system in a paper mill actually retains 3 to 10 times the amount of water sewered. The paper industry is continually looking for new ways to reuse much of this water. In many cases, the mill uses internal treatment to renovate these waters before reuse in process showers or for stock dilution. Also, the effluents are generally treated within the mill before being discharged to the sewer. Most of the usable fiber is scalped from the water by a variety of savealls. The savealls may be screens, filters of various sorts, dissolved air flotation units, or mechanical clarifiers.

When the recycle system is working properly, the treatment plant receives moderate quantities of water containing the fine cellulose particles and dirt, which have been deliberately separated from the paper. However, when the recycle system is out of balance, large quantities of water and cellulose may be sent to the sewer. Most of these cellulose fibers settle readily and will clog screens and block the passage of effluent unless adequate mixing or dilution is maintained in the sewers. When the fibers reach the primary clarifier, they settle readily and increase the load on the rakes. Most clarifiers are designed to handle this load without overloading, but the sludge removal rate is usually increased to keep the torque in control.

2. *FLOW*—Most paper mills operate around the clock. Thus, flows are fairly constant and do not decrease during the night as happens with the flow to municipal plants. Depending on what product is being manufactured, however, discharges high in cellulose content may occur. Large quantities of extra water may also accompany abnormally high fiber discharges. The size of the plant should be adequate to handle these surges. However, it may be necessary to adjust the chemical feed rates and to use standby pumps after hydraulic surges have reached the plant.

3. *SETTLEABILITY*—If the paper plant contains a pulp mill, some further differences from a municipal treatment may be evident. The pulping of wood separates a number of chemicals from the fiber. These chemicals are believed to cause higher SVI levels in biological sludges produced during treatment. In the case of treating 100 percent pulp mill effluent, secondary clarifiers are designed with overflow rates lower than systems processing only domestic wastes in order to accommodate this slower settling floc. Also, the final effluent may contain slightly higher suspended solids concentrations where 100 percent pulp mill wastes are being treated.

4. *COLOR AND TURBIDITY*—The chemicals extracted from wood are colored very much like swamp water, which contains vegetative extracts. Thus, the wastewater will generally appear yellow-brown. A wastewater can be clear in terms of suspended solids but still colored by dissolved solids. Wastewater with a dissolved color will change in color with changes in pH as well as mill operations.

Certain paper mills use dyes or pigments to color the fibers. Tissue paper, construction paper, and a variety of specialty products are examples of grades that are colored. The wastewater associated with these grades will also be colored. Mills with one paper machine produce effluents that change in color every time the papermaker changes the dye. However, the color change is not as noticeable in effluents from mills with many paper machines because the colors blend toward neutral and often appear gray. Most dyes are oxidized and thus disappear during biological treatment.

Mills making filled grades or using starch will produce waters that appear white. Starch will usually degrade in biological treatment and the cloudiness and turbidity will dis-

appear during treatment. Some mills use titanium dioxide and talc to fill the paper sheet. These very fine white particles may not be completely removed during treatment; thus, the final effluent will contain white solids. Also, the return activated sludge may appear white because it contains large quantities of these inorganic pigments. Titanium dioxide pigments are highly refractive, which means they reflect and disperse light very effectively, thus increasing the white appearance of wastewater containing the pigment.

5. *ODOR*—The odor of the water discharged from a pulp mill will reflect the chemicals used and produced in the pulping process. In most cases, these odors are associated with gases that were captured in the process water back in the mill. When these waters arrive at the treatment plant and are aerated, the gases are stripped from the wastewater. The extent to which these gases are present in the wastewater depends on the pulping process and how much gas stripping occurs before the wastewater is discharged to the sewer.

EMERGENCY SYSTEMS

The plant probably has some emergency features that are seldom used. Industrial systems must be able to be started up and shut down in accordance with production variations. In many cases, these are planned far enough in advance to turn down or shut down the plant in an efficient manner. However, occasionally, because of an equipment failure at the plant, personnel have to respond quickly with seldom-used facilities. The operators should run through periodic drills that simulate their reactions to these emergencies so there will be no delay when a real emergency occurs. Examples of seldom-used facilities are standby pumps, diversion valves, spill tanks, and lagoons as well as feed tanks used to supply organic load to build up an activated sludge culture prior to plant start-up.

Spill tanks should normally be operated empty so their capacity is available during an emergency. After the tanks are used to scalp a spill, they should be emptied (usually gradually) at a rate that will allow the plant to treat the wastewater adequately. However, the goal should be to empty them as quickly as possible because the plant is susceptible to upset if another spill should occur that is larger than the reserve spill capacity. The spill tank is often used to store waste during emergency shutdown periods.

2.553 Start-Up and Shutdown Procedures

The paper industry has occasion to shut down and start up an activated sludge plant for several reasons. Holidays, maintenance (preventive and emergency), process upsets, strikes, and scheduled production reductions may require that the activated sludge plant be turned down or shut down. The procedures for shutdown and subsequent start-up vary with the nature and duration of the shutdown, as well as the weather expected during the shutdown. To avoid problems during brief shutdowns, a set of action plans must be developed to achieve some short-term goals. The goals must be selected prior to each planned shutdown.

2.554 Management of Shutdowns and Start-Ups
by W. A. Eberhardt

WHY MANAGE SHUTDOWNS AND START-UPS?

The performance of an activated sludge plant improves with increasing stability in influent loadings (waste composition and volume) and environmental conditions (pH, temperature, DO). Typically, as the steady-state condition is lost, discharges of suspended solids and BOD increase. Also of significance is the fact that once the favorable performance associated with steady-state operation is lost, it is not quickly regained.

Shutdown/start-up conditions have a high potential for producing a loss of biological equilibrium or steady state. In addition, shutdowns/start-ups increase the risks of personal injury and produce abnormally high operating costs due to the new or unusual conditions typically experienced.

As an operator, it is your responsibility to the environment, your employer, and yourself to prevent or minimize the potential adverse impacts of a shutdown/start-up. To adequately meet your responsibilities, you must carefully manage the situation.

WHAT IS A SHUTDOWN/START-UP?

Shutdown/start-up is a situation whereby the normal wastewater feed is interrupted, or one or more of the activated sludge support systems (such as aeration) malfunctions. The result is that the biological process is prone to lose equilibrium. Typical circumstances surrounding a shutdown/start-up are:

1. Interruption of Normal Feed

    a. Manufacturing interruptions (weekends, holidays, annual shutdown, equipment failures)

    b. Manufacturing abnormalities (spills, process changes or tests, product changes)

    c. Feed support loss (failure of a feed pump)

2. Interruption of Biological Support Systems

    a. Mechanical failures or planned outages (recirculation pumps, aerator, chemical feed pump)

    b. Runout of a chemical (nutrient, neutralization, polymer)

HOW CAN I MANAGE SHUTDOWNS/START-UPS?

To successfully manage shutdowns/start-ups, you must have objectives and goals, and a plan for implementation.

1. OBJECTIVES AND GOALS—Several basic objectives to be accomplished during shutdowns/start-ups are:

    a. No personal injuries

    b. Maintain satisfactory and legal process performance

    c. Operate at lowest cost for satisfactory performance

    For each objective, specific goals must be established. Table 2.7 provides some suggested goals. For example, one goal for achieving the process performance objective is to maintain a constant F/M ratio. This would prevent one or more of the potential problems cited for failure to control or change the ratio.

    YOU must customize the suggested objectives and goals to your operation.

2. PLANNING—The shutdown/start-up plan is a strategy to accomplish your selected goals. The plan is necessary for both scheduled and sudden occurrences. The plan may be formally typed in a manual or written in a log book. This plan should include the specific activities and associated timing and responsibilities necessary to overcome all identified barriers to accomplishing each goal. For example, the shutdown might involve discontinuance of wastewater from manufacturing; this will be a barrier to maintaining a constant F/M ratio. Your plan might call for storage of wastewater in a basin in advance of the shutdown with provision to pump it to the process during the outage. (Holding wastewater in a basin at all times might be a continuing plan to cover unscheduled losses of process wastewater.) Finally, your plan must be coordinated with other wastewater treatment and manufacturing activities.

3. IMPLEMENTATION—Successful implementation of your plan requires:

    a. COMMUNICATIONS with involved and affected personnel and organizations before and during implementation

    b. DISCIPLINED EXECUTION of the planned action steps

    c. EVALUATION OF RESULTS during execution with appropriate plan adjustments

    d. REVIEW AND EVALUATION of results and performance after the incident with emphasis on learning for future improvement

4. SUMMARY—The benefits of well-managed shutdowns/start-ups include satisfactory and legal performance, fewer injuries, and reduced costs. To manage an interruption, you must identify its occurrence (actual or potential) and aggressively act to minimize its consequences. Success will follow if objectives and goals are established, plans are developed to achieve them, and plan implementation is coordinated, disciplined, and flexible.

TABLE 2.7   GOALS FOR SHUTDOWNS/START-UPS

Goals	Potential Problems	Action Plans
1. Maintain constant food characteristics	Solids in effluent SVI increase Low BOD removal Lost activated sludge organisms	Normal waste: • Store prior to shutdown • Feed during shutdown Abnormal wastes: • Minimize • Bypass to storage for equalization
2. Maintain constant food/ microorganism (F/M) ratio	Solids in effluent SVI increase	Pre-shutdown: • Store waste • Lower MLSS During shutdown: • Feed stored waste • Reduce or stop wasting
3. Maintain nutrient balance (N, P, metals)	Solids in effluent Bulking Low BOD removal that persists	Ensure in advance: • Adequate supply of nutrients • Provisions for addition of nutrients
4. Maintain constant environment (temperature, pH, DO)	Solids in effluent Low BOD removal	Optimize aeration for: • DO • Heat retention in winter • Heat loss in summer Introduce heat during prolonged winter usage Ensure in advance: • Supply of neutralization chemicals • Provisions for addition of chemicals or stored supply of preneutralized waste
5. Maintain hydraulic loading control	Hydraulic surges Flows during periods requiring a "dry outage" Sludge recycle imbalances	Plan together with manufacturing staff Use surge/storage basins Closed valve operation Gradual increase in loading at start-up
6. Necessary equipment functional	Power outage (lights, equipment) Instrument air outage (valve positioning)	Advance provisions: • Standby generators • Instrument air override • Use storage basins
7. Complete required maintenance on schedule	Less than optimum performance after start-up Production downtime for subsequent repair	Critical path planning Maximum prework Avoid shutdown of units
8. No personnel injuries	Lack of safety equipment Difficult environment (no lighting) Abnormal (nonroutine) tasks Inadequate staffing	Individual job safety: • Analyses with appropriate follow-up • Safety training program • Appropriate safety equipment
9. Cost optimization	Shutdown extension due to: • Inadequate staffing • Undertaking necessary work Provision of unnecessary equipment (generator for aerators)	Question shutdown needs Complete planning
10. Protect receiving stream	Spills Inadequate treatment	Comprehensive planning
11. Meet permit requirements	Inadequate performance or discharge monitoring	Meet all goals in this table

2.555 The Periodic-Feeding (Step-Feed) Technique for Process Start-Up of Activated Sludge Systems [19]
by R. S. Dorr

NEED FOR EFFECTIVE START-UP PROCEDURES

Activated sludge biological treatment systems are being constructed in many locations across the country by pulp and paper mills. Process start-up of these high-rate biological systems presents a complex problem. Experience at the plants currently in operation indicates that given enough time and attention these systems can be successfully started up. Time, however, is costly and often limited when dealing with discharge permit deadlines.

DEVELOPMENT OF THE START-UP PROCEDURE

A periodic-feeding technique has been developed for the start-up of pulp and paper mill activated sludge systems. Several important concepts should be understood before attempting to use this start-up procedure:

1. The concentration of microorganisms in the system influent (the mill sewer) is insignificant in comparison to concentrations found in domestic collection systems.

2. There exists a critical minimum mean cell residence time (MCRT) below which the system cannot function and the microorganisms cannot establish themselves.

3. The degree of flocculation of the microorganisms is influenced by the food-to-microorganism (F/M) ratio. At very high F/M values, microorganisms are completely dispersed and will not settle, rendering the secondary clarifiers nonfunctional.

4. If the secondary clarifiers are nonfunctional, the mean cell residence time is equal to the hydraulic retention time of the aeration basins.

5. Beyond certain limits, increasing the food (soluble BOD) to a bacterial population does not necessarily increase the rate of population growth.

In view of these facts, a process start-up procedure should provide a means of controlling the food-to-microorganism ratio or the mean cell residence time to prevent dispersal and washout of whatever bacteria may be initially present in the system. Neither the soluble BOD concentration in the system influent nor the amount of bacteria in the system can normally be controlled during a process start-up. [20] The hydraulic loading rate, however, can be regulated. This is the basis of the start-up procedure described in this section.

Assume a wastestream of 10 MGD and 100 mg/L soluble BOD were to be treated in an activated sludge plant designed to have a hydraulic retention time of six hours. Flow is introduced to the system along with normal amounts of nitrogen and phosphorus required, and aerators and recycle pumps are started. After the six hours of flow required to fill the basins and clarifiers, the following conditions exist:

1. The soluble BOD level throughout the system is 100 mg/L.

2. The amount of bacteria present is very small, existing only in the form of "seed."

3. The F/M value is very high; therefore, the bacteria are totally dispersed.

At this point, the flow through the system could be allowed to continue at 10 MGD; however, this might be disadvantageous for two reasons. The "seed" microorganisms cannot possibly consume a very significant portion of the soluble BOD during the six hours in which it would pass through the system. Indeed, it may take several days under very similar conditions to consume the 100 mg/L in a BOD bottle. There is no reason to believe that the reaction will take place any faster inside the aeration basin. Therefore, constantly feeding the system a fresh supply of wastewater containing 100 mg/L soluble BOD provides no real advantage.

Additionally, since the bacteria are dispersed, 10 MGD of water, containing whatever concentration of microorganisms that exist in the system, would be flushed over the effluent weir of the secondary clarifier. This loss must be made up for by the rate of growth of the bacterial population. The six-hour residence time could be at or below the minimum mean cell residence time required for growth; therefore, washout of any bacterial growth will occur. In any case, the six-hour residence time would put the microorganisms at an extreme handicap in establishing themselves.

In the periodic-feeding start-up procedure, the influent flow is shut off. The water within the plant is then recirculated between the secondary clarifiers and the aeration basins. With this method, the mean cell residence time may be made as long as necessary simply by retaining the entire mass of microorganisms within the system. The bacteria proceed to consume the soluble BOD present and subsequently increase in concentration, rapidly lowering the F/M value. Eventually, the energy level drops to a point where the bacteria begin to flocculate and the amount of food remaining becomes a limiting factor in further growth.

Now is the time to introduce more food to the microorganism population. Again, it might be disadvantageous to feed the full 10 MGD flow through the system for the same reasons stated previously. Instead, gradually increase the rate to match the available food supply to the growing concentration of microorganisms. In terms of the food-to-microorganism ratio, "F" might be matched to an ever-increasing "M" to achieve an

---

[19] This section was taken from R. S. Dorr, "Development and Testing of the Step-Feed Technique for the Process Start-Up of Activated Sludge Systems," *PROCEEDINGS OF THE 1976 NCASI NORTHEAST REGIONAL MEETING, NCASI Special Report No. 77-03,* May 1977. NOTE: The original paper referred to this procedure as STEP-FEED, but this section uses the term PERIODIC-FEEDING to avoid confusion with the step-feed activated sludge process.

[20] As an exception, bacteria from another similar activated sludge plant may be introduced by "massive seeding" techniques involving many truckloads of thickened sludge.

F/M ratio that would neither inhibit the growth rate nor disperse the bacteria. As an additional advantage, a reduced rate of feed would produce a lower overflow rate in the secondary clarifiers, making them more efficient than normally possible. This fact could be used as justification for allowing higher than normal F/M values during the low flow portions of the start-up.

The value of F/M could ideally be held constant or gradually decreased by carefully regulating the flow rate through the plant. In practice, however, this is not possible due to variations in the actual soluble BOD concentration in the wastestream and also because the flows cannot usually be regulated with such precision. From an operating standpoint, it would be more feasible to keep the F/M ratio within a range that eventually narrows down to the optimum ratio when the system is started. The feed can be increased in a periodic step-wise manner as long as the periodic steps do not cause the F/M ratio to exceed a "safe" level.

Although bacteria are the primary agents in stabilizing the organic wastes, other types of microorganisms play an important role in an activated sludge system. More complex single-cell organisms called protozoa act as scavengers devouring dispersed bacteria and thereby producing a highly clarified effluent. Protozoa also serve as useful indicators of the overall condition of the system. Because of their greater complexity and size, they are more sensitive to changes in their environment and can be more easily observed under a typical compound microscope.

The usefulness of microscopic indicators in this type of start-up scheme is obvious. Following each periodic (step) increase in feed to the system, the F/M ratio will rapidly pass from a high level to one near the normal optimum. Timing the next periodic (step) increase is critical both to the health of the microbial population and to the overall start-up time requirements. The F/M value may be guessed at based on measured MLVSS and either estimated BOD loads or rapid tests that approximate BOD such as COD to TOC analysis. These lab tests are very important during start-up; however, only by microscopic inspection can the system's energy level be immediately and directly confirmed. In addition, this observation may detect nutrient deficiency, low pH, low dissolved oxygen levels, and other problems hindering most start-ups. Microscopic examinations can reveal developing problems that plant instruments may not detect due to inaccurate calibration or laboratory tests with which the technician is unfamiliar.

Besides producing a rapid and controlled growth of active microorganisms, the periodic (step)-feed procedure has several other desirable effects. Using this procedure, a full, active population may be grown from even the smallest amount of seed organisms. A bucket of activated sludge from another plant will contain all the types of microorganisms necessary to start a system and may be used if it is uncertain that they exist normally in the mill wastewater. This small population of microorganisms, when introduced during the zero-flow step, will multiply rapidly and stabilize the F/M value within a matter of several days.

The periodic (step)-feed method minimizes the possibility of sudden kills of the active bacteria mass. The microbial population is extremely fragile during the early stages of its development, but it is also during this time that most of the wastewater flow bypasses the system. This arrangement reduces the possibility that a slug of toxic, inhibitory, or high strength organic material in the wastestream will endanger growth of the microorganisms.

Finally, there is no period of "population stabilization" after the target MLVSS level is attained using the periodic (step)-feed method. Often in the start-up of an activated sludge system by other methods, the types of microorganisms present in the system when the target is reached are not the same ones that will predominate during the normal operation. This is because the population produced by using these methods develops under conditions, such as the energy level, that are far removed from the conditions that will exist after the target is reached. A considerable length of time may be required for a turnover of the types of microorganisms represented in the population. Plant performance can be expected to be somewhat less than optimum during this period. By gradually "zeroing in" on the optimum F/M value, the periodic (step)-feed technique cultivates a highly diverse population of microorganisms that are more suited for survival during normal operation.

2.556 Operation of a Municipal Plant Receiving Paper Mill Wastewater
by Anthony A. Leotta and W. A. Hopsecger

Operating experiences described in this section were obtained from a treatment plant with domestic flows averaging from 2.0 to 2.5 MGD (7,570 to 9,460 m^3/day) and paper mill flows from 0.3 to 0.5 MGD (1,135 to 1,890 m^3/day). Paper mill wastewater averages 600 mg/L suspended solids and 250 mg/L BOD. Domestic wastewater is relatively weak and averages 75 mg/L suspended solids and 50 mg/L BOD.

Nutrient deficiencies are experienced at this plant. Urea is introduced into the aeration tanks to provide a concentration of 5 mg/L nitrogen at the head of the aeration tank. Another satisfactory approach to solving the nitrogen deficiency is to bypass a portion (10 to 20 percent) of the primary influent directly to the aeration tanks. Nutrients that are usually removed by primary treatment are diverted to the aeration tanks to satisfy the nitrogen deficiency.

Mixed liquor suspended solids are adjusted to 2,500 mg/L. Experience seems to indicate that paper waste is difficult to consume by microorganisms and therefore may require more microorganisms than are typically found in most domestic treatment plants to do the job.

DO levels in the aeration tank are maintained between 2 and 3 mg/L throughout the entire tank. Too much air (over 3 mg/L) breaks up the floc and reduces settleability. Too little air kills some of the beneficial activated sludge bacteria and could result in an undesirable explosion of filamentous growths.

Optimum return activated sludge (RAS) flows range from 40 to 50 percent, depending on flow and secondary clarifier conditions. Return rates of 20 to 25 percent do not maintain the proper MLSS in aeration. Rates in excess of 50 percent cause an increase in overflow rates that result in bulking sludge flowing over the clarifier weirs.

Depending upon the type of solids in the return sludge, the wasting schedule is reduced during the autumn and winter season in order to maintain higher MLSS in aeration. Since microorganisms become less active (sluggish) in colder temperatures, MLSS are raised to about 3,000 mg/L during the colder seasons. Closer control of MLSS in relation to suspended solids in the return sludge is required during the colder season. Experience has shown that the ratio between MLSS and return sludge suspended solids should be about 1:3. In other words, with the best range of MLSS from 2,000 to 3,000 mg/L, suspended solids in the return sludge should be maintained between 6,000 to 9,000 mg/L. Wasting schedules are calculated to maintain this ratio.

The operating guidelines are summarized as follows:

1. F/M ratio less than 1.0

2. SVI around 200

3. MLSS of 2,500 mg/L (warm weather)

4. MLVSS, 70 percent of MLSS

5. DO, 2 to 3 mg/L throughout aeration tank

6. Return activated sludge rate, 40 to 50 percent

7. Return activated sludge suspended solids, 7,500 mg/L

8. MLSS/RAS SS, 0.33

2.557   Acknowledgments

Authors in this section on how to treat pulp and paper mill wastes included James J. McKeown, W. A. Eberhardt, R. S. Dorr, Anthony A. Leotta, and W. A. Hopseger. Reviewers of this section included Al Brosig, Anthony A. Leotta, Larry Metzger, David B. Buckley, Ray Pepin, and W. A. Eberhardt.

QUESTIONS

Write your answers in a notebook and then compare your answers with those on page 149.

2.55A   Why must accurate records be kept?

2.55B   Under what conditions might the paper industry shut down an activated sludge process?

2.55C   Why should the effluent from a pulp and paper mill activated sludge process contain around 0.5 mg/L ammonia nitrogen and 0.25 mg/L phosphorus?

2.55D   Why is the periodic-feeding (step-feed) technique for process start-up of activated sludge systems effective?

2.56   Brewery Wastewaters
by Clifford J. Bruell

2.560   Operational Strategy

In brewery wastewater treatment plants, as in other food or industrial treatment plants, there exists the potential of wide fluctuations in the influent wastewater characteristics. This is especially true with respect to the organic content of the influent. A brewery wastewater treatment plant that receives an organic load of 25,000 pounds (11,000 kg) of BOD on one day may experience an organic loading of 50,000 pounds (22,000 kg) of BOD (or more) the following day. Economically, it is impossible to design, construct, and operate a wastewater treatment plant that will, as a matter of routine, treat highly fluctuating loads or the entire potential load from an industrial plant. Therefore, a comprehensive waste load monitoring program to control the source of the wastewaters and to capture for reuse valuable raw materials is essential.

Due to the nature of brewery operations and the characteristics of brewery wastewater, most brewery wastewater treatment plants are small wastewater treatment plants that do the work of large wastewater treatment plants. Therefore, a tight, well-organized operation is essential in order to meet NPDES permit goals. The various operational strategies described in this section are examples of what has worked best at a typical plant. These methods are not the only way to get the job done. Careful experimentation (where only one variable at a time is changed) is necessary to determine the best operating methods at every individual treatment plant. There is no substitute for experience and every operator has experience. The key to successful operation is using existing experience and building on that foundation with new, well-documented experience.

2.561   Sources and Characteristics of Brewery Wastewater

SOURCES OF BREWERY WASTEWATER

In the brewing industry, as in other food processing industries, there is always some minimal loss of raw materials. Brewing ingredients such as malted barley, rice, starch, hops, and yeast often become part of the wastewater. Raw material losses can be controlled by the use of capture systems and recycle loops; however, it is inevitable that some portion of this material will become waste. The malting of barley requires large volumes of water to flush concentrated organic materials from the grain. To maintain sanitary conditions within a food plant, large volumes of cleaning solutions are required. These soapy solutions, which often have a high pH (caustic soda), and their associated rinse water also become part of the brewery wastewater stream. A small portion of the product (beer) and brewery by-products (brewers' yeast) are very concentrated (organically) and can enter the wastestream. On occasion, lubricating oils or other maintenance or utility materials (ammonia, glycol coolant, acids) may accidentally enter the sewer line. Together, these items are the constituents of brewery wastewater.

CHARACTERISTICS OF BREWERY WASTEWATER

Brewery wastewater that contains starches, sugars, and alcohol can be characterized as organically strong. This means that the wastewater may often exert a BOD of 1,200 mg/L or greater (municipal wastewater often has a BOD of 200 mg/L). Much of this BOD is "soluble" or "dissolved" and only 10 to 15 percent of this BOD can be removed by primary clarification. Therefore, a highly efficient secondary treatment process, such as activated sludge, is necessary to successfully treat brewery waste. Brewery wastewaters are often deficient in nutrients such as nitrogen and phosphorus. Textbooks recommend a BOD:nitrogen:phosphorus ratio of 100:5:1 in the influent to the aeration basins to produce desirable activated sludge microorganisms. If wastewater has a very high BOD, due to starches and sugars, then the nitrogen and phosphorus levels considered normal in municipal wastewater would be deficient when compared to the BOD in brewery wastewater. If cleaning solutions (acids and caustics) are released rapidly to a brewery wastewater treatment plant (instead of being slowly metered), drastic pH fluctuations in the influent will result. This pH fluctuation (pH 4 to 10) can cause operational problems if the influent is not neutralized.

2.562 Brewery Wastewater Treatment Plant Tour

OVERALL PICTURE

A typical brewery wastewater treatment plant flow layout is shown on Figure 2.16. Typical data have been included to give you the "feel" of the relative operating conditions of this plant. To provide you with an overall picture of the plant, let us take a walking tour of the plant shown on Figure 2.16. At each location try to picture the equipment that would be seen and the process that is taking place. Use your existing knowledge of wastewater treatment and try to relate it to this brewery wastewater treatment process. Also, much of the other information in the remainder of this section on brewery wastewater treatment will be somewhat specific to the particular typical brewery wastewater treatment plant described.

PRETREATMENT OF WASTEWATER

To begin the tour let us start at the headworks of the plant. At this point, the flow from various portions of the brewery, which is collected by a massive network of piping, valves, floor drains, and sumps, is delivered to the wastewater treatment plant by means of a lift station and several gravity lines. An influent flow of 3.0 MGD (11,350 m³/day) containing 1,200 mg/L BOD (30,000 lbs or 13,600 kg/day) and 250 mg/L SS (6,000 lbs or 2,700 kg/day) would be an average loading to this plant.

The first step of the treatment process is bar screening. At this brewery, the automatically raked bar screen removes can lids, cans, glass, pieces of wood, rags, and balls of "grain and keg wax" that form within the sewer lines. Immediately following the bar screening step is the grit removal chamber or grit channel. Grit at a brewery wastewater treatment plant consists of malted barley and hops. Depending on what raw materials are used, grains such as corn grits or rice might also be captured within a grit chamber. These brewing materials are readily settled and removed from the wastestream. Therefore, this grit material has some limited reuse value; an example of this would be as a compost mix additive. Estimation of the volumes of grit that are generated at various breweries is very difficult because these volumes would depend on whether malting was being done on site and the efficiency of the recapture systems. This particular brewery generates from 0.5 to 1.0 cubic yard of grit per MG (0.1 to 0.2 L/m³) of flow.

A useful operational aid within the area of the bar screen is an in-line continuous monitoring pH meter. A meter at this location, with a low and high pH alarm, can alert an operator to potential pH problems long before they become problems. The size of a pH fluctuation and the duration can alert an operator to the severity of the problem. An automatic/manual acid addition system is used to meter (pump) concentrated H_2SO_4 to neutralize caustic cleaning solutions. An additional recording pH meter is used to control/monitor the acid addition. At this brewery, the acid is added in the grit chamber area; however, a more desirable addition and monitoring point would be downstream of the surge tank or primary clarifier. Other brewers have successfully used automatic caustic addition systems (50 percent NaOH added by pump) to neutralize acid spills. If an automatic caustic addition system is not present, stores of caustic materials such as lime or NaOH should be kept on hand for emergency situations such as large acid spills.

After grit removal, the influent wastewater stream enters a 0.5 MG (2,100 m³) surge tank (also see Section 7.1, "Equalization of Industrial Wastewater Flows," in INDUSTRIAL WASTE TREATMENT, Volume I). Brewery production is often a batch process. A good example of this is the malting process. When a malting tank is drained of liquid during the "steeping out" process, this can immediately exert a high hydraulic load (8 MGD (30,000 m³/day) flow rate) on the wastewater treatment plant. The surge tank is used to catch this peak hydraulic load and then the waste can be metered to the plant at a controlled rate. Pumping from the surge tank at a controlled rate of flow eliminates shock loadings to the aeration basins and prevents disturbing the sludge blankets in the plant clarifiers.

Another function of the surge tank is dilution of high strength organic wastes in the influent. Beer has a very high BOD that can exceed 200,000 mg/L. An accidental beer spill consisting of 10 to 100 barrels of beer (31 gallons/barrel or 117 liters/barrel) would be considered a major overload situation. A load of this size, unchecked, could cause a plant process failure. However, when a surge tank is used, dilution will occur within the tank and the entire waste load can be metered into the aeration system. When the waste material then enters the aeration system at a slightly elevated organic concentration (rather than

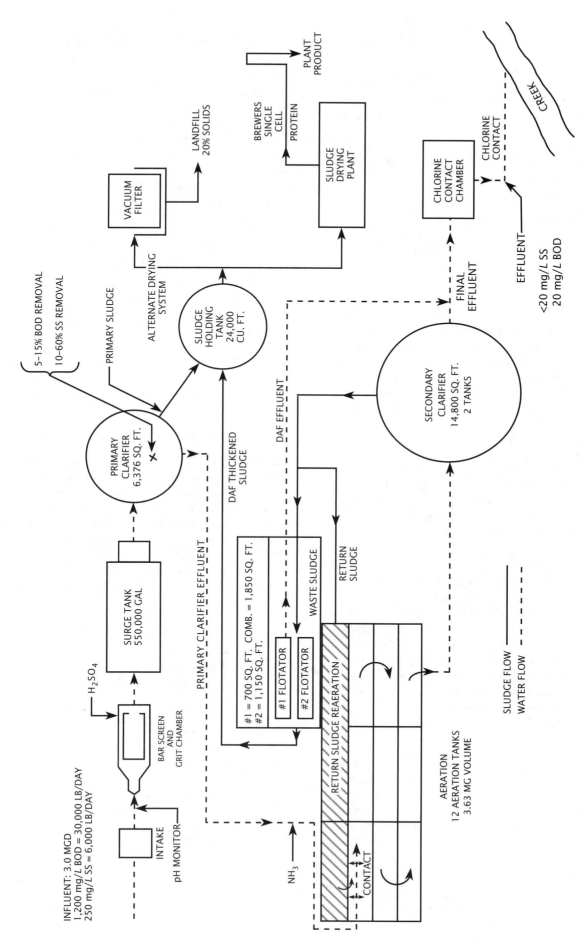

Fig. 2.16   Flow schematic of brewery wastewater treatment plant

greatly elevated), it is possible to meet dissolved oxygen requirements and to form a desirable biomass.

In the surge tank, several floating aerators are used to help mix and dilute the wastestream and to pre-aerate the wastewater to a limited degree. The aerator units rise and fall (float) just as the tank level does. Oils and grease in the wastestream are also "conditioned" by the mixing action of the surge tank. These materials become slightly coagulated (clumped together), which results in an increased removal efficiency of floating oil and grease (FOG) in the primary clarifier.

The surge tank level is manually controlled by the operator (level alarms and overflow provisions are present). Periodically, the operator checks the surge tank level and adjusts the rate of waste pumping to maintain a desired level. From the surge tank, the waste is pumped into the primary clarifier.

PRIMARY CLARIFIER

The primary clarification step typically removes 10 to 60 percent of the influent suspended solids. However, because brewery waste has a highly soluble organic content, a smaller amount of BOD removal is achieved at this point. A 5 to 15 percent BOD removal is typically obtained; this fact also indicates that only a small amount of BOD is associated with the solids removed here. Though oil and grease are not a major portion of the influent stream, some is skimmed from the primary clarifier surface along with small amounts of yeast, hops, and grain hulls that tend to float in the primary clarifier. Sludge solids concentrations of materials removed from the primary clarifier typically range from 3 to 5 percent solids. Sludge blanket depths are measured periodically. Sludge pumping frequency and the rate of removal (pumping rate) are adjusted to maintain a minimum sludge blanket depth in the clarifier without causing CONING.[21]

2.563 Nutrient Addition

Most brewery wastewaters contain too little nitrogen and phosphorus to support microorganism growth. To supplement nitrogen, ammonia gas (or liquid) can be metered into the primary clarifier effluent. Some brewery wastewater treatment plants use phosphoric acid, H_3PO_4, as an additional source of phosphorus. At the brewery wastewater treatment plant under examination here, adequate quantities of phosphorus are present in the influent. This is due to the fact that phosphorus is derived from both malting by-products and phosphorus-based cleaning solutions. However, in this plant, nitrogen is deficient.

A desirable organic strength to nutrient ratio is 100 pounds BOD: 5 pounds N: 1 pound P. In order to calculate the quantities of nutrient addition required, it is first necessary to have some estimate of organic strength of the waste. To run a 5-day BOD test as an estimate of the organic strength is impractical because it takes too long, particularly if nutrient addition rates

have to be adjusted on a daily basis. Therefore, an analysis such as chemical oxygen demand (COD) or total organic carbon (TOC) is used to provide relatively quick results. These tests can be run in several hours and a BOD value can be calculated once a BOD:COD or BOD:TOC ratio has been established. As an indication of the amount of nitrogen that is already present in the influent, NH_3-N analysis is performed on a 24-hour composite sample of the primary clarifier effluent. This same sample is used for other analyses such as TOC analysis. As an estimate of nitrogen content, NH_3-N analysis is used because it is easier to run than TKN (Total Kjeldahl Nitrogen) analysis. This usually ends up in a shift in the BOD:N ratio *FROM* 100 pounds BOD:5 pounds N (TKN) *TO* 100 pounds BOD:2 pounds NH_3-N. (This is true because in this wastewater the ratio of 5 pounds TKN:2 pounds NH_3-N holds true.) The same philosophy applies to phosphorus analysis (it is easier to run ortho-P analyses than total-P analyses).[22]

An example of an NH_3-N addition calculation is provided (see the sample calculation). The basic goals of the addition procedure are to:

1. Estimate the organic strength of the waste (using a quick method such as TOC).

2. Estimate the amount of nutrient already present (using a quick method such as NH_3-N).

3. Estimate the supplemental amount of nutrient required.

The following example shows how to calculate the amount of ammonia that needs to be added to correct a nitrogen deficiency in a brewery wastestream. The sample calculation procedure is designed to allow addition of the nutrient on a continuous basis for 24 hours. Also, when the plant effluent analysis reveals 1.0 mg/L or greater residual of the nutrient that is being supplemented, this is an indication that the addition quantity is sufficient. If the residual is far above or below the 1.0 mg/L concentration, the addition ratio should be adjusted accordingly.

NUTRIENT ADDITION SAMPLE CALCULATION

Sample Data:	Item	Data
	Primary Effluent, 24-hour Composite Total Organic Carbon (TOC)	554 mg/L
	BOD:TOC Ratio	2.2:1
	Calculated Sample BOD 2.2 × 554 mg/L	1,219 mg/L
	Estimated Flow for the Next 24 hours	3.3 MGD
	Final Effluent, 24-hour Composite NH_3-N Concentration	0.88 mg/L
	Primary Effluent, 24-hour Composite NH_3-N Concentration	8.36 mg/L

---

[21] *Coning.* Development of a cone-shaped flow of liquid, like a whirlpool, through sludge. This can occur in a sludge hopper during sludge withdrawal when the sludge becomes too thick. Part of the sludge remains in place while liquid rather than sludge flows out of the hopper. Also called coring.

[22] For a detailed explanation of how to perform the tests mentioned in this paragraph, refer to Chapter 16, "Laboratory Procedures and Chemistry," in *OPERATION OF WASTEWATER TREATMENT PLANTS,* Volume II, in this series of operator training manuals.

SAMPLE AMMONIA ADDITION CALCULATIONS

1. Estimate the present day's BOD loading in pounds of BOD per day.

$$\text{BOD Loading, lbs/day} = \text{Flow, MGD} \times \text{Est BOD} \times 8.34 \text{ lbs/gal}$$

$$= 3.3 \text{ MGD} \times 1{,}219 \text{ mg}/L \times 8.34 \text{ lbs/gal}$$

$$= 33{,}549 \text{ lbs BOD/day}$$

2. Calculate the amount of ammonia (NH_3-N) required in pounds per day. Assume an ammonia requirement of 2 pounds NH_3-N per 100 pounds BOD.

$$\text{NH}_3 \text{ Required, lbs/day} = \text{BOD, lbs/day} \times \frac{\text{NH}_3\text{-N, lbs}}{\text{BOD, lbs}}$$

$$= 33{,}549 \frac{\text{lbs BOD}}{\text{day}} \times \frac{2 \text{ lbs NH}_3\text{-N}}{100 \text{ lbs BOD}}$$

$$= 671 \text{ lbs NH}_3\text{-N/day}$$

3. Estimate the ammonia (NH_3-N) supplied to the aeration basins in the primary clarifier effluent.

$$\text{NH}_3 \text{ Supplied, lbs/day} = \text{Flow, MGD} \times \text{NH}_3\text{-N, mg}/L \times 8.34 \text{ lbs/gal}$$

$$= 3.3 \text{ MGD} \times 8.36 \text{ mg}/L \times 8.34 \text{ lbs/gal}$$

$$= 230 \text{ lbs NH}_3\text{-N/day}$$

4. Determine the amount of ammonia (NH_3-N) that must be added to the primary effluent.

$$\text{NH}_3 \text{ Added, lbs/day} = \text{NH}_3 \text{ Required, lbs/day} - \text{NH}_3 \text{ Supplied, lbs/day}$$

$$= 671 \text{ lbs/day} - 230 \text{ lbs/day}$$

$$= 441 \text{ lbs NH}_3\text{-N/day}$$

NOTE: Effluent NH_3-N concentration is 0.88 mg/L, which is a satisfactory level (possibly a little low).

5. Calculate the rotameter setting. The rotameter constant is 5.45 lbs NH_3-N for a 24-hour period for each one percent.

$$\text{Rotameter Setting, \%} = \frac{\text{NH}_3\text{-N Added, lbs/day}}{5.45 \text{ lbs NH}_3\text{-N/day/\%}}$$

$$= \frac{441 \text{ lbs NH}_3\text{-N/day}}{5.45 \text{ lbs NH}_3\text{-N/day/\%}}$$

$$= 81\%$$

2.564 *Aeration Basin Flow Scheme*

Immediately after nutrient addition, the primary clarifier effluent enters the first aeration basin or "contact cell." Figure 2.17 shows a clear overview of the activated sludge system. The fresh waste is distributed throughout the basin and mixed with return sludge from the return sludge reaeration cell. Together, the return sludge and fresh waste travel in a "plug flow" pattern through nine aeration cells (0.303 MG or 1,150 m³ each) in a serpentine (snake-like) fashion. Assuming an average flow of 3.0 to 3.5 MGD (11,350 to 13,250 m³/day) and a 35 percent return sludge pumping rate, aeration time is approximately 14 to 16 hours.

The mixed liquor leaving the final aeration cell is split and distributed to the two secondary clarifiers. The activated sludge mixed liquor is next separated from the final effluent in the secondary clarifier. The sludge organisms settle in the quiescent (calm) clarifier and are removed from the bottom of the tank while the clear effluent overflows the tank weirs.

From the secondary clarifier, the secondary effluent passes through a chlorine contact chamber and is chlorinated. If necessary, the final effluent is dechlorinated with sodium bisulfite before discharge to the receiving waters.

A small portion of the return sludge is continuously wasted. Return sludge that is not wasted enters the return sludge reaeration basins. Flow of the return sludge in the reaeration system is also in a plug flow pattern. Return sludge in the reaeration phase is allowed to "rest" for 18 to 20 hours before reentering the contact cell of the aeration system.

The trip of a sludge particle that makes the complete loop from the contact cell, through the entire aeration and reaeration system and back to the contact cell, could take 32 to 36 hours under the previously described operating conditions.

2.565 *Activated Sludge System Operation*

Brewery wastewaters, as well as many other food processing wastewaters, are highly bio-oxidizable. This means that when activated sludge organisms come in contact with the fresh wastewater, the organisms will immediately start to use the wastewater as a food source. When this happens, there is a sudden demand for oxygen by the sludge organisms. Dissolved oxygen (DO) uptake rates in excess of 200 mg O_2/hr/L may occur. The aeration basin DO levels might even drop to 0.5 mg/L or less. Some believe that when DO levels greater than 5.0 mg/L are not maintained, this will promote the growth of filamentous organisms, which may cause sludge bulking. However, low DO levels are just one possible cause of filamentous growth. Several design modifications have been made to the aeration basin to avoid DO level sags in the first aeration basin or contact cell and the associated possibility of sludge bulking.

The contact cell aeration mode is shown in Figure 2.18. This is an enlarged view of the first aeration basin that receives the primary effluent. The view shown here represents what would be seen if you were looking down into a nearly empty aeration basin. A jet-type aeration system that uses Venturi nozzles is the type of aeration system used. Two large submerged basin recirculation pumps collect mixed liquor and pump it into the "liquid" header. Air or oxygen is metered into the "gas" header. When the "liquid" and "gas" pass through the aeration Venturi nozzles at the same time, a jet action occurs, which dissolves the gas in the liquid. This type of jet aeration system is used in all aeration basins and it has the capability of using air or oxygen in all of the "gas" headers.

In most basins where DO requirements can be met with air, standard aeration blowers are used to supply air and satisfy

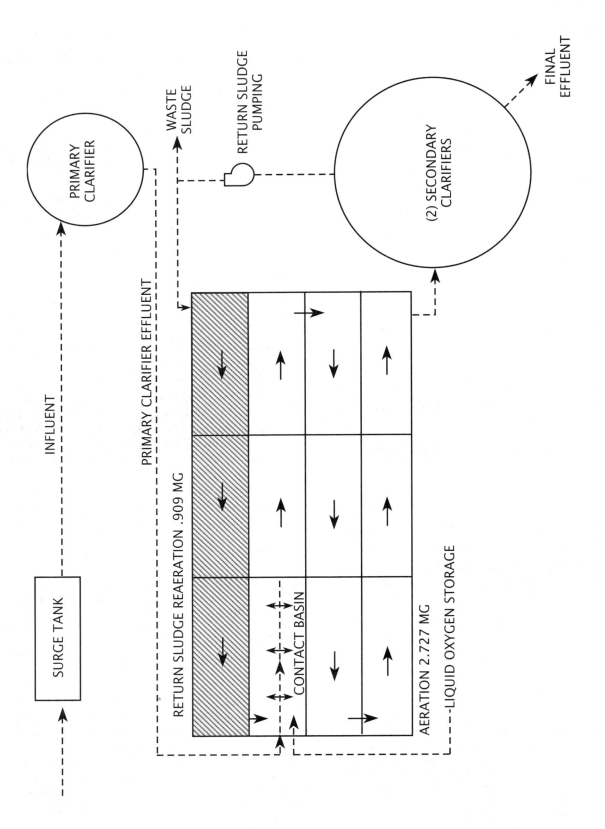

Fig. 2.17   Simplified aeration basin flow schematic

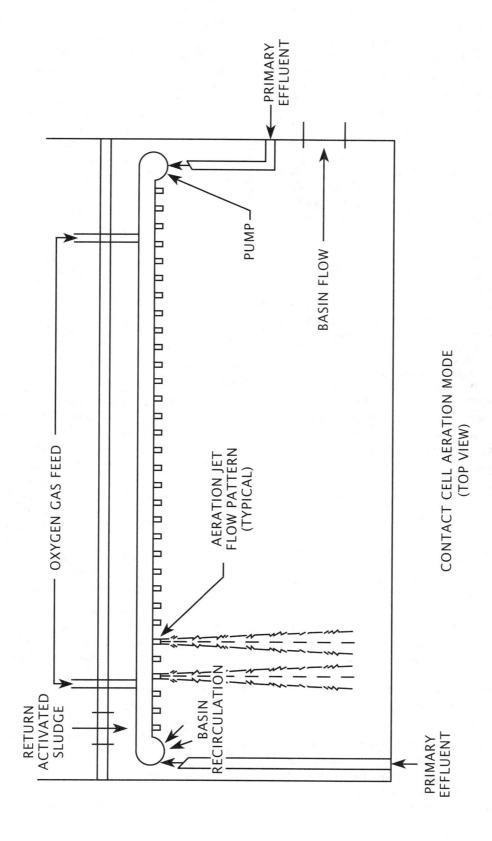

PRIMARY EFFLUENT

PUMP

BASIN FLOW

OXYGEN GAS FEED

AERATION JET
FLOW PATTERN
(TYPICAL)

RETURN
ACTIVATED
SLUDGE

BASIN
RECIRCULATION

PRIMARY
EFFLUENT

CONTACT CELL AERATION MODE
(TOP VIEW)

Fig. 2.18   *Contact cell aeration mode (top view)*

oxygen demands. However, in the contact cell where DO demands are high, oxygen is necessary (pure oxygen is more effective than air because air is only 21 percent O_2). The oxygen is first vaporized from liquid and then metered into the "gas" header within the contact cell. When oxygen is used in the header, no air is used simultaneously.

Within the contact cell, special modifications direct the raw primary clarifier effluent and the return activated sludge from the last reaeration basin to the immediate vicinity of the submerged pump's suction. This is accomplished by piping and duct work that are equipped with flow metering devices. This modification enables the fresh reaerated return sludge and wastewater to be mixed and quickly distributed throughout the contact cell. With the aid of oxygen, it is possible to maintain minimum DO levels. An on-line DO meter and lab meter are used to monitor the basin and frequent oxygen flow rate adjustments are made.

Following "contact," the mixed liquor moves to the aeration basin on the right (in a plug flow fashion) through a submerged flow control gate. The plug flow mode of aeration basin flow is used rather than "complete mix" mode because plug flow helps to prevent the growth of filamentous organisms or dispersed growth. Also, this particular aeration scheme has been the most successful mode at this plant for obtaining the required 20/20 BOD/SS effluent.[23]

Aeration basin MLSS levels are usually between 2,000 to 2,800 mg/L (85 percent VSS). Many factors, including influent organic loadings, dictate the exact MLSS level that is maintained. When very low organic loadings are delivered to the aeration basins, a relatively low MLSS should be maintained to prevent an excessive sludge age from developing. Similarly, high organic loadings call for higher MLSS levels.

Temperature is another factor that greatly influences the MLSS level that should be maintained. The activity (ability to remove carbon) of the sludge organisms within the biomass depends a lot on temperature. An increase in basin temperature of only 10°C (for example, from 20°C to 30°C) can double the activity of the biomass. Therefore, during periods of higher summer temperatures, a lighter biomass MLSS can achieve the same organic removals as twice (2×) the biomass under winter operating conditions (temperatures). As the sludge organisms increase their activity as a result of warmer basin temperatures, the amount of oxygen needed in the system also increases. This means that at high MLSS levels and high basin temperatures, it is often difficult to maintain satisfactory DO levels in the aeration basins. When this happens and aeration capabilities cannot be increased, it often becomes necessary to trim back the MLSS levels until minimum DO levels are achieved again.

The suspended solids levels maintained in the return sludge reaeration basins often range from 6,000 to 10,000 mg/L. The actual concentration depends on the settling characteristics of the sludge and the operation of the secondary clarifiers. Return sludge rates are adjusted to maintain a sludge blanket from 12 to 18 inches deep (30 to 45 cm) in the secondary clarifier. Numerous blanket depth measurements (with a "thief" type sampler) are made to control the blanket depth. The goal is to avoid building a deep blanket that leaves sludge in the clarifiers for a long period of time or pumping the blanket levels too low and then pumping only dilute sludge into the return sludge reaeration basins. An on-line suspended solids meter is used to monitor return sludge suspended solids levels and to aid with this determination. Also, grab samples of the secondary clarifier influent are taken frequently to measure sludge settling rates and sludge volume indices (SVIs). Once again, just as the relationship between temperature and basin DO levels dictates maximum MLSS levels, the settling characteristics of the sludge in the secondary clarifier may also dictate the maximum MLSS level that can be maintained under aeration. To control clarifier sludge blanket depth, return sludge pumping rates can be increased only to a certain point (about 50 percent); after that, further increases in return rates are not beneficial. This is because the additional return that is pumped back through the reaeration and aeration systems ultimately comes back into the clarifier and causes a high hydraulic loading. An increased hydraulic loading on the clarifier can stir up the characteristically light fluffy sludge and possibly cause sludge bulking. If the sludge settling characteristics are poor (high SVI) and increasing the return sludge pumping rates does not control sludge blanket depths, MLSS levels must be trimmed back again.

Other commonly used activated sludge operational control factors such as the food to microorganism (F/M) ratio, mean cell residence time (MCRT), and sludge age are very difficult to use as the sole determining factor to direct sludge wasting on a day-to-day basis. These factors can be very useful as plant design figures or they can be used to determine desirable operating ranges. However, it is nearly impossible to calculate these factors and adjust sludge wasting rates or to attempt to adjust MLSS levels on a daily basis "shooting" at a specific F/M level. When highly fluctuating organic loads enter a plant, it is very difficult to waste or build MLSS rapidly enough (within 24 hours) to meet a specific set point. Instead, select a proper MLSS level that actually produces an F/M, MCRT, or sludge age value that floats in the vicinity of what would approximate a desirable range.

In reality, there are many interrelated and independent factors that influence plant operation. These factors include basin DO levels that can be maintained, sludge settling characteristics, and effluent quality. These and many other factors dictate what MLSS level can be maintained. The proper MLSS level must be determined experimentally by slowly adjusting the MLSS level while carefully observing plant behavior. Ultimately, a proper MLSS set point can be determined and used to control sludge wasting activities. However, you must realize that even this MLSS set point may have to be adjusted when influent characteristics change and when seasonal changes occur.

As previously described, all sludge that is not wasted flows into the reaeration portion of the plant. There are numerous

---

[23] An effluent with less than 20 mg/L of both BOD and suspended solids.

benefits from return sludge reaeration. A list of some of the advantages is as follows:

1. Basic math indicates that if nine aeration basins are maintained at an MLSS concentration of 2,000 mg/L, then the addition of only three return sludge reaeration basins (ALL basins equal 0.303 MG or 1,150 m^3) at an SS level of 6,000 mg/L (return sludge concentration) will *DOUBLE* the available biomass in the activated sludge system (9 aeration × 2,000 mg/L = 3 reaeration × 6,000 mg/L or double initial capacity).

2. During periods of high organic loading, the reaeration phase allows adequate "rest" time for the sludge organisms to metabolize **ad**sorbed and **ab**sorbed BOD.

3. If a toxic substance (heavy metals) should enter the plant or if adverse conditions occur (pH or temperature shift), only a small portion of the biomass will be destroyed. The plant can recover quickly and be back on line achieving satisfactory BOD removal within a short period of time.

4. The "rest" period within reaeration seems to condition the sludge so that it will readily accept new influent loading and consistently obtain high BOD removals (+98 percent).

2.566 *Sludge Wasting*

The following example shows how to calculate sludge wasting rates. This example assumes continuous sludge wasting for a 24-hour period with a desired MLSS set point goal of 2,200 mg/L. Every 24 hours all of the data are reviewed and a new sludge wasting rate is implemented. As a quick indication of influent organic strength, a total organic carbon (TOC) measurement is used to project BOD loading.

Sample Data:	Item	Data
	Final Aeration Cell MLSS	2,420 mg/L
	Primary Effluent, 24-hour Composite Sample, TOC	554 mg/L
	Desired MLSS Set Point	2,200 mg/L
	Return Sludge (waste) SS	7,160 mg/L
	Estimated Flow for the Next 24 hours	3.3 MGD

Volumes and Assumptions

1. Volume of each aeration cell 0.3 MG.

2. Nine aeration cells, total aeration volume 2.7 MG.

3. Solids in secondary clarifiers are equal to 15 percent of the solids within the aeration basins.

4. MLSS of final aeration basin is representative of the MLSS of all the aeration basins. 9 × final aeration basin MLSS = total aeration solids.

5. Yield factor = 0.5 lb MLSS solids produced/lb BOD removed.

6. Estimate primary effluent flow from recent records and actual flow data for first 8 to 10 hours of the day. Take into consideration whether the sludge wasting system is in operation.

7. Conversion factor = Yield × BOD:TOC ratio

$$1.1 = 0.5 \times 2.2$$

SAMPLE WASTING FORMULA CALCULATIONS

1. Calculate the solids in the system in pounds. Since solids in the secondary clarifiers are 15 percent of the solids in the aeration basins, multiply the solids in the aeration basins by 1.15.

$$\text{Solids in System, lbs} = \text{Aeration Volume, MG} \times \text{Final Aeration Cell MLSS, mg/}L \times 8.34 \frac{\text{lbs}}{\text{gal}} \times 1.15$$

$$= 2.7 \text{ M Gal} \times 2,420 \text{ mg/}L \times 8.34 \text{ lbs/gal} \times 1.15$$

$$= 62,668 \text{ lbs}$$

2. Estimate the solids produced in the system in pounds per day. Assume 1.1 pounds of solids are produced per pound of TOC.

$$\text{Solids Produced, lbs/day} = \frac{1.1 \text{ lbs solids/day}}{1 \text{ lb TOC/day}} \times \text{TOC, lbs/day}$$

$$= 1.1 \times \text{Flow, MGD} \times \text{TOC, mg/}L \times 8.34 \text{ lbs/gal}$$

$$= 1.1 \times 3.3 \text{ MGD} \times 554 \text{ mg/}L \times 8.34 \text{ lbs/gal}$$

$$= 16,772 \text{ lbs solids/day}$$

3. Determine the desired pounds of solids in the system based on an MLSS set point of 2,200 mg/L. Assume solids in the secondary clarifiers are 15 percent of the solids in the aeration basins (multiply by 1.15).

$$\text{Desired Solids in System, lbs} = \text{Aeration Volume, MG} \times \text{MLSS Set Point, mg/}L \times 8.34 \frac{\text{lbs}}{\text{gal}} \times 1.15$$

$$= 2.7 \text{ M Gal} \times 2,200 \text{ mg/}L \times 8.34 \text{ lbs/gal} \times 1.15$$

$$= 56,971 \text{ lbs}$$

4. Calculate the sludge wasting amount in pounds per day.

$$\text{Sludge Wasting Amount, lbs/day} = \frac{\left(\text{Solids in System, lbs} - \text{Desired Solids in System, lbs}\right)}{\text{Waste During 1 day}} + \text{Solids Produced, lbs/day}$$

$$= \frac{(62,668 \text{ lbs} - 56,971 \text{ lbs})}{1 \text{ day}} + 16,772 \frac{\text{lbs}}{\text{day}}$$

$$= 5,697 \text{ lbs/day} + 16,772 \text{ lbs/day}$$

$$= 22,469 \text{ lbs/day}$$

NOTE: If the sludge wasting rate is negative, the MLSS is too low. Reduce the existing wasting rate by 10 to 15 percent. Some operators shut off the waste when negative values are obtained, but many operators try to avoid drastic changes in wasting rates by adjusting the existing wasting rate up or down by no more than 10 to 15 percent each day.

5. Determine the sludge wasting rate in MGD and GPM.

$$\text{Sludge Wasting Rate, MGD} = \frac{\text{Sludge Wasting Amount, lbs/day}}{\text{Waste Sludge SS, mg/}L \times 8.34 \text{ lbs/gal}}$$

$$= \frac{22,469 \text{ lbs/day}}{7,160 \text{ mg/}L \times 8.34 \text{ lbs/gal}}$$

$$= 0.376 \text{ M gal/day}$$

$$\text{Sludge Wasting Rate, GPM} = \frac{376,000 \text{ gal/day}}{1,440 \text{ min/day}}$$

$$= 261 \text{ GPM}$$

Waste sludge is first prethickened in dissolved air flotation (DAF) units and mixed with primary sludge in the sludge holding tanks. Water is evaporated from the separated sludge mixture in the sludge drying plant. The product is currently marketed as a high vitamin B-12, high protein, animal feed supplement.

One last item that should be covered along with aeration basin control is the subject of aeration basin foam. At all times, foam is present on the basin surfaces. Foam is a natural part of the biomass. A dark, greasy-looking foam is the sign of an old sludge or long sludge age while a white, clear foam indicates a young sludge or short sludge age. Large quantities of foam can sometimes cause operational problems if the foam does not stay within the basins. Excessive foam can be caused by large quantities of detergents in the wastestream or some brewery materials, such as yeast. Control of this foam can be achieved by: (1) decreasing air flow to the aerators, (2) using chemical surfactants, or (3) using water sprays aimed to physically collapse the foam. A novel approach to foam control involves overflowing the aeration basin into a small side basin, letting the foam collapse and then wasting this foaming material from the plant.

2.567 Filamentous Organisms

In brewery wastewater treatment plants, often the number one operational problem is the control of filamentous organisms (Figure 2.19). Often sludge bulking is related to a filamentous bacteria by the name of *Sphaerotilus natans*. To blame all sludge bulking on this one type of filamentous organism is a misconception. In reality there are about 10 different types of filamentous organisms that may dominate brewery activated sludge and cause sludge settling problems. To a certain degree there is a relationship between the different filamentous organisms and specific operating conditions. Therefore, in some cases, if the type of filamentous organism can be identified, the causative operating condition can also be identified and corrected. However, this approach is usually not practical. To begin with, specific filamentous organism identification requires a 500× to 1,000× power microscope. Also, even if the specific filament can be identified, our understanding of the relationship between specific filament type and particular operating conditions is somewhat limited.

A more successful approach to filamentous organisms management is described in the following paragraphs. (It should be noted that only some of the techniques listed below have been used at the plant described here.)

1. Determine if filamentous organisms are the true culprit. This can usually be done by examining a sludge sample with a relatively inexpensive microscope. The filamentous organisms look like fine hairs or wires extending out of the sludge floc particles or they can be found in the liquor floating free. Some filaments are always present, but if they appear to be 20 percent of the biomass or more, they could be causing settling problems.

2. A short-term corrective step is often necessary to halt immediate bulking. Many process adjustments can be made to bring the filaments under control:

    a. Careful polymer or chemical ($FeCl_3$) dosage to prevent SS loss within the secondary clarifiers.

    b. Excessive sludge wasting to remove filamentous organisms from the system.

    c. Chemical dosage of the return sludge with oxidants such as hydrogen peroxide, H_2O_2 (0.1 lb H_2O_2/1,000 lbs VSS) or chlorine, Cl_2 (3 to 5 lbs Cl_2/1,000 lbs VSS).

3. Develop a long-term control program to correct operational problems and to prevent the recurrence of filamentous organisms. Common causes of filamentous bulking include:

    a. Low DO levels in aeration basins and secondary clarifiers (this is often a *result* of a high F/M ratio, although the F/M ratio alone is not necessarily the problem).

    b. A lack of adequate nutrients (nitrogen or phosphorus) or trace minerals in the influent.

    c. The presence of high levels of sulfur in the influent.

    d. Low F/M levels in the aeration system (less than 0.2).

    e. Large fluctuations in the plant influent organic loading.

Many of these problems can be overcome by making operational changes and by engineering design changes. At the wastewater treatment plant under examination here, many changes have been made to control the growth of filamentous organisms. Experimentation has revealed that the dissolved oxygen concentration (DO) is the most influential factor that affects filamentous organism growth in this plant. Since some filamentous organisms are often desirable (to maintain floc structure), a desired DO level set point is used to adjust filament concentrations. Over time, experience has shown that filamentous organism concentrations can be adjusted by adjusting "contact" cell DO levels. An *increased* DO level (5 to 10 mg/L) will yield a *decrease* in filamentous organisms. A *decreased* DO level (1 to 3 mg/L) will result in an *increased* number of filamentous organisms.

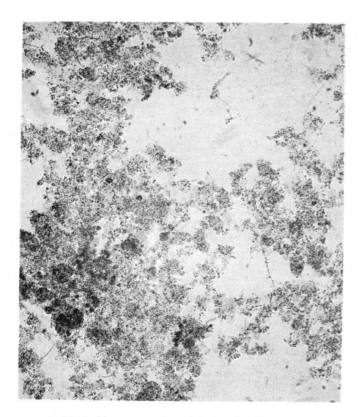

1. Typical brewery activated sludge, few filaments

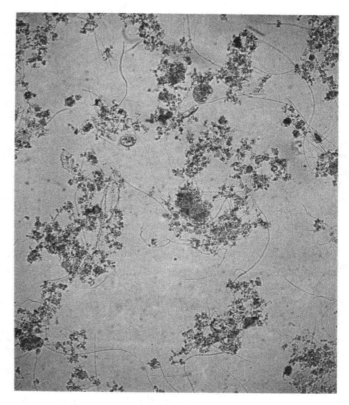

2. Brewery activated sludge, some filaments

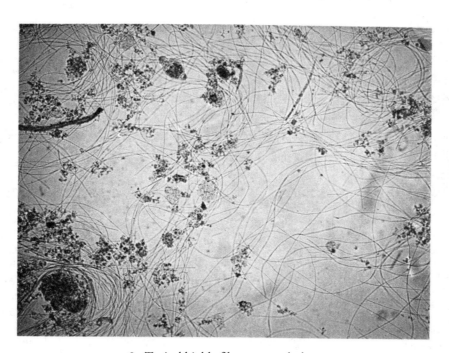

3. Typical highly filamentous sludge

Fig. 2.19   Filamentous organisms (100× magnification)

2.568 Laboratory Testing

Only a small portion of the testing in a brewery wastewater treatment plant is devoted to "permit" monitoring requirements. The majority of laboratory time is spent obtaining test data that are required to assist in making process control adjustments. Numerous MLSS measurements are made to evaluate sludge wasting needs and sludge settling characteristics. Plant loading data are updated daily by the use of total organic carbon (TOC) analysis. Nutrient concentrations are measured at several locations in the plant to assist with nutrient addition calculations. Both 24-hour composite samples and grab samples are used to obtain representative data. Microphotographs (Figure 2.19) of sludge samples from several locations within the aeration system serve as a permanent record of microorganism diversity and relative filamentous organism concentrations. Many additional quantitative and qualitative tests are performed and the results are recorded. These data, along with numerous other measurements and evaluations recorded throughout the wastewater treatment plant, are essential for mapping out operational control strategies.

2.569 Recordkeeping

Every wastewater treatment plant is different and requires different types of testing and recordkeeping. There is no substitute for well-organized laboratory data and operational log sheets. The extreme value of this information cannot be overemphasized. Well-recorded data can serve as a quick update to an operator coming on shift or it may be graphed, charted, or tabulated to indicate more clearly overall trends. Proper use and analysis of records have allowed the operators of plants treating brewery wastes to produce a high-quality effluent.

QUESTIONS

Write your answers in a notebook and then compare your answers with those on pages 149 and 150.

2.56A Why are primary clarifiers not very effective in removing BOD from brewery wastes?

2.56B Where are nutrients added and how are nutrients added when treating brewery wastes?

2.56C What factors influence the MLSS level in the aeration basins?

2.56D How is the sludge wasting rate determined?

2.56E How can filamentous bulking be controlled in activated sludge plants treating brewery wastes?

2.57 Food Processing Wastes

This section discusses how to treat wastewaters from two different types of food processing plants. Treatment of artichoke wastewater and dairy wastes are presented by operators who actually treat these wastes on a day-to-day basis. Almost all of the foods we eat (fruits, vegetables, fish, meats, dairy products) produce wastewaters when they are processed for consumption. Treatment of these wastewaters may be unique for each food, but the basic principles of pretreating the wastewater to produce an environment suitable for activated sludge treatment are similar.

2.570 Treatment of Artichoke Wastewater
by Peter Luthi

Processing of artichokes in California is a year-round operation with a major peak during the months of March, April, and May and a minor peak in September, October, and November. The fact that some wastewater is generated all year long makes the operation of an activated sludge system possible at this pretreatment facility (Figure 2.20). The effluents from the activated sludge plant that treats high-strength wastewater, the flotation unit that treats medium-strength wastewater, and cooling water with a low BOD load are all discharged into the sewer for final treatment by the municipal wastewater treatment plant.

Wastewaters from the artichoke processing plant are segregated according to BOD strength and only the highest BOD portion of the wastewater is treated by the activated sludge system (Figure 2.20). Primary objectives of the activated sludge treatment process are to:

1. Treat the high-strength (high BOD) waste to acceptable levels with a minimum of input (energy, labor, dollars) and a minimum of waste sludge produced.

2. Produce a treated effluent that meets the following discharge limits for discharge to a municipal treatment plant:

   a. BOD less than 500 mg/L.

   b. Suspended solids less than 500 mg/L.

   c. pH within a range of 6 to 9.

BOD strength of the wastewater treated varies between 1,000 and 15,000 mg/L with a pH of around 4.5 (from the use of vinegar and citric acid in the process). The volume of high-strength BOD water ranges from 1,500 gal/week to 15,000 gal/week (5.7 to 57 m³/week) during the peak season.

2.5700 PILOT PROJECT

The unusually high and widely fluctuating BOD levels of the waste required that the feasibility of an activated sludge system be studied on a pilot project. An activated sludge system with extended aeration using mean cell residence times up to 15 days was chosen. As much as 90 to 95 percent BOD removal could be achieved if the influent BOD was kept below 6,000 mg/L. Higher influent BODs resulted in lower effluent quality. Medium-strength wastewater is therefore used to dilute the influent. To accumulate some operating data during the pilot project, the following analyses were performed on a daily basis:

1. Influent and effluent COD and suspended solids

2. Mixed liquor suspended solids

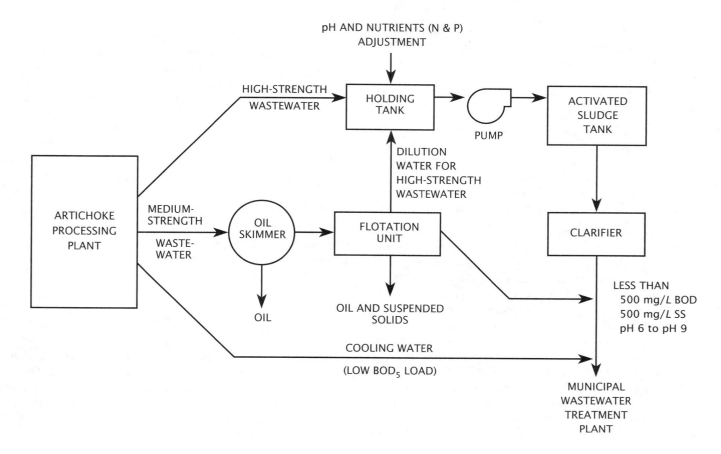

Fig. 2.20   Artichoke activated sludge pretreatment facility

3. pH of influent, mixed liquor, and effluent

4. Dissolved oxygen

5. SVI

COD rather than BOD was chosen for its relatively easy and fast analysis. A graph showing 15-day moving average influent COD, mixed liquor, suspended solids, moving F/M ratio, effluent COD, and suspended solids was drawn and kept up daily to find the best operating guidelines. Fifteen-day moving averages for influent COD and F/M ratio were computed to reduce the effect of fluctuating daily results.

2.5701   DAILY OPERATIONAL PROCEDURES

Through trial and error, and after several upsets over a nine-month period, it was readily visible from the graph and data accumulated that acceptable results (less than 500 mg/L BOD and suspended solids) were obtainable when the F/M ratio was kept between 0.08 and 0.28 pound BOD/day/pound MLVSS.

Daily influent and effluent COD and pH, mixed liquor, and effluent suspended solids analyses are performed. Also, deter-

mining the SVI on a daily basis is helpful in detecting changes in the settling characteristics of the activated sludge. SVIs were, however, generally higher than data given in literature for municipal treatment plants.

Nutrients (N and P, in the form of ammonium phosphate fertilizer) were initially added in batches on a daily basis. Later, however, a continuous addition directly into the in-feed line was used. This arrangement proved much more reliable and provided for a smoother operation. Lack of sufficient nutrients was responsible for bad settling characteristics in several cases. If the SVI increases and lack of nutrients is suspected, take a look at the sludge under the microscope. The presence of filamentous bacteria will confirm the suspicion. Sufficient amounts of nutrients proved to be quite important and upsets caused by their lack took a long time to remedy. No provision for chlorinating the return sludge exists. No adverse effects (other than higher operating costs) were observed to have resulted from the addition of too much nitrogen and phosphorus.

To keep the influent COD concentration as steady as possible on a day-to-day basis, dilution with medium-strength wastewater is used when needed.

The fluctuating concentrations of the influent COD make it necessary to keep the DO level in the aerator between 4.0 and 6.0 mg/L. Rapid changes occur sometimes and DO levels can drop to as low as 2.0 mg/L in a matter of hours. Therefore, keeping levels at 4.0 to 6.0 mg/L ensures that the DO does not drop below the critical 2.0 mg/L.

Effluent pH monitoring shows that when effluent pH drops below 6.5 it is necessary to adjust the influent pH to between 5.5 and 6.5. This is done with granular sodium hydroxide added to the holding tank. At times of low feed rates, however, the system is able to tolerate the normal influent pH of 4.2 to 4.8 and only occasional adjustments are necessary.

Influent levels vary from 4,300 gal/day to 8,600 gal/day (16 to 32 m³/day). Levels higher than 8,600 gal/day (32 m³/day) produce turbulence in the clarifiers with accompanying solids loss in the effluent. Mixed liquor solids vary from 3,000 to 4,500 mg/L. Influent volume is the most important variable used to control the system. By increasing or reducing the volume, the F/M ratio can be increased or decreased more gradually or kept at one level when influent COD concentration changes. In anticipation of heavy production times when larger amounts of wastewater are generated, mixed liquor solids are allowed to build up to 4,500 mg/L. The higher F/M ratio allows treatment of larger volumes with higher concentrations of COD. Sludge is wasted on a daily basis to keep the aerator solids level (MLSS) at 4,500 mg/L. Waste sludge is disposed of on land.

Immediately following heavy production, sludge is wasted at an increased rate to reduce mixed liquor solids to 3,000 mg/L. The goal is to avoid starvation of the system when the COD concentration and waste volume are dropping. During times of low production, the F/M ratio is reduced to around 0.08 lb BOD/day/lb MLVSS. At this level, the microorganisms are just barely able to survive because the influent BOD level and feed rate are at their lowest levels. The effluent produced at this time is of a lower quality. More solids than usual are being carried away with the effluent and sludge wasting only has to be done on an occasional basis. However, the primary objectives of reducing effluent BOD and suspended solids levels below 500 mg/L can still be achieved.

QUESTIONS

Write your answers in a notebook and then compare your answers with those on page 150.

2.57A   How are high strength artichoke waste influents (BOD >6,000 mg/L) adjusted before being treated by the activated sludge process?

2.57B   Why were 15-day moving averages computed and plotted for artichoke influent COD and F/M ratio?

2.57C   What chemical can be used to provide nutrients for the treatment of artichoke wastes?

2.571   Treatment of Dairy Wastes

This section describes treatment of dairy facility wastewater from two perspectives: (1) pretreatment to prepare the wastewater for discharge to a municipal wastewater treatment facility (WWTF), and (2) biological treatment of wastes at the dairy processing facility for discharge either to a municipal treatment system or directly to the environment. Three operators contributed material for Section 2.571: Christine Thompson, Ralph L. Robbins, Jr., and Jon Jewett. Their assistance in describing the treatment of dairy wastewater is greatly appreciated.

2.5710   SOURCES OF WASTEWATER

Dairy wastewaters are generated from milk processing (pasteurization and homogenization), ice cream manufacturing, and butter, cheese, and yogurt production. Wastewaters from these operations are composed of production materials such as lactose, calcium lactate, protein hydrolysates, and fats.

2.5711   VARIABLES AFFECTING TREATMENT OF DAIRY WASTES

Dairy processing wastewater is difficult to treat because of its high strength, variability over time, and basic composition. In addition to high organic loading, dairy wastes may pose additional challenges to the operation of wastewater treatment facilities. These include variable hydraulic loads, changing pH values, nutrient deficiencies, and toxins.

ORGANIC LOADS

The biochemical oxygen demand (BOD) of whole milk is approximately 100,000 mg/L. The oxygen demand of dairy wastewaters is, therefore, substantially higher than the BOD of domestic wastewater. Dairy wastewater BOD typically ranges from 940 to 9,440 mg/L with average values ranging from 940 to 4,790 mg/L.[24] The high oxygen demand exhibited by milk and its by-products comes from its major constituents: lactate, proteins, and fats. Common sources of high BOD wastes include:

- Product spills
- Spoiled batches
- Abnormal losses of product
- Concentrated residuals of product
- Unused by-products (for example, fruit juices)
- Cleanup
- General washdown/sanitizing operations

---

[24] The BOD ranges listed here are based on a survey of 50 dairy facilities by J. H. Martin and R. R. Zall, "Bio-Augmentation in the Treatment of Dairy Processing Wastewaters," 1989.

HYDRAULIC LOADS

Hydraulic loads are generated at various locations throughout the dairy plant and change according to the setup and product line of a particular plant. Common fixed flow sources, those that are not routinely variable in volume, are from truck washdowns, separators, pasteurizers, homogenizers, process silos, line sterilization, filler backwash, and condensate. These flows, though relatively stable in volume, may occur at particular intervals throughout the day and in combination greatly increase flow values. Common variable sources, those that may fluctuate from day to day based on operations, personnel, and equipment in service, are from general washdown hoses, washdown operations, mist sprays, and conveyor lubrication. Washdown operations are often the major contributing source of flows. They represent 50 to 75 percent of total flows from some processing plants.[25]

pH

pH values of dairy plant effluent change according to product line. Cheese or yogurt making operations tend to produce an effluent lower in pH than a milk bottling plant because of the acidic nature of the operations. However, the major cause of pH fluctuations in dairy plant effluent is from cleaning and sanitizing operations. To prevent the buildup of calcium deposits in lines and equipment, which can lead to increased bacteria counts, these industries routinely perform acid or alkaline washes, and rinse with bleaching compounds such as sodium hypochlorite. Where flow equalization is practiced, the equalization tank may act as a neutralization tank. The acid and alkaline washes may come together and react with each other thus maintaining a near neutral pH and preventing major swings in pH. Where effluent discharges are immediate, with no flow equalization, pH values may change frequently. This will create very difficult operating conditions for both the industrial pretreatment operator and the municipal operator. pH values may range from a low of 5.0 to as high as 9.5 putting substantial stress on biological treatment processes.

NUTRIENT DEFICIENCIES

Dairy processing wastewaters contain substantial amounts of nutrients (see Table 2.8). However, these nutrients may not always be in a form that is available to the organisms we wish to have predominate in a biological treatment system. Nutrient deficiencies in the activated sludge process can lead to the development of filamentous organisms that prevent the proper settling of the biomass. The general rule of thumb for activated sludge processes is a BOD:N:P ratio of 100:5:1.

Nutrient deficiencies primarily involve the availability of inorganic nitrogen. Some organisms may more readily use organically bound nitrogen than others. Milk compounds are composed mostly of fats, proteins, carbohydrates, and minerals. Nitrogen compounds are generally in the form of organic nitrogen. The organic nitrogen must be mineralized over time in order to be available to some organisms. However, if the carbonaceous BOD arrives in large volumes over a short period of time, which is common for the dairy industry, the BOD/nutrient ratio may be inadequate. This situation can cause a predominance of unwanted organisms. As a result, some BOD may even pass through the system untreated.

Aerated lagoons tend to handle nutrient deficiencies quite well as long as adequate dissolved oxygen is provided in the system. Lagoons have the advantage of holding a relatively small population of microorganisms in a large volume of wastewater for a relatively long detention time.

As outlined above, the BOD/nutrient ratio is a useful guideline for detecting a nutrient deficiency problem; however, the following factors[26] should also be considered.

- The sources of available nitrogen and phosphorus should be maintained at the appropriate level considered necessary for the complete *METABOLISM*[27] of the carbon source.

- The nutrient supply, if either supplemental or naturally occurring, should never be allowed to be depleted in any section of the bio-reactor/aeration cell as the result of shock loads.

- Because each wastewater and treatment system has its own requirements for nutrients, an influent BOD:N:P ratio should be compared to effluent concentrations of dissolved

TABLE 2.8   DAIRY PROCESSING WASTEWATER CHARACTERISTICS[a]

Characteristic	Range
Biochemical Oxygen Demand (BOD)	15–4,790 mg/L
Total Solids (TS)	135–8,500 mg/L
Total Volatile Solids (TVS)	57–4,700 mg/L
Total Suspended Solids (TSS)	24–5,700 mg/L
Volatile Suspended Solids (VSS)	17–5,260 mg/L
Nitrogen (N)	15–180 mg/L
Phosphorus (P)	11–160 mg/L
pH	5.3–9.4
Temperature, °C	12.8–48.9°C

[a] Harper, Blaisdell, and Grosshopf, 1971, *Dairy Processing Wastewater Characteristics.*

---

[25] N. True and T. Reeves, "Internal Waste Management and Spill Prevention," Dairy Wastewater Operations Seminar, Georgia, VT, May 13, 1992.

[26] Dynamic Corporation, *CAUSES AND CONTROL OF ACTIVATED SLUDGE BULKING AND FOAMING.* July 1987. Environmental Protection Agency. Summary Report.

[27] *Metabolism.* All of the processes or chemical changes in an organism or a single cell by which food is built up (anabolism) into living protoplasm and by which protoplasm is broken down (catabolism) into simpler compounds with the exchange of energy.

phosphate, ammonia, and nitrate. The results of such monitoring should be routinely recorded and used as a guide to future operations. Such monitoring must be performed during good operational periods as well as during poor operational times.

- Ammonia and nitrate are both sources of inorganic nitrogen available for cell growth. Supplemental ammonia may be completely converted to nitrate, but the nitrate will continue to be available as a nutrient source.

As a general rule, the nutrient level of a process is adequate if the effluent concentrations of soluble ammonia nitrogen and phosphate are about 0.5 and 0.25 mg/*L*, respectively. Where nutrient deficiencies are common and supplemental nutrients are added, effluent ammonia and phosphorus concentrations at or slightly below 1.0 mg/*L* indicate adequate available nutrients in the system.

TOXINS

Toxins are often found in dairy plant effluents, primarily detergents, lubricants, sanitizers, and refrigerants. The hazard posed by the normal discharge of these materials is minimized by dilution through the municipal collection/treatment system. It is the industrial operator of a biological treatment system who most often must deal with the effects of toxic substances in the wastestreams.

Toxic wastes entering a treatment system have a number of potentially harmful effects. These include destruction of biomass, pass-through, formation of secondary toxins, and contamination of residual *BIOSOLIDS.*[28]

The best means to avoid the discharge of toxins to the wastewater treatment facility is through awareness of both the appropriate handling techniques and the potential effects of the discharge of toxins to the downstream wastewater treatment system. To prevent toxic dumps and to ensure that treatment plant staff will be notified of dumps when they occur, the wastewater treatment facility (WWTF) operator should assist the industry in the development of a spill prevention and toxins handling program. As a minimum, the program should include the following items:

- Become familiar with the industry's pretreatment or discharge permit.

- Be sure there is a spill containment area for each tank of chemicals.

- See that floor drains are segregated or plugged.

- If a spill has been contained, the operator should control the location and timing of any discharge.

- Make sure that a proper system of labeling is in place.

- Obtain a copy of the corporate emergency response plan.

- Learn the names and responsibilities of key industry staff.

- Be available to answer questions and provide assistance; cooperation goes a long way.

- Educate the industry employees so they are aware of the potential impacts that the discharged toxins may have on your treatment system.

No plan will ever prevent all discharges of toxins to a treatment facility; accidents will still happen. However, cooperation and awareness will help to minimize the occurrences and severity of toxic spills and save the industry money.

QUESTIONS

Write your answers in a notebook and then compare your answers with those on page 150.

2.57D   What wastewater treatment problems do operators have to deal with when treating dairy processing wastewaters?

2.57E   What are common sources of BOD from a milk processing facility?

2.57F   What is the major cause of pH fluctuations in dairy plant effluent?

2.57G   What is the main problem caused by nutrient deficiencies in the activated sludge process when treating dairy processing wastewaters?

2.57H   How can an operator determine if adequate nutrients are available for the activated sludge process?

2.57I   What are sources of toxins found in dairy plant effluent?

2.5712   PRETREATMENT OF DAIRY WASTES

Many industries pretreat their wastewaters before discharge into a wastewater collection system. This wastewater is combined with municipal wastewater and treated by a publicly owned treatment works (POTW). This section covers typical pretreatment practices by the dairy industry and can be used as a guide for pretreatment activities by other industries.

NEED FOR PRETREATMENT PROGRAM

Frequently, dairy industries discharge their treated wastewater to a municipal wastewater treatment facility rather than

---

[28] *Biosolids.*   A primarily organic solid product produced by wastewater treatment processes that can be beneficially recycled. The word biosolids is replacing the word sludge when referring to treated waste.

discharging directly to a surface water. The volume, strength, and quality of the wastewater is usually controlled by a pretreatment permit or the municipality's sewer-use ordinance. Such control over the dairy discharge is necessary in order to prevent process problems from developing at the municipal wastewater treatment facility (WWTF) that could potentially cause effluent violations of their discharge permit.

This section will discuss in more depth how and why limits on pretreatment dairy facilities are set, the potential adverse impacts that a dairy facility can have on a municipal WWTF, and some of the potential corrective actions that can be taken by either the dairy facility or the municipal facility if the dairy wastewater is creating problems for the municipal treatment facility. It is essential that the pretreatment facility operator realize that what leaves the plant will affect the municipal facility at the other end of the pipe. By being aware of this and working with the municipal WWTF operator in preventing or correcting problems caused by the dairy discharge, the industry will be viewed as a productive and desired member of the community rather than the cause of nuisance odors and effluent violations at the community's WWTF.

POTENTIAL ADVERSE IMPACTS OF A DAIRY PRETREATER ON ITS MUNICIPAL WWTF

When a dairy pretreatment facility discharges to a municipal WWTF collection system, the pretreatment and municipal operators should be aware of frequently occurring problems that may develop at the municipal WWTF. It should be noted that these same process problems can also occur at the dairy pretreatment facility itself if it consists of some type of secondary biological treatment process (activated sludge, aerated lagoons). The types of problems encountered will vary depending upon the type and degree of pretreatment the wastewater receives before entering the municipal system.

BLOCKAGES IN THE COLLECTION SYSTEM

If the dairy pretreatment system simply consists of pH adjustment and flow equalization tanks, there is a good chance that the high butterfat content of the wastewater will sooner or later form greasy plugs in the municipal collection system. The most direct way of preventing these plugs would be for the dairy to provide some treatment to its wastewater to lower the butterfat content (see Section 7.22, "Dissolved Air Flotation Thickeners," in Chapter 7, "Residual Solids Management," *INDUSTRIAL WASTE TREATMENT,* Volume II) before discharging it to the municipal facility. The industry might also agree to pay for periodic cleaning of the collection system lines and wet wells where butterfat buildup tends to be a problem as a way of being a "good neighbor" to the municipal WWTF. Chemical additives can be used to emulsify the butterfat to keep it in solution until it arrives at the municipal WWTF. However, this solution only tends to push the butterfat down into the treatment plant where it will probably cause problems by either creating a thick foam or balling up in the aeration tanks/aerated lagoons. Wastewater containing a high butterfat content may also provide an environment that will favor the growth of the troublesome filamen-

tous organism, *Sphaerotilus natans.* Bacterial additives can be applied to the dairy effluent. They can be designed to convert the butterfat into other compounds that will not cause problems either in the sewer or the WWTF.

FLOATING SOLIDS IN THE MUNICIPAL EFFLUENT

This particular problem has been noted at municipal treatment facilities that have a cheese company as a connection. In one case, the cheese company had an extremely lax approach to in-house waste management. Pieces of cheese curd that fell on the floor were simply hosed down the floor drain, went through the company's equalization tanks, and eventually showed up at the municipal activated sludge facility. These particles of curd are basically unaffected by the biological treatment process and pass through the facility and out with the effluent. Attempts by the operators to manually skim the particles from the tank surfaces proved overly time-consuming and basically ineffective. In this case, after discussion with the community, the cheese company installed filter baskets in the floor drains and educated their employees on the need to contain and properly dispose of the spilled curd. This simple process solved the municipal facility's effluent problem.

EFFLUENT BOD VIOLATIONS CAUSED BY ORGANIC OVERLOADS

A dairy industry's pretreatment discharge limits are usually based on the total pounds of BOD and the total flow to be discharged in a day. Permit limits such as these, which do not limit peak hydraulic and BOD loads, tend to create long-term chronic problems at the municipal WWTF. Case studies frequently indicate that as a dairy increases its production, municipal process problems also begin developing even though the dairy industry may be meeting its permit limits and the municipal influent BOD load is within the facility's design numbers. These problems can include poor sludge settling in the secondary clarifier caused by rapid growth of filamentous bacteria, periods of low or zero dissolved oxygen concentrations in the aeration tanks, and odors during periods of prolonged zero dissolved oxygen.

The problem is not the overall BOD load that is coming in over a 24-hour period, but rather the periodic high BOD loads that are reaching the facility throughout the day. These periodic organic overloads are caused by inadequate flow equalization or inadequate flow control out of existing flow equalization tanks at the industry.

High-strength BOD loads discharged over a short time frame can have a significant adverse impact on the receiving activated sludge or RBC process. The sudden rise in oxygen demand as the process's microbes feed on the periodic organic "feasts" causes outstripping of the oxygen input capability of the facility. When the aeration capacity is exceeded, the aeration tank's dissolved oxygen concentration will drop to very low levels or may become nonexistent. Of course, during other times of the day when the dairy's equalization tanks are filling, the dissolved oxygen levels at the municipal facility may return to normal. This fluctuating oxygen concentration can encourage the growth of unwanted filamentous bacteria. Excessive numbers of filamentous bacteria can lead to poor sludge settleability (bulking) with resulting effluent BOD and TSS (total suspended solids) violations. The filaments also tend to predominate in environments having highly variable F/M ratios. Such fluctuating F/Ms will also result from the periodic high-strength BODs that hit a facility throughout the day.

SPILLS OF HIGH-ORGANIC-STRENGTH WASTES/TOXINS

Accidents do happen and so it would be wise for the dairy facility operator and the municipal operator to be prepared for the inevitable. Examples of potential high-organic-strength spills include occasions when milk/whey/sugar solution accidentally flows to a floor drain because the tank's drain valve was left open or the tank overflowed. If chemical storage tanks are not surrounded with controlled drain spill containment berms, a ruptured tank or broken valve can send hundreds of gallons of hypochlorite/hydroxide down the floor drain to the receiving pretreatment or municipal treatment facility as a toxic spill.

Communication between the dairy production staff and the municipal operator is essential. As soon as a spill is discovered, the production staff must notify the pretreatment/municipal operator (or municipal official) as soon as possible about the type and volume of material that was discharged.

RESPONSE BY AERATED LAGOON OPERATOR

In the event of a high-organic-strength spill, an aerated lagoon operator should make sure that all available aeration equipment is turned on. The frequency of DO testing in the lagoon cells should be increased to four times a day, as opposed to the routine once/day measurement, and grab samples for COD analysis should be taken every other day from the end of each aeration cell. The data accumulated can be used to track the passage of the spill through the treatment process. In a lagoon system with three or more aeration cells, recirculation of some of the effluent from the last cell back to the head end of the first cell might be considered if the DO in the first cell drops to less than 1.0 mg/L. If the last cell has not been adversely affected by the spill, this recirculated water will have additional dissolved oxygen in it to assist the treatment process in the first lagoon. The additional hydraulic load of the recirculated water will also help by diluting the incoming waste flow and by reducing the detention time of the high-strength waste in the first aeration cell where the oxygen concentration may have been depleted. This will cause the waste load to be more quickly moved to the second aeration cell where an oxygen concentration exists to provide treatment. Lagoon facilities may have such a recirculation system perma-

nently installed for these situations; otherwise a portable pump and hoses can be used.

Because of the dilution effect provided by the large volume of a lagoon system, a toxic spill usually will not have an overwhelming, long-term impact on the lagoon's final effluent.

RESPONSE BY ACTIVATED SLUDGE/RBC OPERATOR

If the pretreatment facility or municipal facility consists of an activated sludge or RBC process, a spill of high-organic-strength wastes or toxins could kill the microorganisms in the biological process. The high-organic-strength wastes could cause long-term DO depletion, which in turn causes the beneficial microbes to die. In the case of a toxic spill, the toxic chemical itself kills the microorganisms.

If notified of a spill before it arrives at the treatment facility, the operator can set valves at the facility to bypass the influent around the secondary treatment process. Although a substandard quality of effluent will be discharged during the bypass, it will be for a relatively short period of time. If the toxic spill reaches the biological processes, the facility will be discharging substandard effluent until the microbe population has built back up again (a period of a least a week compared against a bypass measured in hours). Any such bypassing and the reason for the bypass must be reported to the regulating authority as soon as possible.

If the microorganisms in an activated sludge facility are killed off by an unannounced spill, the process can be helped back online quicker by hauling in fresh waste activated sludge from a neighboring facility to use as a "seed." The operator in such a situation should also contact the dairy production facility to ask the management to investigate any possible spills as well as to notify management of the problems at the treatment facility. In this way, if an employee refuses to admit to having caused a spill, the employee will be well aware of the adverse impact that the spill had on the downstream treatment facility. As a result, the employee may be more careful in the future.

RESPONSIBILITY FOR CORRECTING PROBLEMS

Before the dairy operator accepts any responsibility for the municipal facility's process problems, it is recommended that a study project be conducted to determine the true cause(s) of the problem.

The operation of the municipal facility should be reviewed. A determination should be made as to whether all equipment is operational and running efficiently and whether the operator's process control scheme is appropriate for running the facility at its optimum capabilities. If it is found that the municipal treatment plant is properly operated and maintained, the study project should be expanded to include time of travel study. To conduct a time of travel study, a fluorometer is installed on the dairy's equalization tank discharge line. A continuously recording DO meter is installed at the municipal aeration tanks. Fluorescent dye is then added to the dairy's wastewater discharge. Data from the fluorometer and the DO meter can be compared to see if periods of reduced DO concentration in the aeration tanks correspond with the arrival of the dairy discharge at the municipal facility. If it is found that a correlation exists, a decision will then have to be made as to how the problem will be resolved. Potential solutions include:

1. At the dairy facility:

   • Reduce production thereby reducing the quantity of wastewater that will be generated and discharged. This will probably not be considered an acceptable alternative by the industry.

   • Arrange to discharge the equalization tanks so that the dairy load will arrive at the municipal facility during the facility's low-flow periods. Such timing may prevent the outstripping of the facility's oxygen input capability. While this alternative may prevent the wildly fluctuating DO concentrations, the variable F/M will still exist, which could still favor the filamentous organisms if the dairy wastewater is discharged in short slugs.

   • Provide adequate flow equalization and positive flow control (pumped discharge) to spread the organic loading of the municipal facility over a longer period of time. By discharging during the municipality's low-flow period, this alternative should solve both the variable DO concentrations and F/M ratios. The dairy industry would incur some capital costs for additional tankage and the necessary pumping system.

2. At the municipal facility:

   • Chlorinate the return activated sludge when the filamentous population begins to cause poor sludge settleability. This "solution" is simply treating the symptoms of the actual problem. When you stop chlorinating, it will only be a matter of time before the filamentous bacteria again predominate. Additionally, there is a risk of accidentally overchlorinating to the point that the beneficial microorganisms are adversely affected.

   • Install more oxygen input capabilities. This is an option that will include capital costs for new equipment and additional energy costs. Both costs could possibly be passed on to some extent through the user fee to the industry.

   • Install variable-frequency drives on existing mechanical aerator systems together with a DO probe control system.

Use of this alternative in conjunction with timing the dairy discharge to arrive during the facility's low-flow period may well prove to be the most economical if the municipal system uses mechanical aerators. The DO probe control system would increase and decrease the aerator speeds in response to the DO concentration in the aeration tanks. During periods of high loadings when the microbes are consuming more oxygen, the dropping DO levels would be measured by the DO probes, which in turn would cause the aerators to speed up. When the microbes are less active during low-load periods, the probes would measure an adequate amount of DO existing in the tank and so keep the aerators on low speed. Frequently, the savings in energy costs using a DO control system provide a quick payback of the money spent to purchase and install the system. Once the payback point is reached, future energy savings are "money in the bank." Keep in mind, however, that increased air input may cause an overall increase in the annual power cost to the municipality.

QUESTIONS

Write your answers in a notebook and then compare your answers with those on page 150.

2.57J   How can a dairy help avoid causing blockages in a wastewater collection system?

2.57K   What problems can a municipal treatment plant experience from organic (BOD) overloads and fluctuating loads?

2.57L   Why would an industry install flow equalization facilities?

2.57M   What should an industrial operator do when a spill of high-strength organic wastes/toxins occurs?

2.5713   TYPICAL METHODS OF TREATING DAIRY WASTEWATERS

Dairy wastewaters may be treated in a variety of ways before discharge to municipal treatment facilities (pretreatment) or directly to the environment. Treatment processes range from simple pH adjustment and flow equalization to more advanced biological treatment and nutrient removal processes. Commonly used treatment processes include:

• Dissolved air flotation

• Conventional aerated lagoons, following pretreatment

• Complete-mix aerated lagoons

• Activated sludge

• Trickling filters

• Rotating biological contactors

• Combined anaerobic/aerobic treatment systems

• Livestock feeding programs

• Land application

The dissolved air flotation (DAF) unit (operation and maintenance details can be found in Chapter 7, "Residual Solids Management," *INDUSTRIAL WASTE TREATMENT,* Volume II), is used in the thickening of conventional secondary biosolids and to remove BOD, total suspended solids (TSS), and fats, oils, and grease (FOG) from dairy wastewaters. One such unit at a Ben & Jerry's ice-cream plant has been reported to remove approximately 50 percent of the BOD and 90 percent of the TSS and FOG. Operators lower the pH to near 3.0 and add carragheenin to enhance coagulation. (Carragheenin is used in the dairy process to increase the *VISCOSITY*[29] of milk products such as chocolate milk. Carragheenin is a natural product derived from seaweed.)

One problem associated with DAF operation is that cleaning solutions, caustics, and any strong *SURFACTANT*[30] can interfere with the coagulation of certain materials. Some cleaners are so effective that the oil, grease, and dirt are solubilized and held in suspension. Even extreme adjustments in pH are not enough to break the strong bond produced by the cleaner.

Complete-mix aerated lagoons, activated sludge, trickling filters, and RBCs are commonly used processes capable of treating or pretreating wastewater flows from various types of dairy industries including ice cream manufacture, milk processing, and cheese manufacture. In Vermont, similar processes consistently pretreat BOD loads from as little as 500 to as high as 8,000 pounds BOD per day.

Combinations of anaerobic/aerobic treatment have been pilot tested in the dairy industry with very good results. One such unit, in Maryland, completed in 1989, had average removal efficiencies of 95, 97, and 85 percent, respectively, for COD, BOD, and TSS.[31]

Other processes that are less complex and have lower capital construction costs are animal feed programs and land application. The cheese and ice-cream industries have been feeding whey and other residual by-products to cattle for many years. Grafton Cheese, a small Vermont manufacturer of cheese products, supplies approximately 800 gallons per day of whey to a nearby farm as cattle feed. The Ben & Jerry's corporation feeds approximately 1,000 gallons of residual ice-cream products to pigs at a nearby central Vermont pig farm. The Vermont whey authority's whey conversion facility, operated by Wyeth Nutritionals, uses high-quality whey, supplied from regional cheese plants, to make protein base for baby formula. There are many options for reusing these valuable food by-products. As cost of treating these residuals increases, simple alternatives such as the ones described here will be more widely practiced throughout the industry.

Dairy processing wastes (residuals) that were discarded as wastes in the past are now considered valuable resources that can be recycled back into the environment. Land application of whey products began in Wisconsin in mid-1970. Vermont issued guidelines in 1990 for the land application of not only whey, but the majority of the common liquid and semisolid dairy residuals. Currently, the Vermont dairy industry land applies approximately 20 million gallons of residuals annually. The residuals are applied by a spray truck or they are mixed with manure and spread using routine agricultural methods.

In order to expand, industries need to handle their waste products. An industry's decision to treat or to pretreat depends on many factors including state and local treatment requirements, cost of the process, ease of operation, and expected success. In many instances, a dairy production facility's ability to maximize production is limited by its ability to meet its pretreatment or discharge standards.

2.5714 EXAMPLE ACTIVATED SLUDGE PLANT TREATING DAIRY WASTEWATERS

PLANT INFLUENT

Dairy wastewaters are composed of production materials such as lactose, calcium lactate, and protein hydrolysates. The sanitary facilities at many plants are separate from the industrial dairy wastes and do not enter the wastewater stream to the treatment facility. Thus chlorination of the effluent is not necessary. The influent in this example facility averages a BOD loading of 3,000 lbs per day (1,360 kg/day) and a total solids level of approximately 3,500 lbs per day (1,590 kg/day). The treatment facility (Figure 2.21) operates throughout the year with little change of temperature in the aeration tanks. The effluent temperature does not fall below 60°F (15°C), thus providing for a stable operation all year.

OPERATION

Aeration tanks are operated in the extended aeration mode. Some dairy wastes require at least 18 hours' aeration time in the tanks. Up to 80 percent of the BOD load can be reduced in the first tank with the other two tanks treating the remaining 20 percent. For this reason, the diffuser capability in the first tank is twice that of the other two tanks.

The mixed liquor suspended solids levels in the aeration tanks are kept at a high level of 1 percent or 10,000 mg/*L*. This high level of MLSS is necessary to prevent process upsets caused by shock organic loadings. Extra air capacity is necessary in case of shock organic loadings in order to keep aeration tank DO levels between 2 and 6 mg/*L*.

---

[29] *Viscosity* (vis-KOSS-uh-tee).   A property of water, or any other fluid, that resists efforts to change its shape or flow. Syrup is more viscous (has a higher viscosity) than water. The viscosity of water increases significantly as temperatures decrease. Motor oil is rated by how thick (viscous) it is; 20 weight oil is considered relatively thin while 50 weight oil is relatively thick or viscous.

[30] *Surfactant* (sir-FAC-tent).   Abbreviation for surface-active agent. The active agent in detergents that possesses a high cleaning ability.

[31] A. A. Cocci, B. F. Burke, R. C. Landine, and D. L. Blickenstaff, "Anaerobic-Aerobic Pretreatment of a Dairy Waste, A Case History," 1990.

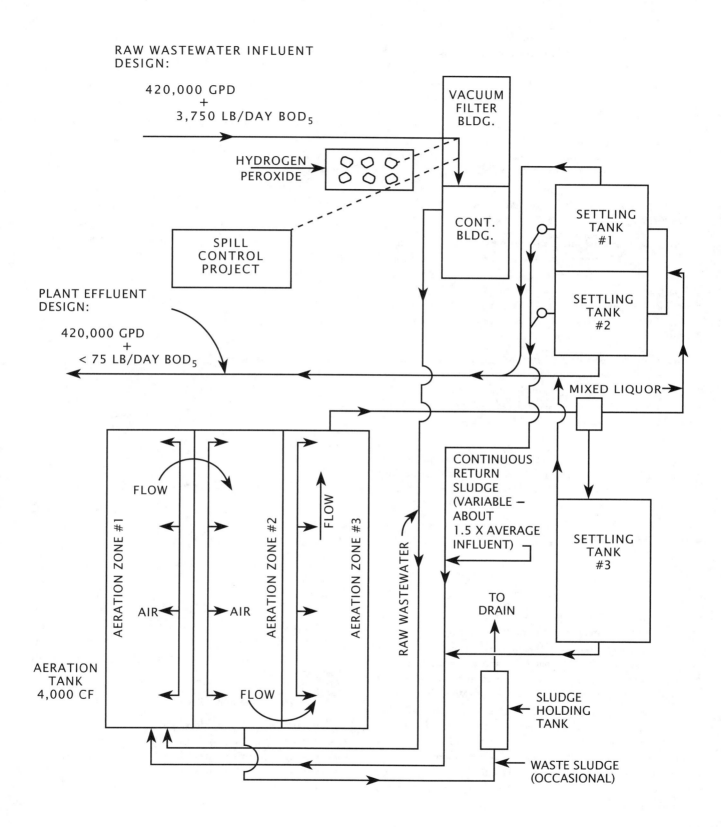

Fig. 2.21   Dairy waste activated sludge treatment facility

Settling tank detention times can be as long as 8 hours. Activated sludge is recirculated to the aeration tanks at a rate of 1½ to 2 times the plant influent rate.

Vacuum filters are used to dewater the sludge. Lime and ferric chloride are added for flocculation of the waste sludge before filtering. The sludge cake runs from 20 to 30 percent dry solids. The waste sludge cake is disposed of at a sanitary landfill site.

PLANT EFFLUENT

Plant effluent is monitored daily by plant personnel. The BOD, SS, COD, phosphorus, nitrate, nitrite, and total nitrogen are monitored by laboratory analysis of samples.

OPERATIONAL TECHNIQUES FOR UPGRADING EFFLUENT

Milk protein is sometimes difficult to break down. The addition of 50 mg/L of anhydrous ammonia can produce a complete breakdown of milk protein in the aeration tanks. Activated carbon is useful in many areas, including odor control and removal of phosphate. Activated carbon can also be used as media for extra bacteria growth in the aeration tanks and as an aid in flocculation and settling in the clarifiers.

A fermenter is used to grow a reserve supply of bacteria for injection into the aeration tanks during times of bacteria kill due to shock loadings from major spills in the production area. Installation of a system for the addition of hydrogen peroxide for odor control and additional oxygen during times of shock loadings can be helpful.

A spill control system has been installed to divert major spills into a pretreatment holding tank. Pretreatment consists of pH adjustment and pre-aeration. The wastewater is then blended into the main aeration tanks.

2.5715   IN-PLANT WASTE REDUCTION

(Also see Chapter 4, "Preventing and Minimizing Wastes at the Source," in Volume I of INDUSTRIAL WASTE TREATMENT in this series of operator training manuals.)

With the implementation of more restrictive discharge requirements, source reduction or pollution prevention can be more cost effective to industry than developing additional treatment alternatives. The operator (municipal or industrial) can perform a number of innovative projects to improve the quality and reduce the volume of wastes produced. Work items for a source reduction program include learning the industrial process, learning the names and functions of key personnel, and performing a walkthrough of the plant to identify key waste production areas. When these first steps are completed, the operator can develop with industry management (including the production manager) a list of attainable goals. The next step is to approach the quick-fix items first:

1. Install automatic hose shutoffs in the production area and other areas that need routine cleaning.

2. Acquire some control over the drainage system, having storage and steering capability.

3. Review transfer operations.

4. Implement daily water meter readings (this is a good way to gauge results).

5. Participate in and encourage management to provide awareness training to plant personnel. Training programs should emphasize water conservation, source recovery, the benefits of waste management, and how wastes affect the downstream treatment facility.

As the operator of the pretreatment or treatment system, it will be your responsibility to see, within reason, that the industry follows through on these programs. A good source reduction program will save money in both production and waste treatment operations.

QUESTIONS

Write your answers in a notebook and then compare your answers with those on page 150.

2.57N   Why might the chlorination of the effluent from a dairy waste treatment plant not be necessary?

2.57O   Why are high levels of MLSS (10,000 mg/L) kept in aeration tanks that treat dairy wastes?

2.58   Petroleum Refinery Wastes
by Cal Davis

2.580   Refinery Wastewater Characteristics

The three main compounds, ammonia, phenols, and sulfide, found in petroleum refining wastewater can be treated very effectively by the activated sludge process.

Wastewater flows from a petroleum refinery can vary rapidly (and without notice) both in rate and contaminant concentration. Waste treatment plants with holding ponds can control hydraulic loadings, but not always BOD loadings. To some extent, hydraulic loading can be used to control BOD loading with the MLVSS remaining constant in the aeration basin. In the event of a shock load, a hydraulic loading change would be the first corrective step.

2.581   Activated Sludge Process

Understanding and monitoring the activated sludge process is important in treating petroleum refining wastewater. Recognizing that each plant operates differently, most petroleum treatment activated sludge units (Figure 2.22) operate in the extended aeration mode with MCRTs up to 30 days. The reason for using extended aeration is to maintain the nitrification

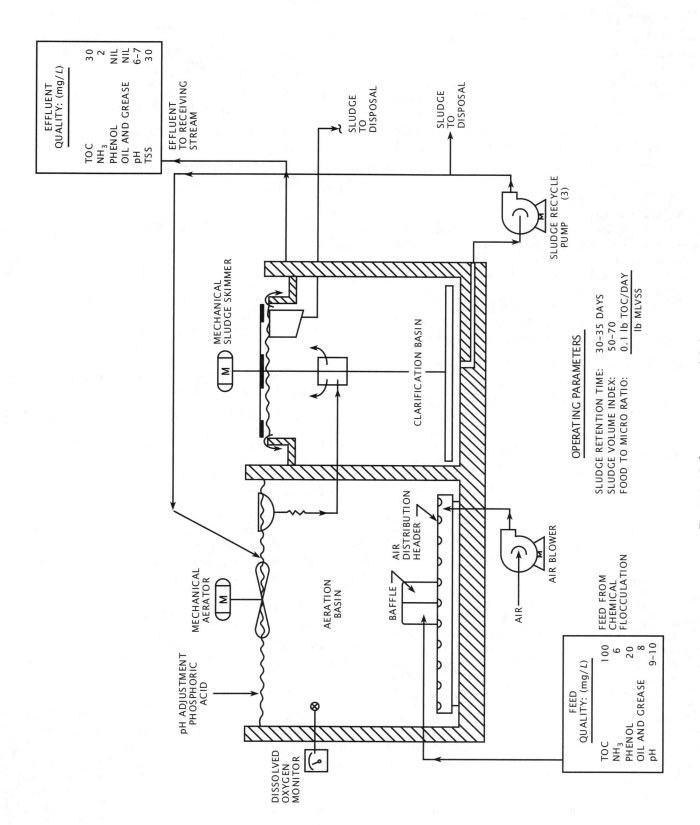

Fig. 2.22   Refinery waste activated sludge process

population necessary to oxidize ammonia. The minimum MCRT for good nitrification seems to be 20 days.

2.582   Frequency of Sampling and Lab Tests

Wastewater from a refinery can change in flow rate and waste concentration very suddenly. Certain tests must be run each shift while others can be run daily. Tests that need to be run each shift at the treatment plant include DO, temperature, pH, sulfide, phenols, ammonia, and 30-minute settleability. During upset conditions these tests need to be run at least twice a shift. Tests that need to be run each day include TOC or COD, BOD, MLTSS, MLVSS, recycle sludge TSS, ammonia, pH, PO_4, and oil.

2.583   Operational Procedures

When you come on duty after a shift change, visually inspect the activated sludge process, and review the log book and lab sheet for any changes in influent rates or concentrations. Check the following items: hydraulic loading, DO, aeration basin color, MLTSS and MLVSS, pH, temperature of influent, sludge recycle rate, and clarifier loading. For comparison purposes, measure sludge settleability and also calculate the SVI and F/M ratio. Under normal conditions, DO levels are 1.5 to 2.0 mg/L, pH from 6.8 to 7.0, and phenols, ammonia, and sulfide are nil.

2.584   Response to Sulfide Shock Load

The symptoms of a sulfide shock to the activated sludge system are DO less than 0.5 mg/L, pH of 5.6, aeration basin turning a light color, and phenols showing in the effluent. Testing may show no phenols in the influent stream since sulfide tends to mask or interfere with the phenol test.

To correct the problem of sulfide shock, reduce the hydraulic loading until DO levels are above 1.5 mg/L and add soda ash to the aeration basin to bring the pH above 6.2 in order to reactivate the phenol-consuming microbes. If phenols in the effluent are near your NPDES limit, adding hydrogen peroxide (H_2O_2) can be beneficial.

2.585   Correcting Excessive Phenols

Phenols in excess may create an odor problem and, in extreme cases of phenol shock, microbes will stop working and DO levels will increase to the saturation limit. To correct this problem, decrease the hydraulic loading. If phenols in the effluent are near the NPDES limit, then add H_2O_2 to help oxidize the phenols.

2.586   Treating Ammonia

Ammonia can be very troublesome since it can be toxic to fish. Ammonia can be difficult to treat biologically because it is difficult to cultivate nitrifying organisms to degrade ammonia. Besides the free ammonia, two other problem compounds that can show up in petroleum refining wastewater are monoethanolamine (MEA) and thiocyanate. Both of these compounds are biologically degraded to ammonia. Lab results must be checked each shift for an indication of a potential increase in ammonia.

Also check the lab results to be sure the MCRT and environmental conditions in the basins are suitable for cultivating nitrifying bacteria.

QUESTIONS

Write your answers in a notebook and then compare your answers with those on pages 150 and 151.

2.58A   What are the three main petroleum refinery waste compounds that can be treated by the activated sludge process?

2.58B   Why are MCRTs as high as 30 days necessary to treat petroleum refinery wastes?

2.58C   How can a shock load of sulfide be treated?

2.59   Summary and Acknowledgments

2.590   Summary

The basic treatment unit in the activated sludge process is a biological reactor (aerated basin or pond). This reactor provides an environment for the conversion of soluble organic material into insoluble microorganism cells. The subsequent unit is a secondary clarifier or pond where the cells are allowed to settle. The settled cells, or sludge, may be either returned to the aeration system, wasted from the system, or stored. As the result of biological growth, large volumes of organic solids are generated in secondary treatment processes.

Although several different activated sludge systems are used to provide secondary treatment (see Sections 11.041 and 11.8 of Volume II of OPERATION OF WASTEWATER TREATMENT PLANTS) for industrial, domestic, and domestic-industrial wastewaters, the control strategies and operating guidelines are essentially the same.

This section has described conditions unique to the treatment of industrial wastewater. Control of your activated sludge system will be enhanced by using the information for the operation and control of your plant contained in this chapter and in Chapters 8 and 11 of Volumes I and II of OPERATION OF WASTEWATER TREATMENT PLANTS.

2.591   Acknowledgments

Portions of Sections 2.50 through 2.54 were taken from "Pollution Abatement in the Fruit and Vegetable Industry" (Volumes 1, 2, and 3), EPA Technology Transfer Seminar Publication, US Environmental Protection Agency.

The representatives of industry who prepared these sections on how to treat industrial wastes are sincerely thanked. Without the contributions from James J. McKeown, Clifford J. Bruell, Peter Luthi, Ralph L. Robbins, Jr., and Cal Davis, this section would not have appeared in this manual. Special thanks also go to Gary Audy, Superintendent, Ben & Jerry's Home Made Inc., Waterbury, Vermont, and William Fletcher Jr., Head Cheese Maker, Grafton Cheese Company, Grafton, Vermont.

2.6 EFFLUENT NITRIFICATION

2.60 Need for Effluent Nitrification

Many activated sludge processes are designed to attain a high degree of nitrification. The degree of nitrification that must be attained is dictated by the maximum allowable limit of ammonia-nitrogen discharged with the final effluent. This limit is usually governed by the NPDES permit issued by state or federal regulatory agencies.

Nitrogenous compounds discharged from wastewater treatment plants can have several harmful effects. These include ammonia toxicity to fish, reduction of chlorine disinfection efficiency, an increase in the dissolved oxygen depletion in receiving waters, adverse public health effects (mainly groundwater), and a reduction in the suitability of the water for reuse.

Nitrogen concentrations in raw municipal wastewaters generally range from 15 to 50 mg/L, of which approximately 60 percent is ammonia-nitrogen, 40 percent is organic nitrogen, and a negligible amount (one percent) is nitrite and nitrate-nitrogen.

2.61 Nitrogen Removal Methods

Ammonia nitrogen can be reduced in concentration or removed from wastewater by several processes. These processes can be divided into two broad categories: physical–chemical methods and biological methods.

This section is devoted mainly to biological nitrification. A brief discussion of some physical–chemical nitrogen removal methods also is included.

2.610 Ammonia Stripping

The ammonia nitrogen that is present in wastewater during conventional biological treatment can be removed by a physical process called desorption (stripping). Simply stated, the wastewater is first made very alkaline by adding lime, and the ammonia is then induced to leave the water phase and enter the gas phase where it is released to the atmosphere. To accomplish this stripping, the wastewater is contacted with a sufficient quantity of ammonia-free air. This contacting with air is done in a slat-filled tower very similar to those used by industry to cool water.

2.611 Ion Exchange

This nitrogen removal process involves passing ammonia-laden wastewater through a series of columns packed with a material called clinoptilolite. The ammonium ion adheres to or is adsorbed by the clinoptilolite. When the first column in a series loses its ammonia adsorptive capacity, it is removed from the treatment scheme and washed with lime water. This step converts the captured ammonium ions to ammonia gas, which is then released to the atmosphere by contacting heated air with the wastewater stream, in much the same manner as described under ammonia stripping.

2.612 Breakpoint Chlorination

Breakpoint chlorination (superchlorination) for nitrogen removal is accomplished by adding chlorine to the wastewater in an amount sufficient to oxidize ammonia-nitrogen to nitrogen gas. After sufficient chlorine has been added to oxidize the organic matter and other readily oxidizable substances present, a stepwise reaction of chlorine with ammonium takes place. This may be the simplest nitrogen removal process, yet it has some disadvantages. In practice, approximately 10 mg/L of chlorine is required for every 1 mg/L of ammonia-nitrogen. In addition, acidity produced by the reaction must be neutralized by the addition of caustic soda or lime, which add greatly to the total dissolved solids in the wastewater.

2.62 Biological Nitrification

The nitrogen present in wastewater predominates as ammonia and organic nitrogen. As the organic matter in the wastewater decomposes, a portion of the organic nitrogen is converted to ammonia-nitrogen. When the wastewater is sufficiently aerated, the nitrite-forming bacteria *(Nitrosomonas)* will oxidize the ammonia-nitrogen to nitrite-nitrogen. The nitrate-forming bacteria *(Nitrobacter)* then oxidize the nitrite-nitrogen to nitrate-nitrogen. Nitrate represents the final form of nitrogen resulting from the oxidation of nitrogenous compounds in the wastewater. The wastewater nitrogen cycle is shown in Figure 2.23.

2.63 Factors Affecting Biological Nitrification

Because of current ammonia removal requirements and anticipation of future "complete nitrogen removal" requirements, you may be required to operate such an activated sludge plant. If you are operating a plant of this type, there are seven principal control guidelines that you must consider to maintain the nitrification process at optimum performance levels:

1. Dissolved oxygen

2. pH

3. Wastewater temperature

4. Nitrogenous food

5. Detention time

6. MCRT, F/M, or sludge age

7. Toxic materials

Each of these guidelines must be properly controlled for the successful operation of a biological nitrification process.

1. Dissolved Oxygen (DO)

    Nitrification exerts a substantial oxygen requirement. Each pound of ammonium-nitrogen that is nitrified requires approximately 4.6 pounds of oxygen (4.6 kg O_2/kg NH_4^+-N).

    Nitrification appears to be uninhibited at DO concentrations of 1 mg/L or more. To ensure adequate nitrification, the DO in the aeration tank must usually be maintained

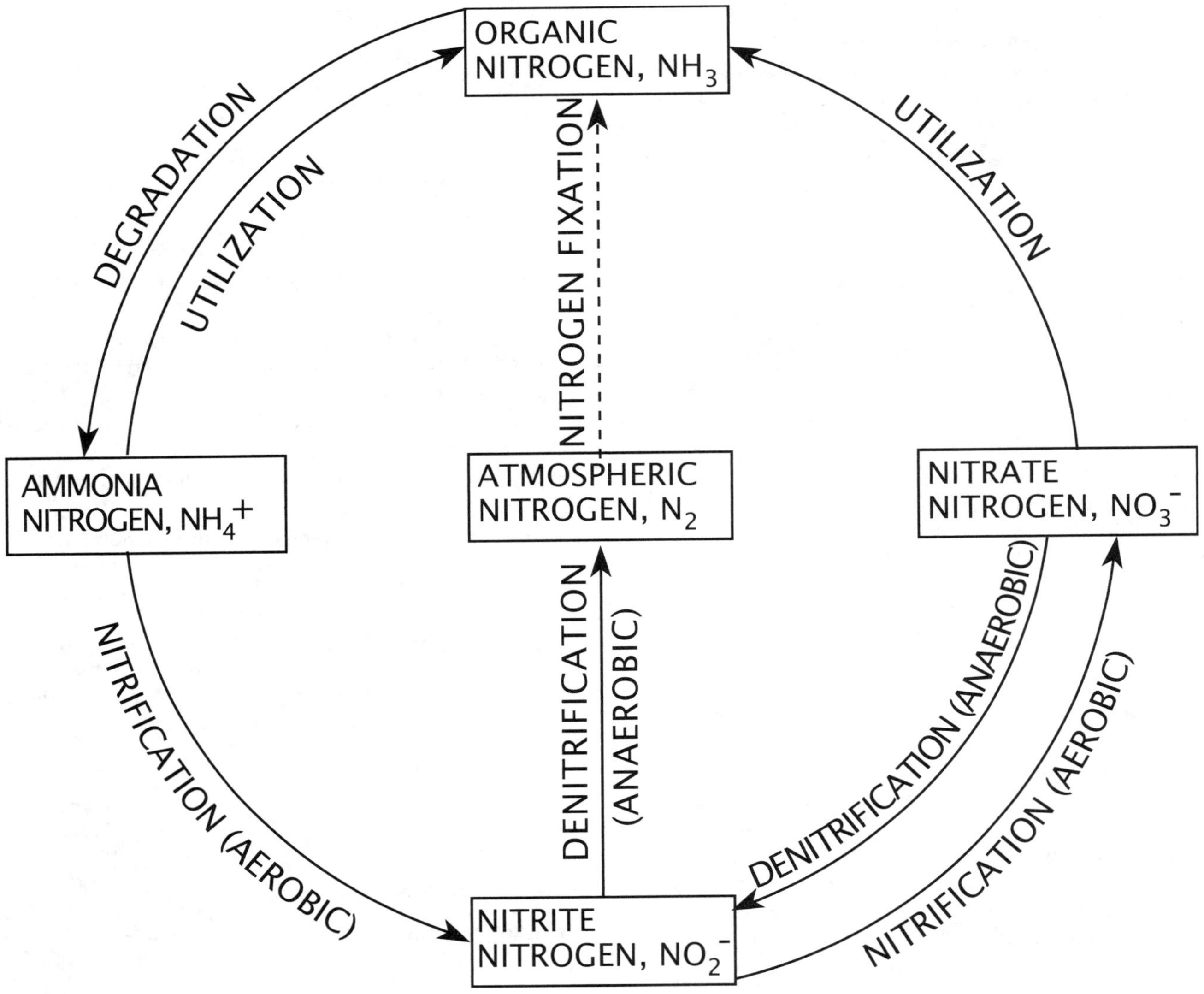

Fig. 2.23 Wastewater nitrogen cycle

between 1.0 and 4.0 mg/L under average loading conditions. This will include a reasonable DO safety factor. Under peak loading, the DO may fall off somewhat, yet should never fall below 1.0 mg/L.

The oxygen requirement may be calculated as shown in the following example.

EXAMPLE 11

Determine the oxygen requirements for the effluent from a 10 MGD activated sludge plant with an average five-day BOD of 30 mg/L and an average ammonium-nitrogen concentration of 15 mg/L.

Known		**Unknown**
Flow, MGD	= 10 MGD	Oxygen Requirement, lbs/day
BOD, mg/L	= 30 mg/L	
NH_4^+-N, mg/L	= 15 mg/L	

1. Calculate the ammonium-nitrogen load in pounds per day.

$$NH_4^+\text{-N Load, lbs/day} = \text{Flow, MGD} \times NH_4^+\text{-N, mg/}L \times 8.34 \text{ lbs/gal}$$

$$= 10 \text{ MGD} \times 15 \text{ mg/}L \times 8.34 \text{ lbs/gal}$$

$$= 1,251 \text{ lbs } NH_4^+\text{-N/day}$$

2. Calculate the BOD load in pounds per day.

$$\text{BOD, lbs/day} = \text{Flow, MGD} \times \text{BOD, mg/}L \times 8.34 \text{ lbs/gal}$$

$$= 10 \text{ MGD} \times 30 \text{ mg/}L \times 8.34 \text{ lbs/gal}$$

$$= 2,502 \text{ lbs BOD/day}$$

3. Calculate the ammonium-nitrogen oxygen requirement (pounds per day of oxygen to oxidize ammonia (NH_3) to nitrate (NO_3^-).

$$\text{Oxygen, lbs/day } (NH_4^+\text{-N}) = NH_4^+\text{-N, } \frac{\text{lbs}}{\text{day}} \times \frac{4.6 \text{ lbs Oxygen}}{\text{lb } NH_4^+\text{-N}}$$

$$= 1,251 \frac{\text{lbs } NH_4^+\text{-N}}{\text{day}} \times \frac{4.6 \text{ lbs Oxygen}}{\text{lb } NH_4^+\text{-N}}$$

$$= 5,755 \text{ lbs Oxygen/day}$$

4. Calculate the BOD oxygen requirement.

$$\text{Oxygen, lbs/day (BOD)} = \text{BOD, } \frac{\text{lbs}}{\text{day}} \times \frac{1.5 \text{ lbs Oxygen}}{\text{lb BOD}}$$

$$= 2,502 \frac{\text{lbs BOD}}{\text{day}} \times \frac{1.5 \text{ lbs Oxygen}}{\text{lb BOD}}$$

$$= 3,753 \text{ lbs Oxygen/day}$$

5. Calculate the total oxygen requirement to properly oxidize ammonium-nitrogen (NH_4^+-N) and biochemical oxygen demand (BOD).

$$\text{Total Oxygen Requirement, lbs/day} = \frac{\text{Oxygen, lbs/day}}{(NH_4^+\text{-N})} + \frac{\text{Oxygen, lbs/day}}{(\text{BOD})}$$

$$= 5,755 \text{ lbs/day} + 3,753 \text{ lbs/day}$$

$$= 9,508 \text{ lbs/day}$$

Because the rate of nitrification will vary significantly with temperature and pH, compensation must be made for these variations. During the summer months, the following methods can be used to match the oxygen requirement to your plant's oxygen capability. These methods attempt to provide more oxygen for nitrification while trying to reduce other oxygen demands.

a. Reduce the aeration system MLSS concentration.

b. Reduce the wastewater pH by reducing chemical addition (if used).

c. Reduce the number of tanks in service while increasing oxygen supply to the tanks remaining in service.

2. pH

In many wastewaters, there is insufficient alkalinity initially present to leave a sufficient residual for buffering the wastewater during the nitrification process. The significance of pH depression in the process is that nitrification rates are rapidly depressed as the pH is reduced below 7.0. Because of the effect of pH on nitrification rate, it is especially important that there be sufficient alkalinity in the wastewater to balance the acid produced by nitrification. A pH between 7.5 and 8.5 is considered optimal. Approximately 7.2 pounds of alkalinity are destroyed per pound (7.2 kg/kg) of ammonia-nitrogen (NH_3-N) oxidized. Caustic or lime addition may be required to supplement moderately alkaline wastewaters.

If it becomes necessary to add chemicals (preferably lime) for pH adjustment, the required quantities of chemical will vary with wastewater temperature, MLVSS concentration, influent ammonia-nitrogen concentration, and the natural alkalinity of the wastewater. As the oxidation of ammonia-nitrogen to nitrate-nitrogen destroys approximately 7.2 pounds of alkalinity per pound (7.2 kg/kg) of ammonia-nitrogen, this loss of alkalinity must be added to the chemical quantity calculated for pH adjustment. For operation under the most adverse temperature and pH conditions, sufficient lime must be added initially to raise the pH into the desired range, and then 5.4 pounds of hydrated lime per pound (5.4 kg/kg) of ammonia-nitrogen will be required to maintain the pH. Sufficient alkalinity should be provided to leave a residual of 30 to 50 mg/L after complete nitrification.

3. Temperature

The optimum wastewater temperature range is between 60 and 95°F (15 and 35°C) for good nitrification operation. Nitrification is inhibited at low wastewater temperatures and up to five times as much detention time may be needed to accomplish "complete nitrification" in the winter as is needed in the summer. The growth rate of nitrifying bacteria increases as the wastewater temperature increases and conversely it decreases as the wastewater temperature decreases. Since there is no control over the wastewater temperature, operating compensations for slower winter growth rates are necessary. Increasing the MLVSS concentration, the MCRT, and adjusting the pH to favorable levels can be expected to provide substantial, if not "complete," oxidation of ammonia-nitrogen compounds. Under summer conditions, operation will be possible at less favorable pH levels and lower MLVSS concentrations.

4. Nitrogenous Food

The growth rate of nitrifying bacteria *(Nitrosomonas* and *Nitrobacter)* is only slightly affected by the organic load applied to the aeration system. However, the population of the nitrifying bacteria will be limited by the amount of nitrogenous food available in the wastewater. Organic nitrogen and phosphorus-containing compounds as well as many trace elements are essential to the growth of microorganisms in the aeration system. The generally recommended ratio of five-day BOD to nitrogen to phosphorus for domestic waste is 100:5:1. Laboratory nitrogen determination (TKN) and phosphorus determination analyses should be performed so that you may add the supplemental phosphorus nutrient if necessary. Phosphorus in the form of phosphate fertilizer may be added and adjusted according to the five-day BOD level and the TKN concentration in the wastewater.

5. Detention Time

The time required for nitrification is directly proportional to the amount of nitrifiers present in the system. Because the rate of oxidation of ammonia-nitrogen is essentially linear or constant, short-circuiting must be prevented. The aeration tank configuration should ensure that flow through the tank follows the plug-flow mixing model as closely as possible and provides a minimum detention time of approximately 4.0 hours. Single-pass tanks may be modified and divided into a series of three compartments with ports between them to prevent short-circuiting. Not all of the various modifications to the activated sludge process are appropriate for nitrification applications, although some may be used where only partial ammonia removal is required.

6. MCRT, F/M, or Sludge Age

To achieve the desired degree of nitrification, the MCRT must be long enough (usually four days plus) to allow the nitrifying bacteria sufficient time to grow. Since the nitrifying bacteria grow much more slowly than the bacteria using the carbonaceous compounds, it is possible to waste the nitrifying bacteria from the system at a higher rate than their growth rate. In simpler terms, this means that nitrification in plants can be maintained only when the rate of growth of nitrifying bacteria is rapid enough to replace organisms lost through sludge wasting. When these bacteria can no longer keep pace, the ability to nitrify decreases and may stop.

When reviewing the performance of your activated sludge process for the selection of an optimum F/M ratio, MCRT, or sludge age, oxygen requirements for bacteria using the carbonaceous compounds should be considered along with nitrification requirements. These guidelines should be selected to provide the degree of nitrification required by the discharge permit. If the ammonia-nitrogen limit is being exceeded, the MCRT or sludge age should be increased. Increasing these guidelines will increase the MLVSS and consequently decrease the F/M ratio. With the other conditions (discussed above) constant, a definite relationship will exist between the weight ratio of the ammonia-nitrogen oxidized per day to the MLVSS under aeration.

The growth of cell mass from the oxidation of ammonia is about 0.05 lb per lb (0.05 kg/kg) of ammonia-nitrogen oxidized. As a result, the degree of nitrification will have little effect on the net sludge yield and WAS rates.

7. Toxic Materials

Various types of toxic materials that will inhibit the nitrification process (in concentrations greater than those indicated) are shown below.

a. Halogen-substituted phenolic compounds, 0.0 mg/L.

b. Thiorea and thiorea derivatives, 0.0 mg/L.

c. Halogenated solvents, 0.0 mg/L.

d. Heavy metals, 10 to 20 mg/L.

e. Cyanides and all compounds from which hydrocyanic acid is liberated on acidification, 1 to 2 mg/L.

f. Phenol and cresol, 20 mg/L.

Pretreatment alternatives provide a degree of removal of the toxicants present in raw wastewater. However, the types of toxicants removed by each pretreatment stage vary among the alternatives. Chemical primary treatment can be used where toxicity from heavy metals is the major problem. Lime primary treatment is one of the most effective processes for removal of a wide range of metals. Chemical treatment is usually not effective for removal of organic toxicants, unless it is coupled with a carbon adsorption step such as in a physical–chemical treatment sequence.

When materials toxic to nitrifiers are present in the raw wastewater on a regular basis, the pretreatment technique most suitable for their removal can be used in the plant to safeguard the nitrifying population. The determination of the most suitable pretreatment process application may be initially developed based on *BENCH-SCALE ANALYSIS* [32] to screen alternatives.

The particular pretreatment technique that is effective may also indicate the type of toxicant that is interfering with nitrification and may permit identification and elimination of the source. Subsequent specific analysis can then be run in the identified category of compounds. If the toxicants cannot be eliminated by a source control program, often a pilot study of the process identified by the bench-scale tests can be justified to confirm the process selection.

2.64   Rising Sludge and the Nitrification Process

Rising sludge caused by unwanted denitrification in the clarifiers may occasionally plague your nitrification operation. Denitrification occurs because the facultative heterotrophic organisms in the biological sludge in the clarifier accomplish nitrate reduction by what is known as a process of nitrate dissimilation. In this process, nitrate and nitrite replace oxygen in the respiratory process of the organisms under oxygen-deficient conditions. This nitrate dissimilation allows bubbles of nitrogen gas and carbon dioxide to adhere to the sludge floc surface resulting in rising sludge.

The degree of stabilization of the sludge in the aeration tank (depending on detention time and DO) also has a profound effect on denitrification in the clarifier. Clarifier sludges containing partially oxidized or unoxidized organics float more readily than well-oxidized sludges. Wastewater temperature is also important as it affects the rate of denitrification and therefore affects the rate of gas and bubble formation (depends on warm temperatures and denitrification rates).

Some considerations for good clarifier operation to prevent denitrification in the clarifier are discussed in the following sections.

1. The settled sludge must be quickly removed from the clarifier to minimize the occurrence and duration of oxygen-deficient conditions. The RAS rate may be proportioned with the aerator influent flow to maintain an essentially "zero" sludge blanket in the clarifier.

2. Since the nitrification sludge is lighter and does not compact as well as carbonaceous sludges, sludges with low SVI values are preferable. They can be withdrawn from the clarifier faster. Since the saturation level of nitrogen is greater in deep tanks than laboratory cylinders, bubbles will form and sludges will float faster in the laboratory than in the field.

3. There is a minimum concentration of nitrate-nitrogen below which there is insufficient nitrogen to cause rising sludge. In weak wastewaters or for those plants in which nitrification is suppressed, rising sludge will not occur.

4. A drop in wastewater temperature will reduce denitrification rates and may render rising sludge a problem only under warmer wastewater conditions.

2.65   Acknowledgments

Major portions of Section 2.6 were taken from the following three publications. *PROCESS CONTROL MANUAL FOR AEROBIC BIOLOGICAL WASTEWATER TREATMENT FACILITIES,* US Environmental Protection Agency. Available from National Technical Information Service (NTIS), 5285 Port Royal Road, Springfield, VA 22161. Order No. PB-279474. EPA No. 430-9-77-006. Price $86.50, plus $5.00 shipping and handling per order. *NITRIFICATION AND DE-NITRIFICATION FACILITIES, WASTEWATER TREATMENT,* EPA Technology Transfer Publication, US Environmental Protection Agency. Available from the NTIS address listed above. Order No. PB-259447. Price $33.50, plus $5.00 shipping and handling per order. *PROCESS DESIGN MANUAL FOR NITROGEN CONTROL,* US Environmental Protection Agency. Available from National Service Center for Environmental Publications (NSCEP), PO Box 42419, Cincinnati, OH 45242-2419. EPA No. 625-R-93-010.

QUESTIONS

Write your answers in a notebook and then compare your answers with those on page 151.

2.6A   List the harmful effects that could result from the discharge of nitrogenous compounds from wastewater treatment plants.

2.6B   What are the principal control guidelines for biological nitrification?

2.6C   How can you control rising sludge that results from unwanted denitrification?

2.7   REVIEW OF PLANS AND SPECIFICATIONS— PURE OXYGEN ACTIVATED SLUDGE SYSTEMS

2.70   Need to Be Familiar With System

The operational staff that reviews the plans and specifications for a pure oxygen plant should be very familiar with the activated sludge process. A tour through an existing pure oxygen system would be extremely helpful. Specific questions regarding the operation and maintenance of the facility can be answered by manufacturers or other treatment plant personnel whose systems use pure oxygen. Also, sources of industrial waste discharges should be identified and investigated for possible toxic wastes, heavy load contributions, and seasonal fluctuations. Be sure your plant has the capacity and flexibility to treat all industrial wastes.

After plans are submitted, the operation and maintenance staff should review all areas of the plans and specifications with special attention directed toward:

1. Physical plant layout

2. Oxygen generation equipment

---

[32] *Bench-Scale Analysis (Test).*   A method of studying different ways or chemical doses for treating water or wastewater and solids on a small scale in a laboratory. Also see JAR TEST.

3. Reactor basins (aeration tanks)

4. Oxygen safety and process instrumentation

5. Preventive maintenance program

2.71   Physical Layout

The pure oxygen generation system should be located near the plant maintenance facilities. If this is not possible, a maintenance area within the system should be included. This will be an aid during major maintenance on the facility. Major pieces of equipment should be easily accessible. A crane or other lifting devices should be provided to lift large pieces of equipment during major overhauls. Road access and loading facilities should also be provided. Sources of noise and vibrations should also be considered. Most equipment in oxygen production systems produce noise similar to air blower systems and, if not properly installed, could produce vibrations that could be transferred through walls or structures adjacent to such facilities. In offices or laboratories, noise considerations should also be reviewed, especially in plants where housing areas are near the oxygen generation site. Silencers are typically provided with the generation equipment. The overall layout of the system should also allow for expansion of the wastewater treatment plant and oxygen production facilities.

2.72   Oxygen Generation Equipment

If the oxygen generation equipment is located within a building, it should be well ventilated. In areas of extremely hot temperatures, vent fans on the roof would be of benefit. A heating system is normally not required if the lowest temperature does not remain below freezing for long periods of time. The compressor equipment generates sufficient heat to heat the building. Systems located in the open must be designed to operate during the most severe weather conditions.

Individual equipment suppliers recommended by the design engineer should be contacted for specific answers to the installation of compressors, valve skids, or oxygen storage facilities. Start-up controls, instrumentation, and safety devices should be carefully reviewed. A vibration shutdown system on compressors should be included. Prior to actual start-up, the manufacturer should run each compressor and check the equipment for proper operation, including excessive vibration. When everything operates in an acceptable manner, calibrate and set the vibration monitor to shut down the equipment if excessive vibration develops. This monitor ensures proper operation and protection of the equipment. If a separate water cooling system is provided, the water used should be treated to avoid scaling and corrosion of the units.

A specific cleaning sequence should be specified in the plans and specifications that would include oxygen pipelines, sample lines, valve skids, and other equipment to ensure equipment protection during start-up. Contamination from dirt, grease, and welding slag should never be allowed in a pure oxygen atmosphere. The specifications should include a requirement that equipment manufacturers be present during the start-up and provide a training program to the operations staff. A recommended spare parts list and names of suppliers or vendors should be provided.

Major oxygen feed lines, valves, sample tubing, and electrical systems must be tagged and indicated in the "as-built" (record) drawings and instruction booklets. Specialized drawings and instruction booklets must include detailed descriptions on preventive maintenance, safety, and operation instructions. At least four copies should be provided. Any modification during start-up should be indicated in each copy of these manuals.

2.73   Reactors (Aeration Tanks)

The location of the reactor should be reviewed with consideration for future expansion. Reactors located above normal ground elevation should have facilities provided to remove equipment located on the deck. Oxygen reactors are usually completely covered and access to each basin must be provided through a sealed and airtight lid or locking manhole. Gas sampling lines or other conduits can be installed within the deck. If they are located on the deck, they should be protected and indicated by safety signs to avoid a tripping hazard.

The deck should be a rough surface such as broomed concrete. A completely smooth deck is a slipping hazard if water is allowed to collect. Warning signs should be provided at each entrance to indicate the presence of oxygen and to warn that no smoking or open flames are allowed. A well-lighted deck is helpful for night shift operators.

Interior metal such as weir plates, mixer blades, and valves should be constructed of stainless steel, carbon steel, or coated carbon steel as dictated by the specific service. A good protective coating should also be provided over any surfaces that may corrode. Control valves should have position indicators and be located in such a manner that preventive maintenance may be performed without draining the reactor. Equipment on the reactor should be tagged with equipment numbers corresponding to electrical control panel facilities.

2.74   Safety and Instrumentation

Operator safety and process safety are both very important in pure oxygen systems. Safety signs, belt guards, temperature shutdown switches, and overload protection devices should be provided and indicated in the specifications by the engineer. A system "emergency trip switch" station should be provided and located away from major electrical controls. If tripped during a major mechanical or electrical equipment failure, the entire oxygen operation shuts down without endangering personnel or equipment.

The specifications should include a requirement that the equipment supplier provide a training class to instruct personnel on operation, maintenance, and safety hazards involved in the pure oxygen generation and waste treatment process.

Process control involves major instrumentation and control systems. Preventive maintenance on such systems is extremely technical in nature and should be completely understood before maintenance personnel attempt to maintain such systems. The design engineer can provide the necessary background to ensure proper training.

One major area of concern is instrumentation sample tubing and heat trace lines. These systems are the main control and safety equipment functions of the entire system. To avoid costly errors, the manufacturer or equipment supplier should be consulted. If sample lines for lower explosive limit (LEL) or system pressure sensing control lines are incorrectly installed, the entire operation could fail. If they must be installed under roadways, they should be protected within rigid, sealed conduit to prevent being crushed or kinked.

2.75   Preventive Maintenance

Most specifications would not include specific requirements for preventive maintenance. The system design should give con-

sideration to the staff size and experience needs of the system. The design engineer could advise key personnel about the needs of the system and requirements. Manufacturers can provide maintenance contracts that provide preventive maintenance as well as major tune-ups. Spare parts can be included but can be purchased separately. The cost of such contracts depends on the needs of the system and the options involved.

End of Lesson 4 of 4 Lessons on ACTIVATED SLUDGE

Please answer the discussion and review questions next.

DISCUSSION AND REVIEW QUESTIONS

Chapter 2.   ACTIVATED SLUDGE

Pure Oxygen Plants and Operational Control Options

(Lesson 4 of 4 Lessons)

Write the answers to these questions in your notebook. The question numbering continues from Lesson 3.

15. Why would an operator want to monitor the influent to a wastewater treatment plant?

16. How can undesirable constituents be detected in a plant influent in addition to the use of automatic monitoring units?

17. Why should an operational strategy be developed before a toxic waste is discovered in the influent to a plant?

18. Why do some industries pretreat industrial process wastewaters?

19. Why does the operator of an industrial waste treatment plant have to know the flow and waste characteristics of the wastewater being treated?

20. How do toxic materials enter the wastestream?

21. Why should grit, soil, grease, and oil be removed from the wastestream?

22. How can the amount of nutrients to be added each day be determined?

23. How does ammonia stripping remove nitrogen from wastewater?

24. How is the nitrification process influenced by temperature?

25. How would you handle materials toxic to a nitrification process?

SUGGESTED ANSWERS

Chapter 2.   ACTIVATED SLUDGE

Pure Oxygen Plants and Operational Control Options

ANSWERS TO QUESTIONS IN LESSON 1

Answers to questions on page 57.

2.0A   Pure oxygen systems dissolve oxygen with a higher efficiency for use by microorganisms treating the wastes. This allows the use of smaller reactor tanks than air activated sludge tanks. The sludge produced has improved settleability and dewaterability.

2.0B   Biological nitrification can remove ammonium from wastewater by converting it to nitrate.

Answers to questions on page 60.

2.1A   Pure oxygen reactors are staged to increase the efficiency of the use of oxygen.

2.1B   As the pure oxygen flows from one stage to the next stage, the oxygen is diluted by the carbon dioxide produced by the microorganisms and the nitrogen stripped from the wastewater being treated.

2.1C   The two methods commonly used to produce pure oxygen are:

1. Pressure Swing Adsorption (PSA) Oxygen Generating System
2. Cryogenic Air Separation Method

Answers to questions on page 64.

2.1D   Cryogenic means very low temperature.

2.1E   Cryogenic plants are usually shut down once a year for approximately five to seven days for maintenance.

Answers to questions on page 66.

2.1F   Special measurements used to control pure oxygen systems include:

1. Reactor vent gas
2. Reactor gas space pressure
3. Dissolved oxygen

2.1G   Hydrocarbons can be detected before they reach the reactor by installing a lower explosive limit (LEL) combustible gas detector in the plant headworks.

ANSWERS TO QUESTIONS IN LESSON 2

Answers to questions on page 70.

2.2A   1. MLSS, Mixed Liquor Suspended Solids
2. MLVSS, Mixed Liquor Volatile Suspended Solids
3. RAS, Return Activated Sludge
4. F/M, Food to Microorganism Ratio

2.2B   The two basic approaches that can be used to control the RAS flow rate are:

1. Controlling the RAS flow rate independently from the influent flow
2. Controlling the RAS flow rate as a constant percentage of influent flow

Answers to questions on page 72.

2.2C   Techniques used to determine the rate of RAS flow include:

1. Monitoring the depth of the sludge blanket
2. Settleability approach
3. SVI approach

2.2D   The sludge blanket depth should be kept to one to three feet (0.3 to 1 m) as measured from the clarifier bottom at the sidewall.

2.2E   The sludge blanket depth should be measured at the same time every day during the period of maximum flow and high solids loading rate so the depth will be measured under the same conditions of flow and solids loading.

Answers to questions on page 73.

2.2F   Sludge is allowed to settle 30 minutes in the settleability test.

2.2G   Calculate the return activated sludge RAS flow rate in MGD when the influent flow is 4 MGD and the sludge settling volume (SV) in 30 minutes is 250 mL/L.

Known		Unknown
Infl Flow, MGD	= 4 MGD	RAS Flow, MGD
Sl Set Vol (SV), mL/L	= 250 mL/L	

1. Calculate RAS flow as a percent of influent flow.

$$\text{RAS Flow, \%} = \frac{\text{SV, m}L/L \times 100\%}{1,000 \text{ m}L/L - \text{SV, m}L/L}$$

$$= \frac{250 \text{ m}L/L \times 100\%}{1,000 \text{ m}L/L - 250 \text{ m}L/L}$$

$$= \frac{250 \text{ m}L/L \times 100\%}{750 \text{ m}L/L}$$

$$= 33\% \text{ of influent flow rate}$$

2. Calculate RAS flow, MGD.

$$\text{RAS Flow, MGD} = \text{RAS Flow, decimal} \times \text{Infl Flow, MGD}$$

$$= 0.33 \times 4 \text{ MGD}$$

$$= 1.32 \text{ MGD}$$

ANSWERS TO QUESTIONS IN LESSON 3

Answers to questions on page 78.

2.3A The objective of wasting activated sludge is to maintain a balance between the microorganisms and the amount of food as measured by the COD or BOD test.

2.3B Wasting of the excess activated sludge is usually achieved by removing a portion of the RAS flow. Sometimes excess sludge is wasted directly from the MLSS in the aeration tank.

2.3C Calculate the change in the waste activated sludge (WAS) flow rate in MGD and GPM for an activated sludge plant that adds 4,750 lbs of solids per day. The solids under aeration are 41,100 pounds and the return activated sludge (RAS) suspended solids concentration is 5,800 mg/L. The desired sludge age is 8 days.

Known		Unknown
Solids Added, lbs/day	= 4,750 lbs/day	WAS Flow, MGD
Solids Aerated, lbs	= 41,100 lbs	and GPM
RAS Susp Sol, mg/L	= 5,800 mg/L	
Desired Sludge Age, days	= 8 days	

1. Calculate the desired pounds of solids under aeration for the desired sludge age of 8 days.

$$\text{Desired Solids, lbs} = \text{Solids Added, lbs/day} \times \text{Sludge Age, days}$$

$$= 4,750 \text{ lbs/day} \times 8 \text{ days}$$

$$= 38,000 \text{ lbs}$$

2. Calculate the WAS flow, MGD and GPM, to maintain the desired sludge age.

$$\text{WAS Flow, MGD and GPM} = \frac{\text{Solids Aerated, lbs} - \text{Desired Solids, lbs*}}{\text{RAS Susp Sol, mg/}L \times 8.34 \text{ lbs/gal}}$$

$$= \frac{(41,100 \text{ lbs} - 38,000 \text{ lbs})/\text{day}}{5,800 \text{ mg/}L \times 8.34 \text{ lbs/gal}}$$

$$= \frac{3,100 \text{ lbs removed per day}}{48,372}$$

$$= 0.064 \text{ MGD} \times 694 \text{ GPM/MGD}$$

$$= 45 \text{ GPM}$$

Increase WAS pumping by 45 GPM.

---

* Difference represents solids wasted in pounds per day.

Answers to questions on page 81.

2.3D Four facts that should be remembered regarding the F/M method of control are:

1. The food concentration is estimated with the COD or BOD tests.
2. The amount of food (COD or BOD) applied is important to calculate the F/M.
3. The quantity of microorganisms can be represented by the quantity of MLVSS.
4. The data obtained to calculate the F/M should be based on the seven-day moving average.

2.3E Calculate the desired pounds of MLSS if the desired F/M ratio is 0.30 lb COD/day/lb MLVSS, 7,000 lbs of COD per day are added, and the volatile matter is 70 percent of the MLSS.

Known		Unknown
Desired F/M, $\frac{\text{lbs COD/day}}{\text{lb MLVSS}}$	$= \frac{0.30 \text{ lb COD/day}}{\text{lb MLVSS}}$	Desired MLSS, lbs
COD Added, lbs/day	= 7,000 lbs COD/day	
Volatile Portion	= 0.70	

1. Determine the desired pounds of MLVSS.

$$\text{Desired MLVSS, lbs} = \frac{\text{COD Applied, lbs/day}}{\text{F/M, lbs COD/day/lb MLVSS}}$$

$$= \frac{7,000 \text{ lbs COD/day}}{0.30 \text{ lb COD/day/lb MLVSS}}$$

$$= 23,333 \text{ lbs MLVSS}$$

2. Determine the desired pounds of MLSS.

$$\text{Desired MLSS, lbs} = \frac{\text{Desired MLVSS, lbs}}{\text{MLSS VM Portion}}$$

$$= \frac{23,333 \text{ lbs MLVSS}}{0.70}$$

$$= 33,333 \text{ lbs}$$

2.3F   The MCRT expresses the average time a microorganism spends in the activated sludge process.

Answers to questions on page 88.

2.3G   Overstabilized organic matter generally produces fast-settling sludge, which may cause high effluent turbidity resulting from pin floc. Understabilized organic matter generally results in a sludge that is slow to settle and hard to keep in the clarifier, and that may even turn septic in the return sludge system.

2.3H   The respiration rate (RR_{AS}) test can be used to alert operators to changes required in waste sludge flow rates, return sludge flow rates, aeration tank detention times, and the quantity of oxygen needed to maintain a healthy activated sludge process.

2.3I   Low respiration rate (RR_{AS}) values may indicate unhealthy or inactive microorganism conditions, or a sludge that settles rapidly in the clarifier yet is associated with a turbid secondary effluent caused by pin floc.

Answers to questions on page 92.

2.3J   You can obtain a fairly good estimate of effluent quality, as well as food quality and strength, by calculating the oxygen uptake rate (OUR) value. Oxygen uptake rate (OUR) values may be correlated with plant influent and secondary effluent BOD_5 values.

2.3K   A high load ratio (LR greater than 5) typically indicates poor organic matter stabilization (anticipate poor sludge settleability), or that the aeration system influent is very strong in terms of organic strength.

2.3L   The organic matter stabilization test is used to determine the best aeration system detention time for complete organic matter stabilization for existing and new industrial waste contributors, for seasonal organic load variations, and for managing the step-feed aeration system process.

Answers to questions on page 95.

2.3M   The microorganisms that are important indicators in the activated sludge process are the protozoa and rotifers.

2.3N   Important activated sludge control tests used in the Al West method to define sludge quality and process status include final clarifier sludge blanket depth, mixed liquor and return sludge concentrations, and sludge settleability.

ANSWERS TO QUESTIONS IN LESSON 4

Answers to questions on page 100.

2.4A   An industrial waste monitoring program is valuable for the operator of an activated sludge plant for the following reasons:

1. To assure regulatory agencies of industrial compliance with discharge requirements and implementation schedules set forth in the discharge permit.
2. To maintain sufficient control of treatment plant operations to prevent NPDES permit violations.
3. To gather necessary data for the future design and operation of the treatment plant.

2.4B   Automatic wastewater monitoring devices are commonly used to monitor:

1. Flow
2. pH
3. Oxidation potential
4. Suspended solids
5. Dissolved oxygen

Other useful automatic devices include wastewater samplers and gas-phase hydrocarbon analyzers.

2.4C   If a high organic waste load enters your plant:

1. Gradually increase the RAS flow
2. Increase the air input to the aeration tanks
3. Add nutrients, if necessary

Answers to questions on page 101.

2.50A   Types of industries whose wastes could be treated by an activated sludge plant include:

1. Fruit and vegetable processing
2. Paper product manufacturing
3. Petroleum refining
4. Dairy products processing

2.51A   The character of industrial wastewater is dependent on the particular production process and the way it is operated.

2.52A   One approach to solving a hydraulic overloading problem is to reduce water consumption and minimize waste generation through proper management of processing and production operations.

2.52B   The pH of an industrial wastewater can be adjusted (1) upward by the addition of caustic, and (2) downward by the addition of sulfuric acid.

2.52C   Chemical oxidation (COD) measures the presence of (1) *CARBON* and (2) *HYDROGEN*, but not (3) *AMINO NITROGEN* in organic materials.

2.52D   The three major nutrients required by microorganisms are carbonaceous organic matter (measured by BOD test), nitrogen (N), and phosphorus (P). Other elements that might be critical include iron, calcium, magnesium, potassium, cobalt, and molybdenum.

2.52E   The most common causes or kinds of toxicity in industrial wastewaters include excessive amounts of free ammonia, residual chlorine, detergents, paints, solvents, and biocides.

Answers to questions on page 102.

2.53A   Samples from industrial wastewaters should be collected during each operating shift and during different stages of the finished product and raw product runs.

2.53B   Large variations in flow can be reduced or smoothed out by routing the flows through a surge tank or storage tank.

2.53C   Discrete waste solids (such as trimmings, rejects, corn meal, and pulp) are effectively separated from the wastewater flow by various types of screens.

2.53D   Screens should be located as close as possible to the process/production activity producing the waste. The longer the solids are in contact with water and the rougher the flow is handled (more turbulent), the more material will pass through the screens and the more the solids will become dissolved.

Answers to questions on page 106.

2.53E   Generally, the most economical location to treat toxic wastes is at the source. If possible, do not allow toxic wastes to enter the plant wastewater.

2.53F   Industrial waste pretreatment facilities are necessary to treat industrial wastes so they can be treated by the activated sludge process. Pretreatment involves pH adjustment, coarse solids removal, flow equalization, nutrient addition, and cooling.

2.53G   Screens are used to remove coarse solids to reduce unnecessary clogging and wear on downstream pipes, pumps, aerators, and clarifier mechanisms. Also, this process will help avoid odor problems that could develop from the settling out of solids in the equalization and emergency basins.

2.53H   Organic loadings can be smoothed out by the use of an equalizing tank and also by keeping the contents of the tank well mixed.

2.53I   Nitrogen is supplied in the form of ammonia (usually as 30 percent aqueous) and phosphorus as phosphoric acid (usually as 75 percent aqueous).

2.53J   "Seed activated sludge" may be obtained from either a nearby municipal or industrial wastewater treatment plant.

2.53K   Initially, industrial wastes must be added gradually to the aeration basin to allow time for the activated sludge microorganisms to adapt or become acclimated to the wastes.

Answers to questions on page 107.

2.54A   Nutrients should be added to wastewater at a point where the incoming wastewater is highly mixed and preferably in the aeration basin.

2.54B   Supplemental nitrogen can be provided by aqueous ammonia or anhydrous ammonia. Supplemental phosphorus can be provided by dissolved triple superphosphate, phosphate fertilizer, or phosphoric acid.

2.54C   Factors or conditions that have been found to cause the disappearance of higher forms of microorganisms from the activated sludge process include:

1. DO levels below 3 to 4 mg/L
2. High organic loadings
3. Toxic substances or nutrient deficiencies
4. Lack of pH control

2.54D   A gradual decrease in percent solids removal from an activated sludge process over a period of time may be the result of grit and silt accumulation in the aeration basin that was not carried out in the effluent and remained in the basin, thus resulting in reduced aeration volume and detention time.

Answers to questions on page 115.

2.55A   Accurate records must be kept for the following reasons:

1. To help troubleshoot problems that arise
2. To account for the operation of the plant on a weekly or monthly basis
3. To serve as legal documents that will protect the company from unjust claims

2.55B   The paper industry might shut down an activated sludge process during holidays, maintenance (preventive and emergency), process upsets, strikes, and scheduled production reductions.

2.55C   Effluent from a pulp and paper mill activated sludge process should contain some nutrients. If low levels of nutrients are present, a nutrient deficiency could exist in the process.

2.55D   The periodic-feeding (step-feed) technique for process start-up of activated sludge systems is effective because it allows organisms time to consume the available food (waste). When they are ready for more food, more is added. This procedure encourages rapid microorganism reproduction.

Answers to questions on page 126.

2.56A   Primary clarifiers are not very effective in removing BOD because most of the BOD is "soluble" or "dissolved" and cannot be removed by sedimentation.

2.56B   When treating brewery wastes, nutrients can be metered into the primary clarifier effluent. To supplement nitrogen, ammonia gas (or liquid) can be added. Phosphoric acid (H_3PO_4) can be used as a source of phosphorus. Sufficient phosphorus may be present in the influent from malting by-products and phosphorus-based cleaning solutions.

2.56C The MLSS level in aeration basins is influenced by:

1. Influent organic loading
2. Temperature

2.56D The sludge wasting rate is determined by attempting to regulate the actual MLSS in the aeration basins as close as practical to the desired MLSS set point.

2.56E Filamentous bulking can be controlled in activated sludge plants treating brewery wastes by:

1. Proper DO levels in aeration basins
2. Providing sufficient nutrients (nitrogen and phosphorus) and trace minerals
3. Proper F/M ratios
4. Minimizing large fluctuations of influent organic loadings

Answers to questions on page 128.

2.57A High strength artichoke waste influents (BOD greater than 6,000 mg/L) are diluted with medium strength wastewater to produce an influent BOD of less than 6,000 mg/L before treatment by the activated sludge process.

2.57B Fifteen-day moving averages were computed and plotted for artichoke influent COD and F/M ratio to reduce the effect of fluctuating daily results.

2.57C Ammonium phosphate can be used to provide both nitrogen and phosphorus for the treatment of artichoke wastes.

Answers to questions on page 130.

2.57D Wastewater treatment problems that operators have to deal with when treating dairy processing wastewaters include high organic loadings, variable hydraulic loads, changing pH values, nutrient deficiencies, and toxins.

2.57E Common sources of BOD from a milk processing facility include product spills, spoiled batches, abnormal losses of product, concentrated residuals of product, unused by-products, cleanup, and washdown/sanitizing operations.

2.57F The major cause of pH fluctuations in dairy plant effluent is from cleaning and sanitizing operations.

2.57G The main problem caused by nutrient deficiencies in the activated sludge process when treating dairy processing wastewaters is the development of filamentous organisms that prevent the proper settling of the biomass.

2.57H An operator can conclude that adequate nutrients are available for the activated sludge process if effluent ammonia and phosphorus concentrations are 1.0 mg/L or just under this value.

2.57I Sources of toxins in dairy plant effluent include detergents, lubricants, sanitizers, and refrigerants.

Answers to questions on page 133.

2.57J The most direct way for a dairy to prevent blockages in a wastewater collection system is to provide some treatment to its wastewater to lower the butterfat content (such as dissolved air flotation).

2.57K Problems a municipal treatment plant can experience from organic (BOD) overloads and fluctuating loads can include poor settling sludge in the secondary clarifier caused by excessive filamentous bacteria, periods of low or zero dissolved oxygen concentrations in the aeration tanks, and, in the case of prolonged zero dissolved oxygen, odors can develop.

2.57L An industry might install flow equalization facilities to spread the organic loading of the municipal facility over a longer period of time. By discharging during the municipality's low-flow period, a constant loading can be applied to the municipal biological treatment processes.

2.57M When a spill of high-strength organic wastes or toxins occurs, the industrial operator should try to contain the spill and also notify the pretreatment or municipal operator. Communication between the dairy production staff and operators is essential.

Answers to questions on page 136.

2.57N If domestic or sanitary wastes are not treated by a dairy waste treatment plant, effluent chlorination may not be necessary.

2.57O High levels of MLSS are necessary to prevent process upsets caused by shock organic loadings.

Answers to questions on page 138.

2.58A The three main petroleum refinery waste compounds that can be treated by the activated sludge process are ammonia, phenols, and sulfide.

2.58B MCRTs as high as 30 days are necessary to treat petroleum refinery wastes in order to maintain the nitrification population necessary to oxidize ammonia.

2.58C   A shock load of sulfide can be treated by decreasing the hydraulic loading and adding soda ash to the aeration basin to reactivate the phenol-consuming microbes. Hydrogen peroxide can be used to help oxidize phenols in the effluent.

Answers to questions on page 143.

2.6A   Harmful effects that could result from the discharge of nitrogenous compounds include:

1. Ammonia toxicity to fish
2. Reduction in effectiveness of chlorine disinfection
3. Increase in DO depletion in receiving waters
4. Adverse public health impact on groundwater
5. Reduction in the suitability of water for reuse

2.6B   The principal control guidelines for biological nitrification are:

1. Dissolved oxygen
2. pH
3. Wastewater temperature
4. Nitrogenous food
5. Detention time
6. MCRT, F/M, or sludge age
7. Toxic materials

2.6C   Rising sludge resulting from unwanted denitrification can be controlled by:

1. Returning settled sludge as quickly as possible and maintaining essentially "zero" sludge blanket in the clarifier.
2. Maintaining sludges with low SVI values.

CHAPTER 3

RESIDUAL SOLIDS MANAGEMENT

by

Liberato D. Tortorici

Boris L. Pastushenko

James F. Stahl

(With Special Sections by Richard Best and William Anderson)

TABLE OF CONTENTS
Chapter 3.   RESIDUAL SOLIDS MANAGEMENT

LESSON 4

LESSON 5

OBJECTIVES

Chapter 3.   RESIDUAL SOLIDS MANAGEMENT

Categories of residual solids management processes contained in this chapter include thickening, stabilization, conditioning, dewatering, volume reduction, and land disposal. Following completion of Chapter 3, with regard to the processes in these solids handling and disposal categories, you should be able to:

1. Explain the purposes of the processes.

2. Properly start up, operate, shut down, and maintain these processes.

3. Develop operating procedures and strategies for both normal and abnormal operating conditions.

4. Identify potential safety hazards and conduct your duties using safe procedures.

5. Troubleshoot when a process does not function properly.

6. Review plans and specifications for the processes.

WORDS
Chapter 3.   RESIDUAL SOLIDS MANAGEMENT

AGRONOMIC RATES AGRONOMIC RATES

Sludge application rates that provide the amount of nitrogen needed by the crop or vegetation grown on the land while minimizing the amount that passes below the root zone.

ANAEROBIC (AN-air-O-bick) ANAEROBIC

A condition in which atmospheric or dissolved oxygen (DO) is *NOT* present in the aquatic (water) environment.

ASPIRATE (AS-per-rate) ASPIRATE

Use of a hydraulic device (aspirator or eductor) to create a negative pressure (suction) by forcing a liquid through a restriction, such as a Venturi tube. An aspirator may be used in the laboratory in place of a vacuum pump; sometimes used instead of a sump pump.

BAFFLE BAFFLE

A flat board or plate, deflector, guide, or similar device constructed or placed in flowing water, wastewater, or slurry systems to cause more uniform flow velocities, to absorb energy, and to divert, guide, or agitate liquids (water, chemical solutions, slurry).

BIOSOLIDS BIOSOLIDS

A primarily organic solid product produced by wastewater treatment processes that can be beneficially recycled. The word biosolids is replacing the word sludge when referring to treated waste.

BLINDING BLINDING

The clogging of the filtering medium of a microscreen or a vacuum filter when the holes or spaces in the media become clogged or sealed off due to a buildup of grease or the material being filtered.

BOUND WATER BOUND WATER

Water contained within the cell mass of sludges or strongly held on the surface of colloidal particles. One of the causes of bulking sludge in the activated sludge process.

BULKING BULKING

Clouds of billowing sludge that occur throughout secondary clarifiers and sludge thickeners when the sludge does not settle properly. In the activated sludge process, bulking is usually caused by filamentous bacteria or bound water.

CAVITATION (kav-uh-TAY-shun) CAVITATION

The formation and collapse of a gas pocket or bubble on the blade of an impeller or the gate of a valve. The collapse of this gas pocket or bubble drives water into the impeller or gate with a terrific force that can knock metal particles off and cause pitting on the impeller or gate surface. Cavitation is accompanied by loud noises that sound like someone is pounding on the impeller or gate with a hammer.

CENTRATE CENTRATE

The water leaving a centrifuge after most of the solids have been removed.

CENTRIFUGE CENTRIFUGE

A mechanical device that uses centrifugal or rotational forces to separate solids from liquids.

COAGULATION (ko-agg-yoo-LAY-shun)   COAGULATION

The clumping together of very fine particles into larger particles (floc) caused by the use of chemicals (coagulants). The chemicals neutralize the electrical charges of the fine particles, allowing them to come closer and form larger clumps.

CONING   CONING

Development of a cone-shaped flow of liquid, like a whirlpool, through sludge. This can occur in a sludge hopper during sludge withdrawal when the sludge becomes too thick. Part of the sludge remains in place while liquid rather than sludge flows out of the hopper. Also called coring.

DENITRIFICATION (dee-NYE-truh-fuh-KAY-shun)   DENITRIFICATION

(1) The anoxic biological reduction of nitrate nitrogen to nitrogen gas.

(2) The removal of some nitrogen from a system.

(3) An anoxic process that occurs when nitrite or nitrate ions are reduced to nitrogen gas and nitrogen bubbles are formed as a result of this process. The bubbles attach to the biological floc and float the floc to the surface of the secondary clarifiers. This condition is often the cause of rising sludge observed in secondary clarifiers or gravity thickeners. Also see NITRIFICATION.

DENSITY   DENSITY

A measure of how heavy a substance (solid, liquid, or gas) is for its size. Density is expressed in terms of weight per unit volume, that is, grams per cubic centimeter or pounds per cubic foot. The density of water (at 4°C or 39°F) is 1.0 gram per cubic centimeter or about 62.4 pounds per cubic foot.

DEWATERABLE   DEWATERABLE

This is a property of sludge related to the ability to separate the liquid portion from the solid, with or without chemical conditioning. A material is considered dewaterable if water will readily drain from it.

EDUCTOR (e-DUCK-ter)   EDUCTOR

A hydraulic device used to create a negative pressure (suction) by forcing a liquid through a restriction, such as a Venturi. An eductor or aspirator (the hydraulic device) may be used in the laboratory in place of a vacuum pump. As an injector, it is used to produce vacuum for chlorinators. Sometimes used instead of a suction pump.

ELUTRIATION (e-LOO-tree-A-shun)   ELUTRIATION

The washing of digested sludge with either fresh water, plant effluent, or other wastewater. The objective is to remove (wash out) fine particulates and/or the alkalinity in sludge. This process reduces the demand for conditioning chemicals and improves settling or filtering characteristics of the solids.

ENDOGENOUS (en-DODGE-en-us) RESPIRATION   ENDOGENOUS RESPIRATION

A situation in which living organisms oxidize some of their own cellular mass instead of new organic matter they adsorb or absorb from their environment.

FILAMENTOUS (fill-uh-MEN-tuss) ORGANISMS   FILAMENTOUS ORGANISMS

Organisms that grow in a thread or filamentous form. Common types are *Thiothrix* and *Actinomycetes*. A common cause of sludge bulking in the activated sludge process.

FLOCCULATION (flock-yoo-LAY-shun)   FLOCCULATION

The gathering together of fine particles after coagulation to form larger particles by a process of gentle mixing. This clumping together makes it easier to separate the solids from the water by settling, skimming, draining, or filtering.

GASIFICATION (gas-uh-fuh-KAY-shun)   GASIFICATION

The conversion of soluble and suspended organic materials into gas during aerobic or anaerobic decomposition. In clarifiers, the resulting gas bubbles can become attached to the settled sludge and cause large clumps of sludge to rise and float on the water surface. In anaerobic sludge digesters, this gas is collected for fuel or disposed of using a waste gas burner.

GROWTH RATE, Y

An experimentally determined constant to estimate the unit growth rate of bacteria while degrading organic wastes.

INCINERATION

The conversion of dewatered wastewater solids by combustion (burning) to ash, carbon dioxide, and water vapor.

INORGANIC WASTE

Waste material such as sand, salt, iron, calcium, and other mineral materials that are only slightly affected by the action of organisms. Inorganic wastes are chemical substances of mineral origin; whereas organic wastes are chemical substances usually of animal or plant origin. Also see NONVOLATILE MATTER, ORGANIC WASTE, and VOLATILE SOLIDS.

LYSIMETER (lie-SIM-uh-ter)

A device containing a mass of soil and designed to permit the measurement of water draining through the soil.

NITRIFICATION (NYE-truh-fuh-KAY-shun)

An aerobic process in which bacteria change the ammonia and organic nitrogen in water or wastewater into oxidized nitrogen (usually nitrate).

NITRIFYING BACTERIA

Bacteria that change ammonia and organic nitrogen into oxidized nitrogen (usually nitrate).

NONVOLATILE MATTER

Material such as sand, salt, iron, calcium, and other mineral materials that are only slightly affected by the actions of organisms and are not lost on ignition of the dry solids at 550°C (1,022°F). Volatile materials are chemical substances usually of animal or plant origin. Also see INORGANIC WASTE and VOLATILE SOLIDS.

POLYELECTROLYTE (POLY-ee-LECK-tro-lite)

A high-molecular-weight (relatively heavy) substance, having points of positive or negative electrical charges, that is formed by either natural or synthetic (manmade) processes. Natural polyelectrolytes may be of biological origin or obtained from starch products or cellulose derivatives. Synthetic polyelectrolytes consist of simple substances that have been made into complex, high-molecular-weight substances. Used with other chemical coagulants to aid in binding small suspended particles to larger chemical flocs for their removal from water. Often called a POLYMER.

POLYMER (POLY-mer)

A long-chain molecule formed by the union of many monomers (molecules of lower molecular weight). Polymers are used with other chemical coagulants to aid in binding small suspended particles to larger chemical flocs for their removal from water. Also see POLYELECTROLYTE.

POLYSACCHARIDE (poly-SAC-uh-ride)

A carbohydrate, such as starch or cellulose, composed of chains of simple sugars.

PRECOAT

Application of a free-draining, noncohesive material, such as diatomaceous earth, to a filtering medium. Precoating reduces the frequency of media washing and facilitates cake discharge.

PROTEINACEOUS (PRO-ten-NAY-shus)

Materials containing proteins, which are organic compounds containing nitrogen.

PUG MILL

A mechanical device with rotating paddles or blades that is used to mix and blend different materials together.

PUTRESCIBLE (pyoo-TRES-uh-bull)

Material that will decompose under anaerobic conditions and produce nuisance odors.

RABBLING RABBLING

The process of moving or plowing the material inside a furnace by using the center shaft and rabble arms.

RISING SLUDGE RISING SLUDGE

Rising sludge occurs in the secondary clarifiers of activated sludge plants when the sludge settles to the bottom of the clarifier, is compacted, and then starts to rise to the surface, usually as a result of denitrification, or anaerobic biological activity that produces carbon dioxide and/or methane.

SCFM SCFM

Standard Cubic Feet per Minute. Cubic feet of air per minute at standard conditions of temperature, pressure, and humidity (0°C, 14.7 psia, and 50 percent relative humidity).

SECONDARY TREATMENT SECONDARY TREATMENT

A wastewater treatment process used to convert dissolved or suspended materials into a form more readily separated from the water being treated. Usually, the process follows primary treatment by sedimentation. The process commonly is a type of biological treatment followed by secondary clarifiers that allow the solids to settle out from the water being treated.

SEPTIC (SEP-tick) or SEPTICITY SEPTIC or SEPTICITY

A condition produced by bacteria when all oxygen supplies are depleted. If severe, the bottom deposits produce hydrogen sulfide, the deposits and water turn black, give off foul odors, and the water has a greatly increased oxygen and chlorine demand.

SHORT-CIRCUITING SHORT-CIRCUITING

A condition that occurs in tanks or basins when some of the flowing water entering a tank or basin flows along a nearly direct pathway from the inlet to the outlet. This is usually undesirable since it may result in shorter contact, reaction, or settling times in comparison with the theoretical (calculated) or presumed detention times.

SLUDGE/VOLUME (S/V) RATIO SLUDGE/VOLUME (S/V) RATIO

The volume of sludge blanket divided by the daily volume of sludge pumped from the thickener.

SLURRY SLURRY

A watery mixture or suspension of insoluble (not dissolved) matter; a thin, watery mud or any substance resembling it (such as a grit slurry or a lime slurry).

SPECIFIC GRAVITY SPECIFIC GRAVITY

(1) Weight of a particle, substance, or chemical solution in relation to the weight of an equal volume of water. Water has a specific gravity of 1.000 at 4°C (39°F). Particulates with specific gravity less than 1.0 float to the surface and particulates with specific gravity greater than 1.0 sink.

(2) Weight of a particular gas in relation to the weight of an equal volume of air at the same temperature and pressure (air has a specific gravity of 1.0). Chlorine gas has a specific gravity of 2.5.

STABILIZATION STABILIZATION

Conversion to a form that resists change. Organic material is stabilized by bacteria that convert the material to gases and other relatively inert substances. Stabilized organic material generally will not give off obnoxious odors.

STRUVITE (STREW-vite) STRUVITE

A deposit or precipitate of magnesium ammonium phosphate hexahydrate found on the rotating components of centrifuges and centrate discharge lines. Struvite can be formed when anaerobic sludge comes in contact with spinning centrifuge components rich in oxygen in the presence of microbial activity. Struvite can also be formed in digested sludge lines and valves in the presence of oxygen and microbial activity. Struvite can form when the pH level is between 5 and 9.

THERMOPHILIC (thur-moe-FILL-ick) BACTERIA THERMOPHILIC BACTERIA

A group of bacteria that grow and thrive in temperatures above 113°F (45°C). The optimum temperature range for these bacteria in anaerobic decomposition is 120°F (49°C) to 135°F (57°C). Aerobic thermophilic bacteria thrive between 120°F (49°C) and 158°F (70°C).

VECTOR

VECTOR

An insect or other organism capable of transmitting germs or other agents of disease.

VOLATILE SOLIDS

VOLATILE SOLIDS

Those solids in water, wastewater, or other liquids that are lost on ignition of the dry solids at 550°C (1,022°F). Also called organic solids and volatile matter.

Y, GROWTH RATE

Y, GROWTH RATE

An experimentally determined constant to estimate the unit growth rate of bacteria while degrading organic wastes.

CHAPTER 3. RESIDUAL SOLIDS MANAGEMENT

(Lesson 1 of 5 Lessons)

3.0 NEED FOR SOLIDS HANDLING AND DISPOSAL

3.00 Sludge Types and Characteristics

Solids removal processes in municipal and industrial wastewater treatment plants generate wastestreams containing grit, screenings, scum (floatable materials), and sludge (Figure 3.1). Of these wastestreams, sludge is by far the largest in volume. The handling and disposal of residual sludge and other residual solids represents one of the most challenging problems wastewater treatment plant operators must solve. Treating and disposing of sludge is complicated by the very nature of sludge: (1) sludge is composed largely of the substances responsible for the offensive character of untreated wastewater, (2) only a small portion of the sludge is solid matter, and (3) the response of similar sludge types to various handling techniques differs from one treatment plant to the next.

Some general statements can be made about the reaction of similar sludge types to a specific unit process, but the operator should be aware that the exact response of a particular sludge depends on the plant location and on a large number of variables.

This chapter is designed to familiarize the operator with: (1) the general characteristics of wastewater treatment plant sludges, (2) the unit processes used to effectively handle sludge, and (3) the operating guidelines necessary for successful unit process performance.

Basically, two types of sludges could be produced at any secondary wastewater treatment facility. They are classified as primary and secondary sludges.

Primary sludge includes all those solids that are physically removed from the wastestream by such processes as primary sedimentation or, in some cases, rotary screening. For the purpose of this chapter, primary sludges are defined as all suspended solids that are removed from the wastestream and that are not a by-product of biological removal of organic matter. Generally speaking, primary sludge solids are usually fairly coarse and fibrous, have *SPECIFIC GRAVITIES*[1] or *DENSITIES*[2] significantly greater than that of water, and could be composed of 40 to 80 percent *VOLATILE* (organic) *SOLIDS*.[3] The remaining 20 to 60 percent of the sludge solids are classified as *NONVOLATILE* (inorganic) *MATTER*.[4]

In the biological treatment processes, new bacterial cells are produced as the bacteria feed on and degrade organic matter. In order to maintain the desired quantity or population of bacteria within the biological treatment system, some of these new bacterial cells have to be removed from the process stream. Usually, these bacterial cells are removed in the secondary clarifier. The biological solids or bacterial cells that are removed are called secondary sludge. Generally, secondary sludges are more flocculant than primary sludge solids, less fibrous, have specific gravities closer to that of water, and consist of 75 to 80 percent volatile (organic) matter and 20 to 25 percent nonvolatile (inorganic) material.

The solids removed from wastewater may either be disposed of or reused. When wastewater treatment processes produce a sludge or solid waste product that is suitable for some further beneficial use, the material is commonly referred to as "biosolids" rather than simply "sludge" or "solids."

3.01 Sludge Quantities

The daily quantity of primary and secondary sludges removed will vary from one treatment plant to the next. Before sludge handling equipment can be designed, purchased, and installed, estimates have to be made to determine the daily quantity of sludge removed from the system. Although the design

---

[1] *Specific Gravity.* (1) Weight of a particle, substance, or chemical solution in relation to the weight of an equal volume of water. Water has a specific gravity of 1.000 at 4°C (39°F). Particulates with specific gravity less than 1.0 float to the surface and particulates with specific gravity greater than 1.0 sink. (2) Weight of a particular gas in relation to the weight of an equal volume of air at the same temperature and pressure (air has a specific gravity of 1.0). Chlorine gas has a specific gravity of 2.5.

[2] *Density.* A measure of how heavy a substance (solid, liquid, or gas) is for its size. Density is expressed in terms of weight per unit volume, that is, grams per cubic centimeter or pounds per cubic foot. The density of water (at 4°C or 39°F) is 1.0 gram per cubic centimeter or about 62.4 pounds per cubic foot.

[3] *Volatile Solids.* Those solids in water, wastewater, or other liquids that are lost on ignition of the dry solids at 550°C (1,022°F). Also called organic solids and volatile matter.

[4] *Nonvolatile Matter.* Material such as sand, salt, iron, calcium, and other mineral materials that are only slightly affected by the actions of organisms and are not lost on ignition of the dry solids at 550°C (1,022°F). Volatile materials are chemical substances usually of animal or plant origin. Also see INORGANIC WASTE and VOLATILE SOLIDS.

TREATMENT PROCESS

FUNCTION

PRELIMINARY TREATMENT

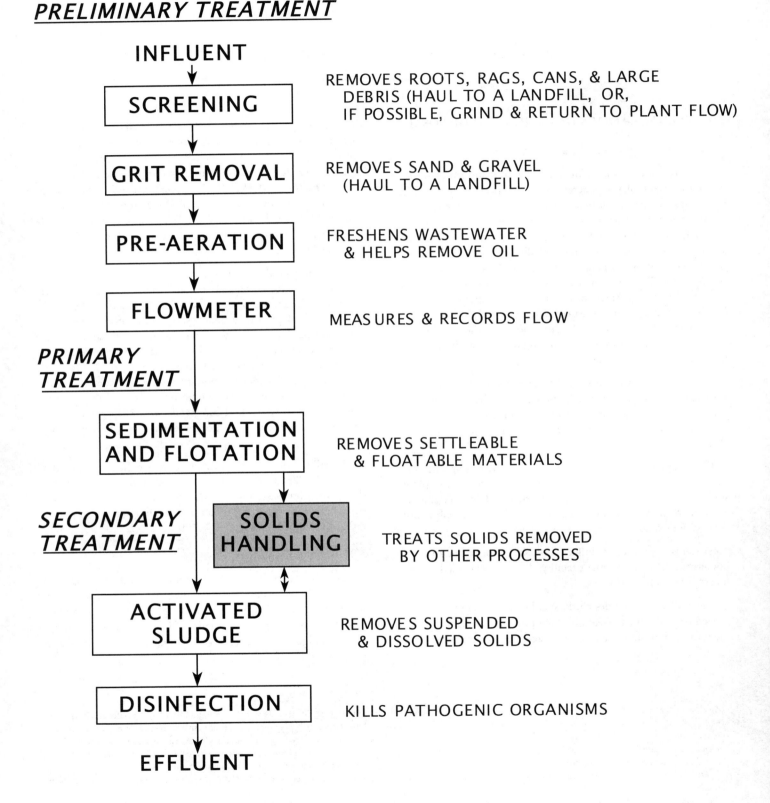

INFLUENT

| SCREENING |
REMOVES ROOTS, RAGS, CANS, & LARGE DEBRIS (HAUL TO A LANDFILL, OR, IF POSSIBLE, GRIND & RETURN TO PLANT FLOW)

| GRIT REMOVAL |
REMOVES SAND & GRAVEL (HAUL TO A LANDFILL)

| PRE-AERATION |
FRESHENS WASTEWATER & HELPS REMOVE OIL

| FLOWMETER |
MEASURES & RECORDS FLOW

PRIMARY TREATMENT

| SEDIMENTATION AND FLOTATION |
REMOVES SETTLEABLE & FLOATABLE MATERIALS

SECONDARY TREATMENT

| SOLIDS HANDLING |
TREATS SOLIDS REMOVED BY OTHER PROCESSES

| ACTIVATED SLUDGE |
REMOVES SUSPENDED & DISSOLVED SOLIDS

| DISINFECTION |
KILLS PATHOGENIC ORGANISMS

EFFLUENT

Fig. 3.1   Typical flow diagram of a wastewater treatment plant and installation of such equipment is the engineer's responsibility, the operator should be aware of how engineers make these estimates.

3.010 Primary Sludge Production

The quantity of primary sludge generated depends on: (1) the influent wastewater flow, (2) the concentration of influent settleable suspended solids, and (3) the *EFFICIENCY*[5] of the primary sedimentation basin. The estimation of primary sludge production is illustrated in the following example.

EXAMPLE 1   PRIMARY SLUDGE PRODUCTION

Given: The influent flow to a primary clarifier is 1.5 MGD. The influent suspended solids concentration is 350 mg/L. The effluent dry suspended solids concentration from the primary clarifier is 150 mg/L.

Find: 1. The total pounds of dry suspended solids entering the plant per day (lbs SS/day).

   2. The total pounds of dry suspended solids leaving the primary clarifier with the primary effluent (lbs SS/day).

   3. The total pounds of dry (no moisture present) sludge solids produced per day.

Solution:

Known	Unknown
Flow, MGD = 1.5 MGD	1. Dry SS Entering Plant, lbs/day
Infl SS, mg/L = 350 mg/L	2. Dry SS Leaving Primary Clarifier, lbs/day
Effl SS, mg/L = 150 mg/L	3. Dry Sludge Solids Produced, lbs/day

1. Calculate the amount of dry influent suspended solids, lbs/day.

$$\text{Infl Susp Sol, lbs/day} = \text{Flow, MGD} \times \text{Susp Sol, mg}/L \times 8.34 \text{ lbs/gal}$$
$$= 1.5 \text{ MGD} \times 350 \text{ mg}/L \times 8.34 \text{ lbs/gal}$$
$$= 4,379 \text{ lbs/day}$$

2. Calculate the amount of dry suspended solids leaving the primary clarifier, lbs/day.

$$\text{Prim Clar Effl Susp Sol, lbs/day} = \text{Flow, MGD} \times \text{Susp Sol, mg}/L \times 8.34 \text{ lbs/gal}$$
$$= 1.5 \text{ MGD} \times 150 \text{ mg}/L \times 8.34 \text{ lbs/gal}$$
$$= 1,877 \text{ lbs/day}$$

3. Calculate the amount of dry primary sludge produced, lbs/day.

$$\text{Primary Sludge, lbs/day} = \text{Infl Susp Sol, lbs/day} - \text{Effl Susp Sol, lbs/day}$$
$$= 4,379 \text{ lbs/day} - 1,877 \text{ lbs/day}$$
$$= 2,502 \text{ lbs/day}$$

OR

$$\text{Primary Sludge, lbs/day} = \text{Flow, MGD} \times (\text{Infl SS, mg}/L - \text{Effl SS, mg}/L) \times 8.34 \text{ lbs/gal}$$
$$= 1.5 \text{ MGD} \times (350 \text{ mg}/L - 150 \text{ mg}/L) \times 8.34 \text{ lbs/gal}$$
$$= 2,502 \text{ lbs/day}$$

NOTE: All answers are in terms of pounds of dry solids per day.

3.011 Secondary Sludge Production

The daily quantity of sludge produced by a secondary treatment system depends on many factors, the most influential of which are: (1) the influent flow to the biological or secondary system, (2) the influent organic load to the biological system, (3) the efficiency of the biological system in removing organic matter, and (4) the growth rate, Y, of the bacteria within the system. The determination of secondary sludge production is rather complicated due to the mathematics and variables involved. The rate of biological growth, Y, is highly dependent on such variables as temperature, nutrient balances, the amount of oxygen supplied to the system, the ratio between the amount of food supplied (BOD) and the mass or quantity of biological cells developed within the system, detention time, and other factors. A detailed discussion of the estimation of growth rates and sludge production is beyond the scope of this chapter. A general rule of thumb that operators may use to estimate secondary sludge production is that for every pound of organic matter (soluble 5-day BOD) used by the bacterial cells, approximately 0.30 to 0.70 pound of new bacterial cells are produced and have to be taken

---

[5] *Efficiency, % =* $\dfrac{(\textit{Influent Suspended Solids, mg/L} - \textit{Effluent Suspended Solids, mg/L})100\%}{\textit{Influent Suspended Solids, mg/L}}$

out of the system. The following example illustrates the estimation of secondary sludge production.

EXAMPLE 2   SECONDARY SLUDGE PRODUCTION

Given: The primary effluent organic content, as measured by the 5-day BOD test, to a secondary treatment facility is 200 mg/L. The secondary effluent 5-day BOD is 30 mg/L. The bacterial growth rate, Y, is 0.50 lb SS/lb BOD removed and the flow rate, Q, is 1.5 MGD.

NOTE: The secondary influent flow in your plant may be different from the actual plant effluent flow due to in-plant recycle uses of effluent or secondary system streams.

Find: 1. The total pounds of BOD entering the secondary system per day.

2. The total pounds of BOD leaving the secondary system with the effluent per day.

3. The total pounds of BOD removed per day by the secondary system.

4. The total dry pounds of secondary sludge produced per day by the secondary system.

Solution:

Known		Unknown
Flow, MGD	= 1.5 MGD	1. BOD Entering, lbs/day
Infl BOD, mg/L	= 200 mg/L	2. BOD Leaving, lbs/day
Effl BOD, mg/L	= 30 mg/L	3. BOD Removed, lbs/day
$\dfrac{\text{Y, lbs Sl Sol Prod}}{\text{lb BOD Rem}}$	= $\dfrac{\text{0.50 lb Sl Sol}}{\text{lb BOD}}$	4. Sludge Prod, lbs/day

1. Determine the 5-day BOD entering the secondary system, lbs BOD/day.

$$\frac{\text{Entering BOD,}}{\text{lbs BOD/day}} = \text{Flow, MGD} \times \text{BOD, mg/}L \times 8.34 \text{ lbs/gal}$$

$$= 1.5 \text{ MGD} \times 200 \text{ mg/}L \times 8.34 \text{ lbs/gal}$$

$$= 2,502 \text{ lbs BOD/day}$$

2. Determine the 5-day BOD leaving the secondary system, lbs BOD/day.

$$\frac{\text{Leaving BOD,}}{\text{lbs BOD/day}} = \text{Flow, MGD} \times \text{BOD, mg/}L \times 8.34 \text{ lbs/gal}$$

$$= 1.5 \text{ MGD} \times 30 \text{ mg/}L \times 8.34 \text{ lbs/gal}$$

$$= 375 \text{ lbs BOD/day}$$

3. Determine the 5-day BOD removed from the secondary system, lbs BOD/day.

$$\frac{\text{BOD Removed,}}{\text{lbs BOD/day}} = \frac{\text{Entering BOD,}}{\text{lbs BOD/day}} - \frac{\text{Leaving BOD,}}{\text{lbs BOD/day}}$$

$$= 2,502 \text{ lbs BOD/day} - 375 \text{ lbs BOD/day}$$

$$= 2,127 \text{ lbs BOD/day}$$

OR

$$\frac{\text{BOD Removed,}}{\text{lbs BOD/day}} = \frac{\text{Flow, MGD} \times (\text{Infl BOD, mg/}L - \text{Effl BOD, mg/}L)}{\times 8.34 \text{ lbs/gal}}$$

$$= 1.5 \text{ MGD} \times (200 \text{ mg/}L - 30 \text{ mg/}L) \times 8.34 \text{ lbs/gal}$$

$$= 2,127 \text{ lbs BOD/day}$$

4. Determine the secondary sludge produced in terms of pounds of dry sludge solids per day.

$$\frac{\text{Sludge Produced,}}{\text{lbs Dry Solids/day}} = \frac{\text{BOD Removed,}}{\text{lbs BOD/day}} \times \text{Y,} \frac{\text{lbs Sl Sol Prod/day}}{\text{lb BOD Rem/day}}$$

$$= 2,127 \text{ lbs BOD/day} \times \frac{0.50 \text{ lb Sl Sol/day}}{1 \text{ lb BOD/day}}$$

$$= 1,064 \text{ lbs Dry Sludge Solids/day}$$

3.02   Sludge Volumes

Examples 1 and 2 illustrated how to estimate the quantity or pounds of primary and secondary sludge solids, respectively. The total volume or gallons of primary and secondary sludges are equally as important for sizing sludge handling equipment. Sludge volumes in gallons are determined by the sludge solids content (% SS) and the pounds of solids in a sludge sample according to the following equation:

$$\boxed{\text{Sludge Volume, gal} = \frac{\text{Sludge Quantity, lbs Dry Solids}}{8.34 \text{ lbs/gal} \times \text{Sludge Solids, \%/100\%}}}$$

If the primary sludge from Example 1 is withdrawn from the primary clarifier at a sludge solids concentration (% Sl Sol) of 5 percent, the daily volume of sludge would be determined as follows:

Solution:

Known		Unknown
Primary Sludge Quantity, lbs/day	= 2,502 lbs/day	Primary Sludge Volume, gal/day
Sludge Solids, %	= 5%	

Determine the daily primary sludge volume in gallons per day.

$$\frac{\text{Primary Sludge Volume,}}{\text{gal/day}} = \frac{\text{Sludge Quantity, lbs Dry Solids/day}}{8.34 \text{ lbs/gal} \times \text{Sludge Solids, \%/100\%}}$$

$$= \frac{2,502 \text{ lbs/day}}{8.34 \text{ lbs/gal} \times 5\%/100\%}$$

$$= \frac{2,502}{8.34 \times 0.05}$$

$$= 6,000 \text{ gal/day}$$

Likewise, if the secondary sludge from Example 2 is withdrawn from the secondary clarifier at a sludge solids concentration (% Sl Sol) of 1.0 percent, the daily volume of sludge would be determined as follows:

Solution:

Known		Unknown
Secondary Sludge Quantity, lbs/day	= 1,064 lbs/day	Secondary Sludge Volume, gal/day
Sludge Solids, %	= 1.0%	

Determine the daily secondary sludge volume in gallons per day.

$$\text{Secondary Sludge Volume, gal/day} = \frac{\text{Sludge Quantity, lbs Dry Solids/day}}{8.34 \text{ lbs/gal} \times \text{Sludge Solids, \%}/100\%}$$

$$= \frac{1,064 \text{ lbs/day}}{8.34 \text{ lbs/gal} \times 1.0\%/100\%}$$

$$= 12,758 \text{ gal/day}$$

3.03 Sludge Handling Alternatives

Depending on the type and quantity of sludge produced, a variety of unit processes and overall sludge handling systems can be used to process the sludge. The chart presented in Table 3.1 illustrates these sludge processing alternatives. The remainder of this chapter will be divided into separate sections on thickening, stabilization, conditioning, dewatering, volume reduction, and ultimate disposal of solids.

QUESTIONS

Write your answers in a notebook and then compare your answers with those on pages 305 and 306.

3.0A List the two types and general characteristics of sludges that are produced at a typical wastewater treatment facility.

3.0B What are biosolids?

3.0C List the variables that govern the quantity of primary sludge production.

3.0D Determine the daily quantity (lbs/day) of primary sludge produced for the following conditions: (1) influent flow of 2.0 MGD, (2) influent suspended solids of 200 mg/L, and (3) primary effluent suspended solids of 120 mg/L.

3.0E List the variables that influence the production of secondary sludges.

3.0F Estimate the daily quantity of secondary sludge produced for the following conditions: (1) influent flow of 2.0 MGD, (2) influent BOD to the secondary system of 180 mg/L and effluent from the secondary system of 30 mg/L, and (3) growth rate coefficient, Y, of 0.50 pound of solids per pound of BOD removed.

3.0G For the conditions given in problem 3.0D, estimate the daily volume (gal/day) of primary sludge if it is withdrawn from the primary clarifier at a sludge solids concentration of 4.0 percent.

3.1 THICKENING

3.10 Purpose of Sludge Thickening

Settled solids removed from the bottom of the primary clarifier (primary sludge) and settled biological solids removed from the bottom of secondary clarifiers (secondary sludge) contain large volumes of water. Typically, primary sludge contains approximately 95 to 97 percent water. For every pound of primary solids, there are 20 to 30 pounds of water and for every pound of secondary solids, approximately 50 to 150 pounds of water are incorporated in the sludge mass. If some of the water is not removed from the sludge mass, the size of subsequent sludge handling equipment (digester, mechanical dewatering equipment, pumps) have to be larger to handle the greater volumes and this would obviously increase equipment costs.

Concentration or thickening is usually the first step in a sludge processing system following initial separation of solids by

TABLE 3.1 SLUDGE PROCESSING ALTERNATIVES

	Thickening	Stabilization	Conditioning	Dewatering	Volume Reduction	Disposal
OBJECTIVE	Remove water from the sludge mass	Convert odor-causing portion of sludge solids to nonodorous end products	Pretreat sludge to facilitate removal of water in subsequent treatment processes	Reduce sludge moisture and volume to allow economical disposal	Reduce sludge mass prior to ultimate disposal	Ultimate disposal
SPECIFIC PROCESSES	1. Gravity 2. Flotation 3. Centrifugation 4. Filtration	1. Digestion 2. Thermal 3. Chemical	1. Chemical 2. Thermal 3. Elutriation	1. Filtration 2. Centrifugation 3. Drying Beds	1. Drying 2. Incineration 3. Composting	1. Land 2. Ocean 3. Air

sedimentation from the wastewater being treated. Maximum sludge thickening should always be attempted in the sedimentation tank before using a separate sludge thickener. The primary function of sludge thickening is to reduce the sludge volume to be handled in subsequent processes. The advantages normally associated with sludge thickening include: (1) improved digester performance due to a smaller volume of sludge, (2) construction cost savings for new digestion facilities due to smaller sludge volumes treated, and (3) a reduction in digester heating requirements because less water has to be heated. Also, reduced sludge volumes result in smaller facilities for storing, blending, dewatering, and incinerating or disposing of the sludge. The following example illustrates the reduction in sludge volume when a sludge is thickened.

EXAMPLE 3

Given: A primary sludge is withdrawn from a primary clarifier at a sludge solids concentration of 3.0 percent. The volume of sludge withdrawn is 2,000 gallons per day.

Find: 1. The amount of primary sludge solids withdrawn in pounds per day.

2. If the sludge is concentrated (thickened) to 5.0 percent sludge solids, find the new sludge volume.

Solution:

Known	Unknown
Sludge Solids, % = 3.0%	1. Amount of Dry Sludge, lbs/day
Sludge Vol, gal/day = 2,000 gal/day	2. Thickened Sludge Volume, gal/day

1. Determine the amount of primary dry sludge withdrawn in pounds per day.

$$\text{Dry Sludge Solids,} \atop \text{lbs/day} = \text{Sludge Vol,} \frac{\text{gal}}{\text{day}} \times \frac{8.34 \text{ lbs}}{\text{gal}} \times \text{Sl Sol,} \frac{\%}{100\%}$$

$$= 2,000 \frac{\text{gal}}{\text{day}} \times \frac{8.34 \text{ lbs}}{\text{gal}} \times \frac{3.0\%}{100\%}$$

$$= 500 \text{ lbs/day}$$

2. Calculate the new thickened sludge volume in gallons per day.

$$\text{Sludge Volume,} \atop \text{gal/day} = \frac{\text{Dry Sludge Solids, lbs/day}}{8.34 \text{ lbs/gal} \times \text{Sl Sol, \%}/100\%}$$

$$= \frac{500 \text{ lbs/day}}{8.34 \text{ lbs/gal} \times 5.0\%/100\%}$$

$$= 1,200 \text{ gal/day}$$

Although the pounds of sludge solids remained constant at 500 lbs/day, the total volume of sludge decreased from 2,000 gal/day to 1,200 gal/day when the sludge was thickened from 3 percent sludge solids to 5 percent sludge solids. Four unit processes commonly used to concentrate wastewater sludges include gravity thickening, dissolved air flotation (DAF) thickeners, centrifuge thickeners, and gravity belt thickeners. A discussion of each process is contained in this section.

QUESTIONS

Write your answers in a notebook and then compare your answers with those on page 306.

3.10A   What is the primary function of sludge thickening?

3.10B   Determine the amount of dry sludge (lbs/day) if 12,000 gal/day of secondary sludge are produced with a solids concentration of 1.0%.

3.10C   For the conditions given in problem 3.10B, determine the secondary sludge volume (gal/day) if the sludge is withdrawn from the secondary clarifier at a solids concentration of 1.5%.

3.11   Gravity Thickening

Gravity thickening of wastewater sludges uses gravity forces to separate solids from the sludges being treated. Those solids that are heavier than the water settle to the bottom of the thickener by virtue of gravity and are then compacted by the weight of the overlying solids. Gravity concentrators or thickeners are typically circular in design and resemble circular clarifiers. The main components of gravity thickeners, as shown in Figure 3.2, are: (1) the inlet and distribution assembly, (2) a sludge rake to move the sludge to a sludge hopper, (3) vertical steel members or "pickets" mounted on the sludge rake, (4) an effluent or overflow weir, and (5) scum removal equipment.

The inlet or distribution assembly usually consists of a circular steel skirt or *BAFFLE*[6] that originates above the water surface and extends downward approximately 2 to 3 feet (0.6 to 0.9 m) below the water surface. The sludge to be thickened enters the assembly and flows downward under the steel skirt and through the tank where the solids settle to the bottom. The inlet assembly provides for an even distribution of sludge throughout the tank and reduces the possibility of *SHORT-CIRCUITING*[7] to the effluent end of the thickener.

---

[6] *Baffle.*   A flat board or plate, deflector, guide, or similar device constructed or placed in flowing water, wastewater, or slurry systems to cause more uniform flow velocities, to absorb energy, and to divert, guide, or agitate liquids (water, chemical solutions, slurry).

[7] *Short-Circuiting.*   A condition that occurs in tanks or basins when some of the flowing water entering a tank or basin flows along a nearly direct pathway from the inlet to the outlet. This is usually undesirable since it may result in shorter contact, reaction, or settling times in comparison with the theoretical (calculated) or presumed detention times.

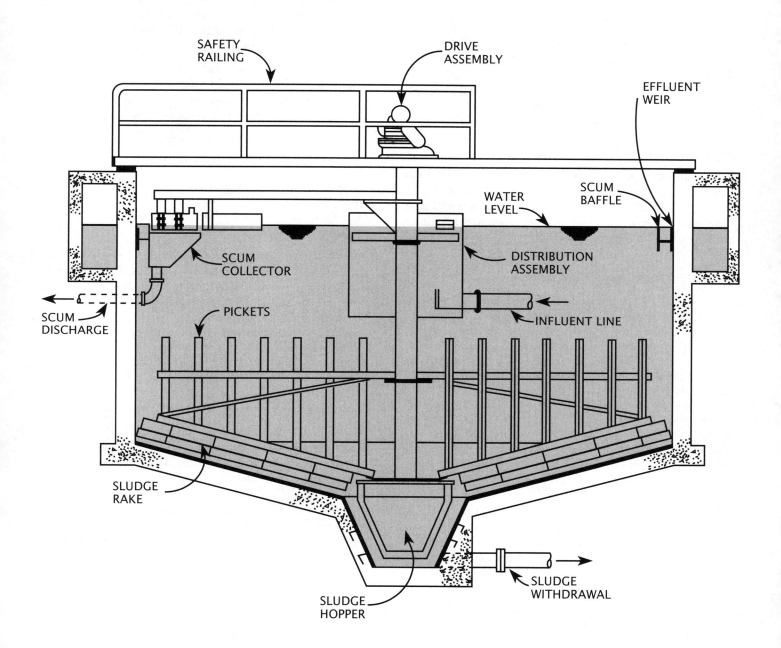

SAFETY RAILING

DRIVE ASSEMBLY

EFFLUENT WEIR

WATER LEVEL

SCUM BAFFLE

SCUM COLLECTOR

DISTRIBUTION ASSEMBLY

SCUM DISCHARGE

PICKETS

INFLUENT LINE

SLUDGE RAKE

SLUDGE HOPPER

SLUDGE WITHDRAWAL

Fig. 3.2   Gravity thickener

The sludge rake provides for movement of the settled (thickening) sludge. As the rake slowly rotates, the settled solids are moved to the center of the tank where they are deposited in a sludge hopper. The tank bottom is usually sloped toward the center to facilitate the movement of sludge to the collection point. Typically, sludge pumps are centrifugal recessed-impeller pumps or positive displacement progressive cavity pumps.

The vertical steel members (pickets) that are usually mounted on the sludge rake assembly provide for gentle stirring or flocculation of the settled sludge as the rake rotates. This gentle stirring action serves two purposes. Trapped gases in the sludge are released to prevent RISING[8] of the solids. Also, stirring prevents the accumulation of a large volume of solids (scum) floating on the thickener surface. Scum must be removed to prevent nuisance and odor problems.

The effluent or thickener overflow flows over a continuous weir located on the periphery (outside) of the thickener. The outlet works usually include an effluent baffle to retain floating debris and a scum scraper and collection system to remove these floatables.

3.110 Factors Affecting Gravity Thickeners

The successful operation of gravity thickeners is dependent on the following factors: (1) type of sludge, (2) age of the feed sludge, (3) sludge temperature, (4) sludge blanket depth, (5) solids and hydraulic detention times, and (6) solids and hydraulic loadings. The first three factors deal with the characteristics of the influent sludge while the remaining three factors deal with operational controls.

Both the type and age of sludge to be thickened can have pronounced effects on the overall performance of gravity thickeners. Fresh primary sludge usually can be concentrated to the highest degree. If the primary sludge is septic or allowed to go ANAEROBIC,[9] hydrogen sulfide (H_2S), methane (CH_4), and carbon dioxide (CO_2) gases may be produced (GASIFICATION[10]). If gas is produced, it will attach to sludge particles and carry these solids to the surface. The net effect(s) of gas production due to anaerobic conditions will be reduced thickener efficiency and lower solids concentration.

Secondary sludges are not as well suited for gravity thickening as primary sludge. Secondary sludges contain large quantities of BOUND WATER,[11] which makes the sludge less dense than primary sludge solids. Biological solids are composed of approximately 85 to 90 percent water by weight within the cell mass. The water contained within the cell wall is referred to as "bound water." The fact that biological solids contain large volumes of cell water and are often smaller or finer than primary sludge solids makes them harder to separate by gravity concentration.

The age of the secondary sludge also plays an important role in the efficiency of gravity thickening processes. In the activated sludge process, ammonia is converted to nitrite and nitrate by bacterial metabolism according to the following equations:

Ammonia (NH_3) + Oxygen → Nitrite (NO_2^-) + H_2O

Nitrite (NO_2^-) + Oxygen → Nitrate (NO_3^-)

The conversion of ammonia to nitrite and then nitrate is called "nitrification." This conversion will occur if sufficient oxygen and aeration time are allotted in the activated sludge process to permit the buildup of NITRIFYING BACTERIA.[12]

In the liquid-solids separation section (secondary sedimentation) of activated sludge wastewater treatment plants, the available oxygen in the settled sludge may be depleted to the point where no dissolved oxygen remains. If the sludge is held too long in the final clarifier or gravity thickener and the dissolved oxygen concentration decreases to zero, the nitrate can be converted by bacterial metabolism to nitrogen gas. The conversion of nitrate to nitrogen gas is called DENITRIFICATION.[13] Rising bubbles of nitrogen gas due to denitrification will carry settled solids to the surface of secondary clarifiers or gravity thickeners and will adversely affect process performance.

Another problem occasionally encountered with activated sludge processes is "sludge bulking." If sufficient oxygen is not available in the aeration basin or nutrient imbalances are present, FILAMENTOUS ORGANISMS[14] may grow in the aeration basins. The predominance of these organisms will decrease the settleability of activated sludge and it will not settle as readily in the secondary clarifiers or compact to its highest degree in gravity thickeners. Greater compaction can be achieved by the addition of chemicals.

---

[8] *Rising Sludge.*   Rising sludge occurs in the secondary clarifiers of activated sludge plants when the sludge settles to the bottom of the clarifier, is compacted, and then starts to rise to the surface, usually as a result of denitrification, or anaerobic biological activity that produces carbon dioxide and/or methane.

[9] *Anaerobic* (AN-air-O-bick).   A condition in which atmospheric or dissolved oxygen (DO) is *NOT* present in the aquatic (water) environment.

[10] *Gasification* (gas-uh-fuh-KAY-shun).   The conversion of soluble and suspended organic materials into gas during aerobic or anaerobic decomposition. In clarifiers, the resulting gas bubbles can become attached to the settled sludge and cause large clumps of sludge to rise and float on the water surface. In anaerobic sludge digesters, this gas is collected for fuel or disposed of using a waste gas burner.

[11] *Bound Water.*   Water contained within the cell mass of sludges or strongly held on the surface of colloidal particles. One of the causes of bulking sludge in the activated sludge process.

[12] *Nitrifying Bacteria.*   Bacteria that change ammonia and organic nitrogen into oxidized nitrogen (usually nitrate).

[13] *Denitrification* (dee-NYE-truh-fuh-KAY-shun).   (1) The anoxic biological reduction of nitrate nitrogen to nitrogen gas. (2) The removal of some nitrogen from a system. (3) An anoxic process that occurs when nitrite or nitrate ions are reduced to nitrogen gas and nitrogen bubbles are formed as a result of this process. The bubbles attach to the biological floc and float the floc to the surface of the secondary clarifiers. This condition is often the cause of rising sludge observed in secondary clarifiers or gravity thickeners. Also see NITRIFICATION.

[14] *Filamentous* (fill-uh-MEN-tuss) *Organisms.*   Organisms that grow in a thread or filamentous form. Common types are *Thiothrix* and *Actinomycetes*. A common cause of sludge bulking in the activated sludge process.

Temperature is another sludge characteristic that affects the degree of sludge thickening. As the temperature of the sludge (primary or secondary) increases, the rate of biological activity is increased and the sludge tends to gasify and rise at a faster rate. During summertime (warm weather) operation, the settled sludge has to be removed at a faster rate from the thickener than during wintertime operation. When the sludge temperature is lower during the winter, biological activity and gas production proceed at a slower rate.

Solids and hydraulic loadings and sludge detention times are discussed in the next section, which reviews operating guidelines.

3.111 Operating Guidelines

The size of gravity thickeners is determined and designed by engineers. The operator controls the solids retention (detention) time within the thickener and achieves peak performance by controlling the speed of the sludge collection mechanism (if possible), adjusting the sludge withdrawal rate, and controlling the sludge blanket depth. Successful operation of gravity concentrators requires that the operator be able to calculate applied loading rates and sludge detention (retention) time, and be aware of the available controls.

3.1110 HYDRAULIC AND SOLIDS LOADINGS

The hydraulic surface loading or overflow rate is defined as the total number of gallons applied per square foot of thickener water surface area per day (GPD/sq ft). To calculate hydraulic surface loading, first determine the total water surface area in square feet. Then divide the number of gallons applied per day by the surface area (sq ft). The gallons applied usually include more than just the sludge pumped, because it is better to keep a good high flow of fresh liquid entering the thickener to prevent septic conditions and odors from developing. To accomplish this higher flow, secondary effluent is usually blended with the sludge feed to the thickener. Typical hydraulic loading rates are from 400 to 800 GPD/sq ft (16 to 32 m^3 per day/sq m). For a very thin mixture or for waste activated sludge (WAS) only, hydraulic loading rates of 100 to 200 GPD/sq ft (4 to 8 m^3 per day/sq m) would be appropriate.

EXAMPLE 4

Given: A 20-foot diameter gravity thickener is used to thicken 20 GPM of primary sludge. 80 GPM of secondary effluent is blended with the raw sludge to prevent septic conditions and odors.

Find: The hydraulic surface loading applied to the gravity thickener in GPD per sq ft.

Solution:

Known	Unknown
Thickener Diameter, ft = 20 ft	Hydraulic Surface Loading, GPD/sq ft
Sludge Flow, GPM = 20 GPM	
Blend Flow, GPM = 80 GPM	
Total Flow, GPM = 100 GPM	

Determine the flow in gallons per day and the water surface area in square feet. Calculate the hydraulic surface loading in gallons per day (GPD) per square foot.

$$\text{Hydraulic Surface Loading, GPD/sq ft} = \frac{\text{Total Flow, gal/min} \times 60 \text{ min/hr} \times 24 \text{ hr/day}}{\frac{\pi}{4} \times (\text{Diameter, ft})^2}$$

$$= \frac{100 \text{ gal/min} \times 60 \text{ min/hr} \times 24 \text{ hr/day}}{0.785 \times (20 \text{ ft})^2}$$

$$= \frac{144,000 \text{ GPD}}{314 \text{ sq ft}}$$

$$= 460 \text{ GPD/sq ft}$$

The solids loading is defined as the total number of pounds of solids applied per square foot of thickener surface area per day. To calculate the solids loading, first find the total surface area (sq ft) of the thickener. Next, using the flow rate and solids concentration, calculate the total pounds of solids applied per day. Finally, divide the total solids (lbs/day) by the total surface area (sq ft) to find the solids loading. The proper operating solids loading will vary with the type of sludge. Typical values are discussed in Section 3.113, "Typical Performance."

EXAMPLE 5

Given: A 45-foot diameter gravity thickener is used to thicken 100 GPM of primary sludge. The primary sludge is applied to the thickener at an initial sludge solids concentration of 3.5 percent.

Find: The solids loading (S.L.) applied to the gravity thickener in lbs/day per sq ft.

Solution:

Known	Unknown
Thickener Diameter, ft = 45 ft	Solids Loading, lbs/day/sq ft
Sludge Flow, GPM = 100 GPM	
Sludge Solids, % = 3.5%	

1. Determine the solids applied to the thickener in pounds per day.

$$\text{Solids Applied, lbs/day} = \text{Flow, GPD} \times 8.34 \text{ lbs/gal} \times \text{Solids,} \frac{\%}{100\%}$$

$$= 100 \text{ gal/min} \times 1,440 \text{ min/day} \times 8.34 \text{ lbs/gal} \times \frac{3.5\%}{100\%}$$

$$= 42,034 \text{ lbs/day}$$

2. Calculate the solids loading.

$$\text{Solids Loading, lbs/day/sq ft} = \frac{\text{Solids Applied, lbs/day}}{\text{Surface Area, sq ft}}$$

$$= \frac{42,034 \text{ lbs/day}}{0.785 (45 \text{ ft})^2}$$

$$= 26 \text{ lbs/day/sq ft}$$

3.1111 SLUDGE DETENTION TIME

The sludge detention time is defined as the length of time the solids remain in the gravity thickener. This time is based on the amount of solids applied, the depth and concentration of the sludge blanket, and the quantity of solids removed from the bottom of the thickener. The operator has the ability to control the solids detention time and the degree of thickening to some extent by controlling the depth of the sludge blanket. If the blanket is maintained at too high a level and the solids detention time is excessive, gasification may develop with subsequent rising sludge and deterioration of effluent quality. The actual response of a particular sludge to gravity thickening depends on the treatment plant. Trial-and-error procedures usually determine the best operation. To aid the operator in controlling the detention time of the solids in the thickener, the sludge/volume (S/V) ratio can be used. S/V ratio is defined as the volume of the sludge blanket divided by the daily volume of sludge pumped from the thickener. The S/V ratio is a relative measure of the average detention time of solids in the thickener and is calculated in days. Typical S/V ratio values are between 0.5 and 2.0 days. The higher S/V values are desirable for a maximum sludge concentration; however, to guard against gasification, the lower S/V values are maintained during warm weather.

3.112 Normal Operating Procedures

Typically, the flow through the thickener is continuous and should be controlled to be as constant as possible. Monitoring of the influent, effluent, and concentrated sludge streams should be done at least once per shift and should include collection of samples for later laboratory analysis.

Under normal operating conditions, water at the surface should be relatively CLEAR and free from solids and gas bubbles. The sludge blanket depth is usually kept around 5 to 8 feet (1.5 to 2.4 m). The speed of the sludge collectors should be fast enough to allow the settled solids to move toward the sludge collection sump. The bottom sludge collectors should not be operated at speeds that will disrupt the settled solids and cause them to float to the surface. Sludge withdrawal rates should be sufficient to maintain a constant blanket level.

Normal procedures for starting up and shutting down a gravity thickener are outlined below.

START-UP

- Fill the unit with fresh water or plant effluent.

- Turn on the sludge collectors and scum collection equipment. Let the equipment run long enough to ensure proper operation prior to introducing any sludge.

- Activate chemical conditioning systems, if used.

- Open all appropriate inlet valves.

- Turn on and adjust, if possible, the influent sludge pump.

- Check the sludge blanket depth.

- Open all appropriate sludge withdrawal valves.

- Set the sludge pump in the automatic ON position.

- Routinely check the blanket depth and thickened sludge concentration and adjust the withdrawal rate, as required.

The thickener should be operated continuously. However, if the thickener is not operated as a continuous process and daily or frequent shutdowns are required, the following procedures should be followed:

SHUTDOWN

- Turn off the influent sludge pump and close appropriate inlet valves.

- Turn off the chemical addition equipment, if chemical conditioning is used.

- Allow the scum collection and sludge collection and removal systems to operate until the water surface is free of floating material and settled sludge has been removed from the thickener bottom. Add fresh water or plant effluent to the unit to maintain the appropriate water surface level to accommodate scum removal.

- Turn off the scum collection, sludge collection, and sludge withdrawal equipment.

- Hose down and clean up the area, as required.

3.113 Typical Performance

Typical loadings and thickener output concentrations for various sludge types are summarized in Table 3.2. Note that the data presented in Table 3.2 are generalized and the actual response of a particular sludge at a particular plant may vary significantly.

TABLE 3.2 OPERATIONAL AND PERFORMANCE GUIDELINES FOR GRAVITY THICKENERS

Sludge Type	Solids Loading, lbs/day/sq ft[a]	Thickened Sludge, %
Separate		
Primary	20–30	8–10
Activated Sludge	5–8	2–4
Trickling Filter	8–10	7–9
Combined		
Primary & Act Sl	6–12	4–9
Primary & Trickling Filter	10–20	7–9

[a] lbs/day/sq ft × 4.883 = kg/day/sq m.

In order to rate the performance of gravity thickeners, the operator must know how to calculate process efficiency. The efficiency of any process in removing a particular constituent is determined by the following equation:

$$\text{Efficiency, \%} = \frac{(\text{Influent} - \text{Effluent})}{\text{Influent}} \times 100\%$$

In the case of gravity thickeners, suspended or sludge solids removal is a key performance factor. One of the goals of the operator should be to remove as much of the influent solids as possible. Usually, the supernatant or overflow from the thickener is returned to the plant headworks. If the solids concentration in this stream is high, then you are recirculating solids and can end up "chasing your tail" (having to treat more and more solids). The following example shows how to calculate sludge solids removal efficiency for a gravity thickener.

EXAMPLE 6

Given: A gravity thickener receives 20 GPM of primary sludge at a concentration of 3.0 percent sludge solids (30,000 mg/L). The effluent from the thickener contains 0.15 percent (1,500 mg/L) of sludge solids.

Find: The efficiency in removing sludge solids.

Solution:

Known	Unknown
Gravity Thickener	Thickener Efficiency, %

Flow, GPM (Primary Sludge)	= 20 GPM
Infl SS, %	= 3.0%
Infl SS, mg/L	= 30,000 mg/L
Effl SS, %	= 0.15%
Effl SS, mg/L	= 1,500 mg/L

Determine the thickener efficiency in removing sludge solids.

$$\text{Efficiency, \%} = \frac{(\text{Infl SS, mg/}L - \text{Effl SS, mg/}L) \times 100\%}{\text{Infl SS, mg/}L}$$

$$= \frac{(30{,}000 \text{ mg/}L - 1{,}500 \text{ mg/}L) \times 100\%}{30{,}000 \text{ mg/}L}$$

$$= 95\%$$

Concentrating the sludge and thereby reducing the volume to be pumped to subsequent processes is the main goal of the operator. A concentration factor should be used to determine the effectiveness of the thickener in concentrating the sludge. The concentration factor (cf) is determined by the following formula:

$$\text{Concentration Factor (cf)} = \frac{\text{Thickened Sludge Conc, \%}}{\text{Influent Sludge Conc, \%}}$$

The following example illustrates the use of the above equation.

EXAMPLE 7

Given: A primary sludge with a concentration of 3.0 percent sludge solids is thickened to a concentration of 7.0 percent sludge solids.

Find: The concentration factor (cf).

Solution:

Known	Unknown
Primary Sludge Conc, % = 3.0%	Concentration Factor
Thickened Sludge Conc, % = 7.0%	

$$\begin{aligned}
\text{Concentration Factor} &= \frac{\text{Thickened Sludge Concentration, \%}}{\text{Influent Sludge Concentration, \%}} \\
&= \frac{7.0\% \text{ Sludge Solids}}{3.0\% \text{ Sludge Solids}} \\
&= 2.33
\end{aligned}$$

The concentration factor determined above means that the influent sludge was thickened to a concentration 2.33 times its initial concentration. For primary sludges, the operator should achieve concentration factors of 2.0 or higher. Concentration factors for secondary sludges should be 3.0 or greater.

QUESTIONS

Write your answers in a notebook and then compare your answers with those on pages 306 and 307.

3.11A List the main components of gravity thickeners.

3.11B Discuss the functions of the inlet baffle, sludge rakes, and vertical pickets.

3.11C Discuss how the age of sludge may affect gravity concentration of primary activated sludge and waste activated sludge.

3.11D How does sludge temperature affect the efficiency of gravity thickeners and what measure should be taken during summertime operation to reduce gas production and rising sludge?

3.11E Determine the hydraulic surface (GPD/sq ft) and solids loading (lbs/day/sq ft) to a 30-ft diameter gravity thickener if 60 GPM of primary sludge at an initial suspended solids concentration of 3.0 percent sludge are applied.

3.11F A gravity thickener is used to concentrate 40 GPM of waste activated sludge at a concentration of 0.9% (9,000 mg/L). The underflow sludge is withdrawn at 3 percent and the effluent suspended solids concentration is 1,800 mg/L. Determine the suspended solids removal efficiency (%) and the concentration factor.

3.114 Troubleshooting

Visual inspection of most wastewater treatment unit processes coupled with an understanding of the expected results (compare design values with operating criteria) from good performance are the keys to successful operation. More often than not, the operator is made aware of equipment malfunctions and decreases in process efficiency by observing such items as liquid surfaces and effluent quality and also being aware of uncharacteristic odors.

The importance of the operator's awareness and ability to recognize signs of trouble cannot be overemphasized. In most instances, the experienced operator has the ability to ward off major operational problems and to maintain efficient operation by careful inspection and operational adjustments.

The specific areas of concern regarding gravity thickeners are: (1) surface and overflow quality, and (2) sludge blanket depth and thickened sludge concentrations.

3.1140 LIQUID SURFACE

As previously discussed, the overflow or effluent should be relatively clear (less than 500 mg/L suspended solids) and the liquid surface should be free of gas bubbles. If you notice an excessively high carryover of suspended solids, attention should immediately focus on the hydraulic loading and signs of gasification. If gas bubbles are evident at the tank surface, the problem may be caused by an excessive sludge detention time and subsequent gasification. The action(s) to be taken should include a determination of sludge blanket depth and a visual estimate of the thickened solids concentration. If the problem is related to an excessive sludge retention (detention) time, the thickened sludge concentration will be thicker than normal and the depth of the sludge blanket will be higher than usual.

To correct the problem, increase the rate of sludge withdrawal from the bottom of the thickener or lower the feed rate, if possible. Once a change of this nature is made, periodically check the condition of the effluent and the thickened (underflow) solids so as not to completely remove the sludge blanket and drastically reduce the thickened sludge concentration.

If the sludge blanket depth and solids concentrations are not high enough to be considered causes of gasification, investigate the speed of the sludge collectors and the influent characteristics. Sludge collection equipment may be equipped with a variable-speed mechanism. If the scrapers are operating at too low a speed, gasification may develop because pickets are not stirring the sludge and releasing the gas. Another common cause of gassing in gravity thickeners is the age of the influent sludge. If the sludge is held too long in the primary or secondary clarifiers, it may be well on its way to releasing gas before it enters the thickener. If gassing in the thickener cannot be attributed to operation of the thickener, observe the influent sludge and adjust the flow rate of sludge to the thickener. Secondary effluent can be recirculated to the thickener to freshen the influent sludge.

If poor effluent quality cannot be attributed to gasification, the problem may be the result of malfunctions in chemical conditioning equipment or increased hydraulic loadings. Chemical conditioning will be covered in Section 3.31, but you should be aware that chemical underdosing or overdosing can lead to equipment inefficiencies and you should inspect the chemical addition equipment. Hydraulic loadings in excess of design values also may lead to solids carryover and decreased efficiency. Check the rate of sludge pumping to the thickener, determine the hydraulic loading according to the calculations presented in Example 4, and adjust the thickener feed rate for successful operation. Coagulating chemicals may be used if the effluent quality needs improvement.

3.1141 THICKENED SLUDGE CONCENTRATION

Even if the thickener appears to be operating effectively, as evidenced by the lack of gas on the surface and solids carryover with the effluent, you should periodically check the thickened sludge concentration and the sludge blanket depth. The main objective of sludge thickening is to produce as concentrated a sludge as possible to effect volume reductions and cost savings in subsequent processes. If the thickened sludge concentration is not as thick as desired, check the blanket depth before making any adjustment to the withdrawal rate. On occasion, sludge in primary sedimentation tanks and gravity thickeners can become very thick and resistant to pumping. If this happens, a "hole" (CONING[15]) can develop in the blanket and liquid from above the blanket can be pulled through the pump. Lowering the rate of sludge withdrawal would increase the amount of solids at the bottom of the thickener and eventually result in SEPTICITY[16] and rising sludge. A hole (cone) in the sludge blanket (indicated by a low thickened sludge concentration and a high blanket level) can best be corrected by: (1) lowering the flow to the affected thickener, (2) increasing the speed of the collectors to keep the sludge at the point of withdrawal, and (3) increasing the rate of thickened sludge pumping. If both the blanket and the thickened sludge solids concentrations are low, you should lower the rate of sludge withdrawal in accordance with the calculations outlined below.

With time and experience, you should be able to roughly estimate the concentration of the influent and thickened sludges. This ability to "eyeball" concentrations, coupled with previous performance data, should enable you to control withdrawal rates.

EXAMPLE 8

Given: A 40-foot diameter by 10-foot SWD (Side Water Depth) gravity thickener is used to concentrate 100 GPM of primary sludge. The primary sludge enters the thickener at approximately 3.5 percent concentration based on the previous week's data. The sludge is withdrawn from the bottom of the thickener at 40 GPM at a concentration of 7.0 percent. The thickener effluent has a suspended solids concentration of 700 mg/L and the sludge blanket is 3 feet thick.

---

[15] *Coning.* Development of a cone-shaped flow of liquid, like a whirlpool, through sludge. This can occur in a sludge hopper during sludge withdrawal when the sludge becomes too thick. Part of the sludge remains in place while liquid rather than sludge flows out of the hopper. Also called coring.

[16] *Septic* (SEP-tick) or *Septicity.* A condition produced by bacteria when all oxygen supplies are depleted. If severe, the bottom deposits produce hydrogen sulfide, the deposits and water turn black, give off foul odors, and the water has a greatly increased oxygen and chlorine demand.

Find: 1. The sludge/volume (S/V) ratio in days.

2. If the present influent and effluent conditions are maintained, will the sludge blanket increase or decrease in depth?

3. What changes should be made if a higher concentration of underflow (thickened sludge) solids is desired?

4. What changes would stop gasification? How would these changes affect thickened sludge concentrations?

Solution:

Known		Unknown
Gravity Thickener		1. Sludge/volume (S/V) ratio, days
Diameter, ft	= 40 ft	
Side Water Depth, ft	= 10 ft	2. Will sludge blanket increase or decrease?
Flow In, GPM (Primary Sludge)	= 100 GPM	3. What changes would increase underflow sludge concentrations?
Sludge Out, GPM	= 40 GPM	
Primary Sludge Conc, %	= 3.5%	4. What changes would stop gasification? How would these changes affect thickened sludge concentrations?
Sludge Out Conc, %	= 7.0%	
Thickener Effluent Susp Sol, mg/L	= 700 mg/L	
, %	= 0.07%	
Sludge Blanket, ft	= 3 ft	

1. Calculate the sludge/volume (S/V) ratio in days.

a. Determine the sludge blanket volume in gallons.

$$\text{Sludge Blanket Volume, gal} = \frac{\pi}{4} \times (\text{Diameter, ft})^2 \times \text{Blanket, ft} \times \frac{7.48 \text{ gal}}{\text{cu ft}}$$

$$= 0.785 \times (40 \text{ ft})^2 \times 3 \text{ ft} \times \frac{7.48 \text{ gal}}{\text{cu ft}}$$

$$= 28,200 \text{ gallons}$$

b. Determine the sludge pumped in gallons per day.

$$\text{Sludge Pumped, gal/day} = \text{Sludge Out, GPM} \times 1,440 \frac{\text{min}}{\text{day}}$$

$$= 40 \frac{\text{gal}}{\text{min}} \times 1,440 \frac{\text{min}}{\text{day}}$$

$$= 57,600 \text{ gal/day}$$

c. Calculate the sludge/volume (S/V) ratio in days.

$$\text{S/V Ratio, days} = \frac{\text{Sludge Blanket Volume, gal}}{\text{Sludge Pumped, gal/day}}$$

$$= \frac{28,200 \text{ gal}}{57,600 \text{ gal/day}}$$

$$= 0.5 \text{ day}$$

2. Will the sludge blanket increase or decrease? If the quantity of solids entering the thickener is greater than the quantity leaving the thickener, then the blanket depth will increase. If the quantity of solids entering the thickener is less than the quantity leaving the thickener, the blanket thickness will decrease. The solution to this problem is based on mass balance calculations, as shown below:

a. Determine the pounds of sludge solids entering the thickener daily.

$$\text{Sludge Solids Entering, lbs/day} = \text{Flow In, GPM} \times 1,440 \frac{\text{min}}{\text{day}} \times 8.34 \frac{\text{lbs}}{\text{gal}} \times \text{Sl In, } \frac{\%}{100\%}$$

$$= 100 \frac{\text{gal}}{\text{min}} \times 1,440 \frac{\text{min}}{\text{day}} \times 8.34 \frac{\text{lbs}}{\text{gal}} \times \frac{3.5\%}{100\%}$$

$$= 42,034 \text{ lbs/day}$$

b. Determine the pounds of sludge solids withdrawn in the thickener underflow daily.

$$\text{Sludge Solids Withdrawn, lbs/day} = \text{Sludge Out, GPM} \times 1,440 \frac{\text{min}}{\text{day}} \times 8.34 \frac{\text{lbs}}{\text{gal}} \times \text{Sl Out, } \frac{\%}{100\%}$$

$$= 40 \frac{\text{gal}}{\text{min}} \times 1,440 \frac{\text{min}}{\text{day}} \times 8.34 \frac{\text{lbs}}{\text{gal}} \times \frac{7.0\%}{100\%}$$

$$= 33,627 \text{ lbs/day}$$

c. Determine the pounds of solids lost in the thickener effluent daily.

$$\text{Solids Lost in Effl, lbs/day} = \text{Flow, GPM} \times 1,440 \frac{\text{min}}{\text{day}} \times 8.34 \frac{\text{lbs}}{\text{gal}} \times \text{Effl, } \frac{\%}{100\%}$$

$$= (100 \text{ GPM} - 40 \text{ GPM}) \times 1,440 \frac{\text{min}}{\text{day}} \times 8.34 \frac{\text{lbs}}{\text{gal}} \times \frac{0.07\%}{100\%}$$

$$= 60 \frac{\text{gal}}{\text{min}} \times 1,440 \frac{\text{min}}{\text{day}} \times 8.34 \frac{\text{lbs}}{\text{gal}} \times \frac{0.07\%}{100\%}$$

$$= 504 \text{ lbs/day}$$

d. Determine total pounds of solids removed daily.

$$\underset{\text{lbs/day}}{\text{Solids Out,}} = \underset{\text{Withdrawn, lbs/day}}{\text{Sludge Solids}} + \underset{\text{Effl, lbs/day}}{\text{Solids Lost in}}$$

$$= 33{,}627 \text{ lbs/day} + 504 \text{ lbs/day}$$

$$= 34{,}131 \text{ lbs/day}$$

e. Compare the sludge solids in with the solids out.

$$\underset{\text{lbs/day}}{\underset{\text{Entering,}}{\text{Sludge Solids}}} = 42{,}034 \text{ lbs/day}$$

$$\underset{\text{lbs/day}}{\text{Solids Out,}} = 34{,}131 \text{ lbs/day}$$

Therefore, since the solids entering (42,034 lbs/day) are greater than the solids out (34,131 lbs/day), the sludge blanket will increase in depth.

3. What changes would increase the thickened sludge concentration? Higher thickened sludge solids concentration will normally result if the depth of the sludge blanket is increased. To increase the blanket depth, the flow rate of the thickened sludge should be decreased somewhat. The thickened sludge is withdrawn at a rate of 40 GPM and the rate should not be changed at increments of greater than 20 percent when steady-state (lbs in = lbs out) conditions exist. Drastic changes should be avoided and a close watch should be kept over the depth of the blanket after such changes are made. To increase the sludge blanket depth and the thickened sludge concentration, the sludge withdrawal rate should be decreased to approximately 40 GPM − (40 GPM × 20%/100%) = 40 GPM − 8.0 GPM = 32 GPM.

Another approach to regulating the sludge blanket depth is to sound (measure) the depth of the sludge blanket. In general, if the depth is greater than 7 feet (2.1 m), increase the underflow withdrawal rate and if the depth is less than 5 feet (1.5 m), decrease the withdrawal rate.

4. What changes would stop gasification? How would these changes affect thickened sludge concentrations?

If gasification develops as a result of excessive sludge retention times, the rate of the sludge withdrawal should be increased so as to lower the sludge blanket depth with subsequent lowering of the sludge retention time. The net effect on thickener performance will be a decrease in thickened sludge concentration and a possible improvement in effluent quality. Another alternative may be to recirculate secondary effluent to freshen the sludge.

EXAMPLE 9

Given: The thickener from Example 8 has just been restarted following routine maintenance shutdown. The influent concentration is "eyeballed" at approximately 3.0 percent sludge solids. The influent flow is 150 GPM and the sludge withdrawal pump is set at 15 GPM. After a few hours of continuous operation, the sludge blanket depth is measured and found to be 2 feet thick. The underflow concentration is estimated to be approximately 6 percent.

Find: Should the operator increase, decrease, or maintain the current rate of withdrawal?

Solution:

Known	**Unknown**
Known information from Example 8	Should rate of sludge withdrawal be increased, decreased, or not changed?
Infl Sl Conc, % = 3.0% Sl Sol	
Infl flow, GPM = 150 GPM	
Sludge Withdrawal Pump, GPM = 15 GPM	
Sludge Blanket Depth, ft = 2 ft	
Thickened Sl Conc, % = 6.0% Sl Sol	

1. Calculate the sludge solids entering in pounds per minute.

$$\underset{\text{lbs/min}}{\text{Solids Entering,}} = \text{Infl Flow, GPM} \times 8.34 \frac{\text{lbs}}{\text{gal}} \times \text{Sl Sol In}, \frac{\%}{100\%}$$

$$= 150 \text{ gal/min} \times 8.34 \frac{\text{lbs}}{\text{gal}} \times \frac{3.0\%}{100\%}$$

$$= 37.5 \text{ lbs/min}$$

2. Calculate the sludge solids leaving the thickener in pounds per minute.

$$\underset{\text{lbs/min}}{\underset{\text{Withdrawn,}}{\text{Solids}}} = \text{Underflow, GPM} \times 8.34 \frac{\text{lbs}}{\text{gal}} \times \text{Unfl Sl}, \frac{\%}{100\%}$$

$$= 15 \frac{\text{gal}}{\text{min}} \times 8.34 \frac{\text{lbs}}{\text{gal}} \times \frac{6.0\%}{100\%}$$

$$= 7.5 \text{ lbs/min}$$

The number of pounds exiting with the effluent can be neglected if the effluent is clear (less than 500 mg/L suspended solids) and little solids carryover is observed.

Based on the visual estimations of sludge concentration and the above calculations, sludge is being stored at the rate of 30.0 lbs/min (lbs enter − lbs exit). The sludge blanket depth is 2 feet, but let us assume typical operation for this thickener indicates that a blanket depth of 5 feet can be maintained. The operator should therefore determine the time required to fill the thickener with 3 additional feet of sludge at the present conditions. The calculations are shown below.

$$\underset{\text{Time, min}}{\text{Storage}} = \frac{\underset{\text{Volume, cu ft}}{\text{Storage}} \times \frac{62.4 \text{ lbs}}{\text{cu ft}} \times \text{Unfl Sl}, \frac{\%}{100\%}}{\text{Sludge Storage Rate, lbs/min}}$$

$$= \frac{\frac{\pi}{4} \times (40 \text{ ft})^2 \times 3 \text{ ft} \times \frac{62.4 \text{ lbs}}{\text{cu ft}} \times \frac{6.0\%}{100\%}}{30 \text{ lbs/min}}$$

$$= 470 \text{ min}$$

$$\text{Storage Time, hr} = \frac{470 \text{ min}}{60 \text{ min/hr}}$$

$$= 7.8 \text{ hours}$$

If the unit is left as is, the blanket will reach a depth of 5 feet in approximately 8 hours. However, at the end of the 8 hours, the operator will again have to adjust the withdrawal rate to avoid even greater buildup of sludge blanket. Drastic changes in withdrawal rates are not desirable and can be avoided by making a slight adjustment at the start of the 8-hour period. This adjustment should be made based on the ratio of volume stored to total storage volume as shown below.

$$\text{Sludge Storage Rate, lbs/min} = \frac{\text{Stored Sludge Height, ft}}{\text{Total Storage Height, ft}} \times \frac{\text{Solids Entering,}}{\text{lbs/min}}$$

$$= \frac{2 \text{ ft}}{5 \text{ ft}} \times 37.5 \text{ lbs/min}$$

$$= 15 \text{ lbs/min}$$

We therefore want to store solids at a rate of 15 lbs/min instead of the current 30 lbs/min. To obtain this storage rate, the desired sludge withdrawal rate must be determined in pounds per minute.

$$\text{Sludge Withdrawal Rate, lbs/min} = \frac{\text{Solids Entering,}}{\text{lbs/min}} - \frac{\text{Sludge Storage}}{\text{Rate, lbs/min}}$$

$$= 37.5 \text{ lbs/min} - 15 \text{ lbs/min}$$

$$= 22.5 \text{ lbs/min}$$

The sludge withdrawal pumping rate must be increased in order to remove underflow solids at a rate of 22.5 pounds per minute.

$$\text{Sludge Withdrawal Pumping Rate, GPM} = \frac{\text{Sludge Withdrawal, lbs/min}}{8.34 \text{ lbs/gal} \times \text{Unfl Sl, \%/100\%}}$$

$$= \frac{22.5 \text{ lbs/min}}{8.34 \text{ lbs/gal} \times \dfrac{6.0\%}{100\%}}$$

$$= 45 \text{ GPM}$$

The sludge withdrawal pumping rate should therefore be increased from 15 GPM to 45 GPM at this time. This change represents a 200 percent increase in withdrawal rate, which is substantially greater than the 20 percent change outlined in Example 8. In Example 8, the thickener was operating under steady-state (lbs in = lbs out) conditions and under such conditions the withdrawal rate should not be changed by increments greater than 20 percent. For this example, the thickener is not at steady state and the formulas outlined above should govern the withdrawal rate changes. Approximately 4 hours after the above change is made, the operator should recheck the blanket depth, sludge concentrations, and effluent quality, rerun the above calculation, and change the withdrawal rate, if required.

Table 3.3 summarizes the operational problems that may develop in gravity thickeners and lists the corrective measures that might be taken.

TABLE 3.3   TROUBLESHOOTING GRAVITY THICKENERS

Operational Problem	Possible Cause	Check or Monitor	Possible Solution
1. Liquid level clear but sludge rising and solids carry over into effluent	1a. Gasification	1a. Sludge blanket and sludge detention	1a. Increase sludge withdrawal rate
	1b. Septic feed	1b. Characteristics of feed	1b. Increase sludge pumping from clarifier
	1c. Blanket disturbances	1c. Sludge collector speed	1c. Lower collector speed[a]
	1d. Chemical inefficiencies	1d. Chemical equipment	1d. Increase chemical feed rate
	1e. Excessive loadings	1e. Hydraulic flow rate	1e. Lower flow, if possible
2. Thin (dilute) underflow sludge and clear effluent	2a. Low blanket	2a. Blanket level	2a. Decrease sludge withdrawal rate
	2b. Sludge withdrawal rate too high		
3. Thin (dilute) underflow sludge, liquid level clear but sludge rising with solids carryover	3a. Collector speed too low or inoperative	3a. Collector mechanism and speed	3a. Turn on or increase collector speed
	3b. Hole or cone in sludge blanket	3b. Blanket level	3b. Increase collector speed and increase withdrawal rate
4. Thin (dilute) underflow sludge, liquid surface laden with solids, and solids carryover	4a. Hydraulic loading high	4a. Loadings. Influent sludge flow	4a. Lower influent sludge
	4b. Chemical system inoperative	4b. Chemical equipment	4b. Increase chemical rate

[a] If solids carryover is caused by gasification, increase collector speed.

QUESTIONS

Write your answers in a notebook and then compare your answers with those on pages 307 and 308.

3.11G   Why should the operator make routine visual checks on gravity thickeners as well as any other equipment?

3.11H   What is the meaning of the term "hole" in the sludge blanket and how can a hole be corrected?

3.11I   A gravity thickener has been operating successfully. On a routine check, the operator notices that solids are rising to the surface. List the possible causes and outline the procedures the operator should follow to correct the problem(s).

3.12   Dissolved Air Flotation (DAF) Thickeners (Figure 3.3)

The objective of flotation thickening is to separate solids from the liquid phase in an upward direction by attaching air bubbles to particles of suspended solids. Four general methods of flotation are commonly employed. These include:

1. Dispersed air flotation where bubbles are generated by mixers or diffused aerators.

2. Biological flotation where gases formed by biological activity are used to float solids.

3. Dissolved air (vacuum) flotation where water is aerated at atmospheric pressure and released under a vacuum.

4. Dissolved air (pressure) flotation where air is put into solution under pressure and released at atmospheric pressure.

Flotation by dissolved air (pressure) is the most commonly used procedure for wastewater sludges and will be the topic of discussion in this section. Flotation units may be either rectangular or circular in design. The dissolved air system uses either a compressed air supply or an *ASPIRATOR-TYPE*[17] air injection assembly to obtain a pressurized air-water solution. The key components of dissolved air flotation thickener units are: (1) air injection equipment (located on and within pressurized retention tank), (2) agitated or unagitated pressurized retention tank, (3) recycle pump, (4) inlet or distribution assembly, (5) sludge scrapers, and (6) an effluent baffle.

The sludge to be thickened is either introduced to the unit at the bottom through a distribution box and blended with a pre-pressurized effluent stream or the influent stream is saturated with air, pressurized, and then released to the inlet distribution assembly. Total wastestream pressurization may shear (break up) flocculated sludges and seriously reduce process efficiency. Direct saturation and pressurization of the sludge stream is not the preferred mode of operation where primary sludges are to be thickened. Primary sludges often contain stringy material that can clog or "rag up" the aeration equipment in a pressurized retention tank. Flotation thickening of excess biological solids may use air saturation and pressurization of the wastestream with less possibility of clogging the air addition and dissolution equipment.

The preferred mode of operation from a maintenance standpoint is the use of a recycle stream to serve as the air carrying medium. Referring to Figure 3.3, the operation of dissolved air flotation (DAF) units that incorporate recycle techniques is as follows. A recycled DAF effluent, primary or secondary effluent stream, is introduced into a retention tank to dissolve air into the liquid. The retention tank is maintained at a pressure of 45 to 70 psig (3.2 to 4.9 kg/sq cm). Compressed air is either introduced into the retention tank directly or at some point upstream of the retention tank or an aspirator assembly is used to draw air into the stream.

The pressurized air-saturated liquid then flows to the distribution or inlet assembly and is released at atmospheric pressure through a back pressure-relief valve. The decrease in pressure causes the air to come out of solution in the form of thousands of minute air bubbles. The bubbles make contact with the influent sludge solids in the distribution box and attach to the solids causing them to rise to the surface. These concentrated solids are then removed from the surface. An effluent baffle is provided to keep the floated solids from floating out with the effluent.

The effluent baffle extends approximately 2 to 3 inches (5.0 to 7.5 cm) above the water surface and down to within 12 to 18 inches (0.3 to 0.45 m) of the bottom of the tank. The effluent baffle is provided to keep the floated solids from contaminating the effluent. Clarified effluent flows under the baffle and leaves the unit through an effluent weir. If air is introduced or aspirated upstream of the retention tank, it is usually done on the suction side of the recycle pump to use the pump as a driving force for dissolving air into the liquid. The main disadvantage associated with introducing air to the suction side of pumps is the possibility of pump *CAVITATION*[18] and the subsequent loss of pump capacity. Systems that add compressed air directly to the retention tank commonly use a float control mechanism to maintain a desired air-liquid balance. A sight glass should be provided to periodically check the level of the air-liquid interface (point of contact) because if the float mechanisms fail, the retention tank may fill completely with either liquid or air. In both cases, the net effect will be a drastic reduction in flotation efficiency.

3.120   *Factors Affecting Dissolved Air Flotation*

The performance of dissolved air flotation units depends on: (1) type of sludge, (2) age of the feed sludge, (3) solids and hydraulic loadings, (4) air to solids (A/S) ratio, (5) recycle rate, and (6) sludge blanket depth.

---

[17] *Aspirate* (AS-per-rate).   Use of a hydraulic device (aspirator or eductor) to create a negative pressure (suction) by forcing a liquid through a restriction, such as a Venturi tube. An aspirator may be used in the laboratory in place of a vacuum pump; sometimes used instead of a sump pump.

[18] *Cavitation* (kav-uh-TAY-shun).   The formation and collapse of a gas pocket or bubble on the blade of an impeller or the gate of a valve. The collapse of this gas pocket or bubble drives water into the impeller or gate with a terrific force that can knock metal particles off and cause pitting on the impeller or gate surface. Cavitation is accompanied by loud noises that sound like someone is pounding on the impeller or gate with a hammer.

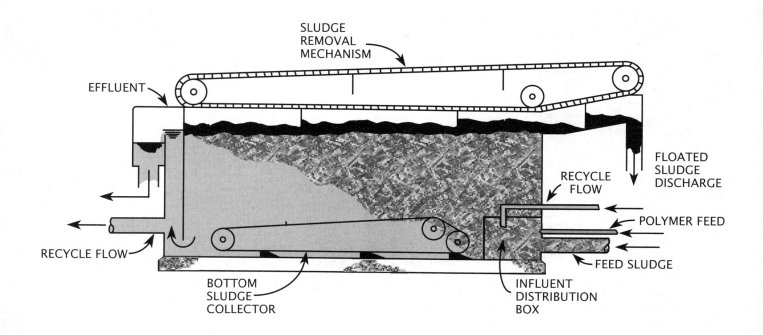

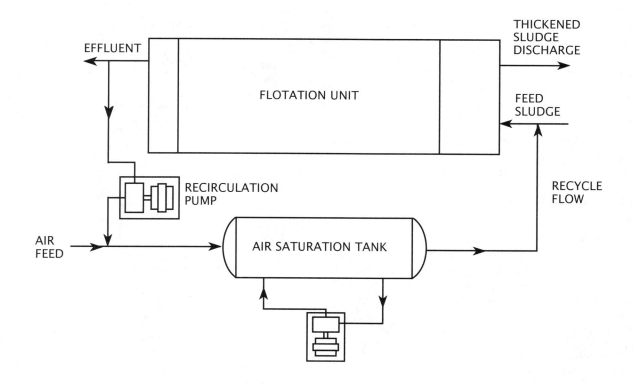

Fig. 3.3   Dissolved air flotation thickener

As is the case with gravity thickeners, the type and age of sludge applied to flotation thickeners will affect the overall performance. Primary sludges are generally heavier than excess biological sludges and are not as easy to treat by flotation concentration. If enough air is introduced to float the sludge mass, the majority of the primary sludge solids will float to the surface and be removed by the skimming mechanisms. Gritty or heavy primary sludge particles will settle and be deposited on the floor of the flotation unit and provisions should be made to remove these settled solids. If a flotation unit is used for primary sludge thickening, the flotation cell is usually equipped with sludge scrapers to push the settled solids to a collection hopper for periodic removal. Problems will arise when concentrating primary sludges or combinations of primary sludge and waste activated sludge if the flotation chamber is not equipped with bottom sludge scrapers and sludge removal equipment. Solids buildup will result in a decrease in flotation volume and a reduction in thickener efficiency.

Excess biological sludges are easier to treat by flotation thickening than primary sludges because they are generally lighter and thus easier to float. Bottom sludge scrapers should still be incorporated in the design of units used solely for biological sludge because a small fraction of solids will settle. These settled solids will eventually become anaerobic and rise due to gasification. If these solids are deposited at the effluent end of the unit, solids may be carried under the effluent baffle and exit the unit with the effluent.

Sludge age usually does not affect flotation performance as drastically as it affects gravity concentrators. A relatively old sludge has a natural tendency to float due to gasification and this natural buoyancy will have little or no negative effect on the operation of flotation thickeners. However, rising sludge does create problems in primary and final sedimentation processes and should be avoided by controlling the sludge withdrawal rate from these unit processes.

Solids and hydraulic loadings, A/S (air to solids) ratios, recycle rate, and sludge blanket depth are normal operational guidelines and are discussed in the following paragraphs.

3.121   Operating Guidelines

The size of dissolved air flotation units is determined by the engineers who design them. The operator has control over A/S ratio, recycle rate, and the blanket thickness and can optimize performance by properly adjusting these variables. Before discussing the control variables, the operator should be familiar with determining applied loading rates.

3.1210   SOLIDS AND HYDRAULIC LOADINGS

Solids and hydraulic loadings for flotation units are based on the same calculations used to determine loading rates for gravity thickeners. If either the solids or hydraulic loading becomes excessive, effluent quality declines and thickened sludge concentrations are reduced. The following example shows how to calculate loading rates.

EXAMPLE 10

Given: A dissolved air flotation unit receives 100 GPM of waste activated sludge with a suspended solids concentration of 8,000 mg/L. The rectangular flotation unit is 40 feet long and 15 feet wide.

Find: The hydraulic loading (GPM/sq ft) and solids loading (lbs/hr/sq ft).

Solution:

Known	**Unknown**
Flow, GPM = 100 GPM	1. Hydraulic Loading, GPM/sq ft
Sus Sol, mg/L = 8,000 mg/L	
, % = 0.8%	2. Solids Loading, lbs/hr/sq ft
Flotation Unit	
Length, ft = 40 ft	
Width, ft = 15 ft	

1. Determine the hydraulic loading, GPM/sq ft.

$$\text{Hydraulic Loading, GPM/sq ft} = \frac{\text{Flow, GPM}}{\text{Liquid Surface Area, sq ft}}$$

$$= \frac{100 \text{ gal/min}}{40 \text{ ft} \times 15 \text{ ft}}$$

$$= \frac{100 \text{ gal/min}}{600 \text{ sq ft}}$$

$$= 0.2 \text{ GPM/sq ft}$$

2. Determine the solids loading, lbs/hr/sq ft.

$$\text{Solids Loading, lbs/hr/sq ft} = \frac{\text{Flow, GPM} \times 60 \frac{\text{min}}{\text{hr}} \times 8.34 \frac{\text{lbs}}{\text{gal}} \times \text{SS}, \frac{\%}{100\%}}{\text{Liquid Surface Area, sq ft}}$$

$$= \frac{100 \frac{\text{gal}}{\text{min}} \times 60 \frac{\text{min}}{\text{hr}} \times 8.34 \frac{\text{lbs}}{\text{gal}} \times \frac{0.8\%}{100\%}}{40 \text{ ft} \times 15 \text{ ft}}$$

$$= \frac{400 \text{ lbs/hr}}{600 \text{ sq ft}}$$

$$= 0.67 \text{ lb/hr/sq ft} = 16 \text{ lbs/day/sq ft}$$

3.1211   AIR TO SOLIDS (A/S) RATIO

The quantity of air introduced and dissolved into the recycle or wastestream is critical to the operation of flotation thickeners. Enough air has to be added and dissolved to float the sludge solids. The most effective method of accomplishing this is to introduce air into a pressurized retention tank along with the wastestream to be thickened or along with a portion of the thickener effluent stream. Air also can be dissolved in primary or secondary effluent, thus avoiding solids recycling in the DAF unit. Mixing of the retention tank contents should also be used to increase the amount of air that can be put into solution. In unmixed pressure retention tanks, only about 50 percent of the injected air will dissolve while 90 percent saturation can be obtained by vigorous agitation of the tank contents. As previously discussed, following a short detention time in the pressurized retention tank, the saturated air/liquid stream is pumped to the inlet side of the flotation unit where it enters a distribution assembly by operation of a back pressure-relief valve. The release of the saturated air stream to atmospheric pressure causes the air to come out of solution in the form of very small bubbles. Thousands of these minute bubbles attach to particles of suspended solids. The solids float to the surface, concentrate, and are removed by the sludge skimming mechanism. The more air you have dissolved in the retention tank, the greater the number of minute air bubbles that will be released in the distribution assembly. And, the more bubbles you produce in the distribution assembly, the more efficient your operation will be.

The amount of air supplied to the unit is usually controlled by an air rotameter and compressor assembly, which are activated by a liquid level indicator in the retention tank. The most important operational concern is to ensure that the air rotameter, compressor, and the float mechanism to actuate air injection are in proper working order.

Example 11 shows how to calculate the quantity of air applied to the system.

EXAMPLE 11

Given: An air rotameter and compressor provide for 10 cubic feet per min (SCFM)[19] of air to be injected into a pressurized retention tank.

Find: The pounds of air applied to the unit per hour (lbs/hr).

Solution:

Known	Unknown
Air Flow, SCFM = 10 SCFM	Air Applied, lbs/hr

Calculate the air applied in pounds of air per hour.

$$\text{Air Applied, lbs/hr} = \text{Air Flow, } \frac{\text{cu ft}}{\text{min}} \times 60 \frac{\text{min}}{\text{hr}} \times 0.075 \frac{\text{lb Air}}{\text{cu ft Air}}$$

$$= 10 \frac{\text{cu ft}}{\text{min}} \times 60 \frac{\text{min}}{\text{hr}} \times 0.075 \frac{\text{lb}}{\text{cu ft}}$$

$$= 45 \text{ lbs/hr}$$

NOTE: The conversion factor of 0.075 pound of air per cubic foot of air will change with temperature and elevation or barometric pressure.

The ratio between air supplied and the quantity of solids applied to the flotation unit is then the air to solids (A/S) ratio. The following example illustrates the determination of air/solids (A/S) ratio.

EXAMPLE 12

Given: A dissolved air flotation unit receives 100 GPM of waste activated sludge at a concentration of 9,000 mg/L (0.9% solids). Air is supplied at a rate of 5.0 cu ft/min.

Find: The air to solids (A/S) ratio.

Solution:

Known	Unknown
Solids Flow, GPM = 100 GPM	Air to Solids (A/S) Ratio
Sl Conc, mg/L = 9,000 mg/L	
, % = 0.9% Solids	
Air, cu ft/min = 5.0 cu ft/min	

Calculate the air to solids (A/S) ratio.

$$\frac{\text{Air, lbs}}{\text{Solids, lbs}} = \frac{\text{Air, cu ft/min} \times 0.075 \text{ lb/cu ft}}{\text{Solids, GPM} \times 8.34 \frac{\text{lbs}}{\text{gal}} \times \text{Sl Conc, } \frac{\%}{100\%}}$$

$$= \frac{5.0 \text{ cu ft/min} \times 0.075 \text{ lb/cu ft}}{100 \frac{\text{gal}}{\text{min}} \times 8.34 \frac{\text{lbs}}{\text{gal}} \times \frac{0.9\%}{100\%}}$$

$$= \frac{0.375 \text{ lb Air}}{7.5 \text{ lbs Solids}}$$

$$= 0.05 \text{ lb Air/lb Solids}$$

3.1212 RECYCLE RATE AND SLUDGE BLANKET

Both the rate of effluent recycle and the thickness of the sludge blanket are operational controls available to optimize DAF performance. Typically, recycle rates of 100 to 200 percent are used. A recycle rate of 100 percent means that for every gallon of influent sludge there is one (1) gallon of DAF effluent recycled to the DAF inlet works. The following example illustrates the determination of recycle rate.

---

[19] SCFM. Standard Cubic Feet per Minute. Cubic feet of air per minute at standard conditions of temperature, pressure, and humidity (0°C, 14.7 psia, and 50 percent relative humidity).

EXAMPLE 13

Given: A dissolved air flotation unit receives waste activated sludge flow of 50 GPM. The recycle rate is set at 75 GPM.

Find: Percentage of recycle.

Solution:

Known	**Unknown**
Waste Flow, GPM = 50 GPM	Percentage of Recycle, %
Recycle Flow, GPM = 75 GPM	

Calculate the percentage of recycle.

$$\text{Recycle, \%} = \frac{\text{Recycle Flow, GPM} \times 100\%}{\text{Waste Flow, GPM}}$$

$$= \frac{75\ \text{GPM}}{50\ \text{GPM}} \times 100\%$$

$$= 150\%$$

The optimum recycle rate for a particular unit will vary from one treatment plant to the next and it is impossible to define that rate for every application. The important point is that the recycle stream carries the air to the inlet of the unit. Obviously, as the rate of recycle increases, the potential to carry more air to the inlet also increases. The term "potential" is used here because the recycle rate and the quantity of air dissolved and released in the inlet assembly are dependent on one another because of what happens in the retention tank. DAF recycle pumps are usually centrifugal pumps, which means that as the pressure upstream (retention tank) increases, the output (flow) decreases. Therefore, the rate of recycle is directly dependent on the pressure maintained within the retention tank. As stated previously, retention tank pressures of 45 to 70 psi (3.2 to 4.9 kg/sq cm) are commonly used. As the pressure within the retention tank is increased or decreased by closing or opening the back pressure-relief valve, the recycle rate will decrease or increase. The optimum recycle rate and retention tank pressure are usually determined by experimentation.

The thickness of the floating sludge blanket can be varied by increasing or decreasing the speed of the surface sludge scrapers. Increasing the sludge scrapers speed usually tends to thin out the floated sludge while decreasing the scrapers speed will generally result in a more concentrated sludge.

3.122   Normal Operating Procedures

Typically, the flow through the thickener is continuous and should be set as constant as possible. Monitoring of the influent, effluent, and thickened sludge streams should be done at least once per shift and composite samples should be taken for later laboratory analysis.

Under normal operating conditions, the effluent stream should be relatively free of solids (less than 100 mg/L suspended solids) and will resemble secondary effluent. The float (thickened sludge) solids will have a consistency resembling cottage cheese. The depth of the float solids should extend approximately 6 to 8 inches (15 to 20 cm) below the surface. The surface farthest from the float solids collection and the discharge point should be scraped clean of floating solids with each pass of the sludge collection scrapers. If the sludge blanket is allowed to build up (too thick) and drop too far below the surface, thickened (floated) solids will be carried under the effluent baffle and contaminate the effluent.

Normal start-up and shutdown procedures are outlined below:

START-UP

- Fill the unit with fresh water, primary effluent, or secondary effluent.

- Open the inlet and discharge valves on the recycle pump and turn on the recycle pump only when thickener is full.

- Adjust the retention tank pressure to the desired pressure (45 to 70 psig or 3.2 to 4.9 kg/sq cm) by opening or closing the pressure regulating valve.

- Open the inlet and discharge valves on the reaeration pump (if provided) and turn on the reaeration pump. If a mechanical mixer is used instead of a reaeration pump, turn on the mixer in the retention tank. Mixing in the retention tank may be accomplished by methods other than the use of mechanical mixers.

- Open the appropriate air injection valves and turn on the air compressor.

- Open the appropriate chemical addition valves and turn on the chemical pump, if chemicals are used.

- Wait until the surface of the DAF thickener is covered by a uniform pattern of small bubbles.

- Open the inlet and discharge valves on the sludge feed pump and start the feed pump.

- Allow floated sludge mat to build up, then turn on sludge collection scrapers.

- Turn on the thickened sludge pump and adjust the withdrawal rate, as required.

If the thickener is not operated in a continuous mode and daily or frequent shutdowns are required, the following procedures should be followed:

SHUTDOWN

- Turn off the sludge inlet pump and close the appropriate valves.

- Turn off the chemical pump and close appropriate valves.

- Turn on the fresh water supply to the unit and allow it to run on fresh water until the surface is free of floating sludge.

- Turn off the air compressor and close appropriate air valves.

- Turn off the reaeration and recycle pumps and close appropriate valves.

- Turn off the sludge collectors.
- Turn off the thickened sludge pump.
- Hose down and clean up, as required.

3.123 Typical Performance

Typical operating guidelines as well as thickened sludge concentration and suspended solids removals for waste activated sludge are presented in Table 3.4.

TABLE 3.4 OPERATIONAL AND PERFORMANCE GUIDELINES FOR FLOTATION THICKENERS

	Without Polymer Addition	With Polymer Addition
Solids Loading,		
lbs/hr/sq ft[a]	0.4–1	1–2
lbs/day/sq ft	9.6–24	24–48
Hydraulic Loading, GPM/sq ft[b]	0.5–1.5	0.5–2.0
Recycle, %	100–200	100–200
Air/Solids, lb/lb	0.01–0.10	0.01–0.10
Minimum Influent Solids Concentration, mg/L	5,000	5,000
Float Solids Concentration, %	2–4	3–5
Solids Recovery, %	50–90	90–98

[a] lbs/hr/sq ft × 4.883 = kg/hr/sq m.
[b] GPM/sq ft × 0.679 = L/sec/sq m.

The determination of solids recovery in the operation of the DAF unit is based on laboratory analysis and the following calculations.

EXAMPLE 14

Given: A 100-foot diameter dissolved air flotation unit receives 750 GPM of waste activated sludge at a concentration of 0.75% (7,500 mg/L) sludge solids. The effluent contains 50 mg/L of suspended solids. The float or thickened sludge is at a concentration of 3.3 percent.

Find: The solids removal efficiency (%) and the concentration factor (cf).

Solution:

Known	**Unknown**
Dissolved Air Flotation Unit	
Infl Solids, mg/L = 7,500 mg/L	1. Solids Removal Efficiency, %
, % = 0.75%	
Effl Solids, mg/L = 50 mg/L	2. Concentration Factor (cf)
Effl Sludge, % = 3.3% (Thickened Sludge)	

1. Determine the solids removal efficiency, %.

$$\text{Solids Removal Efficiency, \%} = \frac{(\text{Infl Solids, mg/}L - \text{Effl Solids, mg/}L)\,100\%}{\text{Infl Solids, mg/}L}$$

$$= \frac{(7,500 \text{ mg/}L - 50 \text{ mg/}L)\,100\%}{7,500 \text{ mg/}L}$$

$$= 99.3\%$$

2. Calculate the concentration factor (cf) for the thickened sludge.

$$\text{Concentration Factor (cf)} = \frac{\text{Thickened Sludge Concentration, \%}}{\text{Influent Sludge Concentration, \%}}$$

$$= \frac{3.3\%}{0.75\%}$$

$$= 4.4$$

QUESTIONS

Write your answers in a notebook and then compare your answers with those on pages 308 and 309.

3.12A List the main components of dissolved air flotation (DAF) units.

3.12B Discuss the function of the distribution box, the retention tank, and the effluent baffle.

3.12C Why should a sight glass be provided on the retention tank?

3.12D List the factors that affect the performance of DAF thickeners.

3.12E What effect does sludge age have on the performance of DAF thickeners?

3.12F Determine the hydraulic loading (GPD/sq ft) for a 20-foot diameter DAF unit. The influent flow is 100 GPM. The formula for surface area of a circular tank is:

$$\text{Area} = \frac{\pi}{4} \text{ Diameter}^2 \text{ or Area} = 0.785 \text{ Diameter}^2$$

3.12G For the above problem, determine the solids loading, A/S ratio, and recycle flow rate (GPM), if the influent sludge has a suspended solids concentration of 0.75% (7,500 mg/L), and is supplied at a rate of 2.5 cu ft/min. Air is supplied at a rate of 0.75 cu ft/min and a recycle ratio of 100 percent is required.

3.12H Determine the suspended solids removal efficiency (%) and the concentration factor (cf) if a DAF unit receives an influent sludge at 1.0 percent (10,000 mg/L) suspended solids. The effluent is at 50 mg/L suspended solids and the float or thickened sludge is at a concentration of 3.8 percent.

3.124   Troubleshooting

VISUAL INSPECTION of the dissolved air flotation unit in conjunction with a working knowledge of the operating techniques is the operator's biggest asset in ensuring efficient operation. The specific areas that the operator should be concerned with are: (1) effluent quality, and (2) thickened sludge (float) characteristics. The effluent from DAF units should be relatively clear (less than 100 mg/L suspended solids). Well-operated units should produce effluents equivalent in appearance to secondary clarifier effluent. If the effluent from the unit contains an unusually high amount of suspended solids, the problem may be related to: (1) sludge blanket thickness, (2) chemical conditioning, (3) A/S ratio, (4) recycle rate, (5) solids or hydraulic loading, or (6) any combination of these factors.

If the float solids appear to be well flocculated and concentrated (resembling cottage cheese), the speed of the sludge scrapers should be increased. Poor effluent quality in conjunction with a concentrated float sludge usually results from allowing the sludge blanket to develop too far below the surface. When this happens, the undermost portions of the blanket will break off and be carried under the effluent baffle. Increasing the sludge collector speed will result in a decrease in blanket thickness and prevent solids from flowing under the baffle.

If the scrapers are already operating at full speed and the blanket level is below the effluent baffle, the unit is probably overloaded with regard to solids. In this case, the influent flow rate and concentration should be checked and the flow rate should be decreased, if possible.

High solids carryover with the effluent, in conjunction with lower than normal float solids concentrations, usually indicates that problems exist with the air system, chemical conditioning system, or the loading rates. The operator should systematically check the retention tank pressure and sight glass, the recycle pump, the air compressor assembly, the reaeration pump, the chemical conditioning equipment, and the influent flow.

Equipment malfunctions are quickly revealed by checking the retention tank pressure and the sight glass. Higher than desired pressures will result in decreased recycle rates and the back pressure-relief valve should be opened somewhat to decrease the pressure and increase the recycle rate. Lower than normal pressures will result in higher recycle rates. In this case, the pressure-relief valve should be closed somewhat to decrease the recycle rate and allow more time for air to dissolve in the retention tank.

Malfunctions in the retention tank liquid level indicator and air compressor activation assembly will also cause drastic decreases in flotation efficiency. The liquid level in the sight glass is the best indicator of this problem.

If the liquid level in the retention tank is lower or higher than normal, either the float mechanism to activate the air inlet valve or control is malfunctioning or the air compressor and solenoid valves are not operating correctly. If the liquid level in the retention tank is not at the desired level, shut off the DAF unit, open the hatch on the retention tank, and clean the liquid level indicator probes.

If everything (air, recycle, and retention pressure) seems to be in proper order, but the DAF effluent is still high in solids and the float solids are at a low concentration, check the retention tank mixer (reaeration pump), the chemical conditioning system, and the loading rates.

If chemical conditioners are used, they must be prepared properly and applied at the desired dosage. Chemical conditioning is discussed in Section 3.31. Proper operation of the chemical conditioning system will greatly help the performance of the DAF unit. The chemical mixing and delivery systems should be carefully watched and calibrated because of the high cost of chemicals.

If all the equipment is operating properly and the problem still exists, check the hydraulic and solids loading according to Example 10 and adjust flow rates as required.

Table 3.5 summarizes problems that may arise when operating a dissolved air flotation system and the corrective measures that might be taken.

TABLE 3.5   TROUBLESHOOTING DISSOLVED AIR FLOTATION

Operational Problem	Possible Cause	Check or Monitor	Possible Solution
1. Solids carry over with effluent but good float (thickened sludge) concentration	1. Float blanket too thick	1a. Flight speed 1b. Solids loading	1a. Increase flight speed 1b. Lower flow rate to unit, if possible
2. Good effluent quality but float thin (dilute)	2. Float blanket too thin	2a. Flight speed 2b. Solids loading	2a. Decrease flight speed 2b. Increase flow rate, if possible
3. Poor effluent quality and thin (dilute) float	3a. A/S low	3a. (1) Air rate (2) Compressor	3a. (1) Increase air input (2) Repair or turn on compressor
	3b. Pressure too low or too high	3b. Pressure gauge	3b. Open or close valve
	3c. Recycle pump inoperative	3c. Pressure gauge and pump	3c. Turn on recycle pump
	3d. Reaeration pump inoperative	3d. Pump pressure	3d. Turn on reaeration pump
	3e. Chemical addition inadequate	3e. Chemical system	3e. Increase chemical dosage
	3f. Loading excessive	3f. Loading rates	3f. Lower flow rate

Write your answer in a notebook and then compare your answer with the one on page 309.

QUESTION

3.12I   On a routine check of a dissolved air flotation unit, the operator notices high suspended solids in the effluent and a thinner than normal sludge. *DISCUSS* the possible causes and solutions to the problem.

3.13   Centrifuge Thickeners

NOTE: Centrifuge popularity depends on costs, energy requirements, performance, and alternative processes. Centrifuge manufacturers are always striving to improve their equipment.

Centrifugal thickening of wastewater sludge results from sedimentation and high centrifugal forces. Sludge is fed at a constant feed rate to a rotating bowl. Solids are separated from the liquid phase by the centrifugal forces and are forced to the bowl wall and compacted. The liquid and fine solids (*CENTRATE* [20]) exit the unit through the effluent line.

Three centrifuge designs are commonly installed today. They are: (1) basket centrifuges, (2) scroll centrifuges, and (3) disc-nozzle centrifuges. The mechanical operation of the three centrifuges varies significantly and separate descriptions of each will follow.

BASKET CENTRIFUGE. The basket centrifuge is a solid bowl that rotates along a vertical axis and operates in a batch manner. A schematic of a typical basket centrifuge is shown in Figure 3.4. Feed material is transported by a pipe through the top and is introduced at the bottom of the unit. This sludge is accelerated radially outward to the basket wall by centrifugal force. Cake continually builds up within the basket until the quality of the centrate, which overflows a weir at the top of the unit, begins to deteriorate. At that point, feed to the unit is stopped and a nozzle skimmer enters the bowl to remove the innermost and wettest portion of the retained solids. The inner solids are generally too wet for conveyor belt transport through the system. Upon completion of the skimming sequence, which takes about one-half minute, deceleration of the bowl takes place followed by knife or plow insertion. As the knife moves toward the bowl wall, retained solids are scraped out and fall through the bottom of the basket for conveyance to a discharge point as cake. Upon retraction of the knife, the solids discharge cycle is completed.

When basket centrifuges are used as thickening devices, full-depth skimming is commonly practiced with the nozzle skimmer while the basket is revolving at full speed, and the deceleration and knife insertion sequences are eliminated from the operation.

SCROLL CENTRIFUGE. The scroll centrifuge is a solid bowl that rotates along a horizontal axis and operates in a continuous manner. This type of centrifuge is illustrated in Figures 3.5 and 3.6. The newest design in scroll centrifuges is a tapered bowl that uses an inner scroll to evenly distribute the feed sludge. Sludge is fed to the unit through a stationary tube along the centerline of the inner screw, which accelerates the sludge and minimizes turbulence. Sludge passes through ports in the inner conveyor shaft and is distributed to the periphery (outer edge) of the bowl. Solids settled through the liquid pool in the separating chamber are compacted by centrifugal force against the bowl and are conveyed by the outer screw conveyor to the discharge point. Separated liquid (centrate) is discharged continuously over an adjustable weir.

DISC-NOZZLE CENTRIFUGE. The disc-nozzle centrifuge is a solid bowl that rotates along a vertical axis and operates in a continuous manner. A schematic of the centrifuge is shown in Figure 3.7. Feed material is introduced at the top of the unit and flows through a set of some 50 conical discs that are used for stratification (separation into layers) of the wastestream to be clarified. The discs are fitted quite closely together and centrifugal force is applied to the relatively thin film of liquid and solids between the discs. This force throws the denser, solid material to the wall of the centrifuge bowl where it is subjected to additional centrifugal force and concentrated before it is discharged through nozzles located on the periphery. Clear liquid continuously flows over a weir at the top of the bowl and exits through the centrate line. The bowl is equipped with 12 nozzle openings, but various numbers and sizes of discharge nozzles can be used, depending on the wastewater characteristics and the desired results. The number and size of discharge nozzles used directly controls final sludge concentration for any given feed condition.

3.130   Factors Affecting Centrifuge Thickeners

The performance of centrifugal thickeners depends on: (1) type of sludge, (2) age of the feed sludge, (3) solids and hydraulic loading, (4) bowl speed and resulting gravitational ("g") forces, (5) pool depth and differential scroll speed for scroll centrifuges, (6) size and number of nozzles for disc centrifuges, and (7) chemical conditioning.

Centrifuges are not commonly used to thicken primary sludges because all three of the designs have sludge inlet assemblies that clog easily. For this reason, there will be no discussion of centrifugal thickening of primary sludge.

Secondary sludges are more suited to centrifugal thickening because their relative lack of stringy and bulky material reduces the potential for plugging. Centrifuges are less affected by sludge characteristics such as age of sludge and bulking or rising conditions due to the high centrifugal forces developed. Usually, centrifugal forces of 600 to 1,400 "g's" or 600 to 1,400 times the force of gravity are developed and fluctuations in sludge thickening characteristics can generally be overcome. However, in all cases, if the sludge is fresh and exhibits good settling characteristics, it would more readily be thickened than an old sludge. Every attempt should be made, regardless of the thickening

---

[20] *Centrate.*   The water leaving a centrifuge after most of the solids have been removed.

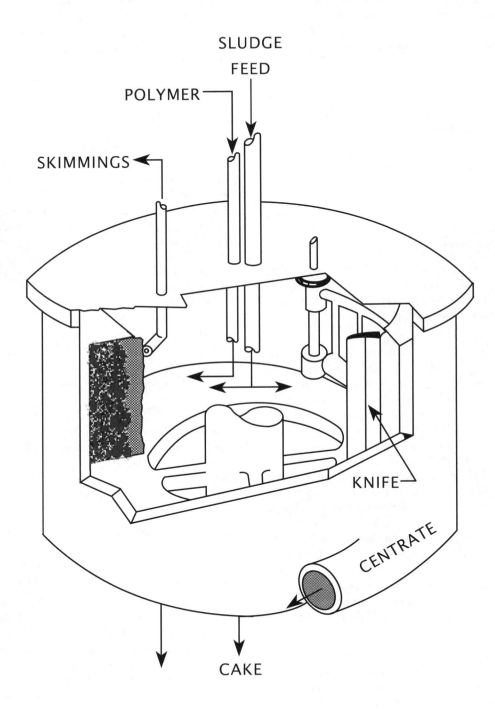

Fig. 3.4   Basket centrifuge

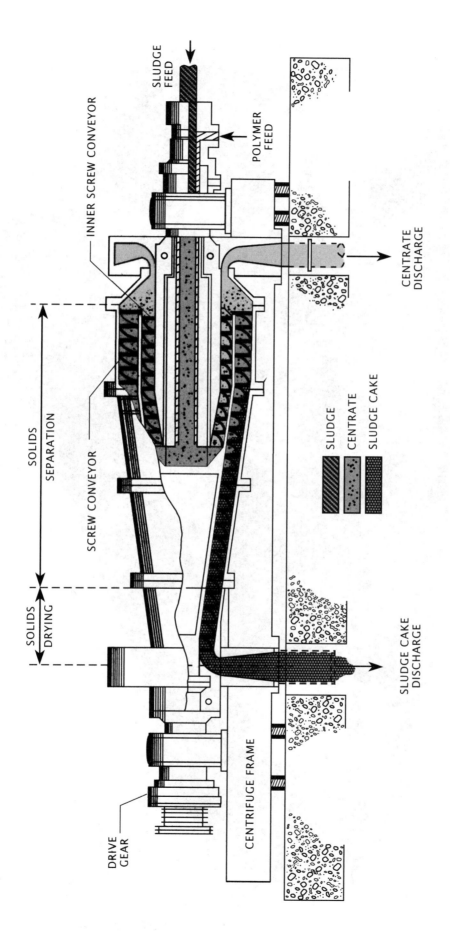

SLUDGE
FEED

POLYMER
FEED

INNER SCREW CONVEYOR

SCREW CONVEYOR

SOLIDS
SEPARATION

SOLIDS
DRYING

DRIVE
GEAR

CENTRIFUGE FRAME

SLUDGE CAKE
DISCHARGE

CENTRATE
DISCHARGE

SLUDGE

CENTRATE

SLUDGE CAKE

Fig. 3.5   Scroll centrifuge (horizontal-tapered bowl)

Fig. 3.6   Photo of scroll centrifuge
(Permission of Dorr-Oliver Incorporated)

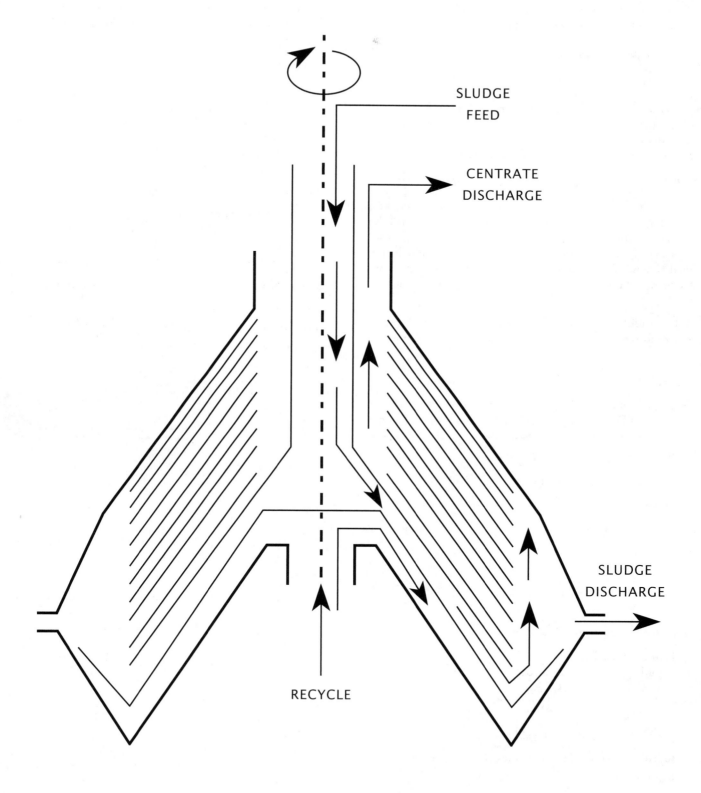

SLUDGE
FEED

CENTRATE
DISCHARGE

SLUDGE
DISCHARGE

RECYCLE

Fig. 3.7   Disc centrifuge

system used, to feed a consistent and fresh sludge to the thickening facility. Solids and hydraulic loading, bowl speed, feed time, differential scroll speed and pool depth, and nozzle size and number will be discussed in the following paragraphs.

3.131 Operating Guidelines

The physical size and number of centrifugal thickeners needed depend on the anticipated sludge volume and its dewatering properties. For a specific plant, the operator usually has a variety of operational controls to optimize centrifuge performance. Prior to a discussion of these control strategies, the operator should be familiar with determining hydraulic and solids loadings.

3.1310 HYDRAULIC AND SOLIDS LOADINGS

Unlike gravity and flotation thickeners, the loading rates for centrifuges are not related to unit areas (GPM/sq ft or GPD/sq ft). The accepted loading unit terminology for centrifuge loadings are gal/hr/unit or lbs/hr/unit. This type of terminology is used because of the various sizes available from different manufacturers and the variations in design from one unit to the next unit.

The loading rates of scroll centrifuges and disc centrifuges are straightforward and illustrated in the following sample calculations.

EXAMPLE 15

Given: A scroll centrifuge was selected to thicken 120,000 GPD of waste activated sludge with an initial sludge solids (SS) concentration of 0.80 percent (8,000 mg/L).

Find: 1. The hydraulic load (gal/hr).

2. The solids load (lbs SS/hr).

Solution:

Known		**Unknown**
Scroll Centrifuge		1. Hydraulic Load, gal/hr
Flow, GPD	= 120,000 GPD	
Sludge Solids, mg/L	= 8,000 mg/L	2. Solids Load, lbs Sl Sol/hr
, %	= 0.80%	

1. Determine the hydraulic load in gallons per hour.

$$\text{Hydraulic Load,}\ \text{gal/hr} = \frac{\text{Flow, GPD}}{24\ \text{hr/day}}$$

$$= \frac{120,000\ \text{gal/day}}{24\ \text{hr/day}}$$

$$= 5,000\ \text{gal/hr}$$

2. Calculate the solids load in pounds of sludge solids per hour.

$$\text{Solids Load,}\ \text{lbs/hr} = \text{Flow,}\ \frac{\text{gal}}{\text{hr}} \times 8.34\ \frac{\text{lbs}}{\text{gal}} \times \text{Sl Sol,}\ \frac{\%}{100\%}$$

$$= 5,000\ \frac{\text{gal}}{\text{hr}} \times 8.34\ \frac{\text{lbs}}{\text{gal}} \times \frac{0.80\%}{100\%}$$

$$= 334\ \text{lbs/hr}$$

These same calculations would apply for disc-nozzle centrifuges. The determination of loading rates for basket centrifuges is more complicated because they operate in a batch manner and the downtime required to remove the thickened solids must be incorporated in the loading calculation as follows.

EXAMPLE 16

Given: A basket centrifuge is fed 50 GPM of waste activated sludge at a sludge solids concentration of 0.80 percent. The basket run time is 20 minutes for the unit to fill completely with solids. After the unit is full, the solids are skimmed out. The skimming operation takes 1½ minutes.

Find: 1. The hydraulic load (gal/hr).

2. The solids load (lbs SS/hr).

Solution:

Known		**Unknown**
Basket Centrifuge		1. Hydraulic Load, gal/hr
Flow, GPM	= 50 GPM	
Sludge Solids, %	= 0.80%	2. Solids Load, lbs Sl Sol/hr
Run Time, min	= 20 min	
Skimming Time, min	= 1.5 min	

1. Determine the hydraulic load in gallons per hour.

$$\text{Hydraulic Load,}\ \text{gal/hr} = \text{Flow,}\ \frac{\text{gal}}{\text{min}} \times \frac{\text{Run Time, min}}{\left(\text{Run Time,}\ \text{min} + \text{Skim Time,}\ \text{min}\right)} \times 60\ \frac{\text{min}}{\text{hr}}$$

$$= 50\ \frac{\text{gal}}{\text{min}} \times \frac{20\ \text{min}}{(20\ \text{min} + 1.5\ \text{min})} \times 60\ \frac{\text{min}}{\text{hr}}$$

$$= 2,790\ \text{gal/hr}$$

If the unit were fed continuously at a rate of 50 GPM, the hydraulic loading rate would be 3,000 gal/hr (50 gal/min × 60 min/hr).

2. Calculate the solids load in pounds of sludge solids per hour.

$$\text{Solids Load,}\ \text{lbs/hr} = \text{Flow,}\ \frac{\text{gal}}{\text{hr}} \times 8.34\ \frac{\text{lbs}}{\text{gal}} \times \text{Sl Sol,}\ \frac{\%}{100\%}$$

$$= 2,790\ \frac{\text{gal}}{\text{hr}} \times 8.34\ \frac{\text{lbs}}{\text{gal}} \times \frac{0.80\%}{100\%}$$

$$= 186\ \text{lbs/hr}$$

3.1311 BOWL SPEED

Regardless of the type of centrifuge (basket, scroll, or disc) used, increasing the bowl speed (RPM) will result in higher gravitational forces and thicker sludge concentration. This is because gravitational forces are directly proportional to the bowl diameter and revolutions per minute. For a given machine, the bowl diameter is fixed and cannot be changed. If more or less "g" force is desired, the bowl speed should be increased or decreased but in no case should the bowl be operated at speeds out of the manufacturer's recommended range. Operation out of the recommended range can lead to bearing failures and costly repairs.

3.1312 FEED TIME

This section deals with the basket-type batch operated centrifuges only. The actual feed time (run time) for basket centrifuges will depend on the influent sludge solids concentration (% SS), the flow rate (GPM), and the average concentration of the solids retained in the basket. The solids storage volume within a basket is fixed. If the feed is shut off prior to filling the storage area with solids, the portion of retained sludge farthest from the basket wall will be extremely wet. The net effect of not filling the basket completely with solids is an overall decrease in the cake solids concentration because large volumes of water are carried out during the skimming sequence.

Conversely, if the feed time exceeds the time required to fill the storage area with solids, the effluent quality will decrease drastically after the bowl is full. This is because once the basket is filled with solids, no more storage area is available for additional incoming solids.

The following example illustrates the feed time required to fill a 48-inch diameter basket centrifuge with concentrated solids. All 48-inch diameter baskets have solids storage volumes of approximately 16 cubic feet.

EXAMPLE 17

Given: A 48-inch diameter basket is used to thicken waste activated sludge at a concentration of 0.75 percent sludge solids. The applied flow rate is 50 GPM and the average concentration of solids within the basket is 7.0 percent.

Find: The time required to fill the storage volume with thickened sludge.

Solution:

Known	**Unknown**
Basket Centrifuge, 48-inch diameter	Time required to fill storage volume, min
Flow, GPM $= 50$ GPM	
Infl Solids, % $= 0.75\%$	
Basket Solids, % $= 7.0\%$	
Solids Storage Vol, cu ft $= 16$ cu ft	

1. Calculate the amount of stored solids in pounds.

$$\text{Solids, lbs} = \text{Storage Vol, cu ft} \times 62.4 \frac{\text{lbs}}{\text{cu ft}} \times \text{Bkt Sol,} \frac{\%}{100\%}$$

$$= 16 \text{ cu ft} \times 62.4 \frac{\text{lbs}}{\text{cu ft}} \times \frac{7.0\%}{100\%}$$

$$= 70 \text{ lbs}$$

Therefore, under these conditions, the centrifuge could store 70 pounds of dry solids.

2. Determine the time required to fill the storage volume or the feed time in minutes.

$$\text{Feed Time, min} = \frac{\text{Stored Solids, lbs}}{\text{Flow, GPM} \times 8.34 \frac{\text{lbs}}{\text{gal}} \times \text{Infl Sol,} \frac{\%}{100\%}}$$

$$= \frac{70 \text{ lbs}}{50 \frac{\text{gal}}{\text{min}} \times 8.34 \frac{\text{lbs}}{\text{gal}} \times \frac{0.75\%}{100\%}}$$

$$= \frac{70 \text{ lbs}}{3.13 \text{ lbs/min}}$$

$$= 22 \text{ minutes}$$

For the conditions given in the above example, feed times less than 22 minutes will result in wetter discharge solids and feed times greater than 22 minutes will result in poorer effluent quality.

3.1313 DIFFERENTIAL SCROLL SPEED AND POOL DEPTH

This section deals with scroll-type centrifuges only. In addition to being able to adjust the bowl speed, the operator can adjust the differential or relative scroll speed and the liquid depth (pool) within the bowl. As previously discussed, scroll centrifuges have an inner screw (scroll) that rotates at a different speed than the bowl. The difference between the bowl speed and the speed of the inner screw is called the "differential" (relative) scroll speed. As the differential scroll speed is increased, the solids that are compacted on the bowl wall are conveyed out of the centrifuge at a faster rate, resulting in a decrease in the concentration of these solids. Lower concentrations result because as the solids are moved out at a faster rate, they are subjected to centrifugal forces for shorter periods of time. Likewise, as the relative scroll speed is decreased, the solids at the bowl wall are moved out at a slower rate and are more compact because they are subjected to the centrifugal forces for longer times.

The liquid depth (pool depth) within the bowl can be varied by adjusting or changing the effluent weirs. As the bowl depth is increased, the effluent quality will also increase because the liquid level and consequently the hydraulic detention time within the bowl increases. Longer retention times result in increased suspended solids capture because these solids have a better opportunity to be thrown to the bowl wall. Conversely, as the pool depth decreases, the suspended solids removal and effluent quality also decrease due to shorter detention times within the bowl. However, changing the pool depth has just the opposite effect on the cake solids. As the pool depth is increased, solids recovery increases but the cake solids get wetter. As the pool depth is decreased and solids recovery decreases, the cake solids get drier. Thus, the operator must adjust the pool depth to get the recovery and cake solids desired, realizing that a high recovery will usually result in the wettest cakes, while dry cakes are normally accompanied with lower solids recovery.

3.1314   NOZZLE SIZE AND NUMBER

This section deals only with disc-nozzle type centrifuges. The degree of sludge thickening can be controlled somewhat by increasing or decreasing both the number of nozzles and nozzle openings. Nozzles are located at the periphery (outer edge) of the disc centrifuge bowl and are used to discharge the thickened sludge from the unit. If the size of the nozzles is increased, the dryness of the compacted sludge will decrease because the sludge will exit the unit at a faster rate and will not concentrate to its highest degree. This principle is much like that of the scroll speed where increasing differential scroll speeds result in wetter sludge. If the nozzle openings are reduced or the number of nozzles is decreased, the sludge will remain subjected to centrifugal forces for longer times and will dry or become thicker.

3.132   Normal Operating Procedures

Typically, the flow through centrifuges is continuous and should be set as constant as possible. Even though the basket centrifuge is a batch process, the flow rate during the feeding time should be as constant as possible. Routine monitoring of the influent, effluent, and thickened sludge streams should be done at least once per shift and samples should be collected for solids analysis. Normal start-up and shutdown procedures vary for the three centrifuge types and each is outlined below.

BASKET CENTRIFUGE

- Retract the skimmer and plow (knife).

- Turn on the drive motor.

- When the centrifuge reaches approximately 80 percent full speed, open appropriate chemical and sludge inlet valves and turn on the pumps.

- When the centrate "breaks" (high solids in effluent), turn off the sludge and chemical pumps.

- While the machine is operating at full speed, activate the skimmer to advance toward the wall and remove solids.

- If full-depth skimming cannot be used (sludge is too thick), retract the skimmer and push the deceleration button.

- When the bowl is rotating at approximately 50 to 70 RPM, activate the plow (knife).

- After all the solids are removed, retract the plow, accelerate the bowl, and proceed as above.

For any prolonged machine shutdown, pump fresh water into the bowl while the knife is inserted to clean the wall. Following clean-out, retract the knife and turn off the drive motor.

SCROLL CENTRIFUGE

- Turn on drive motor.

- When the bowl is at full speed, open appropriate chemical and sludge valves and turn on the respective pumps.

- Adjust differential (relative) scroll speed as required.

- Flush centrifuge after each use to prevent solids from caking on inside of bowl.

For prolonged machine shutdown, turn off the feed and chemical pumps, pump fresh water into the unit for approximately 20 to 30 minutes, and then turn off the drive motor.

DISC-NOZZLE CENTRIFUGE

- Turn on drive motor.

- Activate the pre-screens.

- When the unit is at full speed, open appropriate chemical and sludge valves and turn on the respective pumps.

For prolonged machine shutdown, turn off the feed and chemical pumps, pump water into the centrifuge for 20 to 30 minutes, and then turn off the main drive motor.

3.133   Typical Performance

Typical operating guidelines as well as thickened sludge concentrations and sludge solids removals for various centrifuge types are presented in Table 3.6.

TABLE 3.6   OPERATIONAL AND PERFORMANCE GUIDELINES FOR CENTRIFUGAL THICKENERS TREATING WASTE ACTIVATED SLUDGE

Centrifuge Type	Capacity, GPM[a]	Feed Solids, %	Thickened Sludge, %	Solids Recovery, %
Basket	33–70	0.7	9–10	70–90
Scroll	75–100	0.4–0.7	5–7	80–90
Disc	30–150	0.7–1.0	5–5.5	90+

[a] GPM × 0.063 = L/sec.

The variations in solids loading are due to the many different sizes of centrifuges available from various manufacturers. The performance data reflect no chemical conditioning prior to centrifugation. The addition of POLYMERS [21] normally improves the recovery of suspended solids much more than the recovery

---

[21] *Polymer* (POLY-mer).   A long-chain molecule formed by the union of many monomers (molecules of lower molecular weight). Polymers are used with other chemical coagulants to aid in binding small suspended particles to larger chemical flocs for their removal from water. Also see POLYELECTROLYTE.

of cake solids. For example, look at the solids recovery and cake solids versus polymer dosage curves for both a basket and scroll centrifuge shown in Figures 3.8, 3.9, 3.10, and 3.11. For the basket centrifuge, it can be seen that with no polymer addition, the thickened sludge solids were 4.5 percent and the suspended solids recovery was 78 percent. At a polymer dosage of approximately 5 lbs/ton (2.5 gm/kg), the thickened sludge solids were increased to 6 percent and the solids recovery leveled off at 95 percent. For the scroll centrifuge, the thickened sludge solids remained fairly constant at 7 percent regardless of the polymer dosage. However, with no polymer addition, the solids recovery was at 25 percent and could not reach 90 percent until the polymer dosage exceeded 11 lbs/ton (5.5 gm/kg). In all, the operator must realize that when using polymers, a great deal of experimentation and tinkering with both the dosage and point of application must be done to obtain the best results and minimize chemical costs. A more detailed discussion of the basics of chemical conditioning will be presented in Section 3.31.

The determination of centrifuge performance is based on laboratory solids analysis and the following calculations.

EXAMPLE 18

Given: A 22-inch diameter by 60-inch long scroll centrifuge is used to thicken 80 GPM of waste activated sludge (WAS). The WAS has an initial sludge solids concentration of 0.80 percent (8,000 mg/L). The centrifuge effluent (centrate) has a sludge solids concentration of 0.20 percent (2,000 mg/L).

Find: The sludge solids removal efficiency.

Solution:

Known	**Unknown**
Scroll Centrifuge, 22-inch diameter by 60-inch length	Sludge Solids Removal Efficiency, %

Flow, GPM   = 80 GPM

Infl SS, %   = 0.80%
  , mg/L = 8,000 mg/L

Effl SS, %   = 0.20%
  , mg/L = 2,000 mg/L

Determine the sludge solids removal efficiency as a percent.

$$\text{Efficiency, \%} = \frac{(\text{Infl SS, \%} - \text{Effl SS, \%}) \times 100\%}{\text{Infl SS, \%}}$$

$$= \frac{(0.80\% - 0.20\%) \times 100\%}{0.80\%}$$

$$= \frac{0.60 \times 100\%}{0.8}$$

$$= 75\%$$

The determination of thickened sludge concentrations (% TS) for scroll and disc centrifuges is based on collecting thickened sludge samples and analyzing for total solids content (% TS) according to procedures outlined in Chapter 16, Volume II, *OPERATION OF WASTEWATER TREATMENT PLANTS*. The determination of thickened sludge concentrations of basket centrifuges is more complicated because samples of the skimmed portions and the knifed portion of the retained solids have to be collected and analyzed, and the composite (average solids) has to be calculated based on the relative quantity of skimmed and knifed solids.

EXAMPLE 19

Given: A 48-inch diameter basket centrifuge with a total sludge storage volume of 16 cu ft (120 gal) is used to thicken WAS at an initial suspended solids concentration of 0.80 percent (8,000 mg/L). Approximately 13 cu ft of solids in the basket bowl were skimmed out and the average total solids concentration of the skimmed portion was determined to be 4.0 percent thickened sludge (TS). The remaining 3 cu ft were removed by inserting the knife (plow) and the average concentration of these solids was 7.5 percent total solids.

Find: The average (composite) total solids concentration of the thickened sludge removed from the basket.

Solution:

Known	**Unknown**
Basket Centrifuge, 48-inch diameter	Average Total Thickened Sludge Solids Removed, %

Storage Volume, cu ft   = 16 cu ft
  , gal   = 120 gal

Infl Sl Sol, %   = 0.80%
  , mg/L = 8,000 mg/L

Skimmed Volume, cu ft   = 13 cu ft

Skimmed Sl, %   = 4.0%

Knife Volume, cu ft   = 3 cu ft

Knife Solids, %   = 7.5%

Calculate the average thickened sludge solids as a percent.

$$\text{Thickened Sludge, \%} = \frac{(\text{Sk Vol, cu ft} \times \text{Sk Sl, \%}) + (\text{Kn Vol, cu ft} \times \text{Kn Sol, \%})}{\text{Sk Vol, cu ft} + \text{Kn Vol, cu ft}}$$

$$= \frac{(13 \text{ cu ft} \times 4.0\%) + (3 \text{ cu ft} \times 7.5\%)}{13 \text{ cu ft} + 3 \text{ cu ft}}$$

$$= \frac{52 + 22.5}{16}$$

$$= 4.66\%$$

This mathematical calculation assumes perfect mixing of the skimmed and knifed solids. In actual practice, perfect mixing is very difficult to achieve.

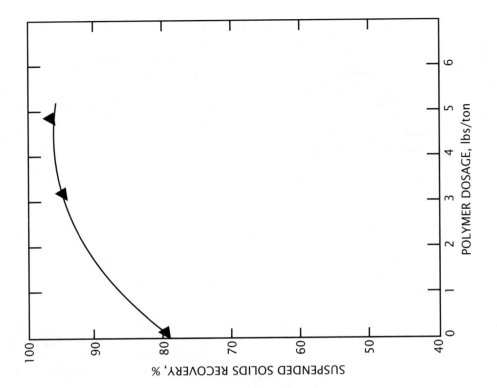

Fig. 3.9   Suspended solids recovery from basket centrifuge thickening

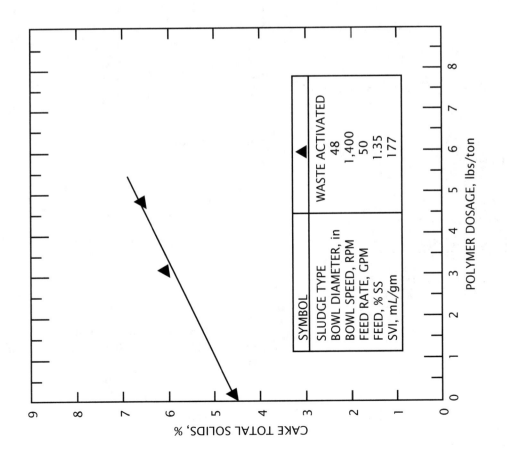

Fig. 3.8   Cake solids from basket centrifuge thickening

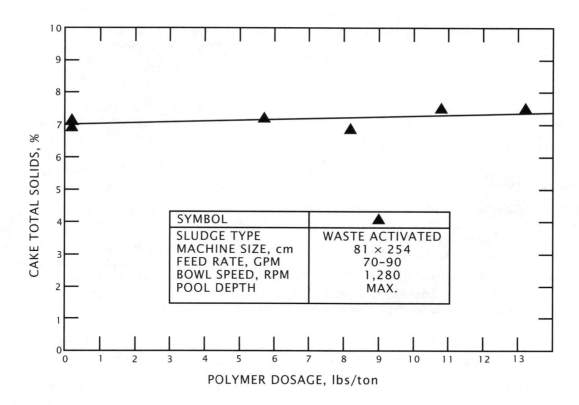

Fig. 3.10   *Cake solids from scroll centrifuge thickening*

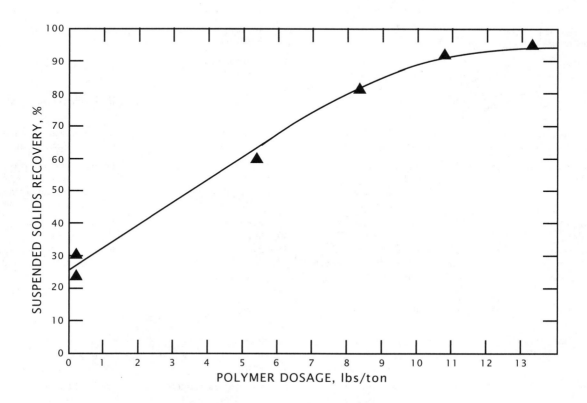

Fig. 3.11   *Suspended solids recovery from scroll centrifuge thickening*

QUESTIONS

Write your answers in a notebook and then compare your answers with those on pages 309 and 310.

3.13A   List the three centrifuge types. Which ones are continuous and which operate in a batch (intermittent feed) mode?

3.13B   List the factors that affect centrifugal thickening.

3.13C   Why are centrifuges not commonly used to thicken primary sludges?

3.13D   Determine the solids and hydraulic loading for a 20-inch by 60-inch scroll centrifuge. The feed rate is 30 GPM and the influent solids concentration is 1.1 percent (11,000 mg/L) suspended solids.

3.13E   Determine the hydraulic and solids loading for a 48-inch diameter basket centrifuge. The feed rate is 30 GPM, the feed time is 25 minutes and 3 minutes are required to receive the solids and restart the feed. The influent solids concentration is 1.1 percent.

3.13F   How does differential scroll speed affect performance of scroll centrifuges?

3.13G   A 20-inch disc centrifuge receives 25 GPM of waste activated sludge with a suspended solids concentration of 0.65 percent. The effluent (centrate) contains 0.03 percent SS (300 mg/L) and the thickened sludge concentration is 4.9 percent. Determine the percent efficiency and the concentration factor (cf).

3.134   Troubleshooting

Since the operating characteristics of the three centrifuges are quite different, the operating problems and corrective measures also are different. Troubleshooting for each of the centrifuges will be discussed in the following paragraphs. Table 3.7 lists operational problems that may develop and corrective measures

TABLE 3.7   TROUBLESHOOTING CENTRIFUGAL THICKENERS

Operational Problem	Possible Cause	Check or Monitor	Possible Solution
Basket			
1. Centrate quality good but discharge solids dilute	1a. Feed time too short 1b. Flow rate too low	1a. Time for centrate to break 1b. Sludge flow rate	1a. Increase feed time 1b. Increase flow rate
2. Centrate quality poor during the end of the run, but discharge solids OK	2a. Feed time too long 2b. Flow rate too high 2c. Incorrect chemical dose	2a. Time for centrate to break 2b. Sludge flow rate 2c. Chemical system	2a. Lower feed time 2b. Lower flow rate 2c. Increase chemical dosage
3. Centrate quality poor and discharge solids dilute	3a. High loadings 3b. Insufficient chemicals	3a. Flow rate and break time 3b. Chemical system	3a. Lower flow rate 3b. Increase chemical dosage
4. Vibrations	4a. Mechanical malfunctions such as bearings, drive unit, or base support 4b. Plugged feed port	4a. Inspect all mechanical equipment	4a. Mechanical repairs 4b. Unplug, as required
Scroll			
1. Centrate quality good but discharge solids dilute	1a. Scroll speed too high 1b. Pool depth too high	1a. Scroll RPM 1b. Pool depth	1a. Decrease scroll speed 1b. Lower pool depth
2. Centrate poor but discharge solids OK	2a. Scroll speed too slow 2b. Hydraulic load too high 2c. Pool depth too low 2d. Incorrect chemical dose	2a. Scroll RPM 2b. Flow rate 2c. Pool depth setting 2d. Chemical system	2a. Increase scroll speed 2b. Decrease flow 2c. Increase pool depth 2d. Increase chemical dosage
3. Centrate poor and discharge solids dilute	3a. Bowl speed too low 3b. Loading too high 3c. Chemical inefficiencies 3d. Scroll speed and pool depth not optimum	3a. Bowl RPM 3b. Flow rate 3c. Chemical system 3d. Scroll RPM and pool depth	3a. Increase bowl speed 3b. Decrease flow rate 3c. Increase chemical dosage 3d. Vary scroll speed and pool depth
Disc-Nozzle			
1. Centrate good but discharge solids dilute	1a. Size and number of nozzles too large	1a. Nozzles	1a. Decrease number or size of nozzles
2. Centrate poor but discharge solids OK	2a. Size and number of nozzles too large 2b. Hydraulic load too high	2a. Nozzles 2b. Flow rate	2a. Increase number or size of nozzles 2b. Decrease flow rate
3. Vibrations	3a. Mechanical malfunctions such as bearings, drive unit, or base support 3b. Plugged nozzle	3a. Inspect all mechanical equipment 3b. Nozzle	3a. Mechanical repairs 3b. Unplug, as required

the operator may take. Remember that close visual monitoring is the operator's best indication of operational problems.

3.1340   BASKET CENTRIFUGE

The operator should be concerned with the concentration of the thickened excess biological sludges. The entire volume of stored sludge can usually be skimmed out without having to use the deceleration and knife insertion sequences when chemical conditioners are not used. If you noticed that the initial skimmings (stored solids farthest from basket wall) contain large quantities of relatively clear water, check the feed time or the influent sludge flow. Short feed times or low flows will result in only partial filling of the storage volume. If the storage volume is not completely filled with solids prior to discharge, the sludge will be thinner than desired because of dilution with the water discharged. Monitor the centrate quality with time for one complete run, then adjust (increase) the feed time or flow rate so that the centrate quality "breaks" when the feed sequence is finished. Conversely, if the thickened sludge concentrations appear to be in a desirable range but the overall centrate quality is poor, monitor the effluent for one complete run and adjust (decrease) the feed time or sludge flow rate. If the feed time or sludge flow exceeds the time required to fill the storage volume, the majority of the solids entering the unit beyond the "break" point will exit with the centrate.

If polymers are added for conditioning, the net effect will be an increase in feed time, suspended solids recovery, and thickened sludge concentrations. If the conditions described above are evident, check the polymer addition system in addition to procedures mentioned. The use of polymers also causes the sludge to thicken to a higher degree and the skimmer may not be able to travel all the way to the basket wall. The skimmer will usually proceed toward the wall until it encounters sludge in excess of approximately 6 percent thickened sludge (TS). At this concentration, thickened biological sludges are usually not fluid enough to flow through the skimmer and flow through downstream piping. To remove the remaining stored solids, the deceleration and knife insertion sequence must be used. The problems that may arise could be plugging of the skimmer, if it proceeds too far into the thickened sludge, or wet and sloppy discharged solids upon deceleration and knife insertion, if the skimmer does not proceed far enough into the sludge. The distance that the skimmer travels is adjustable by set screws on the skimming mechanism. This distance of travel should be set by monitoring a few runs and adjusting as required to obtain a firm and conveyable knifed sludge.

Another problem that may develop with the use of baskets is vibration failures due to plugging of the feed ports or uneven solids distribution. This problem usually develops only when dewatering primary sludges and will be discussed in Section 3.4, "Dewatering."

3.1341   SCROLL CENTRIFUGE

The operational controls for scroll centrifuges on a day-to-day basis usually include relative scroll speed, pool depth, sludge flow, and chemical conditioning when used. Unless the centrifuge is equipped with a hydraulic backdrive, the bowl speed cannot be changed without changing the belt sheaves. In addition, once the optimum bowl speed has been determined and set, there is usually no point in changing it for a given sludge. The same can be said regarding the pool depth because maximum pool depths are commonly used when thickening sludges. This is because in thickening processes it is usually desirable to recover as much of the influent sludge solids as possible.

The performance breakdowns that are commonly encountered are deteriorations in centrate quality and decreases in discharge or cake total solids concentrations. For any given centrifuge, there are upper limits for hydraulic and solids loadings. If these limits are exceeded, both the centrate and cake solids will fall below the desired range. If the centrifuge is operated within its loading limits, the most common problem is a decrease in centrate quality or cake dryness. When this problem is evident by visual observation, adjust the relative scroll speed, monitor the centrate and cake, and readjust the relative scroll speed until the desired results are achieved. If the centrate quality is poor but the cake is within a desired range, increasing the relative scroll speed should result in a cleaner centrate. As the scroll speed is increased, the centrifuged solids are conveyed out at a faster rate and the solids are not given the opportunity to entirely fill the bowl and flow over the effluent weir.

To achieve good solids recovery with a scroll centrifuge, polymers are usually required when thickening biological solids unless the centrifuge is operated well below its loading capacities. Thickened biological solids are plastic in nature and tend to slip within the bowl as the screw or scroll conveyor tries to move them out. In order to successfully move these solids out of the centrifuge and produce a desirable centrate, polymers are often required. Therefore, check the polymer conditioning equipment in conjunction with relative scroll speed to optimize centrate quality and thickened sludge concentrations.

3.1342   DISC-NOZZLE CENTRIFUGE

Disc-nozzle centrifuges operate at higher speeds than other types of centrifuges and usually develop centrifugal forces in excess of 3,000 "g's." Because of these high "g" forces, suspended solids recoveries are almost always in excess of 90 percent. Centrate quality usually poses no operational concerns unless the thickened sludge is not adequately removed. The solids will eventually build up and contaminate the centrate. If this happens, the size or number of the discharge nozzles should be increased to facilitate sludge discharge. The nozzles usually do not have to be changed for day-to-day operation. If the centrate contains a high amount of suspended solids, check the hydraulic flow rate. Operating at loadings in excess of the recommended range will almost always result in less than optimum performance.

One of the major mechanical problems associated with disc-nozzle centrifuges is plugging of the nozzles because of the relatively small openings (0.07 to 0.08 inch or 1.75 to 2.00 mm). When this occurs, pre-screening of the sludge has to be incorporated into the process sequence. Unless the sludge is adequately screened, the nozzles will continuously plug and interrupt operation of the unit. Plugging will be evident by excessive machine vibrations due to an uneven distribution of solids along the bowl wall. All centrifuges are equipped with vibration switches for automatic shutoff in the event of excessive vibration. If a nozzle becomes plugged, disassemble the bowl assembly and remove and clean the nozzles prior to restarting the centrifuge.

3.1343 STRUVITE CONTROL

Struvite (magnesium ammonium phosphate, $MgNH_4PO_4 \cdot 6H_2O$) scales can be formed in anaerobic digesters and also in downstream digested sludge concentration and handling equipment. Anaerobic digestion of sludges converts ammonia (NH_3-N), phosphate (PO_4-P), and magnesium (Mg) to soluble forms. When the concentrations of these constituents of struvite exceed the saturation level, struvite scale deposits will form on piping and equipment. One method used to control struvite is the removal (precipitation) of phosphate (PO_4-P) by the addition of an iron salt (ferric chloride, $FeCl_3$, or ferrous chloride, $FeCl_2$). These iron salts are added either to the wastewater flow or to the digesting sludge.

Studies conducted at the City of San Francisco's Southeast Water Pollution Control Plant determined that the optimum dose of an iron salt for struvite control depends on the amount of soluble phosphorus (P) and soluble magnesium (Mg) available for precipitation as well as the ratio (0.37 for San Francisco) of phosphate (PO_4-P) removed per iron (Fe) added. The optimum doses determined at San Francisco were 2,200 mg/L ferric chloride or 100 kg ferric chloride per ton of total solids under digestion. Specific values can be determined for a particular plant by conducting a series of continuous-flow anaerobic digester experiments at various doses of ferric chloride.

For additional information, see P. Pitt, *et al.,* "Struvite Control Using Ferric Chloride at the Southeast Plant, San Francisco, CA." Paper presented at 65th Annual Conference and Exposition of Water Environment Federation, New Orleans, Louisiana, September 20–24, 1992.

QUESTIONS

Write your answers in a notebook and then compare your answers with those on page 310.

3.13H A scroll centrifuge is used to thicken waste activated sludge. On a routine check, the operator notices a poor centrate quality. The discharge solids are good. What should the operator check and what action should be taken?

3.13I What operational modification must be made to a disc-nozzle centrifuge if thickened sludge solids are building up in the centrifuge?

3.14 Gravity Belt Thickeners

A very effective sludge thickening alternative for secondary sludges is the gravity belt thickener (GBT). A schematic of a typical gravity belt thickener is presented in Figure 3.12. Sludge to be thickened is preconditioned, usually with polymer, then applied to the gravity belt thickener where free water drains through small openings in the belt and is collected in a trough below the belt. After traveling the full length of the upper gravity zone, the thickened sludge is removed from the belt and subsequently transferred to the next sludge treatment process, stabilization. After the thickened sludge is removed from the belt, the belt continues its return travel to the inlet zone and is washed by an internal high-pressure wash water spray system. The ability of gravity belt thickeners to effectively thicken secondary sludge depends on belt type, chemical conditioning, belt speed, and hydraulic and solids loadings.

Belts are available in a variety of materials (nylon, polypropylene), each with various porosities. Porosity is a measure of fiber openings. As the porosity increases, the resistance to flow decreases and larger volumes of water are able to be drained. If the porosity (fiber openings) is too large, sludge solids may pass through the belt and result in poor filtrate quality. If the porosity is too low, the belt may blind or plug, which will produce frequent washouts. Washout occurs when a large quantity of free water is unable to be released in the drainage zone and it travels to the discharge end where it is carried out with the thickened sludge. The excess water will reduce the thickened sludge concentration. The right belt for a particular application is determined by experimentation in conjunction with the manufacturer's recommendations. After a belt is selected and installed, the operator should provide for adequate belt cleaning by maintaining the belt washing equipment in proper working condition and by adjusting the wash water volume, as required.

Polymers are generally used for chemical conditioning in conjunction with gravity belt thickening of sludges. Polymer dosage must be optimum to ensure optimum separation of free moisture from the sludge and subsequent drainage through the belt. With proper operating conditions, GBTs can thicken secondary sludges from concentrations of 0.3 to 0.6 percent suspended solids to concentrations of 4 to 6 percent suspended solids, while achieving a 95 percent or better suspended solids removal efficiency.

3.140 Operating Guidelines

A GBT should produce a high degree of thickening, provided the operator is aware of and exercises proper control of belt speed, hydraulic loading, and solids loading rates.

3.1400 BELT SPEED

The belt speed can be varied from approximately 2 to 10 feet/minute (0.6 to 3 m/min). The speed at which the belt should be operated depends on the sludge flow rate to the belt and the concentration of the influent sludge. As the belt speed is increased, the rate of belt area contacting the influent sludge also increases and allows for greater volumes of water to drain from the sludge. If the belt area is not sufficient to allow the free water to drain, belt washout will cause a reduction in the thickened sludge concentration.

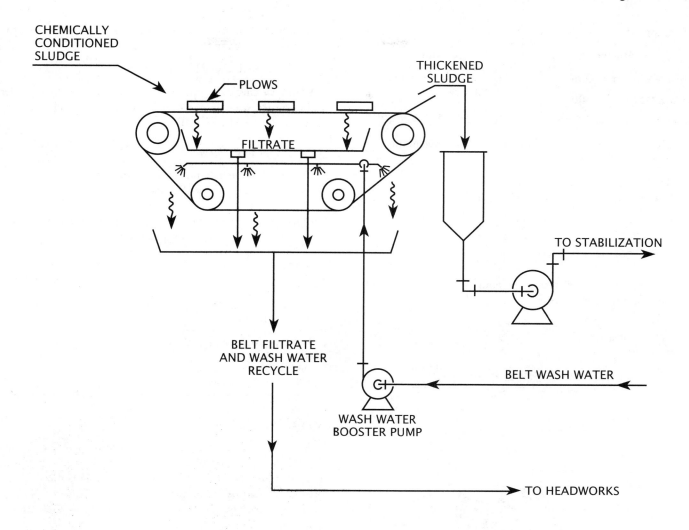

CHEMICALLY
CONDITIONED
SLUDGE

PLOWS

THICKENED
SLUDGE

FILTRATE

TO STABILIZATION

BELT FILTRATE
AND WASH WATER
RECYCLE

BELT WASH WATER

WASH WATER
BOOSTER PUMP

TO HEADWORKS

Fig. 3.12   Gravity belt thickener

The belt speed should be controlled so that the sludge delivered to the discharge zone has a minimum amount of free water. The experienced operator will be able to optimize belt speed by observing the dryness of the solids delivered to the discharge end. Obviously, as the concentration of influent sludge increases, less water is associated with the sludge mass and reduced belt speed can be used. The ideal operating belt speed is the slowest the operator can maintain without washing out the belt.

3.1401   HYDRAULIC AND SOLIDS LOADINGS

Hydraulic and solids loadings are based on the flow applied per unit of belt width and the solids concentration of the influent sludge. GBTs range in width from approximately 3 to 10 feet (1 to 3 m). Manufacturers generally recommend hydraulic loadings of 100 to 250 GPM (6.3 to 16 L/sec) per meter of belt width, and solids loadings of 375 to 600 lbs/hour (170 to 275

kg/hour) of suspended solids per meter of belt width. The ideal loading for a particular belt thickener and a specific sludge should be determined on the basis of thickening performance and consistency of operation. If a GBT is operated close to the upper hydraulic limit, frequent washing out may result, due to slight variations in sludge characteristics or chemical dosages. Example 20 shows how to calculate the hydraulic and solids loadings for a gravity belt thickener.

EXAMPLE 20

Given: A 2-meter wide gravity belt thickener receives 400 GPM of waste activated sludge at a concentration of 0.55 percent (5,500 mg/L).

Find: 1. The hydraulic loading, GPM/m.

2. The suspended solids loading, lbs/hr/m.

Solution:

Known	Unknown
Belt Width, m = 2 m	1. Hydraulic Loading, GPM/m
Flow, GPM = 400 GPM	
SS Conc, % = 0.55%	2. Solids Loading, lbs/hr/m

1. Determine the hydraulic loading in gallons per minute per meter of belt width.

$$\text{Hydraulic Loading, GPM/m} = \frac{\text{Flow, GPM}}{\text{Belt Width, m}}$$

$$= \frac{400 \text{ GPM}}{2 \text{ m}}$$

$$= 200 \text{ GPM/m}$$

2. Determine the solids loading in pounds of suspended solids per hour per meter of belt width.

$$\text{Solids Loading, lbs/hr/m} = \frac{\text{Flow, GPM} \times 60 \frac{\text{min}}{\text{hr}} \times 8.34 \frac{\text{lbs}}{\text{gal}} \times \text{SS Conc}, \frac{\%}{100\%}}{\text{Belt Width, m}}$$

$$= \frac{400 \text{ GPM} \times 60 \frac{\text{min}}{\text{hr}} \times 8.34 \frac{\text{lbs}}{\text{gal}} \times \frac{0.55\%}{100\%}}{2 \text{ m}}$$

$$= \frac{1,100 \text{ lbs/hr}}{2 \text{ m}}$$

$$= 550 \text{ lbs/hr/m}$$

3.141 Normal Operating Procedures

The operation of different gravity belt thickeners will vary somewhat, but the following procedures can be used for most types:

1. Prepare an adequate supply of an appropriate polymer solution.

2. Turn on the wash water sprays and the belt drive.

3. Adjust the belt speed to its maximum setting and allow the wash water to wet the entire belt.

4. Open appropriate sludge and polymer valves.

5. Turn on the polymer pump and adjust the polymer rate as required to achieve the desired dosage.

6. Turn on the sludge feed pump.

7. Lower the belt speed as low as possible without running the risk of washing out.

8. Turn on the thickened sludge pumps.

Routinely check the operation of the gravity belt thickener and make adjustments, as required, to produce a thick discharge sludge and good filtrate quality.

3.142 Typical Performance

Operating guidelines, polymer requirements, and typical performance data are presented in Table 3.8.

TABLE 3.8   TYPICAL PERFORMANCE OF GRAVITY BELT THICKENERS

Sludge Type	Polymer, lbs/ton[a]	Cake, %TS[b]	SS Recovery, %	Hyd. Load, GPM/m[c]
Secondary	3–8	4–7	90–97	130–200

[a] lbs of dry polymer/ton dry sludge solids. lbs/ton × 0.5 = gm polymer/kg dry sludge solids.
[b] Thickened sludge.
[c] GPM/m × 0.0631 = L/sec/m.

3.143 Troubleshooting

A close watch over GBT operation in combination with field experience should enable the operator to optimize performance. In addition to mechanical reliability and maintenance, the operator should be concerned with the filtrate quality and dryness of the thickened sludge. The most frequent problem encountered with gravity belt thickeners is washing out. Usually, this problem is indicated by large volumes of water carrying over with the thickened sludge. When this happens check: (1) the polymer dosage, (2) hydraulic loading, (3) solids loading, (4) belt speed, and (5) belt washing equipment.

If the polymer dosage is too low, the solids will not flocculate and free water will not be released from the sludge mass. If the polymer dosage is adequate, evidenced by large floc particles and free water, increase the belt speed so as to provide more belt surface area for drainage. If the belt is already at its maximum setting, check the flow rate to the belt and reduce it if the rate is higher than normal. If the polymer dosage, belt speed, and hydraulic loading are set properly but washing out still persists, the problem may be related to blinding of the belt. Check the appearance of the belt as it leaves the washing chamber. If the belt appears to be dirtier than normal, increase the wash water rate, turn off the polymer and feed pumps, and allow the belt to be washed until it is clean. Belt blinding often develops because of polymer overdosing. If too much polymer is added, the belts can become coated with a film of excess chemical, which will prevent drainage and result in washing out of the belt. Check the polymer addition rate and adjust the rate, as required, so as not to grossly overdose the sludge. Table 3.9 summarizes typical operational problems and corrective measures that could be taken to optimize GBT performance.

TABLE 3.9 TROUBLESHOOTING GRAVITY BELT THICKENERS

Operational Problem	Possible Cause	Check or Monitor	Possible Solution
1. Washing out of belt	1a. Polymer dosage insufficient 1b. Hydraulic load too high 1c. Belt speed too low 1d. Belt blinding	1a. Polymer flow rate and solution 1b. Sludge flow rate 1c. Belt speed 1d. Washing equipment and polymer overdosage	1a. Increase dose rate 1b. Lower flow 1c. Increase speed 1d. Increase wash water rate. Turn off sludge and polymer and clean belts. Reduce polymer, if overdosing.
2. Cake solids too wet	2a. Belt speed too high 2b. Belt tension too low	2a. Belt speed 2b. Belt tension	2a. Reduce belt speed 2b. Increase belt tension

QUESTIONS

Write your answers in a notebook and then compare your answers with those on pages 310 and 311.

3.14A List the factors affecting gravity belt thickener performance.

3.14B What problem may develop if the belt porosity is too low?

3.14C What does the term "washing out" mean?

3.14D Explain how low belt speed affects GBT performance.

3.14E What is the ideal belt speed?

3.14F What steps should the operator take if washing out of the belt develops?

3.14G How can blinding of the belt be corrected?

3.15 Thickening Summary

The successful operation of any thickening device depends on the operator's knowledge of the operating guidelines, the consistency of the influent sludge, maintaining loading rates within the recommended and design values, and, when used, effective polymer addition. For optimum operation of subsequent sludge processes (stabilization, dewatering), the thickener should be operated so as to produce as thick a sludge as possible with maximum sludge solids recovery.

END OF LESSON 1 OF 5 LESSONS

on

RESIDUAL SOLIDS MANAGEMENT

Please answer the discussion and review questions next.

DISCUSSION AND REVIEW QUESTIONS

Chapter 3. RESIDUAL SOLIDS MANAGEMENT

(Lesson 1 of 5 Lessons)

At the end of each lesson in this chapter you will find some discussion and review questions. The purpose of these questions is to indicate to you how well you understand the material in the lesson. Write the answers to these questions in your notebook.

1. Why is the handling and disposal of sludge such a complicated problem?

2. List the major types of alternatives available for processing sludges.

3. What are the advantages normally associated with sludge thickening?

4. How does temperature affect the operation of gravity thickeners?

5. The performance of dissolved air flotation thickeners depends on what factors?

6. What is the most important operational concern when operating a dissolved air flotation thickener?

7. What is the best way for an operator to detect operational problems in a centrifuge?

8. Describe how a gravity belt thickener works.

CHAPTER 3. RESIDUAL SOLIDS MANAGEMENT

(Lesson 2 of 5 Lessons)

3.2 STABILIZATION [22]

3.20 Purpose of Sludge Stabilization

Prior to the disposal of wastewater treatment plant sludges, federal, state, and local regulatory agencies require that they be stabilized. Stabilization converts the volatile (organic) or odor-causing portion of the sludge solids to nonodorous end products, prevents the breeding of insects upon disposal, reduces the pathogenic (disease-carrying) bacteria content, and improves the sludge dewaterability.

Unit processes commonly used for stabilization of waste-water sludges include: (1) anaerobic digestion, (2) aerobic digestion, and (3) chemical treatment.

For a detailed discussion of anaerobic sludge digestion, including operating procedures for normal and abnormal conditions, see Chapter 12, "Sludge Digestion and Solids Handling," in OPERATION OF WASTEWATER TREATMENT PLANTS, Volume II, in this series of operator training manuals. Only a brief review of the anaerobic digestion process will follow. The remainder of this section will discuss aerobic digestion and chemical treatment methods of sludge stabilization.

3.21 Anaerobic Digestion

The most widely used method of sludge stabilization is anaerobic digestion in which decomposition of organic matter is performed by microorganisms in the absence of oxygen. Anaerobic digestion is a complex biochemical process in which several groups of anaerobic and facultative (survive with or without oxygen) organisms break down organic matter. This process can be considered a two-phase process; in the first phase, facultative, acid-forming organisms convert complex organic matter to volatile (organic) acids. In the second phase, anaerobic methane-forming organisms convert the acids to odorless end products of methane gas and carbon dioxide.

The performance of anaerobic digesters in converting volatile (organic) matter to methane and carbon dioxide depends on: (1) sludge type, (2) digestion time, (3) digestion temperature, and (4) mixing. In general, as the concentration of sludge solids fed to anaerobic digesters increases, the performance or efficiency in converting volatile sludge solids also increases due to lower sludge volumes and longer digestion times. In addition, the methane-forming bacteria are highly sensitive to temperature changes and

anaerobic digesters are usually heated to maintain temperatures of 94 to 97°F (34 to 36°C). If the temperature falls below this range or if the digestion time falls below 15 days, the digester may become upset and require close monitoring and attention.

The presence of certain materials in industrial wastes can severely inhibit or even be toxic to the anaerobic digestion process by affecting primarily the methane formation. Whether a substance is inhibitory or toxic depends on a number of factors, including the substance nature, concentration, pH of the sludge, temperature, presence of other components, and biomass acclimatization.

Table 3.10 presents inhibitory and toxic concentrations of selected inorganic pollutants for anaerobic digestion. Table 3.11 presents inhibitory concentrations of selected organic pollutants for anaerobic digestion.

The anaerobic sludge digestion process has a relatively low ability to adapt to changing conditions compared to other biological processes (for example, the activated sludge process). Under shock load conditions, there is not enough time for the anaerobic microorganisms to adjust to the new environment and the digestion process fails. Additionally, as a result of the chemical and biological factors that affect the inhibitory impact of a pollutant, a given concentration of that pollutant may be inhibitory to a biological process under one set of conditions and noninhibitory under another set of conditions. It is very important to remember that in each specific case a laboratory or pilot study may be required to evaluate the applicability of the anaerobic digestion process for sludge stabilization. In cases where any doubt exists about the effect of a wastewater constituent, laboratory or field prototype studies should be conducted to determine the precise effect on the anaerobic digestion process.

QUESTIONS

Write your answers in a notebook and then compare your answers with those on page 311.

3.20A   What are the goals of stabilization?

3.20B   List the unit processes commonly used for sludge stabilization.

3.21A   Explain the two-step process of anaerobic digestion.

3.21B   List the factors affecting anaerobic digestion.

---

[22] *Stabilization.* Conversion to a form that resists change. Organic material is stabilized by bacteria that convert the material to gases and other relatively inert substances. Stabilized organic material generally will not give off obnoxious odors.

TABLE 3.10   CONCENTRATIONS[a] OF SELECTED INORGANIC POLLUTANTS THAT AFFECT PERFORMANCE OF ANAEROBIC DIGESTERS[b, c, d]

Pollutant	Inhibitory Concentration, mg/L	Toxic Concentration, mg/L	Pollutant	Inhibitory Concentration, mg/L	Toxic Concentration, mg/L
Ammonia	1,500–3,000	3,000	Lead	0.1	10
Arsenic	0.1	1.6	Manganese	10	[e]
Borate (boron)	0.05	2.0	Magnesium	1,000–1,500	3,000
Cadmium	0.02	[e]	Mercury	43	1,300
Calcium	2,500–4,500	8,000	Nickel	2.0	500
Chromium (hexavalent)	3.0	5.0	Potassium	2,500–4,500	12,000
Chromium (trivalent)	50	500	Sodium	3,500–5,500	8,000
Copper	0.5	10	Sulfide	50	200
Cyanide	4	20	Zinc	1.0	5–20
Iron	5	[e]			

[a] Concentrations shown represent influent soluble concentrations (mg/L) to the anaerobic digesters.

[b] J. C. Dyer, A. S. Vernick, & H. D. Feiler, *HANDBOOK OF INDUSTRIAL WASTES PRETREATMENT/WATER MANAGEMENT SERIES,* Garland STPM Press, New York, NY, 1981.

[c] A. W. Obayashi & J. M. Gorgan, *MANAGEMENT OF INDUSTRIAL POLLUTANTS BY ANAEROBIC PROCESSES/INDUSTRIAL WASTE MANAGEMENT SERIES,* Lewis Publishers, Inc., Chelsea, MI, 1985.

[d] *PROCESS DESIGN MANUAL FOR SLUDGE TREATMENT AND DISPOSAL/USEPA TECHNOLOGY TRANSFER,* Municipal Environmental Research Laboratory/Office of Research and Development, Cincinnati, OH, 1979.

[e] Insufficient data.

3.21C   Briefly explain the influence of inorganic and organic materials that can be found in industrial wastes on the performance of anaerobic digestion processes.

3.21D   Describe why laboratory or pilot-scale studies may be required to evaluate whether the anaerobic digestion process is suitable for stabilization of a particular sludge.

3.22   Aerobic Digestion

Aerobic digestion involves the conversion of organic sludge solids to odorless end products of carbon dioxide and water by aerobic microorganisms. This process essentially evolved from the extended aeration version of the activated sludge process. Aerobic digestion may be used for either primary sludge, secondary sludge, or mixtures of the two types of sludges.

Sludge to be stabilized is delivered to the aerobic digester on a continuous or an intermittent basis. A few aerobic digesters are operated on a batch basis. Figure 3.13 shows a typical aerobic sludge digestion process in a schematic fashion. When operated in a continuous mode, thickened sludge is fed continuously to the digester inlet. The digester is equipped with blowers and air diffusion equipment to supply oxygen to the system and to provide for mixing of the digester contents. Digested sludge continuously exits through an effluent line and is either pumped directly to dewatering facilities or may flow to a gravity thickener prior to dewatering. The use of thickening equipment following digestion is to concentrate the sludge and reduce the hydraulic loadings on subsequent dewatering equipment. The overflow (effluent) from the thickener is usually pumped back to the plant headworks for more treatment. The underflow (solids)

from the thickener is pumped to the dewatering facilities. If thickeners are not used, the digested sludge is pumped directly to the dewatering facilities.

In the intermittent or batch mode of operation, the digester receives thickened sludge for a portion of the day. After the digester is fed, the blowers and air diffusion equipment remain in operation until approximately 2 to 3 hours before the next feeding. At 2 to 3 hours before the next feeding, the blowers are turned off and the digester contents are allowed to settle for approximately 1½ to 2½ hours. A portion of the settled solids is then withdrawn and pumped to sludge dewatering facilities. A portion of the supernatant (top liquor) is also withdrawn for either recycle to the plant influent or further treatment with the plant secondary effluent. The blowers are then turned on and thickened sludge is again pumped to the aerobic digester to replace the volume of sludge withdrawn. Automatic controls similar to those used in the sequencing batch reactor activated sludge process variation can be used in order to minimize the number of periodic manual operations.

3.220   Factors Affecting Aerobic Digestion

The operation and performance of aerobic digesters are affected by many variables. These include: (1) sludge type, (2) digestion time, (3) digestion temperatures, (4) volatile solids loading, (5) quantity of air supplied, and (6) dissolved oxygen (DO) concentrations within the digester.

The sludge type deals with the influent characteristics of the wastestream to be stabilized. The operator has little, if any, control over the chemical and biological makeup of the influent sludge. The presence of heavy metals (such as copper, zinc, nickel, lead, cadmium, chromium) or other toxic (including

TABLE 3.11   CONCENTRATIONS[a] OF SELECTED ORGANIC POLLUTANTS INHIBITORY TO ANAEROBIC DIGESTERS[b, c, d]

Pollutant	Inhibitory Concentration, mg/L	Pollutant	Inhibitory Concentration, mg/L
Acetates:		Chlorinated hydrocarbons: *(continued)*	
Ethyl acetate	11[e]	1,1,2-trichloroethane	10–20
Vinyl acetate	11	Trichloroethylene	20
Alcohols:		Trichlorofluoromethane	0.7
Allyl	100	Trichlorotrifluoroethane (Freon)	5
Crotonyl	500	Vinyl chloride	40
Heptyl	500	Chlorobenzenes:	
Hexyl	1,000	Certi-chlor	0.5[f]
Octyl	200	Chlorobenzene	0.9[f]
Propargyl	500	Orthodichlorobenzene	0.7[f]
Aldehydes:		Paradichlorobenzene	1.4[f]
Acrolein	11	Organic acids:	
Acetaldehyde	10[e]	Acrylic acid	12[e]
Crotonaldehyde	50–100	2-chloropropionic acid	8[e]
Formaldehyde	100–200	Lauric acid	2.6[e]
Chlorinated hydrocarbons:		Organic nitrogen compounds:	
Chloroform	10–16	Acrylonitrile	5
1-chloropropane	1.9[e]	Aniline	26[e]
1-chloropropene	0.1[e]	Miscellaneous organic compounds:	
3-chloro-1, 2-propanediol	6[e]	Catechol	26[e]
Carbon tetrachloride	10–20	Benzene	1,000
Methylene chloride	100–500	Benzidine	5
1-2-dichloroethane	1	Ethylbenzene	340
Dichlorophene	1	Nitrobenzene	0.1[e]
Ethylene dichloride	50	Phenol	26[e]
Hexachlorocyclohexane	48	Propanol	90[e]
Pentachlorophenol	0.4	Resorcinol	29[e]
Tetrachloroethylene	20	Toluene	500
1,1,1-trichloroethane	1	Xylene	1,000

[a] Concentrations shown represent influent to the anaerobic digester, mg/L, except as noted.

[b] J. C. Dyer, A. S. Vernick, & H. D. Feiler, *HANDBOOK OF INDUSTRIAL WASTES PRETREATMENT/WATER MANAGEMENT SERIES,* Garland STPM Press, New York, NY, 1981.

[c] A. W. Obayashi & J. M. Gorgan, *MANAGEMENT OF INDUSTRIAL POLLUTANTS BY ANAEROBIC PROCESSES/INDUSTRIAL WASTE MANAGEMENT SERIES,* Lewis Publishers, Inc., Chelsea, MI, 1985.

[d] *PROCESS DESIGN MANUAL FOR SLUDGE TREATMENT AND DISPOSAL/USEPA TECHNOLOGY TRANSFER,* Municipal Environmental Research Laboratory/Office of Research and Development, Cincinnati, OH, 1979.

[e] Concentrations shown represent the sludge in the anaerobic digesters, mM/kg dry sludge solids (millimoles of chemical per kilogram of dry sludge solids).

[f] Concentrations shown represent the sludge in the anaerobic digesters, % wt/wt dry sludge solids.

organic) materials in wastes may inhibit or even upset the aerobic digestion process. It should be noted, though, that many organic chemicals are biodegradable and, therefore, compatible with standard biological treatment processes. Because biological mechanisms of the aerobic digestion process are similar to those of the activated sludge process, the same concerns regarding levels of biologically toxic materials apply. A list of selected pollutants that have an inhibitory effect on the activated sludge process is presented in Table 3.12. The presence of these pollutants may also severely inhibit or upset the aerobic digestion process. The inhibitory and toxicity effects depend on a number of factors, including the nature and concentration of the substance, pH of the sludge, temperature, oxygen uptake rate, presence of other components, and biomass acclimatization. If any of these pollutants are found in the process influent, laboratory or pilot studies should be conducted to determine whether the aerobic digestion process can be used for sludge stabilization.

As stated earlier, aerobic digestion may be used for either primary sludge, secondary sludge, or mixtures of the two. The process has found its widest application with secondary sludges. Secondary sludges are composed primarily of biological cells that are produced in the activated sludge or trickling filter processes as a by-product of degrading organic matter. In simplified terms, by the time secondary sludges leave the biological treatment process, settle in final clarifiers, concentrate in sludge thickening units, and are delivered to aerobic digesters, the quantity of available food (organic matter) is substantially reduced. In the absence of an external food source, these microorganisms enter the *ENDOGENOUS*[23] or death phase of their life cycle. When no

[23] *Endogenous* (en-DODGE-en-us) *Respiration.*   A situation in which living organisms oxidize some of their own cellular mass instead of new organic matter they adsorb or absorb from their environment.

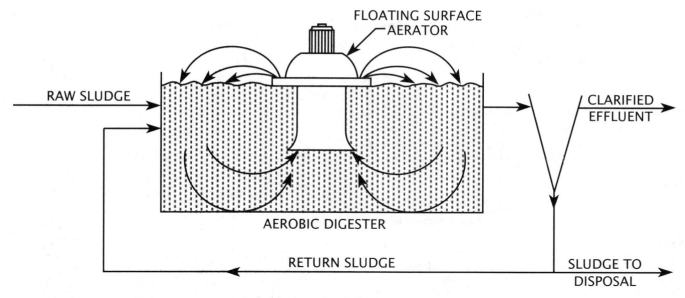

Fig. 3.13 Aerobic sludge digestion process

TABLE 3.12 CONCENTRATIONS OF SELECTED POLLUTANTS INHIBITORY TO ACTIVATED SLUDGE PROCESSES[a, b]

Pollutant	Inhibitory Concentration (influent to the process in dissolved form), mg/L
Inorganic chemicals:	
Aluminum	15–26
Ammonia	480
Arsenic	0.1
Borate (boron)	0.05–100
Cadmium	10–100
Calcium	2,500
Chromium (hexavalent)	1–10
Chromium (trivalent)	50
Copper	1.0
Cyanide	0.1–5
Iron	1,000
Lead	0.1
Manganese	10
Mercury	0.1–5.0
Nickel	1.0–2.5
Silver	5
Zinc	0.08–10
Organic chemicals:	
Benzidine	500
Ceeprine	100
EDTA	25
Nacconol	200
Phenol	200
T-butyl borane	500
Trinitrotoluene (TNT)	20–25

[a] J. C. Dyer, A. S. Vernick, & H. D. Feiler, *HANDBOOK OF INDUSTRIAL WASTES PRETREATMENT/WATER MANAGEMENT SERIES,* Garland STPM Press, New York, NY, 1981.

[b] *WASTEWATER TREATMENT PLANT DESIGN/MANUAL OF PRACTICE NO. 8.* Water Pollution Control Federation, Lancaster Press, Inc., Lancaster, PA, 1982.

food is available (endogenous phase), the biomass begins to self-metabolize (consume its own cellular material), which results in a conversion of the biomass to end products of carbon dioxide and water and a net decrease in the sludge mass.

When primary sludge is introduced to an aerobic digester, food becomes available to the microorganisms. In the presence of an external food source (the primary sludge), the biomass will convert the food to end products of carbon dioxide and water and will function in the growth phase of their life cycle until the food supply is exhausted. During this growth phase, the biomass will reproduce, resulting in a net increase in the sludge mass. Aerobic digestion times are long enough to allow the food to be depleted and the biomass to eventually enter the endogenous or death phase. The main drawback to aerobically digesting primary sludge is that more air has to be supplied to maintain a desirable DO level because the bacteria are more active when food is available.

Digestion time, temperature, volatile solids loading, and air supply are considered operational controls and are discussed below.

3.221 Operating Guidelines

3.2210 DIGESTION TIME

In general, as the digestion time is increased, the efficiency or effectiveness of aerobic digesters in achieving the goals of stabilization is also increased. The physical size of aerobic digesters is determined by engineers and the operator has no control over the digester volume after it is constructed. The operator does have control over the digestion time by controlling the degree of sludge thickening prior to digestion. The digestion time is directly proportional to the thickened sludge flow according to the following equation:

$$\text{Digestion Time, days} = \frac{\text{Digester Volume, gal}}{\text{Sludge Flow, gal/day}}$$

As the flow to the digester increases, the time of digestion decreases; as the flow decreases, the digestion time increases. The following example illustrates the effect of sludge concentration and the resulting sludge volume on digestion time.

EXAMPLE 21

Given: A 40-foot diameter by 10-foot side-wall-depth (SWD) aerobic digester receives 10,000 GPD of thickened secondary sludge. The thickened sludge concentration is 2.5 percent (25,000 mg/L) sludge solids (total solids).

Find: 1. The time of digestion.

2. What effect thickening the sludge to 3.5 percent sludge solids will have on digester performance.

Solution:

Known	**Unknown**
Aerobic Digester	1. Digestion Time, days
Diameter, ft = 40 ft	2. Effect of Increasing Sludge Solids to 3.5%
Depth (SWD), ft = 10 ft	
Flow, GPD = 10,000 GPD	
Thickened Sludge, % = 2.5% Sludge Solids	
, mg/L = 25,000 mg/L	

1. Calculate the digestion time in days.

   a. Calculate the digester volume.

$$\text{Volume, gal} = \frac{\pi}{4} \times (\text{Diameter, ft})^2 \times \text{SWD, ft} \times 7.48 \frac{\text{gal}}{\text{cu ft}}$$

$$= 0.785 \times (40 \text{ ft})^2 \times 10 \text{ ft} \times 7.48 \frac{\text{gal}}{\text{cu ft}}$$

$$= 94,000 \text{ gal (approximate)}$$

   b. Determine the digestion time in days.

$$\text{Digestion Time, days} = \frac{\text{Digester Volume, gallons}}{\text{Flow, GPD}}$$

$$= \frac{94,000 \text{ gal}}{10,000 \text{ gal/day}}$$

$$= 9.4 \text{ days}$$

NOTE: The digestion time in days based on the solids in the digester is more important than the hydraulic digestion time.

$$\text{Digestion Time, days (solids)} = \frac{\text{Digester Solids, lbs}}{\text{Sludge Wasted, lbs/day}}$$

2. Determine the effect of increasing the sludge solids from 2.5 percent to 3.5 percent sludge solids. The total sludge volume pumped to the aerobic digester will be decreased and the digestion time will be increased. Calculate the new digestion time in days.

   a. Determine the new flow to the aerobic digester in gallons per day.

$$\text{Flow, GPD} = \frac{\text{Flow at 2.5\% SS, GPD} \times \text{SS, \%}}{\text{SS, \%}}$$

$$= \frac{10,000 \text{ GPD} \times 2.5\%}{3.5\%}$$

$$= 7,143 \text{ GPD}$$

   b. Calculate the new digestion time.

$$\text{Digestion Time, days} = \frac{\text{Digester Volume, gallons}}{\text{Flow, GPD}}$$

$$= \frac{94,000 \text{ gal}}{7,143 \text{ gal/day}}$$

$$= 13.2 \text{ days}$$

The overall impact of thickening the sludge to 3.5 percent sludge solids is an increase in digestion time and a potential increase in digester efficiency.

The above example illustrates the need to thicken sludge as much as possible prior to stabilization to obtain maximum digestion times.

3.2211   DIGESTION TEMPERATURE

Aerobic digestion, like the activated sludge process, depends on groups of microorganisms performing specific functions. The microorganisms require favorable environments to function properly. An important environmental condition is temperature. As the temperature of the system decreases, the rate of biological activity also decreases. In the case of aerobic digestion, a decrease in biological activity will result in a decreased rate of destruction of the biomass and the potential for unstabilized sludge exiting the digester. Desirable aerobic digestion temperatures are approximately 65 to 80°F (18 to 27°C). In colder climates, provisions may have to be made to heat the digester to maintain temperatures in the above range. In addition, if aerobic digesters are fabricated with steel and erected above ground, sufficient insulation should be provided to prevent excess heat loss and reduce heating costs. Actual temperatures in aerobic digesters depend on the temperature and volume of sludge fed to the digester and also the temperature of the air supplied to the digester.

3.2212   VOLATILE SOLIDS LOADING

Volatile sludge solids loading is an estimate of the quantity of organic matter applied to the digester. Procedures for calculating sludge solids concentration are outlined in Chapter 16, "Laboratory Procedures and Chemistry," *OPERATION OF WASTEWATER TREATMENT PLANTS*, Volume II. The optimum volatile solids loading for aerobic digestion (and anaerobic digestion) depends on the treatment plant and is generally determined by pilot- or full-scale experimentation. In general, volatile sludge solids (VSS) loadings for effective aerobic stabilization vary from 0.07 lb VSS/day/cu ft to 0.20 lb VSS/day/cu ft

(0.7 to 2.0 kg/day/m^3), depending on the temperature, type of sludge, and presence of inhibitory/toxic pollutants as discussed in Section 3.220.

The operator should be familiar with the calculations required to determine volatile suspended solids loading rates. The following example outlines these calculations. The determination of volatile sludge solids loading is identical for aerobic and anaerobic stabilization processes.

EXAMPLE 22

Given: A 40-foot diameter by 10-foot SWD aerobic digester receives 7,140 GPD of secondary sludge. The thickened secondary sludge is at a concentration of 3.5 percent sludge solids and is 75 percent volatile matter.

Find: The volatile sludge solids loading (lbs VSS/day/cu ft).

Solution:

Known		Unknown
Aerobic Digester		Volatile Sludge
Diameter, ft	= 40 ft	Solids Loading,
Depth (SWD), ft	= 10 ft	lbs VSS/day/cu ft
Flow, GPD	= 7,140 GPD	
Sludge Solids, %	= 3.5%	
Volatile Matter, %	= 75%	

Calculate the volatile sludge solids (VSS) loading in pounds of VSS per day per cubic foot of aerobic digester.

1. Determine the digester volume in cubic feet.

$$\text{Volume, cu ft} = \frac{\pi}{4} \times (\text{Diameter, ft})^2 \times \text{Depth, ft}$$

$$= 0.785 \times (40 \text{ ft})^2 \times 10 \text{ ft}$$

$$= 12,560 \text{ cu ft}$$

2. Calculate the volatile sludge solids (VSS) loading in pounds of VSS per day per cubic foot.

$$\text{VSS Loading, lbs VSS/day/cu ft} = \frac{\text{VSS Added, lbs/day}}{\text{Digester Volume, cu ft}}$$

$$= \frac{\text{Flow, GPD} \times 8.34 \frac{\text{lbs}}{\text{gal}} \times \text{SS}, \frac{\%}{100\%} \times \text{VM}, \frac{\%}{100\%}}{\text{Digester Volume, cu ft}}$$

$$= \frac{7,140 \frac{\text{gal}}{\text{day}} \times 8.34 \frac{\text{lbs}}{\text{gal}} \times \frac{3.5\%}{100\%} \times \frac{75\%}{100\%}}{12,560 \text{ cu ft}}$$

$$= \frac{1,563 \text{ lbs VSS/day}}{12,560 \text{ cu ft}}$$

$$= 0.12 \text{ lb VSS/day/cu ft}$$

The VSS loading is affected by the concentration and volume of sludge introduced into the digester and the presence of inhibitory/toxic pollutants as discussed in Section 3.220. In general, the volatile portion of a sludge from a particular plant will not vary from day to day. If the digestion capacity is fixed, the daily sludge flow in combination with the degree of thickening will determine the VSS loading.

3.2213 AIR REQUIREMENTS AND DISSOLVED OXYGEN

Oxygen is supplied to the sludge by using air diffusers or mechanical aerators. Air requirements are usually expressed as CFM (cubic feet per minute) air/1,000 cu ft of aerobic digester capacity for diffuser systems, and as horsepower per 1,000 cubic feet for mechanical aerators. Air requirements also are expressed as 1.5 to 2 pounds of oxygen per pound of volatile sludge solids destroyed. The air requirements stem from a desire to keep the digester solids in suspension (well mixed) and to maintain a dissolved oxygen (DO) concentration of 1 to 2 mg/*L* in the digester.

Depending on the sludge type, temperature, and concentration, and the activity of the biomass in the digester, the quantity of air required to maintain a residual DO of 1 to 2 mg/*L* will vary. As the concentration or activity of the microorganisms increases, more air is required to satisfy the oxygen requirements of the biomass. The residual DO is a measure of the quantity of oxygen supplied beyond that used by the biomass. For example, if the biomass requires 3.0 mg/*L* of oxygen and 5.0 mg/*L* are supplied by the blowers or mechanical aerators, then 2.0 mg/*L* of oxygen are left over. The quantity left over is called the residual DO. The residual DO in the digester should always be greater than 1.0 mg/*L*. If the digester DO falls below 1.0 mg/*L*, filamentous organisms may grow. Filamentous organisms are undesirable because they can lead to sludge bulking or foaming, which will negatively affect digester operation.

Typical air rates required to maintain a residual DO of 1.0 to 2.0 mg/*L* are discussed in Section 3.223, "Typical Performance." You should realize that values in any book are estimates and that the exact air supply requirements are usually determined in the plant by experimentation. The following example illustrates the determination of air rates for aerobic digesters.

EXAMPLE 23

Given: A pilot-scale digestion study showed that 0.040 CFM of air was required per cu ft of digestion capacity to satisfy the biomass oxygen requirements. Based on these pilot studies, a full-scale digester with dimensions of 100 ft long by 25 ft wide by 10 ft SWD has been constructed.

Find: The quantity of air (CFM) to be delivered to the full-scale digester.

Solution:

Known	**Unknown**
Aerobic Digester	Air Rate, CFM

$$\text{Air Required, CFM/cu ft} = \frac{0.040 \text{ CFM Air}}{\text{cu ft Digester}}$$

Length, ft	= 100 ft
Width, ft	= 25 ft
SWD, ft	= 10 ft

Determine the rate of air that must be delivered to the aerobic digester in cubic feet of air per minute (CFM).

1. Calculate the digester volume in cubic feet.

$$\text{Digester Volume, cu ft} = \text{Length, ft} \times \text{Width, ft} \times \text{SWD, ft}$$
$$= 100 \text{ ft} \times 25 \text{ ft} \times 10 \text{ ft}$$
$$= 25,000 \text{ cu ft}$$

2. Determine the rate of air that must be supplied in CFM.

$$\text{Air Rate, CFM} = \frac{\text{Air Required, CFM Air}}{\text{cu ft Digester}} \times \text{Dig Vol, cu ft}$$
$$= \frac{0.040 \text{ CFM Air}}{\text{cu ft Digester}} \times 25,000 \text{ cu ft}$$
$$= 1,000 \text{ CFM Air}$$

3.222   Normal Operating Procedures

The sludge feed to aerobic digesters should be as continuous and consistent as possible. This is best achieved by proper operation of the sludge thickening facilities and by pumping the thickest possible sludge. Aerobic digesters should be routinely checked at least once per shift and composite samples of the influent and effluent should be collected daily for laboratory analysis. Daily laboratory analysis on the influent and effluent streams should include suspended solids, percent volatile matter, and pH measurements. Alkalinity, total and soluble COD, ammonia-nitrogen, nitrite, and nitrate should be determined on the influent and effluent weekly. In addition to these laboratory analyses, the residual dissolved oxygen (DO) in the digester should be measured at least once per shift. Digester temperature should be measured daily.

Digester temperature measurements are straightforward and simply involve the use of a thermometer. The determination of dissolved oxygen (DO) in the digester and oxygen (O_2) uptake rates require the use of a membrane-type electrode commonly called a DO probe. Membrane electrode instruments are commercially available in some variety and it is impossible to formulate detailed operational instructions that would apply to every instrument. Calibration procedures and readout are included in manufacturer's instructions and they should be followed exactly to obtain the guaranteed precision and accuracy.

Residual DO in the digester can simply be measured by lowering the probe into the digester, gently raising and lowering the probe approximately 6 to 12 inches (15 to 30 cm), and recording the readout measurement after the readout has stabilized. Depending on the instrument readout measurement and the temperature of the digester, the DO can be determined in mg/L with the aid of charts supplied by the electrode manufacturer. DO measurements should be obtained at several points within the digester to obtain an average dissolved oxygen concentration. These measurements should be obtained at the influent and effluent ends of the digester and at different tank depths.

Oxygen uptake measurements are an indication of the activity of the aerobic digester biomass. Accurate oxygen (O_2) uptake measurements are of importance to the operator because they will readily indicate if the process is functioning properly or if upset conditions exist. Oxygen uptake measurement requires the use of a sealed container into which a DO probe and a mixer can be inserted to measure the oxygen concentration with time. To conduct this test, collect approximately one liter of digested sludge in a wide-mouth container, seal the top, and vigorously mix the contents. Vigorous mixing will saturate the sample with oxygen. Following approximately ½ to 1 minute of mixing, place the oxygen-saturated sample into the oxygen uptake container, insert the DO probe, and turn on the mixer. Record the DO concentrations with time for 10 to 15 minutes or until zero DO is recorded. Then calculate the uptake measurement according to the following equation:

$$\text{Oxygen Uptake, mg } O_2/L/\text{hr} = \frac{(DO_1, \text{mg}/L - DO_2, \text{mg}/L)}{(\text{Time}_2, \text{min} - \text{Time}_1, \text{min})} \times \frac{60 \text{ min}}{\text{hr}}$$

The following example illustrates the determination of O_2 uptake:

EXAMPLE 24

Given: An operator measures the dissolved oxygen (DO) concentration with time on an air-saturated sample taken from an aerobic digester. The following measurements were recorded:

Time (min)	DO (mg/L)
0	7.1
1	6.0
2	5.2
3	4.5
4	3.9
5	3.2

Find: The oxygen uptake in mg/*L*/hr.

Solution:

Known	**Unknown**
Aerobic Digester	Oxygen Uptake, mg/*L*/hr
DO Measurements with Time for an Air-Saturated Sample	

Calculate the oxygen uptake for the air-saturated sample from an aerobic digester in mg/*L*/hr. The 2-minute DO reading is used to allow the DO probe and the sample time to stabilize. The 5-minute DO reading also is used in the calculation.

$$\text{Oxygen Uptake, mg/}L\text{/hr} = \frac{(DO_1, \text{mg/}L - DO_2, \text{mg/}L)}{(\text{Time}_2, \text{min} - \text{Time}_1, \text{min})} \times \frac{60 \text{ min}}{\text{hr}}$$

$$= \frac{(5.2 \text{ mg/}L - 3.2 \text{ mg/}L)}{(5 \text{ min} - 2 \text{ min})} \times \frac{60 \text{ min}}{\text{hr}}$$

$$= \frac{2.0 \text{ mg/}L}{3 \text{ min}} \times \frac{60 \text{ min}}{\text{hr}}$$

$$= 40 \text{ mg/}L\text{/hr}$$

If the uptake measurement and residual DO measurements are significantly different from the values usually measured, the operator should be aware that something may be wrong (see Section 3.2240). Changes in oxygen uptake rates could indicate the presence of substances capable of inhibiting the activities of the organisms treating the sludge (see Table 3.12, page 209, for a partial list of inhibitory substances).

Like a well-operated activated sludge system, a well-operated aerobic digester should be relatively free of odors. The surface will contain a small accumulation of foam due to the turbulence created by the diffusers or mechanical aerators and the operator should be aware of changes in the physical appearance of the system. The operator soon comes to think of an aerobic digester as a living organism. The organisms thrive and enjoy good health or become upset and refuse to function properly. By combining careful observation with experience, you may determine what is happening and what adjustments, if any, are required. Section 3.224 deals with troubleshooting.

3.223   Typical Performance

The efficiency of aerobic digesters is usually measured by the quantity of suspended and volatile (sludge) solids converted to end products of CO_2 and H_2O.

The following example illustrates how to calculate the efficiency of aerobic digesters.

EXAMPLE 25

Given: An aerobic digester receives 9,000 GPD of secondary sludge at a concentration of 3.6 percent sludge solids (SS) and 74 percent volatile solids (matter). The digester effluent is at a concentration of 2.93 percent sludge solids, 64 percent volatile matter, and a flow of 8,000 GPD.

Find: 1. Pounds of sludge solids (SS) and pounds of volatile sludge solids (VSS) entering the digester.

2. Pounds of sludge solids and pounds of volatile sludge solids exiting the digester.

3. Efficiency of digester in destroying sludge solids, %.

4. Efficiency of digester in destroying volatile sludge solids, %.

Solution:

Known		**Unknown**
Aerobic Digester		1. Sludge Solids In, lbs/day Volatile Solids In, lbs/day
Flow In, GPD	= 9,000 GPD	
Sludge Solids In, %	= 3.6%	
Volatile Solids In, %	= 74%	2. Sludge Solids Out, lbs/day Volatile Solids Out, lbs/day
Flow Out, GPD	= 8,000 GPD	
Sludge Solids Out, %	= 2.93%	
Volatile Solids Out, %	= 64%	3. Sludge Solids Removal Eff, %
		4. Volatile Solids Removal Eff, %

1. Determine the sludge solids and volatile solids entering the aerobic digester in pounds per day.

$$\text{Sludge Solids Entering, lbs/day} = \text{Flow, GPD} \times 8.34 \frac{\text{lbs}}{\text{gal}} \times \text{SS In,} \frac{\%}{100\%}$$

$$= 9,000 \frac{\text{gal}}{\text{day}} \times 8.34 \frac{\text{lbs}}{\text{gal}} \times \frac{3.6\%}{100\%}$$

$$= 2,702 \text{ lbs SS/day}$$

$$\text{Volatile Solids Entering, lbs/day} = \text{Sludge Solids,} \frac{\text{lbs}}{\text{day}} \times \text{VSS,} \frac{\%}{100\%}$$

$$= 2,702 \frac{\text{lbs SS}}{\text{day}} \times \frac{74\%}{100\%}$$

$$= 2,000 \text{ lbs VSS/day}$$

2. Determine the sludge solids and volatile solids exiting the aerobic digester in pounds per day.

$$\text{Sludge Solids Exiting, lbs/day} = \text{Flow, GPD} \times 8.34 \frac{\text{lbs}}{\text{gal}} \times \text{SS Out,} \frac{\%}{100\%}$$

$$= 8,000 \frac{\text{gal}}{\text{day}} \times 8.34 \frac{\text{lbs}}{\text{gal}} \times \frac{2.93\%}{100\%}$$

$$= 1,955 \text{ lbs SS/day}$$

$$\text{Volatile Solids Exiting, lbs/day} = \text{Sludge Solids, } \frac{\text{lbs}}{\text{day}} \times \text{VSS, } \frac{\%}{100\%}$$

$$= 1,955 \frac{\text{lbs SS}}{\text{day}} \times \frac{64\%}{100\%}$$

$$= 1,251 \text{ lbs VSS/day}$$

3. Calculate the efficiency of the sludge solids destruction as a percent.

$$\text{SS Removal Efficiency, \%} = \frac{(\text{SS Entering, lbs/day} - \text{SS Exiting, lbs/day}) \times 100\%}{\text{SS Entering, lbs/day}}$$

$$= \frac{(2,702 \text{ lbs SS/day} - 1,955 \text{ lbs SS/day}) \times 100\%}{2,702 \text{ lbs SS/day}}$$

$$= 27.6\%$$

4. Calculate the efficiency of the volatile sludge solids destruction as a percent.

$$\text{VSS Removal Efficiency, \%} = \frac{(\text{VSS In, lbs/day} - \text{VSS Out, lbs/day}) \times 100\%}{\text{VSS Entering, lbs/day}}$$

$$= \frac{(2,000 \text{ lbs VSS/day} - 1,251 \text{ lbs VSS/day}) \times 100\%}{2,000 \text{ lbs VSS/day}}$$

$$= 37.5\%$$

Table 3.13 shows typical operating guidelines for aerobic digestion and includes a summary of performance to be expected.

TABLE 3.13 OPERATIONAL AND PERFORMANCE GUIDELINES FOR AEROBIC DIGESTION

| Sludge Type | Digestion Time, days | Air Rate | | VSS Load, lbs VSS/day/ cu ft[c] | VSS Destruction, % |
		Diffused Air, CFM/cu ft[a]	Mechanical, HP/1,000 cu ft[b]		
Primary	15–20	0.015–0.06	0.05–1.25	0.08–0.20	25–50
Secondary	10–15	0.015–0.04	0.05–1.25	0.08–0.20	25–40

[a] CFM/cu ft × 1.0 = cu m/min/cu m.
[b] HP/1,000 cu ft × 26.34 = W/cu m.
[c] lbs VSS/day/cu ft × 10.02 = kg/day/cu m.

QUESTIONS

Write your answers in a notebook and then compare your answers with those on pages 311, 312, and 313.

3.22A Briefly explain the aerobic digestion process.

3.22B List the factors affecting aerobic digestion.

3.22C Briefly explain the influence of inorganic and organic materials that can be found in wastes on the performance of the aerobic digestion process.

3.22D Describe why laboratory or pilot-scale studies may be required to evaluate whether aerobic digestion can be used for stabilization of a particular sludge.

3.22E Briefly explain why aerobic digestion is more suitable for treating secondary sludges than for treating primary sludges.

3.22F How can an operator control the digestion time?

3.22G A digester with an active volume of 140,000 cubic feet receives 110,000 GPD of primary sludge. What is the digestion time (days)?

3.22H If the sludge from problem 3.22G is thickened from 2.7 percent to 3.5 percent, what will happen to the digestion time?

3.22I How does temperature affect aerobic digester performance?

3.22J An aerobic digester with dimensions of 120 ft in length, 25 feet wide, and 11 feet SWD receives 24,000 GPD of secondary sludge at a concentration of 3.1 percent and 73 percent volatile matter. What is the digestion time (days) and the VSS loading (lbs/day/cu ft)?

3.22K Why should the DO in aerobic digesters be maintained at concentrations greater than 1.0 mg/L?

3.22L Explain how the DO level is determined in aerobic digesters.

3.22M Determine the O_2 uptake rate (mg/L/hr) for the following field measurements:

Time (min)	DO (mg/L)
0	6.3
1	5.1
2	4.2
3	3.4
4	2.6
5	1.8
6	1.0

3.22N A 1,000,000-gallon aerobic digester receives 91,000 GPD of primary sludge at a concentration of 5.1 percent SS and 76 percent volatile matter. The digester effluent is at a concentration of 3.7 percent SS and 67 percent volatile matter. Determine the digestion time (days), VSS loading (lbs/day/cu ft), and percent VSS destruction.

3.224 Troubleshooting

Aerobic digesters, like all biological systems, are subject to upsets. These upsets may result from equipment malfunctions, changes in the influent characteristics, presence of inhibitory/toxic pollutants (as discussed in Section 3.220), or operation out of the range of recommended operating guidelines.

Even before sample analyses are available from the laboratory, the operator may become aware of process inefficiencies by careful observation of the physical appearance of the digested sludge and routine monitoring of the residual dissolved oxygen.

3.2240 DISSOLVED OXYGEN AND OXYGEN UPTAKE

After an aerobic digester reaches steady state, that is, the concentration of solids within the digester is fairly constant, the O_2 uptake rate and residual DO should be relatively constant from day to day. If the residual DO increases significantly, either the air rate is excessive or the O_2 uptake rate and activity of the biomass have decreased. The O_2 uptake should be checked immediately. If the O_2 uptake is in the range normally encountered, then the biomass is functioning properly and the increase in DO is most likely due to high air rates to the digester. Check the air rate and adjust as required to maintain a DO of 1 to 2 mg/L. Excessive air rates are not desired because they may cause high turbulence within the digester, which may adversely affect sludge settleability and may lead to foaming problems. If the O_2 uptake rate is significantly lower than normal, something may be inhibiting the biomass.

Check the temperature and pH of the digester contents. A significant decrease in the temperature or pH will reduce the activity of the biomass. If the temperature is low, try to determine what could have caused the drop in temperature and how the temperature can be returned to the normal operating range. If the pH is significantly lower than normal (6.8 to 7.3), you may have to add caustic or lime for pH adjustment. A decline in the pH may be caused by nitrification in the digester or by changes in the influent sludge characteristics. If the decline is caused by nitrification, the decrease in pH will be gradual over about a week's time. A review of the daily pH measurements will indicate whether the decline was gradual or sudden. If nitrification is the cause of a decrease in pH, lower the air rate somewhat or decrease the hydraulic detention time somewhat to suppress the growth of nitrifying bacteria. In any case, the pH should not be allowed to drop below 6.0. If the pH is below or close to 6.0, take immediate action to neutralize the digester contents. The safest and easiest way to determine the quantity of lime ($Ca(OH)_2$) or caustic (NaOH) to be added is to conduct jar tests (see page 223, JAR TEST PROCEDURE) on the digested sludge according to the following example.

EXAMPLE 26

Given: The pH of an aerobic digester has decreased to 6.1. The operator wishes to raise the pH to 7.0 with the addition of sodium hydroxide.

Find: How much caustic must be added if the digester volume is 100,000 gallons.

Procedure: The operator should run jar tests on the digested sludge and determine how much caustic must be added to a 1-liter sample to raise the pH to 7.0. Assume the operator determines that 20 mg of caustic added to 1 liter of sludge raises the pH to 7.0. The quantity that must be added to the full-scale plant must be calculated according to the following solution.

Solution:

Known		Unknown
Aerobic Digester		Amount of Caustic (NaOH) to be Added, lbs
Current pH	= 6.1	
Desired pH	= 7.0	
Digester Vol, gal	= 100,000 gal	
Jar Test Results, Caustic Added, mg/L	= 20 mg NaOH/L	

Determine the amount of caustic (NaOH) to be added in pounds.

$$\text{Caustic Added, lbs} = \frac{\text{NaOH to Jar, mg} \times \text{Dig Vol, gal} \times 3.78\ L/\text{gal}}{\text{Sludge Sample Vol, } L \times 454\ \text{gm/lb} \times 1{,}000\ \text{mg/gm}}$$

$$= \frac{20\ \text{mg} \times 100{,}000\ \text{gal} \times 3.78\ L/\text{gal}}{1\ L \times 454\ \text{gm/lb} \times 1{,}000\ \text{mg/gm}}$$

$$= 16.7\ \text{lbs NaOH}$$

If a significant rise in DO along with a decrease of O_2 uptake is definitely not being caused by low temperatures or low pH, check the sludge flow and volatile sludge solids loading to the digester during the previous seven days. Excessive sludge flows will reduce the time of digestion and may increase the volatile sludge solids loading to the point where the digester is operating out of the recommended range of operation. Adjust the flows and volatile sludge solids loading so as to operate within the range normally used for good operation. If the sludge flow and volatile sludge solids loading to the digester are within normal operational range, the sludge should be tested for any inhibitory/toxic substances discussed in Section 3.220.

The discussion thus far has dealt with a residual DO higher than normal, but low DO levels may also occur. If the DO drops significantly, check the air rate and adjust as required to increase the residual DO to 1.0 to 2.0 mg/L. If higher than normal O_2 uptake rates are also noted, the volatile sludge solids loading rate to the digester may be higher than normal. As long as sufficient air capacity exists to meet air requirements at higher loading rates, the system can still operate but you should still check critical operating guidelines such as temperature, pH, and digestion time. If low DO exists and the blower is operating at full capacity, decrease the flow and loading to the digester or obtain additional blower capacity if the loadings cannot be decreased.

3.2241 FOAMING

Aerobic digesters often develop foaming problems. If excessive foam develops, check the air rate and residual DO. If the DO is high and the remaining critical factors (O_2 uptake, pH, temperature) are satisfactory, the problem may be related to excessive turbulence. In this case, you should lower the air rate. If the DO is low, increase the air rate and observe a sample of digested sludge under a microscope. Low DO encourages filamentous growth. If filamentous growths are observed, the problem may be related solely to DO and the predominance of filamentous growths. On occasion, foaming will develop even with high DO. If this occurs, the problem may be related to influent characteristics and you should add defoaming agents to the digester to suppress the foam. Foaming in biological systems can be caused by a variety of conditions and generally indicates a rather complex problem. If the procedures given in this section will not cure a foaming problem, a consultant may be helpful.

3.2242 LOADINGS

Both the digestion time as governed by the hydraulic flow rate (GPD) and the volatile sludge solids loading (lbs VSS/day/cu ft) should be maintained in the ranges summarized in Table 3.13 (page 214). Operation outside of the recommended range may lead to decreased digester efficiency. A review of daily influent and effluent pounds of volatile sludge solids will indicate whether or not the digester is efficiently converting volatile (organic) matter to stabilized end products. A decrease in volatile suspended solids destruction could indicate that digestion times are too short, volatile sludge solids loadings are too high, or the digestion process is affected by inhibitory/toxic pollutants as discussed in Section 3.220.

Table 3.14 summarizes aerobic digester operational problems that may be encountered and corrective measures that might be taken.

QUESTIONS

Write your answers in a notebook and then compare your answers with those on page 313.

3.22O What routine checks can the operator make to detect aerobic digestion process inefficiencies?

3.22P A 15,000-gallon aerobic digester has been operating successfully with a sludge flow of 1,000 GPD. The influent sludge is normally at a concentration of 3.6 percent with a volatile content of 74 percent. The operator determines the residual DO in the digester to be 4.0 mg/L. Normally, the digester operates at a DO of 1.5 mg/L. What should the operator do?

3.22Q List the potential causes of foaming and the corrective measures that should be taken.

TABLE 3.14   TROUBLESHOOTING AEROBIC DIGESTERS

Operational Problem	Possible Cause	Check or Monitor	Possible Solution
1. High residual DO and normal uptake rate	1a. High air rates	1a. Air rate	1a. Lower air rate
2. High residual DO and low uptake rate	2a. Low digester temperature	2a. Temperature and heating equipment	2a. Increase temperature
	2b. Low digester pH	2b. pH. Check for nitrification	2b. Neutralize pH. Lower air rate and digestion time
	2c. VSS loading too high or too low	2c. Flow rate and feed concentration	2c. Adjust to obtain recommended loading
	2d. Digestion time too high or too low	2d. Flow rate	2d. Adjust to obtain recommended detention time
	2e. Toxicity	2e. Toxic trace constituents in the influent sludge	2e. Control industrial waste discharges
3. Foaming	3a. Filamentous growth	3a. Residual DO and microscopic exam	3a. Increase air rate. Add defoamant
	3b. Excessive turbulence	3b. Air rate and residual DO	3b. Lower air rate. Add defoamant
4. Reduced VSS destruction	4a. Low digestion time	4a. Flow and concentration of feed	4a. Decrease flow
	4b. High VSS loading	4b. Same as 4a.	4b. Decrease flow
	4c. Low temperature	4c. Temperature	4c. Adjust temperature
	4d. Low DO	4d. DO	4d. Increase air rate
	4e. Low pH	4e. pH	4e. Neutralize pH
	4f. Toxicity	4f. Toxic trace constituents in the influent sludge	4f. Control industrial waste discharges
5. Poor sludge settling	5a. Digestion time too high or too low	5a. Flow rate and solids wasting	5a. Run jar test to determine optimum dosage of alum, lime, or polymer

3.23 Chemical Stabilization

Sludges that are not biologically digested or thermally stabilized can be made stable by the addition of large doses of lime or chlorine to destroy pathogenic and nonpathogenic organisms. Chemical addition to sludge to prepare it for ultimate disposal is not a common practice. Chemical addition is usually considered to be a *temporary* stabilization process and finds application at overloaded plants or at plants experiencing stabilization facility upsets. The main drawback to chemical stabilization is the cost associated with the large quantities of chemical required.

3.230 Lime Stabilization

Lime stabilization is accomplished by adding sufficient quantities of lime to the sludge to raise the pH to 11.5 to 12.0. Estimated dosages to achieve a pH of 11.5 to 12.0 are generally 200 to 220 pounds of lime per ton for primary sludge solids (100 to 110 grams of lime per kg of solids). Waste activated sludge (WAS) will require doses from 400 to 800 pounds of lime per ton of sludge solids (200 to 400 grams of lime per kg of solids). An indication of the quantity required for a medium-sized wastewater treatment plant is given in the following example.

EXAMPLE 27

Given: A 3.0 MGD treatment plant produces 3,400 lbs/day of primary sludge solids and 1,700 lbs/day of secondary sludge solids. Lime stabilization requires the addition of 210 lbs/ton to raise the pH to 12.0.

Find: The pounds of lime required per day.

Solution:

Known		Unknown
Treatment Plant		Lime Required, lbs/day
Flow, MGD	= 3.0 MGD	
Prim Sol, lbs/day	= 3,400 lbs/day	
Sec Sol, lbs/day	= 1,700 lbs/day	
To increase pH to 12.0, add Lime, lbs/ton	$= \dfrac{210 \text{ lbs Lime}}{\text{ton Sl Sol}}$	

Calculate the amount of lime required in pounds of lime per day.

$$\begin{aligned}
\text{Lime Reqd,} \atop \text{lbs/day} &= \text{Dose,} \frac{\text{lbs Lime}}{\text{ton Sludge}} \times \text{Sludge,} \frac{\text{lbs/day}}{2{,}000 \text{ lbs/ton}} \\[2ex]
&= \frac{210 \text{ lbs Lime}}{\text{ton Sludge}} \times \frac{(3{,}400 \text{ lbs/day} + 1{,}700 \text{ lbs/day})}{2{,}000 \text{ lbs/ton}} \\[2ex]
&= 536 \text{ lbs Lime/day}
\end{aligned}$$

A crucial consideration and drawback of lime stabilization is that the sludge mass is not reduced as with other stabilization processes (digestion, oxidation). In fact, the addition of lime adds to the overall quantity of solids that must be ultimately disposed. The high pH of the stabilized solids may also reduce the range of beneficial reuse opportunities.

NORMAL OPERATING PROCEDURES

Lime arrives from the supplier in powder form and cannot be added directly to sludge. The powdered lime must be made into a *SLURRY*[24] with the addition of water prior to blending with the sludge. Slurry concentrations (lbs lime/gal water) are discussed in Section 3.3, "Conditioning." A lime stabilization system must incorporate the use of a slurry or mixing tank to mix the lime with water, a slurry transfer pump, and a sludge mixing tank to mix the sludge and lime slurry. The process may be either continuous or batch and the slurry-sludge mixing tank must be of sufficient size to allow the mixture to remain at least 30 minutes at a pH of 11.5 to 12.0 before disposal. The pH of the slurry-sludge mixture should be measured either manually or automatically to ensure proper pH adjustment. The process of lime stabilization produces an unfavorable environment and destroys pathogenic and nonpathogenic bacteria. Studies have shown that greater than 99 percent of the fecal coliforms and fecal streptococci can be destroyed. If the pH is not adjusted to the above range, the goals of stabilization will not be achieved.

TROUBLESHOOTING

The problem that is usually encountered with lime stabilization is improper pH adjustment and subsequent disposal of unstabilized sludge. Routine pH measurements on the slurry-sludge mixture will reveal process inefficiencies. If the pH is lower than desired, check the slurry tank, slurry transfer pump, and the flow and concentration of solids to the slurry-sludge mixing tank. If an insufficient amount of lime was slurried or the slurry transfer pump is inoperative, the desired pH rise will not be achieved. If the lime slurry equipment and accessories are operating properly, check the flow rate and solids concentration to the mix tank. If either the flow or concentration is higher than normal, the total pounds of sludge solids also will be higher. In this case, increase the rate of flow of the lime slurry to the mixing tank until the desired pH is obtained.

3.231 Chlorine Stabilization

Chlorine stabilization is accomplished by adding sufficient quantities of gaseous chlorine to the sludge to kill pathogenic and nonpathogenic organisms. Estimated dosages to achieve disinfection are generally 100 to 300 lbs chlorine/ton of sludge solids (50 to 150 gm chlorine/kg sludge solids or about 2,000 mg chlorine per liter depending on percent sludge solids). Waste activated sludge (WAS) requires higher doses than primary sludge. As is the case with lime stabilization, there is very little

---

[24] *Slurry.*   A watery mixture or suspension of insoluble (not dissolved) matter; a thin, watery mud or any substance resembling it (such as a grit slurry or a lime slurry).

reduction of the sludge mass with chlorine stabilization. Therefore, the quantity of solids that remain for ultimate disposal are significantly greater than with digestion processes. The addition of the large quantities of chlorine required for stabilization will result in an acidic (pH less than 3.5) sludge and neutralization with lime or caustic may be required prior to dewatering due to the corrosive condition of the mixture.

NORMAL OPERATING PROCEDURES

Sludge to be treated enters the chlorine-sludge retention tank (Figure 3.14) through a feed or recirculation pump. The retention tank is normally operated at a pressure of 35 to 45 psig (2.5 to 3.2 kg/sq cm) and detention times of 10 to 15 minutes. After

leaving the reactor, the flow splits; about 10 percent is discharged for further solids processing and 90 percent passes through an *EDUCTOR*[25] and is recycled back to the reactor. The passage of the sludge through the eductor creates a vacuum, which causes the chlorine gas to move from the chlorine supply container into the sludge line.

Chlorine stabilization systems are completely automated with shutdown switches that will activate in the event of equipment malfunctions. Refer to manufacturer's literature for routine operating procedures and troubleshooting techniques. In general, since these systems are fully automated, you need only be concerned with maintaining desired flows, replenishing chlorine supplies, as needed, adjusting the chlorine addition rate, and

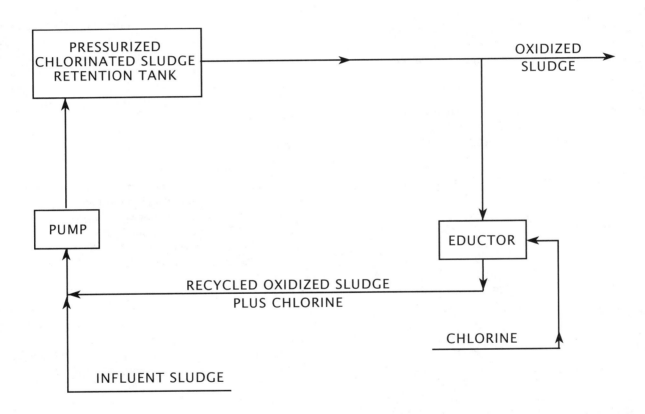

Fig. 3.14   Sludge stabilization with chlorine

---

[25] *Eductor* (e-DUCK-ter).   A hydraulic device used to create a negative pressure (suction) by forcing a liquid through a restriction, such as a Venturi. An eductor or aspirator (the hydraulic device) may be used in the laboratory in place of a vacuum pump. As an injector, it is used to produce vacuum for chlorinators. Sometimes used instead of a suction pump.

checking equipment operation according to manufacturer's recommendations.

QUESTIONS

Write your answers in a notebook and then compare your answers with those on page 314.

3.23A   List two chemicals used to stabilize sludges.

3.23B   Under what circumstances are chemicals most often used to stabilize sludges?

3.23C   What are two major limitations of using chemicals to stabilize sludges?

3.24   Additional Reading

1. *ANAEROBIC SLUDGE DIGESTION* (MOP 16). Obtain from Water Environment Federation (WEF), Publications Order Department, 601 Wythe Street, Alexandria, VA 22314-1994. Order No. M0024. Price to members, $26.75; nonmembers, $36.75; price includes cost of shipping and handling.

END OF LESSON 2 OF 5 LESSONS

on

RESIDUAL SOLIDS MANAGEMENT

Please answer the discussion and review questions next.

DISCUSSION AND REVIEW QUESTIONS

Chapter 3.   RESIDUAL SOLIDS MANAGEMENT

(Lesson 2 of 5 Lessons)

Write the answers to these questions in your notebook. The question numbering continues from Lesson 1.

9. Why do wastewater treatment plant sludges have to be stabilized before disposal?

10. What variables affect the operation and performance of aerobic digesters?

11. What laboratory tests should be performed on aerobic digester influent and effluent samples?

12. What factors could upset an anaerobic or aerobic digester?

13. What problem is often encountered with the lime stabilization process?

CHAPTER 3.   RESIDUAL SOLIDS MANAGEMENT

(Lesson 3 of 5 Lessons)

3.3   CONDITIONING

3.30   Purpose of Conditioning

Conditioning is defined as the pretreatment of sludge to facilitate the removal of water in subsequent treatment processes. Solid particles in sludge usually require conditioning because they are fine in particle size, hydrated (combined with water), and may carry an electrostatic charge. Removal of toxic substances such as heavy metals, as discussed in Sections 3.21 and 3.220, is beyond the scope of this chapter. If these toxic substances are present in the sludge, laboratory or pilot studies should be conducted to identify the proper chemical or physical method of sludge pretreatment, such as chemical *COAGULATION,*[26] *FLOCCULATION,*[27] sedimentation and filtration, air stripping or carbon adsorption, or ion exchange.

Sludge conditioning reduces mutually repelling electrostatic charges on suspended sludge particles, decreases the ability of biological sludges to entrain (hold) water, and promotes coagulation of the sludge solids. Sludge conditioning methods include: (1) chemical treatment, (2) thermal treatment, (3) wet oxidation, (4) *ELUTRIATION,*[28] and (5) other methods such as freezing, electrical treatment, and ultrasonic treatment. Of these, only chemical treatment, thermal treatment, wet oxidation, and elutriation are practiced on a full-scale basis and so the following discussion will focus on these four methods.

3.31   Chemical Conditioning

The most commonly used chemical for sludge conditioning is ferric chloride, either alone or in combination with lime. A group of synthetic organic chemicals known as *POLY-ELECTROLYTES*[29] or polymers have been developed and their use is gaining popularity and acceptance. Polymers are usually classified in three general types: anionic, cationic, or nonionic.

Anionic polymers have a negative charge and are normally used as coagulant aids with positively charged alum and ferric chloride. Cationic polymers are positively charged and can serve as the sole coagulant or can be used in combination with an inorganic coagulant such as alum. The use of cationic polymers is most common in sludge dewatering. Nonionic polymers normally consist of equal portions of cationic and anionic polymers and have a charge that can vary with the pH of the solution. Nonionic, anionic, and cationic polymers are all used as coagulant aids. A seasonal fluctuation has been noted in chemical conditioning requirements so that many plants can successfully condition using cationic polymers during the summer and anionic polymers during the winter.

The primary use for polymers in wastewater treatment is the conditioning of sludge to facilitate removal of water in subsequent treatment processes, such as belt filter presses, centrifuges, gravity belt thickeners, and dissolved air flotation thickeners. Polymers are available in either liquid or dry forms. The type of polymer you use depends upon the application, product performance, volume of use, space availability, and equipment capabilities.

Dry polymers come in various types of powder, crystals, and beads. Most dry polymers have a carrier in them and are about 94 to 96 percent active. The powder form is the most economical, but it may release a fine dust when handled, which can present a safety hazard.

Liquid polymers are available in either solution or emulsion/dispersion types. Both types consist of a polymer material in a medium, either water or oil, respectively. Emulsion/dispersion polymers range from 25 to 50 percent concentration and must be mixed with a large volume of water to achieve full activation before use. When water is added to emulsion/dispersion polymers, there is a dramatic increase in viscosity. Solution polymers

---

[26] *Coagulation* (ko-agg-yoo-LAY-shun).   The clumping together of very fine particles into larger particles (floc) caused by the use of chemicals (coagulants). The chemicals neutralize the electrical charges of the fine particles, allowing them to come closer and form larger clumps.

[27] *Flocculation* (flock-yoo-LAY-shun).   The gathering together of fine particles after coagulation to form larger particles by a process of gentle mixing. This clumping together makes it easier to separate the solids from the water by settling, skimming, draining, or filtering.

[28] *Elutriation* (e-LOO-tree-A-shun).   The washing of digested sludge with either fresh water, plant effluent, or other wastewater. The objective is to remove (wash out) fine particulates and/or the alkalinity in sludge. This process reduces the demand for conditioning chemicals and improves settling or filtering characteristics of the solids.

[29] *Polyelectrolyte* (POLY-ee-LECK-tro-lite).   A high-molecular-weight (relatively heavy) substance, having points of positive or negative electrical charges, that is formed by either natural or synthetic (manmade) processes. Natural polyelectrolytes may be of biological origin or obtained from starch products or cellulose derivatives. Synthetic polyelectrolytes consist of simple substances that have been made into complex, high-molecular-weight substances. Used with other chemical coagulants to aid in binding small suspended particles to larger chemical flocs for their removal from water. Often called a POLYMER.

range from 4 to 50 percent actual polymer and are extremely viscous and difficult to pump if not mixed with water. A widely used solution polymer is the *mannich reaction* polymer, which typically contains 4 to 6 percent active polymer. Mannich reaction polymers are produced by using formaldehyde as a catalyst. Since vapors from these polymers pose a safety hazard (formaldehyde is a carcinogen) they should be stored and used only by trained personnel in well-ventilated areas.

A detailed review of the chemistry involved when chemicals are used for conditioning is beyond the scope of this chapter. Essentially, the addition of chemicals reduces natural repelling forces and allows the solids to come together (coagulate) and gather (flocculate) into a heavier solid mass.

The optimum chemical(s) type and dosage for a particular sludge is highly dependent on the characteristics of that sludge. Calculation of chemical requirements is usually based on on-site experimentation and trial-and-error procedures. Sludge types and characteristics vary from one treatment plant to the next and there is no one chemical or dosage that can be applied to all plants and sludges.

3.310 Polymer Safety

There are a number of safety hazards associated with the use of polymers that require appropriate precautions to prevent injury or illness. The hazards generally fall into two areas, slipping hazards and personal exposure hazards from contact or inhalation:

- Slipping hazards—Polymers have a moisture attracting property. Even a thin film of dry polymer can combine with moisture and become extremely slippery. Liquid polymers are inherently slippery. Polymer spills must be cleaned up immediately. Gather up as much of the spilled material as possible by using an appropriate method, such as gentle sweeping, vacuuming, soaking it up with rags, or using "kitty litter" as an absorbent. Water flushing and the use of household bleach can remove remaining polymer material. The proper disposal of "cleaned up" material should be confirmed with your local environmental regulatory agency.

- Personal exposure hazards—Irritation of skin, eyes, or lungs can result from contact with polymer, polymer dust, or polymer fumes. Personal protective equipment (PPE) such as chemical-resistant gloves, splashproof goggles, and an apron may be required. If exposure to polymer dust or fumes from mannich reaction polymer may occur, use appropriate respiratory protection. Environmental monitoring should be performed to determine the extent of exposure as well as the required level of protection.

Areas where polymers are handled or stored should be equipped with sufficient continuous ventilation and emergency eye wash/deluge showers. *Remember, read and heed the safety requirements identified on your polymer's MSDS!*

3.311 Chemical Requirements

In selection of chemical types and the determination of chemical requirements, it is important that the operator be very familiar with the selection procedures and be able to compare the efficiency and cost of one product or chemical with other products.

Chemical requirements are usually determined by preliminary laboratory jar tests (page 223) followed by pilot- or full-scale trial experiments. Jar tests are effective in indicating the *RELATIVE* quantity of chemical(s) required, but should be followed by on-site dewatering experiments to more accurately determine the required chemical dosage.

Chemicals are available in either liquid or solid (powder, crystals) form and the best way to equate one product to the next is to express the quantity required (lbs) per unit (tons) of dry sludge solids. The quantity required per unit of dry sludge solids (lbs/ton) can then be multiplied by the chemical cost per pound ($/lb) to give you the cost in dollars per ton ($/ton) of sludge processed for each type of chemical.

EXAMPLE 28

Given: Jar tests indicate that a waste activated sludge flow of 30,000 GPD with a solids concentration of 1.5 percent sludge solids will require 18 pounds per day of Polymer A or 165 pounds per day of Polymer B for successful gravity thickening. Polymer A is a dry product and costs $2.00 per dry pound. Polymer B is a liquid product and costs $0.21 per liquid pound.

Find: 1. The polymer dosage in pounds polymer/ton of solids for Polymer A and Polymer B.

2. The unit cost in $/ton.

Solution:

Known	Unknown
Jar Tests on Waste Activated Sludge	1. Polymer dosage in lbs polymer per ton of solids for both Polymer A and B
Flow, GPD = 30,000 GPD	
Sl Sol, % = 1.5%	
Polymer A, lbs/day = 18 lbs/day	
Polymer B, lbs/day = 165 lbs/day	2. Unit cost in dollars per ton for both Polymer A and B
Polymer A, $/lb = $2.00/dry lb	
Polymer B, $/lb = $0.21/liquid lb	

1. Determine the polymer dosage in pounds of polymer per ton of sludge for both Polymer A and B.

a. Calculate the tons of dry sludge solids per day treated by the polymers.

$$\text{Sludge, tons/day} = \frac{\text{Flow, GPD} \times 8.34 \frac{\text{lbs}}{\text{gal}} \times \text{Sl Sol,} \frac{\%}{100\%}}{2{,}000 \text{ lbs/ton}}$$

$$= \frac{30{,}000 \frac{\text{gal}}{\text{day}} \times 8.34 \frac{\text{lbs}}{\text{gal}} \times \frac{1.5\%}{100\%}}{2{,}000 \text{ lbs/ton}}$$

$$= 1.88 \text{ tons/day}$$

b. Calculate the dosage for Polymer A in dry pounds of polymer per ton of sludge solids.

$$\text{Polymer A Dose,} \atop \frac{\text{lbs Polymer}}{\text{ton Sludge}} = \frac{\text{Amount of Polymer A, lbs/day}}{\text{Sludge, tons/day}}$$

$$= \frac{18 \text{ lbs Polymer A/day}}{1.88 \text{ tons/day}}$$

$$= 9.6 \text{ lbs Dry Polymer A/ton Sludge}$$

c. Calculate the dosage for Polymer B in liquid pounds of polymer per ton of sludge solids.

$$\text{Polymer B Dose,} \atop \frac{\text{lbs Polymer}}{\text{ton Sludge}} = \frac{\text{Amount of Polymer B, lbs/day}}{\text{Sludge, tons/day}}$$

$$= \frac{165 \text{ lbs Polymer B/day}}{1.88 \text{ tons/day}}$$

$$= 88 \text{ lbs Liquid Polymer B/ton Sludge}$$

2. Determine the unit cost in dollars per ton for both Polymer A and B.

a. Calculate the unit cost for Polymer A in dollars of polymer per ton of sludge solids treated.

$$\text{Unit Cost, \$/ton} = \text{Dose,} \frac{\text{lbs Polymer}}{\text{ton Sludge}} \times \text{Cost,} \frac{\$}{\text{lb}}$$

$$= 9.6 \frac{\text{lbs Polymer}}{\text{ton Sludge}} \times \frac{\$2.00}{\text{lb Polymer}}$$

$$= \$19.20/\text{ton of Sludge}$$

b. Calculate the unit cost for Polymer B in dollars of polymer per ton of sludge solids treated.

$$\text{Unit Cost, \$/ton} = \text{Dose,} \frac{\text{lbs Polymer}}{\text{ton Sludge}} \times \text{Cost,} \frac{\$}{\text{lb}}$$

$$= 88 \frac{\text{lbs Polymer}}{\text{ton Sludge}} \times \frac{\$0.21}{\text{lb Polymer}}$$

$$= \$18.50/\text{ton of Sludge}$$

This example illustrates the need to compare polymers on a cost per ton of solids (\$/ton) basis rather than on a pound of product per ton of solids (lb/ton) basis. Even though more pounds of Product B were required, it yielded a lower cost than Product A.

Successful jar test, pilot-scale, or full-scale chemical addition studies require that the chemicals be prepared prior to application. Liquid and powder or crystal polymers and lime must be mixed with water before they can be used as sludge conditioners. Ferric chloride, which is usually delivered in bulk tanks or 55-gallon drums, can be added directly to sludge.

Typically, dry polymers are mixed with water to produce a solution strength of 0.05 to 0.25 percent. Liquid polymers are usually diluted to 1 to 10 percent polymer solutions as product, while lime is mixed to create 5 to 30 percent lime solutions. These solution strengths are all based on the ratio of product weight to the weight of water. Sample calculations to determine the pounds of product required per gallon of solution are illustrated below.

EXAMPLE 29

Given: Twenty-five gallons of a 0.1 percent polymer solution is to be prepared by an operator at a wastewater treatment plant.

Find: The pounds of dry polymer to be added to the 25 gallons of water.

Solution:

Known	Unknown
Volume of Solution, gal = 25 gal	Dry Polymer Added, lbs
Polymer Solution, % = 0.1%	

1. Determine the pounds of dry polymer to be added by setting up the problem as a proportion.

$$\frac{\text{Polymer Solution, \%}}{100\%} = \frac{\text{Dry Polymer, lbs}}{\text{Vol of Sol, gal} \times 8.34 \text{ lbs/gal}}$$

$$\frac{0.1\%}{100\%} = \frac{\text{Dry Polymer, lbs}}{25 \text{ gal} \times 8.34 \text{ lbs/gal}}$$

2. Rearrange the terms in the above equation and solve for the pounds of dry polymer.

$$\text{Dry Polymer, lbs} = 25 \text{ gal} \times 8.34 \text{ lbs/gal} \times \frac{0.1\%}{100\%}$$

$$= 0.21 \text{ lb of Dry Polymer}$$

In the above example, if 0.21 lb of dry polymer is mixed with 25 gallons of water, the solution strength would be 0.10 percent.

NOTE: To be theoretically correct, the above equation should be:

$$\frac{\text{Polymer Solution, \%}}{100\%} = \frac{\text{Dry Polymer, lbs}}{\text{Solution, lbs} + \text{Dry Polymer, lbs*}}$$

---

* Dry polymer, lbs, term can be ignored when it is small compared with the solution, lbs, term.

EXAMPLE 30

Given: Six pounds of dry polymer are added to 480 gallons of water.

Find: The strength of the polymer solution.

Solution:

Known	Unknown
Volume of Solution, gal = 480 gal	Strength of Polymer Solution, %
Dry Polymer Added, lbs = 6 lbs	

Calculate the strength of the polymer solution as a percent.

$$\text{Polymer Solution, \%} = \frac{\text{Dry Polymer Added, lbs} \times 100\%}{\text{Vol of Sol, gal} \times 8.34 \text{ lbs/gal} + \text{Dry Pol, lbs}}$$

$$= \frac{6 \text{ lbs Polymer} \times 100\%}{480 \text{ gal} \times 8.34 \text{ lbs/gal} + 6 \text{ lbs Polymer}}$$

$$= 0.15\%$$

EXAMPLE 31

Given: A lime solution is prepared by adding 250 pounds of lime to 100 gallons of water.

Find: The strength of the lime solution as a percent.

Solution:

Known	**Unknown**
Lime Solution	Strength of Lime Solution, %
Lime Dose, lbs = 250 lbs	
Water Volume, gal = 100 gal	

Calculate the strength of the lime solution as a percent.

$$\text{Lime Solution, \%} = \frac{\text{Dry Lime Dose, lbs} \times 100\%}{\text{Volume of Water, gal} \times 8.34 \text{ lbs/gal} + \text{Lime, lbs}}$$

$$= \frac{250 \text{ lbs Lime} \times 100\%}{100 \text{ gal Water} \times 8.34 \text{ lbs/gal} + 250 \text{ lbs Lime}}$$

$$= 23\%$$

EXAMPLE 32

Given: Five gallons of a liquid polymer are added to 395 gallons of water.

Find: The strength of the polymer solution as a percent.

Solution:

Known	**Unknown**
Liquid Polymer, gal = 5 gal	Strength of Polymer Solution, %
Volume Water, gal = 395 gal	

Calculate the strength of the polymer solution as a percent.

$$\text{Polymer Solution, \%} = \frac{\text{Liquid Polymer, gal} \times 8.34 \text{ lbs/gal*} \times 100\%}{\text{Total Volume, gal} \times 8.34 \text{ lbs/gal}}$$

$$= \frac{5 \text{ gal} \times 8.34 \text{ lbs/gal} \times 100\%}{(395 \text{ gal} + 5 \text{ gal}) \times 8.34 \text{ lbs/gal}}$$

$$= 1.25\%$$

---

* Most liquid polymers are heavier than 8.34 lbs/gal.

The procedures for jar tests are outlined below and again it should be noted that these tests only indicate the relative effectiveness of sludge conditioners. Jar tests should be followed by pilot- or full-scale tests to determine the exact chemical requirements.

JAR TEST PROCEDURE

1. Collect approximately one gallon (approximately 4 liters) of a representative sample of sludge to be tested.

2. Prepare chemical solutions according to the manufacturer's recommendation. Only a small amount of chemical solution is needed for the jar test as compared with actual doses for wastewater being treated.

3. Save approximately ½ liter of the sludge sample for the sludge solids determination.

4. Fill a 1-liter beaker up to the 1-liter mark with the sludge to be tested.

5. Pipet a portion of the prepared chemical solution into the beaker containing the sludge. Polymer dosages should be increased by increments of 5 lbs/ton (2.5 gm/kg) or less for dry polymers, 25 lbs/ton (12.5 gm/kg) or less for liquid polymers, 100 lbs/ton (50 gm/kg) or less for ferric chloride, and 200 lbs/ton (100 gm/kg) or less for lime.

6. After the chemical is placed in the beaker containing the sludge sample, the entire contents should be poured *SLOWLY* into a second 1-liter beaker and then poured slowly back to the original 1-liter beaker. This slow pouring action allows the chemical to mix with the sludge and coagulate and flocculate the sludge solids. If the chemical is effective, large floc particles will develop and free water will be observed. If the floc does not develop, add another portion of the chemical solution and slowly pour from one beaker to next. Continue adding portions of the chemical followed by gentle pouring until floc formation and clear water or a supernatant are observed.

7. Instead of pouring the chemical and sludge sample back and forth from one beaker to another, a stirring apparatus can be used as described in Chapter 4, Section 4.120, "The Jar Test." The chemical mixing, flocculation, and settling conditions used in the jar test should be similar to the actual conditions in the treatment plant in order to obtain realistic results.

8. Record the volume of chemical solution required for floc formation.

9. After the solids analysis is performed on the initial sludge sample, determine the chemical dosage and costs.

The following example illustrates the incremental procedure and calculations for jar testing.

EXAMPLE 33

Given: A 1-gallon (4-liter) sample of digested primary sludge is to be collected and jar tests run using Polymer A, a dry polymer, and Polymer B, a liquid product. A ½-liter sample of the digested sludge was analyzed for suspended solids concentration. The sludge solids concentration was found to be 2 percent (20,000 mg/*L*).

Find: The quantity required and approximate cost of Polymer A and Polymer B.

Polymer Preparation: The solution to this problem is a series of jar tests. The first step is to prepare the polymer solutions. Polymer A is a dry polymer, therefore, mix a 0.05 to 0.25 percent solution. For jar tests, a 0.1 percent solution is desirable. Approximately 1 liter of solution should be prepared. The quantity of dry polymer to be added to 1 liter of water is determined based on the calculation in Example 29.

Solution:

Known	**Unknown**
Jar Tests Run on Polymer A and Polymer B	Quantity and Cost of Polymer A and Polymer B
Sludge Solids, % = 2% , mg/*L* = 20,000 mg/*L*	
Polymer A is a dry powder 0.05 to 0.25 percent solution	
Polymer B is a liquid mix 1 to 10 percent solution	

$$1 \text{ liter} = \frac{1 \text{ liter}}{3.78 \text{ } L/\text{gal}} = 0.265 \text{ gal}$$

1. To dose the jars, prepare 1 liter of a 0.10 percent solution of the dry polymer.

   a. Calculate the polymer dose in pounds.

   $$\text{Dry Polymer, lbs} = \text{Volume, gal} \times 8.34 \text{ lbs/gal} \times \text{Solution, } \frac{\%}{100\%}$$

   $$= 1 \, L \times \frac{0.265 \text{ gal}}{L} \times 8.34 \text{ lbs/gal} \times \frac{0.10\%}{100\%}$$

   $$= 0.0022 \text{ lb of Dry Polymer}$$

   b. Convert the dose in pounds to grams.

   $$\text{Dry Polymer, grams} = 0.0022 \text{ lb} \times 454 \text{ grams/lbs}$$

   $$= 1.00 \text{ gram}$$

   OR

   Calculate the polymer dose in grams directly.

   $$\text{Dry Polymer, grams} = \text{Volume, } L \times \frac{1,000 \text{ gm}}{L} \times \text{Solution, } \frac{\%}{100\%}$$

   $$= 1 \, L \times \frac{1,000 \text{ gm}}{L} \times \frac{0.10\%}{100\%}$$

   $$= 1.00 \text{ gram}$$

   Therefore, 1.00 gram of dry polymer mixed with 1 liter of water will produce 0.10 percent polymer solution.

2. To dose the jars, prepare 1 liter of a 2.5 percent solution of the liquid polymer.

   a. Calculate the liquid polymer dose in gallons.

   $$\text{Liquid Polymer, gal} = \text{Volume, gal} \times \text{Solution, } \frac{\%}{100\%}$$

   $$= 1 \, L \times \frac{0.265 \text{ gal}}{L} \times \frac{2.5\%}{100\%}$$

   $$= 0.0066 \text{ gal}$$

   b. Convert the dose in gallons to milliliters.

   $$\text{Liquid Polymer, m}L = 0.0066 \text{ gal} \times \frac{3,780 \text{ m}L}{\text{gal}}$$

   $$= 25 \text{ m}L$$

   OR

   Calculate the liquid polymer dose in milliliters directly.

   $$\text{Liquid Polymer, m}L = \text{Volume, } L \times \frac{1,000 \text{ m}L}{L} \times \text{Solution, } \frac{\%}{100\%}$$

   $$= 1 \, L \times \frac{1,000 \text{ m}L}{L} \times \frac{2.5\%}{100\%}$$

   $$= 25 \text{ m}L$$

3. Determine the amount of water to be mixed with the liquid polymer.

$$\text{Volume Water, m}L = \text{Total Volume, m}L - \text{Liquid Polymer, m}L$$

$$= 1,000 \text{ m}L - 25 \text{ m}L$$

$$= 975 \text{ m}L$$

Therefore, 25 mL of liquid polymer mixed with 975 mL of water will produce a 2.5 percent solution.

Following polymer preparation, conduct the jar tests and record the results in the following format:

	Polymer				
Sludge Type	Product Type	Form	% Solution	mL Added	Observation
Dig. Primary	A	dry	0.10	15	No floc formed
Dig. Primary	A	dry	0.10	30	Small floc formed
Dig. Primary	A	dry	0.10	50	Large floc and clear supernatant
Dig. Primary	B	liquid	2.5	15	No floc formed
Dig. Primary	B	liquid	2.5	30	Small floc formed
Dig. Primary	B	liquid	2.5	50	Large floc and clear supernatant

Based on the amount of polymer added and the observations made for the above tests, approximately 50 mL of Polymer Solution A and 50 mL of Polymer Solution B are required to coagulate and flocculate the solids.

4. Calculate the dosage for Polymer A and Polymer B in lbs/ton.

POLYMER A (DRY)

Determine the dosage of Polymer A in pounds of polymer per ton of sludge solids treated.

$$\text{Dosage, lbs/ton} = \frac{\text{Sol, }\frac{\%}{100\%} \times \frac{\text{Polymer, m}L}{\text{(Added)}} \times \frac{1 \text{ gal}}{3,780 \text{ m}L} \times \frac{8.34 \text{ lbs}}{\text{gal}} \times \frac{2,000 \text{ lbs}}{\text{ton}}}{\text{Sl Vol, }L \times \frac{1 \text{ gal}}{3.78 \text{ }L} \times 8.34 \frac{\text{lbs}}{\text{gal}} \times \text{Sl Sol, }\frac{\%}{100\%}}$$

By cancelling out similar terms, the equation can be reduced to:

$$\text{Dosage, }\frac{\text{lbs}}{\text{ton}} = \frac{\text{Sol, }\% \times \text{Polymer Added, m}L}{\text{Sl Vol, }L \times \text{Sl Sol, }\%} \times 2$$

$$= \frac{0.10 \times 50}{1.0 \times 2.0} \times 2$$

$$= 5 \text{ lbs Dry Polymer/ton Sludge Solids}$$

POLYMER B (LIQUID)

$$\text{Dosage, }\frac{\text{lbs}}{\text{ton}} = \frac{\text{Sol, }\% \times \text{Polymer Added, m}L}{\text{Sl Vol, }L \times \text{Sl Sol, }\%} \times 2$$

$$= \frac{2.5 \times 50}{1.0 \times 2.0} \times 2$$

$$= 125 \text{ lbs Liquid Polymer/ton Sludge Solids}$$

5. Calculate the cost per ton to use Polymer A if the dry polymer costs $2.00 per pound.

$$\text{Cost, }\frac{\$}{\text{ton}} = \frac{5 \text{ lbs Polymer}}{\text{ton Sludge Solids}} \times \frac{\$2.00}{\text{lb Polymer}}$$

$$= \$10/\text{ton Sludge Solids}$$

6. Calculate the cost per ton to use Polymer B if the liquid polymer costs $0.21 per pound.

$$\text{Cost, }\frac{\$}{\text{ton}} = \frac{125 \text{ lbs Polymer}}{\text{ton Sludge Solids}} \times \frac{\$0.21}{\text{lb Polymer}}$$

$$= \$26.25/\text{ton Sludge Solids}$$

Based on these jar tests, Polymer A would cost about one-half as much as Polymer B.

Following jar test experiments, the polymer or any other chemical should be evaluated on pilot- or full-scale equipment. The determination of polymer dosages is identical to calculations used for the jar test examples except that larger values of chemical and sludge are used. The following example illustrates the calculation on a full-scale basis.

EXAMPLE 34

Given: A waste activated sludge flow of 200 GPM at 0.90 percent (9,000 mg/L) solids is to be conditioned with 20 GPM of a 0.05 percent dry polymer solution.

Find: The pounds of dry polymer to be mixed with 5,000 gallons of water to produce a 0.05 percent solution and the resulting dosage in lbs/ton.

Solution:

Known	Unknown
Waste Activated Sludge	1. Pounds of dry polymer to be mixed with 5,000 gallons to produce a 0.05 percent polymer solution
Sludge Flow, GPM = 200 GPM	
Sl Sol, % = 0.90%	
, mg/L = 9,000 mg/L	
Polymer Flow, GPM = 20 GPM	
Polymer Solution, % = 0.05%	2. Dosage in pounds polymer per ton of sludge solids

1. Determine the pounds of dry polymer to be mixed with 5,000 gallons to produce a 0.05 percent polymer solution.

$$\text{Dry Polymer Required, lbs} = \frac{\text{Polymer Sol, }\% \times \text{Vol, gal} \times 8.34 \text{ lbs/gal}}{100\%}$$

$$= \frac{0.05\%}{100\%} \times 5,000 \text{ gal} \times \frac{8.34 \text{ lbs}}{\text{gal}}$$

$$= 20.9 \text{ lbs}$$

2. Calculate the dosage in pounds of polymer per ton of sludge solids.

$$\text{Dosage, } \frac{\text{lbs}}{\text{ton}} = \frac{\text{Sol, } \% \times \text{Polymer Added, GPM}}{\text{Sl Flow, GPM} \times \text{Sl Sol, } \%} \times \frac{2,000 \text{ lbs}}{\text{ton}}$$

$$= \frac{0.05\% \times 20 \text{ GPM}}{200 \text{ GPM} \times 0.9\%} \times \frac{2,000 \text{ lbs}}{\text{ton}}$$

$$= 11.1 \frac{\text{lbs Polymer}}{\text{ton Sludge Solids}}$$

An extensive amount of time and discussion was devoted to the determination of chemical solution requirements and chemical dosages because many times the proper amounts of chemicals are not added in routine operation. In order to chemically condition sludge at the required dosage, the operator must be able to determine the quantity to be prepared and added to the sludge.

QUESTIONS

Write your answers in a notebook and then compare your answers with those on pages 314 and 315.

3.30A   Why do solid particles present in sludge usually require conditioning?

3.30B   List the different types of sludge conditioning methods available.

3.31A   Briefly explain how chemical addition conditions sludge.

3.31B   Explain why chemical types and dosage requirements vary from plant to plant.

3.31C   Briefly explain how chemical requirements are determined for a particular sludge.

3.31D   Three pounds of dry polymer are added to 360 gallons of water. What is the solution strength of the mixture?

3.31E   Ten pounds of lime are added to 100 gallons of water. What is the solution strength of the mixture?

3.31F   Ten gallons of liquid polymer are added to 790 gallons of water. What is the solution strength of the mixture?

3.31G   Five gallons of commercially available ferric chloride are added to 50 gallons of water. What is the solution strength of the mixture?

3.31H   A jar test has been conducted on digested primary sludge. The sludge has a concentration of 3.0 percent SS (30,000 mg/L) and 60 mL of a 0.15 percent solution of polymer was required to flocculate the sludge. Determine the polymer dosage in lbs/ton and the cost in $/ton if the polymer costs $1.50/lb.

3.31I   A polymer solution of 2.5 percent is prepared from a liquid polymer and added at a rate of 3 GPM to a sludge flow of 30 GPM. The sludge has a solids content of 4 percent sludge solids. Determine the dosage (lbs/ton) and the cost ($/ton) if the liquid polymer costs $.20/lb.

3.312   Chemical Solution Preparation

One of the keys to successful chemical conditioning is the preparation of the chemical solution. Depending on the solution strength to be made, the proper amount (lbs) of dry polymer or lime must first be weighed out and then added to a predetermined amount of water and mixed. The weighing container should be dry, and the drums or bulk storage tanks of dry chemicals should not be allowed to absorb moisture. If these chemicals are stored in a dry place, there should be no problems with handling and transferring to a weighing container. If the chemicals absorb moisture or become wet, balls or cakes of chemicals will form and prevent easy handling and transferring.

If the quantity of chemical used exceeds approximately 25 to 50 lbs per day (11 to 22 kg/day), automatic chemical feed systems are commonly used. Such equipment usually includes a storage hopper to hold bulk supplies of the chemical and a screw conveyor system to measure out and transfer the dry chemical to the mixing chamber. These automatic systems are usually activated by liquid-level indicators in the mixing tank. When the liquid level falls below the bottom probe, a signal is automatically sent and water is delivered to the mix tank. After the water level reaches a predetermined point, a second signal activates the screw conveyor system and dry chemical is delivered to the tank. The length of time the screw feeder is operated depends on the number of pounds per minute the feeder can deliver to the mix tank, the solution desired, and the volume of the mix tank.

The most common problems encountered with automatic feed systems are plugging of the screw conveyor and the buildup of chemicals at the discharge side of the screw. These problems can usually be traced back to premature wetting of the chemicals by water sprays coming from the mix tanks or from not having a watertight storage and feed system. If moisture can be prevented from entering the storage hopper and screw conveyor, smooth operation should result. Lime is somewhat easier to put into solution than dry polymers. Automatic dry polymer feed systems are sometimes equipped with wetting mechanisms to prewet the polymer as it falls into the mix tank. If the polymer is not properly wet as it falls into the mix tank, a poor mix will result and will be evident by balls or "fish eyes" of undissolved polymer in the tank. Another method of mixing dry polymers is to use an aspirator or eductor to put the dry polymer into solution.

For smaller systems requiring less than 25 to 50 lbs/day (11 to 22 kg/day), manual batching can successfully be used. The procedure to prepare and apply batch chemicals manually is as follows:

1. Weigh out the desired quantity of dry product in a dry container.

2. Partially fill the mix tank with water until the impellers on the mixer are submerged.

3. Turn on the mixer.

4. Add the premeasured dry product to the mix tank. Lime can be poured directly into the tank. Dry polymers have to be added through an eductor for wetting purposes.

5. Fill the tank to the desired level.

6. Allow tank contents to mix thoroughly before use to sufficiently "cure" the solution.

7. Turn off the mixer.

The curing or mixing time after the dry chemical has been added to the tank should be 45 to 60 minutes for polymers and approximately 30 minutes for lime. If adequate mixing times are not allowed for curing, the chemical will not be as effective as it should be because it will not fully dissolve and chemical requirements for successful conditioning will increase.

The preparation of chemical solutions from liquid polymers and liquid ferric chloride is not as difficult as for dry polymers and lime because the liquids go into solution more rapidly and prewetting is not required.

Automatic batching systems are commonly used for handling quantities in excess of 55 gallons/day (208 liters/day) of product. These systems incorporate the use of bulk storage tanks, bulk solution transfer pumps, and mixing tanks. Manual preparation incorporates the same procedures outlined for dry chemicals except that eductors are not used and the curing time can usually be reduced to approximately 20 to 30 minutes.

After the chemicals are cured, they can either be pumped to another tank or pumped to the sludge stream to be conditioned. The use of a second holding tank provides for a mix tank to be available at any time to prepare another batch of chemicals. Both the mix tank and transfer tank, if used, should be covered and protected from the sun's rays and extreme heat. Covering of the tanks should prevent foreign material from entering and possibly clogging equipment. When polymers are used, covering should be mandatory because the ultraviolet sun rays deteriorate the polymer molecules and can rapidly decrease the effectiveness of the solution. The same is true if the tank contents are allowed to approach temperatures of 120 to 130°F (49 to 54°C). At these temperatures, the polymer molecules can be broken down and the effectiveness of the solution deteriorates. To ensure protection against ultraviolet rays and extreme heat, all chemical tanks should be covered and insulated or painted white to reflect heat.

3.313 Chemical Addition

Once the chemical has been prepared and the approximate dosage and addition rate determined, the solution can be added to the sludge for conditioning. The point(s) of application for the different chemicals will vary depending on the chemical type and the specific mechanical equipment (DAF, centrifuges) used. In general, polymers are added directly into the feed assemblies of the various equipment types. Polymers should not be added to the suction side of sludge feed pumps because the shearing forces through such pumps tend to shear any floc formation. After conferring with the equipment and polymer manufacturers, application points for polymers should be determined by field experimentation. The use of lime and ferric chloride generally requires a blending tank to mix these chemicals with the sludge prior to dewatering. Lime and ferric chloride are generally not used for DAF thickening or centrifugation. Their use is usually limited to gravity thickening and vacuum and pressure filtration. Again, application points and blending requirements can best be determined by field experimentation and discussions with the equipment and chemical manufacturers.

3.314 Typical Chemical Requirements

Table 3.15 summarizes typical chemical dosages required for conditioning various types of sludges. Remember that the actual chemical requirements vary not only with the actual sludge, but also with the dewatering device. The optimum chemical dosage(s) is usually determined by on-site experimentation.

TABLE 3.15   TYPICAL CHEMICAL CONDITIONING REQUIREMENTS[a]

Sludge Type	Ferric Chloride, lbs/ton[b]	Lime, lbs/ton[b]	Polymer, lbs/ton[b]
Primary	20–40	120–200	4–24
Primary WAS	30–50	140–180	10–20
WAS	80–200	—	4–30
Digested Primary	30–100	300–600	5–40
Digested Primary and WAS	30–200	300–600	15–50
Digested WAS	80–200	300–600	15–40
Digested Elutriated Primary	40–80	—	10
Digested Elutriated Primary and WAS	80–125	—	15–30

[a] *SLUDGE PROCESSING AND DISPOSAL: A STATE OF THE ART REVIEW,* LA/OMA Project, County Sanitation Districts of Los Angeles County, Whittier, CA, April 1977.

[b] Expressed as pounds of chemical/ton of dry sludge solids. lbs/ton × 0.5 = gm of chemical/kg of dry sludge solids.

3.315 Troubleshooting

If decreases in thickening or dewatering equipment performance cannot be traced to equipment malfunctions, check the chemical mixing (preparation) and addition equipment. With automatic feeding systems, check: (1) the level of dry product in the storage hopper and replenish, if necessary, (2) the screw conveyor and unplug, if necessary, (3) the quality of the solution (are there large balls of undissolved chemical?), and (4) the chemical addition pump. Many times these chemical pumps are allowed to run dry due to inoperative level indicators and shutoff mechanisms. If pumps are run dry, the interior components may wear and the pump will not deliver at its rated capacity. You should immediately recalibrate the chemical feed pump or repair it, if necessary, because the pump may be the only means of measuring the chemical feed rate. The best indication of a failure in the chemical preparation or the chemical feed rate is to run a jar test on the sludge with a laboratory-prepared batch of the chemical. If the jar tests indicate that substantially less polymer is needed than the amount supposedly being added at full scale, there is usually a problem with the quality of the full-scale solution or the application rate.

Table 3.16 summarizes the problems that may arise during chemical conditioning and the actions that might be taken.

QUESTIONS

Write your answers in a notebook and then compare your answers with those on page 315.

3.31J   Why should dry chemicals be kept in a dry place?

3.31K   What is the purpose of wetting dry polymers?

3.31L   Outline the procedures to prepare a batch solution of dry and liquid chemicals.

3.31M   Why is curing time important?

3.31N   Why should chemical tanks be covered?

3.31O   Why are polymers generally not added to the suction side of sludge pumps?

3.31P   Outline the areas to be checked if sludge thickening or dewatering inefficiencies cannot be traced back to equipment failures.

3.32   Thermal Conditioning

Wastewater sludges, and biological sludges in particular, may have large quantities of bound water associated with them. The cell mass of biological sludges contains water along with other soluble and particulate matter. Outside the cell wall is a gelatinous sheath (cover) composed of PROTEINACEOUS[30] and POLYSACCHARIDE[31] material along with an additional quantity of water referred to as "bound water." Subjecting these sludge particles to extreme heat at elevated pressures hydrolizes (decomposes) the surrounding sheath and bursts the cell wall allowing bound water to escape. The net effect of releasing the cell water is a substantial increase in the dewaterability of the sludge. Thermal conditioning is effective in destroying all pathogens in the sludge, thereby increasing the options available for beneficial reuse of the treated sludge.

When used for conditioning, thermal treatment facilities are usually operated in the heat treatment or low pressure (150 to 300 psi or 1 to 2 kg/sq cm) wet oxidation (LPO) modes. The process descriptions for heat treatment and LPO conditioning are basically identical to that for the wet oxidation process outlined in Section 3.33. The major differences are that: (1) air is not introduced into the reactor for heat treatment conditioning and only a limited or small quantity of air is introduced for LPO conditioning, and (2) the reactor temperatures are lower than those sustained for wet oxidation. Reactor temperatures for heat treatment and LPO conditioning are typically 350 to 400°F (177 to 204°C) with reactor detention times of 20 to 40 minutes.

TABLE 3.16   TROUBLESHOOTING CHEMICAL CONDITIONING PROCESSES

Problem	Possible Cause	Check or Monitor	Possible Solution
Effluent quality or sludge concentrations from thickening or dewatering equipment deteriorating	1. Poor solution mixture	1a. Automatic feed system	1a. Fill storage hoppers and batch tanks
		1b. Mixer operation	1b. Allow for adequate curing time
		1c. Run jar test on sludge with a fresh laboratory solution of chemical	1c. Batch a new supply of chemicals
	2. Chemical dosage inadequate	2a. Chemical feed pump operation	2a. Turn on pump, open appropriate valves. Calibrate pump and increase rate or solution strength of the chemical.

---

[30] Proteinaceous (PRO-ten-NAY-shus).   Materials containing proteins, which are organic compounds containing nitrogen.
[31] Polysaccharide (polly-SAC-uh-ride).   A carbohydrate, such as starch or cellulose, composed of chains of simple sugars.

3.320 Factors Affecting Thermal Conditioning

The performance and efficiency of thermal conditioning systems depend on: (1) the concentration and consistency of the influent sludge, (2) reactor detention times, and (3) reactor temperature and pressure. For conditioning purposes, the introduction of relatively small quantities of air or steam (LPO) results in little, if any, difference in sludge dewaterability. The advantage usually associated with adding air is a reduction in fuel requirements because of increased thermal efficiencies within the reactor. This potential fuel savings may be offset by the power requirements needed to supply the air. For all practical purposes, however, heat treatment and LPO conditioning will be regarded as equivalent in this discussion.

The solids concentration of the influent sludge will have significant effects on the overall heating requirements and the reactor detention times. The physical size of thermal conditioning systems is based on hydraulic and solids loadings. If the concentration of the influent sludge decreases significantly, the volume of water pumped to the reactors will increase. This will cause a decrease in the detention time within the reactor and an increase in the heating requirements due to the increased water volume. The following example illustrates the effects of sludge concentration on the operation of thermal systems.

EXAMPLE 35

Given: A thermal conditioning system is designed to process 200 GPM of waste activated sludge at a concentration of 3.5 percent. The thermal reactor has a volume of 8,000 gallons.

Find: 1. The reactor detention time under design conditions.

2. The reactor detention time if the sludge enters at a concentration of 2.5 percent.

3. The effect of reduced concentration on heat requirements.

Solution:

Known	**Unknown**
Thermal Conditioning System	1. Reactor Detention Time, min
Treat Waste Activated Sludge	
WAS Flow, GPM = 200 GPM	2. Reactor Detention Time if Solids at 2.5%, min
Reactor Vol, gal = 8,000 gal	
Sludge Solids, % = 3.5%	3. Effect of 2.5% Solids on Heat Requirements

1. Calculate the reactor detention time in minutes.

$$\text{Detention Time, min} = \frac{\text{Reactor Volume, gal}}{\text{Flow, GPM}}$$

$$= \frac{8,000 \text{ gallons}}{200 \text{ gal/min}}$$

$$= 40 \text{ min}$$

2. Calculate the reactor detention time if the sludge solids concentration drops from 3.5% to 2.5%. A reduction in solids concentration causes an increase in WAS flow.

$$\text{New Flow, GPM} = \text{Old Flow, GPM} \times \frac{\text{Old Sl Sol, \%}}{\text{New Sl Sol, \%}}$$

$$= 200 \text{ GPM} \times \frac{3.5\%}{2.5\%}$$

$$= 280 \text{ GPM}$$

$$\text{Detention Time, min} = \frac{\text{Reactor Volume, gal}}{\text{Flow, GPM}}$$

$$= \frac{8,000 \text{ gallons}}{280 \text{ gal/min}}$$

$$= 29 \text{ min}$$

3. What is the effect on heat requirements of a decrease in WAS concentration from 3.5% to 2.5% sludge solids?

A specific amount of heat is required to raise a volume of water from one temperature level to the desired level. If the volume of water increases (as it will when WAS concentration drops to 2.5% sludge solids), the amount of heat required to raise the temperature of the increased volume of water also increases.

In the above example, the reactor detention time decreased from 40 minutes to 29 minutes when the sludge concentration decreased from 3.5 to 2.5 percent. This is not necessarily desirable. In general, as the reactor detention time increases from 20 to 40 minutes, the dewaterability of the sludge also increases somewhat and it is important to consistently pump a thickened sludge to the thermal unit to ensure effective and efficient operation.

The temperature and pressure within the reactor also contribute to the degree of conditioning obtainable. As the reactor temperature is increased from 350 to 400°F (177 to 204°C), the general trend is an increase in the dewaterability of the conditioned sludge. Pressure and temperature are dependent upon each other; therefore, pressures should be increased when the temperature is increased to prevent the sludge from boiling.

3.321 Operating Guidelines

The key operating guidelines that the operator has some control over on a day-to-day basis are: (1) inlet sludge flow, (2) reactor temperature and detention time, and (3) sludge withdrawal from the decant tank.

As discussed, the inlet sludge flow and reactor detention time are dictated by the concentration of the thickened feed sludge and the total pounds of solids processed. If you maintain a consistently thick feed by closely monitoring and operating the thickening equipment, the sludge inlet volume will be minimized, reactor detention times will be maximized, and optimum thermal conditioning should result. The control of reactor temperature within the normal range of 350 to 400°F (177 to 204°C) will depend on how the sludge dewaters in subsequent dewatering facilities. In general, the sludge dewaters better following conditioning at higher temperatures but the temperature should be maintained so as to achieve the desired dewatering results. The degree of dewatering required will vary from one treatment plant to the next and you should maintain operating temperatures according to the dewaterability desired and achieved. If it is found that a reactor temperature of 350°F (177°C) provides sufficient conditioning to satisfy the dewatering requirements, do not increase the conditioning temperature. If on the other hand, a temperature of 350°F (177°C) is not adequate from a dewatering standpoint, increase the reactor temperature. Any decisions to vary reactor temperatures or pressures should be based on consultation between process operators and their supervisors.

The operator also has the ability to control the concentration of the underflow solids from the decant tank by controlling the rate of sludge withdrawal. The decant tank is a gravity thickener and the same operating procedures outlined in Section 3.111 should be applied in operating the decant tank. In most instances, gasification is not a problem in thermal conditioning decant tanks because of the lack of biological activity. The high temperatures sustained in the thermal reactors should sterilize the sludge and biological activity and subsequent gas production should not occur.

QUESTIONS

Write your answers in a notebook and then compare your answers with those on pages 315 and 316.

3.32A   Briefly explain how thermal conditioning improves the dewaterability of sludge.

3.32B   List the factors that affect thermal conditioning.

3.32C   Determine the reactor detention time for a reactor volume of 1,000 gallons and a sludge flow of 33 GPM with a concentration of 4.0 percent.

3.32D   If the sludge concentration from problem 3.32C decreases to 2.5 percent, determine the reactor detention time assuming the same total pounds are processed.

3.32E   Briefly discuss the operating controls available to optimize a thermal conditioning process.

3.32F   Why is gasification not usually a problem with gravity thickening of thermally treated sludge?

3.322   Normal Operating Procedures

Thermal conditioning units can be operated in the continuous or batch modes. Continuous operation is the preferred mode because energy is not wasted in allowing the heat exchanger and reactor contents to cool down and be heated back to the desired temperature each time it is operated as a batch process.

To operate a thermal unit in the batch mode:

1. Fill the reactor and heat exchangers with water if the water is drained after the previous day's shutdown.

2. Turn on the boiler makeup water pump and open the valve to the steam boiler.

3. Open the required steam valves to the thermal reactor and start the boiler.

4. After the reactor has reached its desired operating temperature, open the sludge inlet and outlet valves.

5. Turn on the sludge grinder and the stirring mechanism in the decant (gravity thickener) tank.

6. Turn on the vent fan from the decant tank and activate the appropriate odor-control equipment.

7. Turn on the sludge feed pump. If LPO conditioning is used, turn on the air compressor.

These procedures should be followed in reverse for shutdown operation. For continuous operation, these procedures are followed whenever the operation is interrupted for mechanical or routine shutdowns.

Whether operating in the continuous or batch mode, fuel levels for the steam generating system should be routinely checked and replenished, as required, and daily records of the pressure drop across the heat exchangers should be kept. The heat exchangers are subject to clogging due to the formation of scale. Periodic acid flushings are therefore required to remove these deposits and to unplug the heat exchangers. The best indication of scaling and the time at which an acid flushing should be conducted is the pressure drop across the heat exchangers. Pressure drop is determined by measuring the pressures at the heat exchanger inlet and outlet and calculating the pressure differential (Δp) according to the following equation:

$$\Delta p = P \text{ outlet} - P \text{ inlet}$$

When the pressure difference (Δp) reaches a certain magnitude, the system should be taken out of service and an acid flushing should be done. The pressure drop at which an acid flushing is required is determined by the manufacturer and in no case should the pressure differential be allowed to develop beyond the manufacturer's recommended figure. Routine or periodically required acid flushings should be conducted according to the manufacturer's recommended procedure.

3.323   Typical Performance

Typical operating guidelines for thermal conditioning are presented in Table 3.17. The overall evaluation of thermal performance is based on subsequent mechanical dewatering of the conditioned sludge. Performance data for various conditioning and dewatering schemes will be presented in Section 3.4, "Dewatering." The degree of dewatering obtainable is indicated in Table 3.17 from a qualitative standpoint.

TABLE 3.17   DEGREE OF DEWATERING FROM VARIOUS SLUDGE TYPES

| Sludge Type | Thermal Mode | Reactor | | | Dewaterability |
		°F[a]	psig[b]	Detention Time, min	
Primary	LPO or HT	350–400	350–400	20–60	Excellent
Secondary	LPO or HT	350–400	350–400	20–60	Good–Excellent
Dig. Primary	LPO or HT	350–400	350–400	20–60	Good–Excellent
Dig. Primary & Secondary	LPO or HT	350–400	350–400	20–60	Good

[a] $(°F - 32°F) \times 5/9 = °C$.
[b] $psi \times 0.07 = kg/sq\ cm$.

3.324   Troubleshooting

Thermal conditioning systems are high-temperature, high-pressure processes that incorporate the use of sophisticated instrumentation and mechanical equipment. All of the mechanical, electrical, and performance difficulties that might arise cannot be summarized in this section. In the event of complicated mechanical or electrical malfunctions, the operator should not attempt to locate or to correct these problems without the assistance of qualified mechanics, electricians, or instrument personnel. The following discussion will be limited to malfunctions typically encountered on a day-to-day basis.

3.3240   REACTOR TEMPERATURE

If the reactor temperature cannot be maintained at the desired level, check: (1) the fuel supply to the steam boiler, (2) temperature sensor and boiler actuator assembly, and (3) boiler makeup water supply. If the boiler fuel system and makeup water supply are adequate but the reactor temperature fluctuates significantly, the problem may be related to instrumentation malfunctions; seek the assistance of qualified electricians or instrumentation personnel to diagnose and repair the instruments.

3.3241   REACTOR PRESSURE

If the high-pressure feed pump is inoperative, the desired pressure will not be maintained. The usual problem with the feed pump is loss of prime due to a plug on the suction side or clogging of the sludge guide. Loss of prime will result in low or no flow through the system.

3.3242   HEAT EXCHANGER PRESSURE DIFFERENTIAL

Increases in the pressure drop across the heat exchangers indicates the buildup of scale. In the event of an excessive pressure

drop, schedule a shutdown and acid flush the system according to the manufacturer's recommended procedure.

3.3243   SLUDGE DEWATERABILITY

A decrease in the dewaterability of the thermally conditioned sludge may be caused by operational difficulties with the specific dewatering equipment or operation at less than optimum conditions. If the deterioration in dewaterability cannot be attributed to dewatering equipment inefficiencies, check: (1) the flow rate through the thermal system, (2) reactor temperature, and (3) operation of the decant (gravity) thickener.

An increase in the flow rate due to introducing a thin feed sludge will result in decreased reactor detention times and possible decreases in dewaterability.

If the problem is attributed to low reactor detention times, check and optimize the operation of the sludge thickening equipment. Decreases in reactor temperatures will also result in inferior dewatering characteristics and you should check and adjust the temperature, as required.

The operation of the decant tank will also affect dewaterability. In general, you should provide for as thick a decant underflow sludge as possible to the dewatering facility. This is controlled by monitoring and controlling the underflow sludge withdrawal rate. Be careful that the sludge does not become so heavy that it cannot be moved out of the decant tank. Decant stirring plows should be operated continuously. Operation of the decant tank should follow the same procedures outlined for gravity thickeners.

Table 3.18 summarizes operational problems that might be encountered with thermal conditioning processes and some possible solutions.

QUESTIONS

Write your answers in a notebook and then compare your answers with those on page 316.

3.32G   Why is continuous operation of a heat treatment unit desirable?

3.32H   Outline the start-up and shutdown procedures for a heat treatment unit.

3.32I   Why should a log be kept on the pressure drop across the heat exchangers and what action should be taken to correct excessive pressure drops?

3.32J   Over the course of a week, the dewaterability of a thermally conditioned sludge decreases drastically. What operating conditions or equipment should be checked and what corrective measures can be taken?

3.33   Wet Oxidation

Wet oxidation is a thermal treatment process that stabilizes organic matter and results in a net reduction in the sludge mass and a total destruction of pathogenic organisms.

TABLE 3.18 TROUBLESHOOTING THERMAL CONDITIONING PROCESSES

Problem	Possible Cause	Check or Monitor	Possible Solution
1. Reactor temperature not maintained	1a. Fuel exhausted 1b. Temperature sensor and actuators inoperative 1c. Makeup water supply inadequate	1a. Fuel supply and fuel lines 1b. Instrumentation 1c. Water supply	1a. Replenish 1b. Clean and repair or replace 1c. Replenish
2. Reactor pressure not maintained	2a. Feed pump inoperative	2a. Grinder, pump suction and discharge valve, and sludge supply	2a. Unplug grinder and pump suction. Open suction and discharge valves. Provide thickened sludge.
3. Heat exchanger Δp excessive	3a. Scaling	3a. Inlet and outlet pressures	3a. Acid flush exchangers
4. Reduction in sludge dewaterability	4a. Low temperature 4b. Low detention time 4c. Poor operation of decant (gravity) thickener	4a. Reactor temperature 4b. Sludge flow 4c. Thickness of blanket and sludge concentration	4a. Same as 1 4b. Thicken feed sludge 4c. Thicken underflow sludge

Three modes of wet oxidation exist: low-pressure wet oxidation (LPO), intermediate-pressure wet oxidation (IPO), and high-pressure wet oxidation (HPO). Figure 3.15 is a schematic of the process. Sludge to be processed is first passed through a grinder to reduce the particle size of the sludge solids and thereby reduce the potential for clogging inside the wet oxidation unit. The sludge may then be pumped to the oxidation unit by a high-pressure, positive displacement pump along with air supplied by an air compressor. A high-pressure feed pump is used to produce and maintain required pressures in the oxidation unit. Sludge and air are then passed through heat exchangers and delivered to the thermal reactor. Heat is added to the reactor from an external source, usually a steam boiler, to maintain desired reactor temperatures. Stabilization takes place within the reactor. The stabilized sludge leaving the reactor is cooled in the heat exchangers against the entering cold sludge and then released to a decant (gravity) thickener for separation and compaction of the stabilized sludge solids. Off gases from the decant tank are vented to gas scrubbers and carbon adsorbers or to a catalytic combustion unit for odor control. Overflow from the decant tank may be returned to the plant headworks while the underflow (thickened) solids are pumped to subsequent dewatering units. The decant tank overflow may require additional treatment prior to recycling to the plant headworks.

Under LPO conditions, feed sludge is reacted with approximately 15 SCF[32] air/lb solids (0.94 SCM[33] air/kg solids), while temperatures around 400°F (204°C) and pressures of 400 psig (28 kg/sq cm) are maintained. IPO treatment requires the addition of approximately 45 SCF air/lb solids (2.81 SCM air/kg solids) and reactor temperatures and pressures of 450°F (232°C)

and 500 to 600 psig (35 to 42 kg/sq cm), respectively. Under HPO conditions, feed sludge is reacted with approximately 100 SCF air/lb solids (6.24 SCM air/kg solids) while reactor temperatures and pressures approximate 500°F (260°C) and 1,000 to 1,500 psig (70 to 105 kg/sq cm), respectively. For each of the three modes of wet oxidation, reactor detention times usually vary from 20 to 40 minutes.

3.330 Factors Affecting Wet Oxidation

The performance and efficiency of wet oxidation units are dependent on: (1) the concentration and consistency of the feed sludge, (2) reactor detention time, (3) reactor temperature and pressure, and (4) quantity of air supplied.

The effects of feed sludge, reactor temperature and pressure, and reactor detention times are discussed in Section 3.32, "Thermal Conditioning." The major difference between wet oxidation and thermal conditioning is that air is introduced for wet oxidation. As wet oxidation progresses from the LPO to HPO mode of operation, the degree of oxidation or conversion of the sludge solids to volatile gases also increases. Thus, an increase in oxidation is due primarily to reacting the sludge with greater quantities of oxygen at elevated temperatures and pressures.

The operating guidelines, normal operating procedures, and troubleshooting are the same as those discussed in Section 3.32 (3.321, 3.322, and 3.324) for thermal conditioning except that reactor temperatures and pressures are higher for IPO and HPO. In addition, the quantity of air supplied (SCF per pound of sludge solids) is also higher for IPO and HPO when compared to thermal conditioning.

---

[32] SCF. Standard Cubic Feet of air at standard conditions of temperature, pressure, and humidity (0°C, 14.7 psia, and 50 percent relative humidity).

[33] SCM. Standard Cubic Meters of air at standard conditions of temperature, pressure, and humidity.

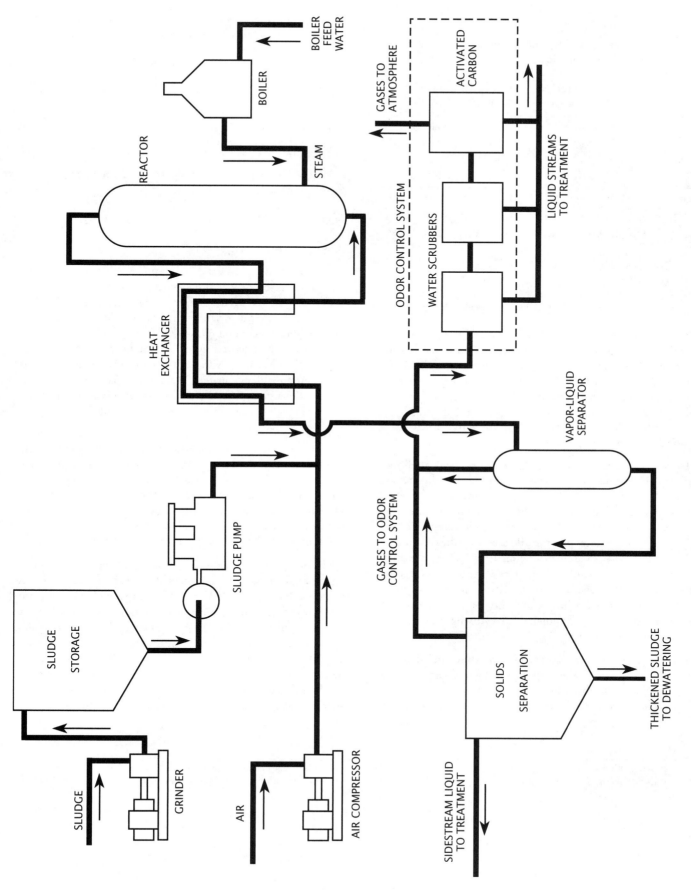

Fig. 3.15   Wet air oxidation system

3.331   Typical Performance

Typical operating guidelines and the degree of oxidation for the three modes of wet oxidation are presented in Table 3.19.

TABLE 3.19   OPERATIONAL AND PERFORMANCE GUIDELINES FOR WET OXIDATION UNITS

Mode of Operation	Reactor				% Reduction		
	°F[a]	psig[b]	Detention Time, min	Air, SCF/lb solids[c]	Total Solids	VSS	Total COD
LPO	350–400	350–400	20–60	15	20–25	25–40	25–40
IPO	450	500–600	20–60	45	30–50	40–60	40–60
HPO	500	1,000–1,500	20–60	100	70–75	75–85	75–85

[a] (°F – 32°F) × 5/9 = °C.
[b] psi × 0.07 = kg/sq cm.
[c] SCF/lb × 0.0624 = SCM/kg.

In addition to reducing the sludge mass and total COD, wet oxidation should generally result in sterilization (total destruction of pathogenic and nonpathogenic organisms) of the sludge because of the elevated temperatures and the reactor detention times used. The thermally oxidized sludge, which is thickened and withdrawn from the bottom of the decant tank, usually exhibits excellent dewatering characteristics as will be discussed in Section 3.4, "Dewatering."

Two of the drawbacks of wet oxidation or thermal conditioning are the production of noxious odors and high-strength liquid sidestreams. The odors closely resemble the smell of burned plastic and are produced by volatilizing or converting the organics in the sludge to complex organic gases. The production of odors requires that the off gases from the decant tank be deodorized prior to atmospheric discharge. Therefore, thermal treatment systems must be equipped with gas scrubbers and carbon adsorbers or catalytic combustion units. The operator should become familiar with the operation and maintenance of the air pollution control equipment by reviewing the manufacturer's literature and Chapter 1, "Odor Control."

The liquid sidestreams (decant tank overflow and dewatering equipment) are extremely high in soluble organics. Recycling these liquids to the treatment plant headworks may result in operational problems in secondary treatment processes. If problems develop because of the recycling of thermal liquors, separate aerobic or anaerobic treatment may be required.

QUESTIONS

Write your answers in a notebook and then compare your answers with those on page 316.

3.33A   Explain the differences between LPO, IPO, and HPO.

3.33B   List the factors affecting wet oxidation.

3.33C   Why is air pollution control equipment required on thermal treatment units?

3.34   Elutriation[34]

3.340   Process Description

Elutriation is basically a washing process that may not actually improve the dewatering characteristics of digested sludge but does reduce chemical conditioning requirements. The reduction in chemical conditioning requirements has often been attributed to a reduction in sludge alkalinity and subsequent reduction in lime requirements for pH adjustment. While dilution of digested sludge with fresh water or plant effluent results in a dilution and an apparent reduction in alkalinity, the major reason for improved dewaterability is most probably the result of washing out of fine, difficult-to-dewater solids.

As discussed in Section 3.0, fine, low-density solids have large surface areas with possibly high electrostatic charges and are more difficult to dewater than larger and coarser solids. If these fine solids are taken out of the sludge mass, the remaining sludge solids would naturally be easier to dewater. The problem with elutriation is that the fine solids removed from the sludge stream are recycled back to the plant headworks. Sometimes these fine solids may pass through the plant and leave in the plant effluent.

Various case histories demonstrate that elutriation lowers chemical demands and improves dewaterability. However, as much as 50 percent of the digested solids may be lost to the plant effluent with the elutriation effluent (elutriate). The loss of these fine solids into the plant effluent will lower the effluent quality while recycling to the plant headworks generally results in operational problems due to the buildup of fine solids throughout the system. In general, elutriation is not a preferred or efficient method of sludge conditioning in light of the ever-increasing federal, state, and local effluent discharge requirements.

---

[34] *Elutriation* (e-LOO-tree-A-shun).   The washing of digested sludge with either fresh water, plant effluent, or other wastewater. The objective is to remove (wash out) fine particulates and/or the alkalinity in sludge. This process reduces the demand for conditioning chemicals and improves settling or filtering characteristics of the solids.

3.341 Operating Guidelines

The simplest and most common method of elutriation is the single-stage method, which uses a single contact between the solids and elutriating liquid (elutriant). In this system, the sludge and elutriant make contact in an elutriating tank and are vigorously mixed for 30 to 60 seconds. The mixer is then turned off and the contents are allowed to settle from 4 to 24 hours under batch operation, or the contents are delivered to a gravity thickening tank under continuous operation. After the sludge and elutriant are vigorously mixed and settled or delivered to a gravity thickener, the operation becomes one of gravity thickening. Refer to Section 3.1 for operating strategies for a gravity thickener.

QUESTION

Write your answer in a notebook and then compare your answer with the one on page 316.

3.34A   How does elutriation improve the dewaterability of sludge? Discuss the problems associated with the process.

END OF LESSON 3 OF 5 LESSONS

on

RESIDUAL SOLIDS MANAGEMENT

Please answer the discussion and review questions next.

DISCUSSION AND REVIEW QUESTIONS

Chapter 3.   RESIDUAL SOLIDS MANAGEMENT

(Lesson 3 of 5 Lessons)

Write the answers to these questions in your notebook. The question numbering continues from Lesson 2.

14. How is the optimum type of chemical and dose to condition a particular sludge determined? Why?

15. What are the most common problems encountered with automatic dry chemical feed systems? List the cause of each problem.

16. How would you attempt to identify the cause of a decrease in the performance of sludge thickening or dewatering processes when the problems appear to be with the chemical conditioning facilities?

17. List the types of problems typically encountered on a day-to-day basis with a thermal conditioning system.

18. How are odors from thermal treatment systems controlled?

CHAPTER 3. RESIDUAL SOLIDS MANAGEMENT

(Lesson 4 of 5 Lessons)

3.4 DEWATERING

3.40 Purpose of Dewatering

Following stabilization, wastewater sludges can be ultimately disposed of by a variety of methods or they can be dewatered prior to further processing and ultimate disposal. In general, it is more economical to dewater sludge before disposal than it is to pump or haul liquid sludge to disposal sites. The primary objective of dewatering is to reduce sludge moisture and consequently sludge volume to a degree that will allow for economical disposal. Unit processes most often used for dewatering are: (1) pressure filtration, (2) vacuum filtration, (3) centrifugation, and (4) sand drying beds.

3.41 Pressure Filtration

Basically, two types of pressure filtration systems are used for sludge dewatering: (1) plate and frame filter press, and (2) belt filter press. The operating mechanics of these two filter press types are totally different and each will be discussed.

3.410 Plate and Frame Filter Press

The plate and frame filter press consists of vertical plates that are held rigidly in a frame and pressed together. A schematic diagram of a typical plate section is shown in Figure 3.16. The plate and frame filter press operates in a batch mode. Sludge is fed into the press through feed holes along the length of the press. A filter cloth is mounted on the face of each individual plate. As filtration proceeds, water (filtrate) passes through the fibers of the cloth, is collected in drainage ports provided at the bottom of each press chamber and is discharged. Sludge solids are retained on the filter cloths and are allowed to build up until the cavities between the plates are completely filled with solids (cake). As the cake builds up between the plates, the resistance to flow increases because the water has to pass through a thicker layer of compacted solids. As the cake builds up and the resistance increases, the volume of sludge fed to the filter and consequently the volume of filtrate decreases. When the filtrate flow is near zero, the feed is shut off and the plates are disengaged. As the plates are pulled away from each other, the retained cakes are discharged by gravity and fall into a hopper or conveyor. The diaphragm press or variable-volume type filter presses have expandable membranes on the plate faces to further dewater the cake and to ease cake removal. After the cakes are discharged, the plates are pulled back together and the feed is restarted.

3.4100 FACTORS AFFECTING PRESSURE FILTRATION PERFORMANCE

The degree of dewatering and sludge solids removal efficiency are affected by: (1) sludge type, (2) conditioning, (3) filter pressure, (4) filtration time, (5) solids loadings, (6) filter cloth type, and (7) PRECOAT.[35]

Both the sludge type and the conditioning methods used will drastically affect the operation and performance of filter presses. In general, primary sludges dewater more readily and require fewer chemical conditioners than secondary sludges. Chemicals used to condition sludge prior to plate and frame filtration generally are lime or ferric chloride. As the quantity of chemical conditioners increases, the dryness of the discharged cake solids and the sludge solids removal efficiency also increases. As discussed in Section 3.3, the optimum chemical dosages are determined by jar test experiments followed by pilot- or full-scale tests. Experience has shown that various combinations of lime, ferric chloride, ash, and polymer can be used to condition sludge prior to plate and frame filtration. If the chemical dosages are less than optimum, the performance of the filter press also will be less than optimum. Thermal conditioning or wet oxidation of wastewater sludges followed by gravity thickening yields a readily dewaterable sludge. The operating criteria maintained in the thermal conditioning system will have definite effects on sludge dewaterability. As discussed in Section 3.32, the thermal conditioning system should be operated so as to obtain the desired degree of dewatering by pressure filtration. The operating guidelines for filter presses and their effects on filter performance are discussed below.

3.4101 OPERATING GUIDELINES

The operator has the ability to control filter press performance to a certain degree by controlling the pressure, time of filtration, and the solids loading. Selection of a particular type of filter cloth for a specific sludge is generally done by pilot- or full-scale testing with various cloth types. It should be noted that the presence of certain inorganic and organic compounds in industrial wastes may degrade or clog the filter cloth. Once a filter cloth is selected and installed, the operator must control the frequency and duration of media cleaning.

---

[35] *Precoat.* Application of a free-draining, noncohesive material, such as diatomaceous earth, to a filtering medium. Precoating reduces the frequency of media washing and facilitates cake discharge.

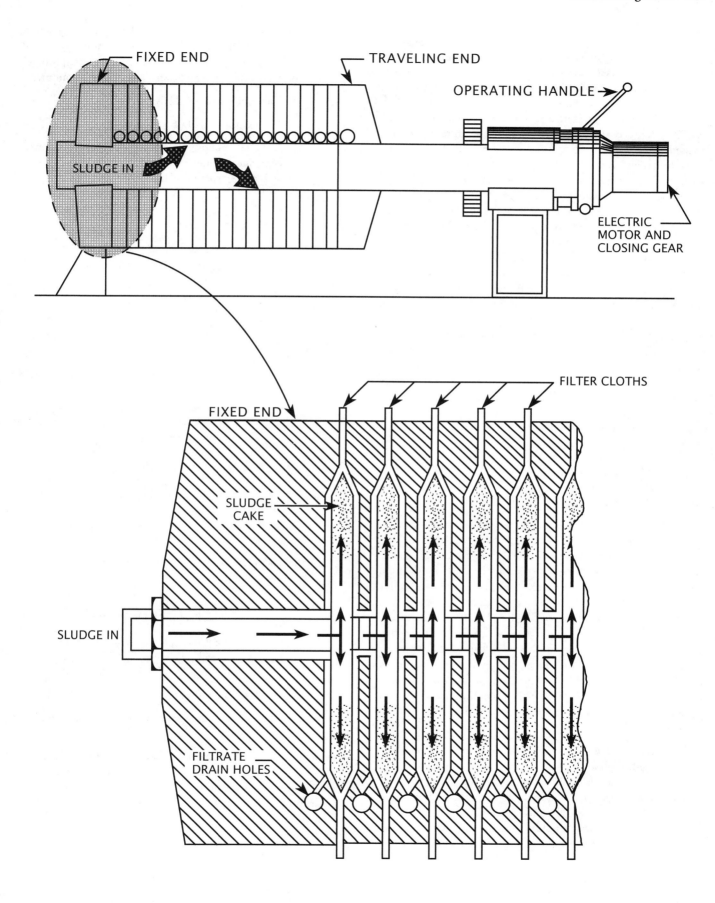

Fig. 3.16 Plate and frame filter press

PRESSURE

The feed to filter presses starts out at low pressure and high flow rates. As the cake builds up, the resistance to flow increases. A pneumatically or hydraulically driven positive displacement feed pump provides increasing pressure as the flow drops off. Generally, the initial pressure is maintained at approximately 25 psi (1.75 kg/sq cm) for 5 to 10 minutes. Pressure is then increased at intervals approximating 5 psi/min (0.35 kg/sq cm/min) until the terminal operating pressure is reached. Final operating pressures usually vary from 100 to 225 psi (7.0 to 15.8 kg/sq cm) depending on the particular type of press. The pressure is applied and maintained for one to three hours. Some presses are designed to operate at 100 to 125 psi (7.0 to 8.8 kg/sq cm). In general, higher pressures should produce a somewhat drier discharge by forcing more water from the sludge mass. The most effective pressure for a particular sludge is determined by experimentation and the operator should be aware that increased cake dryness may result from increased operating pressures. For some sludges, especially secondary sludge, the reverse might happen. That is, as the pressure is increased, the sludge retained on the filtering media may compress to a higher degree and reduce the porosity (openings) of the sludge cake that is formed. If the openings in the sludge formation are reduced, fine, low-density solids may be captured and incorporated in the sludge cake. The inclusion of fine solids generally results in wetter cakes because these solids have large surface areas and relatively large quantities of water associated with them. For each sludge, the optimum pressure should be determined by conducting laboratory or pilot studies and by on-site experimentation.

FILTRATION TIME AND SOLIDS LOADING

The time of filtration is actually controlled by the physical size of the filter and the solids loading rate. When the cavities between the plates are filled with solids and the filtrate flow is almost zero, the filtering sequence is complete. Obviously, for a given cavity volume, the filtration time will vary as the solids loading rate and the dewaterability vary. If the time of filtration is not adequate to completely fill the plate cavities with dewatered solids, large volumes of water will be discharged when the plates are disengaged and the cakes discharged. The operator should control filtration time based on the filtrate flow rate. If, on the other hand, the filtration time exceeds the time required to fill the cavity volume, the cakes will be firm and dry upon discharge but the quantity of solids processed per hour or per day (solids loading) will decrease.

The solids loading is determined by dividing the pounds of solids applied per hour by the surface area of the plates. Since filter presses are batch systems, time is lost in disengaging the plates, discharging the cakes, and re-engaging the plates prior to restarting the feed pump. The incorporation of downtime into the solids loading equation results in a net filter yield. The following example illustrates the determination of solids loading and net filter yield.

EXAMPLE 36

Given: A filter press with a plate surface area of 100 sq ft is used to dewater digested primary sludge. The digested sludge is at a concentration of 3.0 percent sludge solids. The filtration time is 2 hours and the total volume of sludge processed is 700 gallons. The time required to discharge the cakes and restart the feed is 20 minutes.

Find: 1. The solids loading (lbs/hr/sq ft).

2. The net filter yield (lbs/hr/sq ft).

3. If the feed solids concentration decreased to 2 percent sludge solids and the filtration time remained at 2 hours, what problems might develop?

Solution:

Known		Unknown
Plate Area, sq ft	= 100 sq ft	1. Solids Loading, lbs/hr/sq ft
Sl Sol, %	= 3.0%	
Filtration Time, hr	= 2 hr	2. Net Filter Yield, lbs/hr/sq ft
Sludge Volume, gal	= 700 gal	3. If solids drop to 2% Sl Sol, what problems might develop?
Discharge and Restart, min	= 20 min	

1. Calculate the solids loading in pounds per hour per square foot.

$$\text{Solids Loading,} \atop \text{lbs/hr/sq ft} = \frac{\text{Sl Vol, gal} \times 8.34 \frac{\text{lbs}}{\text{gal}} \times \text{Sl Sol,} \frac{\%}{100\%}}{\text{Filt Time, hr} \times \text{Area, sq ft}}$$

$$= \frac{700 \text{ gal} \times 8.34 \frac{\text{lbs}}{\text{gal}} \times \frac{3.0\%}{100\%}}{2 \text{ hr} \times 100 \text{ sq ft}}$$

$$= 0.88 \text{ lb/hr/sq ft}$$

2. Calculate net filter yield in pounds per hour per square foot.

$$\text{Net Filter} \atop \text{Yield,} \atop \text{lbs/hr/sq ft} = \frac{\text{Loading,} \frac{\text{lbs/hr}}{\text{sq ft}} \times \text{Filt Time, min}}{\text{Filt Time, min} + \text{Downtime, min}}$$

$$= \frac{0.88 \text{ lb/hr/sq ft} \times 120 \text{ min}}{120 \text{ min} + 20 \text{ min}}$$

$$= 0.75 \text{ lb/hr/sq ft}$$

3. What would happen if the feed concentration decreased to 2 percent sludge solids?

If the feed solids concentration decreases to 2 percent sludge solids, the cake MAY be wetter upon discharge if the filtration time is not increased. Check the filtrate flow and adjust the filtration time so that the filtrate flow is near zero when the feed pump is turned off.

FILTER CLOTH AND PRECOAT

The selection of a particular cloth type is done by experimentation in conjunction with the manufacturer's recommendation. Once a cloth is selected and installed, the operator must determine the frequency of media cleaning by inspecting the condition of the cloth and monitoring filter performance. After repeated use, the cloth media may *BLIND*[36] and adversely affect filter performance. If the cloth is clogged, water will not drain as readily and the discharged cakes will be wet and sloppy upon release from the plates. Also, as the cloth becomes clogged, a longer time will be required to dry the cake. The operator should be aware that certain chemicals that are present in industrial sludges could degrade the filter cloth and could accelerate the clogging of the cloth. Some presses are furnished with media washing equipment and the media can be cleaned in place. If a washing system is not furnished with the press, the operator will have to remove the filter cloths, wash them according to the manufacturer's recommended procedure, then reinstall them.

To reduce the frequency of media washing and to facilitate cake discharge, a precoat may be applied (this is an optional operation) before each batch is loaded for filtering. A free-draining, noncohesive material such as diatomaceous earth (a fine, siliceous (made of silica) earth composed mainly of the skeletal remains of diatoms) is made into a slurry and is applied to the filter so as to leave a thin layer on the filter cloth. When the sludge is applied, the precoat material prevents the sludge solids from sticking to the filter cloth. The net effect is that when the filter press is opened, the cake will readily discharge and solids remaining on the cloth will be minimized. Using precoats can improve operation and performance of filter presses, but precoating adds to the solids load to be disposed of.

3.4102   NORMAL OPERATING PROCEDURES

The specific operation of different pressure filters will vary somewhat, but the basic operational procedures are similar. The filtration cycle can be divided into various steps: (1) preparation of precoat and chemical conditioners, (2) chemical conditioner and sludge mixing, (3) precoat application, (4) sludge application, and (5) cake discharge.

The procedures for normal operation are outlined below:

1. Slurry (add water to) the precoat mix in a separate precoat tank.

2. Slurry the lime, if used, in a separate lime slurry tank.

3. Transfer the lime to a separate tank containing the sludge to be filtered and provide gentle stirring. Add the appropriate quantity of ferric chloride, if used. Usually, either lime or ferric chloride is used as a chemical conditioner.

4. Apply the precoat material to the filter.

5. Introduce the conditioned sludge to the filter.

6. When the filtrate flow decreases to near zero, turn off the feed pump.

7. Disengage and open the press for cake discharge.

8. Close the press and repeat the above procedures.

Full-scale filter press installations are usually automated or semiautomated so as to reduce operator attention. Even with fully automated systems, the operator should routinely check the operation of the various equipment and should make adjustments, as required.

3.4103   TYPICAL PERFORMANCE

Operating guidelines, conditioner requirements, and filter press performance for various sludge types are summarized in Table 3.20. Remember that the actual cake solids concentration, net filter yield, solids recovery, pressure, chemical requirement, and type of filter cloth and precoat vary with the actual sludge and should be determined by pilot testing whenever possible. Note that when thermal conditioning is used, precoat material and chemical conditioners are often not required.

TABLE 3.20   TYPICAL PERFORMANCE OF PLATE AND FRAME FILTER PRESSES

| Sludge Type | Chemical Conditioners | | Thermal Treat. | Pressure, psig[b] | Yield, lbs/hr/sq ft[c] | Cake Solids, %TS[d] | Solids Recovery, % |
	Lime, lbs/ton[a] or FeCl₃, lbs/ton[a]						
Primary	100–200	100–200	—	100–200	0.5–1.0	40–50	90–99
Primary	—	—	LPO	100–200	0.5–1.2	40–50	90–99
Secondary	200–500	100–400	—	100–200	0.1–0.3	20–30	90–99
Secondary	—	—	LPO	100–200	0.1–0.4	20–40	90–99
Dig. Primary	100–400	100–200	—	100–200	0.5–1.0	40–50	90–99
Dig. Primary	—	—	LPO	100–200	0.5–1.0	40–50	90–99
Dig. Primary	200–600	100–400	—	100–200	0.1–0.3	20–30	90–99
Dig. Secondary	200–600	100–400	—	100–200	0.1–0.3	20–30	90–99
Dig. Secondary	—	—	LPO	100–200	0.1–0.4	20–30	90–99

[a] lbs chemical/ton dry solids. lbs/ton × 0.5 = gm chemical/kg dry solids.

[b] psi × 0.07 = kg/hr/sq m.

[c] lbs/hr/sq ft × 4.883 = kg/hr/sq m.

[d] Thickened sludge.

---

[36] *Blinding.*   The clogging of the filtering medium of a microscreen or a vacuum filter when the holes or spaces in the media become clogged or sealed off due to a buildup of grease or the material being filtered.

QUESTIONS

Write your answers in a notebook and then compare your answers with those on page 317.

3.40A   What is the primary objective of sludge dewatering?

3.40B   What unit processes are most commonly used for sludge dewatering?

3.41A   Why does the flow through plate and frame filter presses decrease with filtration time?

3.41B   List the factors that affect pressure filtration performance.

3.41C   Increasing the operating pressure might result in wetter cakes. Why?

3.41D   The typical performance data presented in Table 3.20 indicate that secondary sludges do not dewater as readily as primary sludges. Why is this so?

3.41E   How should the time of filtration be controlled?

3.41F   What purpose does precoating serve?

3.41G   List the normal operating procedures for filter presses.

3.4104   TROUBLESHOOTING

The operator should be concerned with the characteristics of the cakes discharged at the end of the filtration cycle. Generally, filter presses consistently produce excellent effluents (filtrate) unless the filtering media is torn or not properly installed. Routine monitoring of both the filtrate and cake is required for continued successful operation.

Depending on the operation and performance of filter presses, the discharge cakes will be: (1) firm and dry throughout, (2) firm and dry at the outer sections with wet and sloppy inner sections, or (3) wet and sloppy throughout.

A firm and dry cake indicates good filter press operation and no adjustments are necessary. If the cakes are firm and dry at the ends but are composed of liquid centers, check filtration time and chemical dosages. The filtration time should be checked by monitoring the filtrate flow on a subsequent filter run. If the filtration time is not adequate, the cavities between the plates will not fill completely with solids and the innermost portions of the cakes will be wet. When this occurs, increase the filtration time so as to obtain near zero filtration flow at the end of the feed cycle. If the cakes are wet throughout, either the filtration time should be increased, if necessary, or the pressure should be monitored during a subsequent run. If the desired pressure is not being developed, check the condition and operation of the high-pressure feed pump and the condition and installation of the filter cloths. A tear in the filtering media or misalignment of the cloths will cause a lot of the sludge to pass through the filter without building up between the plates, and will usually result in poor effluent quality. Also, check to determine if the poor effluent quality is related to a clogged or dirty filter cloth. If the pressure is as desired and the filtrate

quality is good, inconsistent and wet cakes could develop from a low chemical dosage or from changes in the incoming sludge characteristics. Check all aspects of the chemical conditioning system and adjust the chemical dosage to achieve the desired degree of dewatering.

The operator has little, if any, control over the characteristics of the incoming sludge and should coordinate with the supervisor any adjustments for upstream processes such as sludge sedimentation, thickening, stabilization, and conditioning. In cases where any doubt exists about the effect of the industrial waste based upon personal experience or literature survey, laboratory or pilot studies should be conducted to determine the precise effect on the sludge dewatering.

If precoats are used to aid in discharge of the dewatered solids, check the precoat system if relatively large quantities of solids remain on the filter cloths upon discharge.

Table 3.21 summarizes potential operational problems and corrective measures to assist in maintaining effective filter press dewatering.

QUESTIONS

Write your answers in a notebook and then compare your answers with those on page 317.

3.41H   What measures should be taken if the discharge cakes from a filter press are wet throughout?

3.41I   Why do solids occasionally cling to the filtering media when the cakes are discharged?

3.411   Belt Filter Press

Belt filter presses consist of two endless belts that travel continuously over a series of rollers. Variations in belt filter designs are available from different manufacturers, but the basic principles are the same for all belt filters. A schematic of a typical belt filter press is presented in Figure 3.17. Sludge to be dewatered is preconditioned, usually with polymers, then applied to the free water drainage zone of the filter belt. This portion of the belt is so named because it allows for most of the free water to drain through the filter and to be collected in a trough on the underside of the belt. The main differences between different brand name filter types are the method of introducing and mixing chemicals with the sludge and the type of drainage zone used. Some presses use in-line polymer mixing where the polymer is added directly to the feed line and mixed with sludge by passing the flow through a Venturi-type restriction to create mild turbulence. With this type of chemical mixing system, the conditioned sludge is applied to a horizontal drainage zone as shown in Figure 3.17.

Mixing chambers also can be used to ensure adequate polymer and sludge contact. Such chambers are cylindrical in design and slowly rotate to allow the polymer to mix with the sludge. Mixing chambers simply replace the Venturi-type restriction for creating mild turbulence. When the conditioned sludge moves out of the mix chamber, it can be applied directly to a horizontal

TABLE 3.21   TROUBLESHOOTING PLATE AND FRAME FILTER PRESSES

Operational Problem	Possible Cause	Check or Monitor	Possible Solution
1. Inner portions of cakes wet and sloppy upon discharge	1a. Low filtration time 1b. Low pressure 1c. Chemical inefficiencies	1a. Filtrate flow for an entire run 1b. Pressure developed 1c. Chemical dosages	1a. Increase filtration time 1b. Repair or unplug feed pump. Align filter media 1c. Increase chemical dosage
2. Cakes wet throughout	2a. Low filtration time 2b. Low pressure 2c. Chemical inefficiencies 2d. Changes in influent characteristics	2a. Filtration flow 2b. Pressure developed. Check media for tears or misalignment 2c. Chemical equipment and dosage 2d. Chemical constituents in the influent sludge	2a. Increase filtration time 2b. Repair feed pump. Replace or realign media 2c. Increase chemical dosage 2d. Control of industrial waste discharges. Adjustments to the upstream processes. Adjustments to the chemical dosage. Laboratory or pilot studies
3. Solids remain on cloth upon discharge	3a. Precoat inefficiencies	3a. Precoat application and dosage	3a. Increase precoat dosage

drainage zone as discussed above or it can be delivered to a cylindrical "reactor chamber." The reactor chamber replaces the horizontal drainage zone and consists of a screen around the outside edge of the chamber, which allows most of the free water to drain out.

Regardless of the type of polymer mixing and "drainage zone" dewatering used, the partially dewatered solids are carried to a point on the unit where they are trapped between two endless belts where the solids are further dewatered as they travel over a series of perforated and unperforated rollers. This zone is known as the "press" or "dewatering zone." In this zone, the entrapped solids are subjected to shearing forces as they proceed over the rollers. Water is forced from between the belts and collected in filtrate trays while the retained solids are scraped from the two belts when they separate at the discharge end of the press. The two endless belts then travel through respective washing chambers for the removal of fine solids to decrease the possibility of plugging.

3.4110   FACTORS AFFECTING BELT PRESSURE FILTRATION

The ability of belt filter presses to dewater sludge and to remove suspended solids is dependent on: (1) sludge type, (2) conditioning, (3) belt tension or pressure, (4) belt speed, (5) hydraulic loading, and (6) belt type.

Sludge type, consistency of the feed, and chemical conditioning will affect the performance of belt filters. Because the mechanisms of the dewatering process (pressure filtration) are similar to those of plate and frame filter presses, the same concerns regarding the presence of chemicals that can inhibit the process apply (see Section 3.4100). In each specific case, a laboratory or pilot study should be conducted to identify the applicability of belt filter presses for dewatering specific sludges. Polymers are generally used for chemical conditioning in conjunction with belt filter operations. Polymer dosages must be optimized to ensure optimum dewatering. Unlike plate and frame filter presses, which can consistently handle secondary sludge if properly conditioned with chemicals, the belt filter might not be able to handle secondary or waste activated sludges consistently. Even with adequate polymer addition to flocculate and to cause a separation of solids from the liquid, the belt press might not be suitable for dewatering secondary sludges. Secondary sludges generally lack fibrous materials and exhibit a plastic or jello-like nature. When these sludges are trapped between the two belts and pressure is applied by the rollers, the solids tend to slip toward the belt sides and eventually squeeze out from between the belts. The net effects are that these solids contaminate the effluent by falling into the filtrate trays and continuous housekeeping is required. If this problem is evident, primary sludge may have to be blended with the secondary sludge to add fibrous material necessary to contain the sludge between the belts. This procedure has produced sludge cakes in the range of 24 to 26 percent solids with the use of polymers.

The operating guidelines available to vary the degree of dewatering are discussed below.

3.4111   OPERATING GUIDELINES

A well-operated belt filter press should result in a high degree of dewatering provided the operator is aware of the important operating controls.

BELT TENSION PRESSURE

The pressure applied to the sludge can be increased or decreased by adjusting the tension rollers to take up slack on the

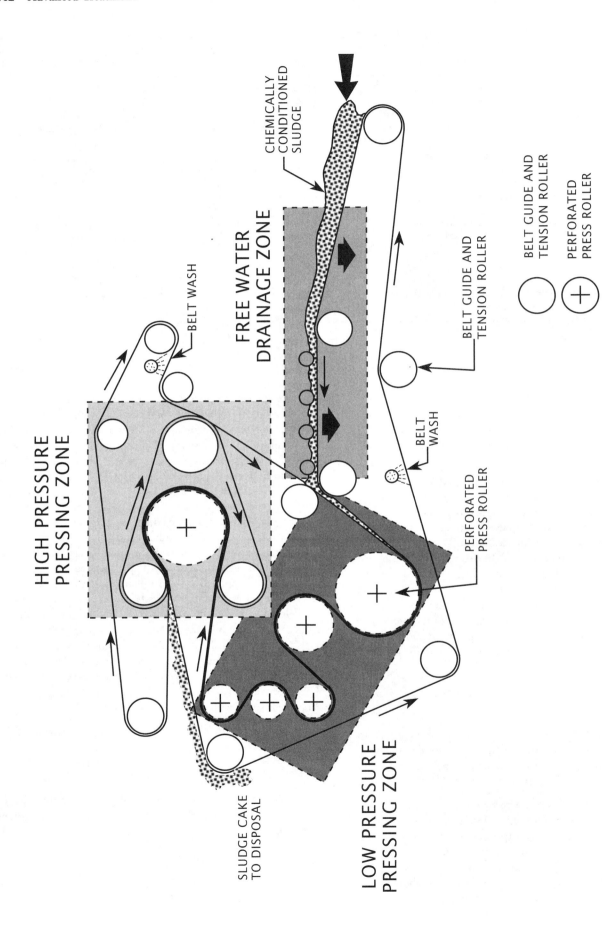

CHEMICALLY CONDITIONED SLUDGE

FREE WATER DRAINAGE ZONE

BELT WASH

HIGH PRESSURE PRESSING ZONE

BELT GUIDE AND TENSION ROLLER

PERFORATED PRESS ROLLER

BELT GUIDE AND TENSION ROLLER

BELT WASH

PERFORATED PRESS ROLLER

SLUDGE CAKE TO DISPOSAL

LOW PRESSURE PRESSING ZONE

Fig. 3.17   Belt filter press

two endless belts. As the belt tension is increased, more water is generally squeezed from the belt resulting in drier cakes. The pressure variations available on each manufacturer's belt press are different and you should consult the manufacturer's literature to determine the range of operating pressures for your equipment. Although pressure increases usually result in drier cakes, some undesirable conditions may develop as a result of increasing the tension between the belts. These are: (1) sludge may be forced from between the belts due to increased shear forces, or (2) sludge may be forced through the belt. Both of these conditions will result in filtrate contamination and increased housekeeping requirements. The optimum operating pressure is the one that produces the driest cake possible while containing the sludge between the belts.

BELT SPEED AND HYDRAULIC LOADING

The belt speed can be varied from approximately 2 to 10 feet/minute (0.6 to 3 m/min). The speed at which the belt should be operated depends on the sludge flow rate to the belt and the concentration of the influent sludge. Since most of the water associated with the sludge is removed in the drainage zone, sufficient belt area has to be provided to allow the water to drain. As the belt speed is increased, the rate of belt area contacting the influent sludge also increases and allows for greater volumes of water to drain from the sludge. If the belt area is not sufficient to allow the free water to drain, a "washing out" of the belt will occur. Washing out means that large quantities of free water unable to be released in the drainage zone will travel to the dewatering zone, flow out from between the belts, and drastically reduce effluent quality.

The belt speed should be controlled so that the sludge delivered to the dewatering zone has a minimum amount of free water. The experienced operator will be able to optimize belt speed by observing the dryness of the solids delivered to the dewatering zone. Obviously, as the concentration of influent sludge increases, less water is associated with the sludge mass and reduced belt speeds can be used. The ideal operating belt speed is the slowest the operator can maintain without washing out the belt. As the belt speed decreases, cake dryness increases because the sludge is subjected to pressure and shearing forces for longer periods of time.

Hydraulic loadings are based on the flow rate applied per unit of belt width. Belt presses range in width from approximately 1.5 feet to 10 feet (0.5 to 3 m). Manufacturers generally recommend loadings of 20 to 50 GPM (1.3 to 3.2 *L*/sec) per meter of belt width. The ideal loading for a particular belt press and a specific sludge should be determined on the basis of dewatering performance and consistency of operation. If a belt press is operated close to the upper hydraulic limit, frequent washing out may result, due to slight variations in sludge characteristics or chemical dosages. An example for the calculation of the hydraulic loading follows.

EXAMPLE 37

Given: A 2-meter wide belt press receives 100 GPM of primary sludge at a concentration of 5 percent.

Find: The hydraulic loading, GPM/m.

Solution:

Known	**Unknown**
Belt Filter Press	Hydraulic Loading, GPM/m
Belt Width, m = 2 m	
Flow, GPM = 100 GPM	
Sl Sol, % = 5%	

Determine the hydraulic loading in gallons per minute per meter.

$$\text{Hydraulic Loading, GPM/m} = \frac{\text{Flow, GPM}}{\text{Belt Width, m}}$$

$$= \frac{100 \text{ GPM}}{2 \text{ m}}$$

$$= 50 \text{ GPM/m}$$

BELT TYPE

Belts are available in a variety of materials (nylon, polypropylene), each with various porosities. Porosity is a measure of fiber openings. As the porosity increases, the resistance to flow decreases and larger volumes of water are able to be drained. If the porosity (fiber openings) is too large, sludge solids may pass through the belt and cause poor filtrate quality. If the porosity is too low, the belt may blind or plug, which will produce frequent washouts. The right belt for a particular application is determined by experimentation in conjunction with manufacturer's recommendations. After a belt is selected and installed, the operator should provide for adequate belt cleaning by maintaining the belt washing equipment in proper working condition and by adjusting the wash water volume as required.

3.4112 *NORMAL OPERATING PROCEDURES*

The operation of different belt filters will vary somewhat, but the following procedures will be applicable for most cases:

1. Prepare an adequate supply of an appropriate polymer solution.
2. Turn on the dewatered sludge conveyor.
3. Turn on the wash water sprays and the belt drive.
4. Adjust the belt speed to its maximum setting and allow the wash water to wet the entire belt.
5. Turn on the mix drum and reactor drum, if applicable.
6. Open appropriate sludge and polymer valves.
7. Turn on the polymer pump and adjust the polymer rate, as required, to achieve the desired dosage.
8. Turn on the sludge feed pump.
9. Lower the belt speed as low as possible without running the risk of washing out.
10. Adjust the belt tension, as required.

Routinely check the operation of the belt press and make adjustments, as required, to produce a dry cake and good filtrate quality.

3.4113 TYPICAL PERFORMANCE

Operating guidelines, polymer requirements, and typical performance data are presented in Table 3.22. Remember that the actual cake solids concentration, hydraulic loading rate, solids recovery, chemical requirement, and type of belt vary with the actual sludge and should be determined by on-site experimentation.

TABLE 3.22   TYPICAL PERFORMANCE OF BELT PRESS FILTRATION UNITS

Sludge Type	Polymer, lbs/ton[a]	Cake, %TS[b]	SS Recovery, %	Hyd. Load, GPM/m[c]
Primary	4–8	25–35	95–99	35–80
Secondary	9–20	17–20	90–99	16–50
Dig. Primary	4–8	25–30	95–99	35–80
Dig. Secondary	15–30	17–20	90–99	16–50

[a] lbs of dry polymer/ton dry sludge solids. lbs/ton × 0.5 = gm polymer/kg dry sludge solids.
[b] Thickened sludge.
[c] GPM/m × 0.0631 = L/sec/m.

QUESTIONS

Write your answers in a notebook and then compare your answers with those on page 317.

3.41J   What purpose does the drainage zone serve?

3.41K   What is the function of mix chambers used on some belt presses?

3.41L   What is the purpose of reactor chambers used on some belt presses?

3.41M   Describe the function of the "press" or "dewatering zone."

3.41N   List the factors affecting belt filter performance.

3.41O   What problems might be expected when using a belt press to dewater secondary sludge?

3.41P   How does pressure affect belt press operation?

3.41Q   What does the term "washing out" mean?

3.41R   What is the ideal belt speed?

3.41S   Explain how low belt speed affects belt press performance.

3.41T   How does belt type affect belt press performance and what problem may develop if the porosity is too low?

3.4114 TROUBLESHOOTING

A close watch over belt press operation in combination with field experience should enable the operator to optimize performance. In addition to mechanical reliability and maintenance, the operator should be concerned with the filtrate quality and dryness of the dewatered sludge. The most frequent problem encountered with belt presses is washing out. Usually, this problem is indicated by large volumes of water carrying onto the dewatering zone and overrunning the sides of the belt. When this happens, check: (1) the polymer dosage, (2) mixing in the reactor, (3) hydraulic loading, (4) drum speed, (5) belt speed, and (6) belt washing equipment.

If the polymer dosage is too low, the solids will not flocculate and free water will not be released from the sludge mass. If the polymer dosage is adequate (as evidenced by large floc particles and free water), increase the belt speed so as to provide more belt surface area for drainage. If the belt is already at its maximum setting, check the flow rate to the press and reduce it if the rate is higher than normal. If the polymer dosage, belt speed, and hydraulic loading are set properly but washing out still persists, the problem may be related to blinding of the filter media. Check the appearance of the belts as they leave their respective washing chambers. If the belts appear to be dirtier than normal, increase the wash water rate, turn off the polymer and feed pumps, and allow the belts to be washed until they are clean. Belt blinding often develops because of polymer overdosing. If too much polymer is added, the belts can become coated with a film of excess chemical, which will prevent drainage and result in washing out of the belt. Check the polymer addition rate and adjust the rate, as required, so as not to grossly overdose the sludge.

Poor effluent quality will generally result from washing out or from sludge being forced either through or from the sides of the belt. If sludge is being forced from the belts, check the belt tension and condition of the belts. Again, if the belts are dirty and water is prevented from draining, the sludge and water will squeeze from the belt sides when they are subjected to pressure in the dewatering zone. Reduce belt tension somewhat and clean the belts to eliminate this problem.

If the unit is working as expected (the sludge is contained between the belts and filtrate quality is good) but the discharge cakes are not as dry as desired, you can reduce the belt speed or increase the tension somewhat to achieve drier cakes. After a change in belt speed is made, you should observe the operation for at least 15 minutes to make certain that a washout will not occur.

A decrease in the cake solids concentration can occur as a result of changes in the characteristics of the incoming sludge. The operator has little, if any, control over the characteristics of the incoming sludge and should coordinate with the supervisor any adjustments for the upstream processes such as sludge sedimentation, thickening, stabilization, and conditioning. A laboratory or pilot study may be required to determine the precise effect of upstream process changes on sludge dewatering.

Table 3.23 summarizes typical operational problems and corrective measures that could be taken to optimize belt press performance.

3.412   Vacuum Filtration *(Figures 3.18, 3.19, and 3.20)*

A vacuum filter consists of a rotating drum that continuously passes through a trough or pan containing the sludge to be dewatered. The cylindrical drum is covered with a filter media that may be a cloth of natural or synthetic fibers, coil springs, or a wire-mesh fabric that is submerged about 20 to 40 percent in the trough. The trough is usually equipped with an agitator to keep the chemically conditioned sludge well mixed and to keep sludge from settling in the trough. The filter drum is divided into compartments. In sequence, each compartment is subjected to vacuums ranging from 15 to 30 inches (38 to 75 cm) of mercury. As the vacuum is applied to the compartment of the drum submerged in the trough, sludge is picked up on the filter media and a sludge mat is formed. This is known as the "mat formation" or "sludge pick-up zone." As the drum rotates out of the trough, the vacuum is decreased slightly and water is drawn from the sludge mat, through the filter media, and discharged through internal pipes to a drainage system. This is known as the "drying zone" of the cycle. Just prior to the point where the media separates from the drum, the vacuum is reduced to zero. The dewatered solids remaining on the filter media are then discharged to a sludge hopper or conveyor by means of scrapers or discharge rollers. After dewatered solids are separated from the belt, the belt enters the "wash zone," where accumulated fines can be flushed from the belt using water to prevent blinding. The cloth then reunites with the drum and is reintroduced into the sludge trough for another cycle.

3.4120   FACTORS AFFECTING VACUUM FILTRATION

The following factors can have pronounced effects on the operation and performance of vacuum filters: (1) sludge type, (2) conditioning, (3) applied vacuum, (4) drum speed or cycle time, (5) depth of submergence, and (6) media type and condition.

The sludge type and the conditioning methods used will drastically affect filter operation. As for plate and frame filter presses and for the belt filter presses, the presence of certain materials in industrial wastes can significantly decrease the dewaterability of the respective sludges. In each specific case, a laboratory or pilot study should be conducted to identify the applicability of the vacuum filtration for sludge dewatering. In general, straight (only) secondary sludges (digested or undigested) are not easily dewatered by vacuum filtration because they do not dewater enough to readily discharge from the belt; however, thermally conditioned secondary sludges can be effectively dewatered with a vacuum filter. Even with extremely high dosages of lime and ferric chloride, secondary sludges that have not been thermally conditioned will generally not dewater effectively. Such sludges usually require blending with primary sludges prior to vacuum filtration for successful dewatering.

TABLE 3.23   TROUBLESHOOTING BELT FILTER PRESSES

Operational Problem	Possible Cause	Check or Monitor	Possible Solution
1. Washing out of belt	1a. Polymer dosage insufficient 1b. Hydraulic load too high 1c. Belt speed too low 1d. Belt blinding	1a. Polymer flow rate and solution 1b. Sludge flow rate 1c. Belt speed 1d. Washing equipment and polymer overdosage	1a. Increase dose rate 1b. Lower flow 1c. Increase speed 1d. Increase wash water rate. Turn off sludge and polymer and clean belts. Reduce polymer, if overdosing
2. Poor filtrate quality	2a. Washing out 2b. Sludge squeezed from belt	2a. Same as 1 2b. Belt tension. Washing equipment	2a. Same as 1 2b. Decrease tension. Wash the belt
3. Cake solids too wet	3a. Belt speed too high 3b. Belt tension too low 3c. Changes in influent characteristics	3a. Belt speed 3b. Belt tension 3c. Chemical constituents in the influent sludge	3a. Reduce belt speed 3b. Increase belt tension 3c. Control of industrial waste discharges. Adjustments to the upstream processes. Adjustments to the chemical dosage. Laboratory or pilot studies

Fig. 3.18   Vacuum filter
(Permission of Eimco)

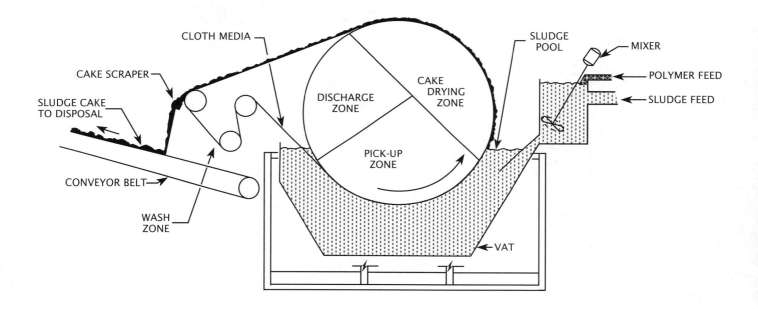

Fig. 3.19 Vacuum filter operating schematic

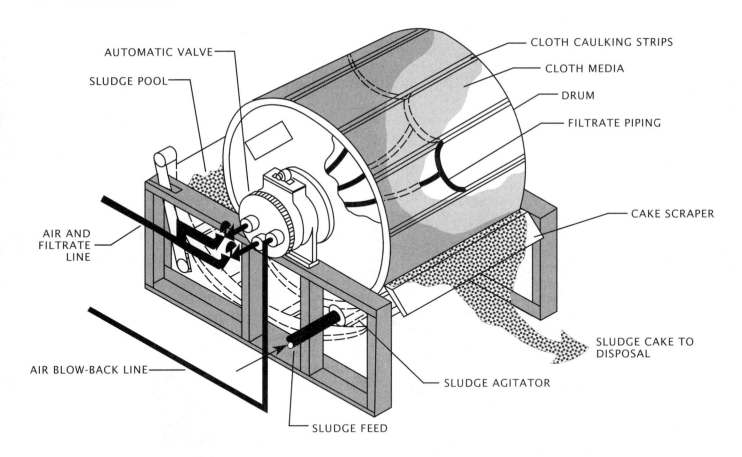

AUTOMATIC VALVE

SLUDGE POOL

AIR AND
FILTRATE
LINE

AIR BLOW-BACK LINE

SLUDGE FEED

SLUDGE AGITATOR

CLOTH CAULKING STRIPS

CLOTH MEDIA

DRUM

FILTRATE PIPING

CAKE SCRAPER

SLUDGE CAKE TO
DISPOSAL

Fig. 3.20   Rotary drum vacuum filter

As discussed in Section 3.3, sludge conditioning must be optimized to achieve the desired degree of dewatering regardless of the sludge type and type of dewatering equipment used. To ensure successful vacuum filtration, follow the procedures outlined in Section 3.3 to condition and prepare the sludge for dewatering.

3.4121 OPERATING GUIDELINES

The applied vacuum will affect the degree and rate of sludge pick-up in the formation zone and will affect the quantity of water withdrawn from the sludge in the drying zone. In general, reduced vacuums produce wetter cakes and less-than-optimum discharge characteristics. The operator should attempt to maintain as high a vacuum as possible to obtain a high degree of dewatering.

The drum speed or the time required to make one complete cycle also controls the degree of dewatering. Typically, cycle times vary from 2 to 6 minutes and the longer the cycle time the higher the degree of dewatering. Cycle time as controlled by

drum speed affects sludge dewatering in two ways. First, it controls the rate of sludge pick-up and the thickness of the sludge mat in the formation zone, and second, it controls the length of time the sludge remains in the drying zone. As the cycle time is increased, the opportunity to pick up sludge from the trough and the time the sludge mat is subjected to a vacuum in the drying zone are increased. This generally results in drier cakes and improved discharge characteristics. The operator should maintain as low a drum speed as possible to obtain the highest degree of dewatering. The cycle time is dependent on the solids loading or net filter yield; the minimum speed that can be used depends on the filter area and the quantity of sludge to be processed. Obviously, if the number or size of the filters is not adequate to handle the entire sludge load at the lowest drum speed, the operator will have to maintain higher drum speeds. In addition to controlling the degree of dewatering, the drum speed also controls the net filter loading. As the drum speed increases, the net filter loading and the total quantity of sludge processed per day increases. The following example illustrates the determination of filter loading and filter yield.

EXAMPLE 38

Given: A 6-foot diameter by 10-foot long vacuum filter with a total surface area of 188 sq ft (3.14 × 6 feet × 10 feet)* dewaters 4,500 lbs/day of primary sludge solids. The filter operates for 7 hours per day at a drum cycle time of 3 minutes and produces a dewatered sludge of 25 percent thickened sludge (TS) with 95 percent solids recovery.

NOTE: Some people use an "effective area." This may be the area under water or the area where the vacuum is applied.

Find: The filter loading in lbs/hr/sq ft and filter yield in lbs/hr/sq ft.

Solution:

Known		**Unknown**
Vacuum Filter		1. Filter Loading, lbs/hr/sq ft
Diameter, ft	= 6 ft	
Length, ft	= 10 ft	2. Filter Yield, lbs/hr/sq ft
Surface Area, sq ft	= 188 sq ft	
Sl Sol Loading, lbs/day	= 4,500 lbs/day	
Filter Operation, hr/day	= 7 hr/day	
Drum Cycle Time, min	= 3 min	
Dewatered Sludge Sol, %	= 25%	
Solids Recovery, %	= 95%	

1. Calculate the filter loading in pounds per hour per square foot.

$$\text{Filter Loading,} \atop \text{lbs/hr/sq ft} = \frac{\text{Sl Sol Loading, lbs/day}}{\text{Fil Operation, hr/day} \times \text{Area, sq ft}}$$

$$= \frac{4,500 \text{ lbs/day}}{7 \text{ hr/day} \times 188 \text{ sq ft}}$$

$$= 3.4 \text{ lbs/hr/sq ft}$$

2. Calculate the filter yield in pounds per hour per square foot.

$$\text{Filter Yield,} \atop \text{lbs/hr/sq ft} = \frac{\text{Sl Sol Loading,} \frac{\text{lbs}}{\text{day}} \times \text{Recov,} \frac{\%}{100\%}}{\text{Fil Op, hr/day} \times \text{Area, sq ft}}$$

$$= \frac{4,500 \frac{\text{lbs}}{\text{day}} \times \frac{95\%}{100\%}}{7 \text{ hr/day} \times 188 \text{ sq ft}}$$

$$= 3.2 \text{ lbs/hr/sq ft}$$

EXAMPLE 39

Given: Based on previous experimentation with the filter and sludge from Example 38, the operator knows that lowering the drum cycle to 2 minutes will produce a 30 percent dewatered sludge with 95 percent solids capture or recovery, but the filter yield will decrease to 1.7 lbs/hr/sq ft.

Find: The time the filter must be operated to process 4,500 pounds of primary sludge solids.

Solution:

Known		**Unknown**
Information given in Example 38		Time filter must be operated in hours per day
If drum cycle time reduced to 2 min,		
Dewatered Sl Sol, %	= 30%	
Solids Recovery, %	= 95%	
Filter Yield, lbs/hr/sq ft	= 1.7 lbs/hr/sq ft	

Calculate the time the filter must be operated to process 4,500 pounds per day of primary sludge solids.

$$\text{Filter Yield, lbs/hr/sq ft} = \frac{\text{Sl Sol Loading,} \frac{\text{lbs}}{\text{day}} \times \text{Recov,} \frac{\%}{100\%}}{\text{Fil Op, hr/day} \times \text{Area, sq ft}}$$

Rearrange the terms.

$$\text{Filter Operation,} \atop \text{hr/day} = \frac{\text{Sl Sol Loading,} \frac{\text{lbs}}{\text{day}} \times \text{Recov,} \frac{\%}{100\%}}{\text{Fil Yield,} \frac{\text{lbs/hr}}{\text{sq ft}} \times \text{Area, sq ft}}$$

$$= \frac{4,500 \frac{\text{lbs}}{\text{day}} \times \frac{95\%}{100\%}}{1.7 \frac{\text{lbs/hr}}{\text{sq ft}} \times 188 \text{ sq ft}}$$

$$= 13.4 \text{ hours/day}$$

Therefore, the filter must be operated for 13.4 hours per day to produce a dewatered sludge cake of 30 percent solids when using a two-minute cycle time.

The exact response (filter yield and dewatered sludge as percent sludge solids) of a particular sludge depends on the treatment plant. Experimentation is required to determine the filter yields and degrees of dewatering at different cycle times.

The depth of submergence of the drum within the trough affects the formation of the sludge mat on the media. In general, as the depth of submergence increases, more sludge solids are picked up and filter yields may increase somewhat. The depth of submergence should always be within the manufacturer's recommended range. In addition, the liquid level in the trough should never drop low enough to cause a loss in vacuum. If the liquid level in the trough is too low, air will be pulled into the vacuum compartments and will result in a loss of vacuum and a subsequent loss of mat formation and sludge dewatering.

The material of construction for the filter media is selected based on experimentation with various cloth types. As the porosity (openings) of the media increases, the ability to capture suspended solids decreases because fine, low-density solids can pass directly through the media. Regardless of the type and size of media selected for a particular application, the operator must ensure adequate media cleaning in the wash water zone. If the media blinds with fine solids or chemical coatings, sludge will not be picked up in the trough (vat) in the sludge blanket formation zone and the vacuum filter will become inoperative.

3.4122 NORMAL OPERATING PROCEDURES

The specific operation of a particular vacuum will vary somewhat, but the following procedures will be applicable for most operations:

1. Prepare chemicals in appropriate chemical tanks.

2. Transfer lime, if used, to a mixing tank to provide contact with the sludge and/or add the required amount of ferric chloride, if used.

3. Turn on the filter drum and the wash water to wet the entire belt.

4. Pull up tension.

5. Fill the trough to the operating level with the conditioned sludge.

6. Turn on the sludge mixer in the trough.

7. Turn on the vacuum pumps and filtrate pumps.

8. Adjust the drum speed and vacuum until the desired results are achieved.

Whenever the unit is shut down, the cloth should be thoroughly cleaned to reduce the possibility of plugging during subsequent operation. After shutdown, release cloth tension.

3.4123 TYPICAL PERFORMANCE

Table 3.24 summarizes typical vacuum filter operation and performance guidelines for various sludge types. Remember that the actual cake solids concentration, solids loading rate, solids recovery, vacuum, chemical requirement, and type of the filter media vary with the actual sludge and should be determined by pilot testing, whenever possible.

QUESTIONS

Write your answers in a notebook and then compare your answers with those on pages 318.

3.41W What is the purpose of the agitator or mixer in the trough of a vacuum filter?

3.41X Explain what happens in the "mat formation" and "drying zones."

3.41Y Why does the filter media pass through a washing zone?

3.41Z List the factors that affect vacuum filtration performance.

3.41AA What is the vacuum range that should be applied to the filter compartments?

3.41BB Explain how cycle time or drum speed affects vacuum filter dewatering.

3.41CC Determine the filter yield (lbs/hr/sq ft) for a vacuum filter with a surface area of 300 sq ft. Digested sludge is applied at a rate of 75 GPM with a suspended solids concentration of 4.7 percent. The filter recovers 93 percent of the applied suspended solids.

3.41DD How does the porosity of the filter media affect vacuum filter performance?

3.4124 TROUBLESHOOTING

Successful vacuum filtration produces a relatively clear filtrate and a relatively thick sludge mat that will readily discharge from the filter media. The most common problems that develop with vacuum filters are deterioration of the filtrate quality and wet cakes that are difficult to discharge from the belt.

If the filtrate quality is noted to contain more than the usual amount of solids, check the vacuum and the condition of the filter media. The filter media can easily move out of alignment and often works its way free of the drum. If this happens, a proper seal will not develop between the media and drum and unfiltered sludge will be sucked from the trough and contaminate the filtrate. If the seal is broken between the media and drum, a loss of vacuum will develop when the unsealed point on the drum is subjected to a vacuum. Regularly check the vacuum gauges for loss of vacuum and realign the filter media,

TABLE 3.24   TYPICAL PERFORMANCE OF VACUUM FILTERS

Sludge Type	Lime, lbs/ton[a]	FeCl₃, lbs/ton[a]	Thermal Conditioning	Vacuum, in Hg[b]	Cycle, min	Yield, lbs Sol/hr/sq ft[c]	Cake, % TS[d]	Solids Recovery, %
Primary	50–150	25–50	—	15–30	2–6	4–12	24–40	85–95
Primary and Trickling Filter	—	—	LPO	15–30	2–6	4–8	30–45	85–95
Primary and Air-Act.	—	—	LPO	15–30	2–6	4–5	25–40	85–95
Digested Primary	200–600	—	—	15–30	2–6	4–8	25–35	85–95
Digested Primary and Trickling Filter	—	—	LPO	15–30	2–6	4–5	25–40	85–95
Digested Primary and Air-Act.	—	—	LPO	15–30	2–6	4–5	25–30	85–90

[a] lbs of chemical/ton of dry sludge solids. lbs/ton × 0.5 = gm of chemical/kg of dry sludge solids.
[b] in Hg (Mercury) × 2.54 = cm Hg.
[c] lbs solids/hr/sq ft × 4.883 = kg/hr/sq m.
[d] Thickened sludge.

as required. If the filter media (cloth blanket) is misaligned, this is a major repair job and will require the unit to be shut down. A tear in the filter media will have the same effect as misalignment and you should repair or replace the media promptly if it is defective.

A substantial reduction in cake dryness or changes in discharge or filtrate characteristics will result from: (1) inadequate chemical conditioning, (2) cycle times too low (drum speed too fast), or (3) media blinding. Check the condition of the media and increase the wash water rate, or turn off the sludge feed and wash the belt if it is still covered with solids as it leaves the washing zone. A clogged media will usually develop when poor mat formation occurs in the pick-up zone. Whatever solids are picked up will not effectively drain in the drying zone and a wet cake will develop.

If the media is clean and a good seal is developed, a decrease in sludge dryness and an increase in filtrate solids could result from high drum speeds or improper conditioning. If possible, the drum speed should be lowered to afford more drying time, and the chemical mixing and addition system should be checked for malfunctions. Drastic changes in the influent sludge characteristics could also affect the chemical dosage and the degree of dewatering. The operator has little, if any, control over the characteristics of the incoming sludge and should coordinate with the supervisor any adjustments for the upstream processes such as sludge sedimentation, thickening, stabilization, and conditioning. Laboratory or pilot studies may be required to measure the precise effect of changing influent characteristics on sludge dewatering. If the influent sludge appears to be thicker than normal, increase the rate of chemical addition to match the increase in influent solids.

Table 3.25 summarizes the most common operational problems and the usual corrective measures taken to maintain good vacuum filter dewatering.

QUESTIONS

Write your answers in a notebook and then compare your answers with those on page 318.

3.41EE   What can cause a loss in vacuum and how will such a loss affect filter performance?

3.41FF   If the sludge is not picked up in the mat formation zone, what should the operator do?

3.41GG   How can the operator increase cake dryness?

3.42   Centrifugation

3.420   Process Description

Centrifuge designs for dewatering sludge include batch-operated basket centrifuges and continuous-flow scroll centrifuges. Factors affecting centrifugation, operating guidelines, and troubleshooting were described in detail in the thickening section (Section 3.13) of this manual. The reader should refer to Section 3.13 for a review of the mechanics of centrifugation. The main differences between centrifugal thickening and centrifugal dewatering are that the concentration of feed sludge is somewhat higher for dewatering than for thickening, and that a drier cake

TABLE 3.25   TROUBLESHOOTING VACUUM FILTERS

Operational Problem	Possible Cause	Check or Monitor	Possible Solution
1. Loss of vacuum	1a. Filter media misaligned 1b. Tear in filter media 1c. Trough empty 1d. Vacuum pumps inoperative	1a. Media alignment 1b. Media condition 1c. Level in trough 1d. Operation of vacuum pumps	1a. Realign media 1b. Repair/replace media 1c. Fill trough 1d. Repair or turn on pumps
2. Poor filtrate quality	2a. Same as 1a and 1b 2b. Insufficient conditioning	2a. Same as 1a and 1b 2b. Conditioning system	2a. Same as 1a and 1b 2b. Increase dosage or degree of conditioning
3. Wet cake and poor discharge	3a. Same as 1a, 1b, and 1d 3b. Drum speed too high 3c. Insufficient conditioning 3d. Changes in influent characteristics	3a. Same as 1 3b. Drum speed 3c. Conditioning system 3d. Chemical constituents in the influent sludge	3a. Same as 1 3b. Lower drum speed 3c. Increase dosage or degree of conditioning 3d. Control of industrial waste discharges. Adjustments to the upstream processes. Adjustments to the chemical dosage. Laboratory or pilot studies.
4. Poor mat formation	4a. Same as 1, 2, and 3 4b. Clogged media	4a. Same as 1, 2, and 3 4b. Media condition	4a. Same as 1, 2, and 3 4b. Clean media

is the usual goal for dewatering. The operator may, therefore, have to maintain a higher differential (relative) scroll speed when dewatering because of the higher solids loading, but the principles of centrifugation remain the same as previously discussed.

3.421   Typical Performance

Table 3.26 summarizes typical centrifuge operating guidelines, conditioning requirements, and performance data for various sludge types. Remember that the actual cake solids concentration, solids and hydraulic loading rates, solids recovery, and chemical requirements vary with the actual sludge and should be determined by pilot testing, whenever possible.

Usually, the physical size of the centrifuge will govern the maximum throughput loading for a particular sludge and, in general, the exact response of a particular sludge depends on the treatment plant.

Troubleshooting techniques for basket and scroll centrifuge dewatering are identical to those already outlined in Section 3.134.

QUESTIONS

Write your answers in a notebook and then compare your answers with those on pages 318 and 319.

3.42A   Why are higher scroll speeds usually required to dewater sludges as compared to sludge thickening?

3.42B   A scroll centrifuge is used to dewater 60 GPM of digested primary sludge at a concentration of 3.0 percent sludge solids. A liquid polymer is used for conditioning. The polymer solution is 2.5 percent and 2 GPM are added to the sludge stream. What is the hydraulic loading (GPH), the solids loading (lbs SS/hr), and the polymer dosage (lbs liq/ton)?

3.42C   A 48-inch diameter basket centrifuge is used to dewater 50 GPM of digested primary sludge. The feed is at a concentration of 2.7 percent sludge solids and polymers are added to achieve 95 percent suspended solids recovery. The average concentration of solids stored within the basket is 23 percent. The feed time is automatically set at 17 minutes. Is the feed time too long, too short, or OK? Assume the basket has a solids storage capacity of 16 cubic feet.

3.42D   A basket centrifuge is used to dewater digested secondary sludge. On a routine check, the operator notices that the centrate quality is poor but the discharge solids are dry. What should the operator check and what action should be taken to produce a clean centrate?

TABLE 3.26   TYPICAL PERFORMANCE OF CENTRIFUGES

Sludge Type	Centrifuge Type	Conditioning		Flow, GPM[b]	Solids Load, lbs/hr[c]	Cake, % TS[d]	Solids Recovery, %
		Polymer, lbs/ton[a]	Thermal				
Primary	Scroll	0–5	—	20–150	200–1,500	25–35	30–95
Primary	Scroll	0–5	LPO	20–150	200–1,500	25–40	50–95
Secondary	Scroll	0–15	—	20–150	100–700	6–9	30–95
Secondary	Scroll	0–7	LPO	20–150	100–700	20–30	70–95
Secondary	Basket	0–10	—	20–70	70–350	6–9	50–95
Secondary	Basket	0–4	LPO	20–70	70–350	20–30	70–95
Dig. Primary	Scroll	0–15	—	20–150	200–1,500	20–25	30–95
Dig. Primary	Scroll	0–5	LPO	20–150	200–1,500	20–30	70–95
Dig. Primary	Basket	0–15	—	35–50	200–500	20–25	70–95
Dig. Secondary	Scroll	0–40	—	20–150	200–1,500	6–12	30–95
Dig. Combined[e]	Scroll	0–20	—	20–150	200–1,500	10–25	30–95
Dig. Combined[e]	Basket	0–20	—	35–50	200–500	10–20	50–95

[a] lbs of dry polymer/ton dry sludge solids. lbs/ton × 0.5 = gm of polymer/kg of dry sludge solids.
[b] GPM × 5.45 = cu m/day.
[c] lbs/hr × 0.454 = kg/hr.
[d] Thickened sludge.
[e] Digested Primary + Digested Combined.

3.43   Sand Drying Beds

There are various types of drying beds. Sand drying beds are the most widely used for sludge dewatering. The features and operating characteristics of other types of drying beds such as paved drying beds, wedge-wire drying beds, and vacuum-assisted drying beds vary significantly. Therefore, it is nearly impossible to provide general operating guidelines for all types of drying beds. This section is limited to a description of widely used sand drying beds. Consult the operation and maintenance manual for your plant to obtain detailed operation and maintenance guidelines for the specific types of drying beds installed at your plant.

The use of sand drying beds to dewater wastewater sludges is usually limited to small or medium-sized plants (less than 5 MGD or 19,000 m³/day) because of land restrictions. Drying beds usually consist of 4 to 9 inches (10 to 23 cm) of sand placed over 8 to 20 inches (20 to 50 cm) of graded gravel or stone. Sludge is placed on the beds in 12- to 18-inch (30- to 45-cm) layers. The sludge is allowed to dewater by drainage through the sludge mass and supporting sand, and by evaporation from the surface exposed to the atmosphere. An underdrain system composed of a lateral network of perforated pipes, vitrified clay pipe with open joints, or trenches can be used to collect the filtrate for subsequent recycle to the plant headworks. If provided, the underdrains are usually spaced 8 to 20 feet apart. After drying, the dewatered sludge is removed by manual shoveling (forks) or a small front end loader for further processing or ultimate disposal. Sludge drying beds are usually not used for sludges that have been stabilized or conditioned by wet oxidation because of the odorous nature of thermally treated sludge.

3.430   Factors Affecting Sand Drying Beds

The design, use, and performance of drying beds are affected by many factors. These factors include: (1) sludge type, (2) conditioning, (3) climatic conditions, (4) sludge application rates and depths, and (5) dewatered sludge removal techniques.

As is the case with mechanical dewatering devices, the type of sludge applied and the effectiveness of chemical conditioners can determine the degree of dewatering and the operation of the beds. Since the majority of dewatering is by drainage through the support sand, the sludge has to be adequately conditioned to flocculate the sludge solids and release free water. Polymer conditioning is often used to enhance the sludge drying process. Care must be taken to prevent chemical overdosing for two reasons: (1) media (sand) blinding with unattached polymer may develop, and (2) large floc particles that settle too rapidly may also blind the media. Whatever the mechanism for blinding, the net result will be the same. The liquid portion of the sludge will be unable to drain through the bed and only dewatering by evaporation will occur. The time required to evaporate water is substantially greater than the time required to remove water by a combination of evaporation and drainage.

With some sludges, a relatively clear supernatant may develop over time, and decant tubes are often provided in drying beds to allow its removal.

Chemical requirements, prescreening requirements, and the response of sludges to sand drying operations are generally determined by laboratory, pilot-, or full-scale experimentations. Sludge from the bottom of an anaerobic digester may require prescreening because greases and hair-like stringy material can clog the sand bed.

Climatic conditions are very important to an efficient sand drying bed operation. After all possible water has been removed by drainage, evaporation is the only mechanism available for further dewatering. As climatic conditions vary from inclement and humid to dry and arid, so does the rate of evaporation and, consequently, the time required to achieve the desired degree of dewatering. In wet or cold climates, the sand beds are usually covered with greenhouse-type enclosures to protect the drying sludge from rain and to reduce the drying period. Such enclosures should be well ventilated to promote evaporation. They also can serve to control odors and insects.

The operator has little, if any, control over the sludge type and characteristics, no control over climatic conditions and physical size of the sand beds, but can control, to some extent, the depth of sludge application and dewatered sludge removal techniques.

3.431   Operating Guidelines

The physical size of sand drying beds is based on the amount of sludge to be dried each year. In general, sand beds are loaded at rates of 10 to 35 lbs of dry sludge solids/yr/sq ft (50 to 170 kg/yr/sq m) for open drying beds and 20 to 45 lbs of dry sludge solids/yr/sq ft (100 to 220 kg/yr/sq m) for covered drying beds. Obviously, as the loading rate decreases, the time required to dewater the sludge decreases as does the potential for blinding of the bed. The operator should control loading rates according to the total area of bed available and the estimated quantity of sludge production. The operator should also provide for a standby capacity of 10 percent additional area, if possible, to meet unexpected increases in sludge loads and to allow for

downtime for operational problems. If sufficient bed area is not available to allow for standby capacity, auxiliary mechanical dewatering equipment or liquid sludge haulers should be available for emergency situations.

EXAMPLE 40

Given: An operator at a treatment plant has two sand beds available for dewatering experiments. Each bed is 200 feet long by 25 feet wide. Sludge is applied to Bed A to a depth of 3 inches and to Bed B to a depth of 9 inches. Bed A requires six days to dry and one day to remove the sludge for another application. Bed B requires 21 days to dry and one day to remove the sludge for another application. The sludge is applied at a concentration of 3 percent sludge solids.

Find: 1. The total gallons and pounds of sludge applied per application for Bed A and Bed B.

2. The loading rates (lbs/yr/sq ft) for Bed A and Bed B assuming repeated applications are made on each bed with no operational or maintenance downtime.

3. Which application depth should be used?

Solution:

Known	Unknown
Two Sand Beds	1. Sludge Applied, gallons/application and pounds/application for both Beds A and B
Length, ft = 200 ft	
Width, ft = 25 ft	
Sl Depth, in = 3 in (Bed A)	
Sl Depth, in = 9 in (Bed B)	2. Loading Rates, lbs/yr/sq ft for both Beds A and B
Drying Time	
Bed A, days = 6 days	
Bed B, days = 21 days	3. Which application depth should be used?
Sludge Removal, days = 1 day	
Sludge Solids, % = 3%	

1. Determine the sludge applied in gallons per application and pounds per application for both Beds A and B.

BED A

$$\text{Sludge Applied, gal/application} = \frac{L, ft \times W, ft \times D, in/appl}{12\ in/ft} \times 7.48\ \frac{gal}{cu\ ft}$$

$$= \frac{200\ ft \times 25\ ft \times 3\ in/appl}{12\ in/ft} \times 7.48\ \frac{gal}{cu\ ft}$$

$$= 9{,}350\ gal/application$$

$$\text{Sludge Applied, lbs/application} = \text{Sl Appl,}\ \frac{gal}{appl} \times 8.34\ \frac{lbs}{gal} \times \text{Sl Sol,}\ \frac{\%}{100\%}$$

$$= 9{,}350\ \frac{gal}{appl} \times 8.34\ \frac{lbs}{gal} \times \frac{3.0\%}{100\%}$$

$$= 2{,}340\ lbs/application$$

BED B

$$\text{Sludge Applied,} \atop \text{gal/application} = \frac{L, \text{ft} \times W, \text{ft} \times D, \text{in/appl}}{12 \text{ in/ft}} \times 7.48 \frac{\text{gal}}{\text{cu ft}}$$

$$= \frac{200 \text{ ft} \times 25 \text{ ft} \times 9 \text{ in/appl}}{12 \text{ in/ft}} \times 7.48 \frac{\text{gal}}{\text{cu ft}}$$

$$= 28{,}050 \text{ gal/application}$$

$$\text{Sludge Applied,} \atop \text{lbs/application} = \text{Sl Appl,} \frac{\text{gal}}{\text{appl}} \times 8.34 \frac{\text{lbs}}{\text{gal}} \times \text{Sl Sol,} \frac{\%}{100\%}$$

$$= 28{,}050 \frac{\text{gal}}{\text{appl}} \times 8.34 \frac{\text{lbs}}{\text{gal}} \times \frac{3.0\%}{100\%}$$

$$= 7{,}018 \text{ lbs/application}$$

2. Determine the loading rates for the sludge applied in pounds per year per square foot for both Beds A and B.

BED A

$$\text{Loading Rate,} \atop \text{lbs/yr/sq ft} = \frac{\text{Sl Appl, lbs/appl} \times 365 \text{ days/yr}}{L, \text{ft} \times W, \text{ft} \times \text{Cycle, days/appl}}$$

$$= \frac{2{,}340 \text{ lbs/appl} \times 365 \text{ days/yr}}{200 \text{ ft} \times 25 \text{ ft} \times (6 \text{ days} + 1 \text{ day})/\text{appl}}$$

$$= 24.4 \text{ lbs/yr/sq ft}$$

BED B

$$\text{Loading Rate,} \atop \text{lbs/yr/sq ft} = \frac{\text{Sl Appl, lbs/appl} \times 365 \text{ days/yr}}{L, \text{ft} \times W, \text{ft} \times \text{Cycle, days/appl}}$$

$$= \frac{7{,}018 \text{ lbs/appl} \times 365 \text{ days/yr}}{200 \text{ ft} \times 25 \text{ ft} \times (21 \text{ days} + 1 \text{ day})/\text{appl}}$$

$$= 23.3 \text{ lbs/yr/sq ft}$$

3. Which application depth should be used?

Based on the data given and the above analysis, there is no substantial difference in the amount of solids that can be applied per year for application depths of 3 inches and 9 inches. The operator should choose the 9-inch application because it will result in less operator time. A 3-inch application would require the operator to refill and possibly remove solids every 7 days while a 9-inch application will require operator attention every 22 days.

The preceding example *only* illustrates the calculations necessary to determine loading rates and it should *not* be misinterpreted that high application depths are more efficient than low application depths as a result of these calculations. The operator should go through the above analysis for a particular sludge and specific data to determine the optimum depth of application. Usually, greater depths of digested sludge are applied and the drying times are longer than used in this example.

Sludge removal from sand drying beds should be done so as to remove as little of the sand media as possible and care should be taken to avoid compacting the sand bed. Heavy equipment should not be allowed on the bed. Provisions should be made to remove dewatered sludge by surface scrapers or collectors that

are mounted on the vertical walls of the bed or on the access road between the beds. Compaction of the bed will result in reduced drainage rates, longer drying times, and may increase the potential for plugging.

3.432 Normal Operating Procedures

The sludge should be applied to the bed as evenly as possible and should be done with minimal disturbance of the bed surface. This is best accomplished by applying the sludge through an inlet distribution assembly, which may consist of troughs and weirs to apply the sludge evenly and with as little turbulence as possible. Be sure to flush the sludge out of the pipe and leave one end open for any gas produced by anaerobic decomposition to escape. Some operators make a large two-wheel "pizza cutter" out of disk harrows, or a tine-type drag device, either of which is dragged across the sludge as soon as it begins to "fetch up." This promotes cracking along these lines, which allows operators to fork out pieces of dried sludge sized for easy handling. Remove the sludge from the bed when it reaches a dryness that will allow for easy removal. Scrape and smooth the surface of the bed with a rake to prepare the drying bed for more sludge.

3.433 Typical Performance

Table 3.27 summarizes typical operating guidelines and performance data for sand drying of wastewater sludges.

TABLE 3.27 TYPICAL PERFORMANCE OF SAND BEDS

Sludge Type	Loading, lbs SS/sq ft/yr[a]		Cake, % TS[b]	Solids Recovery, %
	Open Bed	Covered Bed		
Digested Primary	20–35	20–45	30–70	95–99
Digested Secondary	10–20	10–25	30–50	95–99
Digested Combined	10–25	10–35	30–70	95–99

[a] lbs/sq ft/yr × 4.883 = kg/sq m/yr.
[b] Thickened sludge.

The final concentration to which the sludge can be dried is dependent on the climatic conditions and time the sludge remains in the bed after most of the water has drained through the sand. In general, the sludge is dried to the point where it can easily be removed from the bed. If the sludge is removed when it is relatively wet and sticking to the bed surface, large quantities of sand also will be removed.

Sand bed operations are more of an art than a science because of the large number of uncontrolled variables. Even though sand beds are a common method of sludge dewatering, it is difficult to list typical operation and performance data with a reasonable degree of certainty.

3.434 Troubleshooting

Sand drying beds are relatively simple to operate and the only problem that appears to develop at most installations is plugging of the media (sand) surface. When sludge is applied, the majority of water should drain within the first 3 to 10 days after the application. If poor drainage is evident by small filtrate quantities and a slow rate of drop of the liquid surface, you can assume the media is plugged. If sufficient area or standby capacity is available, the affected bed(s) should be allowed to dry by evaporation. After the sludge is dried and removed, rake the surface of the bed and consider removing the upper 2 to 3 inches (5 to 8 cm) of sand and replenish with fresh sand if excessive blinding is evident. If sufficient capacity is not available to allow time for the sludge in the affected bed to dry by evaporation, pump the sludge out of the bed and clear or replace the upper layer of sand.

QUESTIONS

Write your answers in a notebook and then compare your answers with those on pages 319 and 320.

3.43A   Why are sludge drying beds not commonly used for wet oxidized sludges?

3.43B   List the factors that affect sand drying bed performance.

3.43C   Why should overdosing with chemicals, particularly polymer, be avoided?

3.43D   Why might primary sludge from the bottom of anaerobic digesters require prescreening?

3.43E   Why are drying beds sometimes covered?

3.43F   A 150-foot long by 30-foot wide sand drying bed is loaded at 15 lbs/yr/sq ft. One application of sludge is made per month (12 applications per year). Determine the depth (in) of each application if the sludge has a concentration of 3.0 percent sludge solids.

3.43G   How should an operator determine the most desirable depth of sludge to be applied to sand beds?

3.43H   Why should sand bed compaction be avoided?

3.43I   Why should the sand bed surface be raked after sludge is removed?

3.43J   What determines the final concentration to which sludge is dewatered on sand beds?

3.43K   What is the major problem that is encountered when operating sand drying beds?

3.44   Surfaced Sludge Drying Beds

3.440   Need for Surfaced Drying Beds

Sludge drying beds using gravel and sand with an underlying pipe system for bed drainage have a limitation of manual cleaning with forks or shovels to remove the dried sludge. Equipment such as tractors with front end loaders cannot be used to remove dry sludge from sand beds because the equipment would break the pipe underdrains, compact the sand bed, or mix the sand and gravel layers. Some operators lay planks or mats on the sand bed surface to distribute the weight of cleaning equipment.

After the sludge removal operation is complete, the operator simply removes the planks or mats leaving the sand bed undisturbed. Surfaced drying beds of either blacktop or concrete have been used to facilitate easier sludge removal by the use of skip loaders. One advantage of a surfaced bed is the ability to speed up sludge drying.

3.441   Layout of Surfaced Drying Beds (Figure 3.21)

Surfaced drying beds are designed to allow the use of mechanical sludge removal equipment. Other design considerations include the application of digested sludge to be dried and the drainage of water released from the sludge being dried. Drying beds are rectangularly shaped with widths of 40 to 50 feet (12 to 15 m) and lengths of 100 to 200 feet (30 to 60 m). A two-foot high retaining wall around the outside of the bed contains the sludge.

The actual size of a drying bed depends on the sludge produced by the treatment plant. A plant with small flows, less than 2 MGD (7,570 m³/day), and with secondary treatment may require only four 40- by 100-foot (12- by 30-m) drying beds for sludge handling purposes. However, a 10 MGD (37,850 m³/day) plant with activated sludge and no solids thickening processes before sludge digestion may require 20 drying beds 50 by 200 feet (15 by 60 m) to provide adequate sludge handling capabilities. The drying beds are usually sized so that one bed can be filled with sludge from the digesters to an 18-inch (45-cm) sludge depth in an 8-hour period.

The drying bed is provided with a 4-inch (100-mm) or 6-inch (150-mm) drain line down the center of the bed and 18 inches (45 cm) to 30 inches (75 cm) below the surface of the drying bed. The drain line is either a perforated pipe or a pipe with pulled joints (space between joints) (Figure 3.22). The trench in which the drain pipe is placed is filled with gravel and sand that serve as a media to filter solids out of the drainage water.

The drying bed drain pipe to the plant headworks is equipped with a valve outside of the bed for isolation of the drying bed. At the other end (high end) of the drain pipe, the line is extended to the ground surface (grade) and a cleanout is installed to allow cleaning of the drain line. Digested sludge is usually applied to the drying bed through a 6-inch (150-mm) sludge feed line with a control valve or a box arrangement.

Equipment access to the bed is through a 12-foot (3.6-m) opening in the end wall. The opening is closed off with 2×12-inch (5×30-cm) planks (stop logs) to seal the opening when the drying bed is being filled and in use.

3.442   Operation

When placing a sludge drying bed into operation, the following tasks should be performed:

1. Remove the drain line cleanout cap on the bed to be filled and flush the drain line with water.

2. Close the valve on the drain line discharge after flushing the drain line.

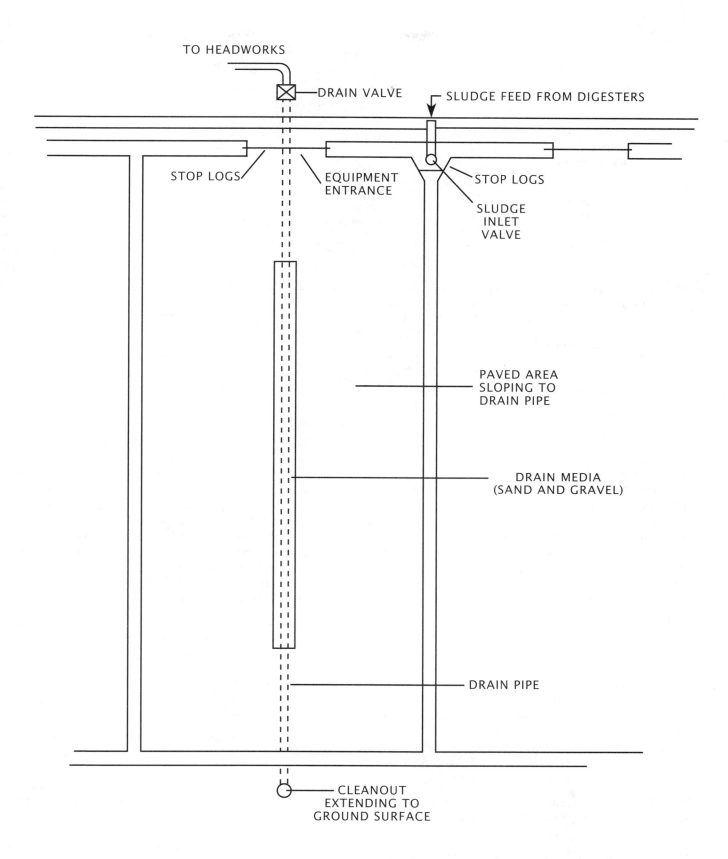

Fig. 3.21   Surfaced sludge drying bed (top view)

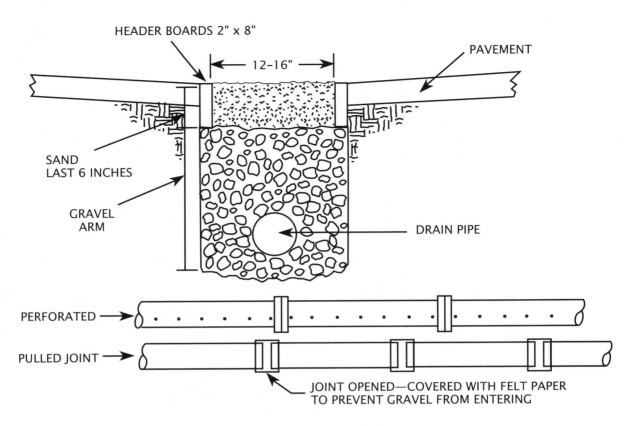

HEADER BOARDS 2" x 8"

PAVEMENT

12–16"

SAND
LAST 6 INCHES

GRAVEL
ARM

DRAIN PIPE

PERFORATED

PULLED JOINT

JOINT OPENED—COVERED WITH FELT PAPER
TO PREVENT GRAVEL FROM ENTERING

Fig. 3.22   Drainage details

3. If the drain media consists only of gravel, fill the drain line and filter area with water until the gravel is flooded. This will prevent sludge from entering the voids (empty spaces) in the gravel over the drain line during filling of the drying bed. Beds with gravel and an overlying sand upper layer do not require flooding before filling with digested sludge. However, the sand layer should be loosened and raked level before applying sludge to the drying bed. After flooding the drain media, make sure the cleanout cap is replaced and secured.

4. Install the 2×12-inch (5×30-cm) stop logs in the equipment entrance opening to the bed. To seal the ends where the stop logs fit into the wall slot, use rags or burlap sacks wrapped around each end of the stop log and along the bottom of the stop log next to the paving. Sealing is necessary to prevent sludge leaking from the bed during the filling operation and the first days of drying. Sand or soil also may be used to form a small dam to prevent leaking. If the drying bed being filled has a gate for entrance to an adjacent bed that is not to be filled, seal the stop logs in this gate to prevent leakage.

WARNING: NEVER smoke near a sludge drying bed that is being filled or has been filled recently because the gases from the sludge could form an explosive mixture.

5. Select the digester that sludge is to be withdrawn from and close feed valves to all drying beds except to the drying bed to be filled. Start sludge flowing from the digester to the drying bed by opening the proper valves.

Control the sludge feed to the drying bed at the drying bed inlet valve. This allows you to observe the flow of sludge and the rate of sludge application. The sludge should not be applied too fast because it may cause coning in the digester. When coning occurs, only the water will be drawn, which will greatly extend the drying time. Eight to 12 hours may be required to fill a bed 40 by 100 feet (12 by 20 m), 18 inches (45 cm) deep at the side walls with sludge having a 6 to 8 percent solids content.

Samples should be taken of the applied sludge 30 minutes after the start of the filling by placing 100 to 200 mL of sludge in a 1,000-mL beaker. An additional 100 to 200 mL sample is placed in the same beaker every hour thereafter while filling the drying bed. Always use the same size sample. Upon completion of filling the bed, the 1,000-mL beaker should be full of sludge from samples taken during the filling time.

When the bed is filled to the desired level (usually 6 to 18 inches (15 to 45 cm) deep at the side walls), close the sludge feed valve at the digester. If no more sludge is to be withdrawn for at least a week, the sludge line should be cleared of sludge and flushed with water to prevent plugging of the line. Always leave the valve open at the filled drying bed, or another empty drying bed, to prevent gas from building up in the line and damaging the pipes, valves, or fittings.

6. Dewatering the drying bed. Mix the contents in the 1,000-mL sample beaker. Remove a 200- to 300-mL portion for laboratory analysis. Leave the remaining portion of sludge in

the beaker and set it on the wall of the sludge drying bed. If the wastewater plant is staffed for 24-hour operation, an operator should check the beaker every four hours looking for signs of water-sludge separation.

Water-sludge separation usually occurs in 12 to 24 hours after the sludge has been applied to the drying bed. The sludge will rise to the surface leaving a 10 to 40 percent portion by volume of water under the sludge. When this occurs, partially open the drying bed drain line valve to drain off the water from the drying bed. The sample beaker can then be returned to the laboratory. The water-sludge separation will not continue for more than several hours because the sludge will resettle to the bottom. If the sludge resettles, a large percentage of the water will move back into the sludge or to the surface of the sludge in pools. These pools will take a considerable amount of time to evaporate from the sludge. To hasten sludge drying, open the drain valve when the sludge sample shows that the water-sludge separation has occurred. This procedure has reduced the sludge volume in a drying bed by as much as 30 percent during the first day of drying.

Sludge drying may be further hastened after a crust has formed on the sludge surface and has started to crack. Mix or break down the sludge by driving equipment through the drying bed to expose new sludge to surface evaporation.

3.443 Cleaning the Drying Bed

When the sludge on the bed has dried, remove the stop logs and use a tractor equipped with a front bucket (skip loader) to scoop up the sludge and remove it from the drying bed. Do not drive on or across the drain trench because you will damage the trench or the drain pipe.

3.45 Dewatering Summary

Successful dewatering requires that: (1) the operator be very familiar with the operation of the particular dewatering device(s) used, (2) sludge conditioning be optimum, and (3) the influent sludge be as thick and consistent as possible. Presence of certain materials in industrial wastes can significantly decrease the dewaterability of the respective sludges. The operator has little, if any, control over the sludge type and characteristics. In each specific case, a laboratory, pilot-, or full-scale study may be required to determine the applicability of the dewatering device(s).

QUESTIONS

Write your answers in a notebook and then compare your answers with those on page 320.

3.44A   What is a limitation of using sludge drying beds?

3.44B   How would you start to fill a surfaced drying bed that has only gravel in the drainage trench?

3.44C   What should be done when water-sludge separation is observed in a beaker containing digested sludge that is sitting on the wall of the sludge drying bed?

3.45A   How can an operator have a successful sludge dewatering program?

3.46 Additional Reading

1. *OPERATION AND MAINTENANCE OF SLUDGE DEWATERING SYSTEMS* (MOP OM-8). Obtain from Water Environment Federation (WEF), Publications Order Department, 601 Wythe Street, Alexandria, VA 22314-1994. Order No. MOM8. Price to members, $26.75; nonmembers, $36.75; price includes cost of shipping and handling.

END OF LESSON 4 OF 5 LESSONS

on

RESIDUAL SOLIDS MANAGEMENT

Please answer the discussion and review questions next.

DISCUSSION AND REVIEW QUESTIONS

Chapter 3.   RESIDUAL SOLIDS MANAGEMENT

(Lesson 4 of 5 Lessons)

Write the answers to these questions in your notebook. The question numbering continues from Lesson 3.

19. The degree of dewatering and the sludge solids removal efficiency for pressure filters are influenced by what factors?

20. How would you determine filtration time for a pressure filter?

21. What could be the cause or problem if most of the sludge passes through the filter without building up between the plates?

22. How would you determine the speed of the belt on a belt filter press?

23. How would you attempt to correct the cause of poor filtrate quality from a vacuum filter?

24. Why must care be taken to avoid chemical overdosing of sludges applied to sand drying beds?

25. Why should smoking not be allowed around drying beds while digested sludge is being applied or shortly thereafter?

CHAPTER 3.   RESIDUAL SOLIDS MANAGEMENT

(Lesson 5 of 5 Lessons)

3.5   VOLUME REDUCTION

3.50   Purpose of Volume Reduction

Drying of wastewater sludges beyond the level attained by normal dewatering methods could result in a product that could be marketed as a soil amendment or fertilizer (subject to meeting applicable regulatory requirements relative to pollutant concentrations and pathogen quality). Drying might be a feasible and required step to produce sludges suitable for landfill disposal, especially those that are difficult to dewater. In addition, drying may improve the economy and overall efficiency of subsequent incineration processes.

Drying and incineration of wastewater sludges result in a net reduction of the sludge mass and are therefore called *VOLUME REDUCTION PROCESSES*. The distinction between drying and incineration is that drying removes water from sludge without the combustion (burning) of solid material.

Volume reduction is accomplished by a variety of methods including composting, direct and indirect heat drying, and incineration. In addition to substantially reducing the volume of the sludge mass, heat drying and incineration processes should result in a complete destruction of pathogenic organisms due to the high temperatures maintained.

3.51   Composting

Composting results in the decomposition of organic matter by the action of *THERMOPHILIC*[37] facultative aerobic microorganisms to sanitary, nuisance-free, humus-like material. Composting is a biological process and requires that a suitable environment be established and maintained to ensure the survival and health of this group of bacteria. In order to create a suitable environment, several guidelines must be met.

First, composting of wastewater sludges requires that these sludges be blended with previously composted material or bulking agents such as sawdust, straw, wood shavings, or rice hulls. The initial solids content of the mixture should be about 35 to 40 percent solids. The blending process should produce a fairly uniform, porous structure in the composting material to improve aeration. Second, aeration must be sufficient to maintain aerobic conditions in the composting material. Third, proper moisture content and temperatures must be maintained. Microorganisms require moisture to function, but too much moisture can cause the process to become anaerobic or reduce the composting temperature below that which is suitable for the bacteria.

Composting generally falls into three categories: (1) windrow, (2) static pile, and (3) mechanical. In a static pile (Figure 3.23) operation, a *PUG MILL*[38] might best be used to achieve a uniform 35 percent solids mixture of dewatered stabilized sludge and bulking material. This mixture is placed in a pile containing 150 to 200 cu yd (115 to 150 m^3). A forced-air system draws air through the pile to a perforated pipe network beneath the pile. Warm, moist air is drawn off and blown through a small pile of previously composted material to reduce odors. The static pile is also covered with a layer of composted material to contain odors. The process takes about three weeks to complete. The compost material is then typically stored a month for curing. Open storage may have a slight musty odor. After curing, the material is ready for bulk use or it can be dried further, pulverized, and then bagged for sale.

A variety of proprietary mechanical reactor composters are currently available and in operation, but the specific features and operating characteristics of the systems vary widely. It is, therefore, difficult to provide general operating guidelines that would apply to all types of mechanical composters. If your plant uses a mechanical composter, consult the plant O & M manual to obtain detailed operation and maintenance guidelines for the specific type of system installed at your plant.

The most common method of sludge composting is by windrow operation (Figure 3.24). Windrow composting is generally limited to digested sludge and consists of collecting dewatered digested sludge, mixing it with previously composted material or bulking agents, and forming windrow piles. Typical windrow stack (pile) dimensions and spacing between stacks are shown in Figure 3.25. Specialized machinery such as "Flow-Boy"-type trailers can form windrows and, at the same time, mix dewatered sludge with compost material or bulking agents.

The initial moisture content of the blend of the dewatered sludge and bulking agents or compost should be approximately 45 to 65 percent moisture, and the carbon to (total) nitrogen ratio should be about 25:1. After formation of the windrows,

---

[37] *Thermophilic* (thur-moe-FILL-ick) *Bacteria.*   A group of bacteria that grow and thrive in temperatures above 113°F (45°C). The optimum temperature range for these bacteria in anaerobic decomposition is 120°F (49°C) to 135°F (57°C). Aerobic thermophilic bacteria thrive between 120°F (49°C) and 158°F (70°C).

[38] *Pug Mill.*   A mechanical device with rotating paddles or blades that is used to mix and blend different materials together.

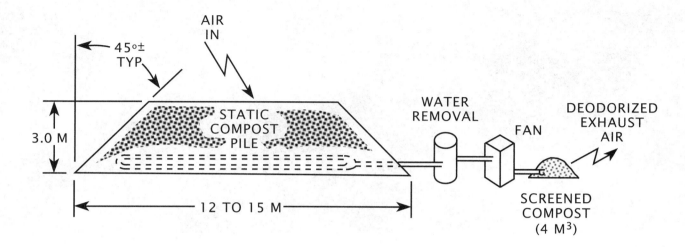

AIR IN

45°± TYP

3.0 M

STATIC COMPOST PILE

12 TO 15 M

WATER REMOVAL

FAN

DEODORIZED EXHAUST AIR

SCREENED COMPOST (4 M³)

GENERAL LAYOUT

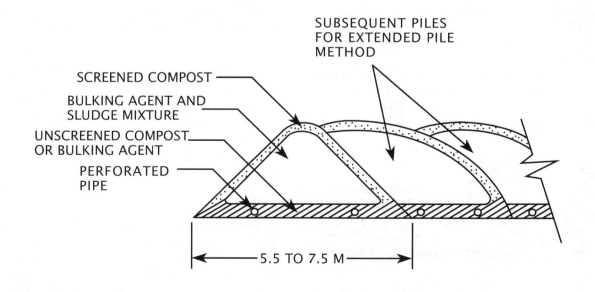

SUBSEQUENT PILES FOR EXTENDED PILE METHOD

SCREENED COMPOST

BULKING AGENT AND SLUDGE MIXTURE

UNSCREENED COMPOST OR BULKING AGENT

PERFORATED PIPE

5.5 TO 7.5 M

CROSS SECTION

Fig. 3.23   Typical static compost pile for 40 cubic meters of dewatered sludge
(from EPA Capsule Report, EPA 625/2-77-014)

Fig. 3.24   Windrow composting

the stacks should be turned once or twice daily during the first five days to begin the compost action and to ensure a uniform mixture. Thereafter, the windrows should be turned anywhere from once per day to once a week to provide aeration and to encourage drying by exposing the compost material to the atmosphere. After the process is complete, the compost product must be loaded onto trucks for disposal or recycling. This can be accomplished by a variety of equipment including skip loaders and specially designed compost loaders.

Chemically stabilized and wet-oxidized sludges are generally not suited for compost operations. Chemical stabilization produces environments that are unsuitable for microorganism survival and usually will not support the life of composting bacteria unless the sludges are neutralized and favorable conditions exist. Sludge that has been stabilized by wet oxidation can be composted, but noxious odors will more than likely develop in and around the compost operation. Unless provisions are made to house the compost area and scrub the exhaust gases, a severe lowering of air quality could develop and lead to numerous odor complaints. The persistence of organic chemicals, pathogenic organisms, or heavy metals in some composted sludges may prevent the use of the material for application to crops for human consumption.

3.510   Factors Affecting Composting

The time required to complete the composting process and the efficiency of the operation depend on many factors. These include: (1) sludge type, (2) initial moisture content and uniformity of the mixture, (3) frequency of aeration or windrow turning, (4) climatic conditions, and (5) desired moisture content of the final product.

The type of digested sludge to be treated in a compost facility can drastically affect performance and operation. Primary and secondary sludges can be treated by compost processes, but the plastic nature of dewatered secondary sludges and increased moisture content make them more difficult to compost than primary sludges. The difficulties arise with secondary sludge because greater efforts have to be put into mixing with compost material or bulking agents to produce an evenly blended, porous mixture. Dewatered secondary sludges tend to clump together and form "balls" when they are blended with compost material. The balls that are formed within the windrows readily dry on the outer surface but remain moist on the inside. Occasionally, this creates anaerobic conditions leading to odor production and a lower composting temperature. The problem is further complicated when large quantities of polymer are used in the dewatering step because of the sticky nature of polymers.

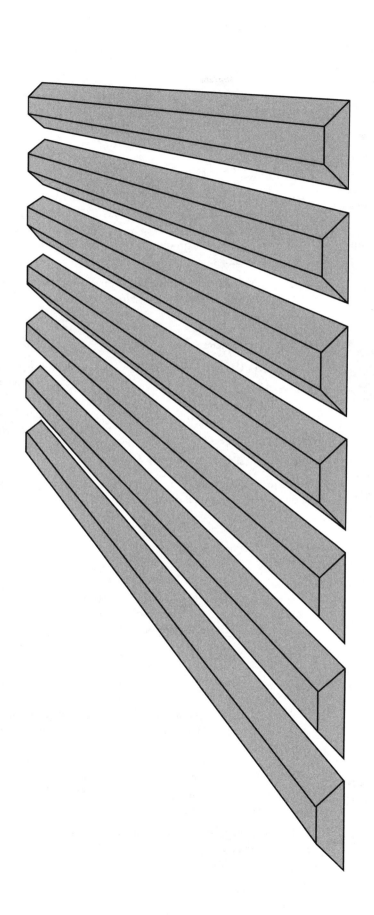

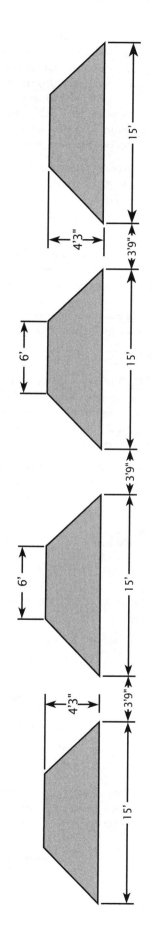

Fig. 3.25   Windrow stack dimensions and spacing between stacks

Composting of the sludges from industrial wastewater treatment plants may be inhibited by the presence of a variety of inorganic and organic chemicals that are toxic to facultative aerobic microorganisms. Some well-digested sludges and industrial sludges do not contain enough nitrogen to compost well and nitrogen supplements might have to be added. In general, special laboratory or pilot studies may be required to verify the applicability of the compost processes for sludge handling.

The initial moisture content and homogeneity (evenness) of the mixture are important considerations in starting the compost process. These factors depend on sludge type, use of polymers, as discussed above, and the effectiveness of the blending operation.

The quantity of compost product that must be blended with dewatered sludge to produce an initial moisture content of 45 to 65 percent depends on the concentration of the dewatered sludge and the moisture content of the recycled compost material. The following examples show how to calculate the amount of compost required for blending purposes.

EXAMPLE 41

Given: A 5 MGD plant produces 4,100 lbs/day of dewatered digested primary sludge. The dewatered sludge is at a concentration of 30 percent thickened sludge (TS). Final compost product at the plant has a moisture content of 30 percent.

Find: The total pounds of compost that must be blended with the dewatered sludge to produce a moisture content of the mixture of 50 percent.

Solution:

Known		**Unknown**
Flow, MGD	= 5 MGD	Pounds of compost blended daily with dewatered sludge to produce a mixture with 50 percent moisture content
Sludge, lbs/day (Dewatered Digested Primary Sludge)	= 4,100 lbs/day	
Dewatered Sl Sol, %	= 30% Solids	
Final Compost, %	= 30% Moisture (70% Solids)	

1. Determine the moisture content of the dewatered sludge.

$$\text{Sludge Moisture, \%} = 100\% - \text{Dewatered Sl Sol, \%}$$
$$= 100\% - 30\%$$
$$= 70\% \text{ Moisture}$$

2. Calculate the pounds of compost that must be blended daily with the dewatered sludge to produce a mixture with 50 percent moisture content.

$$\frac{\text{Mixture}}{\text{Moisture, \%}} = \frac{\text{Sludge, lbs/day} \times \text{Sl Moist, \%} + \text{Comp, lbs/day} \times \text{C Moist, \%}}{\text{Sludge, lbs/day} + \text{Compost, lbs/day}}$$

Rearranging terms:

$$\frac{\text{Compost,}}{\text{lbs/day}} = \frac{(\text{Sludge, lbs/day} \times \text{Sl M, \%}) - (\text{Sludge, lbs/day} \times \text{Mix M, \%})}{\text{Mix M, \%} - \text{Comp M, \%}}$$

$$= \frac{(4{,}100 \text{ lbs/day} \times 70\%) - (4{,}100 \text{ lbs/day} \times 50\%)}{50\% - 30\%}$$

$$= \frac{4{,}100 \text{ lbs/day} (70\% - 50\%)}{50\% - 30\%}$$

$$= 4{,}100 \text{ lbs/day}$$

EXAMPLE 42

Given: A 5 MGD plant produces 2,700 lbs/day of dewatered digested secondary sludge. The dewatered sludge is at a concentration of 17 percent thickened sludge. The final compost product is at a moisture content of 30 percent.

Find: The total pounds of compost that must be recycled to produce an initial mixture moisture of 50 percent.

Solution:

Known		**Unknown**
Flow, MGD	= 5 MGD	Pounds per day of compost to produce an initial mixture of 50 percent
Sludge, lbs/day (Dewatered Digested Secondary Sludge)	= 2,700 lbs/day	
Dewatered Sludge, %	= 17% Solids	
Final Compost, %	= 30% Moisture	

Calculate the pounds per day of compost that must be recycled to produce an initial mixture moisture of 50 percent.

$$\frac{\text{Sludge}}{\text{Moisture, \%}} = 100\% - \text{Dewatered Sl Sol, \%}$$
$$= 100\% - 17\%$$
$$= 83\%$$

$$\frac{\text{Compost,}}{\text{lbs/day}} = \frac{(\text{Sludge, lbs/day} \times \text{Sl M, \%}) - (\text{Sludge, lbs/day} \times \text{Mix M, \%})}{\text{Mix M, \%} - \text{Comp M, \%}}$$

$$= \frac{(2{,}700 \text{ lbs/day} \times 83\%) - (2{,}700 \text{ lbs/day} \times 50\%)}{50\% - 30\%}$$

$$= \frac{2{,}700 \text{ lbs/day} (83\% - 50\%)}{50\% - 30\%}$$

$$= 4{,}455 \text{ lbs/day}$$

The preceding examples illustrate the effect the degree of sludge dewatering has on the quantity of compost that must be recycled to obtain the desired initial moisture content of the mixture. Even though fewer pounds of secondary sludge were produced (Example 42), more pounds of compost had to be recycled than for the primary sludge because of the differences in the degree of dewatering obtained.

The frequency of aeration or turning of the stacks is determined by trial and error. The stacks should be turned frequently enough to prevent anaerobic conditions but not so frequently that excessive heat loss occurs. If the stacks are turned too often, excessive heat will be released and the temperature may drop to a point where it is unfavorable for the thermophilic composting bacteria. To find the right turning balance, closely monitor the windrows and adjust the turning frequency. If the frequency of turning is not optimized as evidenced by anaerobic conditions or low temperatures, the time required to complete the process will increase.

Another reason it is important to optimize turning of the stacks is that in a properly managed windrow composting operation, pathogenic bacteria can be reduced to safe levels provided the temperature in the stack stays high enough for a long enough time. Too frequent turning of the stacks may lower the temperature so much that pathogens will survive the process. When pathogen destruction is a goal, temperature measurements must be taken and recorded.

Climatic conditions play an important role in compost operations. Wet and cold climates generally require longer composting times than hot and arid regions. Wet weather is particularly damaging to windrow composting because the piles can become soaked and lose heat and the area generally becomes inaccessible to heavy equipment.

The desired final moisture content of compost product affects the time of composting because longer times are required for higher degrees of drying. In general, a well-operated windrow compost facility can dry sludge from an initial moisture content of approximately 60 percent to a moisture content of 30 percent in about 15 to 20 days and to a final moisture of 20 percent in approximately 20 to 30 days. The final moisture content at which the compost process is stopped depends on whether the material is used as a fertilizer base and the economics of hauling the compost to final disposal.

3.511 Normal Operating Procedures

The required steps for successful composting are generally the same from one operation to the next, although the mechanisms for blending sludge with bulking agents or compost material and the method of aeration might vary depending on the type of equipment used. The procedures for successful composting are listed below:

1. Dewater sludge to the highest degree economically practical.

2. Blend dewatered sludge with recycled compost or bulking agents to produce a homogeneous (evenly blended) mixture with a moisture content of 45 to 65 percent.

3. Form the windrow piles and turn (aerate) once or twice daily for the first 4 to 5 days after windrow formation.

4. Turn the piles approximately once every two days to once a week to maintain the desired temperature (130 to 140°F or 55 to 60°C) until the process is complete. The temperature of the piles should be routinely monitored during this period.

5. Load the compost onto trucks for disposal or recycle purposes.

3.512 Typical Performance

Typical performance and operational data for windrow composting operations are summarized in Table 3.28. Since climatic conditions play an important role in windrow composting, the data should be viewed with caution because it reflects summertime operation. During wet weather periods and in cold climates, the composting times may double or triple those presented in Table 3.28. The data presented in Table 3.28 reflect the windrow composting of municipal wastewater sludges. It should be noted that the presence of some inorganic or organic chemicals in industrial wastewater sludges may drastically affect the performance of the process.

TABLE 3.28 TYPICAL PERFORMANCE OF WINDROW COMPOSTING

Sludge Type[a]	Blend Material Ratio, lb/lb[b]	Initial Moisture Content, %	Max Compost Temp, °F[c]	Final Moisture Content, %	Compost Time, days
Primary	0.5:1–1:1	45–65	130–140	30–25	8–15
Secondary	1:1–1.5:1	45–65	130–140	30–25	15–25

[a] Sludge is digested and dewatered.
[b] Assuming compost product is used for blending. Ratio is lb Compost/lb Sludge or kg Compost/kg Sludge.
[c] (°F − 32) × 5/9 = °C.

A higher ratio of blend material and longer compost times for secondary sludge are usually required because dewatered secondary sludges are commonly wetter than dewatered primary sludges. As a result, it is more difficult to produce a homogeneous blend when secondary sludges are composted.

3.513 Troubleshooting

Windrow composting is a relatively simple and cost-effective method to further reduce and stabilize sludge, but difficulties can arise that will cause the process to be ineffective and troublesome. Apart from the sludge type and climatic conditions, which the operator has no control over but should plan for, the most common problems that arise are anaerobic conditions and reductions in compost temperatures.

Anaerobic conditions can prevail if the initial moisture content is greater than the optimum range (45 to 65 percent), the stacks are not turned frequently enough, or balling occurs. If anaerobic conditions develop, increase the frequency of aeration and inspect the pile for balling. If balling is not evident and a uniform mixture exists, then more frequent turning should correct the problem. If balling is evident and a nonhomogeneous mixture exists, increasing the frequency of turning may result in a decrease in the compost temperature. You should monitor the temperature and turn the stacks frequently enough to reduce the anaerobic odors while maintaining thermophilic temperatures.

As discussed earlier, balling will occur if the dewatered sludge is too wet or polymer overdosing occurs. Optimizing the performance of the dewatering facilities will help to prevent subsequent compost problems. If balling is not evident and piles are turned frequently, anaerobic conditions might develop if the moisture content of the stack is too high. This condition will usually result in corresponding reduced composting temperatures and can generally be traced back to not blending enough compost or bulking agents with the sludge prior to windrow formation. The stack will eventually recover from too much moisture but if the odors are severe, try remixing the stack with a sufficient quantity of bulking agents or compost to bring the windrow to a desirable moisture content. Balling can also develop when compost temperatures are low and there is a nitrogen deficiency. If the organic matter (carbon) in the C:N relationship is greater than 25, add urea nitrogen (N) to the compost to adjust the C:N relationship to 25:1.

Temperature decreases can be caused by too high a moisture content or too much aeration or turning. If the pile is homogeneous and the moisture content is within the optimum range,

reduce the frequency of turning to maintain thermophilic temperatures.

Table 3.29 summarizes usual operational problems and corrective measures that may alleviate or eliminate inefficiencies.

QUESTIONS

Write your answers in a notebook and then compare your answers with those on pages 320, 321, and 322.

3.50A What is the difference between drying and incineration? What category does composting fall into and why?

3.51A Why does a suitable environment need to be established in compost piles?

3.51B List the guidelines that must be met to create a suitable compost environment.

3.51C Why are chemically stabilized and wet-oxidized sludges generally not composted?

3.51D List the factors that affect compost operations.

3.51E Explain why secondary sludges are not as easy to compost as primary sludges. Include a discussion of the "balling" phenomenon.

3.51F A medium-sized wastewater treatment plant produces 4,700 lbs/day of dewatered digested primary sludge with a solids concentration of 27 percent and 3,300 lbs/day of dewatered secondary sludge with a solids concentration of 15 percent. The sludges are blended together and then composted in windrows. Determine the total pounds of compost that must be recycled and blended with the combined sludge to produce a moisture of 60 percent. The final compost product of the plant has a moisture content of 30 percent.

3.51G List the operational procedures required for windrow composting.

TABLE 3.29   TROUBLESHOOTING WINDROW COMPOSTING

Operational Problem	Possible Cause	Check or Monitor	Possible Solution
1. Anaerobic conditions	1a. Aeration frequency too low	1a. Frequency of aeration and stack temperature	1a. Increase aeration frequency
	1b. Stack moisture too high	1b. Stack moisture and temperature	1b. Blend additional compost or bulking agents
	1c. Balling	1c. See Item 3	1c. See Item 3
2. Low compost temperatures	2a. Aeration frequency too high	2a. Frequency of aeration and stack moisture	2a. Decrease aeration frequency
	2b. Stack moisture	2b. Stack moisture	2b. Blend additional compost or bulking agents
	2c. Balling	2c. See Item 3	2c. See Item 3
	2d. Lack of nitrogen	2d. Total Kjeldahl Nitrogen	2d. Add urea to 25:1 C:N
3. Sludge balling	3a. Dewatered sludge too wet	3a. Dewatering facility	3a. Increase cake dryness
	3b. Polymer overdosing	3b. Dewatering facility	3b. Reduce polymer dosage
	3c. Ineffective blending	3c. Blending operation	3c. Improve blending operations

3.52   Mechanical Drying

Mechanical heat drying is a dehydration process that removes water from sludge without combustion of the solid material. Mechanical drying is either direct or indirect. In an indirect dryer, steam fills the outer shell of a rotating cylinder. Sludge circulates through the inner compartment and is dried by the heat from the steam. Direct dryers use the direct contact of sludge with preheated gases.

The most common types of dryers include direct and indirect rotary dryers, rotary vacuum dryers, and modified multiple-hearth incinerators. Other mechanical dryers include flash dryers, atomized-spray dryers, disk and paddle dryers, and fluidized-bed dryers. Only the direct and indirect dryers will be discussed here. Drying by modified multiple-hearth incinerators will be discussed in Section 3.53.

Rotary dryers or rotary kilns (Figure 3.26) consist of a horizontal cylindrical steel shell with flights (mixing blades) projecting from the inside wall of the shell. The basic difference between direct and indirect rotary dryers is that indirect dryers are equipped with a hollow jacket through which steam is passed while hot gases are passed directly through the dryer for direct drying. In indirect and direct dryers, the hottest gases surround a central shell containing the material, but return through the shell at reduced temperatures.

Rotary kilns may be used to either dry or incinerate sludge. They also have been used for refuse incineration and ore processing. In Figure 3.26, a storage bin is located on the lower left to hold the sludge for processing. The clamshell bucket loads the sludge into the feed hopper. A conveyor moves the sludge to the inlet hopper of the rotary kiln. Inside the kiln the sludge is either dried or incinerated. Dried sludge or ash is removed from the kiln into the hopper in front of the operator. Hot gases flow through the scrubber on the right to remove the particulate matter suspended in the gases. The stack in the middle right serves as an emergency bypass stack in case of system failure. Behind the stack is a clarifier that is used for ash separation in liquid ash systems.

Rotary dryers or rotary vacuum dryers can be operated as either batch or continuous type processes. In the batch mode, the rotary vacuum dryer is charged with the sludge to be dried and then sealed. A vacuum (approximately 26 to 28 inches Hg) is then applied to the internally charged compartment. Steam (50 to 100 psig) from a boiler is passed through a jacketed hollow in the outer shell wall and, in some cases, through the internal central portion of an agitator assembly. Vapor is removed by vacuum pumps and passed through a condenser prior to discharge.

In the continuous mode, no vacuum is applied and the drying process takes place at atmospheric pressure. Material is introduced continuously at one end of the dryer and is discharged continuously at the opposite end. All other process functions are the same except for exhaust gas temperatures, which are much higher in the continuous flow process. This is due to the differences in vaporization temperature requirements at atmospheric pressure versus a vacuum of 26 to 28 inches Hg.

In each case, the dewatered sludge is blended with previously dried material and continuously fed into the dryer. The cylindrical drum rotates about 4 to 8 RPM and the inlet end is slightly higher than the discharge end. As the dryer rotates, the flights projecting from the shell wall elevate, tumble, and mix the material (like a clothes dryer) to provide frequent contact by tumbling the wetted sludge. The rotation of the dryer drum causes the sludge to fall off the walls of the drum near the top or crown portion of the kiln. As the sludge falls it becomes drier and is conveyed toward the outlet end of the drum. Also, the speed of rotation of the drum will affect the moisture content of the sludge being dried.

Dried product is generally blended with the sludge to improve the conveying characteristics of the sludge and to reduce the potential for balling. This blending of wet sludge with a previously dried sludge is accomplished in a pug mill. As is the case with compost operations, the sludge may clump together and form balls that readily dry on the outer surface but remain wet on the inside if the blending is not done properly.

3.520   Factors Affecting Mechanical Drying

The degree of drying obtained and the efficiency of rotary dryers depend on: (1) sludge type, (2) sludge detention time within the dryer or hot gas velocity, (3) temperature, (4) vacuum (for vacuum dryers), and (5) moisture content and ratio of wet-to-dry sludge before being fed to the rotating kiln.

Secondary sludges have a greater tendency to ball and contain more water than primary or digested sludges. Secondary sludges are, therefore, not as well suited to mechanical drying operations as primary sludges. Obviously, any sludge should be dewatered to the highest degree possible by centrifuges, filter presses, or vacuum filters to reduce the volume of water delivered to the dryer and to facilitate the drying process.

The length of time the sludge remains in the dryer and the temperature of the drying gas will affect the degree of drying obtained. As drying time, vacuum, and temperature increase, the moisture content of the dried product will decrease. In the continuous mode of operation, drying time is governed by the size of the dryer, the quantity of sludge applied, and the speed of the drum. As the speed increases, the drying time decreases because the sludge is picked up and tumbled toward the outlet at a faster rate. To increase the drying time, lower the drum speed or reduce the quantity of sludge applied, if possible. Keep in mind, however, that as the drum speed is reduced to increase the drying time, the frequency of contact between wetted sludge particles and the drying medium will also decrease. A fine line exists between operating the drum at a speed to maximize drying time while still providing frequent contact between wet sludge particles and hot gases or heated surfaces. Experiment with different drum speeds to determine the best speed for your situation.

3.521   Normal Operation and Performance

Operational data on heat dryers are scarce since the process is not routinely used at the present time. The use of mechanical dryers is uncommon because this is a very expensive process

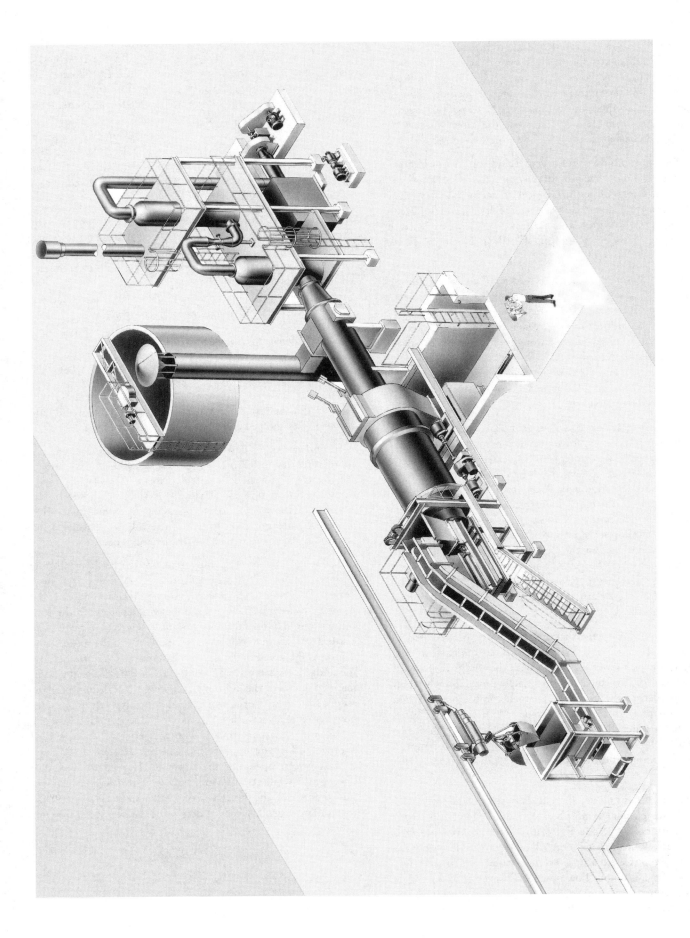

Fig. 3.26   Rotary kiln
(Permission of Envirotech)

principally due to large energy demands. Due to this relative lack of operating data, it is virtually impossible to list typical performance and operating data or to outline specific operational procedures. However, some limited operating data for the indirect and direct rotary dryers and the vacuum rotary dryers published in the literature could be helpful for operator orientation. These data are presented in Tables 3.30, 3.31, and 3.32.

In general, the sludge to be dried should be dewatered to the highest degree possible, blended with previously dried material to produce a homogeneous structure, and adequately mixed or tumbled within the dryer to maximize contact with the drying medium (gases or heated surfaces) and to minimize sludge balling.

TABLE 3.30 LIMITED EXPERIMENTAL DATA FOR DIRECT ROTARY DRYERS

Parameter	Unit	Value
Sludge Type(s)	—	Dewatered Secondary/Primary
Heat Source	—	Products of Natural Gas Combustion
Hot Gas Temperature:		
—Inlet	°F	800–1,200
—Outlet	°F	180–300
Hot Gas Velocity	fps	4–12
Rotary Drum Speed	RPM	5–8
Evaporative Capacity	lbs/hr	4,000–6,000
Dry Solids Concentration:		
—Inlet	%	10–20
—Outlet[a]	%	30–90

[a] For the batch mode of operation, the detention times of 20 to 60 minutes have been used.

TABLE 3.31 LIMITED EXPERIMENTAL DATA FOR INDIRECT ROTARY DRYERS (CONTINUOUS OPERATION)

Parameter	Unit	Value
Sludge Type(s)	—	Dewatered Secondary/Primary
Heat Source	—	Steam
Steam:		
—Pressure	psig	100–140
—Temperature	°F	300–350
Linear Rotor Speed	fps	20–25
Heat Transfer Coefficient	BTU/hr-sf	15–75
Dry Solids Concentration:		
—Inlet	%	10–20
—Outlet	%	30–50

TABLE 3.32 LIMITED EXPERIMENTAL DATA FOR INDIRECT VACUUM ROTARY DRYERS (BATCH OPERATION)[a, b, c, d]

Sludge Type	Inlet Concentration (%)	Outlet Concentration (%)	Heat Capacity (BTU/lb Water Vaporized)	Heat Transfer Coefficient (BTU/hr-sf)
Digested/Dewatered Primary	30–32	95–98	1,300–1,350	8–9
Digested/Dewatered Oxygen Activated Secondary	14–15	33–35	1,500–1,600	7–8
Undigested/Dewatered Oxygen Activated Secondary	10–11	40–42	1,300–1,400	8–9
Digested/Dewatered Air Activated Secondary	9–10	34–35	1,200–1,300	9–10

[a] Size of the drum: 1.5 feet diameter, 3.0 feet length.
[b] Jacket temperature: 297°F.
[c] Drying time: 4 hours.
[d] Applied vacuum: 28 inches Hg.

3.522 Troubleshooting

Successful operation of rotary dryers produces a relatively dry sludge product that will readily discharge from the outlet port. The most common problems that develop with rotary dryers are excessive moisture content and balling of sludge, both of which reduce the dried product quality.

A substantial reduction in the product dryness will result from: (1) decrease in temperature of incoming heating media (hot gases, steam), (2) decrease in the vacuum applied (for the vacuum dryers), (3) decrease in detention time, (4) decrease in frequency (or area) of contact between sludge and drying medium, (5) sludge balling, and (6) deterioration of the incoming sludge quality.

The operator should check and adjust (as required) the temperature of the incoming heating media (hot gases, steam), the respective furnace or boiler, and the temperature sensors. Also, check all fuel supply and heat media supply piping for any signs of cracking or leakage and make repairs, as necessary.

The operator should check the vacuum gauges at the vacuum dryers for loss of vacuum and seal the unit, as required, or check and adjust operation of the respective vacuum pump. The vacuum lines should also be checked for any signs of cracking or leakage and be repaired, as necessary.

As explained in Section 3.520, the operator should optimize the dryer sludge loading and the drum speed in order to provide proper detention time and the required frequency (or area) of contact between the sludge and the drying medium, and to minimize sludge balling. Balling will occur if the incoming dewatered sludge is too wet, blending of the incoming sludge with dried product is ineffective, or polymer overdosing occurs. In this situation, the sludge particles may agglomerate into 2-inch to 4-inch balls within the dryer, preventing maximum sludge-dryer contact. The operator should optimize the performance of the dewatering facilities to eliminate future dryer problems, to provide proper blending of the incoming sludge with dried product, and (if possible) to increase the drum speed.

In addition, proper hot gas flow/velocities (for the direct dryers) should be maintained to prevent excessive amounts of sludge and dust from being carried out with the exhaust gas from the dryers.

Table 3.33 summarizes the most common operational problems and the usual corrective measures taken to maintain good rotary drying.

QUESTIONS

Write your answers in a notebook and then compare your answers with those on page 322.

3.52A   Explain the main difference between direct and indirect drying.

3.52B   What purpose do the flights installed on rotary dryers serve?

3.52C   Why is sludge blended with previously dried material prior to rotary drying?

3.52D   Why should sludge be dewatered to its maximum degree prior to mechanical drying or incineration?

3.52E   What effect does drum speed have on rotary dryer performance?

3.52F   What are the major problems that are encountered when operating rotary dryers?

3.53   Sludge Incineration
by Richard Best

3.530   Process Description

Sludge incineration is defined as the conversion of dewatered cake by combustion to ash, carbon dioxide, and water vapor. As a result of incineration, the volume of sludge is significantly reduced (up to 90 percent by weight). This reduction in volume is caused by the evaporation of the water in the cake and the conversion of the volatile matter in the cake to carbon dioxide and water. The only material remaining after the incineration of a cake is some ash and the inert matter in the cake.

TABLE 3.33   TROUBLESHOOTING ROTARY DRYERS

Operational Problem	Possible Cause	Check or Monitor	Possible Solution
1. Dried product too wet	1a. Temperature of incoming heating media too low	1a. Furnace or boiler. Hot gas or steam lines. Temperature sensors.	1a. Adjust or repair furnace or boiler. Repair fuel supply lines and hot gas or steam lines. Repair temperature sensors.
	1b. Vacuum too low (in vacuum dryers)	1b. Rotary dryer (for gas leaks). Vacuum pump.	1b. Eliminate gas leakage. Adjust or repair vacuum pump. Repair vacuum lines.
	1c. Detention time too low	1c. Sludge flow. Speed of drum.	1c. Decrease sludge feed. Lower drum speed.
	1d. Frequency (or area) of contact between sludge and drying medium too low	1d. Speed of drum rotation. Hot gas flow/velocity (for direct dryers).	1d. Increase drum speed. Increase hot gas flow/velocity (for direct dryers).
	1e. Sludge balling	1e. See Item 2.	1e. See Item 2.
	1f. Incoming sludge too wet	1f. Dewatering facility.	1f. Increase incoming sludge dryness.
2. Sludge balling	2a. Incoming sludge too wet	2a. See Item 1f.	2a. See Item 1f.
	2b. Polymer overdosing	2b. Dewatering facility.	2b. Reduce polymer dosage.
	2c. Ineffective blending	2c. Blending operation.	2c. Improve blending operation.
	2d. Drum speed too low	2d. Drum speed.	2d. Increase drum speed.
3. Excessive sludge/dust in exhaust gas (for direct dryers)	3a. Gas flow/velocity too high	3a. Gas supply blower.	3a. Lower gas flow/velocity.
4. Vibrations, jam, or overload	4a. Mechanical malfunctions such as rotor misalignment, bearings, drive unit, or base support	4a. Inspect all mechanical equipment.	4a. Mechanical repairs.
	4b. Plugged feed port or outlet port	4b. Feed port, outlet port.	4b. Unplug as required.
	4c. Dryer plugged	4c. Dryer.	4c. Clean as required. Reduce sludge loading and increase drum speed.
	4d. Sludge balling	4d. See Item 2.	4d. See Item 2.

The most common type of sludge incinerator is the multiple-hearth furnace (MHF) (Figure 3.27) and it will be the only process discussed in detail in this section. Other incineration designs include fluidized-bed reactors and rotary kilns (see Section 3.52, "Mechanical Drying," and Figure 3.26).

The MHF, in its simplest form, is a steel cylinder lined with refractory (heat-resistant material) and equipped with a centralized shaft and runners (Figure 3.27). Inside the MHF there are a number of levels called hearths. A center shaft passes through the center of the hearths. Rabble arms are attached to this center shaft. Plows or rabble teeth are attached to each arm. The hearths are "sprung" arches and are not connected to the center shaft.

During incineration, sludge cake enters the top of the furnace and is moved back and forth across the hearths by the rabble teeth. The cake drops alternatingly on each hearth through an outer drop opening and then through a center drop and works its way down through the hearths. As the cake reaches the edge of one hearth it drops to the hearth below until it reaches ignition temperature and burns. Ash remaining after the sludge is burned is then rabbled (moved across the hearths) until it reaches the bottom of the MHF and has been cooled.

3.531 Furnace Description

Before attempting to understand the operation of an MHF, the purpose of the various parts of the furnace must be understood.

3.5310 FURNACE REFRACTORY

The furnace shell is insulated to prevent the loss of heat into the atmosphere and to protect the equipment and workers from the high temperatures (1,500 to 1,700°F) found within the furnace. The outer steel shell is protected from the internal heat by 9 to 13 inches (23 to 33 cm) of refractory brick that lines the inside.

The hearths or levels within the furnace are actually self-supported arches. The weight of the arches or the thrust is transmitted to the outside shell. The hearths are made of a specially shaped refractory brick with castable insulation poured into the odd-shaped spaces.

Figures 3.28 and 3.29 illustrate the two different types of hearths installed within the MHF. The hearths on which the cake drops to the next lower hearth from the center are called "in" hearths. The hearths on which the cake drops through holes on the outside of the furnace are called "out" hearths.

Out hearths (those hearths where the cake moves to the outside edge of the hearth) have a circular cap ringing the center shaft at the hearth/shaft meeting point. This is known as the lute cap (Figure 3.30). The purpose of this cap is to prevent air and sludge from passing through the shaft opening rather than through the drop holes.

The holes around the outside of the hearth are called "drop holes." The sludge cake passes to the next lower or in hearth

(those hearths where the waste material moves toward the inside of the furnace) through these drop holes.

When the furnace is in operation, cake enters the furnace through a counterbalanced flap gate. The flap gate prevents the escape of gases from the furnace and limits the flow of cold air into the top of the furnace.

Once the cake goes through the flap gate, it drops onto the top hearth of the furnace. This is called "hearth No. 1." At this point, the cake begins to be moved through the furnace by action of the rabble teeth (Figure 3.30). This process is called "rabbling."

Rabbling is a term used to describe the process of moving or plowing the material inside a furnace by using the center shaft and rabble arms. Rabbling forms spiral ridges of cake on each hearth, which aids with the drying and burning of the cake. The surface area of these ridges varies with the side slope of the cake (or the slope of the sides of the furrows). This angle may vary widely from 20 degrees up to 60 degrees. Rabbling, in addition to exposing the wet sludge cake surface to the furnace gases, helps to break up large cake particles, which increases the surface area of the sludge available for drying.

Because of the ridges formed by rabbling, the surface area of cake exposed to the hot gases is as much as 130 percent greater than the hearth area provided. During rabbling, the cake falls through the in hearth and the out hearth ports. The counter-current flow of hot gases over the cake decreases the drying time.

3.5311 CENTER SHAFT

The rotating shaft to which the rabble arms and teeth are attached is called the "center shaft." The center shaft has seals at the top and bottom called "sand seals" (Figure 3.31). These are stationary cups partially filled with sand that surround the shaft. A cylindrical steel ring is attached to the shaft and extends down into the sand to form the seal. At the bottom of the shaft, the sand cup is attached to the shaft and rotates while the steel ring is fixed to the furnace bottom. These seals are very effective if properly maintained. They prevent the escape of heat and gases from the furnace and the entrance of air at these points. Gases escaping from the furnace could cause potential air pollution problems. Unplanned entrance of air can cause draft changes and false furnace conditions, which will reflect on the control panel instruments. These seals also allow for the differential expansion and contraction of the furnace body due to changes in temperature.

Due to the extremely high temperatures within the MHF, the center shaft and the rabble arms are hollow. This allows a fan, installed at the bottom of the furnace, to blow cool air (ambient air) through the center shaft and rabble arms while the furnace is in operation (Figures 3.30 and 3.32). This fan is called the cooling air fan. The hot air exhausted at the top of the furnace from the shaft is called cooling air. Depending on the furnace design, the cooling air can either be returned to the burning zone of the MHF or vented to the atmosphere.

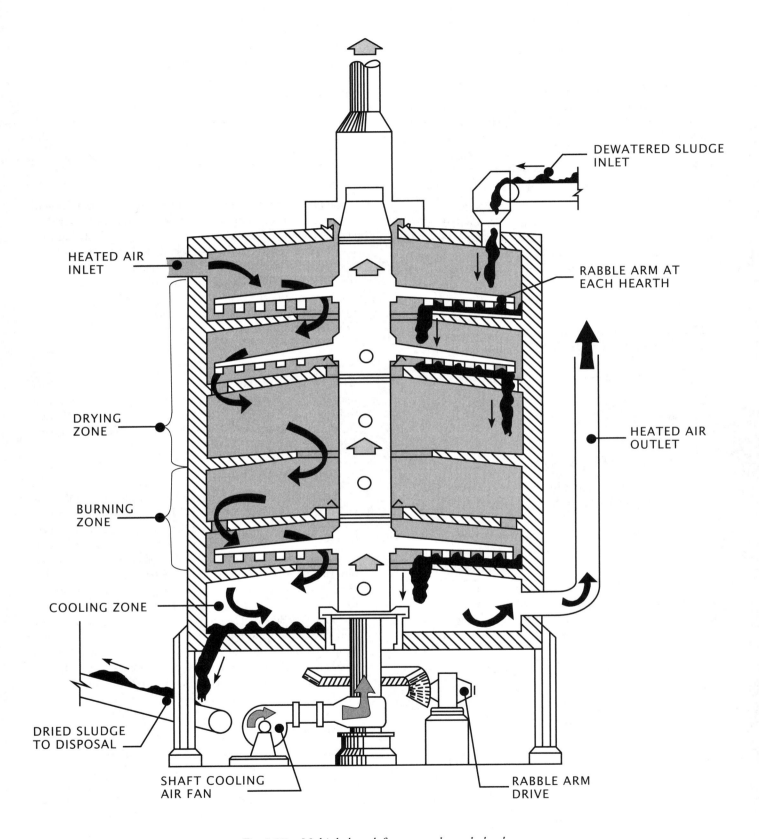

DEWATERED SLUDGE INLET

HEATED AIR INLET

RABBLE ARM AT EACH HEARTH

DRYING ZONE

HEATED AIR OUTLET

BURNING ZONE

COOLING ZONE

DRIED SLUDGE TO DISPOSAL

SHAFT COOLING AIR FAN

RABBLE ARM DRIVE

Fig. 3.27   Multiple-hearth furnace used as a sludge dryer

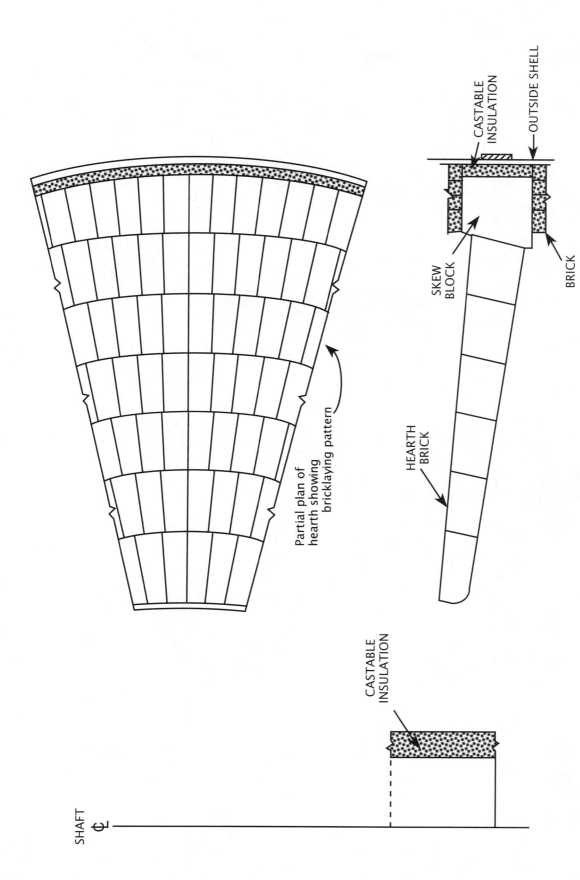

CASTABLE
INSULATION

OUTSIDE SHELL

BRICK

SKEW
BLOCK

HEARTH
BRICK

Partial plan of
hearth showing
bricklaying pattern

CASTABLE
INSULATION

SHAFT
₵

Fig. 3.28   Section through a single "in" hearth

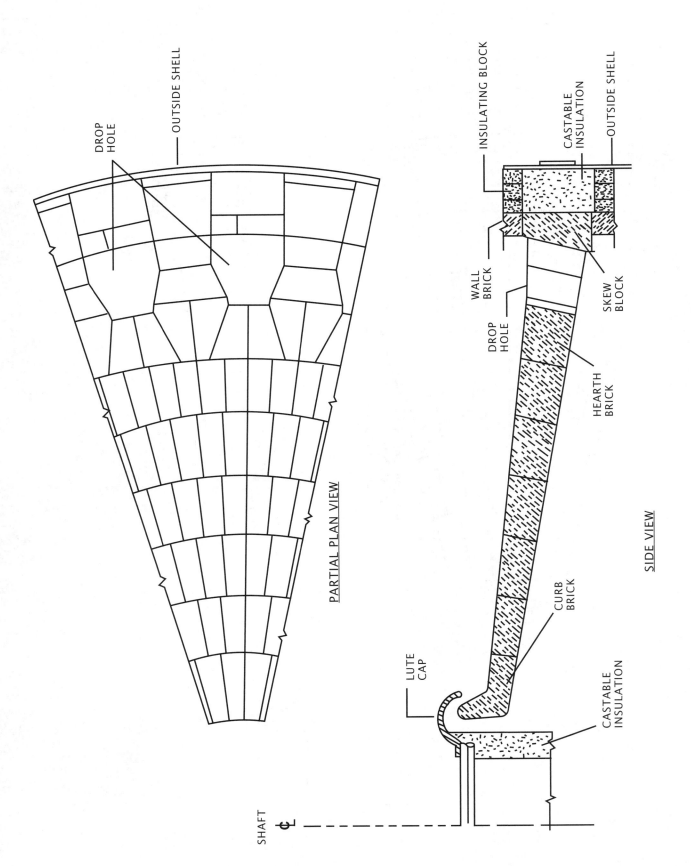

DROP HOLE

OUTSIDE SHELL

PARTIAL PLAN VIEW

INSULATING BLOCK

CASTABLE INSULATION

OUTSIDE SHELL

WALL BRICK

DROP HOLE

SKEW BLOCK

HEARTH BRICK

SIDE VIEW

LUTE CAP

CURB BRICK

CASTABLE INSULATION

SHAFT ℄

Fig. 3.29   Section through a single "out" hearth

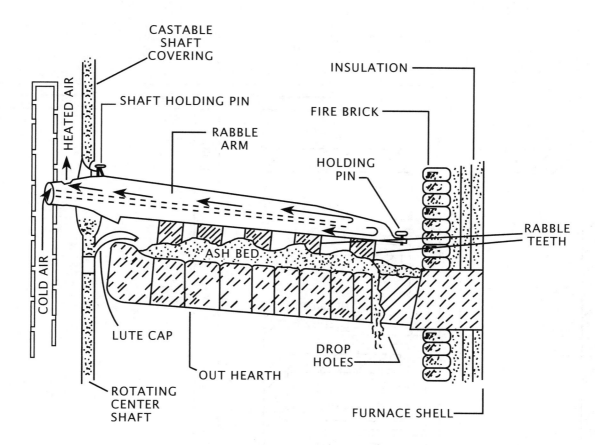

Fig. 3.30   *Action of rabble arm and rabble teeth*

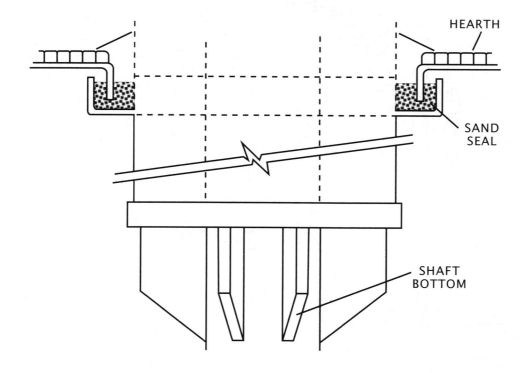

Fig. 3.31   *Lower sand seal schematic*

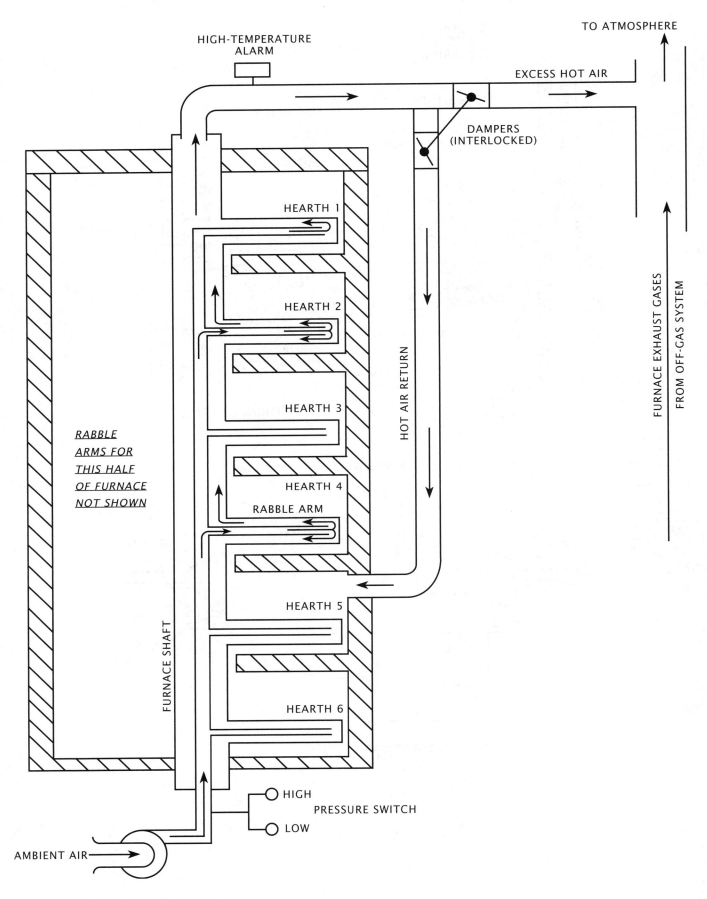

Fig. 3.32   Shaft cooling air system for a six-hearth furnace

3.5312 SHAFT DRIVE

The center shaft drive mechanism on an MHF is usually a combination of an electric variable-speed drive and an independent gear reducer. Occasionally, hydraulic drives are used instead of electric variable-speed drives. Connected to the output shaft of the gear reducer is a pinion gear that drives the large bevel or bell gear attached to the bottom of the furnace shaft.

3.5313 TOP AND LOWER BEARINGS

Each MHF manufacturer uses its own bearing design for the top and lower shaft bearings. The operational principles are the same for all manufacturers.

The lower bearing supports the entire weight of the center shaft. In a larger furnace this could be 60,000 pounds (27,240 kg) or more. The top bearing maintains shaft alignment. The shaft rotates within the bearing and the bearing housing maintains alignment.

QUESTIONS

Write your answers in a notebook and then compare your answers with those on page 322.

3.53A   What is sludge incineration?

3.53B   What is the meaning of the term "refractory" as it relates to a multiple-hearth furnace?

3.53C   What are rabble arms and rabble teeth?

3.53D   The lute cap serves what purpose?

3.53E   What is the purpose of the sand seal?

3.5314 FURNACE OFF-GAS SYSTEM
   (Figure 3.33, page 278)

As the sludge cake is incinerated, hot air and gases must be vented from the MHF. To vent these gases, almost all incinerators are equipped with an emergency bypass damper located on the top of the MHF. The function of the bypass damper is to vent the furnace gases to the atmosphere during emergency conditions. This device protects equipment and operating personnel.

CYCLONE SEPARATOR. Under normal operation of the MHF, the furnace gases are vented into the off-gas system. As the gases leave the furnace, the first unit that the gases may enter is the cyclone separator.

Hot furnace gases, which have fly ash and solid particles in suspension, are drawn through the furnace into the cyclone by the induction draft (ID) fan. The cyclone is constructed so that the gas flow sets up a separating current. This current causes the fly ash and heavy particles to settle out into the cyclone bin at the bottom. The cyclone bin has a flap gate that dumps the ash into the recycle screw. The recycle screw returns the fly ash and heavy particles to the furnace on a middle hearth. This material is then carried out with the ash. A vibrator assists in keeping the fly ash and particles moving downward in the cyclone bin. The hot gas and finer particles are drawn up out of the cyclone and move on to the precooler.

PRECOOLER. The precooler is a section of furnace exhaust ducting in which water is sprayed to cool the furnace exhaust gases to saturation temperature and to wet the small particles of light ash (particulate matter). The precooler lowers the temperature of the exhaust enough to prevent damage to the rest of the off-gas system components.

VENTURI SCRUBBER. Immediately below the precooler is a constant- or variable-throat Venturi scrubber. The Venturi scrubber cleans the particulate matter from the cooled furnace gases. Water is sprayed into the top of the Venturi for even distribution. The water and gases are mixed and accelerated in an adjustable, narrow Venturi throat. As the gases re-expand in the exit portion of the Venturi, the water is split into tiny droplets that trap the particulate matter and remove it from the gas stream.

The ducting directly below the Venturi scrubber usually makes a sharp bend. Because of the high air speed, the water droplets with their collected particulate matter cannot make the turn and instead run into the bottom of the ducting. From there, the water and particulate matter flow to a drain. Following the Venturi scrubber there is usually an impingement scrubber.

IMPINGEMENT SCRUBBER (Figure 3.34). The impingement scrubber consists of three sections:

1. The lower sprays

2. The impingement baffles

3. The mist eliminator

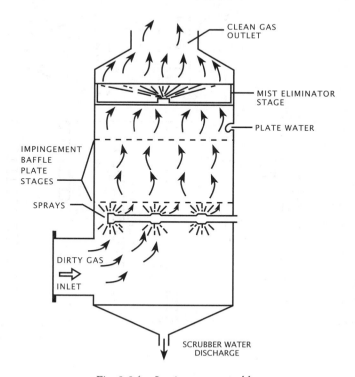

Fig. 3.34   Impingement scrubber

The impingement baffles are level, flat, stainless-steel plates with thousands of tiny holes in them. Water flows continuously across the plates. As the gas is drawn up through the perforated

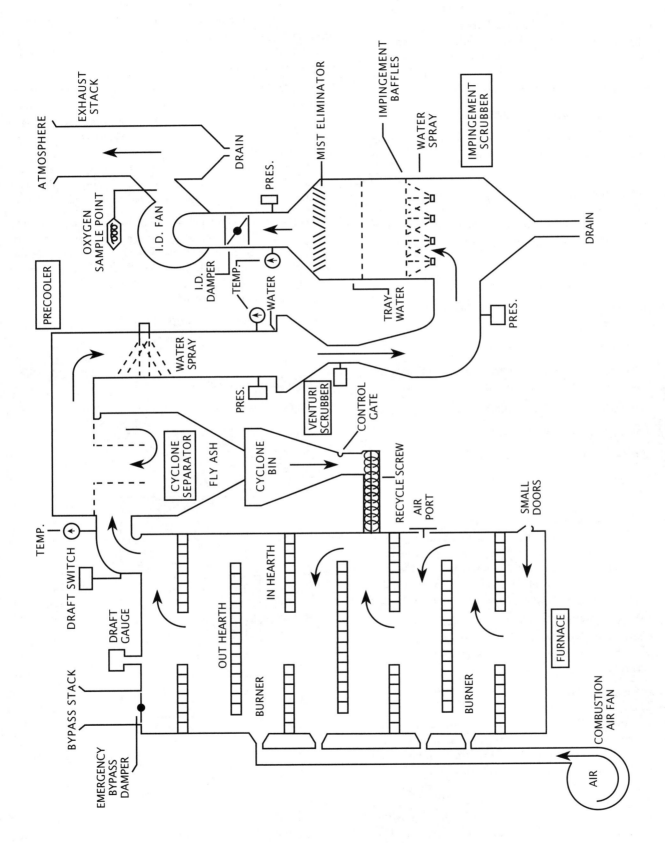

Fig. 3.33   Furnace off-gas system

impingement baffles, it collects water droplets while leaving any remaining particles trapped in the flowing water due to impingement action of the bars just above the perforations. The particle and water slurry overflows the baffles, is collected in the bottom of the scrubber, and is drained out.

The gas, carrying water droplets, is drawn up into the mist eliminator.

A fixed-bladed mist eliminator directs the gas stream to the side of the scrubber shell where the droplets collect by centrifugal action. The collected droplets drain back down to the impingement baffle section for reuse. The remaining cool, clean gas is drawn out the top of the scrubber by the induced draft (ID) fan.

INDUCED DRAFT (ID) FAN AND DAMPER. The ID fan provides the suction or draft necessary to vent the furnace gases and pull them through the off-gas system. Since the quantity of these gases varies with the quantity and type of cake burned in the MHF, a damper is installed immediately before the ID fan. This damper is used to regulate the suction or draft within the MHF.

ASH HANDLING SYSTEM. Ash is the inorganic material left after the sludge cake is burned. This material is disposed of in a variety of methods depending on the MHF installation. The two most common are wet and dry ash systems, which we will discuss briefly. Other types of ash systems include pneumatic ash transport, ash classification, and ash eductors.

The wet ash system is the simplest of all ash handling systems. The ash drops out of the MHF into a mix tank where effluent water is continuously added. This produces an ash slurry that is pumped to an ash lagoon. The ash settles out in the lagoon and the water is returned to the front end for treatment. The ash is left to dry and is ultimately removed from the lagoon for disposal.

In the dry ash system, the ash drops from the MHF onto an ash screw conveyor. This screw conveyor carries the ash to a bucket elevator. The bucket elevator transports the ash to a storage bin where it awaits disposal.

At the bottom of the storage bin is the ash conditioner. This is a screw conveyor equipped with a series of water sprays. The water sprays wet the ash so it does not create dust or blow off a truck during transport.

3.5315 BURNER SYSTEM

Burners are provided on an MHF to supply the necessary heat to ignite the sludge. Prior to discussing the burner system, it is important to understand combustion. In order for any combustion to occur, the three ingredients of the fire triangle (Figure 3.35) must be present.

For complete combustion to occur, there must be a specific ratio between the amount of fuel and the amount of oxygen. The burner system described here is manufactured by the North American Manufacturing Company. This type of system is common to all burner systems supplied on MHFs although component names may be different.

COMBUSTION AIR FAN. This fan supplies the filtered burner/combustion air for the burner system.

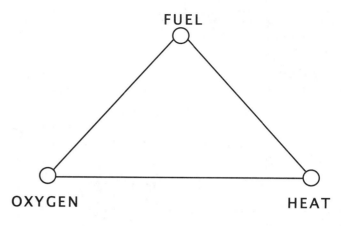

Fig. 3.35   Fire triangle

OIL PUMPS. Positive displacement pumps supply oil to the burners (except for natural gas burners).

SAFETY VALVES. Electric solenoid valves are used to stop the fuel flow during a burner shutdown.

PRESSURE REGULATORS. Standard pressure regulators are used to control the fuel pressure.

AIR/FUEL RATIO REGULATOR. A pressure regulator that maintains a specific ratio between fuel and air.

COMBUSTION-AIR CONTROL VALVE. A butterfly valve that governs the flow of combustion air to the burner.

FLOW-LIMITING VALVE. A metering valve that allows a specific flow of combustion air to the burner.

FUEL. MHF burners are fired by a variety of fuels. Natural gas, number 2 fuel oil, and heavy oil are some examples. The fuel and air are regulated to the proper pressure and injected into the burner. At this point the fuel/air mixture is ignited by the pilot and the resulting flame is sensed by the ultraviolet scanner, which signals the burner control station that a flame-safe condition exists.

The temperature and firing rate of the burner are then controlled by the temperature indicator controller (TIC).

3.532   Controls and Instrumentation

Usually, multiple-hearth furnaces are equipped with the following controls to maintain temperature, draft, and oxygen.

LOW DRAFT SWITCH. This switch shuts down the MHF in the event of an unsafe draft condition.

DRAFT CONTROLLER/INDICATOR. This is a controller that opens and closes the induced draft damper in order to maintain the draft within the MHF.

OXYGEN ANALYZER/CONTROLLER. This instrument measures the percent oxygen in the stack gas, which is an indication of complete combustion. In some cases, a controller is attached to this instrument to control the oxygen level in the furnace by regulating the combustion air.

TEMPERATURE INDICATOR CONTROLLER (TIC). This instrument controls the burner firing rate and the hearth temperature.

TEMPERATURE RECORDING CHART. A strip recorder used to record the temperatures throughout the furnace.

SCRUBBER DIFFERENTIAL-PRESSURE INDICATOR. An instrument that indicates the pressure difference across the scrubber. This pressure difference is the main operating variable on the Venturi scrubbers.

QUESTIONS

Write your answers in a notebook and then compare your answers with those on page 322.

3.53F   List the parts of the furnace off-gas system and the purpose of each part.

3.53G   Why do multiple-hearth furnaces contain burners?

3.53H   What three ingredients are necessary for combustion to occur?

3.533   MHF Operations

Operation of an MHF requires a person knowledgeable enough in furnace theory and operations to keep the fire burning in the desired location and prevent damage to the equipment. More importantly, a furnace operator must be able to look at the instruments, fire, and feed and be able to predict what is going to happen. The operator makes the necessary changes to maintain a stable burn and the most efficient burn possible. The operator's main objective is to operate the furnace at design conditions while keeping operating costs to an absolute minimum.

3.5330   FURNACE ZONES

The furnace is generally considered to be separated into three functional zones: drying, combustion, and cooling. None of these zones are confined to any specific hearth or hearths, but will always operate in this order. The area of each zone is determined by actual conditions in the furnace.

The furnace zones are as follows:

1. *THE DRYING ZONE.* In this area, generally the top one-fourth of the furnace, the sludge is exposed to high temperatures while being continuously turned over by the rabble teeth. The constant turning over of the sludge exposes more surface area to the high-temperature gases flowing over the cake surface and increases the rate at which moisture is driven out of the wet sludge. The wetter the cake entering the furnace and the greater the feed rate, the more hearths will be in the drying zone.

2. *THE COMBUSTION ZONE.* Ideally, this zone is where the actual burning of the volatile materials in the sludge takes place. Usually, the combustion zone will be confined to only one hearth. Ideally, the actual burning should occur approximately at the midpoint of an out hearth (Figure 3.36). At

this location, all gases given off in the final stages of the drying process, just prior to ignition, will be destroyed. This happens because the gases must pass through the flame, with very high temperatures, as the flame goes through the drop hole around the shaft.

3. *THE COOLING ZONE OR AIR PREHEAT ZONE* is where the ash is cooled. Any remaining carbon in the sludge is burned off here before the ash falls into the ash hopper. At the same time as the ash passes down the furnace, the air admitted through the air ports, slide doors, or shaft return-air duct is flowing over the hot ash and being preheated.

3.5331   AUXILIARY FUEL

The amount of fuel used will depend on several factors:

1. Conditions in the furnace

    Items such as shaft speed, number of slide doors, air ports open, or forced air ducted to the furnace, and the amount of shaft cooling air being returned to the furnace influence conditions in the furnace.

    For proper combustion, air should be added low in the furnace. This allows the cool ambient air to pick up a great amount of heat as it passes across the hot ash and also cools the ash before disposal.

    Air that is added in or above the fire zone will cool the heated air already in the furnace. This will cause more fuel consumption. Air should be added at or above the fire only when absolutely necessary to reduce the temperature or quench the fire.

2. The moisture content of the sludge feed to the furnace

    The drier the cake being fed to the furnace, the less fuel will be required to dry the sludge to maintain a good burn. Ideally, the moisture content of the furnace feed should not exceed 75 percent. Every bit of moisture entering the furnace requires a great deal of fuel to be consumed in the process of evaporation.

3. The volatile content of the solids in the sludge feed

    The higher the percent volatile material, the less fuel will be required, assuming, of course, a reasonable percent solids in the sludge.

4. Cake feed rate

    A constant cake feed rate coupled with the items mentioned above helps to reduce the quantity of fuel required.

3.5332   AIR FLOW

The draft or vacuum in the furnace is the direct cause of all air flow within the MHF. The draft within the furnace is caused by the induced draft fan and by the convection flow caused by the temperature differentials between the interior of the furnace and the atmosphere. Convection flows also explain why there is a draft within the MHF when the induced draft fan is off and the bypass damper is open.

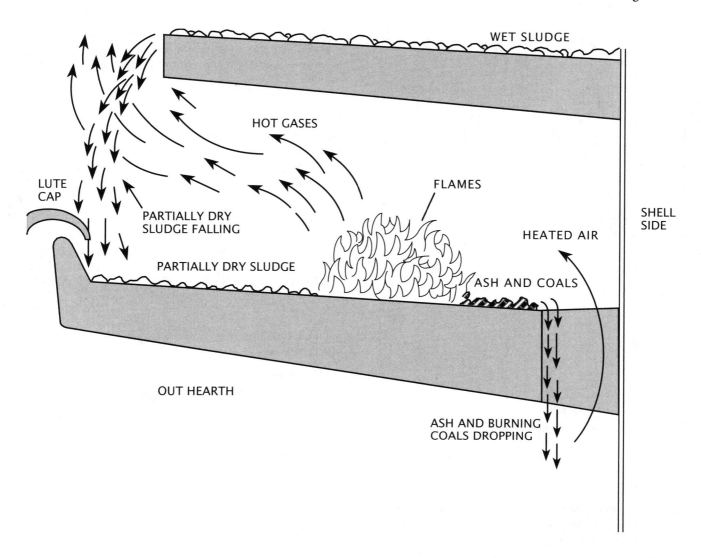

Fig. 3.36 Flames in the middle of an out hearth

Convection air flow develops in the MHF as a result of air being heated by the burners in the MHF (Figure 3.37). The air flow within an MHF is from the bottom to the top. The flow occurs because as the air in the MHF is heated and rises, it creates a vacuum in the bottom of the furnace. This vacuum causes cooler, outside air to flow into the bottom of the furnace where the air is heated and rises. The term used to describe this process is called "convection flow" and "draft" is the measurement of the negative pressure or vacuum created by this flow.

The draft in the furnace is measured either in tenths of inches or millimeters of water column. Relating this height of column to pounds per square inch or kilograms per square centimeter, you must remember that 1 foot (304.8 mm) of water column is equal to 0.434 psi (0.0305 kg/sq cm). Therefore, one inch (25.4 mm) of water column equals 0.0378 psi (0.00266 kg/sq cm)

and one-tenth of one inch (2.54 mm) water column equals 0.0038 psi (0.00027 kg/sq cm), a very small pressure.

We want only enough air in the furnace to allow for the complete combustion of the volatile matter in the sludge cake. Therefore, we must control the draft within the furnace carefully. The normal range of draft within the MHF is from 0.05 to 0.2 inch (1.3 to 5.1 mm) of water.

In operating a multiple-hearth furnace, it is extremely important to remember that an excess amount of air must be available at all times. This excess air ensures that all volatiles (combustibles) can contact sufficient oxygen to ensure complete combustion. If there is inadequate oxygen present, there will be smoke, which means incomplete combustion. Smoke indicates unburned hydrocarbons. While remembering that the human

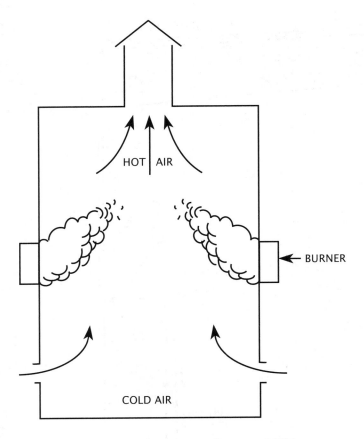

Fig. 3.37   Convection air flow in an MHF

eye cannot detect the ultraviolet spectrum of the flames, a visual inspection of the burning zone can give a good indication of the amount of excess air present.

1. A darker flame, ending in a curlicue of smoke, with a dull smokey atmosphere means there is a lack of air (volatile hydrocarbons are carbonizing).

2. A bright, sharp flame indicates excess air, but not how much. You can decrease the amount of excess air (which will decrease the amount of fuel being burned) as long as the stack plume does not become smoky, black, light blue, or brownish.

3. Short blue flames on the lower hearths indicate burning of fixed carbon, which means there is adequate excess air.

There are two ways of getting excess air into the furnace. These are:

1. By evenly opening the smaller doors and possibly the big doors a little on the bottom hearth, or opening the dampers of the air ducts, the incoming air rushes over the hot ash and is preheated while the ash is being cooled.

2. Opening the center shaft cooling air return damper allows the hot air to be returned, if the furnace is so equipped.

Since air is approximately 21 percent oxygen, an oxygen content of about 8 to 12 percent in the furnace gas indicates that adequate excess air has been added to ensure complete combustion.

The automatic oxygen analyzer is the operator's best tool for measuring the excess air in the furnace. This device draws samples from the exhaust after the induced draft (ID) fan, analyzes the amount of oxygen remaining after combustion, and sends the reading to the control panel. Fluctuation in the excess oxygen level is an indication of change in the furnace. A rapidly decreasing excess oxygen usually means that the fire is growing and is removing the excess oxygen. This is a common occurrence during a burnout.

A rising excess oxygen reading can mean a reduction in fire. By adding air above the fire, a false indication of excess oxygen will be given. Since it is a false reading, a visual inspection of the burn for smoke should be made until the air above the fire can be removed.

General rules for making excess oxygen changes are summarized as follows:

O_2 Change	Cause
Increase O_2	Decreasing fire and/or increasing air above fire zone.
Decrease O_2	Increasing fire and temperatures.

The oxygen demand of the furnace changes with the amount of combustion going on inside the furnace. Too much cool air is a waste of fuel and too little air causes smoke. Control must be maintained to provide the proper amount of air to meet the demands of the furnace.

3.5333   COMBUSTION

As previously explained, combustion is a chemical reaction that requires oxygen, fuel, and heat. In the furnace, air provides oxygen, the primary fuel is sludge, and the heat comes from the burning sludge. Initially, fuel of various kinds (natural gas, number 2 fuel oil, heavy oil, among others) is burned as an auxiliary fuel in the burners to help provide the heat needed to burn the sludge and to preheat the furnace to combustion temperatures.

3.5334   AIR FLOW AND EVAPORATION

Above the fire the furnace has wet, cold sludge in it. As the hot, dry air and combustion gases pass over the upper hearths, the heat is transferred to the sludge. At the same time the moisture in the sludge is evaporated, the dry gases pick up the moisture and carry it out of the furnace.

Return shaft cooling air (Figure 3.38) is the usable by-product of the center shaft cooling air. The main shaft is double-walled

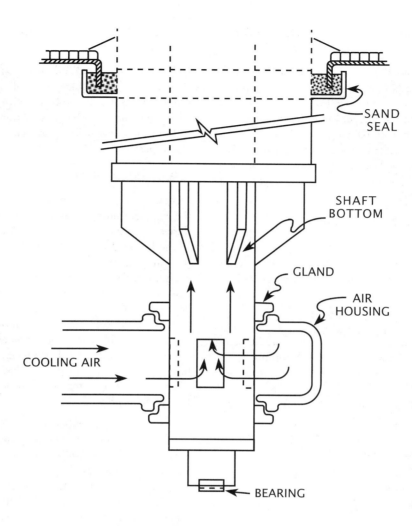

Fig. 3.38   Shaft cooling air return system

and cast in sections. The sections have a tubular inner duct called a cold air tube. The annular (ring shaped) space between the inner tube and the outer shaft wall serves as a passageway for hot air and is referred to as the hot air compartment. The central shaft and rabble arms are cooled by air supplied at a fixed quantity and pressure from a blower that discharges air through a housing into the bottom of the shaft.

Two or more rabble arms are held in shaft sockets above each hearth where the cold air tube as well as the outer shaft wall provide support. Each rabble arm has a central tube that conducts the cold air from the cold air tube to the extreme end of the rabble arm (Figures 3.30 and 3.32). From there, the air goes through the outer air space in the arm, back toward the shaft, and through the openings into the hot air compartments. Unused heated cooling air may be returned to the atmosphere by other means depending on the design.

The hot air compartment may be vented to the atmosphere or the hot shaft air may be returned to the furnace for combustion purposes. Shaft return air is sent below the normal burn zone but still above the final ash cooling hearth.

> THE SHAFT AIR TEMPERATURE SHOULD NOT EXCEED A MAXIMUM TEMPERATURE OF 550°F (290°C)

When the preheated shaft cooling air is not required for furnace operation, it is vented to the induced draft stack. This hot air will prevent steam plume formation. As this heated air exits the furnace at the top of the center shaft, the operator may direct its path. By means of two mechanically linked dampers (called proportional dampers), the operator at the main control panel directs the heated air out into the atmosphere or back to the furnace as hot air return.

Hot air return should be used with some discretion. The hot air return can: (1) provide a fast source of air (oxygen) within the furnace, (2) use less fuel to heat the air, (3) reduce smoking, or (4) increase the drying rate of sludge. Returning too great a volume has the disadvantages of blowing fly ash as well as dumping the return air into the furnace at one point rather than distributing it evenly around the hearths. Remember that as the air passes over the hot coals, some of the oxygen is being removed before it gets to the fire.

3.5335   RECOMMENDED FURNACE OPERATING RANGES

Table 3.34 summarizes the general temperature and pressure ranges maintained on the various hearths when burning wastewater sludge.

TABLE 3.34   MHF OPERATING RANGES

Location	Range	Optimum
Hearth #1 (Gas Exit)	700 to 1,000°F[a]	As low as possible
Burning Hearth	1,300 to 1,700°F[a]	1,600°F[a]
Bottom Hearth	200 to 800°F[a]	200°F[a] or As low as possible
Scrubber Inlet	100 to 300°F[a]	150°F[a]
Furnace Draft	0.05 to 0.2 inch[b] of water	0.1 inch[b] of water
Furnace Oxygen	8 to 12 percent	10 percent

[a]   (°F − 32) 5/9 = °C.
[a]   Inches × 25.4 = millimeters.

3.5336   ALARM SYSTEMS

Almost all MHFs are equipped with an alarm system because of the speed at which the system will react to changes. The alarm system informs the operator of abnormal conditions. The operator should react to an alarm as follows:

1. Understand what malfunction is causing the alarm before pressing the acknowledge button. Many times a group of alarms will go off at one time. Find the alarm that is downstream from the rest of the alarms (for example, the ash bin is downstream from the center shaft).

2. Press the reset button to see if the alarm conditions still exist; if not, proceed with Step 3.

3. Try to restart the equipment that caused the alarm at the location of the equipment rather than from an alarm panel. *Do not force it to run.* If it starts, restart all equipment that was shut down by the alarm. Press reset button.

4. If it does not start, make a brief visual inspection of the malfunctioning equipment.

5. Make a decision as to whether or not you can safely correct the problem.

6. Return to the panel and furnace and check the conditions. If you can fix the problem, you still must control the furnace. If you cannot fix the problem, burn out the remaining sludge or maintain as stable a condition as possible. If you must burn out the remaining sludge, try to control the burnout temperature.

3.5337   BURNOUTS

A burnout occurs when the sludge feed has been stopped and the fire continues to burn. Eventually, the final quantity of sludge is dry enough to burn and does. This final rapid burning of the last sludge can cause high temperatures in the furnace that can potentially cause damage to the furnace and its related equipment. Operators must control the burnout temperatures. The ideal burnout is to have only a 100 degree Fahrenheit (55 degree Celsius) increase in any hearth temperature during a burnout. This requires concentration by the operator on the furnace's changing conditions: fire position, sludge remaining in the furnace, shaft speed, temperatures, excess oxygen, and burner settings. By anticipating what the fire will do long before it happens, the operator can make adjustments that control the fire's final burnout.

The operator may choose to increase the shaft speed in preparation for the burnout. The purpose is to move the last sludge lower in the furnace before it burns. This results in keeping the high temperatures away from the top of the furnace. The operator must be careful not to allow unburned sludge to pass through the furnace. The high shaft speed will also generate a higher furnace temperature because of the increased rate at which fresh fuel is available to the fire.

There are three basic adjustments that control the fire during a burnout:

1. Opening doors at or just above the fire. By opening the doors, a large volume of cold air enters the furnace. This replaces the heat needed for combustion and cools the fire. The draft must be maintained (0.05 to 0.2 inch or 1.3 to 5.1 mm water column) by opening the induced draft damper. To prevent an induced draft fan overload, air ports and doors below the fire and the hot air return can be closed to reduce the total air flow. The air flow removed from underneath can now be added directly at the fire for combustion and also above the fire where it will have a cooling effect.

Cooling air will only provide short-term control. The fire will eventually generate more heat than can be offset by cooling air. Another control must be used *quickly.*

2. Stopping and restarting the center shaft or changing the shaft speed. Stopping the center shaft will remove the fresh fuel for the fire, which will cause the fire to die down. By starting and stopping the shaft, the operator controls the amount of fuel available for the fire to burn. The shaft should be stopped no longer than three minutes. When it is started again, a low shaft speed (0.4 to 0.7 RPM) will help control the fire by

turning over the fresh sludge at a slower rate. When the temperature rises again, the shaft is stopped again. Continue this procedure until the fire does not cause a high temperature jump. Once the temperature starts to drop with the shaft running, close up the furnace doors to control the temperature change. Once the fire has burned out completely, switch from induced draft to natural draft to control the temperature.

> *DO NOT ASSUME THE CAKE HAS BURNED OUT BY LOOKING AT THE TEMPERATURES SHOWN IN THE TEMPERATURE RECORDER. OPEN THE DOOR ON EACH HEARTH AND LOOK IN CAREFULLY.*

3. Reducing or shutting off the burners. During a burnout, the sludge needs little or no heat from the burners. Once the fire starts to increase the temperatures, reduce and then shut off the burners. This will provide two advantages:

   a. Temperatures will be reduced.

   b. Gases from the burner will be removed. This missing gas volume can be replaced by cooling air above the fire.

   Once the temperature starts to fall, the burners can be relit to control the dropping temperatures.

QUESTIONS

Write your answers in a notebook and then compare your answers with those on pages 322 and 323.

3.53I List the three distinct zones in a furnace.

3.53J List the factors that influence the amount of fuel required.

3.53K What three factors are essential for combustion and what is the source of each factor in a furnace?

3.53L What is a burnout?

3.53M What steps should an operator take to control a burnout?

3.534 *General Operational Procedures (Start-Up, Normal Operation, and Shutdown)*

MHF operational procedures are based on consideration of the interrelationships between air flow, shaft speed, and temperature.

The first step in starting an MHF is to set the water flows to the scrubbers and then start the induction draft (ID) fan. Once the ID fan is at the operational speed, set the ID damper to maintain the optimum draft (0.1 inch or 2.54 mm water column). At this point, begin the purge cycle (removing unwanted gases from the furnace). The purge cycle is usually controlled by a timer and will last from 3 to 10 minutes. At the end of the purge cycle (usually indicated by a panel light), start the shaft cooling air fan, the combustion air fan, the ash system, and the center shaft. You are now ready to start heating up the MHF.

> **IMPORTANT**
> *DUE TO THE NATURAL CHARACTERISTICS OF REFRACTORIES (HEARTHS), THE MHF MUST BE HEATED UP AT AN EXTREMELY SLOW RATE. THEREFORE, THE FOLLOWING DRY-OUT MUST BE FOLLOWED EXACTLY.*

The first step in the warm-up of an MHF is to light the pilot lights on the bottom hearth. The temperature is then brought up to 200°F (93°C). Do not allow the temperature to exceed 200°F (93°C) at any point in the furnace. Hold the temperature at 200°F (93°C) until the refractory is dry and warm. To check this, open the door of the MHF and observe the refractory where it meets the steel shell. The refractory and the shell must be dry. If it is dry, carefully feel the refractory where it joins the shell. The refractory at this point should be warm to the touch. The time required for the dry-out/warm-up varies depending on how long the furnace has been out of service. If the temperature of the hearth never dropped below 200°F (93°C), the dry-out/warm-up will not be required, but if the MHF has been off line for an extended period of time, the dry-out/warm-up may take from several days to a week. If there is any question as to whether the furnace is dry or warm, let the furnace warm up for an additional period of time. Remember, this is one of the most critical periods in MHF operation.

Once the refractory is warm, you may proceed to heat up the unit at a rate of 50°F/hr (28°C/hr). This rate of temperature increase must be maintained carefully, even at the expense of adding air for cooling.

Once the temperature on a given hearth reaches 1,000°F (540°C) that hearth's temperature may be raised at a rate of 100°F/hr (56°C/hr). This procedure is followed until the burning zone of the furnace reaches 1,600°F (870°C) at which point the feed to the furnace may be started.

NOTE: The same rate of temperature losses applies when the furnace is taken out of service. Drop the temperature at a rate of 100°F/hr (56°C/hr) to 1,000°F (540°C) and then lower the temperature at 50°F/hr (28°C/hr) when the temperature is below 1,000°F (540°C).

When feed is introduced to the furnace, the temperatures will initially drop. This is due to the cooling effect of the wet cake.

Once the cake reaches the burning zone, however, the cake should start to burn and the temperature profile will even out. Once this profile is established, the profile should be maintained within a 200°F (93°C) range by controlling the shaft speed, return air, and the burners.

Shaft speed controls the cake feed rate, which affects the location of the burning zone in the MHF. Setting the proper shaft speed will establish the sludge cake depth, which will control the drying rate of the cake. Because the cake cannot combust until it reaches a certain moisture content and temperature, the drying rate affects the location of the burning zone. If the shaft speed is too fast, the cake depth is reduced, causing the drying rate to increase and the cake burning to occur sooner—higher in the furnace. If the shaft speed is too slow, the cake depth is increased, causing the drying rate to decrease and the cake burning to occur later—lower in the furnace, and incomplete cake burning can result.

The sludge cake has a very high heating value (approximately 10,000 BTU/lb volatile solids or 23,260 kilojoules/kg volatile solids). In many cases, the volatile content of the sludge cake is high enough that the cake will burn without the additional heat input from the burners. This condition is called an autogenous (aw-TAW-jen-us) burn. To achieve this condition, the sludge cake must generally exceed 25 percent total solids and 70 percent total volatile solids. To maintain an autogenous burn condition, a constant steady-state sludge feed is mandatory. An autogenous burn represents the most economical mode of MHF operation.

Even when an autogenous burn cannot be established, fuel usage is affected by the heat released by the burning volatile material in the sludge cake. The MHF should be operated on a continuous basis when the unit is in operation to take advantage of the heat from the burning sludge. Remember that all the fuel used to maintain the temperature in a furnace at a standby mode represents money added to the total cost of solids disposal.

Another benefit of continuous operation is an increase in refractory life. As the MHF is cycled up and down in temperature, the refractory expands and contracts with the temperature changes. As this expansion and contraction occurs, the joints between the hearth bricks open and close. As the joints open, ash falls into the joint. When the MHF is later reheated, the surrounding brick expands and compresses the ash and a tremendous pressure is exerted on the brick. This process occurs repeatedly until the brick finally breaks.

Ultimately, there is a trade-off between fuel cost to maintain furnace temperatures and the cost of refractory repair. As a general rule, MHF operation is most economical when scheduled on as nearly continuous a basis as possible.

3.535 Common Operating Problems (Troubleshooting)

3.5350 SMOKE

The most common cause of smoke or air emissions from an MHF is low oxygen content. This means there is insufficient oxygen in the furnace to completely burn the hydrocarbons. The solution to this problem is to add air to the furnace.

Air may be added to the furnace through the doors, air ducts, or through the use of the shaft cooling air return. Just as important as the excess air is where you add the air. Air generally should be added at or below the fire.

NEVER ADD AIR ABOVE THE FIRE, UNDER NORMAL CONDITIONS.

3.5351 CLINKERING

Many times hard, rock-like clinkers will form within the furnace. If this situation is allowed to continue, the clinkers may grow to a point where the drop holes plug and the rabble teeth become blinded (plugged). The solution to clinkering lies in understanding how a clinker is formed.

A clinker is nothing more than melted ash that has cooled. The only temperature in the furnace usually high enough to melt the ash is the actual flame temperature of the burners. The flame temperature from the burning sludge is rarely high enough. Therefore, as a general rule, the solution to clinkering lies in distributing the burner input into the MHF. An example of this would be running burners on separate hearths at lower firing rates. If this does not correct the problem, it will be necessary to reduce the feed rate to the furnace and have the cake analyzed for mineral content and for excessive levels of polyelectrolyte.

3.5352 INABILITY TO STABILIZE BURN

If you cannot stabilize the burn, the first place to look for a problem is your feed cake. The MHF loading must be constant with little or no change in moisture or volatile content. If the feed is constant, then you may be making other process changes too quickly. The ultimate effect of any process change to a furnace will show up one hour after the change was made. Therefore, make one change at a time and wait for the results.

3.536 Safety

An MHF presents several safety hazards beyond those in the rest of the treatment plant. These hazards all revolve around the fact that the MHF uses high temperatures to destroy the solids. Therefore, *treat everything as if it were hot.*

Anytime you are in the furnace area, wear protective clothing including heavy leather gloves, face shield, hard hat, long-sleeved shirt, and long pants. Never wear synthetic fabrics. The heat from the furnace can cause synthetic fabrics to ignite and act like napalm on your skin. Wear clothing made of cotton.

Never look directly into a furnace door when the furnace is in operation. Always approach the door from an angle and look in at an angle.

A furnace, when out of operation, is a confined space. Treat it accordingly, checking the atmosphere before entering for oxygen deficiency/enrichment, combustible gases and explosive conditions, and toxic gases (hydrogen sulfide). *NEVER* enter a furnace alone. Follow all confined space entry procedures.

Always check the temperature of the ash bed before entry. Though the furnace may be cold, the ash bed temperature below the surface may still be several hundred degrees.

Always lock out and tag the main fuel control valve and the control power prior to furnace entry.

Check and verify the operation of all safety controls and interlocks on a regular basis.

QUESTIONS

Write your answers in a notebook and then compare your answers with those on page 323.

3.53N   At what rate of temperature increase do you bring a cold furnace up to temperature?

3.53O   What is an autogenous burn?

3.53P   Why is it desirable to operate an MHF on a continuous basis?

3.53Q   What is the cause of smoke from an MHF and how can this problem be corrected?

3.53R   What protective clothing should be worn when in the furnace area?

3.54   Facultative Sludge Storage Lagoons

Facultative sludge storage lagoons can serve three very important purposes:

1. Volume reduction. Volatile solids can be reduced by 45 percent and solids concentrations can be increased from two percent up to eight percent or more. Sludge concentrations as high as 25 percent solids have been obtained in the bottom layers of some lagoons.

2. Storage buffer. Storage is frequently required when sludge production is continuous and land disposal is affected by changing seasonal conditions.

3. Further stabilization. Anaerobically digested sludge is further stabilized in the storage lagoon by continued anaerobic biological activity.

Facultative sludge storage lagoons vary in depth from 10 to 16 feet (3 to 5 m). Surface areas are based on solids loadings of 20 to 50 pounds of volatile sludge solids per day per 1,000 square feet of surface area (0.1 to 0.25 kg VSS per day per square meter). Surface aerators are commonly used to maintain aerobic conditions near the surface in order to avoid odor problems.

There should be enough lagoons to allow each lagoon to be out of service for approximately six months. Stabilized and thickened sludge can be removed from the basins using a mud pump mounted on a floating platform.

QUESTION

Write your answer in a notebook and then compare your answer with the one on page 323.

3.54A   List three purposes of facultative sludge storage lagoons.

3.6   SOLIDS DISPOSAL
by William Anderson

3.60   Regulations Governing Disposal

The methods available for ultimate disposal or reuse of sludge-based products (biosolids) from wastewater treatment plants are dictated by federal, state, and local regulations that have been developed to protect the environment and public health. Complying with these regulations and securing necessary approvals and permits for sludge management projects has become increasingly challenging, and may be more difficult for treatment plants handling wastewater because of the potential for the sludge-based products to be classified as a hazardous material. A listing and discussion of all of the applicable regulations relative to sludge management practices is beyond the scope of this operator training manual. However, the planners and operators of wastewater treatment plants should be familiar with the general procedures that should be followed for implementing a sludge management project, and with the self-implementing EPA sewage sludge regulations discussed below. Some general sludge management procedures are:

Step 1 -   Analyze the sludge or otherwise predict the concentration of pollutants identified by the appropriate state health department and the Environmental Protection Agency (EPA) that could cause the sludge to be classified as a hazardous waste.

Step 2 -   If the concentrations of pollutants exceed established limits, the sludge will be classified as hazardous and can only be disposed of at an approved Class I landfill.

Step 3 -   If the concentrations of all pollutants are lower than the established limits, the sludge may be classified as nonhazardous and may fall under applicable regulations that govern the disposal or reuse of municipal treatment plant sludges.

Step 4 -   If the sludge is classified as nonhazardous by the federal, state, and local regulatory authorities, then it should be analyzed and compared with the requirements of the EPA 40 CFR 257 or 503 regulations and the appropriate state health department guidelines. These regulatory requirements will establish the acceptable alternatives for disposing of sludges.

Step 5 -   Keep in contact with your state regulatory agency for current rules and regulations concerning the management and disposal of sludge from your treatment plant.

On November 25, 1992, EPA signed PART 503 FINAL RULE, "Standards for the Use or Disposal of Sewage Sludge." The compliance date for the Part 503 standards was February 19, 1994 for most facilities. The information in the remainder of this section was obtained from EPA's *SEWAGE SLUDGE USE AND DISPOSAL REGULATION (40 CFR PART 503)— FACT SHEET* dated November 1992. To obtain details of the FINAL RULE, contact your state regulatory agency or your regional EPA office.

The sewage sludge use and disposal rule (40 CFR Part 503) sets national standards for pathogens and 10 heavy metals in domestic sewage sludge.[39] It also defines standards (or management practices) for the safe handling and use of domestic sewage sludge. This rule is designed to protect human health and the environment when sewage sludge is beneficially applied to the land, placed in a surface disposal site, or incinerated. These regulations do not apply to sludge produced from strictly industrial sources but do apply to mixed domestic and industrial sludges. (Refer to 40 CFR Parts 257 and 258 for regulations affecting strictly industrial sludge.)

The scientific research used to develop the 503 rule shows that most sewage sludge can be safely and beneficially used in a wide variety of ways. It can be applied safely to agricultural land, lawns and gardens, golf courses, forests, and parks, and is a valuable resource for land reclamation projects. This rule is designed to protect human health and the environment with a margin of safety equal to any of the unregulated use or disposal practices. It sets standards for pathogens and limits for 12 pollutants that have the potential for adverse effects, and explains why limits are not needed for 61 other pollutants that were considered. Additionally, it contains a comprehensive set of management practices to ensure that sewage sludge is beneficially used or disposed of properly. Table 3.35 summarizes the numerical limits established by the Part 503 Sewage Sludge Regulations.

TABLE 3.35   40 CFR PART 503 SEWAGE SLUDGE RULE LIMITS SUMMARY

	Land Application				Surface Disposal		Incineration
	Ceiling, mg/kg	Cumulative Load, kg/ha	APL[a] (clean sludge), mg/kg	APLR,[b] kg/ha/yr	Unlined, mg/kg	Lined, mg/kg	
Arsenic	75	41	41	2.0	73	—	c
Beryllium	—	—	—	—	—	—	10 gm/24 hr
Cadmium	85	39	39	1.9	—	—	c
Chromium	3,000	3,000	1,200	150	600	—	c
Copper	4,300	1,500	1,500	75	—	—	—
Lead	840	300	300	15	—	—	c
Mercury	57	17	17	0.85	—	—	3,200 gm/24 hr
Molybdenum	75	—	—	—	—	—	—
Nickel	420	420	420	21	420	—	c
Selenium	100	100	36	5.0	—	—	—
Zinc	7,500	2,800	2,800	140	—	—	—
Total Hydrocarbons	—	—	—	—	—	—	100 ppm

[a] APL.   Monthly Average Pollutant Load.
[b] APLR.   Annual Pollutant Loading Rate.
[c] Allowable concentration in sewage sludge is determined site-specifically for each incinerator through performing incinerator testing and emissions dispersion modeling.

[39] The Water Environment Federation (WEF) refers to sludge at wastewater treatment plants as "wastewater solids." After these solids receive further treatment and meet US federal criteria for beneficial use, the solids are called "biosolids."

To convert the information in Table 3.35 from metric units to English units, use the following procedures:

1. Milligrams per kilogram (mg/kg) is the same as parts per million or pounds per million pounds. Therefore, a cadmium limit of 39 mg/kg means 39 milligrams of cadmium per kilogram of dry sludge solids or 39 pounds of cadmium per million pounds of dry sludge solids.

$$\text{Limit,} \frac{\text{lbs Cadmium}}{\text{ton Dry Solids}} = \frac{39 \text{ mg Cadmium}}{1 \text{ kg Dry Solids}}$$

$$= \frac{39 \text{ lbs Cadmium}}{1,000,000 \text{ lbs Dry Sludge Solids}}$$

$$= \frac{(39 \text{ lbs Cadmium})(2,000 \text{ lbs/ton})}{1,000,000 \text{ lbs Dry Sludge Solids}}$$

$$= 0.078 \text{ lb Cadmium/ton Dry Solids}$$

2. Kilograms per hectare per year. This value is often converted to tons per acre per year. Therefore a cadmium Annual Pollutant Loading Rate of 1.9 kg/ha/yr would be converted to pounds or tons per acre per year by the following calculation.

$$\text{Cadmium APLR, lbs/acre/yr} = 1.9 \text{ kg/ha/yr}$$

$$= (1.9 \text{ kg/ha/yr})(2.2 \text{ lbs/kg})(2.47 \text{ ha/ac})$$

$$= 10.3 \text{ lbs Cadmium/acre/yr}$$

$$\text{Cadmium APLR, tons/acre/yr} = \frac{10.3 \text{ lbs Cadmium/acre/yr}}{2,000 \text{ lbs/ton}}$$

$$= 0.0052 \text{ ton Cadmium/acre/yr}$$

The goal of all regulations governing the ultimate disposal of sludge or reuse of biosolids is the protection of public health and the environment. Major areas of concern are the emissions to the atmosphere from furnaces, groundwater and surface water contamination, and the potential health consequences of applying biosolids to land used for food production.

Sludge may be disposed of in a sanitary landfill at the treatment plant site or off site. Under these conditions, surface runoff must be prevented. Also, percolation of leachate to groundwater must be carefully controlled or eliminated.

At dedicated land disposal (DLD) sites, stabilized sludge is applied to the land and then ploughed under. At sites of this type, sludge must be covered the same day it is applied. Public access must be avoided because pathogens or parasites may not have been removed.

Beneficial reuse of biosolids for agricultural purposes is encouraged, and is safe when accomplished in accordance with federal and state guidelines. If biosolids are to be reused for agricultural purposes in areas where groundwater contamination is a concern, monitoring wells should be installed. Where groundwater is not a problem, *LYSIMETERS*[40] placed at several intervals can indicate if nitrate is moving through the soil. Use of biosolids on agricultural land requires close monitoring of nitrogen and heavy metals; cadmium is one metal of concern. Monitoring

for toxic substances and pathogens must also be conducted. (See Section 3.62 for further information about monitoring requirements for sludge disposal operations.)

QUESTIONS

Write your answers in a notebook and then compare your answers with those on page 323.

3.60A What are the major concerns regarding the disposal of sludge?

3.60B Under what conditions may sludges or biosolids be safely applied to food crops?

3.61 Disposal Options

The alternatives for land disposal or reuse of wastewater solids are shown in Figures 3.39, 3.40, and 3.41. The alternatives for sludge are placed in either of two categories based on the process used after stabilization (digestion or chemical stabilization): (1) dewatering to about 20 to 30 percent solids, or (2) concentration during liquid storage to 6 percent solids.

Disposal of sludge or reuse of biosolids following stabilization without additional treatment to reduce water content is to be avoided for the following reasons:

1. Water content (97 to 98 percent) of stabilized sludge is too high to permit landfill or composting operations.

2. Sludge application and surface runoff problems in the wet season are difficult to handle.

3. Land requirements necessary to evaporate the excessive moisture are unreasonable.

The regulations concerning the disposal of sludge usually refer to the following categories of disposal methods:

1. Land application

2. Distribution and marketing (D & M)

3. Monofilling (sludge-only landfills)

4. Surface disposal

5. Incineration

Land application is the disposal of domestic wastewater sludge on agricultural or nonagricultural lands. Regulations describe the requirements for stabilization of sludge to reduce the levels of pathogens and the attraction of vectors (an insect or rodent capable of spreading germs or other agents of disease). Limits are also placed on the concentrations of metals, especially cadmium, lead, and copper, in the sludge applied to land.

Distribution and marketing refers to treated sludge products sold to the general public. This category includes composted sludge and other products produced by high-temperature drying and inactivation processes. Rules are generally more restrictive than the land application regulations.

---

[40] *Lysimeter* (lie-SIM-uh-ter).   A device containing a mass of soil and designed to permit the measurement of water draining through the soil.

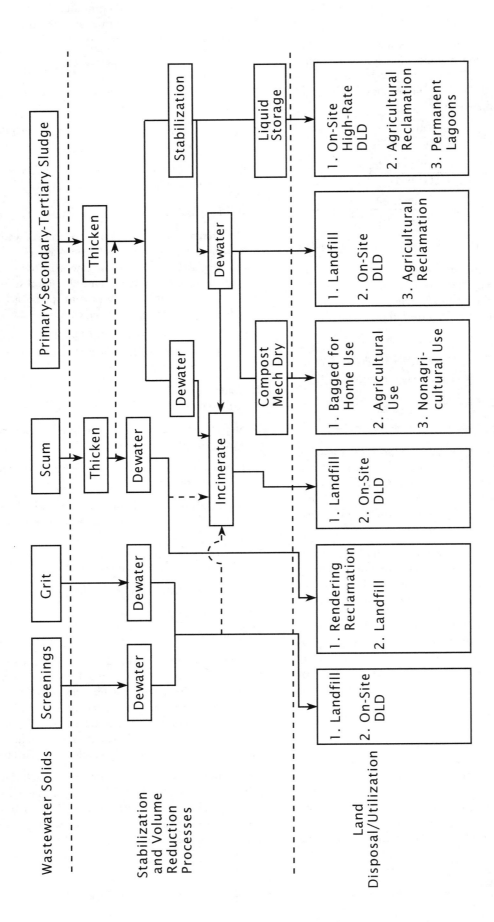

Fig. 3.39 Alternatives for land disposal or utilization

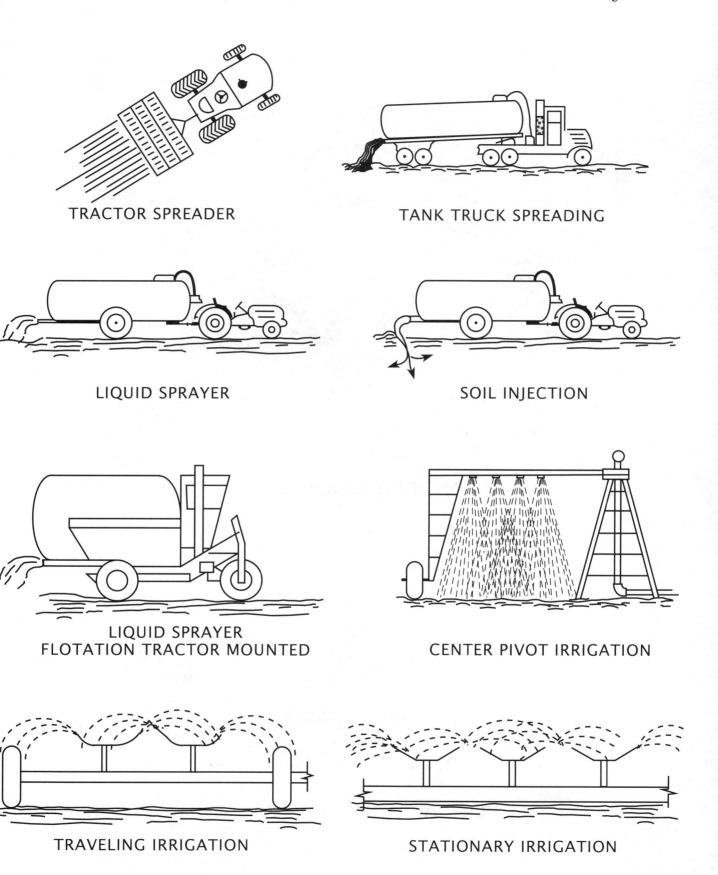

TRACTOR SPREADER

TANK TRUCK SPREADING

LIQUID SPRAYER

SOIL INJECTION

LIQUID SPRAYER
FLOTATION TRACTOR MOUNTED

CENTER PIVOT IRRIGATION

TRAVELING IRRIGATION

STATIONARY IRRIGATION

Fig. 3.40   Typical dewatered sludge application systems on land
(Adapted from "Sludge Processing and Disposal - LA/OMA Project")

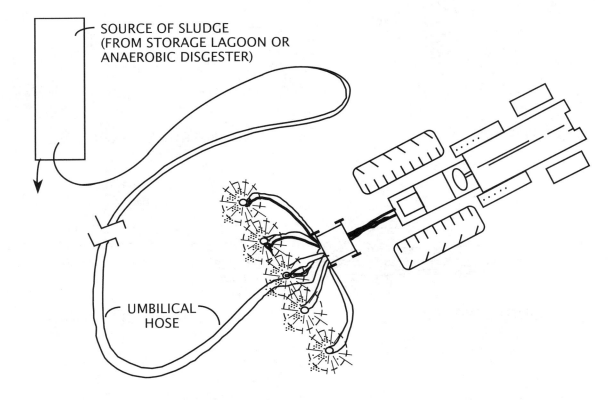

SOURCE OF SLUDGE
(FROM STORAGE LAGOON OR
ANAEROBIC DISGESTER)

UMBILICAL
HOSE

UMBILICAL CORD TRACTOR - SURFACE SPREADING

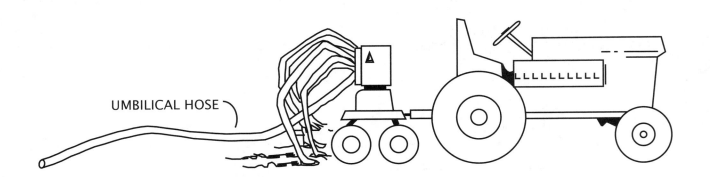

UMBILICAL HOSE

UMBILICAL CORD TRACTOR - SOIL INJECTION

Fig. 3.40   Typical dewatered sludge application systems on land (continued)
(Adapted from "Sludge Processing and Disposal - LA/OMA Project")

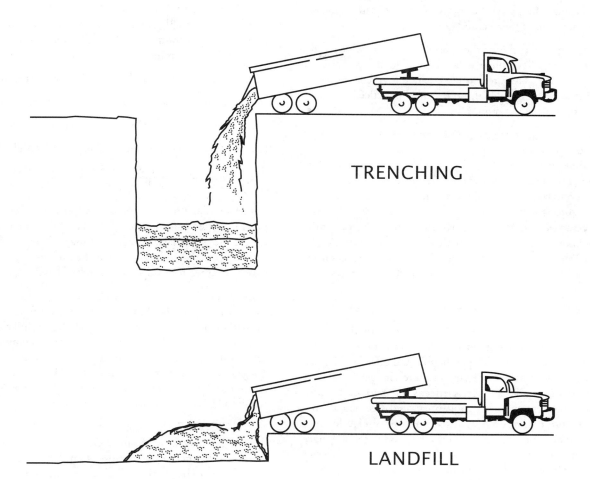

TRENCHING

LANDFILL

MANURE
SPREADER

SURFACE SPREADING

Fig. 3.41   Typical dewatered sludge application systems on land

Monofilling is the practice of burying sludge in a dedicated landfill. Regulations are very restrictive and may result in the ultimate elimination of this method.

Surface disposal includes those applications in which only wastewater sludge is applied to an area of land (see Sections 3.610 and 3.611). The sites do not include a vegetative cover. This is a disposal practice with no intent of beneficial use. The regulations are similar to monofills.

Incineration is the burning (combustion) of sludge at a high temperature in a furnace. Incineration reduces the volume of sludge while producing heat, dry inorganic ash, and gaseous emissions. The rules for incineration deal mainly with gas emissions. Of concern is the fate of metals and total hydrocarbons as they pass through the incineration process and are either destroyed, removed with the ash, or pass through the gaseous emission scrubbing system and exit with the exhaust gases. Scrubbing equipment removes particulates and undesirable gases from gaseous emissions by passing dirty air through water sprays.

3.610 Stabilized Sludge—Dewatered

STORAGE. Storage often must be provided in the sludge treatment system to accommodate differences between disposal rates and production rates. Depending on the ultimate disposal method, it may also be necessary to store sludge during the winter months. Sludge storage is effective as part of a liquid treatment system, such as anaerobic or aerobic digesters. If lime or ferric chloride have been used to condition sludge for mechanical dewatering, the sludge may be stored for a longer period than if polymers were used. Dewatered sludge often can be stored for long periods of time in open stockpiles.

TRANSPORTATION. The number of trucks required to haul the dewatered stabilized sludge cake to the disposal site must be determined. A common average value is 1,600 lbs of dewatered sludge per cubic yard as deposited in a truck. Route possibilities should have been evaluated in the environmental impact report for the project. Round-trip travel, loading, and unloading time need to be estimated. Operating hours for truck transport also must be considered. If the dewatering facility operates during times when transportation is not available, storage will have to be provided at the dewatering facility. Usually, it is not cost-effective to transport sludge cake to a landfill on Saturdays, Sundays, or holidays. However, if the dedicated land disposal (DLD) site is located at the treatment plant site, sludge or sludge cake may be deposited at any time. On-site dedicated land disposal operations use pumping or pneumatic ejection pipelines for transporting material to the disposal site. Agricultural reuse is always seasonal and trucking must always be available to respond to the needs of the farmer, not the operator of the treatment plant.

3.6100 SANITARY LANDFILL DISPOSAL

Sludge cake (20 to 30 percent solids—discharged from the dewatering machines (centrifuges, vacuum filters, or filter presses)) or cake (50 to 60 percent solids) removed from drying beds or drying lagoons (see Chapter 12, Section 12.7, "Digested Sludge Handling," *OPERATION OF WASTEWATER TREATMENT PLANTS*, Volume II) can be transported to a sanitary landfill for disposal with the municipal refuse. The impact on the capacity of the landfill is determined as shown by the following calculation:

Given: Rate of Municipal Refuse Deposit, D tons/day

Rate of Sludge Cake Production, P cu yd/day

Assume: A Compacted Density of 560 lbs/cu yd in the landfill

Find: The percent of increased usage of landfill, U.

$$\text{Usage, \%} = \frac{560 \text{ lbs/cu yd} \times P, \text{Sludge Cake, cu yd/day} \times 100\%}{2{,}000 \text{ lbs/ton} \times D, \text{Refuse Deposit, tons/day}}$$

The additional water associated with the stabilized sludge will provide better compaction in the landfill, which will slightly reduce the percentage of increased usage. The landfill operator must be able and willing to accept the sludge cake. A higher rate may be charged for sludge cake if special handling is needed.

LANDFILL MOISTURE ABSORPTION CAPACITY. The water balance of a landfill is important in minimizing leachate formation. The absorption capacity of the landfill should be investigated when adding materials with large amounts of water such as sludges. A general guideline is that no more than 25 to 40 gallons of water per cubic yard (125 to 200 liters/m^3) of refuse be allowed. Many landfills will no longer accept any materials of less than 50 percent solids content.

PLACEMENT OF SLUDGE IN A LANDFILL. Direct dumping or tailgating of sludge cake on the working face of the landfill is the method used most often. However, the following factors need to be considered for large-scale operations:

1. Operators must work in the immediate area.

2. Compaction must be achieved.

3. The general public discharging trash at the landfill must be protected.

DURING DRY WEATHER, either open-air drying or mixing with soil cover material prior to placement in the landfill would improve the operation.

DURING WET WEATHER, direct disposal in trenches or pits would probably be necessary to avoid surface runoff contamination and extremely unpleasant working conditions.

3.6101 ON-SITE DEDICATED LAND DISPOSAL (DLD)

Sludge (6 to 10 percent solids), sludge cake (20 to 30 percent solids) discharged from the dewatering machines (centrifuges, vacuum filters, or filter presses), and cake (50 to 60 percent solids) from drying beds and drying lagoons can be moved to an area (DLD) on the treatment plant site that has been dedicated to the disposal of sludge, incinerator ash, and dewatered grit and screenings. To avoid the potential problems of having future development immediately adjacent to the sludge disposal areas, buffer land adjacent to these areas should be acquired by the treatment plant.

Surface runoff control facilities must be provided.

1. Flood Protection. The disposal site should be protected from flooding by a continuous dike.

2. Existing Drainage. The existing drainage into the disposal site should be intercepted and directed outside the flood-protection dike.

3. Contaminated Runoff. Runoff from the disposal site should be collected in a detention basin and allowed to evaporate during the summer or recycled to the treatment plant headworks.

PLACEMENT OF SLUDGE CAKE IN A DEDICATED LAND DISPOSAL SITE. Several methods of placement of sludge cake in a DLD site can be used and are described in the following paragraphs.

TRENCHING. This method has been used in some areas for many years.

1. Shallow trenches. Construct a trench about 4 feet (1.2 m) deep. Add sludge to a depth of about 2 feet (0.6 m). Backfill the trench to its original grade. Substantial amounts of land are required. Small treatment plants can use pits instead of trenches.

2. Deep trenches. Construct a trench about 20 feet (6 m) deep. Add sludge cake in 2-foot (0.6-m) lifts with a 1-foot (0.3-m) soil cover over each lift. When the trench is full, place a 5-foot (1.5-m) soil cover on top and sow grass on this cover. An average annual cake production of 50 tons/day (dry) will require about 20 acres/year for disposal.

Trenching of sludge cake prevents rapid decomposition of organic material and rapid removal of water, both of which would reduce the sludge disposal volume. Also, the DLD site is unsuitable for many purposes when trenching operations are complete because the surface may not support much weight. Depending on the area available, a DLD site may not be suitable for re-trenching at the end of 20 years. Additional land would then be required. Also, at the end of 20 years, the ground surface elevation may have been raised about 5 to 7 feet (1.5 to 2.1 m).

Long-term effects of trenching on groundwater are unknown and depend on soil conditions and the level of the groundwater table. Water addition to the trench areas will be about 100 gal/cu yd (500 liters/m^3). This is 2½ times the 40 gal/cu yd (200 liters/m^3) criteria for sanitary landfills. Specific site evaluation is required to satisfy the regulatory agencies and trench liners or leachate control facilities may be necessary.

Trenching operations are most difficult and sometimes impossible during extreme wet periods. No matter how the sludge cake is moved to the DLD site, by truck or by pneumatic ejection pipelines, it must be mixed with soil or buried daily to avoid odor production. Bulldozers, roadgraders, or bucket loaders have been successfully used to mix the materials before placement. To operate in wet weather, either paved areas or a well-prepared gravel base is needed. Also, rainwater that collects in the trenches or other working areas will have to be transported to a detention basin and allowed to evaporate during the summer or be recycled to the treatment plant headworks.

LANDFILLING. Landfilling of sludge cake is an above-ground operation of mixing or interlaying sludge with soil. Usually, several feet of soil is excavated beneath the existing ground surface to obtain sufficient soil for final cover material. The excavated site is used as a cell where the sludge is mixed with the soil on a 1:1 ratio to aid in the placement of the material.

The problems and disadvantages of landfilling are similar to those listed for trenching.

1. Continued use of land, 20 ac/yr (8 hectares/yr), for a 50 ton/day (45,000 kg/day) cake production.

2. Large quantities of water are added to the DLD site creating potential leaching.

3. Large excavation operations could damage existing natural (clay soils) groundwater protection.

4. Operations will raise the ground surface several feet.

5. Wet weather may prevent operations.

The advantages of on-site landfilling of sludge cake only are also similar to those for trenching.

1. Odor problems are minimized if sludge is covered each day.

2. Operations eliminate off-site transport.

3. Operations are fully controlled by the treatment plant.

4. Operations will also provide for on-site disposal of incinerator ash, dewatered grit, and screenings.

INCORPORATION INTO SURFACE SOILS. Mixing sludge cake with surface soils is a method that minimizes the problems of trenching and landfilling while retaining the advantages of on-site disposal. However, it is only recommended for use in dry weather because of the inability to move equipment and the odor-generation potential in wet weather. No application of sludge should be permitted on frozen or snow-covered ground because the sludge cannot be properly mixed into the soil. Without proper mixing, runoff may carry contaminants from the sludge into nearby bodies of water. Extensive storage facilities are required to successfully practice this method of sludge cake disposal.

About 200 dry tons/acre (450,000 dry kg/hectare) could be applied in a 6-month dry season. Sludge could be reincorporated every few weeks after the soil/sludge mixture has dried. The DLD site may have a life of at least 40 years.

Equipment required for this operation consists of dump trucks or manure-spreading equipment to haul dewatered sludge from the dewatering facility to the DLD. Since travel time should be minimal to an on-site DLD, only relatively small trucks are needed. A tractor, plow, and disking equipment also are necessary. Liquid process application techniques are also sometimes used (see Section 3.611).

3.6102   AGRICULTURAL RECLAMATION

The main reasons for using wastewater solids for agriculture are to supply the nutritional requirements of crops and improve the texture of the soil by adding humus without adversely affecting the crop produced, the soil, or the groundwater. Determining safe loadings within the above limitations requires a complete chemical and biological analysis of the wastewater biosolids coupled with an evaluation of soil types, crops, and irrigation practices.

APPLICATION RATE. The controlling factor for the biosolids (sludge) application rate is generally the nitrogen requirement of the crop. Biosolids that are to be land applied should be applied according to the *AGRONOMIC RATES*[41] in accordance with the predicted nitrogen uptake of the crop. In the majority of cases, nitrogen will be the factor that limits the annual application to a particular field. However, in states that have stringent metals limits, metals industries, or aggressive water supplies, the annual application rate may be controlled by the metals concentration of the sludge. An allowable long-term annual cadmium application of 0.5 kg/ha/yr would allow an annual application of about 12.5 tons dry solids/acre (28,000 kg/ha) if the cadmium concentration was 18 mg Cd/kg dry solids. The following assumptions concerning nitrogen requirements and losses result in a 3.3 tons/acre/yr (dry) (7,400 kg/ha/yr) biosolids (sludge) application rate.

1. Nitrogen content of 6 percent (dry weight basis of dewatered biosolids).

2. One application per season before or between plantings (this procedure will require biosolids storage facilities).

3. Rate of nitrogen mineralization is assumed to be an annual percentage of the remaining unmineralized portion. This results in 67 percent of the nitrogen being mineralized in 20 years.

4. Loss of 25 percent of mineralized nitrogen to the atmosphere due to volatization (ammonia release) and denitrification (nitrogen gas release). This loss is appropriate when biosolids are disked into soil after application.

5. Crop demand is 200 pounds nitrogen/acre/yr (255 kg N/ha/yr) which is appropriate for many crops, including field corn.

$$
\begin{aligned}
\text{Average Annual Biosolids Application Rate, dry tons/acre} &= \frac{\text{Annual Crop Nitrogen Demand, lbs/acre}}{\text{N Content, \% × N Mineral, \% × N Remain, \% × 2,000 lbs/ton}} \\
&= \frac{200 \text{ lbs/acre}}{6\%/100\% × 67\%/100\% × 75\%/100\% × 2,000 \text{ lbs/ton}} \\
&= \frac{200 \text{ lbs/acre}}{0.06 × 0.67 × 0.75 × 2,000 \text{ lbs/ton}} \\
&= 3.3 \text{ dry tons Biosolids/acre}
\end{aligned}
$$

This analysis results in a low application rate that optimizes the reuse of the biosolids (sludge) nitrogen. If maximum nitrogen reuse was not an objective, higher application rates (7.5 tons/acre or 16,800 kg/ha) could be made without injuring most crops. Nitrogen not used by the crop is denitrified during the winter when the soil becomes saturated with water. In this manner, nitrogen does not move downward in the soil to pollute groundwater.

Double-cropping in some areas may permit additional biosolids (sludge) application.

A zinc-to-cadmium (Zn/Cd) ratio lower than the EPA's guideline of 1,000:1 may not have an adverse effect on crops, but monitoring of cadmium additions should be done on a regular basis.

METHOD OF APPLICATION. For low rates of biosolids application (3.3 tons/acre/yr or 7,400 kg/ha/yr), good control over spreading methods is necessary. Manure spreaders or similar equipment are recommended for applying the biosolids cake. A plowing or disking operation should follow closely to incorporate the material into the soil and cover it. Application of biosolids to land controlled by the treatment plant affords maximum matching of disposal rates to plant production rates. Treatment plant agreements with individual land owners to spread biosolids at certain times is less desirable because it reduces flexibility and reliability of operation. Monitoring of spreading sites will also be more difficult.

QUESTIONS

Write your answers in a notebook and then compare your answers with those on page 323.

3.61A   What methods are available for the disposal of mechanically dewatered digested sludge?

3.61B   What kinds of surface runoff control facilities must be provided at an on-site dedicated land disposal operation?

3.611   Stabilized Sludge—Liquid Process

Stabilized sludge can be disposed of in a liquid form (less than 10 percent solids) by: (1) high-rate incorporation into the surface soils of a site dedicated to land disposal (DLD); (2) low-rate application to agricultural sites; or (3) confinement in permanent lagoons.

STORAGE. Long-term storage (about 5 years) of sludge in facultative ponds immediately after digestion is recommended in order to accomplish the following objectives:

1. Separate the daily production of sludge by the wastewater treatment process from the final disposal operation, which may be seasonal, sporadic, highly variable in quantity, or subject to changing regulatory requirements.

2. Achieve substantial additional destruction of remaining volatile solids.

---

[41] *Agronomic Rates.*   Sludge application rates that provide the amount of nitrogen needed by the crop or vegetation grown on the land while minimizing the amount that passes below the root zone.

3. Maximize reduction of total volume by evaporation.

4. Consolidate sludge to about 6 to 12 percent solids or greater.

TRANSPORTATION. Liquid transportation of stabilized sludge is usually best accomplished by pipelines. This is especially true when high-rate on-site dedicated land disposal is used. Sludge concentrations much above four to six percent can be difficult to pump if the pipeline designer did not realize that the friction head loss increases when the sludge concentration increases.

For agricultural reuse of liquid stabilized sludges (biosolids), trucks are usually the best means of transportation. The main advantage here is the flexibility of application sites. Sludge is dredged at 8 or 10 percent solids concentration and placed in large, specially designed tanker trucks of about 10,000 gallons (38 m³) capacity. The material is flooded onto the field and disked in. If 10 percent solids concentrations are achieved, smaller open trucks with manure-spreading devices fitted onto them may be used. This allows a less sophisticated operation and makes it easier for farmers to use existing equipment.

3.6110 HIGH-RATE DEDICATED LAND DISPOSAL

This process uses facultative sludge lagoons (FSLs) for storage and further stabilization prior to disposal. Sludge is dredged from the facultative sludge lagoons and transported by pipeline to a dedicated land disposal (DLD) site, which should be located adjacent to the facultative sludge lagoons. The DLD site should be loaded at a rate of about 100 dry tons/acre/yr (224,000 kg/ha/yr) and should be expected to operate for at least 40 years. Several sludge-spreading techniques have been tested for DLD operations, but shallow injection beneath the surface of the soil appears to be the most cost-effective and environmentally acceptable operation at this time.

RATE OF APPLICATION. The disposal system should operate during the months with the greatest potential net evaporation. Experiments indicate that an application of 100 dry tons/acre (224,000 kg/ha) of sludge at six percent solids concentration would be feasible over a four- to six-month period in some areas.

The application rate results in a water loading rate of 1,570 tons/acre (3,520,000 kg/ha) for an average of 14 inches (36 cm) of water that must be evaporated. Despite the fact that evaporation from wet soil will be less than lake evaporation, there is no problem in meeting these evaporation needs in most arid areas.

DISPOSAL TECHNIQUES. Techniques for spreading sludge at six percent solids concentration dredged from the bottom of the facultative sludge lagoons include:

1. *RIDGE AND FURROW.* In theory, this technique could be very cost-effective if sludge could be made to flow down furrows, then, after being applied, be covered by splitting the ridges and throwing the dirt over the top of the sludge in the furrows. However, the system seems to be unmanageable due to:

   a. Difficulty in maintaining the required relationship between the sludge viscosity (percent solids) and slope of the furrows.

   b. Clogging of sludge debris in the individual furrow gates of the distribution pipe.

   c. Cloddy, puddled soil that does not cover the sludge adequately. This may be due to soil characteristics. (However, this type of soil is usually needed for protection of the groundwater.)

2. *FLOODING.* This method consists of spreading the sludge as evenly as possible over the surface of the disposal area and then incorporating it with the soil. Flooding needs to be controlled with low (6 to 8 inch or 15 to 20 cm) borders or dikes running the length of the field in the direction of flow. The borders control the lateral movement of sludge, thus giving more uniform coverage. After spreading to an average depth of 1½ to 2 inches (3.8 to 5.1 cm), the sludge is turned into the soil with a large disk. At this application rate, the result is a well-aerated soil/sludge mixture. At heavier applications, which may occur in some areas due to uneven flooding, the mixture approaches saturation and results in anaerobic conditions and some odors.

   A comparison of flooding with other methods produces the following disadvantages:

   a. Occasional odor problems from areas receiving excessive sludge application.

   b. The difficulty of maintaining the proper relationship between the slope of the land and sludge viscosity makes even or uniform sludge application practically impossible.

   However, the advantages of flooding over subsurface injection are:

   a. Lower labor requirement and total cost.

   b. Lower energy requirements since sludge would not be pumped through a small-diameter hose and an injection system (both have high head losses).

3. *SUBSURFACE INJECTION.* Subsurface injection is presently the preferred disposal technique because of its ability to ensure consistent sludge application rates and to avoid odorous conditions. This system uses an umbilical cord (4-inch-diameter hose), tractor-mounted, subsurface injector (Figure 3.42) to distribute sludge at about 6 to 8 inches (14 to 20 cm) beneath the surface of the soil. The sweep on the injector unit opens a small cavity in the soil into which the sludge flows. After the sweep passes, the soil falls back into place leaving the sludge completely covered.

TRACTOR AND INJECTION UNIT

INJECTION UNIT

Fig. 3.42   Sludge tractor and injector unit

(From Sewage Sludge Management Program, Wastewater Solids Processing and Disposal, Draft EIR, Sacramento Area Consultants, October, 1978)

Sludge is transported from the dredge, operating in the facultative sludge lagoons, through pipes to the DLD site. Booster pumps may be needed at appropriate locations to move sludge the required distance. A pipeline extends into the DLD site, underground, and risers are appropriately located for hookup with the injection system. A 4-inch (10-cm) diameter flexible hose is attached to the riser and the tractor-mounted injection unit. Sludge application is directly beneath the surface of the ground and there is no visible evidence of the sludge after the injection unit moves on.

Problems may include:

a. Insufficient durability of the flexible supply hose.

b. Coordination of dredging and injection. The flow must be stopped or relieved through a booster pump bypass system when the tractor-injector is being turned at the end of a row. This requires good communication between the injector, booster pump, and the dredge operators.

The cleanliness and odor-free operation of subsurface injection make it a desirable disposal technique despite its slightly higher cost.

SITE LAYOUTS

1. Each dedicated land disposal site layout should have a gross area that includes area for drainage, road access, and injector turning.

2. Each site should be approximately 1,300 feet (400 m) wide. This dimension is determined by the 660-foot (220-m) length of the subsurface injector feed hose and the turnaround space required at each side of the field.

3. Each site should be approximately 800 feet (240 m) in length. This dimension is determined by the area required to allow one injector to operate continuously (six hours/day, five days/week) during peak dry months. Consideration also must be given to the net evaporation during the dry months. Peak dry month operation assumes one pass is made over the DLD site each week. During other harvesting months, it is assumed maximum operation will be limited to one injector making one pass every four weeks.

4. Each DLD site should be graded so that runoff drains (flows) from the center of the DLD to ditches on both sides. Runoff from these ditches should collect in a runoff detention basin at the end of the field. To prevent erosion, the maximum field slope should be held to 0.5 percent and the drainage ditches on the sides of each dedicated land disposal site should be designed so that the runoff water velocity does not exceed five feet per second. The drainage ditches should be "V" ditches with minimum side slopes of 4:1. These flat side slopes will permit vehicle access across the ditches during the dry weather for site control activities. The collected runoff must be returned to some point in the liquid treatment process.

5. Each dedicated land disposal site should be surrounded by an isolation berm designed to keep uncontaminated surface runoff out and contaminated DLD site runoff in. The berm should be 15 feet (4.5 m) wide at the top and should be provided with 3:1 side slopes. The top of the berm should be finished with an all-weather gravel road.

6. Each dedicated land disposal site should be provided with an all-weather gravel access road to ensure light truck access at all times to its isolation berm system.

7. Capability to purge sludge from the dedicated land disposal piping without discharging excessive purge water on the dedicated land disposal site must be provided.

SYSTEM OPERATIONAL CRITERIA

1. Dedicated land disposal sites should be loaded at a minimum of 100 tons (dry weight) per acre (224,000 kg/ha) of harvested sludge per year.

2. Dedicated land disposal sites should be disked, as required, to ensure all sludge is covered daily. Intermittent disking is helpful in drying the fields quickly.

3. Harvested sludge piping should be flushed and purged at the end of each day's run. Sludge purged from piping should be injected into the disposal site. When starting each day's operation, liquid within harvested sludge piping should be purged, as much as possible, back to the facultative sludge lagoons.

EQUIPMENT NEEDS. The operation of the dredge is very important to the efficient application of sludge to the DLD site. This operation should be able to average 6 percent solids concentration, although even with a 4½ percent average solids content, the operation would still be quite effective. The greater the solids content, the more efficient the operation becomes since the system is limited by the amount of water that must be evaporated. Problems may be encountered trying to maintain a constant solids content in the dredged material since the sludge mass in the facultative sludge lagoons shifts from time to time.

3.6111 AGRICULTURAL RECLAMATION

This process uses facultative sludge lagoons, as described previously for sludge storage, and further stabilization of the sludge prior to its use on nearby cropland. The purpose of this process is to maximize reuse of the nutrient components of the sludge (called biosolids when usable for some beneficial purpose). Some degree of control must be exercised over the land to be used for biosolids spreading in order to ensure that full sludge utilization is guaranteed each year. Short of such a guarantee, a backup disposal system may be necessary.

APPLICATION RATE. The controlling factor in the rate of biosolids application is the nitrogen requirement of the crop. The same assumptions apply to agricultural reclamation as to liquid sludge application rates. The slightly lower nitrogen content (5.9 percent) allows a slightly higher application rate of 3.4 ton (dry weight)/acre/yr (7,600 kg/ha/yr).

METHOD OF APPLICATION. For these rather low rates of biosolids application, good control over spreading methods is necessary. Several techniques are discussed briefly in the following paragraphs.

1. *SUBSURFACE INJECTION.* The concentration of the biosolids pumped from the storage ponds should be three or four percent to allow pumping of sludge the required distances for umbilical cord injection systems. This also keeps trucking costs down when subsurface injection tractors are used. This would increase costs of operating an injection system. Overall, costs of subsurface injection would be rather high but it would provide a safe, nonodorous application system with excellent control over application rate.

2. *RIDGE AND FURROW OR CONTROLLED FLOODING.* Portable irrigation pipe is often used to take biosolids from fixed sludge feeder mains to the actual spreading areas. A 3.4 ton/acre (7,600 kg/ha) biosolids application at 3 percent solids concentration would be about 1 inch (2.5 cm) of liquid/biosolids mixture. Each spreading area is leveled or managed along contours. Spreading can be done directly from the back of tanker trucks. Sludge (biosolids) is disked into the soil after spreading to minimize odors. Timing and scheduling of biosolids application and crop planting must be properly managed.

3. *BIOSOLIDS MIXED WITH IRRIGATION WATER.* This system requires a tie-in between a crop irrigation system and the biosolids dredging system. The advantage is that a separate biosolids distribution system is unnecessary although there are costs associated with the inter-ties. Sludge (biosolids) application is spread over several irrigation applications depending on the crop type and other factors. This helps to minimize the problem of scheduling biosolids applications and crop planting. Disadvantages include the fact that the biosolids are not disked into the soil immediately but dry as a layer of sludge cake in the furrows. This may produce additional odors.

3.6112   PERMANENT LAGOONS

This process uses facultative sludge lagoons as described previously for further stabilization and volume reduction prior to transfer of the sludge to permanent lagoons for disposal. The land used for lagoons is permanently dedicated to the disposal of sludge. Many agencies operate permanent sludge lagoons, sending anaerobically digested sludge directly to them without the intermediate step of long-term storage and stabilization in a facultative pond. These types of lagoons typically have odor problems. To minimize that problem, the facultative pond is sometimes used as a highly controlled intermediate environment to achieve substantial additional volatile solids destruction. An obvious secondary advantage of facultative ponds is their storage capacity, which allows transferring sludge to lagoons at the most advantageous times.

Permanent lagoons should be approximately 20 feet (6 m) deep with a 15-foot (4.5-m) working depth. Sludge is permanently stored at an average solids concentration of 12.5 percent and an assumed solids loading of 2,900 tons (dry weight)/acre/yr (6,500,000 kg/ha/yr). Lagoon construction is similar to facultative ponds without the mechanical and piping equipment and electrical connections. Such lagoons have to be located on land dedicated in perpetuity (forever) for that purpose and always maintained with a top cap layer of aerobic liquid to minimize nuisance odors and vector problems. Liquid levels are maintained with plant effluent and rainfall. Barriers surrounding the lagoons aid in odor dispersion. Loading is intermittent from facultative ponds.

Permanent lagoons may also generate odors. If odors become a problem with the lagoons, surface aeration equipment could be added; however, the increased costs, both capital and operational, are substantial.

3.612   *Disposal of Reduced Volume Sludge*

3.6120   COMPOSTED MATERIALS

The composting processes described in Section 3.5, "Volume Reduction," are used for both volume reduction and final disposal of stabilized, dewatered sludge. Because of frequent odor problems with the windrow method and the need to use larger amounts of bulking materials to maintain aerobic conditions, many operators prefer the static pile compost method.

Using the static pile method, the composting process takes about three weeks to complete, followed by a curing period of approximately one month. After curing, the material is ready for bulk use as compost or it can be dried further, pulverized, and then bagged for sale as a soil amendment.

3.6121   MECHANICALLY DRIED SOLIDS
(also see Section 3.52)

Mechanically dried sludge solids may be further processed or disposed of in a landfill or DLD site, used for agricultural reclamation, or composted and sold commercially.

3.6122   INCINERATOR ASH

On-site landfilling of incinerator ash is accomplished similar to a sanitary landfill and provides good control over disposal of this material. Ash also is disposed of on a dedicated land disposal site as long as it is turned under the soil quickly and not subject to wind action.

3.6123   BIOSOLIDS REUSE OPTIONS

There are basically three options for use of the composted and mechanically dried materials.

1. Provide a bagged commercial product for sale to the public as a soil amendment.

2. Provide compost for agricultural land use. This would provide more of a marketing problem due to concern over some materials used for bulking: for example, wood chips have nitrogen demand and some tree leaves have toxic effects. Coarse bulking agents, such as wood chips, are usually screened out and reused.

3. Provide compost for nonagricultural land uses. This would include use in parks, golf courses, or other recreational areas.

3.613 Screenings, Grit, and Scum

Screenings, grit, and scum are usually the most difficult solids to handle and dispose of because of odor and vector problems. Be sure that whatever method of disposal your plant uses will not cause any impact on groundwater or surface waters. Contact your regulatory agency to obtain any necessary permits or approvals for disposal procedures.

Dewatered grit and screenings can be placed in a Class III landfill as long as they are buried the same day as produced to avoid odors.

Dewatered scum should be disposed of in a Class III sanitary landfill. Another possibility is the sale of the scum for recovery of grease and other potentially useful by-products.

Process incinerated ash from dewatered raw sludge the same as incinerated ash from grit, scum, and screenings.

QUESTIONS

Write your answers in a notebook and then compare your answers with those on page 323.

3.61C How can liquid digested sludge be disposed of?

3.61D How can liquid sludge be spread over land?

3.61E Why must the surface layer of liquid on permanent lagoons be aerobic?

3.61F How can composted material be disposed of?

3.62 Monitoring Program

The size and nature of disposal facilities require that major attention be directed to proper operation and control. A monitoring program is essential and its elements are described here. An annual report should be issued describing each year's operation and monitoring results. This report is important in communicating with regulatory agencies and local citizens to ensure that problems that develop are being resolved, and in documenting operating results for future enlargements of the system.

The recommended monitoring program for sludge disposal systems is basically a data collection and analysis function aimed at determining if permit conditions, operational and regulatory guidelines, and design objectives are being met.

Listed below is a typical recommended monitoring schedule for sludges and other constituents of a liquid dedicated land disposal system with facultative sludge lagoons. Most of this schedule is similar to monitoring that is routinely conducted by treatment plant staff.

Sample Type	Sampling Frequency	Constituents
Raw and thickened sludge (to digesters)	Daily	Flow, TS, VS
Digested sludge (from digesters)	Daily	Flow, TS, VS
Facultative sludge lagoon stored sludge	Quarterly Profiles	TS, VS
Facultative sludge lagoon supernatant layer (each pond)	Every two days	pH, temp, H_2S, DO
Recycled supernatant (total of all ponds)	Daily when operating	TKN, NH_3-N, PO_4-P, COD, BOD, SS, pH

3.620 Odors

The following monitoring is recommended to prevent occurrences of odors and to assess and correct odor problems.

1. Meteorological monitoring and general operational control includes the following items. (It is expected that this information will be continuously recorded.)

   a. Air temperature measurement at 25 feet (7.5 m) and 5 feet (1.5 m) above the ground. This provides data to calculate ΔT (temperature differential), which indicates the strength of low-level inversion conditions.

   b. Wind direction. This is used primarily to calculate the rate of change of wind direction.

   c. Wind speed measurement.

   These measurements are also necessary to provide a historical record. If and when an odor complaint is recorded at the treatment plant, the meteorology occurring at the time of the problem is available to assist in assessing the problem and correcting it.

2. The best overall odor monitoring program is responding to odor complaints received from nearby residents. It may be difficult to determine from complaints received which facility was the problem. For this reason, a response team should be available to immediately check on all odor complaints, track them back to the source, if possible, and provide a written record. In this manner, problem areas can be identified and solutions undertaken.

   However, in order to avoid problems and determine if the system is operating as intended, odor readings [42] using an

---

[42] For a description of how to measure odors, see Chapter 1, "Odor Control."

olfactometer from the surface of the ponds should be taken periodically, perhaps for three consecutive days during each quarter. Also, odor testing at the ponds should involve qualitative operator judgments recorded daily.

3.621   Sludge/Dedicated Land Disposal Sites

Recommended monitoring of sludge removed from the facultative sludge lagoons is intended mainly for operational purposes, but also to determine if sludge chemical content is compatible with future reuse options. Removed sludge should be monitored for the following:

Sampling Frequency	Constituents
Daily each dredge	TS, Flow
Two composites from each pond dredged/season	Alkalinity, Cl, NH$_3$-N, Soluble SO$_4$, TP, TN, Ca, Mg, K, Na, As, Be, Cd, Cr, Cu, Pb, Hg, Mn, Ni, Se, Ag, Zn, PCB 1242, PCB 1254, Technical Chlordane, DDE, TS, VS, pH, EC

Daily records should be made of the location and quantity of sludge spread on the DLD sites. Each DLD site should be sampled in two locations at the end of each spreading season. Soils should be sampled for pH, nitrogen, and heavy metals.

3.622   Groundwater

Potential degrading of groundwater is associated with sludge disposal due to the substantial amount of water (80 to 98 percent) that would be buried with the sludge. This water may travel laterally (sideways) as well as vertically.

Groundwater monitoring wells should be provided around all the disposal facilities. There should be at least four test wells for a DLD site. Two of these should not extend below the confining soil layer and two should extend below this level. The following sampling program should be used for groundwater monitoring wells.

Type	Sampling Frequency	Constituents
Ponds	Quarterly	Alkalinity (phenolphthalein and methyl orange), Cl, TPO$_4$, hardness (Ca and Mg), pH, TKN, NH$_3$-N, NO$_2$-N, organic-N, COD, pH, EC
DLD/Landfill	Annually	Same as above

3.623   Surface Water Monitoring

Surface water runoff, which needs to be monitored, includes the DLD landfill contaminated runoff, which should be recycled to the plant, as well as any intermittent and continuously flowing surface streams that pass through the plant site. Following the first storm with substantial runoff, a sample of DLD and landfill runoff water should be taken. At two or three other times during the rainy season additional samples should be taken and analyzed. Runoff constituents sampled should include the complete list of items analyzed for the plant influent.

Any surface streams that pass through the site require monitoring. These samples should be taken periodically throughout the year. Constituents sampled should include the complete list of items analyzed for plant effluent.

3.624   Public Health Vectors

A variety of disease-causing organisms can be found in untreated wastewater and wastewater solids. These include:

1. Bacteria such as SALMONELLA, SHIGELLA, STREPTO-COCCUS, and MYCOBACTERIUM, which are responsible for typhoid and paratyphoid fever, shigellosis, scarlet fever, and tuberculosis

2. Protozoans and their cysts such as ENTAMOEBA HISTOLYTICA, which induces amoebic dysentery

3. Helminths and their ova, which include both parasitic and free-living roundworms, flukes, and tapeworms that can infect humans and animals

4. Viral agents that can infect humans and animals

5. Viral agents of human and animal origin that can cause infectious hepatitis, polio, meningitis, and a variety of other diseases

One of the basic goals of wastewater treatment processes is to destroy pathogenic microorganisms. Sludge treatment processes such as anaerobic digestion, incineration, composting, and disinfection have variable effectiveness in reducing pathogenic microorganism concentrations.

Stabilization destroys most of the pathogenic organisms. Further destruction is accomplished with composting and anaerobic digestion in facultative sludge lagoons. When appropriate precautions are taken, plant operators are exposed to an undefinably low level of risk; public exposure and risk are even less.

The potential for the propagation or attraction of birds, rodents, and flies is discussed briefly in the following paragraphs.

1. *BIRDS.* Salmonella bacteria are known to be excreted by birds that have fed on food material found in polluted waters. Gull feces have been found to contain *S. TYPHORA* and *S. TYPNI* even after they have been isolated from access to polluted waters. Gulls have been implicated in *SALMONELLA* contamination of a community surface water supply in Alaska after feeding near a wastewater outfall.

   Several species of birds have been observed feeding on floatable materials on facultative sludge lagoons. This is a common occurrence in most wastewater treatment plants that have open basins in which edible organic materials are concentrated. No cases of disease transmission by birds in wastewater treatment plants have been reported. However, the potential threat of diseases being transmitted by birds is very real and operators should try to minimize any bird contact with any materials or solids that may contain pathogenic organisms. The potential aircraft hazard associated with attracting birds to the facultative sludge lagoons is probably minimal. If you find a dead bird, pick it up with a shovel and bury it. You could become infected with lice if you pick up dead birds with your hands.

2. *RODENTS.* Rats, mice, and other rodents, which can multiply in response to human activities that create a favorable environment for their survival, can serve as vectors for several human diseases including leptospirosis and plague, and for the dwarf tapeworm *HYMENOLEPIS NANA*. To date, no conditions associated with the operation of facultative sludge lagoons have been observed that contribute to the propagation of rodents. Sludge storage and disposal procedures can be conducted so as to prevent the propagation of rodents.

3. *FLIES.* Flies, known nuisances and vectors of disease, can propagate as a result of certain treatment plant operations or disposal practices. Facultative sludge lagoons have not contributed to fly breeding.

   If sanitary landfilling, agricultural reclamation, lagooning, or composting are the disposal or reuse methods, proper management of the processes will minimize fly propagation or attraction. There is no evidence to indicate that anaerobically digested sludge will support fly breeding.

3.63 Acknowledgment

This section was reviewed by Warren Uhte who provided many helpful comments and suggestions.

QUESTIONS

Write your answers in a notebook and then compare your answers with those on page 324.

3.62A What items should be measured in an odor-monitoring program?

3.62B How many groundwater monitoring wells are needed for a dedicated land disposal (DLD) site and where should they be located?

3.7 REVIEW OF PLANS AND SPECIFICATIONS

On occasion, operators will be asked to review and comment on design drawings and specifications. Engineering drawings are usually very detailed and can be rather complicated to understand. When you are asked to review design drawings, you should be concerned with those features that will directly affect day-to-day operation, routine maintenance, and periodic repairs.

Day-to-day operation, maintenance, and repairs are affected by effort(s) required to open and close valves; read meters; lubricate equipment; repair pumps and drive motors; replace equipment such as chains, sprockets, and bearings; unclog and clean pipes and float control mechanisms; and wash down the area.

Regardless of the equipment or process being installed, the operator should be sure that the following provisions are incorporated into the design:

1. All valves are easily accessible and enough area is provided to facilitate turning of the valves.

2. All meters and gauges are easily readable and located where process adjustments can be made if so indicated by the meters and gauges.

3. Sufficient area is provided around pumps and drives to facilitate routine maintenance and repairs.

4. Sufficient area, wash water capacity, and drains are provided around major pieces of equipment such as thickeners and centrifuges to facilitate repairs and cleanup operations.

When reviewing plans and specifications, you must consider what you will do if each particular treatment process breaks down. You have to decide where to store the sludge and how long you might have to store it. For example, if you have centrifuges to thicken the sludge and then burn the sludge in an incinerator, what will you do if the incinerator is out of service? Can the sludge be hauled to a sanitary landfill? How many truckloads would have to be hauled per day? Where would the sludge be stored while the truck is traveling to and from the dump? These are the types of questions you must answer because they are the problems you will have to solve when equipment fails. Simple alterations in the design before facilities are built can make problems caused by equipment failures easier to solve.

Specific areas of concern for the unit processes described in this chapter are summarized in Table 3.36.

QUESTION

Write your answer in a notebook and then compare your answer with the one on page 324.

3.7A List the important design provisions you would look for when reviewing plans and specifications.

TABLE 3.36 SPECIFIC O & M ITEMS CONSIDERED WHEN REVIEWING PLANS AND SPECIFICATIONS

Unit Process	Recommended Provisions	Unit Process	Recommended Provisions
Gravity Thickening	1. Sludge collectors and surface scrapers are equipped with variable-speed motors.	Centrifugation and Filtration *(continued)*	4. Same as Item 4, Gravity Thickening.
	2. Sludge withdrawal line is *AT LEAST 4 INCHES* (10 cm) in diameter and valves are provided on the suction side and discharge side of the sludge withdrawal pump.		5. Same as Item 5, Gravity Thickening.
	3. A tee with a valve is provided between the suction valve described in Item 2 and the sludge outlet to facilitate backflushing in the event of clogging.	Chemical Conditioning	1. Dry chemical feeders are equipped with infrared heating lamps to prevent the absorption of moisture.
			2. An eductor-type dry chemical feeder is provided to serve as standby for automatic feeders.
	4. A valve is provided on the suction side and discharge side of the influent sludge pump.		3. A bulk storage tank is provided for liquid chemicals.
	5. Sample taps and valves are provided on the influent, effluent, and sludge withdrawal lines to facilitate sample collection.		4. Chemical feed pumps are equipped with variable-speed motors.
Dissolved Air Flotation	1. Surface and bottom sludge collectors are equipped with variable-speed motors.		5. The walking areas around chemical systems are coated with a nonskid-type paint.
	2. Same as Item 2, Gravity Thickening.	Thermal Conditioning	1. A water softener is provided where appropriate to remove hardness from the boiler makeup water. Also, provisions to de-aerate the water.
	3. Same as Item 3, Gravity Thickening.		
	4. Same as Item 4, Gravity Thickening.		
	5. Same as Item 5, Gravity Thickening.		2. An acid-resistant flushing system is incorporated for acid washing of reactors and heat exchangers.
	6. A sight glass is provided on the retention tank.		
	7. Level indicators in the retention tank are accessible for cleaning or repairs.		3. Sample taps and valves are provided on the influent, high-rate, and decant overflow and underflow lines.
Centrifugation and Filtration	1. Sludge hoppers are provided with movable catch troughs to collect and divert wash water to a common drain.		4. The decant tank is totally enclosed and equipped with a vent fan.
	2. A wash water supply and drains are provided in the area of sludge conveyors to facilitate cleaning operations.		5. Gas scrubbers or carbon adsorbers are provided to clean the vent gases.
			6. Same as Item 2, Gravity Thickening.
	3. A permanent wash water line and valve are provided on the influent to the unit.		7. Same as Item 3, Gravity Thickening.
			8. Same as Item 4, Gravity Thickening.

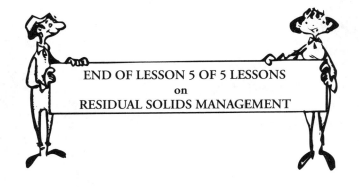

END OF LESSON 5 OF 5 LESSONS
on
RESIDUAL SOLIDS MANAGEMENT

Please answer the discussion and review questions next.

DISCUSSION AND REVIEW QUESTIONS
Chapter 3.   RESIDUAL SOLIDS MANAGEMENT
(Lesson 5 of 5 Lessons)

Write the answers to these questions in your notebook. The question numbering continues from Lesson 4.

26. Why are chemically stabilized and wet-oxidized sludges generally not suited for compost operations?

27. How would you determine the frequency of turning compost stacks?

28. How does incineration reduce the volume of sludge?

29. Where should air be added to a multiple-hearth furnace for proper combustion?

30. How can you tell by looking at the flames in the burning zone of an MHF how much air is present?

31. What happens if the incinerator shaft speed is too high or too low?

32. Why should the disposal or reuse of sludge from digesters without additional treatment be avoided?

33. What are the main reasons for using wastewater solids for agricultural purposes?

34. What are the advantages of using trucks to spread digested sludge in an agricultural reclamation project?

SUGGESTED ANSWERS
Chapter 3.   RESIDUAL SOLIDS MANAGEMENT

ANSWERS TO QUESTIONS IN LESSON 1

Answers to questions on page 171.

3.0A The two types of sludges produced at a wastewater treatment facility are primary sludge and secondary sludge. Primary sludge includes all the solids that settle to the bottom of the primary sedimentation basin and are removed from the wastestream. Primary sludges are usually fairly coarse and fibrous, have specific gravities greater than water, and are composed of 40 to 80 percent volatile matter. Secondary sludge is generated as a by-product of biological degradation of organic wastes. Secondary sludges are finer than primary sludge solids, less fibrous, have specific gravities closer to that of water, and consist of 75 to 80 percent volatile matter.

3.0B Biosolids are wastewater treatment process sludges or solid waste products that are suitable for some further beneficial use.

3.0C Volumes of primary sludge depend on inflow, influent suspended solids, and efficiency of the primary sedimentation basin.

3.0D Determine the daily quantity (lbs/day) of primary sludge produced for the following conditions: (1) influent flow of 2.0 MGD, (2) influent suspended solids of 200 mg/L, and (3) primary effluent suspended solids of 120 mg/L.

Known		Unknown
Flow, MGD	= 2.0 MGD	Sludge, lbs/day
Influent SS, mg/L	= 200 mg/L	
Effluent SS, mg/L	= 120 mg/L	

Determine the quantity of primary sludge in pounds per day.

$$\text{Primary Sludge, lbs/day} = \text{Flow, MGD} \times (\text{Infl SS, mg/}L - \text{Effl SS, mg/}L) \times 8.34 \text{ lbs/gal}$$

$$= 2.0 \text{ MGD} \times (200 \text{ mg/}L - 120 \text{ mg/}L) \times 8.34 \text{ lbs/gal}$$

$$= 2.0 \text{ MGD} \times 80 \text{ mg/}L \times 8.34 \text{ lbs/gal}$$

$$= 1,335 \text{ lbs/day}$$

3.0E Variables that influence the production of secondary sludges include the flow to the biological system, the BOD loading to the biological system, the efficiency of the biological system in removing BOD, and the growth rate of the bacteria in the system.

3.0F Estimate the daily quantity of secondary sludge produced for the following conditions: (1) influent flow of 2.0 MGD, (2) influent BOD to the secondary system of 180 mg/L and effluent from the secondary system of 30 mg/L, and (3) growth rate coefficient, Y, of 0.50 pound of solids per pound of BOD removed.

Known		Unknown
Flow, MGD	= 2.0 MGD	Sec Sludge, lbs/day
Sec Infl BOD, mg/L	= 180 mg/L	
Sec Effl BOD, mg/L	= 30 mg/L	

$$\text{Growth Rate, Y } \frac{\text{lbs Sl}}{\text{lb BOD rem}} = \frac{0.50 \text{ lb Sludge}}{1 \text{ lb BOD removed}}$$

Estimate the quantity of secondary sludge in pounds per day.

$$\frac{\text{Sec Sludge,}}{\text{lbs/day}} = \text{Flow, MGD} \times (\text{Infl BOD, mg/}L - \text{Effl BOD, mg/}L) \propto$$
$$8.34 \text{ lbs/gal} \times \frac{0.50 \text{ lb Sludge}}{1 \text{ lb BOD removed}}$$

$$= 2.0 \text{ MGD} \times (180 \text{ mg/}L - 30 \text{ mg/}L) \times 8.34 \text{ lbs/gal} \times 0.50$$

$$= 1,251 \text{ lbs/day}$$

3.0G For the conditions given in problem 3.0D, estimate the daily volume (gal/day) of primary sludge if it is withdrawn from the primary clarifier at a sludge solids concentration of 4.0 percent.

Known	Unknown
Conditions in Problem 3.0D	Sludge, gal/day
Sludge, lbs/day = 1,335 lbs/day	
Solids Conc, % = 4.0%	

Estimate the primary sludge volume in gallons per day.

$$\frac{\text{Sludge,}}{\text{gal/day}} = \frac{\text{Sludge, lbs/day}}{8.34 \text{ lbs/gal} \times (\text{Solids, \%/100\%})}$$

$$= \frac{1,335 \text{ lbs/day}}{8.34 \text{ lbs/gal} \times (4.0\%/100\%)}$$

$$= 4,002 \text{ gallons/day}$$

Answers to questions on page 172.

3.10A The primary function of sludge thickening is to reduce the sludge volume to be handled in subsequent processes.

3.10B Determine the amount of dry sludge (lbs/day) if 12,000 gal/day of secondary sludge are produced with a solids concentration of 1.0%.

Known	Unknown
Sec Sludge, gal/day = 12,000 gal/day	Dry Sludge, lbs/day
Solids Conc, % = 1.0%	

Determine the amount of dry sludge in pounds per day.

$$\frac{\text{Dry Sludge,}}{\text{lbs/day}} = \text{Sludge, gal/day} \times 8.34 \text{ lbs/gal} \times \text{Solids Conc, } \frac{\%}{100\%}$$

$$= 12,000 \text{ gal/day} \times 8.34 \text{ lbs/gal} \times \frac{1\%}{100\%}$$

$$= 12,000 \text{ gal/day} \times 8.34 \text{ lbs/gal} \times 0.01$$

$$= 1,000 \text{ lbs/day}$$

3.10C For the conditions given in problem 3.10B, determine the secondary sludge volume (gal/day) if the sludge is withdrawn from the secondary clarifier at a solids concentration of 1.5%.

Known	Unknown
Conditions in Problem 3.10B	Sludge Volume, gal/day
Sec Sludge, lbs/day = 1,000 lbs/day	
Solids Conc, % = 1.5%	

Determine the sludge volume in gallons per day.

$$\frac{\text{Sludge Vol,}}{\text{gal/day}} = \frac{\text{Sec Sludge, lbs/day}}{8.34 \text{ lbs/gal} \times \text{Solids Conc, } \frac{\%}{100\%}}$$

$$= \frac{1,000 \text{ lbs/day}}{8.34 \text{ lbs/gal} \times \frac{1.5\%}{100\%}}$$

$$= 8,000 \text{ gal/day}$$

Answers to questions on page 177.

3.11A The main components of gravity thickeners include:

1. Inlet and distribution assembly
2. Sludge rake
3. Vertical steel members or "pickets" mounted on the sludge rake
4. Effluent or overflow weir
5. Scum removal equipment

3.11B 1. The inlet baffle causes the influent to flow downward toward the bottom of the tank where the solids settle. The inlet baffle provides for an even distribution of sludge throughout the tank and reduces the possibility of short-circuiting to the effluent end of the thickener.
2. Sludge rakes cause the settled sludge to move toward the center of the tank to be removed by a sludge pump.
3. The vertical pickets provide for gentle stirring of the settled sludge as the rake rotates. This gentle stirring action opens up channels for the vertical release of entrapped gases and free moisture, which promotes or enhances the concentration of the settled sludge.

3.11C The age of the sludge to be thickened is very important. Fresh primary sludge usually can be concentrated to the highest degree. If gasification occurs due to anaerobic conditions, sludges are difficult to thicken. Secondary sludges are not as well suited for gravity thickening as primary sludge. Secondary sludges contain large quantities of "bound" water, which makes the sludge less dense than primary sludge solids.

3.11D As the temperature of the sludge (primary or secondary) increases, the rate of biological activity increases and the sludge tends to gasify and rise at a higher rate. During summertime operation, the settled sludge has to be removed at a faster rate from the thickener than during wintertime operation when the sludge temperature is lower and biological activity and subsequent gas production proceed at a slower rate.

3.11E Determine the hydraulic surface (GPD/sq ft) and solids loading (lbs/day/sq ft) to a 30-ft diameter gravity thickener if 60 GPM of primary sludge at an initial suspended solids concentration of 3.0 percent sludge are applied.

Known		Unknown
Diameter, ft	= 30 ft	1. Hydraulic Surface Loading, GPD/sq ft
Sludge Solids, %	= 3%	
Flow, GPM	= 60 GPM	2. Solids Loading, lbs/day/sq ft

1. Determine the water surface area.

$$\text{Surface Area, sq ft} = \frac{\pi}{4} (\text{Diameter, ft})^2$$

$$= 0.785 \, (30 \text{ ft})^2$$

$$= 707 \text{ sq ft}$$

2. Calculate the hydraulic surface loading.

$$\text{Hydraulic Surface Loading, GPD/sq ft} = \frac{\text{Flow, GPD}}{\text{Surface Area, sq ft}}$$

$$= \frac{60 \text{ gal/min} \times 1,440 \text{ min/day}}{707 \text{ sq ft}}$$

$$= 122 \text{ GPD/sq ft}$$

3. Determine the solids applied to the thickener in pounds per day.

$$\text{Solids Applied, lbs/day} = \text{Flow, GPD} \times 8.34 \text{ lbs/day} \times \text{Solids, } \frac{\%}{100\%}$$

$$= 60 \, \frac{\text{gal}}{\text{min}} \times 1,440 \, \frac{\text{min}}{\text{day}} \times 8.34 \, \frac{\text{lbs}}{\text{gal}} \times \frac{3\%}{100\%}$$

$$= 21,617 \text{ lbs/day}$$

4. Calculate the solids loading.

$$\text{Solids Loading, lbs/day/sq ft} = \frac{\text{Solids Applied, lbs/day}}{\text{Surface Area, sq ft}}$$

$$= \frac{21,617 \text{ lbs/day}}{707 \text{ sq ft}}$$

$$= 31 \text{ lbs/day/sq ft}$$

3.11F A gravity thickener is used to concentrate 40 GPM of waste activated sludge at a concentration of 0.9% (9,000 mg/L). The underflow sludge is withdrawn at 3 percent and the effluent suspended solids concentration is 1,800 mg/L. Determine the suspended solids removal efficiency (%) and the concentration factor.

Known		Unknown
Flow, GPM	= 40 GPM	1. SS Removal Eff, %
WAS Conc, %	= 0.9%	
, mg/L	= 9,000 mg/L	2. Concentration Factor
Underflow Sludge, %	= 3%	
Effl Susp Sol, mg/L	= 1,800 mg/L	

1. Calculate the suspended solids removal efficiency as a percent.

$$\text{Efficiency, \%} = \frac{(\text{Infl SS, mg/}L - \text{Effl SS, mg/}L)}{\text{Infl SS, mg/}L} \times 100\%$$

$$= \frac{(9,000 \text{ mg/}L - 1,800 \text{ mg/}L)}{9,000 \text{ mg/}L} \times 100\%$$

$$= 80\%$$

2. Determine the concentration factor.

$$\text{Concentration Factor} = \frac{\text{Thickened Sludge Concentration, \%}}{\text{Influent Sludge Concentration, \%}}$$

$$= \frac{3.0\%}{0.9\%}$$

$$= 3.33$$

Answers to questions on page 182.

3.11G Routine visual checks on gravity thickeners as well as other equipment help the operator identify equipment malfunctions and decreases in process efficiency.

3.11H The term "hole" or coning refers to a cone-shaped hole that can develop in the sludge blanket that allows liquid from above the sludge blanket (rather than sludge from the sludge blanket) to be pumped from the thickener. A hole in the sludge blanket can best be corrected by lowering the flow to the thickener, increasing the speed of the collectors to keep the sludge at the point of withdrawal, and increasing the rate of underflow sludge pumping. However, if both the blanket and thickened sludge solids concentrations are low, decrease the rate of sludge withdrawal.

3.11I   Possible causes of solids rising to the surface of a gravity thickener include:

1. Gasification
2. Septic feed
3. Blanket disturbances
4. Chemical inefficiencies
5. Excessive loadings

Procedures to correct the problem(s) include:

1. Increase sludge withdrawal rate
2. Increase sludge pumping from clarifier
3. Lower collector speed
4. Increase chemical feed rate
5. Lower flow, if possible

Answers to questions on page 187.

3.12A   The main components of dissolved air flotation (DAF) units are: (1) air injection equipment, (2) agitated or unagitated pressurized retention tank, (3) recycle pump, (4) inlet or distribution assembly, (5) sludge scrapers, and (6) an effluent baffle.

3.12B   1. The function of the distribution box is to allow the air to come out of solution in the form of minute air bubbles that attach to the solids and cause them to rise to the surface.
2. The retention tank provides a location to dissolve air into the liquid.
3. The effluent baffle is provided to keep the floated solids from contaminating the effluent.

3.12C   A sight glass should be provided on the retention tank to periodically check the level of the air-liquid interface because on occasion the float mechanisms may fail and the retention tank will either fill completely with liquid or fill completely with air.

3.12D   The performance of DAF thickeners depends on: (1) type of sludge, (2) age of the feed sludge, (3) solids and hydraulic loadings, (4) air to solids (A/S) ratio, (5) recycle rate, and (6) sludge blanket depth.

3.12E   The age of the sludge usually does not affect flotation performance as drastically as it affects gravity concentrators. A relatively old sludge has a natural tendency to float due to gasification and this natural buoyancy will have little or no adverse effect on the operation of flotation thickeners.

3.12F   Determine the hydraulic loading (GPD/sq ft) for a 20-foot diameter DAF unit. The influent flow is 100 GPM. The formula for surface area of a circular tank is:

$$\text{Area} = \frac{\pi}{4}\,\text{Diameter}^2 \text{ or Area} = 0.785\,\text{Diameter}^2$$

Known	Unknown
DAF Diameter, ft = 20 ft	Hydraulic Surface
Flow, GPM = 100 GPM	Loading, GPD/sq ft

$$\text{Area} = \frac{\pi}{4}\,\text{Diameter}^2 \text{ or Area} = 0.785\,\text{Diameter}^2$$

1. Determine the liquid surface area, sq ft.

$$\text{Surface Area, sq ft} = \frac{\pi}{4} \times (\text{Diameter, ft})^2$$

$$= 0.785 \times (20 \text{ ft})^2$$

$$= 314 \text{ sq ft}$$

2. Calculate the hydraulic surface loading, GPD/sq ft.

$$\text{Surface Loading, GPD/sq ft} = \frac{\text{Flow, GPD}}{\text{Surface Area, sq ft}}$$

$$= \frac{100 \text{ gal/min} \times 1,440 \text{ min/day}}{314 \text{ sq ft}}$$

$$= 459 \text{ GPD/sq ft}$$

3.12G   For the above problem, determine the solids loading, A/S ratio, and recycle flow rate (GPM), if the influent sludge has a suspended solids concentration of 0.75% (7,500 mg/L), and is supplied at a rate of 2.5 cu ft/min. Air is supplied at a rate of 0.75 cu ft/min and a recycle ratio of 100 percent is required.

Known	Unknown
Conditions in Problem 3.12F	1. Solids Loading, lbs/day/sq ft
Influent Sludge, % = 0.75%	
, mg/L = 7,500 mg/L	
Influent Sludge	2. A/S Ratio, lbs air/lbs solids
Flow, CFM = 2.5 cu ft/min	
Air, CFM = 0.75 cu ft/min	3. Recycle Flow Rate, GPM
Recycle Ratio, % = 100%	

1. Determine solids applied, lbs/day.

$$\text{Solids Applied, lbs/day} = \text{Flow,} \frac{\text{cu ft}}{\text{min}} \times 1,440\,\frac{\text{min}}{\text{day}} \times 62.4\,\frac{\text{lbs}}{\text{cu ft}} \times \text{SS,}\,\frac{\%}{100\%}$$

$$= 2.5\,\frac{\text{cu ft}}{\text{min}} \times 1,440\,\frac{\text{min}}{\text{day}} \times 62.4\,\frac{\text{lbs}}{\text{cu ft}} \times 0.75\,\frac{\%}{100\%}$$

$$= 1,685 \text{ lbs/day}$$

2. Calculate solids loading, lbs/day/sq ft.

$$\text{Solids Loading, lbs/day/sq ft} = \frac{\text{Solids Applied, lbs/day}}{\text{Surface Area, sq ft}}$$

$$= \frac{1,685 \text{ lbs/day}}{314 \text{ sq ft}}$$

$$= 5.4 \text{ lbs/day/sq ft}$$

3. Determine the air supply in pounds per hour.

$$\text{Air Supply, lbs/hr} = \text{Air Flow,}\,\frac{\text{cu ft}}{\text{min}} \times 60\,\frac{\text{min}}{\text{hr}} \times 0.075\,\frac{\text{lb air}}{\text{cu ft}}$$

$$= 0.75\,\frac{\text{cu ft}}{\text{min}} \times 60\,\frac{\text{min}}{\text{hr}} \times 0.075\,\frac{\text{lb}}{\text{cu ft}}$$

$$= 3.375 \text{ lbs/hr}$$

4. Determine the solids applied in pounds per hour.

$$\text{Solids Applied, lbs/hr} = \frac{\text{Solids Applied, lbs/day}}{24 \text{ hr/day}}$$

$$= \frac{1,685 \text{ lbs/day}}{24 \text{ hr/day}}$$

$$= 70.2 \text{ lbs/hr}$$

5. Calculate the pounds of air to pounds of solids (A/S) ratio.

$$\frac{\text{Air, lbs}}{\text{Solids, lbs}} = \frac{\text{Air Supply, lbs/hr}}{\text{Solids Applied, lbs/hr}}$$

$$= \frac{3.375 \text{ lbs Air/hr}}{70.2 \text{ lbs Solids/hr}}$$

$$= 0.05 \text{ lb Air/lb Solids}$$

6. Determine the recycle flow rate, GPM.

$$\text{Recycle Flow Rate, GPM} = \text{Inflow, GPM} \times \text{Recycle Ratio}, \frac{\%}{100\%}$$

$$= 2.5 \frac{\text{cu ft}}{\text{min}} \times 7.5 \frac{\text{gal}}{\text{cu ft}} \times \frac{100\%}{100\%}$$

$$= 18.8 \text{ GPM}$$

$$\text{or} = 19 \text{ GPM for pumping rate}$$

3.12H Determine the suspended solids removal efficiency (%) and the concentration factor (cf) if a DAF unit receives an influent sludge at 1.0 percent (10,000 mg/L) suspended solids. The effluent is at 50 mg/L suspended solids and the float or thickened sludge is at a concentration of 3.8 percent.

Known	Unknown
DAF Unit	1. SS Removal Eff, %
Infl Sludge, % = 1.0%	
, mg/L = 10,000 mg/L	2. Concentration Factor (cf)
Effl Sludge, % (Thickened Sludge) = 3.8%	
Effl Liquid SS, mg/L = 50 mg/L	

1. Determine the suspended solids removal efficiency.

$$\text{SS Efficiency, \%} = \frac{(\text{SS Infl, mg/}L - \text{SS Effl, mg/}L)}{\text{SS Infl, mg/}L} \times 100\%$$

$$= \frac{(10,000 \text{ mg/}L - 50 \text{ mg/}L)}{10,000 \text{ mg/}L} \times 100\%$$

$$= 99.5\%$$

2. Determine the concentration factor for the thickened sludge.

$$\text{Concentration Factor (cf)} = \frac{\text{Thickened Sludge Concentration, \%}}{\text{Influent Sludge Concentration, \%}}$$

$$= \frac{3.8\%}{1.0\%}$$

$$= 3.8$$

Answer to question on page 189.

3.12I *PROBLEM.* Poor effluent quality (high suspended solids) and thinner than normal sludge.

Possible Causes	Possible Solutions
1. A/S low	1. Increase air input. Repair or turn on compressor.
2. Pressure too low or too high	2. Open or close valve.
3. Recycle pump inoperative	3. Turn on recycle pump.
4. Reaeration pump inoperative	4. Turn on reaeration pump.
5. Chemical addition inadequate	5. Increase dosage.
6. Loading excessive	6. Lower flow rate.

Answers to questions on page 200.

3.13A Three centrifuge designs commercially available today are: (1) basket centrifuges, (2) scroll centrifuges, and (3) disc-nozzle type centrifuges. Basket centrifuges operate in a batch mode, while scroll and disc-nozzle types operate continuously.

3.13B Centrifugal thickening is affected by: (1) type and age of the feed sludge, (2) solids and hydraulic loading, (3) bowl speed and resulting gravitational ("g") forces, (4) pool depth and differential scroll speed for scroll centrifuges, (5) size and number of nozzles for disc centrifuges, and (6) chemical conditioning.

3.13C Centrifuges are not commonly used to thicken primary sludges because they have inlet assemblies that clog easily.

3.13D Determine the solids and hydraulic loading for a 20-inch by 60-inch scroll centrifuge. The feed rate is 30 GPM and the influent solids concentration is 1.1 percent (11,000 mg/L) suspended solids.

Known	Unknown
Scroll Centrifuge, 20 in ∞ 60 in	1. Hydraulic Loading, gal/hr
Feed Rate, GPM = 30 GPM	
Infl Solids, % = 1.1 %	2. Solids Loading, lbs/hr
, mg/L = 11,000 mg/L	

1. Calculate the hydraulic loading in gallons per hour.

$$\text{Hydraulic Load, gal/hr} = \text{Flow, GPM} \times 60 \text{ min/hr}$$

$$= 30 \text{ gal/min} \times 60 \text{ min/hr}$$

$$= 1,800 \text{ gal/hr}$$

2. Calculate the solids loading in pounds of solids per hour.

$$\begin{aligned}\text{Solids Load,}\atop\text{lbs/hr} &= \text{Flow, gal/hr} \times 8.34 \text{ lbs/gal} \times \text{SS, } \frac{\%}{100\%} \\ &= 1{,}800 \text{ gal/hr} \times 8.34 \text{ lbs/gal} \times \frac{1.1\%}{100\%} \\ &= 165 \text{ lbs Solids/hr}\end{aligned}$$

3.13E Determine the hydraulic and solids loading for a 48-inch diameter basket centrifuge. The feed rate is 30 GPM, the feed time is 25 minutes and 3 minutes are required to receive the solids and restart the feed. The influent solids concentration is 1.1 percent.

Known	Unknown
Basket Centrifuge, 48-inch diameter	1. Hydraulic Loading, gal/hr
Feed Rate, GPM = 30 GPM	
Infl Solids, % = 1.1%	2. Solids Loading, lbs/hr
, mg/L = 11,000 mg/L	
Feed Time, min = 25 min	
Downtime, min = 3 min	

1. Calculate the hydraulic loading in gallons per hour.

$$\begin{aligned}\text{Hydraulic}\atop\text{Load,}\atop\text{gal/hr} &= \text{Flow, GPM} \times 60 \frac{\text{min}}{\text{hr}} \times \frac{\text{Run Time, min}}{(\text{Run, min} + \text{Down, min})} \\ &= 30 \frac{\text{gal}}{\text{min}} \times 60 \frac{\text{min}}{\text{hr}} \times \frac{25 \text{ min}}{(25 \text{ min} + 3 \text{ min})} \\ &= 30 \times 60 \times \frac{25}{28} \\ &= 1{,}607 \text{ gal/hr}\end{aligned}$$

2. Calculate the solids loading in pounds of solids per hour.

$$\begin{aligned}\text{Solids Load,}\atop\text{lbs/hr} &= \text{Hyd Load, } \frac{\text{gal}}{\text{hr}} \times 8.34 \frac{\text{lbs}}{\text{gal}} \times \text{SS, } \frac{\%}{100\%} \\ &= 1{,}607 \frac{\text{gal}}{\text{hr}} \times 8.34 \frac{\text{lbs}}{\text{gal}} \times \frac{1.1\%}{100\%} \\ &= 147 \text{ lbs Solids/hr}\end{aligned}$$

3.13F As the differential scroll speed is increased, the solids that are compacted on the bowl wall are conveyed out of the centrifuge at a faster rate, resulting in a decrease in the concentration of these solids. Lower concentrations result because as the solids are moved out at a faster rate, they are subjected to centrifugal forces for shorter periods of time.

3.13G A 20-inch disc centrifuge receives 25 GPM of waste activated sludge with a suspended solids concentration of 0.65 percent. The effluent (centrate) contains 0.03 percent SS (300 mg/L) and the thickened sludge concentration is 4.9 percent. Determine the percent efficiency and the concentration factor (cf).

Known	Unknown
Disc Centrifuge, 20-inch diameter	1. Efficiency, %
Flow, GPM = 25 GPM	2. Concentration Factor (cf)
Infl SS, % = 0.65%	
, mg/L = 6,500 mg/L	
Effl SS, % = 0.03%	
, mg/L = 300 mg/L	
Thick Sl, % = 4.9%	
, mg/L = 49,000 mg/L	

1. Calculate the efficiency of the disc centrifuge.

$$\begin{aligned}\text{Efficiency,}\atop\% &= \frac{(\text{Infl SS, mg/}L - \text{Effl SS, mg/}L)}{\text{Infl SS, mg/}L} \times 100\% \\ &= \frac{(6{,}500 \text{ mg/}L - 300 \text{ mg/}L)}{6{,}500 \text{ mg/}L} \times 100\% \\ &= 95.4\%\end{aligned}$$

2. Determine the concentration factor for the thickened sludge.

$$\begin{aligned}\text{Concentration}\atop\text{Factor (cf)} &= \frac{\text{Thickened Sludge Concentration, \%}}{\text{Influent Sludge Concentration, \%}} \\ &= \frac{4.9\%}{0.65\%} \\ &= 7.5\end{aligned}$$

Answers to questions on page 202.

3.13H *PROBLEM:* Scroll centrifuge has poor centrate quality, but discharge solids are good.

Check	Possible Solutions
1. Scroll RPM	1. Increase scroll speed
2. Flow rate	2. Decrease flow
3. Pool depth setting	3. Increase pool depth
4. Chemical system	4. Increase chemical dosage

3.13I Increasing the number and size of the nozzles in a disc-nozzle centrifuge will facilitate sludge discharge from the centrifuge.

Answers to questions on page 205.

3.14A Factors that affect gravity belt thickener performance include belt type, chemical conditioning, belt speed, and hydraulic and solids loadings.

3.14B If belt porosity is too low, the belt may blind or plug, which will produce frequent washouts.

3.14C "Washing out" means sludge with an excess of free water is discharged from a gravity belt thickener.

3.14D Low belt speed reduces the belt area contacting the influent sludge. If the belt area is not sufficient to allow the free water to drain, washout may occur and cause a reduction in the thickened sludge concentration.

3.14E The ideal belt speed is the slowest the operator can maintain without washing out the belt.

3.14F If a gravity belt thickener frequently washes out, the operator should check: (1) the polymer dosage, (2) hydraulic loading, (3) solids loading, (4) belt speed, and (5) belt washing equipment.

3.14G The steps an operator should take to correct blinding of the belt depend on the cause of the problem. If the belt appears dirtier than normal, increase the wash water rate, turn off the polymer and feed pumps, and allow the belt to be washed until it is clean. If polymer overdosing is the cause of blinding, check and adjust the polymer feed rate.

ANSWERS TO QUESTIONS IN LESSON 2

Answers to questions on pages 206 and 207.

3.20A The goals of stabilization are to convert the volatile (organic) or odor-causing portion (fraction) of the sludge solids to nonodorous end products, to prevent the proliferation (breeding) of insects upon disposal, to reduce the pathogenic (disease-carrying) organism content, and to improve sludge dewaterability.

3.20B Unit processes commonly used for sludge stabilization include: (1) anaerobic digestion, (2) aerobic digestion, and (3) chemical treatment.

3.21A In the two-step process of anaerobic digestion, first, facultative acid-forming organisms convert complex organic matter to volatile (organic) acids. Next, the anaerobic methane-forming organisms convert the acids to odorless end products of methane gas and carbon dioxide.

3.21B Factors affecting anaerobic digestion include: (1) sludge type, (2) digestion time, (3) digestion temperature, and (4) mixing.

3.21C Inorganic and organic materials can influence the performance of anaerobic digesters by severely inhibiting or even being toxic to the digestion process by affecting primarily the methane formation.

3.21D Laboratory or pilot-scale studies may be required to identify chemical and biological factors that may affect the inhibitory impact of a pollutant on the anaerobic digestion process for sludge stabilization.

Answers to questions on page 214.

3.22A In the aerobic digestion process, thickened sludge is fed to the digester inlet. Air from blowers is diffused into the digester to provide oxygen for the organisms and to cause mixing of the digester contents. Digested sludge exits through an effluent line and is either pumped directly to dewatering facilities or to a gravity thickener prior to dewatering.

3.22B Factors affecting aerobic digestion include: (1) sludge type, (2) digestion time, (3) digestion temperatures, (4) volatile solids loading, (5) quantity of air supplied, and (6) dissolved oxygen concentrations within the digester.

3.22C The presence of heavy metals or other toxic (including organic) materials in wastestreams may inhibit or even upset the aerobic digestion process.

3.22D Laboratory or pilot studies may be required to identify the impact of pollutants found in a particular process influent on the aerobic digestion process for sludge stabilization.

3.22E Aerobic digestion is more suitable for treating secondary sludges than primary sludges because secondary sludges are composed primarily of biological cells that are produced in the activated sludge or trickling filter processes as a by-product of degrading organic matter. In the absence of food, these microorganisms enter the endogenous or death phase of their life cycle. When no food is available, the biomass begins to self-metabolize, which results in a conversion of the biomass to end products of carbon dioxide and water and a net decrease in the sludge mass.

3.22F The operator can control digestion time by controlling the degree of sludge thickening prior to digestion. The thicker the sludge, the longer the digestion time.

3.22G A digester with an active volume of 140,000 cubic feet receives 110,000 GPD of primary sludge. What is the digestion time (days)?

Known	Unknown
Digester Volume, cu ft $= 140,000$ cu ft	Digestion Time, days
Flow, GPD $= 110,000$ GPD	

1. Calculate the digester volume in gallons.

$$\text{Digester Volume, gal} = \text{Digester Volume, cu ft} \times 7.48 \text{ gal/cu ft}$$
$$= 140,000 \text{ cu ft} \times 7.48 \text{ gal/cu ft}$$
$$= 1,047,200 \text{ gallons}$$

2. Determine the digestion time in days.

$$\text{Digestion Time, days} = \frac{\text{Digester Volume, gallons}}{\text{Flow, GPD}}$$
$$= \frac{1,047,200 \text{ gallons}}{110,000 \text{ GPD}}$$
$$= 9.5 \text{ days}$$

3.22H If the sludge from problem 3.22G is thickened from 2.7 percent to 3.5 percent, what will happen to the digestion time?

Known	Unknown
Conditions in Problem 3.22G	New Digestion Time, days
Increase thickened sludge from 2.7 to 3.5 percent solids	
Digestion Time, 9.5 days	

Calculate the new digestion time in days.

$$\text{Digestion Time, days} = \frac{\text{Old Digestion Time, days} \times \text{New Thickened Solids, \%}}{\text{Old Thickened Solids, \%}}$$

$$= \frac{9.5 \text{ days} \times 3.5\%}{2.7\%}$$

$$= 12.3 \text{ days}$$

3.22I Desirable aerobic digestion temperatures are approximately 65 to 80°F (18 to 27°C). As the temperature decreases from desirable temperatures, the rate of biological activity decreases.

3.22J An aerobic digester with dimensions of 120 ft in length, 25 feet wide, and 11 feet SWD receives 24,000 GPD of secondary sludge at a concentration of 3.1 percent and 73 percent volatile matter. What is the digestion time (days) and the VSS loading (lbs/day/cu ft)?

Known	Unknown
Aerobic Digester Dimensions, ft	1. Digestion Time, days
$\quad$ L $\quad$ = 120 ft	
$\quad$ W $\quad$ = 25 ft	2. VSS Loading, lbs/day/cu ft
$\quad$ SWD = 11 ft	
Flow, GPD $\quad$ = 24,000 GPD	
Sludge Solids, % = 3.1%	
Volatile Matter, % = 73%	

1. Calculate the aerobic digester volume in cubic feet and gallons.

$$\text{Volume, cu ft} = \text{Length, ft} \times \text{Width, ft} \times \text{SWD, ft}$$

$$= 120 \text{ ft} \times 25 \text{ ft} \times 11 \text{ ft}$$

$$= 33,000 \text{ cu ft}$$

$$\text{Volume, gal} = 33,000 \text{ cu ft} \times 7.48 \text{ gal/cu ft}$$

$$= 246,840 \text{ gallons}$$

2. Determine the digestion time in days.

$$\text{Digestion Time, days} = \frac{\text{Digester Volume, gal}}{\text{Flow, GPD}}$$

$$= \frac{246,840 \text{ gallons}}{24,000 \text{ GPD}}$$

$$= 10.3 \text{ days}$$

3. Calculate the VSS applied in pounds of volatile matter per day.

$$\text{VSS Applied, lbs/day} = \text{Flow, GPD} \times \frac{8.34 \text{ lbs}}{\text{gal}} \times \text{SS,} \frac{\%}{100\%} \times \text{VM,} \frac{\%}{100\%}$$

$$= 24,000 \text{ GPD} \times \frac{8.34 \text{ lbs}}{\text{gal}} \times \frac{3.1\%}{100\%} \times \frac{73\%}{100\%}$$

$$= 4,530 \text{ lbs VSS/day}$$

4. Determine the VSS loading in pounds per day per cubic foot.

$$\text{VSS Loading, lbs/day/cu ft} = \frac{\text{VSS Applied, lbs/day}}{\text{Digester Volume, cu ft}}$$

$$= \frac{4,530 \text{ lbs VSS/day}}{33,000 \text{ cu ft}}$$

$$= 0.14 \text{ lb VSS/day/cu ft}$$

3.22K DO in aerobic digesters should be maintained at concentrations greater than 1.0 mg/L to avoid the growth of filamentous organisms, which can lead to sludge bulking or foaming.

3.22L DO is measured in aerobic digesters by lowering a DO probe into the digester, gently raising and lowering the probe 6 to 12 inches, and recording the readout measurement after the readout has stabilized.

3.22M Determine the O_2 uptake rate (mg/L/hr) for the following field measurements:

Time (min)	DO (mg/L)
0	6.3
1	5.1
2	4.2
3	3.4
4	2.6
5	1.8
6	1.0

Known	Unknown
O_2 Uptake Data	O_2 Uptake Rate, mg/L/hr

Calculate the O_2 uptake rate in mg/L/hr.

$$\text{O}_2 \text{ Uptake,} \atop \text{mg/}L\text{/hr} = \frac{(DO_1 - DO_2)}{(Time_2 - Time_1)} \times \frac{60 \text{ min}}{\text{hr}}$$

$$= \frac{(4.2 \text{ mg/}L - 1.8 \text{ mg/}L)}{(5 \text{ min} - 2 \text{ min})} \times \frac{60 \text{ min}}{\text{hr}}$$

$$= 48 \text{ mg/}L\text{/hr}$$

3.22N A 1,000,000-gallon aerobic digester receives 91,000 GPD of primary sludge at a concentration of 5.1 percent SS and 76 percent volatile matter. The digester effluent is at a concentration of 3.7 percent SS and 67 percent volatile matter. Determine the digestion time (days), VSS loading (lbs/day/cu ft), and percent VSS destruction.

Known	Unknown
Digester Vol, gal = 1,000,000 gal	1. Digestion Time, days
Inflow, GPD = 91,000 GPD	2. VSS Loading, lbs/day/cu ft
Infl Sludge	
SS, % = 5.1%	3. VSS Destruction, %
VM, % = 76%	
Effl Sludge	
SS, % = 3.7%	
VM, % = 67%	

1. Calculate the digestion time in days.

$$\text{Digestion Time, days} = \frac{\text{Digester Volume, gal}}{\text{Inflow, GPD}}$$

$$= \frac{1,000,000 \text{ gal}}{91,000 \text{ GPD}}$$

$$= 11.0 \text{ days}$$

2. Calculate the digester volume in cubic feet.

$$\text{Volume, cu ft} = \frac{\text{Volume, gal}}{7.48 \text{ gal/cu ft}}$$

$$= \frac{1,000,000 \text{ gal}}{7.48 \text{ gal/cu ft}}$$

$$= 133,700 \text{ cu ft}$$

3. Determine the volatile suspended solids applied (entering) in pounds per day.

$$\text{VSS Applied,} \atop \text{lbs/day} = \text{Inflow, GPD} \times \frac{8.34 \text{ lbs}}{\text{gal}} \times \text{SS,} \frac{\%}{100\%} \times \text{VM,} \frac{\%}{100\%}$$

$$= 91,000 \text{ GPD} \times \frac{8.34 \text{ lbs}}{\text{gal}} \times \frac{5.1\%}{100\%} \times \frac{76\%}{100\%}$$

$$= 29,400 \text{ lbs/day}$$

4. Calculate the VSS loading in pounds per day per cubic foot.

$$\text{VSS Loading,} \atop \text{lbs/day/cu ft} = \frac{\text{VSS Applied, lbs/day}}{\text{Digester Volume, cu ft}}$$

$$= \frac{29,400 \text{ lbs/day}}{133,700 \text{ cu ft}}$$

$$= 0.22 \text{ lb VSS/day/cu ft}$$

5. Determine the volatile suspended solids exiting in pounds per day.

$$\text{VSS Exiting,} \atop \text{lbs/day} = \text{Inflow, GPD} \times \frac{8.34 \text{ lbs}}{\text{gal}} \times \text{SS,} \frac{\%}{100\%} \times \text{VM,} \frac{\%}{100\%}$$

$$= 91,000 \text{ GPD} \times \frac{8.34 \text{ lbs}}{\text{gal}} \times \frac{3.7\%}{100\%} \times \frac{67\%}{100\%}$$

$$= 18,800 \text{ lbs/day}$$

6. Calculate the VSS destruction as a percent.

$$\text{VSS Destruction,} \atop \% = \frac{(\text{VSS In, lbs/day} - \text{VSS Out, lbs/day})}{\text{VSS Applied, lbs/day}} \times 100\%$$

$$= \frac{(29,400 \text{ lbs/day} - 18,800 \text{ lbs/day})}{29,400 \text{ lbs/day}} \times 100\%$$

$$= 36.1\%$$

Answers to questions on page 216.

3.22O Process inefficiencies can be detected by careful observation of the physical sludge and routine monitoring of the DO and O_2 uptake rates. Laboratory analyses of influent and effluent suspended solids (% solids) and volatile matter content (% volatile matter) also will reveal process inefficiencies.

3.22P Normally, digester DO is 1.5 mg/L. A DO residual of 4.0 mg/L is measured. The operator should verify the DO and check the O_2 uptake rate. If the O_2 uptake rate is normal, the air rate should be lowered. If the O_2 uptake rate is low, the cause should be identified and corrected. Possible causes include low digester temperature, low digester pH, too high or too low a VSS loading, and digestion time too high or too low.

3.22Q Foaming problems.

Potential Causes	Corrective Measures
1. Filamentous growth	1. Increase air rate. Add defoamant.
2. Excessive turbulence	2. Lower air rate. Add defoamant.

Answers to questions on page 219.

3.23A   Two chemicals used to stabilize sludges are lime and chlorine.

3.23B   Chemicals are used as a temporary stabilization process at overloaded plants or at plants experiencing stabilization facility upsets.

3.23C   Major limitations of using chemicals to stabilize sludge include: (1) costs, and (2) the volume of sludge is not reduced.

ANSWERS TO QUESTIONS IN LESSON 3

Answers to questions on page 226.

3.30A   Solid particles present in sludge usually require conditioning in order to separate from wastewater because they are fine in particle size, hydrated (combined with water), and may carry an electrostatic charge.

3.30B   Different types of sludge conditioning methods include: (1) chemical treatment, (2) thermal treatment, (3) wet oxidation, and (4) elutriation. These are the most common. Other types include: freezing, electrical treatment, and ultrasonic treatment.

3.31A   The addition of chemicals to sludge reduces the natural repelling forces and allows the solids to come together (coagulate) and gather (flocculate) into a heavier solid mass.

3.31B   Chemical types and dosage requirements vary from plant to plant because sludge types and characteristics vary from plant to plant.

3.31C   Chemical requirements are determined for a particular sludge by the use of laboratory-scale jar tests. Various amounts of a chemical are added to different jars containing the sludge. The chemical requirements are based on the volume of chemical solution required for floc formation.

3.31D   Three pounds of dry polymer are added to 360 gallons of water. What is the solution strength of the mixture?

Known	Unknown
Polymer Added, lbs = 3 lbs	Strength of Solution, %
Total Volume, gal = 360 gallons	

Determine the strength of the polymer solution in percent.

$$\text{Solution, \%} = \frac{\text{Polymer Added, lbs} \times 100\%}{\text{Total Volume, gal} \times 8.34 \text{ lbs/gal} + \text{Polymer, lbs}}$$

$$= \frac{3 \text{ lbs} \times 100\%}{360 \text{ gallons} \times 8.34 \text{ lbs/gal} + 3 \text{ lbs Polymer}}$$

$$= 0.10\% \text{ Polymer}$$

3.31E   Ten pounds of lime are added to 100 gallons of water. What is the solution strength of the mixture?

Known	Unknown
Lime Added, lbs = 10 lbs	Strength of Solution, %
Total Volume, gal = 100 gallons	

Determine the strength of the lime solution in percent.

$$\text{Solution, \%} = \frac{\text{Chemical Added, lbs} \times 100\%}{\text{Total Volume, gallons} \times 8.34 \text{ lbs/gal} + \text{Lime, lbs}}$$

$$= \frac{10 \text{ lbs} \times 100\%}{100 \text{ gallons} \times 8.34 \text{ lbs/gal} + 10 \text{ lbs Lime}}$$

$$= 1.20\% \text{ Lime}$$

3.31F   Ten gallons of liquid polymer are added to 790 gallons of water. What is the solution strength of the mixture?

Known	Unknown
Liquid Polymer, gal = 10 gal	Strength of Solution, %
Volume Water, gal = 790 gal	

Determine the strength of the polymer solution in percent.

$$\text{Solution, \%} = \frac{\text{Polymer Added, gal} \times 8.34 \text{ lbs/gal} \times 100\%}{\text{Total Volume, gallons} \times 8.34 \text{ lbs/gal}}$$

$$= \frac{10 \text{ gal} \times 8.34 \text{ lbs/gal} \times 100\%}{(790 \text{ gal} + 10 \text{ gal}) \times 8.34 \text{ lbs/gal}}$$

$$= 1.25\% \text{ Polymer}$$

3.31G   Five gallons of commercially available ferric chloride are added to 50 gallons of water. What is the solution strength of the mixture?

Known	Unknown
Ferric Chloride, gal = 5 gal	Strength of Solution, %
Volume Water, gal = 50 gal	

Determine the strength of the ferric chloride solution in percent.

$$\text{Solution, \%} = \frac{\text{Ferric Chloride, gal} \times 8.34 \text{ lbs/gal} \times 100\%}{\text{Total Volume, gallons} \times 8.34 \text{ lbs/gal}}$$

$$= \frac{5 \text{ gal} \times 8.34 \text{ lbs/gal} \times 100\%}{(50 \text{ gal} + 5 \text{ gal}) \times 8.34 \text{ lbs/gal}}$$

$$= 9.1\% \text{ Ferric Chloride}$$

3.31H   A jar test has been conducted on digested primary sludge. The sludge has a concentration of 3.0 percent SS (30,000 mg/L) and 60 mL of a 0.15 percent solution of polymer was required to flocculate the sludge. Determine the polymer dosage in lbs/ton and the cost in $/ton if the polymer costs $1.50/lb.

Known		Unknown
Sludge Conc,		1. Polymer
%	= 3.0%	Dose, lbs/ton
mg/L	= 30,000 mg/L	2. Polymer
Polymer		Cost, $/ton
Volume, mL	= 60 mL	
Strength, %	= 0.15%	
Cost, $/lb	= $1.50/lb	
Sample Vol, L	= 1 liter	
, gal	= 0.265 gal	

1. Determine the dosage in pounds of polymer per ton of sludge.

$$\text{Dosage, lbs/ton} = \frac{\text{Polymer Solution, \%} \times \text{Polymer Added, m}L \times 2}{\text{Sample Volume, }L \times \text{Sludge Conc, \%}}$$

$$= \frac{0.15 \times 60 \times 2}{1 \times 3.0}$$

$$= 6 \text{ lbs Dry Polymer/ton of Sludge}$$

2. Determine the cost in dollars of polymer per ton of sludge.

$$\text{Cost, \$/ton} = \text{Dosage, lbs/ton} \times \text{Polymer Cost, \$/lb}$$

$$= \frac{6 \text{ lbs Polymer}}{\text{ton of Sludge}} \times \frac{\$1.50}{\text{lb Polymer}}$$

$$= \$9.00/\text{ton of Sludge}$$

3.31I   A polymer solution of 2.5 percent is prepared from a liquid polymer and added at a rate of 3 GPM to a sludge flow of 30 GPM. The sludge has a solids content of 4 percent sludge solids. Determine the dosage (lbs/ton) and the cost ($/ton) if the liquid polymer costs $.20/lb.

Known		Unknown
Polymer Solution, %	= 2.5%	1. Dosage,
Polymer Flow Rate, GPM	= 3 GPM	lbs/ton
Sludge Flow, GPM	= 30 GPM	2. Cost,
Sludge Conc, %	= 4%	$/ton
Polymer Cost, $/lb	= $0.20/lb	

1. Determine the dosage of polymer in pounds of polymer per ton of sludge.

$$\text{Dosage, lbs/ton} = \frac{\text{Poly Sol, \%} \times \text{Poly Flow, GPM} \times 2{,}000 \text{ lbs/ton}}{\text{Sludge Flow, GPM} \times \text{Sludge Conc, \%}}$$

$$= \frac{2.5\% \times 3.0 \text{ GPM} \times 2{,}000 \text{ lbs/ton}}{30 \text{ GPM} \times 4\%}$$

$$= 125 \text{ lbs/ton}$$

2. Determine the cost in dollars of polymer per ton of sludge.

$$\text{Cost, \$/ton} = \text{Dosage, lbs/ton} \times \text{Polymer Cost, \$/lb}$$

$$= 125 \text{ lbs/ton} \times \$0.20/\text{lb}$$

$$= \$25/\text{ton of Sludge}$$

Answers to questions on page 228.

3.31J   Dry chemicals should be kept in a dry place to avoid chemical handling and transferring problems. If allowed to get wet, the dry chemicals will not move freely.

3.31K   The purpose of wetting dry polymers is to produce a properly mixed solution that will not have balls of undissolved polymer.

3.31L   Procedures to prepare a batch solution of dry chemicals.

1. Calculate amount of dry chemical needed.
2. Weigh out dry chemical.
3. Partially fill mix tank with water until impellers are submerged.
4. Turn on mixer.
5. Add premeasured dry product to mix tank.
6. Fill tank to desired level.
7. Allow to mix before use to sufficiently cure solution.
8. Turn off the mixer.

   Procedures to prepare a batch solution of liquid chemicals.

1. Calculate the volume of liquid chemical needed.
2. Measure the volume of liquid chemical.
3. Follow Steps 3 through 8.

3.31M   Curing time is important to allow the chemical to fully dissolve and be as effective as possible.

3.31N   Chemical tanks should be covered to prevent foreign material from entering and possibly clogging equipment. Polymers must be covered to protect polymers from ultraviolet sun rays.

3.31O   Polymers should not be added to the suction side of sludge feed pumps because the shearing forces through such pumps tend to shear any floc formation.

3.31P   If sludge thickening or dewatering inefficiencies cannot be traced back to equipment failures, check the chemical mixing (preparation) and addition equipment. With automatic feeding systems, the operator should check: (1) the level of dry product in the storage hopper and replenish, if necessary, (2) the screw conveyor and unplug, if necessary, (3) the quality of the solution, and (4) the chemical addition pump.

Answers to questions on page 230.

3.32A   When sludge particles are exposed to extreme heat at elevated pressures, the surrounding sheath hydrolyzes (decomposes) and ruptures the cell wall allowing bound water to escape.

3.32B   The performance and efficiency of thermal conditioning systems are affected by: (1) the concentration and consistency of the influent sludge, (2) reactor detention times, and (3) reactor temperature and pressure.

3.32C   Determine the reactor detention time for a reactor volume of 1,000 gallons and a sludge flow of 33 GPM with a concentration of 4.0 percent.

Known	Unknown
Reactor Volume, gal = 1,000 gal	Detention Time, min
Sludge Flow, GPM = 33 GPM	
Sludge Conc, % = 4.0%	

Calculate detention time in minutes.

$$\text{Detention Time, min} = \frac{\text{Reactor Volume, gal}}{\text{Flow, GPM}}$$

$$= \frac{1,000 \text{ gal}}{33 \text{ gal/min}}$$

$$= 30 \text{ min}$$

3.32D   If the sludge concentration from problem 3.32C decreases to 2.5 percent, determine the reactor detention time assuming the same total pounds are processed.

Known	Unknown
Information from 3.32C	Reactor Detention Time, min
Sludge Concentration Decreases to 2.5%	

Estimate the reactor detention time in minutes.

$$\text{Detention Time, min} = \frac{\text{Old Detention Time, min} \times \text{New Conc, \%}}{\text{Old Sludge Conc, \%}}$$

$$= 30 \text{ min} \times \frac{2.5\%}{4.0\%}$$

$$= 19 \text{ min}$$

3.32E   Operating controls available to optimize a thermal conditioning process include: (1) inlet sludge flow, (2) reactor temperature and detention time, and (3) sludge withdrawal from the decant tank.

3.32F   Gasification usually is not a problem in gravity thickeners with thermally treated sludge because of the lack of biological activity.

Answers to questions on page 231.

3.32G   Continuous operation of a heat treatment unit is desirable because energy is not wasted in allowing the heat exchanger and reactor contents to cool down and be heated back to the desired temperature each day when operated as a batch process.

3.32H   Start-up procedures for a heat treatment unit:

1. Fill reactor and heat exchangers with water, if necessary.
2. Turn on boiler makeup water pump and open valve to the steam boiler.
3. Open required steam valves and start the boiler.
4. After desired temperature is reached, open sludge inlet and outlet valves.
5. Turn on sludge grinder and stirring mechanisms.
6. Turn on vent fan and activate odor-control equipment.
7. Turn on sludge feed pump.

Reverse these procedures for shutdown.

3.32I   A log of the pressure drop across the heat exchangers must be kept so the operator can determine when the pressure drop is excessive. When the pressure drop becomes excessive, the system should be acid flushed to remove scale deposits and to unplug the heat exchangers.

3.32J   Loss of sludge dewaterability.

Possible Causes	Corrective Measures
1. Low temperature	1. Increase temperature. Check fuel supply, system instrumentation, and makeup water supply.
2. Low or short detention time	2. Check sludge flow. Thicken feed sludge.
3. Poor operation of decant	3. Check thickness of blanket and sludge concentration. Thicken underflow sludge.

Answers to questions on page 234.

3.33A   The major difference between LPO, IPO, and HPO is the pressure (low, 400 psig; intermediate, 500 to 600 psig; high, 1,000 to 1,500 psig) in the reactor. As the pressure increases, the amount of air reacted with the feed sludge and the temperatures also increase.

3.33B   The performance and efficiency of wet oxidation units are dependent on: (1) the concentration and consistency of the feed sludge, (2) reactor detention time, (3) reactor temperature and pressure, and (4) the quantity of air supplied.

3.33C   Air pollution control equipment is required on thermal treatment units due to the production of noxious odors.

Answer to question on page 235.

3.34A   Elutriation improves the dewaterability of sludge by washing out the fine, difficult-to-dewater solids. Problems associated with the elutriation process result from solids lost to the plant effluent with the elutriation effluent (elutriate). The loss of these fine solids into the plant effluent will deteriorate the effluent quality while recycling to the plant headworks generally results in operational problems due to buildup of fine solids throughout the system.

ANSWERS TO QUESTIONS IN LESSON 4

Answers to questions on page 240.

3.40A The primary objective of sludge dewatering is to reduce sludge moisture and consequently sludge volume to a degree that will allow for economical disposal.

3.40B Unit processes most often used for sludge dewatering are: (1) pressure filtration, (2) vacuum filtration, (3) centrifugation, and (4) sand drying beds.

3.41A Flow through plate and frame filter presses decreases with filtration time because as the cake builds up between the plates, the resistance to flow increases because the water must pass through thicker and thicker layers of compacted solids.

3.41B Pressure filtration performance is affected by: (1) sludge type, (2) conditioning, (3) filter pressure, (4) filtration time, (5) solids loadings, (6) filter cloth type, and (7) precoat.

3.41C Increasing the operating pressure might result in wetter cakes when dewatering secondary sludges. As the pressure is increased, the sludge retained on the filtering media may compress to a higher degree and reduce the porosity (openings) of the sludge cake that is formed. If the openings are reduced, fine, low-density solids may be captured, which results in wetter cakes because these solids have large surface areas and relatively large quantities of water associated with them.

3.41D Secondary sludges do not dewater as readily as primary sludges because secondary sludges contain fine, low-density solids that have large surface areas and relatively large quantities of water associated with them.

3.41E The time of filtration depends on the physical size of the filter and applied solids loading rate. The operator controls filtration time on the basis of the actual filtrate flow rate. When the cavities between the plates are filled with solids and the filtrate flow is almost zero, the filtering sequence is complete.

3.41F The purpose of precoating is to reduce the frequency of media washing and to facilitate cake discharge.

3.41G Normal operating procedures for a filter press are as follows:

1. Slurry precoat mix.
2. Slurry the lime, if used.
3. Transfer lime slurry to tank containing sludge and gently stir. Add ferric chloride, if used.
4. Apply the precoat material to the filter.
5. Introduce the conditioned sludge to the filter.
6. When the filtrate flow decreases to near zero, turn off the feed pump.
7. Disengage and open the press for cake discharge.
8. Close the press and repeat the above procedures.

Answers to questions on page 240.

3.41H If discharge cakes from a filter press are wet throughout, try to identify the cause and correct the problem. Causes of wet cakes include: (1) low filtration time, (2) low pressure, (3) chemical inefficiencies, and (4) changes in influent characteristics.

3.41I Solids may cling to filtering media when the cakes are discharged due to precoat inefficiencies.

Answers to questions on page 244.

3.41J The purpose of the drainage zone (portion of the belt) is to allow for most of the free water to drain through the filter and to be collected in a trough on the underside of the belt.

3.41K Mix chambers can be used to ensure adequate polymer and sludge contact.

3.41L Some belt filter presses use a reactor chamber instead of the horizontal drainage zone to allow most of the free water to drain out.

3.41M In the "press" or "dewatering zone," the entrapped solids are subjected to shearing forces created as the two belts travel over rollers that bring them closer and closer together. Water is forced from between the belts and collected in filtrate trays while the retained solids are scraped from the two belts when they separate at the discharge end of the press.

3.41N The ability of belt filter presses to dewater sludge and to remove suspended solids is dependent on: (1) sludge type, (2) conditioning, (3) belt tension or pressure, (4) belt speed, (5) hydraulic loading, and (6) belt type.

3.41O When using a belt press to dewater secondary sludges, the sludges may tend to slip toward the belt sides and eventually squeeze out from between the belts. The net effects are that these solids contaminate the effluent by falling into the filtrate trays and continuous housekeeping is required.

3.41P As the belt tension increases, the pressure on the sludge increases and more water is generally squeezed from the belt, which results in drier cakes.

3.41Q Washing out means that large quantities of free water travel to the dewatering zone and flow out from between the belts, drastically reducing effluent quality.

3.41R The ideal operating belt speed is the slowest the operator can maintain without washing out the belt.

3.41S As the belt speed decreases, cake dryness increases because the sludge is subjected to pressure and shearing forces for longer periods of time.

3.41T Porosity of a belt depends on the belt type. As the porosity increases, the resistance to flow decreases and larger volumes of water are able to be drained. If the porosity is too low, the belt may blind or plug, which will produce frequent washouts.

Answers to questions on page 245.

3.41U   If washing out of the belt develops, check: (1) polymer dosage, (2) mixing in the reactor, (3) hydraulic loading, (4) drum speed, (5) belt speed, and (6) belt washing equipment.

3.41V   Blinding can be corrected by reducing the polymer dosage.

Answers to questions on page 250.

3.41W   The purpose of the agitator in the trough is to keep the chemically conditioned sludge well mixed and to prevent the sludge from settling in the trough.

3.41X   In the "mat formation" or "sludge pick-up zone," the vacuum is applied to the compartment of the drum submerged in the trough. This vacuum causes the sludge to be picked up on the filter media and a sludge mat is formed.

In the "drying zone" of the cycle, the drum rotates out of the trough. When this occurs, the vacuum is decreased slightly and water is drawn from the sludge mat, through the filter media, and discharged through internal pipes to a drainage system.

3.41Y   The filter media passes through a washing zone to remove fine particles from the media and to reduce the possibility of media blinding.

3.41Z   Factors affecting vacuum filtration performance include: (1) sludge type, (2) conditioning, (3) applied vacuum, (4) drum speed or cycle time, (5) depth of submergence, and (6) media type and condition.

3.41AA   A vacuum range of 15 to 30 inches (38 to 75 cm) of mercury should be applied to the filter compartments.

3.41BB   The longer the cycle time the higher the degree of dewatering. Cycle time controls the rate of sludge pick-up and the thickness of the sludge mat in the formation zone. Also, cycle time controls the length of time the sludge remains in the drying zone.

3.41CC   Determine the filter yield (lbs/hr/sq ft) for a vacuum filter with a surface area of 300 sq ft. Digested sludge is applied at a rate of 75 GPM with a suspended solids concentration of 4.7 percent. The filter recovers 93 percent of the applied suspended solids.

Known	Unknown
Vacuum Filter	Filter Yield, lbs/hr/sq ft
Surface Area, sq ft = 300 sq ft	
Sludge Flow, GPM = 75 GPM	
Suspended Solids, % = 4.7%	
Filter Recovery, % = 93%	

Calculate the filter yield in pounds per hour applied per square foot of filter surface area.

$$\text{Filter Yield, lbs/hr/sq ft} = \frac{\text{Flow, GPM} \times 8.34 \text{ lbs/gal} \times 60 \text{ min/hr} \times \text{SS, }\% \times \text{Rec, }\%}{\text{Surface Area, sq ft} \times 100\% \times 100\%}$$

$$= \frac{75 \text{ gal/min} \times 8.34 \text{ lbs/gal} \times 60 \text{ min/hr} \times 4.7\% \times 93\%}{300 \text{ sq ft} \times 100\% \times 100\%}$$

$$= \frac{1,640 \text{ lbs/hr}}{300 \text{ sq ft}}$$

$$= 5.5 \text{ lbs/hr/sq ft}$$

3.41DD   As the porosity (openings) of the media increases, the ability to capture suspended solids decreases because fine, low-density solids can pass directly through the media. If the porosity of the media decreases too much, the media can blind with fine solids or chemical coatings, sludge will not be picked up in the formation zone, and the vacuum filter will become inoperative.

Answers to questions on page 251.

3.41EE   A loss of vacuum can be caused by: (1) filter media misaligned, (2) tear in filter media, (3) trough empty, and (4) vacuum pumps inoperative. A loss of vacuum will result in deterioration of the effluent quality and wet cakes that are difficult to discharge from the belt.

3.41FF   If sludge is not picked up in the mat formation zone, a poor effluent quality will result. To look for the cause of poor effluent quality, look for: (1) a loss of vacuum, or (2) insufficient chemical conditioning.

3.41GG   To increase cake dryness, the operator could: (1) increase vacuum, (2) reduce drum speed, and (3) improve chemical conditioning.

Answers to questions on page 252.

3.42A   Higher scroll speeds are usually required to dewater sludges as compared to sludge thickening because the concentration of feed sludge is somewhat higher for dewatering than for thickening.

3.42B   A scroll centrifuge is used to dewater 60 GPM of digested primary sludge at a concentration of 3.0 percent sludge solids. A liquid polymer is used for conditioning. The polymer solution is 2.5 percent and 2 GPM are added to the sludge stream. What is the hydraulic loading (GPH), the solids loading (lbs SS/hr), and the polymer dosage (lbs liq/ton)?

Known	Unknown
Sludge Flow, GPM = 60 GPM	1. Hydraulic Loading, gal/hr
Sludge Solids, % = 3.0%	2. Solids Loading, lbs Solids/hr
Polymer Solution, % = 2.5%	3. Polymer Dose, lbs Polymer/ton Sludge
Polymer Flow, GPM = 2 GPM	

1. Determine the hydraulic loading in gallons per hour.

$$\text{Hydraulic Loading, gal/hr} = \text{Flow, GPM} \times 60 \text{ min/hr}$$

$$= 60 \text{ gal/min} \times 60 \text{ min/hr}$$

$$= 3,600 \text{ gal/hr}$$

2. Calculate the solids loading in pounds per hour.

$$\text{Solids Loading, lbs/hr} = \text{Flow, GPM} \times 8.34 \frac{\text{lbs}}{\text{gal}} \times 60 \frac{\text{min}}{\text{hr}} \times \text{Solids, } \frac{\%}{100\%}$$

$$= 60 \frac{\text{gal}}{\text{min}} \times 8.34 \frac{\text{lbs}}{\text{gal}} \times 60 \frac{\text{min}}{\text{hr}} \times \frac{3.0\%}{100\%}$$

$$= 900 \text{ pounds Solids/hr}$$

3. Determine the polymer flow in pounds per hour.

$$\text{Polymer Flow, lbs/hr} = \text{Flow, GPM} \times 8.34 \frac{\text{lbs}}{\text{gal}} \times 60 \frac{\text{min}}{\text{hr}} \times \text{Polymer, } \frac{\%}{100\%}$$

$$= 2 \frac{\text{gal}}{\text{min}} \times 8.34 \frac{\text{lbs}}{\text{gal}} \times 60 \frac{\text{min}}{\text{hr}} \times \frac{2.5\%}{100\%}$$

$$= 25 \text{ lbs Polymer/hr}$$

4. Calculate the polymer dose in pounds of polymer applied per ton of sludge treated.

$$\text{Polymer Dose, } \frac{\text{lbs Polymer}}{\text{ton Sludge}} = \frac{\text{Polymer Flow, lbs/hr} \times 2,000 \text{ lbs/ton}}{\text{Solids Loading, lbs/hr}}$$

$$= \frac{25 \text{ lbs Polymer/hr} \times 2,000 \text{ lbs/ton}}{900 \text{ pounds Solids/hr}}$$

$$= 55.6 \text{ pounds Polymer per ton Sludge}$$

3.42C   A 48-inch diameter basket centrifuge is used to dewater 50 GPM of digested primary sludge. The feed is at a concentration of 2.7 percent sludge solids and polymers are added to achieve 95 percent suspended solids recovery. The average concentration of solids stored within the basket is 23 percent. The feed time is automatically set at 17 minutes. Is the feed time too long, too short, or OK? Assume the basket has a solids storage capacity of 16 cubic feet.

Known	Unknown
Basket Centrifuge, 48-in diameter	Is feed time OK?
Sludge Flow, GPM = 50 GPM	
Sludge Feed, % = 2.7% Solids	
Solids Recovery, % = 95%	
Stored Solids, % = 23%	
Feed Time, min = 17 min	
Basket Capacity, cu ft = 16 cu ft	

1. Determine the volume available to store solids in pounds.

$$\text{Solids Stored, lbs} = \text{Vol, cu ft} \times 7.48 \frac{\text{gal}}{\text{cu ft}} \times 8.34 \frac{\text{lbs}}{\text{gal}} \times \text{Solids, } \frac{\%}{100\%}$$

$$= 16 \text{ cu ft} \times 7.48 \frac{\text{gal}}{\text{cu ft}} \times 8.34 \frac{\text{lbs}}{\text{gal}} \times \frac{23\%}{100\%}$$

$$= 230 \text{ lbs Solids}$$

2. Determine solids retained in pounds per minute.

$$\text{Solids Retained, lbs/min} = \text{Flow, GPM} \times 8.34 \frac{\text{lbs}}{\text{gal}} \times \text{Solids, } \frac{\%}{100\%} \times \text{Recovery, } \frac{\%}{100\%}$$

$$= 50 \frac{\text{gal}}{\text{min}} \times 8.34 \frac{\text{lbs}}{\text{gal}} \times \frac{2.7\%}{100\%} \times \frac{95\%}{100\%}$$

$$= 10.7 \text{ lbs Solids/min}$$

3. Calculate feed time in minutes.

$$\text{Feed Time, min} = \frac{\text{Solids Stored, lbs}}{\text{Solids Retained, lbs/min}}$$

$$= \frac{230 \text{ lbs Solids}}{10.7 \text{ lbs Solids/min}}$$

$$= 21.5 \text{ min}$$

The feed time should be increased from 17 to 21 minutes.

3.42D   If the centrate quality is poor, but discharge solids are dry:

Possible Causes	Possible Solutions
1. Feed time too long	1. Lower feed time
2. Flow rate too high	2. Lower flow rate
3. Chemical insufficient	3. Increase chemical dosage

Answers to questions on page 256.

3.43A   Sludge drying beds are usually not used for sludges that have been stabilized by wet oxidation because of the odorous nature of thermally treated sludge.

3.43B   Factors affecting sand drying bed performance include: (1) sludge type, (2) conditioning, (3) climatic conditions, (4) sludge application rates and depths, and (5) dewatered sludge removal techniques.

3.43C   Care must be taken to prevent chemical overdosing for two reasons: (1) media blinding with unattached polymer may develop, and (2) large floc particles that settle too rapidly may also blind the media.

3.43D   Primary sludge from the bottom of an anaerobic digester may require prescreening because greases and hair-like stringy material can clog the sand bed.

3.43E   Drying beds are usually covered in wet or cold climates to protect the drying sludge from rain and to reduce the drying period during cold weather. Covered drying beds should be well ventilated to promote evaporation. The cover also serves to control odors and insects.

3.43F   A 150-foot long by 30-foot wide sand drying bed is loaded at 15 lbs/yr/sq ft. One application of sludge is made per month (12 applications per year). Determine the depth (in) of each application if the sludge has a concentration of 3.0 percent sludge solids.

Known		Unknown
Sand Drying Bed		Application
Length, ft	= 150 ft	Depth, in
Width, ft	= 30 ft	
Loading, lbs/yr/sq ft	= 15 lbs/yr/sq ft	
Applications, no/mo	= 1 per month	
Sludge Solids, %	= 3.0%	

1. Determine the total pounds of sludge that can be applied during the year.

$$\text{Sludge Applied,} \atop \text{lbs/yr} = \text{Loading, lbs/yr/sq ft} \times \text{L, ft} \times \text{W, ft}$$

$$= 15 \text{ lbs/yr/sq ft} \times 150 \text{ ft} \times 30 \text{ ft}$$

$$= 67,500 \text{ lbs/yr}$$

NOTE: Standby area not included in this calculation.

2. Calculate the total volume of sludge applied in gallons per year.

$$\text{Sludge Applied,} \atop \text{gallons/yr} = \frac{\text{Sludge Applied, lbs/yr}}{8.34 \text{ lbs/gal} \times \text{Sl Solids, \%}/100\%}$$

$$= \frac{67,500 \text{ lbs/yr}}{8.34 \text{ lbs/gal} \times 3.0\%/100\%}$$

$$= 269,784 \text{ gal/yr}$$

3. Calculate the depth of sludge in inches per application.

$$\text{Sludge Depth,} \atop \text{in/Appl} = \frac{\text{Sludge Applied, gal/yr} \times 12 \text{ in/ft}}{7.48 \text{ gal/cu ft} \times \text{Length, ft} \times \text{Width, ft} \times \text{Appl/yr}}$$

$$= \frac{269,784 \text{ gal/yr} \times 12 \text{ in/ft}}{7.48 \text{ gal/cu ft} \times 150 \text{ ft} \times 30 \text{ ft} \times 12 \text{ Appl/yr}}$$

$$= 8 \text{ in/Application}$$

3.43G   To determine the optimum depth of sludge to be applied to sand beds, the operator should apply different depths of sludge to different sand beds, allow the sludge to dry and be removed, and then calculate the loading rate in pounds of sludge per year per square foot of drying bed area. The highest loading rate indicates the optimum depth of sludge.

3.43H   Sand bed compaction should be avoided to prevent reduced drainage rates, longer drying times, and an increased potential for plugging.

3.43I   The sand bed surface should be raked after sludge is removed to break up any scum or mat formations.

3.43J   The final concentration to which the sludge can be dried is dependent on the climatic conditions and time the sludge remains in the bed after most of the water has drained through the sand.

3.43K   The only problem that appears to develop at most sand drying bed installations is plugging of the media surface.

Answers to questions on page 259.

3.44A   One limitation of using sludge drying beds is that the dried sludge must be removed manually with forks or shovels.

3.44B   To fill a surfaced drying bed that has only gravel in the drainage trench, start by adding water until the gravel is flooded. Digested sludge may be applied to the drying bed after the gravel is flooded.

3.44C   When water-sludge separation is observed in a beaker containing digested sludge, partially open the drying bed drain line valve to drain off the water from the drying bed.

3.45A   A successful sludge dewatering program requires that: (1) the operator be very familiar with the operation of the particular dewatering device(s) used, (2) sludge conditioning be optimum, and (3) the influent sludge be as thick and consistent as possible.

ANSWERS TO QUESTIONS IN LESSON 5

Answers to questions on page 266.

3.50A   The difference between drying and incineration is that drying removes water from sludge without the combustion of solid material while incineration results in combustion or burning of solid material. Composting is a drying process that removes moisture without the combustion of solid material.

3.51A   A suitable environment must be established in compost piles to ensure the health and growth of the thermophilic facultative aerobic microorganisms.

3.51B   Guidelines that must be met to create a suitable compost environment include:

1. Sludges must be blended with previously composted material or bulking agents such as sawdust, straw, wood shavings, or rice hulls.
2. Aeration must be sufficient to maintain aerobic conditions in the composting material.
3. Proper moisture content and temperatures must be maintained.

3.51C   Chemically stabilized and wet-oxidized sludges are generally not suited for compost operations. Chemical stabilization produces environments that are unsuitable for microorganism survival and will not support life of composting bacteria unless the sludges are neutralized and favorable conditions exist. Sludge that has been stabilized by wet oxidation can be composted, but noxious odors are likely to occur in and around the compost operation.

3.51D Factors affecting composting include: (1) sludge type, (2) initial moisture content and homogeneity of the mixture, (3) frequency of aeration or windrow turning, (4) climatic conditions, and (5) desired moisture content of the final product.

3.51E Secondary sludges are not as easy to compost as primary sludges because of the plastic nature of dewatered secondary sludge and its higher moisture content. Dewatered secondary sludges tend to clump together and form "balls" when they are blended with compost material. The "balls" that are formed within the windrows readily dry on the outer surface but remain moist on the inside. The net effect of this "balling" phenomenon is the occasional creation of anaerobic conditions with odor production and a reduction in composting temperature.

3.51F A medium-sized wastewater treatment plant produces 4,700 lbs/day of dewatered digested primary sludge with a solids concentration of 27 percent and 3,300 lbs/day of dewatered secondary sludge with a solids concentration of 15 percent. The sludges are blended together and then composted in windrows. Determine the total pounds of compost that must be recycled and blended with the combined sludge to produce a moisture of 60 percent. The final compost product of the plant has a moisture content of 30 percent.

Known		Unknown
Dewatered Primary Sludge Solids, %	= 27%	Compost, lbs/day (needed to produce 60% initial moisture content of mixture)
Dewatered Primary Sludge Solids, lbs/day	= 4,700 lbs/day	
Dewatered Secondary Sludge Solids, %	= 15%	
Dewatered Secondary Sludge Solids, lbs/day	= 3,300 lbs/day	
Combined Sludge Moisture, %	= 60%	
Compost Product Moisture, %	= 30%	

1. Determine the moisture content of the dewatered primary and secondary sludges in percent.

$$\text{Primary Sludge Moist, \%} = 100\% - \text{Primary Sludge Solids, \%}$$
$$= 100\% - 27\%$$
$$= 73\%$$

$$\text{Secondary Sludge Moist, \%} = 100\% - \text{Secondary Sludge Solids, \%}$$
$$= 100\% - 15\%$$
$$= 85\%$$

2. Determine the pounds per day of compost products that must be recycled and blended.

$$\text{Mixture Moisture, \%} = \frac{\text{Primary, lbs/day} \times \text{Pri Moist, \%} + \text{Sec, lbs/day} \times \text{Sec Moist, \%} + \text{Compost, lbs/day} \times \text{Mix Moist, \%}}{\text{Primary, lbs/day} + \text{Sec, lbs/day} + \text{Compost, lbs/day}}$$

$$60\% = \frac{4,700 \text{ lbs/day} \times 73\% + 3,300 \text{ lbs/day} \times 85\% + \text{Compost, lbs/day} \times 30\%}{4,700 \text{ lbs/day} + 3,300 \text{ lbs/day} + \text{Compost, lbs/day}}$$

3. Divide both sides of the equation by 100%.

$$0.60 = \frac{3,431 + 2,805 + 0.30 \text{ Compost, lbs/day}}{8,000 + \text{Compost, lbs/day}}$$

$$0.60 \, (8,000 + \text{Compost, lbs/day}) = 3,431 + 2,805 + 0.30 \text{ Compost, lbs/day}$$

$$4,800 + 0.60 \text{ Compost, lbs/day} = 6,236 + 0.30 \text{ Compost, lbs/day}$$

4. Subtract 0.30 Compost, lbs/day and 4,800 from both sides of the equation.

$$0.30 \text{ Compost, lbs/day} = 1,436$$

$$\text{Compost, lbs/day} = \frac{1,436}{0.30}$$

$$= 4,787 \text{ lbs/day}$$

or ALTERNATIVE SOLUTION

1. Determine pounds of moisture in primary sludge.

$$\text{Pri Sludge Moist, lbs} = (\text{Pri Sludge, lbs})(\text{Moist, \%}/100\%)$$
$$= (4,700 \text{ lbs})(73\%/100\%)$$
$$= 3,431 \text{ lbs Moisture}$$

2. Determine pounds of moisture in secondary sludge.

$$\text{Sec Sludge Moist, lbs} = (\text{Sec Sludge, lbs})(\text{Moist, \%}/100\%)$$
$$= (3,300 \text{ lbs})(85\%/100\%)$$
$$= 2,805 \text{ lbs Moisture}$$

3. Determine percent moisture of mixture of primary and secondary sludge.

$$\text{Mix Moist, \%} = \frac{(\text{Pri Moist, lbs} + \text{Sec Moist, lbs})(100\%)}{\text{Pri Sludge, lbs} + \text{Sec Sludge, lbs}}$$

$$= \frac{(3,431 \text{ lbs} + 2,805 \text{ lbs})(100\%)}{4,700 \text{ lbs} + 3,300 \text{ lbs}}$$

$$= 77.95\% \text{ or } 78\%$$

4. Calculate the pounds per day of compost that must be recycled to produce an initial mixture moisture of 60 percent.

$$\text{Compost, lbs/day} = \text{Sludge, lbs/day} \frac{(\text{Sl Mix M, \%} - \text{Comp Mix M, \%})}{(\text{Comp Mix M, \%} - \text{Comp Prod M, \%})}$$

$$= 8,000 \text{ lbs/day} \frac{(78\% - 60\%)}{(60\% - 30\%)}$$

$$= 4,787 \text{ lbs/day}$$

3.51G   Operational procedures for windrow composting.

1. Dewater sludge to highest degree practical.
2. Blend dewatered sludge with recycled compost or bulking agents to a consistency that will stack.
3. Form the windrow piles and turn (aerate) once or twice daily for the first 4 to 5 days after windrow formation.
4. Turn the piles approximately once every two days to once a week until the process is complete.
5. Load the compost onto trucks for disposal or recycle purposes.

Answers to questions on page 270.

3.52A   Indirect dryers use indirect contact of sludge with preheated gases by circulating steam through a jacketed hollow in an outer shell of a rotating cylindrical compartment. Direct dryers use the direct contact of sludge with preheated gases.

3.52B   Flights on rotary dryers elevate and mix the sludge being dried to provide frequent contact of all wetted particles with hot gas streams or heated surfaces for direct or indirect drying, respectively.

3.52C   Blending of the sludge with the dried product is generally practiced to improve the conveying characteristics of the sludge and to reduce the potential for balling.

3.52D   Sludge should be dewatered to the highest degree practical to reduce the volume of water delivered to the dryer and to facilitate the drying process.

3.52E   As the drum speed increases, the drying time decreases because the sludge is picked up and tumbled toward the outlet at a faster rate. Too high a drum speed will not provide time for complete drying of the sludge.

3.52F   The major problems that are encountered when operating rotary dryers include: (1) dried product too wet, (2) sludge balling, (3) excessive sludge or dust in exhaust gas, and (4) vibrations, jam, or overload.

Answers to questions on page 277.

3.53A   Sludge incineration is the conversion of dewatered sludge cake by combustion to ash, carbon dioxide, and water vapor.

3.53B   The refractory is the heat-resistant material that lines the steel cylinder of a multiple-hearth furnace.

3.53C   On the center shaft there are arms to which plows are attached. These arms are called rabble arms and the plows are called rabble teeth.

3.53D   The purpose of the lute cap is to prevent air and sludge from passing through the shaft opening, rather than the drop holes.

3.53E   The purpose of the sand seal is to prevent the escape of heat and gases from the furnace and the entrance of air.

Answers to questions on page 280.

3.53F   Furnace off-gas system.

Part	Purpose
1. Emergency bypass damper	Vents gases to the atmosphere during emergency conditions. This device protects equipment and operating personnel.
2. Cyclone separator	Causes fly ash and heavy particles to settle out into the cyclone bin.
3. Precooler	Cools the furnace exhaust gases to saturation temperature and wets the small particles of light ash.
4. Venturi scrubber	Cleans the particulate matter from the cooled furnace gases.
5. Impingement scrubber	Traps remaining particles in flowing water.
6. Induced draft fan	Pulls gases through the off-gas system and vents gases.
7. Induced draft damper	Regulates suction or draft within the MHF.
8. Ash handling system	Removes ash for ultimate disposal.

3.53G   Burners are provided to supply the necessary heat to ignite the sludge.

3.53H   The three ingredients necessary for combustion to occur are fuel, oxygen, and heat.

Answers to questions on page 285.

3.53I   The three distinct zones in a furnace are the drying, combustion, and cooling zones.

3.53J   The factors that influence the amount of fuel required include:

1. Conditions in the furnace
2. Moisture content of the sludge
3. Volatile content of the solids
4. Feed rate of the cake

3.53K Combustion is a chemical reaction that requires oxygen, fuel, and heat. In the furnace, air provides oxygen, the primary fuel is sludge, and heat comes from the burning sludge.

3.53L A burnout occurs when the sludge feed has been stopped and the fire continues to burn.

3.53M To control a burnout, the operator should take the following steps:

1. Open the doors at or just above the fire
2. Repeatedly stop and restart the center shaft or change the shaft speed
3. Reduce or shut off the burners

Answers to questions on page 287.

3.53N To bring a cold furnace up to temperature:

1. Slowly increase the temperature until the temperature is up to 200°F (93°C) throughout the furnace.
2. Hold the temperature at 200°F (93°C) until the refractory is dry and warm.
3. Increase the temperature at a rate of 50°F/hr (28°C/hr) until the temperature on a given hearth reaches 1,000°F (540°C).
4. Once the temperature of a hearth reaches 1,000°F (540°C), the temperature of the hearth may be increased at a rate of 100°F/hr (56°C/hr) until the burning zone of the furnace reaches 1,600°F (870°C). At this point the feed to the furnace may be started.

3.53O An autogenous burn occurs when the volatile content of the sludge cake is high enough that the cake will burn without the additional heat input from the burners.

3.53P MHFs should be operated on a continuous basis to take advantage of the heat from the burning sludge. In addition, continuous operation extends refractory life.

3.53Q Smoke is caused by too low an oxygen content in the furnace. The solution to this problem is to add air at or below the fire.

3.53R When in the furnace area, wear protective clothing including heavy leather gloves, face shield, hard hat, long-sleeved shirt, and long pants. *ALWAYS* follow confined space entry procedures whenever anyone must enter a furnace.

Answer to question on page 287.

3.54A Three purposes of facultative sludge storage lagoons are to:

1. Reduce the volume of sludge
2. Store sludge
3. Stabilize sludge

Answers to questions on page 289.

3.60A Major areas of concern regarding the disposal of sludge include emissions to the atmosphere from furnaces, groundwater and surface water contamination, and the health aspects of sludge applied to land used for food crops.

3.60B Sludge or biosolids may be safely applied to food crops if they are applied in accordance with federal and state guidelines. Use of biosolids on agricultural land requires monitoring of nitrogen, heavy metals, toxic substances, and pathogens.

Answers to questions on page 296.

3.61A Mechanically dewatered sludge may be disposed of by:

1. Sanitary landfill disposal
2. On-site dedicated land disposal
3. Agricultural reclamation
4. Composting and utilization

3.61B The following types of surface runoff control facilities must be provided at an on-site dedicated land disposal operation:

1. Flood protection. The disposal site should be protected from flooding by a continuous dike.
2. Existing drainage. The existing drainage into the disposal site should be intercepted and directed outside the flood-protection dike.
3. Contaminated runoff. Runoff from the disposal site should be collected in a detention basin and allowed to evaporate during the summer or recycled to the treatment plant headworks.

Answers to questions on page 301.

3.61C Liquid digested sludge can be disposed of by:

1. High-rate incorporation into the surface soils of a site dedicated to land disposal (DLD)
2. Low-rate application to agricultural sites
3. Confinement in permanent lagoons

3.61D Liquid sludge can be spread over land by:

1. Ridge and furrow
2. Flooding
3. Subsurface injection

3.61E The surface layer of liquid on permanent lagoons must be aerobic to minimize nuisance odors and vector problems.

3.61F Composted material can be disposed of by:

1. Providing a bagged commercial product for sale to the public as a soil amendment
2. Providing compost for agricultural land use
3. Providing compost for nonagricultural land uses

Answers to questions on page 303.

3.62A   An odor-monitoring program should monitor:

1. Meteorological conditions (air temperature at 5 and 25 feet above the ground, wind direction, and wind speed)
2. Number of complaints

3.62B   There should be at least four test wells for a DLD site. Two of these should not extend below the confining soil layer and two should extend below this level.

Answer to question on page 303.

3.7A   Important design provisions to look for when reviewing plans and specifications include:

1. Space for operation and maintenance of valves and pumps
2. Meters and gauges easily readable and located for ease in process adjustment, if necessary
3. Sufficient area, wash water capacity, and drains to maintain, repair, and clean up equipment and areas
4. Provision for handling sludge when equipment fails

CHAPTER 4

SOLIDS REMOVAL FROM SECONDARY EFFLUENTS

by

Paul Amodeo

Ross Gudgel

James L. Johnson

Paul J. Kemp

Robert G. Blanck

Francis J. Brady

TABLE OF CONTENTS
Chapter 4. SOLIDS REMOVAL FROM SECONDARY EFFLUENTS

LESSON 6

OBJECTIVES

Chapter 4.   SOLIDS REMOVAL FROM SECONDARY EFFLUENTS

Following completion of Chapter 4, you should be able to:

CHEMICALS

1. Describe the proper procedures for using chemicals to remove solids from your treatment plant's secondary effluent.

2. Operate and maintain chemical feed equipment.

3. Safely store and handle chemicals.

4. Review plans and specifications of chemical feed systems.

5. Start up and shut down a chemical feed system.

6. Perform a jar test.

7. Select the most cost-effective chemicals and determine proper dosage.

8. Troubleshoot a chemical feed system.

9. Develop an operational strategy for a chemical feed system.

FILTRATION

1. Identify and describe the components of gravity and pressure filters.

2. Explain how membrane filters operate.

3. Safely operate and maintain filters.

4. Start up and shut down filters.

5. Troubleshoot a filtration system.

6. Develop operational strategies for inert-media and membrane filtration systems.

7. Review plans and specifications for filter systems.

WORDS

Chapter 4. SOLIDS REMOVAL FROM SECONDARY EFFLUENTS

AGE TANK   AGE TANK

A tank used to store a chemical solution of known concentration for feed to a chemical feeder. It usually stores sufficient chemical solution to properly treat the water being treated for at least one day. Also called a DAY TANK.

AGGLOMERATION (uh-glom-er-A-shun)   AGGLOMERATION

The growing or coming together of small scattered particles into larger flocs or particles, which settle rapidly. Also see FLOC.

AIR BINDING   AIR BINDING

The clogging of a filter, pipe, or pump due to the presence of air released from water. Air entering the filter media is harmful to both the filtration and backwash processes. Air can prevent the passage of water during the filtration process and can cause the loss of filter media during the backwash process.

ALGAE (AL-jee)   ALGAE

Microscopic plants containing chlorophyll that live floating or suspended in water. They also may be attached to structures, rocks, or other submerged surfaces. Excess algal growths can impart tastes and odors to potable water. Algae produce oxygen during sunlight hours and use oxygen during the night hours. Their biological activities appreciably affect the pH, alkalinity, and dissolved oxygen of the water.

ALKALINITY (AL-kuh-LIN-it-tee)   ALKALINITY

The capacity of water or wastewater to neutralize acids. This capacity is caused by the water's content of carbonate, bicarbonate, hydroxide, and occasionally borate, silicate, and phosphate. Alkalinity is expressed in milligrams per liter of equivalent calcium carbonate. Alkalinity is not the same as pH because water does not have to be strongly basic (high pH) to have a high alkalinity. Alkalinity is a measure of how much acid must be added to a liquid to lower the pH to 4.5.

ANHYDROUS (an-HI-drous)   ANHYDROUS

Very dry. No water or dampness is present.

ANION (AN-EYE-en)   ANION

A negatively charged ion in an electrolyte solution, attracted to the anode under the influence of a difference in electrical potential. Chloride ion (Cl^-) is an anion.

ANNULAR (AN-yoo-ler) SPACE   ANNULAR SPACE

A ring-shaped space located between two circular objects. For example, the space between the outside of a pipe liner and the inside of a pipe.

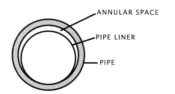

BOD (pronounce as separate letters)   BOD

Biochemical Oxygen Demand. The rate at which organisms use the oxygen in water or wastewater while stabilizing decomposable organic matter under aerobic conditions. In decomposition, organic matter serves as food for the bacteria and energy results from its oxidation. BOD measurements are used as a surrogate measure of the organic strength of wastes in water.

BASE

(1) A substance that takes up or accepts protons.

(2) A substance that dissociates (separates) in aqueous solution to yield hydroxyl ions (OH^-).

(3) A substance containing hydroxyl ions that reacts with an acid to form a salt or that may react with metals to form precipitates.

BATCH PROCESS

A treatment process in which a tank or reactor is filled, the water (or wastewater or other solution) is treated or a chemical solution is prepared, and the tank is emptied. The tank may then be filled and the process repeated. Batch processes are also used to cleanse, stabilize, or condition chemical solutions for use in industrial manufacturing and treatment processes.

BULKING

Clouds of billowing sludge that occur throughout secondary clarifiers and sludge thickeners when the sludge does not settle properly. In the activated sludge process, bulking is usually caused by filamentous bacteria or bound water.

COD (pronounce as separate letters)

Chemical Oxygen Demand. A measure of the oxygen-consuming capacity of organic matter present in wastewater. COD is expressed as the amount of oxygen consumed from a chemical oxidant in mg/L during a specific test. Results are not necessarily related to the biochemical oxygen demand (BOD) because the chemical oxidant may react with substances that bacteria do not stabilize.

COAGULANT (ko-AGG-yoo-lent)

A chemical that causes very fine particles to clump (floc) together into larger particles. This makes it easier to separate the solids from the liquids by settling, skimming, draining, or filtering.

COAGULANT (ko-AGG-yoo-lent) AID

Any chemical or substance used to assist or modify coagulation.

COAGULATION (ko-agg-yoo-LAY-shun)

The clumping together of very fine particles into larger particles (floc) caused by the use of chemicals (coagulants). The chemicals neutralize the electrical charges of the fine particles, allowing them to come closer and form larger clumps.

COLIFORM (KOAL-i-form)

A group of bacteria found in the intestines of warm-blooded animals (including humans) and also in plants, soil, air, and water. The presence of coliform bacteria is an indication that the water is polluted and may contain pathogenic (disease-causing) organisms. Fecal coliforms are those coliforms found in the feces of various warm-blooded animals, whereas the term "coliform" also includes other environmental sources.

COLLOIDS (KALL-loids)

Very small, finely divided solids (particles that do not dissolve) that remain dispersed in a liquid for a long time due to their small size and electrical charge. When most of the particles in water have a negative electrical charge, they tend to repel each other. This repulsion prevents the particles from clumping together, becoming heavier, and settling out.

COMPOSITE (PROPORTIONAL) SAMPLE

A composite sample is a collection of individual samples obtained at regular intervals, usually every one or two hours during a 24-hour time span. Each individual sample is combined with the others in proportion to the rate of flow when the sample was collected. Equal volume individual samples also may be collected at intervals after a specific volume of flow passes the sampling point or after equal time intervals and still be referred to as a composite sample. The resulting mixture (composite sample) forms a representative sample and is analyzed to determine the average conditions during the sampling period.

CONCENTRATION POLARIZATION

(1) A buildup of retained particles on the membrane surface due to dewatering of the feed closest to the membrane. The thickness of the concentration polarization layer is controlled by the flow velocity across the membrane.

(2) Used in corrosion studies to indicate a depletion of ions near an electrode.

(3) The basis for chemical analysis by a polarograph.

CONFINED SPACE CONFINED SPACE

Confined space means a space that:

(1) Is large enough and so configured that an employee can bodily enter and perform assigned work; and

(2) Has limited or restricted means for entry or exit (for example, manholes, tanks, vessels, silos, storage bins, hoppers, vaults, and pits are spaces that may have limited means of entry); and

(3) Is not designed for continuous employee occupancy.

Also see DANGEROUS AIR CONTAMINATION and OXYGEN DEFICIENCY.

CONTINUOUS PROCESS CONTINUOUS PROCESS

A treatment process in which water is treated continuously in a tank or reactor. The water being treated continuously flows into the tank at one end, is treated as it flows through the tank, and flows out the opposite end as treated water.

DALTON DALTON

A unit of mass designated as one-sixteenth the mass of oxygen-16, the lightest and most abundant isotope of oxygen. The dalton is equivalent to one mass unit.

DAY TANK DAY TANK

A tank used to store a chemical solution of known concentration for feed to a chemical feeder. A day tank usually stores sufficient chemical solution to properly treat the water being treated for at least one day. Also called an AGE TANK.

DECANT (de-KANT) DECANT

To draw off the upper layer of liquid (water) after the heavier material (a solid or another liquid) has settled.

DELAMINATION (DEE-lam-uh-NAY-shun) DELAMINATION

Separation of a membrane or other material from the backing material on which it is cast.

DETENTION TIME DETENTION TIME

(1) The time required to fill a tank at a given flow.

(2) The theoretical (calculated) time required for water to pass through a tank at a given rate of flow.

(3) The actual time in hours, minutes, or seconds that a small amount of water is in a settling basin, flocculating basin, or rapid-mix chamber. In septic tanks, detention time will decrease as the volumes of sludge and scum increase. In storage reservoirs, detention time is the length of time entering water will be held before being drafted for use (several weeks to years, several months being typical).

$$\text{Detention Time, hr} = \frac{(\text{Basin Volume, gal})(24 \text{ hr/day})}{\text{Flow, gal/day}}$$

or

$$\text{Detention Time, hr} = \frac{(\text{Basin Volume, m}^3)(24 \text{ hr/day})}{\text{Flow, m}^3/\text{day}}$$

DIAPHRAGM PUMP DIAPHRAGM PUMP

A pump in which a flexible diaphragm, generally of rubber or equally flexible material, is the operating part. It is fastened at the edges in a vertical cylinder. When the diaphragm is raised, suction is exerted, and when it is depressed, the liquid is forced through a discharge valve.

DIATOMS (DYE-uh-toms) DIATOMS

Unicellular (single cell), microscopic algae with a rigid, box-like internal structure consisting mainly of silica.

ELECTROLYTE (ee-LECK-tro-lite) ELECTROLYTE

A substance that dissociates (separates) into two or more ions when it is dissolved in water.

FILTER AID FILTER AID

A chemical (usually a polymer) added to water to help remove fine colloidal suspended solids.

FLOC FLOC

Clumps of bacteria and particles, or coagulants and impurities, that have come together and formed a cluster. Found in flocculation tanks, sedimentation basins, aeration tanks, secondary clarifiers, and chemical precipitation processes.

FLOCCULATION (flock-yoo-LAY-shun) FLOCCULATION

The gathering together of fine particles after coagulation to form larger particles by a process of gentle mixing. This clumping together makes it easier to separate the solids from the water by settling, skimming, draining, or filtering.

FLOW EQUALIZATION SYSTEM FLOW EQUALIZATION SYSTEM

A device or tank designed to hold back or store a portion of peak flows for release during low-flow periods.

FLUIDIZED (FLOO-id-i-zd) FLUIDIZED

A mass of solid particles that is made to flow like a liquid by injection of water or gas is said to have been fluidized. In water and wastewater treatment, a bed of filter media is fluidized by backwashing water through the filter.

HEAD LOSS HEAD LOSS

The head, pressure, or energy (they are the same) lost by water flowing in a pipe or channel as a result of turbulence caused by the velocity of the flowing water and the roughness of the pipe, channel walls, or restrictions caused by fittings. Water flowing in a pipe loses head, pressure, or energy as a result of friction losses. The head loss through a filter is due to friction losses caused by material building up on the surface or in the top part of a filter. Also called FRICTION LOSS.

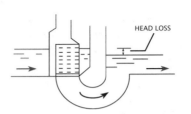

HYDROLYSIS (hi-DROLL-uh-sis) HYDROLYSIS

(1) A chemical reaction in which a compound is converted into another compound by taking up water.

(2) Usually a chemical degradation of organic matter.

INDICATOR INDICATOR

(1) (Chemical indicator) A substance that gives a visible change, usually of color, at a desired point in a chemical reaction, generally at a specified end point.

(2) (Instrument indicator) A device that indicates the result of a measurement, usually using either a fixed scale and movable indicator (pointer), such as a pressure gauge, or a moving chart with a movable pen like those used on a circular flow-recording chart. Also called a RECEIVER.

INTEGRATOR INTEGRATOR

A device or meter that continuously measures and sums a process rate variable in cumulative fashion over a given time period. For example, total flows displayed in gallons per minute, million gallons per day, cubic feet per second, or some other unit of volume per time period. Also called a TOTALIZER.

JAR TEST JAR TEST

A laboratory procedure that simulates coagulation/flocculation with differing chemical doses. The purpose of the procedure is to estimate the minimum coagulant dose required to achieve certain water quality goals. Samples of water to be treated are placed in six jars. Various amounts of chemicals are added to each jar, stirred, and the settling of solids is observed. The lowest dose of chemicals that provides satisfactory settling is the dose used to treat the water.

LAUNDERS LAUNDERS

Sedimentation basin and filter discharge channels consisting of overflow weir plates (in sedimentation basins) and conveying troughs.

LINEAL (LIN-e-ul)   LINEAL

The length in one direction of a line. For example, a board 12 feet (meters) long has 12 lineal feet (meters) in its length.

MATERIAL SAFETY DATA SHEET (MSDS)   MATERIAL SAFETY DATA SHEET (MSDS)

A document that provides pertinent information and a profile of a particular hazardous substance or mixture. An MSDS is normally developed by the manufacturer or formulator of the hazardous substance or mixture. The MSDS is required to be made available to employees and operators or inspectors whenever there is the likelihood of the hazardous substance or mixture being introduced into the workplace. Some manufacturers are preparing MSDSs for products that are not considered to be hazardous to show that the product or substance is not hazardous.

MICRON (MY-kron)   MICRON

μm, Micrometer or Micron. A unit of length. One millionth of a meter or one thousandth of a millimeter. One micron equals 0.00004 of an inch.

MIL   MIL

A unit of length equal to 0.001 of an inch. The diameter of wires and tubing is measured in mils, as is the thickness of plastic sheeting.

MUDBALLS   MUDBALLS

Material, approximately round in shape, that forms in filters and gradually increases in size when not removed by the backwashing process. Mudballs vary from pea-sized up to golf-ball-sized or larger.

NPDES PERMIT   NPDES PERMIT

National Pollutant Discharge Elimination System permit is the regulatory agency document issued by either a federal or state agency that is designed to control all discharges of potential pollutants from point sources and stormwater runoff into US waterways. NPDES permits regulate discharges into US waterways from all point sources of pollution, including industries, municipal wastewater treatment plants, sanitary landfills, large animal feedlots, and return irrigation flows.

NAMEPLATE   NAMEPLATE

A durable, metal plate found on equipment that lists critical operating conditions for the equipment.

NEPHELOMETRIC (neff-el-o-MET-rick)   NEPHELOMETRIC

A means of measuring turbidity in a sample by using an instrument called a nephelometer. A nephelometer passes light through a sample and the amount of light deflected (usually at a 90-degree angle) is then measured.

NEUTRALIZATION (noo-trull-uh-ZAY-shun)   NEUTRALIZATION

Addition of an acid or alkali (base) to a liquid to cause the pH of the liquid to move toward a neutral pH of 7.0.

ORIFICE (OR-uh-fiss)   ORIFICE

An opening (hole) in a plate, wall, or partition. An orifice flange or plate placed in a pipe consists of a slot or a calibrated circular hole smaller than the pipe diameter. The difference in pressure in the pipe above and at the orifice may be used to determine the flow in the pipe. In a trickling filter distributor, the wastewater passes through an orifice to the surface of the filter media.

OVERFLOW RATE   OVERFLOW RATE

One factor of the design flow of settling tanks and clarifiers in treatment plants used by operators to determine if tanks and clarifiers are hydraulically (flow) over- or underloaded. Also called SURFACE LOADING.

$$\text{Overflow Rate, GPD/sq ft} = \frac{\text{Flow, gallons/day}}{\text{Surface Area, sq ft}}$$

or

$$\text{Overflow Rate, } \frac{\text{m}^3/\text{day}}{\text{m}^2} = \frac{\text{Flow, m}^3/\text{day}}{\text{Surface Area, m}^2}$$

PATHOGENIC (path-o-JEN-ick) ORGANISMS

PATHOGENIC ORGANISMS

Organisms, including bacteria, viruses, or cysts, capable of causing diseases (such as giardiasis, cryptosporidiosis, typhoid, cholera, dysentery) in a host (such as a person). Also called PATHOGENS.

POLYELECTROLYTE (POLY-ee-LECK-tro-lite)

POLYELECTROLYTE

A high-molecular-weight (relatively heavy) substance, having points of positive or negative electrical charges, that is formed by either natural or synthetic (manmade) processes. Natural polyelectrolytes may be of biological origin or obtained from starch products or cellulose derivatives. Synthetic polyelectrolytes consist of simple substances that have been made into complex, high-molecular-weight substances. Used with other chemical coagulants to aid in binding small suspended particles to larger chemical flocs for their removal from water. Often called a POLYMER.

POLYMER (POLY-mer)

POLYMER

A long-chain molecule formed by the union of many monomers (molecules of lower molecular weight). Polymers are used with other chemical coagulants to aid in binding small suspended particles to larger chemical flocs for their removal from water. Also see POLYELECTROLYTE.

POTTING COMPOUNDS

POTTING COMPOUNDS

Sealing and holding compounds (such as epoxy) used in electrode probes.

PRECIPITATE (pre-SIP-uh-TATE)

PRECIPITATE

(1)   An insoluble, finely divided substance that is a product of a chemical reaction within a liquid.

(2)   The separation from solution of an insoluble substance.

PROGRAMMABLE LOGIC CONTROLLER (PLC)

PROGRAMMABLE LOGIC CONTROLLER (PLC)

A microcomputer-based control device containing programmable software; used to control process variables.

RECEIVER

RECEIVER

A device that indicates the result of a measurement, usually using either a fixed scale and movable indicator (pointer), such as a pressure gauge, or a moving chart with a movable pen like those used on a circular flow-recording chart. Also called an INDICATOR.

RISING SLUDGE

RISING SLUDGE

Rising sludge occurs in the secondary clarifiers of activated sludge plants when the sludge settles to the bottom of the clarifier, is compacted, and then starts to rise to the surface, usually as a result of denitrification, or anaerobic biological activity that produces carbon dioxide and/or methane.

ROTAMETER (ROTE-uh-ME-ter)

ROTAMETER

A device used to measure the flow rate of gases and liquids. The gas or liquid being measured flows vertically up a tapered, calibrated tube. Inside the tube is a small ball or bullet-shaped float (it may rotate) that rises or falls depending on the flow rate. The flow rate may be read on a scale behind or on the tube by looking at the middle of the ball or at the widest part or top of the float.

ROTARY PUMP

ROTARY PUMP

A type of displacement pump consisting essentially of elements rotating in a close-fitting pump case. The rotation of these elements alternately draws in and discharges the water being pumped. Such pumps act with neither suction nor discharge valves, operate at almost any speed, and do not depend on centrifugal forces to lift the water.

SPC CHART

SPC CHART

Statistical Process Control chart. A plot of daily performance such as a trend chart.

SEPTIC (SEP-tick) or SEPTICITY

SEPTIC or SEPTICITY

A condition produced by bacteria when all oxygen supplies are depleted. If severe, the bottom deposits produce hydrogen sulfide, the deposits and water turn black, give off foul odors, and the water has a greatly increased oxygen and chlorine demand.

SHORT-CIRCUITING SHORT-CIRCUITING

A condition that occurs in tanks or basins when some of the flowing water entering a tank or basin flows along a nearly direct pathway from the inlet to the outlet. This is usually undesirable since it may result in shorter contact, reaction, or settling times in comparison with the theoretical (calculated) or presumed detention times.

SLAKE SLAKE

To mix with water so that a true chemical combination (hydration) takes place, such as in the slaking of lime.

SPECIFIC GRAVITY SPECIFIC GRAVITY

(1) Weight of a particle, substance, or chemical solution in relation to the weight of an equal volume of water. Water has a specific gravity of 1.000 at 4°C (39°F). Particulates with specific gravity less than 1.0 float to the surface and particulates with specific gravity greater than 1.0 sink.

(2) Weight of a particular gas in relation to the weight of an equal volume of air at the same temperature and pressure (air has a specific gravity of 1.0). Chlorine gas has a specific gravity of 2.5.

SURFACE LOADING SURFACE LOADING

One factor of the design flow of settling tanks and clarifiers in treatment plants used by operators to determine if tanks and clarifiers are hydraulically (flow) over- or underloaded. Also called OVERFLOW RATE.

$$\text{Surface Loading, GPD/sq ft} = \frac{\text{Flow, gallons/day}}{\text{Surface Area, sq ft}}$$

or

$$\text{Surface Loading, } \frac{\text{m}^3/\text{day}}{\text{m}^2} = \frac{\text{Flow, m}^3/\text{day}}{\text{Surface Area, m}^2}$$

SURFACTANT (sir-FAC-tent) SURFACTANT

Abbreviation for surface-active agent. The active agent in detergents that possesses a high cleaning ability.

TOTALIZER TOTALIZER

A device or meter that continuously measures and sums a process rate variable in cumulative fashion over a given time period. For example, total flows displayed in gallons per minute, million gallons per day, cubic feet per second, or some other unit of volume per time period. Also called an INTEGRATOR.

TRAMP OIL TRAMP OIL

Oil that comes to the surface of a tank due to natural flotation. Also called free oil.

TRUE COLOR TRUE COLOR

Color of the water from which turbidity has been removed. The turbidity may be removed by double filtering the sample through a Whatman No. 40 filter when using the visual comparison method.

TURBID TURBID

Having a cloudy or muddy appearance.

TURBIDITY (ter-BID-it-tee) TURBIDITY

The cloudy appearance of water caused by the presence of suspended and colloidal matter. In the waterworks field, a turbidity measurement is used to indicate the clarity of water. Technically, turbidity is an optical property of the water based on the amount of light reflected by suspended particles. Turbidity cannot be directly equated to suspended solids because white particles reflect more light than dark-colored particles and many small particles will reflect more light than an equivalent large particle.

TURBIDITY (ter-BID-it-tee) UNITS (TU)

TURBIDITY UNITS (TU)

Turbidity units are a measure of the cloudiness of water. If measured by a nephelometric (deflected light) instrumental procedure, turbidity units are expressed in nephelometric turbidity units (NTU) or simply TU. Those turbidity units obtained by visual methods are expressed in Jackson turbidity units (JTU), which are a measure of the cloudiness of water; they are used to indicate the clarity of water. There is no real connection between NTUs and JTUs. The Jackson turbidimeter is a visual method and the nephelometer is an instrumental method based on deflected light.

VISCOSITY (vis-KOSS-uh-tee)

VISCOSITY

A property of water, or any other fluid, that resists efforts to change its shape or flow. Syrup is more viscous (has a higher viscosity) than water. The viscosity of water increases significantly as temperatures decrease. Motor oil is rated by how thick (viscous) it is; 20 weight oil is considered relatively thin while 50 weight oil is relatively thick or viscous.

WATER HAMMER

WATER HAMMER

The sound like someone hammering on a pipe that occurs when a valve is opened or closed very rapidly. When a valve position is changed quickly, the water pressure in a pipe will increase and decrease back and forth very quickly. This rise and fall in pressures can cause serious damage to the system.

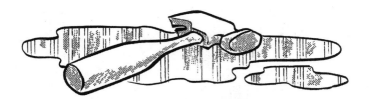

CHAPTER 4.   SOLIDS REMOVAL FROM SECONDARY EFFLUENTS

(Lesson 1 of 6 Lessons)

4.0   NEED TO REMOVE SOLIDS FROM SECONDARY EFFLUENTS

As increasing demands are placed upon our nation's receiving waters, it has often become necessary to set wastewater treatment plant discharge standards at a level that cannot be consistently met by conventional secondary wastewater treatment plants.

Some locations have stringent discharge requirements because more and more wastes are being discharged into the receiving waters and increasing demands are being placed on the waters by water users. At some locations, National Pollutant Discharge Elimination System (NPDES) discharge requirements as stringent as those listed in Table 4.1 are being imposed. To comply with these requirements, the effluent from a standard secondary treatment plant must receive additional or tertiary treatment. Improving solids removal from the effluent of secondary wastewater treatment plants may be accomplished by chemical addition or by several filtration processes. In addition to removing solids, these treatment processes also remove particulates, *BOD*,[1] and *COLIFORMS*.[2] This chapter will review (1) physical–chemical methods presently in use for improving solids removal from the effluent of secondary treatment plants, and (2) four types of filtration systems used for solids removal.

TABLE 4.1   EXAMPLE OF STRICT NPDES REQUIREMENTS

Water Quality Indicator	7-Day Average	30-Day Average	No Sample to Exceed
Biochemical Oxygen Demand (BOD$_5$), mg/L	5	5	7.5
Suspended Solids, mg/L	5	5	7.5
Total Coliform, MPN/100 mL *MEDIAN,* not average	2.2	2.2	240
Chlorine Residual, mg/L *AFTER* dechlorination	0	0	0

QUESTIONS

Write your answers in a notebook and then compare your answers with those on page 459.

4.0A   Why do some locations have stringent discharge requirements?

4.0B   What do the initials NPDES stand for?

4.1   SOLIDS REMOVAL FROM WASTESTREAMS USING CHEMICALS

Physical–chemical treatment is a process that consists of three steps: (1) *COAGULATION*,[3] (2) *FLOCCULATION*,[4] and (3) liquid/solids separation. The three steps must occur in the proper sequence. During the coagulation phase, chemicals are added to the wastewater and rapidly mixed with the process flow. At this time, certain chemical reactions occur quickly, resulting in the formation of very small particles, usually called "pinpoint floc."

Flocculation follows coagulation and consists of gentle mixing of the wastewater. The purpose of the gentle or slow mixing is to produce larger, denser floc particles that will settle rapidly. The liquid is agitated slowly to ensure contact of coagulating chemicals with particles in suspension. Floc growth is accelerated by controlled particle collisions. Suspended particles gather together and form larger particles with higher settling velocities.

---

[1] *BOD* (pronounce as separate letters).   Biochemical Oxygen Demand. The rate at which organisms use the oxygen in water or wastewater while stabilizing decomposable organic matter under aerobic conditions. In decomposition, organic matter serves as food for the bacteria and energy results from its oxidation. BOD measurements are used as a surrogate measure of the organic strength of wastes in water.

[2] *Coliform* (KOAL-i-form).   A group of bacteria found in the intestines of warm-blooded animals (including humans) and also in plants, soil, air, and water. The presence of coliform bacteria is an indication that the water is polluted and may contain pathogenic (disease-causing) organisms. Fecal coliforms are those coliforms found in the feces of various warm-blooded animals, whereas the term "coliform" also includes other environmental sources.

[3] *Coagulation* (ko-agg-yoo-LAY-shun).   The clumping together of very fine particles into larger particles (floc) caused by the use of chemicals (coagulants). The chemicals neutralize the electrical charges of the fine particles, allowing them to come closer and form larger clumps.

[4] *Flocculation* (flock-yoo-LAY-shun).   The gathering together of fine particles after coagulation to form larger particles by a process of gentle mixing. This clumping together makes it easier to separate the solids from the water by settling, skimming, draining, or filtering.

The liquid/solids separation step follows flocculation and is almost always conventional sedimentation by gravity settling, although other processes, such as dissolved air flotation, are used occasionally.

Sedimentation can be defined in a broad sense as those operations performed in which a suspension of particles is separated into a clarified liquid and a more concentrated suspension. Sedimentation can be physically located downstream of any process in which such a suspension is generated (such as a biological treatment process) or it may be autonomous (by itself), representing the bulk of the treatment process of a plant. (The latter is most common in the industrial waste treatment facilities in which metals and large quantities of suspended matter exist prior to any treatment.) In most cases, allowing the solids to settle at their own pace takes too long to be cost-effective. Consequently, chemical coagulation is generally used to enhance the settling quality of the suspension, thereby decreasing the detention time required to achieve the desired liquid clarification.

Care should be taken not to use the words flocculation and coagulation interchangeably. Coagulation is the act of adding and mixing a coagulating chemical to destabilize the suspended particles allowing the particles to collide and "stick" together, forming larger particles. Flocculation is the actual gathering together of smaller suspended particles into flocs, thus forming a more readily settleable mass.

A separate chemical treatment process can be added on to an existing primary or secondary treatment plant as a tertiary treatment process. Chemical treatment performed in this manner requires the construction of additional basins or tanks, which may significantly increase the capital cost of the treatment plant. However, chemical treatment can also be practiced by adding chemicals at specific locations in existing primary or secondary treatment plants (Figure 4.1). This approach is often called chemical addition, and it eliminates the need for constructing additional clarifiers.

Regardless of the form of the chemical treatment process (tertiary or chemical addition), the most important process control guidelines are:

1. Providing enough energy to completely mix the chemicals with the wastewater

2. Controlling the intensity of mixing during flocculation

3. Controlling the chemical(s) dose

Filters often are installed after chemical treatment to produce a highly polished effluent. Also, chemicals may be added to reduce emergency problems such as those created by sludge BULKING[5] in the secondary clarifier, upstream equipment failure, accidental spills entering the plant, and seasonal overloads. Chemicals can be used effectively as a "band-aid" during problem situations with relatively minor capital expense.

Keep in mind that the addition of chemicals is usually meant to capture some additional solids; therefore, more sludge must be handled. Care must be taken in controlling dosage into the secondary system because large chemical additions may be toxic to the organisms in biological treatment processes. This will reduce the activity or even kill the organisms treating the wastes in the system.

Whenever applying chemicals, it is important to always know, understand, and carefully control the dosage. You must understand each chemical's characteristics so the chemical will be properly stored and safely handled. *Many of the chemicals used are harmful, especially to the eyes. Safety is of the utmost importance when chemicals are stored or applied.*

QUESTIONS

Write your answers in a notebook and then compare your answers with those on page 459.

4.1A   What is coagulation?

4.1B   What is flocculation?

4.1C   Why might chemicals be used in a wastewater treatment plant in addition to removing solids from secondary effluents?

4.1D   What precaution must be exercised when adding chemicals upstream from a biological treatment process?

4.10   How Coagulation/Flocculation Works

Efficient removal of solids from a suspension such as wastewater requires use of both chemical (destabilization) and physical (mixing and AGGLOMERATION[6]) processes. Two processes are employed sequentially: coagulation followed by flocculation. The first step, coagulation, is a conditioning step that increases the tendency for very small particles to stick together. The second step, flocculation, collects small particles together, forming larger particles (floc) that can be separated from wastewater using sedimentation or filtration.

---

[5] *Bulking.*   Clouds of billowing sludge that occur throughout secondary clarifiers and sludge thickeners when the sludge does not settle properly. In the activated sludge process, bulking is usually caused by filamentous bacteria or bound water.

[6] *Agglomeration* (uh-glom-er-A-shun).   The growing or coming together of small scattered particles into larger flocs or particles, which settle rapidly. Also see FLOC.

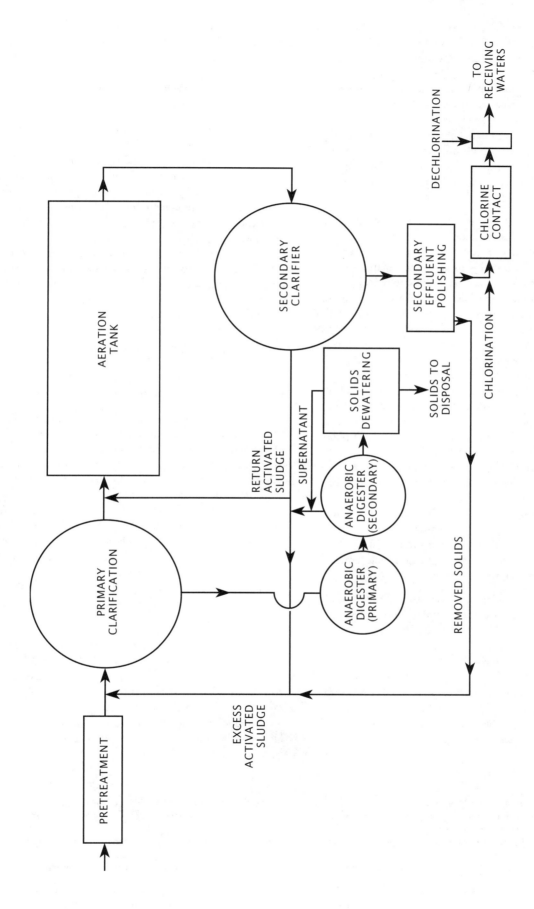

Fig. 4.1   Plan layout of a typical activated sludge plant with secondary effluent polishing process

For suspended solids particles in the ultra-fine or *COLLOI-DAL*[7] size range (1 to 2 *MICRONS*[8]), random physical contact between particles within the suspension would normally cause a "snowball" type of effect commonly known as aggregation. However, fine suspended solids particles tend to accumulate an electrical charge that prevents aggregation. Whether the charge is positive (+) or negative (–) depends on the composition of the particles involved. This accumulated charge on particles in suspension causes electrostatic repulsion between the particles when they approach each other. Electrostatic repulsion is the tendency of particles with the same electrical charge to push away from or repel each other.

In order to get aggregation to proceed, the electrostatic charges of the particles must be modified to reduce their tendency to repel each other, a process called coagulation. This is the process of destabilization by ionic charge neutralization (sometimes also referred to as electrostatic charge reduction). It is accomplished by adding chemicals with a charge opposite to the charge of the suspended particles. Wastewater solids particles tend to have a negative ionic charge and the wastewater tends to have a slightly acidic pH—around pH 6.5. Sometimes, the chemicals added for pH adjustment will produce ions (truly dissolved charged atoms or molecules) having a charge opposite to the one that tends to disperse solids particles. The chemicals used for pH adjustment are commonly a caustic (alkali or base) such as sodium hydroxide, soda ash, or lime, or an acid such as sulfuric or muriatic acid. When the wastewater has a high phosphate content, the addition of lime ($Ca(OH)_2$) may be sufficient to both destabilize and aggregate the solids particles. Lime reacts with phosphate to precipitate calcium phosphate that collects extra calcium ions (Ca^{2+}). The positively charged phosphate precipitate adsorbs onto (sticks on the surface of) the negatively charged solids particles in the wastewater, neutralizing the electrostatic charge(s) and eliminating the repulsion effect. When lime addition is sufficient for coagulation requirements, the low cost of lime makes its use attractive.

Unfortunately (from a chemical conditioning standpoint), high phosphate wastewaters are not common. More often, aluminum (3+) (alum) or iron (3+) (ferric) metal salts are used to coagulate wastewater solids. The cationic (positively charged) metal salts adsorb onto negatively charged wastewater solids and neutralize their negative charge(s). In addition, chemical reactions involving the metal salts produce insoluble (they do not dissolve easily) *PRECIPITATES*[9] that also carry positive charges. The precipitates further assist in the destabilization reactions. This process is called coprecipitation.

Once a colloidal suspension has been destabilized, aggregation of particles can begin. Since this process mainly involves interparticle collisions, energetic mixing or stirring generally enhances the aggregation process. The physical turbulence created by rapid mixing helps collect the colloidal solids in a way that resembles churning butter or rolling up a snowball.

The initial process of rapid mix is important in two regards. First, it ensures that there is a homogeneous mixture (complete mix) of suspended particles and coagulating chemicals. Second, it causes the needed contact between particles. The rapid mix operation is typically a relatively quick process. It should continue only long enough to create a homogeneous mixture because too long a rapid mix may break up and separate the forming floc. Therefore, the speed of the paddles becomes very important. Too rapid a speed may mechanically break up floc. On the other hand, too slow a speed may not provide the needed mixing and may promote dead spots within the tank where mixing does not occur. Floc remaining too long in a dead spot may begin to settle out in the mixing tank before it can be effectively removed by equipment in the downstream processes. The amount of energy applied to optimize coagulation will vary from system to system and may also vary with time in an individual system. Routine testing is the only means available to keep the performance optimized as operations proceed.

Proper coagulation produces agglomerated solids particle units that tend to hold together and do not easily redisperse. Depending on the concentration of suspended solids particles, their composition and the destabilization methods used, coagulation may or may not be sufficient by itself to allow satisfactory solids removal from wastestreams. For example, coagulation alone may be insufficient when suspended solids loadings are high (as with waste activated sludge and anaerobic digester sludge) or when a high-quality effluent must be produced (as in the case of wastewater reuse systems). In these circumstances, it is often desirable or necessary to go one step further using a process called flocculation to precondition the solids particles and enhance the sludge dewatering processes or the tertiary filtration processes.

Flocculation is a process of further collecting together the coagulated solids particles into still larger aggregated units (floc). This process most resembles stringing beads together on a long thread. It is most easily accomplished using synthetic *POLYMERS*,[10] which are, by comparison, much lighter and more fragile than the solids particles (floc) they gather together. Flocculation, therefore, is a more placid process of gentle

---

[7] *Colloids* (KALL-loids).   Very small, finely divided solids (particles that do not dissolve) that remain dispersed in a liquid for a long time due to their small size and electrical charge. When most of the particles in water have a negative electrical charge, they tend to repel each other. This repulsion prevents the particles from clumping together, becoming heavier, and settling out.

[8] *Micron* (MY-kron).   μm, Micrometer or Micron. A unit of length. One millionth of a meter or one thousandth of a millimeter. One micron equals 0.00004 of an inch.

[9] *Precipitate* (pre-SIP-uh-TATE).   (1) An insoluble, finely divided substance that is a product of a chemical reaction within a liquid. (2) The separation from solution of an insoluble substance.

[10] *Polymer* (POLY-mer).   A long-chain molecule formed by the union of many monomers (molecules of lower molecular weight). Polymers are used with other chemical coagulants to aid in binding small suspended particles to larger chemical flocs for their removal from water. Also see POLYELECTROLYTE.

mixing that requires special care. The results, however, can be spectacular. Flocculation can produce floc that is visible to the naked eye from a coagulated mixture that only looked cloudy or turbid before.

Stirring during flocculation is for the purpose of promoting maximum contact between suspended particles so that they will gather together or agglomerate to form a larger floc mass. Paddle configuration (layout) and speed are such that the water and floc are encouraged to move slowly through the tank. The aggregates or flocs have a greater overall density after flocculation and can be more readily separated from the liquid portion. As in the case of the rapid mix, two dangers must be avoided during flocculation. Paddle speed must be sufficient to keep the floc from settling while at the same time it must not be so great as to shear and break up the floc formed.

By far the most precise application of flocculation is in the preparation of secondary effluents for tertiary filtration (often used to "polish" the effluent). Commonly, coagulation alone is insufficient to enlarge the suspended solids particles enough to meet prefiltration conditioning requirements. This is especially true when the concentration of suspended solids in the wastewater leaving the secondary treatment process is very low (15 mg/L or less). Polymers are often used under these circumstances. Selection of the best type of polymer for a specific application depends on factors such as pH, conductivity (dissolved solids content), type and concentration of suspended solids,

particle size ranges, type and amount of coagulant(s) applied, and what the next stage of treatment will be.

Two other important physical–chemical reactions occur during coagulation and flocculation that also contribute to the formation of floc that can be easily removed by sedimentation or other processes. The two reactions are interparticle bridging and physical enmeshment.

Some polymers have the capacity to adsorb (stick) to sites on suspended particles and act as a bridge between them, thus enhancing floc formation in two ways: (1) the positively charged polymer neutralizes the negatively charged wastewater particle, and (2) the polymer acts as a bridge that connects solids particles. The effects of interparticle bridging are: (1) particles are attached into larger floc, and (2) the larger floc traps (physical enmeshment) additional solids within its mass as it collides with them during flocculation or sedimentation. Figure 4.2 illustrates how a positively charged polymer adsorbs onto negatively charged solids particles creating interparticle bridging.

Physical enmeshment also takes place when cationic (positively charged) metal salts of iron, aluminum, or (at high pH) magnesium are used for destabilization. These cations will combine with hydroxyl ions (OH^-) found in water from its natural dissociation as shown below.

$$H_2O \rightleftharpoons H^+ + OH^-$$

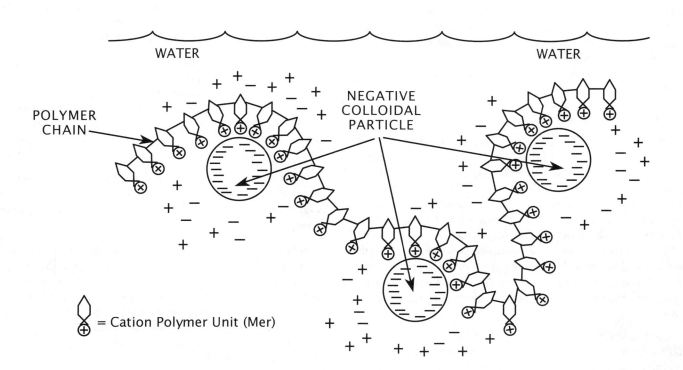

Fig. 4.2 Drawing illustrating the coagulation-flocculation reaction

This natural dissociation of water comes about either as a result of the alkalinity of the water or as a result of increasing the pH by the use of lime or soda ash. The metal ions combine with the hydroxide ions to form precipitates that tend to have an extended open flake form, commonly referred to as gelatinous. This means that they form in soft strings and flakes containing a lot of incorporated water and "void" space. As this gelatinous precipitate forms, it also collects any other solids in the vicinity as a coprecipitate, thus enlarging the forming floc. The gelatinous flake structure can be destroyed by severe turbulence, but vigorous mixing during coagulation produces a more granular gelatinous formation that is more physically stable.

There is no substitute for testing the wastewater to select the right combination of chemicals and the proper dosage rates for effective coagulation and flocculation.

QUESTIONS

Write your answers in a notebook and then compare your answers with those on page 459.

4.10A   What is destabilization and why is it necessary?

4.10B   What is the typical particle charge when a wastewater tends to have a pH on the low (acidic) side of pH 6.5?

4.10C   What is the importance of mixing in the coagulation process?

4.11   Chemicals Used to Improve Settling

Secondary effluent quality may be improved by adding coagulant aids ahead of the secondary settling tanks or filters. Chemicals usually added are alum, ferric chloride, lime, or polyelectrolytes. Other useful chemicals may include sodium aluminate, ferric sulfate, ferrous chloride, and ferrous sulfate. These chemicals may be used alone or in combinations as determined by laboratory testing (Section 4.12, "Selecting Chemicals and Determining Dosages") and the results obtained in actual plant operation.

4.110   Aluminum Sulfate (Dry) (Al$_2$(SO$_4$)$_3$ • 14 H$_2$O)

Alum may be purchased in varying grades identified as lump, ground, rice, and powdered. Lump alum consists of lumps varying in size from 0.8 inch to 8 inches (2 to 20 cm) in diameter and is rarely used due to its irregularity in size and the difficulties of applying and achieving a satisfactory dose. Ground (granulated) alum is a mixture of rice-size material and some fines (very small particles). This form of alum is used by the majority of the water and wastewater plants. Ground alum feeds easily and does not bulk (stick together) in the hoppers if kept free of moisture or water. Also, ground alum does not require special protection of the hopper interiors from corrosion and wear.

Commercial filter alum (ground alum) has a bulk density of 60 to 70 lbs/cu ft (1.0 to 1.1 gm/cm^3) and is shipped in 100-pound (45-kg) bags or in large quantities (20 tons or 18,000 kg) by bulk trucks and railroad hopper cars. Special care should be taken to prevent alum from getting damp or it will cake into a solid lump. All mechanical equipment such as conveyors should be run until well cleaned of all alum before shutting down because the alum can harden and jam the equipment. Keep alum dry by storing it inside a well-ventilated location. Storage bins should have a 60-degree slope to the bottom to ensure complete emptying. Be sure the alum will not get wet when hosing down equipment or washing floors.

Both dry dust and liquid forms of alum are irritating to the skin and mucous membranes and can cause serious eye injury. Wear protective clothing to protect yourself from dust, splashes, or sprays. Proper clothing consists of a face shield, rubber or plastic gloves, rubber shoes, and rubber clothing when working around alum dust. Prevent inhalation by the use of local exhaust or approved respiratory protection.

4.111   Aluminum Sulfate (Liquid)

Alum is also available as a liquid. One gallon weighs about 11 pounds and contains the equivalent of 5.4 pounds (2.45 kg) of dry aluminum sulfate (49% as Al$_2$(SO$_4$)$_3$ • 14 H$_2$O). Obtain a chemical analysis from the supplier for each delivery to determine the exact content. Liquid alum is preferred by operators because of its ease of handling; however, you must pay shipping costs for transporting the water portion.

Liquid alum is shipped in 2,000- to 4,000-gallon tank trucks or 55- to 110-ton railroad tank cars.

Alum becomes very corrosive when mixed with water; therefore, dissolving tanks, pumps, and piping must be protected. Liquid storage tanks must be constructed of corrosion-resistant material such as rubber-lined steel or fiberglass. Bulk liquid alum storage tanks must be protected from extreme cold because normal commercial concentrations will crystallize at temperatures below 32°F (0°C) and freeze at about 18°F (−8°C).

Alum will support a bacterial growth and cause sludge deposits in feed lines if wastewater is used to transport the alum to the point of application. These growths and deposits can completely plug the chemical feed line. This problem can be reduced by maintaining a high velocity to scour the line continuously. Also, a concentrated alum solution will not support the bacterial growth so reducing the amount of carrier water helps.

Alum reduces the alkalinity in the water being treated during the coagulation process. Hydrated lime, soda ash, or caustic soda may be required if there is not enough natural alkalinity present to satisfy the alum dosage.

When added to water, alum is acidic; a 1 percent solution will have a pH of 3.5. Overdosing of alum may depress the pH to a point that it will reduce the biological activity in the secondary system. Also, this lowered pH may allow the chlorine added as a disinfectant to further depress the pH and affect the aquatic life in the receiving waters. This, along with chemical costs, emphasizes the need to maintain proper chemical dosages and closely monitor effluent quality.

Regularly analyze the bulk chemicals to determine if the concentration has changed. If the concentration has changed, you will need to adjust the chemical feed rate. Also, test the effluent quality to determine if sufficient solids are being removed or if an adjustment in the chemical feed rate might be helpful.

Liquid alum can be very hazardous. A face shield and gloves should be worn around leaking equipment. The eyes or skin should be flushed and washed upon contact with liquid alum. Liquid alum becomes very slick upon evaporation and therefore spills and leaks should be avoided.

QUESTIONS

Write your answers in a notebook and then compare your answers with those on page 459.

4.11A What are the four most common chemicals added to improve settling of solids?

4.11B Why should alum be kept dry?

4.11C Why should all mechanical equipment, such as conveyors, be run until well cleaned of all alum before shutting down?

4.112 Ferric Chloride (FeCl₃)

Ferric chloride is available in three forms—*ANHYDROUS*,[11] crystal hydrated, and liquid. The dry forms will absorb enough moisture from the air to quickly form highly corrosive solutions.

Anhydrous ferric chloride is shipped in 150- and 350-pound drums. Once these drums are opened, they should be completely emptied to prevent the formation of corrosive solutions. Care must be taken when making up solutions because the solution temperature will rise as the chemical dissolves.

Crystal ferric chloride is shipped in 100-, 400-, or 450-pound drums. Store the crystals in a cool, dry place and always completely empty any opened containers. The heat rise in dissolving crystal ferric chloride is much lower than that of anhydrous ferric chloride and is not a problem.

Liquid ferric chloride (35 to 45% FeCl₃) is shipped in rubber-lined tank cars or trucks (3,000 to 10,000 gallons). This chemical must be stored in corrosion-resistant tanks. If storage occurs at temperatures below 30°F (–1°C), it may be necessary to provide tank heaters or insulation to prevent crystallization.

Positive displacement metering pumps should be used for accurate measurements. Both the feeder and the lines must be corrosion-resistant.

All forms of ferric chloride will cause bad stains. This staining will occur on almost every material including walls, floors, equipment, and even operators.

Safety precautions required for handling ferric chloride in concentrated forms should be the same as those for acids. Wear protective clothing, chemical goggles, rubber or plastic gloves, and rubber shoes. Flush all splashes off clothing and skin immediately.

4.113 Lime (Ca(OH)₂)

Hydrated lime (calcium hydroxide or Ca(OH)₂) is used to coagulate solids or adjust the pH to improve the coagulation process of other chemicals. Lime may be purchased in 50- or 100-pound bags or in bulk truck or railroad car loads. Lime should be stored in a dry place to avoid absorbing moisture. Bulk bin outlets should be provided with nonflooding rotary feeders. Hopper slopes may vary from 60 to 66 degrees.

Lime also may be purchased as anhydrous or quicklime, but must be *SLAKED*[12] before it can be used. Quicklime is more difficult to store because it will easily absorb moisture and cake into a solid clump. Quicklime is less expensive to purchase than lime; however, the added equipment for slaking and the requirement for increased operational safety must be considered.

Heat is generated when water is added to quicklime. If the controlled water supply fails and the water is shut off while the lime feed continues, boiling temperatures can be reached quickly. If a boiling reaction results, hot lime may cause the slaker to erupt and spew out hot lime. If high temperature controls are properly installed, they should activate an alarm or shut down the unit. Mixers and pumps should be inspected frequently (daily) for wear because the lime slurry will rapidly erode or wear moving parts.

When transporting concentrated lime slurries in pipelines, a scale will build up on the inside of the pipe and eventually plug the line. A 2- to 3-inch (50- to 60-mm) diameter pipe may need replacing every year or two due to this scale. Rubber or flexible piping with easy access and short runs will permit cleaning by squeezing the walls and washing out the broken scale. Standby lines should be provided for use during the cleaning operation.

Lime is irritating to the skin, the eyes, the mucous membranes, and the lungs. Protect your eyes and lungs with approved full-face respiratory protection devices and wear protective clothing when working around lime.

4.114 Polymeric Flocculants

Polymeric flocculants are high-molecular-weight (molar mass) organic compounds with the characteristics of both polymers and *ELECTROLYTES*.[13] They are commonly called *POLYELECTROLYTES*.[14] These flocculants may be of natural or synthetic origin.

---

[11] *Anhydrous* (an-HI-drous).   Very dry. No water or dampness is present.

[12] *Slake.*   To mix with water so that a true chemical combination (hydration) takes place, such as in the slaking of lime.

[13] *Electrolyte* (ee-LECK-tro-lite).   A substance that dissociates (separates) into two or more ions when it is dissolved in water.

[14] *Polyelectrolyte* (POLY-ee-LECK-tro-lite).   A high-molecular-weight (relatively heavy) substance, having points of positive or negative electrical charges, that is formed by either natural or synthetic (manmade) processes. Natural polyelectrolytes may be of biological origin or obtained from starch products or cellulose derivatives. Synthetic polyelectrolytes consist of simple substances that have been made into complex, high-molecular-weight substances. Used with other chemical coagulants to aid in binding small suspended particles to larger chemical flocs for their removal from water. Often called a POLYMER.

Technically speaking, a polymer is any material that is composed of one single base unit that is chemically linked together with many more base units of the same type. In many ways, a polymer's base unit (properly named its "monomer") resembles a link in a chain. When a large number of these "links" are assembled into one single unit, the entire assembly is an entity with entirely different characteristics from the individual monomer "links" in their detached or unassembled form. It is this similarity to a chain with which we commonly have personal experience that has generated much of the polymer terminology, such as references to "polymer chains" when discussing individual polymer molecules or talking about "chain lengths" when describing the size of a polymer molecule.

All synthetic polyelectrolytes are classified on the basis of the type of charge on the polymer chain, the molecular weight (or length) of the polymer chain, and the charge density (spacing of the charges) along the chain. Negative-charge polymers are called "anionic" and positive-charge polymers are called "cationic." Polymers carrying no free electrical charge are "nonionic polyelectrolytes."

The sizes of polyelectrolytes are designated by their molecular weights. This is a chemical term used to describe the length of a polymer chain and has the units of gm/mole or MW/mole (Molecular Weight/mole). A medium-molecular-weight polymer will weigh <100,000 gm/mole. A high-molecular-weight polymer will range from 100,000 to 1,000,000 gm/mole, a very high-molecular-weight polymer will be from 1,000,000 to 10,000,000 gm/mole and an ultrahigh-molecular-weight polymer will be >10,000,000 gm/mole.

Charge densities refer to the relative number of locations along the polymer chain that actually carry a charge. The more precise specification reports the percentage of charged sites out of the maximum number of possible locations. Polymers are also often grouped as low charge (<25% charge), medium charge (26–50% charge), or high charge (>50% charge).

A great assortment of polyelectrolytes are available to the wastewater treatment facility operator. They may be applied alone or in combination with other chemicals to aid coagulation.

With this large selection of polymers available, it is possible to find a beneficial combination for almost all conditions.

Because of the variety of polymers available, a method of cataloging polymers has been developed called a polymer map. This can be in the form of a table or an actual graphical "map" (Figure 4.3) that organizes polymers according to their design or production specifications. The polymer map is actually a logarithmic graph of polymer specifications. This kind of map is particularly useful when an initial set of polymer selection studies is undertaken because it helps in determining which polymers may work well in a particular application.

Across the upper portion of the map (horizontal "x" axis) is shown the range of polymer ionic charges. These charges range from −100% (minus 100 percent) to +100% (plus 100 percent). The definition of a −100% value is that every available site along a polymer chain in this group has an anionic (negative) charge when the polymer is properly dissolved in water. Conversely, a +100% value indicates that every available site along a polymer chain in this group has a cationic (positive) charge when the polymer is properly dissolved in water.

The center of the polymer map (vertical "y" axis) displays the uncharged (0%) designation assigned to nonionic polymers. Nonionic polymers do not develop a "free" ionic charge in solution. However, they have the ability to form bonds with both positively and negatively charged suspended solids particles.

Along the vertical "y" axis are marked the molecular weights (MW) of polymers. The values are presented in orders of magnitude (powers of 10) of the molecular weight (for example $10^4 = 10,000$; $10^5 = 100,000$), thus the scale is logarithmic.

For you to prepare a polymer map, you will need to know both the "charge density" (± % charge) and the molecular weight for each polymer you will be investigating. With such information, each polymer to be tested can be plotted as a point on the map. A completed polymer map will resemble a map of the stars in the sky—a scattering over all the possible range of polymers of interest to you. In Figure 4.3, some of the typical molecular weight "zones" of polymer types suitable for various process applications are identified (the irregularly shaped boxes).

Because of the wide selection of available products and the different and changing chemical characteristics of water being treated, extensive laboratory testing should be conducted before treating the entire plant effluent. The selection of a polymer for any individual application is highly dependent on the characteristics of the wastewater to be treated. Specific influences (factors) that require attention are: pH, conductivity (dissolved solids content), type and concentration of suspended solids, particle size ranges, type and amount of coagulant(s) applied, and what the next stage of treatment will be. Most polyelectrolyte suppliers have field representatives who will assist with the testing of their products at the treatment plant.

Polyelectrolytes are commonly used in very small doses, usually less than 1 mg/L. The effective dosage range is limited. An overdose can be worse than no polymer addition at all.

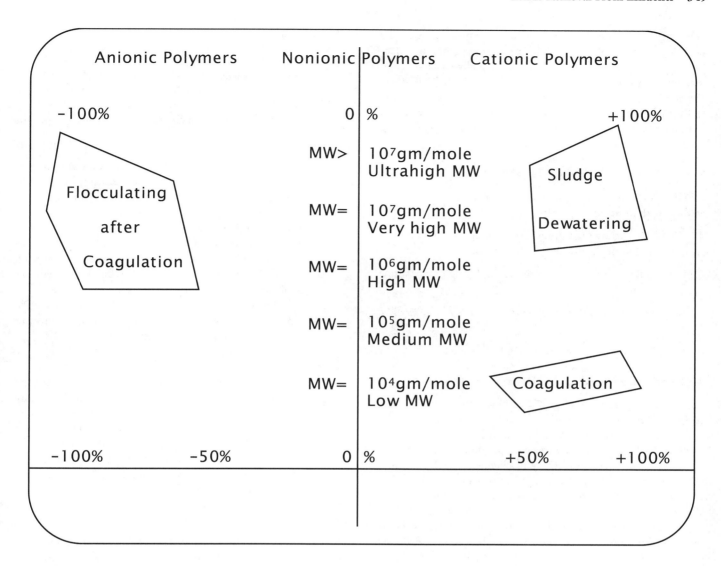

Definitions:

1. MW = Molecular Weight
2. gm/mole = molecular weight of a substance expressed in grams

Fig. 4.3   Polymer map

Polyelectrolytes are available as dry powder, as a liquid suspension, or as a true solution. Care must be taken when storing powders because they may quickly absorb moisture and become ineffective.

Solutions for treating wastewater must be made up in water prior to their use. This step is especially important with high- and ultrahigh-molecular-weight polymers supplied as powders or suspensions. With these materials, a minimum two-hour hydration time is necessary to get the polymer fully dissolved in water. True solution polymers do not require this kind of preparation and they are applied with dilution water in order to disperse or mix them with the wastewater to be treated. When mixing a batch, care is required to add a powder or a suspension liquid slowly while continuously mixing. If care is not taken, useless lumps will form that can clog feed pumps and lines.

Polyelectrolytes are considered nonhazardous to handle; however, good housekeeping must be practiced because polyelectrolytes will create an extremely slippery surface when wet. Clean up spills immediately. Salts, drying agents, and highly concentrated solutions will break down polyelectrolytes and help with a cleanup effort. There are commercially available polymer cleaning solutions based on isopropyl (rubbing) alcohol but a variety of cleaning solutions are equally effective. Polymer manufacturers' Material Safety Data Sheets (MSDSs) recommend using inert absorbent materials to clean up polymer spills, such as sand or earth. Some polyelectrolytes have a low pH and can be corrosive to the makeup day or age tank (tank used to store solution).

QUESTIONS

Write your answers in a notebook and then compare your answers with those on page 459.

4.11D   What safety precautions are required for handling ferric chloride in concentrated solutions?

4.11E   How can the scale of lime that builds up on the inside of pipe be cleared?

4.11F   What problems can be created when a polyelectrolyte is spilled?

4.11G   How would you clean up a polyelectrolyte spill?

4.12   Selecting Chemicals and Determining Dosages

In order to select the chemical products to be used in any treatment process application, a systematic accumulation of data is required. Factors that should be considered when selecting a chemical include minimum dosage, upper dosage limit, volume of sludge, cost, safety, availability of chemical supply, and reliability of chemical supply. When preparing to set up a chemical treatment program, a three-stage series of tests is used. These stages are: (1) preliminary screening, (2) dosage testing, and (3) full-scale trial. The common element through all stages is the jar test procedure, although specific details and objectives are different for each stage. Once the correct combination of chemicals has been selected for actual process use, jar test procedures for system performance optimization (similar to full-scale trial tests) are regularly used by the operator to maintain the best possible chemical use efficiency during routine process operation.

One invariable requirement in all jar test procedures is that the tests only have meaning if the tested wastewater exactly resembles the flow stream that will be ultimately treated by the chemicals being evaluated. The bottom line is that for the tests to be truly valid, they need to be performed at the chemical application location using freshly drawn samples from the actual flow stream. A delay of as little time as 30 minutes between sample collection time and jar testing can significantly affect the test results. It is equally important that all other test conditions match the actual system (plant) conditions.

1.   Preliminary Screening

Preliminary screening tests cover a wide variety of possible chemicals for treatment use and are more qualitative than quantitative. These tests are meant to show obvious differences between potential treatment chemicals and narrow the possibilities down to a few choices.

2.   Dosage Testing

Dosage testing is used to identify (bracket) the minimum-to-maximum amounts of potential treatment chemicals that will be needed. Performance results for a range of application rates are measured. These tests are used to predict the costs associated with the use of each potential treatment chemical.

3.   Full-Scale Trial

Following completion of dosage testing, precise estimates of chemical application guidelines for treating the process flow stream are available and can be used to begin treatment. While the chemicals are being applied, their performance is measured through the actual treatment process, along with a parallel series of jar tests. This allows two essential sets of data to be collected. First, the actual performance of the treatment process can be measured against minor fine-tuning adjustments of the chemical(s) being applied. Second, the actual treatment results will show the operator how closely the jar tests predicted what would happen. Knowing this helps the operator interpret and use future test results.

4.   System Performance Optimization

In order to maintain optimum, cost-effective treatment process performance, regular monitoring is essential. In addition to regularly testing the effluent quality from the treatment process, regular jar testing of the influent just upstream of the chemical application point will enable the operator to confirm chemical treatment dosage rates and provide required dosage rate adjustment information when flow stream conditions change.

4.120   The Jar Test

Probably the single most valuable tool in operating and controlling a chemical treatment process is the variable speed, multiple station (or "gang") jar test unit (Figure 4.4). Various types of chemicals or different doses of a single chemical are added to sample portions of wastewater and all portions of the sample are rapidly mixed. After rapid mixing, the samples are slowly mixed to approximate the conditions in the plant. Mixing is then stopped and the floc formed is allowed to settle. The appearance of the floc, the time required to form a floc, and the settling conditions are recorded. The supernatant (liquid above the settled sludge) is analyzed for turbidity, suspended solids, and pH. With this information, the operator selects the best chemical or best dosage to feed on the basis of clarity of effluent and minimum cost of chemicals.

Jar test units vary somewhat in configuration depending on the manufacturer. These differences, such as the number of test stations, the size and shape of the test jars (round or square), stirrer controls, portability, internal illumination, and method of mixing, do not affect the performance of the device. However, two things must be known before a particular jar test unit is used: (1) the volume of the test jars, and (2) the speed rates of the stirrers. Some jar test units are supplied with a gauge that displays the turning rate of the stirrers in revolutions per minute

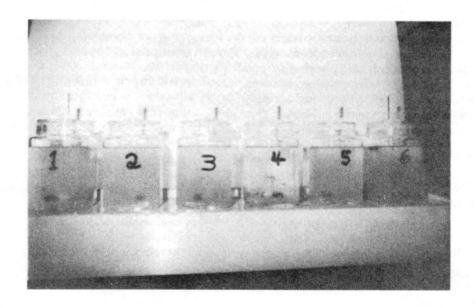

Fig. 4.4   Jar test units with mechanical (top) and magnetic (bottom) stirrers
(Source: EPA *PROCESS DESIGN MANUAL FOR SUSPENDED SOLIDS REMOVAL*)

(RPM). Other units have a control dial calibrated in RPM, and some are marked in percent of maximum speed (0–100%). In the latter case, it will be necessary for the operator to perform a calibration check so that the percentage numbers can be associated with the RPM of the stirrers. When testing flow streams containing low suspended solid concentrations, any type of stirring option is acceptable: rotating paddles, magnetically driven stirring bars, or vertical plungers. For flow streams containing high suspended solid concentrations (for example, waste activated sludge, digested sludge), magnetically driven stirring bars will be unsuitable because the resistance of the solids within the sample can decouple the stirring bar from the drive magnet.

QUESTIONS

Write your answers in a notebook and then compare your answers with those on pages 459 and 460.

4.12A   What are the three stages of testing used to select chemical products for a treatment process?

4.12B   Briefly describe the jar test procedure.

4.12C   What two features must be known about a jar test unit before it is used?

4.121 Jar Testing Procedures

4.1210 PREPARATION

In many cases, the preparation of the chemical(s) to be tested will be virtually the same. In some cases, especially with treatment system performance optimization testing, normal, convenient concentrations of the chemical(s) to be used may be selected. The descriptions that follow are intended as examples and are meant to be typical rather than specific. Always perform jar tests using the same chemicals that will be used to treat the water. Do not use laboratory grade chemicals because they may contain fewer impurities and therefore could produce slightly different results.

BASIC GUIDELINES

1. Jar test volume: 1,000 mL (= 1.0 L = 1,000 gm).

2. Test chemical stock solution strength: 0.5% (= 5,000 mg/L).

3. Working chemical solution strength: 0.1% (= 1,000 mg/L).

4. Rapid mix time at 140 to 160 RPM for 10 to 15 seconds for polymers; 3 to 5 minutes for aluminum or iron metal salts.

5. Slow mix time at 15 to 20 RPM = 1 to 3 minutes for polymers; slow mix is not used for aluminum or iron metal salts alone.[15]

6. One gallon of water weighs 8.34 lbs.

NOTE: The ranges given for items 4 and 5 above represent the ranges of normal operating conditions. The actual values selected for the tests will be fixed at single values based on experience with the type of jar test equipment used and the composition of the flow stream being examined.

4.1211 GENERAL SETUP CONSIDERATIONS

THE JARS

If the jars used in the jar test unit have a calibration mark at the 1,000 mL level, this is usually satisfactory for most testing applications. If there is no such mark on the jars, obtain and use a measuring container that is calibrated for 1,000 mL to fill the test jars. For precision testing, it is preferable to measure sample volumes using a graduated cylinder.

TEST CHEMICALS

To prepare stock solutions of the chemicals to be tested:

1. Prepare clean 250-mL (half-pint) storage bottles with appropriate labels and the date of stock solution preparation.

2. Into each storage bottle, accurately measure 199 mL of deionized or distilled water.

3. Weigh exactly 1.0 gm of the selected chemical(s) into the respective storage bottle(s).[16]

4. Thoroughly mix the polymer and deionized or distilled water with constant stirring for at least 10 minutes. Polymer solutions will require an additional aging period of at least 2 hours with occasional stirring at 10 to 15 minute intervals.[17]

5. Store the chemical stock solutions in a cool place, convenient to the jar testing work area.[18]

To prepare a working solution of a chemical to be tested:

1. Prepare a clean, temporary container and mark it for identification.

2. Into the container, add exactly 20 mL of the chemical stock solution.

---

[15] Slow mix will be used with aluminum or iron metal salts during the second stage of a combination test where a flocculant (polymer) is added as a second chemical after the initial metal salt coagulation period.

[16] When preparing solutions from liquid polymer supplies, a positive displacement measuring device will be needed, such as a hypodermic syringe, or a positive displacement micropipet. For preliminary screening tests, it is adequate to dispense 1.0 mL of the polymer supply into the storage bottle. For more precise testing, it is necessary to correct the volume for the density of the liquid polymer supply. (For example, if a polymer liquid has a density of 8.8 lbs/gallon (consult the polymer MSDS or other data sheet), the accurate volume of 1.0 gm of the polymer can be found using the ratio of the density of water (8.34 lbs/gallon) divided by the density of the polymer (8.8 lbs/gallon) = 0.94 mL.)

[17] When mixing a dry polymer into water, the water must be stirred rapidly while gradually sprinkling the polymer granules or powder over the water surface. Care must be taken to be sure that the granules of powder do not pile up in lumps on the water surface or cling to the sides of the mixing container, and that each granule of polymer is wetted individually.

[18] Generally, stock solutions of polymer are stable for 5 to 7 days and stock solutions should only be used for that length of time. After the designated period, leftover polymer stock solutions should be discarded. Metal salt solutions and soda ash are stable indefinitely and may be kept until the solutions are exhausted.

3. Into the same container, add 80 mL deionized or distilled water.

4. Thoroughly mix the contents of the container for at least 5 minutes, and let the solution stand for at least 5 minutes before use.

5. Discard any unused working solution at the end of the sequence of jar tests.

To calculate trial dosage levels for a jar test a generally applicable formula is:

$$\left(\begin{array}{c}\text{Volume}\\\text{(of Stock}\\\text{Solution)}\end{array}\right)\left(\begin{array}{c}\text{Concentration}\\\text{(of Stock}\\\text{Solution)}\end{array}\right) = \left(\begin{array}{c}\text{Final Volume}\\\text{(of Working}\\\text{Solution)}\end{array}\right)\left(\begin{array}{c}\text{Final Concentration}\\\text{(of Working}\\\text{Solution)}\end{array}\right)$$

Which can be solved to find the final concentration as:

$$\begin{array}{c}\text{Final Concentration}\\\text{(of Working Solution)}\end{array} = \frac{\left(\begin{array}{c}\text{Volume}\\\text{(of Stock Solution)}\end{array}\right)\left(\begin{array}{c}\text{Concentration}\\\text{(of Stock Solution)}\end{array}\right)}{\begin{array}{c}\text{Final Volume}\\\text{(of Working Solution)}\end{array}}$$

Or, which can be solved to find the needed amount of stock solution as:

$$\begin{array}{c}\text{Volume}\\\text{(of Stock Solution)}\end{array} = \frac{\left(\begin{array}{c}\text{Final Volume}\\\text{(of Working Solution)}\end{array}\right)\left(\begin{array}{c}\text{Final Concentration}\\\text{(of Working Solution)}\end{array}\right)}{\begin{array}{c}\text{Concentration}\\\text{(of Stock Solution)}\end{array}}$$

EXAMPLES:

1. Using 1.0 mL of stock solution (5,000 mg/L) in a 1,000-mL jar, what is the dosage?

$$\begin{array}{c}\text{Final Concentration}\\\text{(of Jar Test Solution)}\end{array} = \frac{(1.0\ \text{m}L)(5,000\ \text{mg}/L)}{1,000\ \text{m}L}$$

$$= 5.0\ \text{mg}/L\ \text{(in the Jar)}$$

2. How much working solution is needed to provide 5.0 mg/L in the test jar?

$$\begin{array}{c}\text{Volume}\\\text{(of Working Solution)}\end{array} = \frac{(1,000\ \text{m}L)(5.0\ \text{mg}/L)}{5,000\ \text{mg}/L}$$

$$= 1.0\ \text{m}L\ \text{(into the Jar)}$$

4.122 Example Calculations

When conducting jar tests and setting chemical feed rates, the following standard units of measurement are used:

- Jar test results are expressed in milligrams of chemical per liter of sample (mg/L) or parts chemical per million parts sample (ppm).

- Chemical feed rates are expressed in pounds of chemical delivered per hour (lbs/hr) or pounds of chemical delivered per day (lbs/day).

- Chemical feed pump settings are expressed in gallons of chemical fed per hour (gal/hr).

- Chemical pumping rates are expressed in gallons of chemical fed per hour (gal/hr).

- Chemical application rates are expressed in pounds of chemical applied per hour (lbs/hr) or pounds of chemical applied per day (lbs/day).

- Chemical dosage rates are expressed in milligrams of chemical per liter of wastewater (mg/L), parts of chemical per million parts wastewater (ppm), or pounds of chemical per million pounds of wastewater (lbs/M lbs equivalent process flow).

The basic formula for calculating a chemical feed is:

Chem Feed Rate, lbs/day = (Flow, MGD)(Conc, mg/L)(8.34 lbs/gal)

EXAMPLE 1

A process flow rate is 347 GPM. What is the equivalent process flow rate in M lbs/day?

Known	**Unknown**
Flow Rate, GPM = 347 GPM	Flow Rate, M lbs/day

Convert the flow rate from GPM to M lbs/day.

1. Flow Rate, GPD   = (Flow Rate, GPM)(1,440 min/day)

$$= (347\ \text{GPM})(1,440\ \text{min/day})$$

$$= 499,680\ \text{GPD (or 0.5 MGD)}$$

2. Flow Rate, lbs/day   = (Flow Rate, GPD)(8.34 lbs/gal)

$$= (499,680\ \text{GPD})(8.34\ \text{lbs/gal})$$

$$= 4,167,331\ \text{lbs/day}$$

3. Flow Rate, M lbs/day $= \dfrac{\text{Flow Rate, lbs/day}}{1,000,000/\text{M}}$

$$= \frac{4,167,331\ \text{lbs/day}}{1,000,000/\text{M}}$$

$$= 4.167\ \text{M lbs/day}$$

Therefore, 347 GPM = 4.167 M lbs/day equivalent process flow rate.

EXAMPLE 2

Results of jar testing indicate that a concentration of 5.0 mg/L of chemical should be used to treat a process flow of 0.5 MGD. What is the chemical application rate in lbs/day and lbs/hr? (*NOTE:* For each 1,000,000 lbs of water, it takes 1 lb of chemical to produce a 1 mg/L (1 ppm) chemical concentration in that water.)

Known	**Unknown**
Flow Rate, MGD = 0.5 MGD	Chemical Application
Chem Conc, mg/L = 5.0 mg/L	Rate, lbs/day and lbs/hr

1. Calculate the chemical application rate in lbs/day.

$$\begin{array}{c}\text{Application Rate,}\\\text{lbs/day}\end{array} = (\text{Flow, MGD})(\text{Chem Conc, mg}/L)(8.34\ \text{lbs/gal})$$

$$= (0.5\ \text{MGD})(5.0\ \text{mg}/L)(8.34\ \text{lbs/gal})$$

$$= 20.85\ \text{lbs/day}$$

2. Convert the application rate from lbs/day to lbs/hr.

$$\text{Application Rate, lbs/hr} = \frac{\text{Application Rate, lbs/day}}{24 \text{ hr/day}}$$

$$= \frac{20.85 \text{ lbs/day}}{24 \text{ hr/day}}$$

$$= 0.869 \text{ lb Chemical/hr}$$

Chemicals are supplied or prepared in solutions of various strengths. Polymers are usually limited to 1.0 percent maximum solution strength so that the solution can be pumped (metered) easily.

EXAMPLE 3

Based on jar test results, 5.0 mg/L of chemical is applied to 0.5 MGD equivalent process flow rate (from Example 1 flow of 347 GPM). If the chemical feed solution strength is 0.4 percent, what is the chemical pumping rate in gallons per hour?

Known	Unknown
Chem Dosage, mg/L = 5.0 mg/L	Chem Pumping Rate, gal/hr
Equiv Process Flow Rate, MGD = 0.5 MGD	
Solution Strength, % = 0.4%	

Calculate the chemical pumping rate for a 0.4% solution.

$$\text{Chem Pumping Rate, gal/hr} = \frac{(\text{Flow, MGD})(\text{Dosage, mg/}L)(8.34 \text{ lbs/gal})(100\%)}{(24 \text{ hr/day})(8.34 \text{ lbs/gal})(\text{Sol Strength, }\%)}$$

$$= \frac{(0.5 \text{ MGD})(5.0 \text{ mg/}L)(8.34 \text{ lbs/gal})(100\%)}{(24 \text{ hr/day})(8.34 \text{ lbs/gal})(0.4\%)}$$

$$= 26.05 \text{ gal/hr}$$

4.123 Sampling

Before starting the jar test procedure, outline the objectives of your test and list the information you need to collect to reach your objectives. Choose a sampling point that is as near as possible to the location where treatment chemicals will be or are being fed into the flow stream. When testing a treatment process that has an existing chemical feeding system, take care to obtain the water test samples far enough upstream of the chemical injection point(s) so that there will not be any added chemicals in the test sample.

Next, use a container that will hold enough sample to fill all the jars in the jar test unit. If for some reason you can only collect small portions of the sample, all of the scoops of sample should be placed in a single container and stirred thoroughly before the sample is put into the individual jars. When running jar tests on high solids solutions (for example, waste activated sludge or digested sludge), stirring of the sample must be continuous in the holding container. It may be necessary to work with a partner to accomplish the stirring while the samples are being measured into the jars. Retain a portion of the untreated sample for laboratory analysis uses.

4.124 Addition of Test Chemicals

Performance of the treatment chemicals depends on chemical concentration, method of application, and time of reaction. It is, therefore, important to add the treatment chemicals to all of the test jars as nearly simultaneously as possible. To do this, prefill the chemical measuring device with the required amount of chemical working solution before beginning chemical addition. For VISCOUS[19] polymer solutions, a positive displacement device such as a hypodermic syringe or positive displacement micropipet is needed. For nonviscous solutions such as aluminum or iron metal salts, an ordinary pipet or graduated cylinder is satisfactory.

It is usually preferable to obtain assistance when injecting the chemical working solution into the jars. One person can handle up to two samples simultaneously, so it takes two people for four jars and three people for six jars. Simultaneous injection is most important with flocculant addition where the rapid mixing time is only a few seconds. When all the chemical working solutions are ready for injection, start the jar test unit stirrer at rapid mix speed and proceed with the chemical injection. Some newer jar test equipment allows the chemical working solution to be placed in six dishes that can be simultaneously placed into the jars, and then the unit stirrer can be started. As soon as the chemicals are in the test jars, begin the appropriate timing sequence and data logging.

4.125 Specific Tests for Chemical Selection, Dosage, and Process Control

4.1250 PRELIMINARY SCREENING TEST

Objective: To select a limited number of potentially suitable chemical treatment products from a large number of choices.

In order to properly run a preliminary screening test series on polymer samples, it is essential to make use of the polymer map (Figure 4.3). The polymers to be screened are first marked on the map, and then a number of polymers are chosen with substantially different properties. An initial screening dosage level suitable to the flow stream to be examined is then selected. For secondary effluent flow streams, a dosage of 3.0 mg/L is usually a surplus and makes a suitable initial dosage. This will require application of 3.0 mL of polymer working solution (1,000 mg/L). For sludge samples, dosage ranges of 75 mg/L or higher are not uncommon, for which 15 mL of chemical stock solution (5,000

---

[19] *Viscosity* (vis-KOSS-uh-tee). A property of water, or any other fluid, that resists efforts to change its shape or flow. Syrup is more viscous (has a higher viscosity) than water. The viscosity of water increases significantly as temperatures decrease. Motor oil is rated by how thick (viscous) it is; 20 weight oil is considered relatively thin while 50 weight oil is relatively thick or viscous.

mg/*L*) will be needed. To calculate trial dosage levels for a jar test, refer to *TEST CHEMICALS* in Section 4.1211.

1. Prepare a "jar test bench sheet" (similar to the one shown in Figure 4.5) on which to record specific data including, but not limited to:

   a. Source of water sample

   b. Name of each treatment product to be tested

   c. Dosage of treatment chemicals

   d. Rapid mixing speed and duration

   e. Slow mixing speed and duration

   f. The sequence in which visible flocculation appears (which jar is first, which is second)

   g. The relative apparent size of the floc that forms

   h. How quickly the floc settles to the bottom of the jar after stirring stops

   i. How deep the floc is at the bottom of the jar after settling

   j. The relative clarity of the supernatant water above the settled floc after settling

2. Collect a suitable amount of sample and fill all the jars with sample.

3. Start the jar test unit stirrer and set it at the rapid mixing speed.

4. Fill the chemical measuring device with the appropriate amount of chemical working solution.

5. Add the chemical working solution to the test jars and start the timer. (Coagulation tests skip to step 7.)

6. Reduce the stirring speed after the appropriate interval. (Flocculation tests only.)

7. Record data items 1-f and 1-g during the stirring period(s).

8. At the end of the timing sequence, stop the stirrer and record data items 1-h, 1-i, and 1-j on the jar test bench sheet.

Before going on to additional preliminary screening tests, compare the results obtained so far with the polymer map layout and make a note of what kind of chemical gave the best kind of result. It is often useful to mark the relative performance of each polymer screened by rank (for example: 1 = best, 2 = next best) next to the associated polymer location on the polymer map. This will help you see the comparative effectiveness of the various polymer types used in your investigation. Using this as your guide to predicting results, select another set of test chemicals for the next stage of screening. It is usually a good idea to carry along the chemical having the best performance into the next round of screening to provide a cross reference between tests. Using the newly selected test chemicals, repeat the screening procedure beginning at step 2 above.

Use each set of preliminary screening tests to help you decide which, if any, new chemicals look promising. After a few screenings, most of the ineffective chemicals will be eliminated and four to six chemicals will have begun to show consistently favorable results. If there is a clear performance difference in the last few "best performers," especially if there is a single obvious best choice, the next stage of testing, dosage testing, may be started. If there are two or more apparent best performers, repeat the preliminary screening test with these products at reduced chemical levels. For secondary effluent wastewater, drop the test dosage from 3.0 mg/*L* to 1.0 mg/*L*. For sludges, drop the dosage to 50 mg/*L*. If this reduced dosage does not reveal noticeable differences between the choices, try again at still lower application rates, or go on to dosage testing with all of the "winners" of the screening test.

4.1251 DOSAGE TESTING

Objective: To determine the chemical dosages required to achieve the desired treatment performance.

The information collected in this and subsequent test methods must be more detailed and more precise than for the preliminary screening test procedure above. What is required are reliable numbers that will allow you to set up a dependable chemical addition program at the lowest practical cost. The term practical is used here rather than "lowest possible cost" because as regulatory requirements become more and more stringent, and enforcement penalties become more costly, the value of reliability is substantial.

Along with dosage testing, a detailed analysis of the untreated flow stream is needed. In addition to a number of particular factors that will be measured after chemical treatment (for example, suspended solids, turbidity, and BOD/*COD*[20]), a number of other factors directly influence the performance of the treatment chemicals. These include pH, conductivity (dissolved solids), calcium and magnesium hardness, phosphate, sulfate, sulfide, sodium, potassium, nitrogen, and chloride. For polymers, the most critical of these are conductivity (dissolved solids) and pH. For inorganic coagulants, those two, along with phosphate and sulfate, are also critical. For iron metal salts, nitrogen (particularly ammonia nitrogen) and sulfide are very important as well.

The dosage tests follow the same general procedures as the preliminary screening tests. The main difference is that only one chemical is tested at a time, and a different amount of it is used

---

[20] *COD* (pronounce as separate letters). Chemical Oxygen Demand. A measure of the oxygen-consuming capacity of organic matter present in wastewater. COD is expressed as the amount of oxygen consumed from a chemical oxidant in mg/*L* during a specific test. Results are not necessarily related to the biochemical oxygen demand (BOD) because the chemical oxidant may react with substances that bacteria do not stabilize.

JAR TEST BENCH SHEET

PERFORMED BY: _____

DATE: _____

SAMPLE DATA

SOURCE: _____
TIME TAKEN: _____ AM/PM
FLOW: _____ pH: _____
TEMPERATURE: _____
TURBIDITY: _____
SUSPENDED SOLIDS: _____
COD: _____
CONDUCTIVITY: _____

PROCESS DATA

RAPID MIX
SPEED (RPM): _____
TIME (MINUTES): _____

SLOW MIX
SPEED (RPM): _____
TIME (MINUTES): _____

CHEMICALS USED: _____

| JAR NO. | CHEMICALS (PPM) | | | | | FLOC CHARACTERISTIC AND TIME OF APPEARANCE | ANALYSIS OF SUPERNATANT | | | | | SLUDGE | |
	ALUM	LIME	FERRIC	POLYMER	CAUSTIC		pH	TURB	COD	SUSP SOLIDS	COAG RESIDUAL	VOLUME	SETTLING TIME
UNTREATED													
1													
2													
3													
4													
5													
6													

REMARKS: _____

CHEMICALS(S) AND DOSAGE TO BE USED: _____

Fig. 4.5　Jar test bench sheet

in each jar. Also, some additional tests need to be made on the "clarified" water.

1. Prepare a jar test bench sheet similar to the one shown in Figure 4.5 on which to record specific data including, but not limited to:

   a. Source of water sample

   b. Name of each treatment product to be tested

   c. Dosage of treatment chemicals

   d. Rapid mixing speed and duration

   e. Slow mixing speed and duration

   f. The sequence in which visible flocculation appears (which jar is first, which is second)

   g. The relative apparent size of the floc that forms

   h. How quickly the floc settles to the bottom of the jar after stirring stops

   i. How deep the floc is at the bottom of the jar after settling

   j. The turbidity of the clarified liquid

   k. The suspended solids in the clarified liquid

   l. The *TRUE COLOR*[21] of the clarified liquid

   m. The residual BOD or COD (whichever is part of normal discharge guideline testing) of the clarified liquid

   n. The volatile and nonvolatile suspended solids

2. Collect a suitable amount of flow stream sample. Using a graduated cylinder, accurately fill all the jars with their respective sample volumes.

3. Start the jar test unit stirrer and set it at the rapid mixing speed.

4. Fill the chemical measuring devices with the different dosages of the same chemical working solution.

5. Add the various dosages of chemical working solution to the respective test jars and start the timer. (Coagulation tests skip to step 7.)

6. Reduce the stirring speed after the appropriate interval. (Flocculation tests only.)

7. Record data items 1-f and 1-g during the stirring period(s).

8. At the end of the timing sequence, stop the stirrer and record data items 1-h through 1-n on the jar test bench sheet.[22]

Because of the time required to analyze the results of the chemical treatment on the flow stream samples in each test jar, these jar tests are run "semiblind." That is, variations in performance at near-optimum conditions may not be visible to the eye of the observer (this is the reason for the laboratory analysis). However, the judgment of the testing personnel should be applied wherever there is an obviously inferior test. Discard any such samples and note this on the jar test bench sheet.

Two very important pieces of information should become evident following the dosage testing: (1) there will likely be a minimum dosage that will give suitable performance in meeting treatment specifications and above which better than necessary treatment process performance results can be obtained, and (2) nearly always there will be an upper dosage limit beyond which performance will diminish. (This is one classic example of "if a little bit is good, a lot is not necessarily better.") These two dosage limits are the upper and lower ends of the safe operating dosage range for each chemical product. If more than one treatment chemical is carried through this dosage testing procedure, the test exhibiting the widest safe operating dosage range will most likely handle the greatest variation in treatment process operating conditions. This factor needs to be included in the considerations used to select a treatment chemical. Two other considerations should be evaluated along with volume of sludge, cost, and safe operating range; they are: (1) availability of the chemical supply, and (2) reliability of the chemical supply (which are not necessarily the same issues.)

4.1252 FULL-SCALE TRIAL

Objective: To confirm the performance and fine-tune the dosage rates for the selected chemical(s).

By the time dosage testing has been completed, the performance of one or two chemicals will stand out as clearly superior. Because the decision about which chemical to use has far-reaching effects for both plant operation and future planning, the final stage is to try the selected chemicals in the process flow stream itself. The jar tests that will be run during this process are identical to those for dosage testing.

Use a set of chemical dosages that start below the level being used in the process flow stream and extending well above it. During the full-scale trial period, detailed analyses (as described in dosage testing) for the untreated flow, as well as the chemically treated flow, should be run at various times of the day in order to record any significant changes in flow composition as treatment process flow rates vary. The range of dosages for these jar tests should be run at about the same times that the samples for untreated and chemically treated flows are taken. By doing this, the results from the jar tests and the results in the treatment

---

[21] *True Color.* Color of the water from which turbidity has been removed. The turbidity may be removed by double filtering the sample through a Whatman No. 40 filter when using the visual comparison method.

[22] To complete the data collection for items numbered 1-j through 1-n, it will be necessary to carefully remove the clarified liquid from each test jar using a pipet or other suitable extraction device, taking care not to disturb either the sediment at the bottom of the jar or anything that might be floating on the surface or sticking to the sides.

process flow can be measured and confirmed and any correlations or variations can be noted. When this procedure is followed, the jar tests can then be applied with full confidence to the fourth phase of testing described in the following section.

4.1253 SYSTEM PERFORMANCE OPTIMIZATION

Objective: To keep treatment process performance within the desired limits while minimizing the chemical costs involved.

The required frequency of system performance optimization testing will be determined by a number of factors. These include the stability (or variability) of treatment process flow rates, the reliability of the treatment process itself, and the nature and frequency of process monitoring. A secondary objective of system performance optimization testing is to detect and compensate for potential treatment process upsets before they become an urgent problem. A treatment plant that operates with virtually no variation in effluent quality will need optimization testing very infrequently, perhaps monthly or quarterly and then mainly to ensure that chemical treatment dosages are not excessive. The more common situations, however, involve systems that operate near or at (or even significantly above) their design capacities on a daily basis. In these systems, jar testing should coincide with peak flows in order to obtain the information necessary to keep the treatment process operating satisfactorily.

The system performance optimization test procedure is identical to the ones described above for dosage testing and for full-scale trial testing except that the detailed analysis of the untreated water may be considered a matter of historical information rather than an essential guideline. The second difference is that, where at all possible, an attempt should be made to bracket the "safe operating dosage range" of the chemical(s) in such a way that the limits are apparent to the operator's naked eye. This is so that the time delay between the jar testing and completion of the laboratory analyses does not completely limit the operator's ability to adjust operation of the treatment process to keep it within performance specifications.

A third variation is that the jar test results (all of them), along with chemical system feed rates, should be recorded in a log format that can be used for tracking and projection purposes.

4.126 Procedure for Plants Without Laboratory Facilities

Source: *PROCEDURE FOR DETERMINING THE ALUM DOSAGE. Permission of Industrial Chemicals Division, Allied Chemical Corporation, Solvay, New York*

In case a laboratory is unavailable, it is possible to maintain reasonable process control by conducting coagulation tests using a simple hand-stirring method.

Clear glass fruit jars, one- to two-quart capacity, are a good substitute for beakers and are easily obtainable. If necessary, the local druggist or high-school chemistry teacher will usually assist in preparing alum solutions and lime suspensions. If a pipet is unavailable, approximately 20 drops from a common medicine

dropper is roughly equivalent to one milliliter. A calibrated dropper (1.0 mL) may be obtained from the drugstore.

Procedure:

1. Dissolve 9.46 grams of alum and dilute to 1 liter (8.95 grams of alum to one quart). One mL of this solution will provide a treatment of 10 mg/L of alum when added to one quart (946 mL) of the water sample.

2. With the pipet or medicine dropper, add 1 mL of the alum solution to one quart of the sample and stir rapidly for approximately two minutes. Actual rapid mixing time should be similar to the actual detention time of the water being treated in the flash mixer. This actual detention time may vary from 50 seconds to six minutes depending on design and flows. Then stir gently for at least 15 minutes to permit floc particles to form. Again, this actual gentle mixing time should be similar to the actual process flocculation time. During gentle mixing, the speed of the paddles in the jars should be similar to the speed of the actual flocculator paddles. When running the jar test, try adjusting the paddle speed to produce the best floc particle growth and then operate the flocculator paddles at this speed. Under some conditions, the best floc can be produced by stopping the flocculators. Avoid violent agitation during this floc conditioning stage to prevent the breakup of the floc.

3. Observe the quality of the floc, the rate of settling, and the clarity of the settled water.

4. Repeat the above steps using a higher alum dose until the desired floc and clarity are achieved.

   Helpful conversion factors:

1 liter	= 1,000 mL
1 quart	= 946 mL
1 grain/US gal	= 17.1 mg/L
1 grain/US gal	= 143 lbs/million US gallons
1 mg/L	= 8.34 lbs/million US gallons

4.127 Phosphate Monitoring

When the coagulant is being used to precipitate phosphate as well as to remove solids, coagulant dosage control may be obtained by automatically analyzing the incoming wastewater for soluble orthophosphate. The coagulant feeder is set to maintain a selected ratio of coagulant-to-phosphate either automatically or manually. Equipment is available that will automatically do this type of coagulant feeding.

Coagulant dosages that produce good phosphate removal will generally produce good solids removal if polymers are used to aid in flocculating the fine, precipitated matter. Usually, the polymer feed should be flow-paced; however, the polymer feeder may function with manual adjustments.

4.128 Safe Working Habits

Chemical feed equipment and chemical handling areas have safety hazards that each operator should become aware of for each plant. In addition to the usual electrical and mechanical hazards associated with automatic equipment, chemical treatment hazards include:

1. Strong acids

2. Strong caustics

3. High pressures

4. High temperatures

5. Dust in the air

6. Slippery walk areas

Avoid getting inorganic coagulants on the skin or in the eyes. These chemicals are either corrosive or caustic and cause irritation and possibly permanent damage. Wear protective gloves, chemical goggles, and a face shield, and ensure that an emergency eye wash and deluge shower are immediately available. Use local exhaust or approved respiratory protection to prevent inhalation of mists and dusts. Areas contacted should be flushed immediately with water. If a dry coagulant chemical comes in contact with your skin, attempt to mechanically wipe or brush off as much as possible before flushing. Check the appropriate MATERIAL SAFETY DATA SHEET (MSDS)[23] for specific health and safety information.

Polymers mixed with water are extremely slippery and can cause a fall hazard if left unattended on walkways. Spills should be cleaned immediately, using inert absorbent materials such as sand or earth.

Develop safe working habits by always wearing proper safety equipment and protective clothing. If you do not have the required safety equipment or are not trained in its use, do not use the chemical until you have both the equipment and the training.

Good housekeeping is a part of the total plant operation. Good housekeeping around the chemical feed systems is very important to good operations and safety. A dry chemical feeder that weighs its output will change its feed rate if chemicals are allowed to build up on the scales. Good housekeeping will reduce the hazard of slipping around chemical handling areas and will keep the dust down in work areas. Good housekeeping is a daily duty and must not be neglected.

4.129 Summary

Treatment chemical testing is most effective when it proceeds from a general review (preliminary screening) through more precise analyses as optimum chemical and treatment process operating performance is approached. Evaluation of test results will initially lead to another set of tests, which, in turn, will indicate further testing. Simply running a single grab batch of tests will rarely provide information of sufficient reliability to operate a treatment process. As enough information accumulates, the number of key guidelines that are significant will become evident, and those of less importance can be reduced in test frequency while any that prove to be virtually unimportant can be discontinued. Testing is not only valuable in controlling treatment process performance, but regular review of recorded test, application, and performance data will also serve to evaluate the needed testing as well.

QUESTIONS

Write your answers in a notebook and then compare your answers with those on page 460.

4.12D   How are stock solutions of polymers prepared?

4.12E   What is the maximum solution strength for a polymer?

4.12F   Where should the sampling point be located to collect samples for a jar test?

END OF LESSON 1 OF 6 LESSONS

on

SOLIDS REMOVAL FROM SECONDARY EFFLUENTS

Please answer the discussion and review questions next.

DISCUSSION AND REVIEW QUESTIONS
Chapter 4.   SOLIDS REMOVAL FROM SECONDARY EFFLUENTS
(Lesson 1 of 6 Lessons)

At the end of each lesson in this chapter you will find some discussion and review questions. The purpose of these questions is to indicate to you how well you understand the material in the lesson. Write the answers to these questions in your notebook.

1. What is the difference between coagulation and flocculation?

2. What are the three most important process control guidelines in any chemical treatment process?

3. What is the meaning of "electrostatic repulsion"?

4. How do the precipitates of metal salts assist in the destabilization of suspended particles?

5. Why is paddle speed important during flocculation?

6. What precautions should be taken when using alum?

7. What is a polymer map?

8. Why should extensive laboratory testing be conducted before treating the entire plant effluent with a particular polyelectrolyte?

9. What factors influence the selection of a polymer?

10. What factors should be considered when selecting a chemical?

11. Explain how to run the jar test.

12. What would you do if a dry coagulant chemical came in contact with your skin?

CHAPTER 4.   SOLIDS REMOVAL FROM SECONDARY EFFLUENTS

(Lesson 2 of 6 Lessons)

4.13   Physical–Chemical Treatment Process Equipment

4.130   *Chemical Storage and Mixing Equipment*

Coagulants can be divided into two physical categories, liquid and solid. The physical category will necessarily dictate the type of equipment needed to accommodate it.

Solids are often dissolved before actual use and, therefore, use some equipment similar to that for liquid coagulants. Storage of all inorganic solids should be in dry tanks or bins because almost all inorganic solids exhibit caustic or corrosive properties if they become moist. Bins for powdered solids generally will be provided with a dust collector to keep the material ventilated as well as contained as a precaution against flash combustion. A shaker or vibrator is often installed at the feed system exit to prevent caking of the material. In the case of some limes, a slaker is often found at this point.

Chemical mixing equipment (Figure 4.6) is needed to prepare a solution of known concentration that can be metered (measured) into the water being treated. Polyelectrolytes can be difficult to dissolve. Also, polyelectrolytes may need a period of aging prior to application. A dissolving tank with a mechanical mixer is used to prepare the solution for feeding. The resulting solution is stored in a day tank (holding tank) from which it is metered out at the proper dosage into the water being treated.

4.131   *Chemical Feed Equipment*

Chemical feeders (metering equipment) are required to accurately control the desired dosage. The chemical to be used and the form in which it will be purchased must be determined first because chemicals used for solids removal in wastewater treatment usually can be purchased in either solid or liquid form.

After metering, solid chemicals are generally converted into a solution or a slurry (watery mixture) before being fed to the wastewater stream. Flushing water is often used with both slurries and liquids to rapidly carry the chemical to the point of application. This is especially true with lime (Figure 4.7), caustic (Figure 4.8), alum (Figure 4.9), and organic polymers (Figure 4.10). In such cases, measured amounts of the dry chemical are passed to a makeup or slurry tank in which a desired fluid dilution is made.

Dry chemicals may also be added directly under the assistance of water jets spraying at the points of entry. The latter method is an acceptable means of delivering dry polymer.

4.1310   *TYPES OF CHEMICAL METERING EQUIPMENT*

To maintain accurate feed rates, there cannot be any slippage in the metering equipment; therefore, most liquid feeders are of the positive displacement type. The quality of the water used for both mixing and flushing the polymer system is important. Use of poor-quality plant effluent for either purpose may cause clumps ("fish eyes") to form, which will plug feeders, small orifices (openings), and even piping.

POSITIVE DISPLACEMENT PUMPS

A plunger pump (Figure 4.11) is used for metering chemicals due to the accuracy of the positive displacement stroke and the ease of adjusting the piston stroke to regulate the chemical feed rate. With each stroke, a fixed amount or volume of chemical or solution is discharged. By knowing the amount discharged per stroke and the number of strokes per minute, it is easy to calculate the chemical output.

Other positive displacement pumps besides the plunger pump include the gear pump (Figure 4.11) and the diaphragm pump (Figures 4.12 and 4.13). Each of these pumps will produce a constant chemical flow rate for a specific setting. Another type of positive displacement pump is the progressive cavity pump.

The feed rate for dry chemicals must also be accurately controlled. Typical dry chemical feeders include the screw feeder, vibrating trough, rotating feeder, and belt-type gravimetric feeder.

SCREW FEEDER

A screw feeder unit (Figure 4.14) maintains a desired output by varying the speed or the amount of time the screw rotates as it moves chemicals out the discharge port. Care must be taken that the chemical does not cake up in the hopper and stop feeding

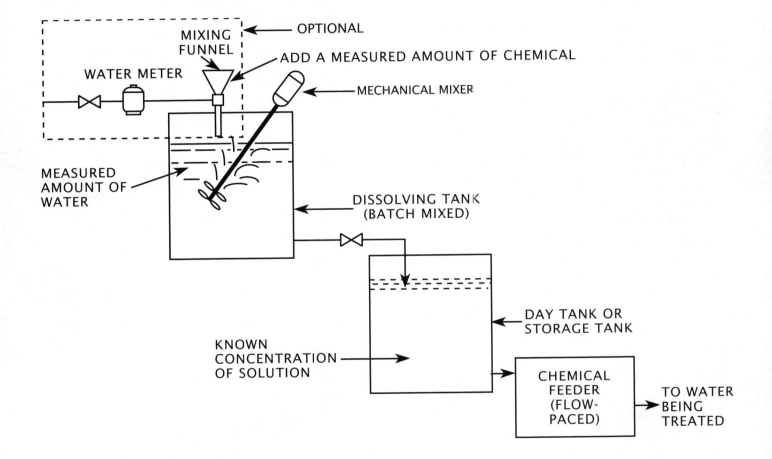

OPTIONAL

MIXING
FUNNEL

ADD A MEASURED AMOUNT OF CHEMICAL

WATER METER

MECHANICAL MIXER

MEASURED
AMOUNT OF
WATER

DISSOLVING TANK
(BATCH MIXED)

DAY TANK OR
STORAGE TANK

KNOWN
CONCENTRATION
OF SOLUTION

CHEMICAL
FEEDER
(FLOW-
PACED)

TO WATER
BEING
TREATED

Fig. 4.6   Dry chemical dissolver, day tank, and feeder

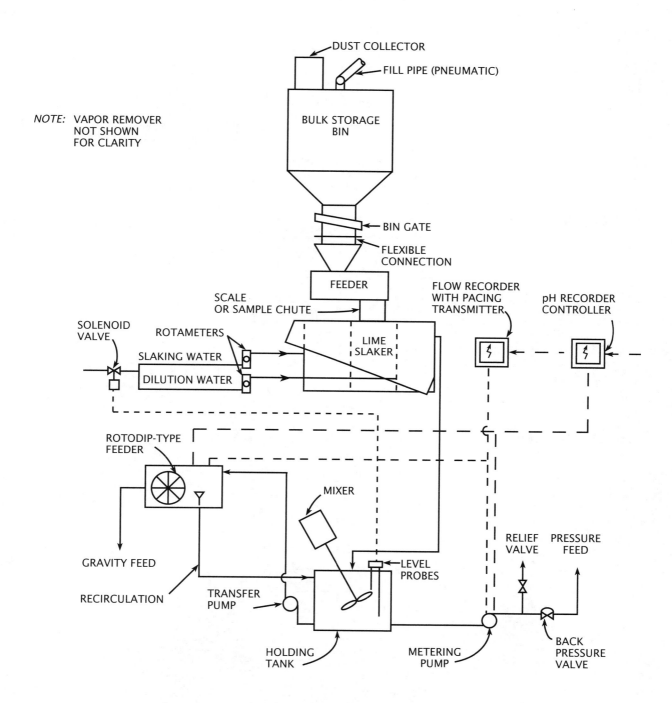

Fig. 4.7   Typical lime feed system

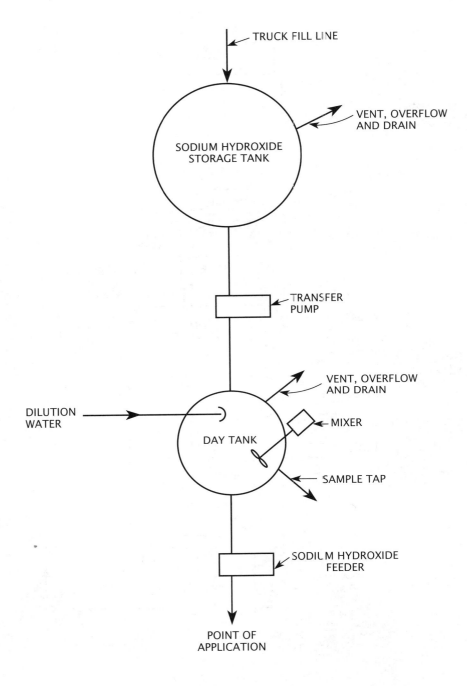

Fig. 4.8   Typical schematic of a caustic soda feed system

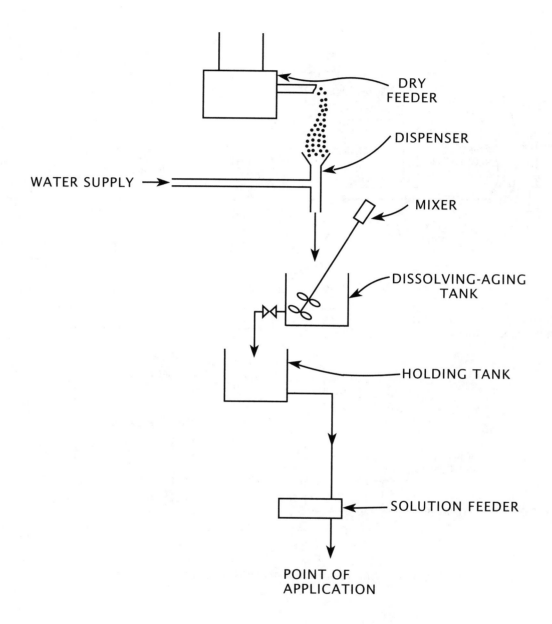

Fig. 4.9   Typical schematic of a dry alum feed system

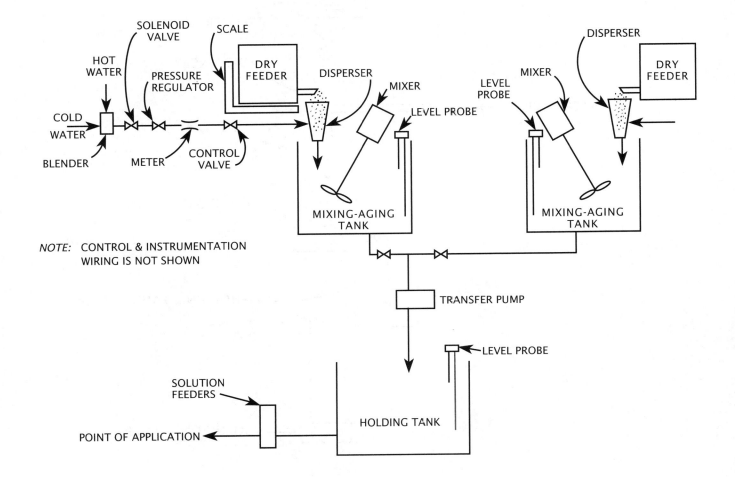

Fig. 4.10   Typical schematic of an automatic dry polymer feed system

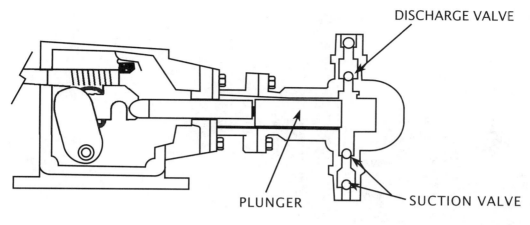

PLUNGER PUMP
(Courtesy of Wallace & Tiernan)

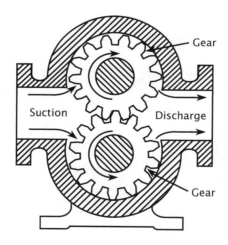

GEAR PUMP
(Courtesy of *CHEMICAL ENGINEERING*, 76 8, 45 (April 1969))

Fig. 4.11   Plunger and gear pumps

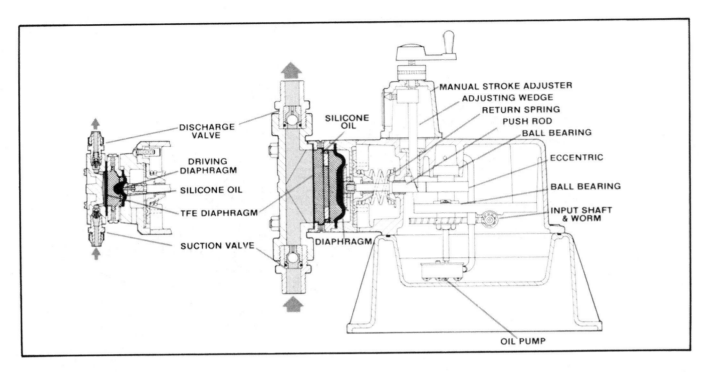

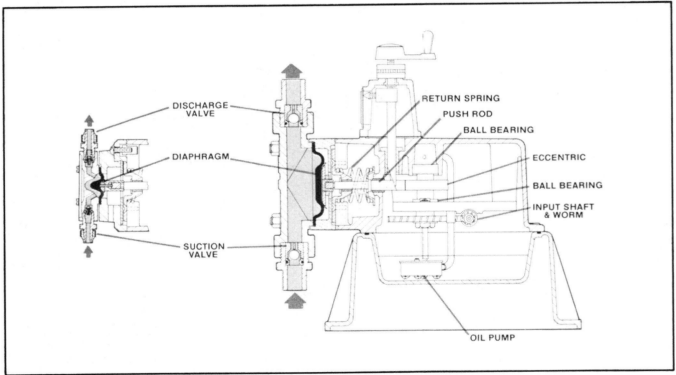

Fig. 4.12   Positive displacement diaphragm pumps

(Permission of Wallace & Tiernan Division, Pennwalt Corporation)

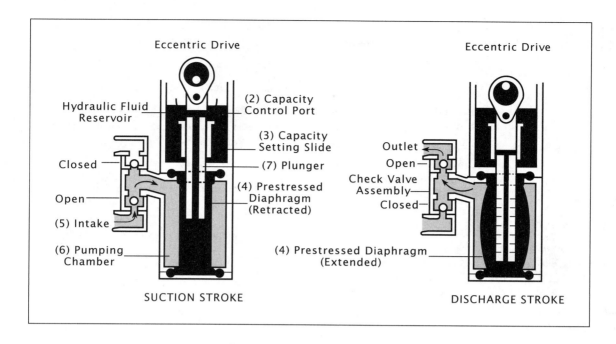

Fig. 4.13   Positive displacement diaphragm metering
(Permission of BIF, a Unit of General Signal)

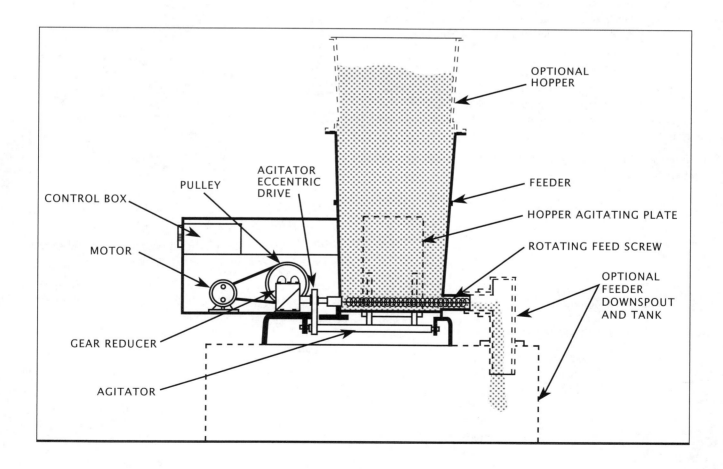

Fig. 4.14   Volumetric screw feeder
(Permission of Wallace & Tiernan Division, Pennwalt Corporation)

the screw. Also, the screw must be kept clean or the amount discharged per revolution will change.

VIBRATING TROUGH FEEDER

The vibrating trough maintains a constant depth of chemical discharged and controls its chemical output by the magnitude and the duration (length of time) of the vibrations.

Care must be taken that the chemical does not cake in the hopper and stop feeding into the trough. Also, caking on the trough will prevent an even flow of chemical, which could change the output volume.

ROTARY FEEDER

Rotary feeders are similar to the positive displacement gear pump because a fixed amount of chemical is discharged from between each tooth (Figure 4.15). The output can be controlled by the speed or running time of the rotor. Care must be taken to maintain the rotor lobes clean and free of buildup that will change the chemical output volume.

BELT-TYPE GRAVIMETRIC FEEDER

A gravimetric belt feeder (Figure 4.16) maintains a constant chemical weight on a revolving belt. This is accomplished using a vibrating trough and a balance system. The chemical output is controlled by the amount of chemical on the belt and the speed and time the belt travels. The amount of chemical is varied by the opening or closing of a feed gate or as a weighing deck moves up or down.

Care must be taken with this unit to ensure that chemicals do not build up on the balance because this will change the chemical output. By catching and weighing the chemical discharged at a constant speed over a measured amount of time, the feeder output can be verified.

Table 4.2 lists various types of chemical feeders, their uses, and limitations.

QUESTIONS

Write your answers in a notebook and then compare your answers with those on page 460.

4.13A   How are chemical solutions prepared for feeding?

4.13B   List the most common types of chemical feeders or metering equipment.

4.1311   SELECTING A CHEMICAL FEEDER

When you must decide which chemical feeder to purchase for your situation, include the following considerations:

1. *TOTAL OPERATING RANGE*

    Will the unit run at today's lowest expected chemical output as well as the future required output?

2. *ACCURACY*

    Will the unit maintain the same feed rate after it has been installed, calibrated, and operated?

3. *REPEATABILITY*

    Can you return to previous settings and obtain the same feed rates as before?

4. *RESISTANCE TO CORROSION*

    Will the equipment, including electrical components, withstand the corrosive environment to which they may be exposed?

5. *DUST CONTROL*

    Is a means provided to control dust, if needed?

6. *AVAILABILITY OF PARTS*

    Are replacement parts readily available?

7. *SAFETY*

    Is the system designed with safety of both operation and maintenance in mind?

8. *ECONOMICS*

    What are the costs of purchase, installation, operation, maintenance, replacement, and energy requirements?

4.1312   REVIEWING CHEMICAL FEED SYSTEM DESIGNS

When reviewing chemical feed system designs and specifications, the operator should check the following items:

1. Review the results of predesign tests to determine the chemical feed rate for both the present and future. The chemical feeders should be sized to handle the full range of chemical doses or provisions should be made for future expansion.

2. Determine if sampling points are provided to measure chemical feeder output.

3. Be sure provisions are made for standby equipment in order to maintain uninterrupted dosages during equipment maintenance.

4. Look for adequate valving to allow bypassing or removing equipment for maintenance without interrupting the chemical dosage.

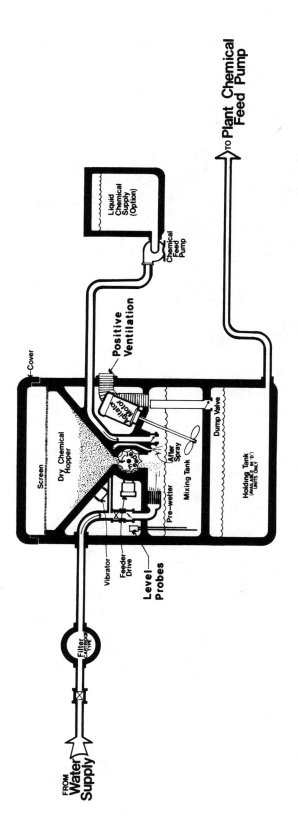

Fig. 4.15   Rotary feeder
(Permission of Neptune Microfloc)

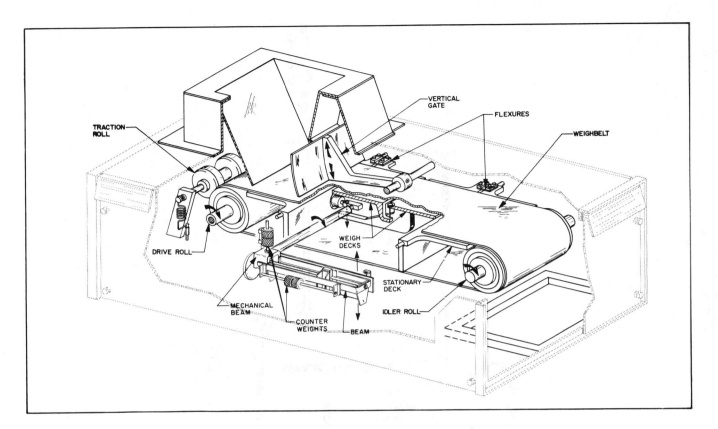

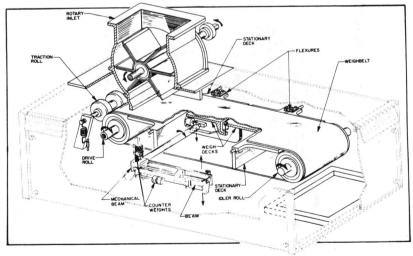

Fig. 4.16   Gravimetric belt feeders

(Permission of Wallace & Tiernan Division, Pennwalt Corporation)

TABLE 4.2 TYPES OF CHEMICAL FEEDERS

(Source: EPA *PROCESS DESIGN MANUAL FOR SUSPENDED SOLIDS REMOVAL*)

Type of Feeder	Use	Limitations General	Capacity, cu ft/hr	Range
Dry feeder:				
Volumetric:				
Oscillating plate	Any material, granules or powder.		0.01 to 35	40 to 1
Oscillating throat (universal)	Any material, any particle size.		0.02 to 100	40 to 1
Rotating disk	Most materials including NaF, granules, or powder.	Use disk unloader for arching.	0.01 to 1.0	20 to 1
Rotating cylinder (star)	Any material, granules or powder.		8 to 2,000 or 7.2 to 300	10 to 1 or 100 to 1
Screw	Dry, free flowing material, powder or granular.		0.05 to 18	20 to 1
Ribbon	Dry, free flowing material, powder, granular, or lumps.		0.002 to 0.16	10 to 1
Belt	Dry, free flowing material up to 1½-inch size, powder or granular.		0.1 to 3,000	10 to 1 or 100 to 1
Gravimetric:				
Continuous-belt and scale	Dry, free-flowing, granular material, or floodable material.	Use hopper agitator to maintain constant density.	0.02 to 2	100 to 1
Loss in weight	Most materials, powder, granular, or lumps.		0.02 to 80	100 to 1
Solution feeder:				
Nonpositive displacement:				
Decanter (lowering pipe)	Most solutions or light slurries.		0.01 to 10	100 to 1
Orifice	Most solutions.	No slurries	0.16 to 5	10 to 1
Rotameter (calibrated valve)	Clear solutions.	No slurries	0.005 to 0.16 or 0.01 to 20	10 to 1
Loss in weight (tank with control valve)	Most solutions.	No slurries	0.002 to 0.20	30 to 1
Positive displacement:				
Rotating dipper	Most solutions or slurries.		0.1 to 30	100 to 1
Proportional pump:				
Diaphragm	Most solutions. Special unit for 5% slurries.[a]		0.004 to 0.15	100 to 1
Piston	Most solutions, light slurries.		0.01 to 170	20 to 1
Gas feeder:				
Solution feed	Chlorine		8,000 lb/day max	20 to 1
	Ammonia		2,000 lb/day max	20 to 1
	Sulfur dioxide		7,600 lb/day max	20 to 1
	Carbon dioxide		6,000 lb/day max	20 to 1
Direct feed	Chlorine		300 lb/day max	10 to 1
	Ammonia		120 lb/day max	7 to 1
	Carbon dioxide		10,000 lb/day max	20 to 1

[a] Use special heads and valves for slurries.

5. Examine plans for valving to allow flushing the system with water before removing from service.

6. Be sure corrosion-resistant drains are provided to prevent chemical leaks from reaching the floor; for example, drips from pump packing.

7. Check for corrosion-resistant pumps, piping, valves, and fittings as needed.

8. Determine the amount of maintenance required. The system should require a minimum of maintenance. Equipment should be standard, with replacement parts readily available.

9. Consider the effect of changing head conditions, both suction and discharge, on the chemical feeder output. Changing head conditions should not affect the output if the proper chemical feeder has been specified.

10. Determine whether locations for monitoring readouts and dosage controls are convenient to the operation center and easy to read and record.

4.1313 CHEMICAL FEEDER START-UP

After the chemical feed system has been purchased and installed, the operator must carefully check it out before starting it up. Even if the contractor who installed the system is responsible for ensuring that the equipment operates as designed, the operation by plant personnel, the functioning of the equipment, and the results from the process are the responsibility of the chief operator. Therefore, before start-up, check the following items:

1. Inspect the electrical system for proper voltage, for properly sized overload protection, for proper operation of control lights on the control panel, for proper safety lockout switches and operation, and for proper equipment rotation.

2. Confirm that the manufacturer's lubrication and start-up procedures are being followed. Equipment may be damaged in minutes if it is run without lubrication.

3. Examine all fittings, inspection plates, and drains to ensure that they will not leak when placed into service.

4. Determine the proper positions for all valves. A positive displacement pump will damage itself or rupture lines in seconds if allowed to run against a closed valve or system.

5. Be sure that the chemical to be fed is available. A progressive cavity pump will be damaged in minutes if it is allowed to run dry.

6. Inspect all equipment for binding or rubbing.

7. Confirm that safety guards are in place.

8. Examine the operation of all auxiliary equipment including the dust collectors, fans, cooling water, mixing water, and safety equipment.

9. Check the operation of alarms and safety shutoffs. If it is possible, operate these devices by manually tripping each one. Examples of these devices are alarms and shutoffs for high water, low water, high temperature, high pressure, and low chemical levels.

10. Be sure that safety equipment, such as emergency eye wash facilities, deluge showers, approved respiratory protection, face shields, gloves, and vent fans, is in place and functional.

11. Record all important NAMEPLATE[24] data and place it in the plant files for future reference.

4.1314 CHEMICAL FEEDER OPERATION

Once the chemical feed equipment is in operation and the major bugs are worked out, the feeder will need to be fine-tuned. To aid in fine-tuning and build confidence in the entire chemical feed system, the operator must maintain accurate records (Figures 4.17 and 4.18). These records will include the flows and characteristics of the wastewater before treatment, the dosage and conditions of the chemical treatment, and the results obtained after treatment. A comment section should be used to note abnormal conditions, such as a feeder plugged for a short time, a sudden change in the characteristics of the influent waste, and related equipment that malfunctions. Daily logs should be summarized into a form that operators can use as a future reference.

4.1315 SHUTTING DOWN CHEMICAL SYSTEMS

If the equipment is going to be shut down for an extended length of time, it should be cleaned out to prevent corrosion or the solidifying of the chemical. Lines and equipment could be damaged when restarted if chemicals left in them solidify. Operators could be seriously injured if they open a chemical line that has not been properly flushed out.

The following items should be included in your checklist for shutting down the chemical system:

1. Shut off the chemical supply.

2. Run chemicals completely out of the equipment and clean the equipment.

3. Flush out all the solution lines.

4. Shut off the electric power.

5. Shut off the water supply and protect from freezing.

6. Drain and clean the mix and feed tanks.

QUESTIONS

Write your answers in a notebook and then compare your answers with those on page 460.

4.13C List the items that should be considered when selecting a chemical feeder.

4.13D What information should be recorded for a chemical feeder operation?

---

[24] *Nameplate.* A durable, metal plate found on equipment that lists critical operating conditions for the equipment.

SODIUM HYDROXIDE LOG

	TANK #1 GAL.	TANK #2 GAL.	TANK #3 GAL.	GAL. #3 BEFORE TRANS.	GAL. #3 AFTER TRANS.	H₂0 TO NAOH	DILUTE GAL. USED	TOTAL GAL. REC.	REMARKS
1	880	-0-	1,280				300		
2	880	-0-	990				290		
3	880	-0-	700				290		
4	880	-0-	580				120		
5	880	-0-	300				280		
6	-0-	-0-	1,840	200	2,000	1-1	260		
7	-0-	-0-	1,600				240		
8	-0-	-0-	1,400				200		
9	-0-	-0-	1,240				160	4,000	
10	1,990	1940	1,050				190		
11	1,990	1940	850				200		
12	1,990	1940	650				200		
13									
14									
15									
16									
17									
18									
19									
20									
21									
22									
23									
24									
25									
26									

Fig. 4.17   Typical record of chemical feeder operation

CHEMICAL FEED RECORD

CHEMICAL *ALUM*   10 GALLONS PER INCH   LOCATION *SECONDARY*

DATE TIME	CHEMICAL TANK LEVEL			TREATED FLOW x 1,000 METER READINGS			PUMP SET A/M	CHEM USED mg/L	OPER.	REMARKS
	PREVIOUS	PRESENT	AMT. USED	PRESENT	PREVIOUS	TOTAL				
5-31-05	—	123"		105,376	—					
6-1/0800	123"	100"	23"	115,376	105,376	10,000	AUTO	23.0	T.J.	OK
6-2/0800	100"	75"	25"	125,026	115,376	9,650	A	25.9	T.S.	OK
6-3/0800	75"	51"	24"	134,574	125,026	9,548	A	25.1	A.L.	OK
6-3/1030	51"	48"	3"	135,976	134,574	1,302	A	23.0	B.L.	REFILLED TANK
6-3/1030		200"								
6-4/0800	200"	184"	16"	144,806	135,876	8,930	M/30%	17.9	B.L.	TREATMENT BETTER
6-5/0800	184"	162"	22"	154,466	144,806	10,660	A	20.6	B.L.	"
6-5/1500	162"	154"	8"	158,140	154,466	3,676	A	21.8	T.J.	CHECKING PUMP
TOTAL MAX. MIN. AVG.								CHEMICAL COST $/MG		

Fig. 4.18   Typical form for chemical feeder operation

4.132 Coagulant Mixing Units

Rapid mixing of coagulants may be accomplished in one of three modes: (1) high-speed mixers (impeller or turbine), (2) in-line blenders and pumps, and (3) baffled compartments or pipes (static mixers). The use of high-speed mechanical mixers is most common. They are often seen in parallel to increase residence time. Static mixers make use of fluid passing through baffled chambers at high velocities to bring about turbulence and mixing. In-line blenders and pumps accomplish the same by virtue of a high velocity through pipelines and pumps.

4.133 Flocculators

Mechanical flocculating units (Figure 4.19) may be rotary, horizontal shaft-reel type, rotary-shaft turbine, or rotary reciprocating. All three rotary systems possess vertical shafts. Standard rotary and rotary reciprocating units use paddle impellers; reciprocating types have two or more shafts rotating in opposite directions.

In all flow-through flocculators, tapered flocculation is found to be most beneficial. By this method, a small, dense floc is formed initially followed by aggregation to form a denser, larger floc. This is accomplished on single shafts by variation of the paddle sizes. On multiple-shaft units, variation of the speed of the individual units or the number of paddles per shaft is effective.

4.134 Clarifiers (Also see Chapter 5, "Sedimentation and Flotation," Volume I, OPERATION OF WASTEWATER TREATMENT PLANTS)

4.1340 TYPES OF CLARIFIERS

Clarifiers can take on two basic configurations based upon the flow character, that is, vertical or horizontal flow. Horizontal flow is the most common in both rectangular (Figure 4.20) and circular (Figure 4.21) clarifiers.

Rectangular clarifiers with horizontal flow have the influent entering at one end. Flow generally hits a baffle and moves by gravity to the opposite end where the effluent overflows the outlet weirs. A surface skimmer made up of flights pushes oil and floating debris to a spiral collector located at one end. Settled sludge is moved by flights along the bottom to a sludge hopper where it is collected and pumped to a dewatering facility.

Circular clarifiers (Figure 4.21) with horizontal flow take on one of three configurations:

1. Center influent with radial effluent

2. Radial influent with center effluent

3. Radial influent and effluent

In each case, sludge is collected at the center of the conical (cone-shaped) base. Oil and scum are skimmed by a radial arm at the surface of the water, which deposits the material into a sump.

Vertical-flow units (Figure 4.22) have the general distinction of the influent flowing along the bottom and rising toward the top to be discharged over the effluent weir. One advantage of vertical-flow clarifiers is that flow can be forced up through the sludge blanket, thus aiding in solids retention and improving flow control. Both rectangular and circular configurations exist.

Circular configurations have solids contact units in which all three activities leading up to and including coagulation and precipitation take place (Figure 4.23). Influent becomes rapidly mixed with the coagulant at the influent discharge and flows down through a center baffle. Flocculation occurs in this zone. Flow proceeds radially upward through the sludge blanket and clarification occurs. Effluent discharge is radial.

Rectangular configurations may have tube and lamella separators in which settling becomes compartmentalized. Tube settlers (Figures 4.24, 4.25, and 4.26) consist of a collection of closely packed, small-diameter tubes placed at an angle. Flow proceeds upward as sludge settles downward. A lamella separator uses parallel plates rather than tubes. The wastewater flows upward and sludge moves downward similar to tube settlers (Figure 4.24). Both configurations depend on the assumption that the paths of all particles with the same settling velocities will be straight, parallel lines. In the above two configurations, the surface area of the clarifier is effectively increased without increasing the actual size.

4.1341 OPERATING GUIDELINES

Four major factors are used to predict the performance of clarifiers: detention time, weir overflow rate, surface loading rate, and solids loading. If any one or more of these factors exceeds design or expected values, we can expect facility performance to deteriorate. Each of these factors is readily calculated and will give the operator an indication of expected clarifier efficiency. These four factors are also used to design clarifiers.

Detention time is the length of time it would take a plug of water to enter a clarifier and to exit in the effluent. This is related to clarifier efficiency in that particles should be allowed ample time to settle out. If the detention time is less than the

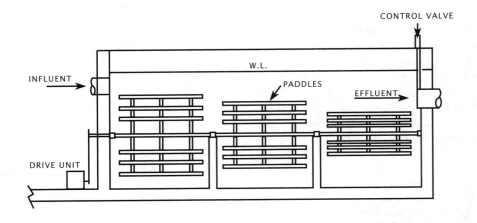

Mechanical Flocculation Basin
Horizontal Shaft—Reel Type

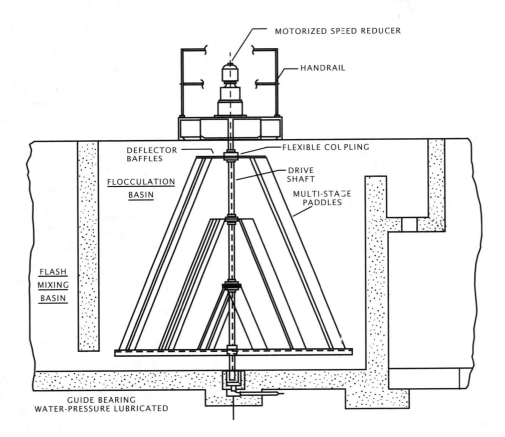

Mechanical Flocculator
Vertical Shaft—Paddle Type

(Courtesy of Ecodyne Corp.)

Fig. 4.19 Mechanical flocculators

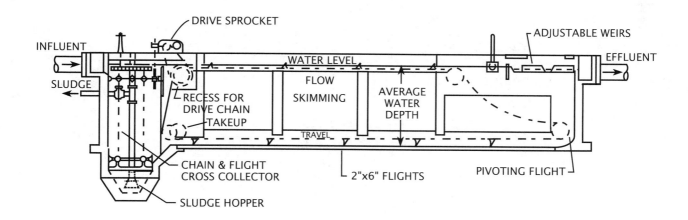

A. WITH CHAIN AND FLIGHT COLLECTOR

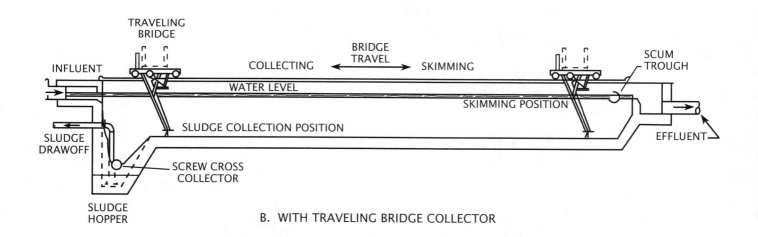

B. WITH TRAVELING BRIDGE COLLECTOR

Fig. 4.20 Rectangular sedimentation tanks
(Courtesy of FMC Corp.)

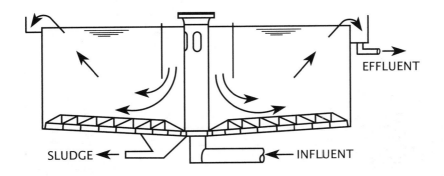

(a) CIRCULAR CENTER-FEED CLARIFIER WITH
A SCRAPER SLUDGE REMOVAL SYSTEM

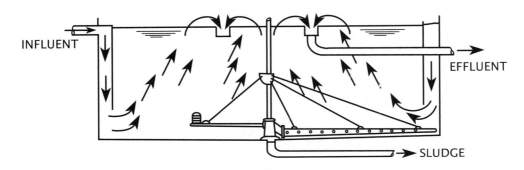

(b) CIRCULAR RIM-FEED, CENTER TAKE-OFF CLARIFIER WITH A
HYDRAULIC SUCTION SLUDGE REMOVAL SYSTEM

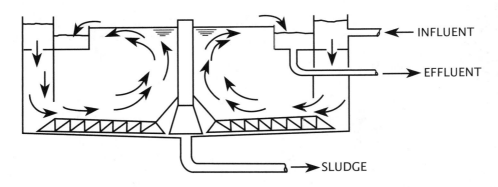

(c) CIRCULAR RIM-FEED, RIM TAKE-OFF CLARIFIER

Fig. 4.21   Typical circular clarifier

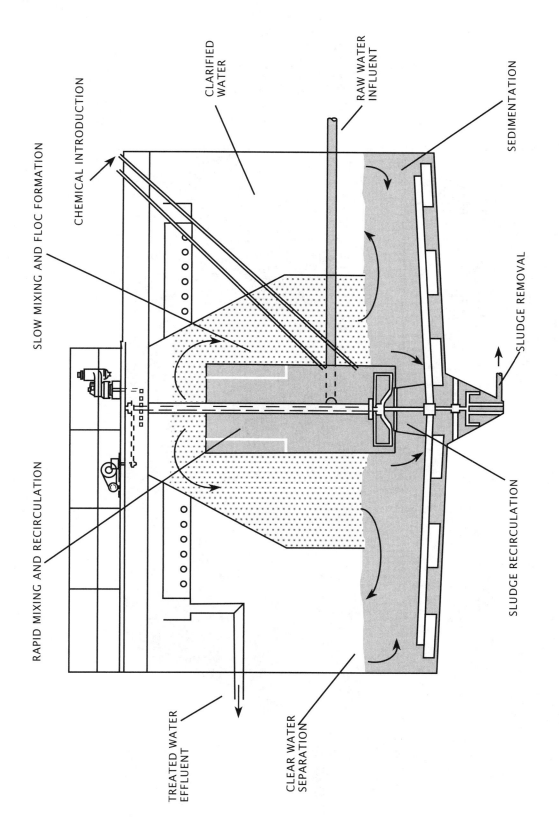

CHEMICAL INTRODUCTION

CLARIFIED WATER

RAW WATER INFLUENT

SLOW MIXING AND FLOC FORMATION

SEDIMENTATION

RAPID MIXING AND RECIRCULATION

SLUDGE REMOVAL

SLUDGE RECIRCULATION

TREATED WATER EFFLUENT

CLEAR WATER SEPARATION

Fig. 4.22   Solids contact clarifier without sludge blanket filtration
(Courtesy of Econodyne Corp.)

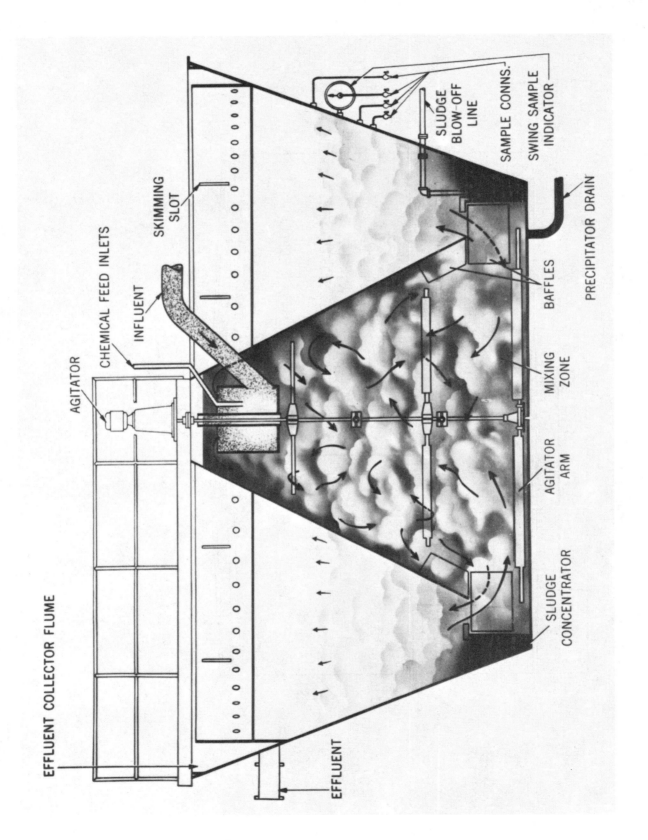

Fig. 4.23   Solids contact clarifier with sludge blanket filtration
(Courtesy of the Permutit Co.)

EFFLUENT COLLECTOR FLUME

AGITATOR

CHEMICAL FEED INLETS

INFLUENT

SKIMMING SLOT

EFFLUENT

SLUDGE CONCENTRATOR

AGITATOR ARM

MIXING ZONE

BAFFLES

PRECIPITATOR DRAIN

SWING SAMPLE INDICATOR

SAMPLE CONNS.

SLUDGE BLOW-OFF LINE

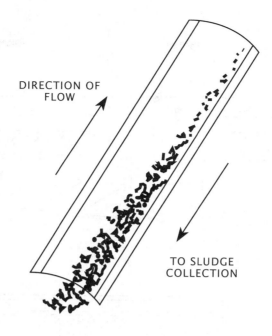

Fig. 4.24   Tube settlers—flow pattern

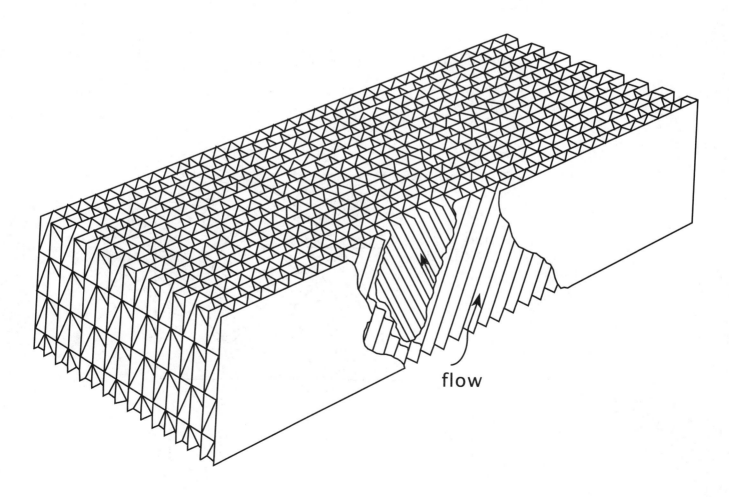

Fig. 4.25   Module of steeply inclined tubes
(Courtesy Neptune Microfloc, Inc.)

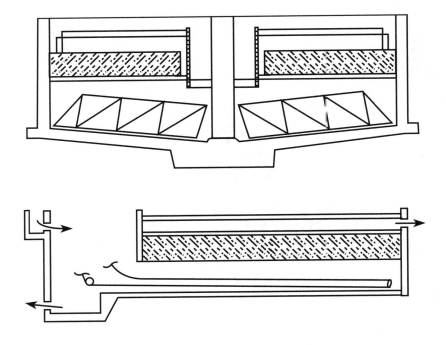

TUBE SETTLERS IN EXISTING CLARIFIER

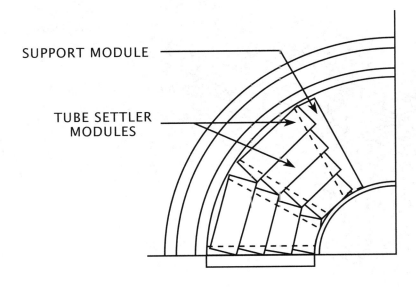

SUPPORT MODULE

TUBE SETTLER
MODULES

Fig. 4.26   Plan view of modified clarifier

settling rate, then there will be a carryover of particles into the effluent. Detention time is calculated knowing the flow and tank dimensions as follows:

Tank Volume, cu ft = Length, ft × Width, ft × Depth, ft

or

Tank Volume, m³ = Length, m × Width, m × Depth, m

and

$$\text{Detention Time, hr} = \frac{\text{Tank Vol, cu ft} \times \frac{7.48 \text{ gal}}{\text{cu ft}} \times \frac{24 \text{ hr}}{\text{day}}}{\text{Flow, gal/day}}$$

or

$$\text{Detention Time, hr} = \frac{\text{Tank Vol, m}^3 \times 24 \text{ hr/day}}{\text{Flow, m}^3/\text{day}}$$

If the detention time proves to be less than the settling rate (as shown by results of laboratory tests), then it may be necessary to increase the clarifier capacity by placing other basins into operation.

The weir overflow rate expresses the quantity of water that passes out of the clarifier in relation to the *LINEAL*[25] feet of weir available. Weir overflow rate is calculated as follows:

$$\text{Weir Overflow Rate, GPD/ft} = \frac{\text{Flow, gal/day}}{\text{Length of Weir, lineal feet}}$$

or

$$\text{Weir Overflow Rate, m}^3/\text{day/m} = \frac{\text{Flow, m}^3/\text{day}}{\text{Length of Weir, m}}$$

Surface loading rate expresses the quantity of water being treated in relation to the available clarifier surface. As previously stated, sedimentation efficiency increases with increased surface area. Surface loading rate may be calculated as follows:

$$\text{Surface Loading Rate, GPD/sq ft} = \frac{\text{Flow, gal/day}}{\text{Clarifier Surface Area, sq ft}}$$

or

$$\text{Surface Loading Rate, m}^3/\text{day/sq m} = \frac{\text{Flow, m}^3/\text{day}}{\text{Clarifier Surface Area, sq m}}$$

Another very important loading guideline for clarifiers is the solids loading. Solids loadings are especially important for industrial waste treatment because the solids carried by industrial wastewater may be significantly different from solids in municipal wastewaters.

$$\text{Solids Loading, lbs/hr/sq ft} = \frac{\text{Solids, lbs/hr}}{\text{Clarifier Surface Area, sq ft}}$$

or

$$\text{Solids Loading, kg/hr/sq m} = \frac{\text{Solids, kg/hr}}{\text{Clarifier Surface Area, sq m}}$$

The above characteristics are useful when compared with the original design considerations and specifications in attempting to diagnose the possible causes for clarifier inefficiency. These, coupled with the results of laboratory tests, such as dye tracer studies, will aid the operator in troubleshooting clarifier performance problems.

Other physical characteristics will also affect clarifier efficiency. One important feature is the settling characteristics of the particles. Settling rate may be affected by many factors including particle size, shape, temperature of the surrounding water, and particle density in relationship to that of the surrounding water. Particles of greater weight and density will settle faster than those of lesser density. The horizontal velocity and the tank depth will also come into play in determining how long it takes a particle to settle.

A particle settling quiescently (in still water) will tend to settle in a path perpendicular to the settling surface. With flow through a clarifier, force is applied to a particle causing it to settle in a plane diagonal to the settling surface. Since it now has a longer path to travel, the particle will take slightly longer to settle.

The ultimate settling velocity of the particle will be affected by the flow rate, the viscosity of the water, and the settling characteristics of the particle. Temperature may play a vital role in the settling characteristics of the particle. With decreasing temperature, water becomes more dense and more viscous. Consequently, the space between the water molecules becomes more restricting on the particle's ability to settle. Increased temperature has the opposite effect. Settling rate is therefore greater at a higher temperature than it is at a lower temperature.

SHORT-CIRCUITING[26] of flow can adversely affect clarifier efficiency. This occurs when the flow is not homogeneous (completely uniform) throughout the tank. That is to say, there exist zones or layers where the flow is either faster or slower than the surrounding areas. If velocity is too high in an area, suspended material may pass out of the clarifier without settling. Too slow a velocity will cause dead spaces and may produce a *SEPTIC*[27] condition if organic or other biologically degradable material is present.

Short-circuiting may be caused by differences in water density due to different temperatures existing at the surface and the bottom of the clarifier. This is especially true in temperate climates during the winter and summer seasons. Density differences may also be caused by a high suspended solids content in the influent. Short-circuiting may be made worse by high inlet

---

[25] *Lineal* (LIN-e-ul).   The length in one direction of a line. For example, a board 12 feet (meters) long has 12 lineal feet (meters) in its length.

[26] *Short-Circuiting.*   A condition that occurs in tanks or basins when some of the flowing water entering a tank or basin flows along a nearly direct pathway from the inlet to the outlet. This is usually undesirable since it may result in shorter contact, reaction, or settling times in comparison with the theoretical (calculated) or presumed detention times.

[27] *Septic* (SEP-tick) or *Septicity.*   A condition produced by bacteria when all oxygen supplies are depleted. If severe, the bottom deposits produce hydrogen sulfide, the deposits and water turn black, give off foul odors, and the water has a greatly increased oxygen and chlorine demand.

velocities, high outlet weir rates, and strong winds blowing along the tank surface. In all cases, the most effective solution is the use of weir plates, baffles, and port openings to produce a flow velocity throughout the clarifier that is as even as possible.

4.1342   CLARIFIER EFFICIENCY

Clarifier efficiency may be defined as the percent of a pollutant removed by the clarifier. Four major factors are used to measure performance or efficiency: flows, suspended solids, settleable solids, and floatable solids. Efficiency should be based upon the analyses of both inlet and outlet *COMPOSITE SAMPLES*[28] over a 24-hour period. Calculation is as follows:

$$\text{Efficiency, } \% = \frac{\text{In} - \text{Out}}{\text{In}} \times 100\%$$

QUESTIONS

Write your answers in a notebook and then compare your answers with those on page 460.

4.13E   List the three common modes of rapid mixing of coagulant chemicals.

4.13F   What is tapered flocculation?

4.13G   What is an advantage of vertical-flow clarifier units?

4.13H   List the four major factors that are used to predict the performance of clarifiers.

4.13I   What happens if the detention time in a clarifier is too short?

4.13J   List the possible causes of short-circuiting in a clarifier.

4.13K   List the four major factors that are used to measure clarifier performance or efficiency.

4.14   Operation, Start-Up, and Maintenance

4.140   Operating Strategy

The development of an operational strategy for a physical–chemical treatment process will prepare you to deal with sudden changes in the water being treated, to train new operators, and to plan for the future. Procedures that should be part of your strategy include:

1. Set up your laboratory so jar tests can be run quickly and easily. The jar test is the most important control test for chemical treatment. Accurate jar tests can result in significant chemical and cost savings. For example, if the dosage of a polyelectrolyte costing $2.00 per pound could be reduced by 0.5 mg/L in a 10 MGD plant, the cost savings would be $83.40 per day.

2. Monitor chemical feeders closely to ensure proper output. On new equipment, measure the actual feed rates and compare them with feed settings at least weekly.

3. Adjust chemical dosages whenever the flow rate changes. Long detention times during low flows may not require as high a chemical dosage as shorter detention times during high flows.

4. Monitor water conditions and quality at least daily for alkalinity, pH, temperature, turbidity, and suspended solids because these water quality indicators may signal a need for a chemical dosage change. If you are removing phosphorus, measure soluble phosphorus also.

5. Consider in-plant conditions when collecting samples for jar tests and adjusting chemical dosages. For example, if one-half of the primary clarifiers in a plant are out of service or if a digester is upset, these situations can affect required chemical doses.

4.141   Start-Up and Maintenance Inspection

The items listed below should be checked during your general start-up inspection of a physical–chemical treatment process. This checklist can also be used as the basis for an overall maintenance inspection. Because of the wide variety of municipal and industrial wastewater treatment plants and the broad differences from one manufacturer's equipment to another, it is suggested that the manufacturer's instructions be consulted for more in-depth maintenance procedures. For simplicity and to avoid duplication, some items in this checklist have been grouped under equipment type rather than process stage.

1. General

   a. Determine that all tanks, basins, and piping are clean and free of debris.

   b. Ensure that all drawings, equipment manufacturer's specifications, and operating manuals are complete, up to date, and available.

---

[28] *Composite (Proportional) Sample.*   A composite sample is a collection of individual samples obtained at regular intervals, usually every one or two hours during a 24-hour time span. Each individual sample is combined with the others in proportion to the rate of flow when the sample was collected. Equal volume individual samples also may be collected at intervals after a specific volume of flow passes the sampling point or after equal time intervals and still be referred to as a composite sample. The resulting mixture (composite sample) forms a representative sample and is analyzed to determine the average conditions during the sampling period.

c. Verify that the correct spare parts are on hand.

d. Ensure the proper operation of all Start-Emergency-Stop controls both on site and at the control panel. Check all electrical connections and power supplies.

e. Ensure that all piping and valves are properly installed and adequately braced.

2. Motors and Drives

a. Ensure that all motors and drives are securely fastened. Check bearing supports, shaft alignments to drive motors, and belts (for both condition and tightness).

b. Ensure proper lubrication of motors, drives, shafts, chains, and bearings according to the manufacturer's specifications.

c. Verify that motors run at the speed and rotational directions prescribed and that all voltage requirements are satisfied.

d. Ensure that chains move freely without binding.

e. Ensure that all chain and other equipment guards are in place.

3. Pumps

a. Piston Pumps. Check ball seatings, packing, shear pin, drive belts, and hydraulic fluid.

b. Centrifugal Pumps. Check impeller for wear or plugging. Check for prime. Check packing.

c. Progressive Cavity Pumps. Check rotor and stator for wear or plugging. Check prime and packing.

d. Diaphragm pumps. Be sure that diaphragm is intact and working properly.

4. Chemical Feed Systems

a. Ensure that all level alarms in tanks are functioning properly.

b. Ensure proper calibration of all flow and metering systems.

c. Ensure that water blenders operate properly and check dilution mixers for proper placement and installation.

d. Ensure proper temperature and pressure for dilution water.

e. Ensure in-line mixers are in place, properly braced with accommodation for proper bypass.

5. Gates for Control of Flow

a. Ensure all gates are properly aligned in angles and for travel clearance.

b. Ensure proper lubrication of wheels and rising stems.

c. Ensure proper operation in both automatic and manual modes.

6. Rapid Mix

a. Inspect impeller conditions. Ensure that impellers are free from obstructions.

b. Check motors and drives (item 2 above).

7. Flocculators

a. Check motors and drives (item 2 above).

b. Ensure baffles are correctly set and securely anchored.

c. Ensure that drive stuffing box is properly placed and grouted.

d. Ensure that drive bearing and sprocket are complete.

e. Ensure that mixers rotate freely before coupling to the gear drive.

f. Ensure that sump pumps are in place and operational.

8. Clarifiers

a. Check motors and drives (item 2 above).

b. Ensure that all sprockets and shafts are in alignment and free for the rotation of the sludge collectors.

c. Lube all rails. Run collectors in empty tank (dry) for two hours before allowing the water into the flocculation tank.

d. Ensure all flights are connected to the chain and that all shoes are attached.

e. Check that all drive sprockets are operational. Ensure that shear pins are installed in all sprockets.

f. Check the operation of limit switches to ensure that they stop the motor.

g. Check cross-collector travel.

9. Scum Collectors

a. Ensure that the collectors and trough are properly secured and aligned.

b. Ensure that the wiper blades on the spiral collectors have proper uniform contact with the breaching plate.

4.142 Actual Start-Up

The following list is a general pattern of procedures to be followed when placing the entire system on line. Sections may be applied to the start-up of an individual part.

1. Chemical Feed System

a. Ensure proper temperature for dilution waters, if applicable.

b. Check that proper chemical strengths are set on automatic feeds.

c. Open all manual valves, as appropriate.

d. Set proper flow rate.

2. Mixing Tank

   a. Open effluent gates.

   b. Allow tanks to fill.

   c. Start mixers and adjust speed.

3. Flocculation Tanks

   a. Open influent gates.

   b. Turn on and adjust paddle/turbine drives.

4. Clarifiers

   a. Turn on cross-collectors and longitudinal collectors.

   b. Start scum collectors.

   c. Allow basins to fill.

4.143   Normal Operation

A general procedure for the normal operation of a chemical coagulation-precipitation system is outlined below.

1. Chemical Feed System

   a. Perform a jar test to determine the proper chemical dosage.

   b. Ensure proper chemical dilution.

   c. Set controls for the proper feed rate.

   d. Report in the operating log the amount of chemical used per unit time.

2. Rapid Mixing Tanks

   a. Ensure proper mixing speed by observing the floc formed.

   b. Check for scum formation. If scum accumulates in the influent, open the scum gates. If scum is floating in the tanks, adjust the mixer speed. If the condition does not improve, open the slide gates and allow the scum to pass through with the effluent.

   c. Only operate mixers when the tank is filled to capacity.

   d. Do not allow mixers to be off for an extended period while material is still in the tank.

3. Flocculation Tanks

   a. Adjust paddle/turbine speed so that particles receive just enough agitation to remain suspended.

   b. Check to see that chemical is being added if no floc forms.

   c. Adjust all paddles to the same speed unless:

      (1) Sludge formation occurs at one point. Increase the paddle speed for this area.

      (2) Coagulant is added at the floc tanks. Increase the tip speed of the mixers immediately preceding the point of discharge.

      (3) Coagulant is added in the influent channel. Increase the tip speed of the first mixer.

4. Clarifiers

   a. If possible, all tanks should be kept in operation since the best sedimentation occurs with the greatest amount of surface area.

   b. Sludge control. Level should be kept at a minimum.

   c. Never store sludge in the clarifiers. Move it to the thickeners for storage.

   d. Scum removal. Check periodically.

      Skimmer. Clean daily.

      Scum pit. Clean after each pumping.

   e. Clean weirs, scum baffles, and launders daily.

4.144   Abnormal Operation

This section contains a list of abnormal conditions that could occur at any time during the operation of a wastewater treatment plant. Included are recommendations that should help you adjust the chemical treatment system in order to maintain a high-quality effluent. Whenever you detect indications of abnormal conditions, increase the frequency of process monitoring.

1. High solids concentrations in effluent leaving the secondary clarifiers due to bulking sludge, rising sludge, or solids washout

   a. Inspect chemical feeders for proper output.

   b. Run jar tests to determine if dosage requirements have changed.

   c. Examine overall plant operations to locate the cause of high solids.

   d. Increase sludge removal rates from clarifiers.

2. Unusually low suspended solids in effluent leaving the secondary clarifiers

   a. Inspect chemical feeders for proper output.

   b. Run jar tests to determine if dosage requirements have changed.

   c. Record in log book conditions and dosage that produced low solids in the effluent. You need to know how you produced a good-quality effluent.

3. High flows passing through the treatment plant

   a. Prepare to feed a greater quantity of chemicals.

   b. Run jar tests to determine the dosage that will produce rapid settling rates when detention times are reduced.

   c. Be sure jar test flash mixing and flocculation times are similar to actual detention times through these units during the high flows.

4. Low flows passing through the treatment plant

   a. Run jar tests to determine optimum dosage because longer detention times may allow a reduction in chemical dosage, which will reduce chemical costs.

   b. Watch for chemical overdoses that could produce toxic conditions in biological treatment processes or in the receiving waters.

5. A change in the pH of the water being treated by one or more units

   a. Inspect chemical feeders to determine if the chemicals being added are causing the pH change.

   b. Run jar tests to determine if chemical feed rates need adjusting.

   c. Extreme pH changes may affect biological activity and effectiveness of disinfection. Try to control chemical feeders to minimize chemical changes.

   d. If existing chemical dosages will not cause coagulation, new chemicals may be required. For example, you may have to switch from one type of polyelectrolyte to another type.

6. A change in water temperature resulting from seasonal weather conditions, groundwater infiltration, or wet weather inflows

   a. Run a jar test to determine if new chemical feed rates are required. Coagulation and settling rates change when the temperature changes.

4.145 Troubleshooting

Two common problems in physical–chemical treatment systems are foaming in the rapid-mix tanks and flocculators and no coagulation of suspended particles. Foaming can be controlled by the use of water spray from hoses or by surface spray nozzles installed on the tanks.

Any number of conditions could lead to a failure of the particles to coagulate properly. First, run a jar test to determine the proper chemical dosage for the wastewater being treated. Be sure the jar test chemicals are the proper strength. If the chemical dosage is correct, inspect the following items:

1. Chemical feed pump operation

2. Chemical supply and valve positions

3. Solution carrier water flow and valve positions

4. Applied water for a significant change

5. Actual feeder output by catching a timed sample

6. Feed chemical strength

Table 4.3 lists several other problems operators should watch for when operating a physical–chemical treatment system and suggests possible causes and solutions.

4.146 Safety

Most of the chemicals used in coagulation and flocculation processes are either caustic or corrosive. Operators who work with these chemicals must make it a habit to use the safe working procedures discussed in Section 4.128, "Safe Working Habits."

In addition to being exposed to dangerous chemicals, operators of coagulation, flocculation, and sedimentation systems also must be aware of the dangers associated with working around large tanks or basins. Drowning in a basin filled with wastewater is not a common occurrence, but it could happen at any time. Spilled polymers tend to be extremely slippery. If not promptly cleaned up, an operator could easily slip and fall into an open tank or basin. Always approach basins through areas that have appropriate walkways and be sure life preservers are readily available in work areas around open tanks or channels. Keep walkways clear of clutter and free of grease, oil, and chemicals. Be sure that approved guardrails are provided around all tanks and channels and make use of handrails in all work locations.

Before working on a piece of mechanical or electrical machinery, be sure all power is turned off, locked out, and properly tagged to prevent accidental start-up. If possible, physically block rotating arms or any part of a machine that could suddenly move or swing free. Avoid placing your arms or legs into

TABLE 4.3　TROUBLESHOOTING A PHYSICAL–CHEMICAL TREATMENT PROCESS

Problem	Cause	Solution
Excess scum buildup	Scum collection device	Inspect scum trough, spiral screw, and scum pumps.
Floc too small	Improper chemical dosage	Check dosage with jar test.
	Low chemical metering	Adjust metering.
	Chemical feed pump adjusted too low	Adjust feed pumps.
	Paddle speed in flocculators or rapid mix too fast	Decrease paddle speed.
	Short-circuiting	Baffling changes, adjust weir plates or port openings.
	Change in pH	Neutralize pH.
Floc too large, settles too soon	Improper chemical dosage	Check dosage with jar test.
	Metering setting	Check and adjust metering.
	Chemical make-up too strong	Check make-up and adjust feed.
	Too little dilution water	Check and adjust metering of dilution water.
	Paddle speed in rapid mix or flocculator too slow	Increase paddle speed.
	Coagulant aid added at wrong point	Optimize point of coagulant aid addition (in rapid mix, before flocculator, in flocculator).
Floating sludge	Sludge collectors not functioning properly	Check motors, chains, drives, and belts for smooth operation.
	Sludge pumping system malfunction	Check sludge pumps for operation, pipes for debris. Switch to standby pump.
	Change in influent character or flow rate	Check dosage with jar test and adjust.
	Excess coagulant aid	Check dosage with jar test and adjust.
Loss of solids over effluent weir	See "floc too small" problem above	
	Improper or misaligned baffling	Adjust baffling at inlet and outlet.
Thin sludge with deep sludge blanket	Cross-collectors not functioning	Check motors, drives, and chains for cross-collectors.
Sludge collector, jerky operation or inoperable	Broken sprocket, chain link, flight, or shear pin	Inspect and repair.
	Sludge blanket too deep	Pump out sludge. May have to drain basin and remove manually.

any moving part of machinery. If you must work on a piece of equipment while it is operating, consider using an extension tool if the work can be done safely using one. Do not remove guards or safety shields around moving equipment unless absolutely necessary and be sure to replace them immediately when the maintenance work or repair is complete.

QUESTIONS

Write your answers in a notebook and then compare your answers with those on pages 460 and 461.

4.14A   What water quality indicators should be monitored when operating a physical–chemical treatment process?

4.14B   List the items you would check during the start-up inspection of a chemical feed system.

4.14C   List the procedures you would follow for the normal operation of a chemical feed system in a coagulation-precipitation system.

4.14D   What abnormal conditions could be encountered in the water being treated when operating a physical–chemical treatment process?

4.14E   What are two common problems that could occur when operating a physical–chemical treatment process?

END OF LESSON 2 OF 6 LESSONS

on

SOLIDS REMOVAL FROM SECONDARY EFFLUENTS

Please answer the discussion and review questions next.

DISCUSSION AND REVIEW QUESTIONS

Chapter 4.   SOLIDS REMOVAL FROM SECONDARY EFFLUENTS

(Lesson 2 of 6 Lessons)

Write the answers to these questions in your notebook. The question numbering continues from Lesson 1.

13.   Why are plunger pumps used for metering chemicals?

14.   What economic factors should be considered when selecting a chemical feeder?

15.   Why should chemical feed equipment be cleaned before being shut down for an extended length of time?

16.   Why are solids loadings especially important in the design of clarifiers treating industrial wastes?

17.   What is the most effective solution for short-circuiting?

18.   Why should you develop an operational strategy for a chemical treatment process?

CHAPTER 4.   SOLIDS REMOVAL FROM SECONDARY EFFLUENTS

(Lesson 3 of 6 Lessons)

4.2   GRAVITY FILTERS
by Ross Gudgel and James L. Johnson

4.20   Use of Filters

The use of gravity filtration is second only to gravity sedimentation for the separation of wastewater solids. This same process, using deep-bed filtration and granular media, has long been used in municipal and industrial water supplies. However, gravity filter systems are more frequently used for domestic water supplies that have much lower suspended solids concentrations than are found in the effluent from secondary wastewater treatment facilities.

The following specific applications have been observed:

1. Removal of residual biological floc in settled effluents from secondary treatment by trickling filters or activated sludge processes

2. Removal of residual chemical–biological floc after alum, iron, or lime precipitation of phosphate in secondary settling tanks of biological treatment processes

3. Removal of solids remaining after the chemical coagulation of wastewaters from tertiary or independent physical–chemical wastewater treatment processes

4.21   Description of Filters

Applied water generally flows through wastewater filters from top to bottom. The applied water is distributed evenly over the surface of the filter media through an inlet distribution system. This may be the same system that is used later to uniformly collect the dirty backwash water.

The water travels through the filter media where the solids are trapped. The filter bed may be made up of one or several materials or several grades of materials. This is determined by the designer based primarily upon the quality of the applied water.

The underdrain system is designed to collect the filtered water uniformly throughout the bed. It also is used to uniformly apply the backwash water during backwashing. The system design must prevent the filter media from passing into the underdrain system, thereby being lost from the bed.

A surface wash system is beneficial during backwashing to scrub the surface mat of accumulated solids, thus breaking it up with minimum amounts of water. In some installations, air is used in place of surface washing as a means of breaking up the accumulated solids.

Valves control the volume, direction, and duration of flows through the unit, and instruments are used to monitor and record the volumes and quality of the water being processed through the filtering system.

4.22   Filtering Process

Water to be treated enters at the top of the filter bed through an inlet valve and is distributed over the entire filter surface. The water passes evenly down through the media (sand) and leaves the solids behind. Filtered water then travels out the bottom of the filter and into the underdrain collection system, which is designed to uniformly collect the flow. Once inside the underdrain collection system, the water passes through a flowmeter and rate-control valve. The rate-control valve maintains the desired flow through the filter and prevents backwash water from mixing with the filtered water during backwashing.

Most gravity filters operate on a batch basis. The filter operates until its capacity to remove solids is nearly reached but before solids break through into the effluent. It is then completely removed from service and cleaned. Other designs are available that filter continuously with a portion of the media always undergoing cleaning. The cleaning of the media may take place either externally or in place.

4.23   Backwashing Process

As suspended solids are removed from water, the filter media becomes clogged. This is indicated by a *HEAD LOSS* [29] reading. Through operating experience, the maximum head loss before backwashing will be determined. The filter should be backwashed after the solids capacity of the media has been reached, but before solids pass through the filter and begin to appear in the effluent (a condition known as breakthrough).

Backwashing consists of closing valves to stop influent flow and to protect the filtered water. Backwash water either flows by gravity or is pumped to the filter. This water flows through the underdrain system and back through the media. As the water flows through the media, the sand particles are lifted and are cleaned by rubbing against each other. The solids retained by the media are washed away with the backwash water and the media has been cleaned.

If the media is cleaned externally, the media is removed from the filter bed, cleaned in a separate system, and recycled back into the bed. In-process cleaning involves washing a small section of the filter bed with a traveling backwash water or air-pulsing system while the remainder of the bed remains in service.

4.24   Methods of Filtration

4.240   Filter Types

Most gravity filters used in wastewater treatment are "rapid sand" filters (Figure 4.27(a)). They also may be called "downflow" (water flows down through the bed) or "static bed" (bed does not move or expand when filtering) filters. These filters operate continuously until they must be shut down for backwashing. Other designs, such as the upflow and the biflow, are on the market. Both of these designs are attempts to use more of the filter media, thereby removing and holding more solids per filter run.

In the upflow filter (Figures 4.27(b) and 4.28), water enters at the bottom of the filter and is removed from the top. The biflow system has water applied at both the top and bottom and water is withdrawn from the interior of the bed. Filters are always backwashed in an upflow direction regardless of the operating flow direction.

4.241   Surface Straining

Downflow filters are designed to remove suspended solids by either the surface-straining method or the depth-filtration method. In surface straining, the filter is designed to remove the solids at the very top of the media. The fine grade-sized media is uniform throughout the bed. Because of this conformity, surface-straining systems will have a rapid head loss buildup, short filter runs, and they must be backwashed frequently. There are, however, no problems with breakthrough of solids. The solids compress into a mat at the surface, which aids in removing solids; however, the mat is difficult to remove during backwashing. Backwashing a surface-straining system, although needed more frequently, requires less water per wash than does a depth-filtration system.

4.242   Depth Filtration

Depth filtration is designed to permit the solids to penetrate deep into the media, thereby capturing the solids within as well as on the surface of the media. Depth filtration will have a slower buildup of head loss, but solids will break through more readily than with surface straining.

To reduce breakthrough, yet retain depth filtering, the multimedia design is used. This combines a fine, denser media (sand) on the bottom with a coarse, lighter media (anthracite coal) on the top (Figures 4.27(d) and 4.29). The coarse media remove large solids that would quickly clog finer media. The fine media will surface strain solids that penetrate the full depth of the coarse media bed thereby preventing a breakthrough of solids. The filter is designed to prevent the fine media from escaping unless the head loss becomes too great.

QUESTIONS

Write your answers in a notebook and then compare your answers with those on page 461.

4.2A   Do most gravity filters operate on a batch or on a continuous basis?

4.2B   When should a gravity filter be cleaned?

4.2C   What is meant by the following terms that are used to describe "rapid sand" filters:

1. Downflow?
2. Static bed?

4.2D   From what part of the filter are solids removed by:

1. Surface straining?
2. Depth filtration?

---

[29] *Head Loss.*   The head, pressure, or energy (they are the same) lost by water flowing in a pipe or channel as a result of turbulence caused by the velocity of the flowing water and the roughness of the pipe, channel walls, or restrictions caused by fittings. Water flowing in a pipe loses head, pressure, or energy as a result of friction losses. The head loss through a filter is due to friction losses caused by material building up on the surface or in the top part of a filter. Also called FRICTION LOSS.

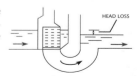

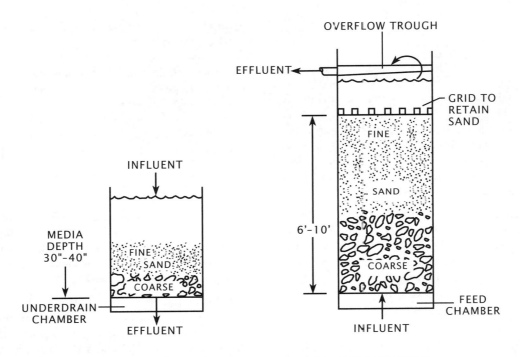

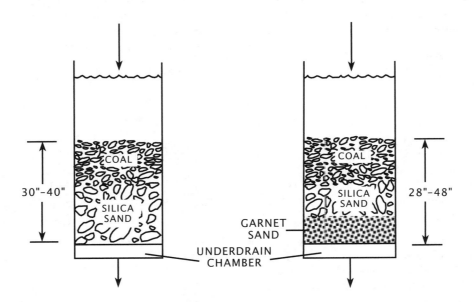

Fig. 4.27 Filter configurations
(Source: EPA *PROCESS DESIGN MANUAL FOR SUSPENDED SOLIDS REMOVAL*)

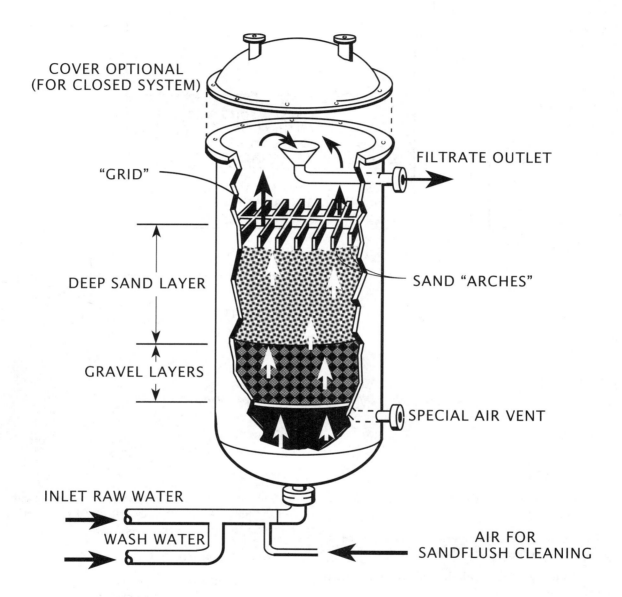

COVER OPTIONAL
(FOR CLOSED SYSTEM)

"GRID"

FILTRATE OUTLET

DEEP SAND LAYER

SAND "ARCHES"

GRAVEL LAYERS

SPECIAL AIR VENT

INLET RAW WATER

WASH WATER

AIR FOR
SANDFLUSH CLEANING

Fig. 4.28   Cross section of upflow filter
(Source: EPA *PROCESS DESIGN MANUAL FOR SUSPENDED SOLIDS REMOVAL*)

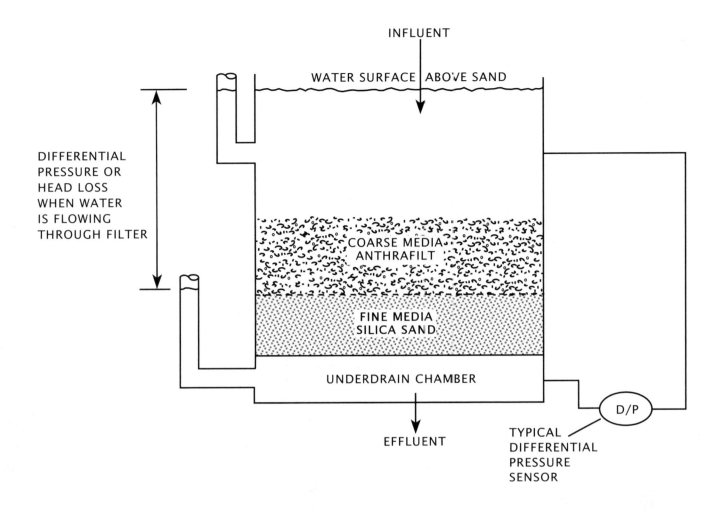

Fig. 4.29 Differential pressure through a sand filter

4.25 Location of Filters in a Treatment System

In wastewater treatment, the filters may be used in the following modes (Figure 4.30):

1. To polish secondary effluent without the addition of chemicals as filter aids just ahead of the filters

2. To polish secondary effluent with the addition of chemicals as filter aids just ahead of the filters

3. To polish secondary effluent that has been chemically pretreated and settled

4. To polish raw wastewater that has undergone coagulation, flocculation, and sedimentation in a physical–chemical treatment system

4.26 Major Parts of a Filtering System (Figure 4.31)

This section describes the major parts of a gravity filtering system and also how each part works or functions during the filtration process.

4.260 Inlet

The filter inlet gate allows the applied water to enter the top of the filter media. When closed, it will permit emptying the filter for backwashing or maintenance.

4.261 Filter Media

The filter media selection is one of the most important design considerations. Filter beds are made up of silica sand, anthracite coal, garnet, or ilmenite. Garnet and ilmenite are commonly used in multi-media beds.

Because of rapid plugging, the conventional single media filter bed commonly used in potable (drinking) water systems is generally unsatisfactory for removing solids from wastewater. To lengthen filter runs and use the full bed depth, the dual- and multi-media filters are used. A layer of coarse media (anthracite) is placed over finer, dense material (sand or garnet). The coarse layer allows deep penetration of the solids into the bed causing a minimum of head loss. The fine material prevents breakthrough of solids into the effluent.

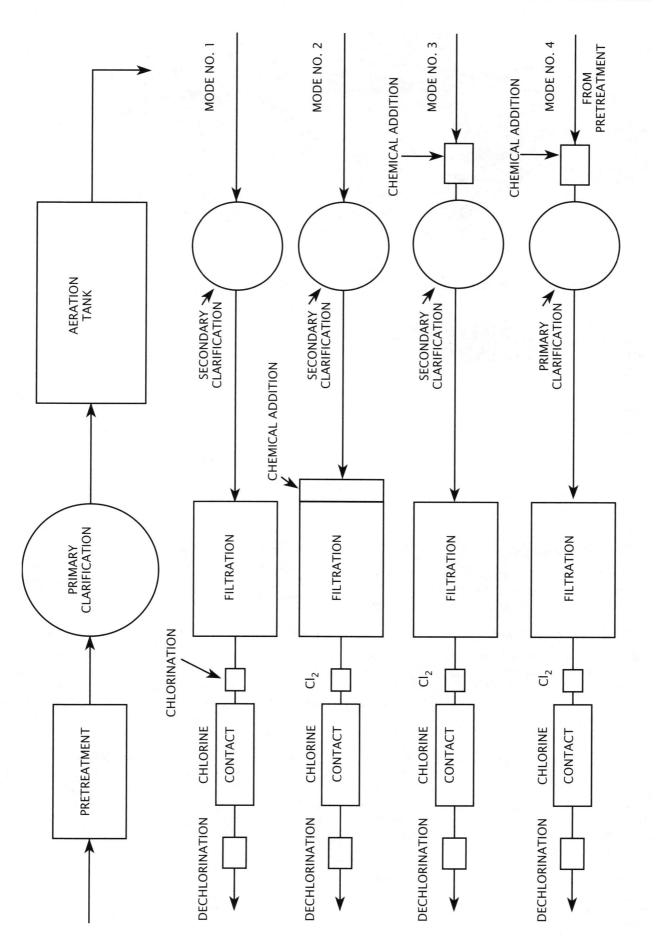

Fig. 4.30   Four possible modes of using filters to remove solids

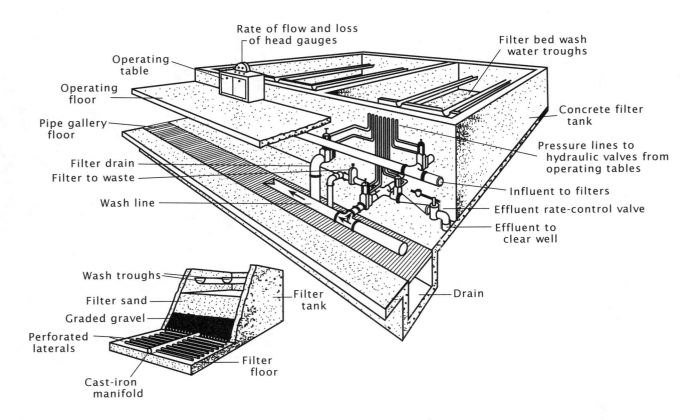

Fig. 4.31 Typical rapid sand filter
(From *WATER SUPPLY AND TREATMENT* by C. P. Hoover, permission of National Lime Association)

4.262 Filter Underdrains (Figure 4.32)

The filter underdrain system is designed to contain the filter media within the bed and to maintain uniform water flows through the entire bed during both filtering and backwashing.

4.263 Filter Media Scouring Systems

If the filter media is not cleaned thoroughly at each backwashing, a buildup of solids will occur. The end result of incomplete cleaning is the formation of *MUDBALLS*[30] within the bed. These mudballs settle to the filter bottom and eventually make it necessary to rebuild the entire bed. "Surface wash" and "air scour" are two systems used to improve cleaning of the media.

The surface-wash system consists of either fixed or rotating nozzles installed just above the media. During a backwash, the sand expansion places the nozzles within the media where high-pressure water jetting out of the nozzles will agitate and clean the surface. Because these wash systems are designed primarily to break up the surface mat, deep filtering beds need nozzles placed deeper within the media.

The air-scour system injects air into the bottom of the bed. This agitates the entire bed, yet requires no additional wash water. Care must be taken to prevent air and water flowing at the same time or the media will be washed out and lost.

4.264 Wash Water Troughs

During backwashing, the accumulated solids strained out by the filter media are carried out of the filter bed by means of the backwash water troughs. The troughs must be level to uniformly collect and withdraw the backwash water. This will help prevent dead spots (areas where no water circulates) during the backwash operation.

A smooth trough surface, such as fiberglass, will reduce routine cleaning; however, fiberglass troughs may be damaged more easily by the weight of the backwash water than steel or concrete troughs. Filter troughs, particularly fiberglass, must be well anchored to ensure that they will not warp or attempt to float during backwashing. They also must be designed to withstand the weight of water if filled when there is no water over the bed.

---

[30] *Mudballs.* Material, approximately round in shape, that forms in filters and gradually increases in size when not removed by the backwashing process. Mudballs vary from pea-sized up to golf-ball-sized or larger.

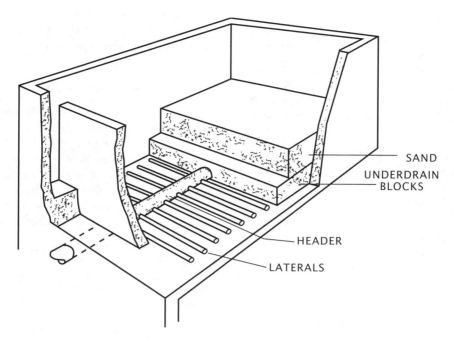

A. HEADER LATERALS
(Courtesy of the AWWA)

SAND
UNDERDRAIN
BLOCKS
HEADER
LATERALS

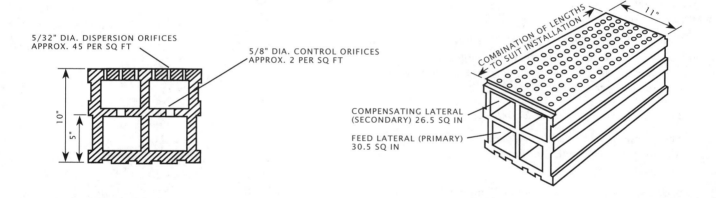

5/32" DIA. DISPERSION ORIFICES
APPROX. 45 PER SQ FT

5/8" DIA. CONTROL ORIFICES
APPROX. 2 PER SQ FT

10"
5"

11"

COMBINATION OF LENGTHS
TO SUIT INSTALLATION

COMPENSATING LATERAL
(SECONDARY) 26.5 SQ IN

FEED LATERAL (PRIMARY)
30.5 SQ IN

B. LEOPOLD BLOCK SYSTEM
(Courtesy of F. B. Leopold Co.)

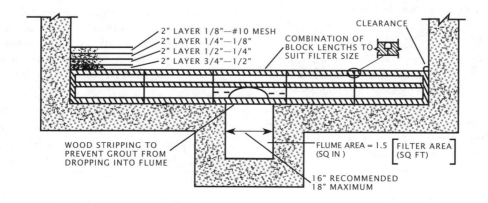

2" LAYER 1/8"—#10 MESH
2" LAYER 1/4"—1/8"
2" LAYER 1/2"—1/4"
2" LAYER 3/4"—1/2"

CLEARANCE

COMBINATION OF
BLOCK LENGTHS TO
SUIT FILTER SIZE

WOOD STRIPPING TO
PREVENT GROUT FROM
DROPPING INTO FLUME

FLUME AREA = 1.5 $\left[\dfrac{\text{FILTER AREA}}{\text{(SQ FT)}}\right]$
(SQ IN)

16" RECOMMENDED
18" MAXIMUM

Fig. 4.32   Underdrains
(Source: EPA *PROCESS DESIGN MANUAL FOR SUSPENDED SOLIDS REMOVAL*)

4.265   Backwash Water Drain

The filter drain allows the backwash water to leave the filter and return to the plant headworks for reprocessing. This drain must be opened before the backwash water flow begins, but not before the filtered water has dropped below the level of the backwash trough. If the drain opens while the filter is still full of applied water, the water above the troughs will needlessly be recycled back through the plant.

The drain should be closed completely before the inlet valve is opened or applied water again will be wasted.

4.266   Backwash Water Supply

The backwash water is usually water that has gone through the complete treatment process and is of the best quality available. If unfiltered water is supplied to the backwash system, clogging of the underdrain system may occur.

Filter backwashes require large volumes of water over a short period of time; therefore, small- to medium-sized plants need a wash water storage reservoir. Water from the chlorine contact tank commonly is used.

Large filters are often split in half to reduce the size of pumps and piping required for backwashing. This also can reduce the water storage requirements because a pause between washing the two halves will provide time to refill the storage reservoir.

Sectional filters (Figure 4.33), designed to backwash one small section at a time, do not require a large backwash water storage supply. These filters use pumped water as it is being filtered through other sections.

4.267   Backwash Water Rate Control

The backwash water may be supplied through pumps or by gravity from a storage tank. Both methods require careful control of the flow rate.

Backwash water supplied through pumps will maintain a more constant flow over the entire wash cycle than wash water from gravity storage. Water supplied from storage tanks may require adjustment of the rate-control valve to maintain constant flows as the storage tank level drops, due to a decrease in the available pressure head on the backwash water.

4.268   Used Backwash Water Holding Tank

The filter backwash water contains solids concentrated from many gallons of applied water. Because of the high solids concentration, this water must be re-treated in the treatment process. Since the backwash flow rates are very high, they must be dampened through a holding tank to avoid hydraulic overloads on the treatment plant. A holding tank is generally provided to equalize flows and prevent plant overloads. The tank is filled during the backwash operation and slowly emptied into the plant headworks between washings. (Some improvement in primary settling efficiency may be noted because of this recycled water.) If these high flows were returned directly to the headworks, a hydraulic overload would occur that would upset the treatment process and flow pacing systems in all but the largest plants. A filter system designed to backwash like the sectional filter may avoid the need for a used backwash water holding tank.

4.269   Effluent Rate-Control Valve

A valve automatically controls the filtered water flow leaving the bed. The effluent rate-control valve is designed to maintain a constant water level in the filter. When operating a clean filter, this valve will be closed down to restrict the flow. As the head loss through the media increases, this valve must open more to maintain a constant flow. The effluent rate-control valve must be closed during filter backwash to prevent backwash water from mixing with previously filtered water.

QUESTIONS

Write your answers in a notebook and then compare your answers with those on page 461.

4.2E   What types of materials are used for filter media?

4.2F   What can happen if the filter media is not thoroughly cleaned during each backwashing?

4.2G   Why should the backwash water be of the best quality available?

4.2H   What is the purpose of a used backwash water holding tank?

4.27   Filter System Instrumentation

Instrumentation is essential for all but the small package plant installations. Instrumentation associated with filtering is used to monitor the plant performance, to operate the plant in the absence of the operator, and to trigger an alarm if abnormal conditions develop. The system may be simple or very complex depending on the facilities; each has its place.

As with any equipment, the usefulness of instrumentation is limited by the quality of the maintenance it receives. Stated another way, if there are intermittent errors in a flowmeter signal and they are not corrected, then the operator cannot trust any of the readings and must disregard all of them. The usefulness of the instrument is then very limited.

Comments regarding instrumentation in the following sections apply to plants of all sizes. (Also see Chapter 9, "Instrumentation," for more detailed information.)

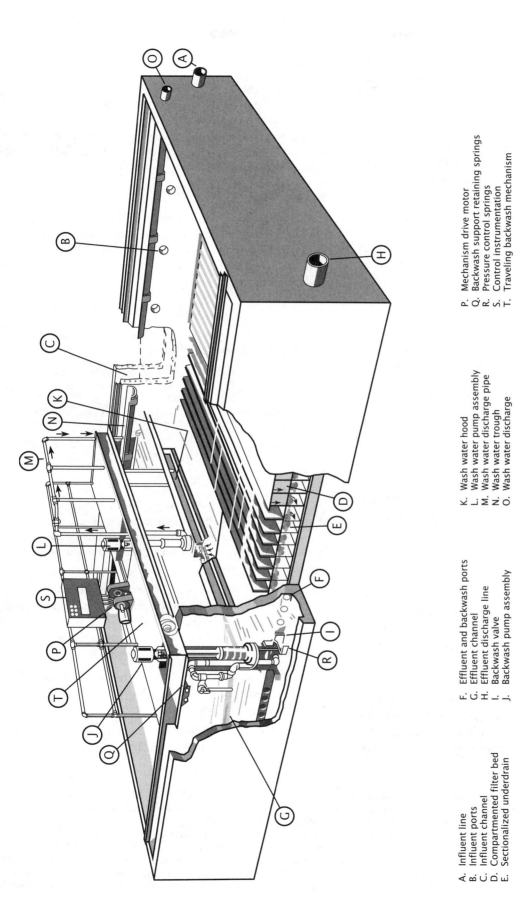

A. Influent line
B. Influent ports
C. Influent channel
D. Compartmented filter bed
E. Sectionalized underdrain

F. Effluent and backwash ports
G. Effluent channel
H. Effluent discharge line
I. Backwash valve
J. Backwash pump assembly

K. Wash water hood
L. Wash water pump assembly
M. Wash water discharge pipe
N. Wash water trough
O. Wash water discharge

P. Mechanism drive motor
Q. Backwash support retaining springs
R. Pressure control springs
S. Control instrumentation
T. Traveling backwash mechanism

Fig. 4.33 Sectional filter
(Permission of Environmental Elements Corporation)

4.270   Head Loss

Head loss is one of the most important control guidelines in the operation of the rapid sand filter. Each filter or filter half requires a head loss (differential pressure) indicator, preferably one with a readout chart. This will indicate the present condition of the bed, its ability to remove solids, and the effectiveness of the backwash operation.

Head loss is determined by measuring the water pressure above and below the filter media (see Figure 4.29, page 396). With the filter out of service, the pressures will be the same (zero difference).

When water flows through the bed of a typical downflow gravity filter, the pressure below the media will be less than the pressure above the media (when the pressure levels are measured or read at the same elevation). Measured in feet (or meters) of water, the difference becomes the head loss.

As the media bed becomes filled with solids, the head loss becomes greater. There is a point at which little or no water can pass through the filter. The operator wants the head loss to always be less than at that point; therefore, the filter backwash control point must be less than the maximum design head loss.

4.271   Filter Flow Rate Indicator [31] and Totalizer [32]

Each filter or filter half requires a flow indicator and totalizer on the filtered water line. This is needed to determine proper filtering rates (gal/min/sq ft or liters/sec/sq meter). Also, knowing the total volume filtered and the volume of backwash water used, the percent of production (filtered) water used for backwashing can be calculated. This is important because excessive wash water usage is costly and must be controlled. The backwash water should average 5 to 10 percent of total water production.

4.272   Applied Turbidity [33]

A continuous-reading turbidimeter with readout chart on the applied water is useful in monitoring the performance of the secondary settling tanks. This readout will alert the operator to developing problems if the turbidity suddenly increases. With experience, chemical dosages can be adjusted as turbidity changes.

4.273   Effluent Turbidity

A continuous reading of turbidity with a chart on the filter effluent will monitor the filter performance. A sudden increase may indicate a filter breakthrough (cracked bed) and may be used or instrumented to set off alarms if specified limits are exceeded. One turbidimeter unit, with proper valving, may be used to monitor more than one filter.

4.274   Indicator Lights

Indicator lights are beneficial to operators in keeping track of the filter system. Lights can easily indicate which filter is in service, out of service, or backwashing. They can indicate if filter pumps, wash water pumps, or air blowers are running, out of service, or on standby and ready to run. Indicator lights are also used with the alarm system to show abnormal conditions.

4.275   Alarms

Alarms needed to alert the operator should include high applied water level, high turbidity, and pump malfunctions. Backwash water supply and holding tanks both need high water level alarms. All alarms should be tested for proper functioning at least every 60 days.

QUESTIONS

Write your answers in a notebook and then compare your answers with those on page 461.

4.2I   How is the head loss through the filter media determined?

4.2J   How often should filter system alarms be tested for proper operation?

4.28   Operation of Gravity Filters

4.280   Pre-Start Checklist

Before starting up any major system, such as gravity filters, a thorough check of each component must be made to prevent damage to the equipment or injury to personnel. The following items should be included in your checklist for starting filtering systems.

1. Be sure all construction debris has been removed. Wood scraps, concrete chips, nails, and other trash can damage equipment such as pumps and valve seats. Trash dropped into the filter media will work its way to the bottom, thus reducing the effective area of the filter.

2. Inspect the electrical installation for completeness. Check safety lockouts, fuse sizes, safety covers, and equipment overload protections.

3. Check motors and drives for proper alignment, for proper safety guards, and for free rotation.

---

[31] *Indicator.*   (1) (Chemical indicator) A substance that gives a visible change, usually of color, at a desired point in a chemical reaction, generally at a specified end point. (2) (Instrument indicator) A device that indicates the result of a measurement, usually using either a fixed scale and movable indicator (pointer), such as a pressure gauge, or a moving chart with a movable pen like those used on a circular flow-recording chart. Also called a RECEIVER.

[32] *Totalizer.*   A device or meter that continuously measures and sums a process rate variable in cumulative fashion over a given time period. For example, total flows displayed in gallons per minute, million gallons per day, cubic feet per second, or some other unit of volume per time period. Also called an INTEGRATOR.

[33] *Turbidity* (ter-BID-it-tee).   The cloudy appearance of water caused by the presence of suspended and colloidal matter. In the waterworks field, a turbidity measurement is used to indicate the clarity of water. Technically, turbidity is an optical property of the water based on the amount of light reflected by suspended particles. Turbidity cannot be directly equated to suspended solids because white particles reflect more light than dark-colored particles and many small particles will reflect more light than an equivalent large particle.

4. Examine motors, drive units, and bearings for proper lubrication.

5. Check motors for proper rotation. (A three-phase motor may run in either direction.)

6. Inspect pumps and motors for excessive vibration.

7. Fill tanks and piping and look for leaks.

8. Open and close valves manually and run each valve through a complete cycle to check limit setting.

9. Put the automatic controls through a "dry run."

10. Inspect the total system for safety hazards.

11. Backwash the media several times. Skim the fines (tiny particles that tend to float) from the surface between each washing prior to placing filter into service. After the final pre-start backwashing sequence, fill the filter with wash water up to the level of the wash water troughs.

4.281   Normal Operation

Since most wastewater gravity filters are deep-bed, downflow, rapid sand-type filters, this section will present information based on them. Nevertheless, most of the information can be applied to other filter designs with some possible modifications.

4.2810   FILTERING

The applied wastewater enters at the top of the filter bed through an inlet valve and is distributed over the entire filter surface. The water passes evenly down through the media and leaves the solids behind. Filtered water then travels out the bottom of the filter and into the underdrain collection system, which is designed to uniformly collect the flow. Once inside the underdrain collection system, the water passes through a flowmeter and rate-control valve. The rate-control valve maintains the desired flow through the filter and prevents backwash water from mixing with the filtered water during backwashing. Successful filter operation depends on effective backwashing of the filter media.

4.2811   BACKWASHING

As suspended solids are removed from wastewater, the filter media becomes clogged. This is indicated by the head loss reading (Figure 4.29, page 396). The filter should be backwashed after the capacity of the media to hold solids is nearly used up, but before solids break through into the effluent. Operating experience will be the best guide to determining the maximum head loss before backwashing is required.

A typical set point to start backwashing is at 7.0 feet (2.0 m) of head loss. If a filter is operating with a 6.0-foot (1.8-m) head loss, the operator knows the filter will need washing soon. If the head loss is 4.0 feet (1.2 m) after washing, this indicates a very poor washing or it may indicate a malfunctioning instrument. After a proper washing, the head loss should be less than 0.5 foot (0.15 m) at start-up. The head loss will then slowly increase to the point where backwashing is required again.

Backwashing a filter manually, although sometimes necessary, is very time-consuming; moreover, manual backwashings are inconsistent. Automatic backwashing, on the other hand, can be a simple procedure that requires a minimum of operator time.

To maintain smooth operations, the automatic backwash cycle should be initiated by the operator. This mode of operation permits the operator to backwash at a convenient time, thereby allowing time for keeping records current and completing the necessary maintenance duties. Automatically starting backwashes, although workable in a large system, can be very inconvenient to the operation of a small system.

At the start of the backwash cycle, the rate-control valve must be opened slowly to a low rate of backwash. This prevents damaging the underdrain system or disturbing the rock and gravel layers of the bed. This damage can occur when an empty bed has high backwash water flows suddenly injected into it or if trapped air in the piping and underdrain system is violently forced into the bed. After the air has been slowly purged and the water level is up to the wash water troughs, the bed can no longer be damaged by high backwash rates.

Some plants use an air-scouring system to clean the filter media. The air-scour system injects air into the bottom of the media bed. This agitates the entire bed, yet requires no additional wash water. Care must be taken to prevent air and water flowing at the same time, or the media will be washed out and lost.

To prevent the loss of filter media into the backwash troughs:

1. Draw the water level in the filter down to within a few inches over the top of the filter media.

2. If air scouring is used, pause a moment after air washing before starting the water wash.

3. Wash with a low water flow rate until the trapped air has escaped the filter media.

4. Never backwash a filter with water containing large quantities of air.

The various types of media become intermixed during the high agitation of air scrubbing or high-rate backwashing. With proper control, however, the media will automatically regrade due to the difference in *SPECIFIC GRAVITIES*[34] of the particles.

By design, the filter media is prevented from escaping into the underdrain system; nevertheless, operational care must be taken to prevent damaging the underdrains while backwashing or the filter media will be lost into the underdrain collection system.

Uniform water flow through the filter bed is important to prevent the breakthrough of solids in the effluent due to localized high velocities. Also, high velocities will cause the media to be disturbed and relocated if the backwash flow is not uniform throughout the bed.

The following situations indicate a disturbed or damaged filter underdrain:

1. Consistently poor-quality effluent (high suspended solids levels) while there is little buildup of the filter head loss.

2. Boiling areas and very quiet ("dead") areas of the filter media during backwashing. This is most noticeable during high wash rates in a nearly clean filter.

3. Filter media in the effluent.

Improper control of the system during backwashing is generally the cause of damaged filter bottoms, providing they were properly installed. Damage to the filter bottom could result if:

1. The maximum backwash rate is allowed to enter an empty filter.

2. A large volume of air preceded the maximum backwash rate causing *WATER HAMMER*.[35]

The only way to correct a damaged filter bottom is to remove the media and rebuild the bed. A bed with the media displaced to a minor extent may be corrected by extended and properly controlled backwashing. This will regrade the media.

After the filter media is clean, slowly reduce the backwash water flow. This permits the media to regrade itself through gravity settling. The heavier particles (gravel, garnet, sand) will settle to the bottom first. Then, as the uplift velocities decline, the lighter particles (anthracite coal) will settle, thereby regrading the filter bed back to its original placement. This regrading must occur at the end of each backwash cycle.

After backwashing, the filter normally contains water up to the sides of the troughs. To fill the remaining portion of the filter, open the inlet valve. Be sure to waste some of the filtered water at the start until completely filtered and clear water is leaving the filter.

If the filter has been drained for maintenance, fill the filter as if you were starting to backwash, up to the top of the sides of the troughs, and then fill the filter using the inlet valve. An empty filter should not be filled through the inlet valve because the water falling onto the media will disturb the bed and result in uneven filtering. Also, filling the backwash troughs with water in an empty filter will place an unnecessary load (weight of water) on the troughs.

When the used backwash water holding tank is empty, the tank should be inspected. An observation of the solids settled on the bottom of the empty holding tank will alert the operator to any loss of filter media due to improper backwash procedures, such as an excessively high flow rate or short-circuiting.

By analyzing the records and observing the complete wash cycle, the operator can determine if the backwashing sequence is adequate. If highly turbid water is still in the bed at the end of the cycle, experiment with one or all of the following:

1. Adjust the media scouring time.

2. Adjust the low wash rate.

3. Adjust the high wash rate.

4. Adjust the time of regrading the media.

5. Backwash more frequently by beginning to wash at a lower head loss.

QUESTIONS

Write your answers in a notebook and then compare your answers with those on page 461.

4.2K   Why should a pre-start check be conducted before starting filtering systems?

4.2L   What is the purpose of the rate-control valve?

4.2M   When should a filter be backwashed?

4.282   Abnormal Operations

Following is a list of conditions that are not normally found in the day-to-day operation of filtration systems; however, these conditions could occur at almost any time. Recommendations are added to aid you in adjusting for the situation.

---

[34] *Specific Gravity.*   (1) Weight of a particle, substance, or chemical solution in relation to the weight of an equal volume of water. Water has a specific gravity of 1.000 at 4°C (39°F). Particulates with specific gravity less than 1.0 float to the surface and particulates with specific gravity greater than 1.0 sink. (2) Weight of a particular gas in relation to the weight of an equal volume of air at the same temperature and pressure (air has a specific gravity of 1.0). Chlorine gas has a specific gravity of 2.5.

[35] *Water Hammer.*   The sound like someone hammering on a pipe that occurs when a valve is opened or closed very rapidly. When a valve position is changed quickly, the water pressure in a pipe will increase and decrease back and forth very quickly. This rise and fall in pressures can cause serious damage to the system.

1. High solids in the applied water due to bulking sludge, *RISING SLUDGE*,[36] or solids washout in the secondary clarifier

    a. Run jar tests and adjust chemical dosage, as needed. (Jar test procedures are described in Section 4.12, "Selecting Chemicals and Determining Dosages.")

    b. Place more filters in service to prevent breakthrough.

    c. Prepare to backwash more frequently.

2. Low suspended solids in applied water; however, solids pass through filter

    a. Run jar tests and adjust chemical dosage as needed. Test a combination of chemicals and polyelectrolytes.

    b. Place more filters in service to reduce velocity through the media.

    c. Backwash filter and precoat clean filter with *FILTER AID*.[37]

3. Loss of filter aid chemical feed

    a. Place more filters in service to reduce velocity through media.

    b. Backwash more frequently.

    c. Precoat clean filters by hand feeding chemicals into them when first placed into service.

4. High wet weather peak flows

    a. Place more filters in service.

    b. Run jar tests and adjust chemical dosage, as needed.

    c. Prepare for peak daily flows by backwashing early.

5. Low applied water flows

    a. Reduce number of filters in service. Run one-half of a filter at a time.

    b. Prepare to take one filter out of service and backwash when flow or head loss increases, thereby preventing breakthrough.

6. High color loading

    a. Run jar tests and adjust chemical dosage, as needed.

    b. Add chlorine to applied water.

    c. Usually, color cannot be removed with filtration; consequently the problem must be corrected at the source.

7. High water temperature

    a. Run jar tests and adjust chemical dosage, as needed.

    b. Prepare for *AIR BINDING*[38] of filters because water will release gases more readily at higher temperatures.

    c. Place more filters in service to reduce head loss through the media.

    d. Increase backwash water flow rates to obtain the same bed expansion as used when backwashing with colder water.

8. Low water temperature

    a. Run jar tests and adjust chemical dosage, as needed.

    b. Prepare for air binding of filters as cold water will carry more gases to the filters. Backwash more frequently if air binding occurs.

    c. Place more filters in service to reduce head loss through filter media.

    d. Reduce backwash water flow rates to obtain the same bed expansion as used when backwashing with warmer water.

9. Air binding

    a. Backwash at a lower head loss.

    b. Place more filters on line to reduce head loss through media.

    c. Take filter out of service and allow air to escape to the atmosphere. This will reduce head loss; however, if placed back into service without backwashing, solids will likely be drawn through the media and into the effluent. These solids may or may not cause a problem.

10. Negative pressure in the filter

    a. Reduce flow through the filter by adding additional units.

    b. Backwash at a lower head loss.

    c. Skim surface of media (about one-half inch or 1.3 cm) to remove fines.

    d. Prevent filter from running at a low filtration rate. This builds a mat on the media surface and then sharply increases the head loss if a higher rate of water flows through the filter.

    e. A negative pressure within the filter will cause a false reading from the differential pressure sensor.

---

[36] *Rising Sludge.* Rising sludge occurs in the secondary clarifiers of activated sludge plants when the sludge settles to the bottom of the clarifier, is compacted, and then starts to rise to the surface, usually as a result of denitrification, or anaerobic biological activity that produces carbon dioxide and/or methane.

[37] *Filter Aid.* A chemical (usually a polymer) added to water to help remove fine colloidal suspended solids.

[38] *Air Binding.* The clogging of a filter, pipe, or pump due to the presence of air released from water. Air entering the filter media is harmful to both the filtration and backwash processes. Air can prevent the passage of water during the filtration process and can cause the loss of filter media during the backwash process.

11. High BOD and COD

   a. Handle same as high suspended solids.

   b. Chlorinate applied water.

12. High coliform-group bacteria levels

   a. Chlorinate applied water to increase contact time.

   b. Place additional units in service to increase contact time.

   c. Run jar tests and adjust chemical dosage, as needed, to reduce suspended solids.

13. Chlorine in the applied water

   a. Discontinue adding polyelectrolytes as chlorine will interfere with them.

   b. Run jar tests and adjust chemical dosage, as needed.

14. pH change in the applied water

   a. Run jar tests and adjust chemical dosage, as needed.

   b. Change type of filter aid, if necessary.

15. High quantities of grease and oil in the applied water

   a. If in solution, they will pass through media.

   b. If not in solution, they will be trapped in the bed, thus requiring extra hosedown during each backwash.

QUESTIONS

Write your answers in a notebook and then compare your answers with those on page 461.

4.2N   List at least five of the various types of abnormal operating conditions that could occur while operating a filtration system.

4.2O   How would you adjust for a situation in which you were treating a high solids content in the water applied to a filter?

4.283   Operational Strategy

The development of an operational strategy for the filtration of wastewater will aid in dealing with situations such as sudden changes in applied water, in training new operators, or in planning for the future. Following are points to consider when developing or reviewing your plans.

1. Maintain the filtering rates within the design limits. Add units or remove them from service, as needed. Very low filtering rates will produce matting on the surface. This matting will cause breakthroughs if the flows are increased sharply. Excessively high rates will pull the solids through the filter and into the effluent.

2. Each backwash must be a complete cleaning of the media or solids will build up and form mudballs, or cause the media to crack.

3. To remove mudballs, first backwash thoroughly. Then, superchlorinate manually and draw the chlorinated water into the filter media. Allow this chlorinated water to stand for 24 or 48 hours to soak the mudballs and then backwash thoroughly again.

4. Run jar tests to maintain optimum chemical dosages. As the applied water quality changes (solids, alkalinity, temperature), the filter aid requirements will change. The operator must be aware of the changes and the effectiveness of the chemicals applied.

5. With complete backwashing, a high-quality effluent can be maintained without filtering to waste before placing the filter back into service.

6. If the effluent turbidity reaches 3 to 4 TURBIDITY UNITS,[39] a change should be made to correct the problem. Either adjust chemical dosage, adjust flow rate, or backwash the filter.

7. Filter walls that are constructed with a smooth surface (sacked) or painted with a good sealant are easy to keep clean. A rough surface provides an excellent area for ALGAE[40] and slimes to grow.

8. Controls and instrumentation must be protected from the elements. Cabinets that are opened to adjust instruments must be out of the rain, dust, and extreme heat.

9. Air used to operate instruments or transmit signals must be cleaned and dried to prevent damaging the equipment.

10. Every three or four months, measure and record the freeboard to the filter media surface (Figure 4.34). A small amount of media loss is normal, but an excessive amount (2 to 3 inches or 5 to 7 centimeters) indicates operational problems.

11. After the filters have been in service for some time, obtain a profile of the media to determine if it is being displaced. A plug sample (Figure 4.34) will show if the media are being regraded after each backwash.

12. When landscaping around uncovered filters, keep trees and shrubs that will drop leaves into the bed away from the filter because leaves are very difficult to backwash out of the media.

---

[39] *Turbidity* (ter-BID-it-tee) *Units (TU).*   Turbidity units are a measure of the cloudiness of water. If measured by a nephelometric (deflected light) instrumental procedure, turbidity units are expressed in nephelometric turbidity units (NTU) or simply TU. Those turbidity units obtained by visual methods are expressed in Jackson turbidity units (JTU), which are a measure of the cloudiness of water; they are used to indicate the clarity of water. There is no real connection between NTUs and JTUs. The Jackson turbidimeter is a visual method and the nephelometer is an instrumental method based on deflected light.

[40] *Algae* (AL-jee).   Microscopic plants containing chlorophyll that live floating or suspended in water. They also may be attached to structures, rocks, or other submerged surfaces. Excess algal growths can impart tastes and odors to potable water. Algae produce oxygen during sunlight hours and use oxygen during the night hours. Their biological activities appreciably affect the pH, alkalinity, and dissolved oxygen of the water.

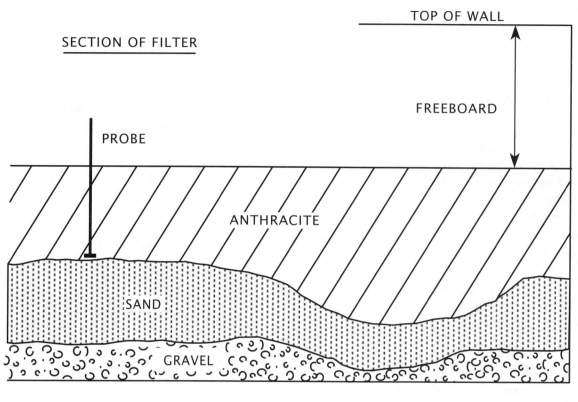

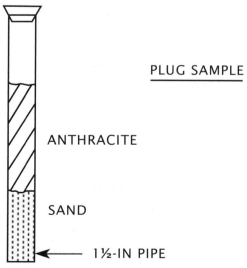

Fig. 4.34   Section of filter and plug sample

13. Never throw trash such as cigarette butts into the filter media. Trash may not backwash out and instead may work its way deep into the media.

14. Occasionally, chlorinate ahead of the filters to control algal and slime growths on the walls and within the media. There will be a short period of discolored effluent after the initial application, but the water will run clear in a short while.

15. Calculate the unit cost to treat wastewater. Apply this cost to the volume of water used per backwash. Inform all operators of this because it may easily cost in excess of $100 per filter wash.

16. Always fill an empty filter bed through the backwash system to prevent disturbing the media surface. If an empty filter is filled through the influent valve, water will flow into the wash water troughs, over the top edges, and onto the top of the media. The force of this falling water will disturb the media.

17. Allow dry filter media to soak several hours before backwashing. Dry media will tend to float out with the backwash water.

18. Maintain a log (Figure 4.35) of the filtering operation that includes the following:

    a. Time filter was placed into service and total hours run between washings

    b. Volume of water processed between washings

    c. Applied water rate at start and end of filter run

    d. Head loss at start and end of filter run

    e. Applied suspended solids and BOD

    f. Effluent suspended solids and BOD

    g. Percent removal of suspended solids and BOD

    h. Chemicals added as filter aids, mg/L

    i. Chlorine added to applied water, mg/L

    j. Remarks of special observations and maintenance

    k. Backwash water flow rates and duration

    l. Surface wash flow rate and duration

    m. Influent and effluent turbidity

4.284   Shutdown of a Gravity Filter

If the filter is to be out of service more than a week, it should be backwashed, dewatered, and air dried. This will help control slime and algal growth on the walls, troughs, and within the media. Dried algae can be hosed from the walls prior to backwashing and returning to service.

To remove a filter from service, first backwash the filter. Then, switch controls to the manual mode of operation, close the influent valve, filter all the water possible through the rate-control valve, and open the drain valve.

Hose down and backwash the filter again before returning it to service.

4.29   Troubleshooting

PROBLEM: HIGH TURBIDITY AND SUSPENDED SOLIDS IN THE EFFLUENT

1. Check for excessive head loss. Breakthrough will occur at a high head loss.

2. Look for fluctuating flows. Widely varying flows will cause breakthrough.

3. Determine filter aid dosages.

4. Examine backwash cycle for complete wash.

5. Inspect for damaged bed due to backwashing.

PROBLEM: RAPID BUILDUP OF HEAD LOSS

1. Check applied water suspended solids.

2. Check filter aid dosage.

3. Determine applied water flow rate.

FILTER LOG

MONTH JAN . 20 05 FILTER NUMBER 1

START FILTER						STOP FILTER								FILTER WASH			
DATE	TIME 6	FILTER RATE MGD A	FILTER RATE MGD B	HEAD LOSS FEET A	HEAD LOSS FEET B	DATE	TIME	FILTER RATE MGD A	FILTER RATE MGD B	HEAD LOSS FEET A	HEAD LOSS FEET B	HRS.	NTU	DATE	TIME	REMARKS	OPER.
						1-3	0100	–	–	–	–	–	–	–	–	BACKWASHED SEMI-AUTOMATIC	A.B
1-4	1500	1.5	1.5	–	–	1-6	2000	1.5	1.5	7	7			1-6	2000	MANUAL BACKWASH	R.M.M.
1-6	2100	1.0	1.0	–	–	1-9	1300	1.1	1.1	6.5	6.5	64		1/9	1300	AUTO	B.T.
1-9	1500					1-11	1100	2.2	2.4	6.0	9.0	44		1-11	1300	AUTO	L.S.
1-12	0900	1.4	1.4	0.4	0.4												
1-14	1300	1.4	1.4	0.4	0.4	1-16	1700	1.7	1.7	8.8	8.6	52		1-16	1700	MANUAL	RICK
1-16	1800	1.8	1.8	.2	.2	1-17	1700	2.0	2.0	7.5	6.5	23	.8	1-17	1700	MANUAL	RICK
1-18	1100	1.3	1.3	–	–	1-20	0300	1.9	1.9	7.0	6.5	60	.7	1-20	0500	AUTO	L.S.
1-25	2000	1.2	1.2	–	–	1-27	2100	.8	1.0	6.9	7.4	49	.5				

Fig. 4.35 Log of filter operation

4. Check backwash cycle for complete wash.

5. Inspect head loss differential pressure sensor for air in one side. This will give a false reading.

PROBLEM: INSIGNIFICANT BUILDUP OF HEAD LOSS

1. Check applied water suspended solids.
2. Check applied water flow rate.
3. Determine filter aid dosages.
4. Check head loss differential pressure sensor for air in one side. This will give a false reading.
5. Examine filter effluent for suspended solids going out (filter breakthrough).
6. Inspect for damaged bed due to backwashing.
7. Backwash and check for complete cycle.

PROBLEM: RAPID LOSS OF FILTER MEDIA

1. Look for washout during backwash cycle.
2. Examine for media in effluent indicating a damaged filter underdrain.
3. Check for excessive scouring during backwash cycle time. Excessive scouring will grind the media into fines.

PROBLEM: HIGH HEAD LOSS THROUGH CLEANED FILTER

1. Inspect differential pressure sensor for air in one side.
2. Check for incomplete backwash cycle.
3. Look for mudballs in filter media. Take a plug sample (Figure 4.34, page 407).

PROBLEM: FLOW INDICATED WHEN EFFLUENT VALVE IS CLOSED

1. Inspect differential pressure sensor for air in one side.
2. Check instrumentation loop for calibration.
3. Examine valve for proper position.

PROBLEM: BACKWASH STOPS BEFORE COMPLETING CYCLE

1. Look for sticking valve.
2. Inspect for electrical relay hang-up.
3. Check for timer out of sequence.
4. Examine backwash water supply.
5. Inspect electrical control (pump lockout).

4.210 Safety

Always think safety when working around moving equipment and motors with automatic controls. Filtration systems have electrical, chemical, and mechanical safety hazards. Operators are usually well protected from electrical hazards; however, there are times when opening a panel to look for trouble or to adjust a timer may expose you to electrical hazards. Always use safety equipment (rubber electrical gloves, multimeters, fuse pullers, and lockout switches) and approved safety procedures when working with electricity. *REMEMBER: Only trained and qualified individuals should be allowed to work on electrical systems.*

Chemical hazards include chemical burns and skin irritation from direct contact with chemicals. Also, there is a hazard of slipping and falling caused by chemical spills. Review appropriate Material Safety Data Sheets (MSDSs) if any doubt exists as to the safety precautions required when dealing with specific chemicals. Good housekeeping will reduce the safety hazards caused by chemicals.

Mechanical hazards associated with filters are similar to those found throughout the treatment plant. Safety guards must be in place, equipment operated automatically must be identified by warning signs, and work areas should be well lighted.

4.211 Review of Plans and Specifications

While reviewing the plans and specifications of a gravity filtration system, you should consider the items listed in this section.

1. Filters require regular servicing; therefore, provisions must be made to handle the normal flows during periods of servicing. Regular maintenance includes servicing of valves, instruments, and filter media.
2. The quality of the water applied to the filters must be considered when a filtering media is specified. A high suspended solids content in the applied water will quickly plug a fine-media filter, thereby requiring frequent backwashing.
3. Install sufficient instrumentation to adequately monitor the process and to determine operating efficiencies. Include instrumentation to measure and record applied flows, backwash flows, head loss, and water quality before and after filtration.
4. Be sure that each step in the automatic system is complete before the following or next step can begin.
5. Provide a means to reset the automatic system if the backwash cycle is interrupted.
6. Keep the automatic backwashing system uncomplicated, especially in small plants. The operator should be on hand at the start of a filter backwash cycle. Housekeeping chores can be performed while keeping an eye on the filter washing process.
7. Install instruments out of the weather and well protected from the weather. Even weather-proof cabinets must be opened during the maintenance and servicing of instruments and equipment.

8. Install the instruments' readout meters, charts, and gauges in a convenient and centralized location.

9. Separate and shield instrumentation signals from all voltages 110 volts and higher and from other equipment noise that may be picked up by the instruments as a false signal.

10. Provide adequate storage for chemicals. A minimum supply of chemicals must be on hand even while waiting for a full shipment.

11. Provide adequate storage for both the backwash water supply and the used backwash water.

12. When reviewing designs for the future, keep today's flows in mind. Equipment operating below 10 percent capacity may be useless for years.

13. Visit similarly designed plants that are currently in operation and talk to the operators regarding possible design improvements.

QUESTIONS

Write your answers in a notebook and then compare your answers with those on pages 461 and 462.

4.2R What are the three main types of safety hazards around filtration systems?

4.2S When reviewing plans and specifications for a filtration system, instrumentation should be available to measure and record what items?

4.2T Where should the instruments' readout meters, charts, and gauges be installed?

END OF LESSON 3 OF 6 LESSONS

on

SOLIDS REMOVAL FROM SECONDARY EFFLUENTS

Please answer the discussion and review questions next.

DISCUSSION AND REVIEW QUESTIONS
Chapter 4. SOLIDS REMOVAL FROM SECONDARY EFFLUENTS
(Lesson 3 of 6 Lessons)

Write the answers to these questions in your notebook. The question numbering continues from Lesson 2.

19. Why are multi-media filters used?

20. What is the purpose of instrumentation used with a filter system?

21. How does a rapid sand filter work?

22. How can a filter bottom be damaged?

23. Why should you attempt to maintain filtering rates within the design limits?

24. How would you remove a gravity filter from service?

CHAPTER 4.   SOLIDS REMOVAL FROM SECONDARY EFFLUENTS

(Lesson 4 of 6 Lessons)

4.3   INERT-MEDIA PRESSURE FILTERS
by Ross Gudgel

The pressure filters described in this section are similar in many ways to the gravity filters discussed earlier in this chapter. In both types of filtration systems, the filter media consists of one or more inert materials, that is, materials that do not react chemically with the wastewater being filtered. Silica sand, anthracite coal, and garnet sand are three types of inert media widely used in both gravity and pressure filters.

Inert-media gravity and pressure filters both remove suspended particles in basically the same way. Solids are trapped in the spaces between media particles and on the surface of the media bed. When the filter approaches the limit of its capacity to remove suspended solids, both types of filters must be backwashed.

As the name implies, pressure is the main difference between an inert-media pressure filter and an inert-media gravity filter. The pressure filter uses a closed tank or vessel that allows the operator to apply pressure to force the wastewater through the media at a faster rate than would be achieved by gravity flow.

4.30   Use of Inert-Media Pressure Filters

Inert-media pressure filters remove suspended solids, associated BOD, and turbidity from the secondary effluent after the addition of chemical coagulants such as a polymer or alum. The filtration process is used to meet waste discharge requirements for final effluent suspended solids and turbidity limits established by an NPDES permit when these limits cannot be met by secondary treatment processes. Filtration also will have a direct bearing on the disinfection of the final effluent by the removal of more solids from the water to be disinfected. Fewer solids will reduce the amount of disinfectant necessary to meet the NPDES permit bacteriological requirements.

The filter system usually consists of:

1. A holding tank or wet well for secondary effluent storage

2. Filter feed pumps that pump the secondary effluent from the holding tank to the filters

3. A chemical coagulant feed pump system that injects the necessary coagulants into the influent line to the filters

4. Single-, dual-, or multi-media filters that trap the suspended solids and remove the turbidity

5. A filter backwash wet well for clean backwash water storage

6. Filter backwash pumps that pump clean water back through the filter to remove the trapped suspended solids

7. A *DECANT*[41] tank that provides for holding the spent backwash water to allow the suspended solids to settle or rise while the clarified water is either directly recycled to the filters or is returned to the headworks

Figure 4.36 shows a schematic view of the items outlined above and they are further discussed in the following sections.

QUESTIONS

Write your answers in a notebook and then compare your answers with those on page 462.

4.3A   What is the purpose of the inert-media pressure filter?

4.3B   What chemicals are commonly used with the filtration process and why?

4.3C   List the major components of a pressure filter system.

4.31   Pressure Filter Facilities

The following sections describe facilities that are typical for a filter plant with a capacity of 5 MGD. Facilities at larger or smaller plants would be quite similar but might differ significantly in the numbers and sizes of the various components.

4.310   Holding Tank (Wet Well)

Secondary effluent from the treatment plant's secondary sedimentation tanks is conducted through a channel or pipe to a holding tank. The purpose of this tank is to store water and to allow additional settling of the suspended solids before the water is applied to the filters. Most tanks of this type are very similar to secondary clarifiers. They have flights or scrapers to move the settled solids toward a sludge hopper for return to the solids handling facility.

A bypass structure should be provided to permit secondary effluent to bypass the pressure filters during emergency conditions, such as equipment failures or clogged filters. Bypassed

---

[41] *Decant* (de-KANT).   To draw off the upper layer of liquid (water) after the heavier material (a solid or another liquid) has settled.

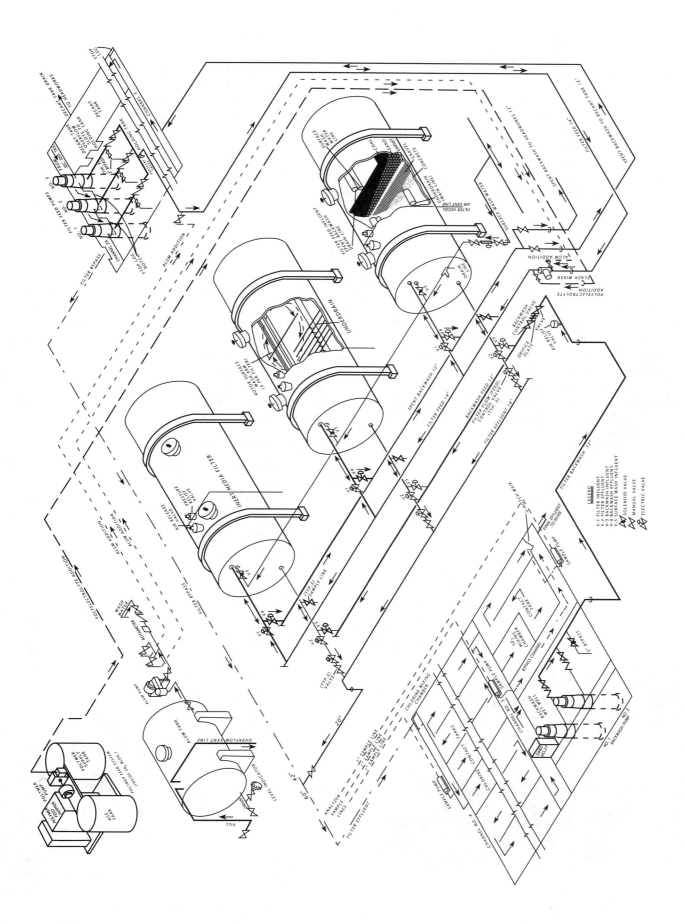

Fig. 4.36   Schematic view of pressure filter system

flows should go into emergency holding basins or into the chlorine contact tank for final treatment before discharge. An alternative emergency storage procedure would be to divert secondary effluent into the decant tank.

Spent backwash water may also be returned to the decant tank. Both flows receive some settling and the clarified effluent then overflows into the holding tank through weir slots between the two tanks for recycle to the filters. In either method of operation, the floatable materials in the holding tank are collected, using a pan-type skimmer, and discharged to the solids handling section of the plant for disposal. For additional information on clarifier operation and maintenance, see Section 4.134, "Clarifiers," in this chapter and Chapter 5, "Sedimentation and Flotation," OPERATION OF WASTEWATER TREATMENT PLANTS, Volume I.

4.311   Filter Feed Pumps (Figure 4.37)

Filter feed pumps lift the secondary effluent from the holding tank and pump it through the filters. Generally, they are of the vertical-turbine wet-pit type pump with either a closed or semi-open impeller. The pumps are driven by either fixed-speed, multi-speed (two speed), variable-speed motors, or a combination of these. Each pump should be equipped with a manually adjusted bypass valve to avoid the possibility of the system operating at the shutoff pressure of the pumps. If this happens, the pumps could be damaged because no water would flow through the pumps. Each valve should be adjusted to allow a given bypass flow as recommended by the manufacturer.

The water level in the holding tank may be sensed by a level transmitter. The transmitter produces a signal to start and stop the pumps and a set point signal for the controller that controls filter flow.

Starting and stopping of the pumps is controlled by a HAND-OFF-AUTO (HOA) switch for each pump. Another switch is used to select the sequence of automatic starting (lead or lag). Normal, automatic start/stop control of the pumps may be by means of current trips using the signal representing the water level in the holding tank. A low-water probe in the holding tank will stop all pumps if the water level drops below a preset elevation. For additional information on the operation and maintenance of pumps, see Chapter 15, "Maintenance," in OPERATION OF WASTEWATER TREATMENT PLANTS, Volume II.

4.312   Chemical Feed Systems

As previously described in Section 4.1, various types of chemicals may be added to the filter influent flow to ensure coagulation and flocculation of the suspended material. This coagulation and flocculation aids the filtering process by joining many of the finely divided and colloidal suspended solids into a floc mass that is easily trapped on or in the filter media, thus allowing clear water to pass through the filter. Alum and polymers are the chemicals most often used as filter aids. The following paragraphs briefly describe the reasons for using chemicals to aid filtration and the methods used to apply the chemicals to water.

4.3120   ALUM (ALUMINUM SULFATE) (Figure 4.38)

Alum is a coagulant that produces a hydrated (containing water) oxide floc. This floc causes suspended material to stick together by electrostatic or interionic force when contact of the

Fig. 4.37   Filter feed pumps

Fig. 4.38   Alum storage tank and feed pump

chemical and a solid particle is made in the filter influent flow. The effective performance of alum is critically affected by the pH of the water in which it is used. The ideal condition is to maintain the value of the pH between 6.5 and 7.5. Economics may dictate that the working range is as much as 1.5 pH units above or below this "perfect" range, depending on the "natural" (or unregulated) pH of the system and the cost of pH control.

The alum may be pumped by a mechanical diaphragm, positive displacement pump. The dosage is manually adjusted by adjustment of the pump stroke length. Motor speed may be controlled by a silicon controlled rectifier (SCR) drive unit that uses a filter flow signal to pace the pump in the automatic mode. The SCR drive is also equipped with a manual potentiometer for manual speed control and a meter indicating percentage of total or full motor speed.

The pump discharge check valve has a built-in back pressure device to prevent siphoning and to prevent uneven delivery due to low discharge pressure. All wetted parts of the pump are selected for their chemical resistance.

4.3121   POLYMERS (POLYELECTROLYTES)

Polymers are flocculation aids that are classified on the basis of the type of electrical charge on the polymer chain, the molecular weight (or length) of the polymer chain, and the charge density (spacing of the charges) along the chain.

Polymers possessing negative charges are called "anionic," positive-charge polymers are called "cationic," and polymers that carry no free electric charge are called "nonionic." Polymers cause the suspended material to stick together by chemical bridging or chemical enmeshment when contact is made in the filter influent flow.

Anionic polymers are most commonly used with the application of alum but systems' performances vary. The optimum alum–polymer combination can only be selected by testing the system water under actual system conditions.

Polymer usually is injected into the influent line of the filters far enough downstream from the point of the alum injection for the alum floc to form properly and far enough upstream of the filters for final coagulation to become complete.

The polymer may be prepared for use (dilution and aging) by a polyelectrolyte mixer unit (Figure 4.39). This unit consists of a dry polymer feeder with storage hopper, a solution water flowmeter with regulating valve, pressure regulating valve, pressure gauge, solenoid valves, dilution water flowmeter with regulating valve, polymer wetting cones, a mixing/aging tank, slow-speed mixer, transfer pump, metering/storage tank, and a metering pump with SCR drive.

A variable-area flowmeter (*ROTAMETER*[42]) is provided to indicate flow of dilution water to the metered polymer. The polymer is mixed automatically by the polyelectrolyte mixer. The polymer feeder is calibrated to dispense a metered quantity of dry polymer to obtain a desired solution concentration. The polymer drops from the feeder hopper to the wetting cones where it is spread on a high-velocity water surface and the individual grains or droplets of polymer are wetted to form a polymer solution. This solution then flows to the aging tank where it is mixed and aged. On completion of the aging cycle, the polymer solution is pumped to the metering/storage tank by the transfer pump. When the aging tank empties, the polymer preparation and mixing cycle begins again. The metering pump, calibrated to deliver a desired dosage, draws the polymer solution from the metering/storage tank and delivers it to the influent line of the filters.

The dry feeder is a screw-type feeder capable of metering dry polymer of densities ranging from 14 to 42 lbs/cu ft (225 to 675 kg/m^3) at an adjustable rate to the wetting cones in order to prepare various solution concentrations.

Application of liquid polymers is essentially the same. Instead of the dry screw feeder, a high-viscosity chemical metering pump is used. The pump may be connected directly to the containers in which the polymer is supplied or to an intermediate holding tank (in place of the hopper).

4.3122   MIXING AND ADDING CHEMICALS

The mixing/aging tank (day tank) and the metering/storage tank are sized based on the projected use of polymer. The tanks may be of molded polyethylene, fiberglass reinforced polyester, stainless steel, mild steel with a plastic liner or, in some circumstances, unlined mild steel. The mixer, a low-shear type (to avoid breaking up floc), is fitted with a stainless-steel shaft and impellers.

The slow-speed transfer pump (either a progressive cavity or a gear type is recommended) conveys the mixed polymer solution from the mixing/aging tank to the metering/storage tank with minimal polymer shear. The metering pump is capable of delivering various amounts of polymers at various percent solutions. Flow pacing, adjustment of the dosage rate, and operation of the

---

[42] *Rotameter* (ROTE-uh-ME-ter).   A device used to measure the flow rate of gases and liquids. The gas or liquid being measured flows vertically up a tapered, calibrated tube. Inside the tube is a small ball or bullet-shaped float (it may rotate) that rises or falls depending on the flow rate. The flow rate may be read on a scale behind or on the tube by looking at the middle of the ball or at the widest part or top of the float.

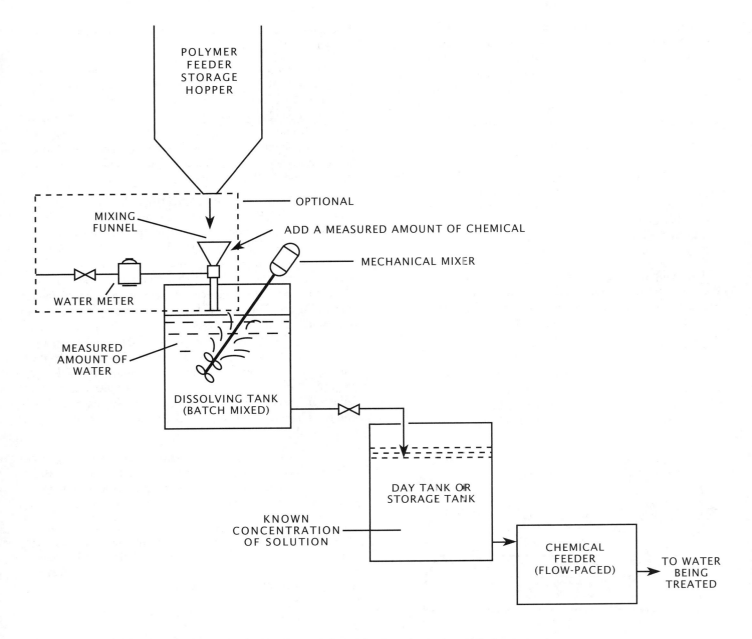

Fig. 4.39   Polymer dissolver, day tank, and feeder

polymer metering pump are the same as for the alum feed pump.

Steps to calculate polymer and alum dosage are outlined in the following examples. Information concerning the concentration of chemical (lbs/gal) delivered to your plant may be obtained from the chemical manufacturer or supplier.

EXAMPLE 4

Known		**Unknown**
Polymer Delivered, lbs/day	= 72 lbs Polymer/day	Polymer Dose, mg/L
Flow Through Filter, GPM	= 6,000 GPM	

Determine the pounds of polymer used per million pounds of water, which is the same as mg per million mg or mg/L.

$$\text{Polymer Dose, mg/}L = \frac{\text{Polymer Delivered, lbs Polymer/day}}{\text{Flow Through Filter, M lbs Water/day}}$$

$$= \frac{72 \text{ lbs Polymer/day}}{6,000 \text{ gal/min} \times 8.34 \text{ lbs/gal} \times 60 \text{ min/hr} \times 24 \text{ hr/day}}$$

$$= \frac{72 \text{ lbs Polymer/day}}{72,057,600 \text{ lbs Water/day}}$$

$$= \frac{72 \text{ lbs Polymer/day}}{72 \text{ M lbs Water/day}}$$

$$= 1 \text{ mg Polymer/liter Water}$$

$$= 1 \text{ mg/}L$$

EXAMPLE 5

Polymer is supplied to your plant at a concentration of 0.5 pound polymer per gallon (60 gm/L or 60 kg/m³). The polymer feed pump delivers a flow of 0.10 GPM (0.0063 L/sec) and the flow to the filters is 3,000 GPM (190 L/sec). Calculate the concentration or dose of polymer in mg/L in the water applied to the filter.

Known	**Unknown**
Polymer Conc, lbs/gal = 0.5 lb/gal	Polymer Dose, mg/L
Polymer Pump, GPM = 0.1 GPM	
Flow to Filter, GPM = 3,000 GPM	

ENGLISH

Calculate the polymer dose, mg/L.

$$\text{Dose, mg/}L = \frac{\text{Pump, gal/min} \times \text{Conc, lbs Polymer/gal}}{\text{Flow, gal/min} \times 8.34 \text{ lbs Water/gal}}$$

$$= \frac{0.1 \text{ gal/min} \times 0.5 \text{ lb Polymer/gal}}{3,000 \text{ gal/min} \times 8.34 \text{ lbs Water/gal}}$$

$$= \frac{0.05 \text{ lb Polymer}}{25,020 \text{ lbs Water}} \times \frac{1,000,000^*}{1 \text{ M}}$$

$$= \frac{2.0 \text{ lbs Polymer}}{1 \text{ M lbs Water}}$$

$$= \frac{2.0 \text{ mg Polymer}}{1 \text{ M mg Water}^{**}}$$

$$= 2.0 \text{ mg/}L$$

---

* We multiplied the top and the bottom by the same number, 1,000,000 or 1 M. This is similar to multiplying the top and bottom by 1, you do not change the equation, only the units.

** 1 M mg Water = 1 liter.

METRIC

$$\text{Dose, mg/}L = \frac{\text{Flow, }L/\text{sec} \times \text{Conc, gm Polymer/}L \times 1,000 \text{ mg/gm}}{\text{Flow, }L/\text{sec}}$$

$$= \frac{0.0063 \text{ }L/\text{sec} \times 60 \text{ gm Polymer/}L \times 1,000 \text{ mg/gm}}{190 \text{ }L/\text{sec}}$$

$$= 2 \text{ mg/}L$$

HOW MUCH?

EXAMPLE 6

Liquid alum usually is supplied at a concentration of 5.4 pounds alum per gallon (650 gm/L or 650 kg/m³). In this example, the alum feed pump delivers 88 mL per minute and the flow to the filter is 3,000 GPM (190 L/sec). Calculate the concentration or dose of alum in mg/L in the water applied to the filter.

Known	**Unknown**
Alum Conc, lbs/gal = 5.4 lbs/gal	Alum Dose, mg/L
Alum Pump, mL/min = 88 mL/min	
Flow to Filter, GPM = 3,000 GPM	

ENGLISH

Calculate the alum dose, mg/L.

$$\text{Dose, mg/}L = \frac{\text{Pump, m}L/\text{min} \times \text{Conc, lbs Alum/gal} \times 0.00026 \text{ gal/m}L^*}{\text{Flow, gal/min} \times 8.34 \text{ lbs Water/gal}}$$

$$= \frac{88 \text{ m}L/\text{min} \times 5.4 \text{ lbs Alum/gal} \times 0.00026 \text{ gal/m}L}{3,000 \text{ gal/min} \times 8.34 \text{ lbs Water/gal}}$$

$$= \frac{0.124 \text{ lb Alum}}{25,020 \text{ lbs Water}} \times \frac{1,000,000}{1 \text{ M}}$$

$$= \frac{5 \text{ lbs Alum}}{1 \text{ M lbs Water}}$$

$$= 5 \text{ mg/}L$$

---

* Conversion factor: 1 mL = 0.00026 gallon

METRIC

$$\text{Dose, mg/}L = \frac{\text{Pump, m}L/\text{min} \times \text{Conc, gm Alum/}L \times 1,000 \text{ mg/gm}}{\text{Flow, }L/\text{sec} \times 60 \text{ sec/min} \times 1,000 \text{ m}L/L}$$

$$= \frac{88 \text{ m}L/\text{min} \times 650 \text{ gm Alum/}L \times 1,000 \text{ mg/gm}}{190 \text{ }L/\text{sec} \times 60 \text{ sec/min} \times 1,000 \text{ m}L/L}$$

$$= \frac{950 \text{ mg/sec}}{190 \text{ }L/\text{sec}}$$

$$= 5 \text{ mg/}L$$

QUESTIONS

Write your answers in a notebook and then compare your answers with those on page 462.

4.3D What is the purpose of the holding tank located just ahead of a filter?

4.3E Choose the correct word within each set of parentheses in order to make the statement correct.

Alum is used for *(COAGULATION* or *FLOCCULATION)* while polymers are used for *(COAGULATION* or *FLOCCULATION)*.

4.3F Polymer is supplied at a concentration of 0.6 pound polymer per gallon (72 gm/L or 72 kg/m³). The polymer feed pump delivers a flow of 0.15 GPM (0.0095 L/sec) and the flow to the filters is 5,000 GPM (315 L/sec). Calculate the concentration or dose of polymer in mg/L in the water applied to the filter.

4.313   Filters (See Figure 4.36, page 413)

This section discusses the parts of a pressure filter and the purpose of each part.

4.3130   VESSELS (Figures 4.40 and 4.41)

Each pressure vessel containing filter media consists of a cylindrical shell closed at both ends. Accessways are provided to allow entry to the vessel for media installation and maintenance work. Pressure gauges are attached to the accessway covers to facilitate monitoring of the vessel pressure.

A direct, spring-loaded pressure relief valve is installed on top of the filter and is set to release at a preset pressure. The relief valve is provided to prevent vessel rupture in case effluent flow is restricted or stopped while influent flow continues.

A combination-type air release valve with a large orifice (opening) is installed on top of the filter to permit air to exhaust when the filter vessel is charged with water and to allow air to re-enter when the filter vessel is drained. A small orifice is also provided to exhaust small pockets of air that may collect during operation of the filter.

4.3131   INTERIOR PIPING (Figure 4.36, page 413)

Interior vessel surfaces, influent and effluent headers, and supports are painted with a protective coating to inhibit corrosion. The influent header is suspended and supported from the upper side of the vessel by lugs. Each filter is equipped with rotary surface wash arms that are installed and supported just beneath the influent header. These are self-propelling, revolving, "straight line" wash arms.

The surface wash piping consists of an influent water line, solenoid valve, a central bearing of all-bronze construction, a bronze tee with a water nozzle to spray water directly below and from the center of the tee, and arms extending laterally from the tee. The lateral arms are fitted with numerous brass nozzles located at double-angle positions to most effectively cover the area of the filter bed to be cleaned. Each nozzle is fitted with a synthetic rubber cap, slit or grooved, to act as a check valve to keep filter media away from the nozzle.

Water to the wash arms is supplied from an external source, usually from the treatment plant wash water system. Water from the surface wash arms quickly breaks up the mat of suspended material that has accumulated in and on the top layer of filter media. This occurs during the first portion of the backwash cycle.

The effluent header is encased in concrete fill in the lower section of the filter. PVC underdrain laterals are attached to the effluent header. Each lateral has numerous small-diameter holes facing toward the bottom of the filter. The ends of the laterals are capped. The filtered water is collected by the underdrain laterals that pass the water to the effluent header for discharge from the filter.

4.3132   UNDERDRAIN GRAVEL (SUPPORT MEDIA)

The inert filtering media is supported by underdrain gravel consisting of specifically sized, hard, durable, rounded stones with an average specific gravity of not less than 2.5. The gravel is placed in the filter in many specific layers starting with the larger stones (2-inch or 5-cm diameter) on the bottom and progressing to the smallest stones (¼-inch or 0.7-cm diameter) on top. The depth of each layer, specific stone sizes, and overall gravel depth will depend on the application, type, and quantity of inert media that will be used in a filter.

4.3133   INERT MEDIA

Granular filter media commonly used in wastewater filtration include anthracite coal, silica sand, and garnet sand. These

Fig. 4.40   Filter vessels

Fig. 4.41   Filter vessels

filter media range in size from 0.20 mm to 1.20 mm and specific gravities range from 1.35 to 4.5. The largest media, anthracite coal, has the lowest specific gravity. Conversely, the smallest media, garnet sand, has the highest specific gravity.

Inert-media filter configurations vary according to the specific characteristics of the water to be filtered. The common applications use either silica sand or garnet sand in a single-media filter; anthracite coal and silica sand or garnet sand in dual-media filters; and anthracite coal, silica sand, and garnet sand in multi-media or mixed media filters.

In most filter applications, the various types of media are placed in the filter in the following order and proportions:

1. The smaller size, higher specific gravity media is placed on the bottom of the filter first and makes up about 10 percent of the total media depth.

2. Next, the medium size, medium density media material is added on top of the small, dense layer. This middle layer equals approximately 30 percent of the total media depth.

3. The top and final layer is made up of the larger size, lower specific gravity media material. This layer equals about 60 percent of the total media depth. (*NOTE:* Total filter media depth varies with the application.)

Due to the size and density ratio of the media and its placement in the filter, the larger size, lower specific gravity media stay at the top and the smaller size, higher specific gravity media remain at the bottom. Most dual-media filters are designed to keep the media separated after backwashing.

4.3134 FLOW CONTROL METHOD (Figure 4.42)

In filter operation, the rate of flow through a filter is expressed in gallons per minute per square foot:

$$\text{Rate of Flow,} \atop \text{GPM/sq ft} \sim \frac{\text{Driving Force}}{\text{Filter Resistance}} \sim \frac{\text{Total Available Head}}{\text{Total Head Loss}}$$

Therefore, as the total head loss increases, the rate of flow decreases. The driving force refers to the pressure drop across the filter that is available to force the water through the filter. At the start of the filter run, the filter is clean and the driving force only needs to overcome the resistance of the clean filter media. As filtration continues, the suspended solids removed by the filter collect on the media surface or in the filter media, or both, and the driving force must overcome the combined resistance of the filter media and the solids removed by the filter.

The filter resistance (head loss) refers to the resistance of the filter media to the passage of water. The head loss increases during a filter run because of the accumulation of solids removed by the filter. The head loss increases rapidly as the pressure drop across the suspended solids mat increases because the suspended solids already removed compress and become more resistant to flow. As the head loss increases, the driving force across the filter must increase proportionally to maintain a constant rate of flow.

The constant rate method of filtration is commonly used for pressure filters. In this method, a constant pressure is supplied to the filter and the filtration rate is then held constant by the

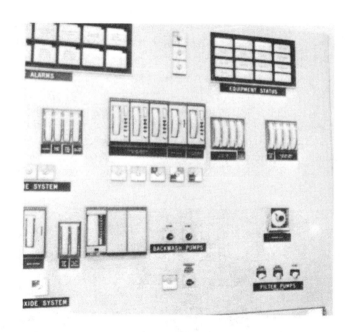

Fig. 4.42   Filter controls

action of a manually or automatically operated filter rate-of-flow controller (RFC). At the beginning of the filter run, the filter is clean and has little resistance. If the maximum available water pressure was applied to the filter, and the effluent flow was not restricted, the flow rate would be very high. To maintain a constant flow rate, the available pressure is dampened or reduced by the RFC. At the start of the filter run, the RFC is nearly closed to provide the additional head loss needed to maintain the desired flow rate. As filtration continues, the filter gradually becomes clogged with suspended solids and the RFC opens proportionally. When the valve is fully opened, any further increase in the head loss will not be balanced by a corresponding decrease in the head loss of the RFC. Thus, the ratio of pressure to filter resistance will decrease, and the flow rate will decrease. This action is also known as filter differential pressure. When the flow rate decreases, filter differential pressure increases and this is an indication that the filter run must be terminated and a filter backwash should be initiated.

In general, a pressure filter should run for a minimum of 6 to 8 hours during peak load conditions before backwashing is required. Under average flow conditions, this will mean a filter run length of about 24 hours.

4.314   Backwash System

As the suspended material accumulates on the filter media surface, or in the filter media bed, or both, the differential pressure across the filter increases, flow through the filter decreases, and filter effluent quality deteriorates. The filter backwash cycle removes the suspended solids accumulation from the filter, thus restoring the filter efficiency.

4.3140   WET WELL

The backwash wet well is used to store a large volume of filtered or chlorinated wastewater to backwash the filters. The water, usually from the chlorine contact tank, flows to the wet well until it is filled up and then it flows to further final treatment processes. This flow method ensures a continuous water supply to the wet well.

4.3141   PUMPS (Figure 4.43)

The filter backwash pumps lift the filtered or chlorinated wastewater from the backwash wet well and pump it through the filters to remove the trapped suspended material. The pumps are generally of the vertical-turbine wet-pit type with either a closed or semi-open impeller. The pumps may be driven by fixed-speed motors.

Fig. 4.43   Backwash pumps

The pump system is equipped with a solenoid-operated bypass valve installed on the main discharge line to prevent the possibility of the system operating at the shutoff pressure of the pumps. A pressure switch with an adjustable operating range (psi or kPa) will cause the valve to open in response to rising pressure. The bypass flow is returned to the backwash wet well. The main discharge line is also equipped with an air relief valve that purges air from the system to prevent air slugs from disturbing the filter media bed.

Normal starting and stopping of the pumps may be controlled by means of current switches in the backwash program unit. Lead-lag position selector switches provide the means of selecting the sequence of starting for the pumps. A low-water probe in the backwash wet well will stop the pumps in the event that the water level drops below a preset elevation.

The backwash pump main discharge line is provided with an *ORIFICE*[43] plate and flow-control valve. Flow control is accomplished by means of a cascade control system using a cam programmer to provide a set point signal. A cam is cut so as to gradually introduce the backwash flow to the filters, thereby avoiding sudden disturbance or uneven expansion of the filter media bed.

4.3142   BACKWASH CYCLE

Whenever possible, the filters should be backwashed during the plant's low-flow hours when the full capacity of the filters is not needed. The backwash cycle may be activated either manually, automatically by a preset filter differential-pressure level, or automatically by a programmed timer. In the manual mode, one or more filters may be backwashed, as needed. In the automatic mode, all filters in the system that are on line will be washed when the differential pressure reaches the preset level. Upon completion of backwash of one filter, the next filter on line will begin to backwash.

The total backwash duration per filter usually is adjustable. The total backwash flow and duration should be adequate to *FLUIDIZE*[44] and expand the media bed. The largest media size and the warmest expected water temperature will dictate the maximum filter backwash rates required.

When the backwash cycle is manually or automatically activated, the following sequence occurs:

1. Filter influent valve (V-1) and effluent valve (V-2) close to terminate filter feed flow. (See Figure 4.36, page 413, for locations of valves.)

2. Backwash influent valve (V-3) and effluent valve (V-4) open to allow backwash flow into and out of the filter.

3. The surface wash arms' influent water line solenoid valve (V-5) opens allowing the wash arms to function in initially breaking up the mat of suspended material that has accumulated on the top layer of filter media.

4. The backwash pumps start pumping against a closed backwash control valve. The backwash flow rate is brought up to full rate in one to two minutes as determined by the cam programmer transmitting a gradual open signal to the backwash flow-control valve operator.

As the surface wash continues to operate, the backwash flow gradually enters the filter from the bottom. As the flow increases, the bed fluidizes and expands upward (about 20 percent of the total media depth) allowing a uniform rolling action of the filter media bed, which results in cleaning of the media due to the hydrodynamic shear (water causes grains to clean each other) that occurs. The media bed expands upward

---

[43] *Orifice* (OR-uh-fiss).   An opening (hole) in a plate, wall, or partition. An orifice flange or plate placed in a pipe consists of a slot or a calibrated circular hole smaller than the pipe diameter. The difference in pressure in the pipe above and at the orifice may be used to determine the flow in the pipe. In a trickling filter distributor, the wastewater passes through an orifice to the surface of the filter media.

[44] *Fluidized* (FLOO-id-i-zd).   A mass of solid particles that is made to flow like a liquid by injection of water or gas is said to have been fluidized. In water and wastewater treatment, a bed of filter media is fluidized by backwashing water through the filter.

and into the rotating surface wash arms. The arms now aid in breaking up the suspended material and mudballs that have accumulated in the top section of the media.

5. After two to five minutes of surface wash, the surface wash influent water line solenoid valve closes. Surface wash is discontinued two to ten minutes before the backwash ends so that the surface of the filter media will be smooth and level at the beginning of the cleaned filter run cycle.

6. After seven to twenty minutes of backwash, the backwash flow-control valve gradually begins to close. Shortly after the backwash flow-control valve is fully closed, the backwash pumps stop.

7. Backwash influent valve (V-3) and effluent valve (V-4) close.

8. Filter influent valve (V-1) and effluent valve (V-2) open.

NOTE: When the backwash cycle is activated, the filter flow-control valve fully closes. Upon completion of the cycle, the valve opens slightly.

The two valve sequences indicated in items 1 and 2, and items 7 and 8 occur simultaneously to ensure that the filter does not become "air bound" (clogged by air released from water). Air binding will reduce or block filter influent flow and create media bed disturbance when filter backwash begins.

4.315 Decant Tank (Backwash Recovery)
(Figures 4.44 and 4.45)

Most decant tanks are very similar to secondary clarifiers because they have flights or scrapers to collect settled material toward a sludge hopper.

Filter backwash effluent leaves the filter and may be discharged to the decant tank. The suspended material in the backwash water is dosed with a cationic polymer and allowed to rise to the surface. The clarified effluent overflows to the holding tank through weir slots between the two tanks for recycle to the filters. The floating solids are collected toward a skimmings trough and the settled material is collected toward a hopper in the tank for periodic discharge to the solids handling facility.

If poor solids capture occurs in the decant tank, all spent backwash flow may be returned to the plant headworks through the tank drain line. The drain line may be equipped with a propeller meter and a motor-operated butterfly valve. The opening limit of the valve should be set to discharge tank flow at a rate that will not hydraulically overload the plant.

The tank may be equipped with high water level probes that will open the motor-operated valve fully to allow a predetermined volume of water to leave the tank rapidly. This may be necessary if the tank becomes surcharged (overloaded) due to frequent filter backwashes.

QUESTIONS

Write your answers in a notebook and then compare your answers with those on page 462.

4.3G List the major components of pressure filters.

4.3H How is the mat of suspended material on the media surface initially broken up during a backwash?

4.3I What is the source of water used to backwash the filter?

4.3J What is the purpose of the decant tank?

Fig. 4.44   Decant tank drain line automatic valves

Fig. 4.45   Backwash recovery tank and feed pumps

4.32   Operation

4.320   *Operational Strategy*

This lesson has covered some of the basic concepts of inert-media pressure filters used to remove suspended solids, associated BOD, and turbidity from secondary effluents before chlorination.

If the filters become overloaded due to high suspended solids concentrations, excessive plant flows, high chemical concentrations, or exposure to very cold temperatures, be prepared for the problems discussed in this section.

1. High suspended solids concentrations will cause a filter to plug up fast, thus requiring very frequent filter backwashes. This will result in high recycle flow rates through the plant and eventually the filters. This problem may be eliminated or reduced by having adequate spent backwash storage capacity or by having a closed filter system that will allow for clarification and reuse of spent backwash water for subsequent backwashes.

2. If no backwash storage or a closed system is provided, hydraulic surcharge (overload) on the upstream side of the filters will result. Provisions must be made for filter bypass or storage. If bypass is the only alternative, you should anticipate increased disinfection demands at the disinfection injector point as a result of the increase in unfiltered suspended material. Adjust the disinfection rate to compensate for the greater demands.

3. By allowing suspended material to bypass the filters and enter chlorine contact tanks, more frequent cleaning of these tanks will be required.

4. If higher than normal plant flows can be anticipated (rain), it would be a good idea to operate the filter holding tank/wet well at a lower water level to provide for additional water storage. This action will reduce the surcharge possibility on the upstream side of the filters. This preventive action should also be used if a filter must be taken out of service for repairs or inspection.

5. If liquid alum is used and it is exposed to cold temperatures, the liquid alum will start to crystallize and the delivery of alum to the filter influent flow will be seriously impaired. If climatic conditions of this type are common in your area, consider storing the alum in an enclosed, warm space or providing insulated storage tanks and heat-traced piping to prevent the alum from crystallizing.

6. If chemical feed pump check valves or antisiphon devices fail, large quantities of chemical will be drawn into the filters. This will result in short filter run times due to increased differential pressure across the filter when a polymer is the chemical involved. When excessive alum concentrations are involved, the alum will pass through the filter media and filter effluent turbidity and suspended solids values will increase due to the alum breakthrough.

7. Filter flow and differential pressure valves may be sensed by differential pressure cells. These cells are water activated and are fed through small-diameter piping. During periods of extremely cold weather, these cells could freeze and prevent proper functioning of the filter and control instruments. Heavy insulation or heat tape will prevent the water in the cell piping from freezing.

4.321   *Abnormal Operation*

Efficient filter operation is essential if your final effluent quality is to comply with the waste discharge requirements established for your plant. Table 4.4 lists a few of the more common pressure filter operational problems and suggestions on how to correct them.

QUESTIONS

Write your answers in a notebook and then compare your answers with those on page 463.

4.3K   What happens when large quantities of alum or polymer accidentally reach the filter?

4.3L   What precautions should be taken in regions where freezing temperatures occur?

4.3M   What could cause high operating filter differential pressures?

4.33   Maintenance

A comprehensive preventive maintenance program is an essential part of plant operations. Good maintenance will ensure longer and better equipment performance. The following may be used as a guideline in performing the required maintenance on the pressure filter system.

A filtration system performance test should be conducted monthly. This test will enable you to evaluate performance and determine if the pumps, valves, filters, and control instruments are functioning properly. If they are not, the proper corrective action must be taken. A sample performance test chart for three filters is shown in Figure 4.46.

Filter media and interior vessel surfaces should be inspected quarterly. The filter should be backwashed just prior to the inspection. Some of the items to look for are:

1. Is the media surface fairly flat and level? If not, the surface wash time should be reduced.

2. Are there mudball formations on or in the media? If so, an increased surface wash time in conjunction with a lower backwash rate should bring the media back to a clean condition.

3. Are very small quantities of mid-filter media particles visible on the top layer media surface? This condition is normal.

TABLE 4.4   ABNORMAL PRESSURE FILTER OPERATION

Abnormal Condition	Possible Cause	Operator Response
PUMPS Do not meet pumping requirements.	Inappropriate valve positioning.	Adjust valves to proper position.
	Insufficient motor speed.	Install higher RPM motors.
	Pump impeller improperly set in bowl of pump.	Set impeller as per manufacturer's instructions.
	Excessive filter system head losses.	Air in filter system. Analyze problem and take corrective action such as install higher RPM motors, redesign pump station, redesign force main, redesign orifice plates.
	Broken pump shaft.	Replace.
FILTERS (GENERAL) High operating filter differential pressure.	Filled with suspended material.	Backwash filter at least once every 24 hours.
	Excessive chemical feed "blinding" media.	Evaluate and reduce dosage. Backwash filter.
Water discharges from pressure-relief valve.	Effluent valve(s) blocked or closed.	Investigate and correct valve problem.
	Foreign object lodged between valve and seat.	Secure filter and clean valve seat.
Water discharges from air-relief valve.	Air pocket in underdrain system (most common after a backwash).	Secure filter for 2 to 3 minutes to allow vessel water level to stabilize, return filter to service. Adjust filter feed and backwash valves to open and close simultaneously to keep vessel full of water.
MEDIA Support media upset.	Air slug forced out by the backwash flow.	Install air-relief valve on backwash influent line.
	Backwash flow pumped too suddenly.	Install flow-control valve for regulated flow rates.
	Backwash flow rate too high.	Install valve stops or limiting orifices.
Mudball formation. Media surface cracks.	Inadequate surface wash time.	Increase surface wash time. Check to ensure arms are operating.
Backwash water dirty at end of wash cycle.	Insufficient backwash time or flow rate.	Increase time or flow rate until clean water appears.
Media surface uneven after backwash.	Surface wash too long.	Decrease surface wash time.
Algal growth in media bed.	Nutrient-rich water being filtered.	Prechlorinate continuously at low rates.

4. Do the surface wash arms rotate freely and in the proper direction? If not, the trouble could be a defective central bearing and tee. Are any nozzles plugged? If so, they must be cleaned.

CAUTION: Wear goggles when observing the operation of the surface wash arms. The velocity of water produced from the wash arms is great and will kick up surface media.

5. Is there a small amount of foreign matter on the media surface (plastic, cigarette butts)? This condition is fairly common and most often occurs at the extreme effluent end of the backwash effluent header. The foreign matter is carried away by the backwash water during the next backwash cycle. If a large amount of foreign matter accumulates, it will have to be removed by manual means.

6. Inspect all interior metal surfaces to ensure that the corrosion-inhibitive protective coating is in good condition. If not, prepare the affected surface and reapply the proper coating. An epoxy tar is frequently used for this purpose.

At least once monthly, the backwash rate should be observed to ensure that the flow rate is correct as specified by the manufacturer's backwash rate/flow curve and that the backwash flow is allowed to enter the filter at a regulated rate. Observe the backwash effluent flow. The water should be clear at the end of the wash cycle. If it is not, an increase in the wash time is indicated.

The flow rates for the chemical feed pumps should be checked at least every two weeks. Corrective adjustments should be made to maintain the proper flow rates.

VALENCIA WRP						FILTER AND FILTER FEED PERFORMANCE TESTS						DATE: 7-21-05 BY: REDNER & KETTLE						
Test	Pumps Running			Filter On			Discharge Through Filters, GPM				Holding Tank Level	Pressure at Pump Discharge	Filter Differential Pressure			Filter Effluent Flow Control Valve, % Open		
#	1	2	3	1	2	3	1	2	3	TOTAL	FEET	PSI	1	2	3	1	2	3
1	on	off	off	on	off	off	1400			1400	7.8	1.5	6.5			100	0	0
2	off	on	off	on	off	off	1100			1100	8.3	3.0	8.3			100	0	0
3	off	off	on	on	off	off	970			970	8.5	3.5	9.5			100	0	0
4	on	on	off	on	on	off	850	1400		2250	8.5	—	9.0	5.0		100	100	0
5	on	on	off	on	on	on	725	1224	1000	2950	8.5	5.0	8.5	5.0	5.0	100	100	100
6	on	off	on	on	on	on	700	1200	1000	2900	8.5	5.2	8.5	5.0	5.0	100	100	100
7	off	off	on	on	on	on	625	1075	925	2625	8.5	3.0	8.5	5.5	5.5	100	100	100
8	on	on	on	on	on	on	900	1275	1150	3325	7.7	8.0	9.5	6.0	6.0	100	100	100
9	on	off	off	off	off	on			1150	1150	7.7	3.0			7.0	0	0	100
10	on	off	off	off	on	off		1300		1300	7.9	3.0		6.5		0	100	0
11	on	off	off	on	off	off	975			975	8.0	3.2	10.0			100	0	0

Note: The filter effluent flow control valve should be 100 percent open during the monthly test.

REMARKS:

Fig. 4.46 Filtration system performance test chart

4.34 Safety

Safety precautions for working in *CONFINED SPACES*[45] and working around mechanical equipment should be observed when operating and maintaining this equipment in the pressure filter system. Confined space safety precautions are discussed in detail in Chapter 14, "Plant Safety," in *OPERATION OF WASTEWATER TREATMENT PLANTS,* Volume II.

In addition, the following safety precautions should be observed:

1. Wear safety goggles and gloves when working with alum or polymers. Wear approved respiratory protection if inhalation of dry alum or polymer is possible. Flush away any alum or polymer that comes in contact with your skin with cool water for a few minutes.

2. Be very careful when walking in an area where polymer mixing takes place. When a polymer is wet, it is very slippery. Clean up polymer spills with an inert, absorbent material such as sand or earth.

3. When inspecting the interior of a filter vessel, *ALWAYS* follow confined space entry procedures, and:

   a. Ensure that all flow-control instruments are in the OFF position and that all valves are in the MANUAL or OFF position. Position all valves to prevent flow from entering the filter. Lock out and tag any controls or valves as well as the power supply to the equipment to prevent accidental start-up.

   b. Always ventilate vessel. Open two access covers. Install and start an exhaust blower in one of the two openings to provide fresh air circulation before entering the filter.

   c. Check vessel atmosphere for oxygen deficiency/enrichment, combustible gases and explosive conditions, and toxic gases (hydrogen sulfide).

   d. Entering a filter vessel is a three-person operation. Two must be outside the vessel whenever one person is inside.

   e. Wear a hard hat when working inside a filter or around the filter vessel piping to protect your head from injury.

---

[45] *Confined Space.*   Confined space means a space that: (1) Is large enough and so configured that an employee can bodily enter and perform assigned work; and (2) Has limited or restricted means for entry or exit (for example, manholes, tanks, vessels, silos, storage bins, hoppers, vaults, and pits are spaces that may have limited means of entry); and (3) Is not designed for continuous employee occupancy. Also see DANGEROUS AIR CONTAMINATION and OXYGEN DEFICIENCY.

4.35 Review of Plans and Specifications

As an operator, you can be very helpful to design engineers in pointing out some design features that would make your job easier and safer. This section attempts to point out some of the items that you should look for when reviewing plans and specifications for expansion of existing facilities or construction of new pressure filter systems.

1. The variable hydraulic and suspended solids load in secondary effluents must be considered in the design to avoid short filter runs and excessive backwash water requirements.

2. A filter that allows penetration of suspended solids (a coarse-to-fine filtration system) is essential to obtain reasonable filter run lengths. The filter media on the influent side should be at least 1 to 1.2 mm in diameter.

3. Auxiliary agitation of the media is essential for proper backwashing. Surface washers should be installed.

4. The effect of recycling used backwash water through the plant on the filtration rate and filter operation must be considered in predicting peak loads on the filters and resulting run lengths.

5. The filtration rate and head loss should be selected to achieve a minimum filter run length of 6 to 8 hours during peak load conditions. This requirement will mean an average filter run length of 24 hours. Estimates of head loss development and filtrate quality should be based on pilot-scale observations of the proposed facility conducted at the treatment plant before the full-scale facility is designed.

6. Accessways should be sized large enough to allow operators and equipment ease of entering and leaving the filter.

7. A media core sample port(s) should be provided to allow evaluation of the entire media depth.

8. Ladders and walkways should be provided to allow easy access to vessels, pipes, and valves.

9. Filter flow charts should be provided to aid in monitoring filter performance.

4.36 Acknowledgments

1. County Sanitation Districts of Los Angeles, Valencia Water Reclamation Plant.

2. Jerry Schmitz, Draftsman, County Sanitation Districts of Los Angeles.

4.37 Additional Reading

The two EPA publications listed below are available from the National Technical Information Service (NTIS), 5285 Port Royal Road, Springfield, VA 22161.

1. *WASTEWATER FILTRATION DESIGN CONSIDERATIONS,* Technology Transfer, July 1974, US Environmental Protection Agency. Order No. PB-259448. Price, $33.50, plus $5.00 shipping and handling per order.

2. *PROCESS DESIGN MANUAL FOR SUSPENDED SOLIDS REMOVAL,* Technology Transfer, US Environmental Protection Agency. Order No. PB-259147. Price, $78.50, plus $5.00 shipping and handling per order.

QUESTIONS

Write your answers in a notebook and then compare your answers with those on page 463.

4.3N   How frequently should a filter system performance test be conducted?

4.3O   What caution should be exercised when observing the operation of the surface wash arms?

4.3P   What safety precautions should be taken when working with alum or polymers?

END OF LESSON 4 OF 6 LESSONS

on

SOLIDS REMOVAL FROM SECONDARY EFFLUENTS

Please answer the discussion and review questions next.

DISCUSSION AND REVIEW QUESTIONS

Chapter 4.   SOLIDS REMOVAL FROM SECONDARY EFFLUENTS

(Lesson 4 of 6 Lessons)

Write the answers to these questions in your notebook. The question numbering continues from Lesson 3.

25. How are floatable and settleable solids removed from a holding tank?

26. How would you attempt to control corrosion of the interior surfaces of a pressure filter vessel?

27. How can a constant rate of filtration be maintained in a pressure filter?

28. During what time of the day should the filters be backwashed?

29. What would be the impact of allowing suspended material to bypass the filters and enter chlorine contact tanks?

CHAPTER 4.   SOLIDS REMOVAL FROM SECONDARY EFFLUENTS

(Lesson 5 of 6 Lessons)

4.4   CONTINUOUS BACKWASH, UPFLOW, DEEP-BED SILICA SAND MEDIA FILTERS
by Ross Gudgel

4.40   Use of Continuous Backwash, Upflow, Deep-Bed Silica Sand Media Filters

As water demands for agricultural, industrial, and municipal use increase, wastewater reuse has become more important, especially as our nation implements essential water conservation programs. In many parts of the country, a high level of water quality is required and specific treatment criteria must be met before treated wastewaters can be reused. Continuous backwash, upflow, deep-bed silica sand media filters consistently produce high-quality effluents that do not exceed average operating values of 2 turbidity units (NTUs). This high level of water quality significantly reduces demands on the disinfection process because *PATHOGENS*[46] and viruses can be removed or inactivated much more effectively and at lower cost when turbidity levels are low.

The actual filtering process, as with most conventional granular media type filters, involves the capture of previously coagulated and flocculated suspended solids within the voids (spaces) between the granules of media that make up the filter bed. Continuous backwash, upflow, deep-bed silica sand media filters continuously remove solids from the influent while at the same time cleaning and recycling the filter media internally through an airlift pipe and sand washer. The cleaned sand is redistributed on top of the filter media bed, which allows for uninterrupted flow of filtered effluent and backwash reject water. Therefore, this type of filter has two distinct advantages over other types of granular media filters.

First, these filters do not need to be taken out of service for backwashing. This is of particular operational importance with regard to the addition of coagulants during periods of low filter influent suspended solids. In most conventional downflow filters, coagulants added to the filter influent flow when the suspended solids are low tend to accumulate and create a mat of coagulant on top of and within the uppermost portions of the granular media. When the filter influent suspended solids subsequently increase, virtually all of the incoming suspended solids are captured on top of and within the coagulant mat. The filter

then blinds, filter head loss increases dramatically, and the filter must be taken out of service for backwashing.

Second, because these filters do not need to be taken out of service for backwashing, full design flow processing capacity can be provided with fewer filters or smaller-sized filters.

Continuous backwash, upflow, deep-bed silica sand media filters are supplied in two basic styles, bottom feed cylindrical (Figure 4.47) and bottom feed concrete basin (Figure 4.48). The internal design and the operation of the two styles are virtually identical. The main difference between the two types is the outer housing of the filter. Therefore, the following section will use the bottom feed cylindrical filter to explain this type of filtration. Figure 4.49 shows a schematic view of the filter components described.

The filter system usually consists of:

1. A channel or pipe from the upstream treatment process to conduct flow to the filtration system. Alternatively, a holding tank or wet well may be provided for storage of treatment process effluent. Filter feed pumps may also be provided that pump the effluent from the holding tank or wet well to the filtration system.

2. Filter influent and effluent turbidity metering to continuously monitor and record filter influent and effluent water quality.

3. Filter influent and effluent flow metering to quantify and record flow values and provide dosage control signals to metering pumps and disinfection chemical addition control systems.

4. A chemical coagulant (ferric chloride or alum) dosage control and metering pump system that injects coagulant solution into the filter influent flow.

5. A chemical flocculant (polyelectrolyte) mixing, dosage control, and metering pump system that injects flocculant solution into the filter influent flow.

6. A flocculation tank to enhance the development of large floc particles (the sticking together of chemically treated suspended solid and dissolved materials).

---

[46] *Pathogenic* (path-o-JEN-ick) *Organisms.*   Organisms, including bacteria, viruses, or cysts, capable of causing diseases (such as giardiasis, cryptosporidiosis, typhoid, cholera, dysentery) in a host (such as a person). Also called PATHOGENS.

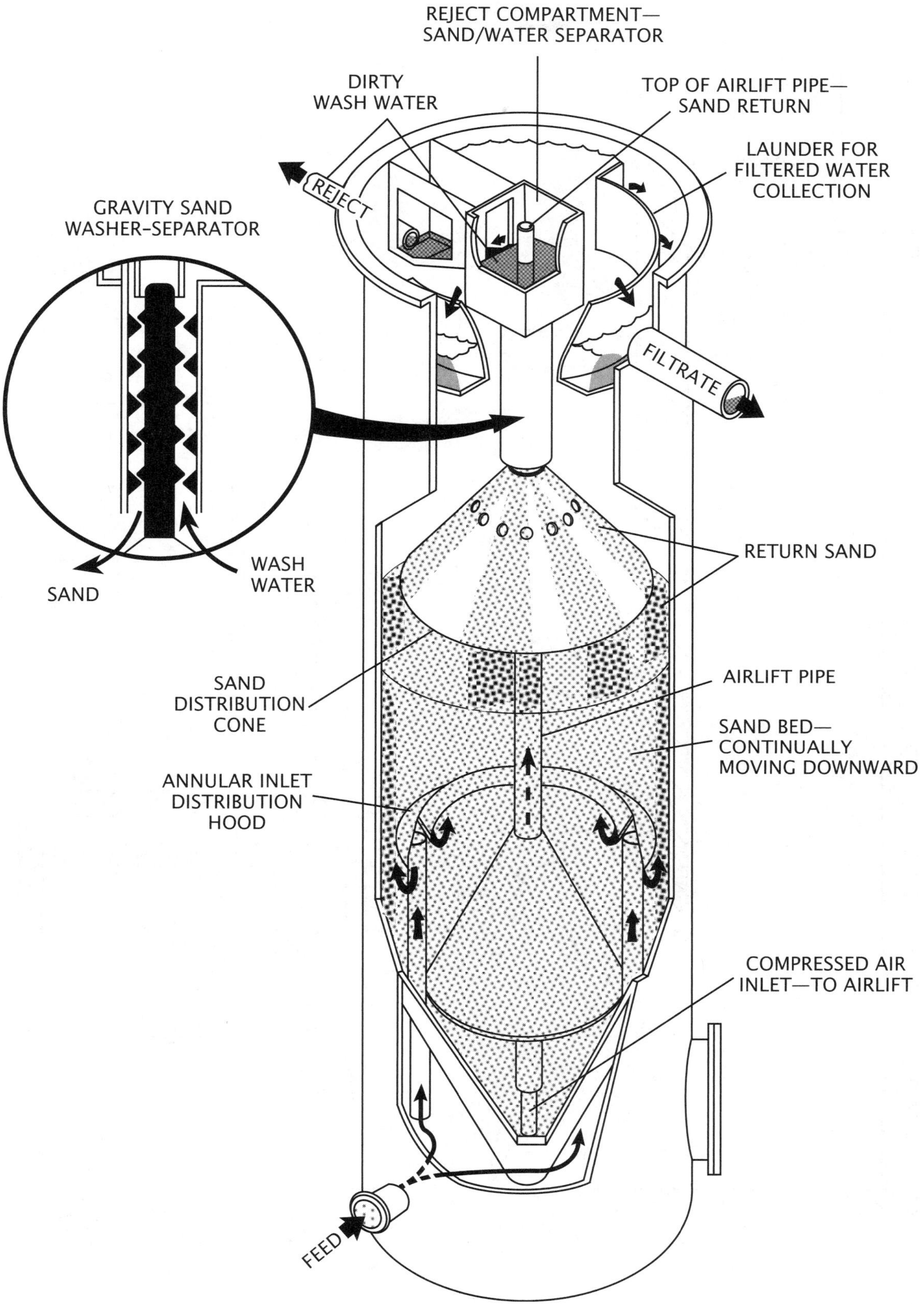

Fig. 4.47 Bottom feed cylindrical filter

(Reproduced by permission. DynaSand® Filter is a registered
trademark of Parkson Corporation)

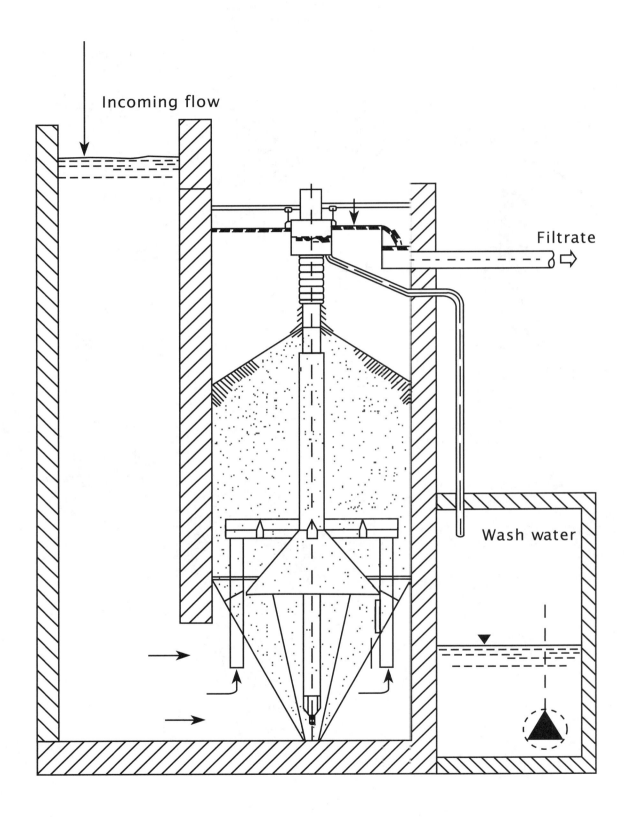

Incoming flow

Filtrate

Wash water

Fig. 4.48   Bottom feed concrete basin filter
(Reproduced by permission. DynaSand® Filter is a registered
trademark of Parkson Corporation)

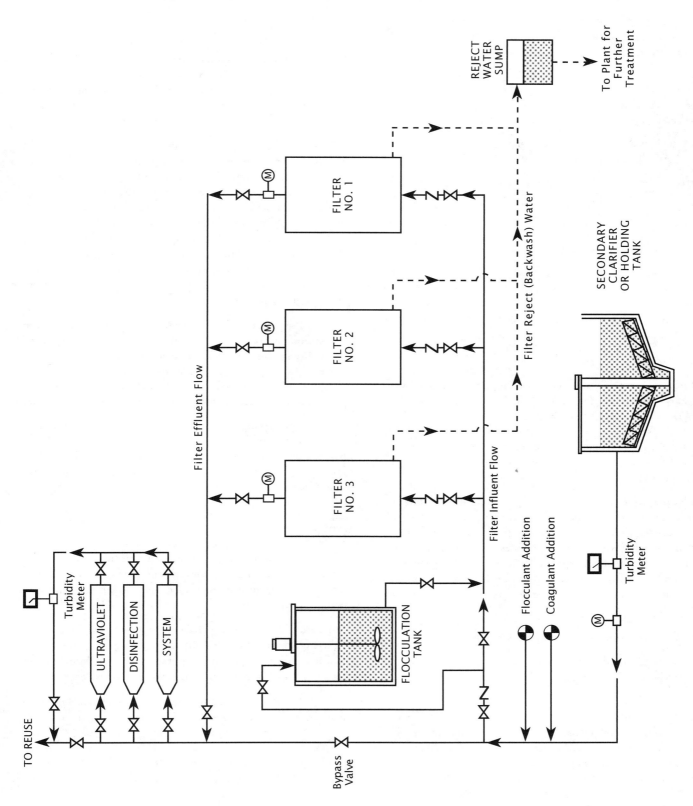

Fig. 4.49 Schematic view of continuous backwash, upflow, deep-bed silica sand media filter system

7. Continuous backwash, upflow, deep-bed silica sand media filters that trap suspended solids, associated BOD, and turbidity and remove them from the treatment flow stream.

8. A filter reject (backwash) water sump or wet well equipped with pumps to collect and return suspended solids and backwash water to the plant for further treatment.

QUESTIONS

Write your answers in a notebook and then compare your answers with those on page 463.

4.4A   How does a continuous backwash, upflow, deep-bed silica sand media filter work?

4.4B   What is the purpose of the filter influent and effluent turbidity meters?

4.4C   Why are the influent and effluent flows metered?

4.41   Auxiliary Equipment

The following sections describe facilities that are typical for a filtration system plant using bottom feed cylindrical filters with a rated flow capacity of 0.5 MGD. Facilities at larger plants would be quite similar but would differ significantly in the numbers and sizes of the various components.

4.410   Channel or Piping

Effluent, normally from the treatment plant's secondary sedimentation tanks, is typically conducted to the filtration system facilities through a covered channel or through piping. Where channels are used, covered channels are preferred in order to limit the risk of having foreign objects get into the flow stream. Such objects can damage flocculation equipment or plug the airlift pump in a filter.

Features to bypass the filtration system facilities should be provided to permit diverting treatment process effluent to holding or equalization basins during emergency conditions, such as extremely high flows, equipment failures, or clogged filters. Following resolution of the emergency, flow can then be regulated back into the filter influent flow stream from the holding or equalization basin for normal and complete treatment through the filtration system.

4.411   Filter Influent and Effluent Turbidity Metering

Low-range turbidity meters are normally used to continuously analyze and record the turbidity (clarity) of the filter influent and effluent flows.

Portions of the respective flow streams are continuously withdrawn from the flow channel or pipe by a sample pump. The sample portions first pass through a bubble trap to remove the air bubbles (air bubbles interfere with turbidity measurement). The bubble trap also serves as a head regulator to dampen fluctuations in flow due to pulsations from the sample pump.

The wastewater being sampled enters the bottom of the bubble trap assembly and rises toward the top. As the flow rises, most of the air bubbles are released and vented to the atmosphere. The wastewater flows out of the trap by gravity through an overflow discharge pipe that is fitted with a shutoff/flow control valve. The valve allows the operator to regulate flow to the turbidity meter; a typical flow range is 3.5 to 11.5 gallons per hour (13.2 to 43.5 liters per hour). Additional flow control/head regulation of the flow fed to the turbidity meter can be obtained by raising or lowering the bubble trap assembly within two pipe hanger straps that support the assembly.

Wastewater entering the turbidity meter (Figure 4.50) passes through another bubble trap baffle network that forces a downward flow of the water. This downward water flow (relatively slow) allows any remaining air bubbles to rise and either cling to surfaces of the baffle or rise to the surface and vent to the atmosphere. At the bottom of the bubble trap baffle network, flow enters a center column and rises up into the turbidity measuring chamber head where turbidity is measured. The flow then spills over a weir contained within the turbidity measuring chamber head and is returned to the process flow stream through an overflow discharge pipe.

Turbidity is measured by directing a strong beam of light from the optical section of the turbidity measuring chamber head down through the surface of the passing water flow. Light is scattered at a 90-degree angle when it strikes suspended solids particles (turbidity) in the water. The scattered light is detected by a photocell located just below the surface of the passing water flow.

The amount of light scattered is proportional to the turbidity of the water. If the turbidity of the water is low, little light will be scattered to the photocell and the turbidity reading will be low. High turbidity, on the other hand, will cause a high degree of light scattering and will result in a high turbidity reading.

A control unit for the turbidity meter has a keyboard, microprocessor board, and power supply components. Operating controls and indicators located on the control unit keyboard are used to program the turbidity meter for recorder output minimums and maximums and for turbidity level alarm set points as well as to perform a number of diagnostic self-tests and programming operations.

Turbidity levels are displayed continuously by a four-digit, light emitting diode (LED) display during normal operations. Alarm conditions and certain critical system malfunctions or impending malfunctions are also indicated on the keyboard.

Programmable alarm circuits provide relay closures, both normally open and normally closed, for two turbidity level set points (high turbidity and low turbidity). Alarm set points can be programmed by the operator anywhere within the overall turbidity measuring range of 0.001 to 100.0 *NEPHELOMETRIC* [47] Turbidity Units (NTUs). High turbidity alarms are transmitted to an alarm monitoring center in the treatment

---

[47] *Nephelometric* (neff-el-o-MET-rick).   A means of measuring turbidity in a sample by using an instrument called a nephelometer. A nephelometer passes light through a sample and the amount of light deflected (usually at a 90-degree angle) is then measured.

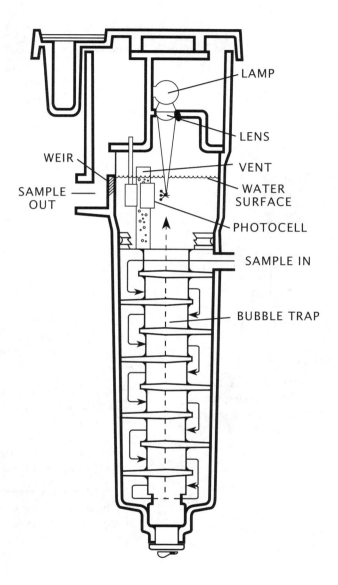

LAMP

LENS

WEIR

VENT

SAMPLE OUT

WATER SURFACE

PHOTOCELL

SAMPLE IN

BUBBLE TRAP

Fig. 4.50   Turbidity meter

(Reproduced by permission of HACH Company)

plant, which, in turn, notifies the operator that a high turbidity condition exists.

NTU values are continuously transmitted from the control unit to a circular chart recorder. The recorder may be a single pen, 10-inch round chart recorder. Chart rotation can be selected for 24-hour or 7-day rotation. Recorder span settings (minimum and maximum values) in NTUs are programmed by the operator at the turbidity meter control unit keyboard.

Most of the information used to develop this section was obtained from the "Hach Model 1720C Low Range Process Turbidimeter Instruction Manual."

4.412 Filter Influent and Effluent Flow Metering

Various types of flow measuring devices suitable for use in these applications are described in Chapter 9, "Instrumentation." Typical flowmeters include the Venturi and orifice types.

4.413 Coagulation and Flocculation Equipment

The suspended solids removal rates of most types of filters, including continuous backwash filters, can be greatly improved by the addition of chemicals before filtration. Coagulants such as ferric chloride and alum cause the suspended and colloidal (finely divided solids that will not settle by gravity) particles to clump together into bulkier hydroxy precipitates.

To ensure maximum solids removal during filtration, the wastewater is next treated with another type of chemical, usually a polymer, in a process called flocculation. This process gathers still more of the suspended and colloidal solids into larger (visible to the naked eye), stronger clumps (floc), which can be efficiently removed by filtration.

Figure 4.51 shows one example of the types of equipment used for flocculation by polymer addition. Section 4.1 of this chapter describes the range of equipment and chemicals available for

coagulation and flocculation processes. In addition, the section explains how to perform a jar test to select the proper chemicals and determine dosages. That information will not be repeated here, so please refer to Section 4.1 to learn about chemical addition and the coagulation and flocculation processes.

4.42 Operation of Continuous Backwash, Upflow, Deep-Bed Silica Sand Media Filters

4.420 Normal Operation

As shown in Figures 4.47 and 4.52, Part A, filter influent flow (feed) is introduced into the bottom of the filter through a feed inlet where it enters a section of the filter called the plenum area. It flows upward through a series of riser tubes and is evenly distributed into the sand bed through the open bottom of the ANNULAR[48] inlet distribution hood. The rate of flow through the filter is expressed as the surface loading rate, in gallons per minute per square foot of filter surface area:

$$\text{Loading Rate, GPM/sq ft} = \frac{\text{Flow, GPM}}{\text{Filter Surface Area, square feet}}$$

The wastewater flows upward through the downward-moving sand bed. The coagulated solids in the wastewater are trapped in the spaces between the granules of sand. A head loss gauge indicates the water pressure (head in inches of water) on the influent side of the filter relative to the atmospheric pressure at the LAUNDER[49] for filtered water collection.

The now cleaned (suspended solids free) influent continues to pass through the sand bed, overflows a launder for filtered water collection, and is discharged from the filter as filtrate (tertiary treated effluent).

Filtrate leaving each filter flows through an effluent pipe and a filtrate flowmeter. The meter indicates and totalizes flow locally and also transmits a flow signal to a filter effluent flow metering system. The flow then passes through an effluent gate valve into a larger discharge pipe that directs flow from each filter to the disinfection system.

As the filter is removing the coagulated solids from the influent flow, the silica sand media is continuously cleaned by recycling the sand internally through an airlift pipe and gravity sand washer-separator. The sand bed, along with the solids trapped among the sand particles, is drawn downward into the suction to the airlift pipe (Figure 4.52, Part B), which is positioned at the bottom and center of the filter. A moderate volume of compressed air (supplied by a compressed air system) is introduced into the bottom of the airlift pipe by an air control panel located

---

[48] *Annular* (AN-yoo-ler) *Space.*   A ring-shaped space located between two circular objects. For example, the space between the outside of a pipe liner and the inside of a pipe.

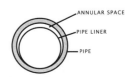

[49] *Launders.*   Sedimentation basin and filter discharge channels consisting of overflow weir plates (in sedimentation basins) and conveying troughs.

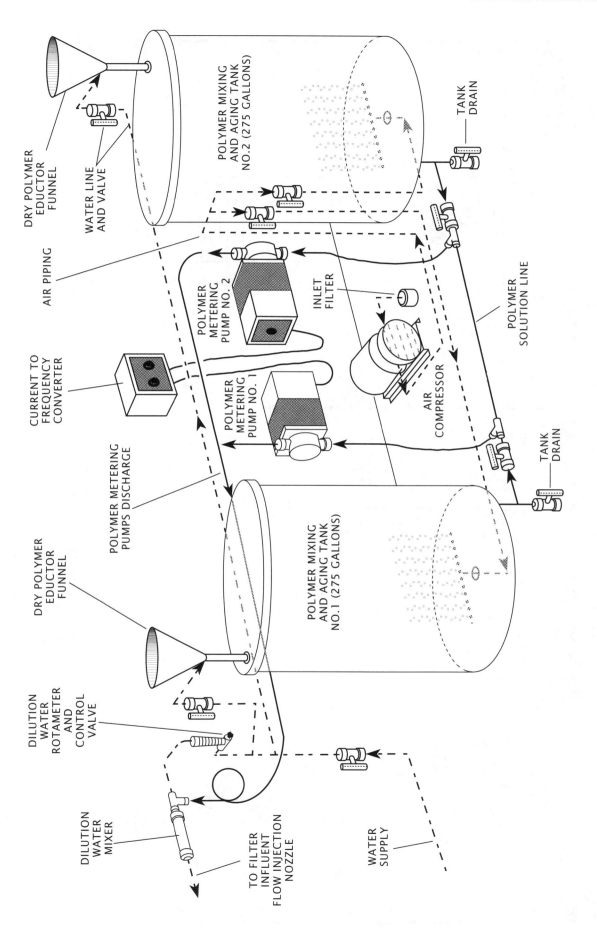

DRY POLYMER EDUCTOR FUNNEL

WATER LINE AND VALVE

POLYMER MIXING AND AGING TANK NO.2 (275 GALLONS)

TANK DRAIN

AIR PIPING

POLYMER METERING PUMP NO. 2

INLET FILTER

CURRENT TO FREQUENCY CONVERTER

POLYMER METERING PUMP NO. 1

AIR COMPRESSOR

POLYMER SOLUTION LINE

POLYMER METERING PUMPS DISCHARGE

DRY POLYMER EDUCTOR FUNNEL

TANK DRAIN

DILUTION WATER ROTAMETER AND CONTROL VALVE

POLYMER MIXING AND AGING TANK NO.1 (275 GALLONS)

DILUTION WATER MIXER

TO FILTER INFLUENT FLOW INJECTION NOZZLE

WATER SUPPLY

Fig. 4.51   Polymer mixing and aging system

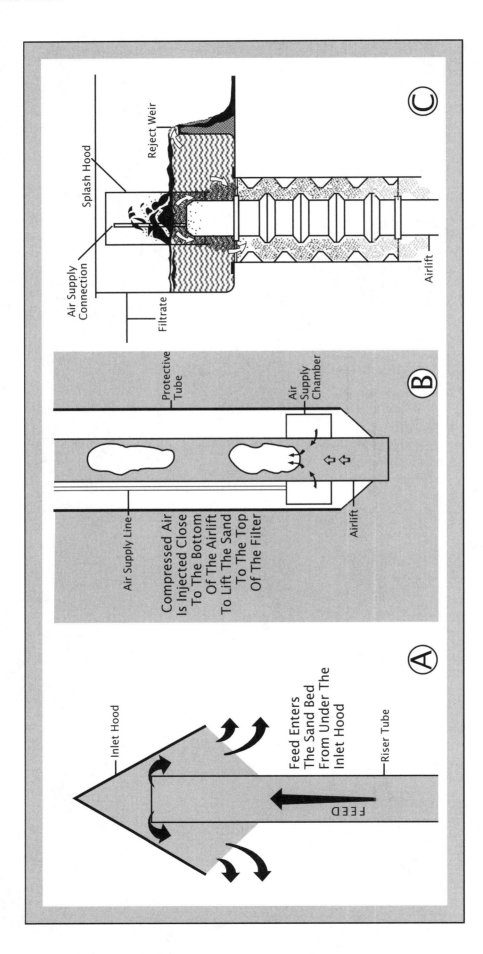

Fig. 4.52 Internal detail of continuous backwash, upflow, deep-bed silica sand media filter
(Reproduced by permission. DynaSand® Filter is a registered trademark of Parkson Corporation)

at the top of each filter. The air lifts the sand and trapped solids up the airlift pipe into the gravity sand washer-separator. The violently turbulent flow, which is created as the sand and trapped solids move up the airlift pipe and into the gravity sand washer-separator, scours the trapped solids loose from the sand particles. A typical air flow rate is 100 to 150 standard cubic feet of air per minute per square foot (SCFM/sq ft) (30 to 45 SCMM/sq m).

As the sand reaches the top of the airlift pipe (Figure 4.52, Part C), the loosened solids and dirty wash water (reject water) spill over into the reject compartment-sand/water separator. The sand passes downward through the gravity sand washer-separator (Figure 4.47) while a small volume of the filtrate is passing up through the gravity sand washer-separator. This counterflow of rising filtrate and falling sand washes any remaining solids away from the sand.

The cleaned sand is returned to the sand bed through the gravity sand washer-separator and is redistributed over the top of the sand bed by means of gravity, in conjunction with the sand distribution cone.

The small volume of filtrate water, which has now become dirty wash water (reject water) in the gravity sand washer-separator, carries the remaining solids out of the top of the gravity sand washer-separator (Figure 4.52, Part C) into the reject compartment-sand/water separator. From there, the reject water with solids joins with the other reject water and solids exiting at the top of the airlift pipe.

The combined reject waters and solids pass over an adjustable reject weir at a rate of 10 to 15 gallons per minute (GPM) (38 to 57 liters per minute). The reject water flow rate leaving the filter is one of the most important determinants for controlling filter operation.

Reject waters and solids leaving the filters are directed to a reject water sump or wet well. A float switch-activated pump returns the reject water and solids to the treatment plant for further treatment.

QUESTIONS

Write your answers in a notebook and then compare your answers with those on page 463.

4.4D Why should features to bypass filtration system facilities be provided?

4.4E What is the purpose of adding chemical coagulants to the filter influent?

4.4F What is the purpose of adding flocculant solution to the filter influent?

4.421 Operating Strategy

Normal operation should require very little attention from the operator. This section describes basic requirements for developing an operational strategy. If these requirements are not met, the filters may not consistently perform optimally.

1. When influent water or supplemental water is being processed through the filter, the airlift must be pumping sand. If not, the sand bed will not be cleaned and will eventually become plugged with suspended solids and chemicals.

2. When influent water or supplemental wash water is not being processed through the filter, the airlift must not be operating. If the airlift is allowed to operate without water flow to the filter, no water will be available to the gravity sand washer-separator. This condition will allow dirty sand to be returned to the top of the sand bed. When normal flow and operating conditions are resumed, it will take many hours before clean filtrate is produced.

3. Whenever the filter is operating, there must be sufficient reject flow going out over the reject weir (10 to 15 GPM or a crest of water over the top weir plate of ¼ to ½ inch in depth). Without this reject water flow rate, the sand will not be washed away fully and properly. Instead, it will be returned to the top of the sand bed only partially clean and the result will be a low-quality filtrate.

4. There is a minimum (100 SCFM) and maximum (150 SCFM) value for the rate at which sand can be pumped through the airlift. Between these values the rate can be adjusted to match the typical range at which influent suspended solids will be optimally removed. If the sand pumping rate is too low, then the sand will only be partially cleaned. This will result in decreased flow of the influent water through the filter (as the result of increased head loss) and associated poor filtrate quality.

5. The sand pumping rate should be adjusted so that the head loss gauge reading remains stable. The air rate should be high enough to cope with all normal operating flow and suspended solids ranges so that constant air adjustment is not required. Specific head loss is determined by the influent flow fed to the filter and the amount of suspended solids in that flow. Since flow and solids guidelines vary with each filter installation, the equipment manufacturer will provide the specific operating head loss values for your particular filtration system.

   An operating log should be maintained so that head loss readings, influent flow rates, suspended solids values, air flow rates, and back pressure rates for different operating conditions can be recorded. This will allow the operator to evaluate the various filter operating conditions and determine the optimum air rate setting for each filter.

6. If there are problems with filter performance, investigate these questions first:

   a. Is sand being pumped to the top of the airlift and at the proper rate?

   b. Is there enough reject water flow?

c. Does the influent flow rate exceed the design flow and loading rate?

d. Are the chemical feed systems functioning properly and are the chemical dosage rates being applied at the proper concentrations based on jar tests?

e. Do influent suspended solids exceed the design solids loading rate?

7. It is very important not to drop anything into the filter. If this should happen, immediately stop the air flow to the airlift and remove the object from the top of the sand bed. If not removed, the object will move downward with continually moving sand. It will eventually block the bottom of the airlift and prevent the filter from operating properly.

8. Calcium scaling can occur within the filter and on the surface of the sand media granules when calcium ions and sulfate or carbonate ions are present in the filter influent flow. Usually, this happens when lime or sulfuric acid are used for pH control. Prevention or control of scaling can be accomplished in several ways:

a. Use caustic in place of lime, or hydrochloric acid in place of sulfuric acid in neutralization.

b. Use scale retardant/dispersant available from water treatment chemical suppliers.

c. Keep the sand bed moving whenever the influent flow is off by introducing clean water, or wash the sand with clean water for four hours before securing the filter.

4.422 *Abnormal Operation*

Efficient filter operation is essential if your final effluent quality is to comply consistently with effluent water quality criteria established for your plant. Table 4.5 lists some of the more common filter operational problems and suggestions on how to correct them.

4.43 Maintenance

A comprehensive preventive maintenance program is an essential part of plant operations. Good maintenance will ensure longer and better equipment performance. The following procedures may be used as a guideline in performing the required maintenance on the continuous backwash, upflow, deep-bed silica sand media filters.

1. A filter sand movement evaluation test should be done monthly. The test will enable you to evaluate and determine if the sand granules have "cemented" together, if the entire sand bed is moving downward equally throughout the inner diameter of the filter, and if the rate of downward sand movement is in accordance with manufacturer's specifications.

a. Make a sand measuring device using a length of ½-inch schedule 80 PVC pipe. Glue caps on both ends and mark one-inch intervals using tape or a permanent-ink marking pen.

b. Place the sand measuring device on top of the filter sand bed at location "A" shown on Figure 4.53.

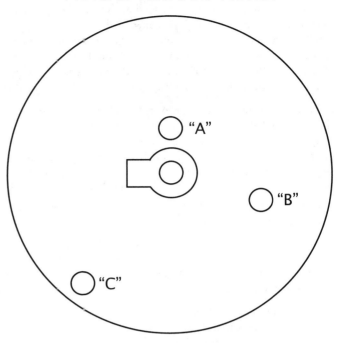

TYPICAL TERTIARY FILTER

NOTE: Measuring locations shown are approximate.

Fig. 4.53 Sand movement evaluation test locations (typical)

c. Ensure that the end of the PVC pipe sand measuring device only contacts the filter sand bed surface and not any supporting structure within the filter.

d. Enter the "Start Time" and "Start Inches" on the "Filter Sand Movement Evaluation" form shown as Figure 4.54.

e. Leave the measuring device in place for 15 minutes with the filter in full operation.

f. At 5-minute intervals during the 15-minute evaluation period, record (on a separate piece of paper) air flow, filter effluent flow, influent turbidity, effluent turbidity, and head loss values associated with the filter being evaluated.

g. At the end of the 15-minute evaluation period, enter the "End Time," "End Inches," and "Total Inches" values on the "Filter Sand Movement Evaluation" form.

h. Calculate the "Inches/Hour" value by multiplying the "Total Inches" value by 4. Sand movement should be 12 to 16 inches per hour. This represents 7 to 10 sand bed turnovers per day on a 24-hour per day basis.

i. Calculate the "Average Air Flow," "Average Flow," "Average Turbidities," and "Average Head Loss" readings using the data you recorded at 5-minute intervals during the 15-minute evaluation (item "f" above). Enter these average values in the appropriate columns on the "Filter Sand Movement Evaluation" form.

TABLE 4.5   TROUBLESHOOTING GUIDE FOR
CONTINUOUS BACKWASH, UPFLOW, DEEP-BED SILICA SAND MEDIA FILTER

Abnormal Condition	Possible Cause	Operator Response
Poor Filtrate Quality	Reject water flow rate too low	Ensure that reject water flow rate is 10 to 15 GPM (reject water crest over the top weir plate is ¼- to ½-inch).
	Influent flow rate or influent solids exceed design range	Calculate the loading rate (GPM/sq ft) and adjust flow to within design range. Evaluate solids concentration in the influent flow and (a) correct upstream process performance, or (b) adjust chemical addition process following performance of jar test. Inspect chemical addition system for proper equipment operation.
	Filter has just experienced solids loading in excess of maximum design loading	After a major solids overload incident, it may take a few hours before filtrate quality returns to normal. Use clean supplemental backwash water, if available, to speed up the sand media washing process.
High or Increasing Head Loss Reading	Sand wash (airlift) air rate too low	Adjust the air rate to clean the sand bed at a higher rate for an hour or two while reducing the filter influent flow rate until filtrate quality returns to normal.
	Air bubbles or foreign matter in or obstructing head loss gauge	Check air relief vent pipe for proper operation. Check head loss gauge inlet for obstruction.
	Airlift is not pumping sand	Check and determine if air is being supplied to the airlift and that all pneumatic systems and control instruments are operating properly. If pneumatic systems are OK, secure filter and (a) check airlift intake for foreign object, or (b) secure filter and remove airlift and clean screen in airlift.
	Excessive chemical application dosages	Adjust chemical addition process following performance of jar test. Inspect chemical addition system for proper equipment operation.
Airlift Functioning Incorrectly	Top of the airlift and the bottom of the splash hood are piled up with sand:	
	1. Sand bed has become plugged with solids due to overloading or improper filter operation, or	1. Reduce or stop filter influent flow until the airlift is operating properly (use clean, supplemental backwash water, if provided).
	2. Airlift lifts only water or just a small amount of sand, or	2. Check that the required air flow is getting through to the airlift. Wearing eye protection, disconnect the air hose from the top of the airlift to determine if the problem is in the airlift itself or the air supply to it. Next, turn off the air and "bump" the airlift by suddenly turning the air on at maximum flow. Try this repeatedly. Wait for one minute between "bumps." Continue with this procedure for some time before moving on to the next troubleshooting step (if your filter is supplied with an "air burst" feature, use it to get maximum air flow for the "bumps").
	3. Airlift is plugged with solids, or	3. Make up an air lance from two sections of ½-inch CPVC pipe. Glue a female threaded fitting to one end of one pipe section, and a male and female threaded fitting on the respective ends of the other pipe section. This will provide you with a two-part air lance. Next, gather together the appropriate pipe fittings that will enable you to connect an air line to the female threaded fitting on the pipe section that also has the male threaded fitting attached. Put on safety goggles or other eye protection. Using both pipe sections screwed together, insert the air lance down into the airlift. With a strong volume of air blowing through the air lance, push the lance down into the airlift as far as it will go. By working the lance up and down, it should be possible to reach all the way down to the bottom of the airlift. Secure the air flow to the lance, remove the lance, and try to get the airlift operating again. If sand pumps for a while and then stops, repeat the air lancing procedure.
	4. Airlift screen is plugged, or	4. Secure the filter and remove the airlift to inspect, clean, repair, or replace the screen as required.
	5. Airlift is plugged	5. If none of the above procedures corrects the problem, it is likely that an object has gotten into the filter and is blocking the bottom of the airlift. Secure and drain the filter. When the filter is drained of water, remove the accessway cover to the plenum area and the inner cone inspection port plate. *WARNING!* The plenum area of the filter is a confined space. Confined space procedures must be followed prior to entry while working with and when exiting the plenum area.

**TABLE 4.5   TROUBLESHOOTING GUIDE FOR
CONTINUOUS BACKWASH, UPFLOW, DEEP-BED SILICA SAND MEDIA FILTER** (continued)

Abnormal Condition	Possible Cause	Operator Response
Airlift Functioning Incorrectly (continued)	5. Airlift is plugged (continued)	DANGER! Before entering the plenum area, check the inner cone for signs of damage. In no event should the plenum area be entered if damage to the inner cone is suspected without first removing all sand from the filter. If sand backflow has occurred as the result of improper filter operation, sand accumulation piles will usually be noted just below each vertical riser tube in the plenum area. Sand accumulation in other areas of the plenum may indicate a damaged inner cone.
		Once the inner cone inspection port plate has been removed, it is possible to dig through the sand with one hand to reach the area around the bottom of the airlift pipe (wear leather gloves). Because the sand will be falling through the inspection port as fast as you dig, it is best to make a sleeve out of a piece of thin-wall PVC pipe that will fit through the inspection port opening and through which you can slide your hand and arm. Cut one end of the pipe at a 45-degree angle and slide this end through the inspection port. Using this device will help keep the weight of the sand off your arm and will make it possible to reach all around the bottom of the airlift to search for the foreign object. This job will take a great deal of determination to do properly and safely.
Plugged Sand Washer	Washer plugged with trash	Because it may be difficult to see down into the sand washer to view possible trash accumulations, use a water hose with significant water pressure to clear the obstruction. Wear eye protection. Insert water hose down into the washer, turn water supply on, and flush the area around the full circumference of the washer to dislodge the obstruction.
	Washer assembly has been displaced	Check to see that the wash rings and spider rings (consult manufacturer's drawings) for the sand washer are in place.
	Sand bed level too high	Check to see if the sand bed level has reached the underside of the washer. If it has, it is most likely that the sand bed is expanding due to calcium scaling. Correction of the scaling problem can be accomplished in one of two ways: (1) while following proper safety precautions, acid wash the sand by maintaining the pH of the water in the filter between 3.0 and 4.0 with inhibited hydrochloric or muriatic acid while the sand is circulated with the airlift. This process will take 8 to 12 hours. Progress can be monitored by probing the sand bed with CPVC pipe and air lancing as necessary. Upon completion of acid cleaning, the filter should be flushed with clean water until the reject water is clean, or (2) remove the contaminated sand from the filter following the manufacturer's instructions and discard the sand in a legal and environmentally approved manner.
Filter Flow Capacity Greatly Reduced	Sand has backflowed into the plenum area and is blocking the riser tubes or the influent pipe	Check the height of the sand bed and compare with height values recorded at initial start-up to determine if sand bed has dropped. It may also be necessary to secure and drain the filter, open the accessway, and inspect for and remove sand accumulation in the plenum area. Follow the safety precautions and procedures described on the previous page (Airlift Functioning Incorrectly, response item #5) before proceeding with the tasks described here. Wear eye protection, use a water hose with significant water pressure, insert water hose up into each riser tube, turn water on, and flush each riser tube to clear any sand from tube. Return removed sand to the filter sand bed.
	Sand bed is plugged due to severe solids overloading or due to operating the filter without operating the airlift	Using the air lance described on the previous page (Airlift Functioning Incorrectly, response item #3), wear eye protection and air scour the sand bed. Secure influent flow to the filter and apply clean supplemental backwash water. As the supplemental backwash water is applied, air lance the sand bed by working the lance up and down through the sand bed from the top of the filter for 15 to 30 minutes. Work the air lance through the lower part of the sand bed as well as the upper part. The reject water will be very dirty after this operation so it is advisable to continue applying the supplemental backwash water until the reject water runs clean. CAUTION! Upper surfaces of the filter unit for which catwalks or walkways have not been constructed are not safe to walk on or work from. Obtain and install proper temporary safety planking to walk on and work from, wear a safety harness, and have an assistant tend the safety line, which is attached to the harness you are wearing.
High or Increasing Back Pressure at the Air Control Panel	Screen in airlift is becoming restricted with solids, trash, or scale	Secure filter, remove airlift, and clean screen.

TERTIARY FILTERS SAND MOVEMENT EVALUATION

Evaluator: _____   Evaluation Date: ____ / ____ / ____

D S F Model: 64,   Serial No: D S F - - 1970,   Sand Size: 0.9 to 1.0 mm,   Sand Depth: 40 inches

Filter No: _____

Test Point "A"

Start Time	Start Inches	End Time	End Inches	Total Inches	Factor	Inches/Hour	Average Air Flow, SCFH	Average Flow, GPM	Average Loading Rate, GPM/SQ FT	Average Influent Turbidity, NTU	Average Effluent Turbidity, NTU	Average Head Loss Reading, Inches
					X-4							
					X-4							
					X-4							
					X-4							

Test Point "B"

Start Time	Start Inches	End Time	End Inches	Total Inches	Factor	Inches/Hour	Average Air Flow, SCFH	Average Flow, GPM	Average Loading Rate, GPM/SQ FT	Average Influent Turbidity, NTU	Average Effluent Turbidity, NTU	Average Head Loss Reading, Inches
					X-4							
					X-4							
					X-4							
					X-4							

Fig. 4.54 Filter sand movement evaluation form

j. Using the "Average Flow" value, calculate the "Average Loading Rate" value as follows:

$$\text{Average Loading Rate,} \atop \text{GPM/sq ft} = \frac{\text{Average Flow, GPM}}{\text{Filter Surface Area, sq ft}}$$

k. Repeat steps "b" through "j" for locations "B" and "C" as shown on Figure 4.53.

2. Clean the head loss measuring tube and the wetted areas around the top of the filter, as necessary, to remove algal growth.

3. In some applications, sludge can accumulate in the plenum area of the filter. If necessary, remove the filter from service, drain the filter, and remove the accessway cover. Hose out the plenum area to remove accumulations. Because the filter plenum drain outlet is located slightly above the absolute bottom of the filter, an industrial vacuum cleaner designed for both wet and dry work may be used to remove all accumulations and water.

4. The airlift should be removed from the filter once a year for inspection and cleaning. The air chamber at the bottom of the airlift should be disassembled and the screen removed, inspected, cleaned, or replaced. This action is of particular importance if ferric chloride (which is very corrosive to most metals), lime, or sulfuric acid (which causes calcium scaling) are used in your process.

4.44 Safety

Safety precautions described in Chapters 5, 14, and 15 *(OPERATION OF WASTEWATER TREATMENT PLANTS)* for sedimentation tanks and pumps should be observed when operating and maintaining this equipment in the continuous backwash, upflow, deep-bed silica sand media filter system.

In addition, the following safety precautions should be observed:

1. Great care must be taken to avoid the contact of ferric chloride with any part of the body, *especially* with eyes. Ferric chloride must be handled with the same care as acid solutions, since ferric chloride causes burns similar to those caused by acids. If ferric chloride comes in contact with any part of the body, flush the body area with large amounts of cool water for at least 15 minutes. Seek medical attention immediately.

   a. Before handling ferric chloride solutions or any system components, operators must wear all of the following personal protective equipment. *REMEMBER, no condition or situation involving ferric chloride is more important than your personal safety.*

      • Safety goggles.

      • Face shield with hard hat.

      • Rubber apron properly secured to the body.

      • Rubber gloves of the full gauntlet (flared cuff) type.

      • Rubber boots. Alternatively, regular work shoes with rubber soles and heels and waterproofed leather uppers may be worn (ferric chloride deteriorates leather rapidly).

2. Wear safety goggles and gloves when working with alum or polymers. Wear approved respiratory protection if inhalation of dry alum or polymer is possible. Flush away any alum or polymer that comes in contact with your skin using cool water for a few minutes.

3. Be very careful when walking in an area where polymer is mixed or used. When polymer is wet, it is very slippery. Clean up polymer spills using a highly concentrated salt water solution made with rock salt or use inert absorbent material such as earth or sand.

4. The plenum area of a filter is a *CONFINED SPACE*.[50] Confined space procedures must be followed prior to entry, while working within, and when exiting the plenum area. A confined space permit may be required and anyone working in or around confined spaces must be trained to recognize safety hazards and use safe working procedures. For a description of the dangerous conditions associated with confined spaces as well as recommended procedures for confined space work see Chapter 14, "Plant Safety," in *OPERATION OF WASTEWATER TREATMENT PLANTS,* Volume II.

5. Before entering the plenum area of a filter, check the inner cone for signs of damage. In no event should the plenum area be entered if damage to the inner cone is suspected without first removing all sand from the filter. The weight of sand in a filter is 20 tons or greater. If sand backflow has occurred as the result of improper filter operation, sand accumulation piles will usually be noted just below each vertical riser tube in the plenum area. Sand in other areas of the plenum may indicate a damaged inner cone.

4.45 Review of Plans and Specifications

As an operator, you can be very helpful to design engineers by suggesting some design features that would make your job easier. This section attempts to point out some of the items that you should look for when reviewing plans and specifications for expansion of existing facilities or construction of new continuous backwash, upflow, deep-bed silica sand media filters.

1. The effect of recycling filter reject (backwash) water through the treatment plant on the filtration rate and filter operation must be considered in predicting peak loads on the filters.

2. Estimates of solids loading, chemical conditioning, head loss development, and filtrate quality should be based on bench-scale tests before the full-scale facility is designed. Bench-scale testing may be performed by the filtration equipment manufacturer.

---

[50] *Confined Space.*   Confined space means a space that: (1) Is large enough and so configured that an employee can bodily enter and perform assigned work; and (2) Has limited or restricted means for entry or exit (for example, manholes, tanks, vessels, silos, storage bins, hoppers, vaults, and pits are spaces that may have limited means of entry); and (3) Is not designed for continuous employee occupancy. Also see DANGEROUS AIR CONTAMINATION and OXYGEN DEFICIENCY.

3. Ladders and walkways should be provided to allow easy access to filters, pipes, and valves.

4. Adequate quantities of spare parts for the filtration system as a whole should be provided for in the construction contract.

5. A source of clean, supplemental water should be piped into the influent line (immediately upstream of the filter) for bottom feed cylindrical filters. For bottom feed concrete basin filters, supplemental water piping should be routed into the top of the filter. In each case, solenoid valve controls on the supplemental water piping may be interlocked with filter influent feed controls (if provided) so that supplemental water will flow to the filter to provide reject water whenever filter influent flow is stopped.

6. Sand backflow from the filter will occur if water within the filter flows backward. To prevent this from happening, check valves must be installed on the influent feed or bypass piping to each filter, or the feed and bypass piping should be

installed at an elevation above the level of the launder for filtered water collection.

7. Influent flow should be delivered to the filter at a pressure equal to 3 to 4 feet (0.9 to 1.2 m) of head.

8. Piping, pumps, valves, and other ferric chloride handling equipment should be made with, lined with, or coated with Kynar® vinylidene plastic, polyvinyl chloride, rubber, glass, Bakelite, Haveg, ceramic materials, or various other plastics that have given good service with all concentrations of ferric chloride at normal temperatures. The only metals suitable for use with ferric chloride are titanium and tantalum. Additional materials will become available over time. Consult your chemical supplier for up-to-date recommendations.

QUESTIONS

Write your answers in a notebook and then compare your answers with those on page 463.

4.4G What happens if the airlift is allowed to operate without water flow to the filter?

4.4H What should be done if a tool is dropped into the filter?

4.4I Why must ferric chloride be handled with care?

4.4J What safety precautions should be taken before entering the plenum area of a filter?

END OF LESSON 5 OF 6 LESSONS

on

SOLIDS REMOVAL FROM SECONDARY EFFLUENTS

Please answer the discussion and review questions next.

DISCUSSION AND REVIEW QUESTIONS

Chapter 4.   SOLIDS REMOVAL FROM SECONDARY EFFLUENTS

(Lesson 5 of 6 Lessons)

Write the answers to these questions in your notebook. The question numbering continues from Lesson 4.

30. What are the advantages of continuous backwash, upflow, deep-bed silica sand media filters over other types of granular media filters?

31. How is the silica sand media cleaned?

32. Why should a filter sand movement evaluation test be performed monthly?

33. What conditions might be the cause of high or increasing head loss readings?

CHAPTER 4.   SOLIDS REMOVAL FROM SECONDARY EFFLUENTS

(Lesson 6 of 6 Lessons)

4.5   CROSS FLOW MEMBRANE FILTRATION
by Robert G. Blanck and Francis J. Brady

"Cross flow" or "tangential flow" membrane filtration is an effective unit operation for treating industrial, municipal, and food processing wastes. Like all filtration processes, cross flow filtration separates the components of a wastestream. Unlike the upflow and downflow filters described earlier in this chapter, wastewater flows across the surface of a membrane rather than through a granular media. The membrane permits water (permeate) to pass through the membrane to be discharged or recycled, but prevents the passage of solids particles (reject or retentate). The concentration of solids in the retentate, therefore, increases as the wastewater progresses through the membrane filtration system. A major advantage of membrane filters over granular media filters is their relatively long cycle time of 5 to 7 days compared to 8 to 24 hours for granular media filters. Membranes are also easy to clean in a very short period of time (2 to 4 hours) and they are reusable.

Membrane filtration processes are classified on the basis of the size of particle they separate from the wastestream. These processes are: microfiltration (MF); ultrafiltration (UF); nanofiltration (NF); and reverse osmosis (RO). All are rapidly gaining acceptance as alternatives to dissolved air flotation (DAF), biological treatment, chemical treatment, settling ponds, and other conventional techniques. Common commercial membrane applications areas include:

Food processing wastes
Metal working wastes
Primary metal wastes
Parts washers
Pulp and paper wastes
Landfill leachates
Plating wastes
Truck wash water waste
Die casting wastes
Corrugated box plant wastes
Flexographic printing wastes
Laundry wastes
Textile plant wastes
Chemical process plant wastes

The separation technique involves a thin, semipermeable membrane that acts as a selective barrier that separates particles on the basis of molecular size. The membrane blocks passage of large particles and molecules, while letting water and smaller constituents pass through the membrane. Membranes made of a variety of materials are available with various pore sizes. Membranes are selected for particular applications on the basis of their differential separation properties. Figure 4.55 shows the particle size and molecular weight retention capacities of one manufacturer's line of membranes.

The driving force of the filtration process is a pressure differential between the wastewater side of the membrane and the effluent side of the membrane. The process differs from conventional filtration in that the waste is pumped at high flow velocities along the surface of the membrane preventing a cake buildup and minimizing a loss of filter capacity.

4.50   Types of Membrane Filtration Processes

4.500   Microfiltration

Microfiltration (MF) membranes have pores ranging from 0.1 to 2.0 microns. MF processes are less common with waste treatment processes because permeate from a microfilter is generally unacceptable for discharge. In some cases, this membrane process may be used in conjunction with settling agents, polymers, activated carbon, and other chemicals that assist in the retention of waste constituents. When using this membrane type, care must be taken to prevent membrane pore blockage by wastestream components by selecting the proper membrane type for the specific plant wastes.

4.501   Ultrafiltration

The process of ultrafiltration (UF) is the most common membrane-based wastewater treatment process; it uses a membrane with pore sizes ranging from 0.005 to 0.1 micron. Particles larger than the pores in the membrane, such as emulsified oils, metal hydroxides, proteins, starches, and suspended solids, are retained on the feed side of the membrane. Molecules smaller than the pores in the membrane, such as water, alcohols, salts, and sugars pass through the membrane. This filtrate (treated water passing through filter) is often referred to as permeate.

Ultrafiltration membranes are sometimes rated on the basis of molecular weight cut-off (MWCO) and range from 1,000 to 500,000 MWCO. For example, a membrane rated at 100,000 MWCO will retain most molecules 100,000 *DALTONS*[51] or

---

[51] *Dalton.*   A unit of mass designated as one-sixteenth the mass of oxygen-16, the lightest and most abundant isotope of oxygen. The dalton is equivalent to one mass unit.

PARTIAL LIST OF STANDARD MEMBRANES

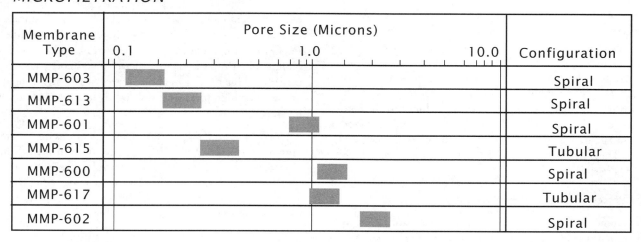

REVERSE OSMOSIS

Membrane Type	Molecular Weight Cut-Off (Daltons)				Configuration
	10	100	1,000	10,000	
KMS-CA		■			Tubular

ULTRAFILTRATION

Membrane Type	Molecular Weight Cut-Off (Daltons)				Configuration
	1,000	10,000	100,000	1,000,000	
HFK-328	■				Spiral
HFK-434		■			Spiral
HFK-131		■			Spiral
HFM-100			■		Spiral, Tubular
HFM-116			■		Spiral
HFM-251			■		Spiral, Tubular
HFP-276				■	Tubular
HFM-183			■		Spiral, Tubular

MICROFILTRATION

Membrane Type	Pore Size (Microns)			Configuration
	0.1	1.0	10.0	
MMP-603	■			Spiral
MMP-613	■			Spiral
MMP-601		■		Spiral
MMP-615	■			Tubular
MMP-600		■		Spiral
MMP-617		■		Tubular
MMP-602		■		Spiral

Fig. 4.55   Cross flow membrane filtration
(Source: Koch Membrane Systems, Inc.)

larger. The use of MWCO is only an approximate indication of membrane retention capabilities and should be used with guidance of the membrane manufacturer.

4.502 Nanofiltration

Nanofiltration (NF) uses a membrane pore size between UF and RO. These membranes are effective in removing salts from a wastestream by allowing them to pass into the permeate while concentrating other components such as sugars, nitrogen components, and other waste constituents causing high BOD/COD in wastestreams.

4.503 Reverse Osmosis

Reverse osmosis (RO) is the tightest membrane process in that it allows only water to pass through the membrane, retaining salts and higher molecular weight components. RO membranes are used for tertiary treatment producing water with low BOD/COD and of near-potable water quality. RO permeate may be recycled throughout the plant and reused for various plant processes. RO is normally used as a post-treatment process following coarser filtration processes such as ultrafiltration.

4.51 Membrane Materials

Membranes are made from durable polymers. They are reusable after regular cleaning cycles with cleaning chemicals designed to remove foulants from the membrane surface without damaging the membrane. Typical polymers are polysulfone, polyvinylidene fluoride, polyacrylonitrile, and polyamides. The membrane polymer is chosen for its compatibility with the wastestream, ability to retain particles, and ability to show reproducible performance over a long life. Other membrane types are manufactured from ceramics, sintered stainless steel, and carbon. They are used for specialty applications such as high-temperature streams.

4.52 Process Time and Membrane Life

Membranes are designed to process waste and then be cleaned on a periodic basis. Typical membrane processes operate over 5- to 7-day cycles, depending on the membrane type and process design of the system, followed by short cleaning cycles. Membrane filters have a usable life of 1 to 5 years when in daily operation. Most membrane systems are designed to operate 24 hours per day, 7 days per week. Shorter duty cycles may be designed, depending on plant requirements.

4.53 Membrane Configurations

Membranes are housed in various types of modular units. The basic types of membrane configurations are:

Tubular
Hollow Fiber
Spiral
Plate and Frame
Ceramic Tube or Monolith

4.530 Tubular Membranes

Tubular membranes are available in either single- or multi-tube configurations of various diameters. Common tube diameters are 1 inch and ½ inch. Figure 4.56 illustrates a 1-inch tube where the membrane is cast on a rigid, porous tube and mounted in a PVC, CPVC, or stainless-steel support shell. Figure 4.57 shows a ½-inch multitube configuration mounted in a similar PVC or CPVC housing. Tubular membranes are a nominal 10 feet long and are connected together by U-bends to form a number of tubes in series. The number of tubes in series ranges from 4 to 16, depending on the system design. A number of "series passes" may be present on a commercial system.

Tubular membranes offer advantages in their ability to concentrate to high solids levels without plugging and their ability to operate at pressures as high as 90 psi. Some tubular membranes, such as one-inch diameter tubes, have the ability to be mechanically cleaned with spongeballs. The spongeballs are recirculated through the membrane system along with cleaning solutions to mechanically scour the membrane surface.

4.531 Hollow Fiber Membranes

Hollow fiber membranes, shown in Figure 4.58, are thin tubules of membrane polymer, with the membrane surface normally on the inside of the hollow fiber. The hollow fibers are *POTTED*[52] in a plastic support structure called a cartridge. Hollow fibers offer a high density of membrane packed into a cartridge. This type of membrane produces a high rate of permeate flow (high process fluxes), but is limited to about 30 psi pressure.

Hollow fiber cartridges are available in various fiber diameters ranging from 20 to 106 *MILS*.[53] For waste treatment, common fiber diameters are 43, 75, and 106 mils. The larger diameter fibers are used for concentrating waste to higher solids levels and are more resistant to plugging. The smaller fibers are used for lower concentrations and reduced suspended solids levels.

The hollow fiber cartridge housing is usually constructed of polysulfone, PVC, or other plastic material and may be found in diameters of 3 inches or 5 inches. The length of a hollow fiber typically is approximately 36 inches.

---

[52] *Potting Compounds.*   Sealing and holding compounds (such as epoxy) used in electrode probes.
[53] *Mil.*   A unit of length equal to 0.001 of an inch. The diameter of wires and tubing is measured in mils, as is the thickness of plastic sheeting.

The Koch one-inch-diameter "FEG" tube is one of the most rugged membrane configurations. This, coupled with the ability to mechanically clean the tubes with spongeballs, makes the FEG tube the membrane configuration that is the most forgiving to system upsets. Wastestreams that would cause severe fouling or plugging of other membrane configurations can be processed with the FEG tube.

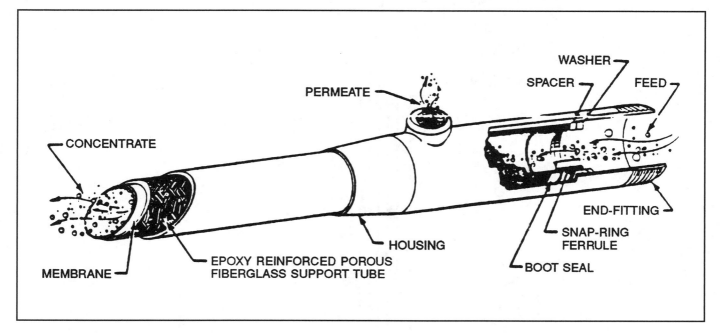

Fig. 4.56   FEG one-inch tube
(Source: Koch Membrane Systems, Inc.)

The Koch ULTRA-COR® VII module packs more membrane area per module than the FEG tube. However, the ULTRA-COR® VII modules are more limited in achieving maximum yields, are less durable and forgiving to upsets, and cannot be mechanically cleaned with spongeballs.

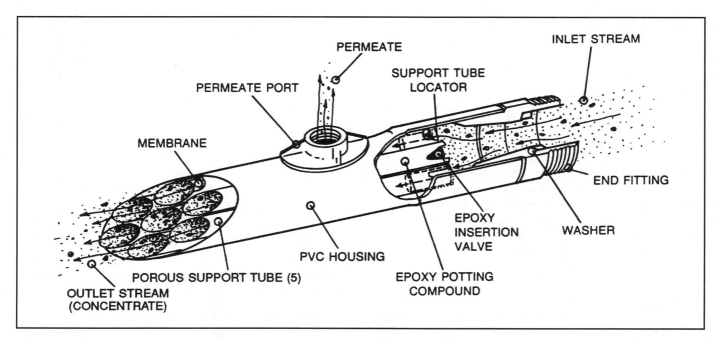

Fig. 4.57   ULTRA-COR® VII ½-inch tubules module
(Source: Koch Membrane Systems, Inc.)

The self-supported hollow fiber membranes are cast in a spinning operation. Since the membrane has no support backing, the cartridges can be backflushed to provide for longer process runs in some applications. With proper operation and waste pretreatment, this configuration can be the most economical choice for many applications.

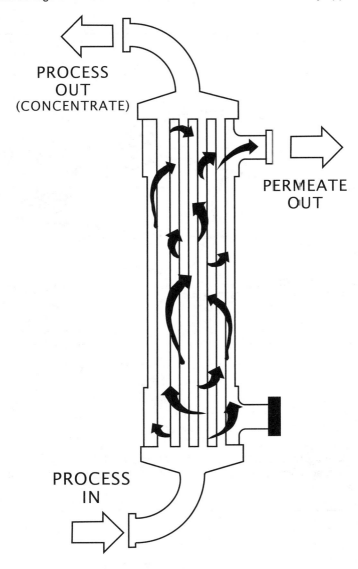

Fig. 4.58   Hollow fiber cartridge
(Source: Koch Membrane Systems, Inc.)

A major advantage of a hollow fiber cartridge is its ability to be backwashed or backflushed without damage to the membrane. In this operation, the permeate is pressurized and made to flow in the opposite direction through the membrane. This technique pushes fouling material off the membrane surface, resulting in increased flow of permeate.

4.532   Spiral Membranes

Membranes can be cast on materials such as nonwoven polyester and manufactured into spiral modules. Spirals also offer a high packing of membrane area into a compact cartridge. They may operate at pressures up to 150 psi for UF and as high as 1,200 psi for RO. A spiral module, shown in Figure 4.59, is composed of the flat sheet membrane, a permeate carrier, a feed spacer and glue to isolate the feed from the permeate side.

Spiral modules are available with various feed spacer thicknesses that serve to maintain a distance between the membrane sheets. Selection of the correct spacer for a particular feed stream is critical for proper filter operation. In general, high solids wastewaters are more easily processed with thicker feed spacers. Spacer thicknesses range from 28 to 80 mils.

Spiral modules are available in nominal diameters of 2, 4, 6, and 8 inches. Large-diameter spiral modules are used for high-volume wastestreams because they provide large membrane surface areas for a given system size.

Spiral membrane modules are capable of packing a large surface area of membrane into a compact design. For proper operation, these modules require the highest degree of pretreatment of all membrane configurations.

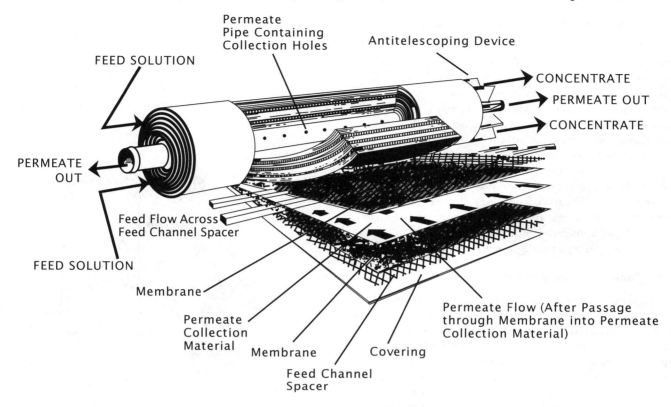

Fig. 4.59   Spiral wound module
(Source: Koch Membrane Systems, Inc.)

The spiral is mounted into either a plastic or stainless-steel shell. This may be replaceable or permanently mounted in the shell. Figure 4.60 illustrates a permanently mounted SPIRAPAK™ module used for wastewater treatment. Most wastewater RO applications are spiral modules.

While offering a large membrane surface area, spiral modules are limited in their ability to process wastes with large amounts of suspended solids and particulates.

4.533   Plate and Frame Membranes

Plate and frame configurations, less common in wastewater treatment, are a series of flat sheet membranes mounted on a frame. These membrane systems resemble a pressure leaf filter, with the membranes fitted in a series of plates. Permeate is diverted out of the plates by a series of small tubes.

QUESTIONS

Write your answers in a notebook and then compare your answers with those on page 463.

4.5A   How do cross flow filtration processes differ from conventional filtration?

4.5B   How are membrane processes classified?

4.5C   List the basic membrane configurations.

Fig. 4.60   SPIRAPAK™ module
(Source: Koch Membrane Systems, Inc.)

4.6   BASIC ELEMENTS OF A MEMBRANE FILTRATION PROCESS

4.60   Concentrating Components in the Wastewater

Membrane systems concentrate components in the wastestream such as oil, grease, fats, BOD, COD, metal hydroxides, suspended solids, particulates, starches, *SURFACTANTS,*[54] bacteria, and other microorganisms. The concentrated feed (known as the retentate or concentrate) may be reduced in volume 20- to 100-fold compared to the original feed volume. The level of concentration depends on the original concentration of components in the waste feed. Retentate from an ultrafilter may be as high as 60 percent total solids, depending on the nature of the incoming feed. The level of concentration is also determined by the choice of membrane configuration. Tubular membranes may concentrate to 50 to 60 percent total solids, while spiral and hollow fiber membrane configurations may be limited to 10 to 30 percent, again depending on the type of wastewater.

4.61   Permeate Discharge to Drain

Permeate is made up of the solvent and low-molecular-weight constituents that pass through the membrane. In some plants, the permeate from an ultrafiltration system meets local and state discharge requirements and can be sent directly to the drain. More stringent discharge requirements may require post-treatment of the UF permeate by processes such as pH adjustment, carbon filtration, ozone treatment, and reverse osmosis. The permeate from an ultrafilter is typically clear (nonturbid) and without suspended solids. In some cases, dye or color bodies will tint the permeate. The amount of dissolved substances in the permeate will depend on the retention properties of the membrane.

4.62   Membrane Flux

The rate of mass flow of permeate passing through the membrane is called flux. This measurement is the basis for sizing membrane equipment and is an essential value to evaluate membrane performance. Flux (J) is expressed on the basis of membrane area:

$$\text{Flux, J} = \frac{\text{Permeate Flow Rate}}{\text{Membrane Area}}$$

Units for flux are "gallons/sq ft-day" (GFD) or "liters/sq m-hour" (LMH). Flux values for typical ultrafiltration waste treatment processes are approximately 25 to 50 GFD (43 to 85 LMH). Reverse osmosis and nanofiltration fluxes typically range from 10 to 30 GFD (17 to 51 LMH). Conversion of GFD to LMH is $1.7 \times \text{GFD} = \text{LMH}$.

The amount of flux across a membrane is dependent on a number of factors:

- Transmembrane pressure (driving force)

- Flow rate across the membrane surface (turbulence on the membrane)

- Concentration of waste material

- Temperature of concentrate

- Viscosity of concentrate

- Cleanliness of the membrane surface

To calculate membrane flux of a commercial system, one needs to know the permeate flow rate measured with a flowmeter, rotameter, or bucket and stopwatch, and the membrane area of the system.

EXAMPLE 7

A membrane filtration system is operating with 1,200 square feet of membrane area and the permeate flow is 25 gallons per minute. What is the flux (J) in GFD (gallons/sq ft-day)?

Known	Unknown
Membrane Area, sq ft = 1,200 sq ft	J (Flux), GFD
Flow, gal/min = 25 gal/min	

Calculate the flux in GFD.

$$\text{J, GFD} = \frac{(\text{Flow, gal/min})(60 \text{ min/hr})(24 \text{ hr/day})}{\text{Membrane Area, sq ft}}$$

$$= \frac{25 \text{ gal/min} \times 60 \text{ min/hr} \times 24 \text{ hr/day}}{1,200 \text{ sq ft}}$$

$$= 30 \text{ GFD}$$

4.63   Retention

A membrane filter achieves a separation of various components in the wastewater stream. The ability of a membrane to retain particles is known as retention or rejection. A membrane's retention coefficient (R) is a numerical expression of the concentration of a component in the retentate (concentrate) relative to its concentration in the permeate. The retention coefficient is:

$$\text{Retention, R} = 1 - \left(\frac{\text{Permeate}}{\text{Concentrate}} \right)$$

---

[54] *Surfactant* (sir-FAC-tent).   Abbreviation for surface-active agent. The active agent in detergents that possesses a high cleaning ability.

Retention is often expressed as a percent by multiplying by 100%.

Retention, % = Retention Coefficient × 100%

Retention ranges from 0 to 100 percent, depending on the molecular or particle size of the wastewater component being measured. Retentions may be obtained on commercial systems by sampling the concentrate and permeate, and submitting samples to a local lab for component analysis. For most waste-streams, analysis is suggested for freon-extractable oil and grease (*STANDARD METHODS FOR THE EXAMINATION OF WATER AND WASTEWATER, METHOD 5520B*, "Liquid-Liquid, Partition-Gravimetric Method," 21st Edition, 2005), true oil and grease (*STANDARD METHOD 5520F*, "Hydro-carbons"), total soluble solids, total suspended solids, heavy metals, BOD, and COD. Other components analyses for fats, proteins, and sugars may be required for food processing wastes.

EXAMPLE 8

Samples of retentate and permeate from an ultrafiltration system were sent to an analytical lab for oil and grease analysis. The retentate has a concentration of 5,000 mg/L oil and grease and the permeate has a concentration of 25 mg/L oil and grease. What is the percent retention of oil and grease?

Known	Unknown
Permeate Conc, mg/L = 25 mg/L	Retention, %
Retentate Conc, mg/L = 5,000 mg/L	

Calculate the percent retention.

$$R, \% = \left(1.00 - \frac{(25 \text{ mg}/L)}{(5,000 \text{ mg}/L)}\right)(100\%)$$

$$= (1.00 - 0.005)(100\%)$$

$$= 99.5\%$$

4.64   Calculating the Concentration of Waste Components

The wastestream is split into the retentate stream and the permeate stream. A mass balance around the membrane system gives us the volumes and concentrations of permeate and concentrate.

Feed Volume = Permeate Volume + Concentrate Volume

or

Vf = Vp + Vc

Membrane filtration data are often presented in terms of the volumetric concentration factor (CF). This value is the amount the waste material is concentrated and can also be expressed as "X." For example, a system concentrating waste 50 times has a CF = 50×. The concentration factor of a system is calculated by the following equation:

$$CF = \frac{\text{Feed Volume}}{\text{Concentrate Volume}}$$

or

$$CF = Vf/Vc$$

EXAMPLE 9

A waste treatment plant has an original waste volume of 20,000 gallons. After ultrafiltration, a concentrate volume of 400 gallons remains. What is the volumetric concentration factor?

$$CF = \frac{20,000 \text{ gallons}}{400 \text{ gallons}} = 50\times$$

Another way to express the volume balance of a membrane system is "percent volume reduction" or "percent water removed."

Percent volume reduction is determined by:

$$\text{Volume Reduction, \%} = \left(\frac{\text{Feed Volume} - \text{Concentrate Volume}}{\text{Feed Volume}}\right)(100\%)$$

or

$$= \left(\frac{Vf - Vc}{Vf}\right)(100\%)$$

or

$$= \frac{Vp}{Vf}(100\%)$$

Where Vp = Permeate Volume

EXAMPLE 10

From above, 20,000 gallons of waste are reduced to 400 gallons of concentrate. What is the volumetric reduction of the wastewater feed?

Known	Unknown
Feed Vol, gal = 20,000 gal	Volume Reduction, %
Concentrate Vol, gal = 400 gal	

Calculate the percent volume reduction.

$$\text{Volume Reduction, \%} = \left(\frac{Vf - Vc}{Vf}\right)(100\%)$$

$$= \left(\frac{20,000 \text{ gal} - 400 \text{ gal}}{20,000 \text{ gal}}\right)(100\%)$$

$$= 98\%$$

The concentration of components in the concentrated waste-water can be determined by:

Concentration in Retentate = Original Feed Concentration × (CF)^R

where R = Retention Coefficient

EXAMPLE 11

A concentrate and permeate sample were collected from an ultrafilter and analyzed for BOD. The retention coefficient was 0.70 (R, % = 70.0%). The original feed BOD was 5,000 mg/*L* and the concentration factor of the UF system was 50×. What is the concentration of BOD at the end of the batch?

Known	Unknown
BOD Feed, mg/*L* = 5,000 mg/*L*	Ending BOD Conc, mg/*L*
CF (Concentration = 50× Factor)	
BOD Retention, R = 0.70 or % = 70%	

Calculate the ending BOD concentration in mg/*L*.

$$\text{BOD Conc, mg/}L = (\text{BOD Feed, mg/}L)(\text{CF})^R$$
$$= (5{,}000 \text{ mg/}L)(50)^{0.7}$$
$$= 77{,}312 \text{ mg/}L$$

For UF and tighter membranes, components like true oil and grease, suspended solids, proteins, and fats have a retention close to 1.0. There is little or no passage of these components into the permeate. This means that the concentration of components becomes:

$$\text{Retentate Concentration} = \text{Original Feed Concentration} \times \text{CF}$$

4.65 Transmembrane Pressure

The permeate rate of a membrane system is dependent on a pressure differential between the feed side and the permeate side of the membrane. This pressure, known as transmembrane pressure, is the driving force of the separation process. Transmembrane pressure typically ranges from 20 to 80 psi (140 to 550 kPa) for ultrafiltration, 300 to 500 psi (2,070 to 3,450 kPa) for nanofiltration, and 400 to 1,000 psi (2,760 to 6,890 kPa) for reverse osmosis. Process fluxes are a function of pressure; however, as concentration of waste increases in the retentate, fluxes will become independent of pressure. At this point, increasing transmembrane pressures show no increase in process flux. This is due to formation of a resistant gel layer on the membrane surface. Once the gel layer forms, the mass flow of particles toward the membrane surface is equal to the diffusion of particles away from the membrane. Fluxes are most pressure-dependent on water feed rate or when low concentrations of waste are present in the feed.

The average transmembrane pressure (Ptm) is a measure of the driving force of the membrane process exactly in the middle of the system. One can calculate the average transmembrane pressure by knowing the inlet pressure (Pin) and outlet pressures (Pout) of the waste feed recirculating in the membrane system and the permeate pressure.

The inlet pressure is measured on the discharge side of the recirculation pump, just prior to the entrance to the modules or tubes. The outlet pressure is measured at the point where the waste leaves the bank of modules. The permeate pressure may be measured on the permeate manifold, but in most cases is assumed to be zero.

To calculate the average transmembrane pressure of a system:

$$\text{Ptm} = \frac{\text{Pin} + \text{Pout}}{2} - P_{\text{permeate}}$$

EXAMPLE 12

The inlet pressure of a commercial system is 64 psi and the outlet pressure is 20 psi. The permeate pressure is zero. What is the average transmembrane pressure?

Known	Unknown
Inlet Pressure, psi = 64 psi	Average Transmembrane Pressure (Ptm), psi
Outlet Pressure, psi = 20 psi	
Permeate Pressure, psi = 0 psi	

Calculate the transmembrane pressure in psi.

$$\text{Ptm, psi} = \frac{\text{Pin} + \text{Pout}}{2} - P_{\text{permeate}}$$
$$= \frac{(64 \text{ psi} + 20 \text{ psi})}{2} - 0$$
$$= 42 \text{ psi}$$

4.66 Recirculation Flow

All cross flow membrane systems recirculate the waste feed across the membrane to prevent buildup of particles on the surface. This tangential flow or cross flow creates high turbulence and a sweeping effect on the membrane and enhances the movement of molecules away from the membrane surface. Increasing the recirculation rate usually increases the flux rate, especially when wastes are highly concentrated. Most systems are designed at flow rates that optimize flux and choice of hardware (pumps and pipework).

Tangential flow across the membrane surface is created by recirculation pumps on the system. The amount of recirculation flow is selected by the membrane manufacturer based on the unit's design.

Recirculation flow is normally constant throughout the process run and must not be confused with permeate flow. Usually, recirculation flow is not directly measured in commercial systems. An indirect measurement of recirculation flow is the drop in pressure between the inlet (Pin) and outlet (Pout) of the membrane system. Delta pressure (dP) is measured as:

Delta pressure dP = 64 psi – 20 psi = 44 psid or psi delta pressure

Delta pressure is an important guideline to monitor in a system because it indicates if the membrane modules have sufficient cross flow velocity to meet design fluxes or are becoming plugged with waste. Normally, delta pressure is relatively constant throughout the process run. The delta pressure might increase when the viscosity of the feed is increasing or when material is accumulating in the membrane modules. (One can imagine two pressure gauges on either side of a valve. When the valve closes, the dP across the valve increases.) Operators should use delta pressure as an indicator of major system problems. Major changes in dP signal the operator to investigate problems in prefiltration or overconcentration in the membrane system. When the delta pressure increases to the shutoff pressure of the pump, there is little or no flow across the surface of the membrane. This can result in significant fouling and eventually will permanently plug the modules.

4.67 Temperature

Increasing the temperature of the feed increases flux through the membrane. This is mainly due to a reduction in viscosity of the waste feed. Membranes have maximum temperature limits based on their materials of construction. The design temperature of a membrane system should be optimized for flux performance and membrane temperature compatibility. Always refer to the system operating manual to determine the temperature limits of the membrane.

Daily temperature variations in the waste feed will cause the membrane system to vary in productivity. For example, the flux at 50 degrees F (8 degrees C) will be approximately one-half of the flux at 100 degrees F (38 degrees C).

4.68 Concentration-Dependent Flux

As materials such as oils, grease, fats, particulate matter, and proteins are concentrated on the feed side of the membrane, the flux of the membrane will drop. This is caused by the formation of the gel layer or *CONCENTRATION POLARIZATION*[55] layer at the membrane surface. Initially, when wastewater concentrations are low, the flux is high. But, while concentrating over the course of a process run, the productivity of permeate will progressively decline as the concentration increases. Eventually, the increasing concentration will drive the membrane flux to zero, although a properly operated membrane system is stopped before the maximum concentration is reached. The typical maximum oil and grease concentration in the retentate on a tubular commercial system is about 50 percent.

4.69 Membrane Fouling

Under normal operation, the surface of the membrane becomes fouled with oil, grease, and other foulants. This fouling results in a decline in the flux of the membrane, however, it is different from the concentration-dependent flux decline described above. Fouling is a molecular attraction between the components in the feed stream and the membrane material.

Two kinds of membrane fouling occur: reversible and irreversible. Reversible fouling is expected. It occurs in all membrane systems and is cleanable with cleaning chemicals. A regular cleaning with caustic, acids, surfactants, chlorine, or hydrogen peroxide (as recommended by the membrane manufacturer) returns the membrane flux very nearly to original values. Irreversible fouling is rare and is a result of incompatible components in the feed stream that cannot be cleaned with normal cleaning chemicals. An example of a common irreversible foulant is silicone. The operator must be aware of potential irreversible foulants in the wastestream and isolate them from the waste at their point sources.

QUESTIONS

Write your answers in a notebook and then compare your answers with those on page 464.

4.6A The amount of flux across a membrane is dependent upon what factors?

4.6B How do membrane systems prevent the buildup of particles on the surface?

4.6C Under what conditions might the delta pressure increase?

4.6D What causes fouling of the membrane surface?

4.7 OPERATION OF A CROSS FLOW MEMBRANE SYSTEM

4.70 Filter Staging

Membrane modules or tubes are mounted on a frame and fitted with a pump, valves, and instrumentation. This functional unit is termed a membrane stage. A membrane system includes a membrane stage or stages, a feed pump to bring waste to the stage, a feed tank, a cleaning station including pump and tank, and control panel. Prefiltration (pretreatment) may be required on some systems.

The amount of membrane area on a stage is a function of the plant effluent requirements and may range from as little as 4 to 1,200 sq ft (0.4 to 110 sq m) or more. Pump and line sizes on the stage increase correspondingly when plant capacities increase. Membranes are placed on the stage in a series and parallel arrangement to maximize the efficiency of the design. For example, a tubular system with 248 tubes on the stage may have them arranged 8-in-series with 31 parallel passes. Figure 4.61 shows the series arrangement of tubes.

4.71 Operating Modes

Most waste treatment systems are designed to operate in a batch or modified batch process mode. A third, less frequently used process mode is a continuous, stages-in-series design. The choice of process mode depends on the volume of waste to be processed and specific plant requirements.

A batch process is the most efficient mode in terms of membrane area requirements and system design. It requires only a feed tank and the membrane stage. As seen in Figure 4.62, the waste feed from the batch tank is recirculated through the membrane

---

[55] *Concentration Polarization.* (1) A buildup of retained particles on the membrane surface due to dewatering of the feed closest to the membrane. The thickness of the concentration polarization layer is controlled by the flow velocity across the membrane. (2) Used in corrosion studies to indicate a depletion of ions near an electrode. (3) The basis for chemical analysis by a polarograph.

CROSS FLOW MEMBRANE FILTRATION

Industrial Wastewater Ultrafiltration 8-in-Series One-Inch Tubular Arrangement

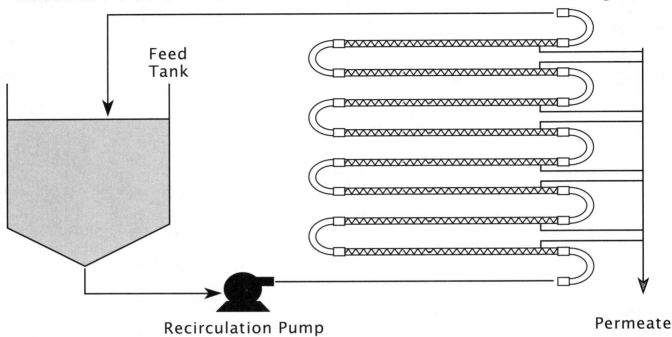

Fig. 4.61 Series tube arrangement
(Source: Koch Membrane Systems, Inc.)

CROSS FLOW MEMBRANE FILTRATION

Batch Process

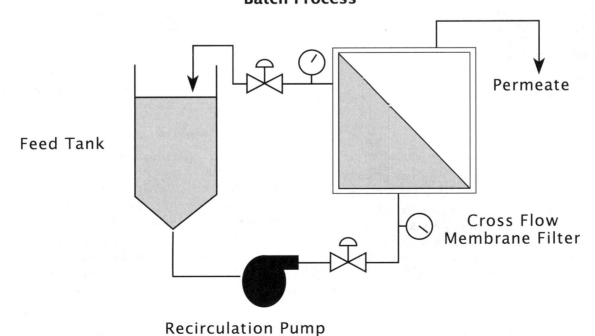

Fig. 4.62 Batch mode
(Source: Koch Membrane Systems, Inc.)

stage and back to the feed tank. Permeate is discharged from the membranes and is sent to drain or to further processing. The recirculation rate across the membrane surface is very large compared to the permeate flow off the system. For example, on a tubular design, the recirculation flow is approximately 100 times the permeate rate. The feed material as it exits the stage and returns back to the feed tank is slightly concentrated. The feed is continuously recirculated back through the membranes until the desired concentration of waste is achieved in the feed tank. Since waste treatment batches run for as long as 4 to 7 days, the feed tank size must be large enough to accommodate all the waste over this time period. A disadvantage of a batch design is the large tank requirements for large waste volumes.

In a modified batch mode (Figure 4.63), a smaller feed tank is used and the waste is continuously fed into the feed tank at a rate equal to the permeate production. Material is concentrated in the tank over a 4- to 7-day period, followed by turning off the feed into the tank and batching down the waste volume in the feed tank to its final desired concentration. The final batch-down normally takes 2 to 8 hours, depending on the size of the feed tank. This modified batch mode with end-of-run batch-down is the most common operating mode found in wastewater treatment applications.

Continuous processing uses multiple stages placed in series with each other, as illustrated in Figure 4.64. Feed from a small feed tank enters the first stage of the system where it is slightly concentrated. The waste then enters a second stage where further concentration takes place. After successive concentration in multiple stages (usually 2 to 6), the concentrated waste is bled off the system at a fraction of the feed rate. The system is continuous in that it produces concentrate at maximum concentration during the length of the process run. This differs from a batch system where concentrate reaches maximum concentrations at the end of the batch. Continuous stages-in-series processing is used for RO plants and larger UF plants.

4.72 Feed Pretreatment

Membrane filtration requires proper pretreatment of the feed to optimize the process flux and prevent fouling or plugging of the membrane. For ultrafiltration systems processing primarily industrial waste, the following pretreatment processes are common (see Figure 4.65):

1. Equalization of wastes from throughout the plant combined with surge capacity

2. Prescreening (prefiltration) to remove large suspended material

3. Free oil (*TRAMP OIL*[56]) and settleable solids removal

4. pH treatment (to alkaline conditions) to stabilize the oil/water emulsion

Equalization of wastes is necessary to level out differences in waste concentrations from different parts of the plant. The equalization tank will also allow for some settling of sludge that can be removed on a periodic basis from the bottom of the tank. A typical equalization tank detention time is one day's waste volume. An alternative to using a large equalization tank is the addition of chemicals to enhance the formation of easily settled floc.

The prefiltration requirements depend on the membrane configuration used and the particle size in the wastestream. Tubular systems require loose prefiltration (5/64th screen), whereas hollow fibers and spirals require prefilter pore sizes as low as 10 microns. Operators must always maintain the prefilter and never bypass it because irreversible membrane plugging or other membrane damage may occur.

Free oil removal is essential for proper membrane operation. Free oil will cause fouling of the membrane surface and reduce the capacity of the membrane system. Oil skimmers are normally placed on the equalization tank to remove free oil from the feed.

In an oily waste system, pH control at the ultrafilter is important for two reasons. Flux rates are improved at the ultrafilter when the wastewater is pH 9 to 10.5 and fouling of the membrane by oil is minimized by stabilizing the oil emulsion. Lower pH values make the oil emulsion unstable and break off oil in the feed tank. Free oil will coat the membrane and reduce process flux. Operators should look for free oil on the surface of the feed tank and, if present, improve the skimming of surface oil on the equalization tank.

4.73 System Operation

Membrane system operation may be manual or controlled by a *PLC*.[57] The steps listed below are a typical operating sequence for processing wastewater in a clean membrane system operating in a batch mode.

---

[56] *Tramp Oil.* Oil that comes to the surface of a tank due to natural flotation. Also called free oil.
[57] *Programmable Logic Controller (PLC).* A microcomputer-based control device containing programmable software; used to control process variables.

CROSS FLOW MEMBRANE FILTRATION
Modified Batch Process

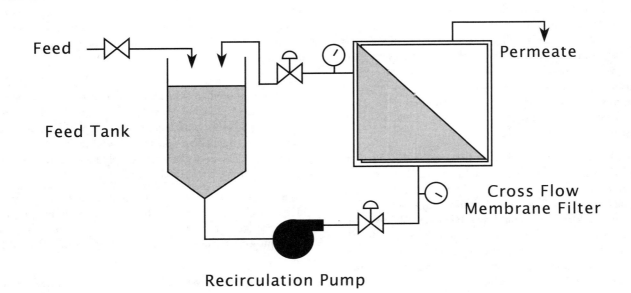

Feed

Feed Tank

Permeate

Cross Flow
Membrane Filter

Recirculation Pump

Fig. 4.63   Modified batch mode
(Source: Koch Membrane Systems, Inc.)

CROSS FLOW MEMBRANE FILTRATION
Stages-in-Series Process

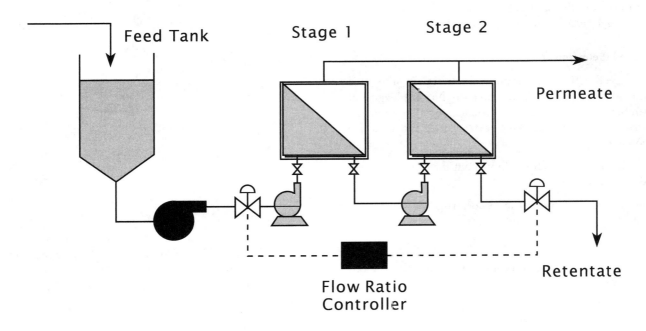

Feed Tank

Stage 1

Stage 2

Permeate

Retentate

Flow Ratio
Controller

Fig. 4.64   Continuous stages-in-series
(Source: Koch Membrane Systems, Inc.)

CROSS FLOW MEMBRANE FILTRATION
Wastewater Process Schematic

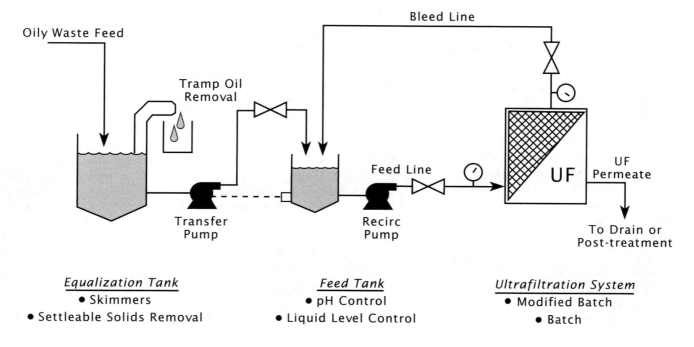

Fig. 4.65 Typical ultrafiltration wastewater treatment flow schematic
(Source: Koch Membrane Systems, Inc.)

START-UP

1. Pretreated waste is introduced into the feed tank.

2. The return valve to the feed tank is opened fully.

3. The recirculation pump is turned on.

4. The discharge valve on the pump is opened slowly.

5. Inlet and outlet pressures are adjusted to the proper delta pressure.

6. System heat exchangers are turned on (if present).

7. For modified batch systems, the rate of feed into the feed tank is adjusted to equal the permeate flow rate.

8. The system is now at steady state.

SHUTDOWN

During batchdown, the following sequence will normally take place:

1. The wastewater to the feed tank is shut off and the UF continues to operate.

2. The feed tank is concentrated down to the final desired concentration factor and the system is stopped.

3. The concentrate in the system is displaced with water.

4. The waste concentrate is transferred to a sludge tank.

5. Cleaning of the system is initiated.

During normal, steady-state operation, conditions should be monitored on a regular basis over the course of a process run. Data collected every few hours is usually adequate to ensure proper system operation.

4.74 Cleaning Procedures

Membrane systems are cleaned on a regular basis with chemicals on cycles prescribed by the membrane supplier, typically every 4 to 7 days on wastewater treatment systems. The cleaning cycle usually takes approximately 2 to 4 hours, depending on the required wash cycles. A typical cleaning will include the following cycles:

WASH CYCLES

1. Displacement of waste from the system with water.

2. Washing the membranes with a caustic and surfactant to remove oils and grease. Typical wash cycles are performed at 0.1 percent surfactant, pH 10.5, for 1 hour at 100 to 120 degrees F (40 to 50°C).

3. Flushing the surfactant from the membrane with warm water.

4. Washing the membrane with an acid cleaner to remove salt buildup. This cycle is typically 1 hour at 100 to 120 degrees F (40 to 50°C).

5. Flushing the acid from the membrane with warm water.

6. Recording clean membrane water flux as a check of cleaning effectiveness.

All wash cycles are performed with transmembrane pressures and recirculation flow across the membrane similar to process conditions.

For some tubular membrane systems, the chemical cleaning may be supplemented with spongeballs circulated through the tubes. Spongeballs are sometimes required when membrane foulants are difficult to remove, such as with metal hydroxides or waste latex. Spongeballs may also be used to speed up the cleaning procedure in some plants.

Some membrane configurations, such as hollow fibers, may be backflushed to assist in removal of particles from the membrane surface. Backflushing is the reverse flow of water or permeate from the permeate side of the membrane to the feed side. This technique is most common during cleaning; however, it may also be used during the treatment process on high-fouling wastewater streams. Most tubular and spiral membranes cannot be backflushed due to potential *DELAMINATION*[58] of the membrane.

Exposure to aggressive cleaning chemicals decreases the life of the membrane and is the principal reason membranes must be replaced periodically. Overuse of chemicals accelerates the failure of membranes and operators must take care not to overclean membranes

4.75 Water Flux Measurements

After cleaning, the rate of water passage through a membrane in flux, or permeate flow, Qp, is measured to assess the effectiveness of the cleaning cycle. To measure water flux or permeate flow, all dirty wash water is drained from the system. Clean water is placed in the cleaning tank and then pumped to the membrane system. Permeate flow is measured under standard pressure conditions. Typical transmembrane conditions for UF are 20 or 50 psi (depending on the membrane type) at 77 degrees F (25°C). Permeate flow, Qp, taken at other temperature and pressure conditions must be corrected to these standard conditions. Correction is performed as follows:

$$Q_{P\,CORRECTED} = \frac{Q_{P\,ACTUAL} \times 50 \text{ psi}}{P_{tm_{avg}}} \times F$$

where F = Temperature correction from standard 77 degrees F (see Table 4.6).

TABLE 4.6   WATER FLUX TEMPERATURE CORRECTION FACTOR (F)[a]

Flux (25°C) = Flux (T°C) × F
or Flux (77°F) = Flux (T°F) × F

Temp (°F)	Temp (°C)	F	Temp (°F)	Temp (°C)	F	Temp (°F)	Temp (°C)	F
125.6	52	0.595	96.8	36	0.793	68.0	20	1.125
123.8	51	0.605	95.0	35	0.808	66.2	19	1.152
122.0	50	0.615	93.5	34	0.825	64.4	18	1.181
120.2	49	0.625	91.4	33	0.842	62.6	17	1.212
118.4	48	0.636	89.6	32	0.859	60.8	16	1.243
116.6	47	0.647	87.8	31	0.877	59.0	15	1.276
114.8	46	0.658	86.0	30	0.896	57.2	14	1.320
113.0	45	0.670	84.2	29	0.915	55.4	13	1.346
111.2	44	0.682	82.4	28	0.935	53.6	12	1.383
109.4	43	0.694	80.6	27	0.956	51.8	11	1.422
107.6	42	0.707	78.8	26	0.978	50.0	10	1.463
105.8	41	0.720	77.0	25	1.000	48.2	9	1.506
104.0	40	0.734	75.2	24	1.023	46.4	8	1.551
102.2	39	0.748	73.4	23	1.047	44.6	7	1.598
100.4	38	0.762	71.6	22	1.072	42.8	6	1.648
98.6	37	0.777	69.8	21	1.098	41.0	5	1.699

[a] Based on water fluidity relative to 25°C (77°F) fluidity value.

F = $(\mu_{T°C/25°C})$, or

F = $(\mu_{T°F/77°F})$

---

[58] *Delamination* (DEE-lam-uh-NAY-shun).   Separation of a membrane or other material from the backing material on which it is cast.

EXAMPLE 13

The clean water productivity of a system is 50 GPM at 100 degrees F, with an average transmembrane pressure of 40 psi. What is the corrected permeate flow, Qp, at 50 psi, 77 degrees F (standard conditions)?

Known	Unknown
Actual Qp, GPM = 50 GPM	Corrected Qp, GPM
Avg Pressure (Ptm$_{avg}$), psi = 40 psi	
Temp, degrees F = 100°F	

Calculate the corrected permeate flow in gallons per minute.

$$Q_{P\text{CORRECTED}}, \text{GPM} = \frac{Q_{P\text{ACTUAL}} \times 50 \text{ psi}}{Ptm_{avg}} \times F$$

$$= \frac{(50 \text{ GPM})(50 \text{ psi})(0.762)}{40 \text{ psi}}$$

$$= 47.6 \text{ GPM}$$

To calculate clean membrane water productivity in terms of flux, J, simply calculate the permeate flow on a daily basis and divide by the system's membrane area.

EXAMPLE 14

If the membrane system in Example 13 has 545 sq ft of membrane area, what is the system's clean water productivity in GFD (gallons/sq ft-day)?

Known	Unknown
Q$_{P\text{CORRECTED}}$, GPM = 47.6 GPM	Flux, J, GFD
Membrane Area, sq ft = 545 sq ft	

Calculate the corrected flux, J, in GFD.

$$J_{\text{CORRECTED}}, \text{GFD} = \frac{Q_{P\text{DAILY CORRECTED}}}{\text{Membrane Area, sq ft}}$$

$$= \frac{(47.6 \text{ GPM})(1,440 \text{ min/day})}{545 \text{ sq ft}}$$

$$= 126 \text{ GFD}$$

4.76 Sampling

Samples of feed, concentrate, and permeate should be collected from the membrane system on a regular basis to assess system performance and discharge levels. For most industrial wastestreams, these analyses are recommended:

Total Solids

Total Soluble Solids

Suspended Solids

True Hydrocarbon Oil and Grease
   (*STANDARD METHOD 5520F*, "Hydrocarbons")

Freon-Extractable Oil and Grease (Includes fats, oils, grease surfactants, and fatty acids)
   (*STANDARD METHOD 5520B*, "Liquid-Liquid, Partition-Gravimetric Method")

pH

BOD

COD

Metals

Color

For food processing wastewater, these analyses are recommended:

Total Solids

Total Soluble Solids

Total Suspended Solids

pH

Freon-Extractable Oil and Grease (Includes fats, oils, grease surfactants, and fatty acids)
   (*STANDARD METHOD 5520B*, "Liquid-Liquid, Partition-Gravimetric Method")

COD

BOD

TKN[59] Protein

True Protein

Additional analyses may be required when specific discharge limits must be met.

The analysis of permeate on a regular basis allows the operator to monitor membrane life. As membranes degrade, the retention values decrease as more of the retained particles pass into the permeate. Concentrate analysis and the permeate values allow the operator to calculate retention levels with the formulas presented earlier. Feed analysis is recommended on a regular basis to assess variations in the feed, which may alter the performance of the membrane filter. Frequency of analysis must be determined case by case. Good documentation of analyses in the form of *SPC CHARTS*[60] and log sheets is recommended.

4.77 Recordkeeping

Recording of data is important for system performance evaluation. A daily log sheet should be set up for recording process and analytical data. This log should include the following information:

1. Date

2. Operator's name

3. Feed description

4. Upstream conditions within the plant

5. Time of data collection

6. Flow to the unit

---

[59] Total Kjeldahl Nitrogen

[60] *SPC Chart.*   Statistical Process Control chart. A plot of daily performance such as a trend chart.

7. Permeate flow from the unit

8. Tank levels

9. Inlet and outlet pressures

10. Feed temperature

11. pH of the feed

12. Sample ID numbers of feed, permeate, and concentrate

13. Comment section to record problems

A separate cleaning data sheet is suggested:

1. Cleaning chemical concentrations (pH, amounts of chemicals added, measured chlorine concentration)

2. Cleaning cycle times

3. Inlet and outlet pressures

4. Fluxes during cleaning

5. Water flux after cleaning, corrected to standard conditions

6. Pressure and temperature conditions of water flux

7. Comment section to record problems

Charts or diagrams of the membrane installation should be readily available. Membrane serial numbers are found on individual modules for identification purposes. All maintenance and repairs should also be recorded.

4.8 SAFETY PRECAUTIONS WITH MEMBRANE SYSTEMS

Each manufacturer will provide a list of safety precautions with the equipment to protect the membrane and ensure safe operation to the plant operators and other employees. Basic safety precautions include:

1. Never expose the membrane to unknown or incompatible materials or chemicals. Examples are silicone, polar hydrocarbons, and organic solvents. Consult the membrane supplier for chemical compatibility.

2. Do not run the system at temperatures, pressures, or flow rates exceeding the manufacturer's recommendations.

3. Do not overconcentrate the waste feed beyond the design of the system.

4. Do not overclean the membrane system; this will shorten the membrane life.

5. Do not modify or bypass prefiltration of the feed.

4.9 REFERENCE

STANDARD METHODS FOR THE EXAMINATION OF WATER AND WASTEWATER, ed. Eaton, A. D., Clesceri, L. S., Rice, E. W., 21st Edition, 2005. Obtain from Water Environment Federation (WEF), Publications Order Department, 601 Wythe Street, Alexandria, VA 22314-1994. Order No. S82011. Price to members, $194.75; nonmembers, $259.75; price includes cost of shipping and handling.

QUESTIONS

Write your answers in a notebook and then compare your answers with those on page 464.

4.7A List the common pretreatment processes used for ultrafiltration systems processing primarily industrial waste.

4.7B In an oily waste system, why is pH control at the ultrafilter important?

4.7C List the steps for cleaning (washing) a membrane.

4.7D When would spongeballs be used to remove membrane foulants?

4.7E How is the effectiveness of a cleaning cycle assessed?

END OF LESSON 6 OF 6 LESSONS

on

SOLIDS REMOVAL FROM SECONDARY EFFLUENTS

Please answer the discussion and review questions next.

DISCUSSION AND REVIEW QUESTIONS

Chapter 4. SOLIDS REMOVAL FROM SECONDARY EFFLUENTS

(Lesson 6 of 6 Lessons)

Write the answers to these questions in your notebook. The question numbering continues from Lesson 5.

34. Membrane filtration processes are gaining acceptance as alternatives for what other treatment processes?

35. How does the membrane process differ from conventional filtration?

36. Stringent discharge requirements may require post-treatment of the ultrafiltration (UF) permeate by which treatment processes?

37. What major membrane system problems can be revealed by analyzing delta pressure measurements?

38. How can fouled membranes be cleaned?

39. Why does membrane filtration require proper pretreatment of the feed?

40. Why must membranes be replaced periodically?

SUGGESTED ANSWERS

Chapter 4.   SOLIDS REMOVAL FROM SECONDARY EFFLUENTS

ANSWERS TO QUESTIONS IN LESSON 1

Answers to questions on page 341.

4.0A   Some locations have stringent discharge requirements because more and more wastes are being discharged into the receiving waters and increasing demands are being placed on the waters by water users.

4.0B   NPDES stands for National Pollutant Discharge Elimination System.

Answers to questions on page 342.

4.1A   Coagulation is the act of adding and mixing a coagulating chemical to destabilize the suspended particles allowing the particles to collide and "stick" together, forming larger particles.

4.1B   Flocculation is the actual gathering together of smaller suspended particles into flocs, thus forming a more readily settleable mass.

4.1C   Chemicals may be added to improve filter performance as well as to reduce emergency problems such as those created by sludge bulking in the secondary clarifier, upstream equipment failure, accidental spills entering the plant, and seasonal overloads.

4.1D   When adding chemicals upstream from a biological treatment process, be sure that the chemical or its concentration is not toxic to the organisms treating the wastewater in the biological process.

Answers to questions on page 346.

4.10A   When most of the solids particles in a wastestream have the same electrical charge (usually negative), the particles repel each other. The electrostatic repulsion prevents the particles from clumping into floc large enough and dense enough to settle out. Destabilization is the addition of chemicals to change or neutralize the charge of the particles so they will coagulate and can be separated from the wastestream.

4.10B   Particles tend to have a negative ionic charge when wastewater has a pH on the low (acidic) side of pH 6.5.

4.10C   Vigorous mixing during coagulation is important to ensure that the coagulants are thoroughly mixed into the wastestream and that the solids particles make physical contact with each other.

Answers to questions on page 347.

4.11A   The four most common chemicals added to improve settling are alum, ferric chloride, lime, and polyelectrolytes (polymers).

4.11B   Alum should be kept dry to prevent it from caking into a solid lump.

4.11C   All mechanical equipment such as conveyors should be run until well cleaned of all alum before shutting down because the alum can harden and jam the equipment.

Answers to questions on page 350.

4.11D   Safety precautions required for handling ferric chloride in concentrated forms should be the same as those for acids. Wear protective clothing, chemical goggles, rubber or plastic gloves, and rubber shoes. Flush all splashes off clothing and skin immediately.

4.11E   Rubber or flexible piping with easy access and short runs will permit cleaning by squeezing the walls and washing out the broken scale. Solid piping that is plugged by scale usually requires replacement.

4.11F   Polyelectrolytes will create an extremely slippery surface when wet. Clean up spills immediately.

4.11G   Manufacturers' MSDSs recommend the use of inert absorbent material to clean polymer spills, such as sand or earth.

Answers to questions on page 352.

4.12A   The three stages of testing used to select chemical products for a treatment process are: (1) preliminary screening, (2) dosage testing, and (3) full-scale trial.

4.12B   Various types of chemicals or different doses of a single chemical are added to sample portions of wastewater in a jar test unit and all portions of the sample are rapidly mixed. After rapid mixing, the samples are slowly mixed to approximate the conditions in the plant. Mixing is stopped and the floc formed is allowed to settle. The appearance of the floc, the time required to form a floc, and the settling conditions are recorded. The supernatant is analyzed for turbidity, suspended solids, and pH. With this information, the operator selects the best chemical or best dosage to feed on the basis of clarity of effluent and minimum cost of chemicals.

4.12C   Before a jar test unit is used, it is essential to know the volume of the test jars and the speed rates of the stirrers.

Answers to questions on page 359.

4.12D   When preparing stock solutions of polymers, carefully mix the dry polymer into water. The water must be stirred rapidly while gradually sprinkling the polymer granules or powder over the water surface. Care must be taken to be sure that the granules of powder do not pile up in lumps on the water surface or cling to the sides of the mixing container, and that each granule of polymer is wetted individually.

4.12E   Polymers are usually limited to 1.0 percent maximum solution strength so that the solution can be pumped (metered) easily.

4.12F   The sampling point for a jar test should be located as near as possible to the location where treatment chemicals will be or are being fed into the flow stream.

ANSWERS TO QUESTIONS IN LESSON 2

Answers to questions on page 370.

4.13A   Chemical solutions are prepared for feeding by mixing known amounts of chemicals and water together using a mechanical mixer. The resulting solution is stored in a day tank (holding tank) from which it is metered out at the proper dosage into the water being treated.

4.13B   Common types of chemical feeders or metering equipment include:

1. Positive displacement pumps such as the plunger pump, gear pump, diaphragm pump, and progressive cavity pump
2. Screw feeder
3. Vibrating trough feeder
4. Rotary feeder
5. Belt-type gravimetric feeder

Answers to questions on page 374.

4.13C   Items that should be considered when selecting a chemical feeder include:

1. Total operating range
2. Accuracy
3. Repeatability
4. Resistance to corrosion
5. Dust control
6. Availability of parts
7. Safety
8. Economics

4.13D   The following information regarding a chemical feeder operation should be recorded:

1. Flows
2. Characteristics of wastewater before and after treatment

3. Dosage and conditions of chemical treatment
4. Results after treatment
5. Abnormal conditions observed during treatment

Answers to questions on page 386.

4.13E   Rapid mixing of coagulant chemicals may be accomplished in any of three modes: (1) high-speed mixers (impeller or turbine), (2) in-line blenders and pumps, and (3) baffled compartments or pipes (static mixers).

4.13F   In tapered flocculation, a small, dense floc is formed initially followed by aggregation to form a denser, larger floc.

4.13G   An advantage of vertical-flow clarifier units is that flow is forced up through a sludge blanket, thus aiding in solids retention and improving flow control.

4.13H   The four major factors used to predict the performance of clarifiers include: detention time, weir overflow rate, surface loading rate, and solids loading.

4.13I   If the detention time in a clarifier is too short, there will be a carryover of particles into the effluent.

4.13J   Short-circuiting in a clarifier may be caused by:

1. Differences in water density due to different temperatures existing at the surface and bottom of the clarifier
2. Density differences due to suspended solids
3. High inlet and outlet velocities
4. Strong winds blowing along the surface of the tank

4.13K   The four major factors used to measure clarifier performance or efficiency include: flows, suspended solids, settleable solids, and floatable solids.

Answers to questions on page 391.

4.14A   Water quality indicators that should be monitored when operating a physical–chemical treatment process include alkalinity, pH, temperature, turbidity, and suspended solids.

4.14B   Items to be checked during the start-up inspection of a chemical feed system include:

1 Level alarms in tanks
2 Calibration of flow and metering systems
3 Water blenders and dilution mixers
4 Temperature and pressure of dilution water
5 In-line mixers

4.14C   Procedures for the normal operation of a chemical feed system in a coagulation-precipitation system include:

1 Perform a jar test to determine the proper chemical dosage.
2 Ensure proper chemical dilution.
3 Set controls for the proper feed rate.
4 Report in the operating log the amount of chemicals used per unit time.

4.14D Abnormal conditions that could be encountered in the water being treated when operating a physical–chemical treatment process include high solids, high or low flows, and changes in pH and temperature.

4.14E Two common problems that could occur when operating a physical–chemical treatment process are foaming in the rapid-mix tanks and flocculators and no coagulation of suspended particles.

ANSWERS TO QUESTIONS IN LESSON 3

Answers to questions on page 393.

4.2A Most filters operate on a batch basis. The filter operates until its capacity to remove solids is nearly reached. At this time it is completely removed from service and cleaned.

4.2B A gravity filter should be cleaned when the solids capacity of the media has nearly been reached but before solids break through into the effluent.

4.2C Meanings of the following terms are:

1. Downflow. Water flows down through the bed.
2. Static bed. Bed does not move or expand while water is being filtered.

4.2D 1. In surface straining, the filter is designed to remove the solids at the very top of the media.
2. Depth filtration is designed to permit the solids to penetrate deep into the media, thereby capturing the solids within as well as on the surface of the media.

Answers to questions on page 400.

4.2E Materials used for filter media include silica sand, anthracite coal, garnet, or ilmenite. Garnet and ilmenite are commonly used in multi-media beds.

4.2F If the filter media is not thoroughly cleaned during each backwashing, a buildup of solids will occur. The end result of incomplete cleaning is the formation of mudballs within the bed.

4.2G If unfiltered water is supplied to the backwash system, clogging of the underdrain system may occur.

4.2H Used backwash water holding tanks are needed to equalize flows and prevent hydraulically overloading the treatment plant when backwash waters are returned to the headworks.

Answers to questions on page 402.

4.2I The head loss through the filter media is determined by measuring the water pressure above and below the filter media. When water flows through the media, the pressure below the media will be less than the pressure above the media (when the pressure levels are measured or read at the same elevation). The difference between the two readings is the head loss.

4.2J Filter system alarms should be tested for proper functioning at least every 60 days.

Answers to questions on page 404.

4.2K A pre-start check should be conducted before starting filtering systems to prevent damage to the equipment or injury to personnel.

4.2L The purpose of the rate-control valve is to maintain the desired flow through the filter and prevent the backwash water from mixing with the filtered water during backwashing.

4.2M A filter should be backwashed after the capacity of the media to hold solids is nearly used up, but before solids break through into the effluent.

Answers to questions on page 406.

4.2N Abnormal operating conditions include:

1. High solids in applied water due to bulking sludge, rising sludge, or solids washout in the secondary clarifier
2. Low suspended solids in applied water; however, solids pass through filter
3. Loss of filter aid chemical feed
4. High wet weather peak flows
5. Low applied water flows
6. High color loading
7. High water temperature
8. Low water temperature
9. Air binding
10. Negative pressure in filter
11. High BOD and COD
12. High coliform group bacteria levels
13. Chlorine in applied water
14. pH change in applied water
15. High grease and oil in applied water

4.2O To treat a high solids content in the water applied to a filter:

1. Run jar tests and adjust chemical dosage, as needed.
2. Place more filters in service to prevent breakthrough.
3. Prepare to backwash more frequently.

Answers to questions on page 408.

4.2P To determine if media are being lost, every three or four months, measure and record the freeboard to the filter media surface. A small amount of media loss is normal, but an excessive amount (2 to 3 inches or 5 to 7 centimeters) indicates operational problems.

4.2Q Trees and shrubs should be kept away from uncovered filters because leaves will drop into the filter and they are very difficult to backwash out of the media.

Answers to questions on page 411.

4.2R The three main types of safety hazards around filtration systems are electrical, chemical, and mechanical.

4.2S Filtration instrumentation should measure and record applied flows, backwash flows, head loss, and water quality before and after filtration.

4.2T Install all readout meters, charts, and gauges of instruments in a convenient and centralized location.

ANSWERS TO QUESTIONS IN LESSON 4

Answers to questions on page 412.

4.3A The purpose of the inert-media pressure filter is to remove suspended solids, associated BOD, and turbidity from secondary effluents to meet waste discharge requirements established by NPDES permits.

4.3B Chemicals commonly used with the filtration process are polymers and alum. The chemicals are used as coagulants for the solids and turbidity to aid in their removal by filtration.

4.3C Major components of a pressure filtration system include:

1. A holding tank or wet well
2. Filter feed pumps
3. Chemical coagulant feed pump system
4. Filters
5. Filter backwash wet well
6. Filter backwash pumps
7. Decant tank

Answers to questions on page 417.

4.3D The purpose of the holding tank is to store water and to allow additional settling of the suspended solids before the water is applied to the filters.

4.3E Alum is used for *COAGULATION* while polymers are used for *FLOCCULATION*.

4.3F Polymer is supplied at a concentration of 0.6 pound polymer per gallon (72 gm/L or 72 kg/m³). The polymer feed pump delivers a flow of 0.15 GPM (0.0095 L/sec) and the flow to the filters is 5,000 GPM (315 L/sec). Calculate the concentration or dose of polymer in mg/L in the water applied to the filter.

Known	Unknown
Polymer Conc, lbs/gal = 0.6 lb/gal	Polymer Dose, mg/L
Polymer Pump, GPM = 0.15 GPM	
Flow to Filter, GPM = 5,000 GPM	

ENGLISH

Calculate the polymer dose, mg/L.

$$\text{Dose, mg/}L = \frac{\text{Flow, gal/min} \times \text{Conc, lbs Polymer/gal}}{\text{Flow, gal/min} \times 8.34 \text{ lbs Water/gal}}$$

$$= \frac{0.15 \text{ gal/min} \times 0.6 \text{ lb Polymer/gal}}{5,000 \text{ gal/min} \times 8.34 \text{ lbs Water/gal}}$$

$$= \frac{0.09 \text{ lb Polymer}}{41,700 \text{ lbs Water}} \times \frac{1,000,000}{1 \text{ M}}$$

$$= 2.2 \text{ mg/}L$$

METRIC

$$\text{Dose, mg/}L = \frac{\text{Flow, }L\text{/sec} \times \text{Conc, gm Polymer/}L \times 1,000 \text{ mg/gm}}{\text{Flow, }L\text{/sec}}$$

$$= \frac{0.0095 \, L\text{/sec} \times 72 \text{ gm Polymer/}L \times 1,000 \text{ mg/gm}}{315 \, L\text{/sec}}$$

$$= 2.2 \text{ mg/}L$$

Answers to questions on page 421.

4.3G Major components of pressure filters include:

1. Vessels
2. Interior piping
3. Underdrain gravel (support media)
4. Inert media
5. Flow controls

4.3H Water from the surface wash arms initially breaks up the mat of suspended material on the media surface.

4.3I The water used to backwash the filter comes from the chlorine contact tank (filtered or chlorinated) to the backwash wet well before it is used for backwashing.

4.3J The decant tank receives the backwash water from the filters. The backwash water is allowed to settle and the clarified effluent is recycled to the filters. The settled material is collected and discharged to the solids handling facility.

Answers to questions on page 422.

4.3K When a large quantity of polymer reaches a filter, short filter run times will result due to increased differential pressure across the filter. When excessive alum concentrations are involved, the alum will pass through the filter media and filter effluent turbidity and suspended solids will increase due to alum breakthrough.

4.3L In areas where freezing temperatures occur, heavy insulation or heat tape will prevent the water in the cell piping from freezing. Also, liquid alum should be stored in an enclosed, warm space.

4.3M High operating filter differential pressures could occur if either: (1) the media is filled with suspended material, or (2) excessive chemical feed is "blinding" the media.

Answers to questions on page 425.

4.3N Filter system performance tests should be conducted monthly.

4.3O Wear goggles when observing the operation of the surface wash arms.

4.3P Safety precautions that should be taken when working with alum or polymers include:

1. Wear safety goggles and gloves. Wear approved respiratory protection if inhalation of dry alum or polymer is possible. Flush away any alum or polymer that comes in contact with your skin with cool water for a few minutes.
2. Be very careful when walking in an area where polymer mixing takes place. When a polymer is wet, it is very slippery. Clean up polymer spills with an inert, absorbent material such as sand or earth.

ANSWERS TO QUESTIONS IN LESSON 5

Answers to questions on page 430.

4.4A Continuous backwash, upflow, deep-bed silica sand media filters continuously and simultaneously filter solids from the influent flow, while cleaning and recycling the filter media internally through an airlift pipe and sand washer. The cleaned sand is redistributed on top of the filter media bed, which allows for continuous uninterrupted flow of filtered effluent and backwash reject water.

4.4B Filter influent and effluent turbidity meters continuously monitor and record filter influent and effluent water quality.

4.4C Filter influent and effluent flows are metered to quantify and record flows and provide dosage control signals to metering pumps and disinfection chemical addition control systems.

Answers to questions on page 435.

4.4D Features to bypass the filtration system facilities should be provided to permit diverting treatment process effluent to holding or equalization basins during emergency conditions, such as extremely high flows, equipment failures, or clogged filters.

4.4E The purpose of adding chemical coagulants to the filter influent is to cause the suspended and colloidal (finely divided solids that will not settle by gravity) particles to clump together into bulkier hydroxyl precipitates.

4.4F The purpose of adding flocculant solution to the filter influent is to gather still more of the suspended and colloidal solids into larger (visible to the naked eye), stronger clumps (floc), which can be efficiently removed by filtration.

Answers to questions on page 441.

4.4G If the airlift is allowed to operate without water flow to the filter, no water will be available to the gravity sand washer-separator. This condition will allow dirty sand to be returned to the top of the sand bed.

4.4H If a tool is dropped into the filter, *IMMEDIATELY* stop the air flow to the airlift and remove the tool from the top of the sand bed.

4.4I Ferric chloride must be handled with the same care as acid solutions since ferric chloride causes burns similar to those caused by acids.

4.4J Before entering the plenum area of a filter, observe all confined space procedures and check for possible damage to the inner cone. In no event should the plenum area be entered if damage to the inner cone is suspected without first removing all sand from the filter.

ANSWERS TO QUESTIONS IN LESSON 6

Answers to questions on page 447.

4.5A In cross flow filtration, wastewater flows across the surface of a membrane rather than through a bed of granular media. The membrane permits water to pass through but blocks the passage of particles. Other differences include the length of the filter run and the ease of cleaning the membranes.

4.5B Membrane processes are classified on the basis of the size of particle they separate from the wastestream.

4.5C The basic membrane configurations are tubular, hollow fiber, spiral, plate and frame, and ceramic tube or monolith.

Answers to questions on page 451.

4.6A   The amount of flux across a membrane is dependent upon transmembrane pressure (driving force), flow rate across the membrane surface (turbulence on the membrane), concentration of waste material, temperature, viscosity, and cleanliness of the membrane surface.

4.6B   All membrane systems recirculate the waste feed across the membrane to prevent buildup of particles on the surface.

4.6C   The delta pressure might increase when the viscosity of the feed is increasing or when material is accumulating in the membrane modules.

4.6D   The surface of the membrane can become fouled with oil, grease, and other foulants.

Answers to questions on page 458.

4.7A   For ultrafiltration systems processing primarily industrial waste, the following pretreatment processes are common:

1. Equalization of wastes
2. Prescreening to remove large suspended material
3. Free oil and settleable solids removal
4. pH treatment (to alkaline conditions) to stabilize the oil/water emulsion

4.7B   In an oily waste system, pH control at the ultrafilter is important for two reasons. Flux rates are improved at the ultrafilter when the wastewater pH is 9 to 10.5 and fouling of the membrane by oil is minimized by stabilizing the oil emulsion.

4.7C   The steps for cleaning (washing) a membrane are as follows:

1. Displacement of waste from the system with water
2. Washing the membranes with a caustic and surfactant to remove oils and grease
3. Flushing the surfactant from the membrane with warm water
4. Washing the membrane with an acid cleaner to remove salt buildup
5. Flushing the acid from the membrane with warm water
6. Recording clean membrane water flux as a check on cleaning effectiveness

4.7D   Spongeballs are sometimes required when membrane foulants are difficult to remove, such as with metal hydroxides or waste latex. Spongeballs may also be used to speed up the cleaning procedure in some plants.

4.7E   After cleaning, a water flux is measured to assess the effectiveness of the cleaning cycle.

CHAPTER 5

PHOSPHORUS REMOVAL

by

John G. M. Gonzales

TABLE OF CONTENTS
Chapter 5.   PHOSPHORUS REMOVAL

OBJECTIVES

Chapter 5.   PHOSPHORUS REMOVAL

Following completion of Chapter 5, you should be able to:

1. Explain the need for phosphorus removal and describe some of the different systems used for this purpose at various treatment plants.

2. Place a phosphorus removal system into service.

3. Schedule and safely conduct operation and maintenance duties.

4. Sample influent and effluent, interpret lab results, and make appropriate adjustments in the treatment process.

5. Recognize abnormal operating conditions, understand the cause, and take corrective action to ensure proper phosphorus removal.

6. Inspect a newly installed phosphorus removal facility to determine if installation has been proper.

7. Review plans and specifications for a phosphorus removal system.

WORDS
Chapter 5.   PHOSPHORUS REMOVAL

AEROBIC (air-O-bick)

AEROBIC

A condition in which atmospheric or dissolved oxygen is present in the aquatic (water) environment.

AGGLOMERATION (uh-glom-er-A-shun)

AGGLOMERATION

The growing or coming together of small scattered particles into larger flocs or particles, which settle rapidly. Also see FLOC.

ALKALINITY (AL-kuh-LIN-it-tee)

ALKALINITY

The capacity of water or wastewater to neutralize acids. This capacity is caused by the water's content of carbonate, bicarbonate, hydroxide, and occasionally borate, silicate, and phosphate. Alkalinity is expressed in milligrams per liter of equivalent calcium carbonate. Alkalinity is not the same as pH because water does not have to be strongly basic (high pH) to have a high alkalinity. Alkalinity is a measure of how much acid must be added to a liquid to lower the pH to 4.5.

ANAEROBIC (AN-air-O-bick)

ANAEROBIC

A condition in which atmospheric or dissolved oxygen (DO) is *NOT* present in the aquatic (water) environment.

ANAEROBIC (AN-air-O-bick) SELECTOR

ANAEROBIC SELECTOR

Anaerobic refers to the practical absence of dissolved and chemically bound oxygen Selector refers to a reactor or basin and environmental conditions (food, lack of DO) intended to favor the growth of certain organisms over others. Also see SELECTOR.

ANOXIC (an-OX-ick)

ANOXIC

A condition in which the aquatic (water) environment does not contain dissolved oxygen (DO), which is called an oxygen deficient condition. Generally refers to an environment in which chemically bound oxygen, such as in nitrate, is present. The term is similar to ANAEROBIC.

ARCH

ARCH

(1)   The curved top of a sewer pipe or conduit.

(2)   A bridge or arch of hardened or caked chemical that will prevent the flow of the chemical.

AUTOTROPHIC (auto-TROF-ick)

AUTOTROPHIC

Describes organisms (plants and some bacteria) that use inorganic materials for energy and growth.

BIOMASS (BUY-o-mass)

BIOMASS

A mass or clump of organic material consisting of living organisms feeding on wastes, dead organisms, and other debris. Also see ZOOGLEAL MASS and ZOOGLEAL MAT (FILM).

BOUND WATER

BOUND WATER

Water contained within the cell mass of sludges or strongly held on the surface of colloidal particles. One of the causes of bulking sludge in the activated sludge process.

BUFFER CAPACITY

BUFFER CAPACITY

A measure of the capacity of a solution or liquid to neutralize acids or bases. This is a measure of the capacity of water or wastewater for offering a resistance to changes in pH.

CENTRATE CENTRATE

The water leaving a centrifuge after most of the solids have been removed.

COAGULATION (ko-agg-yoo-LAY-shun) COAGULATION

The clumping together of very fine particles into larger particles (floc) caused by the use of chemicals (coagulants). The chemicals neutralize the electrical charges of the fine particles, allowing them to come closer and form larger clumps.

COLLOIDS (KALL-loids) COLLOIDS

Very small, finely divided solids (particles that do not dissolve) that remain dispersed in a liquid for a long time due to their small size and electrical charge. When most of the particles in water have a negative electrical charge, they tend to repel each other. This repulsion prevents the particles from clumping together, becoming heavier, and settling out.

ENDOGENOUS (en-DODGE-en-us) RESPIRATION ENDOGENOUS RESPIRATION

A situation in which living organisms oxidize some of their own cellular mass instead of new organic matter they adsorb or absorb from their environment.

FILAMENTOUS (fill-uh-MEN-tuss) ORGANISMS FILAMENTOUS ORGANISMS

Organisms that grow in a thread or filamentous form. Common types are *Thiothrix* and *Actinomycetes*. A common cause of sludge bulking in the activated sludge process.

FLOC FLOC

Clumps of bacteria and particles, or coagulants and impurities, that have come together and formed a cluster. Found in flocculation tanks, sedimentation basins, aeration tanks, secondary clarifiers, and chemical precipitation processes.

FLOCCULATION (flock-yoo-LAY-shun) FLOCCULATION

The gathering together of fine particles after coagulation to form larger particles by a process of gentle mixing. This clumping together makes it easier to separate the solids from the water by settling, skimming, draining, or filtering.

HETEROTROPHIC (HET-er-o-TROF-ick) HETEROTROPHIC

Describes organisms that use organic matter for energy and growth. Animals, fungi, and most bacteria are heterotrophs.

HINDERED SOLIDS SEPARATION HINDERED SOLIDS SEPARATION

Solids settling in a thickening rather than in a clarifying mode. Suspended solids settling (clarifying) velocities are strongly influenced by the applied solids concentration. The greater the applied solids concentration, the greater the opportunity for thickening (hindered) solids settling.

HYDROLYSIS (hi-DROLL-uh-sis) HYDROLYSIS

(1) A chemical reaction in which a compound is converted into another compound by taking up water.

(2) Usually a chemical degradation of organic matter.

MCRT MCRT

Mean Cell Residence Time. An expression of the average time (days) that a microorganism will spend in the activated sludge process.

$$\text{MCRT, days} = \frac{\text{Total Suspended Solids in Activated Sludge Process, lbs}}{\text{Total Suspended Solids Removed From Process, lbs/day}}$$

or

$$\text{MCRT, days} = \frac{\text{Total Suspended Solids in Activated Sludge Process, kg}}{\text{Total Suspended Solids Removed From Process, kg/day}}$$

NOTE: Operators at different plants calculate the Total Suspended Solids (TSS) in the Activated Sludge Process, lbs (kg), by three different methods:

1.   TSS in the Aeration Basin or Reactor Zone, lbs (kg)

2.   TSS in the Aeration Basin and Secondary Clarifier, lbs (kg)

3.   TSS in the Aeration Basin and Secondary Clarifier Sludge Blanket, lbs (kg)

These three different methods make it difficult to compare MCRTs in days among different plants unless everyone uses the same method.

MLSS MLSS

Mixed Liquor Suspended Solids. The amount (mg/L) of suspended solids in the mixed liquor of an aeration tank.

MLVSS MLVSS

Mixed Liquor Volatile Suspended Solids. The amount (mg/L) of organic or volatile suspended solids in the mixed liquor of an aeration tank. This volatile portion is used as a measure or indication of the microorganisms present.

METABOLISM METABOLISM

All of the processes or chemical changes in an organism or a single cell by which food is built up (anabolism) into living protoplasm and by which protoplasm is broken down (catabolism) into simpler compounds with the exchange of energy.

OBLIGATE AEROBES OBLIGATE AEROBES

Bacteria that must have atmospheric or dissolved molecular oxygen to live and reproduce.

POLYELECTROLYTE (POLY-ee-LECK-tro-lite) POLYELECTROLYTE

A high-molecular-weight (relatively heavy) substance, having points of positive or negative electrical charges, that is formed by either natural or synthetic (manmade) processes. Natural polyelectrolytes may be of biological origin or obtained from starch products or cellulose derivatives. Synthetic polyelectrolytes consist of simple substances that have been made into complex, high-molecular-weight substances. Used with other chemical coagulants to aid in binding small suspended particles to larger chemical flocs for their removal from water. Often called a POLYMER.

POLYMER (POLY-mer) POLYMER

A long-chain molecule formed by the union of many monomers (molecules of lower molecular weight). Polymers are used with other chemical coagulants to aid in binding small suspended particles to larger chemical flocs for their removal from water. Also see POLY-ELECTROLYTE.

PRECIPITATE (pre-SIP-uh-TATE) PRECIPITATE

(1)   An insoluble, finely divided substance that is a product of a chemical reaction within a liquid.

(2)   The separation from solution of an insoluble substance.

PROTOPLASM PROTOPLASM

A complex substance (typically colorless and semifluid) regarded as the physical basis of life, having the power of spontaneous motion and reproduction; the living matter of all plant and animal cells and tissues.

RECALCINATION (re-kal-sin-NAY-shun)

RECALCINATION

A lime recovery process in which the calcium carbonate in sludge is converted to lime by heating at 1,800°F (980°C).

RECARBONATION (re-kar-bun-NAY-shun)

RECARBONATION

A process in which carbon dioxide is bubbled into the water being treated to lower the pH.

RESPIRATION

RESPIRATION

The process in which an organism takes in oxygen for its life processes and gives off carbon dioxide.

SELECTOR

SELECTOR

A reactor or basin in which baffles or other devices create a series of compartments. The environment and the resulting microbial population within each compartment can be controlled to some extent by the operator. The environmental conditions (food, lack of dissolved oxygen) that develop are intended to favor the growth of certain organisms over others. The conditions thereby select certain organisms.

SELECTOR RECYCLE

SELECTOR RECYCLE

The recycling of return sludge or oxidized nitrogen to provide desired environmental conditions for microorganisms to perform a desired function.

SLAKE

SLAKE

To mix with water so that a true chemical combination (hydration) takes place, such as in the slaking of lime.

SLURRY

SLURRY

A watery mixture or suspension of insoluble (not dissolved) matter; a thin, watery mud or any substance resembling it (such as a grit slurry or a lime slurry).

VISCOSITY (vis-KOSS-uh-tee)

VISCOSITY

A property of water, or any other fluid, that resists efforts to change its shape or flow. Syrup is more viscous (has a higher viscosity) than water. The viscosity of water increases significantly as temperatures decrease. Motor oil is rated by how thick (viscous) it is; 20 weight oil is considered relatively thin while 50 weight oil is relatively thick or viscous.

VOLATILE ACIDS

VOLATILE ACIDS

Fatty acids produced during digestion that are soluble in water and can be steam-distilled at atmospheric pressure. Also called organic acids. Volatile acids are commonly reported as equivalent to acetic acid.

ZOOGLEAL (ZOE-uh-glee-ul) MASS

ZOOGLEAL MASS

Jelly-like masses of bacteria found in both the trickling filter and activated sludge processes. These masses may be formed for or function as the protection against predators and for storage of food supplies. Also see BIOMASS.

CHAPTER 5.   PHOSPHORUS REMOVAL

(Lesson 1 of 2 Lessons)

5.0   WHY IS PHOSPHORUS REMOVED FROM WASTEWATER?

5.00   Phosphorus as a Nutrient

Phosphorus provides a nutrient or food source for algae. Phosphorus combined with inorganic nitrogen poses serious pollution threats to receiving waters because of high algal growths that result from the presence of the two nutrients in water. Algae in water are considered unsightly and can cause tastes and odors in drinking water supplies. Dead and decaying algae can cause serious oxygen depletion problems in receiving streams, which, in turn, can kill fish and other aquatic wildlife.

By removing phosphorus in the effluent of a wastewater treatment plant, the lake or river that the treatment plant discharges into will have one less nutrient that is essential for algal growth. This reduction in an essential nutrient reduces the growth of the algae.

5.01   Need for Phosphorus Removal

The US Environmental Protection Agency and other water quality regulating agencies recognize the need to protect rivers and lakes from excessive growths of algae. Because of this, the agencies are requiring that wastewater treatment plants remove phosphorus in the effluent in order to protect the receiving waters by eliminating a nutrient that can cause algal growth.

5.1   TYPES OF PHOSPHORUS REMOVAL SYSTEMS

Biological phosphorus removal, lime precipitation, and filtration following aluminum sulfate flocculation are the most common types of phosphorus removal systems. Each of these systems will be described in this chapter.

A variety of other systems for the removal of nutrients from wastewater are also available. One such process, the Bardenpho process, removes both nitrogen and phosphorus using a modification of the activated sludge process. The Bardenpho process is described in Chapter 7, Section 7.20, "The Bardenpho Process."

5.10   Biological Phosphorus Removal

Microorganisms normally found in a conventional activated sludge process use phosphorus within the makeup of the cell structure that forms the organism. When the microorganisms are in a state of *ENDOGENOUS RESPIRATION*[1] or are very hungry and need food and oxygen, they tend to absorb phosphorus quite freely. This process is called "luxury uptake" in which the microorganisms take excess phosphorus into their bodies due to the stimulation of being placed in a proper environment containing food and oxygen. When these same organisms are placed in an environment where there is no oxygen (anaerobic), the first element that is released by the microorganisms as they begin to die is phosphorus. As the phosphorus is released, it can be drawn off and removed from the wastewater stream.

5.11   Lime Precipitation

When lime (calcium hydroxide ($Ca(OH)_2$) is mixed with effluent from a wastewater treatment plant in sufficient concentration to bring about high pH in the water, a chemical compound is formed consisting of phosphorus, calcium, and the hydroxyl (OH^-) ion. This compound can be *FLOCCULATED*[2] or combined in such a way as to form heavier solids that can settle in a clarifier for phosphorus removal. A substantial amount of the lime reacts with the alkalinity of the wastewater to form a calcium carbonate *PRECIPITATE,*[3] which also settles

---

[1] *Endogenous* (en-DODGE-en-us) *Respiration.*   A situation in which living organisms oxidize some of their own cellular mass instead of new organic matter they adsorb or absorb from their environment.

[2] *Flocculation* (flock-yoo-LAY-shun).   The gathering together of fine particles after coagulation to form larger particles by a process of gentle mixing. This clumping together makes it easier to separate the solids from the water by settling, skimming, draining, or filtering.

[3] *Precipitate* (pre-SIP-uh-TATE).   (1) An insoluble, finely divided substance that is a product of a chemical reaction within a liquid. (2) The separation from solution of an insoluble substance.

out with the phosphorus sludge. This calcium carbonate precipitate can be separated out of the sludge and *RECALCINED*[4] in a furnace to convert the calcium back to lime for reuse.

5.12 Aluminum Sulfate Flocculation and Precipitation (Sedimentation)

Aluminum sulfate (alum) in combination with wastewater also can flocculate phosphorus in much the same way as lime precipitation. The flocculation that happens with aluminum sulfate addition is the formation of aluminum phosphate particles that attach themselves to one another and become heavy and settle to the bottom of a clarifier. The aluminum sulfate and phosphorus mixture can then be withdrawn, thereby removing the phosphate or phosphorus from the wastewater flow. This alum *FLOC*[5] is difficult to settle out in a clarifier. Therefore, a pressure filter or a sand or mixed-media filter is usually installed after the clarifier to remove the remaining floc.

QUESTIONS

Write your answers in a notebook and then compare your answers with those on page 505.

5.0A   Why is phosphorus removed from wastewater?

5.1A   List the three major types of systems used to remove phosphorus from wastewater.

5.2 BIOLOGICAL PHOSPHORUS REMOVAL

Phosphorus can be removed from wastewater both biologically and chemically. A well-designed facility will always provide for chemical phosphorus removal, whether or not biological phosphorus removal is used, to ensure that the desired phosphorus removal can be achieved consistently. This section describes biological processes for removing phosphorus.

5.20 Definition of Terms

The two elements all biological treatment systems have in common are (1) a microbial population, the *BIOMASS*,[6] contained in some form of a reactor[7] or attached to inert media, and (2) some means of liquid-solids separation. The characteristics of the wastewater being treated and the environmental and physical conditions (detention time) within the reactor have a significant influence on the biomass. As the biomass adapts (acclimates) to reactor and wastewater conditions, the character of

the microbial population changes. That is, the types and numbers of organisms as well as solids separation and settling characteristics change as the biomass adapts to changing conditions.

Control of oxygen is the key to operating biological nutrient removal processes. The microorganisms always require some type of oxygen to support their growth and reproduction as they break down (degrade) the organic wastes to simpler substances such as carbon dioxide and water. The oxygen may be readily available as dissolved molecular oxygen in the wastewater or it may be chemically bound up in other substances such as nitrate, nitrite, or sulfate ions.

Since oxygen is such a key element of any biological process, it is logical that the terms we use to describe the environmental conditions in a reactor and the resulting activities of the microorganisms are defined in terms of the presence and form of oxygen. The environmental conditions and reactions in a process tank or reactor may be described as aerobic, anoxic, or anaerobic. The generally accepted definitions of these terms are as follows:

AEROBIC (or OXIC): A condition in which the aquatic (water) environment contains dissolved molecular oxygen (DO).

ANOXIC: A condition in which the aquatic (water) environment does not contain dissolved molecular oxygen (DO), which is called an oxygen deficient condition. Generally refers to an environment in which chemically bound oxygen, such as in nitrate, is present.

ANAEROBIC: A condition in which the aquatic (water) environment does not contain dissolved molecular oxygen (DO) or chemically bound oxygen (such as oxygen in the sulfate and nitrate ions). Stabilization of wastes (reduction of BOD) under anaerobic conditions typically progresses through three phases: *HYDROLYSIS,*[8] acid formation, and methane production. (Sulfate reduction to sulfide, which only occurs after the oxidized nitrogen is depleted, is usually grouped with the anaerobic reactions even though it is technically an anoxic reaction by the definition given here.)

There is sometimes an overlap between aerobic, anoxic, and anaerobic conditions. For example, anoxic reactions can occur within the floc that make up the mixed liquor suspended solids even when there is sufficient dissolved oxygen in the rest of the

---

[4] *Recalcination* (re-kal-sin-NAY-shun).   A lime recovery process in which the calcium carbonate in sludge is converted to lime by heating at 1,800°F (980°C).

[5] *Floc.*   Clumps of bacteria and particles, or coagulants and impurities, that have come together and formed a cluster. Found in flocculation tanks, sedimentation basins, aeration tanks, secondary clarifiers, and chemical precipitation processes.

[6] *Biomass* (BUY-o-mass).   A mass or clump of organic material consisting of living organisms feeding on wastes, dead organisms, and other debris. Also see ZOOGLEAL MASS and ZOOGLEAL MAT (FILM).

[7] A reactor is nothing more than a tank or basin or a specific portion of a tank or basin.

[8] *Hydrolysis* (hi-DROLL-uh-sis).   (1) A chemical reaction in which a compound is converted into another compound by taking up water. (2) Usually a chemical degradation of organic matter.

mixed liquor. Also, some anaerobic reactions will occur with anoxic reactions when there is little or no dissolved oxygen in the wastewater being treated.

Two other important terms you will need to understand are selector and MCRT; they are defined as follows:

SELECTOR: A selector is a reactor or basin in which baffles or other devices create a series of compartments. The environment and the resulting microbial population within each compartment can be controlled to some extent by the operator. The environmental conditions (food, lack of dissolved oxygen) that develop are intended to favor the growth of certain organisms over others. The conditions thereby *SELECT* certain organisms.

MCRT: Mean Cell Residence Time is an expression of the average time (days) that a microorganism will spend in the activated sludge process or specific process phase. To calculate the MCRT, the mass of suspended solids (SS) contained in a process is divided by the mass of solids removed from the process per day. The solids removed from the process per day in the activated sludge process include both the SS mass removed as waste activated sludge and the SS discharged from the plant with the effluent. In practical system operation, however, the SS discharged with the plant effluent may be ignored if they are less than ten percent of the total SS removed from the system.

$$\text{MCRT, days} = \frac{\text{Total Suspended Solids in Activated Sludge Process, lbs}}{\text{Total Suspended Solids Removed From Process, lbs/day}}$$

$$= \frac{\text{(Process Volume, M Gal)(MLSS, mg/}L\text{)(8.34 lbs/gal)}}{\text{(WAS Flow, MGD)(WAS MLSS, mg/}L\text{)(8.34 lbs/gal)} + \text{(Effl Flow, MGD)(Effl SS, mg/}L\text{)(8.34 lbs/gal)}}$$

NOTE: Operators at different plants calculate the Total Suspended Solids (TSS) in the Activated Sludge Process, lbs (kg), by three different methods:

1. TSS in the Aeration Basin or Reactor Zone, lbs (kg)

2. TSS in the Aeration Basin and Secondary Clarifier, lbs (kg)

3. TSS in the Aeration Basin and Secondary Clarifier Sludge Blanket, lbs (kg)

These three different methods make it difficult to compare MCRTs in days among different plants unless everyone uses the same method.

In practice, MCRT is used interchangeably with the solids retention time (SRT), which is a measure of sludge age. SRT is usually defined as the mass of solids in the reactor divided by the mass of solids in the process influent. In most municipal and many industrial wastewater treatment plants, however, differences between the MCRT and SRT are not significant. MCRT, defined in terms of Mixed Liquor Volatile Suspended Solids (MLVSS) instead of MLSS, is sometimes used, but this is not usually necessary for routine operational control.

5.21 Types of Microorganisms

Two general types of microorganisms make up the microbial population that stabilizes (treats) the waste materials: *AUTOTROPHS,*[9] which use inorganic carbon materials as their food source, and *HETEROTROPHS,*[10] which use organic carbon materials as their food source. Autotrophs and heterotrophs are further classified on the basis of their oxygen needs. Both types of organisms can be *OBLIGATE AEROBES,*[11] obligate anaerobes, or facultative organisms (able to perform oxic, anoxic, and anaerobic reactions).

5.22 Microbial Population Selection

In discussing biological treatment systems, we often talk about operators "selecting" certain types of microorganisms to accomplish a specific objective. What this means is that the operator creates an environment that provides the right conditions so that the desired type of organism will grow and reproduce, in manageable numbers, and achieve the intended processing objective. Lack of a proper environment for obligate organisms does not simply reduce the organisms' ability to perform their desired function. Without the appropriate environment, the organisms will not be able to reproduce at all or to perform their desired function at any level. Therefore, control of the reactor environment, particularly with regard to oxygen supplies, is a critical element in "selecting" the desired type of organisms.

QUESTIONS

Write your answers in a notebook and then compare your answers with those on page 505.

5.2A List the two elements all biological treatment systems have in common.

5.2B What factors have a significant influence on the biomass in a reactor?

5.2C Briefly define the following terms:

1. Aerobic or oxic
2. Anoxic
3. Anaerobic

5.2D How does the operator "select" the organisms needed to meet a particular processing objective?

---

[9] *Autotrophic* (auto-TROF-ick). Describes organisms (plants and some bacteria) that use inorganic materials for energy and growth.

[10] *Heterotrophic* (HET-er-o-TROF-ick). Describes organisms that use organic matter for energy and growth. Animals, fungi, and most bacteria are heterotrophs.

[11] *Obligate Aerobes.* Bacteria that must have atmospheric or dissolved molecular oxygen to live and reproduce.

5.23   Types of Process Layouts

Either of two general process layouts (Figure 5.1) may be used for biological phosphorus removal. The first, called mainstream, requires the use of an anaerobic selector at the beginning of the processing sequence (this tank or portion of a tank may also be anoxic for part of the reaction time). The second process layout makes use of a sidestream anaerobic stripper and a phosphorus extractor (this is the lime precipitation unit in Figure 5.1) or a clarifier. In both mainstream and sidestream processes, the way in which microorganisms remove phosphorus from the wastewater is the same. The advantages of the mainstream biological phosphorus removal layout are:

1. It offers complete and thorough mixing of the entire wastestream with the biomass in the anaerobic contactor.

2. It is a relatively simple design compared to the sidestream process sequence, which is somewhat more difficult to balance and operate.

The advantages of the sidestream biological phosphorus control layout are:

1. Hydrolysis and *VOLATILE ACID*[12] formation may be superior (however, this advantage may be offset by the incomplete mixing of the organisms with the raw wastewater).

2. It provides for routine phosphorus extraction, making it less dependent on sludge production and metabolic limits to achieve the required phosphorus removal.

Another possible advantage of the sidestream system is that it can potentially achieve lower effluent phosphorus levels. However, the higher phosphorus removal potential declines in importance as influent phosphorus to sludge production ratios decline. This factor could be important in areas where regional phosphate detergent bans are in place. (The sidestream biological phosphorus control system was first advanced nearly 40 years ago when phosphate levels in detergents and wastewaters were much higher than they are today.)

This lesson describes a sidestream process using the Phostrip® method as an example. The Phostrip® system is based on the luxury uptake of phosphorus by activated sludge microorganisms and precipitation of the stripped phosphorus using lime. (Chapter 7, Section 7.20, "The Bardenpho Process," contains additional information about luxury uptake as it is used for removal of both nitrogen and phosphorus.)

5.24   Luxury Uptake of Phosphorus

Luxury uptake of phosphorus is a modification of the basic activated sludge treatment process. The microorganisms usually found in activated sludge routinely remove some phosphorus for use in their own life processes as they break down the wastes in wastewater. Higher phosphorus removals can be achieved by setting up conditions that will first cause the microorganisms to take up and store in their cells[13] more phosphorus than they actually need. This phase of the process takes place under aerobic conditions. The phosphorus is stored as polyphosphate, that is, a *POLYMER*[14] of phosphorus (Figure 5.2). Once the organisms have stored the maximum amount of phosphorus in their cells, they are transferred to an anaerobic environment. To survive under anaerobic conditions, the microorganisms must chemically convert some of the carbon materials in their cells to get the oxygen they need for *METABOLISM*.[15] The energy used in this chemical reaction comes from the polyphosphate stored in the organisms' cells. As a result of the chemical reaction, phosphorus is released from the organisms' cells.

After releasing their phosphorus, the microorganisms are returned to the aeration tank where food, oxygen, and phosphorus are plentiful. Since the organisms used up the phosphorus in their cells just staying alive in the anaerobic environment, the first thing they do upon return to aerobic conditions is take up and store large quantities of phosphorus. In fact, they store far more than they need for their life processes. (The process is called "luxury" uptake for this very reason—the organisms take up more phosphorus than they actually need.)

The microorganisms remain in the aerobic phase until they have completely revived. Then, the sequence is repeated; the organisms are transferred to the anaerobic stripping tank to release their stored phosphorus.

In the Phostrip® sidestream process, lime is added to the supernatant from the phosphorus stripping tank to cause the phosphorus to precipitate out in a clarifier (Figures 5.3 and 5.4). Polymers may also be added to improve coagulation and flocculation of the solid precipitates so they will settle rapidly and can be removed. Sludge from the stripping tank, which contains the microorganisms, is returned to the aerobic reactor. In contrast, a mainstream luxury uptake process provides aerobic conditions where the organisms take up phosphorus in concentrated amounts but then the microorganisms are removed entirely from the wastestream, thus removing the phosphorus.

---

[12] *Volatile Acids.*   Fatty acids produced during digestion that are soluble in water and can be steam-distilled at atmospheric pressure. Also called organic acids. Volatile acids are commonly reported as equivalent to acetic acid.

[13] It is not clear if the phosphorus (PO_4) is stored within the cell or adsorbed on the organisms' cell walls, or a combination of both occurs. In any case, the organisms are removing phosphorus from the wastestream.

[14] *Polymer* (POLY-mer).   A long-chain molecule formed by the union of many monomers (molecules of lower molecular weight). Polymers are used with other chemical coagulants to aid in binding small suspended particles to larger chemical flocs for their removal from water. Also see POLYELECTROLYTE.

[15] *Metabolism.*   All of the processes or chemical changes in an organism or a single cell by which food is built up (anabolism) into living protoplasm and by which protoplasm is broken down (catabolism) into simpler compounds with the exchange of energy.

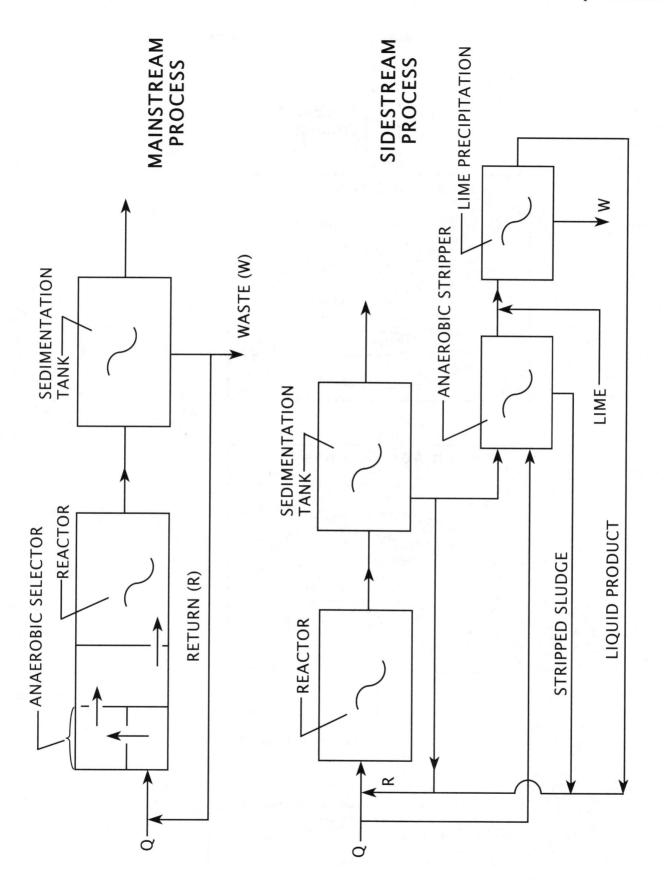

MAINSTREAM PROCESS

ANAEROBIC SELECTOR

REACTOR

SEDIMENTATION TANK

WASTE (W)

RETURN (R)

Q

SIDESTREAM PROCESS

REACTOR

SEDIMENTATION TANK

ANAEROBIC STRIPPER

LIME PRECIPITATION

W

LIME

STRIPPED SLUDGE

LIQUID PRODUCT

R

Q

Fig. 5.1   Biological phosphorus removal process layouts

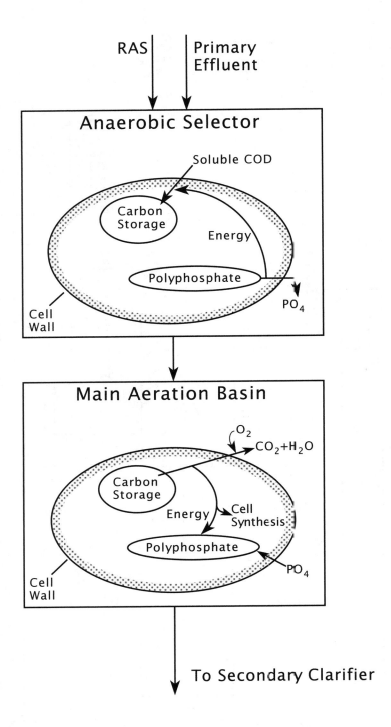

Fig. 5.2   Microorganism cell reactions during phosphorus release and luxury uptake processes

(Provided by Krishna Pagilla, Sacramento Regional Wastewater Treatment Plant)

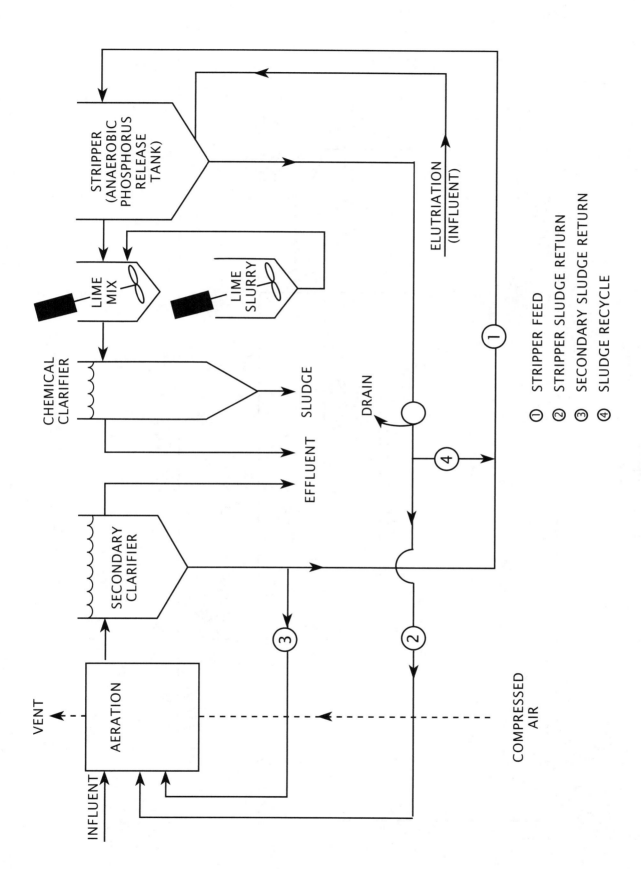

Fig. 5.3 Luxury uptake of phosphorus (elevation flow diagram)

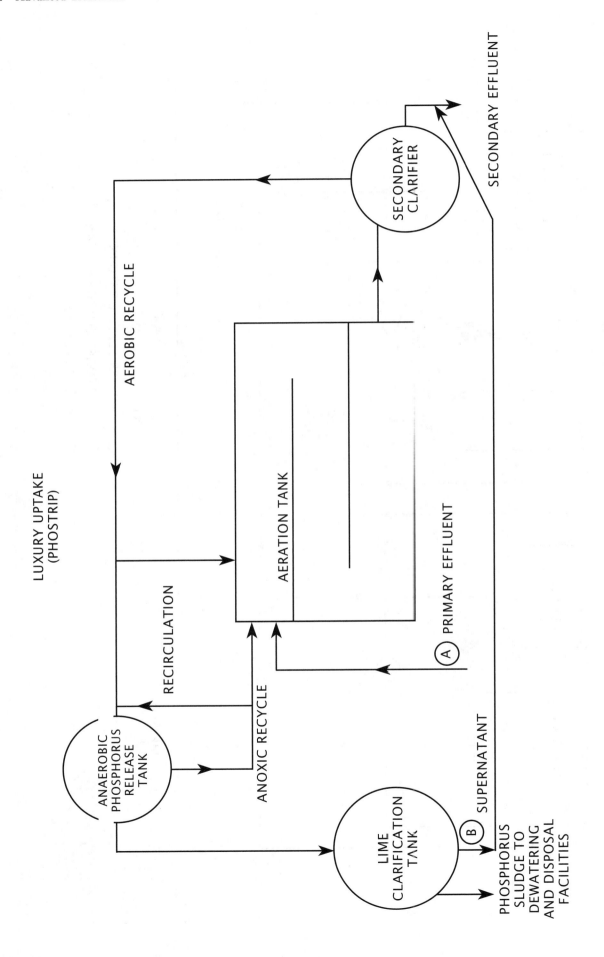

LUXURY UPTAKE (PHOSTRIP)

AEROBIC RECYCLE

SECONDARY CLARIFIER

SECONDARY EFFLUENT

AERATION TANK

RECIRCULATION

ANAEROBIC PHOSPHORUS RELEASE TANK

ANOXIC RECYCLE

(A) PRIMARY EFFLUENT

(B) SUPERNATANT

LIME CLARIFICATION TANK

PHOSPHORUS SLUDGE TO DEWATERING AND DISPOSAL FACILITIES

Fig. 5.4   Luxury uptake of phosphorus (plan flow diagram)

5.2E   What are the two process layouts commonly used for biological phosphorus removal?

5.2F   What is luxury uptake of phosphorus?

5.2G   Where and under what conditions do microorganisms release phosphorus from their cell structure in the luxury uptake process?

5.240   Basic Principles of the Luxury Uptake Process

Because luxury uptake can only take place in a very controlled environment, the microorganisms cannot be exposed to any condition that would prevent them from either taking up phosphorus or releasing it at the proper time. The basic operation requires operators to remove the activated sludge from the secondary clarifier and provide enough detention time in the anaerobic stripping tank to cause the microorganisms to release their stored phosphorus. Therefore, strict anaerobic conditions must be maintained in the stripping tank at all times. In addition, the operator must carefully regulate the detention time so that it is long enough to remove as much phosphorus as possible but not so long that the organisms die of starvation.

Once the microorganisms have released phosphorus from their cells, the activated sludge, which at this point is basically anaerobic, must be quickly returned to the aeration tank. Therefore, the sludge recycle timing is very important and a great deal of care must be taken to ensure that the sludge return is neither too fast nor too slow. Operators should be very careful of this system to ensure proper compatibility with the activated sludge portion of the plant (proper solids levels) and to maintain the highest efficiency of phosphorus removal using the luxury uptake principle.

In the Phostrip® process, the liquid from the anaerobic phosphorus release tank flows into a chemical clarification unit where lime is added to coagulate the phosphorus to enhance settling. The chemical reaction of lime with the phosphorus in the phosphorus release tank effluent is as follows:

$$5Ca^{2+} + 3PO_4^{3-} + OH^- \rightarrow Ca_5(PO_4)_3(OH)\downarrow \text{ Sludge}$$

Unlike other lime clarification processes for phosphorus removal, the luxury uptake and phosphorus stripping process only requires addition of lime to approximately ten percent of the entire plant flow stream. Once the phosphorus has been removed from the phosphorus stripper (release tank) effluent in the clarifier, the remaining liquid is combined with the secondary effluent from the plant for further (tertiary) treatment or final disposal.

5.241   Equipment (Refer to Figure 5.3)

The luxury uptake process uses the standard activated sludge treatment units plus a relatively deep anaerobic detention basin. Another unit commonly found in the luxury uptake and removal system is a lime clarification tank (clarifier), which is usually sized to treat about 10 percent of the overall plant influent. Return pumps and piping move the activated sludge from the anaerobic phosphorus release tank to the aeration tank for the luxury uptake process to begin again.

Lime precipitation for phosphorus removal requires lime feeding systems, mixing and flocculation areas, chemical clarifiers for sedimentation, and the proper pumps and piping for removal of lime-phosphorus sludge. Other equipment includes facilities for pH adjustment of the effluent, recovery of the lime, and disposal of the phosphorus sludge.

LIME FEED EQUIPMENT

Lime is usually added to the wastewater by a slaker, which is equipment that mixes dry powdered lime with water to obtain a SLURRY.[16] The slurry is then fed to the mixing basin for coagulation and flocculation of phosphorus.

MIXING BASIN

In the mixing basin, a high-speed mixer called a "flash mixer" blends the lime slurry as rapidly as possible with the phosphorus release tank effluent. Following this instant mixing (coagulation) of the lime slurry and wastewater, a slower mixing process called flocculation forms floc consisting of suspended and COLLOIDAL[17] matter, including the phosphorus precipitate.

PUMPS AND PIPING FOR LIME-PHOSPHORUS SLUDGE REMOVAL

After the lime-phosphorus mixture settles to the bottom of the chemical clarifier, the sludge is withdrawn and pumped to a thickening process for dewatering and disposal.

5.242   Safety

Operators working with any form of lime are exposed to a number of very serious hazards. Whenever dry lime (CaO) is mixed with water, for example, to prepare a lime slurry, the chemical reaction that occurs can produce very high temperatures, steam, and sometimes splattering of the mixture. Because slaked lime (Ca(OH)₂) has an extremely high pH, it is a very caustic solution and can cause eye irritation and skin irritation upon contact. Wear goggles and approved respiratory protection to prevent the lime from entering your eyes or lungs. Wear protective clothing to prevent skin contact. If any part of your body is exposed to lime, be sure to rinse thoroughly with water. In the case of severe burns, rinse thoroughly with copious

---

[16] *Slurry.*   A watery mixture or suspension of insoluble (not dissolved) matter; a thin, watery mud or any substance resembling it (such as a grit slurry or a lime slurry).

[17] *Colloids* (KALL-loids).   Very small, finely divided solids (particles that do not dissolve) that remain dispersed in a liquid for a long time due to their small size and electrical charge. When most of the particles in water have a negative electrical charge, they tend to repel each other. This repulsion prevents the particles from clumping together, becoming heavier, and settling out.

amounts of water and see a physician immediately. An emergency eye wash facility and deluge shower should be readily available in lime handling and storage areas and wherever contact could occur.

Polymers are often used to help flocculate colloidal particles of lime and phosphorus to provide faster sedimentation in a chemical clarification unit. Wet polymers are extremely slippery. Use inert absorbent materials such as sand or earth or a strong saltwater solution to clean up spilled polymers. Be very careful when walking on or near surfaces that have been exposed to any kind of polymer. Install and use guardrails on stairs and around concrete areas where polymers may accumulate or be spilled on the surface.

Anaerobic phosphate stripping tank gases are another potential safety hazard. As with any unit of wastewater treatment, certain gases may be released from the tank whenever anaerobic conditions prevail. Operators and other personnel must understand the potential hazards and be very cautious around any tank where anaerobic conditions exist and where toxic or explosive gases may accumulate. Just as it is important not to smoke around an anaerobic digester, it is also important not to smoke around an anaerobic phosphorus stripping tank. The gases given off could include methane, which could cause explosions or a fire. Furthermore, oxygen deficiencies can exist within these tanks near the water surface. Oxygen level testing should be done before entering an operating or recently drained tank to prevent suffocation. Appropriate signs must be posted in conspicuous places warning of specific hazards in each area.

QUESTIONS

Write your answers in a notebook and then compare your answers with those on page 506.

5.2H   List the equipment used for lime clarification in the luxury uptake process.

5.2I   How can lime be harmful to your body?

5.2J   Why should you not smoke around a phosphorus stripping tank?

5.243   Start-Up, Operation, and Shutdown

5.2430   PRE-START-UP

1. Lime. Lime (calcium oxide) is usually purchased in dry form and must be mixed with water to form a slurry in order to be fed to a wastewater treatment process. The chemical mixture is:

Calcium Oxide + Water → Calcium Hydroxide

$$CaO + H_2O \rightarrow Ca(OH)_2$$

Before start-up, check the calcium oxide strength to be sure that a high concentration of lime is available for the chemical reaction necessary to precipitate phosphorus. Chemical strength can be tested by determining the percentage of available calcium oxide in the dry lime. A concentration of at least 90 percent calcium oxide is needed to ensure a highly reactive slurry for proper precipitation of phosphorus. (See Chapter 16, "Laboratory Procedures and Chemistry," Section 16.47, "Lime Analysis," Volume II, *OPERATION OF WASTEWATER TREATMENT PLANTS*, for the testing procedure.)

2. Lime Feeding Equipment. A routine check of lime feeding equipment is necessary before start-up and several times during each work shift. Since most dry lime contains a certain amount of grit, rocks, and sand, a grit removal system associated with the slaker or lime mixing and feed system is important to prevent plugging and equipment wear.

A rock-hard lime precipitate called calcium carbonate (limestone) will form when lime combines with carbon dioxide. This rock-hard substance will attach itself to almost anything, including slaking mechanisms and piping. To operate properly, lime slaking equipment must be kept free from a serious buildup of calcium carbonate. For instructions on removing calcium carbonate, see Section 5.244, "Maintenance."

3. Pumps, Valves, and Piping. The most serious maintenance problem in a lime system is the formation of limestone. All pumps, valves, and piping must be regularly checked and cleaned to prevent a buildup of limestone scale, which can cause plugging and malfunction. This applies to both the lime feed system, which carries lime to the mixing chamber, and to the chemical clarification unit. Similar maintenance procedures also apply to the lime slurry return system, which takes settled lime-phosphorus sludge from the bottom of the clarifier and conveys the sludge to further dewatering and disposal processes.

4. Clarifier Mechanism. The lime-phosphorus unit functions the same way as a secondary clarifier. Before start-up, check to be sure the clarifier sweep arm moves at a slow, steady rate to collect the settled lime-phosphorus sludge at the bottom of the tank. Also, check the design specifications for the chemical clarifier to be certain that the overflow rates (weir loading rates and hydraulic loading rates) are not exceeded when starting up the chemical clarifier. The clarifier operates best at or below the overflow rate for which it was designed.

5.2431 START-UP

As you start up the phosphorus removal process, check the pH adjustment in the rapid-mix basin to be certain that the pH of the combined wastewater and lime slurry is 11 or above. If the pH falls below 11, phosphorus removal efficiency could be reduced. The first 75 percent of phosphorus precipitation usually occurs before the pH drops below 10. It may be necessary to increase the lime dosage to keep the pH near 11 in order to reduce phosphorus to very low levels. Measuring the pH manually on a routine basis is a good way to double-check the automatic pH adjustment and recording mechanism installed in the rapid-mix basin area.

5.2432 MONITORING DAILY OPERATION

To ensure smooth, efficient operation of a biological phosphorus removal system, the operator will need to carefully monitor the lime feed and mixing systems, the pH level in the chemical clarifier, the state of anaerobic conditions in the phosphorus stripping tank, and dissolved oxygen levels in the aeration basin.

The lime feed system must operate very reliably. Make frequent checks (several times each shift) of the automatic dry lime feed system, the mixing of dry lime and water, the slurry transfer to the rapid-mix basin, and the grit removal system that removes sand from the lime slurry. As lime is fed into the phosphorus stripping tank effluent flow stream, it is very important that the mixing time and lime feed rate (as well as polymer dose, if used) be correct. You must keep the pH above 11 to promote substantial floc formation and encourage rapid settling of the largest possible floc.

If the lime clarification process is automatically controlled, check the clarifier influent flow with a pH meter every 8 hours to be sure the pH is correct and that the automatic feed control is functioning properly. Daily tests for removal of phosphate compounds through the chemical clarification system (grab samples of influent and effluent) should be run to determine the precise pH setting that works best for your treatment plant and the operating conditions at your facility. (The phosphate tests are also required by state regulatory agencies that monitor phosphate levels in the final effluent from the wastewater treatment facility.)

Lower a dissolved oxygen probe into the anaerobic phosphate stripping tank on a regular basis to ensure that no dissolved oxygen is present. If dissolved oxygen is detected, it may be a sign that sludge is being fed into the tank too fast or that sludge is being withdrawn too quickly from the stripping unit.

The entire process of biological phosphorus removal depends on maintaining a healthy mass of microorganisms. Therefore, detention time in the anaerobic phosphorus stripping tank must be carefully controlled. The organisms must remain there long enough to release the maximum amount of phosphorus from their cells, but must not be left in the anaerobic environment so long that they die off. The detention time can be calculated using the following formula:

$$\text{Detention Time, hours} = \frac{(\text{Tank Volume, gallons})(24 \text{ hr/day})}{\text{Flow, gallons/day}}$$

When the activated sludge from the phosphorus stripping tank is returned to the aeration tank, it is important that there be sufficient food, nutrients, and oxygen available to revive the microorganisms. Periodically, check the conditions at the site where the sludge enters the aeration tanks to be sure adequate dissolved oxygen is present (2 to 4 mg/L DO). If the dissolved oxygen level drops substantially, it may be necessary to increase the air supply to that section of the aeration tank where the activated sludge is returned from the stripping unit.

Lime-phosphorus sludge must not be allowed to build up on the bottom of the clarifier. Excessive sludge accumulations will reduce the effective capacity of the clarifier by reducing the detention time, overload sludge removal equipment, and adversely affect settling conditions. Check the sludge pumping equipment regularly to be sure it is functioning properly and that the lime-phosphorus sludge is not building up in the clarifier.

Improving the performance of biological phosphorus removal systems (both mainstream and sidestream types) usually involves taking one or more of the following steps:

- Achieving more hydrolysis and volatile fatty acid formation in the anaerobic stage

- Avoiding or minimizing the introduction of dissolved and chemically bound oxygen (nitrate nitrogen) in the anaerobic stage

- Avoiding or minimizing the opportunities for phosphorus resolubilization due to periods of extended aeration or anaerobic holding

- Stabilizing and smoothing out any uneven return flows of solubilized phosphorus

- Increasing the extracted phosphorus removal rate through increased sludge production

QUESTIONS

Write your answers in a notebook and then compare your answers with those on page 506.

5.2K What routine checks should an operator make before starting up lime feeding equipment?

5.2L What is calcium carbonate and what effect does it have on equipment?

5.2M How often should the pH be tested manually in the lime clarification tank to ensure that the automatic controls are functioning properly?

5.2433 LOADING GUIDELINES

1. Typical Loading Rates. The loading rates for various units will depend on the design and the operation of that specific treatment facility.

2. Hydraulic Loading for Phosphorus Stripper. The hydraulic loading for an anaerobic phosphorus stripper depends on the dissolved oxygen of the activated sludge when it enters the anaerobic stripper. The hydraulic loading also depends on

the ability of the anaerobic phosphorus stripper to remain anaerobic at all times during the release of phosphate from the cell structure of the microorganisms.

$$\text{Hydraulic Loading, days} = \frac{\text{Stripper Solids, lbs}}{\text{Solids Leaving Stripper, lbs/day}}$$

3. Hydraulic Loading Rate for Chemical Clarifier. The typical loading rate for a chemical clarification unit is normally 800 gallons per day per square foot (32 m³ per day/sq m) of surface area to 1,500 gallons per day per square foot (60 m³ per day/sq m) of surface area.

The operator must know two things to calculate the hydraulic loading rate (also called the overflow rate) for a chemical clarification unit: (1) the flow into the unit, and (2) the surface area of the clarifier. To calculate the hydraulic loading rate, divide the average gallons per day (cubic meters per day) of flow to the clarifier by the square feet (square meters) of surface area.

$$\text{Hydraulic Loading, GPD/sq ft} = \frac{\text{Flow, GPD}}{\text{Surface Area, sq ft}}$$

$$\text{Hydraulic Loading, m}^3/\text{day/sq m} = \frac{\text{Flow, m}^3/\text{day}}{\text{Surface Area, sq m}}$$

5.2434   SHUTDOWN

If the lime clarification system for phosphorus precipitation must be shut down, take the following steps in the order listed:

1. Shut down the activated sludge feed from the secondary clarifier to the anaerobic phosphorus release tank. (See Figure 5.3, page 481.)

2. Remove sludge from the anaerobic phosphorus release tank.

3. Shut down the lime feed equipment.

4. Close the influent valve to the chemical clarifier.

5. Pump the settled lime sludge from the chemical clarifier.

6. If necessary, pump any remaining liquid from the chemical clarifier into a *RECARBONATION*[18] basin.

7. Inspect the empty clarifier and other tanks and perform any repairs that are necessary.

Also, be aware that shutting down a sidestream phosphorus removal process may cause operational problems in the activated sludge treatment processes. These problems could include: (1) the appearance of straggler floc, or (2) some decrease in the dissolved oxygen uptake rate. Both of these conditions require operator attention. Refer to Chapters 8 and 11 of Volumes I and II, *OPERATION OF WASTEWATER TREATMENT PLANTS,* and Chapter 2 in this manual on activated sludge in order to ensure that the treatment facility is operating in the proper mode.

5.2435   ABNORMAL AND EMERGENCY OPERATING CONDITIONS

Lower- and higher-than-normal flow conditions require operators to be alert for necessary changes in sidestream processes. If the lower or higher flows are temporary, then minor adjustments may be appropriate. If the flow changes will continue for longer than one day, then adjustments for the long term are necessary.

During low-flow conditions, check the automatic lime feed system to be sure it has adjusted to the reduced flows. Excessive lime feed rates into the clarifier causes a waste of chemicals and increase in costs. If the automatic lime feed system cannot be throttled down far enough, manual feed control is necessary.

High flows can pose a more serious problem to lime clarification processes. First, high flows may lower the pH in the treatment basins if the lime feed system is not properly pacing the flow, that is, the lime feed system is not adding more lime when the high flow conditions occur. Secondly, high flows may hydraulically overload the chemical clarification units thereby decreasing the phosphorus removal efficiency of the clarification unit.

5.2436   SOLIDS MANAGEMENT

Ultimately, effluent quality depends on the ability of the plant and the operator to properly handle solids. Specifically, the operator must understand and manage the impact of solids in recycle flows and must remove the sludge solids from the plant as early in the process as possible.

Solids management begins at the sedimentation tank. The ideal situation is unhindered[19] settling velocities and no sludge blanket that could be resuspended in the tank effluent when higher flow conditions occur. It is important not to overload the secondary clarifiers. Many plants try to maintain a constant ratio of the return sludge rate to the inflow rate. When proper conditions are achieved, using a constant sludge return-to-influent ratio, the return sludge concentration can vary to accommodate the daily ups and downs and the higher solids loadings that occasionally occur. Once the system operation is stable, it is only necessary to set solids wasting rates to achieve your desired overall system MCRTs, review performance, and occasionally make any slight adjustments that may be needed. Oxygen supplies should be controlled in a similar manner. Determine and maintain appropriate dissolved oxygen levels at critical process points. Minor adjustments may be necessary to meet effluent objectives when influent loads change.

Solids separation from the wastewater should emphasize the capture and removal of the solids, not solids concentration and storage in a treatment process. Once solids are removed by any process, avoid or minimize returning the solids in a recycle flow to an upstream process. (An obvious exception to this rule is the

---

[18] *Recarbonation* (re-kar-bun-NAY-shun).   A process in which carbon dioxide is bubbled into the water being treated to lower the pH.

[19] Unhindered liquid-solids separation occurs when lower suspended solids concentrations allow free-falling (settling) characteristics. Hindered settling is caused by higher SS concentrations blocking free-falling (settling) characteristics.

recycle of activated sludge organisms, which are deliberately transferred between treatment phases in some systems.) Be aware that the processing system can be upset at any point where solids may accumulate (aeration basin, gravity thickener, or storage) because the solids could wash out and return to the wastewater being treated at any time.

Biological treatment processes become more easily upset when operations are intermittent or irregular. If industrial discharges make up a significant portion of plant influent, manufacturing production changes, shift cleanup, or weekend slowdowns or shutdowns could cause intermittent flow volumes or pollutant concentrations. When intermittent operations are planned or expected, reduce the solids inventories to give yourself reserve process capacity well beyond what you would need with a continuous processing system. One way to do this is to recycle flows and use solids equalization tanks. Then use slow bleedback from the equalization tank, especially when the pollutants are dissolved in the wastestream. This approach is particularly important when influent flows are intermittent and as effluent standards become more stringent.

Do not simply react to daily events. Study your plant's operating history and then plan and implement a sludge removal schedule. Do not attempt to store solids in tanks that are intended to separate or concentrate solids unless there is no other choice, and then do it only on an emergency basis.

5.2437 OPERATING STRATEGY

Many operators (and managers) have a tendency to overreact to daily events. Take note of daily process changes but wait for sustained changes and monitor process trends. Trust your own common sense and base your decisions on sound fundamentals. When making a process change, remember that all processes reflect changing influences of process recycle flows as well as immediate changes associated with influent conditions or chemical reactions. Your operating strategy needs to include a fallback position you can use if performance fails to meet your expectations.

The Mean Cell Residence Time (MCRT) is the key to understanding how well the microorganisms are adapting and how stable the process is. Do not attempt to interpret the success or failure of a new processing strategy until at least two or three times the overall operating MCRT of the system has elapsed and, thereafter, evaluate its success over at least one or more MCRT intervals. For example, if the desired MCRT is three days, wait six to nine days before deciding whether the change was beneficial or before making further changes in the process.

5.2438 SLUDGE DISPOSAL

Methods of treatment and disposal of the lime-phosphorus sludge withdrawn from the chemical clarification tank vary from facility to facility. Some treatment plants use sludge drying beds to dry the lime sludge. In some cases, the dried solids receive further processing to recover a portion of the lime. In other treatment facilities, the dry solids are simply disposed of in an approved landfill or by some other appropriate means.

To recover lime for reuse, many treatment facilities use centrifuges (usually two in series) to recover as much of the lime as possible. By operating the first centrifuge at a low solids removal efficiency, most of the phosphorus sludge is discharged with the CENTRATE[20] while most of the calcium carbonate is removed from the centrifuge as a cake ready for the lime recalcining recovery process through a multiple-hearth furnace. Calcium carbonate can make up about three-quarters of the mass of the sludge. The centrate from the first centrifuge (containing most of the phosphorus sludge) is passed through the second centrifuge for further dewatering of the phosphorus cake. The centrate from the second centrifuge is usually returned to the primary sedimentation system and the dewatered phosphorus sludge (cake) is disposed of in a landfill or by some other appropriate sludge disposal mechanism.

QUESTIONS

Write your answers in a notebook and then compare your answers with those on pages 506.

5.2N   What information is needed to calculate the hydraulic loading on a chemical clarifier?

5.2O   What are ideal conditions in the sedimentation tank?

5.2P   What should be the emphasis of solids separation from the wastewater being treated?

5.2Q   How should an operator attempt to respond to intermittent or irregular operations?

5.2R   How can the lime in lime-phosphorus sludge be recovered for reuse?

5.244 Maintenance

5.2440 PIPING

The piping of a phosphorus stripping process using luxury uptake will be similar to an activated sludge plant except that three additional tanks are provided. These tanks are the stripper

---

[20] *Centrate.*   The water leaving a centrifuge after most of the solids have been removed.

(the anaerobic phosphorus release tank), the rapid-mix tank, and the chemical clarification unit. The piping that will need the most care and attention is the lime clarification process piping. Because of the tendency of lime to form scale, frequent maintenance is necessary. Lime builds up on the interior walls of pipes and may completely plug the pipes at a rapid rate. It is necessary to flush and scour pipes periodically in order to maintain the full carrying capacity of the pipelines. Hot water or steam is very effective in dislodging limestone buildup within pipes or pumps. The hot water softens the calcium carbonate scale enough that it can be chipped or scoured away relatively easily.

5.2441   PUMPS AND EQUIPMENT

The system for pumping the activated sludge into and out of the anaerobic phosphorus release tank (stripper) should be equipped with variable-speed controls so that the operator can adjust the feed rate into and out of the stripping unit. The pumps and equipment must be well maintained because the phosphorus stripping process depends heavily on fairly precise timing at each step. For example, the microorganisms in the activated sludge must not release phosphorus before they reach the anaerobic stripping tank. Also, the activated sludge that has undergone anaerobic conditions must be returned to the aeration tank as quickly as possible. Both of these operations require pumps and equipment that operate dependably with a minimum of downtime.

5.2442   LIME SLAKING MECHANISM

The lime slaking mechanism must be kept in good operating order to provide the correct amount of lime slurry for the lime feed station. Grit must be removed from the slaker mechanism and the mixing arms must be kept free of excessive buildups of lime. The slaking process generates a considerable amount of heat. Be sure to check the slaker manufacturer's recommendation on temperature. Frequently, inspect the water sprays used for condensing the steam and for controlling dust. The lime slaking mechanism should also be cleaned frequently to prevent lime buildup. *CAUTION: Whenever you are working with lime, particularly around lime slaking equipment, wear appropriate personal protective equipment and use proper procedures. Lime is extremely hazardous.*

5.2443   LIME SLURRY AND MIXING OPERATION

Lime has a tendency to build up on the surface of the mixing paddles in a rapid-mix chamber. Be sure that no excessive amounts of lime are allowed to gather on any mixing paddles. This could cause heavily weighted sides that set the mechanism off balance. The mixing paddles should be cleaned frequently to prevent lime buildup.

5.2444   SLUDGE WITHDRAWAL AND DISPOSAL PUMPS

The lime sludge pumps (usually centrifugal) must be kept clean and free of scale buildup. Use hot water or steam to soften the calcium carbonate scale that tends to form a crust on the impeller vanes. Then, chip or scour the softened scale from the affected surfaces and flush it away.

5.245   Calculating Process Efficiency

Phosphorus removal efficiency for a sidestream luxury uptake system can be calculated by comparing the phosphorus level in the primary effluent (point A on Figure 5.4) with the phosphorus level in the effluent from the chemical clarification tank (point B on Figure 5.4).

$$\text{Process Treatment Efficiency, \%} = \frac{(\text{Phos Conc In, mg}/L) - (\text{Phos Conc Out, mg}/L)}{\text{Phos Conc In, mg}/L} \times 100\%$$

In addition to removing phosphorus, a chemical clarifier also removes turbidity from the wastewater being treated. The operator should check both the turbidity removal efficiency of the clarifier and the phosphorus removal efficiency. Clarifier removal efficiencies are determined by collecting and analyzing clarifier influent and effluent samples. Turbidity is a measure of the clarity of the effluent; therefore, turbidity removal efficiency measures the effectiveness or performance of the process.

5.246   Review of Plans and Specifications

LIME STORAGE AND UNLOADING FACILITIES

If lime is used to coagulate and settle phosphorus released from the wastewater in the phosphorus stripping tank, the operator can provide useful information to the design engineer and should have the opportunity to review the plans and make comments. The following are some of the items that should be reviewed by operators to aid the design engineer during the facility's design for the lime portion of biological phosphorus removal:

1. Quicklime (powdered calcium hydroxide) is abrasive. Therefore, pipelines used to transfer dry lime should be made of glass-lined pipe and should be designed with large sweeping curves. The sweeping curves will minimize the centrifugal force and velocity of the lime particles and thus decrease erosion of the interior of the pipe wall on the curves while transferring lime from the hauling vehicle to the storage bins.

2. Lime-water generates steam. The quick chemical reaction of lime with water produces a great deal of heat and generates a certain amount of steam. The steam can cause lime in the storage bin to clump into large particles that will interfere with the feed mechanism. Clumping can cause bridging (also

called *ARCHING*[21]); the resulting erratic flow of lime into the chemical slakers will result in uneven concentrations of lime solution. Provisions should be made for water sprays to control dust and to condense vapors so they will not rise into the storage bin.

3. Lime feed piping needs to be accessible. Since lime tends to build up on metal and other types of surfaces, it is important to be able to clean lime feed pipes. If the lime mixing and slaking area is a long distance from the rapid-mix basin, several cleanouts should be installed along the pipe route to ensure easy access for rodding.

4. Piping arrangement should be flexible. Pumps and pipes plug up frequently in a lime sludge pump station; therefore, alternate piping and valving should be provided so that lime sludge removal can continue while cleaning and repairs are made on the affected equipment.

5. Handling of dry lime from the unloading dock. The trucks hauling dry lime to storage bins at the treatment plant must have enough room to maneuver and park in the unloading area. Usually, the trucks have their own pneumatic exhaust system to transfer the lime from the truck to the storage bins. If the bins are taller than the truck, it is critical that the feed line first go straight up from the unloading vehicle. Lime will be prevented from depositing on the bottom of the pipe with the air flowing over the top. This arrangement will allow much faster unloading time for the vehicles.

6. Dust control must be provided. Because lime is extremely hazardous to human health, it is important that proper dust control be provided. Exhaust fans and lime filter covers must be provided to keep lime dust from spreading throughout the lime feeding building or into the operational area causing operators breathing problems and other unpleasant conditions. Approved respiratory and eye protection should be worn during unloading operations.

7. Safety around lime storage bins. Pneumatic feeding mechanisms are the most commonly used devices for transferring lime to the storage bins. To prevent the release of airborne lime particles, storage bins must be properly vented and dust collectors should be installed at each vent.

8. The chemical storage and supply building should be equipped with emergency eye wash facilities and deluge showers and floor washdown equipment. Electrical control panels and chemical feed pumps should be protected from accidental spills or splashing during washdown. If polymers are used for coagulation, all possible locations of polymer spillage should have a false grated floor, with washdown drain, to allow the spilled material to pass to a point where its *VISCOUS*,[22] slippery nature is not a threat to operator safety.

The ultimate destination of floor drains should always comply with sewer-use ordinances or discharge permits.

MAINTAINING PROPER DISSOLVED OXYGEN LEVELS

The operator may want to insist on installation of automatic dissolved oxygen probes in the anaerobic stripping tank. Warning signals and devices can be included to ensure that the operator is notified if oxygen levels in the stripping tank rise above zero. Automatic dissolved oxygen meters and controls can also be very useful in maintaining adequate dissolved oxygen in the aeration tank after the anaerobically treated sludge has been returned to the aeration tanks.

SOLIDS AND SLUDGE HANDLING FACILITIES

Biological phosphorus removal requires facilities to handle at least one-half the amount of sludge that would be produced by chemical methods of phosphorus removal. In some areas with stringent discharge limits, the additional solids produced by phosphorus removal strategies may make it necessary to filter the secondary treatment process effluent. For example, filtration would probably be necessary where the routine effluent SS limit is 30 mg/L or less or the effluent phosphorus limit is 1 mg/L or less.

5.247 References

1. *BIOLOGICAL-CHEMICAL PROCESS FOR REMOVING PHOSPHORUS* by Union Carbide Corporation under a grant from the US Environmental Protection Agency, Cincinnati, Ohio.

2. *THE PHOSTRIP PROCESS* by Union Carbide Corporation, Tonawanda, New York.

QUESTIONS

Write your answers in a notebook and then compare your answers with those on page 506.

5.2S How can limestone scale be removed from equipment and piping?

5.2T What general types of items should an operator look for when reviewing plans and specifications for a biological phosphorus removal system?

5.2U Why might an operator insist on installation of automatic dissolved oxygen probes in the phosphorus stripping tank?

END OF LESSON 1 OF 2 LESSONS

on

PHOSPHORUS REMOVAL

Please answer the discussion and review questions next.

---

[21] *Arch.* (1) The curved top of a sewer pipe or conduit. (2) A bridge or arch of hardened or caked chemical that will prevent the flow of the chemical.

[22] *Viscosity* (vis-KOSS-uh-tee). A property of water, or any other fluid, that resists efforts to change its shape or flow. Syrup is more viscous (has a higher viscosity) than water. The viscosity of water increases significantly as temperatures decrease. Motor oil is rated by how thick (viscous) it is; 20 weight oil is considered relatively thin while 50 weight oil is relatively thick or viscous.

DISCUSSION AND REVIEW QUESTIONS
Chapter 5.   PHOSPHORUS REMOVAL
(Lesson 1 of 2 Lessons)

At the end of each lesson in this chapter you will find some discussion and review questions. The purpose of these questions is to indicate to you how well you understand the material in the lesson. Write the answers to these questions in your notebook.

1. Why is oxygen so important in biological treatment processes?

2. What is the luxury uptake of phosphorus?

3. In the luxury uptake process, what happens if the sludge feed rate is (a) too high, or (b) too low into or out of the phosphorus stripping tank?

4. Why is it important to closely control the detention time in the anaerobic phosphorus stripping tank?

5. Why should sludge solids usually be captured and removed from the treatment system rather than being recycled or allowed to accumulate?

6. The phosphorus stripping process using luxury uptake is similar to an activated sludge plant with the exception of what tanks?

7. Why should the system that pumps the activated sludge into and out of the anaerobic phosphorus stripping tank be equipped with variable-speed controls?

8. What safety precautions should you take when working with or near lime?

9. Why must the lime slurry and mixing equipment and the lime sludge pumps be kept clean at all times?

10. If polymers are used at a plant, what safety provisions are necessary?

CHAPTER 5. PHOSPHORUS REMOVAL

(Lesson 2 of 2 Lessons)

Many aspects of phosphorus removal by lime precipitation were described earlier in Lesson 1 of this chapter where lime precipitation was a component of the biological process of phosphorus removal. In the biological phosphorus removal process, activated sludge microorganisms take up the phosphorus into their cells and later release it in a phosphorus stripping tank. The phosphorus is then removed from the stripping tank sludge by lime precipitation or some other chemical precipitation process. However, many wastewater treatment facilities use only chemical means to remove phosphorus. The remainder of this chapter explains how to remove phosphorus from wastestreams by chemical means alone, using lime or alum to precipitate the phosphorus so that it can be removed from the wastestream.

5.3 LIME PRECIPITATION

5.30 How the Lime Precipitation Process Removes Phosphorus

There are three general physical or chemical reactions that take place during lime precipitation for phosphorus removal (Figure 5.5).

1. Coagulation. When chemicals are added to wastewater, the result may be a reduction in the electrostatic charges that tend to keep suspended particles apart. After chemical addition, the electrical charge on the particles is altered so that the suspended particles containing phosphorus tend to come together rather than remain apart.

2. Flocculation. Flocculation occurs after coagulation and consists of the collection or AGGLOMERATION[23] of the suspended material into larger particles. Gravity causes these larger particles to settle.

3. Sedimentation. As discussed in previous chapters on primary and secondary clarification methods, sedimentation is simply the settling of heavy suspended solid material in the wastewater due to gravity. The suspended solids that settle to the bottom of clarifiers can then be removed by pumping and other collection mechanisms.

For a more detailed description of the physical and chemical reactions involved in coagulation and flocculation, refer to Chapter 4, Section 4.10, "How Coagulation/Flocculation Works."

5.31 Equipment

Lime precipitation for phosphorus removal requires lime feeding systems, mixing and flocculation areas, chemical clarifiers for sedimentation, and the proper pumps and piping for removal of lime-phosphorus sludge. Other equipment includes facilities for pH adjustment of the effluent, recovery of the lime, and disposal of the phosphorus sludge. More specifically, the equipment needed for precipitation includes the following:

1. Lime Feed Equipment. Lime usually comes in a dry form (calcium oxide (CaO)) and must be mixed with water to form a slurry (calcium hydroxide ($Ca(OH)_2$)) in order to be fed to a wastewater treatment process to produce the required results.

Calcium Oxide + Water → Calcium Hydroxide

$$CaO + H_2O \rightarrow Ca(OH)_2$$

Since most dry lime contains a certain amount of grit, rocks, and sand, a grit removal system associated with the slaker or lime mixing and feed system is important to prevent plugging and equipment wear.

2. Mixing Chamber. A basin in which the lime slurry is blended with the wastewater as rapidly as possible with the use of a high-speed mixer called a "flash mixer." After this instant mixing of the lime slurry and wastewater, a slower mixing process called flocculation follows to allow the formation of floc. This floc consists of suspended and colloidal matter, including the phosphorus precipitate.

3. Clarification Process. Clarification is used to allow the floc to settle out of the wastewater being treated. In order to settle lime-phosphorus sludges, the velocity of the flowing wastewater must be slowed down sufficiently to allow for sedimentation. Because the coagulation and flocculation processes produce heavier particles, a clarifier similar to a secondary sedimentation clarifier may be used in the process to settle lime-phosphorus sludges.

4. Pumps and Piping for Lime-Phosphorus Removal Process. After the lime-phosphorus mixture has been settled on the bottom of the chemical clarifier, pipes and pumps are used to transport the sludge to a thickening process for further dewatering and disposal.

---

[23] Agglomeration (uh-glom-er-A-shun).   The growing or coming together of small scattered particles into larger flocs or particles, which settle rapidly. Also see FLOC.

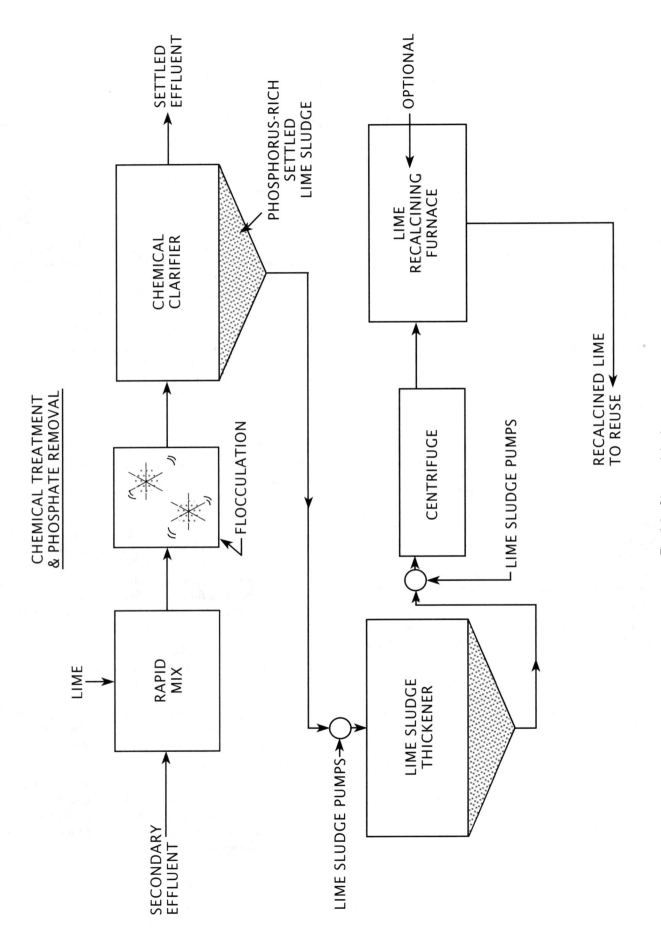

Fig. 5.5   Lime precipitation process

5.32 Operation

5.320 Start-Up

For a description of how to start up a lime precipitation process for phosphorus removal, please refer to the information presented earlier in this chapter in Section 5.2430, "Pre-Start-Up," and Section 5.2431, "Start-Up."

5.321 Operating Procedures

1. Flow Rate into Chemical Clarifier. Check the design specifications for the chemical clarifier to be certain that the overflow rates (weir loading rates and hydraulic loading rates) are not exceeded. The clarifier operates best at or below the overflow rate that was designed into the facility.

2. Lime Feed for pH Control. pH adjustment for phosphorus removal means raising the pH to a very highly alkaline state (pH of 11 or higher) so that phosphorus and calcium hydroxide bond together forming a heavier substance that will settle out in a clarification process. Adding lime (calcium hydroxide) to the wastewater being treated will produce a pH high enough to allow the formation of this lime-phosphorus precipitate. The pH must be maintained above 11.0 in order to achieve the highest possible phosphorus removal; however, a large fraction of the phosphorus can be removed at a pH as low as 9.0.

   Check the pH adjustment in the rapid-mix basin to be certain that the pH of the combined wastewater and lime slurry is 11 or above. If pH falls below 11, phosphorus removal efficiencies could be reduced. The first 75 percent of phosphorus reduction usually occurs before a pH of 10. In order to reduce phosphorus to very low levels, the pH must be higher and overall lime demands become much higher. Measuring the pH on a regular and routine basis with a pH meter will provide a double-check on the automatic pH adjustment and recording mechanism installed in the rapid-mix basin area.

3. Clarification and Settling Process. After mixing, flocculation, and coagulation, the lime-phosphorus precipitate is ready for settling in a clarification tank. The hydraulic loading rates must be adjusted to prevent short-circuiting or hydraulic washout of the floc prior to its complete settling to the bottom of the clarifier. A well-operated chemical clarifier will be very clear and you should be able to see down into the clarifier at least 10 feet (3 m).

4. Pumping and Disposal of Lime Precipitate. Once the lime precipitate containing phosphorus is removed from the wastewater stream, it is important not to allow the same phosphorus to be recycled back through the treatment plant. When pumping lime precipitate from the clarifiers, adjust pumping rates so that all of the lime sludge is removed. Without proper pump regulation and adequate pumping times, heavy accumulations of lime could build up within the clarifier. Poor pump regulation could also lead to the pumping of only a very thin slurry composed mainly of water and little lime sludge to the disposal facilities.

Two methods are commonly used to dispose of lime-phosphorus sludge. In the first method, a centrifuge is used to remove the phosphorus from the lime mud. The remaining lime sludge can be further processed to recover the lime. In the second method, the phosphorus-lime sludge is simply pumped to an appropriate disposal site.

Some treatment plants use sludge drying beds to dry the lime sludge prior to final disposal by landfilling or other means. Other treatment plants recover as much of the lime as possible for reuse. These treatment plants must use two centrifuges. By operating the first centrifuge at a low removal efficiency, most of the phosphorus sludge is discharged with the centrate while most of the calcium carbonate is removed from the centrifuge as a cake ready for the lime recovery process in a multiple-hearth furnace. Calcium carbonate can make up about three-quarters of the mass of the sludge. The centrate containing most of the phosphorus sludge is passed through the second centrifuge for separation of the phosphorus compounds as a cake. The centrate is usually returned to the primary sedimentation system and the dewatered phosphorus sludge (cake) goes to a landfill for ultimate disposal.

If the lime sludge is to be dewatered, check to be sure that the percent concentration of solids remaining after the dewatering process will provide the most efficient lime recovery operation.

The residual solids management portion (Chapter 3) of this manual discusses thickening and centrifugation processes in more depth. This chapter is very closely related to solids handling and you will need to understand the solids handling chapter thoroughly before you can reach a thorough working knowledge of the phosphorus removal system using lime precipitation.

5. Daily Maintenance of Pumps, Piping, and Other Equipment to Prevent Lime Scale Plugging. Lime (calcium hydroxide) and carbon dioxide form what is known as limestone or calcium carbonate. As explained previously, calcium carbonate is a very stubborn substance that sticks to all types of surfaces. This scaling ability of the calcium carbonate causes pumps, piping, and other equipment to scale very readily and they must be cleaned to prevent plugging problems. Daily maintenance will ensure that pumps operate properly and that pipes do not become completely plugged. Hot water or steam is very effective in dislodging limestone buildup within pipes or pumps. The hot water makes the calcium carbonate scale (limestone) soft and it can be chipped or scoured away relatively easily.

5.322 Shutdown

If the lime clarification system for phosphorus precipitation must be shut down, take the following steps in the order listed:

1. Shut off valve to clarifier basin stopping the secondary effluent flow into the chemical clarifier.

2. Shut down the lime feed equipment.

3. Bypass chemical clarifier by opening proper valves beyond secondary clarifier.

4. Pump the settled lime sludge from the clarifier basin.

5. If necessary, pump any remaining liquid from the chemical clarifier into a recarbonation basin.

6. Inspect empty clarifier and perform any repairs that are necessary.

7. Flush equipment and chemical lines with water.

5.323   Sampling and Analysis

1. Phosphorus Removal Efficiencies.   The purpose of lime precipitation of phosphorus is to reduce the phosphorus level in the effluent of the wastewater treatment plant and thereby remove a nutrient source for algae in the receiving waters. Daily phosphorus tests should be run on composite samples of chemical clarifier effluent and also secondary clarifier effluent to see a comparison of results for determining three factors:

a. Which pH setting works best for this treatment plant and these operating conditions?

b. Does the treatment plant meet effluent discharge requirements for phosphorus?

c. Is the chemical clarification lime precipitation process adequately performing at the efficiencies desired?

Along with phosphorus removal, a chemical clarifier is capable of removing turbidity from the secondary effluent at a wastewater treatment plant. Check the turbidity removal efficiency through the chemical clarifier as well as checking for phosphorus removal efficiency.

Consult the chemical analysis portion of Chapter 16, Volume II, of *OPERATION OF WASTEWATER TREATMENT PLANTS,* to understand the laboratory testing process used for phosphorus and turbidity analyses and interpretation of results.

2. Calcium Oxide Content of Lime Feed.   A calcium oxide content of at least 90 percent available calcium oxide is needed in dry lime (quicklime) to bring the pH of the secondary treated water up to at least 11.0. Therefore, you should check the concentration of any lime purchased. Consult the laboratory analysis section (Section 16.47) of Chapter 16 in *OPERATION OF WASTEWATER TREATMENT PLANTS,* Volume II, to determine how to run a calcium oxide test and interpret the results. These test results will ensure the use of high-grade lime as well as inform the operator about the reliability of the supplier.

Calcium oxide content is also very important for treatment plants that recalcine their own lime. If the calcium oxide content drops in the recalcined lime, it is most likely due to higher concentrations of phosphorus within the lime sludge. When higher concentrations of phosphorus are returned to the system, lime clarification of phosphorus precipitation becomes less efficient. If calcium oxide levels drop too low in recalcined lime (less than 70 percent oxide), the recalcined lime should be wasted or disposed of rather than reused within the system.

3. Jar Tests to Determine Flocculation Efficiency.   An efficiently operated chemical clarification unit will allow for proper settling of as large and as heavy a floc as possible. *JAR TESTS* [24] can be used to determine what pH levels form the largest floc possible and allow the fastest settling of the floc formed. *POLYELECTROLYTES* [25] have been used with lime precipitation for phosphorus removal and are added after the fast-mix reaction. The jar tests are very good indicators of the concentration of polymers that produces optimum floc formation and sedimentation of the calcium hydroxide-phosphate combination floc particles.

5.324   Abnormal and Emergency Conditions

1. Changing Flow Conditions.   During low-flow conditions, check the automatic lime feed system to be sure it has adjusted to the reduced flows. Excessive lime feed rates into the clarifier waste the chemicals and increase the costs. If the automatic lime feed system cannot be throttled down far enough, manual feed control is necessary.

High flows can pose a more serious problem to lime clarification processes. First, high flows may cause a lowering of pH if the lime feed system is not properly pacing the flow by adding more lime when the high flow conditions occur. Secondly, high flow can mean hydraulic overloading of chemical clarification units thereby causing a decrease in the efficiency of phosphorus removal in the clarification unit.

2. Factors Affecting Phosphorus Removal Efficiency

a. Short-Circuiting.   Short-circuiting can be caused by too high a flow within the chemical clarification unit. A high flow will not allow adequate detention time. When short-circuiting occurs, the flocculated particles do not settle properly and are washed over the effluent weirs.

b. Changes in pH.   Fluctuating pH levels may have the effect of causing cloudy conditions in certain portions of the clarification tank. The most efficient manner in which to operate a chemical clarification unit is to maintain a constant pH above 11.0. When the pH drops below 11.0 for even a short period of time, floc for that

---

[24] See Chapter 4, Section 4.120, "The Jar Test," for details on how to run a jar test.

[25] *Polyelectrolyte* (POLY-ee-LECK-tro-lite).   A high-molecular-weight (relatively heavy) substance, having points of positive or negative electrical charges, that is formed by either natural or synthetic (manmade) processes. Natural polyelectrolytes may be of biological origin or obtained from starch products or cellulose derivatives. Synthetic polyelectrolytes consist of simple substances that have been made into complex, high-molecular-weight substances. Used with other chemical coagulants to aid in binding small suspended particles to larger chemical flocs for their removal from water. Often called a POLYMER.

flow may not be as large and a cloud of suspended particles may appear within the clarification unit.

c. Solids Loading.   Since most chemical clarification units for phosphorus precipitation follow a secondary clarifier, it is important that the secondary clarifier run as efficiently as possible. If solids are not settled properly within the secondary clarification unit causing high solids to appear in the chemical clarifier phosphorus precipitation units, then removal efficiencies for phosphorus will decrease. Under these conditions, it will be more difficult to maintain a high pH in the clarifier and clarity will be impaired substantially.

d. Small Straggler Floc.   Usually, a small floc will occur because of improper pH or because the polyelectrolyte being used is either at too high a dose or too low a dose for proper control of the flocculation process. The straggler floc will not settle as readily as the large floc and, therefore, efficiency in the phosphorus removal process may be drastically reduced.

e. Stormwater.   If substantial high flows result because of high stormwater runoff into the sewer system, phosphorus removal efficiencies will be substantially reduced because of the short-circuiting due to lack of detention time within the clarification basin. Be sure to adjust (close down) the influent valves if you expect a high-flow condition resulting from storm conditions. High flows through the chemical clarifier will cause settled floc to rise and will prevent sedimentation of the particles.

f. Industrial Dischargers.   Industrial dischargers who ignore discharge restrictions can cause serious problems for the chemical clarification unit. Discharge of toxic wastes can destroy secondary biological systems, thereby reducing the secondary clarification efficiency ahead of the phosphorus removal system.

g. Plugged Pumps or Piping.   Lime-phosphorus sludge must be pumped from the bottom of a chemical clarifier as it accumulates. If plugging problems occur because of a calcium carbonate buildup, they should be corrected immediately. If piping or pump problems are allowed to continue, they could result in an excess buildup of solids within the chemical clarifier and seriously reduce phosphorus removal efficiency.

h. Lime Feed Equipment to Maintain Adequate Lime Supply.   The most important phase of the chemical clarification process that requires close control is the lime feed process. If there is a breakdown in the lime feed operation, phosphorus removal cannot take place. Lime feeding must be continuous and have an adequate feed supply program. The lime slakers or mixers should be regularly inspected. All piping should be examined and deposited scale should be chipped off the pipes. The lime feed should also be using a high concentration of calcium oxide.

If the treatment plant has a recalcining furnace in which lime is reclaimed, be sure that the ratio of new lime to recalcined lime is adjusted so that the quality of lime fed to the mixing chamber is high enough to provide adequate pH control and phosphorus removal. This ratio is computed by using the known factors of calcium oxide content in both the recalcined or reclaimed lime and the new lime.

i. Operational Problems with Upstream or Downstream Treatment Processes.   The most serious effect from an upstream treatment process on chemical clarification and phosphorus removal is an upset condition in the secondary treatment process of the treatment plant. The upset condition can cause a lowering of the pH, too many suspended solids in the chemical clarifier, and a substantial problem for removal efficiencies for not only phosphorus but turbidity and suspended solids. The solids carryover from a secondary clarifier also interferes with any recalcining operation that may be used at the treatment plant if lime is reclaimed.

Downstream treatment processes will usually include a recarbonation basin to bring the pH back to a neutral point. Carbon dioxide from the recalcining process is commonly used in the recarbonation basin. If the carbon dioxide feed is not adequate, a high pH will result in the remainder of the treatment plant. This condition may be in violation of the discharge permit for the treatment facility. Most states require a pH in the relatively neutral range before discharge to any body of water or even for land disposal. The operator should be sure that the recarbonation process is working properly and that an adequate supply of carbon dioxide is being fed to bring the pH back down to a range within the plant's effluent discharge permit requirements.

5.325 Recarbonation for pH Control and Calcium Carbonate Recapture

Effluent from a high pH chemical clarifier used for phosphorus reduction will usually have a pH of at least 11. Use of carbon dioxide (CO_2) is the most common method of neutralizing the pH (bringing the pH of the water down to almost 7). A byproduct of the lowered pH is the formation of settleable calcium carbonate that can be recalcined for reuse in the lime treatment procedure. The process can be accomplished by either using a single- or a two-stage recarbonation and settling process.

Single-stage recarbonation is shown in Figure 5.6. Carbon dioxide (CO_2) gas is bubbled into the effluent stream from a chemical clarifier to allow calcium carbonate to form. As a byproduct of the calcium carbonate formation, pH is reduced.

The calcium carbonate precipitate formed is captured on filters, which usually follow a chemical clarification process. The calcium carbonate captured on the filter media must be settled following a filter backwashing procedure.

The two-stage recarbonation and settling process shown in Figure 5.7 is a more effective method to reduce wastewater pH and recapture calcium carbonate. Carbon dioxide gas is bubbled

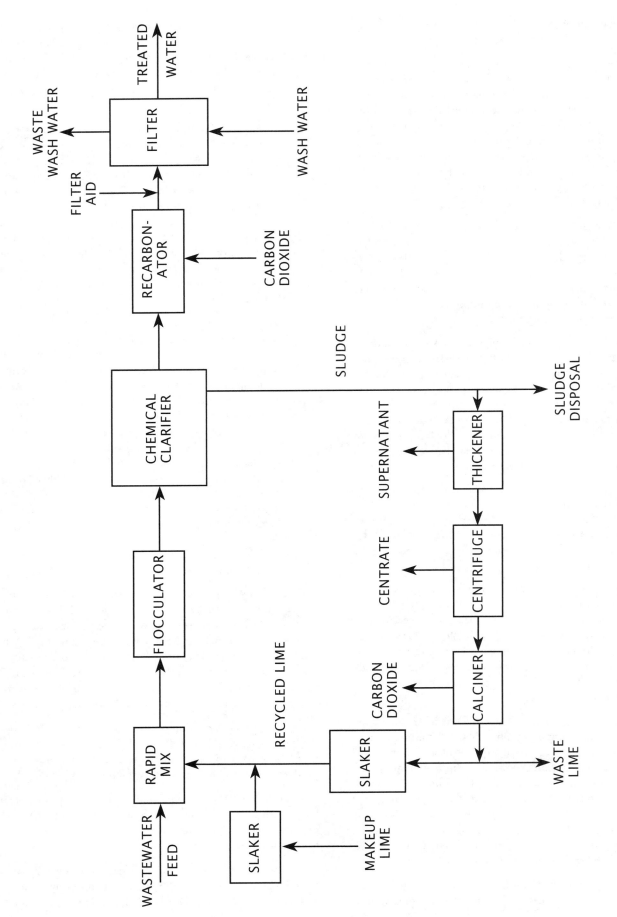

Fig. 5.6   Single-stage lime recarbonation process

(Source: *PROCESS DESIGN MANUAL FOR PHOSPHORUS REMOVAL*, EPA 625/1-76-001A)

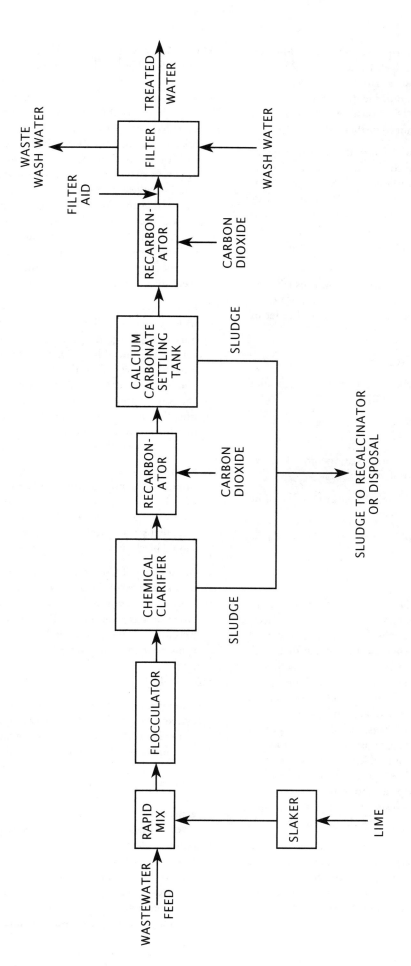

Fig. 5.7   Two-stage lime recarbonation process
(Source: *PROCESS DESIGN MANUAL FOR PHOSPHORUS REMOVAL*, EPA 625/1-76-001A)

into a basin just after the chemical clarification process. However, unlike single-stage recarbonation, the calcium carbonate precipitate formed is allowed to settle in a basin or tank. The settled calcium carbonate is collected and pumped to dewatering and recalcination or hauled to a landfill. Carbon dioxide gas is again bubbled into the wastewater stream to further reduce the pH.

Usually, pH is reduced to around 8.0 to 8.5 in first-stage recarbonation and further reduced to 7.0 in second-stage recarbonation. After second-stage recarbonation, the wastewater is treated by filtration. An additional advantage of two-stage recarbonation over single-stage recarbonation is that the effluent contains lower quantities of calcium carbonate. Therefore, filters that follow the recarbonation process are less likely to become plugged.

The stack gases from the recalcination furnace usually are used as a source of carbon dioxide gas. Additional carbon dioxide may be required to be produced by an auxiliary burner or from commercially tanked carbon dioxide.

QUESTIONS

Write your answers in a notebook and then compare your answers with those on pages 506 and 507.

5.3A   Why must the slaker or lime mix feed system have a grit removal system?

5.3B   Daily operation of a lime precipitation process to remove phosphorus consists of what tasks?

5.3C   Why are low-flow conditions of concern for the lime precipitation process?

5.3D   What factors affect phosphorus removal efficiency in the lime precipitation process?

5.33   Maintenance

1. Pumps and Seals.   As indicated in previous sections, pumps are a very important and major part of any wastewater treatment facility. Pumps and seals must be properly maintained in order that the pumps can work and function at their peak efficiency without needing major repair over the period of use. Seals and packing must be kept in good condition so that the pump can run cool and efficiently at all times. Lime is both caustic and abrasive and, therefore, can wear equipment at a rapid rate. Special attention must be paid to any pumps that handle lime sludges in order to be certain that plugging and excessive wear are not major problems in the pump operation. Pumps that operate at slower-than-normal feed rates or that will not pump are a strong indication that the pump or the sludge lines are plugged. Observation of pump pressure gauges will indicate if a pump is not working properly.

2. Piping.   Because of the tendency of lime to form scale, piping must be kept clean at all times. Lime builds up on the interior walls of pipes and can plug the pipes at a rapid rate. Pipes must be flushed and scoured periodically to ensure that the lime sludge is being moved to other parts of the treatment plant as expected.

3. Clarifier Mechanism.   Make sure the sludge scraper mechanisms for chemical clarification units are operating in good condition before allowing any flow to enter the sedimentation basin. Since the clarifier rotating arm moves continuously, it is very important that the proper oil and grease be provided for lubrication. Be certain that the bearings are lubricated and that no obstructions cause jamming or excessive wear on the equipment. A high torque level on the rotation of the clarifier drive mechanism provides a warning that the rotating collection arms of the chemical clarifier have a problem or that the unit is binding. Be sure to check the manufacturer's recommendations for the operation of the clarifier mechanism. All internal parts must be sealed and preventive maintenance should ensure that lime and weather elements do not affect the operation of the working gears within the mechanism drive.

4. Lime Slaking Mechanism.   The lime slaking mechanism, which converts calcium oxide to a calcium hydroxide slurry, must be kept in good operating order so that it will provide the correct amount of lime slurry for the lime feed station. Grit must be removed from the slaker mechanism and the mixing arms must be kept free of excessive buildups of lime. Be sure to check the manufacturer's recommendation on slake temperature. Frequently, inspect the water sprays used for condensing the steam and for dust control. The lime slaking mechanism should be cleaned frequently to prevent lime buildup.

5. Flash-Mix Basin.   Lime has a tendency to build up on the surface of the mixing paddles in a rapid-mixing chamber. Be sure that no excessive amounts of lime are allowed to gather on any mixing paddles. This could cause heavily weighted sides that set the mechanism off balance. The mixing paddles should be cleaned frequently to prevent lime buildup.

6. Automatic pH Control.   The pH probe that helps the lime feed system work in automatic mode must be cleaned on a daily basis to ensure that a lime scale buildup does not cause false readings. The automatic recording station to control the pH must be calibrated periodically to ensure that the mechanism is functioning properly.

5.34   Safety

1. Lime is a Powerful Caustic Solution.   Because lime has an extremely high pH, it is a very caustic solution and can cause eye irritation and skin irritation when it comes in contact with operators. Be very careful when using lime. Wear goggles and approved respiratory protection to prevent the lime from entering eyes or lungs. Wear protective clothing to prevent skin contact. If your eyes or other parts of your body are exposed to lime, be sure to rinse thoroughly with water. In the case of severe burns, be certain to see a physician immediately. A mild solution of boric acid may be kept on hand to help flush eyes in case of a severe lime burn or exposure of the eyes to dry lime. Because lime is a very caustic solution, an emergency eye wash station and deluge shower should be readily available in areas where personal contact may occur through handling or through a reasonably foreseeable emergency.

2. Polymers Can Cause Slick Surfaces. Polymers are often used in lime processes to help form colloidal particles of lime and phosphorus to provide faster sedimentation in a lime clarification unit. When wet, polymers are extremely slippery. Use inert absorbent materials such as sand or earth or a strong saltwater solution to clean up spilled polymers. Be very careful when walking near surfaces that have been exposed to any kind of polyelectrolyte or polymer. Use guard railings on stairs or near concrete areas that may have polyelectrolytes on the surface.

5.35 Loading Guidelines

1. Typical Loading Rates. The typical loading rate for a chemical clarification unit is normally 800 gallons per day per square foot (32 m³ per day/sq m) of surface area to 1,500 gallons per day per square foot (60 m³ per day/sq m) of surface area.

2. Hydraulic Loading Computation. In order to calculate the hydraulic loading rate for a lime clarification unit, the operator must know two things: (1) the flow into the lime clarification unit, and (2) the surface area of the clarifier. To calculate the hydraulic loading rate, divide the average gallons per day (cubic meters per day) of flow to the clarifier by the square feet (square meters) of surface area. This will give the overflow rate or hydraulic loading rate of the lime clarification unit.

$$\text{Hydraulic Loading Rate, GPD/sq ft} = \frac{\text{Flow, GPD}}{\text{Surface Area, sq ft}}$$

or

$$\text{Hydraulic Loading Rate, m}^3\text{/day/sq m} = \frac{\text{Flow, m}^3\text{/day}}{\text{Surface Area, sq m}}$$

3. Phosphate Loading Computation. Phosphate loading is normally described as pounds per day (kilograms per day) of phosphorus to be treated. This loading can either be a total loading rate of phosphate into the lime clarification unit or it can be the phosphorus removed from the lime clarification unit. To calculate the phosphate loading, the operator needs to know: (1) the gallons per day (cubic meters per day) of flow entering the lime clarification system, (2) the milligrams per liter of phosphate in the secondary effluent, and (3) the milligrams per liter of phosphate in the chemical clarification effluent. In order to compute the phosphate[26] loading, use the following equations:

$$\text{Phosphate Loading, lbs/day} = \text{Flow, MGD} \times \text{Phosphate, mg/}L \times 8.34 \text{ lbs/gal}$$

$$\text{Phosphate Loading, kg/day} = \text{Flow, }\frac{\text{m}^3}{\text{day}} \times \text{Phosphate, }\frac{\text{mg}}{L} \times \frac{1 \text{ kg}}{1{,}000{,}000 \text{ mg}} \times \frac{1{,}000\ L}{\text{m}^3}$$

To determine the efficiency of the phosphate removal process by lime clarification, use the following equations:

$$\text{Phosphate Removal Efficiency, \%} = \frac{\left(\begin{array}{c}\text{Influent Phosphate,} \\ \text{lbs/day}\end{array} - \begin{array}{c}\text{Effluent Phosphate,} \\ \text{lbs/day}\end{array}\right)}{\text{Influent Phosphate, lbs/day}} \times 100\%$$

or

$$\text{Phosphate Removal Efficiency, \%} = \frac{\left(\begin{array}{c}\text{Influent Phosphate,} \\ \text{kg/day}\end{array} - \begin{array}{c}\text{Effluent Phosphate,} \\ \text{kg/day}\end{array}\right)}{\text{Influent Phosphate, kg/day}} \times 100\%$$

or

$$\text{Phosphate Removal Efficiency, \%} = \frac{\left(\begin{array}{c}\text{Influent Phosphate,} \\ \text{mg/}L\end{array} - \begin{array}{c}\text{Effluent Phosphate,} \\ \text{mg/}L\end{array}\right)}{\text{Influent Phosphate, mg/}L} \times 100\%$$

5.36 Review of Plans and Specifications

In many instances, it is very beneficial to the design engineer, to the operator, and to the facility for the operator to review the plans and specifications of an expanded treatment plant or new treatment facility prior to the completion of the plans and specifications. The design review helps the design engineer to know what details to look for to make operations easier and to anticipate problems that might otherwise require design modifications after construction is completed. Without the operator's assistance, modification might later be necessary because someone forgot or did not have the knowledge to recommend specific details for better operational control of the phosphorus removal process. Refer to Section 5.246 on page 488 for a description of some of the items that can be reviewed by the operators to aid the design engineer during the facility's design for phosphorus removal.

5.37 Additional Reading

1. *HANDBOOK OF PUBLIC WATER SYSTEMS*, Second Edition (2001). Obtain from John Wiley & Sons, Inc., Customer Care Center (Consumer Accounts), 10475 Crosspoint Boulevard, Indianapolis, IN 46256. ISBN 0-471-29211-7. Price, $199.00, plus $5.00 shipping and handling.

2. *DESIGN MANUAL: PHOSPHORUS REMOVAL*, US Environmental Protection Agency. Obtain from National Technical Information Service (NTIS), 5285 Port Royal Road, Springfield, VA 22161. Order No. PB95-232914. EPA No. 625-1-87-001. Price, $47.50, plus $5.00 shipping and handling per order.

[26] Phosphate in these equations usually refers to total phosphate and includes orthophosphates, polyphosphates, and organic phosphorus. Both poly and organic forms of phosphorus must be converted to orthophosphate for measurement.

QUESTIONS

Write your answers in a notebook and then compare your answers with those on page 507.

5.3E   What is the purpose of the lime slaking mechanism?

5.3F   Why might a lime process also use a polymer?

5.3G   What forms of phosphorus are included in the total phosphate measurement?

5.3H   What provisions can be made when a facility is designed to reduce problems that will arise when pumps or pipes become plugged with lime?

5.4 PHOSPHORUS REMOVAL BY ALUM FLOCCULATION

5.40   Variations in the Alum Flocculation Process

5.400   Alum Flocculation as Used in a Clarification Process (Figure 5.8)

Aluminum sulfate (alum) can be used in the same manner as lime for precipitation of phosphorus in a clarifier. The same principles of coagulation, flocculation, and sedimentation apply when using alum for the removal of phosphorus in effluent from a secondary treatment facility. However, because of the difference in cost between aluminum sulfate and lime, lime is more commonly used for the precipitation of phosphorus.

When alum is used for phosphorus removal, two general reactions occur. In the first reaction, alum reacts with the alkalinity of the wastewater to form an aluminum hydroxide floc.

Alum	+	Alkalinity	→	Aluminum Hydroxide Floc	+	Sulfate	+	Carbon Dioxide
$Al_2(SO_4)_3$	+	$6HCO_3^-$	→	$2Al(OH)_3\downarrow$	+	$3SO_4^{2-}$	+	CO_2

The alum also reacts with the phosphate present.

Alum	+	Phosphate	→	Aluminum Phosphate	+	Sulfate
$Al_2(SO_4)_3$	+	$2PO_4^{3-}$	→	$2Al\,PO_4\downarrow$	+	$3SO_4^{2-}$

Phosphorus removal is achieved by the formation of an insoluble complex precipitate and by adsorption on the aluminum hydroxide floc. Depending on the alkalinity of the wastewater, dosages of 200 to 400 mg/L of alum are commonly required to reduce phosphorus in the effluent down to 0 to 0.5 mg/L. Typical alum dosage requirements to obtain a 95 percent reduction will range from 2.1:1 to 2.6:1 (Al:P molar ratio). Optimum phosphorus removal is usually achieved around a pH of 6.0. Alum feed is frequently controlled by automatic pH equipment, which doses according to the pH set point (the more alum, the lower the pH). Jar tests can be used to determine the optimum pH set point and alum dosage rate.

If it is necessary to achieve low effluent phosphorus residuals (less than 1.0 mg/L), the chemical clarifier is usually followed by either a pressure filter or a multi-media gravity filter. Phosphorus sludge from the clarifier goes to dewatering and disposal facilities. At present, there are no economical methods available for alum recovery.

5.401   Alum Flocculation as Used in Conjunction With Filtering of Suspended Solids (Figure 5.9)

1. Aluminum Sulfate (Alum) as a Coagulant.   Because of the proven ability of alum to coagulate and flocculate suspended particles from water and wastewater, the use of alum as a filtering aid has been common for many years. When alum is added to the wastewater entering a filtration unit, electrostatic forces are established on the filter media that allow the trapping of suspended solids and substantially improve the quality of effluent.

   Treatment plants that have used aluminum sulfate as a filtering aid have also experienced a reduction in the phosphorus as the wastewater flows through the filtration units. Although filtration is not usually considered an efficient method to remove phosphorus, the filtering with alum addition has provided a reduction of phosphorus at the same time that BOD and suspended solids are being reduced from the wastewater.

2. Expected Efficiencies of Phosphorus Removal Using Alum in Conjunction With Filtration.   Most advanced wastewater treatment facilities that use a phosphorus removal system followed by filtration can expect that low levels of phosphorus would enter the filter unit. Experiences at various wastewater treatment facilities have indicated that total phosphorus removal through filters using alum as a filtering aid achieved 70 to 95 percent phosphorus removal efficiencies. Influent data indicated a total phosphorus, however, of less than one milligram per liter. Dissolved phosphorus can be reduced 65 to 90 percent assuming that the incoming phosphorus loading would be less than 0.5 mg/L. Particulate phosphorus can be removed up to 100 percent because this phosphorus is usually attached to suspended solids, which are almost totally removed from the wastewater as it passes through a properly operated filter using aluminum sulfate for coagulation.

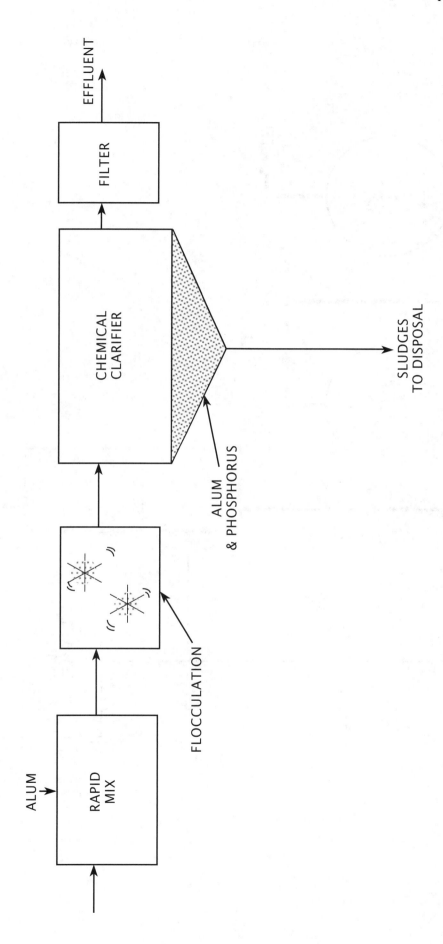

CHEMICAL TREATMENT
& PHOSPHATE REMOVAL

EFFLUENT

FILTER

CHEMICAL
CLARIFIER

SLUDGES
TO DISPOSAL

ALUM
& PHOSPHORUS

FLOCCULATION

ALUM

RAPID
MIX

Fig. 5.8   Alum flocculation as used in a clarification process

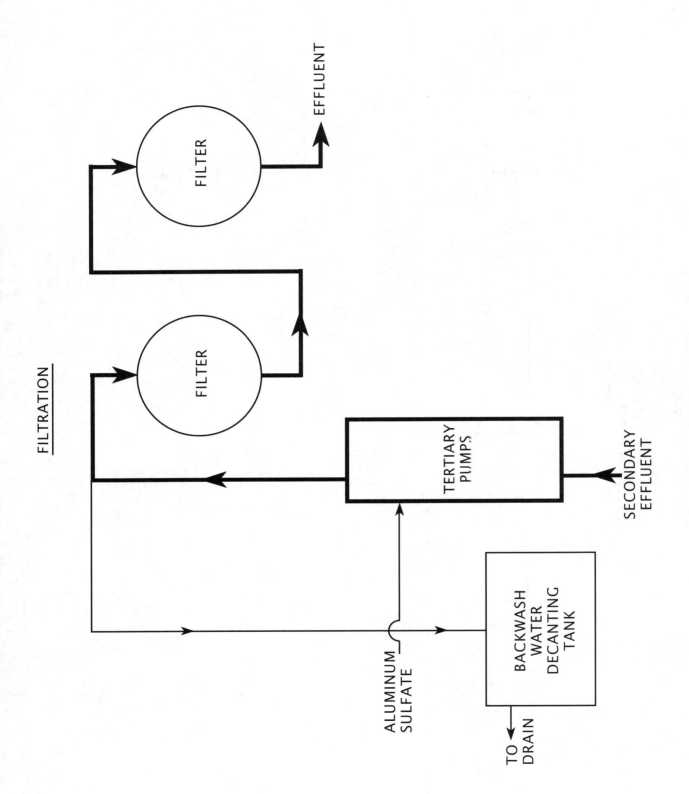

Fig. 5.9   *Alum flocculation as used in conjunction with filtering of suspended solids*

5.41 Maintenance of Alum Feeding Pumps and Associated Equipment

1. Pump Plugging Problems. Aluminum sulfate usually is delivered in a dry powder form or as a liquid; however, it is used as a liquid. The operators of a treatment facility must mix the dry alum with water to obtain a solution for feed to the water before entrance to the filtration units. Because of the chemical nature of aluminum sulfate, it sticks to surfaces very easily. This has caused pump plugging problems for many treatment plants. Most of the pumps that feed aluminum sulfate are small, metered, chemical feed pumps. Alum has a tendency to plug these pumps very rapidly and, therefore, the pumps require a considerable amount of care and maintenance to ensure that the proper dosage of alum is being fed to the filtration units. If plugging problems occur, the accuracy of the feed amount is questionable.

2. Pipe Plugging and Deterioration. Most pipes used for the transport of aluminum sulfate are either plastic or small glass tubing. Alum can stick to most surfaces and can deteriorate metal due to the chemical reaction formed when the aluminum sulfate and ferrous metal react with one another. Once liquid aluminum sulfate dries, plugging is a certain result. Pipes may become permanently plugged with aluminum sulfate and require replacement. The operators should keep pumps and pipes clean and should remember to run clear water through an alum feedline periodically to prevent plugging problems.

5.42 Operation of Alum Flocculation for Phosphorus and Suspended Solids Removal

5.420 Daily Operating Procedures

When used as a filtration aid, aluminum sulfate dosages must be precise. Even slight over- or underdoses may cause reduced efficiency. The operator must rely on jar testing and sampling of influent and effluent from a filter bed to be certain that the feed rate or dosage of alum in milligrams per liter is correct for the optimum phosphorus and suspended solids removal efficiency. The operator should perform the following steps to be certain that aluminum sulfate feed is at the correct ratio:

1. Check the results of jar tests to be certain the dosage of alum will form a good floc.

2. Check laboratory results to be certain that the alum is not overdosed or underdosed so that phosphorus also is removed as efficiently as possible through the filtration unit.

3. Check chemical feed pumps and alum feed piping system to be certain that the setting on the pump actually corresponds to the desired feed rate for efficient operation.

5.421 Abnormal Conditions

1. Pump or Piping Plugging Problems. As mentioned previously, the most important maintenance item for any alum feeding system is keeping the aluminum sulfate from plugging either piping or the chemical feed pump used for the feed of the alum (aluminum sulfate).

2. Operational Problems With Upstream or Downstream Treatment Processes. Tertiary filtration depends to a great extent on the upstream treatment units functioning properly. The operator should be certain of the quality of the wastewater entering the filtration unit. This quality can change daily depending on the upstream processes and the efficiency of their performance. The operator must check the various guidelines including suspended solids loading to calculate and to adjust the dosage of chemicals needed in the filtration unit. Filters also require backwashing to keep them clean and in proper operation. If upstream treatment units are not functioning properly, backwashing cycles may be needed more frequently than normal.

3. Alum Overdoses. One of the most common problems with using chemicals as filtering aids is the possibility of overdosing the filters with the chemical. Aluminum sulfate will react in a very negative way when it is overdosed into a wastewater system. The result is a lowering of the pH; this will hinder the ability of alum to coagulate the suspended solids. The low pH causes a cloudy condition that is visible to the operator in the form of substantial turbidity and suspended solids. If you observe a cloudy appearance in the effluent from the filtration unit, first check to make sure that the alum feed is not in an overdose condition. Underfeeding alum sulfate is better than overdosing with the chemical.

4. Suspended Solids Interference. If treatment units upstream are not operating properly, a substantial amount of suspended solids will load up the filter units and interfere with their ability to remove both suspended solids and phosphorus. From the standpoint of efficient operation, overloading of suspended solids onto the filtration units must not occur. If this occurs, frequent backwashing will be required and an increased quantity of aluminum sulfate or polymer may have to be added to overcome the additional load placed on the filter units.

5.43   Safety

1. Aluminum Sulfate Mixed With Water.   When aluminum sulfate mixes with water, a very slippery combination occurs. Operators should be very cautious around any floor that is wet with a spill of aluminum sulfate. Any surface continually exposed to slippery aluminum sulfate should be roughed up to prevent slipping and to avoid injury. Safety railings should be provided near any containers or working areas in which aluminum sulfate can be found.

2. Materials Handling Precautions.   Aluminum sulfate usually is delivered in a dry powder form. The operator should be very careful when mixing the powder with water that the powder does not get into the operator's respiratory system or eyes. Protective goggles and an approved respirator should be worn to protect your eyes and respiratory system. Use fans and filters to provide a safe air for breathing in the work area. Also, be careful not to allow powdered aluminum sulfate to fall on a wet surface, thereby creating a slippery condition.

5.44   Loading Guidelines

1. Alum Feed Rates for Effective Suspended Solids Removal. Aluminum sulfate is usually added as a filtering aid to remove suspended solids. The dosage rates at various treatment facilities range from 1 to 20 milligrams per liter depending on the incoming suspended solids and wastewater quality. Some phosphorus also will be removed with the suspended solids, but this procedure and dosage do not produce substantial phosphorus removal. See Section 5.400 for information on how to remove phosphorus by the use of alum. To determine the best operating dosage of aluminum sulfate for your treatment facility and type of wastewater, you should rely on jar tests and other laboratory results. The feed rates of aluminum sulfate are usually low; therefore, be certain that the chemical feed pump is operating properly to provide an accurate chemical to flow ratio.

2. Hydraulic Loading on Filtration Process Using Aluminum Sulfate as a Filtering Aid.   The standard design rate for a pressure filtration unit is 5 gallons per minute per square foot (3.4 liters per second per square meter). Loading rates can vary depending on the type of filter system used. The most commonly used type is the pressure filter. The operator should calculate the hydraulic loading based on the flow of wastewater distributed over the surface area of the filtration unit. If a gravity filter unit is used, the operator should obtain information from the manufacturer or design engineer on the proper application rate of wastewater for the unit.

5.45   Review of Plans and Specifications

1. Storage of Aluminum Sulfate.   Because aluminum sulfate comes in a powdered form, it is important to store the product in a dry environment, preferably inside a building. Usually, operators make up a large batch of aluminum sulfate liquid at one time in order to have enough on hand to last several days. The tanks holding the liquid aluminum sulfate should be fiberglass, nonferrous material, or they should be rubber-lined to protect the metal from the corrosive effects of the aluminum sulfate.

2. Piping and Pump Diagrams.   Because aluminum sulfate has the capability of sticking to many surfaces, it is important for the design of a treatment facility using alum to have flexible piping so that if a feedline becomes plugged or a pump is out of service for maintenance, the alum feed can continue without interruption. A pipe chase will enable operators to get to the alum piping in order to make repairs or replace the pipe, if needed. A system of flushing the pumps and piping is necessary to help prevent any plugging problems, which can plague the operation of a facility.

5.46   Additional Reading

1. *HANDBOOK OF PUBLIC WATER SYSTEMS,* Second Edition (2001). Obtain from John Wiley & Sons, Inc., Customer Care Center (Consumer Accounts), 10475 Crosspoint Boulevard, Indianapolis, IN 46256. ISBN 0-471-29211-7. Price, $199.00, plus $5.00 shipping and handling.

QUESTIONS

Write your answers in a notebook and then compare your answers with those on page 507.

5.4A   What is a major difference between the use of lime and the use of alum for precipitation of phosphorus-rich particles?

5.4B   How would you determine the optimum alum dosage?

5.4C   If upstream treatment units are not functioning properly, what happens to filter backwashing cycles?

5.4D   What would you do if you observed a cloudy appearance in the effluent from a filtration unit?

5.4E   What should be done to a floor that is continually exposed to slippery alum?

END OF LESSON 2 OF 2 LESSONS

on

PHOSPHORUS REMOVAL

Please answer the discussion and review questions next.

DISCUSSION AND REVIEW QUESTIONS

Chapter 5.   PHOSPHORUS REMOVAL

(Lesson 2 of 2 Lessons)

Write the answers to these questions in your notebook. The question numbering continues from Lesson 1.

11. What is the most serious problem in a lime system and how can this problem be avoided or corrected?

12. Why would you perform a jar test when removing phosphorus by the lime precipitation process?

13. Why must lime be kept from gathering on the mixing paddles in the flash-mix basin?

14. Alum has proven to be an effective coagulant for removing what pollutants from wastewater?

15. Why do pumps moving alum solutions clog?

16. How would you keep alum feed lines from plugging?

SUGGESTED ANSWERS

Chapter 5.   PHOSPHORUS REMOVAL

ANSWERS TO QUESTIONS IN LESSON 1

Answers to questions on page 476.

5.0A   Phosphorus is removed from wastewater because it provides a nutrient or food source for algae. Dead algae can cause serious oxygen depletion problems in receiving streams, which, in turn, can kill fish and other aquatic life. Also, algae can cause taste and odor problems in drinking water supplies.

5.1A   The three major types of systems used to remove phosphorus from wastewater are:

1. Biological phosphorus removal (luxury uptake)
2. Lime precipitation
3. Aluminum sulfate flocculation and precipitation (sedimentation)

Answers to questions on page 477.

5.2A   The two elements all biological treatment systems have in common are a microbial population (biomass) contained in some form of a reactor or attached to inert media, and some means of liquid-solids separation.

5.2B   The factors that have a significant influence on the biomass in a reactor include the characteristics of the wastewater being treated and the environmental and physical conditions (detention time) within the reactor.

5.2C   Briefly define the following terms:

1. Aerobic or oxic—A condition in which the aquatic (water) environment contains dissolved molecular oxygen (DO).

2. Anoxic—A condition in which the aquatic (water) environment does not contain dissolved molecular oxygen (DO), which is called an oxygen deficient condition. Generally refers to an environment in which chemically bound oxygen, such as in nitrate, is present.

3. Anaerobic—A condition in which the aquatic (water) environment does not contain dissolved molecular oxygen (DO) or chemically bound oxygen (such as oxygen in the sulfate and nitrate ions).

5.2D   To "select" a particular type of organism, the operator creates an environment that provides the right conditions so that the desired type of organism will grow and reproduce, in manageable numbers, and achieve the intended processing objective.

Answers to questions on page 483.

5.2E   The two process layouts commonly used for biological phosphorus removal are sidestream and mainstream.

5.2F   Luxury uptake of phosphorus is a biological process in which the microorganisms normally found in the activated sludge treatment portion of the secondary wastewater treatment plant are withdrawn to an environment without oxygen (anaerobic) for release of phosphorus. When the microorganisms are returned to an ideal environment, the first thing they take in is phosphorus. This phosphorus take-up is known as luxury uptake.

5.2G   In the luxury uptake process, microorganisms release phosphorus from their cell structure in the phosphorus release tank under anaerobic conditions.

Answers to questions on page 484.

5.2H   The equipment used for lime clarification in the luxury uptake process includes:

1. Lime feeding system
2. Lime slaking equipment
3. Lime clarification unit
4. Pumps and piping for lime-phosphorus sludge removal

5.2I   Lime is a very strong base and can cause serious burns and other injuries to your body.

5.2J   You should not smoke around a phosphorus stripping tank because the methane gas produced by the anaerobic conditions can create explosive conditions.

Answers to questions on page 485.

5.2K   Before starting up lime feeding equipment, the operator should routinely check the grit removal system and look for evidence of calcium carbonate buildup on the equipment or in the piping.

5.2L   Calcium carbonate is a rock-hard precipitate formed by the reaction of lime with carbon dioxide. Calcium carbonate scale will stick to almost any surface. A buildup of calcium carbonate scale will quickly plug pipelines and damage pumps or other equipment.

5.2M   The pH in the lime clarification tank should be tested manually every 8 hours to ensure that the automatic controls are functioning properly.

Answers to questions on page 487.

5.2N   Two pieces of information are needed to calculate the hydraulic loading on a chemical clarifier: the flow rate into the unit and the surface area of the clarifier.

5.2O   Ideal conditions in the sedimentation tank include unhindered settling velocities and a sludge blanket low enough to ensure that no solids could be resuspended in the tank effluent when higher flow conditions occur. Applied MLSS and return sludge concentrations within design limits are the keys to operating success.

5.2P   Solids separation from the wastewater being treated should emphasize capture and removal, not solids concentration and storage in a treatment process.

5.2Q   When discontinuous operations are planned or expected, reduce the solids inventories to give yourself reserve process capacity or space well beyond what you would need with a continuous processing system. Recycle flow, use solids equalization tanks, and use slow bleedback from the equalization tank, especially when pollutants are dissolved in the wastestream.

5.2R   Lime can be recovered from lime-phosphorus sludge using centrifuges (usually two in series) to dewater and separate the lime from the sludge. The dewatered solids are treated in a multiple-hearth furnace in a process known as recalcination to extract the lime.

Answers to questions on page 489.

5.2S   To remove limestone scale, use hot water or steam to soften the scale, chip or scour it from the surface of the equipment or pipe, and then flush it away.

5.2T   Items that an operator should check when reviewing plans and specifications for a biological phosphorus removal system include:

1. Lime transfer pipeline material and design
2. Dust and vapor control provisions in the lime slaking area
3. Accessibility of piping for easy maintenance
4. Availability of alternate piping
5. Provisions for unloading dry lime
6. Dust control provisions in lime transfer and storage areas
7. Design features to eliminate other known hazards associated with the use of lime and polymers

Also, you may want to insist that automatic measuring devices be installed for measuring dissolved oxygen levels.

5.2U   An operator may insist on installation of automatic dissolved oxygen probes to verify that continuous anaerobic conditions exist in the phosphorus stripping tank. Anaerobic conditions are necessary for phosphate to be released from the cell structure of the microorganisms.

ANSWERS TO QUESTIONS IN LESSON 2

Answers to questions on page 498.

5.3A   The slaker or lime mix feed system must have a grit removal system because most dry lime has a certain amount of grit, rocks, and sand in the mixture. This material must be removed to prevent plugging and equipment wear.

5.3B   Daily operation of a lime precipitation process to remove phosphorus consists of:

1. Routine pH monitoring to check automatic feed
2. Routine phosphate test for removal efficiencies
3. Verification of calcium oxide content of lime feed
4. Daily maintenance of pumps, piping, and other equipment to prevent plugging by lime scale

5.3C   Low-flow conditions are of concern to avoid the possibility of feeding excess lime and thereby wasting lime and money.

5.3D   Phosphorus removal efficiency may be affected by:

1. Short-circuiting
2. Changes in pH
3. Solids loading
4. Small straggler floc
5. Stormwater
6. Industrial dischargers
7. Plugged pumps or piping
8. Inadequate lime supply
9. Operational problems with upstream or downstream treatment processes

Answers to questions on page 500.

5.3E   The purpose of the lime slaking mechanism is to convert calcium oxide to calcium hydroxide in a slurry form.

5.3F   Many times a lime process uses polymers to help form colloidal particles of lime and phosphorus to provide faster sedimentation in a lime clarification unit.

5.3G   The forms of phosphorus in the total phosphate measurement include orthophosphates, polyphosphates, and organic phosphorus.

5.3H   To reduce problems that will arise when pumps or pipes become plugged with lime, alternate piping and valving should be provided so that while repairing or cleaning one pipe system or pump, continued operation can be provided.

Answers to questions on page 504.

5.4A   A major difference between the use of lime and the use of alum for precipitation of phosphorus-rich particles is that alum is more expensive than lime. Also, there is no economical method available for alum recovery.

5.4B   The optimum alum dosage can be determined by the jar test. Add varying amounts of alum to each jar containing the water being treated. The jar that produces the best clarification with the minimum amount of alum indicates the optimum alum dosage.

5.4C   If upstream treatment units are not functioning properly, filter backwashing cycles may be needed more frequently than normal.

5.4D   If you observe a cloudy appearance in the effluent from a filtration unit, first check to make sure that the alum feed is not in an overdose condition.

5.4E   Any floor that is continually exposed to slippery alum should be roughed up to prevent slipping and to avoid injury.

CHAPTER 6

NITROGEN REMOVAL

by

John G. M. Gonzales

TABLE OF CONTENTS
Chapter 6. NITROGEN REMOVAL

OBJECTIVES

Chapter 6. NITROGEN REMOVAL

Following completion of Chapter 6, you should be able to:

1. Explain why nitrogen is removed from wastewater.

2. Identify the types of nitrogen removal systems.

3. Describe nitrification and denitrification processes.

4. Operate nitrification and denitrification processes.

5. Describe the differences between suspended growth and fixed film reactors.

6. Explain how ammonia stripping, breakpoint chlorination, and ion exchange processes remove nitrogen.

WORDS
Chapter 6.   NITROGEN REMOVAL

ALKALINITY (AL-kuh-LIN-it-tee)   ALKALINITY

The capacity of water or wastewater to neutralize acids. This capacity is caused by the water's content of carbonate, bicarbonate, hydroxide, and occasionally borate, silicate, and phosphate. Alkalinity is expressed in milligrams per liter of equivalent calcium carbonate. Alkalinity is not the same as pH because water does not have to be strongly basic (high pH) to have a high alkalinity. Alkalinity is a measure of how much acid must be added to a liquid to lower the pH to 4.5.

ANAEROBIC (AN-air-O-bick)   ANAEROBIC

A condition in which atmospheric or dissolved oxygen (DO) is *NOT* present in the aquatic (water) environment.

ANOXIC (an-OX-ick)   ANOXIC

A condition in which the aquatic (water) environment does not contain dissolved oxygen (DO), which is called an oxygen deficient condition. Generally refers to an environment in which chemically bound oxygen, such as in nitrate, is present. The term is similar to ANAEROBIC.

ANOXIC DENITRIFICATION (dee-NYE-truh-fuh-KAY-shun)   ANOXIC DENITRIFICATION

A biological nitrogen removal process in which nitrate nitrogen is converted by microorganisms to nitrogen gas in the absence of dissolved oxygen.

ATTACHED GROWTH PROCESSES   ATTACHED GROWTH PROCESSES

Wastewater treatment processes in which the microorganisms and bacteria treating the wastes are attached to the media in the reactor. The wastes being treated flow over the media. Trickling filters and rotating biological contactors are attached growth reactors. These reactors can be used for BOD removal, nitrification, and denitrification.

AUTOTROPHIC (auto-TROF-ick)   AUTOTROPHIC

Describes organisms (plants and some bacteria) that use inorganic materials for energy and growth.

BIOMASS (BUY-o-mass)   BIOMASS

A mass or clump of organic material consisting of living organisms feeding on wastes, dead organisms, and other debris. Also see ZOOGLEAL MASS and ZOOGLEAL MAT (FILM).

BUFFER CAPACITY   BUFFER CAPACITY

A measure of the capacity of a solution or liquid to neutralize acids or bases. This is a measure of the capacity of water or wastewater for offering a resistance to changes in pH.

DENITRIFICATION (dee-NYE-truh-fuh-KAY-shun)   DENITRIFICATION

(1)   The anoxic biological reduction of nitrate nitrogen to nitrogen gas.

(2)   The removal of some nitrogen from a system.

(3)   An anoxic process that occurs when nitrite or nitrate ions are reduced to nitrogen gas and nitrogen bubbles are formed as a result of this process. The bubbles attach to the biological floc and float the floc to the surface of the secondary clarifiers. This condition is often the cause of rising sludge observed in secondary clarifiers or gravity thickeners. Also see NITRIFICATION.

ENDOGENOUS (en-DODGE-en-us) RESPIRATION

ENDOGENOUS RESPIRATION

A situation in which living organisms oxidize some of their own cellular mass instead of new organic matter they adsorb or absorb from their environment.

FIXED FILM

FIXED FILM

Fixed film denitrification is the common name for attached growth anaerobic treatment processes used to achieve denitrification.

HETEROTROPHIC (HET-er-o-TROF-ick)

HETEROTROPHIC

Describes organisms that use organic matter for energy and growth. Animals, fungi, and most bacteria are heterotrophs.

METABOLISM

METABOLISM

All of the processes or chemical changes in an organism or a single cell by which food is built up (anabolism) into living protoplasm and by which protoplasm is broken down (catabolism) into simpler compounds with the exchange of energy.

NITRIFICATION (NYE-truh-fuh-KAY-shun)

NITRIFICATION

An aerobic process in which bacteria change the ammonia and organic nitrogen in water or wastewater into oxidized nitrogen (usually nitrate).

PROTOPLASM

PROTOPLASM

A complex substance (typically colorless and semifluid) regarded as the physical basis of life, having the power of spontaneous motion and reproduction; the living matter of all plant and animal cells and tissues.

SUBSTRATE

SUBSTRATE

(1) The base on which an organism lives. The soil is the substrate of most seed plants; rocks, soil, water, or other plants or animals are substrates for other organisms.

(2) Chemical used by an organism to support growth. The organic matter in wastewater is a substrate for the organisms in activated sludge.

SUSPENDED GROWTH PROCESSES

SUSPENDED GROWTH PROCESSES

Wastewater treatment processes in which the microorganisms and bacteria treating the wastes are suspended in the wastewater being treated. The wastes flow around and through the suspended growths. The various modes of the activated sludge process make use of suspended growth reactors. These reactors can be used for BOD (biochemical oxygen demand) removal, nitrification, and denitrification.

ZOOGLEAL (ZOE-uh-glee-ul) MASS

ZOOGLEAL MASS

Jelly-like masses of bacteria found in both the trickling filter and activated sludge processes. These masses may be formed for or function as the protection against predators and for storage of food supplies. Also see BIOMASS.

ZOE-uh-glee-ul

CHAPTER 6.   NITROGEN REMOVAL

NOTE: For a review of previous discussions, read Chapter 2, "Activated Sludge," Section 2.6, "Effluent Nitrification," pages 139 to 143.

6.0   WHY IS NITROGEN REMOVED FROM WASTEWATER?

6.00   Nitrogen as a Nutrient

Inorganic nitrogen provides a nutrient or food source for algae and a combination of nitrogen and phosphorus in receiving waters can cause algal growths to multiply rapidly. Algae in water are unsightly and cause tastes and odors in a drinking water supply. Dead and decaying algae can result in oxygen depletion problems in receiving waters, which, in turn, adversely affect aquatic life. Fish kills can result from an oxygen deficiency or ammonia toxicity in receiving waters.

By removing nitrogen from the effluent of the wastewater treatment facility, the lake or river that a treatment plant discharges into will not contain one of the nutrients essential for algal growth. The reduction in the nitrogen nutrient thus reduces the growth of algae. However, in most receiving waters, phosphorus is the limiting nutrient; therefore, nitrogen removal will have little impact on algal growth.

6.01   Need for Nitrogen Removal

The United States Environmental Protection Agency and state water pollution control regulatory agencies recognize the need to protect receiving streams from problems that occur because of the growth of algae. The regulatory agencies are requiring that wastewater treatment plants discharging to sensitive receiving waters remove nitrogen in the effluent to protect a river, stream, or lake by eliminating that nutrient from the algal food chain. Nitrogenous compounds must also be controlled in plant effluents to prevent adverse impacts from ammonia toxicity to fish life, reduction of chlorine disinfection efficiency, an increase in the dissolved oxygen depletion in receiving waters, adverse public health effects (mainly in groundwater used for drinking where high nitrate levels can interfere with oxygen utilization in newborn babies), and a reduction in the water's suitability for reuse.

6.1   TYPES OF NITROGEN REMOVAL SYSTEMS

Nitrogen removal from wastewater can be accomplished by a variety of physical, chemical, and biological processes. Nitrification/denitrification, ammonia stripping, and breakpoint chlorination are the most common types of nitrogen removal systems. Ion exchange is also sometimes used. Overland flow systems can remove nitrogen from secondary effluents. Table 6.1 is a summary of types of nitrogen removal systems and operational considerations associated with each process.

TABLE 6.1   TYPES OF NITROGEN REMOVAL SYSTEMS

System	Operational Considerations
1. PHYSICAL TREATMENT METHODS A. Sedimentation B. Gas Stripping	1. Expensive
2. CHEMICAL TREATMENT METHODS A. Breakpoint Chlorination B. Ion Exchange	2. Expensive
3. BIOLOGICAL TREATMENT METHODS A. Activated Sludge Processes B. Trickling Filter Processes C. Rotating Biological Contactor Processes D. Oxidation Pond Processes	3. A–D Operational control. Additional costs for oxygen to produce nitrification.
E. Land Treatment Processes (Overland Flow) F. Wetland Treatment Systems	3. E & F Land requirements. Suitable temperatures. Control of plants.

6.10 Biological Nutrient Removal

6.100 Nitrification

The conversion of ammonia to nitrate requires significant amounts of oxygen. Thus, a wastewater treatment plant effluent high in ammonia could cause oxygen depletion in the receiving waters. Therefore, removal of ammonia by nitrification is an extremely important treatment process used to reduce nitrogen oxygen demand (NOD) on receiving waters. For nutrients (including nitrogen) to be removed from an effluent, thereby reducing the nutrient loading on receiving waters, the process must consist of both nitrification and denitrification.

Nitrification is a biological process accomplished primarily by two types of microorganisms: *Nitrosomonas* and *Nitrobacter*. Unlike most of the common organisms found in a wastewater treatment facility, these microorganisms are *AUTOTROPHS*[1]; that is, they derive carbon for their cellular growth from inorganic sources such as carbon dioxide (CO_2) and bicarbonate *ALKALINITY*[2] (HCO_3^-). The first step in the nitrification process is the conversion of ammonia (NH_3) or ammonium (NH_4^+) to nitrite (NO_2^-) by *Nitrosomonas* bacteria. The second step is conversion of nitrite to nitrate (NO_3^-) by *Nitrobacter* bacteria. The proper conditions must exist for *Nitrosomonas* to be able to separate the nitrogen from the hydrogen in the ammonium molecule and replace the hydrogen with oxygen molecules. Sufficient oxygen and the appropriate temperature and microbiological food must be present to accomplish this process. *Nitrobacter* also rely on oxygen to complete the stabilization of the nitrite molecule into the more stable nitrate substance. Again, adequate amounts of free oxygen and food as well as optimal temperature, pH, and microorganism population along with other conditions are required to complete this reaction.

6.101 Denitrification

Biological denitrification is the process by which microorganisms reduce nitrate (NO_3^-) to nitrogen gas that is released to the atmosphere. A number of bacterial species that naturally occur in wastewater accomplish denitrification. In all cases, the microorganisms are *HETEROTROPHIC*[3] since they can metabolize complex organic substances. Placed in an environment having no dissolved molecular oxygen but containing a carbon food source,[4] denitrifying microorganisms will reduce nitrate to nitrogen gas by breaking down the nitrate molecule to obtain the oxygen they need for cell metabolism. This reduction (breaking down) of nitrate to nitrogen gas by bacteria is called microorganism dissimilation.

Denitrification takes place in two steps. First, nitrate is reduced to nitrite. Next, nitrite is reduced by the microorganism dissimilation process to gaseous forms of nitrogen: nitric oxide (NO), nitrous oxide (N_2O), or nitrogen gas (N_2).

The waste products from cell metabolism are nitrogen gas (primarily), with some minor amounts of nitrous oxide or nitric oxide. Gaseous nitrogen is not removed because it cannot easily be used by microorganisms for biological growth. However, when the denitrified wastewater is later transferred to an aeration or reaeration tank, the nitrogen gases are released. Thus, denitrification converts nitrogen that is in an objectionable form (as a nutrient, nitrate) to one that has no significant impact on environmental quality.

Denitrification is accomplished using attached growth reactors (fixed film denitrification such as the trickling filter or rotating biological contactor (RBC) treatment processes), or within a modified activated sludge process (also known as suspended growth denitrification).

6.102 Modified Activated Sludge Systems

Biological nutrient removal processes using activated sludge systems are popular because they are effective in removing both nitrogen and phosphorus, they are relatively inexpensive processes, and they have proven to be reliable and effective. Typically, nitrogen and phosphorus are removed by modifying the activated sludge process to include separate tanks (or zones within tanks) with and without aeration to produce a sequence of anaerobic, *ANOXIC*,[5] and aerobic zones. The goal is to create favorable environments in which the appropriate microorganisms convert the nitrogen and phosphorus into forms that can be removed from the wastewater being treated. Many different combinations of zones or process sequences have been used successfully and each system is usually known by its proprietary name.[6] One biological nutrient removal system of this type, the Bardenpho process, is described in Chapter 7, Section 7.20.

6.103 Overland Flow

Overland flow systems are capable of producing a very high-quality effluent. The wastewater to be treated is applied to

---

[1] *Autotrophic* (auto-TROF-ick).   Describes organisms (plants and some bacteria) that use inorganic materials for energy and growth.

[2] *Alkalinity* (AL-kuh-LIN-it-tee).   The capacity of water or wastewater to neutralize acids. This capacity is caused by the water's content of carbonate, bicarbonate, hydroxide, and occasionally borate, silicate, and phosphate. Alkalinity is expressed in milligrams per liter of equivalent calcium carbonate. Alkalinity is not the same as pH because water does not have to be strongly basic (high pH) to have a high alkalinity. Alkalinity is a measure of how much acid must be added to a liquid to lower the pH to 4.5.

[3] *Heterotrophic* (HET-er-o-TROF-ick).   Describes organisms that use organic matter for energy and growth. Animals, fungi, and most bacteria are heterotrophs.

[4] If the denitrification reactor is located after the BOD removal process, the carbon food source, typically methanol, must be added by the operator.

[5] *Anoxic* (an-OX-ick).   A condition in which the aquatic (water) environment does not contain dissolved oxygen (DO), which is called an oxygen deficient condition. Generally refers to an environment in which chemically bound oxygen, such as in nitrate, is present. The term is similar to ANAEROBIC.

[6] For a detailed summary of the different types of proprietary processes, see *WASTEWATER ENGINEERING: TREATMENT AND REUSE*, Fourth Edition, Chapter 11, "Advanced Wastewater Treatment." Metcalf & Eddy, Inc., published by the McGraw-Hill Companies, Order Services, PO Box 182604, Columbus, OH 43272-3031. ISBN 0-07-041878-0. Price, $155.63, plus nine percent of order total for shipping and handling.

grass-covered slips by sprinklers or to the land surface through ports in pipes at evenly spaced intervals. The water flows or trickles as sheet flow through water-tolerant grasses to a collection ditch. The microorganisms that live in the grass at the soil surface effectively reduce BOD and cause both nitrification and denitrification to occur. Heavy rainfall and runoff can cause an increase in solids in the effluent during the initial runoff from a storm.

6.11 Physical Nitrogen Removal by Ammonia Stripping

Ammonia nitrogen in the gaseous ammonia (NH_3) form (and only in the gaseous form) has a natural tendency to leave the wastewater and enter the atmosphere. Therefore, for ammonia stripping to work efficiently, the bulk of the ammonium (NH_4^+) (the natural form of "ammonia" in wastewater) must be first converted to the gaseous ammonia (NH_3) form by adding chemicals to increase the pH level of the wastewater up to the 10.5 to 11.5 range. At 25°C and pH 11, the percentage of ammonia is about 98 percent. Then, when this pH level is achieved, a mixture of wastewater droplets and air will result in the ammonia being stripped (removed) from the droplets by the air. This mixing of wastewater and air is accomplished in an ammonia stripping tower where the high pH wastewater usually falls over fixed media (or "packing") while a blower continuously forces fresh air into the mixture to strip off more ammonia.

6.12 Chemical Nitrogen Removal Processes

6.120 Breakpoint Chlorination

Ammonia nitrogen can be oxidized to nitrogen gas (N_2) through the use of chlorine. Breakpoint chlorination is the term used to describe this process. In the breakpoint chlorination process, chlorine is added until the ammonia nitrogen has been oxidized to nitrogen gas. This point is achieved when further additions of chlorine result in an increase in the chlorine residual of the water being treated.

Breakpoint chlorination requires relatively large amounts of chlorine per unit of ammonia removed. The expense and danger of removing high concentrations of ammonia nitrogen by breakpoint chlorination are prohibitive. Consequently, breakpoint chlorination is used primarily to remove small amounts of ammonia nitrogen remaining after wastewater has been treated with other nitrogen removal processes. Polishing treatment plant effluent by decreasing the small amounts of remaining ammonia nitrogen is the recommended use for the breakpoint chlorination process.

The chlorine to ammonia nitrogen ratio is normally around 10:1. In other words, for every one mg/L of ammonia nitrogen, ten mg/L of chlorine are necessary to oxidize the ammonia to nitrogen gas.

6.121 Ion Exchange

The ion exchange process is used to remove undesirable ions from water and wastewater. The nitrogen removal process involves passing ammonia-laden wastewater downward through a series of columns packed with natural or synthetic ion exchange resins. A naturally occurring resin or zeolite called clinoptilolite is commonly used. The clinoptilolite beds or columns are usually four or five feet deep and are packed with 20- × 50-mesh particles. The ammonium ion adheres to or is adsorbed by the clinoptilolite. When the first column in a series loses its ammonium ion adsorptive capacity, it is removed from the treatment scheme and washed with lime water. This step converts the captured ammonium ions to ammonia gas, which is then released to the atmosphere by contacting heated air with the wastewater stream, in much the same manner as described under ammonia stripping. The advantage of this system is that there is no process waste containing ammonia for which ultimate disposal must be provided.

The clinoptilolite may also be regenerated by passing a brine or salt solution through the exchange bed. The sodium in the salt solution exchanges with the ammonium nitrogen. By removing the ammonium from the spent regenerant brine solution, the regenerant may be reused, thus eliminating the difficult problem of brine disposal.

Because of its eutrophication potential, spent nitrate regenerant brine cannot be discharged into rivers and lakes even if it is slowly metered into the receiving water. Also, the high sodium concentration prevents disposal of spent regenerant onto land where nitrogen could serve as a fertilizer. Spent regenerant may be metered into the sanitary sewer for biological treatment at the wastewater treatment plant, if approved by the treatment plant. Small, concentrated volumes of spent regenerant may be disposed of in approved landfills.

QUESTIONS

Write your answers in a notebook and then compare your answers with those on page 542.

6.0A Algal growths are caused mainly by what two nutrients in receiving waters?

6.1A List as many biological nitrogen removal methods as you can recall.

6.1B How is nitrogen removed by the nitrification/denitrification treatment process?

6.1C What is the recommended use of breakpoint chlorination for removing nitrogen?

6.1D How is ammonium nitrogen removed by the ion exchange process?

6.2   BIOLOGICAL NITROGEN REMOVAL

6.20   How Nitrification/Denitrification Is Accomplished

There are two microbial processes or steps, which may occur simultaneously, that produce the end result of nitrification.

1. *Conversion of ammonium to nitrite.* The first step in nitrification is the conversion of ammonium to nitrite. This is accomplished by microorganisms known as *Nitrosomonas. Nitrosomonas* use dissolved molecular oxygen (O_2) in this conversion and, therefore, function only in an aerobic environment. The equation describing *Nitrosomonas'* use of oxygen to break down ammonium (NH_4^+) into nitrite (NO_2^-) is:

$$NH_4^+ + 1.5O_2 \rightarrow 2H^+ + H_2O + NO_2^-$$

Note the hydrogen ion (H^+) production. The importance of this product will be discussed later in Section 6.2210, "Alkalinity and pH."

2. *Conversion of nitrite to nitrate.* The second step in nitrification is the conversion of nitrite to nitrate. This is accomplished by *Nitrobacter* bacteria. Completion of this second step of the nitrification reaction requires adequate amounts of free oxygen and food as well as an optimal temperature, pH, and microorganism population. The carbon food source for the autotrophic nitrifiers is bicarbonate alkalinity. The oxidation of nitrite to nitrate can be represented as follows:

$$NO_2^- + 0.5O_2 \rightarrow NO_3^-$$

The overall oxidation of ammonium by both groups of microorganisms can be represented as follows:

$$NH_4^+ + 2O_2 \rightarrow NO_3^- + 2H^+ + H_2O$$

6.21   Equipment

The nitrification (ammonia conversion) portion of a nitrogen removal process can be accomplished in either suspended growth reactors or attached growth (fixed film) reactors. In either case, a sufficient quantity of oxygen is needed and an ample amount of time is required for the process to develop the proper microbial population age. Both suspended growth reactors and fixed film reactors must be operated so that complete nitrification is achieved; that is, ammonium is converted to nitrate. If the conversion only goes as far as nitrite, disinfection of the effluent can be very difficult and expensive because nitrite (NO_2^-) reacts readily with chlorine, thus making it hard to maintain a chlorine residual to provide adequate disinfection.

1. *Suspended growth reactors.* A suspended growth reactor is normally an aeration basin. Aeration basins must be sized to provide the appropriate volume of mixed liquor, suspended solids mass, and sufficient influent wastewater flow detention time. The basin must be large enough and the MCRT long enough (usually four days plus) (longer in cold weather) to allow the slow-growing nitrifying bacteria sufficient time to grow. Detention times must be at least four hours and preferably eight hours. Sufficient air supply must be available to maintain a dissolved oxygen level in the aeration basins

between 1.5 and 4.0 mg/*L*. Determining the necessary microbial age can sometimes be difficult in suspended growth reactors.

2. *Fixed film reactors.* The two most common types of fixed film reactors are trickling filters and rotating biological contactors (RBCs). In each case, microorganisms grow on the media over which the wastewater is passed to accomplish the desired treatment. Oxygen is supplied to the microorganisms by either natural or forced draft air movement through the media. Proper operation requires sufficient media surface area to allow adequate contact time between the microorganisms and the wastewater to achieve the desired level of treatment. Proper treatment also requires an adequate supply of oxygen and sufficient pretreatment to reduce the applied BOD level. Recycling or recirculation of the reactor effluent is also important to achieve the desired degree of treatment.

6.22   Nitrification Using Suspended Growth Reactors

6.220   *Process Modes*

Activated sludge treatment processes can be operated in a variety of modes, but not all process modes are suitable for biological nitrification.

CONVENTIONAL OR PLUG FLOW AERATION SYSTEM

This type of facility lends itself well to the nitrification process because of the plug flow configuration and the detention time it provides as the wastewater flows through a long, narrow aeration tank. However, pH levels may drop during this detention time because nitrification destroys alkalinity. (See Section 6.221, "Daily Operation.")

COMPLETE MIX ACTIVATED SLUDGE PROCESS

A complete mix design provides a uniform mixture throughout the entire reactor so the problem of uneven distribution of dissolved oxygen within the reactor is lessened. However, this type of reactor may be more sensitive to a drop in alkalinity as ammonia is oxidized to nitrate. (See Section 6.221, "Daily Operation.")

CONTACT STABILIZATION

Because of the separately aerated return activated sludge used in contact stabilization, an insufficient number of nitrifying bacteria may be left in the biomass. Also, the reactor in the main flow stream is too small to be used for nitrification purposes. In a small reactor, the contact time for the wastewater being treated with the mixed liquor is not long enough to produce a nitrified effluent. Therefore, contact stabilization plants are not ideal for the operation of a nitrification facility using a suspended growth reactor.

EXTENDED AERATION

Extended aeration facilities are well suited for use in nitrification due to the long aeration time for the mixed liquor and the long sludge age maintained.

STEP-FEED AERATION

Step-feed aeration can be used to accomplish partial nitrification. Because of the addition of influent wastewater at several points along the aeration basin, the contact time for nitrogen conversion is generally too short to achieve complete nitrification.

6.221   Daily Operation

With adequate alkalinity, dissolved oxygen levels, and MCRT, nitrification systems are stable and relatively easy to operate. Control of the nitrification process in a suspended growth reactor relies primarily on: (1) sludge age or MCRT, and (2) available oxygen. Other important factors include maintaining a low BOD and developing the proper microorganism population. Detention time must be at least four hours and preferably eight hours to achieve complete nitrification. (This is why the step-feed aeration and contact stabilization process modes are not generally suitable for nitrification, although such process modes can nitrify an effluent under certain conditions.) Sufficient air supply must also be available to maintain an average dissolved oxygen level of 2.0 to 4.0 mg/L in the aeration basins. A lower dissolved oxygen level (1.0 to 1.5 mg/L) works well in a stable, high MCRT (low F/M) nitrification process.

6.2210   ALKALINITY AND pH

Alkalinity changes in the nitrification process stream and the final effluent reliably indicate what is happening in the nitrification plant. Therefore, alkalinity determinations (along with dissolved oxygen measurements) are one of the operator's best means of monitoring day-to-day operations.

Alkalinity changes may accompany changes in the wastewater pH. When aerated and exposed to the atmosphere, all water systems will want to come to equilibrium with the carbon dioxide (CO_2) in the atmosphere and the water's pH will tend to move toward a pH of about 8.3. With no change in the accompanying alkalinity, CO_2 will be stripped from the water when the pH is below 8.3 and retained when the pH is above 8.3.

In air activated sludge systems, the rate of change toward a pH of 8.3 will be determined by the rate of the CO_2 production by the microorganisms and the intensity and duration of aeration. In enclosed, less intensely mixed pure oxygen activated sludge systems, CO_2 will accumulate and cause a pH decline. In either system, alkalinity will buffer the tendency of the system to exhibit a pH change.

Losses of alkalinity due to nitrification will generally not cause any apparent significant decline of the pH, due to the counteracting effect of CO_2 stripping. The hydrogen ions produced in such reactors may reduce the BUFFERING CAPACITY[7] of the alkalinity in the wastewater to such an extent that a significant lowering of pH may occur. Lowered pH

could affect the operation of downstream treatment processes or violate effluent discharge requirements. Domestic wastewater usually contains enough alkalinity that nitrification does not create pH problems. However, some industrial wastewaters may lack sufficient alkalinity.

Effluent alkalinity of less than 50 mg/L as $CaCO_3$ is a borderline condition that requires increased operator attention. Sudden declines in the effluent alkalinity may indicate that acid dumps or spills have reached the plant. Sudden increases may indicate a nitrifier upset.

When the alkalinity is depleted to levels of around 50 mg/L as $CaCO_3$, pH levels may decline. At a pH of 7, nitrification may be inhibited by the low pH since little buffer capacity remains. Whether or not the nitrification microorganisms can adjust to this or slightly lower pH values depends on whether or not the lower pH conditions are continuous or change over time. In either case, when the pH reaches 6.5 (as it typically does when alkalinity is 25 mg/L as $CaCO_3$ or less), nitrification is very likely to be inhibited.

If you suspect that nitrification is being inhibited, check carefully before taking action. Remember that 24-hour composite samples often mask fluctuating alkalinity and pH drops. Also, consider the fact that when the alkalinity is low, pH increases can often result from the losses in CO_2 that occur when water flows over final sedimentation tank weirs and from the vigorous mixing that often accompanies standard laboratory sample preparation procedures.

If the alkalinity drops so low that the pH drops low enough to interfere with nitrification, it will be necessary to add chemicals to increase the pH. Therefore, the operator should keep calcium oxide (lime), soda ash, or other chemicals on hand for the purpose of increasing pH at the proper location in a nitrifying aeration process. Table 6.2 shows how much alkalinity is supplied by various chemicals and the alkalinity added by the denitrification process or consumed in the nitrification process.

---

[7] *Buffer Capacity.*   A measure of the capacity of a solution or liquid to neutralize acids or bases. This is a measure of the capacity of water or wastewater for offering a resistance to changes in pH.

TABLE 6.2   SUMMARY OF BIOLOGICAL NITROGEN CONTROL REACTIONS[a]

Biochemical Nitrogen Removal by Microorganisms (microbial uptake)

- 0.075 to 0.10 mg N removed per mg net volatile suspended solids produced by microorganisms

Biochemical Nitrogen Oxidation (Nitrification)

- 4.6 mg oxygen required per mg nitrogen oxidized
- 7.1 mg $CaCO_3$ alkalinity depleted per mg nitrogen oxidized
- approximately 0.1 mg net volatile suspended solids formed per mg nitrogen oxidized

Biochemical Oxidized Nitrogen Removal (Denitrification)

- 2.9 mg oxygen released per mg oxidized nitrogen removed
- 1.5 mg COD per mg methanol (CH_3OH)

   ... 1.9 mg methanol required per mg oxidized nitrogen removed

   ... 0.7 mg methanol required per mg dissolved oxygen removed

   NOTE: Sufficient substrate (COD) must be added to satisfy nitrogen reduction and microorganism needs; typically add 1.5 times theoretical predictions.[b]

- 3.6 mg $CaCO_3$ alkalinity recovered (added to system) per mg oxidized nitrogen removed
- Same to slightly lower net volatile suspended solids per COD (or BOD_5) removed

   NOTE: COD removed is the total amount of COD oxidized by microorganisms.

Chlorine Demand Due to Nitrite Nitrogen (associated with incomplete nitrification or denitrification)

- 2.5 mg chlorine per mg nitrite nitrogen, yielding nitrate nitrogen

---

[a] Prepared by Mike Mulbarger, Paladin Enterprises, Sedona, AZ.
[b] Theoretical prediction. The quantity calculated on the basis of the theoretical amounts involved in chemical reactions.

Be careful in adding chemicals since overdosing with alkaline materials can lead to nitrification inhibition due to ammonia toxicity. High pH levels from the overdosing of alkaline materials shift the ammonium (NH_4^+) - ammonia (NH_3) equilibrium toward converting ammonium to ammonia. Thus, at a high pH, high concentrations of ammonia are produced, which inhibit the activity of nitrification microorganisms.

6.2211   DISSOLVED OXYGEN

Dissolved oxygen levels in the process stream and the final effluent are another reliable indicator of the conditions in a nitrification system. High dissolved oxygen levels (preferably 2.0 to 4.0 mg/L) must be maintained throughout the aeration and nitrification processes. If the dissolved oxygen falls to 0.2 mg/L or less, nitrification may be significantly reduced due to oxygen starvation of the nitrifying bacteria.

6.2212   NITROGEN LEVELS

Nitrogen compounds should be measured throughout the aeration tank, particularly at various key points along the reactor's length if a plug flow mode of operation is used. Ammonia nitrogen (NH_3-N) must be tested along with nitrite (NO_2^-) and nitrate (NO_3^-). These tests will enable the operator to determine the effectiveness of the nitrification process and will help to determine whether detention times and oxygen levels are adequate.

If the ammonia nitrogen levels in the plant effluent are too high and there appears to be no logical reason, you can increase the dissolved oxygen level. Increasing the dissolved oxygen from 1 to 2 mg/L up to 3 to 4 mg/L may be all that is needed to achieve operating success. This type of adjustment may be especially important in the winter due to low temperatures when MCRT conditions are most restrictive. There is less flexibility to increase MCRT when necessary.

Alternatively, to correct operating problems that appear to be related to nitrification MCRT, try increasing the MCRT (and MLSS concentration) (provided the secondary clarifiers are not overloaded).

A short-term remedy may be to begin or to increase chemical applications for enhanced suspended solids capture in the primary sedimentation tank, if available. Metal salt applications with polymer achieve excellent primary effluent suspended solids control. Such strategies will make it possible to achieve an older MCRT at the same MLSS concentration as before chemical applications.

6.2213   TEMPERATURE

The optimum wastewater temperature range is between 60 and 95 degrees F (15 and 30°C) for good nitrification. Nitrification is inhibited at low wastewater temperatures; up to five times as much detention time may be needed to accomplish complete nitrification in the winter as is needed in the summer. The growth rate of nitrifying microorganisms increases as the wastewater temperature increases, and, conversely, it decreases as the wastewater temperature decreases. Since there is no way to control the wastewater temperature, the operator must adjust other process variables to compensate for slower winter growth rates. Increasing the MLVSS concentration, increasing the MCRT, and adjusting the pH to higher levels can be expected to provide substantial, if not complete, oxidation of ammonia-nitrogen compounds. Under summer conditions, successful nitrification will be possible at lower pH levels and lower MLVSS concentrations.

6.2214   NITROGENOUS FOOD

The growth rate of nitrifying microorganisms (*Nitrosomonas* and *Nitrobacter*) is affected very little by the organic load applied to the aeration system. However, the population of the nitrifying bacteria will be limited by the amount of ammonia (NH_3) available in the wastewater. Organic nitrogen, phosphorus-containing compounds, and many trace elements are essential to the growth of microorganisms in the aeration system. The generally recommended ratio of five-day BOD to nitrogen to phosphorus for treating wastes is 100:5:1 for BOD reduction. However, other ratios may be appropriate for nitrogen removal. For example, a ratio of 10:20:0.2 is used for nitrogen removal in a biological tower at Reno-Sparks, Nevada, to produce an effluent ammonia-nitrogen level of less than 0.02 mg/L. Laboratory tests for nitrogen (TKN—Total Kjeldahl Nitrogen analyses) and phosphorus[8] should be performed so that you may add the supplemental phosphorus nutrient, if necessary. Phosphorus, in the form of phosphate fertilizer, may be added and adjusted according to the five-day BOD level and the TKN concentration in the wastewater.

EXAMPLE 1

A nitrification process is treating a flow of 2 MGD from a primary clarifier with an influent BOD of 120 mg/L, total nitrogen of 10 mg/L, and phosphorus of 0.5 mg/L. What is the BOD to nitrogen to phosphorus ratio? Is there a limiting nutrient? The recommended BOD:N:P is 100:5:1.

Known	Unknown
Flow, MGD = 2 MGD	1. BOD:N:P
BOD, mg/L = 120 mg/L	2. Is there a limiting nutrient?
N, mg/L = 10 mg/L	
P, mg/L = 0.5 mg/L	

1. Convert BOD to 100 and adjust N and P by this factor.

$$\text{Adjustment Factor} = \frac{\text{BOD, mg/}L}{100}$$

$$= \frac{120 \text{ mg/}L}{100}$$

$$= 1.2$$

2. Apply the adjustment factor of 1.2 to the nitrogen and phosphorus values to find the N and P values in the ratio.

$$\text{N Ratio Value} = \frac{\text{N, mg/}L}{\text{Adjustment Factor}}$$

$$= \frac{10 \text{ mg/}L}{1.2}$$

$$= 8.3$$

$$\text{P Ratio Value} = \frac{\text{P, mg/}L}{\text{Adjustment Factor}}$$

$$= \frac{0.5 \text{ mg/}L}{1.2}$$

$$= 0.4$$

3. Determine actual BOD:N:P ratio and compare with desired ratio of 100:5:1.

BOD:N:P = 100:8.3:0.4

Nitrogen, N, is not limiting because 8.3 is greater than 5.

Phosphorus, P, is limiting because 0.4 is less than 1. The wastewater being treated is too low in phosphorus for good microorganism performance in treating the wastes.

EXAMPLE 2

Using the information provided in Example 1, determine the amount of phosphorus needed for the optimum BOD:N:P ratio for the microorganisms. To determine the amount of phosphorus needed, calculate the BOD in pounds per day, the needed P in pounds per day, and the available P in pounds per day.

1. Calculate the BOD in lbs per day.

$$\text{BOD, lbs/day} = (\text{Flow, MGD})(\text{BOD, mg/}L)(8.34 \text{ lbs/gal})$$

$$= (2 \text{ MGD})(120 \text{ mg/}L)(8.34 \text{ lbs/gal})$$

$$= 2,000 \text{ lbs BOD/day}$$

2. Determine the amount of phosphorus needed, lbs/day.

$$\text{Needed P, lbs/day} = \frac{(\text{BOD, lbs/day})(\text{P in Ratio})}{\text{BOD in Ratio}}$$

$$= \frac{(2,000 \text{ lbs/day})(1)}{100}$$

$$= 20 \text{ lbs P/day}$$

3. Determine the actual amount of phosphorus available, lbs/day.

$$\text{Actual P, lbs/day} = (\text{Flow, MGD})(\text{P, mg/}L)(8.34 \text{ lbs/gal})$$

$$= (2 \text{ MGD})(0.5 \text{ mg/}L)(8.34 \text{ lbs/gal})$$

$$= 8 \text{ lbs P/day}$$

4. Calculate how much additional phosphorus is needed, lbs/day.

$$\text{Additional P Needed, lbs/day} = \text{Needed P, lbs/day} - \text{Actual P, lbs/day}$$

$$= 20 \text{ lbs P/day} - 8 \text{ lbs P/day}$$

$$= 12 \text{ lbs P/day (This additional phosphate should be added at the beginning of the process.)}$$

EXAMPLE 3

The phosphorus in Example 2 was added using a dry chemical fertilizer. Obtain the pounds of P per pound of fertilizer from the label on the bag or container (for example, 5 lbs of P per 100 lbs of fertilizer). Determine the setting for the fertilizer using a dry chemical feeder.

$$\text{Chemical Feeder, lbs/day} = (\text{Additional P Needed, lbs/day})\frac{(100 \text{ lbs Fertilizer})}{(5 \text{ lbs P})}$$

$$= (12 \text{ lbs P/day})\frac{(100 \text{ lbs Fertilizer})}{(5 \text{ lbs P})}$$

$$= 240 \text{ lbs Fertilizer/day}$$

---

[8] Procedures for conducting nitrogen and phosphorus analyses are presented in *OPERATION OF WASTEWATER TREATMENT PLANTS*, Volume II, in this series of operator training manuals. See Chapter 16, "Laboratory Procedures and Chemistry."

Why is phosphorus added when a phosphorus removal process is being used? The reason is that *healthy* microorganisms are needed for nitrification, denitrification, *and* phosphorus removal.

6.2215 *EXAMPLE FACILITY*

The following operating strategy has been successful for one plant at which an aeration tank and clarifier for biological nitrification are located after a conventional activated sludge process (Figure 6.1). The nitrification aeration tank is operated as a conventional plug flow system. MCRT values range from 14 to 18 days. *NOTE:* The DO and MCRT numbers in this section differ slightly from recommended values in other sections of the chapter. Operators must always develop optimum process control guidelines for their own facilities.

Dissolved oxygen levels are monitored closely to conserve energy and avoid wasting any DO. Target DO levels are 0.5 mg/L in the first five percent of the aeration tank length, 0.5 to 1.0 mg/L from five percent to 80 percent of the tank length, and greater than 2.0 mg/L in the aeration tank effluent.

To produce an effluent that settles well, approximately 15 to 30 percent of the primary effluent (BOD around 40 to 50 mg/L) is bypassed to the nitrification aeration tank. This procedure not only helps with solids settling, but also facilitates sludge wasting. Before this procedure was instituted, the solids were difficult to settle and it was difficult to waste sludge properly because any wasting error (not enough bugs remaining or too many remaining) could be disastrous.

6.23 Nitrification Using Attached Growth Reactors

Attached growth nitrification processes include the low-rate trickling filter process, rotating biological contactors, and packed bed or packed tower reactors. In all of these processes, aerobic conditions are essential for successful nitrification. Sufficient time is also necessary for the nitrifying bacteria to convert ammonia to nitrate.

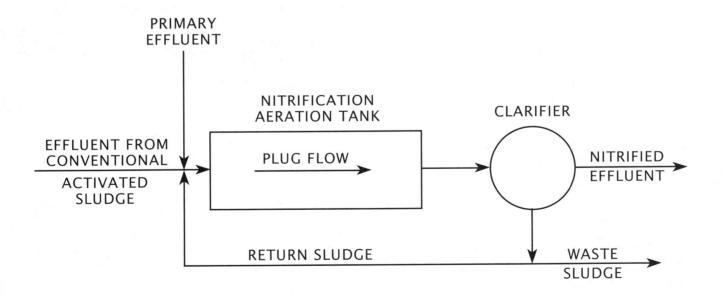

Fig. 6.1 Biological nitrification process layout

6.230 Equipment

6.2300 TRICKLING FILTERS

An attached growth (fixed film) reactor constructed in the shape of a trickling filter must have vessel containment and the appropriate media surface area for *ZOOGLEAL MASS*[9] and bacteriological action to take place. Modern reactors use plastic media (Figure 6.2) in place of rocks or wood slats. Plastic media can provide a great deal of surface area for interaction of bacteria, wastewater, and air, which results in the conversion of ammonium to nitrate.

The trickling filter type of reactor can be provided with either natural or forced ventilation. Blowers or fans can provide the forced ventilation from the bottom to the top of the reactor. As the liquid flows through the media from the top, air is forced up through the media and water trickling downward, thereby causing a cross-flow of wastewater and oxygen.

For trickling filters to nitrify the wastewater being treated, the BOD must be very low. This is achieved by using two-stage trickling filters or rotating biological contactors operated in series with nitrification occurring in the last contactors.

6.2301 PACKED TOWERS (Figures 6.3, 6.4, and 6.5)

The success of attached growth or packed tower (tall trickling filters with synthetic media) nitrification relies on a constant supply of influent and recycled wastewater trickling over the bacteriological growths on the fixed media. Wastewater is pumped to the top of the nitrification towers where distribution arms allow an equal or even distribution of flow over the entire surface of the media. Normally, nitrification towers using the fixed media concept are circular, thereby providing for continuous clockwise motion of distribution arms using the water pressure developed by pumps. Recirculation is provided to allow for a constant and even distribution of the wastewater over the entire surface media. Recirculation also dilutes the primary effluent applied to the filter, returns microorganisms to treat wastes, equalizes food (BOD) loading to microorganisms on media, and provides oxygen for the microorganisms. During low-flow conditions, the recycle will have the beneficial effect of allowing for a constant flow and organic loading capacity. Normally, a wet well is provided from which the effluent is pumped to the top of the nitrification towers to the distributors. These wet wells provide constant head conditions for the recirculation pumps.

6.2302 ROTATING BIOLOGICAL CONTACTORS (Figure 6.6)

Rotating biological contactors (RBCs) have a rotating "shaft" surrounded by plastic disks called the "media." The shaft and media are called the "drum." The plastic-disk media are made of high-density plastic circular sheets usually 12 feet in diameter. These sheets are bonded and assembled onto horizontal shafts up to 25 feet in length. Spacing between the sheets provides the hollow (void) space for distribution of wastewater and air.

A biological slime (zoogleal mass) grows on the media when conditions are suitable. This process is very similar to a trickling filter where the biological slime grows on rock or other media and settled wastewater (primary effluent) is applied over the media. With rotating biological contactors, the biological slime grows on the surface of the plastic-disk media. The slime is rotated into the settled wastewater and then into the atmosphere to provide oxygen for the microorganisms.

Rotating biological contactors provide a surface for the growth of bacteria and use of food in the wastewater for conversion of ammonium nitrogen to nitrate. Oxygen is supplied using natural air convection or ventilation.

Nitrification using a rotating biological contactor is monitored in much the same way as was discussed in the previous section on suspended growth reactors. It is important to monitor oxygen levels, nitrogen levels, and alkalinity throughout the flow stream through the reactors. Nitrification takes place in the last stage or final contactor in a series of contactors.

6.231 Operation

6.2310 WASTEWATER FLOW

The wastewater application rate to the surface of an attached growth nitrification process must not exceed the design specifications in terms of gallons per day per square foot of media. A flow in excess of design guidelines may cause a hydraulic sloughing or

---

[9] *Zoogleal* (ZOE-uh-glee-ul) *Mass.* Jelly-like masses of bacteria found in both the trickling filter and activated sludge processes. These masses may be formed for or function as the protection against predators and for storage of food supplies. Also see BIOMASS.

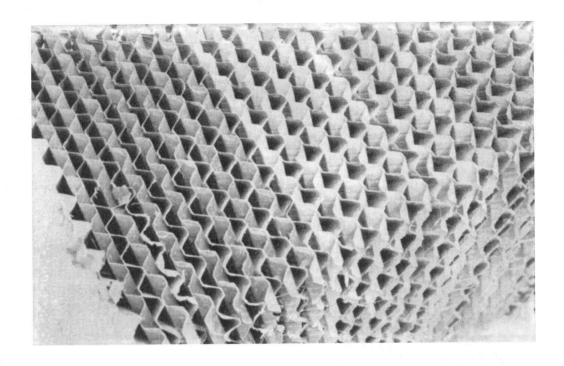

Fig. 6.2   Plastic media used for "trickling filter"
type nitrification

Fig. 6.3   Packed tower for nitrification process

Fig. 6.4   Blowers for forced ventilation into packed towers

Fig. 6.5   Fully packed tower of plastic media used for nitrification

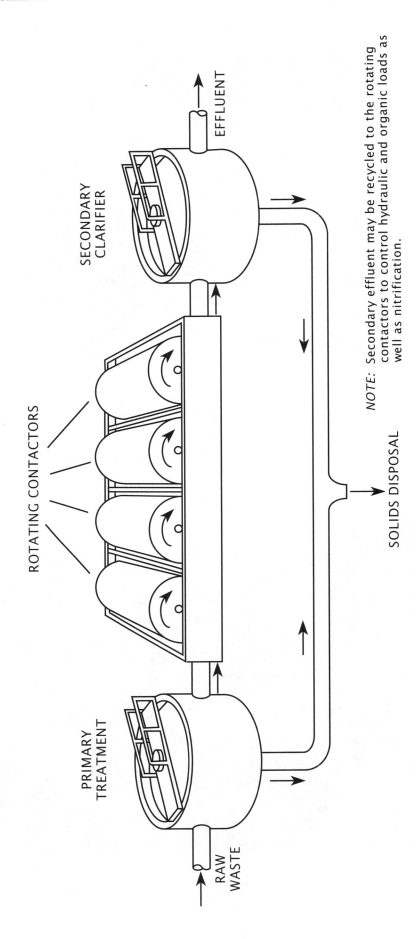

EFFLUENT

SECONDARY CLARIFIER

ROTATING CONTACTORS

PRIMARY TREATMENT

RAW WASTE

SOLIDS DISPOSAL

NOTE: Secondary effluent may be recycled to the rotating contactors to control hydraulic and organic loads as well as nitrification.

Fig. 6.6   Rotating biological contactors capable of providing for conversion of ammonia nitrogen to nitrate nitrogen washing off of the film with the valuable zoogleal growth that provides for the desired treatment and nitrification. If flow is substantially less than recommended for the application rates of wastewater to the surface of the reactor, death or dormancy of organisms may result due to lack of food, oxygen, or drying out of the growth, thus hindering treatment and causing an incomplete nitrification reaction.

6.2311 WASTE LOADS

Waste loads should also be within the range of design guidelines. BOD loadings must be fairly uniform over the surface for good performance. To achieve the desired level of nitrification, ammonia loadings must be within expected levels. The ratio of five-day BOD to nitrogen to phosphorus should be approximately 100:5:1. Laboratory nitrogen determination (TKN) and phosphorus determination analyses should be performed so that supplemental phosphorus nutrient may be added, if necessary.

6.2312 OXYGEN TRANSFER

Oxygen transfer is important in order to accomplish the desired nitrification results. The oxygen must be measured in the wastewater applied to and flowing from attached growth nitrification processes. At least 1 to 2 mg/L of oxygen must be in solution at all times. If oxygen levels become too low, there will not be sufficient oxygen for conversion of ammonium nitrogen to nitrate by the bacteria on the media. In addition, low dissolved oxygen levels may result in the slowing down or dormancy of the organisms, thereby reducing process efficiency.

6.2313 TEMPERATURE VARIATIONS

Fixed film reactors can be adversely affected by low temperatures causing freezing or other process efficiency complications. If the ambient air is too cold, frost or ice may develop on the media, thereby reducing the efficiency of the process and potentially causing mechanical problems should the ice load become heavy or cause additional weight on one side of the reactor.

Nitrification bacteria tend to slow down and become inactive when wastewater temperatures drop below 50°F (10°C). Wastewater temperature may be controlled by adjusting the recycling rate.

6.2314 SLOUGHING OF ORGANISMS

A trickling filter-type attached growth nitrification process can allow for a certain amount of sloughing of zoogleal mass. These microorganisms are dead or flushed off the media and then become part of the effluent, thus increasing the effluent suspended solids. These additional suspended solids must be removed by the treatment plant's final clarifiers prior to any discharge to receiving waters.

6.2315 OXYGEN AND FLOW VARIATIONS

Since the success of nitrification through an attached growth nitrification process is dependent on a constant wastewater flow and the appropriate oxygen levels, it is important that these two elements be monitored closely. In situations where large peak to average flow ratios are encountered, flow equalization basins before the nitrification step may be appropriate. Other important factors requiring monitoring include influent BOD (should be low) and ammonia and also proper recirculation or recycle rates.

6.24 Common Operating Problems

A common problem associated with the standard five-day BOD (BOD$_5$) determination is nitrification in the BOD bottle. This is most likely to be noticeable when a plant is achieving partial nitrification and is releasing effluent ammonia nitrogen levels of 4 to 8 mg/L. Conditions of high effluent suspended solids often mask the significance of nitrification in the BOD bottle. A good rule of thumb is that nitrification (producing high BOD values) is occurring in the BOD test when the effluent BOD is reported as higher than the effluent suspended solids and the effluent suspended solids are between 5 and 15 mg/L. To avoid this problem, always do carbonaceous (nitrification inhibited) BOD$_5$ determinations on any biologically stabilized effluent that may reflect partial nitrification.

In treatment plants that nitrify during the warm summer season, high chlorine demands in warm wastewaters are a common problem. This problem is associated with the presence of elevated nitrite nitrogen levels in the plant's effluent. A high chlorine demand could be an indication that the system's *Nitrobacter* (nitrite to nitrate) population has not yet achieved a balance with the *Nitrosomonas* (ammonia to nitrite) population, a problem that must be controlled quickly. If the plant is required to nitrify, then reduce sludge wasting to the fullest extent possible until the condition passes. If the plant is not required to nitrify, then try increasing the sludge wasting rate to establish a lower MCRT, shutting down dedicated aerators, or dropping the operating level of the dissolved oxygen in the aerator. This latter approach is only safe with an upstream selector (reactor, baffles, aeration, feed, and recycle locations); without such a system, severe bulking may result. An upstream selector allows the operator to adjust the sizes of the reactors (baffles), level of DO in different sections of the aeration basin, and locations where wastewater (feed) and recycle flows are applied to the process.

High chlorine demands can also result from the reaction of free chlorine residuals with organic compounds if nitrification is complete.

A highly nitrified effluent (less than 0.1 mg/L NH$_3$, no chloramines) can be difficult and expensive to disinfect, as previously mentioned. If the nitrate level keeps changing, the chlorine demand can change considerably during a very short time period. Also, the biological nitrification process is too sensitive and difficult to control to try to regulate it so that you will still have some ammonia in the effluent as a result of incomplete nitrification. Therefore, a supplemental source of ammonia may be necessary for effective disinfection. Approximately 1.5 mg/L of ammonia can be added to the effluent leaving the nitrification process. To disinfect the effluent, a chlorine dose of 4 to 6 mg/L of chlorine for each mg/L of ammonia is suggested. The ammonia will react with chlorine to form chloramines, which are an effective disinfectant.

6.25   Troubleshooting

Table 6.3 lists four nitrification process conditions that require operator attention. The probable cause of each condition is explained and some possible solutions are suggested.

QUESTIONS

Write your answers in a notebook and then compare your answers with those on page 542.

6.2A   Nitrification can be accomplished by the use of what two types of biological growth reactors?

6.2B   What can an operator do to maintain sufficient alkalinity in a nitrification process?

6.2C   What tests must be conducted to monitor nitrogen levels in the reactors during the nitrification process?

6.2D   How is oxygen provided in attached growth nitrification processes?

6.2E   Why must the wastewater flow be distributed over the surface of a nitrification tower at an optimum rate?

6.2F   What two items must be monitored closely by operators of attached growth nitrification processes?

6.26   Denitrification

6.260   How Denitrification Is Accomplished in Wastewater

Nitrate will be reduced to gaseous nitrogen by a broad range of microorganisms under suitable environmental conditions. This process, known as "denitrification," is accomplished by several species of heterotrophic bacteria metabolizing the nitrate to obtain oxygen, thus releasing nitrogen gas (primarily), nitrous oxide, or nitric oxide as a waste product. Using methanol as the carbon source, the overall energy reaction can be represented as follows:

$$6NO_3^- + 5CH_3OH \rightarrow 5CO_2 + 3N_2 + 7H_2O + 6OH^-$$

Methanol is closely monitored to maintain a methanol to nitrogen ratio of 3:1. Gaseous nitrogen is relatively unavailable for biological growth; thus, as noted earlier, denitrification converts nitrogen that is in an objectionable form (as a nutrient) to one that has no significant impact on environmental quality.

6.261   Equipment

Bacterial metabolic denitrification requires an environment void of free oxygen. In the absence of free oxygen, the micro-

organisms will be forced to obtain the oxygen necessary for cell metabolism from the nitrate ion.

1. Attached Growth (Fixed Film) Reactors

Most denitrification processes use submerged media columns wherein the voids are filled with the wastewater being denitrified. The varieties of media upon which the denitrifying organisms are attached include packed beds with high-porosity corrugated sheet modules or low-porosity fine media and fluidized beds with high-porosity fine media, such as sand. In a fluidized bed, sand becomes the fixed film on which organisms can attach.

2. Suspended Growth Reactors

Under ideal circumstances a suspended growth reactor similar to an aeration basin used in the activated sludge process can be allowed to operate without oxygen or air being introduced. The suspended growth reactor may be a portion of an existing aeration basin or a separate reaction chamber set aside for the purpose of denitrification. An anoxic environment can be produced to provide the conditions suitable for reduction of nitrate to nitrogen gas by microorganisms.

6.2610   DENITRIFICATION USING A FIXED FILM REACTOR

In the biological fluidized bed process (Figures 6.7 and 6.8), wastewater flows upward through a bed of fine sand at a velocity sufficient for the sand to float or "fluidize." This action allows for the entire surface area of the sand to be available for biological growth. As the wastewater flows into the reactor and the sand is fluidized, oxygen levels are carefully monitored to be certain that no free oxygen exists within the reaction chamber.

In the fixed film reactor, an organic carbon food source is added to promote biological metabolism. This food source can be methanol (wood alcohol), methane gas, primary effluent, or other sources of high-carbon food to stimulate biological growth. The bacteria (which are attached to the fixed media) feed on the organic carbon and must get oxygen from the nitrate ion to break down the food and absorb it into their cells. The bacteria will break down the nitrate ion using the chemically bound oxygen in the metabolic process and releasing the nitrogen as a gas. If primary effluent is used as a carbon source, any organic nitrogen and ammonia nitrogen in the primary effluent will pass through the process without being removed.

6.2611   DENITRIFICATION USING A SUSPENDED GROWTH REACTOR (Figures 6.9 and 6.10)

A suspended growth reactor, when operated under anoxic conditions, will provide an environment where anaerobic bacteria will break down nitrate, thus releasing nitrogen gas. Nitrified wastewater is allowed to flow through a vessel without free oxygen but with a carbon food source. Just as in the operation of a fixed film reactor, microorganisms feed on the organic carbon. If no free oxygen is available, the microorganisms will break down the nitrate ion to obtain the oxygen portion of that nitrate ion and release the nitrogen as a gas.

TABLE 6.3 NITRIFICATION TROUBLESHOOTING GUIDE

(Adapted from *PERFORMANCE EVALUATION AND TROUBLESHOOTING AT MUNICIPAL WASTEWATER TREATMENT FACILITIES*, Office of Water Program Operations, US EPA, Washington, DC.)

Indicator/Observation	Probable Cause	Check or Monitor	Solution
1. Decrease in nitrification unit pH with loss of nitrification.	1a. Need more alkalinity to offset nitrification acidic effects.	1a. Alkalinity in effluent from nitrification unit.	1a. If alkalinity is less than 10 mg/L, start addition of lime or sodium hydroxide to nitrification unit.
	1b. Addition of acidic wastes to sewer system.	1b. Raw waste pH and alkalinity.	1b. Initiate source control.
2. Inability to completely nitrify.	2a. Oxygen concentrations are limiting nitrification.	2a. Minimum DO in nitrification unit should be 1 mg/L or more.	2a. Increase aeration supply or decrease organic (BOD) loading on nitrification unit.
	2b. Cold temperatures are limiting nitrification.	2b. Temperatures.	2b. Decrease organic loading on nitrification unit or increase biological population in nitrification unit. (Increase MCRT.)
	2c. Increases in total daily influent nitrogen loads have occurred.	2c. Current influent nitrogen concentrations.	2c. Place added nitrification units in service or modify pretreatment to remove more nitrogen.
	2d. Biological solids too low in nitrification unit.	2d. MCRT should be greater than 10 days; in cold temperatures, it may need to be greater than 15 days.	2d. (1) Decrease organic loading on nitrification unit and decrease wasting or loss of sludge from nitrification unit. (2) Add settled raw wastewater (primary effluent) to nitrification unit to generate biological solids.
	2e. Peak hourly ammonium concentrations exceed available oxygen supply.	2e. Ammonium concentrations.	2e. Install flow equalization system to minimize peak concentrations or increase oxygen supply.
3. In two-stage activated sludge system, SVI (see Chapter 7, Section 7.3) of nitrification sludge is very high (greater than 250).	3. Nitrification is occurring in first stage.	3. Nitrate in first stage effluent.	3. Transfer sludge from first stage to second and maintain lower MCRT in first stage.
4. Loss of solids from final clarifier.	4. See activated sludge and sedimentation/flotation chapters, *OPERATION OF WASTEWATER TREATMENT PLANTS*, Volumes I and II.		
5. Loss of solids from trickling filter or RBC.	5. See trickling filter and RBC chapters, *OPERATION OF WASTEWATER TREATMENT PLANTS*, Volume I.		

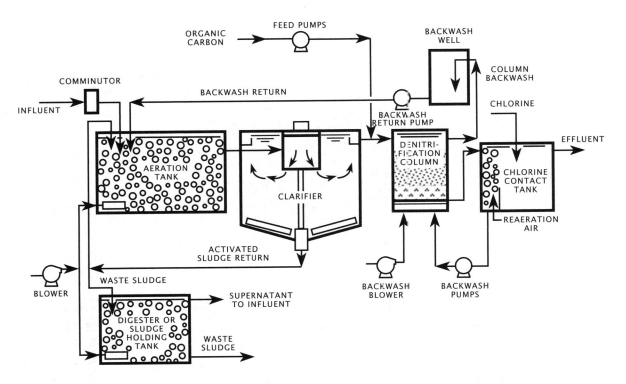

Fig. 6.7   Nitrification-denitrification flow sheet using low-porosity fine media in denitrification column
(from *PROCESS DESIGN MANUAL FOR NITROGEN CONTROL*, US Environmental Protection Agency)

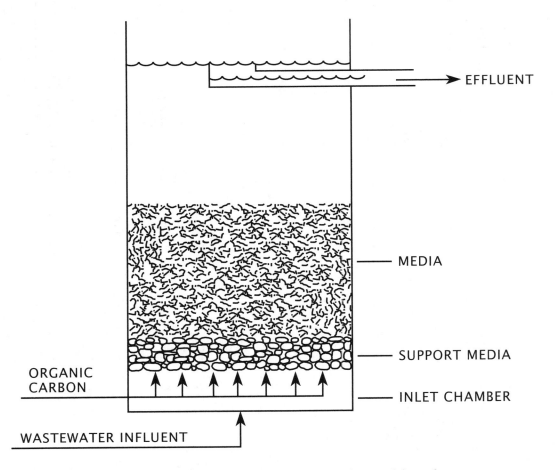

Fig. 6.8   Schematic diagram of fluidized bed used for biological denitrification

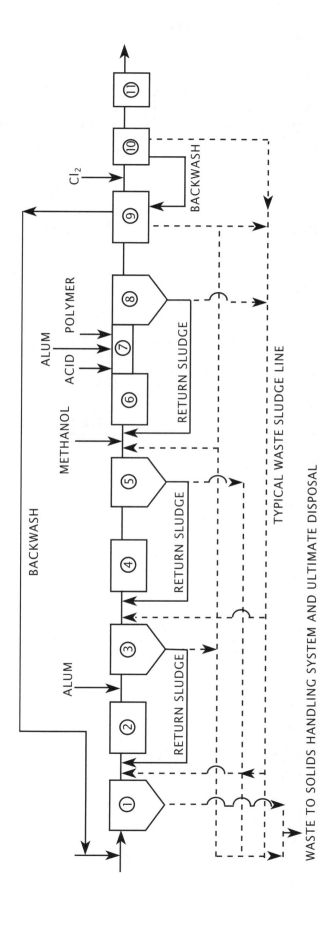

WASTE TO SOLIDS HANDLING SYSTEM AND ULTIMATE DISPOSAL

PRIMARY TREATMENT	HIGH RATE ACTIVATED SLUDGE	NITRIFYING ACTIVATED SLUDGE	DENITRIFYING ACTIVATED SLUDGE	POST TREATMENT
① SEDIMENTATION TANK	② AERATION TANK	④ AERATION TANK	⑥ ANOXIC REACTORS*	⑨ MIXED MEDIA FILTERS
	③ SEDIMENTATION TANK	⑤ SEDIMENTATION TANK	⑦ AERATED CHANNEL	⑩ CHLORINE CONTACT
			⑧ SEDIMENTATION TANK	⑪ POST AERATION

*ANOXIC: Oxygen deficient or lacking sufficient dissolved molecular oxygen.

Fig. 6.9   Nitrification-denitrification flow sheet using modifications
of the activated sludge process

(from PROCESS DESIGN MANUAL FOR NITROGEN CONTROL,
US Environmental Protection Agency)

SEPARATE SLUDGE POST-DN

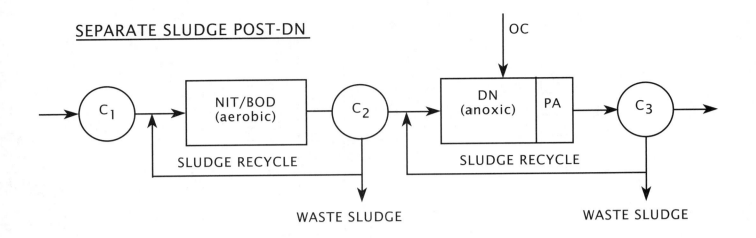

SINGLE SLUDGE POST-DN

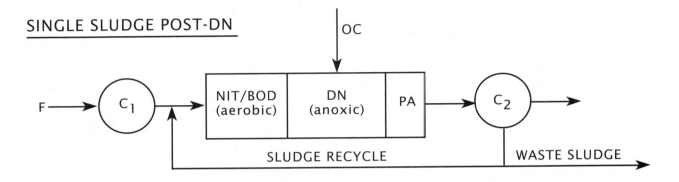

SINGLE SLUDGE PRE-DN

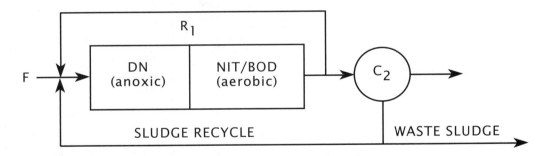

Legend

C_1	Clarifier 1 or Primary	PA	Post Aeration
C_2	Clarifier 2 or Secondary	R_1	Recirculation
C_3	Clarifier 3	OC	Organic Carbon
NIT/BOD	Nitrification and BOD Removal	DN	Denitrification

Fig. 6.10   Nitrification and denitrification using suspended growth reactors

Figure 6.10 shows three types of nitrification/denitrification suspended growth reactors.

SEPARATE SLUDGE POST-DENITRIFICATION

BOD removal and nitrification occur in the first aeration basin/clarifier system. An organic carbon source is added to the second reactor/clarifier system where denitrification takes place. The mixed liquor and return sludge from the first system are kept separate from the mixed liquor and return sludge from the second system.

SINGLE SLUDGE POST-DENITRIFICATION

BOD removal and nitrification occur in the first (aerated) reactor and the mixed liquor flows directly into the second (anoxic) reactor for denitrification without an intermediate clarifier.

SINGLE SLUDGE PRE-DENITRIFICATION

A carbon source is added to the anoxic first reactor where denitrification occurs. The mixed liquor then flows directly to the next reactor (an aeration basin) for nitrification without an intermediate clarifier. The recirculated flow returns most of the nitrate ions to the influent of the first reactor for denitrification.

6.262 Daily Operation

1. Carbon Source Feed Control

The carbon source, particularly in the case of fixed film reactors, must be carefully regulated. The microorganisms require an adequate supply of food to complete their metabolic processes as they break down the nitrate into usable oxygen and nitrogen gas. A typical methanol to nitrogen ratio is 3:1.

The capability to add methanol to the points of denitrification is suggested whenever the total effluent nitrogen limit is less than about 7.5 mg/L. If the effluent limit is less than 5 mg/L, the addition of methanol is essential. If reliable, low-level total nitrogen residuals, for example, 3 mg/L or less, are desired on an ongoing basis (such as average monthly as opposed to average annual or summer season), you will most likely need to add methanol to the process. A system that does not have the capability to add methanol may not be able to meet its nitrogen control objective if denitrification rates are slower than expected.

If too much organic carbon (methanol and primary effluent have been reported as the only effective food sources for microorganisms that convert nitrate to nitrogen gas) is fed into the process, the result can be an increase in the final effluent BOD, COD, or TOC levels (Biochemical Oxygen Demand, Chemical Oxygen Demand, or Total Organic Carbon). This result is not desirable and may lead to violations of BOD or COD effluent requirements.

2. Control of Free Oxygen

The availability of free oxygen in either a suspended growth denitrification reactor or a fixed film denitrification reactor will reduce the efficiency of the process. Special care must be taken to avoid agitation of the wastewater or any other condition allowing oxygen to enter the process flow stream.

3. Control of Effluent Suspended Solids

Suspended solids in the effluent from suspended growth denitrifying reactors and sedimentation tanks can be removed by mixed media filters, if necessary. See Chapter 4, "Solids Removal From Secondary Effluents," for process descriptions and O & M procedures for removal of suspended solids.

6.263 Troubleshooting

Table 6.4 lists four problems that could develop in the operation of a biological denitrification process. For each type of problem, the table suggests one or more probable causes and some possible solutions you may be able to implement.

QUESTIONS

Write your answers in a notebook and then compare your answers with those on pages 542 and 543.

6.2G What kind of environment is necessary for denitrification?

6.2H What is the purpose of sand in a fluidized bed reactor?

6.2I List the two kinds of reactors used in denitrification and give an example of a common type of each reactor.

6.2J What will happen if free oxygen is present in the denitrification process?

6.3 AMMONIA STRIPPING

6.30 How Ammonia Is Stripped From Wastewater

The ammonia stripping process is a reliable means of ammonia removal under suitable environmental conditions. The equilibrium equation for ammonia in water is as follows:

$$NH_4^+ \quad \rightleftarrows \quad NH_3^0 \quad + \quad H^+$$
$$\text{Ammonium Ion} \quad \text{Ammonia Gas} \quad \text{Hydrogen Ion}$$

At normal temperatures and a pH of 7, the reaction is shifted almost completely to the left. Therefore, only ammonium ions are present and virtually no dissolved ammonia gas. When the pH increases above 7 (Figure 6.11), the reaction shifts to the right and the portion of dissolved ammonia gas increases until, at pH levels of 10.8 to 11.5, almost all of the ammonium ion is converted to dissolved ammonia gas. This ammonia gas may be removed by the ammonia stripping process.

The ammonia stripping process (Figure 6.12) requires the pH of the wastewater to be raised to a level of 10.8 to 11.5, the breaking up of water droplets in the stripping process to release the ammonia gas, and the removal of the ammonia gas from the stripping tower by the movement of large quantities of air through the tower.

TABLE 6.4 DENITRIFICATION TROUBLESHOOTING GUIDE

(Adapted from *PERFORMANCE EVALUATION AND TROUBLESHOOTING AT MUNICIPAL WASTEWATER TREATMENT FACILITIES*, Office of Water Program Operations, US EPA, Washington, DC.)

Indicator/Observation	Probable Cause	Check or Monitor	Solution
1. Effluent COD shows sudden increase.	1. Excessive addition of methanol (or other oxygen-demanding material used).	1. Methanol dose.	1a. Reduce methanol addition. 1b. Install automated methanol feed system. 1c. Install aerated stabilization unit for removal of excess methanol.
2. Effluent nitrate shows sudden increase.	2a. Inadequate methanol addition. 2b. pH has drifted outside 7–7.5 range due to low pH in nitrification stage. 2c. Loss of solids from denitrifier due to failure of sludge return. 2d. Excessive DO.	2a. Methanol feed system malfunction. 2b. Alkalinity. 2c. Denitrifier unit solids and clarifier unit. 2d. Denitrifier DO should be as near zero as possible (less than 0.5 mg/L).	2a. Correct malfunction. 2b. Correct pH with addition of lime or soda ash to raise pH to 7–7.5 range. 2c. Increase sludge return; decrease sludge wasting; transfer sludge from carbonaceous unit to denitrifier. 2d. Reduce DO level. Turn some mixers off or reduce speed of blowers.
3. High head loss across packed bed or fluidized bed denitrification units.	3a. Excessive solids accumulation in filter. 3b. Nitrogen gas accumulating in filter.	3a. Length of filter run—if 12 hours or more, this is the probable cause. 3b. Run times of less than 12 hours indicate this may be the cause.	3a. Initiate full backwash cycle. 3b. Backwash bed for 1–2 minutes and return to service.
4. Packed bed or fluidized bed denitrifier that has been out of service blinds immediately upon start-up.	4. Solids have floated to top of bed and blind filter surface.	4. Solids on filter surface.	4. Backwash beds before removing them from service and immediately before starting them.

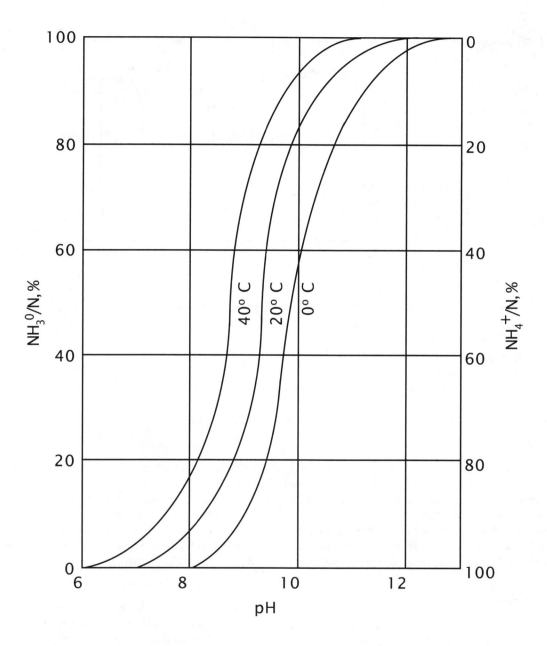

*Fig. 6.11   Effects of pH and temperature on equilibrium
between ammonium ion (NH₄⁺) and ammonia gas (NH₃⁰)*

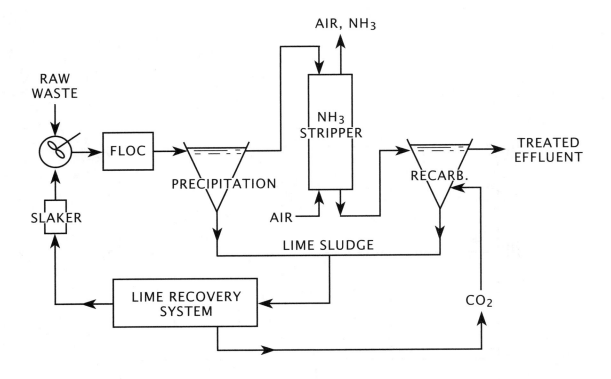

Fig. 6.12   *Schematic of ammonia stripping process with lime recovery*

6.31   Equipment

1. A High pH Source

In order to increase the pH of the wastewater, a chemical such as calcium oxide (lime) must be added to the wastewater. Lime silos, slaking equipment, lime feeders, and flash mixing chambers are all necessary to allow for the introduction of lime into the wastewater and for the mixing to properly increase the pH to between 10.8 and 11.5. (See equipment listing and operation in Chapter 5, "Phosphorus Removal," for additional information.)

2. Pumps and Piping

Most air stripping takes place using packed media towers and, therefore, high pH wastewater must be pumped to the top of these towers to allow for the water to fall, the droplets to break up, and for air contact with the wastewater.

3. Tower and Media

A tower filled with wood or plastic media or a tank filled with media must be provided for the wastewater (at a high pH) to flow or trickle through. The splashing or breaking up of the water droplets allows for the release of the ammonia gas and this gas is removed from the tower by large quantities of air moving through the tower.

6.32   Operation

1. Lime Feed (Figure 6.12)

Lime feed is accomplished normally by mixing quicklime (calcium oxide) with water and feeding the solution to the wastewater. A sufficient amount of lime must be applied to allow for the wastewater pH to be increased to at least 10.8 and preferably above 11.2. By increasing the pH to these levels, ammonium will be converted to ammonia gas, which can then be removed by the air stripping process.

2. Pumping of Wastewater

The more contact the wastewater has with the air after the pH of the wastewater has been increased, the higher the removal efficiency of ammonia gas. Large-capacity pumps are necessary to recycle the high pH wastewater through the packed tower media.

3. Calcium Carbonate Scale

Calcium carbonate is formed when the carbon dioxide in the atmosphere combines with calcium oxide to produce limestone (calcium carbonate scale). The scale can adversely affect the pumping capabilities and can close off the interior walls of pipes, pumps, and channels. Calcium carbonate can fill the voids of packed media, thereby reducing the efficiency of the ammonia stripping process. Scale must be cleaned with muriatic acid or hot water on a routine basis to maintain pumps, piping, and other stripping equipment including media packaging. Scaling potential can be minimized by maximizing the extent of completion of the calcium carbonate reaction in the lime treatment step. Using a high level of solids recycle in the clarification step will ensure a more complete reaction.

4. Freezing

Because of the high air flow and the amount of water pumped for contact with the air stripping process, the cooling effect can cause freezing temperatures within the packed media tower thereby reducing the efficiency of the process. Cold temperatures also reduce the capability of the reaction to convert ammonium ion to ammonia gas. As cold temperatures reduce the gas production capability, additional calcium oxide must be added to compensate. This further complicates problems from the standpoint of calcium carbonate scale formation.

6.33   Troubleshooting

Table 6.5 lists five problems that could develop in the operation of an ammonia stripping process. For each type of problem, the table lists some probable causes as well as some possible solutions to the problem.

TABLE 6.5   AMMONIA STRIPPING PROCESS TROUBLESHOOTING GUIDE

(Adapted from *PERFORMANCE EVALUATION AND TROUBLESHOOTING AT MUNICIPAL WASTEWATER TREATMENT FACILITIES,* Office of Water Program Operations, US EPA, Washington, DC.)

Indicator/Observation	Probable Cause	Check or Monitor	Solution
1. Scale buildup on ammonia stripping packing.	1. Insufficient cleaning of tower (removal of scale).	1. White coating accumulating on packing and reduction of air flow through the tower.	1a. Clean by hosing with a spray of water. 1b. Add a descaling polymer to the tower influent. 1c. Clean with a mixture of muriatic acid and an organic dispersant.
2. Scale buildup on pumping units feeding water into the tower.	2. Insufficient backwashing of pumping units.	2. Backwashing frequency.	2. Backwash pumps 2 or 3 times per day.
3. Ice buildup on inside of tower.	3. Freezing weather.	3. Air temperature.	3. Reverse draft fan to blow warm inside air to the frozen area to melt the ice.
4. Loss of ammonia removal efficiency.	4a. Scale buildup on fill material. 4b. pH of tower influent too low.	4a. See Item 1. 4b. pH.	4a. See Item 1. 4b. Increase pH to at least 10.8 by lime addition in chemical clarifier.
	4c. Excessive hydraulic loading.	4c. Sheets or streams of water flowing through tower rather than in droplets.	4c. See Item 5.
	4d. Insufficient air flow through tower.	4d. (1) Air flow rate.	4d. (1) Increase air flow rate by operating fans at higher speed, or recycle air flow back through the tower.
		(2) Fan inoperable.	(2) Electrical trip out—reset. Loose or damaged blade—tighten or replace. Drive bearing overheats—lube or replace.
5. Sheets or streams of water rather than droplets flowing through tower.	5a. Excessive hydraulic loading rate.	5a. Flow rate should be less than 2 GPM/sq ft.	5a. (1) Decrease the flow rate, or (2) Increase the number of units in service.
	5b. Nonuniform distribution.	5b. Spray inlet nozzle adjustment.	5b. Adjust or clean spray inlet nozzles to provide even flow distribution.
	5c. Scale buildup may be blocking a portion of the fill.	5c. See Item 1.	5c. See Item 1.

QUESTIONS

Write your answers in a notebook and then compare your answers with those on page 543.

6.3A What environmental conditions are important for a successful ammonia stripping process?

6.3B List two operating problems of the air stripping process.

6.3C How is calcium carbonate scale formed during the air stripping process?

6.3D How can calcium carbonate scale be removed?

6.4 BREAKPOINT CHLORINATION (Figure 6.13)

6.40 How Does Breakpoint Chlorination Remove Nitrogen?

When chlorine is added to water, the chlorine first reacts with the inorganic reducing materials such as hydrogen sulfide. These reactions with inorganic reducing materials occur before any chlorine residual occurs. Ferrous iron, manganese, and nitrite are examples of other inorganic reducing agents that react with chlorine and reduce the chlorine to chloride.

When chlorine is added to waters containing ammonia, the ammonia reacts with hypochlorous acid (HOCl) to form monochloramine, dichloramine, and trichloramine (Zone 1 in Figure 6.13). The formation of these chloramines depends on the pH of the solution and the initial chlorine-ammonia ratio. If enough chlorine is added to react with the inorganic compounds and nitrogenous compounds, then this chlorine will react with organic matter to produce chlororganic compounds and other combined forms of chlorine.

Addition of more chlorine results in the destruction of chloramines and chlororganic compounds. The oxidation of these compounds produces nitrous oxide (N_2O), nitrogen gas (N_2), and chlorine (Zone 2 in Figure 6.13). Therefore, if enough chlorine is added to wastewater containing ammonia nitrogen, the complete oxidation of ammonia nitrogen will occur at the "breakpoint." Then, if any additional chlorine is added beyond the breakpoint, the chlorine will exist as free available chlorine (Zone 3 in Figure 6.13).

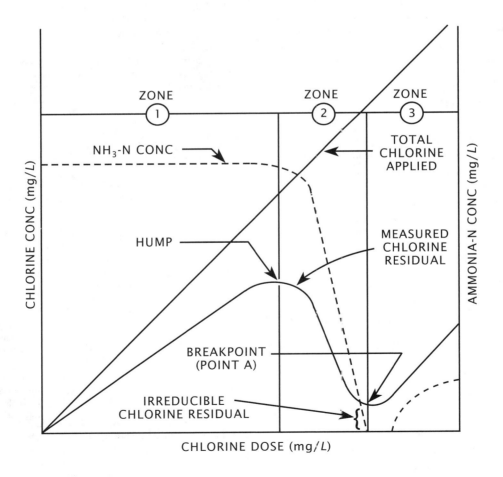

Fig. 6.13 Typical breakpoint chlorination reaction curve illustrating destruction of the ammonia molecule

(From *PROCESS DESIGN MANUAL FOR NITROGEN CONTROL*, US Environmental Protection Agency)

6.41 Equipment

1. Chlorine Feed Equipment

Adequate chlorine feed equipment is necessary to provide the appropriate quantity of chlorine in relationship to the amount of ammonia nitrogen to be reduced in the effluent under peak flow conditions. Approximately 10 mg/L of chlorine is required per mg/L of ammonium plus additional chlorine to react with the inorganic reducing materials and organic compounds.

2. Reaction Chamber

Although the reaction between chlorine and ammonia nitrogen is quite fast, the appropriate flash mixing chamber and detention time must be allowed for in a reaction vessel dedicated to breakpoint chlorination.

6.42 Operation

1. Chlorine Feed and Dose

Chlorine is fed to wastewater containing ammonium nitrogen at a sufficient rate to provide enough chlorine to oxidize the desired amount of ammonia nitrogen. A reaction requirement of 10 parts of chlorine to 1 part of ammonium nitrogen is considered necessary to reach or pass the "breakpoint."

2. Secondary or Filtered Effluent

The breakpoint chlorination process is most efficient in treating an effluent containing low suspended solids, low BOD, and a low chlorine demand. Using secondary or a filtered effluent, the chlorine can work on the ammonia nitrogen instead of being tied up in oxidation reactions with organic matter, reducing materials, and other chemicals. Filtered waters are more desirable than secondary effluent for breakpoint chlorination.

3. Final Cleanup

Breakpoint chlorination is frequently used as final cleanup following other nitrogen removal processes. When treating a high-quality effluent, relatively small amounts of chlorine are needed to remove the remaining nitrogen.

6.43 Daily Operation

1. Contact Time

The appropriate contact time required to oxidize ammonia nitrogen with chlorine varies anywhere from 5 to 15 minutes depending on the wastewater characteristics. The higher the quality of wastewater, the shorter the required contact time.

2. Flash Mixing

It is important that chlorine mixing with the wastewater be instantaneous and complete. A flash mix or other rapid mixing basin must be used to accomplish complete mixing and, thereby, maximize process efficiency.

6.44 Careful Control of Chlorine Feed

All wastewaters vary; however, an approximate ratio of 10 parts of chlorine to 1 part of ammonia nitrogen is normally required for a complete breakpoint reaction to occur. If less chlorine is used, free residual chlorine and chloramine or dichloramine compounds will remain in solution and ammonia nitrogen will not be properly removed. If there is excess chlorine in the plant effluent, this residual can be hazardous to aquatic life. Effluents should be dechlorinated by the use of sulfur dioxide to protect aquatic life. Also, if too much chlorine is added, the pH could be too low and require the addition of a basic chemical to raise the pH to an acceptable discharge level.

QUESTIONS

Write your answers in a notebook and then compare your answers with those on page 543.

6.4A Describe the breakpoint chlorination process.

6.4B What is the appropriate application for breakpoint chlorination?

6.5 LEMNA DUCKWEED SYSTEM

Lemna is a proprietary process that uses aquatic duckweed plants for wastewater treatment. The Lemna Duckweed System is used effectively as a polishing pond after a conventional wastewater treatment pond. The duckweed cover the polishing pond's surface, which prevents the penetration of sunlight and causes the algae to die and settle out of the wastewater being treated. The duckweed are capable of removing phosphorus and nitrogen from the water. Rectangular plastic grids (approximately 10 ft by 10 ft square) are placed on the surface of the pond to prevent the wind from blowing all the duckweed to one side of the pond. The population of duckweed within each grid reproduces and must be harvested on a regular basis for the system to be effective.

Duckweed needs water temperatures of 50°F (10°C) or greater to be effective. If the water temperatures drop below 50°F, duckweed will recover when the water temperatures increase to 50°F.

Please answer the discussion and review questions next.

DISCUSSION AND REVIEW QUESTIONS

Chapter 6. NITROGEN REMOVAL

The purpose of these questions is to indicate to you how well you understand the material in the chapter. Write the answers to these questions in your notebook.

1. Why are algal growths undesirable in receiving waters?

2. List the limitations of the various modes of the activated sludge process in the nitrification process.

3. How can an operator use the results of the ammonia nitrogen (NH_3-N), nitrite (NO_2^-), and nitrate (NO_3^-) tests to control the nitrification process?

4. Why do attached growth reactors use recirculation?

5. Why must the carbon feed source be carefully controlled in an attached growth reactor?

6. How does air come in contact with the wastewater in the air stripping process?

7. Why is calcium carbonate scale a problem in the air stripping process?

8. Why is secondary effluent or filtered effluent the appropriate application for breakpoint chlorination?

SUGGESTED ANSWERS

Chapter 6. NITROGEN REMOVAL

Answers to questions on page 519.

6.0A The two main nutrients that cause algal growths in receiving waters are nitrogen and phosphorus.

6.1A Biological nitrogen removal methods include:

1. Activated sludge processes
2. Trickling filter processes
3. Rotating biological contactor processes
4. Oxidation pond processes
5. Land treatment processes (overland flow)
6. Wetland treatment systems

6.1B Nitrogen is removed by the nitrification/denitrification process. First, microorganisms convert the ammonium ion to the nitrite ion and the nitrite ion to the nitrate ion. Then, in the denitrification process, microorganisms convert the nitrate ion to nitrogen gas and gaseous nitrous oxide, both of which can be removed as gases from the wastewater being treated.

6.1C The recommended use of breakpoint chlorination is to polish treatment plant effluent by decreasing the small amounts of remaining ammonia nitrogen.

6.1D Ammonium nitrogen is removed by the ion exchange process when the ammonium ion is exchanged with another ion on the ion exchange media (resin).

Answers to questions on page 530.

6.2A Nitrification can be accomplished using suspended growth reactors or attached growth (fixed film) reactors.

6.2B Sufficient alkalinity can be maintained in a nitrification process by adding calcium oxide (lime) or soda ash.

6.2C Nitrogen tests that must be performed at various key points along the reactors during the nitrification process include ammonia nitrogen (NH_3-N), nitrite (NO_2^-), and nitrate (NO_3^-).

6.2D Oxygen is provided in attached growth nitrification processes by natural or forced ventilation. Blowers or fans can provide the forced ventilation from the bottom to the top of the reactor.

6.2E The wastewater flow distributed over the surface of a nitrification tower must be at an optimum rate because excessive flow may cause a sloughing or washing off of the film with the valuable zoogleal growth that provides for the desired nitrification. If flow is insufficient, the death of organisms due to drying out, insufficient oxygen, or lack of food may result in an incomplete nitrification reaction.

6.2F A constant wastewater flow and the appropriate oxygen levels must be monitored closely when operating attached growth nitrification processes. Other factors requiring monitoring include influent BOD (should be low) and ammonia and also proper recirculation or recycle rates.

Answers to questions on page 535.

6.2G Denitrification relies on an environment for the wastewater and bacteria that is void of free oxygen. The bacteria must be forced to acquire the necessary oxygen for cell metabolism from the nitrate ion.

6.2H In a fluidized bed reactor, sand is the fixed film on which organisms can attach.

6.2I The two kinds of denitrification reactors are (1) attached growth reactors (fluidized bed reaction vessel), and (2) suspended growth reactor (aeration basin operated without oxygen or air being introduced).

6.2J If free oxygen is present in the denitrification process, there will be a reduction in the process efficiency.

Answers to questions on page 540.

6.3A The environmental conditions important for a successful ammonia stripping process are wastewater pH, temperature, and air movement.

6.3B Two operating problems of the air stripping process are (1) calcium carbonate scale, and (2) freezing.

6.3C Calcium carbonate scale is formed during the air stripping process when the carbon dioxide in the atmosphere combines with calcium oxide (used to increase the pH) and produces limestone (calcium carbonate scale).

6.3D Calcium carbonate scale can be removed with acid or hot water on a routine basis to maintain pumps, piping, and other stripping equipment, including media packing.

Answers to questions on page 541.

6.4A By adding sufficient quantities of chlorine to wastewater containing ammonia nitrogen, the complete oxidation of the ammonia nitrogen takes place at a level of chlorine addition normally referred to as the "breakpoint."

6.4B Secondary or filtered effluent is the appropriate application for breakpoint chlorination. Breakpoint chlorination also is frequently used as final cleanup following other nitrogen removal processes.

CHAPTER 7

ENHANCED BIOLOGICAL (NUTRIENT) CONTROL

by

John G. M. Gonzales

and

Ross Gudgel

TABLE OF CONTENTS

Chapter 7.   ENHANCED BIOLOGICAL (NUTRIENT) CONTROL

OBJECTIVES

Chapter 7.   ENHANCED BIOLOGICAL (NUTRIENT) CONTROL

Following completion of Chapter 7, you should be able to:

1. Explain how an enhanced biological treatment system can be used to improve biological treatment process control and performance.

2. Set priorities when more than one treatment objective must be met.

3. Operate and maintain enhanced nitrogen and phosphorus removal systems.

4. Operate and maintain enhanced SVI control to prevent sludge bulking.

5. Operate and maintain the Bardenpho process.

6. Review plans and specifications for an enhanced biological treatment system.

WORDS
Chapter 7.   ENHANCED BIOLOGICAL (NUTRIENT) CONTROL

AEROBIC (air-O-bick)   AEROBIC

A condition in which atmospheric or dissolved oxygen is present in the aquatic (water) environment.

ALKALINITY (AL-kuh-LIN-it-tee)   ALKALINITY

The capacity of water or wastewater to neutralize acids. This capacity is caused by the water's content of carbonate, bicarbonate, hydroxide, and occasionally borate, silicate, and phosphate. Alkalinity is expressed in milligrams per liter of equivalent calcium carbonate. Alkalinity is not the same as pH because water does not have to be strongly basic (high pH) to have a high alkalinity. Alkalinity is a measure of how much acid must be added to a liquid to lower the pH to 4.5.

ANAEROBIC (AN-air-O-bick)   ANAEROBIC

A condition in which atmospheric or dissolved oxygen (DO) is *NOT* present in the aquatic (water) environment.

ANAEROBIC (AN-air-O-bick) SELECTOR   ANAEROBIC SELECTOR

Anaerobic refers to the practical absence of dissolved and chemically bound oxygen. Selector refers to a reactor or basin and environmental conditions (food, lack of DO) intended to favor the growth of certain organisms over others. Also see SELECTOR.

ANOXIC (an-OX-ick)   ANOXIC

A condition in which the aquatic (water) environment does not contain dissolved oxygen (DO), which is called an oxygen deficient condition. Generally refers to an environment in which chemically bound oxygen, such as in nitrate, is present. The term is similar to ANAEROBIC.

ANOXIC DENITRIFICATION (dee-NYE-truh-fuh-KAY-shun)   ANOXIC DENITRIFICATION

A biological nitrogen removal process in which nitrate nitrogen is converted by microorganisms to nitrogen gas in the absence of dissolved oxygen.

BULKING   BULKING

Clouds of billowing sludge that occur throughout secondary clarifiers and sludge thickeners when the sludge does not settle properly. In the activated sludge process, bulking is usually caused by filamentous bacteria or bound water.

DENITRIFICATION (dee-NYE-truh-fuh-KAY-shun)   DENITRIFICATION

(1)   The anoxic biological reduction of nitrate nitrogen to nitrogen gas.

(2)   The removal of some nitrogen from a system.

(3)   An anoxic process that occurs when nitrite or nitrate ions are reduced to nitrogen gas and nitrogen bubbles are formed as a result of this process. The bubbles attach to the biological floc and float the floc to the surface of the secondary clarifiers. This condition is often the cause of rising sludge observed in secondary clarifiers or gravity thickeners. Also see NITRIFICATION.

ENTRAIN   ENTRAIN

To trap bubbles in water either mechanically through turbulence or chemically through a reaction.

FILAMENTOUS (fill-uh-MEN-tuss) ORGANISMS   FILAMENTOUS ORGANISMS

Organisms that grow in a thread or filamentous form. Common types are *Thiothrix* and *Actinomycetes*. A common cause of sludge bulking in the activated sludge process.

MCRT   MCRT

Mean Cell Residence Time. An expression of the average time (days) that a microorganism will spend in the activated sludge process.

$$\text{MCRT, days} = \frac{\text{Total Suspended Solids in Activated Sludge Process, lbs}}{\text{Total Suspended Solids Removed From Process, lbs/day}}$$

or

$$\text{MCRT, days} = \frac{\text{Total Suspended Solids in Activated Sludge Process, kg}}{\text{Total Suspended Solids Removed From Process, kg/day}}$$

NOTE: Operators at different plants calculate the Total Suspended Solids (TSS) in the Activated Sludge Process, lbs (kg), by three different methods:

1. TSS in the Aeration Basin or Reactor Zone, lbs (kg)

2. TSS in the Aeration Basin and Secondary Clarifier, lbs (kg)

3. TSS in the Aeration Basin and Secondary Clarifier Sludge Blanket, lbs (kg)

These three different methods make it difficult to compare MCRTs in days among different plants unless everyone uses the same method.

MLSS   MLSS

Mixed Liquor Suspended Solids. The amount (mg/L) of suspended solids in the mixed liquor of an aeration tank.

MLVSS   MLVSS

Mixed Liquor Volatile Suspended Solids. The amount (mg/L) of organic or volatile suspended solids in the mixed liquor of an aeration tank. This volatile portion is used as a measure or indication of the microorganisms present.

NITRIFICATION (NYE-truh-fuh-KAY-shun)   NITRIFICATION

An aerobic process in which bacteria change the ammonia and organic nitrogen in water or wastewater into oxidized nitrogen (usually nitrate).

OBLIGATE AEROBES   OBLIGATE AEROBES

Bacteria that must have atmospheric or dissolved molecular oxygen to live and reproduce.

SVI   SVI

Sludge Volume Index. A calculation that indicates the tendency of activated sludge solids (aerated solids) to thicken or to become concentrated during the sedimentation/thickening process. SVI is calculated in the following manner: (1) allow a mixed liquor sample from the aeration basin to settle for 30 minutes; (2) determine the suspended solids concentration for a sample of the same mixed liquor; (3) calculate SVI by dividing the measured (or observed) wet volume (mL/L) of the settled sludge by the dry weight concentration of MLSS in grams/L.

$$\text{SVI, mL/gm} = \frac{\text{Settled Sludge Volume/Sample Volume, mL/L}}{\text{Suspended Solids Concentration, mg/L}} \times \frac{1,000 \text{ mg}}{\text{gram}}$$

SELECTOR   SELECTOR

A reactor or basin in which baffles or other devices create a series of compartments. The environment and the resulting microbial population within each compartment can be controlled to some extent by the operator. The environmental conditions (food, lack of dissolved oxygen) that develop are intended to favor the growth of certain organisms over others. The conditions thereby select certain organisms.

SELECTOR RECYCLE   SELECTOR RECYCLE

The recycling of return sludge or oxidized nitrogen to provide desired environmental conditions for microorganisms to perform a desired function.

SUBSTRATE SUBSTRATE

(1) The base on which an organism lives. The soil is the substrate of most seed plants; rocks, soil, water, or other plants or animals are substrates for other organisms.

(2) Chemical used by an organism to support growth. The organic matter in wastewater is a substrate for the organisms in activated sludge.

VOLATILE ACIDS VOLATILE ACIDS

Fatty acids produced during digestion that are soluble in water and can be steam-distilled at atmospheric pressure. Also called organic acids. Volatile acids are commonly reported as equivalent to acetic acid.

CHAPTER 7. ENHANCED BIOLOGICAL (NUTRIENT) CONTROL

7.0 WHAT IS ENHANCED BIOLOGICAL (NUTRIENT) CONTROL?

This chapter will describe some process modifications the operator can use to improve or enhance the performance of the activated sludge process. In the past, some of these strategies were referred to as "nutrient control" or "nutrient removal" strategies because their primary purpose was the removal of nitrogen and phosphorus. Today, the modified activated sludge processes are sometimes also described as "enhanced biological control." Using the concepts described in this chapter, operators are not only able to control nutrient removal, but also sludge settleability, *FILAMENTOUS ORGANISM*[1] growth, and effluent BOD and suspended solids.

Dozens of wastewater and solids flow schemes have been developed for enhanced biological control processes. Each flow scheme reflects the special objectives, technologies, and physical equipment being used at a particular treatment plant. Many systems are known by their proprietary name,[2] that is, the name of the individual or company that developed the process. One biological nutrient removal system of this type, the Bardenpho process, is described in Section 7.20. Another proprietary process, the Phostrip® process for phosphorus removal, is described in Chapter 5, Section 5.2.

Modified activated sludge systems are popular because they are effective in removing both nitrogen and phosphorus and they are relatively inexpensive processes when compared with other chemical and physical nutrient removal processes. In addition, modified activated sludge techniques have proven to be a reliable and effective way to achieve other process objectives such as controlling sludge settleability. Modified activated sludge systems all depend on creating favorable environments in which the appropriate microorganisms will convert the nitrogen and phosphorus into forms that can be removed from the wastewater.

When enhanced biological control processes are used, the operator is usually trying to achieve more than one objective. This chapter will explain how to set priorities to accomplish multiple objectives such as nutrient removal, control of filamentous bacteria, and improving effluent quality. In addition, operator strategies for resolving conflicts between objectives are described. Information is also provided that can assist operators in reviewing plans and specifications for plant modifications or new plants to ensure that the plant has the necessary operating flexibility to achieve enhanced biological control.

NOTE: In order to understand the material presented in this chapter on the operation of enhanced biological treatment processes, you will first need a thorough understanding of how the activated sludge process works. Chapter 2 of this manual, "Activated Sludge (Pure Oxygen Plants and Operational Control Options)," and the following portions of other operator training manuals in this series may be helpful if you wish to refresh your knowledge of the activated sludge process:

1. *OPERATION OF WASTEWATER TREATMENT PLANTS,* Volume I, Chapter 8, "Activated Sludge (Package Plants and Oxidation Ditches)"

2. *OPERATION OF WASTEWATER TREATMENT PLANTS,* Volume II, Chapter 11, "Activated Sludge (Operation of Conventional Activated Sludge Plants)"

You will also need a basic understanding of the methods commonly used to remove phosphorus and nitrogen from wastestreams. That information is presented in Chapter 5, "Phosphorus Removal," and Chapter 6, "Nitrogen Removal," and will not be repeated here.

7.1 ACHIEVING MULTIPLE PROCESSING OBJECTIVES

7.10 Setting Priorities

In most wastewater treatment plants where enhanced biological treatment processes are used, the operator must meet a variety of effluent standards. Given the equipment one has at hand, it is important to learn techniques that will optimize the performance of each phase of the treatment sequence. By making various technical process adjustments, it is often possible to achieve more than one processing objective. However, the objectives must be prioritized and one controlling variable must be chosen. Table 7.1 lists some common process control objectives for enhanced biological treatment. For each objective (or phase),

---

[1] *Filamentous* (fill-uh-MEN-tuss) *Organisms.* Organisms that grow in a thread or filamentous form. Common types are *Thiothrix* and *Actinomycetes*. A common cause of sludge bulking in the activated sludge process.

[2] For a detailed summary of the different types of proprietary processes, see *WASTEWATER ENGINEERING: TREATMENT AND REUSE,* Fourth Edition, Chapter 11, "Advanced Wastewater Treatment." Metcalf & Eddy, Inc., published by the McGraw-Hill Companies, Order Services, PO Box 182604, Columbus, OH 43272-3031. ISBN 0-07-041878-0. Price, $155.63, plus nine percent of order total for shipping and handling.

TABLE 7.1 WASTEWATER TREATMENT: PROCESS CONTROL OBJECTIVES, PRIMARY AND SECONDARY CONTROLLING VARIABLES, AND POSSIBLE FALLBACK POSITIONS[a]

Process Control Objective	Controlling Variables for Desired Population Selection — Primary	Controlling Variables for Desired Population Selection — Secondary	Phase-Specific Minimum MCRT Days, Dissolved Oxygen (DO), mg/L	Possible Fallback Position
Common to All Activated Sludge Systems				
1. Effluent SS	MLSS concentration (based on target F/M or MCRT)	• SVI-dependent settling velocity • Settled sludge blanket	Uncertain, but likely ≥ 1.5 days and < 10 days with DO ≥ 0.5 mg/L	None, except for additives for floc weight and agglomeration
2. Effluent Soluble BOD_5	Soluble BOD_5	• Form and magnitude • Release patterns • DO	3 to 7 days; DO ≥ 0.5 mg/L at end of reactor	None, except for discharge control
Common to Enhanced Biological Activated Sludge Bulking Control Systems				
3. Bulking Activated Sludge	DO	• Soluble BOD_5	0.5 to 1.5 days with DO ≥ 0.5 mg/L or 1 to 3 days with DO ≤ 0.1 mg/L at beginning of reactor	Return activated sludge chlorination (Cl_2)
Common to Enhanced Biological Nutrient Control Activated Sludge Systems				
4. Nitrogen Oxidation (Nitrification)	DO	• Temperature, pH, and alkalinity	Temperature and DO (≥ 2.0 mg/L) specific	None, breakpoint Cl_2 for NH_4N < 1.5 mg/L
5. Nitrogen Removal (Denitrification)	Oxidized nitrogen and BOD_5, magnitude and form	• DO	DO ≤ 0.1 mg/L	External substrate additions with acclimation
6. Phosphorus Removal (Luxury Uptake)	DO and oxidized nitrogen	• Volatile fatty acids • Resolubilization • Sludge production • Soluble BOD_5 • Phosphate	1 to 3 days at beginning of reactor, anaerobic pathway mandatory, for example, DO ≤ 0.1 mg/L	Chemical phosphate precipitants (also provide weighting and agglomeration enhancement)

[a] Prepared by Mike Mulbarger, Paladin Enterprises, Sedona, AZ.

the table indicates primary and secondary controlling variables for the desired microbial population, the Mean Cell Residence Time (MCRT) required to meet the objective, and a possible fallback position for remedial action.

If nitrification is required, it has to be the first priority since there is no backup for this process. Thereafter, the terms of the discharge permit will determine whether phosphorus removal or nitrogen removal is the highest priority. The end result is the same: additional chemicals will be needed to remove (precipitate) phosphorus or to encourage the microorganisms to convert nitrate to nitrogen gas to achieve nitrogen removal.

If the next priority is anoxic denitrification and an *ANAEROBIC SELECTOR*[3] is available, then additional reactor contact time can be gained by changing the anaerobic selector to an anoxic one; this will make the *SUBSTRATE*,[4] formerly scavenged by the enhanced phosphorus removal organisms, available to the denitrifying organisms for additional nitrate nitrogen removal. This arrangement provides a nitrogen removal benefit by providing additional contact time and substrate. Chemical additions can fully and immediately offset declining biological phosphorus removal but will also increase solids precipitation.

In contrast, if the next priority is to maximize biological phosphorus removal, two control options are available: (1) try to increase the anaerobic MCRT by incorporating the first of at least three subsequent anoxic stages in the process train, and/or (2) use the optional *SELECTOR RECYCLE*,[5] if provided, with proper delivery of the return sludge and oxidized nitrogen recycle. With selector recycle, higher applied wastewater loadings on the recycled solids may give you higher production of volatile fatty acids. The net impact of these attempts will further compromise the denitrification process in some way. The only fallback position is to increase the methanol additions, which increases the oxidized nitrogen removal and results in faster denitrification rates after the biomass becomes accustomed to the additional methanol. A similar acclimation response is possible

for the enhanced biological phosphorus removal organisms. In any event, if the overall operating MCRT is ten days, it will take at least three weeks (two MCRTs) of continued operation before the full benefit of the changeover is realized.

7.11 Process Control

To measure what is happening in a reactor as the microbial population adapts and treats the wastewater, operators usually measure the system loading or the system response. System loading is the pollutant mass applied per unit time per some physical measure of the reactor, such as lbs COD/day/cu ft of basin or lbs COD/day/sq ft of media. System response can be measured as the biomass in the reactor per biomass yield, such as lbs MLSS/lb SS produced. Enhanced biological control also requires measurement of the existing environmental and physical conditions in the reactor, such as temperature, DO, and pH.

Liquid-solids separation is essential for successful biological process control. This is what determines the desired reactor effluent Mixed Liquor Suspended Solids (MLSS) concentration and return sludge rate. In all biological systems, there is a characteristic MCRT for each process phase. The operator is usually working with a fixed reactor volume and will need to determine the desired MLSS concentration and overall MCRT to meet one or more treatment objectives.

The Mean Cell Residence Time (MCRT) is the key to understanding how well the microorganisms are adapting and how stable the process is. Do not attempt to interpret the success or failure of a new processing strategy until at least two or three times the overall operating MCRT of the system has elapsed and, thereafter, evaluate its success over at least one or more MCRT intervals. For example, if the desired MCRT is three days, wait six to nine days before deciding whether the change was beneficial or before making further changes in the process.

The correct MCRT definition must be used for effective operation of enhanced biological control systems. To satisfy the desired MCRT objective for BOD removal, use the combined MCRT of the anaerobic, anoxic, and aerobic treatment processes for phosphorus and nitrogen removal. However, the optimum MCRT for the nitrification process must take into account the ammonia nitrogen concentrations, DO level, pH, and water temperature.

7.12 System Flexibility

Enhanced biological treatment systems should have enough flexibility to allow the operator to switch to an alternate (fallback) operating position if problems develop or the results are different than anticipated. A well-designed process will have the ability to respond to a broad range of possible operating conditions. This means it can be operated and maintained under a

---

[3] *Anaerobic* (AN-air-O-bick) *Selector.*   Anaerobic refers to the practical absence of dissolved and chemically bound oxygen. Selector refers to a reactor or basin and environmental conditions (food, lack of DO) intended to favor the growth of certain organisms over others. Also see SELECTOR.

[4] *Substrate.*   (1) The base on which an organism lives. The soil is the substrate of most seed plants; rocks, soil, water, or other plants or animals are substrates for other organisms. (2) Chemical used by an organism to support growth. The organic matter in wastewater is a substrate for the organisms in activated sludge.

[5] *Selector Recycle.*   The recycling of return sludge or oxidized nitrogen to provide desired environmental conditions for microorganisms to perform a desired function.

variety of possible flows and loads and still meet effluent objectives with a minimum of separate unit operations.

The operator of an existing treatment plant usually must work with the processes that are currently available. It is seldom possible to undertake major design modifications. However, many modifications of the activated sludge processes can be made without a major overhaul of the treatment plant. Even relatively modest changes, such as creation of separate aerobic, anoxic, and anaerobic zones in an aeration basin, can give the operator greater control over effluent quality.

Remember, flexibility is not achieved by installing a large number of treatment processes. Flexibility stems from the ability to make process changes (or alternate uses) within reactor volumes and to deliver applied and return flows to different locations. In a flexible system, auxiliary equipment such as the air supply, return sludge pumps, and chemical metering equipment have large enough operating ranges to respond smoothly and keep pace with changing flows and changes in influent quality. Ideally, plants should have excess capacity to provide sufficient detention time for each process, when necessary. For the most part, a well-designed process will be a self-regulating operation that does not require the operator to make a lot of daily or even monthly adjustments.

7.13   Operating Strategy

Many operators (and managers) have a tendency to overreact to daily events. Take note of daily process changes but wait for sustained changes and monitor process trends. Trust your own common sense and base your decisions on sound fundamentals. When making a process change, remember that all processes reflect changing influences of process recycle flows as well as immediate changes associated with influent conditions or chemical reactions. Your operating strategy needs to include a fallback position you can use if performance fails to meet your expectations.

QUESTIONS

Write your answers in a notebook and then compare your answers with those on page 572.

Write your answers in a notebook and then compare your answers with those on page 572.

7.0A   Why are modified activated sludge systems popular?

7.1A   What has to be the first priority for enhanced biological nutrient (nitrogen, phosphorus) control systems?

7.1B   Why is liquid-solids separation the key to successful biological process control?

7.1C   What factors determine the optimum MCRT for the nitrification process?

7.1D   Why is system flexibility important in biological treatment processes?

7.2   ENHANCED NITROGEN AND PHOSPHORUS REMOVAL

7.20   The Bardenpho Process

The Bardenpho process described in this section is an example of a proprietary biological nutrient removal system based on the continuous-flow activated sludge process. The four-stage Bardenpho process uses nitrification and denitrification to remove nitrogen; the five-stage Bardenpho process adds an additional stage to remove phosphorus by the biological luxury uptake process described earlier in Chapter 5, Section 5.24.

7.200   Process Description

The four-stage Bardenpho process (Figure 7.1) removes between 90 and 95 percent of all the nitrogen present in the raw wastewater by recycling nitrate-rich mixed liquor from the aeration basin to an anoxic zone[6] located ahead of the aeration basin. Denitrification of the recycled nitrate takes place in the anoxic zone in the absence of dissolved oxygen. Ammonia in the influent is converted to nitrate in the aeration basin and then recycled to the anoxic zone for denitrification. Further denitrification may be obtained by adding a second anoxic basin for the removal of nitrate remaining after recycling.

The degree of nitrate removal depends on the rate of recycling of the mixed liquor from the aeration basin to the first anoxic tank. Some plants have three recycle pumps that allow pumping of two, four, or six times the average dry weather flow back to the first anoxic zone. Usually, pumping four times the average dry weather flow is sufficient to achieve satisfactory nitrate removal. The correct recycle flow can be determined by monitoring the nitrate level in the effluent of the first anoxic basin. If the nitrate concentration in the effluent rises above about 1 mg/L, the recycle rate is too high because not enough detention time is provided in the first anoxic zone for denitrification to occur.

The five-stage Bardenpho process (Figure 7.2) is used to remove both nitrogen and phosphorus. To remove phosphorus, an anaerobic zone (fermentation tank) is added before the first anoxic zone in the nitrogen removal process (Figure 7.1). The return activated sludge is mixed with the influent to produce an organism stress condition in the absence of dissolved oxygen and nitrate. Under stress conditions, the bacteria release phosphorus from their cell structure in large quantities. After the bacteria

---

[6] This zone used to be called the anaerobic denitrification zone. Since the biochemical processes involved are not anaerobic, but a modification of aerobic processes, the term anoxic is considered a more accurate description of the processes. Aeration blower requirements usually are reduced in aeration basins (reactors) downstream from anoxic selectors because the anoxic selector removes surfactants and other chemicals that inhibit oxygen transfer from the air bubble to the water. Therefore, anoxic selectors allow much greater oxygen transfer from air bubbles to water. Also, some oxygen is provided by the nitrate in the influent to an anoxic selector.

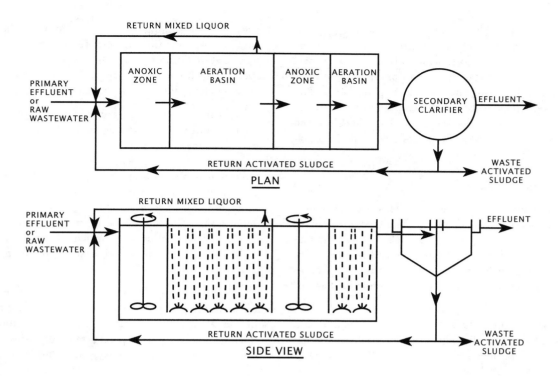

Fig. 7.1 Four-stage Bardenpho process (nitrogen removal)

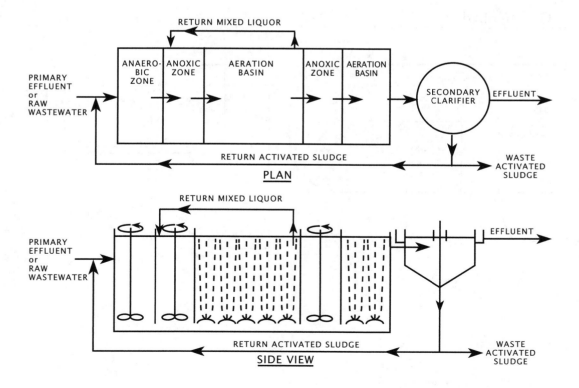

Fig. 7.2 Five-stage Bardenpho process (phosphorus and nitrogen removal)

have released their phosphorus and have gone through the anoxic denitrification zone, they are placed in an ideal environment with oxygen and food (first aeration basin). In this environment, since the bacteria are lacking in phosphorus in their cell structure, the first thing they take in is phosphorus. The phosphorus taken up by the bacteria is removed from the wastestream as waste activated sludge. Figures 7.3 and 7.4 show the four-stage and five-stage Bardenpho processes in oxidation ditch activated sludge plants.

7.201   Operating Procedures

Biological nutrient removal systems can be very effective provided the operators continually monitor the processes and make necessary adjustments, when appropriate. The operating guidelines and troubleshooting suggestions presented in Chapter 6, Section 6.2, "Biological Nitrogen Removal," can be applied to most biological nutrient removal systems, including the Bardenpho process.

Dissolved oxygen should be zero in the first anaerobic and anoxic zones. This may be difficult to achieve if the raw wastewater is fresh and the temperatures are low. Try holding the

wastewater n the collection system to allow the microorganisms in the wastewater to remove any dissolved oxygen. You will need to monitor the conditions in the collection system to prevent corrosion of the pipelines or development of serious odor problems due to septic wastewater.

Biological nitrification (conversion of ammonia to nitrate by microorganisms) theoretically requires 4.6 milligrams of oxygen per milligram of ammonia nitrogen converted to nitrate nitrogen. Sufficient air (oxygen) must be provided to the aeration tanks to maintain a dissolved oxygen (DO) level of 2 to 4 mg/L. During this reaction in an aeration tank, 7.1 milligrams of alkalinity as $CaCO_3$ is consumed per milligram of ammonia nitrogen converted to nitrate nitrogen. There must be sufficient alkalinity in the influent to the aeration tank for the ammonia to be converted to nitrate while still maintaining near-neutral pH.

NPDES permits will specify process and effluent water quality monitoring requirements for nitrogen and phosphorus removal. Water quality monitoring for process control at the influent and effluent of the process tanks or basins should include pH, temperature, dissolved oxygen, alkalinity, suspended solids,

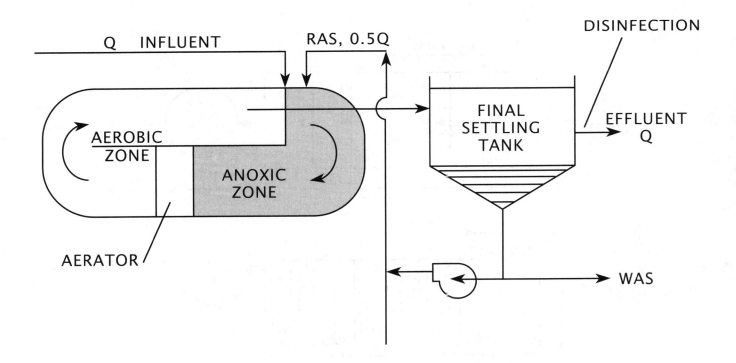

Fig. 7.3   Four-stage Bardenpho oxidation ditch (nitrogen removal)
(Provided by Krishna Pagilla, Sacramento Regional Wastewater Treatment Plant)

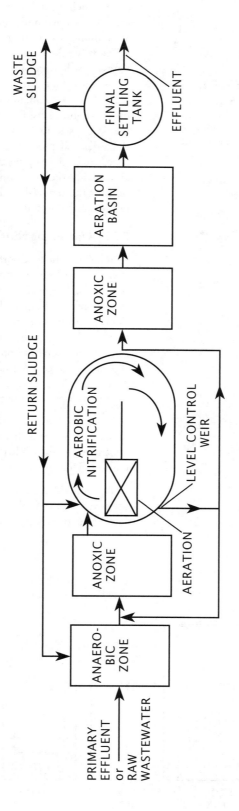

Fig. 7.4   Five-stage Bardenpho process with oxidation ditch (phosphorus and nitrogen removal)

phosphorus, ammonia nitrogen, nitrate nitrogen, total Kjeldahl nitrogen, and chemical oxygen demand (COD) to estimate biochemical oxygen demand (BOD). Typical operating guidelines for biological nutrient removal are summarized in Table 7.2.

TABLE 7.2 OPERATING GUIDELINES FOR BIOLOGICAL NUTRIENT REMOVAL[a]

Operating Guideline	Units	Bardenpho (4-Stage N)	Bardenpho (5-Stage N & P)
Food/Microorganism	$\frac{\text{lbs BOD/day}}{\text{lbs MLVSS}}$	0.1–0.2	0.1–0.2
Mean Cell Residence Time	days	20–30	10–40
MLSS	mg/L	2,000–4,000	2,000–4,000
Hydraulic Detention Time	hours		
Anaerobic Zone			1–2
Anoxic Zone 1		2–4	2–4
Aerobic Zone 1		4–12	4–12
Anoxic Zone 2		2–4	2–4
Aerobic Zone 2		0.5–1	0.5–1
Return Activated Sludge (RAS)	% of influent	50–100	50–100
Nitrate MLSS Recycle	% of influent	400	400

[a] Adapted from *WASTEWATER ENGINEERING: TREATMENT AND REUSE,* Fourth Edition. Metcalf & Eddy, Inc., published by the McGraw-Hill Companies, Order Services, PO Box 182604, Columbus, OH 43272-3031. ISBN 0-07-041878-0. Price $155.63, plus nine percent of order total for shipping and handling.

7.21 Sequencing Batch Reactor (SBR)

Sequencing batch reactors (Figure 7.5) have been used to biologically remove both nitrogen and phosphorus. Two reactors are required to continuously treat a wastestream. One tank is filling while the other tank is operated to treat the wastes. Both tanks must have sufficient capacity to handle or store incoming flows until the treatment processes are completed in the other tank. The approximate treatment times for nitrogen and phosphorus removal using sequencing batch reactors are similar to the times listed in Table 7.2 for the 4-stage N and 5-stage N and P Bardenpho processes. The operator regulates the treatment processes by developing anaerobic, anoxic, or aerobic conditions and mixing the tank contents or allowing liquid-solids separation for the removal of the treated wastewater and the disposal of the remaining sludge.

7.22 Optimizing Nitrogen and Phosphorus Removal

Achieving the best possible performance with a biological nutrient removal system requires the operator to prioritize process objectives. The facility's discharge permit will usually specify effluent nutrient limits, which may largely determine the priorities for you. In some cases, however, you may have the flexibility to set your own priorities. It is important, therefore, to understand how changes in one aspect of system operation can affect the performance of other treatment units or processes. This lesson describes some of the considerations associated with an attempt to optimize nitrogen and phosphorus removal.

Figure 7.6 illustrates the effects of pH on biological activity. What it shows us is that any attempt for optimized phosphorus control is incompatible with any attempt for optimized nitrogen control. Thus, practical operation is a matter of compromise.

The anoxic microorganisms and the microorganisms that achieve enhanced phosphorus removal are more tolerant of lower pH conditions than the nitrifiers. Low pH conditions are

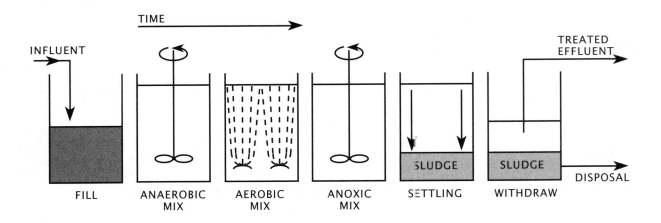

Fig. 7.5 Sequencing batch reactor (SBR) used for nitrogen and phosphorus removal

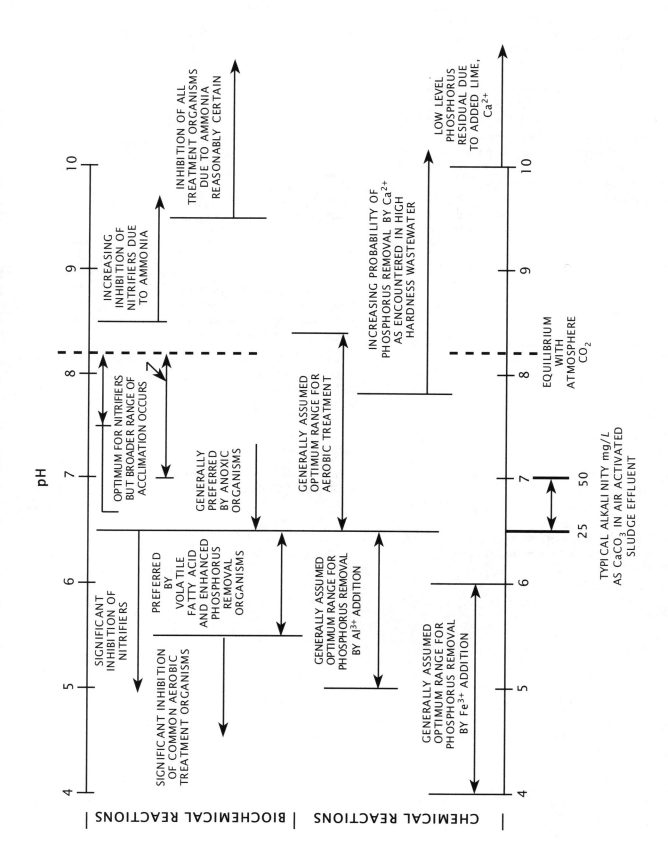

Fig. 7.6 Alkalinity and pH considerations for enhanced biological and chemical nutrient control

(Prepared by Mike Mulbarger, Paladin Enterprises, Sedona, AZ) clearly preferred by the microorganisms responsible for hydrolysis and volatile fatty acid formation; however, these microorganisms can also coexist with the methane formers in anaerobic digesters where a pH of 7 (neutrality) is desired for successful overall solids stabilization.

Anoxic selectors have been used to control filamentous bacteria, to lower SVI, and to reduce nitrate nitrogen concentrations in the effluent from an oxidation ditch. Two or three times per day (cycles per day) the ditch aeration system is turned off. A mixer is turned on to maintain a flow velocity of one ft/sec in the ditch. When the aeration system is off, anoxic conditions develop in the ditch. The microorganisms treating the influent wastes (BOD) use oxygen from the nitrate (because there is no DO), which can result in nitrogen removals as high as 80 to 85 percent. Under these anoxic conditions, filamentous bacteria are controlled and effluent SVI values will be in the 50 to 100 range.

The key to successful operation is to take an ORP probe measurement every minute and to have a transmitter connected to a computer and a programmable logic controller (PLC). Based on changes in the slope of the ORP versus time curve, the computer determines when to initiate aeration after the start of an anoxic period. If the aeration off period continues past the time of nitrate depletion, the overall treatment plant performance drops. If the aeration off period is stopped before all of the nitrate is depleted, maximum nitrogen removal and filamentous bacteria control is not achieved.

Figure 7.6 also shows the pH dependency of the metal ions used for optimum enhanced chemical phosphorus removal.[7] As shown, the optimum range for using trivalent aluminum ion (Al^{3+}) is reached (as pH decreases) about a half pH unit above the optimum range for using ferric ion (Fe^{3+}). Experience suggests that chemical usage and soluble phosphorus residuals can be minimized by operating within or very near the specific pH range cited for each chemical. Thus, in terms of phosphorus removal, the proper application point for the addition of metal salts is after the alkalinity depletion associated with nitrification, that is, to the reactor effluent as it flows to the secondary sedimentation tank. Similarly, because of the gain in alkalinity that occurs during the anoxic reduction of nitrate nitrogen in the denitrification reaction, the optimum point for application of metal salts is before denitrification.

Raw wastewaters generally have the lowest pH. This condition is most favored by the volatile fatty acid formers and enhanced biological phosphorus removal organisms; therefore, this is the most practical place to put the anaerobic reactor for phosphorus removal. If metal salts are also used at this point in the system, or if the wastewater contains metal salts, chemically enhanced phosphorus removals may also occur.

Table 7.3 identifies important reactions that occur in various methods of phosphorus control and shows how alkalinity is added to the process. Table 7.4 summarizes the alkalinity and pH interrelationships during nitrogen removal processes.

TABLE 7.3   SUMMARY OF PHOSPHORUS CONTROL REACTIONS[a]

Biochemical Phosphorus Removal by Microorganisms (microbial uptake)
- 0.015 to 0.02 mg P per mg net volatile suspended solids produced by microorganisms

Enhanced Biological Removal (luxury uptake)
- 3.6 mg P stored by microorganisms per mg P removed under normal background conditions

Enhanced Chemical Removal: Lime
- 5.4 mg $Ca_5OH(PO_4)_3$ per mg P removed plus precipitated $CaCO_3$, which depends on lime dose (influenced by alkalinity) and final pH

Biochemical Anoxic (Anaerobic) Reduction of Sulfate (SO_4)
- 3.1 mg $CaCO_3$ alkalinity added per mg sulfate sulfur reduced to sulfide sulfur

External Additives
- 1.8 mg $CaCO_3$ alkalinity added per mg CaO (quicklime) added
- 1.4 mg $CaCO_3$ alkalinity added per mg $Ca(OH)_2$ (slaked lime) added
- 1.2 mg $CaCO_3$ alkalinity added per mg NaOH (caustic) added
- 0.8 mg $CaCO_3$ alkalinity added per mg Na_2CO_3 (soda) added

[a] Prepared by Mike Mulbarger, Paladin Enterprises, Sedona, AZ.

TABLE 7.4   ALKALINITY AND pH INTERRELATIONSHIPS IN ENHANCED NITROGEN REMOVAL PROCESSES[a]

- The pH can generally be expected to rise throughout most aeration systems, regardless of the alkalinity, since the biochemically produced CO_2 is stripped from the wastewater and wastewater pH wants to come to equilibrium with the CO_2 in the atmosphere. The more vigorous the aeration, the greater the pH rise. Under these conditions, a residual alkalinity of 30 to 50 mg/L will generally result in a pH in excess of 6.8 to 7.0, and a continuing nitrifying reaction can be assumed.
- In contrast, the closed reactor condition of a pure oxygen system will result in a pH decline due to an accumulation of CO_2, regardless of the alkalinity. Under these conditions, a significant decline in nitrification performance may be expected if the pH drops below 6.5.

[a] Prepared by Mike Mulbarger, Paladin Enterprises, Sedona, AZ.

---

[7] Refer to Chapter 5, "Phosphorus Removal," for information about chemical precipitation of phosphorus.

Write your answers in a notebook and then compare your answers with those on page 572.

7.2A How does the operator of the Bardenpho process regulate the degree of nitrate removal?

7.2B Which water quality indicators are monitored for process control of the Bardenpho process?

7.3 ENHANCED SVI[8] CONTROL TO PREVENT SLUDGE BULKING

7.30 Process Types

The sludge volume index (SVI) is a measure of sludge settleability. In the activated sludge process, a condition known as sludge bulking sometimes occurs when filamentous bacteria dominate the population of microorganisms in the process. Sludge bulking is characterized by clouds of billowing sludge that will not settle properly. SVI control selectors can be used to control the growth of filamentous bacteria by preventing the development of oxygen-deficient conditions, which encourage the growth of filamentous bacteria.

SVI control selectors can be aerobic, anaerobic, or anoxic. Figure 7.7 is a schematic representation of the modifications of the activated sludge process used for SVI control. All may use an optional internal process recycle to achieve more operational flexibility. When and if enhanced biological phosphorus removal is desired, this recycle also serves to protect the desired organisms from inhibition by nitrate nitrogen.

7.31 Operations and Troubleshooting

The operation of SVI selectors can be influenced by the actual loading in both aerobic and anaerobic applications. Dropping the MLSS concentration may be desirable, but this can also be done by the compartmentalization illustrated in Figure 7.7, which shows three stages at 25, 25, and 50 percent of the selector volume. Higher MLSS concentrations may be desirable for anoxic selectors; however, staging of the process by compartmentalization is still a good arrangement.

The desired DO range in the aerobic selector or the initial stages of the aerobic system is 1.0 to 2.0 mg/L. The DO in the anoxic selector should be less than 0.3 mg/L and the desired DO in the anaerobic selector should be less than 0.1 mg/L.

Troubleshooting the process is done by looking at the soluble carbonaceous BOD$_5$ (CBOD$_5$) leaving the selector and the reactor SVI. Soluble CBOD$_5$ compartment effluent targets are not firmly fixed; they are influenced by the downstream MLSS concentration and the presence or absence of nitrifiers. The presence of nitrifiers also contributes to high MLSS oxygen uptake rates. Lower CBOD$_5$ values (say 20 to 30 mg/L) would be preferred if nitrifiers are present while higher values (say 40 to 50 mg/L) may be acceptable without their presence. Acceptable SVI target levels are no greater than 150 mL/gm and ideally less than 100 mL/gm.

If high SVIs are encountered, remember that the dominant filamentous microorganisms are not necessarily associated only with low dissolved oxygen levels. Bulking conditions in municipal wastewaters have also been reported due to low BOD loading rates and septic wastewaters or high levels of sulfide. High SVIs are also associated with young sludge (low but stable MCRT or rapid solids buildup in the system) or, in some industrial wastewaters, with nutrient deficiencies.

FOAM CONTROL

Operators may have to control different types of foam. The foam may be unstable and easy to control or the foam could be persistent and difficult to control.

Unstable foam may be caused by nutrient deficiencies or solids from dewatering processes (recycled solids). Polymer overdosing can be a cause of foam. Floating sludges and floating scum also are types of foams. These unstable foams are usually kept down using water sprays.

Persistent foams are often called filamentous or *Nocardia* foam and are difficult to control. These foams are brown, stable, viscous, and usually scum-like in appearance. These foams are often associated with high MCRT values. The higher the concentration of filaments, the greater the tendency for foaming. Also, the higher the MLSS concentration, the more susceptible an aeration basin is to foaming. The aeration rate directly influences foaming and the height of foam.

Filamentous growth rates tend to increase with temperature. Many plants have experienced foaming problems during seasonal temperature changes in the spring and fall. Apparently the optimum pH for filamentous growth is around a pH of 6.5.

Filamentous foam has been controlled by MCRT control, RAS/MLSS chlorination, direct foam chlorination, selective foam wasting, and selector technology. An MCRT of less than six days has been effective. MCRT can be reduced by slowly increasing the wasting rate with care to remain in compliance. Chlorination of RAS/MLSS or return activated sludge or both has been effective in controlling filamentous foam. If a stable foam has already formed, direct foam chlorination is the most effective method of killing foam-forming microorganisms in the foam. A foam trap is installed in the mixed liquor effluent or aeration tank and a highly concentrated chlorine spray is applied directly to the foam-forming microorganisms. Periodic

---

[8] *SVI.* Sludge Volume Index. A calculation that indicates the tendency of activated sludge solids (aerated solids) to thicken or to become concentrated during the sedimentation/thickening process. SVI is calculated in the following manner: (1) allow a mixed liquor sample from the aeration basin to settle for 30 minutes; (2) determine the suspended solids concentration for a sample of the same mixed liquor; (3) calculate SVI by dividing the measured (or observed) wet volume (mL/L) of the settled sludge by the dry weight concentration of MLSS in grams/L.

$$\text{SVI, mL/gm} = \frac{\text{Settled Sludge Volume/Sample Volume, mL/}L}{\text{Suspended Solids Concentration, mg/}L} \times \frac{1{,}000 \text{ mg}}{\text{gram}}$$

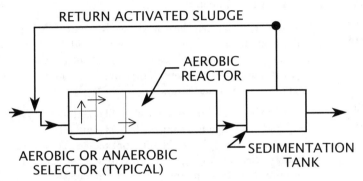

Classical Activated Sludge System

- Not consciously designed for nitrification.

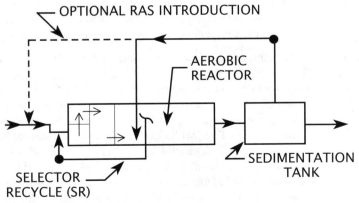

Selector Upgrade

- Aerobic—No biological phosphorus removal.

- Anaerobic—Biological phosphorus removal. (Note: May be anoxic if plant seasonally nitrifies; if so, anaerobic conditions are likely in second and third stages.)

Optional Selector Recycle Enhancement

- Aerobic—Achieves independent control of selector MLSS and reduces dependency on compartmentalization. (Note: Splitting the return sludge achieves the same function.)

- Anaerobic—May be attractive to protect preferred phosphorus removal organisms from nitrate nitrogen if maximum biological phosphorus removal potential is desired. (Note: Maximization of biologically enhanced phosphorus removal may minimize the natural anoxic denitrification potential of a subsequent anoxic system.)

Note: Selector upgrade is achieved within overall reactor or MCRT commitment for carbon stabilization (BOD reduction).

Fig. 7.7   Phases of enhanced biological SVI control and phosphorus removal chlorination of return sludge or MLSS can be useful as a preventive measure to control the number of filaments below the foaming threshold.

Aerobic, anoxic, and anaerobic selectors have been used to prevent the growth of filamentous foam microorganisms by creating an environment in which they are at a competitive disadvantage to nonfoam-forming organisms. Selectors are small reactors usually immediately upstream of the aeration tank. Aerobic selectors operate at high F:M ratios, at dissolved oxygen levels from 2 to 5 mg/L, and at an MCRT of less than 5 days. Anoxic selectors have little or no dissolved oxygen (less than 0.5 mg/L), contain recycled nitrate nitrogen, and operate at high MCRTs.

Anoxic selector zones may be very difficult to maintain during periods of high inflow/infiltration (I/I), low temperatures, diluted wastewaters, or high dissolved oxygen (DO) in influent due to these conditions.

QUESTIONS

Write your answers in a notebook and then compare your answers with those on page 572.

7.3A List the types of SVI control selectors and describe their main purposes.

7.3B What are the desired DO levels for an aerobic selector, anoxic selector, and anaerobic selector?

7.3C What are acceptable SVI target levels?

7.3D What causes elevated SVIs?

7.4 REVIEW OF PLANS AND SPECIFICATIONS

7.40 General Considerations

If possible, the operator of an enhanced biological control activated sludge system should participate to the greatest extent possible in the design of the plant, up to and including the review of plans and specifications. Throughout this process, remember that there is no such thing as a bad question. As an operator, you will have to live with and maintain the facility, daily, through its operating life.

If you are reviewing a processing scheme for possible use, try to find out why all of the particular phases or the particular approaches are proposed. A review of the operating principles described earlier in this chapter may be helpful. Remember, process needs vary as a function of the treatment objective. Some processes may offer advantages that are of no particular value in some applications. Also, some processes may offer equipment that, from a processing point of view, may be a liability compared to other available options.

When you are reviewing plans and specifications, try to anticipate all the things that can go wrong, everywhere. Experience suggests that the preliminary treatment works always require a greater share of the maintenance effort. Check carefully to be sure the design provides ready access for operators and equipment and protection from safety hazards. Often, the biological treatment system is the least demanding of all of the facilities at

the plant, but you should look for potential trouble spots, hazards, and any obstacle that would make your job more difficult or dangerous.

7.41 Overall Facility

Proper design of enhanced biological nutrient control facilities is critical to cost-effective operation, as well as to successful operation, maintenance, and performance of the facility. In many instances, it is very beneficial to both the design engineer and to you, the operator, for you to review and provide input to the plans and specifications beginning at the 50 percent design stage for a new or retrofitted facility. The design review helps the design engineer to know what details to look for to make operations and maintenance easier and to anticipate problems that might otherwise require design modifications after construction is completed. Without your assistance, modification of the facility might later be necessary because someone forgot or did not have the knowledge to recommend specific details for better operational control and ease of maintenance.

The following are some of the items that can be reviewed by you to aid the design engineer during the design stages:

1. Typical enhanced biological control systems normally use common basin construction with anaerobic, aerobic, and anoxic zones located in various portions of the same basin structure, separated by walls and gates. This reduces construction costs through common-wall construction and reduced structural requirements since the dividing walls within a tank are not required to bear the amount of water pressure an exterior tank wall must support. Compartmentalization also conserves site area, which is often critical when retrofitting an existing facility on a site with limited available land.

2. The arrangement of the various reactor zones and interconnecting gates and channels should be such that operational flexibility and component bypassing can be achieved.

3. The design of the aerobic zone is similar to that for a typical activated sludge system. In fact, it may be viewed simply as an aeration basin. The system may be designed as either a plug flow or complete mix basin configuration. Improved performance will result from a plug flow configuration. However, the higher oxygen requirements of nitrification

can create loading problems at the head end of a plug flow system, and this factor should be considered.

4. Three different classes of aeration equipment are typically used in activated sludge aeration systems: (1) mechanical surface aerators, (2) fine or coarse bubble diffused air systems, and (3) submerged turbine aerators. These systems each have different associated oxygen transfer efficiencies, although other operational and maintenance characteristics may be of greater concern than the efficiency factors.

   a. Although they require little maintenance, mechanical surface aerators may not be the system of choice due to their limited turndown capability and high heat loss in cold weather applications. This is a disadvantage, since wide variations in oxygen requirements of the process result from diurnal (daily) and seasonal changes. However, if the aeration system can be adjusted to more closely match those varying needs, the opportunity exists for energy savings.

   b. Diffused air systems are particularly well suited to nitrification systems since diffused air has a much wider turndown range. It is also easier to provide tapered aeration for a plug flow configuration with a diffused air system than with mechanical surface aerators. Due to the relatively high aeration requirements for nitrification, fine bubble diffused aeration (with its higher oxygen transfer efficiency) is preferable over coarse bubble diffusion. However, this higher efficiency comes with greater maintenance and additional chemical costs for diffuser cleaning due to diffuser fouling.

   c. Submerged turbine aerators have the advantages of diffused air in terms of turndown capability, although the energy drawn by the mixer portion of the aerator is essentially fixed with the turndown savings being in the air flow to the diffuser. This type of aerator has the additional advantage of being easily converted to a mixer by simply shutting off the air flow. This can provide additional system flexibility in a plug flow basin configuration by allowing adjustment of the aerobic and anoxic zones.

5. For any aeration system, a dissolved oxygen monitoring/aeration control system is important. The savings in aeration energy resulting from turndown during diurnal periods of low air demand can be substantial and can easily offset the additional capital cost of the control system.

   Typically, the aeration system consists of one or more dissolved oxygen sensors positioned in the aeration basin and coupled to a control system that either adjusts air delivery (for diffused air systems and submerged turbine aerators) or basin level/aerator speed (for mechanical surface aerators). A small, programmable controller is well suited for this control scheme as it can allow more complex time- and dissolved oxygen-related control decisions than typical hardware-based logic systems using relays, timers, and analog controllers. The primary drawback to this type of control system is that the dissolved oxygen monitoring device positioned in the aeration basin, which is central to the control system, requires significant attention in terms of maintenance and calibration to ensure a representative measurement of basin conditions.

6. The anoxic zones have two required features: (1) a basin, or walled-off segment of a basin of sufficient volume, and (2) sufficient mixing of the contents to maintain the microbial solids in suspension without transferring oxygen to the biomass. Submerged propeller (Figure 7.8) or turbine mixers are typically used for this latter purpose. These devices mix without breaking the water surface. They are capable of maintaining biological solids in suspension at minimal energy inputs. While energy input is an important variable in maintaining solids suspension, the number and placement of the mixers are more important factors.

7. Baffles used to define anoxic zones also should allow floating solids to exit the zones. Designs that trap floating solids can cause significant accumulations of scum, leading to odors and other operating problems. The use of submerged baffles is encouraged (Figure 7.9). In such a design, floating solids can pass from one zone to another, finally exiting the aeration basin where they can be collected from the secondary clarifiers. Collected solids should be wasted to the solids handling system, not recycled to the head of the treatment facility. These simple details can significantly reduce the accumulation of floating solids and the associated problems.

8. The recycle of nitrate-containing mixed liquor to the first-stage anoxic zone is generally accomplished by pumping. Since the liquid level in the two zones is virtually the same, the only pumping head results from pipe friction and fitting losses. However, offsetting the low head requirement is the high pumping volume required. The typical recycle ratio (with respect to plant flow) ranges from 1:1 to 4:1 but ratios as high as 6:1 may be required in some cases, particularly with a higher strength influent wastewater.

   a. Rather than using separate dry pit pumping facilities, low-head, submersible, nonclog wastewater pumps, propeller pumps, or nonclog vertical turbine pumps are generally mounted directly in the first-stage aerobic basin. The pumps should be located near the downstream end of a plug flow aerobic basin.

   b. Regardless of the type of aerobic zone, however, the pumps should not be located right next to an aeration device. Using this approach, the amount of dissolved oxygen returned with the mixed liquor will be minimized. Flow should be conveyed in a pipe rather than a channel to avoid *ENTRAINMENT*[9] of dissolved oxygen. The discharge to the first-stage anoxic zone should be submerged for the same reason.

---

[9] *Entrain.* To trap bubbles in water either mechanically through turbulence or chemically through a reaction.

Fig. 7.8   Submerged propeller mixers

c. Another consideration in the design of recycle pumping facilities is the variation of the recycle pumping rate. Unlike a raw wastewater pump station where it is necessary to match a varying influent flow rate, the nitrate-containing mixed liquor recycle flow must only be within a specified range based on a specific facility flow. Hence, there is no need to specifically match the varying facility flow, and constant-speed pumps may be used. However, to accommodate seasonal variations in nitrogen loading and wastewater temperature, it is desirable to have a sufficient number of recycle pumps so that the flow can be stepped up or down to optimize the process and avoid excess energy usage.

9. Some biological nutrient removal systems have a tendency to develop a troublesome scum that can cause odor problems and reduce the plant effluent quality. As discussed above, the basin should be designed to allow floating solids to pass to the secondary clarifiers. Consequently, design of the secondary clarifier scum removal and handling facilities to deal effectively with the potential of excessive scum development is critical for enhanced biological control facilities.

10. The primary additional component in all of the enhanced phosphorus control systems is the anaerobic zone. However, even this component is virtually identical to an anoxic zone in terms of facilities design, since both include mixing without aeration. The mixing energy input should be similar to that for an anoxic basin, or approximately 50 horsepower per million gallons (MG) of volume. The same choices of mixers (propeller type or submerged turbine) are also appropriate for the anaerobic zone. An important consideration in the design of the anaerobic zone is the discharge of the influent and recycle flows; discharge points should always be submerged to avoid entraining air into the basin contents.

11. Other considerations in the design of the anaerobic zone are the detention time (volume), and whether to provide a single basin or multiple tanks in series. Tanks in series may provide improved phosphorus uptake due to the reactions involved in the process. In general, the phosphorus removal process benefits from a higher BOD level (less likely to have any DO and more food for microorganisms) in the first stage and a higher corresponding reaction rate. Offsetting this benefit, however, are the increased construction costs for multiple basins.

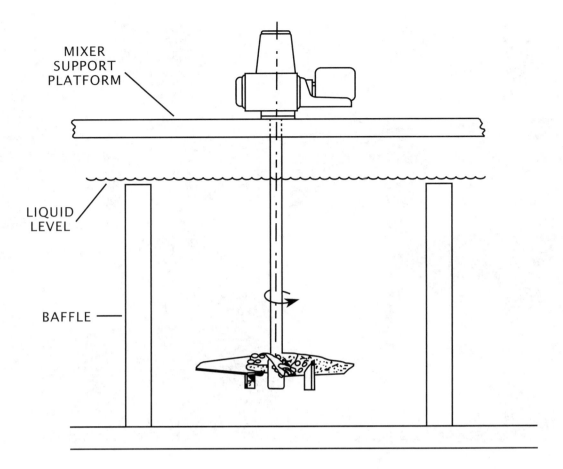

MIXER
SUPPORT
PLATFORM

LIQUID
LEVEL

BAFFLE

Fig. 7.9   Anoxic zone with submerged baffles and turbine mixer

12. Chemical storage tanks should be constructed within diked chemical containment areas that can hold the contents of the largest tank if the tank should rupture. Extreme caution must be observed in storing different chemicals within a common containment area to ensure that the chemicals are compatible in their concentrated forms.

13. Chemical piping should be suitable for the solution being conveyed over the range of anticipated operating temperatures and pressures. Shields or covers over chemical piping joints and valves should be provided to protect personnel from leakage.

14. Chemical metering pumps of the double diaphragm type are preferable due to the higher level of reliability. Two ball check valves in series should be used to protect concentrated chemical solutions from contamination when injecting chemicals into the wastewater flow stream.

15. Portable positive displacement-type pumps should be provided for use in containment area sumps to pump out spilled chemicals and washdown water. Portable pumps will lend some degree of control to the proper disposal of spilled chemicals. Additionally, pump life will be extended if the

pump is not continuously exposed to spilled chemicals contained in a sump.

16. Concrete containment areas should be coated with a corrosion-resistant, nonslip coating rated for the stored chemicals to prevent discoloration and deterioration of the concrete and supporting steel.

17. Safety equipment must also be provided. Eye wash facilities and safety showers should be easily accessible. A locker stocked with protective gear should be located near the chemical storage and metering areas. Safety signs should be prevalent and presented in languages understood by all facility personnel.

QUESTIONS

Write your answers in a notebook and then compare your answers with those on page 572.

7.4A   What are the advantages and limitations of fine bubble diffused aeration?

7.4B   What are the two basic required features of anoxic zones?

7.4C   Why is the use of submerged baffles encouraged?

7.42 Enhanced Phosphorus Control

The two general types of phosphorus removal systems are mainstream processes and sidestream processes (see Chapter 5). The items of most importance when reviewing plans for a mainstream enhanced biological control system have been covered in the previous section. If a sidestream phosphorus removal process is used, the considerations (solids loadings) found in gravity sludge thickening must apply. If you expect to use enhanced biological phosphorus removal, you will need facilities to handle at least one-half the amount of sludge that would be produced by chemical methods of phosphorus removal.

Assuming phosphorus removal by metal salts, as opposed to lime, the chemical storage, handling, and application equipment should be compatible for either alum or ferric chloride usage. The chemical feed pumps should have a reliable backup, and separate metering pumps are preferred for each point of application. Application lines should be heat traced where freezing is possible and they should be designed for easy cleaning (flush to drain) when not in use.

Chemical storage systems should consist of at least two tanks, each able to receive about 1.5 truckloads of chemical. The tanks should be surrounded by a capture basin to contain the chemical in case of a tank rupture or overflow. The storage tank vents must be large enough to allow easy air displacement while the tank is filling. The storage capacity should be at least equivalent to one to two weeks of supply. The chemicals should be stored in such an arrangement as to prevent freezing, and temperatures above 45°F (9°C) are preferred to prevent cold temperature crystallization.

Easy chemical truck access and turnaround space should be provided. Quick connect and disconnect couplings are recommended to transfer the chemical. Bumper posts should be provided to protect water hose bibs, hydrants, and buildings from accidental truck damage.

Building layout or eventual build-out planning should consider the future eventuality of additional chemicals if this seems possible. Additional redundancy in terms of multiple type metal salt additions might be attractive for larger facilities that may have primary treatment works or anticipate the need to provide ferrous salts for sulfide control.

In general, the point of application of the oxidized nitrogen return (ONR) should follow the point where the oxygen demanding substances are expected to be highest. When an anaerobic reactor phase is provided to promote biologically enhanced phosphorus removal, the ability to deliver the ONR after the anaerobic selector is preferred. A flexible design would provide the ability to deliver the return to the last stage of the anaerobic system to achieve more contact time.

Similarly, a flexible design would allow the first stage of the anoxic system to be dedicated to anaerobic processing to achieve more contact time or a higher MCRT for the enhanced biological phosphorus removal system. In this case, the ONR would be applied to the second stage of the planned anoxic reactor phase.

7.43 Enhanced Nitrogen Oxidation

The review of design documents for enhanced nitrogen oxidation should emphasize anticipated problems with *Nocardia* scum collection and management. Verify that the MCRT and dissolved oxygen conditions will allow for successful nitrification. Alkalinity is considered the best water quality indicator to monitor; therefore, be sure to check for adequate alkalinities and provide for supplemental chemical additions.

Look for reasonable predictions of solids production. Remember that point aeration systems do not achieve nonlimiting MCRT for dissolved oxygen in the same dedicated volume as diffused aeration systems. If biological phosphorus removal requirements also must be met, more solids must be considered in determining the mass of solids that must be maintained to achieve the required nonlimiting MCRT for a given DO. Do not rely on biological enhancement for complete phosphorus removal.

Make sure that the designer has adequately anticipated system needs in the early years of operation. Some bubble diffusion systems create too much turbulence; make sure that the blower can be turned down to limit turbulence. Check to be sure that the plant design has the reactors separated in terms of dedicated sedimentation tanks and that the design allows removal of reactors or sedimentation tanks from service without loss of sedimentation or reactor performance.

Check the sludge thickeners to be sure they can handle the anticipated volume of waste secondary sludge. You will need to take into account: (1) high sludge wasting rates during the months with maximum solids loadings, (2) unanticipated recycles, and (3) solids concentrations lower than those anticipated in the design. If at all possible, try to achieve a system that provides for continuous sludge wasting for ease of operation instead of intermittent operations. Similarly, if the wastewaters from sludge processing are generated intermittently (such as sludge feed to the digester with simple volumetric displacement of supernatant, or dewatering operations), look for some way to store these wastewaters temporarily so they can be recycled in a continuous flow throughout the day, 24 hours a day, 7 days a week.

If the plant has an effluent filter that is backwashed intermittently, make sure there are provisions for equalizing the recycle flow of backwash water. (See Chapter 9, "Filtration," in *INDUSTRIAL WASTE TREATMENT*, Volume I, in this series of manuals.) Regardless of the form of filtration, make sure that the backwash can be delivered to the inlet of the primary sedimentation tank or to the headworks of the plant.

If the reactor is a typical rectangular activated sludge aeration basin, explore the possibility of return sludge aeration or step-feed processing modifications to give you additional flexibility. These features would allow you to establish a higher MCRT (or an additional reservoir of nitrifiers) or lower the applied MLSS to the sedimentation tanks, if needed.

Consider using interstage baffles to improve process performance and operational flexibility. These cells need not be water-tight but they should avoid entrapment of surface debris.

7.44   Enhanced Nitrogen Removal

A review of the design drawings and specifications for enhanced nitrogen removal involves many of the same considerations that have been described with the other treatment processes. These common considerations include the use of non-watertight baffles, avoidance of surface debris entrapment, mixing with minimal water surface disruption, and provisions for the use of bulk chemicals. Check to be sure the system has the flexibility to achieve the desired anoxic staging.

Methanol is a hazardous chemical; it is volatile and readily combustible. Adequate ventilation must be provided in areas where methanol is stored or used, and precautions must be taken to prevent fires and explosions.

Methanol application points should be at any place that a continuous or intermittent anoxic reaction is planned. Multi-stage application through the three stages of a planned anoxic reactor may allow lower overall chemical use and higher overall performance. A minimum of two application points is recommended.

If oxidized nitrogen return (ONR) is part of the processing scheme, special care should be taken in locating its point of collection and application to preserve system flexibility, now and in the future. The end of a large nitrification reactor may be inappropriate as an ONR collection point if there is a possibility that more stringent nitrogen removal standards may be enforced in the future, since this end of the reactor may be converted to an anoxic reactor for polishing denitrification. Consider locating the ONR collection point about one-third of the way up from the effluent end of the nitrification reactor phase.

In general, the point of application of the ONR should follow the point where the oxygen demanding substances are expected to be highest. When an anaerobic reactor phase is provided to promote biologically enhanced phosphorus removal, the ability

to deliver the ONR after the anaerobic selector is preferred. A flexible design would provide the ability to deliver the return to the last stage of the anaerobic system to achieve more contact time.

Similarly, a flexible design would allow the first stage of the anoxic system to be dedicated to anaerobic processing to achieve more contact time or a higher MCRT for the enhanced biological phosphorus removal system. In this case, the ONR would be applied to the second stage of the planned anoxic reactor phase.

7.45   Enhanced SVI Control

A flexible selector design (reactors, baffles, and also aeration, feed, and recycle locations) that can be adjusted by the operator is highly desirable. A design that allows independent aeration (intermittent and continuous) and mixing, and the ability to introduce the oxidized nitrogen return (if nitrogen removal is an objective) at more than one point allows easy experimentation. Such a design will also permit future modification as our understanding of the technology develops or as treatment needs change. At the same time, a flexible system enables the operator to control the nuisance organisms that may be encountered.

Selector compartment air supplies can reflect the average values that are found throughout the reactor with a 50 percent turndown to 100 percent expansion over the average. To give you additional processing flexibility, it is useful to have an air supply that can provide enough air to operate the selector as an aerobic reactor. Mixing should be determined by velocity gradient terms (looking for 40 to 60 feet per second per foot (fps/ft)) or less precisely by such rule-of-thumb values as 40 to 60 horsepower per million gallons (HP/MG) (which neglect depth to volume relationships). Select a mixer that will not create surface turbulence.

The baffles between stages do not need to be watertight but should avoid entrapment of surface debris. If mixed liquor solids are recycled, a variability of plus or minus 50 percent of the plant's annual average daily design flow is adequate.

It is unlikely that the anaerobic selector will have significant odor problems so there is no need to cover it or provide for off-gas collection and deodorization. However, the potential for odors does exist; therefore, the site build-out plan should take into account the possible need to retrofit for odor containment and control at the selectors, if it is found to be necessary in the future.

QUESTIONS

Write your answers in a notebook and then compare your answers with those on page 572.

7.4D   What should be the emphasis for the review of design documents for enhanced nitrogen oxidation?

7.4E   If a plant has an effluent filter that is backwashed intermittently, what provisions are important to verify when reviewing plans and specifications?

7.4F   What hazardous chemical is used with enhanced nitrogen removal?

7.5 SUMMARY

Successful operation of enhanced biological nutrient control processes requires operators to understand what is happening and what should be happening in each reactor. Collect and analyze samples to determine the existing situation and any apparent trends. If the processes are not performing as desired or trends indicate the processing is moving in the wrong direction, make gradual changes and *wait* for the results before making additional changes.

7.6 ADDITIONAL READING

1. *OPERATION OF MUNICIPAL WASTEWATER TREATMENT PLANTS* (MOP 11), Chapters 5, 9, 17, 19, and 22,* Water Environment Federation (1996). Obtain from Water Environment Federation (WEF), Publications Order Department, 601 Wythe Street, Alexandria, VA 22314-1994. Order No. M05110. Price to members, $129.75; nonmembers, $157.75; price includes cost of shipping and handling.

2. *DESIGN MANUAL: PHOSPHORUS REMOVAL,* US Environmental Protection Agency. Obtain from National Technical Information Service (NTIS), 5285 Port Royal Road, Springfield, VA 22161. Order No. PB95-232914. EPA No. 625-1-87-001. Price, $47.50, plus $5.00 shipping and handling per order.

3. *HANDBOOK: RETROFITTING POTWs FOR PHOSPHORUS REMOVAL IN THE CHESAPEAKE BAY DRAINAGE BASIN,* US Environmental Protection Agency.

Obtain from National Service Center for Environmental Publications (NSCEP), PO Box 42419, Cincinnati, OH 45242-2419. EPA No. 625-6-87-017.

4. *PROCESS DESIGN MANUAL: NITROGEN CONTROL,* US Environmental Protection Agency. Obtain from National Service Center for Environmental Publications (NSCEP), PO Box 42419, Cincinnati, OH 45242-2419. EPA No. 625-R-93-010.

* Depends on edition.

7.7 ACKNOWLEDGMENTS

Several people contributed to the development of this chapter. Their efforts are greatly appreciated. Mike Mulbarger provided information about enhanced biological control. Ross Gudgel contributed a portion of the information on the review of plans and specifications. Andy Weist, Chief, Compliance Section, New York Department of Environmental Conservation, and Terry Poxon, Operator, Sacramento Regional Wastewater Treatment Plant, contributed significantly to the review process.

The authors wish to acknowledge and thank the operators who provided operating information on their wastewater treatment plants: Terry Anderson, Pinery Water and Wastewater District, Parker, Colorado; Richard Houben, Eastern Water Reclamation Facility, Orlando, Florida; and Mike Luker, Eastern Municipal Water District, San Jacinto, California.

Please answer the discussion and review questions next.

DISCUSSION AND REVIEW QUESTIONS
Chapter 7. ENHANCED BIOLOGICAL (NUTRIENT) CONTROL

The purpose of these questions is to indicate to you how well you understand the material in the chapter. Write the answers to these questions in your notebook.

1. What is the purpose of "enhanced biological control"?

2. Why is liquid-solids separation essential for successful biological process control?

3. How is system flexibility achieved?

4. How does the operator of a sequencing batch reactor regulate the treatment process?

5. List some of the causes of high SVIs.

SUGGESTED ANSWERS

Chapter 7.   ENHANCED BIOLOGICAL (NUTRIENT) CONTROL

Answers to questions on page 556.

7.0A   Modified activated sludge systems are popular because they are effective in removing both nitrogen and phosphorus, they are relatively inexpensive processes, and they have proven to be a reliable and effective way to achieve other process objectives.

7.1A   Nitrification has to be the first priority for enhanced biological nutrient (nitrogen, phosphorus) control systems since there is no backup for this process.

7.1B   Liquid-solids separation is the key to successful biological process control because this is what determines the desired reactor effluent MLSS concentration and return sludge rate.

7.1C   The optimum MCRT for the nitrification process depends on the ammonia nitrogen concentration, pH, dissolved oxygen level, and water temperature.

7.1D   Biological treatment system flexibility is important because it allows the operator to switch to an alternate (fallback) operating position if problems develop or the results are different than anticipated.

Answers to questions on page 563.

7.2A   In the Bardenpho process, the degree of nitrate removal depends on the rate of recycling the mixed liquor from the aeration basin. The correct recycle flow rate can be determined by monitoring the nitrate level in the effluent of the first anoxic basin.

7.2B   For the Bardenpho process, water quality monitoring for process control at the influent and effluent of the process tanks or basins should include pH, temperature, dissolved oxygen, alkalinity, suspended solids, phosphorus, ammonia nitrogen, nitrate nitrogen, total Kjeldahl nitrogen (TKN), and chemical oxygen demand (COD) to estimate biochemical oxygen demand (BOD).

Answers to questions on page 565.

7.3A   SVI control selectors can be aerobic, anaerobic, or anoxic. All are used to reduce the immediate biodegradability of the applied wastewater so as to avoid the immediate onset of aerobic oxygen-limiting conditions, which select for (encourage the growth of) filamentous organisms.

7.3B   The desired DO range in the aerobic selector or the initial stages of the aerobic system is 1.0 to 2.0 mg/L. The DO in the anoxic selector should be less than 0.3 mg/L and the desired DO in the anaerobic selector should be less than 0.1 mg/L.

7.3C   Acceptable SVI target levels are no greater than 150 mL/gm and ideally less than 100 mL/gm.

7.3D   Elevated SVIs can be caused by a dominance of filamentous organisms.

Answers to questions on page 568.

7.4A   Fine bubble diffused aeration has a high oxygen transfer efficiency. However, this higher efficiency comes with greater maintenance and additional chemical costs for diffuser cleaning due to diffuser fouling.

7.4B   The two basic required features of anoxic zones are: (1) a basin or walled-off segment of a basin of sufficient volume, and (2) sufficient mixing of the basin contents to maintain the microbial solids in suspension without transferring oxygen to the biomass.

7.4C   The use of submerged baffles is encouraged because floating solids can pass from one zone to another, finally exiting the aeration basin where they can be collected from the secondary clarifier.

Answers to questions on page 570.

7.4D   The review of design documents for enhanced nitrogen oxidation should emphasize anticipated problems with *Nocardia* scum collection and management. Verify that the MCRT and dissolved oxygen conditions will allow for successful nitrification, and be sure to check for adequate alkalinities and provide for supplemental chemical additions.

7.4E   If a plant has an effluent filter that is backwashed intermittently, make sure there are provisions for equalizing the recycle flow of backwash water. Also, regardless of the form of filtration, make sure that the backwash can be delivered to the inlet of the primary sedimentation tank or to the headworks of the plant.

7.4F   Methanol is a hazardous chemical used with enhanced nitrogen removal.

CHAPTER 8

WASTEWATER RECLAMATION AND REUSE

by

Daniel J. Hinrichs

TABLE OF CONTENTS

Chapter 8.   WASTEWATER RECLAMATION AND REUSE

LESSON 2

EFFLUENT DISPOSAL ON LAND

OBJECTIVES
Chapter 8.   WASTEWATER RECLAMATION AND REUSE

Following completion of Chapter 8, you should be able to:

1. Describe the various methods of wastewater reclamation and reuse.

2. Develop operational strategies for wastewater reclamation and reuse facilities.

3. Safely operate and maintain a wastewater reclamation and reuse facility.

4. Monitor a wastewater reclamation and reuse program and make appropriate adjustments in treatment processes.

5. Review the plans and specifications for a wastewater reclamation and reuse facility.

WORDS

Chapter 8.   WASTEWATER RECLAMATION AND REUSE

ABS   ABS

Alkyl Benzene Sulfonate. A type of surfactant, or surface active agent, present in synthetic detergents in the United States before 1965. ABS was especially troublesome because it caused foaming and resisted breakdown by biological treatment processes. ABS has been replaced in detergents by linear alkyl sulfonate (LAS), which is biodegradable.

CATION (KAT-EYE-en) EXCHANGE CAPACITY   CATION EXCHANGE CAPACITY

The ability of a soil or other solid to exchange cations (positive ions such as calcium, Ca^{2+}) with a liquid.

DRAIN TILE SYSTEM   DRAIN TILE SYSTEM

A system of tile pipes buried under agricultural fields that collect percolated waters and keep the groundwater table below the ground surface to prevent ponding.

DRAINAGE WELLS   DRAINAGE WELLS

Wells that can be pumped to lower the groundwater table and prevent ponding.

EVAPOTRANSPIRATION (ee-VAP-o-TRANS-purr-A-shun)   EVAPOTRANSPIRATION

(1)   The process by which water vapor is released to the atmosphere from living plans. Also called TRANSPIRATION.

(2)   The total water removed from an area by transpiration (plants) and by evaporation from soil, snow, and water surfaces.

HYDROLOGIC (HI-dro-LOJ-ick) CYCLE   HYDROLOGIC CYCLE

The process of evaporation of water into the air and its return to earth by precipitation (rain or snow). This process also includes transpiration from plants, groundwater movement, and runoff into rivers, streams, and the ocean. Also called the WATER CYCLE.

RECHARGE RATE   RECHARGE RATE

Rate at which water is added beneath the ground surface to replenish or recharge groundwater.

RECLAMATION   RECLAMATION

The operation or process of changing the condition or characteristics of water so that improved uses can be achieved.

RECYCLE

The use of water or wastewater within (internally) a facility before it is discharged to a treatment system. Also see REUSE.

REUSE

The use of water or wastewater after it has been discharged and then withdrawn by another user. Also see RECYCLE.

SAR

Sodium Adsorption Ratio. This ratio expresses the relative activity of sodium ions in the exchange reactions with soil. The ratio is defined as follows:

$$SAR = \frac{Na}{[\frac{1}{2}(Ca + Mg)]^{\frac{1}{2}}}$$

where Na, Ca, and Mg are concentrations of the respective ions in milliequivalents per liter of water.

$$Na, meq/L = \frac{Na, mg/L}{23.0 \ mg/meq} \qquad Ca, meq/L = \frac{Ca, mg/L}{20.0 \ mg/meq} \qquad Mg, meq/L = \frac{Mg, mg/L}{12.15 \ mg/meq}$$

SIDESTREAM

Wastewater flows that develop from other storage or treatment facilities. This wastewater may or may not need additional treatment.

CHAPTER 8. WASTEWATER RECLAMATION AND REUSE

(Lesson 1 of 2 Lessons)

DIRECT REUSE OF EFFLUENT

8.0 USES OF RECLAIMED WASTEWATER

8.00 Direct Reuse of Effluent

Why might the effluent from your wastewater treatment plant be reused directly by someone? The main reason is that someone needs water and the effluent from your wastewater treatment plant is of an acceptable quality to meet their needs. Effluent reuse is considered when: (1) the volume of municipal water needed is not available, (2) the cost of purchasing available treated water is too expensive, (3) surface waters are not available or the cost of treatment is excessive, and (4) groundwaters are either not available or the costs of pumping and any treatment are prohibitive. In the future, as discharge requirements become more and more stringent, treatment plant effluents will become more and more attractive as the best available source of water. For these reasons, you must be able to produce an effluent that can be used either directly or reclaimed for beneficial uses. Effluent from wastewater may be reclaimed for the uses listed in Table 8.1. Table 8.2 lists the treatment levels necessary for various beneficial uses.

Irrigation is discussed in Lesson 2, "Effluent Disposal on Land." Indirect reuse is the same as disposal by dilution. Groundwater recharge by spreading basins is included with the section on irrigation. This section includes direct reuse by industry, deep well injection, and recreation (Figures 8.1, 8.2 and 8.3). Generally, the operation of these systems is similar.

If water or wastewater is used again within a facility before it is discharged to a treatment system, this water is considered *RECYCLED*. If this water is discharged and then withdrawn by another user, the water is *REUSED*. *RECLAMATION* is the operation or process of changing the condition or characteristics of water so that improved uses can be achieved.

8.01 Equipment Requirements

Equipment used in wastewater reclamation plants is very similar to equipment used in most conventional wastewater treatment plants. Additional equipment requirements consist of a transmission system of pipes, ditches, or canals to transport water to the user's location. Metering and control systems are necessary for monitoring flows and water quality delivered to users.

Deep well injection systems include pipelines, pumps, and wells along with meters and control equipment. Often, treated

TABLE 8.1 TREATMENT LEVELS REQUIRED FOR VARIOUS BENEFICIAL USES[a]

Beneficial Use	Treatment Level[b]
Agricultural Irrigation—Forage Crops	1
Agricultural Irrigation—Truck Crops	5
Urban Irrigation—Landscape	4
Livestock and Wildlife Watering	1
Power Plant and Industrial Cooling	
Once-Through	1
Recirculation	1
Industrial Boiler Makeup	
Low Pressure	6
Intermediate Pressure	10
Industrial Water Supply	
Petroleum and Coal Products	3a or 4
Primary Metals	1
Paper and Allied Products	5c or 8
Chemicals and Allied Products	7 to 11
Food Products	7 to 11
Fisheries	
Warm Water	6
Cold Water	6
Recreation	
Secondary Contact	4
Primary Contact	5 or 6
Public Water Supply	
Groundwater, Spreading Basins	13[b]
Groundwater, Injection	11
Surface Water	9, 10, or 12
Direct Potable	11

[a] From Culp/Wesner/Culp, *WATER REUSE AND RECYCLING,* Volume 2, *EVALUATION OF TREATMENT TECHNOLOGY,* Department of Interior, Office of Water Research and Technology, OWRT/RU-7911, Washington, DC 1979.

[b] See Table 8.2 for an explanation of processes that produce the desired treatment levels.

TABLE 8.2   TREATMENT LEVELS[a]

Treatment Level	Treatment System
1a	Activated sludge
1b	Trickling filter
1c	Rotating biological contactors
2a	2-Stage nitrification
2b	Rotating biological contactors
2c	Extended aeration
3a	Nitrification–Denitrification
3b	Selective ion exchange
4	Filtration of secondary effluent
5a	Alum added to aeration basin
5b	Ferric chloride added to primary
5c	Tertiary lime treatment
6a	Tertiary lime, nitrified effluent
6b	Tertiary lime plus ion exchange
7	Carbon adsorption, filtered secondary effluent
8	Carbon, tertiary lime effluent
9	Carbon, tertiary lime, nitrified effluent
10	Carbon, tertiary lime, ion exchange
11	Reverse osmosis of AWT effluent
12a	Physical–Chemical system, lime
12b	Physical–Chemical system, ferric chloride
13a	Irrigation
13b	Infiltration–Percolation
13c	Overland flow

[a] From Culp/Wesner/Culp, *WATER REUSE AND RECYCLING*, Volume 2, *EVALUATION OF TREATMENT TECHNOLOGY*, Department of Interior, Office of Water Research and Technology, OWRT/RU-7911, Washington, DC 1979.

effluent (the reclaimed wastewater) is mixed with fresh water prior to injection to dilute the treated effluent. In these instances, a blending tank is required. Effluent and fresh water are pumped to the blending tank and the mixture is pumped to the injection wells. Most injection systems are used in coastal areas to serve as barriers for preventing contamination of groundwater by salt water. There may be other uses for injection wells such as injection into oil wells to aid oil pumping operations. Oil that was not removed by previous pumping efforts will float on top of the water and be easier to pump from underground areas.

The equipment requirements for disposal to a recreation lake are basically the same as disposal by dilution (Chapter 13, Volume II, *OPERATION OF WASTEWATER TREATMENT PLANTS*).

There are several variations possible with these systems, but each must provide a means for delivering water to the point of use. In some instances, storage or blending with fresh water may be desired, so Figures 8.1 and 8.2 show provisions for storage.

8.02   Limitations of Direct Reuse

Industrial reuse of wastewater will increase in the future. Water quality requirements for industrial use vary considerably. For example, cooling and washing waters do not require as high or as consistent a quality as water used in food processing or manufacturing processes. For these reasons, large industries often have their own water treatment plants on the site to provide the degree of treatment required for production processes. Under these conditions, reclaimed effluent may be used directly for washing purposes or serve as influent to a specialized water treatment plant. In either case, industries desire as consistent a quality as possible from your wastewater reclamation plant.

Regardless of the use of treated effluent or reclaimed wastewater, the user of the water will expect the water to meet specific water quality guidelines. These guidelines are just as important as NPDES permit requirements. If water quality guidelines are not met, operators must have a plan to notify users, or to store or dispose of the inadequately treated wastewater.

Industrial reuse requires a fairly uniform quality of water. If any water quality indicator fluctuates, notify the industrial user. Water quality indicators of concern include, but are not limited to, temperature, pH, color, and hardness or scale-forming minerals such as calcium, magnesium, and iron.

Deep well injection must follow procedures developed to maintain the *RECHARGE RATES*.[1] Important considerations include type of pumps, pump discharge pressures, and maintenance of any venting systems. Slime growths caused by organisms in the presence of proper temperature, food, and nutrients can reduce recharge rates. Care must be taken to avoid contamination of groundwaters used as a drinking water supply. Groundwaters can be contaminated by the discharge of excessive amounts of trace organics, minerals, nutrients, or toxic materials.

Water reclaimed for recreation use must consider protection of public health and aesthetics. As a minimum standard, public health considerations require the absence of pathogens as measured by the coliform group bacteria test. Aesthetics are evaluated by the lack of floatables and scums. The clearer a body of water, the more pleasing the appearance. Nutrients can contribute to algal growths, which reduce the aesthetic value of water used for recreation. Inadequate removal of BOD can result in the depletion of oxygen in the receiving waters, which may cause fish kills.

---

[1] *Recharge Rate.*   Rate at which water is added beneath the ground surface to replenish or recharge groundwater.

Fig. 8.1   Industrial reuse

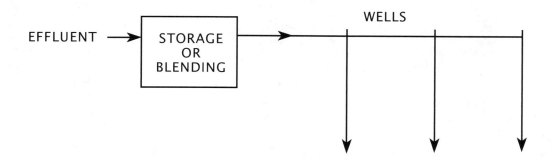

Fig. 8.2   Deep well injection

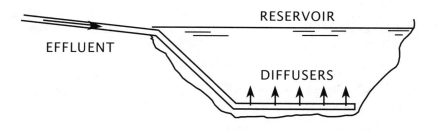

Fig. 8.3   Recreation use

Table 8.1 lists the various beneficial uses of water that are quite likely to require reclaimed water now and in the future. These different uses require different treatment processes and degrees of treatment (treatment levels) to produce a reclaimed effluent suitable for direct reuse by industry or other types of water users. These treatment levels are described in terms of treatment processes or treatment systems in Table 8.2. To determine the effluent quality you should expect from a wastewater reclamation plant, Table 8.3 lists the expected effluent quality from various combinations of treatment processes used to reclaim wastewater. This table also could be used to determine whether the effluent could be used for certain beneficial uses.

QUESTIONS

Write your answers in a notebook and then compare your answers with those on page 613.

8.0A   List the possible uses of reclaimed wastewater.

8.0B   How can deep well injection aid oil pumping operations?

8.03   Case Histories

The following case histories were developed at actual operating wastewater reclamation facilities. The data shown for the

TABLE 8.3   ANTICIPATED PERFORMANCE OF VARIOUS UNIT PROCESS COMBINATIONS[a]

		Estimated AWT Process Effluent Quality						
AWT Pretreatment	AWT Process[b]	BOD (mg/L)	COD (mg/L)	Turb. (JTU)	PO$_4$ (mg/L)	SS (mg/L)	Color (units)	NH$_3$-N (mg/L)
Preliminary[c]	C,S	50–100	80–180	5–20	2–4	10–30	30–60	20–30
	C,S,F	30–70	50–150	1–2	0.5–2	2–4	30–60	20–30
	C,S,F,AC	5–10	25–45	1–2	0.5–2	2–4	5–20	20–30
	C,S,NS,F,AC	5–10	25–45	1–2	0.5–2	2–4	5–20	1–10
Primary	C,S	50–100	80–180	5–15	2–4	10–25	30–60	20–30
	C,S,F	30–70	50–150	1–2	0.5–2	2–4	30–60	20–30
	C,S,F,AC	5–10	25–45	1–2	0.5–2	2–4	5–20	20–30
	C,S,NS,F,AC	5–10	25–45	1–2	0.5–2	2–4	5–20	1–10
High-Rate Trickling Filter	F	10–20	35–60	6–15	20–30	10–20	30–45	20–30
	C,S	10–15	35–55	2–9	1–3	4–12	25–40	20–30
	C,S,F	7–12	30–50	0.1–1	0.1–1	0–1	25–40	20–30
	C,S,F,AC	1–2	10–25	0.1–1	0.1–1	0–1	0–15	20–30
	C,S,NS,F,AC	1–2	10–25	0.1–1	0.1–1	0–1	0–15	1–10
Conventional Activated Sludge	F	3–7	30–50	2–8	20–30	3–12	25–50	20–30
	C,S	3–7	30–50	2–7	1–3	3–10	20–40	20–30
	C,S,F	1–2	25–45	0.1–1	0.1–1	0–1	20–40	20–30
	C,S,F,AC	0–1	5–15	0.1–1	0.1–1	0–1	0–15	20–30
	C,S,NS,F,AC	0–1	5–15	0.1–1	0.1–1	0–1	0–15	1–10

[a] From Russell L. Culp, George M. Wesner, and Gordon L. Culp, *HANDBOOK OF ADVANCED WASTEWATER TREATMENT,* Second Edition, Copyright 1978 by Litton Educational Publishing, Inc. Reprinted by permission of Van Nostrand Reinhold.

[b] C,S—Coagulation and sedimentation; F—mixed-media filtration; AC—activated carbon adsorption; NS—ammonia stripping (lower effluent NH$_3$ value at 18°C). For details on C and S, see Chapter 8; F, Chapter 9; and AC, Chapter 10 in *INDUSTRIAL WASTE TREATMENT,* Volume I, in this series of manuals. For NS (ammonia removal by nitrification), see Chapters 6 and 7 of this manual, *ADVANCED WASTE TREATMENT.*

[c] Preliminary treatment—grit removal, screen chamber, Parshall flume, overflow.

first two case histories were taken from *HEALTH ASPECTS OF WASTEWATER RECHARGE.* This publication was prepared by a consulting panel for the California State Water Resources Control Board, Department of Water Resources, and Department of Health.

8.030   South Lake Tahoe Public Utility District, California (Figure 8.4)

The original advanced waste treatment system operated at the South Lake Tahoe Public Utility District has changed over the years. In 1988, several processes were eliminated since the recreational lake use is no longer required. The current system provides secondary biological treatment followed by filtration and chlorination. However, the original system is described here as an example of advanced waste treatment.

This system was developed to treat wastewater for reuse in a recreation lake and for crop irrigation. The unit processes include primary sedimentation, activated sludge, lime addition and chemical clarification, ammonia stripping, filtration, activated carbon adsorption, and chlorination. The major function of the primary sedimentation process is to remove suspended solids. The activated sludge process removes suspended solids and BOD. Lime addition and chemical clarification remove

phosphorus. Ammonia removal was originally accomplished by ammonia stripping towers and later by biological nitrification or breakpoint chlorination. Filters are provided for suspended solids and turbidity removal. The activated carbon adsorption process removes COD, BOD, and surfactants. The chlorination process is provided for pathogen reduction and, during cold weather, is used for nitrogen removal by breakpoint chlorination.

These processes remove BOD, COD, suspended solids, turbidity, nitrogen, phosphorus, and pathogens from the water being treated. These constituents are removed to make the recreation lake safe for human contact (pathogen removal), prevent unsightly algal growth (nitrogen and phosphorus removal), and provide a pleasant appearance (suspended solids and turbidity removal). Table 8.4 shows water quality following each of these unit processes.

8.031   Deep Well Injection (Figure 8.5)

Wastewater is treated in ponds and then reclaimed and blended with other water for deep well injection. The processes consist of recarbonation of pond effluent followed by algae flotation, chemical clarification, breakpoint chlorination, filtration, and carbon adsorption. Recarbonation lowers the pH.

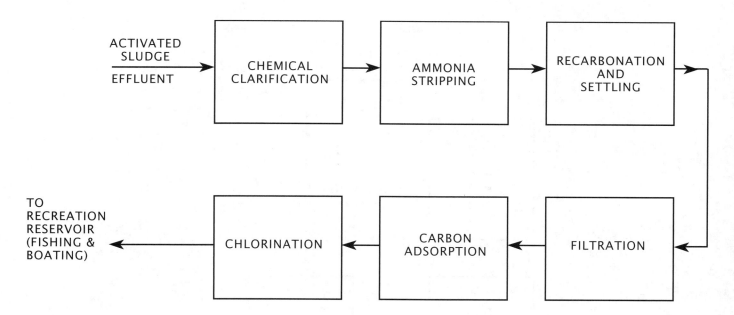

SOUTH LAKE TAHOE, CALIFORNIA

Fig. 8.4   Wastewater reclamation processes previously used at South Lake Tahoe, California

TABLE 8.4   LEVELS OF WATER QUALITY INDICATORS AFTER EACH UNIT PROCESS

Water Quality Indicator	Raw Wastewater	Unit Processes						
		Primary	Secondary	Chemical Clarification	Ammonia Stripping	Filtration	Carbon Adsorption	Chlorination
BOD (mg/L)	140	100	30	30	30	3	1	0.7
COD (mg/L)	280	220	70	70	70	25	10	10
SS (mg/L)	230	100	26	10	10	0	0	0
Turbidity (JTU)	250	150	15	10	10	0.3	0.3	0.3
MBAS (mg/L)	7	6	2	2	2	0.5	0.1	0.1
Ammonia (mg/L)	20	20	15	15	1	1	1	1
Phosphorus (mg/L)	12	9	6	0.7	0.7	0.1	0.1	0.1
Coliform (MPN/100 mL)	50	15	25	50	50	50	50	<2.2

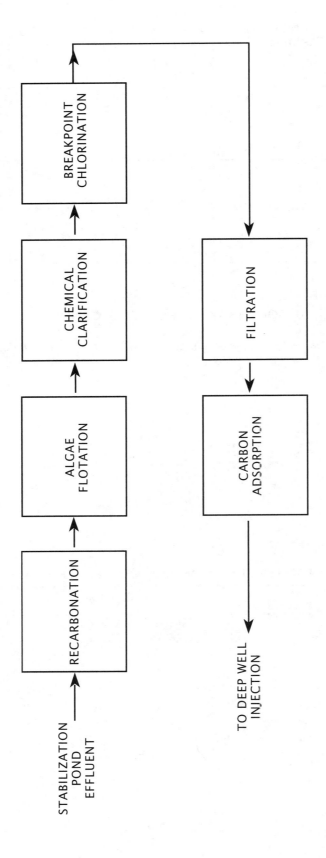

Fig. 8.5   Wastewater reclamation processes used prior to deep well injection

Algae flotation removes algae. Chemical clarification removes phosphorus and turbidity. Breakpoint chlorination destroys pathogens and reduces ammonia levels. Suspended solids and COD are removed by filtration and carbon adsorption. Treatment results are shown in Table 8.5.

8.032 Nuclear Generating Station (Figure 8.6)

Secondary effluent is pumped 40 miles (64 km) through a pipeline to a nuclear generating station. Advanced waste treatment processes produce a highly treated water that is used for makeup water in the cooling towers. There are no liquid discharges from the process and the reclamation plant meets the requirement of "no discharge."

The function of the water reclamation facility is to supply suitable feedwaters for all of the nuclear generating station site uses. The largest water use by far is the circulating water cooling tower makeup of up to 60,000 GPM (327,000 m^3/day) for all three units. This flow rate is based on 15 cycles of concentration in the cooling towers. Since some of the constituents in this water will cause problems of corrosion or deposition, the water is chemically treated before use (Table 8.6). Ammonia causes corrosion of copper and contributes to biological growths in the storage reservoir. Alkalinity contributes to scale formation. Biochemical oxygen demand (BOD) is a measure of the organic materials in the water and contributes to organic materials that cause fouling in condenser tubes. Calcium and magnesium contribute to scale formation as do silica and sulfate. Phosphorus also accelerates biological growth in the storage reservoir. Suspended solids cause sludge formation or deposits.

Upon arrival at the nuclear power site, ammonia is converted to nitrate through biological nitrification (Figure 8.6) by the use of plastic media trickling filters. An additional advantage of this process is a reduction in alkalinity with a subsequent 50 percent reduction in lime demand and sludge handling. Fifty percent of

TABLE 8.6 POTENTIAL PROBLEM CONSTITUENTS IN RECLAIMED WASTEWATER, mg/L

Constituent	Effluent from City Treatment Plant	Target for Reclaimed Water
Ammonia[a]	24–40	5
Alkalinity[b]	216–283	100
BOD	6–42	10
Calcium[b]	110–195	70
Magnesium[b]	60–116	8
Phosphorus[c]	14–41	0.5
Silica[d]	25–34	10
Sulfate	73–90	200
Suspended Solids	20–60	10

[a] mg/L as N.
[b] mg/L as CaCO$_3$.
[c] mg/L as P.
[d] mg/L as SiO$_2$.

the water going through the nitrification process is recycled to stabilize the results.

Lime (Ca(OH)$_2$) is added to the effluent from the trickling filters to increase the pH to 11.3. Lime addition is used to precipitate solids and also to reduce the magnesium (90 percent) and silica (75 percent) content in order to control scaling problems. Lime reduces the phosphate content by 95 percent. Sodium carbonate (Na$_2$CO$_3$) is added to precipitate calcium and soften the water. After lime and sodium carbonate are added and mixed with the water, solids-contact clarifiers (Figure 8.7) are used to allow the solids and precipitates to settle out.

TABLE 8.5 TREATMENT RESULTS, DEEP WELL INJECTION

Water Quality Indicator	Effluent	Flotation	Chemical Clarification and Chlorination	Filtration	Activated Carbon
Total N, mg/L	35	32	15	14	13
Organic N, mg/L	3.2	1.3	0.9	0.7	—
Ammonia N, mg/L	14.9	14	0.2	0.3	0.1
Nitrate, mg/L	17	17	14	13	13
Phosphate, mg/L	10	—	—	—	—
ABS,[a] mg/L	8	7	4	4	0.7
BOD, mg/L	30	4	1	1	0.3
pH	8.5	7.1	8.0	8.0	8.0

[a] ABS. Alkyl Benzene Sulfonate. A type of surfactant, or surface active agent, present in synthetic detergents in the United States before 1965. ABS was especially troublesome because it caused foaming and resisted breakdown by biological treatment processes. ABS has been replaced in detergents by linear alkyl sulfonate (LAS), which is biodegradable.

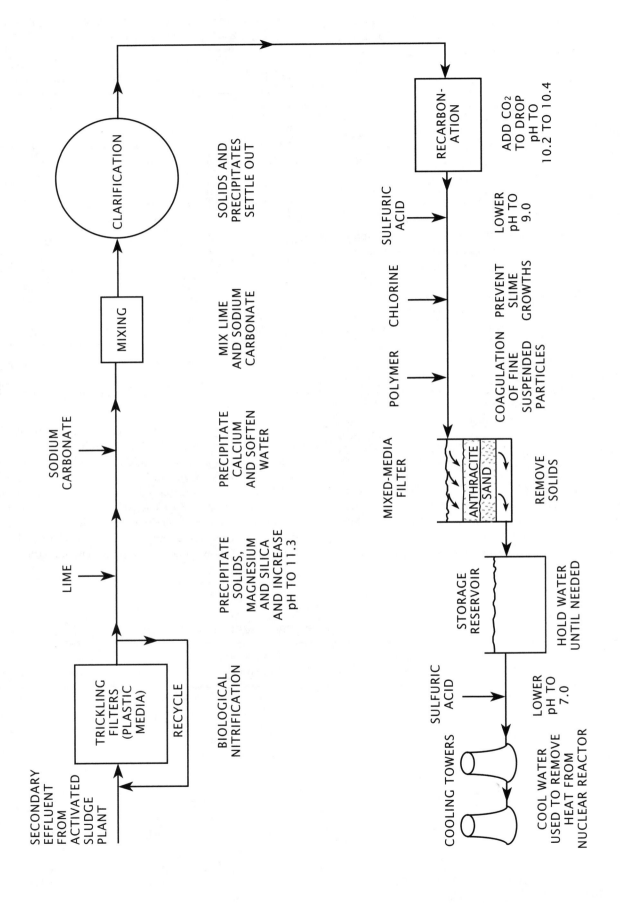

Fig. 8.6   Wastewater reclamation processes used for cooling towers at a nuclear generating station

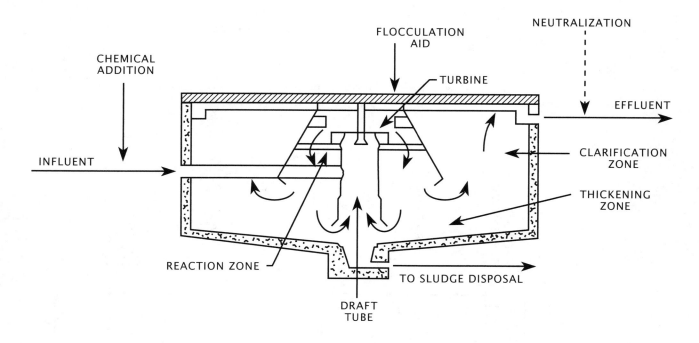

Fig. 8.7 Typical solids-contact clarifier

After clarification, carbon dioxide is used to lower the pH to between 10.2 and 10.4. Sulfuric acid is used to drop the pH to 9 and to stop the formation of any more calcium carbonate ($CaCO_3$). Chlorine is added to prevent slime growth in the filters, and polymer is used to facilitate coagulation of fine suspended particles. The water then passes through mixed-media gravity filters containing anthracite and sand to remove any fine or light solids remaining in suspension in the water. The filters are backwashed with water from the storage reservoir and the backwash water is recycled to the headworks of the trickling filters.

Effluent from the gravity filters is held in a reservoir until needed for cooling tower makeup water. When the water is pumped to the cooling towers, the pH is lowered to 7.0 by the use of sulfuric acid to control scaling problems in the cooling towers.

Polymers are added to the solids from the clarifiers before treatment by a classification centrifuge to separate the calcium carbonate. Centrate (water from centrifuge) is recycled back to the trickling filter headworks. The calcium carbonate is passed through a multiple-hearth furnace to recover the lime for reuse. All residues not recycled from the centrifuge and ash from the furnace are disposed of in a landfill.

8.033 Specialty Steel Mill (Figure 8.8)

Wastewater reclamation and reuse at steel mills requires the identification of sources of wastewater and a determination of the best means of collecting, treating, and recycling or disposing of the wastewater. Table 8.7 summarizes sources of wastewater and treatment methods for a specialty steel mill.

TABLE 8.7   STEEL MILL WASTEWATER SOURCES AND TREATMENT

Source	Treatment
1. Cooling tower blowdown	Discharge into municipal collection system.
2. Rolling mill wastewaters	Collect, provide chemical treatment and recycle within mill.
3. Pickling rinse waters	Collect, provide chemical treatment and recycle within mill.
4. Spent pickling and plating baths	Haul to approved disposal site.
5. Sanitary wastewaters	Collect and discharge into municipal collection system.

Process wastewaters that receive chemical treatment are collected from the various sources in the mill and ultimately reach the aeration tank at the treatment plant (Figure 8.8). Wastewater is aerated to convert soluble iron to an insoluble iron precipitate by chemical oxidation and also to cool the water.

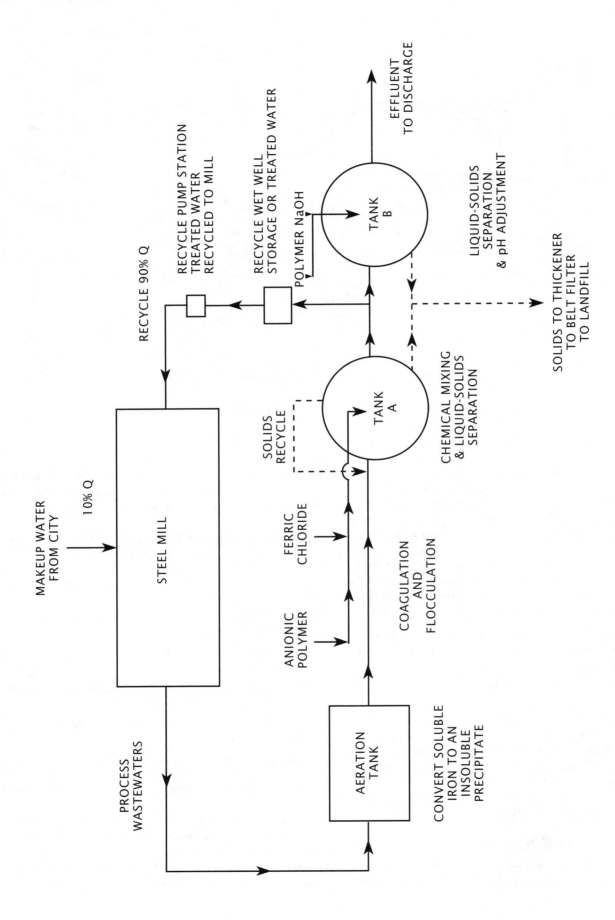

Fig. 8.8   Wastewater reclamation processes for a specialty steel mill

Chemicals are added to the effluent from the aeration tank in Tank A (Figure 8.8). An anionic polymer and ferric chloride provide for chemical coagulation and flocculated solids are recycled within the clarifier (Tank A) in order to provide a prolonged contact between the entering wastewater and previously formed solids. This recycling process increases absorption of the pollutants into the floc particles. Clarified effluent flows over V-notch weirs and is reused in mill operations. The weirs are baffled to hold back floating materials. Water not being recycled flows to Tank B (similar to Figure 8.7) for additional treatment prior to discharge. Tables 8.8 and 8.9 show the removal efficiencies for Tanks (clarifiers) A and B.

TABLE 8.8   REMOVAL EFFICIENCIES FOR TANK A, AVERAGE DAILY VALUES

Water Quality Indicator	Influent, mg/L	Effluent, mg/L	Removal, %
TSS	87.0	21.8	75.9
TOC	17.0	13.3	21.8
Cyanides—Total	0.06	0.049	18.3
Cyanides—Oxid.	.027	0.03	0
Nitrate	21.1	22.5	0
Sulfate	609.0	633.0	0
Chloride	106.0	113.0	0
Cadmium—Total	<0.01	<0.01	ND[a]
Cadmium—Soluble	<0.01	<0.01	ND
Chromium—Hex.	0.15	0.02	86.7
Chromium—Total	3.80	0.75	80
Chromium—Soluble	0.83	0.14	83
Copper—Total	0.48	0.41	14.6
Copper—Soluble	0.10	0.35	0
Iron—Total	27.5	9.0	67.3
Iron—Soluble	4.2	1.95	53.6
Zinc—Total	0.64	0.56	12.5
Zinc—Soluble	0.32	0.51	0
Fluoride	13.9	16.8	0

[a] ND means Not Determined.

8.1   OPERATING PROCEDURES

Operating procedures are generally the same for each type of wastewater reclamation system with variations depending on the processes used. The procedures listed in this section apply mainly to the uses shown in Figures 8.1, 8.2, and 8.3.

8.10   Pre-Start Inspection

1. Examine most recent lab test results.

2. Read preceding day's log for special instructions.

8.11   Start-Up

1. Determine quantity of water required.

2. Read meter totalizer and log.

3. Open valves.

4. Open valves to blending tank from freshwater supply (if blending tank is used).

5. Start the pumps.

8.12   Normal Operation

1. Make one inspection per shift of each pump, check oil levels and packing, and clean area. Listen for any unusual sounds.

2. Lubricate equipment, where necessary.

3. Record flow near the end of the shift.

4. Visually inspect product water every two hours. Look for unusual amounts of solids, floatables, or colors, and try to detect any odors.

5. Collect samples and analyze samples immediately in accordance with Section 8.2, "Monitoring Program."

8.13   Shutdown

1. Turn off pumps.

2. Close all valves.

3. Record meter totalizer reading.

4. Make entry in log.

8.14   Operational Strategy

System flows are controlled by pump run times and valve adjustments. If blending is used, then desired water quality constituent values are reached by adding fresh water. For example, if the plant effluent ammonia concentration is 20 mg/L, the desired concentration is 10 mg/L, and the freshwater ammonia concentration is zero, the delivered water should be slightly less than 50 percent plant effluent and slightly more than 50 percent fresh water.

TABLE 8.9   REMOVAL EFFICIENCIES FOR TANK B, AVERAGE DAILY VALUES

Water Quality Indicator	mg/L Concentration			Kilograms Per 4-Hour Period		
	Influent	Effluent	% Removal	Influent	Effluent	% Removal
TSS	21.8	9.0	58.7	4.45	1.99	55.3
TOC	13.3	9.8	26.3	2.72	1.95	28.3
Cyanides—Total	0.049	.062	0	0.016	0.02	0
Cyanides—Oxid.	0.03	.032	0	0.011	0.0059	46.4
Nitrate	22.5	21.8	3.1	4.54	5.9	0
Sulfate	633	654	0	129.3	127	1.8
Chloride	113	114	0	23	2.22	90.3
Cadmium—Total	ND[a]	ND	0	ND	ND	0
Cadmium—Soluble	ND	ND	0	ND	ND	0
Chromium—Hex.	0.2	0.022	0	0.0041	0.0045	—
Chromium—Total	0.75	0.245	67.3	0.15	0.045	70
Chromium—Soluble	0.14	0.03	78.6	0.03	0.006	80.0
Copper—Total	0.41	0.063	84.6	0.08	0.012	85.0
Copper—Soluble	0.35	0.017	95.1	0.07	0.002	97.1
Iron—Total	9.0	2.30	74.4	3.5	0.45	87.1
Iron—Soluble	1.95	0.028	98.6	0.41	0.008	98.0
Zinc—Total	0.56	0.13	76.8	0.113	0.027	76.1
Zinc—Soluble	0.51	0.06	88.2	0.1	0.01	90
Fluoride	16.8	14.8	11.9	3.58	3.0	16.2

[a] ND means Not Determined.

The process control consists of techniques to blend flows or provide further treatment. Blending procedures are described in the previous paragraph. Further treatment can be accomplished by adding chemicals such as chlorine to kill coliform and pathogenic bacteria or aeration to increase dissolved oxygen. Unacceptable effluent could be returned to the treatment processes for further treatment.

Important sensory observations include detection of odors and colors. Greases and oils also can be seen. The presence of any of these pollutants in the plant effluent means the treatment process is probably upset.

8.15   Emergency Operating Procedures

If one of the treatment processes fails and effluent quality standards cannot be met, immediately stop flow to the water users and send flow to an emergency holding pond or tank, if one is available. Otherwise, reroute the flow in accordance with established procedures to an acceptable means of disposal.

If the power is off, the flow must be contained in the emergency holding area.

8.16   Troubleshooting Guide (Table 8.10)

If your plant is not meeting the water quality requirements of the water users, try to identify the cause of the problem and to select the proper corrective action. Solutions to the problems listed in this section have been covered in more detail in previous chapters. The chapters listed in Table 8.10 refer to *OPERATION OF WASTEWATER TREATMENT PLANTS,* Volumes I and II.

QUESTIONS

Write your answers in a notebook and then compare your answers with those on page 613.

8.1A   How can coliform and pathogenic bacteria be killed in reclaimed wastewater?

8.1B   Why is a "blend" water sometimes mixed with plant effluent?

8.1C   What could be the probable causes of a wastewater reclamation plant being unable to maintain a chlorine residual?

TABLE 8.10 WASTEWATER RECLAMATION SYSTEMS TROUBLESHOOTING GUIDE

Indicator/Observation	Probable Cause	Check or Monitor	Possible Solution
PONDS			Review *OWTP*,[a] Volume I, Chapter 9
1. Floatables in effluent	1a. Outlet baffle not in proper location	1a. Visually inspect outlet baffle	1a. Adjust outlet baffle.
	1b. Excessive floatables and scum on pond surface	1b. Visually inspect pond surface	1b. Remove floatables from pond surface using hand rakes or skimmers. Scum can be broken up using jets of water or a motor boat. Broken scum often sinks.
2. Excessive algae in effluent	2. Temperature or weather conditions may favor a particular species of algae	2. Visually observe effluent or run suspended solids test	2. Operate ponds in series. Draw off effluent from below pond surface by use of a good baffle arrangement.
3. Excessive BOD in effluent	3. Detention time too short, hydraulic or organic overload, poor inlet or outlet arrangements, and possible toxic discharges	3a. Influent flows	3a. Inspect collection systems for infiltration and correct at source.
		3b. Calculate organic loading	3b. Use pumps to recirculate pond contents.
		3c. Observe flow through inlet and outlet structures	3c. Rearrange inlets and outlets or install additional ones.
		3d. Dead algae in effluent	3d. Prevent toxic discharges.
SECONDARY CLARIFIERS FOR TRICKLING FILTERS, ROTATING BIOLOGICAL CONTACTORS, OR ACTIVATED SLUDGE			Review *OWTP*, Volume I, Chapters 6 & 7, or *OWTP*, Volumes I & II, Chapters 8 & 11, and Chapter 2 in this manual.
1. Floatables to effluent	1a. Clarifiers hydraulically overloaded	1a. Visually observe effluent or calculate hydraulic loadings	1a. Install hardware cloth or similar screening device in effluent channels. Review *OWTP*, Volume I, Chapter 5
	1b. Skimmers not operating properly	1b. Observe skimmer movement at beaching plate	1b. Lower skimmer arm or replace neoprene. Review *OWTP*, Volume I, Chapter 5
2. High suspended solids or BOD_5 in effluent	2a. Clarifiers hydraulically overloaded	2a. See 1a above	2a. See 1a above and review operation of biological treatment process.
	2b. Biological treatment process organically overloaded	2b. Calculate BOD or organic loading	2b. Review operation of biological treatment process.
DISINFECTION			Review *OWTP*, Volume I, Chapter 10
1. Unable to maintain chlorine residual	1a. Chlorinator not working properly	1a. Inspect chlorinator	1a. Repair chlorinator.
	1b. Increase in chlorine demand	1b. Run chlorine demand tests	1b. Increase chlorine dose or identify and correct cause of increase in demand.
2. Unable to meet coliform requirements	2a. Chlorine residual too low	2a. See 1 above	2a. See 1 above. Also try increasing chlorine feed rate.
	2b. Chlorine contact time too short	2b. Measure time for dye to pass through contact basin	2b. Improve baffling arrangement.
	2c. Solids in effluent	2c. Observe solids or run suspended solids test	2c. Install hardware cloth or similar screening device in effluent channels. Review operation of biological treatment process.
	2d. Sludge in contact basin	2d. Look for sludge deposits in contact basin	2d. Drain and clean contact basin. Review operation of biological treatment process.
	2e. Diffuser not properly discharging chlorine	2e. Lower tank water level and inspect	2e. Clean diffuser.
	2f. Mixing inadequate	2f. Add dye at diffuser	2f. Add mechanical mixer or move diffuser.
	2g. Incomplete nitrification	2g. Nitrite concentration	2g. Adjust nitrification process.

[a] *OWTP* means *OPERATION OF WASTEWATER TREATMENT PLANTS.*

8.2   MONITORING PROGRAM

8.20   Monitoring Schedule

The monitoring system may vary depending on the type of wastewater reuse and local conditions. The requirements for individual reclamation systems must be provided by the user. Table 8.11 shows a typical monitoring schedule that could be used for any of the three reclamation systems shown in Figures 8.1, 8.2, and 8.3. Samples for these tests should be taken from valves in the effluent pipeline before the effluent leaves the plant. Proper sample containers must be used (for example, sterilized bottles for pathogens) as provided by the laboratory. Temperature, dissolved oxygen, pH, conductivity, and turbidity all may be monitored continuously with the results plotted by recorders.

TABLE 8.11   WATER QUALITY MONITORING SCHEDULE

Grab Sample, Daily	24-Hour Composite Sample, Weekly	Grab Sample, Weekly	24-Hour Composite Sample, Monthly
Alkalinity	Ammonia Nitrogen	Coliforms	Arsenic
Bicarbonate	Bicarbonate	Color	Barium
COD	Boron	Odor	Copper
Calcium	COD		Lead
Chloride	Cadmium		Mercury
Dissolved Oxygen	Calcium		Silver
Electrical Conductivity	Chloride		
Magnesium	Chromium		
	Cyanide		
	Fluoride		
	Iron		
	MBAS		
	Magnesium		
	Manganese		
	Nitrate Nitrogen		
	Organic Nitrogen		
	Phenol		
	Phosphorus		
	Selenium		
	Sodium		
	Sulfate		

In addition to monitoring effluent water quality, flow rates also must be recorded. Hydraulic loading rates on wells in terms of gallons per day per well are very important. Any loss of loading capacity by any well must be investigated immediately. If a recharge well is losing its recharge ability, try to identify the cause of the problem and select appropriate corrective action.

Clogging of the well may be caused by slimes and can be corrected by applying chlorine (10 to 15 mg/L in the well) to kill the slimes or by allowing the well to rest, which can dry out or starve some slimes. If activated carbon is used in a treatment process, clogging may be caused by carbon fines (very tiny pieces of carbon). These fines may be removed by passing the water through a sand/anthracite filter prior to transmission to the well field.

8.21   Interpretation of Test Results and Follow-Up Actions

If pH, dissolved oxygen, chemical oxygen demand, nitrogen compounds, phosphorus, coliforms, or odor standards are not met, then adjustments are needed in the treatment processes. If the concentrations of other water quality indicators exceed standards and cannot be removed by treatment, industrial discharge sites should be tested to see if someone is discharging excessive quantities into the wastewater collection system. Whenever one of the standards is not met, notify the user immediately. If a blending tank is available, add more fresh water to dilute the effluent.

8.3   SAFETY

Always work with another operator when collecting samples or working around storage reservoirs or blending tanks so help will be available if you fall into the water. Take necessary precautions to avoid slipping or falling into the water. Wear approved flotation devices when working around bodies of water without guardrails. Pump station safety is discussed in Chapter 14, "Plant Safety" Volume II, *OPERATION OF WASTEWATER TREATMENT PLANTS*.

A major safety consideration is the health of persons coming in contact with reclaimed water. Be sure the effluent from your wastewater reclamation facility is adequately disinfected at all times. If the effluent ever presents a potential threat to the public's health, *notify the users immediately.*

8.4 MAINTENANCE

There are very few maintenance considerations except for pumps, which are covered in Chapter 15, "Maintenance," Volume II, *OPERATION OF WASTEWATER TREATMENT PLANTS.* Metering system maintenance consists mainly of cleaning and visual inspections. Otherwise, maintenance means good housekeeping.

8.5 REVIEW OF PLANS AND SPECIFICATIONS

Plans and specifications should be reviewed to ensure a piping system with alternate flow paths. If problems develop, there must be a way to pipe the effluent to temporary storage or back to treatment. In other words, there must be an alternate route for the flow.

QUESTIONS

Write your answers in a notebook and then compare your answers with those on page 613.

8.2A List possible causes of clogging in a recharge well and possible cures for each cause.

8.2B What would you do if reclaimed effluent was being used by an industry and one of the water quality standards was not being met?

8.3A Why should you always work with another operator when working around storage reservoirs or blending tanks?

END OF LESSON 1 OF 2 LESSONS

on

WASTEWATER RECLAMATION AND REUSE

Please answer the discussion and review questions next.

DISCUSSION AND REVIEW QUESTIONS

Chapter 8. WASTEWATER RECLAMATION AND REUSE

(Lesson 1 of 2 Lessons)

At the end of each lesson in this chapter you will find some discussion and review questions. The purpose of these questions is to indicate to you how well you understand the material in the lesson. Write the answers to these questions in your notebook.

1. What are some of the limitations of or precautions for direct use of reclaimed wastewater?

2. Why should you be concerned about protecting the public health when operating a wastewater reclamation facility?

3. What would you do if one of your treatment processes failed and you were unable to meet effluent quality standards?

4. What is the purpose of a monitoring program?

5. Why should you look for alternate flow paths when reviewing the plans and specifications for a wastewater reclamation facility?

CHAPTER 8.   WASTEWATER RECLAMATION AND REUSE

(Lesson 2 of 2 Lessons)

EFFLUENT DISPOSAL ON LAND

8.6   LAND TREATMENT SYSTEMS

8.60   Description of Treatment Systems

When a high-quality effluent is required or no discharge is permitted, land treatment offers a means of wastewater reclamation or ultimate disposal (Figure 8.9). Land treatment systems use soil, plants, and bacteria to treat and reclaim wastewaters. They can be designed and operated for the sole purpose of wastewater disposal, for crop production, or for both purposes. In land treatment, effluent is pretreated and applied to land by conventional irrigation methods. When systems are designed for crop production, the wastewater and its nutrients (nitrogen and phosphorus) are used as a resource. This system is then comparable to the reuse systems described in Lesson 1. With either approach, treatment is provided by natural processes (physical, chemical, and biological) as the effluent flows through the soil and plants. Part of the wastewater is lost by *EVAPOTRANSPIRATION*[2] and the rest goes back to the *HYDROLOGIC CYCLE*[3] through surface runoff or percolation to groundwater. Land disposal of wastewater may be done by one of the following methods shown in Figure 8.10.

1. Irrigation

2. Infiltration–percolation

3. Overland flow

The method of irrigation depends on the type of crop being grown. Irrigation methods include traveling sprinklers, fixed sprinklers, furrow, and flooding. Infiltration–percolation systems are not suited for crop growth. Overland flow systems are similar to other treatment processes and have runoff that must either be discharged or recycled in the system. The other systems usually do not have a significant surface runoff. Typical loading rates for these systems are shown in Table 8.12.

Land application systems include the following parts:

1. Treatment before application

2. Transmission to the land treatment site

3. Storage

4. Distribution over site

5. Runoff recovery system (if needed)

6. Crop systems

8.61   Equipment Requirements

Irrigation systems apply water by sprinkling or by surface spreading (Figure 8.11). Sprinkler systems may be fixed or movable. Fixed systems are permanently installed on the ground or buried with sprinklers set on risers that are spaced along pipelines. These systems have been used in all types of terrain. Movable systems include hand-move, center pivot, side wheel roll, and traveling gun sprinklers.

Surface irrigation systems include flooding, border-check, and ridge and furrow systems. Flooding systems are very similar to overland flow systems except the slopes are nearly level. Border-check systems are simply a controlled flooding system. Ridge and furrow systems are used for row crops such as corn where the water flows through furrows between the rows and seeps into the root zone of the crop.

An overland flow system consists of effluent being sprayed or diverted over sloping terraces where it flows down the hill and through the vegetation. The vegetation provides a filtering action, thus removing suspended solids and insoluble BOD. This

---

[2] *Evapotranspiration* (ee-VAP-o-TRANS-purr-A-shun).   (1) The process by which water vapor is released to the atmosphere from living plants. Also called TRANSPIRATION. (2) The total water removed from an area by transpiration (plants) and by evaporation from soil, snow, and water surfaces.

[3] *Hydrologic* (HI-dro-LOJ-ick) *Cycle.*   The process of evaporation of water into the air and its return to earth by precipitation (rain or snow). This process also includes transpiration from plants, groundwater movement, and runoff into rivers, streams, and the ocean. Also called the WATER CYCLE.

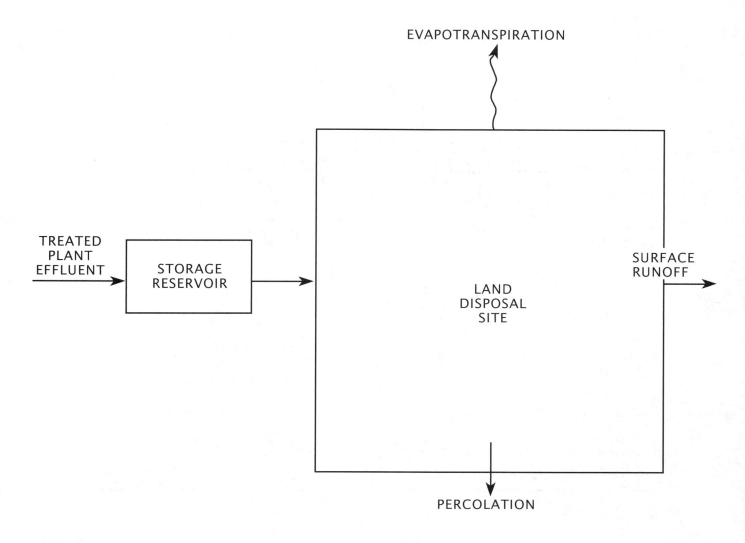

Fig. 8.9   Land disposal system schematic

system can be operated as a treatment system or, with a recycle system included, it can be operated as a disposal system. When operated as a disposal system, the overland flow process is nearly the same as flood irrigation so it will be included with the flood systems in the following paragraphs.

8.62   Sidestreams[4] and Their Treatment

There are two possible sidestreams with land disposal systems. Unlined storage reservoirs will result in wastewater percolating down to groundwater. If the water stored in reservoirs is the final effluent from a treatment plant, percolation down to the groundwater probably is acceptable. However, if the water is untreated or partially treated (primary effluent), the reservoir should be lined or an underground collection system should be installed to collect any percolation or seepage from the reservoir. In some areas, percolation may cause a rise in the area's

groundwater level. To prevent groundwater problems, a seepage ditch may be built around the outside of the reservoir. This ditch should be located somewhat lower than the bottom of the reservoir. Wastewater that percolates through the reservoir bottom is collected in the ditches. This water can be pumped back into the reservoir. The groundwater table also could be lowered by a series of shallow wells with water being pumped out, as necessary. Lowering of the groundwater could result in increased percolation rates.

The other major sidestream is surface runoff. Runoff quantities vary depending on the type of irrigation system used. In all systems, provisions should be made for collecting runoff and returning this flow to be reapplied. In some locations, runoff water can be discharged to surface water. Discharge is the preferred approach due to cost savings and minimizing operational problems.

---

[4] *Sidestream.*   Wastewater flows that develop from other storage or treatment facilities. This wastewater may or may not need additional treatment.

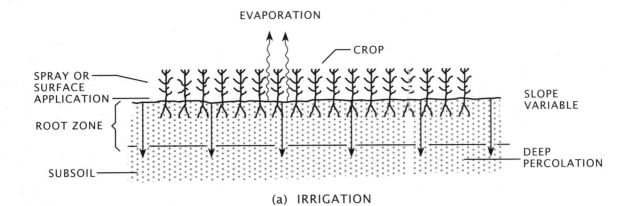

(a) IRRIGATION

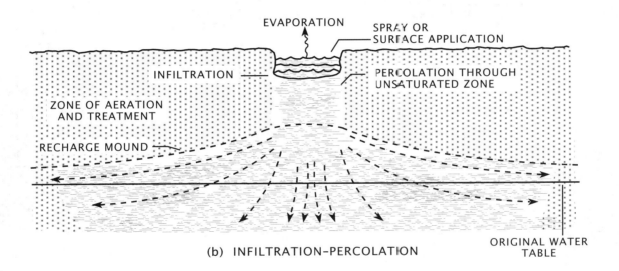

(b) INFILTRATION-PERCOLATION

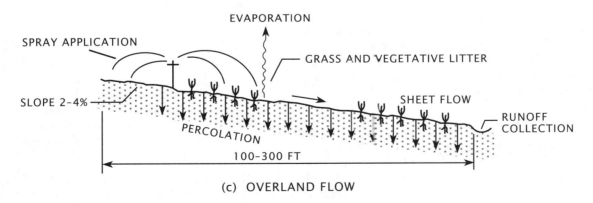

(c) OVERLAND FLOW

Fig. 8.10   Methods of land application

(From C. E. Pound, et al., *COSTS OF WASTEWATER TREATMENT BY LAND APPLICATION,*
US Environmental Protection Agency, Washington, DC 20460, EPA No. 430-9-75-003, June 1975)

TABLE 8.12   TYPICAL LOADINGS FOR IRRIGATION, INFILTRATION–PERCOLATION, AND OVERLAND FLOW SYSTEMS[a]

Factor	Irrigation Low-Rate	High-Rate	Infiltration–Percolation	Overland Flow
Liquid loading rate, in/wk[b]	0.5 to 1.5	1.5 to 4.0	4 to 120	2 to 9
Annual application, ft/yr[c]	2 to 4	4 to 18	18 to 500	8 to 40
Land required for 1 MGD flow rate, acres [d, e]	280 to 560	62 to 280	6 to 62	28 to 140
Application techniques	Spray or surface		Usually surface	Usually surface
Vegetation required	Yes	Yes	No	Yes
Crop production	Excellent	Good/fair	Poor/none	Fair/poor
Soils	Moderately permeable soils with good productivity when irrigated		Rapidly permeable soils such as sands, loamy sands, and sandy loams	Slowly permeable soils such as clay loams and clays
Climatic constraints	Storage often needed		Reduce loadings in freezing weather	Storage often needed
Wastewater lost to:	Evaporation and percolation		Percolation	Surface runoff and evaporation with some percolation
Expected treatment performance				
BOD and SS removal	98+%		85 to 99%	92+%
Nitrogen removal	85+%[d]		0 to 50%	70 to 90%
Phosphorus removal	80 to 99%		60 to 95%	40 to 80%

[a] C. E. Pound, et al., *COSTS OF WASTEWATER TREATMENT BY LAND APPLICATION,* US Environmental Protection Agency, Washington, DC 20460, EPA No. 430-9-75-003, June 1975.
[b] in/wk × 2.54 = cm/wk.
[c] ft/yr × 0.3 = m/yr.
[d] Dependent on crop uptake.
[e] acres × 0.00107 = hectares for 1 m^3/day
   or
   acres × 9.24 = hectares for 1 m^3/sec.

8.63   Limitations of Land Treatment

Problems encountered using land treatment systems usually involve soil problems and weather conditions. If proper care is not taken, the soils can lose their ability to percolate applied water. During the cold winter season, plants and crops will not grow. Under these conditions, no water will be treated by transpiration processes. Also, precipitation can soak the soil so no wastewater can be treated. Provisions must be made to store wastewater during cold and wet weather.

One of the most common land treatment problems is the sealing (water will not percolate) of the soil by suspended solids in the final effluent. These solids are deposited on the surface of the soil and form a mat, which prevents the percolation of water down through the soil. There are three possible solutions to this problem:

1. Remove the suspended solids from the effluent.

2. Apply water intermittently and allow a long enough drying period for the solids mat to dry and crack.

3. Disk or plow the field to break the mat of solids.

Another serious problem is the buildup of salts in the soil. If the effluent has a high chloride content, there can be enough salts in the soil within one year to create a toxic condition to most grasses and plants. To overcome salinity problems:

1. Leach out the salts by applying fresh water (not effluent).

2. Rip the field to a depth of 4 to 5 feet (1.2 to 1.5 meters) to encourage more percolation.

The severity of both soil sealing due to suspended solids and salinity problems due to dissolved solids depends on the type of soil in the disposal area. These problems are more common and more difficult to correct in clay soils than in sandy soils.

Excessive nitrate ions can reach groundwater if irrigation or rapid infiltration systems are overloaded or not properly operated. Crop irrigation systems should be designed so that crop nitrogen uptake is not exceeded. Nitrogen in wastewater will be in the organic, ammonium, or nitrate form. The organic nitrogen is mineralized to form ammonia or ammonium in water. With adequate oxygen and alkalinity, the ammonia will nitrify to form nitrite, which is usually quickly converted to nitrate.

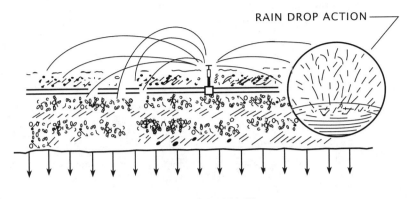

(a) SPRINKLER

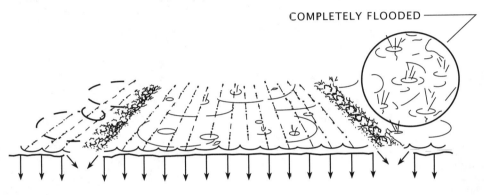

(b) FLOODING

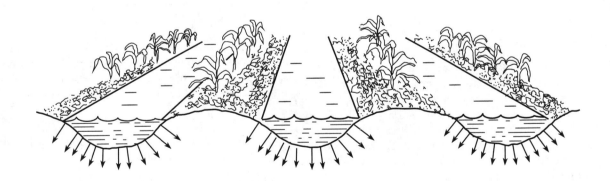

(c) RIDGE AND FURROW

Fig. 8.11   Irrigation techniques

(From C. E. Pound, et al., *EVALUATION OF LAND APPLICATION SYSTEMS,* US Environmental
Protection Agency, Washington, DC 20460, EPA No. 430-9-74-015, September 1974)

Crops use nitrogen in the ammonia or nitrate forms only. If excessive nitrate is present, nitrate will percolate into the groundwater. Contact your local county farm advisor for nitrogen uptake rates for the crops grown.

Rapid infiltration systems have no crops to remove nitrogen so they must be operated in wet/dry cycles in order to first nitrify the ammonium (dry) and then denitrify the nitrate (wet) to form nitrogen gas (N_2), which is released to the atmosphere.

QUESTIONS

Write your answers in a notebook and then compare your answers with those on pages 613 and 614.

8.6A Define the terms evapotranspiration and hydrologic cycle.

8.6B List the three methods by which land disposal of wastewater is accomplished.

8.6C What are the major parts of a land application system?

8.6D A plot of land 2,000 feet long by 1,000 feet wide is used for a land disposal system. If 1 million gallons of water is applied to the land during a 24-hour period, calculate the hydraulic loading in:

1. MGD per acre

2. Inches per day

8.7 OPERATING PROCEDURES

The operating procedures described below apply to a spray irrigation system. This system is for an area where crops are to be grown. The operating procedures for this type of system are more complex than the other systems. The procedures are explained first and then an example is presented to show how to use the procedures.

8.70 Pre-Start Checklist

Table 8.13 consists of a list of items that should be checked before starting a land disposal system.

8.71 Start-Up

1. Determine need to irrigate. The amount of water that can be applied depends on the type of crop. Some crops require a lot of water while other crops will not tolerate any excess water. The procedures in this section are prepared to help you determine when you should irrigate and how much water should be applied. These procedures are based on soil conditions and may require adjustment based on the type of crop being irrigated. The chart shown on Table 8.14 will aid in determining if you need to irrigate based on soil conditions. To use this table, you must first determine the type of soil you are irrigating. Walk around the field and try to identify the different types, if more than one type of soil is present. Pick up a handful of soil and examine the grains or particles. Small, hard, tiny particles indicate a sandy soil. Very fine, soft, smooth particles signify a clay soil. Once you have identified the type of soil, the moisture content can be estimated by squeezing the soil together in your hand. The wetter the soil, the more likely the soil will stick or cling together. By the way the soil sticks together and the type of soil, you can estimate when the available moisture content drops to 50 percent or less. Irrigate when the moisture content is 50 percent or less.

Soils vary throughout an area of land and some areas need irrigating before others. Many farmers have specific areas in their fields that they call "hot spots." Due to soil characteristics or other reasons, these areas dry out faster than the rest of the field. Whenever the soil will not form a ball when squeezed together or crops start to show signs of stress due to lack of water, this is the time to irrigate.

2. Determine amount of moisture to apply using Table 8.15. To use this table, determine the soil type by examining a handful of soil as described in Step 1. The root zone depth is determined by the type of crop. If necessary, dig down to determine how far down the roots are growing. By the use of Table 8.14, you can estimate the percent of available moisture remaining in the soil before irrigation. Knowing the percent available moisture in the soil before irrigation, you can determine the net inches of water to apply from Table 8.15.

3. Determine the time required to apply one inch (2.5 cm) of water from Figures 8.12 or 8.13. Figure 8.12 is used for small areas (up to 60 acres) and Figure 8.13 is used for larger areas (over 60 acres). To use the tables, determine: (1) the area of land you wish to irrigate, and (2) the capacity of your irrigation system in gallons per minute. By starting at the bottom of the figure with the known area, draw a line vertically upward. Next, draw a line from the system capacity on the left, horizontally to the right. Where these two lines intersect is the time required to apply one inch of water to the area being irrigated.

4. Multiply time required to apply one inch (2.5 cm) of water by the inches of moisture to apply to determine the total run time for the irrigation system. After you have determined the net inches of water you wish to apply from Table 8.15 (Step 2), multiply the net inches times the time required (Figures 8.12 and 8.13) to determine the total irrigation time.

If you need help determining soil types, irrigation requirements, soil moisture content, types of crops to plant, salt tolerance of plants, depth of root zone, and any fertilizer needs, contact your local farm adviser. In many areas, an expert adviser is available free of charge through some agency of the federal, state, or local government.

5. Check pump discharge check valve and suction screen for foreign matter.

6. Check pipeline to see that all couplings are still fastened, blocks or pipe supports have not fallen over, and gates and valves are open.

7. Start pump.

8. Inspect the irrigation system to be sure everything is working properly.

TABLE 8.13   SPRAY IRRIGATION PRE-START CHECKLIST [a]

Check Every Time Pump is Started	Check at the Beginning of the Season	
		Electric Motors
	_____	Replace winter lubricant.
	_____	Oil bath bearings. Drain oil and replace with proper weight of clean oil.
	_____	Grease lube bearings. If grease gun is used, be sure old grease is purged through outlet hole.
_____		Before turning on switch, have power company check voltage.
	_____	Check for proper rotation of motor and pump.
_____	_____	Check fuses to make sure they are still good.
_____	_____	Check electrical contact points for excessive corrosion.
	_____	Physically inspect for rodent and insect invasion.
		Pumps
	_____	Replace oil or grease with proper weight bearing lubricant.
	_____	Tighten packing gland to proper setting.
_____	_____	Check discharge head, discharge check valve, and suction screen thoroughly for foreign matter.
_____	_____	Pump shaft should turn freely without noticeable dragging.
		Aluminum Pipe
	_____	If you did not properly handle and store aluminum pipe or tubing last fall, make a mental note to do that at the end of this growing season.
		Always carefully drain aluminum tubing or pipe when you are finished using it. Aluminum pipe bends very easily. If you pick up a length of pipe that is half-full of water and has one end plugged, you will bend the pipe. Flush out the pipe to clean it, drain the pipe, place the pipe on a long-bed trailer for transport, and then store the pipe on racks until you are ready to use the pipe next season.
	_____	Inspect pipe ends to make certain that no damage has occurred. Ends should be round for best operation. A slightly tapered wooden plug of the proper diameter can be used to round out the ends. The diameter of aluminum pipe varies from 2 to 12 inches (50 to 300 mm).

Check Every Time Pump is Started	Check at the Beginning of the Season	
		Aluminum Pipe (continued)
	_____	Check pipe for pit corrosion. If spots are in evidence, contact the aluminum pipe supplier for advice.
	_____	Inspect pipe gaskets, couplers (irrigation pipe couplings), and gates to find those in need of replacement.
_____	_____	Check pipeline to see that all couplers are still fastened, pipe supports have not fallen over, and gates and valves are still open.
_____	_____	Pipe makes an excellent nesting area for small animals. Flush out the pipeline before installing the end plug. Make sure you are away from power lines when you raise the pipe to drain water from the pipe.
		Sprinkler Systems
	_____	Sprinkler bearing washers should be replaced if there is indication of wear.
_____	_____	Visually check all moving parts, seals, bearings, and flexible hose for replacement or repair.
_____	_____	Check to see that hose is laid out straight or on a long radius for turns. Be sure there are no kinks in the hose. There should be sufficient hose at the end of the sprinkler to act as a brake or to hold back the sprinkler system initially as it drags the hose through the field. Also check earth anchors.
_____	_____	If possible, operate the system to check speed adjustment, alignment, and safety switch mechanisms.
_____	_____	Check sprinkler oscillating arm for proper adjustment. If damage has occurred to the sprinkler oscillating arm, the arm should be replaced or bent back to the correct angle. Your dealer can help in correcting a damaged arm. The angle of water-contact surface, if not correct, will change the turning characteristics of the sprinkler. Excessive wear of sprinkler nozzles can be checked with proper size drill bit.

REMEMBER, inspection and corrective maintenance now may save considerable time and money later.

[a] From *WASTEWATER RESOURCES MANUAL,* Edward Norum, ed., by permission of The Irrigation Association.

TABLE 8.14   FEEL AND APPEARANCE GUIDE FOR DETERMINING SOIL MOISTURE[a]

The chart below is very useful in estimating how much available moisture is in your soil. Although the plant's daily moisture use may range from 0.1 inch to 0.4 inch per day (0.25 to 1.0 cm), it will average about 0.20 inch (0.5 cm) per day, and 0.25 inch (0.6 cm) per day during hot days.

Moisture Condition	Percent of Available Moisture Remaining in Soil, %	Soil Texture		
		Sands–Sandy Loams	Loams–Silt Loams	Clay Loams–Clay
Dry	Wilting point	Dry, loose, flows through fingers.	Powdery, sometimes slightly crusted but easily broken down into powdery condition.	Hard, baked, cracked; difficult to break down into powdery condition.
Low	50% or less		Will form a weak ball when squeezed but will not stick to tools.	Pliable, but not slick; will ball under pressure—sticks to tools.
Time to Irrigate When Available Moisture is 50 Percent or Less				
Fair	50 to 75%	Tends to ball under pressure but seldom will hold together when bounced in the hand.	Forms a ball somewhat plastic, will stick slightly with pressure. Does not stick to tools.	Forms a ball, will ribbon out between thumb and forefinger, has a slick feeling.
Good	75 to 100%	Forms a weak ball, breaks easily when bounced in the hand, can feel moistness in soil.	Forms a ball, very pliable, sticks readily, clings slightly to tools.	Easily ribbons out between thumb and forefinger, has a slick feeling, very sticky.
Ideal	Field capacity 100%	Soil mass will cling together. Upon squeezing, outline of ball is left on hand.	Wet outline of ball is left on hand when soil is squeezed. Sticks to tools.	Wet outline of ball is left on hand when soil is squeezed. Sticky enough to cling to fingers.

[a] From *WASTEWATER RESOURCES MANUAL,* Edward Norum, ed., by permission of The Irrigation Association.

TABLE 8.15   AMOUNT OF MOISTURE TO APPLY TO VARIOUS SOILS UNDER DIFFERENT MOISTURE RETENTION CONDITIONS[a]

Soil Type	Root Zone Depth	Available Moisture Plants Will Use	Net Inches to Apply Per Irrigation with Various Percents Available Moisture Retained in the Soil Before Irrigation		
			Percent Available Moisture Before Irrigation		
	Feet	Inches	67%	50%	33%
Light Sandy	1	1.00	0.33	0.50	0.67
	1½	1.50	0.50	0.75	1.00
	2	2.00	0.56	1.00	1.33
	2½	2.50	0.83	1.25	1.67
	3	3.00	0.99	1.50	2.00
Medium	1	1.69	0.57	0.85	1.13
	1½	2.53	0.84	1.26	1.70
	2	3.38	1.11	1.69	2.26
	2½	4.21	1.39	2.11	2.82
	3	5.06	1.67	2.53	3.38
Heavy	1	2.39	0.79	1.20	1.59
	1½	3.58	1.18	1.79	2.38
	2	4.78	1.58	2.39	3.25
	2½	5.97	1.97	2.98	3.97
	3	7.17	2.36	3.58	4.77

[a] From *WASTEWATER RESOURCES MANUAL,* Edward Norum, ed., by permission of The Irrigation Association.

EXAMPLE

Using the procedures outlined in this section, this example shows how to determine the total time to irrigate.

1. Determine need to irrigate. A loamy soil formed a weak ball when squeezed. Table 8.14 indicates the moisture condition is low (50% or less) and that it is time to irrigate.

2. The loamy (light sandy) soil has a crop with a root zone depth of 3 feet (0.9 m) and 50 percent of the available moisture is retained in the soil at irrigation. Table 8.15 indicates that 1.5 inches (3.8 cm) of water should be applied to the soil.

3. The land to be irrigated is a 40-acre plot and the pump has a capacity of 1,000 GPM. From Figure 8.12, find 40 acres across the bottom and draw a line vertically upward to the top. Find the system capacity of 1,000 GPM along the left side and draw a line horizontally to the right. These lines intersect between the diagonal lines labeled 15 hours and 20 hours at approximately 18 hours. Therefore, we should irrigate for 18 hours to apply one inch of water.

4. Determine the total time to irrigate.

Time, hours = Time, (hr) to irrigate one inch × Amount to Apply, in

= 18 hours/inch × 1.5 inches

= 27 hours

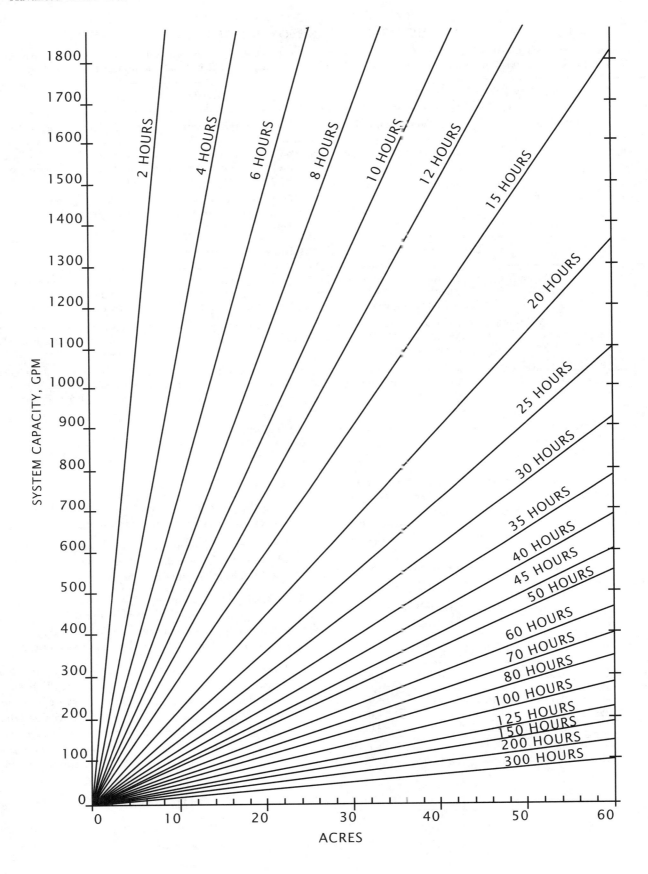

Fig. 8.12   *Time required to apply one inch (2.5 cm) of water on small acreages*

(From *WASTEWATER RESOURCES MANUAL*, Edward Norum, ed.,
by permission of The Irrigation Association)

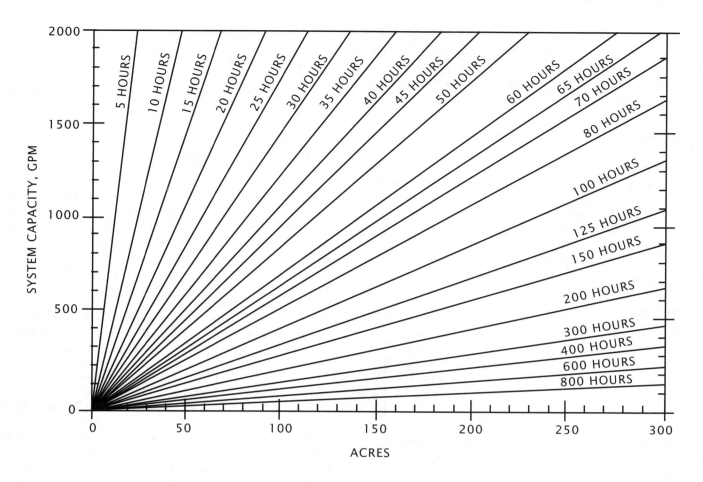

Fig. 8.13   Time required to apply one inch (2.5 cm) of water on large acreages
(From *WASTEWATER RESOURCES MANUAL,* Edward Norum, ed.,
by permission of The Irrigation Association)

8.72   Normal Operation

1. Run the pump or pumps the time determined by Step 4 of the start-up procedure and as shown in the example.

2. Turn the pump off earlier if water begins to pond on the fields.

8.73   Shutdown

This shutdown procedure applies to the end of season shutdown.

1. Drain all lines.

2. Plug open ends of pipelines.

3. Lubricate motors and pumps for winter.

4. Store small movable materials and equipment.

5. Store aluminum tubing or piping.

QUESTIONS

Write your answers in a notebook and then compare your answers with those on page 614.

8.7A   List the major items of equipment that should be inspected before starting a spray irrigation system.

8.7B   Determine the time required to irrigate 30 acres of a medium-type soil where the root zone depth is 2 feet and 50 percent of the available moisture is retained in the soil at irrigation. Pump capacity is 1,200 GPM. Use the figures and tables in this lesson to answer this question.

8.74   Operational Strategy

Physical Control.   The main objectives of a land disposal system are to dispose of effluent without harming surface waters and without creating nuisance conditions. An irrigation system can be designed and operated to produce a crop. The sale of this crop then helps reduce treatment costs. Physical control is then used to dispose of effluent at the highest rate possible without damaging the crop. For all types of land disposal systems, physical controls consist of valves or gates, which are used to direct treated effluent to different disposal areas.

Process Control.   There are three areas of process control: storage reservoirs, runoff and seepage water recycle systems, and systems where crops are grown. The first is the storage reservoir. The reservoir usually will have been equipped with aeration or mixing devices, which are used to maintain aerobic conditions to prevent odor problems. These devices may be operated full-time or for a limited time each day by using timers. The correct time to operate is determined by measuring the dissolved oxygen (DO) content at several points in the reservoir. A minimum of four points should be sampled. An example showing six sampling locations is illustrated in Figure 8.14.

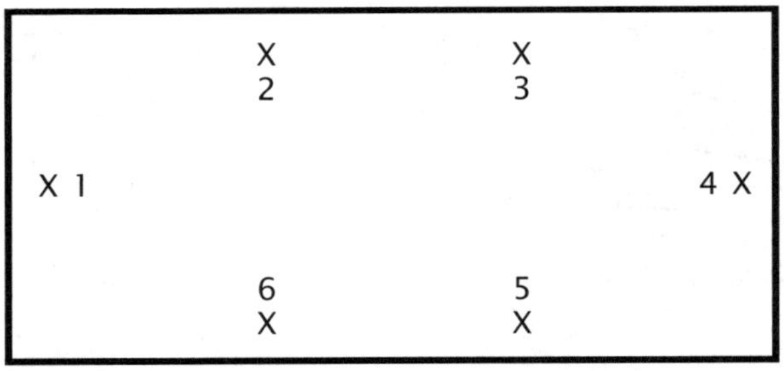

Fig. 8.14   Possible reservoir sampling locations

Each sample should have at least 0.4 mg/L of DO and the average of all samples should be at least 0.8 mg/L. Two sample sets of test results are shown below:

Set 1

Sample No.	mg/L DO
1	1.2
2	1.8
3	2.0
4	1.6
5	0.2
6	0.4

$$\text{Average} = \frac{7.2}{6} = 1.2 \text{ mg/}L$$

Set 2

Sample No.	mg/L DO
1	1.0
2	0.4
3	0.8
4	1.4
5	0.6
6	1.2

$$\text{Average} = \frac{5.4}{6} = 0.90 \text{ mg/}L$$

Set 1 does not meet the requirements. The average DO is 1.2 mg/L, which is good, but one sample (#5) is 0.2 mg/L, which does not meet the minimum requirement. This result indicates that there is adequate aeration but either a portion of the system is not operating properly or the reservoir is not being adequately mixed.

Even though Set 2 has a lower average DO than Set 1, the average is greater than 0.8 and all samples are 0.4 mg/L or greater. Therefore, Set 2 is acceptable.

The second area of process control is the runoff and seepage water recycle systems. These recycle systems may not be necessary, depending on the particular application. For example, the storage reservoir may be lined so there would be no seepage water to recycle. Sprinkler systems that are carefully controlled will have no significant runoff. Your goal is to dispose of as much water as possible without causing runoff and seepage from the disposal area. This is done to reduce recycle pumping costs. The reduction in seepage and runoff is accomplished by taking more care in applying effluent and turning off sprinklers or closing gates when water begins to stand in the field.

The third area of process control applies to those systems where crops are grown. In dry climates such as those found in the southwestern states and western mountain states, farmers who irrigate are concerned with saline and alkaline soils. Some minerals found in the effluent may cause a decrease in crop production in soils of this type. Analyses and irrigation practices for these areas are described in detail in the US Department of Agriculture Handbook No. 60, *SALINE AND ALKALINE SOILS*.[5] A diagram for the classification of irrigation waters (taken from Handbook No. 60) is shown on Figure 8.15.

A simplified version of this diagram, with other critical constituents (substances), is shown in Table 8.16.

Your local US Department of Agriculture, Soil Conservation Service office can provide a list of crops that can be irrigated with Class I and Class III water.

There is nothing that can be done that is economically feasible to control the concentration of these constituents. If one of the constituents exceeds the Class I limit, the crop should be changed to one that is more tolerant. Special agricultural practices can be used to minimize the effects of these constituents. These practices vary from one local area to another. The local soil conservation service office and farm adviser can assist by providing information appropriate to the area.

Important observations and interpretations were discussed earlier for determining when to irrigate. The most important observations in a system where crops are grown and surface runoff is not allowed are observing ponding or runoff. When these occur, either the application amount was excessive or the rate of application was greater than the soil infiltration rate.

---

[5] *DIAGNOSIS AND IMPROVEMENT OF SALINE AND ALKALINE SOILS*, L. A. Richards, ed., Agricultural Handbook No. 60, US Department of Agriculture, Washington, DC.

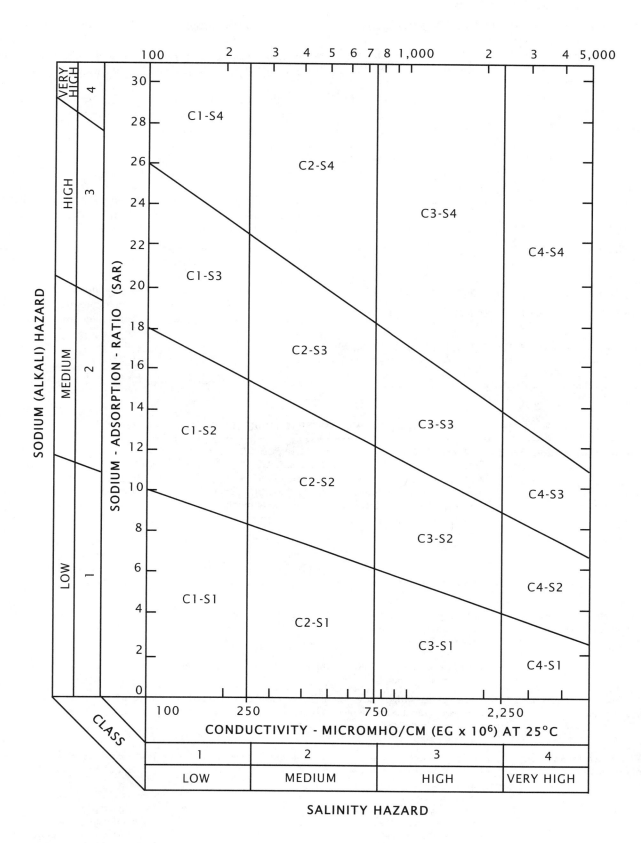

Fig. 8.15   Diagram for the classification of irrigation waters
(From *DIAGNOSIS AND IMPROVEMENT OF SALINE AND ALKALINE SOILS,* L. A. Richards, ed.,
Agricultural Handbook No. 60, US Department of Agriculture, Washington, DC)

TABLE 8.16   CLASSIFICATION OF IRRIGATION WATERS[a]

Chemical Properties	Class I Excellent to Good "Suitable under most conditions"	Class II Good to Injurious "Suitability dependent on soil crop, climate, and other factors"	Class III Injurious to Unsatisfactory "Unsuitable under most conditions"
Total dissolved solids (mg/L)	Less than 700	700–2,000	More than 2,000
Chloride (mg/L)	Less than 175	175–350	More than 350
Sodium (percent of base constituents)	Less than 60	60–75	More than 75
Boron (mg/L)	Less than 0.5	0.5–2.0	More than 2.0

[a] From *DIAGNOSIS AND IMPROVEMENT OF SALINE AND ALKALINE SOILS*, L. A. Richards, ed., Agricultural Handbook No. 60, US Department of Agriculture, Washington, DC.

Visual appearance of the crop being grown is extremely important. Discoloration in plant leaves can indicate excess water (poor drainage) or a nutrient or mineral deficiency. Your local farm adviser can assist in diagnosing the problems.

Observations are critical in storage reservoirs. Odors can result if effluent treatment was inadequate or if insufficient aeration is provided to the reservoir.

8.75   Emergency Operating Procedures

Loss of power will disrupt the sprinkler systems the most since pumping is required (assuming electric motors for pumps). If power is lost, the effluent is retained in the storage reservoir. Gravity-flow flood irrigation systems will not be affected during power outages.

Loss of other treatment units is generally not a problem for a few days. Longer downtimes may result in an overloaded and an odorous storage reservoir and possible odors at the disposal area.

8.76   Troubleshooting Guide (Table 8.17)

TABLE 8.17   SPRAY IRRIGATION SYSTEM TROUBLESHOOTING GUIDE

Indicator/Observation	Probable Cause	Check or Monitor	Solution
1. Water ponding in irrigated area where ponding normally has not been observed.	1a. Application rate is excessive. 1b. If application rate is normal, drainage may be inadequate.	1a. Application rate 1b. (1) Seasonal variation in groundwater level. (2) Operability of any drainage wells. (3) Condition of drain tiles.	1a. Reduce rate to normal value. 1b. (1) Irrigate portions of the site where groundwater is not a problem or store wastewater until level has dropped. (2) Repair drainage wells or increase pumping rate. (3) Repair drain tiles.
	1c. Broken pipe in distribution system.	1c. Leaks in system.	1c. Repair pipe.
2. Lateral aluminum distribution piping deteriorating.	2a. Effluent permitted to remain in aluminum pipe too long causing electrochemical corrosion.	2a. Operating techniques.	2a. Drain aluminum lateral lines except when in use.
	2b. Dissimilar metals (steel valves and aluminum pipe).	2b. Pipe and valve specifications.	2b. Coat steel valves or install cathodic or anodic protection.
3. No flow from some sprinkler nozzles.	3. Nozzle clogged with particles from wastewater due to lack of screening at inlet side of irrigation pumps.	3. Screen may have developed hole due to partial plugging of screen.	3. Repair or replace screen.
4. Wastewater is running off of irrigated area.	4a. Sodium adsorption ratio (SAR) of wastewater is too high and has caused clay soil to become impermeable.	4a. SAR should be less than 9.	4a. Feed calcium and magnesium to adjust SAR.
	4b. Soil surface sealed by solids.	4b. Soil surface.	4b. Strip crop area.
	4c. Application rate exceeds infiltration rate of soil.	4c. Application rate.	4c. Reduce application rate until compatible with infiltration rate.
	4d. Break in distribution piping.	4d. Leaks in distribution piping.	4d. Repair breaks.
	4e. Soil permeability has decreased due to continuous application of wastewater.	4e. Duration of continuous operation on the given area.	4e. Each area should be allowed to rest (2–3 days) between applications of wastewater to allow soil to drain.
	4f. Rain has saturated soil.	4f. Rainfall records.	4f. Store wastewater until soil has drained.

TABLE 8.17 SPRAY IRRIGATION SYSTEM TROUBLESHOOTING GUIDE *(continued)*

Indicator/Observation	Probable Cause	Check or Monitor	Solution
5. Irrigated crop is dead.	5a. Too much (or not enough) water has been applied.	5a. Water needs of specific crop versus application rate.	5a. Reduce (or increase) application rate.
	5b. Wastewater contains excessive amount of toxic elements.	5b. Analyze wastewater and consult with county agricultural agent.	5b. Eliminate industrial discharges of toxic materials.
	5c. Too much insecticide or weed killer applied.	5c. Application of insecticide or weed killer.	5c. Proper control of application of insecticide or weed killer.
	5d. Inadequate drainage has flooded root zone of crop.	5d. Water ponding.	5d. See Item 1.
6. Growth of irrigated crop is poor.	6a. Too little nitrogen (N) or phosphorus (P) applied.	6a. N and P quantities applied—check with county agricultural agent.	6a. If increased wastewater application rates are not practical, supplement wastewater N or P with commercial fertilizer.
	6b. Timing of nutrient application not consistent with crop needs. Also, see 5a–5c.	6b. Consult with county agricultural agent.	6b. Adjust application schedule to meet crop needs.
7. Irrigation pumping station shows normal pressure but above-normal flow.	7a. Broken main, lateral, riser, or gasket.	7a. Inspect distribution system for leaks.	7a. Repair leak.
	7b. Missing sprinkler head or end plug.	7b. Inspect distribution system for leaks.	7b. Repair leak.
	7c. Too many laterals on at one time.	7c. Number of laterals in service.	7c. Make appropriate valving changes.
8. Irrigation pumping station shows above-average pressure but below-average flow.	8. Blockage in distribution system due to plugging sprinklers, valves, screens, or frozen water.		8. Locate blockage and eliminate.
9. Irrigation pumping station shows below-normal flow and pressure.	9a. Pump impeller is worn.	9a. Pump impeller.	9a. Replace impeller. See Section 8.11, "Review of Plans and Specifications," No. 6.
	9b. Partially clogged inlet screen.	9b. Screen.	9b. Clean screen.
10. Excessive erosion occurring.	10a. Excessive application rates.	10a. Application rate.	10a. Reduce application rate.
	10b. Inadequate crop cover.	10b. Condition of crop cover.	10b. See Items 5 and 6.
11. Odor complaints.	11a. Wastewater turning septic during transmission to irrigated site and odors being released as it is discharged to pretreatment.	11a. Sample wastewater as it leaves transmission system.	11a. Contain and treat off-gases from discharge point of transmission system by covering inlet with building and passing off-gases through deodorizing system.
	11b. Odors from storage reservoirs.	11b. DO in storage reservoirs.	11b. Improve pretreatment or aerate reservoirs.
12. Center pivot irrigation rigs stuck in mud.	12a. Excessive application rates.		12a. Reduce application rates.
	12b. Improper tires or rigs.		12b. Install tires with higher flotation capabilities.
	12c. Poor drainage.		12c. Improve drainage. See Item 1b.
13. Nitrate concentration of groundwater in vicinity of irrigation site is increasing.	13a. Application of nitrogen is not in balance with crop needs.	13a. Check lbs/acre/yr of nitrogen being applied with needs of crops.	13a. Change crop to one with higher nitrogen needs.
	13b. Nitrogen being applied during periods when crops are dormant.	13b. Application schedules	13b. Apply wastewater only during periods of active crop growth.
	13c. Crop is not being harvested and removed.	13c. Farming management.	13c. Harvest and remove crop.

QUESTIONS

Write your answers in a notebook and then compare your answers with those on page 614.

8.7C   What are the main objectives of a land disposal system?

8.7D   List the three main areas of process control in a land disposal system.

8.7E   How many points in a storage reservoir should be sampled for DO?

8.7F   What are the minimum recommended DO requirements for a storage reservoir?

8.7G   What are the probable causes of water ponding in an irrigated area where ponding normally has not been observed?

8.8   MONITORING PROGRAM

8.80   Monitoring Schedule

The four monitoring areas for an irrigation system where crops are grown are: effluent, vegetation, soils, and groundwater (or collected seepage). This is reduced to the two areas of effluent and groundwater for systems where crops are not grown. Wells should be monitored to identify any adverse effects on groundwaters. Testing requirements and frequencies are shown in Table 8.18.

Sampling wells should be located within the irrigation site as well as near the site and on all sides to identify any changes or trends in water quality. Typical tests and frequencies are listed in Table 8.19.

TABLE 8.18   TESTING REQUIREMENTS

Area	Test	Frequency
Effluent and groundwater or seepage	BOD	Two times per week
	Fecal coliform	Weekly
	Total coliform	Weekly
	Flow	Continuous
	Nitrogen	Weekly
	Phosphorus	Weekly
	Suspended solids	Two times per week
	pH	Daily
	Total dissolved solids (TDS)	Monthly
	Boron	Monthly
	Chloride	Monthly
Vegetation	—variable depending on crop—	
Soils	Conductivity	Two times per month
	pH	Two times per month
	Cation exchange capacity[a]	Two times per month

[a] Cation (KAT-EYE-en) Exchange Capacity. The ability of a soil or other solid to exchange cations (positive ions such as calcium, Ca^{2+}) with a liquid.

TABLE 8.19   WELL MONITORING PROGRAM

Area	Test	Frequency
Wells	Salinity	
	Conductivity	Monthly
	Chloride	Quarterly
	TDS	Quarterly
	Chemical Buildup	
	Nitrate	Monthly
	Calcium	Semiannually
	Magnesium	Semiannually
	Toxicity (Heavy Metals)	
	Cadmium	Monthly
	Lead	Annually
	Zinc	Annually
	Mercury	Annually
	Molybdenum	Annually
	Selenium	Annually
	Organics	
	Trihalomethanes	Quarterly
	Pesticides (depends on local application)	Quarterly

8.81   Interpretation of Test Results and Follow-Up Actions

Excessive levels and concentrations greater than desired for effluent BOD, fecal and total coliforms, nitrogen, phosphorus, and suspended solids are not a concern for crop-growing operations. The total dissolved solids (TDS), boron, chloride, and pH are important during long periods of land treatment, but not for times less than two to three weeks. Excessive nitrogen is a potential problem in spreading basins since nitrate in water supplies can be harmful to infants. If TDS, boron, or chloride levels increase and do not return to previous levels, a change in farming practices may be necessary.

Increased levels in any of the constituents in the groundwater are unacceptable. Most likely, the only constituent that will increase is nitrate-nitrogen. If this occurs, then a nitrogen removal system (partial or complete) should be added to the treatment plant.

8.9   SAFETY

Safe operating procedures should be practiced in all undertakings. The operation of a sprinkler irrigation system has caused fatalities among operating personnel. Many of the fatalities have resulted from contact with electricity used either to power the pumping plant or to transmit electricity associated with the area being irrigated.

Moving of portable sprinkler lateral pipelines is an extremely hazardous activity. Raising a pipeline into the air to dislodge a small animal or debris and contacting overhead electrical transmission lines has resulted in severe electric shock or death to the person holding the pipe.

A sprinkler throwing a stream of water into a power line has shorted the power to ground through the sprinkler system and resulted in severe electrical injuries to anyone touching the sprinkler system parts.

Always have the electric motor well bonded to a good ground with suitable-size conductors. Injuries have occurred from touching an ungrounded motor or pump frame having shorted electrical windings in electrically powered pumping plants.

Electric shocks have occurred from faulty starting equipment and from working on energized circuits. Always pull the line disconnect switch, lock out, and tag it when making repairs or checks on electrical equipment of any kind.

Look over each sprinkler system and mark the potential safety hazards, then avoid the hazards.

Surface spreading systems can be hazardous due to wet surfaces and muddy areas.

8.10 MAINTENANCE

Maintenance of land treatment systems requires keeping the wastewater distribution piping, valves, and sprinklers in good working condition. Pump and valve maintenance is discussed in Chapter 15, "Maintenance," Volume II; storage reservoir maintenance is similar to pond maintenance outlined in Chapter 9, Volume I, *OPERATION OF WASTEWATER TREATMENT PLANTS.*

8.11 REVIEW OF PLANS AND SPECIFICATIONS

Many operational and maintenance problems can be avoided by a careful review of the plans and specifications for a land treatment system. Check the items listed in this section.

1. Ponding

    Ponding problems can be avoided if the proper site is selected and provided with proper drainage. Soils at the site must be suitable for percolation and for planned crops. Adequate drainage (no ponding) can be provided by leveling or sloping of the land surface so the water will flow evenly over all of the land. *DRAINAGE WELLS*[6] or *DRAIN TILE SYSTEMS*[7] may be necessary to remove excess water and prevent ponding.

2. Plastic pipe laterals

    Plastic pipe laterals installed above ground may break because of cold weather or deteriorate due to sunlight. Install plastic pipe laterals below ground.

3. Screens

    Install screens on the inlet side of irrigation pumps to prevent spray nozzles from becoming plugged.

4. Buffer area

    Be sure sufficient buffer area is provided around spray areas to prevent mist from drifting onto nearby homes and yards. If necessary, do not schedule spraying during days when the wind is blowing toward neighbors.

5. Odor

    If odors may be a problem, consider furrow or flood irrigation rather than spraying. Spraying can cause odor problems by releasing odors to the atmosphere.

6. Protection of pumps

    Excessive wear on pumps can result from sand in the water being pumped. If sand is a problem, improve pretreatment or install a sand trap ahead of the pumps. Remember to drain out-of-service pumps before freezing weather occurs in the fall or winter.

7. Alternate place to pump effluent

    An alternate location to pump or dispose of effluent is very important in case of system failure.

8.12 REFERENCES AND ADDITIONAL READING

8.120 References

1. Pound, C. E., et al., *COSTS OF WASTEWATER TREATMENT BY LAND APPLICATION,* Environmental Protection Agency, EPA No. 430-9-75-003, June 1975.

2. Pound, C. E., et al., *EVALUATION OF LAND APPLICATION SYSTEMS,* Environmental Protection Agency, EPA No. 430-9-74-015, September 1974.

3. Richards, L. A. (ed.), *DIAGNOSIS AND IMPROVEMENT OF SALINE AND ALKALINE SOILS,* Agricultural Handbook No. 60, US Department of Agriculture, August 1969.

8.121 Additional Reading

1. *MANUAL OF WASTEWATER TREATMENT,* Chapters 3 and 22,* prepared by Texas Water Utilities Association, 1106 Clayton Lane, Suite 101 East, Austin, TX 78723-1093. Price to members, $25.00; nonmembers, $35.00; plus $3.50 shipping and handling.

2. *WATER QUALITY CRITERIA,* Second Edition, McKee, J. E. and Wolf, H. W., report to California State Water Resources Control Board, SWRCB Publication 3A, Sacramento, CA (1963). Available from National Technical Information Service (NTIS), 5285 Port Royal Road, Springfield, VA 22161. Order No. PB82-188244. Price, $117.00, plus $5.00 shipping and handling per order.

---

[6] *Drainage Wells.* Wells that can be pumped to lower the groundwater table and prevent ponding.

[7] *Drain Tile System.* A system of tile pipes buried under agricultural fields that collect percolated waters and keep the groundwater table below the ground surface to prevent ponding.

3. *WESTERN FERTILIZER HANDBOOK,* California Fertilizer Association, available through Pearson Education Order Department, PO Box 11073, Des Moines, IA 50336-1073. ISBN 0-8134-3146-8. Price, $40.20.

4. *GOOD UNTIL THE LAST DROP: A PRACTITIONER'S GUIDE TO WATER REUSE.* Obtain from American Public Works Association (APWA), Bookstore, PO Box 802296, Kansas City, MO 64180-2296. Order No. PB.A515. Price to members, $54.00; nonmembers, $65.00; price includes cost of shipping and handling. Phone (800) 848-2792 or visit www.apwa.net/bookstore/.

* Depends on edition.

QUESTIONS

Write your answers in a notebook and then compare your answers with those on page 614.

8.8A   What are the four monitoring areas for an irrigation system where crops are grown?

8.9A   What is the major cause of accidents to operators while working with sprinkler irrigation systems?

8.10A   What equipment needs to be maintained in a land treatment system?

8.11A   List the items that should be examined when reviewing plans and specifications for a land disposal system.

END OF LESSON 2 OF 2 LESSONS

on

WASTEWATER RECLAMATION AND REUSE

Please answer the discussion and review questions next.

DISCUSSION AND REVIEW QUESTIONS

Chapter 8.   WASTEWATER RECLAMATION AND REUSE

(Lesson 2 of 2 Lessons)

Write the answers to these questions in your notebook. The question numbering continues from Lesson 1.

6. How does land treatment work?

7. What should be done with wastewater that seeps out of storage reservoirs and runs off from a land treatment system?

8. What are the limitations of land treatment systems?

9. How long should the pumps be run while irrigating a plot of land?

10. What water quality indicators should be monitored to ensure that a land disposal system does not adversely affect a groundwater supply?

11. How can safety hazards be avoided while operating a sprinkler irrigation system?

SUGGESTED ANSWERS

Chapter 8. WASTEWATER RECLAMATION AND REUSE

ANSWERS TO QUESTIONS IN LESSON 1

Answers to questions on page 583.

8.0A Uses of reclaimed wastewater include:

1. Irrigation for crop or plant growth
2. Indirect reuse by downstream users
3. Direct reuse by industry
4. Use as a freshwater barrier to prevent saltwater intrusion by deep well injection
5. Groundwater recharge by spreading basins
6. Reservoirs for recreation

8.0B Oil that was not removed by previous pumping efforts will float on top of water supplied by deep well injection. The oil is then easier to pump to the surface from underground areas.

Answers to questions on page 592.

8.1A Coliforms and pathogenic bacteria can be killed by chlorination.

8.1B "Blend" water is sometimes mixed with plant effluent because this may be the best (most economical) means of achieving the water quality desired by the water users.

8.1C Probable causes of a wastewater reclamation plant being unable to maintain a chlorine residual include:

1. Chlorinator not working properly
2. An increase in the chlorine demand

Answers to questions on page 595.

8.2A

Possible causes of clogging	Possible cures for cause
1. Slimes	1. Chlorination or allow well to rest.
2. Carbon fines	2. Remove fines by passing the water through a sand/anthracite filter.

8.2B If reclaimed effluent was being used by an industry and one of the water quality standards was not being met, notify the industry immediately.

8.3A Always work with another operator when working around storage reservoirs or blending tanks so help will be available if you fall into the water.

ANSWERS TO QUESTIONS IN LESSON 2

Answers to questions on page 601.

8.6A *EVAPOTRANSPIRATION.* (1) The process by which water vapor is released to the atmosphere from living plants. Also called transpiration. (2) The total water removed from an area by transpiration (plants) and by evaporation from soil, snow, and water surfaces. *HYDROLOGIC CYCLE.* The process of evaporation of water into the air and its return to earth by precipitation (rain or snow). This process also includes transpiration from plants, groundwater movement, and runoff into rivers, streams, and the ocean. Also called the water cycle.

8.6B Land disposal of wastewater is accomplished by irrigation, infiltration–percolation, and overland flow.

8.6C The major parts of land application systems include:

1. Preapplication treatment
2. Transmission to the land site
3. Storage
4. Distribution over site
5. Runoff recovery system (if needed)
6. Crop systems

8.6D A plot of land 2,000 feet long by 1,000 feet wide is used for a land disposal system. If 1 million gallons of water is applied to the land during a 24-hour period, calculate the hydraulic loading in:

1. MGD per acre
2. Inches per day

Known	Unknown
Length, ft = 2,000 ft	Hydraulic Loading,
Width, ft = 1,000 ft	1. MGD/acre
Flow, MGD = 1 MGD	2. inches/day

1. Determine surface area in acres.

$$\text{Area, acres} = \frac{\text{Length, ft} \times \text{Width, ft}}{43,560 \text{ sq ft/acre}}$$

$$= \frac{2,000 \text{ ft} \times 1,000 \text{ ft}}{43,560 \text{ sq ft/acre}}$$

$$= 45.9 \text{ acres}$$

2. Determine hydraulic loading, MGD/acre.

$$\text{Loading, MGD/ac} = \frac{\text{Flow, MGD}}{\text{Area, acres}}$$

$$= \frac{1 \text{ MGD}}{45.9 \text{ acres}}$$

$$= 0.02 \text{ MGD/ac}$$

3. Determine hydraulic loading, inches/day.

$$\text{Loading, in/day} = \frac{\text{Flow, MGD} \times 1,000,000/M \times 12 \text{ in/ft}}{\text{Length, ft} \times \text{Width, ft} \times 7.48 \text{ gal/cu ft}}$$

$$= \frac{1 \text{ MGD} \times 1,000,000/M \times 12 \text{ in/ft}}{2,000 \text{ ft} \times 1,000 \text{ ft} \times 7.48 \text{ gal/cu ft}}$$

$$= 0.8 \text{ in/day}$$

Answers to questions on page 605.

8.7A The major items of equipment that should be inspected before starting a spray irrigation system include:

1. Electric motors
2. Pumps
3. Aluminum pipes
4. Sprinkler systems

8.7B Determine the time required to irrigate 30 acres of a medium-type soil where the root zone depth is 2 feet and 50 percent of the available moisture is retained in the soil at irrigation. Pump capacity is 1,200 GPM. Use the figures and tables in this lesson to answer this question.

Known	Unknown
Area, ac = 30 acres	Time to Irrigate, hr
Soil Type = Medium	
Root Zone, ft = 2 ft deep	
Moisture = 50% retention	
Pump, GPM = 1,200 GPM	

1. Determine inches of water to be applied.

From Table 8.15
Application, in = 1.69 inches

2. Determine time to irrigate 30 acres to apply 1 inch with a 1,200 GPM pumping system capacity.

From Figure 8.12
Time to irrigate 1 inch = 11 hours

3. Determine total time to irrigate in hours.

Time, hr = Time (hr) to Irrigate × Amount to Apply, in
= 11 hours/inch × 1.69 inches
= 18.6 hours

Answers to questions on page 610.

8.7C The main objectives of a land disposal system are to dispose of effluent without harming surface waters and without creating nuisance conditions.

8.7D The three main areas of process control in a land disposal system are:

1. Storage reservoirs
2. Runoff and seepage water recycle systems
3. Systems where crops are grown

8.7E A minimum of four points in a storage reservoir should be sampled for DO.

8.7F The minimum DO requirements for a storage reservoir are a minimum DO of 0.4 mg/L for all samples and the average of all samples should be at least 0.8 mg/L.

8.7G Probable causes of ponding include:

1. Application rate is excessive
2. If application rate is normal, drainage may be inadequate
3. A broken pipe in the distribution system

Answers to questions on page 612.

8.8A The four monitoring areas for an irrigation system where crops are grown are effluent, vegetation, soils, and groundwater.

8.9A The major cause of accidents to operators while working with sprinkler irrigation systems is contact with electricity used either to power the pumping plant or to transmit electricity associated with the area being irrigated.

8.10A Equipment requiring maintenance in a land treatment system includes distribution piping, pumps, valves, and sprinklers.

8.11A Items to be examined when reviewing the plans and specifications for a land disposal system include:

1. Ponding
2. Plastic pipe laterals
3. Screens
4. Buffer area
5. Potential for odor problems
6. Protection of pumps
7. Alternate disposal site for emergencies

CHAPTER 9

INSTRUMENTATION AND CONTROL SYSTEMS

by

Leonard Ainsworth

Revised by

William H. Hendrix

TABLE OF CONTENTS

Chapter 9. INSTRUMENTATION AND CONTROL SYSTEMS

OBJECTIVES

Chapter 9.   INSTRUMENTATION AND CONTROL SYSTEMS

Following completion of Chapter 9, you should be able to:

1. Explain the purpose and nature of instrumentation and control systems.

2. Identify, avoid, and correct safety hazards associated with instrumentation work.

3. Recognize various types of sensors and transducers.

4. Read instruments and make proper adjustments in the operation of wastewater treatment facilities.

5. Identify symptoms of measurement and control system problems.

WORDS

Chapter 9.   INSTRUMENTATION AND CONTROL SYSTEMS

ACCURACY
ACCURACY

How closely an instrument measures the true or actual value of the process variable being measured or sensed.

ALARM CONTACT
ALARM CONTACT

A switch that operates when some preset low, high, or abnormal condition exists.

ANALOG
ANALOG

The continuously variable signal type sent to an analog instrument (for example, 4–20 mA).

ANALOG READOUT
ANALOG READOUT

The readout of an instrument by a pointer (or other indicating means) against a dial or scale. Also see DIGITAL READOUT.

ANALYZER
ANALYZER

A device that conducts periodic or continuous measurement of some factor, such as chlorine or fluoride concentration, or turbidity. Analyzers operate by any of several methods including photocells, conductivity, or complex instrumentation.

CALIBRATION
CALIBRATION

A procedure that checks or adjusts an instrument's accuracy by comparison with a standard or reference.

CONTACTOR
CONTACTOR

An electric switch, usually magnetically operated.

CONTROL LOOP
CONTROL LOOP

The combination of one or more interconnected instrumentation devices that are arranged to measure, display, and control a process variable. Also called a loop.

CONTROL SYSTEM
CONTROL SYSTEM

An instrumentation system that senses and controls its own operation on a close, continuous basis in what is called proportional (or modulating) control.

CONTROLLER CONTROLLER

A device that controls the starting, stopping, or operation of a device or piece of equipment.

DANGEROUS AIR CONTAMINATION DANGEROUS AIR CONTAMINATION

An atmosphere presenting a threat of causing death, injury, acute illness, or disablement due to the presence of flammable and/or explosive, toxic, or otherwise injurious or incapacitating substances.

(1) Dangerous air contamination due to the flammability of a gas, vapor, or mist is defined as an atmosphere containing the gas, vapor, or mist at a concentration greater than 10 percent of its lower explosive (lower flammable) limit (LEL).

(2) Dangerous air contamination due to a combustible particulate is defined as a concentration that meets or exceeds the particulate's lower explosive limit (LEL).

(3) Dangerous air contamination due to the toxicity of a substance is defined as the atmospheric concentration that could result in employee exposure in excess of the substance's permissible exposure limit (PEL).

NOTE: A dangerous situation also occurs when the oxygen level is less than 19.5 percent by volume (OXYGEN DEFICIENCY) or more than 23.5 percent by volume (OXYGEN ENRICHMENT).

DESICCANT (DESS-uh-kant) DESICCANT

A drying agent that is capable of removing or absorbing moisture from the atmosphere in a small enclosure.

DIGITAL DIGITAL

The encoding of information that uses binary numbers (ones and zeros) for input, processing, transmission, storage, or display, rather than a continuous spectrum of values (an analog system) or non-numeric symbols such as letters or icons.

DIGITAL READOUT DIGITAL READOUT

The readout of an instrument by a direct, numerical reading of the measured value or variable.

DISCRETE CONTROL DISCRETE CONTROL

ON/OFF control; one of the two output values is equal to zero.

DISCRETE I/O (INPUT/OUTPUT) DISCRETE I/O (INPUT/OUTPUT)

A digital signal that senses or sends either ON or OFF signals. For example, a discrete input would sense the position of a switch; a discrete output would turn on a pump or light.

DISTRIBUTED CONTROL SYSTEM (DCS) DISTRIBUTED CONTROL SYSTEM (DCS)

A computer control system having multiple microprocessors to distribute the functions performing process control, thereby distributing the risk from component failure. The distributed components (input/output devices, control devices, and operator interface devices) are all connected by communications links and permit the transmission of control, measurement, and operating information to and from many locations.

EFFECTIVE RANGE EFFECTIVE RANGE

That portion of the design range (usually from 10 to 90+ percent) in which an instrument has acceptable accuracy. Also see RANGE and SPAN.

FAIL-SAFE FAIL-SAFE

Design and operation of a process control system whereby failure of the power system or any component does not result in process failure or equipment damage.

FEEDBACK FEEDBACK

The circulating action between a sensor measuring a process variable and the controller that controls or adjusts the process variable.

HERTZ (Hz) HERTZ (Hz)

The number of complete electromagnetic cycles or waves in one second of an electric or electronic circuit. Also called the frequency of the current.

HUMAN MACHINE INTERFACE (HMI)   HUMAN MACHINE INTERFACE (HMI)

The device at which the operator interacts with the control system. This may be an individual instrumentation and control device or the graphic screen of a computer control system. Also called MAN MACHINE INTERFACE (MMI) and OPERATOR INTERFACE.

INTEGRATOR   INTEGRATOR

A device or meter that continuously measures and sums a process rate variable in cumulative fashion over a given time period. For example, total flows displayed in gallons per minute, million gallons per day, cubic feet per second, or some other unit of volume per time period. Also called a TOTALIZER.

INTERLOCK   INTERLOCK

A physical device, equipment, or software routine that prevents an operation from beginning or changing function until some condition or set of conditions is fulfilled. An example would be a switch that prevents a piece of equipment from operating when a hazard exists.

LAG TIME   LAG TIME

The time period between the moment a process change is made and the moment such a change is finally sensed by the associated measuring instrument.

LINEARITY (LYNN-ee-AIR-it-ee)   LINEARITY

How closely an instrument measures actual values of a variable through its effective range.

MAN MACHINE INTERFACE (MMI)   MAN MACHINE INTERFACE (MMI)

The device at which the operator interacts with the control system. This may be an individual instrumentation and control device or the graphic screen of a computer control system. Also called HUMAN MACHINE INTERFACE (HMI) and OPERATOR INTERFACE.

MEASURED VARIABLE   MEASURED VARIABLE

A factor (flow, temperature) that is sensed and quantified (reduced to a reading of some kind) by a primary element or sensor.

OPERATOR INTERFACE   OPERATOR INTERFACE

The device at which the operator interacts with the control system. This may be an individual instrumentation and control device or the graphic screen of a computer control system. Also called HUMAN MACHINE INTERFACE (HMI) and MAN MACHINE INTERFACE (MMI).

ORIFICE (OR-uh-fiss)   ORIFICE

An opening (hole) in a plate, wall, or partition. An orifice flange or plate placed in a pipe consists of a slot or a calibrated circular hole smaller than the pipe diameter. The difference in pressure in the pipe above and at the orifice may be used to determine the flow in the pipe. In a trickling filter distributor, the wastewater passes through an orifice to the surface of the filter media.

PRECISION   PRECISION

The ability of an instrument to measure a process variable and repeatedly obtain the same result. The ability of an instrument to reproduce the same results.

PRIMARY ELEMENT   PRIMARY ELEMENT

(1) A device that measures (senses) a physical condition or variable of interest. Floats and thermocouples are examples of primary elements. Also called a SENSOR.

(2) The hydraulic structure used to measure flows. In open channels, weirs and flumes are primary elements or devices. Venturi meters and orifice plates are the primary elements in pipes or pressure conduits.

PROCESS VARIABLE   PROCESS VARIABLE

A physical or chemical quantity that is usually measured and controlled in the operation of a water, wastewater, or industrial treatment plant. Common process variables are flow, level, pressure, temperature, turbidity, chlorine, and oxygen levels.

PROGRAMMABLE LOGIC CONTROLLER (PLC)

A microcomputer-based control device containing programmable software; used to control process variables.

RANGE

The spread from minimum to maximum values that an instrument is designed to measure. Also see EFFECTIVE RANGE and SPAN.

READOUT

The reading of the value of a process variable from an indicator or recorder or on a computer screen.

RECEIVER

A device that indicates the result of a measurement, usually using either a fixed scale and movable indicator (pointer), such as a pressure gauge, or a moving chart with a movable pen like those used on a circular flow-recording chart. Also called an INDICATOR.

RECORDER

A device that creates a permanent record, on a paper chart, magnetic tape, or in a computer, of the changes in a measured variable.

REFERENCE

A physical or chemical quantity whose value is known exactly, and thus is used to calibrate instruments or standardize measurements. Also called a STANDARD.

ROTAMETER (RODE-uh-ME-ter)

A device used to measure the flow rate of gases and liquids. The gas or liquid being measured flows vertically up a tapered, calibrated tube. Inside the tube is a small ball or bullet-shaped float (it may rotate) that rises or falls depending on the flow rate. The flow rate may be read on a scale behind or on the tube by looking at the middle of the ball or at the widest part or top of the float.

SCADA (SKAY-dah) SYSTEM

Supervisory Control And Data Acquisition system. A computer-monitored alarm, response, control, and data acquisition system used to monitor and adjust treatment processes and facilities.

SCALE

(1) A combination of mineral salts and bacterial accumulation that sticks to the inside of a collection pipe under certain conditions. Scale, in extreme growth circumstances, creates additional friction loss to the flow of water. Scale may also accumulate on surfaces other than pipes.

(2) The marked plate against which an indicator or recorder reads, usually the same as the range of the measuring system. See RANGE.

SENSITIVITY

The smallest change in a process variable that an instrument can sense.

SENSOR

A device that measures (senses) a physical condition or variable of interest. Floats and thermocouples are examples of sensors. Also called a PRIMARY ELEMENT.

SET POINT

The position at which the control or controller is set. This is the same as the desired value of the process variable. For example, a thermostat is set to maintain a desired temperature.

SOFTWARE PROGRAM

Computer program; the list of instructions that tell a computer how to perform a given task or tasks. Some software programs are designed and written to monitor and control treatment processes.

SOLENOID (SO-luh-noid)

A magnetically operated mechanical device (electric coil). Solenoids can operate small valves or electric switches.

SPAN   SPAN

The scale or range of values an instrument is designed to measure. Also see RANGE.

STANDARD   STANDARD

A physical or chemical quantity whose value is known exactly, and thus is used to calibrate instruments or standardize measurements. Also called a REFERENCE.

STANDARDIZE   STANDARDIZE

To compare with a standard.

(1) In wet chemistry, to find out the exact strength of a solution by comparing it with a standard of known strength. This information is used to adjust the strength by adding more water or more of the substance dissolved.

(2) To set up an instrument or device to read a standard. This allows you to adjust the instrument so that it reads accurately, or enables you to apply a correction factor to the readings.

STARTERS (MOTOR)   STARTERS (MOTOR)

Devices used to start up large motors gradually to avoid severe mechanical shock to a driven machine and to prevent disturbance to the electrical lines (causing dimming and flickering of lights).

TELEMETRY (tel-LEM-uh-tree)   TELEMETRY

The electrical link between a field transmitter and the receiver. Telephone lines are commonly used to serve as the electrical line.

THERMOCOUPLE   THERMOCOUPLE

A heat-sensing device made of two conductors of different metals joined together. An electric current is produced when there is a difference in temperature between the ends.

TIMER   TIMER

A device for automatically starting or stopping a machine or other device at a given time.

TOTALIZER   TOTALIZER

A device or meter that continuously measures and sums a process rate variable in cumulative fashion over a given time period. For example, total flows displayed in gallons per minute, million gallons per day, cubic feet per second, or some other unit of volume per time period. Also called an INTEGRATOR.

TRANSDUCER (trans-DUE-sir)   TRANSDUCER

A device that senses some varying condition measured by a primary sensor and converts it to an electrical or other signal for transmission to some other device (a receiver) for processing or decision making.

VARIABLE, MEASURED   VARIABLE, MEASURED

A factor (flow, temperature) that is sensed and quantified (reduced to a reading of some kind) by a primary element or sensor.

VARIABLE, PROCESS   VARIABLE, PROCESS

A physical or chemical quantity that is usually measured and controlled in the operation of a water, wastewater, or industrial treatment plant.

ABBREVIATIONS AND SYMBOLS

Chapter 9.   INSTRUMENTATION AND CONTROL SYSTEMS

Special symbols are used for simplicity and clarity on circuit drawings for instruments. Usually, instrument manufacturers and design engineers provide lists of symbols they use with an explanation of the meaning of each symbol. This section contains a list of typical instrumentation abbreviations and symbols used in this chapter and also used by the waterworks profession.

ABBREVIATIONS

A — Analyzer, such as a device used to measure a water quality indicator (pH, temperature).

C — Controller, such as a device used to start, operate, or stop a pump.

D — Differential, such as a "differential pressure" (DP) cell used with a flowmeter.

E — Electrical or Voltage.
    Element, such as a primary element.

F — Flow rate (*NOT* total flow).

H — Hand (manual operation).
    High as in high-level.

I — Indicator, such as the indicator on a flow recording chart.
    I = E/R where I is the electric current in amps.

L — Level, such as the level of water in a tank.
    Low, as in a low-level switch.
    Light, as in indicator light.

M — Motor.
    Middle, as in a mid-level switch.

P — Pressure (or vacuum).
    Pump.
    Program, as in a software program.

Q — Quantity, such as a totalized volume (Σ for summation is also used).

R — Recorder (or printer), such as a chart recorder.
    Receiver.
    Relay.

S — Switch.
    Speed, such as the RPM (revolutions per minute) of a motor.
    Starter, such as a motor starter.
    Solenoid.

T — Transmitter.
    Temperature.
    Tone.

V — Valve.
    Voltage.

W —Weight.
    Watt.

X — Special or unclassified variable.

Y — Computing function, such as a square root ($\sqrt{}$) extraction.

Z — Position, such as a percent valve opening.

TYPICAL PROCESS AND ELECTRICAL SYMBOLS

1. 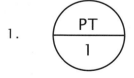   Pressure transmitter #1

2.    Level indicator/recorder #2

3. 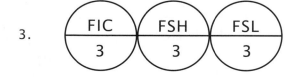   Flow indicator/controller #3 with high-low control switches in the same instrument

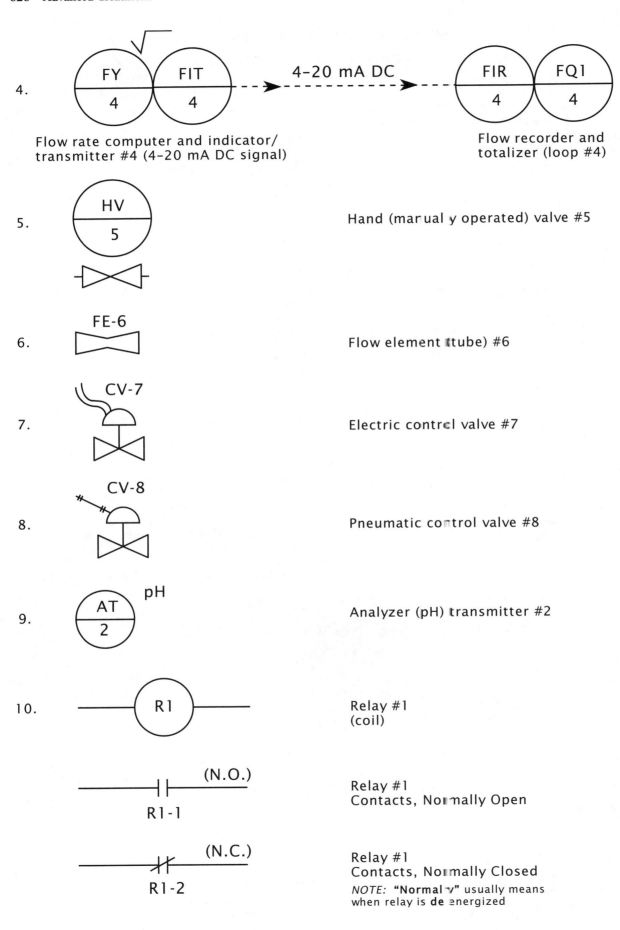

4.   **FY 4 / FIT 4**   4-20 mA DC →   **FIR 4 / FQ1 4**

Flow rate computer and indicator/
transmitter #4 (4-20 mA DC signal)

Flow recorder and
totalizer (loop #4)

5.   **HV 5**   Hand (manually operated) valve #5

6.   **FE-6**   Flow element (tube) #6

7.   **CV-7**   Electric control valve #7

8.   **CV-8**   Pneumatic control valve #8

9.   **AT 2   pH**   Analyzer (pH) transmitter #2

10.   **R1**   Relay #1
(coil)

**(N.O.)
R1-1**   Relay #1
Contacts, Normally Open

**(N.C.)
R1-2**   Relay #1
Contacts, Normally Closed

NOTE: **"Normally"** usually means
when relay is **de**energized

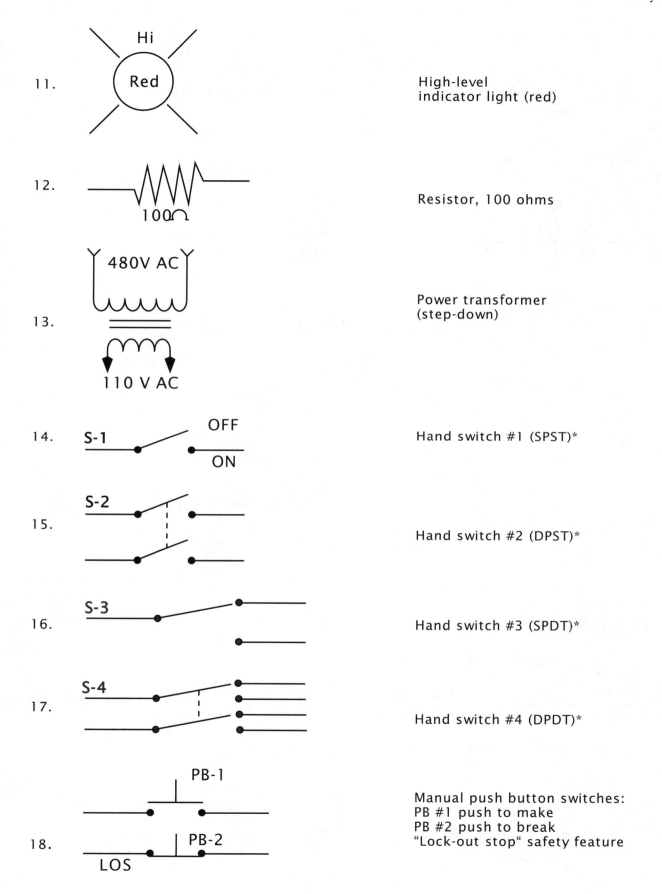

11. High-level
indicator light (red)

12. Resistor, 100 ohms

13. Power transformer
(step-down)

14. Hand switch #1 (SPST)*

15. Hand switch #2 (DPST)*

16. Hand switch #3 (SPDT)*

17. Hand switch #4 (DPDT)*

18. Manual push button switches:
PB #1 push to make
PB #2 push to break
"Lock-out stop" safety feature

* SPST means single-pole, single-throw; DPST means double-pole, single-throw; SPDT means single-pole, double-throw; DPDT means double-pole, double throw

19. Fuses and Circuit Breakers

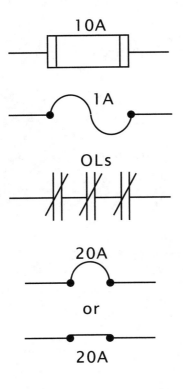

10-amp cartridge fuse

1-amp in-line fuse

Thermal overload contacts (motor

20-amp circuit breaker

20.

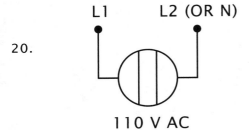

Line 1 and Line 2 (neutral) to standard duplex wall plug outlet

21. H - O - A Function Switch

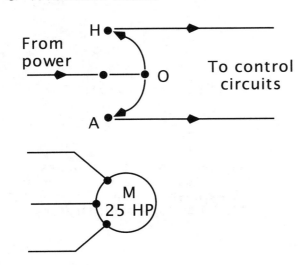

(Hand - Off - Automatic)

22.

Electric motor, 3-phase power, 25 horsepower

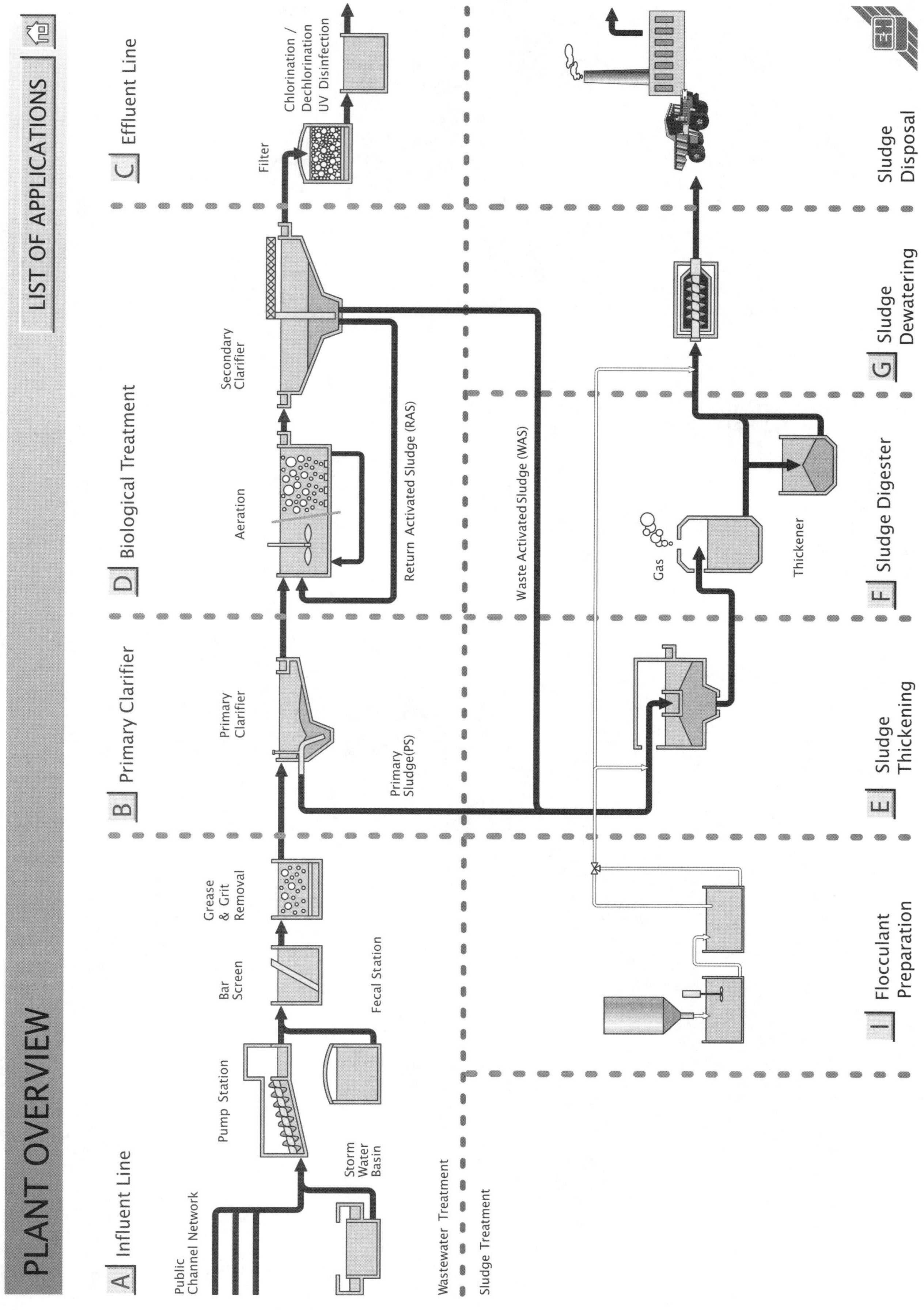

PLANT OVERVIEW

LIST OF APPLICATIONS

A | Influent Line

B | Primary Clarifier

C | Effluent Line

D | Biological Treatment

E | Sludge Thickening

F | Sludge Digester

G | Sludge Dewatering

I | Flocculant Preparation

Public Channel Network

Pump Station

Storm Water Basin

Bar Screen

Grease & Grit Removal

Fecal Station

Primary Clarifier

Aeration

Secondary Clarifier

Filter

Chlorination / Dechlorination UV Disinfection

Return Activated Sludge (RAS)

Waste Activated Sludge (WAS)

Primary Sludge (PS)

Gas

Thickener

Wastewater Treatment

Sludge Treatment

Sludge Disposal

Sludge Dewatering

(Courtesy Endress+Hauser)

CHAPTER 9. INSTRUMENTATION AND CONTROL SYSTEMS

(Lesson 1 of 2 Lessons)

9.0 INSTRUMENTATION AND CONTROL SYSTEMS

9.00 Importance and Nature of Instrumentation and Control Systems

The plant process instrumentation and control system can include manual controls, remote controls, and central computer processing units in any combination. The instrumentation and control system may be computerized or manual: both types provide similar functionality in operating the plant. Today, most facilities are operated through computerized control systems. Classic instrumentation LOOP[1] controllers are still used in new facilities for specific purposes and are in general use in older process facilities. The operator's interface with the control system is a small window into the system. In this chapter, you will learn instrumentation and control system concepts and practices that apply to computerized and manual control systems.

Instrumentation is used to sense process variables that can and should be measured. The treatment plant operator must have a good working familiarity with instrumentation and control systems to properly monitor and control the treatment processes. Operators' primary, and sometimes only, links to the process are the instruments and automatic controls in the treatment plant. The range of instrumentation and control system capabilities is extremely large and is beyond what you will cover in this chapter. Here, you will learn general instrumentation functions and control system capabilities. Once the treatment plant is built, the main processes will be permanently in place. However, by using instrumentation and controls, these processes can be manipulated to achieve and maintain maximum process efficiency. The operator's knowledge and use of instrumentation and control will benefit plant operations.

Your knowledge about instrumentation and control systems can only enhance the effective and efficient operation of your facility. Specifically, if you can recognize that instrumentation and control equipment is not operating properly, your process decisions can be based on that knowledge rather than blind faith. This is especially true for computer automation systems, which tend to be trusted implicitly. The automated instrumentation and control is your main link to the process. Some systems provide backup controls to continue critical operation when the computer control system is out of service. Familiarize yourself with these backup controls and how to operate them for your facility. Instrumentation and control systems have self-diagnostic capabilities that will inform the operator of failures and malfunctions. Again, you will want to familiarize yourself with the control systems to understand the capabilities of their diagnostic features.

Basic process variable measurement functions and concepts are the same, whether applied as a single-loop instrumentation and control system or a fully automated computer control system. Even the most modern plants have many instruments that require the operator to perform some minor preventive maintenance. The operator who understands how to recognize failures, troubleshoot, and make adjustments and minor repairs will be more comfortable in operating the plant without the need to call out maintenance personnel. Thus, the more you know about your plant's instrumentation and control systems, the better operator you become.

9.01 Importance to the Wastewater Treatment Operator

Instrumentation and control systems are essential components in treatment facility operation, providing the operator the ability to monitor and manipulate plant processes remotely.

In a very real sense, measurement instruments can be considered extensions of and improvements to your senses of vision, touch, hearing, and even smell. Not only can instruments provide continuous and simultaneous monitoring of the many process variables throughout the plant, they do so in a more precise and consistent manner than the human senses. In addition, instruments often provide a permanent record of measurements taken. The automated control systems, in turn, provide operators with far-reaching and powerful hands to manipulate switches, valves, motors, and pumps in specific ways. In effect, the control and instrumentation system provides you with a staff of obedient and hard-working assistants, always on the job to help you operate your process more easily and efficiently. To appreciate these great advantages of automation, consider what a treatment plant would be like without modern controls. In the not-too-distant past, and even in some treatment operations today, the situation described in the paragraph below was and is the daily *modus operandi* for some operations:

The plant experiences a power failure in the main and backup electrical systems supplying the main automated

---

[1] *Control Loop.* The combination of one or more interconnected instrumentation devices that are arranged to measure, display, and control a process variable. Also called a loop.

control system; other plant power and process equipment continue operating. The equipment failure is such that the operator cannot repair or reset to restore power. As the operator, you must try to keep the plant on line manually by attempting to control all process variables without any of the normal data indications of the control system. That is, flows and levels have to be estimated visually, valves turned manually, and pumps started and stopped on "HAND" to control them. Additionally, you are forced to observe and try to regulate the other process variables using only your eyes, ears, sense of touch, and probably, in time, even your sense of smell.

Impossible, you say? No, it can be, has been, and is done here and there in old or poorly maintained plants. But, even if you could do it (and on some shifts it does seem like this *is* what you have to do), you certainly could not exercise close control or do it for any extended period. Operating a larger or sophisticated plant would be impossible without its process instrumentation and controls, even with a sizeable crew. Your plant emergency systems and operator knowledge and skills are the final line of defense in such a case as described.

The plant instrumentation and control system is designed with safeguards that nearly eliminate such failures. However, infrequent failures do occur, even in the best-designed instrumentation and control system. Operators who develop their knowledge of the benefits and limitations of the instrumentation and control systems will have enhanced skills during such upset conditions of operation.

QUESTIONS

Write your answers in a notebook and then compare your answers with those on page 680.

9.0A   Why must a treatment plant operator have a good working familiarity with instrumentation and control systems?

9.0B   What kind of information is presented in this chapter with regard to instrumentation?

9.0C   Measurement instruments can serve as an extension of and improvement to which of the operator's physical senses?

9.02   Nature of the Measurement Process

Our senses provide us with qualitative (for example, color, sound, odor) and relatively short-term quantitative (for example, brighter, louder, stronger) information; instrumentation measurements provide us with exact quantitative data. That is to say, measurements give us numbers. Without this objective quantification of our environment (that is, the numbers we use to describe it), we could accomplish very little—just as primitive tribes with no real number systems have no technology worthy of the term. Their processing of natural materials is very limited because the information they use consists only of the relative terms "more than" or "less than." Without objective measurements, they have little control over their environment. Modern

wastewater treatment plants, on the other hand, depend absolutely upon accurate and reproducible measurements of chemical and physical processes, which is what good instrumentation provides.

A measurement is, by definition, the comparison of a quantity with a standard unit of measure. Thus, a tape measure is marked in feet or meters, standard units of length; clocks are marked in hours, minutes, and seconds as the standard units of time; and a small container may have fluid ounces marked on its side. These units, usually of a convenient magnitude (size), have been agreed upon internationally to serve as the accepted standard units. The primary standard for mass (weight) is the weight of an actual physical object, which is located in Paris, France; a standard kilogram is defined as the weight of this one specific object. The primary standards for time are defined in wavelengths of light, and the primary standard for length is the wavelength of a certain spectrum line of the element cesium. Other primary standards can be set up for certain quantities—for example, solution strengths—but generally, industrial measurements depend on secondary standards that are based on more or less exact comparisons to a primary standard. A machinist's caliper is calibrated against precision measuring blocks, which are such secondary standards.

Of course, all the measurements used in water and wastewater treatment ultimately refer back to a few primary and secondary standards. Measurements of length, and the related calculations for area and volume, are in effect comparisons to the standard foot (meter). Weights of chemicals are traceable to the standard pound (kilogram). Timepieces are designed around the standard second. All the other measurements such as flow rate (volume per unit of time), pressure (weight per unit of area), chemical concentration/dosage (weight of chemical per unit of liquid volume), chemical feed (weight of chemical per unit of time), and others, are derived from the fundamental units of measure.

Some important terms directly relating to the measurement process also need to be defined: a *PROCESS VARIABLE* is a physical or chemical quantity, such as flow rate or pH, that is measured or controlled in a wastewater treatment process. *ACCURACY* refers to how closely an instrument measures the true or actual value of a process variable. Accuracy is usually expressed as within plus or minus a given small percent of the true value, for example, ±2 percent is typical instrument accuracy of measurement. Accuracy depends partly upon the *PRECISION* of an instrument, which amounts to how closely the device can reproduce a given reading (or measurement) time after time (a good example of a *precise* but *inaccurate* instrument is a precision electrical gauge—with a bent needle). *SENSITIVITY* refers to the smallest change in a process variable that an instrument can sense. It is usually expressed as a percent of the full-scale value of the instrument. Typical values are ±0.1 percent of full scale for a high-sensitivity instrument and ±2 percent of full scale for an average-sensitivity instrument. *CALIBRATION* is the complete test of an instrument by measuring several (secondary) standards and its complete adjustment (if required) to read the standard values at several points in its range. *STANDARDIZATION* is an abbreviated calibration where only one standard is measured and the instrument is adjusted to read the proper value, such as standardizing a pH meter with pH 7 buffer. The *RANGE* of an instrument is the spread between the minimum and maximum values it is designed to measure, most often from zero to a maximum value, as with, say, a 0–100 psi pressure gauge. And the *EFFECTIVE RANGE* is that portion of an instrument's complete range within which the instrument has acceptable accuracy, commonly between 10 percent and 90+ percent of its design range. *SPAN* is a closely related term to express that procedure in an instrument's calibration that adjusts it to read both low and high standards accurately; an improperly spanned 0–100 psi pressure gauge might read 10 psi "right on" but then indicate a true 80 psi as 70 psi. *LINEARITY* is the term to describe how well a device tracks a series of standards throughout its effective range—the preceding poorly spanned gauge is "off" in linearity by –12.5 percent at 80 percent of scale (off 10 psi at 80 psi). It is evident that to be truly accurate, an instrument must have acceptable precision, be in calibration, and be designed with an effective range that is wide enough to measure the range of the treatment plant's process variables.

Finally, two other common terms should be defined. *ANALOG* displays show a reading as a pointer (or other cursor) against a marked scale, such as simple pressure or level gauges. Numeric *DIGITAL*[2] displays provide direct, numerical readouts, as with most modern instrumentation. The best example of both is a watch that has both readouts; the hands give a quick analog reference and the digital time display gives a more precise reading. Analog process variable values may be transmitted by a continuously variable 4–20 mA (milliamp) signal or a digital-data signal (not to be confused with a numeric digital *display*). Figure 9.1 illustrates the difference between the connections required for each. Computer readouts are analog input signals that have been encoded as digital data. The *OPERATOR INTERFACE*[3] on the computer graphic may be programmed to display the value as an analog or digital readout.

An instrument loop encompasses all devices from the process sensing element to the indicators on a control panel or control system computer screen. Typically, loop components are connected together by a 4–20 mA signal or digital-data link (Figure 9.2).

QUESTIONS

Write your answers in a notebook and then compare your answers with those on page 680.

9.0D   What is the main difference between measurements by our senses and measurements by instruments?

9.0E   What is the definition of a measurement?

9.0F   Explain the difference between accuracy and precision.

9.0G   What is the difference between analog and digital displays?

9.03   Explanation of Control Systems

Control systems are the means by which such process variables as pressure, level, weight, or flow are controlled. The terms "controller" and "control systems" are used to refer to two different types of instrumentation control systems: (1) *MODULATING SYSTEMS,* which sense and control their own operation on a close, continuous basis, and (2) *MOTOR CONTROL STATIONS,* which control only the ON/OFF operation of motors and other devices.

9.030   Modulating Control Systems

The technically proper use of the terms "controller" and "control system" refers to those systems that provide modulating control of process variables. Examples of modulating control systems include: (1) chlorine residual analyzers/controllers; (2) flow-paced (open-loop) chemical feeders; (3) pressure- or flow-regulating valves; (4) continuous level control of process basins; and (5) variable-speed pumping systems for flow/level control.

In order for a process variable, whether pressure, level, weight, or flow, to be closely controlled, it must be measured precisely and continuously. In a modulating control system, the measuring device or primary element sends an electrical or pneumatic signal, proportional to the value of the variable, to the actual system controller.

Within the controller, the signal is compared to the set-point value. A difference between the actual and set-point values results in the controller sending out a command signal to the controlled element, usually a valve, pump, or chemical feeder. Such an error signal produces an adjustment in the system that causes a corresponding change in the original measured variable, making it more closely match the set point. This continuous "cut-and-try" process can result in very fine, ongoing control of variables requiring constant values, such as some flow rates, pressures, levels, or chemical feed rates. The term applied to this circulating action of the variable in such a controller is *FEEDBACK.* The path through the control system is the *CONTROL LOOP.* A diagram of such a control loop, measuring the process variable of wastewater flow, is shown in Figure 9.3.

---

[2] *Digital.*   The encoding of information that uses binary numbers (ones and zeros), for input, processing, transmission, storage, or display, rather than a continuous spectrum of values (an analog system) or non-numeric symbols such as letters or icons.

[3] *Operator Interface.*   The device at which the operator interacts with the control system. This may be an individual instrumentation and control device or the graphic screen of a computer control system. Also called HUMAN MACHINE INTERFACE (HMI) and MAN MACHINE INTERFACE (MMI).

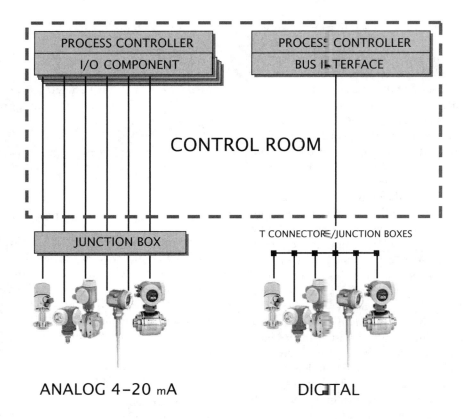

Fig. 9.1   *Connections required for analog and digital transmissions*
(Adapted from figure provided by courtesy of Endress–Hauser)

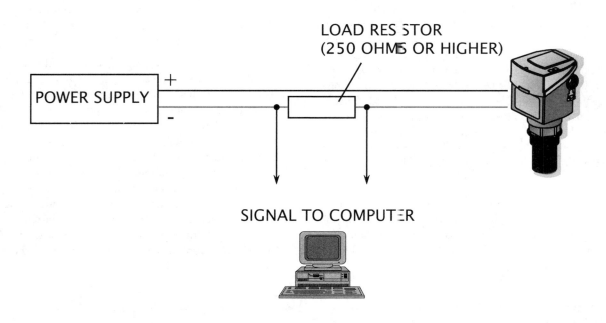

Fig. 9.2   *Typical 4–20 mA instrument loop*
(Adapted from figure provided by courtesy of Endress–Hauser)

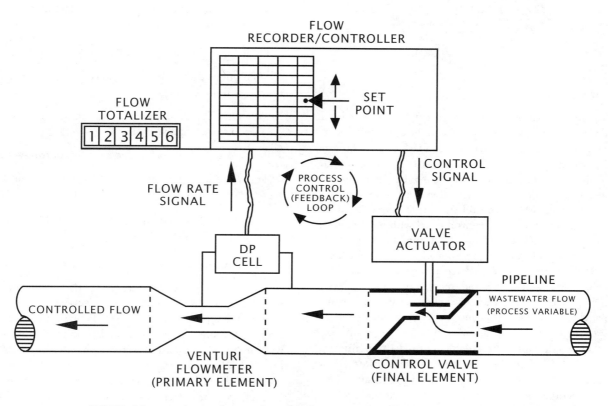

NOTE: Measurement and control system instruments may be electric or pneumatic.

Fig. 9.3   Automatic control system diagram (closed-loop modulating control)

The internal settings of the controller can be quite critical, since close control depends on sensitive adjustments. Thus, you should not try to adjust any such control system unless you know exactly what you are doing. Many plant and system operations have been drastically upset due to such efforts, however well intentioned, by unqualified personnel.

9.031   Motor Control Stations

A motor control station or panel (Figure 9.4) and the related circuitry (Figure 9.5) essentially provide only ON/OFF operation of an electric motor or other device. The electric motor or device, in turn, might power a pump, control valve, or chemical feeder. With this system, control could be manual or automatic, with a switch responding to a preset value of level, pressure, or some other variable. Motor control stations are typically made up of a standard electric power control panel with manual push buttons, overload relays, and Hand-Off-Automatic (H-O-A) or similar On-Off-Remote switch. Additionally, they may include, in good electrical FAIL-SAFE[4] design practice, provisions for power failure or loss-of-phase, and such protective devices as high or low pressure/temperature/level cutoff switches. For this type of panel to be considered a controller (within our secondary meaning of the term), its operation must be directly controlled by the value or values of some variable, not merely by a device such as an ON/OFF timer. In other words, it must be turned on and off as a result of a measurement of a level, pressure, flow, chemical concentration, or other variable as it reaches a predetermined setting or settings. In the automatic mode (A on the H-O-A switch), its operation thus is, in fact, automatic in the sense that the variable is automatically controlled, even though the limits of its value may be quite wide compared to those attainable with a modulating controller as previously described. While a basin level modulating controller may allow only an inch (centimeter) or so of water level change, an ON/OFF system might operate within a few feet (meters) of level difference. In many applications, however, such wide control is of no particular disadvantage, and sometimes is even desirable, as with a tank level where regular exchange of contents is desirable.

Terms used in control practice can now be defined operationally: FEEDBACK and CONTROL LOOP have been mentioned previously, however the term CONTROL LOOP needs some qualification. An OPEN-LOOP control system controls one variable on the basis of another. A good example of this is a chlorinator "paced" by process flow signals (rather than by a chlorine residual analyzer). CLOSED-LOOP control remains as discussed previously, the true control system with measurement and control of the same variable (and feedback). PROPORTIONAL-BAND, RESET, and DERIVATIVE actions are adjustments of a controller that relate to the effectiveness and speed of its control action. OFFSET is the difference between the desired value of the variable (the SET POINT) and the controlled (actual) value. LAG TIME, common to chlorinator control systems, refers to the time period between the moment a process change is made and the moment such a change is finally sensed by the associated measuring instrument. Long lag times can result in unstable processes and poor control.

QUESTIONS

Write your answers in a notebook and then compare your answers with those on page 680.

9.0H   What kinds of wastewater treatment process variables do control systems control?

9.0I   List three examples of modulating control systems.

9.0J   What is the essential purpose of a motor control station or panel?

9.1   SAFETY HAZARDS OF INSTRUMENTATION AND CONTROL SYSTEMS

9.10   General Precautions

The general principles for safe performance on the job, summed up as always avoiding unsafe acts and correcting unsafe conditions, apply as much to instrumentation work as to other plant operations. There are several specific dangers associated with instrument systems that are worth repeating in this section for the sake of safe practice. These include electrical hazards, mechanical and pneumatic hazards, confined spaces, oxygen deficiency and enrichment, explosive gas mixtures, and falls and associated hazards.

9.11   Electrical Hazards

A hidden aspect of energized electrical equipment is that it looks normal; that is, there are no moving parts or other obvious signs that tend to discourage one from touching its components. In fact, there seems to be a peculiar fascination to "see if it is really live" by quickly touching circuit components with a tool, often a screwdriver (usually having an insulated handle, fortunately) but sometimes even with one's finger. Only training (coupled with bad experience, at times) is effective in squelching this morbid curiosity. Even so, most practicing electricians' tools still have the "arc-mark trademark" or two, evidence of the need

---

[4] Fail-Safe.   Design and operation of a process control system whereby failure of the power system or any component does not result in process failure or equipment damage.

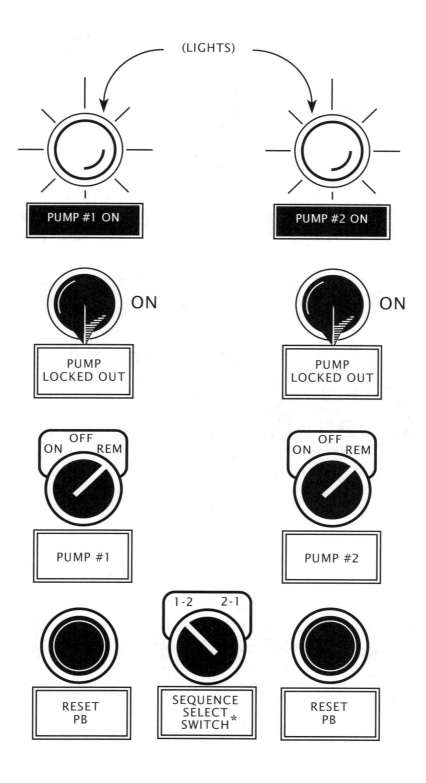

(LIGHTS)

PUMP #1 ON

PUMP #2 ON

ON

ON

PUMP
LOCKED OUT

PUMP
LOCKED OUT

OFF
ON   REM

OFF
ON   REM

PUMP #1

PUMP #2

1-2   2-1

RESET
PB

SEQUENCE
SELECT
SWITCH *

RESET
PB

Note: "REM" on switch above means "remote," which
is automatic operation in this application

*Determines which is "lead" (first to come on) and "lag"
(next to come on) pump in automatic operation

Fig. 9.4   Typical duplex (two-pump) motor control panel

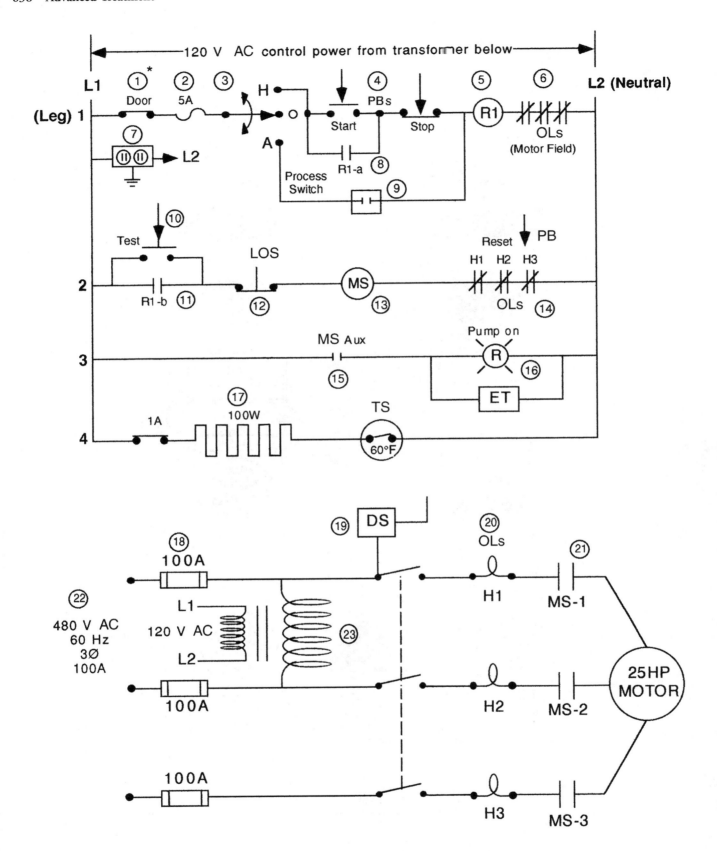

Fig. 9.5　Typical motor control elementary diagram circuitry (do not use for design)
(simplified, self-explanatory ladder diagram)
(see page 639 for parts legend)

FIGURE 9.5 PARTS LEGEND

1. Door or "interlock" switch to kill control power if panel door is opened.

2. Control circuit fuse, 5 amp.

3. Hand-Off-Auto (H-O-A) function switch for manual or automatic operation (shown Off).

4. Momentary contact (spring-loaded) push buttons (Start-Stop), "PB station."

5. Control relay coil (other relays, if used, would be shown as R2, R3, etc.)

6. Motor thermal protection elements embedded in 3 phase motor field ("stator") coils.

7. Duplex 120 V AC outlet within panel (for test equipment usage).

8. "Holding contacts" of control relay; start PB energizes relay coil to close contacts, which remain closed until coil is de-energized.

9. Process control contacts (ON/OFF control); for example, level, pressure, or analytical process variable.

10. Push-to-test motor starter relay circuit leg (if MS coil and OLs OK, pump will start); used for troubleshooting purposes.

11. Main control contacts of relay R1, energizes MS coil to start pump.

12. "Lock-Out Stop" switch to prevent motor operation while being serviced, key lockable in open position.

13. Motor Starter ("Mag starter") relay coil.

14. Manually (push button) resettable overload relay contacts, opened by "heaters" (see 20).

15. Auxiliary contacts in MS relay.

16. Pump running light (red) and elapsed time (total hours run) meter.

17. Enclosure heater in panel (keeps components dry) with 1 amp circuit breaker and thermostat switch.

18. 100 amp 480 V AC cartridge-type fuses.

19. Main 480 V AC manual disconnect switch ("Main switch") local at the motor.

20. Motor circuit thermal overload "heaters" (set for maximum load motor current).

21. Motor starter relay contactors (large contacts).

22. Motor power, 480 volts AC, 60 cycle, 3 phase, 100 amp service.

23. Control power transformer (for control circuit above), 120 volts AC.

DESCRIPTION OF TYPICAL MOTOR CONTROL CIRCUIT OPERATION
(Refer to Figure 9.5)

Control circuit is shown Off; the H-O-A function switch is at "O." To start motor manually, place H-O-A in "H" and depress "Start" push button (PB) to energize main control relay R1. With power through coil, "holding contacts" R1-a close around the start PB to keep R1 energized (and motor running) when the operator's finger is removed from the button (such spring-loaded PBs are termed "momentary" contacts).

When R1 is energized, it also "pulls-in" (closes) contacts R1-b to energize motor starter relay MS in turn. With control power to the MS coil, its contactors MS-1, 2, and 3 "make" (close) to supply 480 V AC to the motor—assuming resettable thermal overloads H1, H2, and H3 are all closed (and the main disconnect is closed and 100A fuses are good, of course).

To stop the motor manually, with the H-O-A still in the "H" position, one pushes the "Stop" PB to break the circuit to the R1 coil, allowing contacts R1-a to "fall out" (open) so the control relay is de-energized (even after this momentary open PB is released).

For automatic operation, H-O-A switch is placed in "A" so motor will start and stop with the open and close, respectively, of the process controller contacts. Note that the PBs cannot start or stop the motor with H-O-A switch in the Auto position.

The "Test" PB, in Leg 1 of the schematic, will start stopped pump as long as it is depressed (being a momentary close PB), by bypassing the control contacts R1-b. This feature permits isolation of a no-start problem to Leg 1 circuitry, in that the motor will start if Leg 2 is OK.

In Leg 3, the "motor on" light and "elapsed time" meter are turned on when the MS relay is energized, through its auxiliary contacts (integral to the relay). Leg 4 consists of a small fused and thermostated space heater to keep all electrical panel components dry.

120 V AC control power for the control circuit (L1 and L2) is transformed from the main 480 V AC 3Ø motor power service. A main disconnect switch kills all power to the motor, for servicing, and all three phases are protected with 100-amp cartridge fuses.

This typical motor control circuitry has two types of motor overheating protection:

1. In-line heater elements H1, H2, and H3, which open any or all of their respective contacts above when the motor draws higher currents for a longer time than it is designed for. These interrupt the power to the MS relay (Leg 2).

2. Small, heat-sensitive contacts embedded in all three stator motor windings, which open at a predetermined maximum temperature to interrupt the control circuit in Leg 1, stopping the motor before it gets too hot.

Only the former (overload heaters) are manually resettable using the PB shown; the latter (stator-imbedded) must be allowed to cool before they remake contact.

for continuing self-discipline in this area. Though such mention may conjure up humorous images of the maintenance person's surprise and shock upon such an incident, one only need consider that electric shock can and does regularly cause serious burns and even death (by asphyxiation due to paralysis of the muscles used in breathing) to bring the problem into sober perspective. Also, the expense and effort caused by a needless shorting-out of an electrical device could be significant. The point is, resist the urge to test any electrical device with a hand tool or part of your body.

If there is *any* doubt in your mind that *all* sources of voltage (not merely the local switch) to a device have been switched off or disconnected, then *DO NOT TOUCH,* except possibly with the insulated probes of a test meter. Remember, you cannot see even the highest voltage, and assuming that a circuit is dead can be very hazardous, maybe even deadly.

Do not simulate an electrical action, for example by pressing down a relay armature, within an electrical panel without a positive understanding of the circuitry. Your innocent action may cause an electrical "explosion" to shower you with molten metal, or startle you into a bumped head or elbow, or cause a bad fall. Remember, *WHEN IN DOUBT—DO NOT* when it comes to electricity.

Usually, a plant operator does not have the test equipment or the technical knowledge to correct an electrical malfunction, other than possibly resetting a circuit breaker, regardless of how critical the device's function is to plant operations. In addition to the shock hazards of motor control centers, there also may be a shock hazard within measurement instrument cases. Expensive instrument components can easily be damaged or destroyed when a foolhardy operator uses the tool-touch-test method.

Most panels have an *INTERLOCK* on the door that interrupts all (local) power to a panel or device when the door is opened or the circuit is exposed for service. Never disconnect or disable interlocks that interrupt control power. Warning labels, insulating covers (over "hot" terminals), safety switches, lockouts, and other safety provisions on electrical equipment must remain functional at all times. Your attention to this crucial aspect of your workplace may save a life, and, as the saying goes, it could be your own.

Operators often use hand power tools around electrical equipment. All such power tools (even double-insulated ones) can present a shock hazard, treatment plants being damp places, at times. Power tools can be a mechanical hazard as well. For your own safety, use power tools only when you can have an observer

on hand in case of an accident. *NEVER* stand in water with a power tool, even when it is turned off. Brace yourself, if necessary, in such a way that electric current cannot flow from arm to arm in case of a faulty power tool. Shocks through the upper body could involve your heart or your head, whose importance to you is self-evident.

QUESTIONS

Write your answers in a notebook and then compare your answers with those on page 680.

9.1A   What are the general principles for safe performance on the job?

9.1B   How can electric shock cause death?

9.1C   What could happen to you as a result of an electrical "explosion"?

9.1D   Why should you brace yourself when operating power equipment so that electric current cannot flow from arm to arm in case of a faulty tool?

9.12   Mechanical and Pneumatic Hazards

There exists a special danger when working around powered mechanical equipment, such as electric motors, valve operators, and chemical feeders that are operated remotely or by an automatic control system. Directly stated, the machinery may suddenly start or move when you are not expecting it. Most devices are powered by motors with enough torque or RPM to severely injure anyone in contact with a moving part. Even when the exposed rotating or meshing elements are fitted with guards in compliance with safety regulations, a danger may exist. A motor started remotely may catch a shirt sleeve, finger, or tool hanging near a loose or poorly fitted shaft or gear train guard.

The sudden automatic operation of equipment, even when half-expected, may startle a nearby operator into a fall or slip. Signs indicating that "This Equipment May Start At Any Time" tend to be ignored after a while. Accordingly, you must stay alert to the fact that any automatic device may begin to operate at any time. Thus, you must stay well clear of active automatic equipment, especially when it is not operating.

Lockout devices on electric switches must be respected at all times. The electrician who attaches a lockout device to physically prevent the operation of an electric circuit is, in effect, trusting his or her life and health to the device. Once the lockout device is attached to the switch (whether the switch is tagged off or actually locked with lock and key), the electrician will consider the circuit and its connected equipment de-energized and safe and will feel free to work on it. Consider the potential consequences, then, of a careless operator who removes a lockout to place needed equipment back into service, *presuming* the electrician is finished (as might occur after several hours). The point cannot be overstressed:

RESPECT ALL LOCKOUT DEVICES AND ALL TAGGED-OFF EQUIPMENT AS IF A LIFE IS ENTRUSTED TO YOU ... IT MAY WELL BE.

PNEUMATIC HAZARDS

Working with and around pneumatic instrumentation presents an additional hazard associated with high-pressure air. Pneumatic devices are powered by air at supply pressures from about 50 to 100 psi (350 to 700 kPa). Serious injury can be caused by the high air (and trapped particle) velocities that can be produced. A pressure regulator normally reduces the high air pressures to a safer 30 psi (210 kPa) or so for each device, but even these lower pressures can cause injury if directed toward the eye or other delicate tissues.

Before disconnecting any air line, put on your safety goggles (not just glasses, but wraparound goggles). Then valve the line off, from both directions ideally, and crack (open slightly) a fitting on the pipe or tubing to permit the pressurized air to bleed out slowly. If it is necessary to purge a line of moisture after inspection or repair, do so through the filter trap valve (pointing down, if standard) on each of a supply line's connected components. If a very dirty, oily, or moist line must be purged with supply air, set up a temporary purge line to a low area and tape a sock or closed rag to the open end to trap particles and minimize dirt pickup by the exhausting air. *NEVER* direct a pressurized air stream toward any part of your body or anyone else's body. High-velocity air can easily penetrate the body's tissues, even the skin of the hands. The tiny particles that are always present, even in filtered air (picked up from the inside of piping), can enter the eye's delicate tissues and cause an irritation at the least, or corneal damage and infection at worst. Breathing oil-laden air (many compressors seal with oil) from extensive line and filter purging can cause "chemical pneumonia" or even bacterial lung infections in sensitive individuals. Wearing a simple painter's (filter) mask will keep the suspended oil droplets and other particulates out of your respiratory system.

Though it is recognized that plant operators do little, if any, corrective maintenance around instrumentation and control components, it seems there is always some exposure to electrical, mechanical, and pneumatic hazards. Operators often do preventive maintenance tasks, and even minor fixing of many devices. Therefore, always play it smart and safe around all instrumentation.

9.13 Confined Spaces

Many measurement and control systems include remotely installed sensors and control valves. Quite often, these are found in vaults or other closed concrete structures. This section outlines procedures for preventing personal exposure to dangerous air contamination and/or oxygen deficiency/enrichment when working within such spaces. If you enter confined spaces, you must develop and implement written, understandable procedures in compliance with OSHA standards and you must provide training in the use of these procedures for all persons whose duties may involve confined space entry. *The procedures presented here are intended as guidelines. Exact procedures for work in confined spaces may vary with different agencies and geographical locations and must be confirmed with the appropriate regulatory safety agency.*

A confined space may be defined as any space that: (1) is large enough and so configured that an employee can bodily enter and perform assigned work; and (2) has limited or restricted means for entry or exit (for example, manholes, tanks, vessels, silos, storage bins, hoppers, vaults, and pits are spaces that may have limited means of entry); and (3) is not designed for continuous employee occupancy. One easy way to identify a confined space is by whether or not you can enter it by simply walking while standing fully upright. In general, if you must duck, crawl, climb, or squeeze into the space, it is considered a confined space.

A major concern in confined spaces is whether the existing ventilation is capable of removing *DANGEROUS AIR CONTAMINATION*[5] and/or oxygen deficiency/enrichment that

---

[5] *Dangerous Air Contamination.* An atmosphere presenting a threat of causing death, injury, acute illness, or disablement due to the presence of flammable and/or explosive, toxic, or otherwise injurious or incapacitating substances.

    (1) Dangerous air contamination due to the flammability of a gas, vapor, or mist is defined as an atmosphere containing the gas, vapor, or mist at a concentration greater than 10 percent of its lower explosive (lower flammable) limit (LEL).

    (2) Dangerous air contamination due to a combustible particulate is defined as a concentration that meets or exceeds the particulate's lower explosive limit (LEL).

    (3) Dangerous air contamination due to the toxicity of a substance is defined as the atmospheric concentration that could result in employee exposure in excess of the substance's permissible exposure limit (PEL).

    NOTE: A dangerous situation also occurs when the oxygen level is less than 19.5 percent by volume (OXYGEN DEFICIENCY) or more than 23.5 percent by volume (OXYGEN ENRICHMENT).

may exist or develop. In wastewater treatment, we are concerned primarily with oxygen deficiency (less than 19.5 percent oxygen by volume), oxygen enrichment (greater than 23.5 percent oxygen by volume), methane (explosive), hydrogen sulfide (toxic), and other gases as identified in Table 9.1.

The potential for buildup of toxic or explosive gas mixtures and/or oxygen deficiency/enrichment exists in all confined spaces. The atmosphere must be checked with reliable, calibrated instruments before every entry. When testing the atmosphere, first test for oxygen deficiency/enrichment, then combustible gases and vapors, and then toxic gases and vapors. The oxygen concentration in normal breathing air is 20.9 percent. The atmosphere in the confined space must not fall below 19.5 percent or exceed 23.5 percent oxygen. Engineering controls are required to prevent low or high oxygen levels. However, personal protective equipment is necessary if engineering controls are not possible. In atmospheres where the oxygen content is less than 19.5 percent, supplied air or self-contained breathing apparatus (SCBA) is required. SCBAs are sometimes referred to as scuba gear because they look and work much like the air tanks used by divers.

Entry into confined spaces is never permitted until the space has been properly ventilated using specially designed forced-air ventilators. These blowers force all the existing air out of the space, replacing it with fresh air from outside. This crucial step must *ALWAYS* be taken even if atmospheric monitoring instruments show the atmosphere to be safe. Because some of the gases likely to be encountered in a confined space are combustible or explosive, the blowers must be specially designed so that the blower itself will not create a source of ignition that could cause an explosion.

There are two general classifications of confined spaces: (1) non-permit confined spaces, and (2) permit-required confined spaces (permit spaces).

A *NON-PERMIT CONFINED SPACE* is a confined space that does not contain or, with respect to atmospheric hazards, have the potential to contain any hazard capable of causing death or serious physical harm. The following steps are recommended *PRIOR* to entry into *ANY* confined space:

1. Ensure that all employees involved in confined space work have been effectively trained.

2. Identify and close off or reroute any lines that may carry harmful substance(s) to, or through, the work area.

3. Empty, flush, or purge the space of any harmful substance(s) to the extent possible.

4. Monitor the atmosphere at the work site and within the space to determine if dangerous air contamination and/or oxygen deficiency/enrichment exists.

5. Record the atmospheric test results and keep them at the site throughout the work period.

6. If the space is interconnected with another space, each space must be tested and the most hazardous conditions found must govern subsequent steps for entry into the space.

7. If an atmospheric hazard is noted, use portable blowers to further ventilate the area; retest the atmosphere after a suitable period of time. Do not place the blowers inside the confined space.

8. If the *ONLY* hazard posed by the space is an actual or potential hazardous atmosphere and the preliminary ventilation has eliminated the atmospheric hazard or continuous forced ventilation *ALONE* can maintain the space safe for entry, entry into the area may proceed.

A *PERMIT-REQUIRED CONFINED SPACE* (permit space) is a confined space that has one or more of the following characteristics:

1. Contains or has the potential to contain a hazardous atmosphere

2. Contains a material that has the potential for engulfing an entrant

3. Has an internal configuration such that an entrant could be trapped or asphyxiated by inwardly converging walls or by a floor that slopes downward and tapers to a smaller cross section

4. Contains any other recognized serious safety or health hazard

OSHA regulations require that a confined space entry permit be completed for each permit-required confined space entry (Figure 9.6). The permit must be renewed each time the space is left and re-entered, even if only for a break or lunch, or to go get a tool. The confined space entry permit is "an authorization and approval in writing that specifies the location and type of work to be done, certifies that all existing hazards have been evaluated by a competent person, and that necessary protective measures have been taken to ensure the safety of each worker." A competent person, in this case, is a person designated in writing as capable, either through education or specialized training, of anticipating, recognizing, and evaluating employee exposure to hazardous substances or other unsafe conditions in a confined space. This person is authorized to specify control procedures and protective actions necessary to ensure worker safety.

The following procedures must be observed before entry into a permit-required confined space:

1. Ensure that personnel are effectively trained.

2. If the confined space has both side and top openings, enter through the side opening if it is within 3½ feet (1.1 meters) of the bottom.

3. Wear appropriate, approved, respiratory protective equipment.

4. Ensure that written operating and rescue procedures are at the entry site.

5. Wear an approved harness with an attached line. The free end of the line must be secured outside the entry point.

6. Test for atmospheric hazards as often as necessary to determine that acceptable entry conditions are being maintained.

TABLE 9.1   COMMON DANGEROUS GASES ENCOUNTERED IN WASTEWATER COLLECTION SYSTEMS AND AT WASTEWATER TREATMENT PLANTS[a]

Name of Gas and Chemical Formula	8TWA PEL[b]	Specific Gravity or Vapor Density[c] (Air = 1)	Explosive Range (% by volume in air) Lower Limit	Explosive Range (% by volume in air) Upper Limit	Common Properties (Percentages below are percent in air by volume)	Physiological Effects (Percentages below are percent in air by volume)	Most Common Sources in Sewers	Method of Testing[d]
Oxygen, O_2 (in Air)		1.11	Not flammable		Colorless, odorless, tasteless, nonpoisonous gas. Supports combustion.	Normal air contains 20.93% of O_2. If O_2 is less than 19.5%, do not enter space without respiratory protection.	Oxygen depletion from poor ventilation and absorption or chemical consumption of available O_2.	Oxygen monitor.
Gasoline Vapor, C_5H_{12} to C_9H_{20}	300	3.0 to 4.0	1.3	7.0	Colorless, odor noticeable in 0.03%. Flammable. Explosive.	Anesthetic effects when inhaled. 2.43% rapidly fatal. 1.1% to 2.2% dangerous for even short exposure.	Leaking storage tanks, discharges from garages, and commercial or home dry-cleaning operations.	Combustible gas monitor.
Carbon Monoxide, CO	50	0.97	12.5	74.2	Colorless, odorless, nonirritating. Tasteless, Flammable. Explosive.	Hemoglobin of blood has strong affinity for gas causing oxygen starvation. 0.2 to 0.25% causes unconsciousness in 30 minutes.	Manufactured fuel gas.	1. CO monitor. 2. CO tubes.
Hydrogen, H_2		0.07	4.0	74.2	Colorless, odorless, tasteless, nonpoisonous, flammable. Explosive. Propagates flame rapidly; very dangerous.	Acts mechanically to deprive tissues of oxygen. Does not support life. A simple asphyxiant.	Manufactured fuel gas.	Combustible gas monitor.
Methane, CH_4		0.55	5.0	15.0	Colorless, tasteless, odorless, nonpoisonous. Flammable. Explosive.	See hydrogen.	Natural gas, marsh gas, manufactured fuel gas, gas found in sewers.	Combustible gas monitor.
Hydrogen Sulfide, H_2S	10	1.19	4.3	46.0	Rotten egg odor in small concentrations, but sense of smell rapidly impaired. Odor not evident at high concentrations. Colorless. Flammable. Explosive. Poisonous.	Death in a few minutes at 0.2%. Paralyzes respiratory center.	Petroleum fumes, from blasting, gas found in sewers.	1. H_2S monitor. 2. H_2S tubes.
Carbon Dioxide, CO_2	5,000	1.53	Not flammable		Colorless, odorless, nonflammable. Not generally present in dangerous amounts unless there is already a deficiency of oxygen.	10% cannot be endured for more than a few minutes. Acts on nerves of respiration.	Issues from carbonaceous strata. Gas found in sewers.	Carbon dioxide monitor.
Ethane, C_2H_4		1.05	3.1	15.0	Colorless, tasteless, odorless, nonpoisonous. Flammable. Explosive.	See hydrogen.	Natural gas.	Combustible gas monitor.
Chlorine, Cl_2	0.5	2.5	Not flammable Not explosive		Greenish yellow gas, or amble color liquid under pressure. Highly irritating and penetrating odor. Highly corrosive in presence of moisture.	Respiratory irritant, irritating to eyes and mucous membranes. 30 ppm causes coughing. 40–60 ppm dangerous in 30 minutes. 1,000 ppm apt to be fatal in a few breaths.	Leaking pipe connections. Overdosage.	1. Chlorine monitor. 2. Strong ammonia on swab gives off white fumes.
Sulfur Dioxide, SO_2	2	2.3	Not flammable Not explosive		Colorless compressed liquified gas with a pungent odor. Highly corrosive in presence of moisture.	Respiratory irritant, irritating to eyes, skin, and mucous membranes.	Leaking pipes and connections.	1. Sulfur dioxide monitor. 2. Strong ammonia on swab gives off white fumes.

[a] Originally printed in Water and Sewage Works, August 1953. Adapted from "Manual of Instruction for Sewage Treatment Plant Operators," State of New York.

[b] 8TWA PEL is the Time Weighted Average permissible exposure limit, in parts per million, for a normal 8-hour workday and a 40-hour workweek to which nearly all workers may be repeatedly exposed, day after day, without adverse effect.

[c] Gases with a specific gravity less than 1.0 are lighter than air; those more than 1.0, heavier than air.

[d] The first method given is the preferable testing procedure.

Confined Space Pre-Entry Checklist/Confined Space Entry Permit

Date and Time Issued: _____ Date and Time Expires: _____ Job Site/Space I.D.: _____

Job Supervisor: _____ Equipment to be worked on: _____ Work to be performed: _____

Standby personnel: _____   _____   _____

1. Atmospheric Checks: Time _____ Oxygen _____ %   Toxic _____ ppm

   Explosive _____ % LEL Carbon Monoxide _____ ppm

2. Tester's signature: _____

3. Source isolation: (No Entry)   N/A   Yes   No

   Pumps or lines blinded,
      disconnected, or blocked   ( )   ( )   ( )

4. Ventilation Modification: N/A   Yes   No

   Mechanical   ( )   ( )   ( )

   Natural ventilation only   ( )   ( )   ( )

5. Atmospheric check after isolation and ventilation: Time _____

   Oxygen _____ % > 19.5% < 23.5%   Toxic _____ ppm < 10 ppm H_2S

   Explosive _____ % LEL   < 10%   Carbon Monoxide _____ ppm < 35 ppm CO

Tester's signature: _____

6. Communication procedures: _____

7. Rescue procedures: _____

   _____

8. Entry, standby, and backup persons   Yes   No

   Successfully completed required training?   ( )   ( )

   Is training current?   ( )   ( )

9. Equipment: N/A   Yes   No

   Direct reading gas monitor tested   ( )   ( )   ( )

   Safety harnesses and lifelines for entry and standby persons   ( )   ( )   ( )

   Hoisting equipment   ( )   ( )   ( )

   Powered communications   ( )   ( )   ( )

   SCBAs for entry and standby persons   ( )   ( )   ( )

   Protective clothing   ( )   ( )   ( )

   All electric equipment listed for Class I, Division I,
      Groups A, B, C, and D, and nonsparking tools   ( )   ( )   ( )

10. Periodic atmospheric tests:

   Oxygen: ___% Time ___;   ___% Time ___;   ___% Time ___;   ___% Time ___;

   Explosive: ___% Time ___;   ___% Time ___;   ___% Time ___;   ___% Time ___;

   Toxic: ___ppm Time ___;   ___ppm Time ___;   ___ppm Time ___;   ___ppm Time ___;

   Carbon Monoxide: ___ppm Time ___;   ___ppm Time ___;   ___ppm Time ___;   ___ppm Time ___;

We have reviewed the work authorized by this permit and the information contained herein. Written instructions and safety procedures have been received and are understood. Entry cannot be approved if any brackets ( ) are marked in the "No" column. This permit is not valid unless all appropriate items are completed.

Permit Prepared By: (Supervisor) _____   Approved By: Unit Supervisor) _____

Reviewed By: (CS Operations Personnel) _____

(Entrant)   (Attendant)   (Entry Supervisor)

This permit to be kept at job site. Return job site copy to Safety Office following job completion.

Fig. 9.6   Confined space pre-entry checklist/confined space entry permit

7. Station at least one person to stand by on the outside of the confined space and at least one additional person within sight or call of the standby person.

8. Maintain effective communication between the standby person and the entry person.

9. The standby person, equipped with appropriate respiratory protection, should only enter the confined space in case of emergency.

10. If the entry is made through a top opening, use a hoisting device with a harness that suspends a person in an upright position. A mechanical device must be available to retrieve personnel from vertical spaces more than five feet (1.5 meters) deep.

11. If the space contains, or is likely to develop, flammable or explosive atmospheric conditions, do not use any tools or equipment (including electrical) that may provide a source of ignition.

12. Wear appropriate protective clothing when entering a confined space that contains corrosive substances or other substances harmful to the skin.

13. At least one person trained in first aid and cardiopulmonary resuscitation (CPR) should be immediately available during any confined space job.

Individuals designated to provide first aid or CPR should be included in a Bloodborne Pathogens (BBP) program. These employees may be exposed to contact with blood or other potentially infectious materials from the performance of their duties. The BBP program includes training in exposure potential determination, engineering and work practice controls, personal protective equipment (PPE), and the availability of the hepatitis B vaccination series (29 CFR 1910.1030). If the operators must enter confined spaces to perform rescue services, they must be trained specifically to perform the assigned rescue duty and to use required personal protective equipment (PPE) and rescue equipment. Rescue practice sessions must be held at least once every 12 months.

If you arrange to have a contractor perform work in confined spaces at your facility or within your collection system, you must inform the contractor:

• That the contractor must comply with confined space regulations

• Of hazards that you have identified and your experience with the space(s)

• Of precautions or procedures you have implemented for the protection of employees in or near the space where the contractor's personnel will be working

• That a debriefing must occur at the conclusion of the entry operations regarding the confined space program followed and any hazards encountered or created during the entry operations

To enhance safety, communications, and coordination of confined space activities, the contractor is also required to obtain available information from you and to inform you of the confined space program the contractor will follow. This exchange of information must occur before the confined space is entered by any operator.

Confined space work can present serious hazards if you are uninformed or untrained. The procedures presented are only guidelines and exact requirements for confined space work for your locale may vary. Contact your local regulatory safety agency for specific requirements.

9.14 Oxygen Deficiency or Enrichment

Low oxygen levels may exist in any poorly ventilated, low-lying structure where gases such as hydrogen sulfide, gasoline vapor, carbon dioxide, or chlorine may be produced or may accumulate (see Table 9.1, "Common Dangerous Gases Encountered in Wastewater Collection Systems and at Wastewater Treatment Plants"). Oxygen in a concentration above 23.5 percent (oxygen enrichment) also can be dangerous because it speeds up combustion.

Oxygen deficiency is most likely to occur when structures or channels are installed below grade (ground level). Several gases (including hydrogen sulfide and chlorine) have a tendency to collect in low places because they are heavier than air. The specific gravity of a gas indicates its weight as compared to an equal volume of air. Since air has a specific gravity of exactly 1.0, any gas with a specific gravity greater than 1.0 may sink to low-lying areas and displace the air from that area or structure. (On the other hand, methane may rise out of a manhole because it has a specific gravity of less than 1.0, which means that it is lighter than air.) You should never rely solely on the specific gravity of a gas to tell you where it is. Air movement or temperature differences within a confined space may affect the location of atmospheric hazards. The only effective way of ensuring safe atmospheric conditions prior to entering a confined space is to test the atmosphere with an appropriate monitor(s) at various levels and locations throughout the space.

When oxygen deficiency or enrichment is discovered, the area should be ventilated with fans or blowers and checked again for oxygen deficiency/enrichment before anyone enters the area to work. Ventilation may be provided by fans or blowers. Follow confined space procedures before entering and during occupancy of any suspect area. *ALWAYS* get air into the confined space *BEFORE* you enter to work and maintain the ventilation until you have left the space. Equipment is available to measure oxygen concentration as well as toxic and combustible atmospheric conditions. You must use this equipment whenever you encounter a potential confined space situation. Ask your local safety regulatory agency or wastewater association about sources of this type of equipment in your area.

9.15 Explosive Gas Mixtures

Explosive gas mixtures may develop in many areas of a treatment facility from mixtures of air and methane, natural gas, manufactured fuel gas, hydrogen, or gasoline vapors. Table 9.1 lists the common dangerous gases that may be encountered in a treatment facility and identifies their explosive range where appropriate. The upper explosive limit (UEL) and lower explosive limit (LEL) indicate the range of concentrations at which combustible or explosive gases will ignite when an ignition source is present at ambient temperature. No explosion or ignition occurs when the concentration is outside these ranges. Gas concentrations below the LEL are too lean to ignite; there is not enough flammable gas or vapor to support combustion. Gas concentrations higher than the UEL are too rich to ignite; there is too much flammable gas or vapor and not enough oxygen to support combustion (Figure 9.7).

Explosive ranges can be measured by using a combustible gas detector calibrated for the gas of concern. Do not rely on your nose to detect gases. The sense of smell is absolutely unreliable for evaluating the presence of dangerous gases. Some gases have no smell and hydrogen sulfide can paralyze the sense of smell.

Avoid explosions by eliminating all sources of ignition in areas potentially capable of developing explosive mixtures. Only explosion-proof electrical equipment and fixtures should be used in these areas (influent/bar screen rooms, gas compressor areas, digesters, battery charging stations). Provide adequate ventilation in all areas that have the potential to develop an explosive atmosphere.

The National Fire Protection Association Standard 820 (NFPA 820), *FIRE PROTECTION IN WASTEWATER TREATMENT AND COLLECTION FACILITIES,* lists requirements for electrical classifications, ventilation, gas detection, and fire control methods in various wastewater treatment and collection system areas. (For ordering information, see Section 9.5, "Additional Reading," item 5.) Comparing these requirements

with your plant's existing design and equipment may indicate deficiencies, which should be remedied to minimize potential hazards.

9.16   Falls and Associated Hazards

Falls from or into tanks, wet wells, catwalks, or conveyors can be disabling or deadly. Most of these can be avoided by the proper use of ladders, hand tools, and safety equipment, and by following established safety procedures. Strains and sprains are probably the most frequent injuries in wastewater treatment facilities. Keep in mind that an electric shock of even minor intensity can result in a serious fall. When working above ground on a ladder, use the proper, nonconductive type of ladder (fiberglass rails) and position it safely. Even an alert, safety-conscious operator could be knocked off balance by a slight electric shock. When required to do preventive maintenance from a ladder, turn off the power to the equipment being serviced, if at all possible. If this is not feasible, take special care to stay out of contact with any component inside the enclosure of an operating mechanism, and well away from terminal strips, unconduited wiring, and electrical black boxes. Though not commonly considered essential, wearing thin rubber or plastic gloves can greatly reduce your chances of electric shock (whether on a ladder or off).

Make provisions for carrying tools or other required objects on an electrician's belt rather than in your hands when climbing up or down ladders. Finally, never leave tools or any object on a step or platform of the ladder when you climb down, even temporarily. You might be the one upon whom they fall if the ladder is moved or even steadied from below. In this regard, it is always a good idea (even if not required) for preventive maintenance personnel to wear a hard hat whenever working on or near equipment, especially when a ladder must be used.

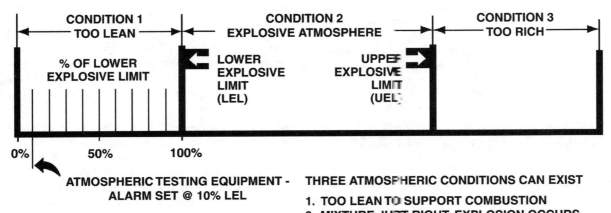

Fig. 9.7   Relationship between the lower explosive limit (LEL) and the upper explosive limit (UEL) of a mixture of air and gas

9.1E   Why should operators be especially careful when working around powered mechanical equipment?

9.1F   What is the purpose of an electrical lockout device?

9.1G   When testing the atmosphere before entry in any confined space, what procedure should be used?

9.1H   What do the initials UEL and LEL stand for?

9.1I   What kind of specific protective clothing could be worn to protect you from electric shock?

9.2 MEASURED VARIABLES AND TYPES OF SENSORS

9.20 General Principles of Sensors

A measured variable is any factor that is sensed and quantified by the primary element or sensor. In wastewater treatment, the variables of pressure, level, and flow are the most common ones measured; at times, chemical feed rates and some physical or chemical process characteristics are also sensed.

The primary element performs the initial conversion of measurement energy to a signal. An example is a pressure gauge that converts the hydraulic action of the water into a mechanical motion to drive a meter indicator, or produces an electrical signal for a remote readout device. For transmitters not having a local readout, the sensing portion is the primary element. If such a signal is produced, be it electric or pneumatic, the sensor is also then considered a transmitter.

The signal produced is not necessarily a continuous one proportional to the variable (that is, an analog signal), but often merely a switch set to detect when the variable goes above or below preset limits. In this type of ON/OFF control, the predetermined setting is called a DISCRETE[6] point. This distinction between continuous and ON/OFF operation relates to the two types of controllers discussed previously.

Process measuring instruments have a common goal: to detect a variable and create an output for use by the control system or the operator. There are two basic types of instrument signals: (1) continuously variable analog, and (2) discrete ON/OFF. Signals to and from a control device are termed as input/output (I/O). The signals can further be classified as analog input (AI), analog output (AO), DISCRETE INPUT (DI), and DISCRETE OUTPUT (DO).[7] Computer control system inputs are converted from the analog and discrete signals and coded into digital-data representations of the inputs. Digital-data output from the computer control system is converted to analog and discrete outputs.

9.21 Pressure Measurements

Since pressure is defined as a force per unit of area (pound per square inch, or kiloPascal), you might expect that sensing pressure would thus entail the small movement of some flexible element subjected to a force. In fact, that is how pressure is measured in practice. There are many classes and brands of sensors, but the most common types contain mechanically deformable components such as the Bourdon tube (Figures 9.8), bellows, or diaphragm arrangements (Figure 9.9). The slight motion each exhibits, directly proportional to the applied force, is then amplified mechanically by levers or gears to position a pointer on a scale or provide an input for an associated transmitter. A blind transmitter for pressure, or any variable, has no local readout.

Some pressure sensors are fitted with surge or overrange protection (snubbers) to limit the effect pressure spikes or water hammer have on the instrument. In most cases, such protection devices function by restricting flow into the sensing element. Surge protection devices thus prevent sudden pressure surges, which can easily damage most pressure sensors.

A snubber (Figure 9.10) consists of a restriction through which the pressure-producing fluid must flow. A more elaborate mechanical snubber responds to surges by moving a piston or plunger that effectively controls the size of an orifice. Some snubbers are subject to clogging or being adjusted so tight as to prevent any response at all to pressure changes. If a pressure sensor is not performing properly, look first for such clogging or adjustment that has become too restrictive.

Another dampening device is an air cushion chamber (Figure 9.11), which is simply constructed yet very effective. The top part of the chamber contains air; water flows into the bottom part. A sudden change in water pressure compresses the air within the chamber, taking the shock. The rate of response can be further dampened by also placing a snubber into the air chamber line.

9.2A   What is a primary element or sensor?

9.2B   How is pressure measured?

9.2C   Why are some pressure sensors fitted with surge or overrange protection?

9.22 Level Measurements

Systems for sensing the level of water or any other liquid, either continuously or at a single point, are very common in municipal and industrial wastewater treatment plants. Pumps are controlled, filters operated, basins and tanks filled and

---

[6] Discrete Control.   ON/OFF control; one of the two output values is equal to zero.
[7] Discrete I/O (Input/Output).   A digital signal that senses or sends either ON or OFF signals. For example, a discrete input would sense the position of a switch; a discrete output would turn on a pump or light.

emptied, chemicals fed and ordered, sumps emptied, and many other variables controlled on the basis of liquid levels. Fortunately, level sensors usually are simple devices. A float, for example, can be a very reliable liquid level sensor both for single-point and continuous-level sensing. Other types of liquid level-sensing devices include displacers, electrical probes, direct hydrostatic pressure, pneumatic bubbler tubes, and ultrasonic (sonar) devices.

Single-point detection of level is very common for levels controlled by pumps or valves within fairly wide limits. Single-point

ball float levels are very simple devices and have proven reliable in applications where set-point changes are rarely necessary. Single-point detectors mount in a sump or tank to sense a fixed elevation. When the water level rises, the ball floats and rolls onto its side actuating an internal mercury switch. The switch provides a discrete ON/OFF signal used to control equipment or send level data to the control system.

An alternative to a float for single-point sensing, is the use of a probe to sense liquid level (Figure 9.12). Only single-point determination can be made this way, though several probes or

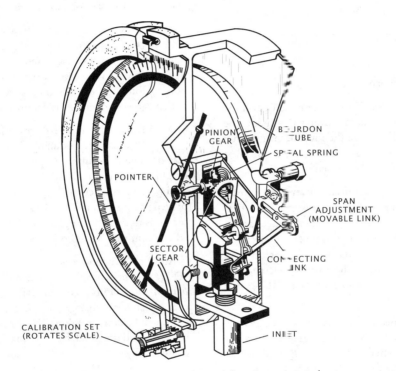

A precision industrial pressure gauge with a Bourdon-tube pressure element.

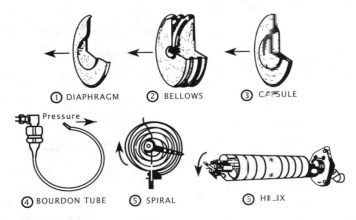

Elastic deformation elements. Pressure tends to expand or unroll elements to indicate pressure as shown by arrows.

Fig. 9.8   Bourdon tube and other pressure-sensing elements

(Permission of Heise Gauge, Dresser Industries)

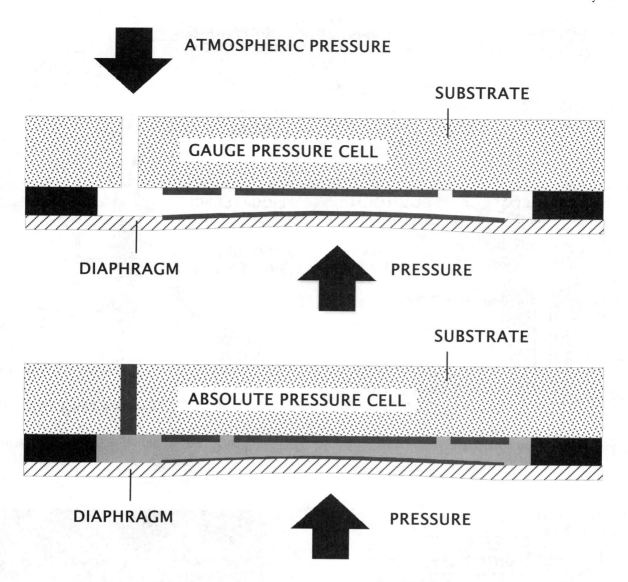

Fig. 9.9   Ceramic diaphragms to measure atmospheric pressure

one with several sensors along its length can be set up to detect several different levels. Level probes are used where a mechanical system is impractical, such as within sealed or pressurized tanks or with chemically active liquids.

The probes can be small-diameter rods of electrically conductive material inserted into a tank through a fitting, usually through the top but at times in the side of a vessel (Figure 9.13). Each rod is cut to length corresponding to a specific liquid level in the case of the top-entering probes; in the side-entering setup, a short rod merely enters the vessel at the appropriate height or depth. One problem encountered with probes is the accumulation of scum or caking (such as grease) on the surface of the rods.

A number of methods are used to detect whether the probe is immersed in liquid. A simple method for conductive liquids is to apply a small voltage to the probe(s) by the system's power supply, with current flowing only when the probe just becomes immersed in the liquid. When immersion is detected, a switch activates a pump/valve control or alarm at as many places as necessary through a discrete output (DO) from the control system. Though at times only a single probe is used, with the metal tank completing the circuit as a ground, usually at least two probes are found. The ground probe extends to within a short distance of the bottom of the tank so as to be in constant contact with the electrically conducting liquid (a liquid ground, as it were). Nonconductive liquids must use other methods to detect immersion of the probe in the liquid.

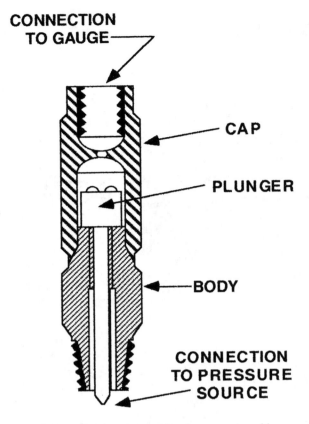

Fig. 9.10   *Internal parts of a plunger-type snubber for surge protection*

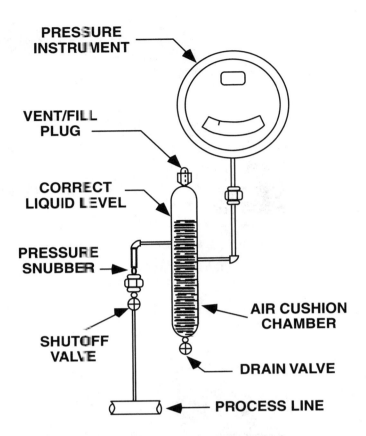

Fig. 9.11   *Air cushion chamber and snubber for surge protection of pressure gauge*

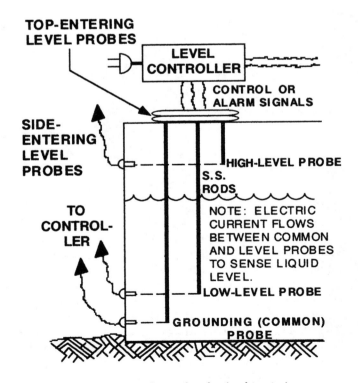

Fig. 9.12   *Electrical probes (multi-point)*

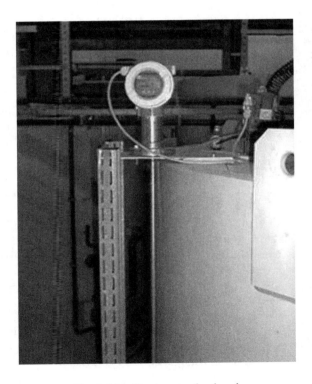

Fig. 9.13   *Single-point level probe*
(Courtesy Endress+Hauser)

Levels can be sensed continuously by measurement of liquid hydrostatic pressure near the bottom of a vessel or basin. The pressure elements used for level sensing must be quite sensitive to the low pressures created by liquid level (23 feet of water column equals only 10 psi, or 70 kPa). Therefore, simple off-the-shelf pressure gauges such as are found on pumps are not used to measure water levels. Instead, very sensitive water level sensors are used to measure levels of filter basins or in process tanks where control or monitoring must be close, continuous, and positive. Rather than being calibrated in units of pressure (psi or kPa), these gauges read directly in units of liquid level (feet or meters). One or more single-point control/alarm contacts can be made a part of this, or any, continuous type of level-sensing system.

Ultrasonic, microwave, and TDR (Time Domain Reflectometry) level sensors have proven very reliable for a wide variety of applications. Typically, the instruments in this group detect level by emitting a pulse that is reflected back to the receiver. The sensor periodically generates a pulse of waves that bounce off the liquid surface and reflect back. The sensor detects the reflected pulse. Based on the speed of the pulse, the travel time between sending and receiving is measured and the distance is calculated and converted into a level measurement. The instruments have many display options: digital readout, analog 4–20 mA, discrete ON/OFF, and digital data. Any combination of one or more may be available on a single device. They are very versatile and reliable and can be used to detect levels of liquid and solid materials.

A very precise method of measuring liquid level is the bubbler tube with its associated pneumatic instrumentation (Figure 9.14). The pressure created by the liquid level is sensed, but not directly as with a liquid pressure element. A bubbler measures the level of a liquid by sensing the air pressure necessary to cause bubbles to just flow out the end of a submerged tube. Air pressure is created in a bubbler tube to just match the pressure applied by the liquid above the open end of the tube when it is immersed to a precisely determined depth in a tank or basin. The air pressure in the tube is then measured as proportional to the liquid level *above* the end of the tube. This indirect determination of level using air permits the placement of the instrumentation anywhere above or below the liquid's surface, whereas direct pressure-to-level gauges must be installed at the very point (or very close to) where liquid pressure must be sensed.

Bubbler systems are adjusted so air *just begins* to bubble slowly out of the submerged end of the sensing tube. They automatically compensate for changes in liquid level by providing a small, constant flow of air through the bubbler tube by means of the constant air flow (also called constant differential) regulator. There is no advantage to turning up the amount of air to create more intense bubbling because the back pressure will still mainly depend on the water level. In fact, increasing the air flow may create a sizeable measuring error in the system, so any air flow changes should be left to qualified instrument service personnel.

Bubbler tube systems are common in basin-level controllers that must maintain water levels within a range of a few inches (centimeters) and for measurement of flow in open channels or over weirs. Usually, the level transmitter for basin-level controllers is blind since it only controls liquid level and need not provide a local readout of the level.

QUESTIONS

Write your answers in a notebook and then compare your answers with those on page 681.

9.2D   List the different types of liquid level-sensing devices.

9.2E   How does a single-point ball float level generate a level signal?

9.2F   Under what circumstances are probes used to measure liquid level?

9.2G   How does an ultrasonic sensor measure the level of a liquid?

9.23   Flow (Rate of Flow and Total Flow)

The two basic types of flow readings are *RATE OF FLOW,* such as MGD, CFS, and GPM or m^3/sec or liters/sec (volume per unit of time), or *TOTAL FLOW,* in simple units of volume, such as the corresponding million gallons, cubic feet, or gallons (liters or cubic meters). Total flow volumes are usually obtained as a running total, with a comparatively long time period for the flow delivery, such as a day or month. The flow instrument may

LEVEL
READOUT
(FEET)

AIR
ROTAMETER

CONTROL

INSTRUMENT
AIR SUPPLY

PRESSURE
REGULATOR
AND
FILTER

CONSTANT
AIR FLOW
REGULATOR

(BACK-)
PRESSURE
SENSOR

BUBBLER
TUBE

LOWEST
LEVEL
SENSED

PURGE VALVE*

*FOR APPLICATIONS WITH CLOGGING TENDENCY
(CHEMICALS, SLURRIES, WASTEWATER)

TANK OR BASIN
(ABOVE, BELOW, OR LEVEL
WITH INSTRUMENTATION)

FUNCTIONAL DIAGRAM

Constant-flow regulator and rotameter on left, back-pressure sensor (DP cell) to right.

Fig. 9.14   Diagram and photo of a bubbler tube system for measuring liquid level provide both the rate of flow and total flow values. Flow instruments applied to computer control systems typically only provide flow rate; total flow is calculated by accumulating the flow rate over some period, such as a day, month, or year.

While it is possible, in principle, to measure process flows directly, it is quite impractical. Direct measurement would involve the constant filling and emptying of, say, a gallon container with water flowing from a pipe on a timed basis. This method is obviously not practical. Therefore, sensing of process flows in waterworks practice is done inferentially, that is by inferring what the flow is from the observation of some associated hydraulic action or effect of the water. Figure 9.15 illustrates measuring level in a flume as a process flow instrument, inferring flow from the level measured. The inferential flow-sensing techniques that are used in flow measurement are: (1) velocity, (2) differential pressure, (3) magnetic, and (4) ultrasonic. First, let us look at one device, the rotameter, used in flow sensing for some specialized applications before studying the devices for process stream flows.

ROTAMETERS (Figures 9.16 and 9.17) are transparent (usually) tubes with a tapered bore containing a ball or float. The float rises up within the tube to a point corresponding to a particular rate of flow. The rotameter tube is set against, or has etched upon it, marks calibrated in whatever flow rate unit is appropriate. Rotameters are used to indicate approximate liquid or gas flow. For example, the readout for a gas chlorinator could be a rotameter. Sometimes a simple rotameter is installed merely to indicate a flow or no-flow condition in a pipe, for example, on a chlorinator injector supply line.

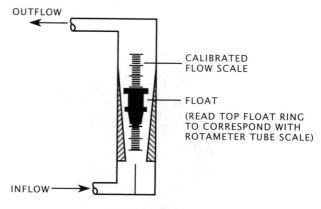

Fig. 9.16   Rotameter

For process flows, as mentioned, *VELOCITY-SENSING* meters measure water speed within a pipeline or channel. One way of doing this is by sensing the rate of rotation of a special impeller (Figure 9.18) placed within the flowing stream; the rate

LEVEL SENSOR MEASURES DEPTH TO INFER FLOW RATE

Fig. 9.15   *Inferring flow ultrasonically in a flume*
(Courtesy Endress+Hauser)

Rotameters (gas **rate**-of-flow)

Fig. 9.17   Flow-sensing devices for fluids

of flow is directly proportional to impeller RPM (within certain limits). Since normal water velocities in pipes and channels are under 10 feet per second (about 7 MPH or 3 m/sec), the impeller turns rather slowly. This rotary motion drives a train of gears, which indicates *RATE* of flow, in the same way a speedometer indicates the rate of travel for a car. *TOTAL* flow then appears as the cumulative number of revolutions, denoting total volume flowed, like the odometer on your car's speedometer indicates total mileage traveled.

Rotation of the velocity-sensing element is not always transferred by gears, but may be picked up as magnetic or electric pulses by a transducer. Nor is velocity always sensed mechanically; it may also be detected or measured purely electrically (the thermistor type), or hydraulically (the pitot tube). In each case, the principle of equating water velocity with rate of flow (within a constant flow area) is the same. Of course, all such flowmeters are calibrated to read out in an appropriate unit of flow rate, rather than velocity units.

Typically, a velocity-sensing element transmits its reading to a remote site as electric pulses, although other devices can be used in order to convert to any standard electrical or pneumatic signal. Figure 9.19 illustrates an electrical method of sensing velocity to measure flow.

Preventive maintenance of impeller-type flowmeters centers around regular lubrication of rotating parts, at least for the older types. Propeller meters, as they are called, have a long history of reliability and acceptable accuracy in both municipal and industrial applications. When propeller meters become old, they can become susceptible to underregistration (read low) due to bearing wear and gear train friction. Accordingly, annual teardown for inspection is indicated. Overregistration is rare, but a partially full pipeline, wrong gears installed, or a malfunctioning transmitter or receiver can cause high readings.

DIFFERENTIAL PRESSURE-SENSING TUBES (Figures 9.20 and 9.21), also called Venturi or differential meters (or flow tubes) depend for their operation upon a basic principle of

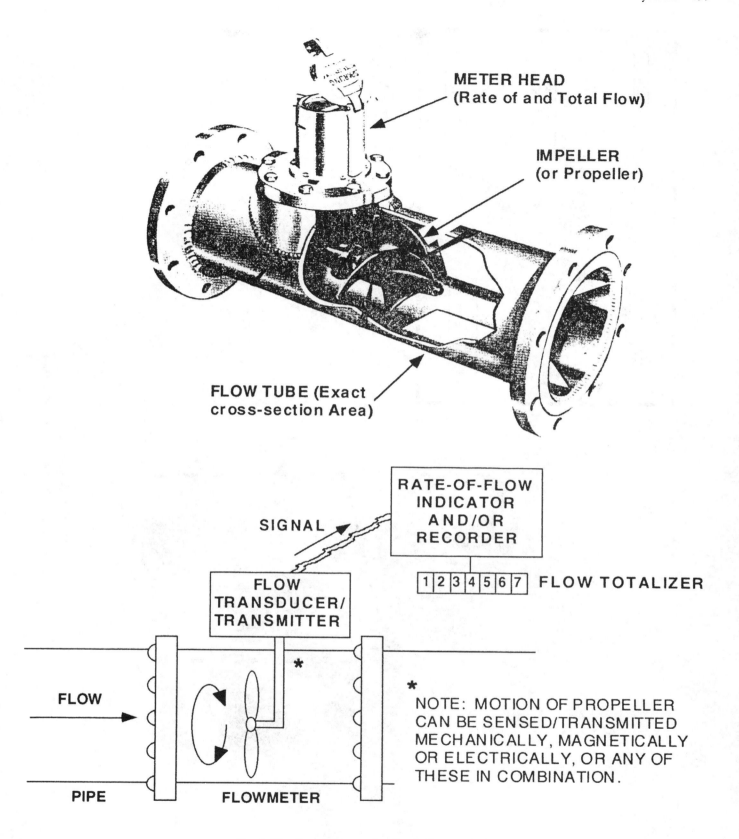

METER HEAD
(Rate of and Total Flow)

IMPELLER
(or Propeller)

FLOW TUBE (Exact
cross-section Area)

RATE-OF-FLOW
INDICATOR
AND/OR
RECORDER

SIGNAL

| 1 | 2 | 3 | 4 | 5 | 6 | 7 | **FLOW TOTALIZER**

FLOW
TRANSDUCER/
TRANSMITTER

*

FLOW

PIPE   FLOWMETER

*
NOTE: MOTION OF PROPELLER
CAN BE SENSED/TRANSMITTED
MECHANICALLY, MAGNETICALLY
OR ELECTRICALLY, OR ANY OF
THESE IN COMBINATION.

Fig. 9.18   Propeller meter (a type of velocity meter)

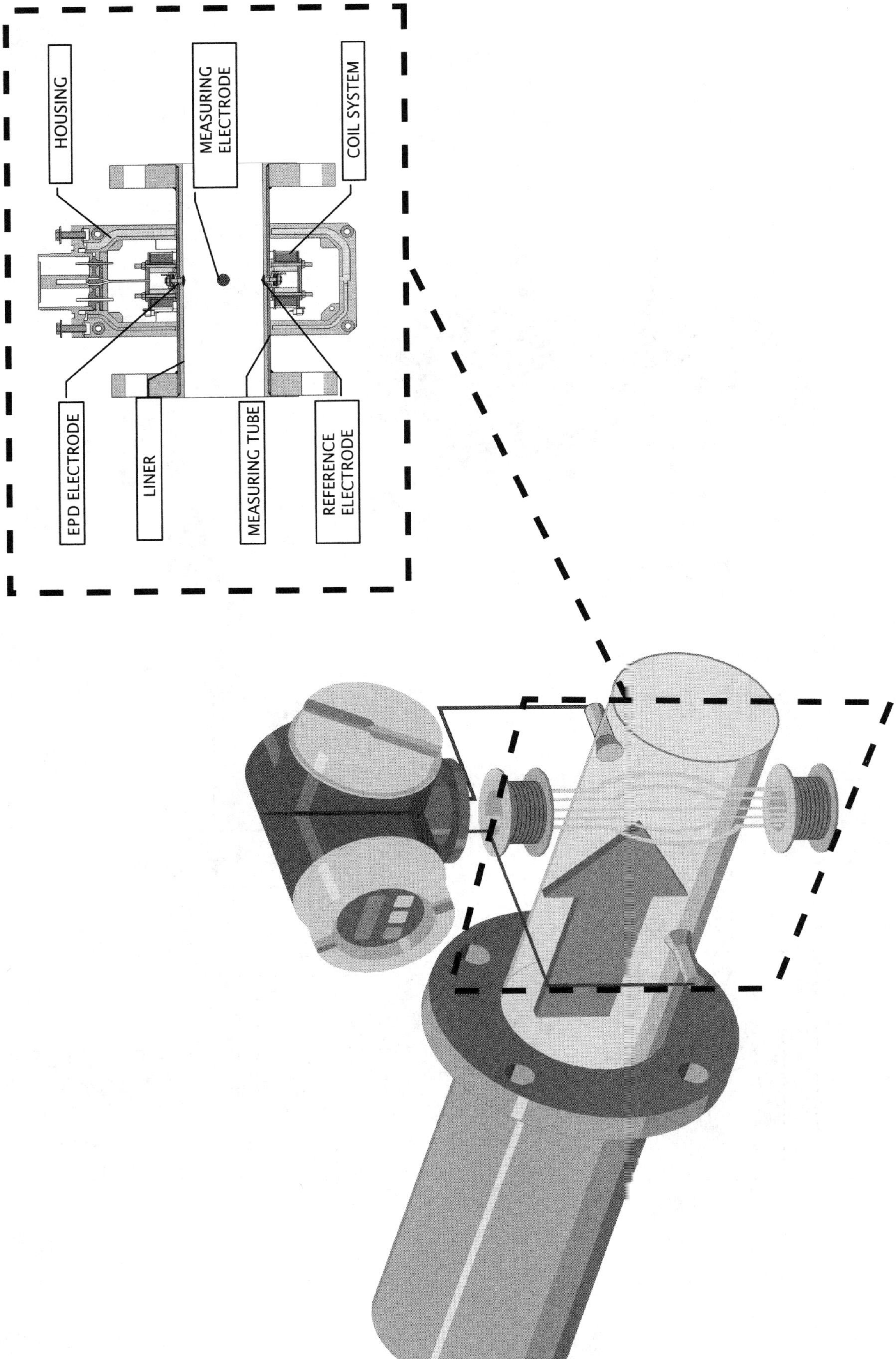

HOUSING

MEASURING ELECTRODE

COIL SYSTEM

EPD ELECTRODE

LINER

MEASURING TUBE

REFERENCE ELECTRODE

Fig. 9.19 Magnetic flowmeter
(Adapted from figure provided by courtesy of Endress+Hauser)

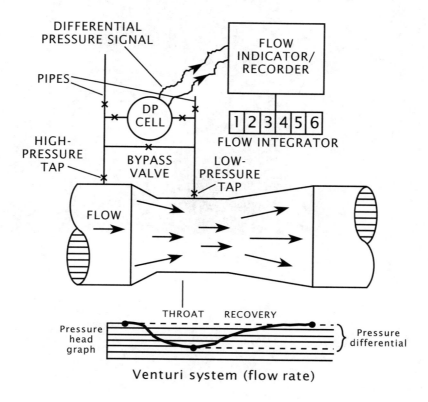

DIFFERENTIAL
PRESSURE SIGNAL

FLOW
INDICATOR/
RECORDER

PIPES

DP
CELL

| 1 | 2 | 3 | 4 | 5 | 6 |

FLOW INTEGRATOR

HIGH-
PRESSURE
TAP

BYPASS
VALVE

LOW-
PRESSURE
TAP

FLOW

THROAT   RECOVERY

Pressure
head
graph

Pressure
differential

Venturi system (flow rate)

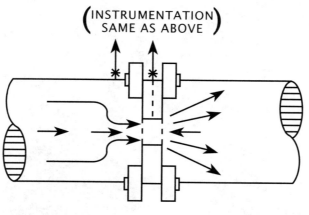

$$\left(\begin{array}{c}\text{INSTRUMENTATION}\\\text{SAME AS ABOVE}\end{array}\right)$$

Orifice plate installation (flow rate)

Fig. 9.20   Schematic diagrams of differential pressure flow measuring devices

54-inch Venturi tube

24-inch orifice plate

Fig. 9.21   Photos of differential pressure flow tubes

hydraulics, the Bernoulli Effect: When a liquid is forced to go faster in a pipe or channel, its internal pressure drops. If a carefully sized restriction is placed within the pipe or flow channel, the flowing water must speed up to get through it. In doing so, its pressure drops a little, and it drops an exact amount for a given flow rate. This small pressure drop, the "pressure differential," is the difference between the water pressure *before* the restriction and *within* the restriction. This difference is proportional in a certain way (but not directly proportional) to the rate of flow. The difference in pressure is measured very precisely by the instrumentation associated with the particular flow tube installed. Typically, a difference of only a few psi (kPa) is required. This small value of pressure difference is often described in inches (centimeters) of water (head).

Measuring flow by the differential pressure method removes a little hydraulic energy from the water. However, the modern flow tube, with its carefully tapered form, allows recovery of well over 95 percent of the original pressure throughout its range of flows. Other ways of constricting the flow, such as installation of orifice plates, do not allow such high recoveries of pressure or the accuracy possible with other modern flow tubes.

An orifice plate (Figures 9.20 and 9.21) is a steel plate with a precisely sized hole (orifice) in it. The plate is inserted between flanges in a pipe. The pressure drop is sensed right at the orifice, or immediately downstream, to yield a less accurate flow indication than a Venturi meter. This drop in pressure is not recovered; that is, a permanent pressure loss occurs with orifice plate installations, unlike Venturi flow tubes.

Differential devices require little, if any, preventive maintenance by the operator since there are no moving parts. Occasional flushing of the hydraulic sensing lines is good practice. However, flushing should only be done by a qualified person. When dealing with an instrument sensitive to fractions of a psi or pascals, opening the wrong valve can instantly damage the internal parts severely. Also, if an older DP (differential pressure) cell containing mercury is used, this toxic (and expensive) metal can easily be blown out of the device and into the process pipeline. Thus, all valve manipulations must be understood and done deliberately after careful planning by a qualified person.

In nearly all cases, the signals from larger flow tubes are transmitted to a remote readout station. Local readout is also provided (sometimes inside the case only) for purposes of calibration. Differential pressure transmitters may be electrical or pneumatic types. The signal transmitted is proportional to the square root of the differential pressure.

Venturi meters have been in use for many decades and can produce very close accuracies year after year. Older flow tubes are quite long physically (to yield maximum accuracy and pressure recovery). Newer units are much shorter but have very good pressure recovery and even better accuracy. With no moving parts, the Venturi-type meter is not subject to mechanical failure as is the comparable propeller meter. Flow tubes, however, must be kept internally clean and without obstructions upstream (and even downstream) to provide the designed accuracies.

The piping design for flowmeters must provide for adequate lengths of straight pipe runs upstream and downstream of the meter. Flowmeters in pipes will produce accurate flowmeter readings when the meter is located at least five pipe diameters distance downstream from any pipe bends, elbows, or valves and also at least two pipe diameters distance upstream from any pipe bends, elbows, or valves. Flowmeters also should be calibrated in place to ensure accurate flow measurements.

In summary, all of the flowmeters described in this section provide rate of flow indication. The rate of flow can also be (and usually is) continuously totalized to give a reading of total flow past the measuring point. Total flow is usually indicated at the readout instrument in units of gallons or cubic feet (liters or cubic meters).

QUESTIONS

Write your answers in a notebook and then compare your answers with those on page 681.

9.2H   What are two basic types of flow readings?

9.2I   List the inferential flow-sensing techniques that are used in flow measurement.

9.2J   How do velocity-sensing meters measure process flows?

9.2K   Flows measured with Venturi meters take advantage of what hydraulic principle?

9.24   Chemical Feed Rate

Chemical feed rate indicators are often an integral part of a particular chemical feed system and thus are usually not considered instrumentation as such. For example, a dry feeder for lime may be provided with an indicator for feed rate in units of weight per time, such as pounds/hour or grams/minute. In a fluid (liquid or gas) feeder, the indication of quantity per unit of time, such as gallons/hour or pounds/day (liters/hour or kilograms/day), may be provided by use of a rotameter (Figure 9.16) or built-in calibrated pump with indicated output settings.

9.25   Process Instrumentation

Process instrumentation, by definition, provides for continuous analysis of physical or chemical indicators of process variables in a municipal or industrial plant. This does not include laboratory "bench" instruments (unless set up to measure sample water continuously), although the operating principles are usually quite similar. The process variables of dissolved oxygen and pH are often monitored closely in a wastewater treatment plant (Figure 9.22). Very frequently, chlorine residuals are also continuously measured and controlled. These variables are usually measured at several locations. Additionally, other indicators of process status may be sensed on a continuous basis, such as electrical conductivity (Total Dissolved Solids, TDS), Oxidation-Reduction Potential (ORP), water alkalinity, turbidity, and temperature. In every case, the instrumentation is specific as to operating principle, standardization procedures, preventive maintenance, and operational checks. The manufacturer's technical manual describes routine procedures to check and operate this sensitive type of equipment.

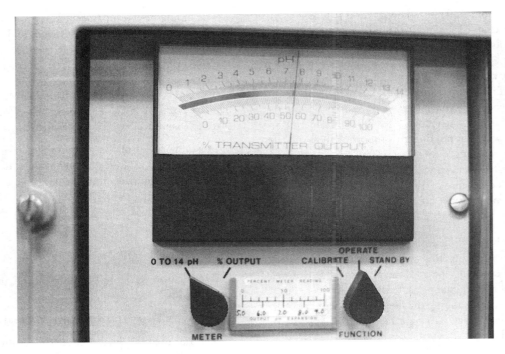

pH meter

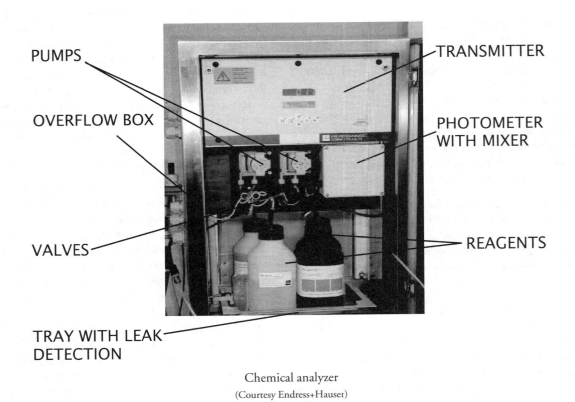

Chemical analyzer
(Courtesy Endress+Hauser)

Fig. 9.22   Wastewater treatment plant analytical process instrumentation

Operators must realize that most process instrumentation is quite delicate and thus requires careful handling and special training to service. No adjustments should be made without a true understanding of the specific device. Generally speaking, this category of instrumentation must be maintained by the plant's instrument specialist, or the factory representative, rather than by an operator (unless specially instructed).

9.26 Signal Transmitters/Transducers

Common system installations measure a variable at one location and provide a readout of the value at a remote location, such as a main control room. Except in the case of a blind transmitter, local indication is provided at the field site as well as being presented at the remote site.

In order to transmit a measured value to a remote location for readout, it is necessary to generate an analog or digital-data signal directly proportional to the value measured. This signal is then transmitted to a remote receiver, which provides a reading based on the signal. Also, a controller may use the signal to control the measured variable and a totalizer to integrate it.

Presently, two general systems for transmission of signals are used in most wastewater treatment facilities: analog or digital data. Analog 4–20 mA transmission has a limited range, typically a few hundred feet or meters, and is being replaced by computer digital-data transmission. For transmission over longer distances, the transmitter output is converted to a digital-data representation of the process variable's measured value. The digital-data transmission simplifies the wired connections that are required for signal transmission.

A power supply to generate the required electrical energy may be located at the analog 4–20 mA transmitter, at the receiver, or at another location in the control instrument loop. The transmitter may be an integral part of the measurement or readout transducer, or it may be housed separately. In either case, the transmitter adjusts the signal to a corresponding value of the measured variable, and the receiver, in turn, converts this signal to a visible indication, the readout. Digital-data transmission distance is limited only by the computer network capability and reliability.

Instrumentation power is required for all components in the instrument loop, for both analog and digital devices. A reliable source of power such as a battery-backed power supply is used to ensure availability during power interruptions. Digital devices have lower power requirements, allowing for smaller and simpler backup power systems.

QUESTIONS

Write your answers in a notebook and then compare your answers with those on page 681.

9.2L What is one way to measure liquid chemical feed rates?

9.2M What process variables are commonly monitored or controlled by process instrumentation?

9.2N What are the two general systems for transmission of measurement signals?

END OF LESSON 1 OF 2 LESSONS

on

INSTRUMENTATION AND CONTROL SYSTEMS

Please answer the discussion and review questions next.

DISCUSSION AND REVIEW QUESTIONS

Chapter 9. INSTRUMENTATION AND CONTROL SYSTEMS

(Lesson 1 of 2 Lessons)

At the end of each lesson in this chapter you will find some discussion and review questions. The purpose of these questions is to indicate to you how well you understand the material in the lesson. Write the answers to these questions in your notebook.

1. Why should operators understand instrumentation and control systems?

2. How can control and instrumentation systems make an operator's job easier?

3. What is the difference between accuracy and precision?

4. Why should a screwdriver not be used to test an electric circuit?

5. What should you do when you discover an area with an oxygen deficiency or enrichment?

6. How can water levels be measured?

7. What problems can develop with propeller meters when they become old or worn?

CHAPTER 9.   INSTRUMENTATION AND CONTROL SYSTEMS

(Lesson 2 of 2 Lessons)

9.3   CATEGORIES OF INSTRUMENTATION

9.30   Primary Elements

The first system element that responds quantitatively to the measured variable is the primary element or sensor. Transducers convert the sensor's minute actions to a usable indication or a signal. If remote transmission of the value is required, a transmitter may be part of the transducer. An illustrative example of these three components is the typical Venturi meter (shown in Figure 9.20, page 657): (1) the flow tube is the primary element, (2) the differential pressure-sensing device (DP cell) is the transducer, and (3) the signal-producing component is the transmitter. An understanding of the separate functions of each section of such a flowmeter is important to the proper understanding of equipment problems.

9.31   Panel Instruments

Process variable readouts are provided locally with indicators and repeated in the control panel and the process computer. These particular components are important to the operator and, hence, to plant operation itself, because they display or control the variable directly. The main control panel devices can also produce alarm signals to indicate if a variable is outside its range of expected values. The controllers are often installed on (or behind) the main panel along with the operating buttons, switches, and indicator lights for the plant's equipment. The controller in an instrument loop control system produces the alarm and control signals.

9.310   Indicators

Indicators give a visual representation of a variable's present value, either as an analog or digital display (Figures 9.23 and 9.24). The analog display uses some type of pointer or graphical display against a scale. A digital display is a direct numerical readout. Chart recorders, which by nature also serve as indicators, give a permanent record of how the variable changes with time by way of a moving chart. These historical records are now largely stored electronically as part of the computer control system. They may be displayed on a computer screen or printed out. Indicators out in the plant or field provide operators with local readouts of the process variables.

Indicators with digital readouts may be devices in a 4–20 mA instrument loop or may receive process variable data from a control system communication network. Digital readouts may be read more quickly and precisely from a longer distance, and can respond virtually instantly to variable changes. But analog indicators are cheaper, more rugged, and may not even require electrical power, an advantage during a power failure or in

Fig. 9.23   Analog chlorine residual indicator

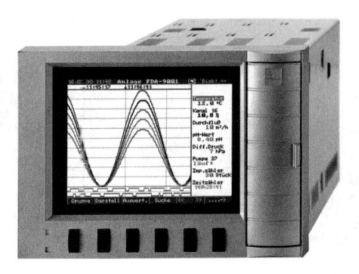

Fig. 9.24   Paperless chart recorder

(Courtesy Endress+Hauser)

hazardous environments. Another advantage of the analog display is that an incorrect readout may be more recognizable than with a digital readout or computer control system, and also is more easily corrected by the operator. For example, the pointer on a flowmeter gauge may merely be stuck, as evidenced by a perfectly constant reading.

Computer control systems provide diagnostic tools to monitor the functionality of instrumentation and troubleshoot process instrumentation loops. Because the capabilities of these systems vary widely, operators must learn the specifics of the system installed. Learning to use the diagnostic tools allows the operator to determine the reliability of the control system.

With all-electronic instrumentation, as advantageous as it may seem from a technical and economic standpoint, the operator has little recourse in case of malfunction of critical instrumentation. Temporary power failures, tripped panel circuit breakers, voltage surges resulting in blown fuses, static electricity, and excessive heat can disrupt the instrument and control systems. Electromechanical or pneumatic instruments may keep operating, or recover operation readily, after such power or heat problems. Computer control systems are typically designed so that the operator can, with training, easily restart the control system and initiate plant operation. Accordingly, the operator should insist upon some input into the design phase of instrument and control systems to ensure that the plant is still operable during power outages, hot weather, and other contingencies. Standby power generators and battery-backed power supplies are used to keep plants operating during commercial power outages, but even they have shortcomings if plant operations depend entirely on electrical power.

QUESTIONS

Write your answers in a notebook and then compare your answers with those on page 681.

9.3A What is the purpose of instrumentation indicators?

9.3B Describe an analog display.

9.3C What is the purpose of chart recorders?

9.3D What factors can disrupt instrument and control systems?

9.311 Recorders

Chart recorder functions have been incorporated into computer control systems (Figure 9.24). However, local recorders are still necessary at some locations throughout a treatment plant to ensure reliable, continuous recording of critical process variables. Recorders are indicators designed to show how the value of the variable has changed with time (Figure 9.25). Usually, this is done by attaching a pen (or stylus) to an indicator's arm, which then marks or scribes the value of the variable onto a continuously moving chart. The chart is marked on a horizontal, vertical, or circular scale in time units. Chart records are also stored on a computer disk or in memory for download from the recorders.

There are two main types of chart recorders: the strip-chart type and the circular-chart type. The strip-chart type carries its chart on a roll or as folded stock, with typically several weeks' supply of chart available. Several hours of charted data are usually visible, or easily available, for the operator to read. On a circular recorder, the chart makes one revolution every day, week, or month, with the advantage that the record of the entire elapsed time period is visible at any time.

Changing of charts is usually the operator's duty. It is easier with circular recorders, though not that difficult with most strip-chart units with some practice.

Chart recorders are typically electrical, powered by the instrumentation and control system electrical supply. Battery-backed power for the recorders may be used, if required, as a backup power source. Recorders are most commonly described by the nominal size of the strip-chart width or circular-chart diameter (for example, a 4-inch (100-mm) strip-chart, or a 10-inch (250-mm) circular-chart recorder). Figure 9.25 shows a combination indicator/recorder and Figure 9.26 presents two models of recorders.

9.312 Totalizers

Rate of flow, as a variable, is a time rate; that is, it involves time directly, such as in gallons per minute, or million gallons per day, or cubic meters per second. Flow rate units become units of volume with the passage of time. For example, flow in gallons per minute accumulates as total gallons during an hour or day. The process of calculating and presenting an ongoing running total of flow volumes passing through a meter is termed "integration" or totalizing.

The totalized flow functions are commonly incorporated into a computer control system, which performs the calculations and stores the historical record of the totals. The historical data may be displayed on the operator interface in trend displays, which are graphical displays similar to strip-chart recorder outputs. The data can be displayed on the computer screen, printed, or sent electronically for analysis by charting and graphing programs.

Large quantities of water (or liquid chemical) are commonly read out in units of hundreds or thousands of gallons (liters). On the face of a totalizer you may find a multiplier such as × 100 or × 1,000. This indicates that the reading is to be multiplied by this factor to yield the full amount of gallons or cubic meters. If the readout uses a large unit, such as mil gal, a decimal will appear between appropriate numbers on the display, or a fractional multiplier (× 0.001, for example) may appear on the face of the totalizer.

Every operator should be able to calculate total flow for a given time period in order to verify that the totalizer is actually producing the correct value. Accuracy to one or two parts in a hundred (1 or 2 percent) is usually acceptable in a totalizer. There are methods to integrate (add up) the area under the flow-rate curve on a recorder chart, to check for long-term accuracy of total flow calculations, but it is cumbersome and rarely necessary.

Fig. 9.25   Chart recorder with digital indicator

Strip-chart recorders

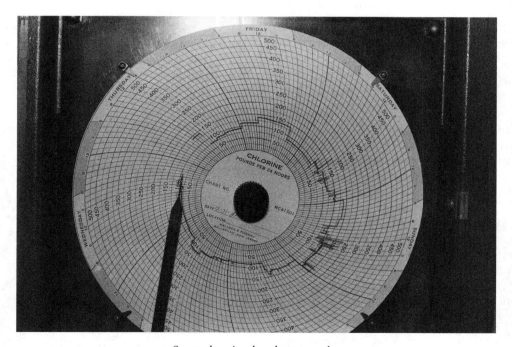

Seven-day circular-chart recorder

Fig. 9.26 Recorders, strip-chart and circular-chart

9.313   *Alarms*

Alarms are visual or audible signals that a variable is out of bounds, or that a condition exists in the plant requiring the operator's attention. Computer control systems have incorporated the alarm annunciator functions into the human machine interface (HMI). Only conditions that require operator attention need alarm signals. For noncritical conditions, a change in color on the operator interface graphic is sufficient notice. For more important variables or conditions, and especially when no computer system is available, an attention-getting annunciator panel (Figure 9.27) with flashing lights and an unmistakable and penetrating alarm horn is commonly used.

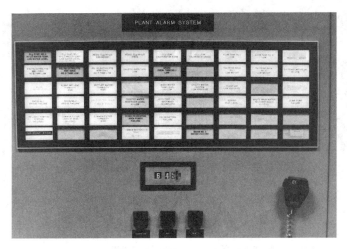

Fig. 9.27   Annunciator (alarm) panel
(each rectangle represents a monitored location)

Operator interface and annunciator panels should all have "acknowledge" and "reset" features to allow the operator to squelch the alarm sound (leaving the visible indication alone), and then to reset the system after the alarm condition is corrected. Annunciator panels should also have a "test" button so that an operator can confirm that no alarm lamps have burned out. The alarm contacts that activate annunciator panel alarms are within the field instrumentation devices. Computer control system alarms are activated by the software's alarm set points and may be programmable from the operator interface. The operator may be responsible for setting the alarm set point and must use judgment as to the actual limits of the particular variables that will ensure meeting proper operational goals. Each system is different, so no attempt will be made here to instruct operating personnel in alarm resetting procedures.

Sometimes operators fail to reset alarm limits as conditions and judgments change in the plant. It is not uncommon to see a plant's operator ignoring alarm conditions on an annunciator panel as normal practice. Such practice is not advised because a true alarm condition requiring immediate operator attention may be lost in the resulting general indifference to the alarm system. For some operators, acknowledging an alarm sound to get rid of the noise is second nature, without due attention to each

and every activating condition. All alarm contact limits should be reset (or deactivated), as necessary, to ensure that the operator is as attentive to the alarm system as necessary to handle real emergencies. The system design should include provisions for disabling alarms when the associated instrument or device is out of service for long periods of time.

QUESTIONS

Write your answers in a notebook and then compare your answers with those on page 681.

9.3E   What is the purpose of recorders?

9.3F   What are the two main types of chart recorders?

9.3G   What are typical power sources for chart recorders?

9.3H   List the two kinds of warning signals that are produced by alarms.

9.32   Automatic Controllers

Section 9.0 explains the nature of control systems as they are used in wastewater treatment plant operations. Automatic controllers may be individual instrumentation devices or a function programmed into a computer control system. Indications of proper and improper control need to be recognized by operators, but adjustment of the controller is left to a qualified instrument technician. By shifting to the manual mode, the operator can bypass the operation of any controller, whether electric or pneumatic. Learn how to shift all your controllers to manual operation. This will allow you to take over control of a critical system, when necessary, in an emergency, as well as at any other time it suits your purposes. For example, you may be able to quickly correct a cycling or sluggish variable by using manual control rather than waiting for the controller to correct the condition in time (if it ever does).

A controller is limited in its capability. It can only do what it has been programmed to do. You, as the operator, can exercise judgment based on your experience and observations, so do not hesitate to intervene if a controller is not exercising control within sensible limits. Of course, you must be sure of your conclusions and competent to take over control if you decide to operate manually.

To repeat a few of the more important operational control considerations, remember that ON/OFF control is quite different in operation from proportional control. Both methods can exercise close control of a variable; however, proportional control is better suited for close control. Attempting to set up an ON/OFF control system to maintain a variable within too close a tolerance may result in rapid ON/OFF operation of equipment. Such operation can damage both the equipment and the switching devices. Therefore, the basic principle to guide you in setting up the frequency of ON/OFF operation of a piece of equipment is to set the ON/OFF controls to operate or cycle associated equipment on and off no more often than actually necessary for plant operation. A level controller, for instance, should be set to cycle a pump or valve only as often as needed.

In the case of a modulating controller, it too may begin to cycle its final control element (pump or valve) through a wide range if any of the internal settings, namely proportional band or reset, are adjusted so as to attempt closer control of the variable than is reasonable. Accordingly, it may be better to accommodate to a small offset (difference between set point and control point) than risk an upset in control by attempting too close control.

9.33 Pump Controllers

Control of pumping systems can be achieved, as we have seen, by an ON/OFF type of controller starting and stopping pumps according to a level, pressure, or flow measurement.

Usually, an ON/OFF pump control system responds to level changes in a tank of some type. Water level can be sensed directly with a float or by a pressure change at the tank or pump site. The pump is thus turned off or on as the tank level rises above or falls below predetermined level or pressure limits. Control is rather simple in this case.

However, such systems may include several extra electrical control features to ensure fail-safe operation. To prevent the pump from running after a loss of level signal, electrical circuitry should be designed so the pump will turn *OFF* on an open signal circuit and *ON* only with a closed circuit. (Ideally, the controller would be able to distinguish between an open or closed remote level/pressure contact and an open or shorted signal line.) Larger pump systems also will often have a low-pressure cutoff switch on the suction side to prevent the pump from running when no water is available, such as with an empty tank or closed suction valve.

Controllers may also protect against overheating a pump (as happens when continuing to pump against a closed discharge valve) by a high-pressure (or low-flow) cutoff switch on the discharge piping. Both the high- and low-pressure switches should shut off a pump through a time delay circuit so that short-term pressure surges (dips and spikes) in the pump's piping can be tolerated. Ideally, the low- or high-pressure switches also key

alarms to notify the operator of the condition. For remote stations, a plant's main panel may include indicator lights to show the pumps' operating conditions. Figure 9.28 shows a simplified diagram of pump control circuitry.

Pump control panels (Figure 9.29) may also include automatic or manual alternators (two pumps) or sequencers (more than two pumps). This provision allows the total pump operating time required for the particular system to be distributed equally among all the pumps at a pump station. A manual switch for a two-pump station, for example, may read "1-2" in one position and "2-1" in the other position. In the first position, pump #1 is the lead pump (which runs most of the time) and #2, the lag pump (which runs less). When the operator changes the switch to "2-1," the lead-lag order of pump operation is reversed, as it should be periodically, to keep the running time (as read on the elapsed time meters, or as estimated) of both pumps to about the same number of total hours. In a station with multiple pumping units, an automatic alternator or sequencer regularly changes the order of the pumps' start-up to maintain similar operating times for all pumps. In order to protect a pump's electric motor from overheating, level controls should be set so that the pump starts no more often than six times per hour (average one start every 10 minutes).

QUESTIONS

Write your answers in a notebook and then compare your answers with those on page 681.

9.3I   Under what conditions might an operator decide to by-pass the operation of a controller? How could this be done?

9.3J   What basic principle should guide you in setting up the frequency of ON/OFF operation of a piece of equipment?

9.3K   How can pumps be prevented from running after a loss of level signal?

9.3L   What provision allows the total pump operating time required for a particular system to be distributed equally among all the pumps at a pump station?

9.34 Air Supply Systems

Pneumatic instrumentation depends on a constant source of clean, dry, pressurized air for reliable operation. Given a quality air supply, pneumatic devices can operate for long periods without significant problems. Without a quality air supply, operational problems can be frequent. The operator of a plant is usually assigned the task of ensuring that the instrument air is always available and dry, so operators must learn how to accomplish this; it cannot be assumed that clean air is there automatically.

The plant's instrument air supply system consists of a compressor with its own controls, master air pressure regulator, air filter, and air dryer, as well as the individual pressure regulator/filters in the line at each pneumatic plant instrument (Figure 9.30). Only the instrument air is filtered and dried; the plant air

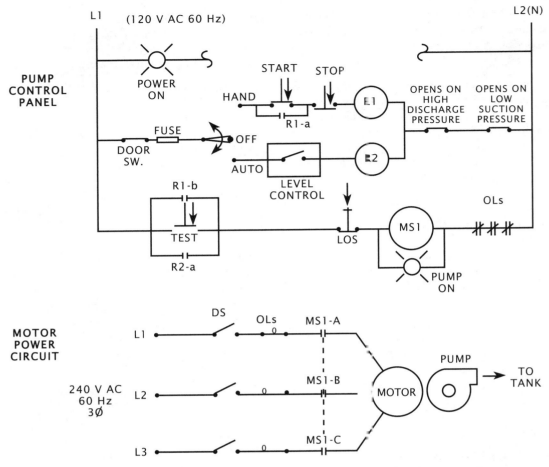

PUMP CONTROL PANEL

MOTOR POWER CIRCUIT

REFER TO FIGURE 9.5 (PAGE 638), AND PAGE 639 FOR PARTS LEGEND

Fig. 9.28 Pump control station ladder diagram (ON/OFF control)
(simplified schematic)

Fig. 9.29 Photo of pump control station

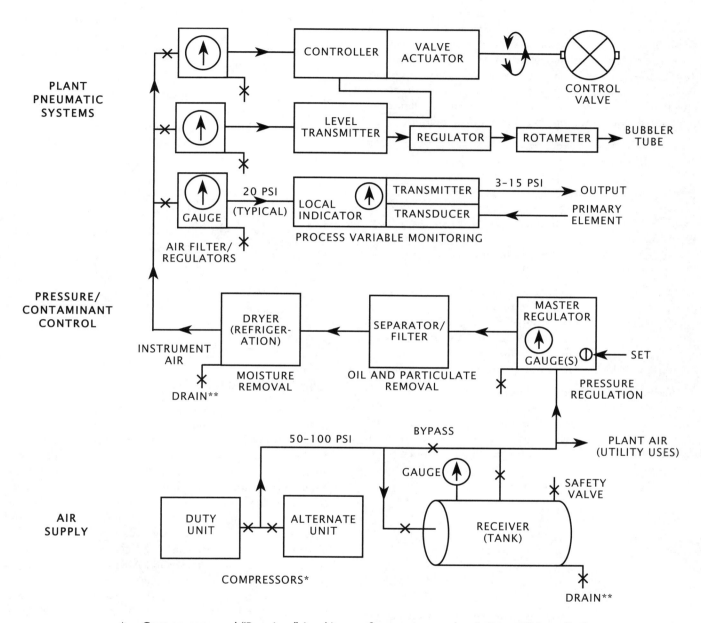

PLANT PNEUMATIC SYSTEMS

CONTROLLER | VALVE ACTUATOR → CONTROL VALVE

LEVEL TRANSMITTER → REGULATOR → ROTAMETER → BUBBLER TUBE

GAUGE   20 PSI (TYPICAL)   LOCAL INDICATOR | TRANSMITTER → 3-15 PSI → OUTPUT

TRANSDUCER ← PRIMARY ELEMENT

AIR FILTER/ REGULATORS

PROCESS VARIABLE MONITORING

PRESSURE/ CONTAMINANT CONTROL

DRYER (REFRIGER-ATION) ← SEPARATOR/ FILTER ← MASTER REGULATOR   GAUGE(S) ← SET

INSTRUMENT AIR

DRAIN**   MOISTURE REMOVAL   OIL AND PARTICULATE REMOVAL   PRESSURE REGULATION

AIR SUPPLY

50-100 PSI   BYPASS   PLANT AIR (UTILITY USES)

GAUGE

DUTY UNIT   ALTERNATE UNIT   RECEIVER (TANK)   SAFETY VALVE

COMPRESSORS*

DRAIN**

*   Compressor and "Receiver" (tank) are often an integral unit in small installations.
** Drains may be automatic type (on timer).

*Fig. 9.30   Typical plant instrument air system functional diagram
(simplified, not all valves and piping shown)*

usually does not require such measures since it is being used only for other purposes.

As air passes through a compressor, it not only can pick up some oil but the air's moisture content is concentrated by the compression process. Special measures must be taken to remove both of the liquids. Oil is removed by filtering the air through special oil-absorbent elements. Water is removed by passing the moisture-laden air through an air-drying system consisting of a moisture separation column and air dryers. Air-drying systems are typically provided in pairs to allow one dryer to regenerate while the other is in service. You must recognize that the capacity of any of these systems of oil or water removal is limited to amounts of liquid encountered under normal conditions.

If the compressor is worn so as to pass more oil than usual, the oil separation process may permit troublesome amounts of oil to pass into the air supply. If the air source contains excessive humidity (due to a rainy day, perhaps, or a wet compressor room), the air-drying system may not handle the excess moisture. Learn enough about the instrument air system to be able to open the drain valves, cycle the air dryer, or even bypass the tank, in order to prevent instrumentation problems due to an oily or moisture-saturated air supply.

Operators should regularly crack the regulator/filter drain valves at each of the plant's process instruments. An unusual quantity of liquid drainage may indicate an overloading or failure in the instrument air filter/drying parts. Also, pneumatic indicators/recorders should be watched for erratic pointer/pen movements, which usually are indicative of air quality problems.

The seriousness of plant power or compressor failures can be lessened if you temporarily turn off all nonessential usages of compressed air in the plant. The air storage tank is usually sized so that there is enough air on hand to last for several hours, if conserved. Knowing this, you may be able to wait out a power failure without undue drastic action by conserving the remaining available pressure of the air supply.

QUESTIONS

Write your answers in a notebook and then compare your answers with those on pages 681 and 682.

9.3M   What are the essential qualities of the air supply needed for reliable operation of pneumatic instrumentation?

9.3N   How are moisture and oil removed from instrument air?

9.35   Laboratory Instruments

This category of instrumentation includes those analytical units typically found in wastewater treatment plant laboratories (Figure 9.31). Some examples are turbidimeters, colorimeters and comparators, pH, conductivity (TDS), and dissolved oxygen (DO) meters. We have already seen that process models of each of the units monitor these same variables out in the plant. The models used in the laboratory are usually referred to as "bench" models rather than "process" instrumentation.

Operators often are required to make periodic readings from laboratory instruments, and periodic standardizing of particular instruments is often required before determinations are made. Preventive, and certainly corrective, maintenance is handled by the laboratory staff, factory representative, or instrument technician since each unit can be quite complex. Some of these countertop instruments or devices are very delicate and replacement parts, such as glass vessels or pH electrodes, are quite expensive. Moreover, the use of some of these instruments requires the regular handling of laboratory glassware and other breakable items. The operator who, through carelessness, lack of knowledge, or simple hurrying, consistently breaks glassware or "finds the darn meter broken again" does not become popular with the chemist, supervisor, or other operators. The byword in the laboratory is *WORK WITH CAUTION*. Protect valuable and essential instrumentation and supplies.

9.36   Test and Calibration Equipment

Plant measuring systems must be periodically calibrated to ensure accurate measurements. In most larger wastewater treatment facility operations, the plant operating staff has little occasion to use testing and calibration devices on the plant instrumentation systems. A trained technician will usually be responsible for using such equipment. There are, however, some general considerations the operator should understand concerning the testing and calibration of plant measuring and control systems. With this basic knowledge, you may be able to discuss needed repairs or adjustments with an instrument technician and perhaps assist with that work. A better understanding of your plant's instrument systems may also enable you to analyze the effects of instrument problems on continued plant operation, and to handle emergency situations created by instrument failure. Your skills in instrument testing and calibration may even eventually result in a job promotion or pay raise.

The most useful piece of general electrical test equipment is the V-O-M, that is the Volt-Ohm-Milliammeter, commonly referred to as a multimeter (Figure 9.32). To use this instrument, you will need a basic understanding of electricity, but once you learn to use it, the V-O-M has potential for universal usage in instrument and general electrical work. Local colleges and other educational institutions may offer courses in basic electricity, which undoubtedly include practice with a V-O-M. You, as a

Fig. 9.31 Industrial waste treatment plant laboratory

SPECIFICATIONS

Voltage Ranges

0–199.9/750 V AC   15 kV AC
0–1.999/19.99/199.9/1,000 V DC   15 kV DC
0–1,999 mV DC

Resistance Ranges

0–199.9/1,999 ohms
0–19.99/199.9/1,999 K ohms

Current Ranges

0–1,999 µA DC
0–19.99/199.9/1,999 mA DC
0–10 Amps DC
0–19.99/199.9/1,999 mA AC
0–10 Amps AC
NOTE: AC accuracy may be affected by outside interference.

Accuracy

DC V: ± 0.5% of rdg ± 2LSD
AC V: ± 1.5% of rdg ± 2LSD
DC Amps: All ranges ± 1.0% of rdg ± 2LSD except 10 Amp range, which is ± 1.5% of rdg ± 3LSD.
AC Amps: All ranges ± 1.5% of rdg ± 2LSD except 10 Amp range, which is ± 2.0% of rdg ± 3LSD.
Ohms: All ranges ± 0.75% of rdg ± 2LSD except 2 megohm range, which is ± 1% of rdg ± 2LSD.
15 kV AC/DC high voltage probe: add up to ± 2% of rdg.

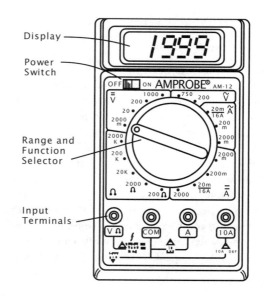

Fig. 9.32 Digital multimeter (V-O-M)

professional plant operator, are unlikely to find technical training of greater practical value than this type of course or program. Your future use of test and calibration equipment, in general, certainly should be preceded by instruction in the fundamentals of electricity.

QUESTIONS

Write your answers in a notebook and then compare your answers with those on page 682.

9.3O   Why should an operator be especially careful when working with laboratory instruments?

9.3P   Why should an operator become familiar with the testing and calibration of plant measuring and control systems?

9.3Q   What is a V-O-M?

9.37   Process Computer Control System

9.370   Computer Control Systems

The computer control system is a computer-monitored alarm, response, control, and data acquisition system used by operators to monitor and adjust their treatment processes and facilities. Computer control systems used for process control may be classified by two commonly used terms: the *DISTRIBUTED CONTROL SYSTEM (DCS)*[8] and the *SUPERVISORY CONTROL AND DATA ACQUISITION (SCADA)*[9] system. The DCS and the SCADA system perform the same functions in different settings. DCSs are typically used to control and monitor processes in treatment plants. SCADA systems are most commonly used to control and monitor collection system facilities that are widely separated geographically. Larger wastewater facilities may have both DCSs and SCADA systems to

provide treatment process control and collection system controls. Smaller utilities often combine all the controls necessary into a SCADA system. In both large and small utilities, the operator interface for both collections and plant processes are provided in a single control room.

The computer control system collects, stores, and analyzes information about all aspects of operation and maintenance, transmits alarm signals, when necessary, and allows fingertip control of alarms, equipment, and processes. The computer control system provides the information that operators need to solve minor problems before they become major incidents. As the nerve center at the treatment plant, the system allows operators to enhance the efficiency of their facility by keeping them fully informed and fully in control. Figure 9.33 shows the basic interaction cycle of the operator with the control systems and plant treatment processes.

The five basic components of the computer control system (Figure 9.34) are as follows:

- Process instrumentation and control devices sense process variables in the field and actuate equipment.

- The input/output (I/O) interface sends and receives data with the process instrumentation and control devices.

- The central processing unit (CPU) is the system component that contains the program instructions for the control system. These instructions are programmed to react based on a control strategy. The CPU gathers data from the various interfaces and sends commands to field devices to operate the plant processes.

- The communication interfaces provide the means for the computer control system to send data to and from outside computer systems, business systems, other process control systems, and equipment.

- The *HUMAN MACHINE INTERFACE (HMI)*[10] is commonly a computer workstation that is running the computer control system software that provides the plant data to the operator on the workstation screen.

These components are the means by which the control system gathers and distributes information for the human operator and the process instrumentation and other equipment.

The computer control systems may be used in various capacities, from data collection and storage only, to total data analysis, interpretation, and process control.

Computer control systems monitor levels, pressures, and flows and also operate pumps, valves, and alarms. They monitor

---

[8] *Distributed Control System (DCS).*   A computer control system having multiple microprocessors to distribute the functions performing process control, thereby distributing the risk from component failure. The distributed components (input/output devices, control devices, and operator interface devices) are all connected by communications links and permit the transmission of control, measurement, and operating information to and from many locations.

[9] *SCADA (SKAY-dah) System.*   Supervisory Control And Data Acquisition system. A computer-monitored alarm, response, control, and data acquisition system used to monitor and adjust treatment processes and facilities.

[10] *Human Machine Interface (HMI).*   The device at which the operator interacts with the control system. This may be an individual instrumentation and control device or the graphic screen of a computer control system. Also called MAN MACHINE INTERFACE (MMI) and OPERATOR INTERFACE.

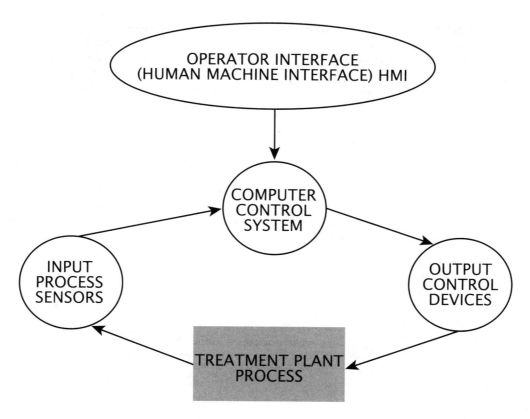

Fig. 9.33   Operator interaction with plant processes

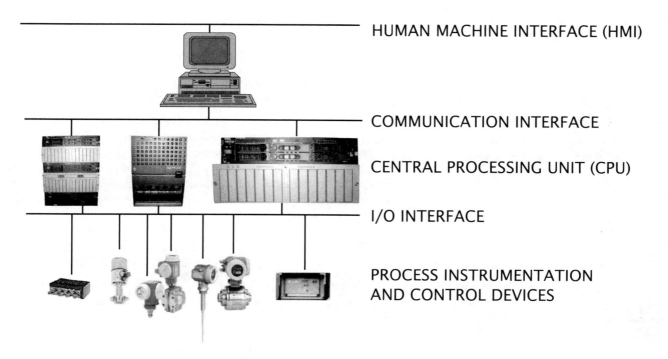

Fig. 9.34   Computer control system components

temperatures, speeds, motor currents, pH, dissolved oxygen levels, and other operating parameters, and provide control, as necessary. Computer control systems provide a log of historical data for events, analog signal trends, and equipment operating time for maintenance purposes. The information collected may be read by an operator on computer screen readouts or analyzed and plotted by the computer as trend charts.

Computer control systems provide a picture of the plant's overall status on a computer screen. In addition, detailed pictures of specific portions of the system can be examined by the operator through the computer workstation. The graphical displays on the computer screens can include current operating information, which the operator can use to determine if the guidelines are within acceptable operating ranges or if any adjustments are necessary.

Computer control systems are capable of analyzing data and providing operating, maintenance, regulatory, and annual reports. Operation and maintenance personnel rely on a computer control system to help them prepare daily, weekly, and monthly maintenance schedules, monitor the spare parts inventory status, order additional spare parts, when necessary, print out work orders, and record completed work assignments.

Computer control systems can also be used to enhance energy conservation programs. For example, operators can develop energy management control strategies that allow for both maximum energy savings and maximum treatment flow prior to peak-flow periods. In this type of system, power meters are used to accurately measure and record power consumption. The information can then be reviewed by operators to watch for changes that may indicate equipment problems.

Emergency response procedures can also be programmed into a computer control system. Operator responses can be provided for different operational scenarios that might be encountered as a result of adverse weather changes, fires, earthquakes, or other emergency situations.

QUESTIONS

Write your answers in a notebook and then compare your answers with those on page 682.

9.3R Computer control systems used for process control are classified by which two commonly used terms?

9.3S What are the five basic components of the computer control system?

9.3T How do computer control systems help operation and maintenance personnel?

9.371 Typical Computer Control System Functions

Computer control systems for wastewater treatment plants and collection systems are usually operated together, with the controls located at the treatment plant. Information that historically was recorded on paper strip-charts is now being recorded and stored (archived) by computers. This information can be retrieved and reviewed easily by the operator, whereas before,

years of strip-chart records would need to be examined to find needed information. Therefore, computer control systems are more efficient in providing operators with the information they need to make informed and timely decisions. Figure 9.35 is a typical modern wastewater treatment plant control room.

Computer control systems give the treatment plant operator the tools to optimize plant processes based on current and historical operating information. The treatment plant influent and effluent are monitored continuously for many process variables such as flow, turbidity, pH, ammonium, chlorine, and nitrogen. If these indicators change significantly or exceed predetermined levels, the computer control system alerts the operator or changes the process based on a preprogrammed control strategy defined by the operator.

Historical operating data stored in a computer control system is readily available at any time. The computer control system can be queried to identify, for instance, when peak plant influent flow was greater than 50 MGD during the previous two years. Plant performance under these conditions can be recalled using the computer control system, analyzed by the operators, and the results used to operate the plant accordingly.

Electrical energy consumption can be optimized by the use of computer control systems. Most power companies are eager to help operators save money by structuring their rates to encourage electrical energy consumption when demands for power are low and to discourage consumption when demands for power are high. Computer controls can be programmed with a control strategy to reduce energy costs by automatically operating equipment when demands for power are low.

Computer control systems are being continually improved to help operators do a better job. Operators can create their own display screens, their own graphics, and show whatever operating characteristics they wish to display. The main screen could be a flow diagram from influent to effluent showing the main treatment and auxiliary process areas. Critical operating information could be displayed for the main treatment flow path and process area, with navigation capabilities to easily access detailed screens for each piece of equipment.

Information on the screen should be color coded to indicate if a pump is running, ready, unavailable, or failed, or if a valve is open, closed, moving, unavailable, or failed. A "failed" signal is used by the computer to inform the operator that something is wrong with the information or the signal it is receiving or is being instructed to display. For example, if there is no power to a motor, then the motor cannot be running even though the computer is receiving a signal that indicates it is running. In this case, the computer would send a "failed" signal, indicating that the information it is receiving is not consistent with the rest of the information available.

The operator can request a computer to display a summary of all alarm conditions in a plant, a particular plant area, or a process system. A blinking alarm signal indicates that the alarm condition has not yet been acknowledged by the operator. A steady alarm signal, one that is not blinking, indicates that the alarm

Fig. 9.35   Control room

has been acknowledged but the condition causing it has not yet been fixed. Also, the screen could be set up to automatically designate certain alarm conditions as *PRIORITY* alarms, requiring immediate operator attention.

With proper security implementation, computer control systems allow operators to have remote access to plant controls from anywhere using a laptop or remote workstation. This provides the flexibility for off-duty staff to help on-duty operators solve operational problems. Computer networking systems allow operators at terminals in offices, in plants, and in the field to work together and use the same information or whatever information they need from one central computer database.

A drawback of some computer control systems is that when the system "goes down" due to a power failure, the numbers displayed will be the numbers that were registered immediately before the failure, not the current numbers. The operator may therefore experience a period of time when accurate, current information about the system is not immediately available.

Customer satisfaction with the performance of a wastewater utility can be enhanced by the use of an effective computer control system. Coordination of the treatment facility control and the collection system control is used to avoid overflows and backups.

When operators decide to initiate or expand a computer control system for their plant process system, the first step is to decide what the computer control system should do to make the operators' jobs easier, more efficient, and safer, and to make their facilities' performance more reliable and cost effective. Cost savings associated with the use of a computer control system frequently include reduced labor costs for operation, maintenance, and monitoring functions that were formerly performed manually. Precise control of chemical feed rates by a computer control system eliminates wasteful overdosing. Preventive maintenance monitoring can save on equipment and repair costs, and, as previously noted, energy savings may result from off-peak (reduced) electrical power rates. Operators should visit facilities with computer control systems and talk to the operators about what they find beneficial, and also detrimental, with regard to computer control systems and how the systems contribute to their performance as operators.

The greatest challenge for operators using computer control systems is to realize that just because a computer says something (a pump is operating as expected), *this does not mean that the computer is always correct*. Also, when the system fails due to a power failure or for any other reason, operators will be required to operate the plant manually and without critical information. Could you do this?

Operators will always be needed to question and analyze the results from computer control systems. They will be needed to *see* if the effluent looks OK, to *listen* to a pump to be sure it sounds right, and to *smell* the process and the equipment to determine if unexpected or unidentified changes are occurring. Treatment plants and collection systems will always need alert, knowledgeable, and experienced operators who have a "feel" for their plant and their collection system.

QUESTIONS

Write your answers in a notebook and then compare your answers with those on page 682.

9.3U   What do computer control systems do with information that historically was recorded on paper strip-charts?

9.3V   Cost savings associated with the use of a computer control system frequently include which items?

9.3W   What is the greatest challenge for operators using computer control systems?

9.4   OPERATION AND PREVENTIVE MAINTENANCE

9.40   Proper Care of Instruments

Usually, instrumentation systems are remarkably reliable year after year, assuming proper application, setup, operation, and maintenance. Reliable measurement systems, though outdated by today's standards, are still found in regular service at some plants up to 50 years after installation. To a certain extent, good design and application account for such long service life, but most important is the careful operation and regular maintenance of the instruments' components. The key to proper operation and maintenance (O & M) is the operator's practical understanding of the system. Operators must know how to: (1) recognize malfunctioning instruments so as to prevent prolonged damaging operation, (2) shut down and prepare devices for seasonal or other long-term nonoperation, and (3) perform preventive (and minor corrective) maintenance tasks to ensure proper operation in the long term. A sensitive instrumentation system can be ruined in short order with neglect in any one of these three areas.

Operators should be familiar with the Technical Manual (also called the Instruction Book or Operating Manual) of each piece of equipment and instrument encountered in a plant. Each manual will have a section devoted to the operation of a certain component of a complete measuring or control system (although frequently not for the entire system). Detailed descriptions of maintenance tasks and operating checks will usually be found in the manual. Depending on the general type of instrument (electromechanical, pneumatic, or electronic), the suggested frequency of the operation and maintenance/checking tasks can range from none to monthly. Accordingly, this section of the course only describes those common and general tasks an operator might be expected to perform to operate and maintain instrumentation systems. From an operations standpoint, these tasks include learning, and constant attention to, what constitutes normal function. From a maintenance standpoint, they include ensuring proper and continuing protection and care of each component.

9.41   Indications of Proper Function

The usual pattern of day-to-day operation of every measuring and control system in a plant should become so familiar to operators that they almost unconsciously sense any significant change. This will be especially evident and true for systems with recorders where the pen trace is visible. An operator should thus watch indicators and controllers for their characteristic actions and pay close attention to trend records. Using the trend capability of the computer control system provides a method to analyze the reaction of one process variable to a change in another or other process variables.

Two of the surest indications of a serious electrical problem in instruments or power circuits are, of course, smoke or a burning odor. Such signs of a problem should *never* be ignored. Smoke/odor means heat, and no device can operate long at unduly high temperatures. Any electrical equipment that begins to show signs of excessive heating must be shut down *immediately*, regardless of how critical it is to plant operation. Overheated equipment will very likely fail very soon anyway, with the damage being aggravated by continued usage. Fuses and circuit breakers do not always de-energize circuits before damage occurs, so they cannot be relied upon to do so.

Finally, operators frequently forget to reset an individual alarm, either after an actual occurrence or after a system test. This is especially prevalent when an annunciator panel is allowed to operate day after day with lit-up alarm indicators (contrary to good practice) and one light (or more) is not easily noticeable. Also, when a plant operator must be away from the main duty station, the system may be set so the audible part of the alarm system is temporarily squelched. When the operator returns, the audible system may inadvertently not be reactivated. In both instances (individual or collective loss of audible alarm), the consequences of such inattention can be serious. Therefore, develop the habit of checking your annunciator system often.

QUESTIONS

Write your answers in a notebook and then compare your answers with those on page 682.

9.4A   List the three areas of operator responsibility that are key to proper instrument operation and maintenance (O & M).

9.4B   What general tasks are expected of an operator of instrumentation systems from an operations standpoint and from a maintenance standpoint?

9.4C   What are two of the surest indications of a serious electrical problem in instruments or power circuits?

9.42   Start-Up/Shutdown Considerations

The start-up and periodic or prolonged shutdown of instrumentation equipment usually require very little extra work by the operator. Start-up is limited mainly to undoing or reversing the shutdown measures taken.

When shutting down any pressure-, flow-, or level-measuring system, valve off the liquid to the measuring element. Exercise particular care, as explained previously, regarding the order in which the valves are manipulated for any flow tube installation. Also, the power source of some instruments may be shut off, unless the judgment is made that keeping an instrument case warm (and thus dry) is in order. Constantly moving parts, such as chart drives, should be turned off. With an electrical panel room containing instrumentation, it is good practice to leave some power components on (such as a power transformer) to provide space heat for moisture control. In a known moist environment, sealed instrument cases may be protected for a while with a container of *DESICCANT*[11] (indicating silica gel, which is blue if OK and pink when the moisture-absorbent capacity is exhausted).

Although preventing the access of insects and rodents into any area is difficult, general cleanliness seems to help considerably. Rodenticides are available to control rodents; this is good preventive maintenance practice in any electrical space. Rats and mice will chew off wire and transformer insulation, and may urinate on other insulator material, leading to serious problems for equipment (not to mention the rodent).

Nest-building activities of some birds can also be a problem. Screening some buildings and equipment against entry by birds has become a design practice of necessity. Insects and spiders are not generally known to cause specific functional problems, but start-up and operation of systems invaded by ants, bees, or spiders should await cleanup of each such component of the system. All of these pests can bite or sting, so take care.

With pneumatic instrumentation, it is desirable to purge each device with dry air before shutdown. This measure helps rid the individual parts of residual oil and moisture to minimize internal sticking and corrosion while standing idle. As before, periodic blow-off of air receivers and filters keeps these liquids out of the instruments to a large degree. Before shutdown, however, extra attention should be paid to instrument air quality for purging. Before start-up, each filter/receiver should again be purged.

Finally, pay attention to the pens and chart drives of recorders upon shutdown. Ink containers (capsules) may be removed if deemed necessary, and chart drives turned off. A dry pen bearing against one track (such as zero) of a chart for weeks on end is an invitation to start-up problems. Re-inking and chart

replacement at start-up are easy if the proper shutdown procedures were followed.

QUESTIONS

Write your answers in a notebook and then compare your answers with those on page 682.

9.4D   How can moisture be controlled in an electrical panel room containing instrumentation or in sealed instrument cases?

9.4E   Why should pneumatic instrumentation be purged with dry air before shutdown?

9.43   Preventive Maintenance

Preventive maintenance (PM) means that attention is given periodically to equipment in order to prevent future malfunctions. Corrective maintenance involves actual, significant repairs, which are beyond the scope of this chapter and, in most cases, are not the responsibility of the operator. Routine operational checks are part of all PM programs in that a potential problem may be discovered and thereby corrected before it becomes serious.

PM duties for instrumentation should be included in the plant's general PM program. If your plant has no formal, routine PM program, it should have. Such a program must be set up on paper (or on a computer program). That is, the regular duties required are printed on forms or cards (or appear on a computer screen) that the operator (or technician) uses as a reminder, guide, and record of PM tasks performed. Without such explicit measures, experience shows that preventive maintenance will almost surely be put off indefinitely. Eventually, the press of critical corrective maintenance (often due to lack of preventive maintenance) and even equipment replacement projects may well eliminate forever any hope of a regular PM program. The fact that instrumentation is usually very reliable (being of quality design) may keep it running long after pumps and other equipment have failed. Nevertheless, instrumentation does require proper attention periodically to maximize its effective life. PM tasks and checks on modern instrument systems are quite minimal (even virtually nonexistent on some), so there are no valid reasons for ever failing to perform these tasks.

The technical manual for each item of instrumentation in your plant should be available so you can refer to it for O & M purposes. When a manual cannot be located, contact the manufacturer of the unit. Be sure to give all relevant serial/ model numbers in your request for the manual. Request two manuals, one to use and one to put in reserve. All equipment manuals should be kept in one protected location and signed out, as needed. Become familiar with the sections of these manuals related to O & M, and follow their procedures and recommendations closely.

A good practice is to have on hand any supplies and spare parts that are or may be necessary for instrument operation (such as charts) or service (such as pens, pen cleaners, and ink).

---

[11] *Desiccant* (DESS-uh-kant).   A drying agent that is capable of removing or absorbing moisture from the atmosphere in a small enclosure.

Some technical manuals contain a list of recommended spare parts that you could use as a guide. Try to obtain these supplies/parts for your equipment. A new pen on hand for a critical recorder can be a lifesaver, at times.

Since PM measures can be so diverse for different types, brands, and ages of instrumentation, only the few general considerations applicable to all will be covered in this section.

1. Protect all instrumentation from moisture (except as needed by design), vibration, mechanical shock, vandalism (a very real problem in the field), and unauthorized access.

2. Keep instrument components clean on the outside, and closed/sealed against inside contamination (for example, spider webs and rodent wastes).

3. *DO NOT* presume to lubricate, tweak, fix, calibrate, free up, or modify any component of a system arbitrarily. If you are not qualified to take any of these measures, then do not do it.

4. *DO* keep recorder pens and charts functioning as designed by frequent checking and service, bleed pneumatic systems regularly, as instructed, ensure continuity of power for electrical devices, and do not neglect routine analytical instrument cleanings and standardizing duties as required by your plant's established procedures.

As a final note, it is a good idea to get to know and cooperate fully with your plant's instrument service person. Good communication between this person and the operating staff can only result in better all-around operation. If your agency is too small to staff such a specialist, it may be a good idea to enter into an instrumentation service contract with an established company or possibly even with the manufacturer of the majority of the components. With rare exceptions, general maintenance persons (even journeyman electricians) are *not* qualified to perform extensive maintenance on modern instrumentation. Be sure that someone takes good care of your instruments and they will take good care of you.

QUESTIONS

Write your answers in a notebook and then compare your answers with those on page 683.

9.4F   Why should regular preventive maintenance duties be printed on forms or cards?

9.4G   How can the technical manual for an instrument be obtained if the only copy in a plant is lost?

9.4H   What instrument supplies and spare parts should always be available at your plant?

9.44   Operational Checks

Computer control systems provide the best tools for observing the operating functions of plant process systems. The operator interface (also called the human machine interface (HMI)) provides the ability to view all areas of plant operation. Most systems provide the ability to display trends of multiple process variables on the same graphical display. These trends are tools you will learn to use extensively to monitor and control your facility's processes.

Operational checks are most efficiently made by always observing each system for its continuing signs of normal operation. However, some measuring systems may be cycled within their range of action as a check on the responsiveness of components. For instance, if a pressure-sensing system indicates only one pressure for months on end, and some doubt arises as to whether it is working or not, the operator may bleed off a little pressure at the primary element to produce a small fluctuation. Or, if a flow has appeared constant for an overly long period, the bypass valve in the DP (differential pressure) cell piping may be cracked open briefly to cause a drop in the reading.[12] Be sure you crack the *bypass* valve, not one of the others on the piping. If you open the wrong valve, the pressure may be excessive and be beyond the range of the DP cell, which could cause problems. A float for a level recorder suspected of being stuck (very constant level indication) may be freed by jiggling its cable, or by taking other measures to cause a slight fluctuation in the reading.

Whenever an operator or a technician disturbs normal operation during checking or for any reason, plant process operating personnel must be informed—ideally prior to the disturbance. If a recorder trace is altered from its usual pattern in the process, the person causing the upset should initial the chart appropriately and note the time. Some plants require operators to mark or date each chart at midnight (or noon) of each day for easy reference and filing.

In the case where a pen or pointer is thought to be stuck mechanically, that is, it does not respond at all to simulated or actual change in the measured variable, it is normally permissible to open an instrument's case and try to move the pointer or pen, but only to the minimum extent necessary to free it. Further deflection may well bend or break the device's linkage. A dead pen often is due only to loss of power or air to the readout mechanism. Any hard or repeated striking of an instrument to make it work identifies the striker as ignorant of good operational practice and can ruin the equipment. Insertion of tools into an instrument case in a random fix-it attempt can easily damage an instrument. Generally speaking, any extensive operating check of instrumentation should be performed by the instrument technician during routine PM program activities.

QUESTIONS

Write your answers in a notebook and then compare your answers with those on page 683.

9.4I   How are operational checks performed on instrumentation equipment?

9.4J   What should be done if a recorder trace is altered from its usual pattern during the process of checking an instrument?

---

[12] There is no similar easy way to check a propeller meter's response.

9.5 ADDITIONAL READING

1. *INSTRUMENTATION—HANDBOOK FOR WATER AND WASTEWATER TREATMENT PLANTS,* Robert G. Skrentner. Obtain from CRC Press LLC, Attn: Order Entry, 6000 Broken Sound Parkway, NW, Suite 300, Boca Raton, FL 33487. ISBN 0873711262. Order No. L126. Price, $189.95, includes shipping and handling.

2. *INSTRUMENTATION IN WASTEWATER TREATMENT FACILITIES* (MOP 21). Obtain from Water Environment Federation (WEF), Publications Order Department, 601 Wythe Street, Alexandria, VA 22314-1994. Order No. MO2021. Price to members, $41.75; nonmembers, $61.75; price includes cost of shipping and handling.

3. *HANDS-ON WATER AND WASTEWATER EQUIPMENT MAINTENANCE.* Obtain from American Public Works Association (APWA) Bookstore, PO Box 802296, Kansas City, MO 64180-2296. Order No. PB.XHOW. Price to members, $110.00; nonmembers, $115.00; plus shipping and handling.

4. *OPERATION OF MUNICIPAL WASTEWATER TREATMENT PLANTS* (MOP 11). Obtain from Water Environment Federation (WEF), Publications Order Department, 601 Wythe Street, Alexandria, VA 22314-1994. Order No. M05110. Price to members, $129.75; nonmembers, $157.75; price includes cost of shipping and handling.

5. National Fire Protection Association Standard 820 (NFPA 820), *FIRE PROTECTION IN WASTEWATER TREATMENT AND COLLECTION FACILITIES.* Obtain from National Fire Protection Association (NFPA), 11 Tracy Drive, Avon, MA 02322. Item No. 82003. Price to members, $31.50; nonmembers, $35.00; plus $7.95 shipping and handling.

6. *INSTRUMENTATION AND CONTROL* (M2). Obtain from American Water Works Association (AWWA), Bookstore, 6666 West Quincy Avenue, Denver, CO 80235. Order No. 30002. Price to members, $87.50; nonmembers, $129.50; price includes cost of shipping and handling.

7. *SUCCESSFUL INSTRUMENTATION AND CONTROL SYSTEMS DESIGN,* Michael D. Whitt. Obtain from ISA—The Instrumentation, Systems, and Automation Society, PO Box 3561, Durham, NC 27702-3561. ISBN 1-55617-844-1. Price to members, $89.00; nonmembers, $99.00.

8. *SCADA SUPERVISORY CONTROL AND DATA ACQUISITION,* Stuart A Boyer. Obtain from ISA—The Instrumentation, Systems, and Automation Society, PO Box 3561, Durham, NC 27702-3561. ISBN 1-55617-877-8. Price to members, $69.00; nonmembers, $79.00.

END OF LESSON 2 OF 2 LESSONS

on

INSTRUMENTATION AND CONTROL SYSTEMS

Please answer the discussion and review questions next.

DISCUSSION AND REVIEW QUESTIONS

Chapter 9.   INSTRUMENTATION AND CONTROL SYSTEMS

(Lesson 2 of 2 Lessons)

Write the answers to these questions in your notebook. The question numbering continues from Lesson 1.

8. What are the advantages and limitations of analog versus digital indicators?

9. Why is it poor practice to ignore the lamps that are lit up (alarm conditions) on an annunciator panel?

10. How should the constantly lit-up lamps (alarm conditions) on an annunciator panel be handled?

11. What electrical control features are available to protect pumps from damage?

12. What problems are created by oil and moisture in instrument air, and how can these contaminants be removed?

13. Why should plant measuring systems be periodically calibrated?

14. Why should rodents be kept out of instrumentation equipment?

15. How could you tell if a float for a level recorder might be stuck, and how would you determine if it was actually stuck?

SUGGESTED ANSWERS

Chapter 9.   INSTRUMENTATION AND CONTROL SYSTEMS

ANSWERS TO QUESTIONS IN LESSON 1

Answers to questions on page 632.

9.0A   A treatment plant operator must have a good working familiarity with instrumentation and control systems to properly monitor and control the treatment processes.

9.0B   This chapter contains information on general instrumentation functions and control system capabilities.

9.0C   Measurement instruments can serve as an extension of and improvement to an operator's senses of vision, touch, hearing, and even smell.

Answers to questions on page 633.

9.0D   Our senses provide us with qualitative and relatively short-term quantitative information; instrumentation measurements provide us with exact quantitative data.

9.0E   A measurement is, by definition, the comparison of a quantity with a standard unit of measure.

9.0F   *ACCURACY* refers to how closely an instrument measures the true or actual value of a process variable, while *PRECISION* refers to how closely the device can reproduce a given reading (or measurement) time after time.

9.0G   *ANALOG* displays show a reading as a pointer against a marked scale and *DIGITAL* displays provide direct, numerical readouts.

Answers to questions on page 636.

9.0H   Control systems are the means by which such process variables as pressure, level, weight, or flow are controlled.

9.0I   Examples of modulating control systems include: (1) chlorine residual analyzers/controllers; (2) flow-paced (open-loop) chemical feeders; (3) pressure- or flow-regulating valves; (4) continuous level control of process basins; and (5) variable-speed pumping systems for flow/level control.

9.0J   A motor control station or panel and the related circuitry essentially provide ON/OFF operation of an electric motor or other device.

Answers to questions on page 640.

9.1A   The general principles for safe performance on the job are to always avoid unsafe acts and correct unsafe conditions immediately.

9.1B   Electric shock can cause serious burns and even death (by asphyxiation due to paralysis of the muscles used in breathing).

9.1C   An electrical "explosion" could shower you with molten metal, startle you into a bumped head or elbow, or cause a bad fall.

9.1D   If electric current flows through your upper body, electric shock could harm your heart or your head.

Answers to questions on page 647.

9.1E   Operators should be especially careful when working around powered mechanical equipment because the equipment could start unexpectedly and cause serious injury.

9.1F   The purpose of an electrical lockout device is to physically prevent the operation of an electric circuit, or to de-energize the circuit temporarily.

9.1G   When testing the atmosphere before entry in any confined space, first test for oxygen deficiency/enrichment, then combustible gases and vapors, and then toxic gases and vapors.

9.1H   UEL stands for upper explosive limit and LEL stands for lower explosive limit. UEL and LEL indicate the range of concentrations at which combustible or explosive gases will ignite when an ignition source is present at ambient temperature.

9.1I   Wearing thin rubber or plastic gloves can greatly reduce your chances of electric shock.

Answers to questions on page 647.

9.2A   A primary element or sensor senses and quantifies measured variables such as pressure, level, and flow; at times, chemical feed rates and some physical or chemical process characteristics are also sensed. The primary element performs the initial conversion of measurement energy to a signal.

9.2B   Pressure is measured by the movement of a flexible element or a mechanically deformable device subjected to the force of the pressure being measured.

9.2C   Some pressure sensors are fitted with surge or overrange protection to limit the effect of pressure spikes or water hammer on the device.

Answers to questions on page 651.

9.2D Different types of liquid level-sensing devices include floats, displacers, electrical probes, direct hydrostatic pressure, pneumatic bubbler tubes, and ultrasonic (sonar) devices.

9.2E Single-point ball float levels mount in a sump or tank to sense a fixed elevation. When the water level rises, the ball floats and rolls onto its side, actuating an internal mercury switch. The switch provides a discrete ON/OFF signal used to control equipment or send level data to the control system.

9.2F Probes are used to measure liquid levels where mechanical systems are impractical, such as within sealed or pressurized tanks or with chemically active liquids.

9.2G Ultrasonic sensors measure the level of a liquid by emitting a pulse that is reflected back to the receiver. The sensor periodically generates a pulse of waves that bounce off the liquid surface and reflect back. The sensor detects the reflected pulse. Based on the speed of the pulse, the travel time between sending and receiving is measured and the distance is calculated and converted into a level measurement.

Answers to questions on page 659.

9.2H The two basic types of flow readings are rate of flow (volume per unit of time) and total flow (volume units).

9.2I The inferential flow-sensing techniques that are used in flow measurement are: (1) velocity, (2) differential pressure, (3) magnetic, and (4) ultrasonic.

9.2J Velocity-sensing meters measure process flows by measuring water speed within a pipeline or channel. One way of doing this is by sensing the rate of rotation of a special impeller placed within the flowing stream; the rate of flow is directly proportional to impeller RPM (within certain limits).

9.2K Flows measured with Venturi meters take advantage of a basic principle of hydraulics, the Bernoulli Effect: When a liquid is forced to go faster in a pipe or channel, its internal pressure drops.

Answers to questions on page 661.

9.2L In a fluid (liquid or gas) chemical feeder, the indication of quantity per unit of time, or liquid chemical feed rate, may be provided by use of a rotameter or built-in calibrated pump with indicated output settings.

9.2M Process variables commonly monitored or controlled by process instrumentation include DO, pH, chlorine residuals, electrical conductivity, ORP, alkalinity, turbidity, and temperature.

9.2N The two general systems for transmission of measurement signals are analog (4–20 mA) or digital data.

ANSWERS TO QUESTIONS IN LESSON 2

Answers to questions on page 663.

9.3A The purpose of instrumentation indicators is to give a visual representation of a variable's present value, either as an analog or digital display.

9.3B An analog display uses some type of pointer or graphical display against a scale.

9.3C Chart recorders, which by nature also serve as indicators, give a permanent record of how a variable changes with time by way of a moving chart.

9.3D Factors that can disrupt instrument and control systems include temporary power failures, tripped panel circuit breakers, voltage surges resulting in blown fuses, static electricity, and excessive heat.

Answers to questions on page 666.

9.3E Recorders are indicators designed to show how the value of the variable has changed with time.

9.3F The two main types of chart recorders are the strip-chart type and the circular-chart type.

9.3G Chart recorders are typically electrical, powered by the instrumentation and control system electrical supply. Battery-backed power for the recorders may be used, if required, as a backup power source.

9.3H Alarms may produce visual or audible warning signals.

Answers to questions on page 667.

9.3I An operator might bypass the operation of a controller in an emergency or when, in the judgment of the operator, the controller is not exercising control within sensible limits. To bypass a controller, switch to the manual mode of operation.

9.3J The basic principle to guide you in setting up the frequency of ON/OFF operation of a piece of equipment is to set the ON/OFF controls to operate or cycle associated equipment on and off no more often than actually necessary for plant operation.

9.3K Pumps can be prevented from running after a loss of level signal by electrical circuitry designed so the pump will turn *OFF* on an open signal circuit and *ON* only with a closed circuit.

9.3L Pump control panels may include automatic or manual alternators (two pumps) or sequencers (more than two pumps). This provision allows the total pump operating time required for a particular system to be distributed equally among all the pumps at a pump station.

Answers to questions on page 670.

9.3M Pneumatic instrumentation depends on a constant source of clean, dry, pressurized air for reliable operation.

9.3N   Oil is removed from instrument air by filtering the air through special oil-absorbent elements. Water is removed from the air by passing the moisture-laden air through an air-drying system consisting of a moisture separation column and air dryers.

Answers to questions on page 672.

9.3O   An operator should be especially careful when working with laboratory instruments because some of these instruments are very delicate and replacement parts, such as glass vessels or pH electrodes, are quite expensive. Moreover, the use of some of these instruments requires the regular handling of laboratory glassware and other breakable items.

9.3P   Operators should become familiar with the testing and calibration of plant measuring and control systems in order to discuss needed repairs or adjustments with an instrument technician and perhaps assist with that work. This knowledge also enables operators to analyze the effects of instrument problems on continued plant operation and to handle emergency situations created by instrument failure.

9.3Q   The most useful piece of general electrical test equipment is the V-O-M, that is the Volt-Ohm-Milliammeter, commonly referred to as a multimeter.

Answers to questions on page 674.

9.3R   Computer control systems used for process control are classified by two commonly used terms: the Distributed Control System (DCS) and the Supervisory Control And Data Acquisition (SCADA) system.

9.3S   The five basic components of the computer control system are: (1) the process instrumentation and control devices, (2) the input/output (I/O) interface, (3) the central processing unit (CPU), (4) the communication interface, and (5) the human machine interface (HMI). These components are the means by which the control system gathers and distributes information for the human operator and the process instrumentation and equipment.

9.3T   Computer control systems help operation and maintenance personnel in preparing daily, weekly, and monthly maintenance schedules, monitoring the spare parts inventory status, and ordering additional spare parts, when necessary, printing out work orders, and recording completed work assignments.

Answers to questions on page 676.

9.3U   Computer control systems record and store (archive) information that historically was recorded on paper strip-charts.

9.3V   Cost savings associated with the use of a computer control system frequently include reduced labor costs for operation, maintenance, and monitoring functions that were formerly performed manually. Precise control of chemical feed rates by a computer control system eliminates wasteful overdosing. Preventive maintenance monitoring can save on equipment and repair costs, and, as previously noted, energy savings may result from off-peak (reduced) electrical power rates.

9.3W   The greatest challenge for operators using computer control systems is to realize that just because a computer says something, this does not mean that the computer is always correct. Operators will always be needed to question and analyze the results from computer control systems.

Answers to questions on page 676.

9.4A   The key to proper instrument operation and maintenance (O & M) is the operator's practical understanding of the system. Operators must know how to: (1) recognize malfunctioning instruments so as to prevent prolonged damaging operation, (2) shut down and prepare devices for seasonal or other long-term nonoperation, and (3) perform preventive (and minor corrective) maintenance tasks to ensure proper operation in the long term.

9.4B   General tasks expected of operators of instrumentation systems can be summed up, from an operations standpoint, as learning, and constant attention to, what constitutes normal function, and, from a maintenance standpoint, as ensuring proper and continuing protection and care of each component.

9.4C   Two of the surest indications of a serious electrical problem in instruments or power circuits are smoke or a burning odor.

Answers to questions on page 677.

9.4D   Moisture can be controlled in an electrical panel room containing instrumentation by leaving some power components on (such as a power transformer) to provide space heat. In a known moist environment, sealed instrument cases may be protected for a while with a container of desiccant.

9.4E   Pneumatic instrumentation should be purged with dry air before shutdown to rid the individual parts of residual oil and moisture to minimize internal sticking and corrosion while standing idle.

Answers to questions on page 678.

9.4F   Regular preventive maintenance duties should be printed on forms or cards (or appear on a computer screen) for use by operators as a reminder, guide, and record of preventive maintenance tasks performed.

9.4G   To obtain a technical manual for an instrument, write to the manufacturer. Be sure to provide all relevant serial/model numbers in your request to the manufacturer for a manual.

9.4H   Instrument supplies and spare parts that should always be available include charts, pens, pen cleaners, and ink, and any other parts listed in the technical manual as necessary for instrument operation or service.

Answers to questions on page 678.

9.4I   Operational checks on instrumentation equipment are performed by always observing each system for its continuing signs of normal operation, and cycling some indicators by certain test methods.

9.4J   If a recorder trace is altered from its usual pattern during the process of checking an instrument, the operator causing the upset should initial the chart appropriately and note the time.

APPENDIX

ADVANCED WASTE TREATMENT

Comprehensive Review Questions and Suggested Answers

Waste Treatment Words

Subject Index

COMPREHENSIVE REVIEW QUESTIONS

This section was prepared to help you review the material in this manual. The questions are divided into five types:

1. True-False

2. Best Answer

3. Multiple Choice

4. Short Answer

5. Problems

To work this section:

1. Write the answer to each question in your notebook.

2. After you have worked a group of questions (you decide how many), check your answers with the suggested answers at the end of this section.

3. If you missed a question and do not understand why, reread the material in the manual.

You may wish to use this section for review purposes when preparing for civil service and certification examinations.

Since you have already completed this course, please *DO NOT SEND* your answers to California State University, Sacramento.

True-False

1. Odor control in wastewater collection systems and at wastewater treatment plants is very important.

   1. True
   2. False

2. Each person has a different tolerance level for various odors.

   1. True
   2. False

3. Odors are always removed by chlorine.

   1. True
   2. False

4. Sampling is the only time operators should not handle liquid oxygen.

   1. True
   2. False

5. In the sludge reaeration (contact stabilization) variation of the activated sludge process, the return sludge rate is very significant.

   1. True
   2. False

6. The character of any industrial wastewater depends on the particular production process and the way it is operated.

   1. True
   2. False

7. Excessive use of defoamers in pulp mill wastewater may also add an undesirable oxygen demand on the system.

   1. True
   2. False

8. The solids removed from wastewater may either be disposed of or reused.

   1. True
   2. False

9. Centrifuges are commonly used to thicken primary sludges.

   1. True
   2. False

10. A seasonal fluctuation has been noted in chemical conditioning requirements so that many plants can successfully condition using cationic polymers during the summer and anionic polymers during the winter.

    1. True
    2. False

11. The plate and frame filter press operates in a continuous mode.

    1. True
    2. False

12. Chemically stabilized and wet-oxidized sludges are suited for compost operations.

    1. True
    2. False

13. Proper coagulation produces agglomerated solids particle units that tend to hold together and do not easily redisperse.

    1. True
    2. False

14. Polymers carrying free electrical charges are "nonionic polyelectrolytes."

    1. True
    2. False

15. A delay of as little time as 30 minutes between sample collection time and jar testing can significantly affect the test results.

    1. True
    2. False

16. The suspended solids removal rates of most types of filters can be greatly improved by the addition of chemicals before filtration.

    1. True
    2. False

17. Under normal operation, the surface of the membrane becomes fouled with oil, grease, and other foulants.

    1. True
    2. False

18. Phosphorus can be removed from wastewater both biologically and chemically.

    1. True
    2. False

19. Wet polymers are extremely slippery.

    1. True
    2. False

20. Once the lime precipitate containing phosphorus is removed from the wastewater stream, it is important to allow the same phosphorus to be recycled back through the treatment plant.

    1. True
    2. False

21. Biological denitrification is the process by which microorganisms reduce nitrate to nitrogen gas that is released to the atmosphere.

    1. True
    2. False

22. The population of the nitrifying bacteria will be limited by the amount of ammonia available in the wastewater.

    1. True
    2. False

23. Denitrification converts nitrogen that is in an objectionable form (as a nutrient) to one that has no significant impact on environmental quality.

    1. True
    2. False

24. In all biological systems, there is a characteristic MCRT for each process phase.

    1. True
    2. False

25. High pH conditions are preferred by the microorganisms responsible for hydrolysis and volatile fatty acid formation.

    1. True
    2. False

26. The more vigorous the aeration, the greater the pH drop.

    1. True
    2. False

27. Wastewater reclamation and recycling at steel mills requires the identification of sources of wastewater.

    1. True
    2. False

28. Maintenance means good housekeeping.

    1. True
    2. False

29. Infiltration–percolation systems are suited for crop growth.

    1. True
    2. False

30. The more you know about your plant's instrumentation, the better operator you become.

    1. True
    2. False

31. Digital displays show a reading as a pointer against a marked scale.

    1. True
    2. False

32. Differential pressure flow measuring devices require significant preventive maintenance because of all the moving parts.

    1. True
    2. False

33. The computer control system can alert the operator or change a plant process based on a preprogrammed control strategy defined by the operator.

    1. True
    2. False

Best Answer (Select only the closest or best answer.)

1. What form(s) of oxygen can be used by facultative microorganisms?

    1. Molecular oxygen and combined (or bound) sources of oxygen
    2. Only atmospheric air oxygen
    3. Only bound sources of oxygen
    4. Only molecular oxygen

2. What is the most common method used to evaluate odor nuisances?

    1. The intensity of neighborhood committees
    2. The number of odor complaints
    3. The odor panel
    4. The use of atmospheric odor analyses equipment

3. Why is soft water supplied to the nozzles in the chemical mist odor control system?

1. To increase odor removal
2. To prevent the creation of side stream odors
3. To produce an acceptable odor
4. To reduce calcium buildup

4. Treatment plant areas exposed to odorous air must be made of what type of materials?

1. Corrosion-resistant materials
2. Easily painted materials
3. Easily polished materials
4. Moisture-resistant materials

5. What is the most effective way of removing ammonia from plant effluents?

1. Biological nitrification (convert ammonium to nitrate)
2. Chemical precipitation
3. pH adjustment
4. Physical screening

6. What is the objective of wasting activated sludge?

1. To allow plant effluent to minimize suspended solids
2. To keep the solids loading in the aeration basin constant at all times
3. To maintain a balance between the microorganisms under aeration and the amount of incoming food as measured by the COD test
4. To prevent the waste activated sludge from flowing over the aeration basin (reactor) walls

7. What is the first indication that there is a toxic waste load within the treatment plant?

1. Denitrification will be observed in the final clarifier
2. The aeration basin will release obnoxious odors
3. The DO residual in the aeration basin will decrease significantly
4. The DO residual in the aeration basin will increase significantly

8. Which item could hinder microorganism activity in the BOD test in some industrial wastes?

1. Additional nutrients
2. Higher levels of coliforms
3. Increase in organic matter
4. Presence of toxic wastes

9. What is denitrification?

1. The conversion of ammonia to nitrate
2. The conversion of nitrate to nitrogen gas
3. The conversion of organic nitrogen to ammonia
4. The conversion of organic nitrogen to nitrate

10. Why are secondary sludges suitable for centrifugal thickening?

1. Because the biological nature of the sludges thickens readily
2. Because the composition of the sludge allows for the release of bound water
3. Because the relative lack of stringy and bulky material reduces the potential for plugging
4. Because the sources of secondary sludges provide for easy transport of sludge to the centrifuge

11. When does washout occur in a gravity belt thickener? When a large quantity of free water is

1. Mixed with the polymer to thicken the sludge
2. Unable to be released in the drainage zone
3. Used to wash the belt during maintenance
4. Used to wash the thickened sludge

12. Why should dry chemicals be stored in a dry place and not allowed to absorb moisture?

1. If chemicals become wet, balls or cakes of chemicals will form and prevent easy handling
2. If chemicals become wet, corrosive conditions could be produced
3. If chemicals become wet, fire hazards could develop
4. If chemicals become wet, they will lose their conditioning strength

13. When disposing of sludge, what is a vector?

1. A device that measures and tracks contaminant movement toward groundwater
2. A type of sludge pump that removes accumulated sludges to prevent odors
3. An insect or rodent capable of spreading germs or other agents of disease
4. The predominant direction wind blows from the disposal site

14. What happens during the coagulation phase of physical–chemical treatment?

1. Chemicals are added to the wastewater and rapidly mixed with the process flow
2. Gentle mixing of the wastewater
3. Media filtration
4. Sedimentation by gravity settling

15. What problem can be created in the coagulation process by too long a rapid mix?

1. The breakup and separation of the forming floc
2. The creation of bulking sludge
3. The formation of "pinpoint floc"
4. The gathering together of large clumps of floc

16. Why is a surface wash system beneficial during the back-washing of gravity filters?

    1. To cleanse mudballs before filtration
    2. To increase the frequency of backwashing
    3. To recover backwash water for future filtration
    4. To scrub the surface mat of accumulated solids

17. What is the permeate?

    1. Solidified solids that gather inside a membrane
    2. Solids that are rejected by a membrane
    3. The thin film layer of solids on the surface of a membrane
    4. Water that passes through a membrane

18. How is the effectiveness of a membrane cleaning procedure assessed?

    1. By measuring the length of membrane filter run
    2. By measuring the particle removal efficiency of the membrane
    3. By measuring the rate of water passage through a membrane in flux
    4. By measuring the retentate flow past the membrane

19. Why is phosphorus removed from wastewater?

    1. Phosphorus encourages the growth of crops
    2. Phosphorus is toxic to fish
    3. Phosphorus produces odors in drinking water
    4. Phosphorus provides a nutrient or food source for algae

20. The flow rate into the chemical clarifier depends on which factor?

    1. CT value
    2. Depth over weir
    3. Detention time
    4. Overflow rates

21. How can an operator determine what pH levels form the largest floc possible and allow the fastest settling of the floc formed?

    1. By the use of chemical analysis tests
    2. By the use of floc density tests
    3. By the use of flocculation tests
    4. By the use of jar tests

22. Which problem can be created in overland flow systems removing nitrogen during the initial runoff from a storm?

    1. A drop in effluent temperature
    2. A "first flush" of toxic chemicals
    3. An increase in solids in the effluent
    4. An increase in the rate of denitrification

23. What problem is created in the attached growth nitrification process when the flow is in excess of design guidelines?

    1. Excessively high removal efficiencies
    2. Hydraulic sloughing or washing off of the film on the media
    3. Odors will be released due to excessive turbulence
    4. Pumps will be operating at optimum efficiencies

24. How is denitrification accomplished in wastewater?

    1. Ammonia is oxidized to nitrate by a broad range of microorganisms under suitable environmental conditions
    2. Nitrate is reduced to gaseous nitrogen by a broad range of microorganisms under suitable environmental conditions
    3. Nitrogen stripping towers are used to accomplish denitrification
    4. Physical–chemical treatment processes are used to accomplish denitrification

25. How is system loading described in enhanced biological (nutrient) control?

    1. Daily pollutant mass applied
    2. Flow volume per unit time
    3. Pollutant mass applied per unit time per some physical measure of the reactor
    4. Toxicity level per unit time

26. Why is the point where raw wastewater enters the plant the most practical place to put the anaerobic reactor for phosphorus removal?

    1. Because converted nitrogen compounds, which can interfere with phosphorus removal, are absent
    2. Because raw wastewaters contain larger portions of solids that are favored by phosphorus removal organisms
    3. Because raw wastewaters generally have the lowest pH and this condition is most favored by the volatile fatty acid formers and enhanced biological phosphorus removal organisms
    4. Because the location is at the highest point and all wastewater being treated is flowing by gravity through the plant

27. What is recycled water?

    1. The O & M procedures used to achieve compliance
    2. The operation or process of changing the condition or characteristics of water so that improved uses can be achieved
    3. Water that is discharged and then withdrawn by another user
    4. Water that is used within a facility before it is discharged to a treatment system

28. What is an important task when reviewing plans and specifications for wastewater reclamation systems?

    1. To avoid fire hazards
    2. To eliminate confined spaces
    3. To ensure a piping system with alternate flow paths
    4. To prevent the emission of toxic gases

29. Why should wells be monitored near a land treatment irrigation system?

    1. To document slope of the groundwater surface
    2. To identify any adverse effects on groundwaters
    3. To monitor any depletion of the groundwater table
    4. To prevent slime growths on well screens

30. Which instrumentation and control system tasks should operators be able to perform in operating a treatment plant?

1. Recognize failures, troubleshoot, and make adjustments and minor repairs
2. Repair and replace defective instruments
3. Specify and prepare plans for new instrumentation systems
4. Write computer programs for instrumentation controls

31. What is a process variable?

1. A physical or chemical quantity that is measured or controlled in a wastewater treatment process
2. How closely a device can reproduce a given reading (or measurement) time after time
3. How closely an instrument measures the true or actual value of a process variable
4. The smallest change in a process variable that an instrument can sense

32. Why are some pressure sensors fitted with surge or over-range protection (snubbers)?

1. To identify the source of the surge and take preventive action
2. To limit the effect pressure spikes or water hammer have on the instrument
3. To protect the pipe from bursting under excessive pressures
4. To record instantaneous peak pressure and alert the operator to inspect for leaks

33. Why should operators learn how to shift all controllers to manual operation?

1. In order to learn the value of controllers
2. In order to observe changes in downstream processes
3. In order to prepare operator training programs
4. In order to take over control of a critical system, when necessary, in an emergency

Multiple Choice (Select all correct answers.)

1. Which gases give off the most offensive odors?

1. Ammonia
2. Carbon dioxide
3. Carbon monoxide
4. Hydrogen sulfide
5. Methane

2. Which items should be included on an "Odor Complaint Form"?

1. Action taken
2. Complaint information
3. Description of odor
4. Investigator information
5. Strength of odor

3. What are the advantages of a packed bed scrubber to remove odors from air?

1. Ability to effectively handle changes in odorous compound concentrations
2. Economical treatment of high gas flows
3. High mass transfer efficiency
4. High oxidant regenerant usage
5. Recycling of odorous compounds

4. Which items are very important factors that operators must control for successful performance of the activated sludge process?

1. Influent flows
2. Influent waste characteristics
3. Plant laboratory performance
4. Return activated sludge (RAS) rate
5. Waste activated sludge (WAS) rate

5. What are the advantages of the constant percentage RAS flow approach?

1. Maximum solids loading on the clarifier occurs at the start of peak flow periods
2. Requires less operational time
3. Simplicity
4. The MLSS will remain in the clarifier for shorter time periods, which may reduce the possibility of denitrification in the clarifier
5. Variations in the MLSS concentration are reduced and the F/M ratio varies less

6. Why is an industrial waste monitoring program valuable?

1. To assure regulatory agencies of industrial compliance with discharge requirements
2. To assure regulatory agencies of industrial implementation schedules set forth in the discharge permit
3. To gather necessary data for the future design and operation of the treatment plant
4. To maintain sufficient control of treatment plant operations to prevent NPDES permit violations
5. To warn industries that noncompliance is unacceptable

7. What are the basic reasons for keeping accurate records for an industrial wastewater treatment plant?

1. To account for the operation of the plant on a weekly or monthly basis
2. To ensure that all slots on recordkeeping forms are filled in
3. To keep a history that helps troubleshoot problems that arise
4. To keep operators busy when they are not performing O & M tasks
5. To prepare records that can serve as legal documents that will protect the company from unjust claims

8. Nitrogenous compounds discharged from wastewater treatment plants can have which harmful effects?

   1. Adverse public health effects (mainly groundwater)
   2. Ammonia toxicity to fish
   3. Increase in the dissolved oxygen depletion in receiving waters
   4. Reduction in chlorine disinfection efficiency
   5. Reduction in the suitability of the water for reuse

9. Which principal control guidelines must an operator consider to operate and maintain the biological nitrification process at optimum performance levels?

   1. Detention time
   2. Dissolved oxygen
   3. Nitrogenous food
   4. pH
   5. Toxic materials

10. The daily quantity of primary sludge generated by a wastewater treatment plant depends on which factors?

    1. Concentration of influent settleable suspended solids
    2. Efficiency of the biological system in removing organic matter
    3. Efficiency of the primary sedimentation basin
    4. Growth rate, Y, of the bacteria within the system
    5. Influent wastewater flow

11. How does an operator achieve peak performance of a gravity thickener?

    1. By adjusting the influent BOD
    2. By adjusting the sludge withdrawal rate
    3. By controlling the sludge blanket depth
    4. By controlling the speed of the sludge collection mechanism
    5. By optimizing the DO level

12. Which factors determine the air requirements for an aerobic digester?

    1. Desire to control effluent BOD
    2. Desire to keep the digester solids in suspension (well mixed)
    3. Desire to maintain a dissolved oxygen (DO) concentration of 1 to 2 mg/L within the digester
    4. Desire to minimize odor production from the process
    5. Desire to optimize methane production for heating purposes

13. In order to create a suitable environment for composting, which guidelines must be met?

    1. Aeration must be sufficient to maintain aerobic conditions in the composting material
    2. Combustible gases must be adequately mixed with air
    3. Odors must be properly controlled
    4. Proper moisture content and temperatures must be maintained
    5. The wastewater sludges must be blended with previously composted material or bulking agents such as sawdust, straw, wood shavings, or rice hulls

14. Physical–chemical treatment consists of which processes?

    1. Coagulation
    2. Comminution
    3. Conditioning
    4. Flocculation
    5. Liquid/solids separation

15. Which items are chemical treatment hazards?

    1. Dust in the air
    2. High temperatures
    3. Slippery walk areas
    4. Strong acids
    5. Strong caustics

16. Which items should be checked before chemical feeder start-up?

    1. Be sure that the chemical to be fed is available
    2. Confirm that the manufacturer's lubrication and start-up procedures are being followed
    3. Determine the proper positions for all valves
    4. Examine all fittings, inspection plates, and drains to ensure that they will not leak when placed into service
    5. Inspect the electrical system for proper voltage, for properly sized overload protection, and for proper operation of control lights on the control panel

17. Which items should be checked before starting a filtering system?

    1. Check motors and drives for proper alignment, for proper safety guards, and for free rotation
    2. Examine motors, drive units, and bearings for proper lubrication
    3. Fill tanks and piping and look for leaks
    4. Inspect pumps and motors for excessive vibration
    5. Inspect the total system for safety hazards

18. What happens when suspended material accumulates on the pressure filter media surface or in the filter media bed?

    1. The differential pressure across the filter increases
    2. The filter effluent quality deteriorates
    3. The flow through the filter decreases
    4. The suspended material deteriorates and releases objectionable odors
    5. The suspended material produces a high chlorine demand

19. Which factors are considered when selecting a membrane material (durable polymer)?

    1. Ability to retain particles
    2. Ability to show reproducible performance over a long life
    3. Color of the durable polymer
    4. Compatibility with the wastestream
    5. Flexibility of the durable polymer

20. What are the most common types of phosphorus removal systems?

1. Biological phosphorus removal
2. Enhanced coagulation
3. Filtration following aluminum sulfate flocculation
4. Iron co-precipitation
5. Lime precipitation

21. To ensure smooth, efficient operation of a biological phosphorus removal system, the operator will need to carefully monitor which items on a daily basis?

1. Coliform levels in the plant effluent
2. Dissolved oxygen levels in the aeration basin
3. Lime feed and mixing systems
4. pH level in the chemical clarifier
5. State of anaerobic conditions in the phosphorus stripping tank

22. Which factors affect phosphorus removal efficiency?

1. Changes in pH
2. Short-circuiting
3. Small straggler floc
4. Solids loading
5. Stormwater

23. Which conditions are essential for the nitrification process?

1. Adequate amounts of food
2. Adequate amounts of free oxygen
3. Optimal microorganism population
4. Optimal pH
5. Optimal temperature

24. Which process variables can an operator adjust to compensate for colder winter conditions when trying to achieve nitrification?

1. Adjust the pH to higher levels
2. Adjust the pH to lower levels
3. Decrease the MCRT
4. Increase the MCRT
5. Increase the MLVSS concentration

25. Which problems could an operator encounter when troubleshooting a nitrification process?

1. Decrease in nitrification unit pH with a loss of nitrification
2. In a two-stage activated sludge system, SVI of nitrification sludge is very high
3. Inability to completely nitrify
4. Loss of solids from the final clarifier
5. Loss of solids from the trickling filter or RBC

26. Enhanced biological control requires measurement of existing environmental and physical conditions in the reactor, such as which items?

1. Coliforms
2. DO
3. Hardness
4. pH
5. Temperature

27. Water quality monitoring for process control at the influent and effluent of the process tanks or basins of biological nutrient removal systems should include which items?

1. Alkalinity
2. Dissolved oxygen
3. pH
4. Suspended solids
5. Temperature

28. Which items should be reviewed by operators to aid the design engineer during the design stages of an enhanced biological control activated sludge system?

1. Chemical storage tanks should be constructed within diked chemical containment areas that can hold the contents of the largest tank if the tank should rupture
2. Design of the secondary clarifier scum removal and handling facilities must be able to deal effectively with the potential of excessive scum development
3. Regardless of the type of aerobic zone, the recycle pumps of the nitrate-containing mixed liquor should not be located next to an aeration device
4. The discharge points of influent and recycle flow into the anaerobic zone should always be submerged to avoid entraining air into the basin contents
5. The recycle pumps of the nitrate-containing mixed liquor should be located near the downstream end of a plug flow aerobic basin

29. Under what conditions is effluent reuse considered?

1. When federal regulations require reuse
2. When groundwaters are either not available or the costs of pumping and any treatment are prohibitive
3. When surface waters are not available or the cost of surface water treatment is excessive
4. When the cost of purchasing available treated water is too expensive
5. When the volume of municipal water needed is not available

30. Water reclaimed for recreation must consider which items?

1. Aesthetics
2. Hardness
3. Photosynthesis
4. Protection of public health
5. THMs

31. How does vegetation treat wastewater in an overland flow system?

    1. By decaying and providing food for organisms
    2. By providing fodder for livestock
    3. By removing carbon dioxide from the atmosphere
    4. By removing insoluble BOD
    5. By removing suspended solids

32. What are possible solutions to sealing of the soil caused by solids forming a mat on the soil surface?

    1. Apply water intermittently and allow a drying period
    2. Disk or plow the field to break the mat of solids
    3. Introduce microorganisms that will degrade the mat
    4. Plant vegetation that thrives on solids in the mat
    5. Remove suspended solids from the effluent

33. Which minor corrective maintenance or preventive maintenance instrumentation tasks should operators be able to perform?

    1. Make adjustments
    2. Make minor repairs
    3. Properly tap an errant gauge
    4. Recognize failures
    5. Troubleshoot

34. What are the most common variables that are measured at wastewater treatment plants?

    1. Flow
    2. Level
    3. Number of dischargers
    4. Pressure
    5. Weather

35. Which factors could cause high readings with a propeller meter?

    1. Bearing wear and gear train friction
    2. Malfunctioning receiver
    3. Malfunctioning transmitter
    4. Partially full pipeline
    5. Wrong gears installed

36. ON/OFF types of controllers are used to start and stop pumps according to which measurements?

    1. Dissolved oxygen
    2. Electrical power supply
    3. Flow
    4. Level
    5. Pressure

Short Answer

1. How should an odor complaint be handled?

2. How can odors in air be treated?

3. Why does the pure oxygen process normally use sealed reactors?

4. Different RAS flow rates will be required as the result of what two activated sludge conditions?

5. What items are affected by the amount of waste activated sludge (WAS) removed from the process?

6. How will you know when you have established the best mode of process control for your plant?

7. Why would an operator want to monitor the influent to a wastewater treatment plant?

8. Why should an operational strategy be developed before a toxic waste is discovered in the influent to a plant?

9. How can the amount of nutrients to be added each day be determined?

10. Why is the handling and disposal of sludge such a complicated problem?

11. What is the most important operational concern when operating a dissolved air flotation thickener?

12. Why do wastewater treatment plant sludges have to be stabilized before disposal?

13. How is the optimum type of chemical and dose to condition a particular sludge determined? Why?

14. How would you determine filtration time for a pressure filter?

15. Why must care be taken to avoid chemical overdosing of sludges applied to sand drying beds?

16. Why are chemically stabilized and wet-oxidized sludges generally not suited for compost operations?

17. What are the main reasons for using wastewater solids for agricultural purposes?

18. What are the three most important process control guidelines in any chemical treatment process?

19. What factors influence the selection of a polymer?

20. Explain how to run the jar test.

21. Why are plunger pumps used for metering chemicals?

22. What is the purpose of instrumentation used with a filter system?

23. How can a constant rate of filtration be maintained in a pressure filter?

24. What are the advantages of continuous backwash, upflow, deep-bed silica sand media filters over other types of granular media filters?

25. How can fouled membranes be cleaned?

26. Why is oxygen so important in biological treatment processes?

27. Why should the system that pumps the activated sludge into and out of the anaerobic phosphorus stripping tank be equipped with variable-speed controls?

28. Why would you perform a jar test when removing phosphorus by the lime precipitation process?

29. Why do pumps moving alum solutions clog?

30. Why do attached growth reactors use recirculation?

31. How does air come in contact with the wastewater in the air stripping process?

32. What is the purpose of "enhanced biological control"?

33. How is system flexibility achieved?

34. What are some of the limitations of or precautions for direct use of reclaimed wastewater?

35. What is the purpose of a monitoring program?

36. How does land treatment work?

37. How long should the pumps be run while irrigating a plot of land?

38. Why should operators understand instrumentation and control systems?

39. What is the difference between accuracy and precision?

40. Why is it poor practice to ignore the lamps that are lit up (alarm conditions) on an annunciator panel?

41. Why should plant measuring systems be periodically calibrated?

Problems

1. An aeration tank has a volume of 1.6 million gallons. The MLSS are 2,400 mg/L and the volatile portion is 0.75. How many pounds of mixed liquor volatile suspended solids (MLVSS) are under aeration?

2. An aerator contains 24,250 lbs of solids, the RAS suspended solids concentration is 7,100 mg/L, and the current WAS flow is 0.02 MGD. The desired sludge age is 6 days. If the plant adds 3,800 lbs of solids per day, what is the total waste activated sludge (WAS) flow rate in GPM?

3. Seven thousand five hundred gallons of sludge with a sludge solids concentration of 1.5 percent is thickened to a sludge solids concentration of 5 percent. What is the reduction in volume of sludge? *HINT:* How many gallons of water were removed by the thickening process?

4. A waste activated sludge is pumped at 50 GPM with a sludge solids concentration of 3.0 percent sludge solids. Jar tests indicate that 8 pounds per day of dry polymer are necessary for successful gravity thickening. What is the polymer dosage in pounds of polymer per ton of dry sludge solids?

SUGGESTED ANSWERS
TO
COMPREHENSIVE REVIEW QUESTIONS

True-False

1. True — Odor control in wastewater collection systems and at wastewater treatment plants is very important.

2. True — Each person has a different tolerance level for various odors.

3. False — Odors are NOT always removed by chlorine.

4. False — Sampling is the only time operators need to handle (NOT should not handle) liquid oxygen.

5. True — In the sludge reaeration (contact stabilization) variation of the activated sludge process, the return sludge rate is very significant.

6. True — The character of any industrial wastewater depends on the particular production process and the way it is operated.

7. True — Excessive use of defoamers in pulp mill wastewater may also add an undesirable oxygen demand on the system.

8. True — The solids removed from wastewater may either be disposed of or reused.

9. False — Centrifuges are NOT commonly used to thicken primary sludges because all three centrifuge designs have sludge inlet assemblies that clog easily.

10. True — A seasonal fluctuation has been noted in chemical conditioning requirements so that many plants can use cationic polymers during the summer and anionic polymers during the winter.

11. False — The plate and frame filter press operates in a batch (NOT continuous) mode.

12. False — Chemically stabilized and wet-oxidized sludges are NOT suited for compost operations.

13. True — Proper coagulation produces agglomerated solids particle units that tend to hold together and do not easily redisperse.

14. False — Polymers carrying NO free electrical charge are "nonionic polyelectrolytes."

15. True — A delay of as little time as 30 minutes between sample collection time and jar testing can significantly affect the test results.

16. True — The suspended solids removal rates of most types of filters can be greatly improved by the addition of chemicals before filtration.

17. True — Under normal operation, the surface of the membrane becomes fouled with oil, grease, and other foulants.

18. True — Phosphorus can be removed from wastewater both biologically and chemically.

19. True — Wet polymers are extremely slippery.

20. False — Once the lime precipitate containing phosphorus is removed from the wastewater stream, it is important NOT to allow the same phosphorus to be recycled back through the treatment plant.

21. True — Biological denitrification is the process by which microorganisms reduce nitrate to nitrogen gas that is released to the atmosphere.

22. True — The population of the nitrifying bacteria will be limited by the amount of ammonia available in the wastewater.

23. True — Denitrification converts nitrogen that is in an objectionable form (as a nutrient) to one that has no significant impact on environmental quality.

24. True — In all biological systems, there is a characteristic MCRT for each process phase.

25. False — Low (NOT high) pH conditions are preferred by the microorganisms responsible for hydrolysis and volatile fatty acid formation.

26. False — The more vigorous the aeration, the greater the pH rise (NOT drop).

27. True — Wastewater reclamation and recycling at steel mills requires the identification of sources of wastewater.

28. True — Maintenance means good housekeeping.

29. False — Infiltration–percolation systems are NOT suited for crop growth.

30. True — The more you know about your plant's instrumentation, the better operator you become.

31. False — Analog (NOT digital) displays show a reading as a pointer against a marked scale.

32. False   Differential pressure flow measuring devices require little (NOT significant) preventive maintenance because there are NO moving parts.

33. True   The computer control system can alert the operator or change a plant process based on a preprogrammed control strategy defined by the operator.

Best Answer

1. 1   Molecular oxygen and combined (or bound) sources of oxygen can be used by facultative microorganisms.

2. 3   The odor panel is the most common method used to evaluate odor nuisances.

3. 4   Soft water is supplied to the nozzles in the chemical mist odor control system to reduce calcium buildup.

4. 1   Treatment plant areas exposed to odorous air must be made of corrosion-resistant materials.

5. 1   Biological nitrification (convert ammonium to nitrate) is the most effective way of removing ammonia from plant effluents.

6. 3   The objective of wasting activated sludge is to maintain a balance between the microorganisms under aeration and the amount of incoming food as measured by the COD test.

7. 4   The first indication that there is a toxic waste load within the treatment plant is that the DO residual in the aeration basin will increase significantly.

8. 4   The presence of toxic wastes could hinder microorganism activity in the BOD test in some industrial wastes.

9. 2   Denitrification is the conversion of nitrate to nitrogen gas.

10. 3   Secondary sludges are suitable for centrifugal thickening because the relative lack of stringy and bulky material reduces the potential for plugging.

11. 2   Washout occurs in a gravity belt thickener when a large quantity of free water is unable to be released in the drainage zone.

12. 1   Dry chemicals should be stored in a dry place and not allowed to absorb moisture because if chemicals become wet, balls or cakes of chemicals will form and prevent easy handling.

13. 3   When disposing of sludge, a vector is an insect or rodent capable of spreading germs or other agents of disease.

14. 1   During the coagulation phase of physical–chemical treatment, chemicals are added to the wastewater and rapidly mixed with the process flow.

15. 1   A problem created in the coagulation process by too long a rapid mix is the breakup and separation of the forming floc.

16. 4   Surface wash systems are used during the backwashing of gravity filters to scrub the surface mat of accumulated solids.

17. 4   The permeate is the water that passes through a membrane.

18. 3   The effectiveness of a membrane cleaning procedure is assessed by measuring the rate of water passage through a membrane in flux.

19. 4   Phosphorus is removed from wastewater because it provides a nutrient or food source for algae.

20. 4   The flow rate into the chemical clarifier depends on overflow rates.

21. 4   Determine the pH levels that form the largest floc possible and allow the fastest settling of the floc formed by the use of jar tests.

22. 3   The initial runoff from a storm can cause an increase in solids in the effluent in overland flow systems removing nitrogen.

23. 2   In the attached growth nitrification process, a flow in excess of design guidelines may cause hydraulic sloughing or washing off of the film on the media.

24. 2   Denitrification is accomplished in wastewater by nitrate being reduced to gaseous nitrogen by a broad range of microorganisms under suitable environmental conditions.

25. 3   System loading in enhanced biological (nutrient) control is described in terms of pollutant mass applied per unit time per some physical measure of the reactor.

26. 3   The most practical place to put the anaerobic reactor for phosphorus removal is the point where raw wastewater enters the plant because raw wastewaters generally have the lowest pH and this condition is most favored by the volatile fatty acid formers and enhanced biological phosphorus removal organisms.

27. 4   Recycled water is water that is used within a facility before it is discharged to a treatment system.

28. 3   The most important task when reviewing plans and specifications for wastewater reclamation systems is to ensure a piping system with alternate flow paths.

29. 2   Wells should be monitored near land treatment irrigation systems to identify any adverse effects on groundwaters.

30. 1   Instrumentation and control system tasks operators should be able to perform include recognizing failures, troubleshooting, and making adjustments and minor repairs.

31. 1   A process variable is a physical or chemical quantity that is measured or controlled in a wastewater treatment process.

32. 2   Some pressure sensors are fitted with surge or overrange protection to limit the effect pressure spikes or water hammer have on the instrument.

33. 4   Operators should learn how to shift all controllers to manual operation in order to take over control of a critical system, when necessary, in an emergency.

Multiple Choice

1. 1, 4   Ammonia and hydrogen sulfide gases give off the most offensive odors.

2. 1, 2, 3, 4, 5   An "Odor Complaint Form" should include the following items: (1) action taken, (2) complaint information, (3) description of odor, (4) investigator information, and (5) strength of odor.

3. 1, 2, 3   The advantages of a packed bed scrubber to remove odors from air include the ability to effectively handle changes in odorous compound concentrations, the economical treatment of high gas flows, and high mass transfer efficiency.

4. 4, 5   Return activated sludge (RAS) rate and waste activated sludge (WAS) rate are two very important factors that must be controlled for successful performance of the activated sludge process.

5. 4, 5   The advantages of the constant percentage RAS flow approach are that the MLSS will remain in the clarifier for shorter time periods, which may reduce the possibility of denitrification in the clarifier, and that variations in the MLSS concentration are reduced and the F/M ratio varies less.

6. 1, 2, 3, 4   An industrial waste monitoring program is valuable to assure regulatory agencies of industrial compliance with discharge requirements, to assure regulatory agencies of industrial implementation schedules set forth in the discharge permit, to gather necessary data for the future design and operation of the treatment plant, and to maintain sufficient control of treatment plant operations to prevent NPDES permit violations.

7. 1, 3, 5   The basic reasons for keeping accurate records for an industrial wastewater treatment plant are to account for the operation of the plant on a weekly or monthly basis, to keep a history that helps troubleshoot problems that arise, and so the records can serve as legal documents that will protect the company from unjust claims.

8. 1, 2, 3, 4, 5   Nitrogenous compounds discharged from wastewater treatment plants can have adverse public health effects (mainly groundwater), can cause ammonia toxicity to fish, can cause an increase in the dissolved oxygen depletion in receiving waters, can cause a reduction in chlorine disinfection efficiency, and can cause a reduction in the suitability of the water for reuse.

9. 1, 2, 3, 4, 5   To operate and maintain the biological nitrification process at optimum performance levels, an operator must consider the following principal control guidelines: (1) detention time, (2) dissolved oxygen, (3) nitrogenous food, (4) pH, and (5) toxic materials.

10. 1, 3, 5   The daily quantity of primary sludge generated by a wastewater treatment plant depends on the concentration of influent settleable suspended solids, the efficiency of the primary sedimentation basin, and the influent wastewater flow.

11. 2, 3, 4   The peak performance of a gravity thickener can be achieved by adjusting the sludge withdrawal rate, by controlling the sludge blanket depth, and by controlling the speed of the sludge collection mechanism.

12. 2, 3   The factors that determine the air requirements for an aerobic digester include the desire to keep the digester solids in suspension (well mixed) and the desire to maintain a dissolved oxygen (DO) concentration of 1 to 2 mg/L within the digester.

13. 1, 4, 5   In order to create a suitable environment for composting, aeration must be sufficient to maintain aerobic conditions in the composting material, proper moisture content and temperatures must be maintained, and the wastewater sludges must be blended with previously composted material or bulking agents such as sawdust, straw, wood shavings, or rice hulls.

14. 1, 4, 5   Physical–chemical treatment consists of coagulation, flocculation, and liquid/solids separation.

15. 1, 2, 3, 4, 5   Chemical treatment hazards include dust in the air, high temperatures, slippery walk areas, strong acids, and strong caustics.

16. 1, 2, 3, 4, 5   The following items should be checked before chemical feeder start-up: (1) be sure that the chemical to be fed is available; (2) confirm that the manufacturer's lubrication and start-up procedures are being followed; (3) determine the proper positions for all valves; (4) examine all fittings, inspection plates, and drains to ensure that they will not leak when placed into service; and (5) inspect the electrical system for proper voltage, for properly sized overload protection, and for proper operation of control lights on the control panel.

17. 1, 2, 3, 4, 5   The following items should be checked before starting a filtering system: (1) check motors and drives for proper alignment, for proper safety guards, and for free rotation; (2) examine motors, drive units, and bearings for proper lubrication; (3) fill tanks and piping and look for leaks; (4) inspect pumps and motors for excessive vibration; and (5) inspect the total system for safety hazards.

18. 1, 2, 3   When suspended material accumulates on the pressure filter media surface or in the filter media bed, the differential pressure across the filter increases, the filter effluent quality deteriorates, and the flow through the filter decreases.

19. 1, 2, 4   When selecting a membrane material (durable polymer), factors to consider include the material's ability to retain particles, ability to show reproducible performance over a long life, and compatibility with the wastestream.

20. 1, 3, 5   The most common types of phosphorus removal systems include biological phosphorus removal, filtration following aluminum sulfate flocculation, and lime precipitation.

21. 2, 3, 4, 5   To ensure smooth, efficient operation of a biological phosphorus removal system, the operator will need to carefully monitor the following items on a daily basis: dissolved oxygen levels in the aeration basin, lime feed and mixing systems, pH level in the chemical clarifier, and the state of anaerobic conditions in the phosphorus stripping tank.

22. 1, 2, 3, 4, 5   Phosphorus removal efficiency is affected by changes in pH, short-circuiting, small straggler floc, solids loading, and stormwater.

23. 1, 2, 3, 4, 5   Conditions essential for the nitrification process include adequate amounts of food, adequate amounts of free oxygen, optimal microorganism population, optimal pH, and optimal temperature.

24. 1, 4, 5   To compensate for colder winter conditions when trying to achieve nitrification, an operator could adjust the pH to higher levels, increase the MCRT, or increase the MLVSS concentration.

25. 1, 2, 3, 4, 5   An operator could encounter the following problems when troubleshooting a nitrification process: (1) a decrease in nitrification unit pH with a loss of nitrification, (2) in a two-stage activated sludge system, SVI of nitrification sludge is very high, (3) an inability to completely nitrify, (4) a loss of solids from the final clarifier, or (5) a loss of solids from the trickling filter or RBC.

26. 2, 4, 5   Enhanced biological control requires measurement of existing environmental and physical conditions in the reactor, such as DO, pH, or temperature.

27. 1, 2, 3, 4, 5   Water quality monitoring for process control at the influent and effluent of the process tanks or basins of biological nutrient removal systems should include alkalinity, dissolved oxygen, pH, suspended solids, and temperature.

28. 1, 2, 3, 4, 5   To aid the design engineer during the design stages of an enhanced biological control activated sludge system, operators should review the following items: (1) chemical storage tanks should be constructed within diked chemical containment areas that can hold the contents of the largest tank if the tank should rupture; (2) design of the secondary clarifier scum removal and handling facilities must be able to deal effectively with the potential of excessive scum development; (3) regardless of the type of aerobic zone, the recycle pumps of the nitrate-containing mixed liquor should not be located next to an aeration device; (4) the discharge points of influent and recycle flow into the anaerobic zone should always be submerged to avoid entraining air into the basin contents; and (5) the recycle pumps of the nitrate-containing mixed liquor should be located near the downstream end of a plug flow aerobic basin.

29. 2, 3, 4, 5   Effluent reuse is considered when groundwaters are either not available or the costs of pumping and any treatment are prohibitive, when surface waters are not available or the cost of surface water treatment is excessive, when the cost of purchasing available treated water is too expensive, or when the volume of municipal water needed is not available.

30. 1, 4   Water reclaimed for recreation must consider aesthetics and protection of public health.

31. 4, 5   Vegetation treats wastewater in an overland flow system by removing insoluble BOD and by removing suspended solids

32. 1, 2, 5   Possible solutions to sealing of the soil caused by solids forming a mat on the soil surface include applying water intermittently and allowing a drying period, disking or plowing the field to break the mat of solids, and removing suspended solids from the effluent.

33. 1, 2, 4, 5   Minor corrective maintenance or preventive maintenance instrumentation tasks that operators should be able to perform include making adjustments, making minor repairs, recognizing failures, and troubleshooting.

34. 1, 2, 4   The most common variables that are measured at wastewater treatment plants are flow, level, and pressure.

35. 2, 3, 4, 5   Factors that could cause high readings with a propeller meter include a malfunctioning receiver, a malfunctioning transmitter, a partially full pipeline, or wrong gears installed.

36. 3, 4, 5   ON/OFF types of controllers are used to start and stop pumps according to measurements of flow, level, and pressure.

Short Answer

1. Answer all odor complaints promptly and courteously and maintain a positive attitude. Make a record of the visit and try to locate the source of the odor.

2. Odors in air can be treated by masking and counteraction, combustion, absorption, activated carbon adsorption, and ozonation.

3. The pure oxygen process uses sealed reactors for the following reasons:

   a. To maximize oxygen utilization.
   b. The slight positive pressure allows control of the flow of oxygen into the reactors.
   c. The oxygen that is not dissolved into the water in the first stage can be used again.
   d. If pure oxygen was released to the atmosphere, an explosive atmosphere could develop.

4. Different RAS flow rates will be required as a result of changes in the activated sludge quality and changes in the settling characteristics of the sludge.

5. The following items are affected by the amount of waste activated sludge (WAS) removed from the process:

   a. Effluent quality
   b. Growth rate of the microorganisms
   c. Oxygen consumption
   d. Mixed liquor settleability
   e. Nutrient quantities needed
   f. Occurrence of foaming/frothing
   g. Possibility of nitrifying

6. The best mode of process control will produce a high-quality effluent that meets NPDES permit requirements with consistent treatment results at a minimal cost.

7. Influent to a wastewater treatment plant should be monitored to provide a warning of undesirable constituents so that the operator can start corrective operational measures where possible.

8. An operational strategy developed before a toxic waste is discovered can significantly reduce adverse effects caused by the waste when the waste is discovered.

9. The amount of nutrients to be added each day can be determined on the basis of the quantity of sludge produced each day. The pounds of nitrogen required per day is equal to 10 percent of the volatile solids (on a dry weight basis) produced each day. Phosphorus requirements are one-fifth of the nitrogen requirements. The amount of nutrients that have to be added each day is determined by the difference between the quantity required and the quantity available in the wastes.

10. The handling and disposal of sludge is a complicated problem due to the following factors:

    1. Sludge is composed largely of the substances responsible for the offensive character of untreated wastewater.
    2. Only a small portion of the sludge is solid matter.
    3. The response of similar-type sludges to various handling techniques differs from one treatment plant to the next.

11. The most important operational concern when operating a dissolved air flotation thickener is to ensure that the air rotameter, compressor, and the float mechanism to actuate air injection are in proper working order.

12. Wastewater treatment plant sludges must be stabilized before disposal to avoid odor problems, to prevent the breeding of insects, to reduce the number of pathogenic organisms, and to improve sludge dewaterability.

13. The optimum chemical type and dose for a particular sludge is highly dependent on the characteristics of that sludge. Determination of chemical dose is usually based on on-site experimentation and trial-and-error procedures. This is because sludge types and characteristics vary from one treatment plant to the next and there is no one chemical or dosage that can be applied to all plants and sludges.

14. Filtration time for a pressure filter should be just long enough to fill the filter cavity. If the time of filtration is not adequate to completely fill the plate cavities with dewatered solids, large volumes of water will be discharged when the plates are disengaged and the cakes discharged. If the filtration time exceeds the time required to fill the cavity volume, the cakes will be firm and dry upon discharge but the quantity of solids processed per hour or per day (solids unloading) will decrease. By watching the condition of the cake when the plates are disengaged, the optimum time of filtration can be determined.

15. Care must be taken to prevent chemical overdosing for two reasons: (1) media (sand) blinding with unattached polymer may develop, and (2) large floc particles that settle too rapidly may also blind the media. When blinding occurs, the liquid portion of the sludge will be unable to drain through the bed and dewatering will be by evaporation only. The time required to evaporate water is substantially greater than the time required to remove water by a combination of evaporation and drainage.

16. Chemical stabilization produces environments that are unsuitable for microorganism survival and will not support the life of composting bacteria unless the sludges are neutralized and favorable conditions exist. Sludge that has been stabilized by wet oxidation can be composted, but noxious odors will more than likely develop in and around the compost operation.

17. The main reasons for using wastewater solids for agricultural purposes are to supply the nutritional requirements of crops and improve the texture of the soil by adding humus without adversely affecting the crop produced, the soil, or the groundwater.

18. The three most important process control guidelines for any chemical treatment process are:

    1. Providing enough energy to completely mix the chemicals with the wastewater
    2. Controlling the intensity of mixing during flocculation
    3. Controlling the chemical(s) dose

19. The selection of a polymer for any individual application is highly dependent on the characteristics of the wastewater to be treated. Specific influences (factors) that require attention are: pH, conductivity (dissolved solids content), type and concentration of suspended solids, particle size ranges, type and amount of coagulant(s) applied, and what the next stage of treatment will be.

20. To run the jar test, various types of chemicals or different doses of a single chemical are added to sample portions of wastewater and all portions of the sample are rapidly mixed. After rapid mixing, the samples are slowly mixed to approximate the conditions in the plant. Mixing is then stopped and the floc formed is allowed to settle. The appearance of the floc, the time required to form a floc, and the settling conditions are recorded. The supernatant (liquid above the settled sludge) is analyzed for turbidity, suspended solids, and pH. With this information, the operator selects the best chemical or best dosage to feed on the basis of clarity of effluent and minimum cost of chemicals.

21. Plunger pumps are used for metering chemicals due to the accuracy of the positive displacement stroke and the ease of adjusting the piston stroke to regulate the chemical feed rate.

22. Instrumentation associated with a filter system is used to monitor the plant performance, to operate the plant in the absence of the operator, and to trigger an alarm if abnormal conditions develop.

23. To achieve a constant rate of filtration with a pressure filter, a constant pressure is supplied to the filter. An automatic or manually operated rate-of-flow controller then holds the filtration rate constant by adjusting the available pressure to correspond with changes in head loss across the filter.

24. Continuous backwash, upflow, deep-bed silica sand media filters have two main advantages over other types of granular media filters:

    1. They do not need to be taken out of service for backwashing.
    2. They operate continuously so fewer filters or smaller-sized filters may be enough to provide full design flow processing capacity.

25. Fouled membranes can be cleaned with caustic, acids, surfactants, chlorine, or hydrogen peroxide, and cleaning will return the membrane flux very nearly to original values.

26. Oxygen is important in biological treatment processes because the microorganisms always need some type of oxygen to support their life processes. The presence or absence of oxygen is mainly what determines which types of microorganisms will survive in various phases of treatment. Availability of oxygen also influences the chemical reactions that occur as the microorganisms break down and thereby treat the wastes in wastewater.

27. The system that pumps the activated sludge into and out of the anaerobic phosphorus stripping tank must be equipped with variable-speed controls so that the operator can adjust the feed rate into and out of the stripping tank.

28. Jar tests are run with the lime precipitation process to determine optimum pH levels and polymer dosages to obtain the largest floc possible for the best and fastest settling of the floc formation.

29. Pumps moving alum solutions clog because alum is very sticky.

30. Attached growth (fixed film) reactors use recirculation to allow for a constant and even distribution of the wastewater over the entire media surface. During low-flow conditions, the recycling of wastewater will have the beneficial effect of allowing for a constant loading.

31. Air comes in contact with the wastewater in the air stripping process when the wastewater is pumped to the top of a packed tower and allowed to trickle down over the media.

32. "Enhanced biological control" is a modification of the activated sludge process that allows operators to control nutrient removal, sludge settleability, filamentous organism growth, and effluent BOD and suspended solids.

33. System flexibility stems from the ability to make process changes (or alternate uses) within reactor volumes and to deliver applied and return flows to different locations.

34. Limitations of direct use of reclaimed wastewater include:

 1. Industrial reuse requires a fairly uniform quality of water.
 2. Deep well injection must follow procedures developed to maintain the recharge rates.
 3. Care must be taken to avoid contaminating groundwater drinking supplies.
 4. Recreation use must consider protection of public health and aesthetics.

35. The purpose of a monitoring program is to be sure the effluent meets the water quality standards of the user.

36. In land treatment, effluent is pretreated and applied to land by conventional irrigation methods. Treatment is provided by natural processes as the effluent flows through the soil and plants. Part of the wastewater is lost by evapotranspiration and the rest goes back to the hydrologic cycle.

37. While irrigating a plot of land, run the pumps the time determined in the start-up procedure or turn the pumps off earlier if water begins to pond on the fields.

38. Operators must have a good working familiarity with instrumentation and control systems to properly monitor and control the treatment processes.

39. *ACCURACY* refers to how closely an instrument measures the true or actual value of a process variable and *PRECISION* refers to how closely the device (instrument) can reproduce a given reading (or measurement) time after time.

40. It is poor practice to ignore the lamps that are lit up (alarm conditions) on an annunciator panel because a true alarm condition requiring immediate operator attention may be lost in the resulting general indifference to the alarm system.

41. Plant measuring systems must be periodically calibrated to ensure accurate measurements.

Problems

1. An aeration tank has a volume of 1.6 million gallons. The MLSS are 2,400 mg/L and the volatile portion is 0.75. How many pounds of mixed liquor volatile suspended solids (MLVSS) are under aeration?

Known	Unknown
Tank Vol, MG = 1.6 MG	MLVSS, lbs
MLSS, mg/L = 2,400 mg/L	
Volatile Portion = 0.75	

Determine mixed liquor volatile suspended solids (MLVSS) in pounds.

$$\text{MLVSS, lbs} = \text{Tank Vol, MG} \times \text{MLSS, mg/}L \times \text{Volatile} \times 8.34 \text{ lbs/gal}$$
$$= 1.6 \text{ MG} \times 2,400 \text{ mg/}L \times 0.75 \times 8.34 \text{ lbs/gal}$$
$$= 24,020 \text{ lbs}$$

2. An aerator contains 24,250 lbs of solids, the RAS suspended solids concentration is 7,100 mg/L, and the current WAS flow is 0.02 MGD. The desired sludge age is 6 days. If the plant adds 3,800 lbs of solids per day, what is the total waste activated sludge (WAS) flow rate in GPM?

Known	Unknown
Solids Added, lbs/day = 3,800 lbs/day	WAS Flow, GPM
Solids Aerated, lbs = 24,250 lbs	
RAS Susp Sol, mg/L = 7,100 mg/L	
Desired Sludge Age, days = 6 days	
Current WAS, MGD = 0.02 MGD	

 1. Calculate the desired pounds of solids under aeration for a sludge age of 6 days.

$$\text{Desired Solids Under Aeration, lbs} = \text{Solids Added, lbs/day} \times \text{Sludge Age, days}$$
$$= 3,800 \text{ lbs/day} \times 6 \text{ days}$$
$$= 22,800 \text{ lbs}$$

2. Calculate the additional WAS flow to maintain the desired sludge age in MGD and convert to GPM.

$$\frac{\text{Additional WAS}}{\text{Flow, GPM}} = \frac{\text{Solids Aerated, lbs} - \text{Desired Solids, lbs*}}{\text{RAS Susp Sol, mg}/L \times 8.34 \text{ lbs/gal}}$$

$$= \frac{24{,}250 \text{ lbs} - 22{,}800 \text{ lbs}}{7{,}100 \text{ mg}/L \times 8.34 \text{ lbs/gal}}$$

$$= 0.024 \text{ MGD} \times 694 \text{ GPM/MGD}$$

$$= 17.0 \text{ GPM}$$

3. Calculate total WAS flow in GPM.

$$\frac{\text{Total WAS}}{\text{Flow, GPM}} = \text{Current WAS, MGD} + \text{Additional WAS, MGD}$$

$$= 0.02 \text{ MGD} + 0.024 \text{ MGD}$$

$$= 0.044 \text{ MGD} \times 694 \text{ GPM/MGD}$$

$$= 30.5 \text{ GPM}$$

* Difference represents solids wasted in pounds per day.

3. Seven thousand five hundred gallons of sludge with a sludge solids concentration of 1.5 percent is thickened to a sludge solids concentration of 5 percent. What is the reduction in volume of sludge? *HINT:* How many gallons of water were removed by the thickening process?

Known		Unknown
Sludge Vol, gal	= 7,500 gal	Reduction in Sludge Volume, gal
Initial Sl Conc, %	= 1.5% Solids	
Final Sl Conc, %	= 5% Solids	

1. Determine the pounds of sludge solids.

$$\frac{\text{Solids,}}{\text{lbs}} = (\text{Sludge Vol, gal})(8.34 \text{ lbs/gal})(\text{Sl Conc, } \frac{\%}{100\%})$$

$$= 7{,}500 \text{ gal} \times 8.34 \text{ lbs/gal} \times 1.5\%/100\%$$

$$= 938 \text{ lbs}$$

2. Determine the volume of thickened sludge.

$$\text{Sludge Vol, gal} = \frac{\text{Solids, lbs}}{(8.34 \text{ lbs/gal}) \times (\text{Sl Conc, }\%/100\%)}$$

$$= \frac{938 \text{ lbs}}{(8.34 \text{ lbs/gal})(5\%/100\%)}$$

$$= 2{,}250 \text{ gal}$$

3. Determine the reduction in sludge volume in gallons.

$$\frac{\text{Reduction,}}{\text{gal}} = \text{Original Sl Vol, gal} - \text{Thickened Sl Vol, gal}$$

$$= 7{,}500 \text{ gal} - 2{,}250 \text{ gal}$$

$$= 5{,}250 \text{ gal}$$

4. A waste activated sludge is pumped at 50 GPM with a sludge solids concentration of 3.0 percent sludge solids. Jar tests indicate that 48 pounds per day of dry polymer are necessary for successful gravity thickening. What is the polymer dosage in pounds of polymer per ton of dry sludge solids?

Known		Unknown
Flow, GPM	= 50 GPM	Polymer Dosage in lbs of Polymer per ton of Dry Sludge Solids
Sl Sol, %	= 3.0%	
Polymer, lbs/day	= 48 lbs/day	

1. Calculate the tons of dry sludge treated per day by the polymer.

$$\frac{\text{Sludge,}}{\text{tons/day}} = \frac{(\text{Flow, GPM})(1{,}440 \text{ min/day})(8.34 \text{ lbs/gal})(\text{Sl Sol, }\%/100\%)}{2{,}000 \text{ lbs/ton}}$$

$$= \frac{(50 \text{ gal/min})(1{,}440 \text{ min/day})(8.34 \text{ lbs/gal})(3.0\%/100\%)}{2{,}000 \text{ lbs/ton}}$$

$$= 9.0 \text{ tons/day}$$

2. Calculate the polymer dosage in pounds of polymer per ton of dry sludge solids.

$$\frac{\text{Polymer Dose,}}{\text{lbs Poly/ton Sludge}} = \frac{\text{Amount of Polymer, lbs/day}}{\text{Sludge, tons/day}}$$

$$= \frac{48 \text{ lbs Polymer/day}}{9 \text{ tons/day}}$$

$$= 5.3 \text{ lbs Polymer/ton Sludge}$$

WASTE TREATMENT WORDS

A Summary of the Words Defined

in

ADVANCED WASTE TREATMENT

PROJECT PRONUNCIATION KEY

by Warren L. Prentice

The Project Pronunciation Key is designed to aid you in the pronunciation of new words. While this key is based primarily on familiar sounds, it does not attempt to follow any particular pronunciation guide. This key is designed solely to aid operators in this program.

You may find it helpful to refer to other available sources for pronunciation help. Each current standard dictionary contains a guide to its own pronunciation key. Each key will be different from each other and from this key. Examples of the difference between the key used in this program and the *WEBSTER'S NEW WORLD COLLEGE DICTIONARY*[1] "Key" are shown below.

In using this key, you should accent (say louder) the syllable that appears in capital letters. The following chart is presented to give examples of how to pronounce words using the Project Key.

WORD	SYLLABLE				
	1st	2nd	3rd	4th	5th
acid	AS	id			
coliform	KOAL	i	form		
biological	BUY	o	LODGE	ik	cull

The first word, *ACID*, has its first syllable accented. The second word, *COLIFORM*, has its first syllable accented. The third word, *BIOLOGICAL*, has its first and third syllables accented.

We hope you will find the key useful in unlocking the pronunciation of any new word.

Term	Project Key	Webster Key
acid	AS-id	aś id
coliform	KOAL-i-form	kō′ lə fôrm
biological	BUY-o-LODGE-ik-cull	bī ə läj′ i kəl

---

[1] The *WEBSTER'S NEW WORLD COLLEGE DICTIONARY*, Fourth Edition, 1999, was chosen rather than an unabridged dictionary because of its availability to the operator.

WASTE TREATMENT WORDS

>GREATER THAN >GREATER THAN

DO >5 mg/*L* would be read as DO GREATER THAN 5 mg/*L*.

<LESS THAN <LESS THAN

DO <5 mg/*L* would be read as DO LESS THAN 5 mg/*L*.

A

ABS ABS

Alkyl Benzene Sulfonate. A type of surfactant, or surface active agent, present in synthetic detergents in the United States before 1965. ABS was especially troublesome because it caused foaming and resisted breakdown by biological treatment processes. ABS has been replaced in detergents by linear alkyl sulfonate (LAS), which is biodegradable.

ACEOPS ACEOPS

See ALLIANCE OF CERTIFIED OPERATORS, LAB ANALYSTS, INSPECTORS, AND SPECIALISTS.

atm atm

The abbreviation for atmosphere. One atmosphere is equal to 14.7 psi or 100 kPa.

ABSORPTION (ab-SORP-shun) ABSORPTION

The taking in or soaking up of one substance into the body of another by molecular or chemical action (as tree roots absorb dissolved nutrients in the soil).

ACCURACY ACCURACY

How closely an instrument measures the true or actual value of the process variable being measured or sensed.

ACID ACID

(1) A substance that tends to lose a proton.

(2) A substance that dissolves in water and releases hydrogen ions.

(3) A substance containing hydrogen ions that may be replaced by metals to form salts.

(4) A substance that is corrosive.

ACIDITY ACIDITY

The capacity of water or wastewater to neutralize bases. Acidity is expressed in milligrams per liter of equivalent calcium carbonate. Acidity is not the same as pH because water does not have to be strongly acidic (low pH) to have a high acidity. Acidity is a measure of how much base must be added to a liquid to raise the pH to 8.2.

ACTIVATED SLUDGE ACTIVATED SLUDGE

Sludge particles produced in raw or settled wastewater (primary effluent) by the growth of organisms (including zoogleal bacteria) in aeration tanks in the presence of dissolved oxygen. The term "activated" comes from the fact that the particles are teeming with bacteria, fungi, and protozoa. Activated sludge is different from primary sludge in that the sludge particles contain many living organisms that can feed on the incoming wastewater.

ACTIVATED SLUDGE PROCESS ACTIVATED SLUDGE PROCESS

A biological wastewater treatment process that speeds up the decomposition of wastes in the wastewater being treated. Activated sludge is added to wastewater and the mixture (mixed liquor) is aerated and agitated. After some time in the aeration tank, the activated sludge is allowed to settle out by sedimentation and is disposed of (wasted) or reused (returned to the aeration tank) as needed. The remaining wastewater then undergoes more treatment.

ACUTE HEALTH EFFECT ACUTE HEALTH EFFECT

An adverse effect on a human or animal body, with symptoms developing rapidly.

ADSORPTION (add-SORP-shun) ADSORPTION

The gathering of a gas, liquid, or dissolved substance on the surface or interface zone of another material.

ADVANCED WASTE TREATMENT ADVANCED WASTE TREATMENT

Any process of water renovation that upgrades treated wastewater to meet specific reuse requirements. May include general cleanup of water or removal of specific parts of wastes insufficiently removed by conventional treatment processes. Typical processes include chemical treatment and pressure filtration. Also called TERTIARY TREATMENT.

AERATION (air-A-shun) AERATION

The process of adding air to water. Air can be added to water by either passing air through water or passing water through air.

AERATION (air-A-shun) LIQUOR AERATION LIQUOR

Mixed liquor. The contents of the aeration tank, including living organisms and material carried into the tank by either untreated wastewater or primary effluent.

AERATION (air-A-shun) TANK AERATION TANK

The tank where raw or settled wastewater is mixed with return sludge and aerated. The same as aeration bay, aerator, or reactor.

AEROBES AEROBES

Bacteria that must have dissolved oxygen (DO) to survive. Aerobes are aerobic bacteria.

AEROBIC (air-O-bick) AEROBIC

A condition in which atmospheric or dissolved oxygen is present in the aquatic (water) environment.

AEROBIC BACTERIA (air-O-bick back-TEER-e-uh) AEROBIC BACTERIA

Bacteria that will live and reproduce only in an environment containing oxygen that is available for their respiration (breathing), namely atmospheric oxygen or oxygen dissolved in water. Oxygen combined chemically, such as in water molecules (H_2O), cannot be used for respiration by aerobic bacteria.

AEROBIC (air-O-bick) DECOMPOSITION AEROBIC DECOMPOSITION

The decay or breaking down of organic material in the presence of free or dissolved oxygen.

AEROBIC (air-O-bick) DIGESTION AEROBIC DIGESTION

The breakdown of wastes by microorganisms in the presence of dissolved oxygen. This digestion process may be used to treat only waste activated sludge, or trickling filter sludge and primary (raw) sludge, or waste sludge from activated sludge treatment plants designed without primary settling. The sludge to be treated is placed in a large aerated tank where aerobic microorganisms decompose the organic matter in the sludge. This is an extension of the activated sludge process.

AEROBIC (air-O-bick) PROCESS AEROBIC PROCESS

A waste treatment process conducted under aerobic (in the presence of free or dissolved oxygen) conditions.

AGE TANK   AGE TANK

A tank used to store a chemical solution of known concentration for feed to a chemical feeder. It usually stores sufficient chemical solution to properly treat the water being treated for at least one day. Also called a DAY TANK.

AGGLOMERATION (uh-glom-er-A-shun)   AGGLOMERATION

The growing or coming together of small scattered particles into larger flocs or particles, which settle rapidly. Also see FLOC.

AGRONOMIC RATES   AGRONOMIC RATES

Sludge application rates that provide the amount of nitrogen needed by the crop or vegetation grown on the land while minimizing the amount that passes below the root zone.

AIR BINDING   AIR BINDING

The clogging of a filter, pipe, or pump due to the presence of air released from water. Air entering the filter media is harmful to both the filtration and backwash processes. Air can prevent the passage of water during the filtration process and can cause the loss of filter media during the backwash process.

AIR GAP   AIR GAP

An open, vertical drop, or vertical empty space, between a drinking (potable) water supply and potentially contaminated water. This gap prevents the contamination of drinking water by backsiphonage because there is no way potentially contaminated water can reach the drinking water supply.

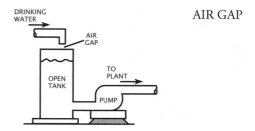

AIR LIFT PUMP   AIR LIFT PUMP

A special type of pump consisting of a vertical riser pipe submerged in the wastewater or sludge to be pumped. Compressed air is injected into a tail piece at the bottom of the pipe. Fine air bubbles mix with the wastewater or sludge to form a mixture lighter than the surrounding water, which causes the mixture to rise in the discharge pipe to the outlet.

AIR PADDING   AIR PADDING

Pumping dry air (dew point −40°F (−40°C)) into a container to assist with the withdrawal of a liquid or to force a liquified gas such as chlorine or sulfur dioxide out of a container.

ALARM CONTACT   ALARM CONTACT

A switch that operates when some preset low, high, or abnormal condition exists.

ALGAE (AL-jee)   ALGAE

Microscopic plants containing chlorophyll that live floating or suspended in water. They also may be attached to structures, rocks, or other submerged surfaces. Excess algal growths can impart tastes and odors to potable water. Algae produce oxygen during sunlight hours and use oxygen during the night hours. Their biological activities appreciably affect the pH, alkalinity, and dissolved oxygen of the water.

ALIQUOT (AL-uh-kwot)   ALIQUOT

Representative portion of a sample. Often, an equally divided portion of a sample.

ALKALI (AL-kuh-lie)   ALKALI

Any of certain soluble salts, principally of sodium, potassium, magnesium, and calcium, that have the property of combining with acids to form neutral salts and may be used in chemical processes such as water or wastewater treatment.

ALKALINITY (AL-kuh-LIN-it-tee)   ALKALINITY

The capacity of water or wastewater to neutralize acids. This capacity is caused by the water's content of carbonate, bicarbonate, hydroxide, and occasionally borate, silicate, and phosphate. Alkalinity is expressed in milligrams per liter of equivalent calcium carbonate. Alkalinity is not the same as pH because water does not have to be strongly basic (high pH) to have a high alkalinity. Alkalinity is a measure of how much acid must be added to a liquid to lower the pH to 4.5.

ALLIANCE OF CERTIFIED OPERATORS,   ALLIANCE OF CERTIFIED OPERATORS,
    LAB ANALYSTS, INSPECTORS,   LAB ANALYSTS, INSPECTORS,
    AND SPECIALISTS (ACEOPS)   AND SPECIALISTS (ACEOPS)

A professional organization for operators, lab analysts, inspectors, and specialists dedicated to improving professionalism; expanding training, certification, and job opportunities; increasing information exchange; and advocating the importance of certified operators, lab analysts, inspectors, and specialists. For information on membership, contact ACEOPS, PO Box 934, Dakota City, NE 68731-0934, phone (402) 698-2330, or e-mail: Info@aceops.org.

AMBIENT (AM-bee-ent) TEMPERATURE   AMBIENT TEMPERATURE

Temperature of the surroundings.

AMPEROMETRIC (am-purr-o-MET-rick)   AMPEROMETRIC

A method of measurement that records electric current flowing or generated, rather than recording voltage. Amperometric titration is a means of measuring concentrations of certain substances in water.

ANAEROBES   ANAEROBES

Bacteria that do not need dissolved oxygen (DO) to survive.

ANAEROBIC (AN-air-O-bick)   ANAEROBIC

A condition in which atmospheric or dissolved oxygen (DO) is *NOT* present in the aquatic (water) environment.

ANAEROBIC BACTERIA (AN-air-O-bick back-TEER-e-uh)   ANAEROBIC BACTERIA

Bacteria that live and reproduce in an environment containing no free or dissolved oxygen. Anaerobic bacteria obtain their oxygen supply by breaking down chemical compounds that contain oxygen, such as sulfate (SO_4^{2-}).

ANAEROBIC (AN-air-O-bick) DECOMPOSITION   ANAEROBIC DECOMPOSITION

The decay or breaking down of organic material in an environment containing no free or dissolved oxygen.

ANAEROBIC (AN-air-O-bick) DIGESTION   ANAEROBIC DIGESTION

A treatment process in which wastewater solids and water (about 5 percent solids, 95 percent water) are placed in a large tank (the digester) where bacteria decompose the solids in the absence of dissolved oxygen. At least two general groups of bacteria act in balance: (1) saprophytic bacteria break down complex solids to volatile acids, the most common of which are acetic and propionic acids; and (2) methane fermenters break down the acids to methane, carbon dioxide, and water.

ANAEROBIC (AN-air-O-bick) SELECTOR   ANAEROBIC SELECTOR

Anaerobic refers to the practical absence of dissolved and chemically bound oxygen. Selector refers to a reactor or basin and environmental conditions (food, lack of DO) intended to favor the growth of certain organisms over others. Also see SELECTOR.

ANALOG   ANALOG

The continuously variable signal type sent to an analog instrument (for example, 4–20 mA).

ANALOG READOUT   ANALOG READOUT

The readout of an instrument by a pointer (or other indicating means) against a dial or scale. Also see DIGITAL READOUT.

ANALYZER

A device that conducts a periodic or continuous measurement of turbidity or some factor such as chlorine or fluoride concentration. Analyzers operate by any of several methods including photocells, conductivity, or complex instrumentation.

ANHYDROUS (an-HI-drous)

Very dry. No water or dampness is present.

ANION (AN-EYE-en)

A negatively charged ion in an electrolyte solution, attracted to the anode under the influence of a difference in electrical potential. Chloride ion (Cl^-) is an anion.

ANNULAR (AN-yoo-ler) SPACE

A ring-shaped space located between two circular objects. For example, the space between the outside of a pipe liner and the inside of a pipe.

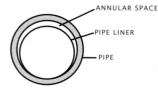

ANOXIC (an-OX-ick)

A condition in which the aquatic (water) environment does not contain dissolved oxygen (DO), which is called an oxygen deficient condition. Generally refers to an environment in which chemically bound oxygen, such as in nitrate, is present. The term is similar to ANAEROBIC.

ANOXIC DENITRIFICATION (dee-NYE-truh-fuh-KAY-shun)

A biological nitrogen removal process in which nitrate nitrogen is converted by microorganisms to nitrogen gas in the absence of dissolved oxygen.

ARCH

(1) The curved top of a sewer pipe or conduit.

(2) A bridge or arch of hardened or caked chemical that will prevent the flow of the chemical.

ASEPTIC (a-SEP-tick)

Free from the living germs of disease, fermentation, or putrefaction. Sterile.

ASPIRATE (AS-per-rate)

Use of a hydraulic device (aspirator or eductor) to create a negative pressure (suction) by forcing a liquid through a restriction, such as a Venturi tube. An aspirator may be used in the laboratory in place of a vacuum pump; sometimes used instead of a sump pump.

ATTACHED GROWTH PROCESSES

Wastewater treatment processes in which the microorganisms and bacteria treating the wastes are attached to the media in the reactor. The wastes being treated flow over the media. Trickling filters and rotating biological contactors are attached growth reactors. These reactors can be used for BOD removal, nitrification, and denitrification.

AUTOTROPHIC (auto-TROF-ick)

Describes organisms (plants and some bacteria) that use inorganic materials for energy and growth.

AVAILABLE EXPANSION

The vertical distance from the sand surface to the underside of a trough in a sand filter. This distance is also called FREEBOARD.

B

BOD (pronounce as separate letters)   BOD

Biochemical Oxygen Demand. The rate at which organisms use the oxygen in water or wastewater while stabilizing decomposable organic matter under aerobic conditions. In decomposition, organic matter serves as food for the bacteria and energy results from its oxidation. BOD measurements are used as a surrogate measure of the organic strength of wastes in water.

BTU (pronounce as separate letters)   BTU

British Thermal Unit. The amount of heat required to raise the temperature of one pound of water one degree Fahrenheit. Also see CALORIE.

BACTERIA (back-TEER-e-uh)   BACTERIA

Bacteria are living organisms, microscopic in size, that usually consist of a single cell. Most bacteria use organic matter for their food and produce waste products as a result of their life processes.

BACTERIAL (back-TEER-e-ul) CULTURE   BACTERIAL CULTURE

In the case of activated sludge, the bacterial culture refers to the group of bacteria classified as AEROBES and FACULTATIVE BACTERIA, which covers a wide range of organisms. Most treatment processes in the United States grow facultative bacteria that use the carbonaceous (carbon compounds) BOD. Facultative bacteria can live when oxygen resources are low. When nitrification is required, the nitrifying organisms are obligate aerobes (require oxygen) and must have at least 0.5 mg/L of dissolved oxygen throughout the whole system to function properly.

BAFFLE   BAFFLE

A flat board or plate, deflector, guide, or similar device constructed or placed in flowing water, wastewater, or slurry systems to cause more uniform flow velocities, to absorb energy, and to divert, guide, or agitate liquids (water, chemical solutions, slurry).

BASE   BASE

(1) A substance that takes up or accepts protons.

(2) A substance that dissociates (separates) in aqueous solution to yield hydroxyl ions (OH^-).

(3) A substance containing hydroxyl ions that reacts with an acid to form a salt or that may react with metals to form precipitates.

BATCH PROCESS   BATCH PROCESS

A treatment process in which a tank or reactor is filled, the water (or wastewater or other solution) is treated or a chemical solution is prepared, and the tank is emptied. The tank may then be filled and the process repeated. Batch processes are also used to cleanse, stabilize, or condition chemical solutions for use in industrial manufacturing and treatment processes.

BENCH-SCALE ANALYSIS (TEST)   BENCH-SCALE ANALYSIS (TEST)

A method of studying different ways or chemical doses for treating water or wastewater and solids on a small scale in a laboratory. Also see JAR TEST.

BENZENE   BENZENE

An aromatic hydrocarbon (C_6H_6) that is a colorless, volatile, flammable liquid. Benzene is obtained chiefly from coal tar and is used as a solvent for resins and fats and in the manufacture of dyes. Benzene has been found to cause cancer in humans.

BIOASSAY (BUY-o-AS-say)   BIOASSAY

(1) A way of showing or measuring the effect of biological treatment on a particular substance or waste.

(2) A method of determining the relative toxicity of a test sample of industrial wastes or other wastes by using live test organisms, such as fish.

BIOCHEMICAL OXYGEN DEMAND (BOD)   BIOCHEMICAL OXYGEN DEMAND (BOD)

See BOD.

BIOCHEMICAL OXYGEN DEMAND (BOD) TEST

BIOCHEMICAL OXYGEN DEMAND (BOD) TEST

A procedure that measures the rate of oxygen use under controlled conditions of time and temperature. Standard test conditions include dark incubation at 20°C for a specified time (usually five days).

BIODEGRADABLE (BUY-o-dee-GRADE-able)

BIODEGRADABLE

Organic matter that can be broken down by bacteria to more stable forms that will not create a nuisance or give off foul odors is considered biodegradable.

BIODEGRADATION (BUY-o-deh-grah-DAY-shun)

BIODEGRADATION

The breakdown of organic matter by bacteria to more stable forms that will not create a nuisance or give off foul odors.

BIOFLOCCULATION (BUY-o-flock-yoo-LAY-shun)

BIOFLOCCULATION

The clumping together of fine, dispersed organic particles by the action of certain bacteria and algae. This results in faster and more complete settling of the organic solids in wastewater.

BIOMASS (BUY-o-mass)

BIOMASS

A mass or clump of organic material consisting of living organisms feeding on wastes, dead organisms, and other debris. Also see ZOOGLEAL MASS and ZOOGLEAL MAT (FILM).

BIOMONITORING

BIOMONITORING

A term used to describe methods of evaluating or measuring the effects of toxic substances in effluents on aquatic organisms in receiving waters. There are two types of biomonitoring, the BIOASSAY and the BIOSURVEY.

BIOSOLIDS

BIOSOLIDS

A primarily organic solid product produced by wastewater treatment processes that can be beneficially recycled. The word biosolids is replacing the word sludge when referring to treated waste.

BIOSURVEY

BIOSURVEY

A survey of the types and numbers of organisms naturally present in the receiving waters upstream and downstream from plant effluents. Comparisons are made between the aquatic organisms upstream and those organisms downstream of the discharge.

BLANK

BLANK

A bottle containing only dilution water or distilled water; the sample being tested is not added. Tests are frequently run on a sample and a blank and the differences are compared. The procedure helps to eliminate or reduce test result errors that could be caused when the dilution water or distilled water used is contaminated.

BLINDING

BLINDING

The clogging of the filtering medium of a microscreen or a vacuum filter when the holes or spaces in the media become clogged or sealed off due to a buildup of grease or the material being filtered.

BLOCKOUT

BLOCKOUT

The physical prevention of the operation of equipment.

BOUND WATER

BOUND WATER

Water contained within the cell mass of sludges or strongly held on the surface of colloidal particles. One of the causes of bulking sludge in the activated sludge process.

BREAKOUT OF CHLORINE (CHLORINE BREAKAWAY)

BREAKOUT OF CHLORINE (CHLORINE BREAKAWAY)

A point at which chlorine leaves solution as a gas because the chlorine feed rate is too high. The solution is saturated and cannot dissolve any more chlorine. The maximum strength a chlorine solution can attain is approximately 3,500 mg/L. Beyond this concentration molecular chlorine will break out of solution and cause off-gassing at the point of application.

BREAKPOINT CHLORINATION

BREAKPOINT CHLORINATION

Addition of chlorine to water or wastewater until the chlorine demand has been satisfied. At this point, further additions of chlorine will result in a free chlorine residual that is directly proportional to the amount of chlorine added beyond the breakpoint.

BUFFER

BUFFER

A solution or liquid whose chemical makeup neutralizes acids or bases without a great change in pH.

BUFFER ACTION

BUFFER ACTION

The action of certain ions in solution in opposing a change in hydrogen ion concentration.

BUFFER CAPACITY

BUFFER CAPACITY

A measure of the capacity of a solution or liquid to neutralize acids or bases. This is a measure of the capacity of water or wastewater for offering a resistance to changes in pH.

BUFFER SOLUTION

BUFFER SOLUTION

A solution containing two or more substances that, in combination, resist any marked change in pH following addition of moderate amounts of either strong acid or base.

BULKING

BULKING

Clouds of billowing sludge that occur throughout secondary clarifiers and sludge thickeners when the sludge does not settle properly. In the activated sludge process, bulking is usually caused by filamentous bacteria or bound water.

C

COD (pronounce as separate letters)

COD

Chemical Oxygen Demand. A measure of the oxygen-consuming capacity of organic matter present in wastewater. COD is expressed as the amount of oxygen consumed from a chemical oxidant in mg/L during a specific test. Results are not necessarily related to the biochemical oxygen demand (BOD) because the chemical oxidant may react with substances that bacteria do not stabilize.

CALIBRATION

CALIBRATION

A procedure that checks or adjusts an instrument's accuracy by comparison with a standard or reference.

CALORIE (KAL-o-ree)

CALORIE

The amount of heat required to raise the temperature of one gram of water one degree Celsius. Also see BTU.

CARBONACEOUS (car-bun-NAY-shus) STAGE

CARBONACEOUS STAGE

A stage of decomposition that occurs in biological treatment processes when aerobic bacteria, using dissolved oxygen, change carbon compounds to carbon dioxide. Sometimes referred to as first-stage BOD because the microorganisms attack organic or carbon compounds first and nitrogen compounds later. Also see NITRIFICATION STAGE.

CATHODIC (ka-THOD-ick) PROTECTION

CATHODIC PROTECTION

An electrical system for prevention of rust, corrosion, and pitting of metal surfaces that are in contact with water, wastewater, or soil. A low-voltage current is made to flow through a liquid (water) or a soil in contact with the metal in such a manner that the external electromotive force renders the metal structure cathodic. This concentrates corrosion on auxiliary anodic parts, which are deliberately allowed to corrode instead of letting the structure corrode.

CATION (KAT-EYE-en) EXCHANGE CAPACITY

CATION EXCHANGE CAPACITY

The ability of a soil or other solid to exchange cations (positive ions such as calcium, Ca^{2+}) with a liquid.

CAUTION

CAUTION

This word warns against potential hazards or cautions against unsafe practices. Also see DANGER, NOTICE, and WARNING.

CAVITATION (kav-uh-TAY-shun)

CAVITATION

The formation and collapse of a gas pocket or bubble on the blade of an impeller or the gate of a valve. The collapse of this gas pocket or bubble drives water into the impeller or gate with a terrific force that can knock metal particles off and cause pitting on the impeller or gate surface. Cavitation is accompanied by loud noises that sound like someone is pounding on the impeller or gate with a hammer.

CENTRATE

CENTRATE

The water leaving a centrifuge after most of the solids have been removed.

CENTRIFUGE

CENTRIFUGE

A mechanical device that uses centrifugal or rotational forces to separate solids from liquids.

CERTIFIED OPERATOR

CERTIFIED OPERATOR

A person who has the education and experience required to operate a specific class of treatment facility as indicated by possessing a certificate of professional competence given by a state agency or professional association.

CHEMICAL EQUIVALENT

CHEMICAL EQUIVALENT

The weight in grams of a substance that combines with or displaces one gram of hydrogen. Chemical equivalents usually are found by dividing the formula weight by its valence.

CHEMICAL OXYGEN DEMAND (COD)

CHEMICAL OXYGEN DEMAND (COD)

A measure of the oxygen-consuming capacity of organic matter present in wastewater. COD is expressed as the amount of oxygen consumed from a chemical oxidant in mg/L during a specific test. Results are not necessarily related to the biochemical oxygen demand (BOD) because the chemical oxidant may react with substances that bacteria do not stabilize.

CHEMICAL PRECIPITATION

CHEMICAL PRECIPITATION

(1) Precipitation induced by the addition of chemicals.

(2) The process of softening water by the addition of lime or lime and soda ash as the precipitants.

CHLORAMINES (KLOR-uh-means)

CHLORAMINES

Compounds formed by the reaction of hypochlorous acid (or aqueous chlorine) with ammonia.

CHLORINATION (klor-uh-NAY-shun)

CHLORINATION

The application of chlorine to water or wastewater, generally for the purpose of disinfection, but frequently for accomplishing other biological or chemical results—aiding coagulation and controlling tastes and odors in drinking water, or controlling odors or sludge bulking in wastewater.

CHLORINE BREAKAWAY

CHLORINE BREAKAWAY

See BREAKOUT OF CHLORINE.

CHLORINE DEMAND

CHLORINE DEMAND

Chlorine demand is the difference between the amount of chlorine added to water or wastewater and the amount of residual chlorine remaining after a given contact time. Chlorine demand may change with dosage, time, temperature, pH, and nature and amount of the impurities in the water.

Chlorine Demand, mg/L = Chlorine Applied, mg/L – Chlorine Residual, mg/L

CHLORINE REQUIREMENT CHLORINE REQUIREMENT

The amount of chlorine that is needed for a particular purpose. Some reasons for adding chlorine are reducing the MPN (Most Probable Number) of coliform bacteria, obtaining a particular chlorine residual, or oxidizing some substance in the water. In each case, a definite dosage of chlorine will be necessary. This dosage is the chlorine requirement.

CHLORINE RESIDUAL CHLORINE RESIDUAL

The concentration of chlorine present in water after the chlorine demand has been satisfied. The concentration is expressed in terms of the total chlorine residual, which includes both the free and combined or chemically bound chlorine residuals. Also called RESIDUAL CHLORINE.

CHLORORGANIC (klor-or-GAN-ick) CHLORORGANIC

Organic compounds combined with chlorine. These compounds generally originate from, or are associated with, living or dead organic materials, such as algae in water.

CHRONIC HEALTH EFFECT CHRONIC HEALTH EFFECT

An adverse effect on a human or animal body with symptoms that develop slowly over a long period of time or that recur frequently.

CILIATES (SILLY-ates) CILIATES

A class of protozoans distinguished by short hairs on all or part of their bodies.

CLARIFICATION (klair-uh-fuh-KAY-shun) CLARIFICATION

Any process or combination of processes the main purpose of which is to reduce the concentration of suspended matter in a liquid.

CLARIFIER (KLAIR-uh-fire) CLARIFIER

A tank or basin in which water or wastewater is held for a period of time during which the heavier solids settle to the bottom and the lighter materials float to the surface. Also called settling tank or SEDIMENTATION BASIN.

COAGULANT (ko-AGG-yoo-lent) COAGULANT

A chemical that causes very fine particles to clump (floc) together into larger particles. This makes it easier to separate the solids from the liquids by settling, skimming, draining, or filtering.

COAGULANT (ko-AGG-yoo-lent) AID COAGULANT AID

Any chemical or substance used to assist or modify coagulation.

COAGULATION (ko-agg-yoo-LAY-shun) COAGULATION

The clumping together of very fine particles into larger particles (floc) caused by the use of chemicals (coagulants). The chemicals neutralize the electrical charges of the fine particles, allowing them to come closer and form larger clumps.

COLIFORM (KOAL-i-form) COLIFORM

A group of bacteria found in the intestines of warm-blooded animals (including humans) and also in plants, soil, air, and water. The presence of coliform bacteria is an indication that the water is polluted and may contain pathogenic (disease-causing) organisms. Fecal coliforms are those coliforms found in the feces of various warm-blooded animals, whereas the term "coliform" also includes other environmental sources.

COLLOIDS (KALL-loids) COLLOIDS

Very small, finely divided solids (particles that do not dissolve) that remain dispersed in a liquid for a long time due to their small size and electrical charge. When most of the particles in water have a negative electrical charge, they tend to repel each other. This repulsion prevents the particles from clumping together, becoming heavier, and settling out.

COLORIMETRIC MEASUREMENT

COLORIMETRIC MEASUREMENT

A means of measuring unknown chemical concentrations in water by measuring a sample's color intensity. The specific color of the sample, developed by addition of chemical reagents, is measured with a photoelectric colorimeter or is compared with color standards using, or corresponding with, known concentrations of the chemical.

COMBINED AVAILABLE CHLORINE

COMBINED AVAILABLE CHLORINE

The total chlorine, present as chloramine or other derivatives, that is present in a water and is still available for disinfection and for oxidation of organic matter. The combined chlorine compounds are more stable than free chlorine forms, but they are somewhat slower in disinfection action.

COMBINED AVAILABLE CHLORINE RESIDUAL

COMBINED AVAILABLE CHLORINE RESIDUAL

The concentration of residual chlorine that is combined with ammonia, organic nitrogen, or both in water as a chloramine (or other chloro derivative) and yet is still available to oxidize organic matter and help kill bacteria.

COMBINED CHLORINE

COMBINED CHLORINE

The sum of the chlorine species composed of free chlorine and ammonia, including monochloramine, dichloramine, and trichloramine (nitrogen trichloride). Dichloramine is the strongest disinfectant of these chlorine species, but it has less oxidative capacity than free chlorine.

COMBINED RESIDUAL CHLORINATION

COMBINED RESIDUAL CHLORINATION

The application of chlorine to water or wastewater to produce a combined available chlorine residual. The residual may consist of chlorine compounds formed by the reaction of chlorine with natural or added ammonia (NH_3) or with certain organic nitrogen compounds.

COMBINED SEWER

COMBINED SEWER

A sewer designed to carry both sanitary wastewaters and stormwater or surface water runoff.

COMMINUTION (kom-mih-NEW-shun)

COMMINUTION

A mechanical treatment process that cuts large pieces of wastes into smaller pieces so they will not plug pipes or damage equipment. Comminution and SHREDDING usually mean the same thing.

COMMINUTOR (kom-mih-NEW-ter)

COMMINUTOR

A device used to reduce the size of the solid materials in wastewater by shredding (comminution). The shredding action is like many scissors cutting to shreds all the large solids in the wastewater.

COMPOSITE (PROPORTIONAL) SAMPLE

COMPOSITE (PROPORTIONAL) SAMPLE

A composite sample is a collection of individual samples obtained at regular intervals, usually every one or two hours during a 24-hour time span. Each individual sample is combined with the others in proportion to the rate of flow when the sample was collected. Equal volume individual samples also may be collected at intervals after a specific volume of flow passes the sampling point or after equal time intervals and still be referred to as a composite sample. The resulting mixture (composite sample) forms a representative sample and is analyzed to determine the average conditions during the sampling period.

COMPOUND

COMPOUND

A pure substance composed of two or more elements whose composition is constant. For example, table salt (sodium chloride, NaCl) is a compound.

CONCENTRATION POLARIZATION

CONCENTRATION POLARIZATION

(1) A buildup of retained particles on the membrane surface due to dewatering of the feed closest to the membrane. The thickness of the concentration polarization layer is controlled by the flow velocity across the membrane.

(2) Used in corrosion studies to indicate a depletion of ions near an electrode.

(3) The basis for chemical analysis by a polarograph.

CONFINED SPACE

CONFINED SPACE

Confined space means a space that:

(1) Is large enough and so configured that an employee can bodily enter and perform assigned work; and

(2) Has limited or restricted means for entry or exit (for example, manholes, tanks, vessels, silos, storage bins, hoppers, vaults, and pits are spaces that may have limited means of entry); and

(3) Is not designed for continuous employee occupancy.

Also see DANGEROUS AIR CONTAMINATION and OXYGEN DEFICIENCY.

CONING

CONING

Development of a cone-shaped flow of liquid, like a whirlpool, through sludge. This can occur in a sludge hopper during sludge withdrawal when the sludge becomes too thick. Part of the sludge remains in place while liquid rather than sludge flows out of the hopper. Also called coring.

CONTACT STABILIZATION

CONTACT STABILIZATION

Contact stabilization is a modification of the conventional activated sludge process. In contact stabilization, two aeration tanks are used. One tank is for separate reaeration of the return sludge for at least four hours before it is permitted to flow into the other aeration tank to be mixed with the primary effluent requiring treatment. The process may also occur in one long tank.

CONTACTOR

CONTACTOR

An electric switch, usually magnetically operated.

CONTINUOUS PROCESS

CONTINUOUS PROCESS

A treatment process in which water is treated continuously in a tank or reactor. The water being treated continuously flows into the tank at one end, is treated as it flows through the tank, and flows out the opposite end as treated water.

CONTROL LOOP

CONTROL LOOP

The combination of one or more interconnected instrumentation devices that are arranged to measure, display, and control a process variable. Also called a loop.

CONTROL SYSTEM

CONTROL SYSTEM

An instrumentation system that senses and controls its own operation on a close, continuous basis in what is called proportional (or modulating) control.

CONTROLLER

CONTROLLER

A device that controls the starting, stopping, or operation of a device or piece of equipment.

CONVENTIONAL TREATMENT

CONVENTIONAL TREATMENT

(1) The common wastewater treatment processes such as preliminary treatment, sedimentation, flotation, trickling filter, rotating biological contactor, activated sludge, and chlorination wastewater treatment processes used by POTWs.

(2) The hydroxide precipitation of metals processes used by pretreatment facilities.

CROSS CONNECTION

CROSS CONNECTION

(1) A connection between drinking (potable) water and an unapproved water supply.

(2) A connection between a storm drain system and a sanitary collection system.

(3) Less frequently used to mean a connection between two sections of a collection system to handle anticipated overloads of one system.

CRYOGENIC (KRY-o-JEN-nick)

CRYOGENIC

Very low temperature. Associated with liquified gases (liquid oxygen).

D

DO (pronounce as separate letters)   DO

Dissolved Oxygen. DO is the molecular oxygen dissolved in water or wastewater.

DALTON   DALTON

A unit of mass designated as one-sixteenth the mass of oxygen-16, the lightest and most abundant isotope of oxygen. The dalton is equivalent to one mass unit.

DANGER   DANGER

The word *DANGER* is used where an immediate hazard presents a threat of death or serious injury to employees. Also see CAUTION, NOTICE, and WARNING.

DANGEROUS AIR CONTAMINATION   DANGEROUS AIR CONTAMINATION

An atmosphere presenting a threat of causing death, injury, acute illness, or disablement due to the presence of flammable and/or explosive, toxic, or otherwise injurious or incapacitating substances.

(1) Dangerous air contamination due to the flammability of a gas, vapor, or mist is defined as an atmosphere containing the gas, vapor, or mist at a concentration greater than 10 percent of its lower explosive (lower flammable) limit (LEL).

(2) Dangerous air contamination due to a combustible particulate is defined as a concentration that meets or exceeds the particulate's lower explosive limit (LEL).

(3) Dangerous air contamination due to the toxicity of a substance is defined as the atmospheric concentration that could result in employee exposure in excess of the substance's permissible exposure limit (PEL).

NOTE: A dangerous situation also occurs when the oxygen level is less than 19.5 percent by volume (OXYGEN DEFICIENCY) or more than 23.5 percent by volume (OXYGEN ENRICHMENT).

DATEOMETER (day-TOM-uh-ter)   DATEOMETER

A small calendar disk attached to motors and equipment to indicate the year in which the last maintenance service was performed.

DAY TANK   DAY TANK

A tank used to store a chemical solution of known concentration for feed to a chemical feeder. A day tank usually stores sufficient chemical solution to properly treat the water being treated for at least one day. Also called an AGE TANK.

DECANT (de-KANT)   DECANT

To draw off the upper layer of liquid (water) after the heavier material (a solid or another liquid) has settled.

DECHLORINATION (DEE-klor-uh-NAY-shun)   DECHLORINATION

The deliberate removal of chlorine from water. The partial or complete reduction of residual chlorine by any chemical or physical process.

DECIBEL (DES-uh-bull)   DECIBEL

A unit for expressing the relative intensity of sounds on a scale from zero for the average least perceptible sound to about 130 for the average level at which sound causes pain to humans. Abbreviated dB.

DECOMPOSITION or DECAY   DECOMPOSITION or DECAY

The conversion of chemically unstable materials to more stable forms by chemical or biological action.

DEFINING   DEFINING

A process that arranges the activated carbon particles according to size. This process is also used to remove small particles from granular contactors to prevent excessive head loss.

DEGRADATION (deh-gruh-DAY-shun)   DEGRADATION

The conversion or breakdown of a substance to simpler compounds, for example, the degradation of organic matter to carbon dioxide and water.

DELAMINATION (DEE-lam-uh-NAY-shun)   DELAMINATION

Separation of a membrane or other material from the backing material on which it is cast.

DENITRIFICATION (dee-NYE-truh-fuh-KAY-shun)   DENITRIFICATION

(1) The anoxic biological reduction of nitrate nitrogen to nitrogen gas.

(2) The removal of some nitrogen from a system.

(3) An anoxic process that occurs when nitrite or nitrate ions are reduced to nitrogen gas and nitrogen bubbles are formed as a result of this process. The bubbles attach to the biological floc and float the floc to the surface of the secondary clarifiers. This condition is often the cause of rising sludge observed in secondary clarifiers or gravity thickeners. Also see NITRIFICATION.

DENSITY   DENSITY

A measure of how heavy a substance (solid, liquid, or gas) is for its size. Density is expressed in terms of weight per unit volume, that is, grams per cubic centimeter or pounds per cubic foot. The density of water (at 4°C or 39°F) is 1.0 gram per cubic centimeter or about 62.4 pounds per cubic foot.

DESICCANT (DESS-uh-kant)   DESICCANT

A drying agent that is capable of removing or absorbing moisture from the atmosphere in a small enclosure.

DESICCATION (dess-uh-KAY-shun)   DESICCATION

A process used to thoroughly dry air; to remove virtually all moisture from air.

DESICCATOR (DESS-uh-kay-tor)   DESICCATOR

A closed container into which heated weighing or drying dishes are placed to cool in a dry environment in preparation for weighing. The dishes may be empty or they may contain a sample. Desiccators contain a substance (DESICCANT), such as anhydrous calcium chloride, that absorbs moisture and keeps the relative humidity near zero so that the dish or sample will not gain weight from absorbed moisture.

DETENTION TIME   DETENTION TIME

(1) The time required to fill a tank at a given flow.

(2) The theoretical (calculated) time required for water to pass through a tank at a given rate of flow.

(3) The actual time in hours, minutes, or seconds that a small amount of water is in a settling basin, flocculating basin, or rapid-mix chamber. In septic tanks, detention time will decrease as the volumes of sludge and scum increase. In storage reservoirs, detention time is the length of time entering water will be held before being drafted for use (several weeks to years, several months being typical).

$$\text{Detention Time, hr} = \frac{(\text{Basin Volume, gal})(24 \text{ hr/day})}{\text{Flow, gal/day}}$$

or

$$\text{Detention Time, hr} = \frac{(\text{Basin Volume, m}^3)(24 \text{ hr/day})}{\text{Flow, m}^3/\text{day}}$$

DETRITUS (dee-TRY-tus)   DETRITUS

The heavy mineral material present in wastewater such as sand, coffee grounds, eggshells, gravel, and cinders. Also called GRIT.

DEW POINT   DEW POINT

The temperature to which air with a given quantity of water vapor must be cooled to cause condensation of the vapor in the air.

DEWATER

DEWATER

(1)   To remove or separate a portion of the water present in a sludge or slurry. To dry sludge so it can be handled and disposed of.

(2)   To remove or drain the water from a tank or a trench. A structure may be dewatered so that it can be inspected or repaired.

DEWATERABLE

DEWATERABLE

This is a property of sludge related to the ability to separate the liquid portion from the solid, with or without chemical conditioning. A material is considered dewaterable if water will readily drain from it.

DIAPHRAGM PUMP

DIAPHRAGM PUMP

A pump in which a flexible diaphragm, generally of rubber or equally flexible material, is the operating part. It is fastened at the edges in a vertical cylinder. When the diaphragm is raised, suction is exerted, and when it is depressed, the liquid is forced through a discharge valve.

DIATOMACEOUS (DYE-uh-toe-MAY-shus) EARTH

DIATOMACEOUS EARTH

A fine, siliceous (made of silica) earth composed mainly of the skeletal remains of diatoms.

DIATOMS (DYE-uh-toms)

DIATOMS

Unicellular (single cell), microscopic algae with a rigid, box-like internal structure consisting mainly of silica.

DIFFUSED-AIR AERATION

DIFFUSED-AIR AERATION

A diffused-air activated sludge plant takes air, compresses it, and then discharges the air below the water surface of the aerator through some type of air diffusion device.

DIFFUSER

DIFFUSER

A device (porous plate, tube, bag) used to break the air stream from the blower system into fine bubbles in an aeration tank or reactor.

DIGESTER (dye-JEST-er)

DIGESTER

A tank in which sludge is placed to allow decomposition by microorganisms. Digestion may occur under anaerobic (more common) or aerobic conditions.

DIGITAL

DIGITAL

The encoding of information that uses binary numbers (ones and zeros) for input, processing, transmission, storage, or display, rather than a continuous spectrum of values (an analog system) or non-numeric symbols such as letters or icons.

DIGITAL READOUT

DIGITAL READOUT

The readout of an instrument by a direct, numerical reading of the measured value or variable.

DISCHARGE HEAD

DISCHARGE HEAD

The pressure (in pounds per square inch (psi) or kilopascals (kPa)) measured at the centerline of a pump discharge and very close to the discharge flange, converted into feet or meters. The pressure is measured from the centerline of the pump to the hydraulic grade line of the water in the discharge pipe.

$$\text{Discharge Head, ft} = (\text{Discharge Pressure, psi})(2.31 \text{ ft/psi})$$

or

$$\text{Discharge Head, m} = (\text{Discharge Pressure, kPa})(1 \text{ m/9.8 kPa})$$

DISCRETE CONTROL

DISCRETE CONTROL

ON/OFF control; one of the two output values is equal to zero.

DISCRETE I/O (INPUT/OUTPUT)   DISCRETE I/O (INPUT/OUTPUT)

A digital signal that senses or sends either ON or OFF signals. For example, a discrete input would sense the position of a switch; a discrete output would turn on a pump or light.

DISINFECTION (dis-in-FECT-shun)   DISINFECTION

The process designed to kill or inactivate most microorganisms in water or wastewater, including essentially all pathogenic (disease-causing) bacteria. There are several ways to disinfect, with chlorination being the most frequently used in water and wastewater treatment plants. Compare with STERILIZATION.

DISSOLVED OXYGEN   DISSOLVED OXYGEN

Molecular oxygen dissolved in water or wastewater, usually abbreviated DO.

DISTILLATE (DIS-tuh-late)   DISTILLATE

In the distillation of a sample, a portion is collected by evaporation and recondensation; the part that is recondensed is the distillate.

DISTRIBUTED CONTROL SYSTEM (DCS)   DISTRIBUTED CONTROL SYSTEM (DCS)

A computer control system having multiple microprocessors to distribute the functions performing process control, thereby distributing the risk from component failure. The distributed components (input/output devices, control devices, and operator interface devices) are all connected by communications links and permit the transmission of control, measurement, and operating information to and from many locations.

DISTRIBUTOR   DISTRIBUTOR

The rotating mechanism that distributes the wastewater evenly over the surface of a trickling filter or other process unit. Also see FIXED SPRAY NOZZLE.

DOCTOR BLADE   DOCTOR BLADE

A blade used to remove any excess solids that may cling to the outside of a rotating screen.

DRAIN TILE SYSTEM   DRAIN TILE SYSTEM

A system of tile pipes buried under agricultural fields that collect percolated waters and keep the groundwater table below the ground surface to prevent ponding.

DRAINAGE WELLS   DRAINAGE WELLS

Wells that can be pumped to lower the groundwater table and prevent ponding.

DRIFT   DRIFT

The difference between the actual value and the desired value (or set point); characteristic of proportional controllers that do not incorporate reset action. Also called OFFSET.

DUCKWEED   DUCKWEED

A small, green, cloverleaf-shaped floating plant, about one-quarter inch (6 mm) across, which appears as a grainy layer on the surface of a pond.

DYNAMIC HEAD   DYNAMIC HEAD

When a pump is operating, the vertical distance (in feet or meters) from a point to the energy grade line. Also see ENERGY GRADE LINE, STATIC HEAD, and TOTAL DYNAMIC HEAD.

DYNAMIC PRESSURE   DYNAMIC PRESSURE

When a pump is operating, pressure resulting from the dynamic head.

$$\text{Dynamic Pressure, psi} = (\text{Dynamic Head, ft})(0.433 \text{ psi/ft})$$

or

$$\text{Dynamic Pressure, kPa} = (\text{Dynamic Head, m})(9.8 \text{ kPa/m})$$

E

EGL
EGL

See ENERGY GRADE LINE.

EDUCTOR (e-DUCK-ter)
EDUCTOR

A hydraulic device used to create a negative pressure (suction) by forcing a liquid through a restriction, such as a Venturi. An eductor or aspirator (the hydraulic device) may be used in the laboratory in place of a vacuum pump. As an injector, it is used to produce vacuum for chlorinators. Sometimes used instead of a suction pump.

EFFECTIVE RANGE
EFFECTIVE RANGE

That portion of the design range (usually from 10 to 90+ percent) in which an instrument has acceptable accuracy. Also see RANGE and SPAN.

EFFLORESCENCE (EF-low-RESS-ense)
EFFLORESCENCE

The powder or crust formed on a substance when moisture is given off upon exposure to the atmosphere.

EFFLUENT (EF-loo-ent)
EFFLUENT

Water or other liquid—raw (untreated), partially treated, or completely treated—flowing *FROM* a reservoir, basin, treatment process, or treatment plant.

ELECTROCHEMICAL CORROSION
ELECTROCHEMICAL CORROSION

The decomposition of a material by: (1) stray current electrolysis, (2) galvanic corrosion caused by dissimilar metals, or (3) galvanic corrosion caused by differential electrolysis.

ELECTROCHEMICAL PROCESS
ELECTROCHEMICAL PROCESS

A process that causes the deposition or formation of a seal or coating of a chemical element or compound by the use of electricity.

ELECTROLYSIS (ee-leck-TRAWL-uh-sis)
ELECTROLYSIS

The decomposition of material by an outside electric current.

ELECTROLYTE (ee-LECK-tro-lite)
ELECTROLYTE

A substance that dissociates (separates) into two or more ions when it is dissolved in water.

ELECTROLYTIC (ee-LECK-tro-LIT-ick) PROCESS
ELECTROLYTIC PROCESS

A process that causes the decomposition of a chemical compound by the use of electricity.

ELECTROMAGNETIC FORCES
ELECTROMAGNETIC FORCES

Forces resulting from electrical charges that either attract or repel particles. Particles with opposite charges are attracted to each other, while particles with similar charges repel each other. For example, a particle with a positive charge is attracted to a particle with a negative charge but is repelled by another particle with a positive charge.

ELECTRON
ELECTRON

(1) A very small, negatively charged particle that is practically weightless. According to the electron theory, all electrical and electronic effects are caused either by the movement of electrons from place to place or because there is an excess or lack of electrons at a particular place.

(2) The part of an atom that determines its chemical properties.

ELEMENT
ELEMENT

A substance that cannot be separated into its constituent parts and still retain its chemical identity. For example, sodium (Na) is an element.

ELUTRIATION (e-LOO-tree-A-shun)   ELUTRIATION

The washing of digested sludge with either fresh water, plant effluent, or other wastewater. The objective is to remove (wash out) fine particulates and/or the alkalinity in sludge. This process reduces the demand for conditioning chemicals and improves settling or filtering characteristics of the solids.

EMULSION (e-MULL-shun)   EMULSION

A liquid mixture of two or more liquid substances not normally dissolved in one another; one liquid is held in suspension in the other.

ENCLOSED SPACE   ENCLOSED SPACE

See CONFINED SPACE.

END POINT   END POINT

The completion of a desired chemical reaction. Samples of water or wastewater are titrated to the end point. This means that a chemical is added, drop by drop, to a sample until a certain color change (blue to clear, for example) occurs. This is called the end point of the titration. In addition to a color change, an end point may be reached by the formation of a precipitate or the reaching of a specified pH. An end point may be detected by the use of an electronic device, such as a pH meter.

ENDOGENOUS (en-DODGE-en-us) RESPIRATION   ENDOGENOUS RESPIRATION

A situation in which living organisms oxidize some of their own cellular mass instead of new organic matter they adsorb or absorb from their environment.

ENERGY GRADE LINE (EGL)   ENERGY GRADE LINE (EGL)

A line that represents the elevation of energy head (in feet or meters) of water flowing in a pipe, conduit, or channel. The line is drawn above the hydraulic grade line (gradient) a distance equal to the velocity head ($V^2/2g$) of the water flowing at each section or point along the pipe or channel. Also see HYDRAULIC GRADE LINE.

[SEE DRAWING ON PAGE 723]

ENTERIC   ENTERIC

Of intestinal origin, especially applied to wastes or bacteria.

ENTRAIN   ENTRAIN

To trap bubbles in water either mechanically through turbulence or chemically through a reaction.

ENZYMES (EN-zimes)   ENZYMES

Organic or biochemical substances that cause or speed up chemical reactions.

EQUALIZING BASIN   EQUALIZING BASIN

A holding basin in which variations in flow and composition of a liquid are averaged. Such basins are used to provide a flow of reasonably uniform volume and composition to a treatment unit. Also called a balancing reservoir.

ESTUARIES (ES-chew-wear-eez)   ESTUARIES

Bodies of water that are located at the lower end of a river and are subject to tidal fluctuations.

EVAPOTRANSPIRATION (ee-VAP-o-TRANS-purr-A-shun)   EVAPOTRANSPIRATION

(1)   The process by which water vapor is released to the atmosphere from living plants. Also called TRANSPIRATION.

(2)   The total water removed from an area by transpiration (plants) and by evaporation from soil, snow, and water surfaces.

EXPLOSIMETER   EXPLOSIMETER

An instrument used to detect explosive atmospheres. When the lower explosive limit (LEL) of an atmosphere is exceeded, an alarm signal on the instrument is activated. Also called a combustible gas detector.

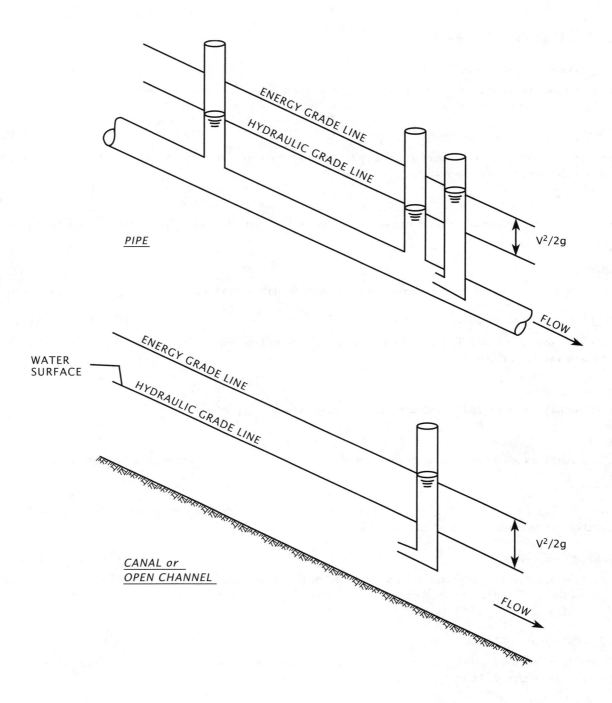

PIPE

WATER
SURFACE

CANAL or
OPEN CHANNEL

$V^2/2g$ = Velocity Head

Energy grade line and hydraulic grade line

F

F/M RATIO

See FOOD/MICROORGANISM RATIO.

FACULTATIVE (FACK-ul-tay-tive) BACTERIA

Facultative bacteria can use either dissolved oxygen or oxygen obtained from food materials such as sulfate or nitrate ions. In other words, facultative bacteria can live under aerobic, anoxic, or anaerobic conditions.

FACULTATIVE (FACK-ul-tay-tive) POND

The most common type of pond in current use. The upper portion (supernatant) is aerobic, while the bottom layer is anaerobic. Algae supply most of the oxygen to the supernatant.

FAIL-SAFE

Design and operation of a process control system whereby failure of the power system or any component does not result in process failure or equipment damage.

FEEDBACK

The circulating action between a sensor measuring a process variable and the controller that controls or adjusts the process variable.

FILAMENTOUS (fill-uh-MEN-tuss) ORGANISMS

Organisms that grow in a thread or filamentous form. Common types are *Thiothrix* and *Actinomycetes*. A common cause of sludge bulking in the activated sludge process.

FILTER AID

A chemical (usually a polymer) added to water to help remove fine colloidal suspended solids.

FIXED FILM

Fixed film denitrification is the common name for attached growth anaerobic treatment processes used to achieve denitrification.

FIXED SAMPLE

A sample is fixed in the field by adding chemicals that prevent the water quality indicators of interest in the sample from changing before final measurements are performed later in the laboratory.

FIXED SPRAY NOZZLE

Cone-shaped spray nozzle used to distribute wastewater over the filter media, similar to a lawn sprinkling system. A deflector or steel ball is mounted within the cone to spread the flow of wastewater through the cone, thus causing a spraying action. Also see DISTRIBUTOR.

FLAME POLISHED

Melted by a flame to smooth out irregularities. Sharp or broken edges of glass (such as the end of a glass tube) are rotated in a flame until the edge melts slightly and becomes smooth.

FLIGHTS

Scraper boards, made from redwood or other rot-resistant woods or plastic, used to collect and move settled sludge or floating scum.

FLOC

Clumps of bacteria and particles, or coagulants and impurities, that have come together and formed a cluster. Found in flocculation tanks, sedimentation basins, aeration tanks, secondary clarifiers, and chemical precipitation processes.

FLOCCULATION (flock-yoo-LAY-shun)

FLOCCULATION

The gathering together of fine particles after coagulation to form larger particles by a process of gentle mixing. This clumping together makes it easier to separate the solids from the water by settling, skimming, draining, or filtering.

FLOW EQUALIZATION SYSTEM

FLOW EQUALIZATION SYSTEM

A device or tank designed to hold back or store a portion of peak flows for release during low-flow periods.

FLUIDIZED (FLOO-id-i-zd)

FLUIDIZED

A mass of solid particles that is made to flow like a liquid by injection of water or gas is said to have been fluidized. In water and wastewater treatment, a bed of filter media is fluidized by backwashing water through the filter.

FOOD/MICROORGANISM (F/M) RATIO

FOOD/MICROORGANISM (F/M) RATIO

Food to microorganism ratio. A measure of food provided to bacteria in an aeration tank.

$$\frac{\text{Food}}{\text{Microorganisms}} = \frac{\text{BOD, lbs/day}}{\text{MLVSS, lbs}}$$

$$= \frac{\text{Flow, MGD} \times \text{BOD, mg}/L \times 8.34 \text{ lbs/gal}}{\text{Volume, MG} \times \text{MLVSS, mg}/L \times 8.34 \text{ lbs/gal}}$$

or by calculator math system

$$= \text{Flow, MGD} \times \text{BOD, mg}/L \div \text{Volume, MG} \div \text{MLVSS, mg}/L$$

or metric

$$= \frac{\text{BOD, kg/day}}{\text{MLVSS, kg}}$$

$$= \frac{\text{Flow, M}L/\text{day} \times \text{BOD, mg}/L \times 1 \text{ kg/M mg}}{\text{Volume, M}L \times \text{MLVSS, mg}/L \times 1 \text{ kg/M mg}}$$

FORCE MAIN

FORCE MAIN

A pipe that carries wastewater under pressure from the discharge side of a pump to a point of gravity flow downstream.

FREE AVAILABLE CHLORINE RESIDUAL

FREE AVAILABLE CHLORINE RESIDUAL

That portion of the total available chlorine residual composed of dissolved chlorine gas (Cl_2), hypochlorous acid (HOCl), and/or hypochlorite ion (OCl^-) remaining in water after chlorination. This does not include chlorine that has combined with ammonia, nitrogen, or other compounds. Also called free available residual chlorine.

FREE CHLORINE

FREE CHLORINE

Free chlorine is chlorine (Cl_2) in a liquid or gaseous form. Free chlorine combines with water to form hypochlorous (HOCl) and hydrochloric (HCl) acids. In wastewater, free chlorine usually combines with an amine (ammonia or nitrogen) or other organic compounds to form combined chlorine compounds.

FREE OXYGEN

FREE OXYGEN

Molecular oxygen available for respiration by organisms. Molecular oxygen is the oxygen molecule, O_2, that is not combined with another element to form a compound.

FREEBOARD

FREEBOARD

(1) The vertical distance from the normal water surface to the top of the confining wall.

(2) The vertical distance from the sand surface to the underside of a trough in a sand filter. This distance is also called AVAILABLE EXPANSION.

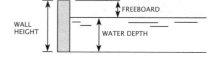

FRICTION LOSS

FRICTION LOSS

The head, pressure, or energy (they are the same) lost by water flowing in a pipe or channel as a result of turbulence caused by the velocity of the flowing water and the roughness of the pipe, channel walls, or restrictions caused by fittings. Water flowing in a pipe loses head, pressure, or energy as a result of friction losses. Also called HEAD LOSS.

G

GIS

GIS

See GEOGRAPHIC INFORMATION SYSTEM.

GASIFICATION (gas-uh-fuh-KAY-shun)

GASIFICATION

The conversion of soluble and suspended organic materials into gas during aerobic or anaerobic decomposition. In clarifiers, the resulting gas bubbles can become attached to the settled sludge and cause large clumps of sludge to rise and float on the water surface. In anaerobic sludge digesters, this gas is collected for fuel or disposed of using a waste gas burner.

GATE

GATE

(1) A movable, watertight barrier for the control of a liquid in a waterway.

(2) A descriptive term used on irrigation distribution piping systems instead of the word "valve." Gates cover outlet ports in the pipe segments. Water flows are regulated or distributed by opening the gates by either sliding the gate up or down or by swinging the gate to one side and uncovering an individual port to permit water flow to be discharged or regulated from the pipe at that particular point.

GEOGRAPHIC INFORMATION SYSTEM (GIS)

GEOGRAPHIC INFORMATION SYSTEM (GIS)

A computer program that combines mapping with detailed information about the physical locations of structures, such as pipes, valves, and manholes, within geographic areas. The system is used to help operators and maintenance personnel locate utility system features or structures and to assist with the scheduling and performance of maintenance activities.

GRAB SAMPLE

GRAB SAMPLE

A single sample of water collected at a particular time and place that represents the composition of the water only at that time and place.

GRAVIMETRIC

GRAVIMETRIC

A means of measuring unknown concentrations of water quality indicators in a sample by weighing a precipitate or residue of the sample.

GRIT

GRIT

The heavy mineral material present in wastewater such as sand, coffee grounds, eggshells, gravel, and cinders. Also called DETRITUS.

GRIT REMOVAL

GRIT REMOVAL

Grit removal is accomplished by providing an enlarged channel or chamber that causes the flow velocity to be reduced and allows the heavier grit to settle to the bottom of the channel where it can be removed.

GROWTH RATE, Y

GROWTH RATE, Y

An experimentally determined constant to estimate the unit growth rate of bacteria while degrading organic wastes.

H

HGL

HGL

See HYDRAULIC GRADE LINE.

HARMFUL PHYSICAL AGENT
  or TOXIC SUBSTANCE

HARMFUL PHYSICAL AGENT
  or TOXIC SUBSTANCE

Any chemical substance, biological agent (bacteria, virus, or fungus), or physical stress (noise, heat, cold, vibration, repetitive motion, ionizing and non-ionizing radiation, hypo- or hyperbaric pressure) that:

(1) Is regulated by any state or federal law or rule due to a hazard to health

(2) Is listed in the latest printed edition of the National Institute of Occupational Safety and Health (NIOSH) Registry of Toxic Effects of Chemical Substances (RTECS)

(3) Has yielded positive evidence of an acute or chronic health hazard in human, animal, or other biological testing conducted by, or known to, the employer

(4) Is described by a Material Safety Data Sheet (MSDS) available to the employer that indicates that the material may pose a hazard to human health

Also see ACUTE HEALTH EFFECT and CHRONIC HEALTH EFFECT.

HEAD   HEAD

The vertical distance, height, or energy of water above a reference point. A head of water may be measured in either height (feet or meters) or pressure (pounds per square inch or kilograms per square centimeter). Also see DISCHARGE HEAD, DYNAMIC HEAD, STATIC HEAD, SUCTION HEAD, SUCTION LIFT, and VELOCITY HEAD.

HEAD LOSS   HEAD LOSS

The head, pressure, or energy (they are the same) lost by water flowing in a pipe or channel as a result of turbulence caused by the velocity of the flowing water and the roughness of the pipe, channel walls, or restrictions caused by fittings. Water flowing in a pipe loses head, pressure, or energy as a result of friction losses. The head loss through a filter is due to friction losses caused by material building up on the surface or in the top part of a filter. Also called FRICTION LOSS.

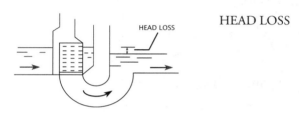

HEADER   HEADER

A large pipe to which the ends of a series of smaller pipes are connected. Also called a MANIFOLD.

HEADWORKS   HEADWORKS

The facilities where wastewater enters a wastewater treatment plant. The headworks may consist of bar screens, comminutors, a wet well, and pumps.

HEPATITIS (HEP-uh-TIE-tis)   HEPATITIS

Hepatitis is an inflammation of the liver caused by an acute viral infection. Yellow jaundice is one symptom of hepatitis.

HERTZ (Hz)   HERTZ (Hz)

The number of complete electromagnetic cycles or waves in one second of an electric or electronic circuit. Also called the frequency of the current.

HETEROTROPHIC (HET-er-o-TROF-ick)   HETEROTROPHIC

Describes organisms that use organic matter for energy and growth. Animals, fungi, and most bacteria are heterotrophs.

HINDERED SOLIDS SEPARATION   HINDERED SOLIDS SEPARATION

Solids settling in a thickening rather than in a clarifying mode. Suspended solids settling (clarifying) velocities are strongly influenced by the applied solids concentration. The greater the applied solids concentration, the greater the opportunity for thickening (hindered) solids settling.

HUMAN MACHINE INTERFACE (HMI)

HUMAN MACHINE INTERFACE (HMI)

The device at which the operator interacts with the control system. This may be an individual instrumentation and control device or the graphic screen of a computer control system. Also called MAN MACHINE INTERFACE (MMI) and OPERATOR INTERFACE.

HUMUS SLUDGE

HUMUS SLUDGE

The sloughed particles of biomass from trickling filter media that are removed from the water being treated in secondary clarifiers.

HYDRAULIC GRADE LINE (HGL)

HYDRAULIC GRADE LINE (HGL)

The surface or profile of water flowing in an open channel or a pipe flowing partially full. If a pipe is under pressure, the hydraulic grade line is that level water would rise to in a small, vertical tube connected to the pipe. Also see ENERGY GRADE LINE.

[SEE DRAWING ON PAGE 723]

HYDRAULIC LOADING

HYDRAULIC LOADING

Hydraulic loading refers to the flows (MGD or m^3/day) to a treatment plant or treatment process. Detention times, surface loadings, and weir overflow rates are directly influenced by flows.

HYDROGEN ION CONCENTRATION [H$^+$]

HYDROGEN ION CONCENTRATION [H$^+$]

The weight of hydrogen ion in moles per liter of solution. Commonly expressed as the pH value, which is the logarithm of the reciprocal of the hydrogen ion concentration.

$$pH = \log \frac{1}{[H^+]}$$

HYDROGEN SULFIDE GAS (H$_2$S)

HYDROGEN SULFIDE GAS (H$_2$S)

Hydrogen sulfide is a gas with a rotten egg odor, produced under anaerobic conditions. Hydrogen sulfide gas is particularly dangerous because it dulls the sense of smell, becoming unnoticeable after you have been around it for a while; in high concentrations, it is only noticeable for a very short time before it dulls the sense of smell. The gas is very poisonous to the respiratory system, explosive, flammable, colorless, and heavier than air.

HYDROLOGIC (HI-dro-LOJ-ick) CYCLE

HYDROLOGIC CYCLE

The process of evaporation of water into the air and its return to earth by precipitation (rain or snow). This process also includes transpiration from plants, groundwater movement, and runoff into rivers, streams, and the ocean. Also called the WATER CYCLE.

HYDROLYSIS (hi-DROLL-uh-sis)

HYDROLYSIS

(1) A chemical reaction in which a compound is converted into another compound by taking up water.

(2) Usually a chemical degradation of organic matter.

HYGROSCOPIC (hi-grow-SKOP-ick)

HYGROSCOPIC

Absorbing or attracting moisture from the air.

HYPOCHLORINATION (HI-poe-klor-uh-NAY-shun)

HYPOCHLORINATION

The application of hypochlorite compounds to water or wastewater for the purpose of disinfection.

HYPOCHLORINATORS (HI-poe-KLOR-uh-nay-tors)

HYPOCHLORINATORS

Chlorine pumps, chemical feed pumps, or devices used to dispense chlorine solutions made from hypochlorites, such as bleach (sodium hypochlorite) or calcium hypochlorite into the water being treated.

HYPOCHLORITE (HI-poe-KLOR-ite)

HYPOCHLORITE

Chemical compounds containing available chlorine; used for disinfection. They are available as liquids (bleach) or solids (powder, granules, and pellets) in barrels, drums, and cans. Salts of hypochlorous acid.

I

IMHOFF CONE

IMHOFF CONE

A clear, cone-shaped container marked with graduations. The cone is used to measure the volume of settleable solids in a specific volume (usually one liter) of water or wastewater.

IMPELLER

IMPELLER

A rotating set of vanes in a pump or compressor designed to pump or move water or air.

IMPELLER PUMP

IMPELLER PUMP

Any pump in which the water is moved by the continuous application of power to a rotating set of vanes from some rotating mechanical source.

INCINERATION

INCINERATION

The conversion of dewatered wastewater solids by combustion (burning) to ash, carbon dioxide, and water vapor.

INDICATOR

INDICATOR

(1) (Chemical indicator) A substance that gives a visible change, usually of color, at a desired point in a chemical reaction, generally at a specified end point.

(2) (Instrument indicator) A device that indicates the result of a measurement, usually using either a fixed scale and movable indicator (pointer), such as a pressure gauge, or a moving chart with a movable pen like those used on a circular flow-recording chart. Also called a RECEIVER.

INDOLE (IN-dole)

INDOLE

An organic compound (C_8H_7N) containing nitrogen that has an ammonia odor.

INFILTRATION (in-fill-TRAY-shun)

INFILTRATION

The seepage of groundwater into a sewer system, including service connections. Seepage frequently occurs through defective or cracked pipes, pipe joints and connections, interceptor access risers and covers, or manhole walls.

INFLOW

INFLOW

Water discharged into a sewer system and service connections from sources other than regular connections. This includes flow from yard drains, foundations, and around access and manhole covers. Inflow differs from infiltration in that it is a direct discharge into the sewer rather than a leak in the sewer itself.

INFLUENT

INFLUENT

Water or other liquid—raw (untreated) or partially treated—flowing *INTO* a reservoir, basin, treatment process, or treatment plant.

INHIBITORY SUBSTANCES

INHIBITORY SUBSTANCES

Materials that kill or restrict the ability of organisms to treat wastes.

INOCULATE (in-NOCK-yoo-late)

INOCULATE

To introduce a seed culture into a system.

INORGANIC WASTE

INORGANIC WASTE

Waste material such as sand, salt, iron, calcium, and other mineral materials that are only slightly affected by the action of organisms. Inorganic wastes are chemical substances of mineral origin; whereas organic wastes are chemical substances usually of animal or plant origin. Also see NONVOLATILE MATTER, ORGANIC WASTE, and VOLATILE SOLIDS.

INTEGRATOR   INTEGRATOR

A device or meter that continuously measures and sums a process rate variable in cumulative fashion over a given time period. For example, total flows displayed in gallons per minute, million gallons per day, cubic feet per second, or some other unit of volume per time period. Also called a TOTALIZER.

INTERFACE   INTERFACE

The common boundary layer between two substances, such as water and a solid (metal); or between two fluids, such as water and a gas (air); or between a liquid (water) and another liquid (oil).

INTERLOCK   INTERLOCK

A physical device, equipment, or software routine that prevents an operation from beginning or changing function until some condition or set of conditions is fulfilled. An example would be a switch that prevents a piece of equipment from operating when a hazard exists.

IONIC CONCENTRATION   IONIC CONCENTRATION

The concentration of any ion in solution, usually expressed in moles per liter.

IONIZATION (EYE-on-uh-ZAY-shun)   IONIZATION

(1)   The splitting or dissociation (separation) of molecules into negatively and positively charged ions.

(2)   The process of adding electrons to, or removing electrons from, atoms or molecules, thereby creating ions. High temperatures, electrical discharges, and nuclear radiation can cause ionization.

J

JAR TEST   JAR TEST

A laboratory procedure that simulates coagulation/flocculation with differing chemical doses. The purpose of the procedure is to estimate the minimum coagulant dose required to achieve certain water quality goals. Samples of water to be treated are placed in six jars. Various amounts of chemicals are added to each jar, stirred, and the settling of solids is observed. The lowest dose of chemicals that provides satisfactory settling is the dose used to treat the water.

JOGGING   JOGGING

The frequent starting and stopping of an electric motor.

JOULE (JOOL)   JOULE

A measure of energy, work, or quantity of heat. One joule is the work done when the point of application of a force of one newton is displaced a distance of one meter in the direction of the force. Approximately equal to 0.7375 ft-lbs (0.1022 m-kg).

K

KJELDAHL (KELL-doll) NITROGEN   KJELDAHL NITROGEN

Nitrogen in the form of organic proteins or their decomposition product ammonia, as measured by the Kjeldahl Method.

L

LAG TIME   LAG TIME

The time period between the moment a process change is made and the moment such a change is finally sensed by the associated measuring instrument.

LAUNDERS

Sedimentation basin and filter discharge channels consisting of overflow weir plates (in sedimentation basins) and conveying troughs.

LAUNDERS

LIMIT SWITCH

A device that regulates or controls the travel distance of a chain or cable.

LIMIT SWITCH

LINEAL (LIN-e-ul)

The length in one direction of a line. For example, a board 12 feet (meters) long has 12 lineal feet (meters) in its length.

LINEAL

LINEARITY (lin-ee-AIR-it-ee)

How closely an instrument measures actual values of a variable through its effective range.

LINEARITY

LIQUEFACTION (lick-we-FACK-shun)

The conversion of large, solid particles of sludge into very fine particles that either dissolve or remain suspended in wastewater.

LIQUEFACTION

LOADING

Quantity of material applied to a device at one time.

LOADING

LYSIMETER (lie-SIM-uh-ter)

A device containing a mass of soil and designed to permit the measurement of water draining through the soil.

LYSIMETER

M

M or MOLAR

A molar solution consists of one gram molecular weight of a compound dissolved in enough water to make one liter of solution. A gram molecular weight is the molecular weight of a compound in grams. For example, the molecular weight of sulfuric acid (H_2SO_4) is 98. A one *M* solution of sulfuric acid would consist of 98 grams of H_2SO_4 dissolved in enough distilled water to make one liter of solution.

M or MOLAR

MBAS

Methylene Blue Active Substance. Another name for surfactants or surface active agents. The determination of surfactants is accomplished by measuring the color change in a standard solution of methylene blue dye.

MBAS

MCRT

Mean Cell Residence Time. An expression of the average time (days) that a microorganism will spend in the activated sludge process.

MCRT

$$\text{MCRT, days} = \frac{\text{Total Suspended Solids in Activated Sludge Process, lbs}}{\text{Total Suspended Solids Removed From Process, lbs/day}}$$

or

$$\text{MCRT, days} = \frac{\text{Total Suspended Solids in Activated Sludge Process, kg}}{\text{Total Suspended Solids Removed From Process, kg/day}}$$

NOTE: Operators at different plants calculate the Total Suspended Solids (TSS) in the Activated Sludge Process, lbs (kg), by three different methods:

1. TSS in the Aeration Basin or Reactor Zone, lbs (kg)

2. TSS in the Aeration Basin and Secondary Clarifier, lbs (kg)

3. TSS in the Aeration Basin and Secondary Clarifier Sludge Blanket, lbs (kg)

These three different methods make it difficult to compare MCRTs in days among different plants unless everyone uses the same method.

mg/*L*   mg/*L*

See MILLIGRAMS PER LITER, mg/*L*.

MLSS   MLSS

Mixed Liquor Suspended Solids. The amount (mg/*L*) of suspended solids in the mixed liquor of an aeration tank.

MLVSS   MLVSS

Mixed Liquor Volatile Suspended Solids. The amount (mg/*L*) of organic or volatile suspended solids in the mixed liquor of an aeration tank. This volatile portion is used as a measure or indication of the microorganisms present.

MPN   MPN

MPN is the Most Probable Number of coliform-group organisms per unit volume of sample water. Expressed as a density or population of organisms per 100 m*L* of sample water.

MSDS   MSDS

See MATERIAL SAFETY DATA SHEET.

MAN MACHINE INTERFACE (MMI)   MAN MACHINE INTERFACE (MMI)

The device at which the operator interacts with the control system. This may be an individual instrumentation and control device or the graphic screen of a computer control system. Also called HUMAN MACHINE INTERFACE (HMI) and OPERATOR INTERFACE.

MANIFOLD   MANIFOLD

A large pipe to which the ends of a series of smaller pipes are connected. Also called a HEADER.

MANOMETER (man-NAH-mut-ter)   MANOMETER

An instrument for measuring pressure. Usually, a manometer is a glass tube filled with a liquid that is used to measure the difference in pressure across a flow measuring device, such as an orifice or a Venturi meter. The instrument used to measure blood pressure is a type of manometer.

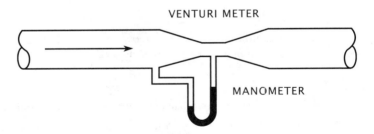

MASKING AGENTS   MASKING AGENTS

Substances used to cover up or disguise unpleasant odors. Liquid masking agents are dripped into the wastewater, sprayed into the air, or evaporated (using heat) with the unpleasant fumes or odors and then discharged into the air by blowers to make an undesirable odor less noticeable.

MATERIAL SAFETY DATA SHEET (MSDS)   MATERIAL SAFETY DATA SHEET (MSDS)

A document that provides pertinent information and a profile of a particular hazardous substance or mixture. An MSDS is normally developed by the manufacturer or formulator of the hazardous substance or mixture. The MSDS is required to be made available to employees and operators or inspectors whenever there is the likelihood of the hazardous substance or mixture being introduced into the workplace. Some manufacturers are preparing MSDSs for products that are not considered to be hazardous to show that the product or substance is not hazardous.

MEAN CELL RESIDENCE TIME (MCRT)   MEAN CELL RESIDENCE TIME (MCRT)

See MCRT.

MEASURED VARIABLE

A factor (flow, temperature) that is sensed and quantified (reduced to a reading of some kind) by a primary element or sensor.

MECHANICAL AERATION

The use of machinery to mix air and water so that oxygen can be absorbed into the water. Some examples are: paddle wheels, mixers, or rotating brushes to agitate the surface of an aeration tank; pumps to create fountains; and pumps to discharge water down a series of steps forming falls or cascades.

MEDIA

The material in a trickling filter on which slime accumulates and organisms grow. As settled wastewater trickles over the media, organisms in the slime remove certain types of wastes, thereby partially treating the wastewater. Also, the material in a rotating biological contactor or in a gravity or pressure filter.

MEDIAN

The middle measurement or value. When several measurements are ranked by magnitude (largest to smallest), half of the measurements will be larger and half will be smaller.

MENISCUS (meh-NIS-cuss)

The curved surface of a column of liquid (water, oil, mercury) in a small tube. When the liquid wets the sides of the container (as with water), the curve forms a valley. When the confining sides are not wetted (as with mercury), the curve forms a hill or upward bulge.

MERCAPTANS (mer-CAP-tans)

Compounds containing sulfur that have an extremely offensive skunk-like odor; also sometimes described as smelling like garlic or onions.

MESOPHILIC (MESS-o-FILL-ick) BACTERIA

Medium temperature bacteria. A group of bacteria that grow and thrive in a moderate temperature range between 68°F (20°C) and 113°F (45°C). The optimum temperature range for these bacteria in anaerobic digestion is 85°F (30°C) to 100°F (38°C).

METABOLISM

All of the processes or chemical changes in an organism or a single cell by which food is built up (anabolism) into living protoplasm and by which protoplasm is broken down (catabolism) into simpler compounds with the exchange of energy.

MICRON (MY-kron)

μm, Micrometer or Micron. A unit of length. One millionth of a meter or one thousandth of a millimeter. One micron equals 0.00004 of an inch.

MICROORGANISMS (MY-crow-OR-gan-is-ums)

Living organisms that can be seen individually only with the aid of a microscope.

MICROSCREEN

A device with a fabric straining medium with openings usually between 20 and 60 microns. The fabric is wrapped around the outside of a rotating drum. Wastewater enters the open end of the drum and flows out through the rotating screen cloth. At the highest point of the drum, the collected solids are backwashed by high-pressure water jets into a trough located within the drum.

MIL MIL

A unit of length equal to 0.001 of an inch. The diameter of wires and tubing is measured in mils, as is the thickness of plastic sheeting.

MILLIGRAMS PER LITER, mg/L MILLIGRAMS PER LITER, mg/L

A measure of the concentration by weight of a substance per unit volume in water or wastewater. In reporting the results of water and wastewater analysis, mg/L is preferred to the unit parts per million (ppm), to which it is approximately equivalent.

MILLIMICRON (MILL-uh-MY-kron) MILLIMICRON

A unit of length equal to $10^{-3}\mu$ (one thousandth of a micron), 10^{-6} millimeters, or 10^{-9} meters; correctly called a nanometer, nm.

MIXED LIQUOR MIXED LIQUOR

When the activated sludge in an aeration tank is mixed with primary effluent or the raw wastewater and return sludge, this mixture is then referred to as mixed liquor as long as it is in the aeration tank. Mixed liquor also may refer to the contents of mixed aerobic or anaerobic digesters.

MIXED LIQUOR SUSPENDED SOLIDS (MLSS) MIXED LIQUOR SUSPENDED SOLIDS (MLSS)

The amount (mg/L) of suspended solids in the mixed liquor of an aeration tank.

MIXED LIQUOR VOLATILE SUSPENDED SOLIDS (MLVSS) MIXED LIQUOR VOLATILE SUSPENDED SOLIDS (MLVSS)

The amount (mg/L) of organic or volatile suspended solids in the mixed liquor of an aeration tank. This volatile portion is used as a measure or indication of the microorganisms present.

MOLE MOLE

The name for a number (6.02×10^{23}) of atoms or molecules. See MOLECULAR WEIGHT.

MOLECULAR OXYGEN MOLECULAR OXYGEN

The oxygen molecule, O_2, that is not combined with another element to form a compound.

MOLECULAR WEIGHT MOLECULAR WEIGHT

The molecular weight of a compound in grams per mole is the sum of the atomic weights of the elements in the compound. The molecular weight of sulfuric acid (H_2SO_4) in grams is 98.

Element	Atomic Weight	Number of Atoms	Molecular Weight
H	1	2	2
S	32	1	32
O	16	4	64
			98

MOLECULE MOLECULE

The smallest division of a compound that still retains or exhibits all the properties of the substance.

MOST PROBABLE NUMBER (MPN) MOST PROBABLE NUMBER (MPN)

See MPN.

MOTILE (MO-till) MOTILE

Capable of self-propelled movement. A term that is sometimes used to distinguish between certain types of organisms found in water.

MOVING AVERAGE MOVING AVERAGE

To calculate the moving average for the last 7 days, add up the values for the last 7 days and divide by 7. Each day add the most recent day's value to the sum of values and subtract the oldest value. By using the 7-day moving average, each day of the week is always represented in the calculations.

MUDBALLS

Material, approximately round in shape, that forms in filters and gradually increases in size when not removed by the backwashing process. Mudballs vary from pea-sized up to golf-ball-sized or larger.

MUFFLE FURNACE

A small oven capable of reaching temperatures up to 600°C (1,112°F). Muffle furnaces are used in laboratories for burning or incinerating samples to determine the amounts of volatile solids and/or fixed solids in samples of wastewater.

MULTISTAGE PUMP

A pump that has more than one impeller. A single-stage pump has one impeller.

N

N or NORMAL

A normal solution contains one gram equivalent weight of reactant (compound) per liter of solution. The equivalent weight of an acid is that weight which contains one gram atom of ionizable hydrogen or its chemical equivalent. For example, the equivalent weight of sulfuric acid (H_2SO_4) is 49 (98 divided by 2 because there are two replaceable hydrogen ions). A one *N* solution of sulfuric acid would consist of 49 grams of H_2SO_4 dissolved in enough water to make one liter of solution.

NPDES PERMIT

National Pollutant Discharge Elimination System permit is the regulatory agency document issued by either a federal or state agency that is designed to control all discharges of potential pollutants from point sources and stormwater runoff into US waterways. NPDES permits regulate discharges into US waterways from all point sources of pollution, including industries, municipal wastewater treatment plants, sanitary landfills, large animal feedlots, and return irrigation flows.

NAMEPLATE

A durable, metal plate found on equipment that lists critical operating conditions for the equipment.

NEPHELOMETRIC (neff-el-o-MET-rick)

A means of measuring turbidity in a sample by using an instrument called a nephelometer. A nephelometer passes light through a sample and the amount of light deflected (usually at a 90-degree angle) is then measured.

NEUTRALIZATION (noo-trull-uh-ZAY-shun)

Addition of an acid or alkali (base) to a liquid to cause the pH of the liquid to move toward a neutral pH of 7.0.

NITRIFICATION (NYE-truh-fuh-KAY-shun)

An aerobic process in which bacteria change the ammonia and organic nitrogen in water or wastewater into oxidized nitrogen (usually nitrate).

NITRIFICATION (NYE-truh-fuh-KAY-shun) STAGE

A stage of decomposition that occurs in biological treatment processes when aerobic bacteria, using dissolved oxygen, change nitrogen compounds (ammonia and organic nitrogen) into oxidized nitrogen (usually nitrate). The second-stage BOD is sometimes referred to as the nitrification stage (first-stage BOD is called the carbonaceous stage).

NITRIFYING BACTERIA

Bacteria that change ammonia and organic nitrogen into oxidized nitrogen (usually nitrate).

NITROGENOUS (nye-TRAH-jen-us)

A term used to describe chemical compounds (usually organic) containing nitrogen in combined forms. Proteins and nitrate are nitrogenous compounds.

MUDBALLS

MUFFLE FURNACE

MULTISTAGE PUMP

N or NORMAL

NPDES PERMIT

NAMEPLATE

NEPHELOMETRIC

NEUTRALIZATION

NITRIFICATION

NITRIFICATION STAGE

NITRIFYING BACTERIA

NITROGENOUS

NONCORRODIBLE NONCORRODIBLE

A material that resists corrosion and will not be eaten away by wastewater or chemicals in wastewater.

NONSPARKING TOOLS NONSPARKING TOOLS

These tools will not produce a spark during use. They are made of a nonferrous material, usually a copper-beryllium alloy.

NONVOLATILE MATTER NONVOLATILE MATTER

Material such as sand, salt, iron, calcium, and other mineral materials that are only slightly affected by the actions of organisms and are not lost on ignition of the dry solids at 550°C (1,022°F). Volatile materials are chemical substances usually of animal or plant origin. Also see INORGANIC WASTE and VOLATILE SOLIDS.

NORMAL NORMAL

See *N* or NORMAL.

NOTICE NOTICE

This word calls attention to information that is especially significant in understanding and operating equipment or processes safely. Also see CAUTION, DANGER, and WARNING.

NUTRIENT NUTRIENT

Any substance that is assimilated (taken in) by organisms and promotes growth. Nitrogen and phosphorus are nutrients that promote the growth of algae. There are other essential and trace elements that are also considered nutrients. Also see NUTRIENT CYCLE.

NUTRIENT CYCLE NUTRIENT CYCLE

The transformation or change of a nutrient from one form to another until the nutrient has returned to the original form, thus completing the cycle. The cycle may take place under either aerobic or anaerobic conditions.

O

O & M MANUAL O & M MANUAL

Operation and Maintenance Manual. A manual that describes detailed procedures for operators to follow to operate and maintain a specific treatment plant and the equipment of that plant.

OSHA (O-shuh) OSHA

The Williams-Steiger Occupational Safety and Health Act of 1970 (OSHA) is a federal law designed to protect the health and safety of workers, including the operators of water supply and treatment systems and wastewater collection and treatment systems. The Act regulates the design, construction, operation, and maintenance of water and wastewater systems. OSHA regulations require employers to obtain and make available to workers the Material Safety Data Sheets (MSDSs) for chemicals used at industrial facilities and treatment plants. OSHA also refers to the federal and state agencies that administer the OSHA regulations.

OBLIGATE AEROBES OBLIGATE AEROBES

Bacteria that must have atmospheric or dissolved molecular oxygen to live and reproduce.

OCCUPATIONAL SAFETY AND
    HEALTH ACT OF 1970 (OSHA)
OCCUPATIONAL SAFETY AND
    HEALTH ACT OF 1970 (OSHA)

See OSHA.

ODOR PANEL ODOR PANEL

A group of people used to measure odors.

ODOR THRESHOLD ODOR THRESHOLD

The minimum odor of a gas or water sample that can just be detected after successive dilutions with odorless gas or water. Also called THRESHOLD ODOR.

OFFSET OFFSET

(1) The difference between the actual value and the desired value (or set point); characteristic of proportional controllers that do not incorporate reset action. Also called DRIFT.

(2) A pipe fitting in the approximate form of a reverse curve or other combination of elbows or bends that brings one section of a line of pipe out of line with, but into a line parallel with, another section.

(3) A pipe joint that has lost its bedding support, causing one of the pipe sections to drop or slip, thus creating a condition where the pipes no longer line up properly.

OLFACTOMETER (ALL-fak-TOM-utter) OLFACTOMETER

A device used to measure odors in the field by diluting odors with odor-free air.

OPERATOR INTERFACE OPERATOR INTERFACE

The device at which the operator interacts with the control system. This may be an individual instrumentation and control device or the graphic screen of a computer control system. Also called HUMAN MACHINE INTERFACE (HMI) and MAN MACHINE INTERFACE (MMI).

ORGANIC WASTE ORGANIC WASTE

Waste material that may come from animal or plant sources. Natural organic wastes generally can be consumed by bacteria and other small organisms. Manufactured or synthetic organic wastes from metal finishing, chemical manufacturing, and petroleum industries may not normally be consumed by bacteria and other organisms. Also see INORGANIC WASTE and VOLATILE SOLIDS.

ORGANISM ORGANISM

Any form of animal or plant life. Also see BACTERIA.

ORIFICE (OR-uh-fiss) ORIFICE

An opening (hole) in a plate, wall, or partition. An orifice flange or plate placed in a pipe consists of a slot or a calibrated circular hole smaller than the pipe diameter. The difference in pressure in the pipe above and at the orifice may be used to determine the flow in the pipe. In a trickling filter distributor, the wastewater passes through an orifice to the surface of the filter media.

ORTHOTOLIDINE (or-tho-TOL-uh-dine) ORTHOTOLIDINE

Orthotolidine is a colorimetric indicator of chlorine residual. If chlorine is present, a yellow-colored compound is produced. This reagent is no longer approved for chemical analysis to determine chlorine residual.

OVERFLOW RATE OVERFLOW RATE

One factor of the design flow of settling tanks and clarifiers in treatment plants used by operators to determine if tanks and clarifiers are hydraulically (flow) over- or underloaded. Also called SURFACE LOADING.

$$\text{Overflow Rate, GPD/sq ft} = \frac{\text{Flow, gallons/day}}{\text{Surface Area, sq ft}}$$

or

$$\text{Overflow Rate, } \frac{\text{m}^3\text{/day}}{\text{m}^2} = \frac{\text{Flow, m}^3\text{/day}}{\text{Surface Area, m}^2}$$

OXIDATION

OXIDATION

Oxidation is the addition of oxygen, removal of hydrogen, or the removal of electrons from an element or compound; in the environment and in wastewater treatment processes, organic matter is oxidized to more stable substances. The opposite of REDUCTION.

OXIDATION-REDUCTION POTENTIAL (ORP)

OXIDATION-REDUCTION POTENTIAL (ORP)

The electrical potential required to transfer electrons from one compound or element (the oxidant) to another compound or element (the reductant); used as a qualitative measure of the state of oxidation in water and wastewater treatment systems. ORP is measured in millivolts, with negative values indicating a tendency to reduce compounds or elements and positive values indicating a tendency to oxidize compounds or elements.

OXIDIZED ORGANICS

OXIDIZED ORGANICS

Organic materials that have been broken down in a biological process. Examples of these materials are carbohydrates and proteins that are broken down to simple sugars.

OXIDIZING AGENT

OXIDIZING AGENT

Any substance, such as oxygen (O_2) or chlorine (Cl_2), that will readily add (take on) electrons. When oxygen or chlorine is added to water or wastewater, organic substances are oxidized. These oxidized organic substances are more stable and less likely to give off odors or to contain disease-causing bacteria. The opposite is a REDUCING AGENT.

OXYGEN DEFICIENCY

OXYGEN DEFICIENCY

An atmosphere containing oxygen at a concentration of less than 19.5 percent by volume.

OXYGEN ENRICHMENT

OXYGEN ENRICHMENT

An atmosphere containing oxygen at a concentration of more than 23.5 percent by volume.

OZONATION (O-zoe-NAY-shun)

OZONATION

The application of ozone to water, wastewater, or air, generally for the purposes of disinfection or odor control.

P

POTW

POTW

Publicly Owned Treatment Works. A treatment works that is owned by a state, municipality, city, town, special sewer district, or other publicly owned and financed entity as opposed to a privately (industrial) owned treatment facility. This definition includes any devices and systems used in the storage, treatment, recycling, and reclamation of municipal sewage (wastewater) or industrial wastes of a liquid nature. It also includes sewers, pipes, and other conveyances only if they carry wastewater to a POTW treatment plant. The term also means the municipality (public entity) that has jurisdiction over the indirect discharges to and the discharges from such a treatment works.

PACKAGE TREATMENT PLANT

PACKAGE TREATMENT PLANT

A small wastewater treatment plant often fabricated at the manufacturer's factory, hauled to the site, and installed as one facility. The package may be either a small primary or a secondary wastewater treatment plant.

PARALLEL OPERATION

PARALLEL OPERATION

Wastewater being treated is split and a portion flows to one treatment unit while the remainder flows to another similar treatment unit. Also see SERIES OPERATION.

PARASITIC (pair-uh-SIT-tick) BACTERIA

PARASITIC BACTERIA

Parasitic bacteria are those bacteria that normally live off another living organism, known as the host.

PATHOGENIC (path-o-JEN-ick) ORGANISMS   PATHOGENIC ORGANISMS

Organisms, including bacteria, viruses, or cysts, capable of causing diseases (such as giardiasis, cryptosporidiosis, typhoid, cholera, dysentery) in a host (such as a person). Also called PATHOGENS.

PATHOGENS (PATH-o-jens)   PATHOGENS

See PATHOGENIC ORGANISMS.

PERCENT SATURATION   PERCENT SATURATION

The amount of a substance that is dissolved in a solution compared with the amount dissolved in the solution at saturation, expressed as a percent.

$$\text{Percent Saturation, \%} = \frac{\text{Amount of Substance That Is Dissolved} \times 100\%}{\text{Amount Dissolved in Solution at Saturation}}$$

PERCOLATION (purr-ko-LAY-shun)   PERCOLATION

The slow passage of water through a filter medium; or, the gradual penetration of soil and rocks by water.

PERISTALTIC (PAIR-uh-STALL-tick) PUMP   PERISTALTIC PUMP

A type of positive displacement pump.

pH (pronounce as separate letters)   pH

pH is an expression of the intensity of the basic or acidic condition of a liquid. Mathematically, pH is the logarithm (base 10) of the reciprocal of the hydrogen ion activity. If $\{H^+\} = 10^{-6.5}$, then pH = 6.5. The pH may range from 0 to 14, where 0 is most acidic, 14 most basic, and 7 neutral.

PHENOLIC (fee-NO-lick) COMPOUNDS   PHENOLIC COMPOUNDS

Organic compounds that are derivatives of benzene. Also called phenols (FEE-nolls).

PHENOLPHTHALEIN (FEE-nol-THAY-leen)   PHENOLPHTHALEIN
    ALKALINITY   ALKALINITY

The alkalinity in a water sample measured by the amount of standard acid required to lower the pH to a level of 8.3, as indicated by the change in color of phenolphthalein from pink to clear. Phenolphthalein alkalinity is expressed as milligrams per liter of equivalent calcium carbonate.

PHOTOSYNTHESIS (foe-toe-SIN-thuh-sis)   PHOTOSYNTHESIS

A process in which organisms, with the aid of chlorophyll (green plant enzyme), convert carbon dioxide and inorganic substances into oxygen and additional plant material, using sunlight for energy. All green plants grow by this process.

PHYSICAL WASTE TREATMENT PROCESS   PHYSICAL WASTE TREATMENT PROCESS

Physical wastewater treatment processes include use of racks, screens, comminutors, clarifiers (sedimentation and flotation), and filtration. Chemical or biological reactions are important treatment processes, but not part of a physical treatment process.

PLUG FLOW   PLUG FLOW

A type of flow that occurs in tanks, basins, or reactors when a slug of water or wastewater moves through a tank without ever dispersing or mixing with the rest of the water or wastewater flowing through the tank.

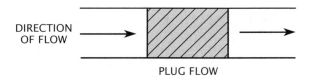

DIRECTION
OF FLOW

PLUG FLOW

POLLUTION   POLLUTION

The impairment (reduction) of water quality by agricultural, domestic, or industrial wastes (including thermal and radioactive wastes) to a degree that the natural water quality is changed to hinder any beneficial use of the water or render it offensive to the senses of sight, taste, or smell or when sufficient amounts of wastes create or pose a potential threat to human health or the environment.

POLYELECTROLYTE (POLY-ee-LECK-tro-lite)   POLYELECTROLYTE

A high-molecular-weight (relatively heavy) substance, having points of positive or negative electrical charges, that is formed by either natural or synthetic (manmade) processes. Natural polyelectrolytes may be of biological origin or obtained from starch products or cellulose derivatives. Synthetic polyelectrolytes consist of simple substances that have been made into complex, high-molecular-weight substances. Used with other chemical coagulants to aid in binding small suspended particles to larger chemical flocs for their removal from water. Often called a POLYMER.

POLYMER (POLY-mer)   POLYMER

A long-chain molecule formed by the union of many monomers (molecules of lower molecular weight). Polymers are used with other chemical coagulants to aid in binding small suspended particles to larger chemical flocs for their removal from water. Also see POLY-ELECTROLYTE.

POLYSACCHARIDE (poly-SAC-uh-ride)   POLYSACCHARIDE

A carbohydrate, such as starch or cellulose, composed of chains of simple sugars.

PONDING   PONDING

A condition occurring on trickling filters when the hollow spaces (voids) become plugged to the extent that water passage through the filter is inadequate. Ponding may be the result of excessive slime growths, trash, or media breakdown.

POPULATION EQUIVALENT   POPULATION EQUIVALENT

A means of expressing the strength of organic material in wastewater. In a domestic wastewater system, microorganisms use up about 0.2 pound (90 grams) of oxygen per day for each person using the system (as measured by the standard BOD test). May also be expressed as flow (100 gallons (378 liters)/day/person) or suspended solids (0.2 lb (90 grams) SS/day/person).

$$\text{Population Equivalent, persons} = \frac{\text{Flow, MGD} \times \text{BOD, mg}/L \times 8.34 \text{ lbs/gal}}{0.2 \text{ lb BOD/day/person}}$$

or

$$\text{Population Equivalent, persons} = \frac{\text{Flow, m}^3/\text{day} \times \text{BOD, mg}/L \times 10^6 \ L/\text{m}^3}{90,000 \text{ mg BOD/day/person}}$$

POSTCHLORINATION   POSTCHLORINATION

The addition of chlorine to the plant discharge or effluent, following plant treatment, for disinfection purposes.

POTABLE (POE-tuh-bull) WATER   POTABLE WATER

Water that does not contain objectionable pollution, contamination, minerals, or infective agents and is considered satisfactory for drinking.

POTTING COMPOUNDS   POTTING COMPOUNDS

Sealing and holding compounds (such as epoxy) used in electrode probes.

PRE-AERATION   PRE-AERATION

The addition of air at the initial stages of treatment to freshen the wastewater, remove gases, add oxygen, promote flotation of grease, and aid coagulation.

PRECHLORINATION

The addition of chlorine in the collection system serving the plant or at the headworks of the plant prior to other treatment processes.

(1) For drinking water, used mainly for disinfection, control of tastes, odors, and aquatic growths, and to aid in coagulation and settling.

(2) For wastewater, used mainly for control of odors, corrosion, and foaming, and for BOD reduction and oil removal.

PRECIPITATE (pre-SIP-uh-TATE)

(1) An insoluble, finely divided substance that is a product of a chemical reaction within a liquid.

(2) The separation from solution of an insoluble substance.

PRECISION

The ability of an instrument to measure a process variable and repeatedly obtain the same result. The ability of an instrument to reproduce the same results.

PRECOAT

Application of a free-draining, noncohesive material, such as diatomaceous earth, to a filtering medium. Precoating reduces the frequency of media washing and facilitates cake discharge.

PRELIMINARY TREATMENT

The removal of metal, rocks, rags, sand, eggshells, and similar materials that may hinder the operation of a treatment plant. Preliminary treatment is accomplished by using equipment such as racks, bar screens, comminutors, and grit removal systems.

PRESSURE MAIN

See FORCE MAIN.

PRETREATMENT FACILITY

Industrial wastewater treatment plant consisting of one or more treatment devices designed to remove sufficient pollutants from wastewaters to allow an industry to comply with effluent limits established by the US EPA General and Categorical Pretreatment Regulations or locally derived prohibited discharge requirements and local effluent limits. Compliance with effluent limits allows for a legal discharge to a POTW.

PRIMARY ELEMENT

(1) A device that measures (senses) a physical condition or variable of interest. Floats and thermocouples are examples of primary elements. Also called a SENSOR.

(2) The hydraulic structure used to measure flows. In open channels, weirs and flumes are primary elements or devices. Venturi meters and orifice plates are the primary elements in pipes or pressure conduits.

PRIMARY TREATMENT

A wastewater treatment process that takes place in a rectangular or circular tank and allows those substances in wastewater that readily settle or float to be separated from the wastewater being treated. A septic tank is also considered primary treatment.

PROCESS VARIABLE

A physical or chemical quantity that is usually measured and controlled in the operation of a water, wastewater, or industrial treatment plant. Common process variables are flow, level, pressure, temperature, turbidity, chlorine, and oxygen levels.

PROGRAMMABLE LOGIC CONTROLLER (PLC)

A microcomputer-based control device containing programmable software; used to control process variables.

PROPORTIONAL WEIR (WEER)   PROPORTIONAL WEIR

A specially shaped weir in which the flow through the weir is directly proportional to the head.

PROTEINACEOUS (PRO-ten-NAY-shus)   PROTEINACEOUS

Materials containing proteins, which are organic compounds containing nitrogen.

PROTOPLASM   PROTOPLASM

A complex substance (typically colorless and semifluid) regarded as the physical basis of life, having the power of spontaneous motion and reproduction; the living matter of all plant and animal cells and tissues.

PROTOZOA (pro-toe-ZOE-ah)   PROTOZOA

A group of motile, microscopic organisms (usually single-celled and aerobic) that sometimes cluster into colonies and generally consume bacteria as an energy source.

PRUSSIAN BLUE   PRUSSIAN BLUE

A blue paste or liquid (often on a paper like carbon paper) used to show a contact area. Used to determine if gate valve seats fit properly.

PSYCHROPHILIC (sy-kro-FILL-ick) BACTERIA   PSYCHROPHILIC BACTERIA

Cold temperature bacteria. A group of bacteria that grow and thrive in temperatures below 68°F (20°C).

PUG MILL   PUG MILL

A mechanical device with rotating paddles or blades that is used to mix and blend different materials together.

PURGE   PURGE

To remove a gas or vapor from a vessel, reactor, or confined space, usually by displacement or dilution.

PUTREFACTION (PYOO-truh-FACK-shun)   PUTREFACTION

Biological decomposition of organic matter, with the production of foul-smelling and -tasting products, associated with anaerobic (no oxygen present) conditions.

PUTRESCIBLE (pyoo-TRES-uh-bull)   PUTRESCIBLE

Material that will decompose under anaerobic conditions and produce nuisance odors.

PYROMETER (pie-ROM-uh-ter)   PYROMETER

An apparatus used to measure high temperatures.

Q

(NO LISTINGS)

R

RAS (pronounce as separate letters, or RAZZ)   RAS

Return Activated Sludge. Settled activated sludge that is collected in the secondary clarifier and returned to the aeration basin to mix with incoming raw or primary settled wastewater.

RABBLING

The process of moving or plowing the material inside a furnace by using the center shaft and rabble arms.

RACK

Evenly spaced, parallel metal bars or rods located in the influent channel to remove rags, rocks, and cans from wastewater.

RANGE

The spread from minimum to maximum values that an instrument is designed to measure. Also see EFFECTIVE RANGE and SPAN.

RAW WASTEWATER

Plant influent or wastewater before any treatment.

READOUT

The reading of the value of a process variable from an indicator or recorder or on a computer screen.

REAGENT (re-A-gent)

A pure, chemical substance that is used to make new products or is used in chemical tests to measure, detect, or examine other substances.

RECALCINATION (re-kal-sin-NAY-shun)

A lime recovery process in which the calcium carbonate in sludge is converted to lime by heating at 1,800°F (980°C).

RECARBONATION (re-kar-bun-NAY-shun)

A process in which carbon dioxide is bubbled into the water being treated to lower the pH.

RECEIVER

A device that indicates the result of a measurement, usually using either a fixed scale and movable indicator (pointer), such as a pressure gauge, or a moving chart with a movable pen like those used on a circular flow-recording chart. Also called an INDICATOR.

RECEIVING WATER

A stream, river, lake, ocean, or other surface or groundwaters into which treated or untreated wastewater is discharged.

RECHARGE RATE

Rate at which water is added beneath the ground surface to replenish or recharge groundwater.

RECIRCULATION

The return of part of the effluent from a treatment process to the incoming flow.

RECLAMATION

The operation or process of changing the condition or characteristics of water so that improved uses can be achieved.

RECORDER

A device that creates a permanent record, on a paper chart, magnetic tape, or in a computer, of the changes in a measured variable.

RECYCLE

The use of water or wastewater within (internally) a facility before it is discharged to a treatment system. Also see REUSE.

RABBLING

RACK

RANGE

RAW WASTEWATER

READOUT

REAGENT

RECALCINATION

RECARBONATION

RECEIVER

RECEIVING WATER

RECHARGE RATE

RECIRCULATION

RECLAMATION

RECORDER

RECYCLE

REDUCING AGENT REDUCING AGENT

Any substance, such as base metal (iron) or the sulfide ion (S^{2-}), that will readily donate (give up) electrons. The opposite is an OXIDIZING AGENT.

REDUCTION (re-DUCK-shun) REDUCTION

Reduction is the addition of hydrogen, removal of oxygen, or the addition of electrons to an element or compound. Under anaerobic conditions (no dissolved oxygen present), sulfur compounds are reduced to odor-producing hydrogen sulfide (H_2S) and other compounds. In the treatment of metal finishing wastewaters, hexavalent chromium (Cr^{6+}) is reduced to the trivalent form (Cr^{3+}). The opposite of OXIDATION.

REFERENCE REFERENCE

A physical or chemical quantity whose value is known exactly, and thus is used to calibrate instruments or standardize measurements. Also called a STANDARD.

REFRACTORY (re-FRACK-toe-ree) MATERIALS REFRACTORY MATERIALS

Materials difficult to remove entirely from wastewater, such as nutrients, color, taste- and odor-producing substances, and some toxic materials.

RELIQUEFACTION (re-lick-we-FACK-shun) RELIQUEFACTION

The return of a gas to the liquid state; for example, a condensation of chlorine gas to return it to its liquid form by cooling.

REPRESENTATIVE SAMPLE REPRESENTATIVE SAMPLE

A sample portion of material, water, or wastestream that is as nearly identical in content and consistency as possible to that in the larger body being sampled.

RESIDUAL CHLORINE RESIDUAL CHLORINE

The concentration of chlorine present in water after the chlorine demand has been satisfied. The concentration is expressed in terms of the total chlorine residual, which includes both the free and combined or chemically bound chlorine residuals. Also called CHLORINE RESIDUAL.

RESPIRATION RESPIRATION

The process in which an organism takes in oxygen for its life processes and gives off carbon dioxide.

RETENTION TIME RETENTION TIME

The length of time water, sludge, or solids are retained or held in a clarifier or sedimentation tank. Also see DETENTION TIME.

RETURN ACTIVATED SLUDGE (RAS) RETURN ACTIVATED SLUDGE (RAS)

Settled activated sludge that is collected in the secondary clarifier and returned to the aeration basin to mix with incoming raw or primary settled wastewater.

REUSE REUSE

The use of water or wastewater after it has been discharged and then withdrawn by another user. Also see RECYCLE.

RIPRAP RIPRAP

Broken stones, boulders, or other materials placed compactly or irregularly on levees or dikes for the protection of earth surfaces against the erosive action of waves.

RISING SLUDGE RISING SLUDGE

Rising sludge occurs in the secondary clarifiers of activated sludge plants when the sludge settles to the bottom of the clarifier, is compacted, and then starts to rise to the surface, usually as a result of denitrification, or anaerobic biological activity that produces carbon dioxide and/or methane.

ROTAMETER (ROTE-uh-ME-ter)

ROTAMETER

A device used to measure the flow rate of gases and liquids. The gas or liquid being measured flows vertically up a tapered, calibrated tube. Inside the tube is a small ball or bullet-shaped float (it may rotate) that rises or falls depending on the flow rate. The flow rate may be read on a scale behind or on the tube by looking at the middle of the ball or at the widest part or top of the float.

ROTARY PUMP

ROTARY PUMP

A type of displacement pump consisting essentially of elements rotating in a close-fitting pump case. The rotation of these elements alternately draws in and discharges the water being pumped. Such pumps act with neither suction nor discharge valves, operate at almost any speed, and do not depend on centrifugal forces to lift the water.

ROTIFERS (ROTE-uh-fers)

ROTIFERS

Microscopic animals characterized by short hairs on their front ends.

ROTOR

ROTOR

The rotating part of a machine. The rotor is surrounded by the stationary (nonmoving) parts (stator) of the machine.

S

SAR

SAR

Sodium Adsorption Ratio. This ratio expresses the relative activity of sodium ions in the exchange reactions with soil. The ratio is defined as follows:

$$SAR = \frac{Na}{[\frac{1}{2}(Ca + Mg)]^{\frac{1}{2}}}$$

where Na, Ca, and Mg are concentrations of the respective ions in milliequivalents per liter of water.

$$Na, meq/L = \frac{Na, mg/L}{23.0\ mg/meq} \qquad Ca, meq/L = \frac{Ca, mg/L}{20.0\ mg/meq} \qquad Mg, meq/L = \frac{Mg, mg/L}{12.15\ mg/meq}$$

SCADA (SKAY-dah) SYSTEM

SCADA SYSTEM

Supervisory Control And Data Acquisition system. A computer-monitored alarm, response, control, and data acquisition system used to monitor and adjust treatment processes and facilities.

SCFM

SCFM

Standard Cubic Feet per Minute. Cubic feet of air per minute at standard conditions of temperature, pressure, and humidity (0°C, 14.7 psia, and 50 percent relative humidity).

SPC CHART

SPC CHART

Statistical Process Control chart. A plot of daily performance such as a trend chart.

SVI

SVI

Sludge Volume Index. A calculation that indicates the tendency of activated sludge solids (aerated solids) to thicken or to become concentrated during the sedimentation/thickening process. SVI is calculated in the following manner: (1) allow a mixed liquor sample from the aeration basin to settle for 30 minutes; (2) determine the suspended solids concentration for a sample of the same mixed liquor; (3) calculate SVI by dividing the measured (or observed) wet volume (mL/L) of the settled sludge by the dry weight concentration of MLSS in grams/L.

$$SVI, mL/gm = \frac{Settled\ Sludge\ Volume/Sample\ Volume, mL/L}{Suspended\ Solids\ Concentration, mg/L} \times \frac{1,000\ mg}{gram}$$

SANITARY SEWER SANITARY SEWER

A pipe or conduit (sewer) intended to carry wastewater or waterborne wastes from homes, businesses, and industries to the treatment works. Stormwater runoff or unpolluted water should be collected and transported in a separate system of pipes or conduits (storm sewers) to natural watercourses.

SAPROPHYTES (SAP-row-fights) SAPROPHYTES

Organisms living on dead or decaying organic matter. They help natural decomposition of organic matter in water or wastewater.

SCALE SCALE

(1) A combination of mineral salts and bacterial accumulation that sticks to the inside of a collection pipe under certain conditions. Scale, in extreme growth circumstances, creates additional friction loss to the flow of water. Scale may also accumulate on surfaces other than pipes.

(2) The marked plate against which an indicator or recorder reads, usually the same as the range of the measuring system. See RANGE.

SCREEN SCREEN

A device used to retain or remove suspended or floating objects in wastewater. The screen has openings that are generally uniform in size. It retains or removes objects larger than the openings. A screen may consist of bars, rods, wires, gratings, wire mesh, or perforated plates.

SEALING WATER SEALING WATER

Water used to prevent wastewater or dirt from reaching moving parts. Sealing water is at a higher pressure than the wastewater it is keeping out of a mechanical device.

SECCHI (SECK-key) DISK SECCHI DISK

A flat, white disk lowered into the water by a rope until it is just barely visible. At this point, the depth of the disk from the water surface is the recorded Secchi disk transparency.

SECONDARY TREATMENT SECONDARY TREATMENT

A wastewater treatment process used to convert dissolved or suspended materials into a form more readily separated from the water being treated. Usually, the process follows primary treatment by sedimentation. The process commonly is a type of biological treatment followed by secondary clarifiers that allow the solids to settle out from the water being treated.

SEDIMENTATION (SED-uh-men-TAY-shun) BASIN SEDIMENTATION BASIN

A tank or basin in which water or wastewater is held for a period of time during which the heavier solids settle to the bottom and the lighter materials float to the surface. Also called settling tank or CLARIFIER.

SEED SLUDGE SEED SLUDGE

In wastewater treatment, seed, seed culture, or seed sludge refer to a mass of sludge that contains populations of microorganisms. When a seed sludge is mixed with wastewater or sludge being treated, the process of biological decomposition takes place more rapidly.

SEIZING or SEIZE UP SEIZING or SEIZE UP

Seizing occurs when an engine overheats and a part expands to the point where the engine will not run. Also called freezing.

SELECTOR SELECTOR

A reactor or basin in which baffles or other devices create a series of compartments. The environment and the resulting microbial population within each compartment can be controlled to some extent by the operator. The environmental conditions (food, lack of dissolved oxygen) that develop are intended to favor the growth of certain organisms over others. The conditions thereby select certain organisms.

SELECTOR RECYCLE SELECTOR RECYCLE

The recycling of return sludge or oxidized nitrogen to provide desired environmental conditions for microorganisms to perform a desired function.

SENSITIVITY

SENSITIVITY

The smallest change in a process variable that an instrument can sense.

SENSOR

SENSOR

A device that measures (senses) a physical condition or variable of interest. Floats and thermocouples are examples of sensors. Also called a PRIMARY ELEMENT.

SEPTIC (SEP-tick) or SEPTICITY

SEPTIC or SEPTICITY

A condition produced by bacteria when all oxygen supplies are depleted. If severe, the bottom deposits produce hydrogen sulfide, the deposits and water turn black, give off foul odors, and the water has a greatly increased oxygen and chlorine demand.

SERIES OPERATION

SERIES OPERATION

Wastewater being treated flows through one treatment unit and then flows through another similar treatment unit. Also see PARALLEL OPERATION.

SET POINT

SET POINT

The position at which the control or controller is set. This is the same as the desired value of the process variable. For example, a thermostat is set to maintain a desired temperature.

SEWAGE

SEWAGE

The used household water and water-carried solids that flow in sewers to a wastewater treatment plant. The preferred term is WASTEWATER.

SHEAR PIN

SHEAR PIN

A straight pin that will fail (break) when a certain load or stress is exceeded. The purpose of the pin is to protect equipment from damage due to excessive loads or stresses.

SHOCK LOAD

SHOCK LOAD

The arrival at a treatment process of water or wastewater containing unusually high concentrations of contaminants in sufficient quantity or strength to cause operating problems. Organic or hydraulic overloads also can cause a shock load.

(1) For activated sludge, possible problems include odors and bulking sludge, which will result in a high loss of solids from the secondary clarifiers into the plant effluent and a biological process upset that may require several days to a week to recover.

(2) For trickling filters, possible problems include odors and sloughing off of the growth or slime on the trickling filter media.

(3) For drinking water treatment, possible problems include filter blinding and product water with taste and odor, color, or turbidity problems.

SHORT-CIRCUITING

SHORT-CIRCUITING

A condition that occurs in tanks or basins when some of the flowing water entering a tank or basin flows along a nearly direct pathway from the inlet to the outlet. This is usually undesirable since it may result in shorter contact, reaction, or settling times in comparison with the theoretical (calculated) or presumed detention times.

SHREDDING

SHREDDING

A mechanical treatment process that cuts large pieces of wastes into smaller pieces so they will not plug pipes or damage equipment. Shredding and COMMINUTION usually mean the same thing.

SIDESTREAM

SIDESTREAM

Wastewater flows that develop from other storage or treatment facilities. This wastewater may or may not need additional treatment.

SIGNIFICANT FIGURE

SIGNIFICANT FIGURE

The number of accurate numbers in a measurement. If the distance between two points is measured to the nearest hundredth and recorded as 238.41 feet (or meters), the measurement has five significant figures.

SINGLE-STAGE PUMP

SINGLE-STAGE PUMP

A pump that has only one impeller. A multistage pump has more than one impeller.

SKATOLE (SKAY-tole)

SKATOLE

An organic compound (C_9H_9N) that contains nitrogen and has a fecal odor.

SLAKE

SLAKE

To mix with water so that a true chemical combination (hydration) takes place, such as in the slaking of lime.

SLIME GROWTH

SLIME GROWTH

See ZOOGLEAL MAT (FILM).

SLOUGHINGS (SLUFF-ings)

SLOUGHINGS

Trickling filter slimes that have been washed off the filter media. They are generally quite high in BOD and will lower effluent quality unless removed.

SLUDGE (SLUJ)

SLUDGE

(1) The settleable solids separated from liquids during processing.

(2) The deposits of foreign materials on the bottoms of streams or other bodies of water or on the bottoms and edges of wastewater collection lines and appurtenances.

SLUDGE AGE

SLUDGE AGE

A measure of the length of time a particle of suspended solids has been retained in the activated sludge process.

$$\text{Sludge Age, days} = \frac{\text{Suspended Solids Under Aeration, lbs or kg}}{\text{Suspended Solids Added, lbs/day or kg/day}}$$

SLUDGE DENSITY INDEX (SDI)

SLUDGE DENSITY INDEX (SDI)

This calculation is used in a way similar to the Sludge Volume Index (SVI) to indicate the settleability of a sludge in a secondary clarifier or effluent. The weight in grams of one milliliter of sludge after settling for 30 minutes. SDI = 100/SVI. Also see SLUDGE VOLUME INDEX.

SLUDGE DIGESTION

SLUDGE DIGESTION

The process of changing organic matter in sludge into a gas or a liquid or a more stable solid form. These changes take place as microorganisms feed on sludge in anaerobic (more common) or aerobic digesters.

SLUDGE GASIFICATION

SLUDGE GASIFICATION

A process in which soluble and suspended organic matter are converted into gas by anaerobic decomposition. The resulting gas bubbles can become attached to the settled sludge and cause large clumps of sludge to rise and float on the water surface.

SLUDGE VOLUME INDEX (SVI)

SLUDGE VOLUME INDEX (SVI)

A calculation that indicates the tendency of activated sludge solids (aerated solids) to thicken or to become concentrated during the sedimentation/thickening process. SVI is calculated in the following manner: (1) allow a mixed liquor sample from the aeration basin to settle for 30 minutes; (2) determine the suspended solids concentration for a sample of the same mixed liquor; (3) calculate SVI by dividing the measured (or observed) wet volume (mL/L) of the settled sludge by the dry weight concentration of MLSS in grams/L.

$$\text{SVI, } mL/\text{gm} = \frac{\text{Settled Sludge Volume/Sample Volume, } mL/L}{\text{Suspended Solids Concentration, mg/}L} \times \frac{1,000 \text{ mg}}{\text{gram}}$$

SLUDGE/VOLUME (S/V) RATIO

SLUDGE/VOLUME (S/V) RATIO

The volume of sludge blanket divided by the daily volume of sludge pumped from the thickener.

SLUG

SLUG

Intermittent release or discharge of wastewater or industrial wastes.

SLURRY   SLURRY

A watery mixture or suspension of insoluble (not dissolved) matter; a thin, watery mud or any substance resembling it (such as a grit slurry or a lime slurry).

SODIUM ADSORPTION RATIO (SAR)   SODIUM ADSORPTION RATIO (SAR)

See SAR.

SOFTWARE PROGRAM   SOFTWARE PROGRAM

Computer program; the list of instructions that tell a computer how to perform a given task or tasks. Some software programs are designed and written to monitor and control treatment processes.

SOLENOID (SO-luh-noid)   SOLENOID

A magnetically operated mechanical device (electric coil). Solenoids can operate small valves or electric switches.

SOLUBLE BOD   SOLUBLE BOD

Soluble BOD is the BOD of water that has been filtered in the standard suspended solids test. The soluble BOD is a measure of food for microorganisms that is dissolved in the water being treated.

SOLUTE   SOLUTE

The substance dissolved in a solution. A solution is made up of the solvent and the solute.

SOLUTION   SOLUTION

A liquid mixture of dissolved substances. In a solution it is impossible to see all the separate parts.

SPAN   SPAN

The scale or range of values an instrument is designed to measure. Also see RANGE.

SPECIFIC GRAVITY   SPECIFIC GRAVITY

(1) Weight of a particle, substance, or chemical solution in relation to the weight of an equal volume of water. Water has a specific gravity of 1.000 at 4°C (39°F). Particulates with specific gravity less than 1.0 float to the surface and particulates with specific gravity greater than 1.0 sink.

(2) Weight of a particular gas in relation to the weight of an equal volume of air at the same temperature and pressure (air has a specific gravity of 1.0). Chlorine gas has a specific gravity of 2.5.

SPLASH PAD   SPLASH PAD

A structure made of concrete or other durable material to protect bare soil from erosion by splashing or falling water.

STABILIZATION   STABILIZATION

Conversion to a form that resists change. Organic material is stabilized by bacteria that convert the material to gases and other relatively inert substances. Stabilized organic material generally will not give off obnoxious odors.

STABILIZED WASTE   STABILIZED WASTE

A waste that has been treated or decomposed to the extent that, if discharged or released, its rate and state of decomposition would be such that the waste would not cause a nuisance or odors in the receiving water.

STANDARD   STANDARD

A physical or chemical quantity whose value is known exactly, and thus is used to calibrate instruments or standardize measurements. Also called a REFERENCE.

STANDARD SOLUTION   STANDARD SOLUTION

A solution in which the exact concentration of a chemical or compound is known.

STANDARDIZE

STANDARDIZE

To compare with a standard.

(1)   In wet chemistry, to find out the exact strength of a solution by comparing it with a standard of known strength. This information is used to adjust the strength by adding more water or more of the substance dissolved.

(2)   To set up an instrument or device to read a standard. This allows you to adjust the instrument so that it reads accurately, or enables you to apply a correction factor to the readings.

STARTERS (MOTOR)

STARTERS (MOTOR)

Devices used to start up large motors gradually to avoid severe mechanical shock to a driven machine and to prevent disturbance to the electrical lines (causing dimming and flickering of lights).

STASIS (STAY-sis)

STASIS

Stagnation or inactivity of the life processes within organisms.

STATIC HEAD

STATIC HEAD

When water is not moving, the vertical distance (in feet or meters) from a reference point to the water surface is the static head. Also see DYNAMIC HEAD, DYNAMIC PRESSURE, and STATIC PRESSURE.

STATIC PRESSURE

STATIC PRESSURE

When water is not moving, the vertical distance (in feet or meters) from a specific point to the water surface is the static head. The static pressure in psi (or kPa) is the static head in feet times 0.433 psi/ft (or meters × 9.81 kPa/m). Also see DYNAMIC HEAD, DYNAMIC PRESSURE, and STATIC HEAD.

STATOR

STATOR

That portion of a machine that contains the stationary (nonmoving) parts that surround the moving parts (rotor).

STEP-FEED AERATION

STEP-FEED AERATION

Step-feed aeration is a modification of the conventional activated sludge process. In step-feed aeration, primary effluent enters the aeration tank at several points along the length of the tank, rather than at the beginning or head of the tank and flowing through the entire tank in a plug flow mode.

STERILIZATION (STAIR-uh-luh-ZAY-shun)

STERILIZATION

The removal or destruction of all microorganisms, including pathogens and other bacteria, vegetative forms, and spores. Compare with DISINFECTION.

STETHOSCOPE

STETHOSCOPE

An instrument used to magnify sounds and carry them to the ear.

STOP LOG

STOP LOG

A log or board in an outlet box or device used to control the water level in ponds and also the flow from one pond to another pond or system.

STORM SEWER

STORM SEWER

A separate pipe, conduit, or open channel (sewer) that carries runoff from storms, surface drainage, and street wash, but does not include domestic and industrial wastes. Storm sewers are often the recipients of hazardous or toxic substances due to the illegal dumping of hazardous wastes or spills caused by accidents involving vehicles transporting these substances. Also see SANITARY SEWER.

STRIPPED GASES

STRIPPED GASES

Gases that are released from a liquid by bubbling air through the liquid or by allowing the liquid to be sprayed or tumbled over media.

STRIPPED ODORS

STRIPPED ODORS

Odors that are released from a liquid by bubbling air through the liquid or by allowing the liquid to be sprayed or tumbled over media.

STRUVITE (STREW-vite)

STRUVITE

A deposit or precipitate of magnesium ammonium phosphate hexahydrate found on the rotating components of centrifuges and centrate discharge lines. Struvite can be formed when anaerobic sludge comes in contact with spinning centrifuge components rich in oxygen in the presence of microbial activity. Struvite can also be formed in digested sludge lines and valves in the presence of oxygen and microbial activity. Struvite can form when the pH level is between 5 and 9.

STUCK DIGESTER

STUCK DIGESTER

A stuck digester does not decompose organic matter properly. The digester is characterized by low gas production, high volatile acid/alkalinity relationship, and poor liquid-solids separation. A digester in a stuck condition is sometimes called a sour or UPSET DIGESTER.

SUBSTRATE

SUBSTRATE

(1)   The base on which an organism lives. The soil is the substrate of most seed plants; rocks, soil, water, or other plants or animals are substrates for other organisms.

(2)   Chemical used by an organism to support growth. The organic matter in wastewater is a substrate for the organisms in activated sludge.

SUCTION HEAD

SUCTION HEAD

The positive pressure [in feet (meters) of water or pounds per square inch (kilograms per square centimeter) of mercury vacuum] on the suction side of a pump. The pressure can be measured from the centerline of the pump up to the elevation of the hydraulic grade line on the suction side of the pump.

SUCTION LIFT

SUCTION LIFT

The negative pressure [in feet (meters) of water or inches (centimeters) of mercury vacuum] on the suction side of a pump. The pressure can be measured from the centerline of the pump down to (lift) the elevation of the hydraulic grade line on the suction side of the pump.

SUPERNATANT (soo-per-NAY-tent)

SUPERNATANT

Liquid removed from settled sludge. Supernatant commonly refers to the liquid between the sludge on the bottom and the scum on the surface.

SURFACE LOADING

SURFACE LOADING

One factor of the design flow of settling tanks and clarifiers in treatment plants used by operators to determine if tanks and clarifiers are hydraulically (flow) over- or underloaded. Also called OVERFLOW RATE.

$$\text{Surface Loading, GPD/sq ft} = \frac{\text{Flow, gallons/day}}{\text{Surface Area, sq ft}}$$

or

$$\text{Surface Loading, } \frac{\text{m}^3/\text{day}}{\text{m}^2} = \frac{\text{Flow, m}^3/\text{day}}{\text{Surface Area, m}^2}$$

SURFACTANT (sir-FAC-tent)

SURFACTANT

Abbreviation for surface-active agent. The active agent in detergents that possesses a high cleaning ability.

SUSPENDED GROWTH PROCESSES

SUSPENDED GROWTH PROCESSES

Wastewater treatment processes in which the microorganisms and bacteria treating the wastes are suspended in the wastewater being treated. The wastes flow around and through the suspended growths. The various modes of the activated sludge process make use of suspended growth reactors. These reactors can be used for BOD (biochemical oxygen demand) removal, nitrification, and denitrification.

SUSPENDED SOLIDS   SUSPENDED SOLIDS

(1) Solids that either float on the surface or are suspended in water, wastewater, or other liquids, and that are largely removable by laboratory filtering.

(2) The quantity of material removed from water or wastewater in a laboratory test, as prescribed in *STANDARD METHODS FOR THE EXAMINATION OF WATER AND WASTEWATER,* and referred to as Total Suspended Solids Dried at 103–105°C.

T

TOC (pronounce as separate letters)   TOC

Total Organic Carbon. TOC measures the amount of organic carbon in water.

TARE WEIGHT   TARE WEIGHT

The weight of an empty weighing dish or container.

TELEMETRY (tel-LEM-uh-tree)   TELEMETRY

The electrical link between a field transmitter and the receiver. Telephone lines are commonly used to serve as the electrical line.

TERTIARY (TER-she-air-ee) TREATMENT   TERTIARY TREATMENT

Any process of water renovation that upgrades treated wastewater to meet specific reuse requirements. May include general cleanup of water or removal of specific parts of wastes insufficiently removed by conventional treatment processes. Typical processes include chemical treatment and pressure filtration. Also called ADVANCED WASTE TREATMENT.

THERMOCOUPLE   THERMOCOUPLE

A heat-sensing device made of two conductors of different metals joined together. An electric current is produced when there is a difference in temperature between the ends.

THERMOPHILIC (thur-moe-FILL-ick) BACTERIA   THERMOPHILIC BACTERIA

A group of bacteria that grow and thrive in temperatures above 113°F (45°C). The optimum temperature range for these bacteria in anaerobic decomposition is 120°F (49°C) to 135°F (57°C). Aerobic thermophilic bacteria thrive between 120°F (49°C) and 158°F (70°C).

THIEF HOLE   THIEF HOLE

A digester sampling well that allows sampling of the digester contents without venting digester gas.

THRESHOLD ODOR   THRESHOLD ODOR

The minimum odor of a gas or water sample that can just be detected after successive dilutions with odorless gas or water. Also called ODOR THRESHOLD.

TIMER   TIMER

A device for automatically starting or stopping a machine or other device at a given time.

TITRATE (TIE-trate)   TITRATE

To titrate a sample, a chemical solution of known strength is added drop by drop until a certain color change, precipitate, or pH change in the sample is observed (end point). Titration is the process of adding the chemical reagent in small increments (0.1–1.0 milliliter) until completion of the reaction, as signaled by the end point.

TOTAL CHLORINE   TOTAL CHLORINE

The total concentration of chlorine in water, including the combined chlorine (such as inorganic and organic chloramines) and the free available chlorine.

TOTAL CHLORINE RESIDUAL

TOTAL CHLORINE RESIDUAL

The total amount of chlorine residual (including both free chlorine and chemically bound chlorine) present in a water sample after a given contact time.

TOTAL DYNAMIC HEAD (TDH)

TOTAL DYNAMIC HEAD (TDH)

When a pump is lifting or pumping water, the vertical distance (in feet or meters) from the elevation of the energy grade line on the suction side of the pump to the elevation of the energy grade line on the discharge side of the pump. The total dynamic head is the static head plus pipe friction losses.

TOTAL ORGANIC CARBON (TOC)

TOTAL ORGANIC CARBON (TOC)

TOC is a measure of the amount of organic carbon in water.

TOTALIZER

TOTALIZER

A device or meter that continuously measures and sums a process rate variable in cumulative fashion over a given time period. For example, total flows displayed in gallons per minute, million gallons per day, cubic feet per second, or some other unit of volume per time period. Also called an INTEGRATOR.

TOXIC

TOXIC

A substance that is poisonous to a living organism. Toxic substances may be classified in terms of their physiological action, such as irritants, asphyxiants, systemic poisons, and anesthetics and narcotics. Irritants are corrosive substances that attack the mucous membrane surfaces of the body. Asphyxiants interfere with breathing. Systemic poisons are hazardous substances that injure or destroy internal organs of the body. Anesthetics and narcotics are hazardous substances that depress the central nervous system and lead to unconsciousness.

TOXIC SUBSTANCE

TOXIC SUBSTANCE

See HARMFUL PHYSICAL AGENT and TOXIC.

TOXICITY (tox-IS-it-tee)

TOXICITY

The relative degree of being poisonous or toxic. A condition that may exist in wastes and will inhibit or destroy the growth or function of certain organisms.

TRAMP OIL

TRAMP OIL

Oil that comes to the surface of a tank due to natural flotation. Also called free oil.

TRANSDUCER (trans-DUE-sir)

TRANSDUCER

A device that senses some varying condition measured by a primary sensor and converts it to an electrical or other signal for transmission to some other device (a receiver) for processing or decision making.

TRANSPIRATION (TRAN-spur-RAY-shun)

TRANSPIRATION

The process by which water vapor is released to the atmosphere by living plants. This process is similar to people sweating. Also see EVAPOTRANSPIRATION.

TRICKLING FILTER

TRICKLING FILTER

A treatment process in which wastewater trickling over media enables the formation of slimes or biomass, which contain organisms that feed upon and remove wastes from the water being treated.

TRICKLING FILTER MEDIA

TRICKLING FILTER MEDIA

Rocks or other durable materials that make up the body of the filter. Synthetic (manufactured) media have also been used successfully.

TRUE COLOR

TRUE COLOR

Color of the water from which turbidity has been removed. The turbidity may be removed by double filtering the sample through a Whatman No. 40 filter when using the visual comparison method.

TRUNK SEWER

A sewer line that receives wastewater from many tributary branches and sewer lines and serves as an outlet for a large territory or is used to feed an intercepting sewer. Also called MAIN SEWER.

TURBID

Having a cloudy or muddy appearance.

TURBIDIMETER

See TURBIDITY METER.

TURBIDITY (ter-BID-it-tee)

The cloudy appearance of water caused by the presence of suspended and colloidal matter. In the waterworks field, a turbidity measurement is used to indicate the clarity of water. Technically, turbidity is an optical property of the water based on the amount of light reflected by suspended particles. Turbidity cannot be directly equated to suspended solids because white particles reflect more light than dark-colored particles and many small particles will reflect more light than an equivalent large particle.

TURBIDITY (ter-BID-it-tee) METER

An instrument for measuring and comparing the turbidity of liquids by passing light through them and determining how much light is reflected by the particles in the liquid. The normal measuring range is 0 to 100 and is expressed as nephelometric turbidity units (NTUs).

TURBIDITY (ter-BID-it-tee) UNITS (TU)

Turbidity units are a measure of the cloudiness of water. If measured by a nephelometric (deflected light) instrumental procedure, turbidity units are expressed in nephelometric turbidity units (NTU) or simply TU. Those turbidity units obtained by visual methods are expressed in Jackson turbidity units (JTU), which are a measure of the cloudiness of water; they are used to indicate the clarity of water. There is no real connection between NTUs and JTUs. The Jackson turbidimeter is a visual method and the nephelometer is an instrumental method based on deflected light.

TURBULENT MIXERS

Devices that mix air bubbles and water and cause turbulence to dissolve oxygen in the water.

TWO-STAGE FILTERS

Two filters are used. Effluent from the first filter goes to the second filter, either directly or after passing through a clarifier.

U

ULTRAFILTRATION

A membrane filter process used for the removal of some organic compounds in an aqueous (watery) solution.

UPSET DIGESTER

An upset digester does not decompose organic matter properly. The digester is characterized by low gas production, high volatile acid/alkalinity relationship, and poor liquid-solids separation. A digester in an upset condition is sometimes called a sour or STUCK DIGESTER.

V

VARIABLE, MEASURED

A factor (flow, temperature) that is sensed and quantified (reduced to a reading of some kind) by a primary element or sensor.

TRUNK SEWER

TURBID

TURBIDIMETER

TURBIDITY

TURBIDITY METER

TURBIDITY UNITS (TU)

TURBULENT MIXERS

TWO-STAGE FILTERS

ULTRAFILTRATION

UPSET DIGESTER

VARIABLE, MEASURED

VARIABLE, PROCESS

VARIABLE, PROCESS

A physical or chemical quantity that is usually measured and controlled in the operation of a water, wastewater, or industrial treatment plant.

VECTOR

VECTOR

An insect or other organism capable of transmitting germs or other agents of disease.

VELOCITY HEAD

VELOCITY HEAD

The energy in flowing water as determined by a vertical height (in feet or meters) equal to the square of the velocity of flowing water divided by twice the acceleration due to gravity ($V^2/2g$).

VISCOSITY (vis-KOSS-uh-tee)

VISCOSITY

A property of water, or any other fluid, that resists efforts to change its shape or flow. Syrup is more viscous (has a higher viscosity) than water. The viscosity of water increases significantly as temperatures decrease. Motor oil is rated by how thick (viscous) it is; 20 weight oil is considered relatively thin while 50 weight oil is relatively thick or viscous.

VOLATILE (VOL-uh-tull)

VOLATILE

(1) A volatile substance is one that is capable of being evaporated or changed to a vapor at relatively low temperatures. Volatile substances can be partially removed from water or wastewater by the air stripping process.

(2) In terms of solids analysis, volatile refers to materials lost (including most organic matter) upon ignition in a muffle furnace for 60 minutes at 550°C (1,022°F). Natural volatile materials are chemical substances usually of animal or plant origin. Manufactured or synthetic volatile materials, such as plastics, ether, acetone, and carbon tetrachloride, are highly volatile and not of plant or animal origin. Also see NONVOLATILE MATTER.

VOLATILE ACIDS

VOLATILE ACIDS

Fatty acids produced during digestion that are soluble in water and can be steam-distilled at atmospheric pressure. Also called organic acids. Volatile acids are commonly reported as equivalent to acetic acid.

VOLATILE LIQUIDS

VOLATILE LIQUIDS

Liquids that easily vaporize or evaporate at room temperature.

VOLATILE SOLIDS

VOLATILE SOLIDS

Those solids in water, wastewater, or other liquids that are lost on ignition of the dry solids at 550°C (1,022°F). Also called organic solids and volatile matter.

VOLUMETRIC

VOLUMETRIC

A measurement based on the volume of some factor. Volumetric titration is a means of measuring unknown concentrations of water quality indicators in a sample by determining the volume of titrant or liquid reagent needed to complete particular reactions.

VOLUTE (vol-LOOT)

VOLUTE

The spiral-shaped casing that surrounds a pump, blower, or turbine impeller and collects the liquid or gas discharged by the impeller.

W

WAS

WAS

See Waste Activated Sludge.

WARNING

WARNING

The word *WARNING* is used to indicate a hazard level between *CAUTION* and *DANGER*. Also see CAUTION, DANGER, and NOTICE.

WASTE ACTIVATED SLUDGE (WAS)　　　　　　　WASTE ACTIVATED SLUDGE (WAS)

The excess quantity (mg/*L*) of microorganisms that must be removed from the process to keep the biological system in balance.

WASTEWATER　　　　　　　WASTEWATER

A community's used water and water-carried solids (including used water from industrial processes) that flow to a treatment plant. Stormwater, surface water, and groundwater infiltration also may be included in the wastewater that enters a wastewater treatment plant. The term sewage usually refers to household wastes, but this word is being replaced by the term wastewater.

WATER CYCLE　　　　　　　WATER CYCLE

The process of evaporation of water into the air and its return to earth by precipitation (rain or snow). This process also includes transpiration from plants, groundwater movement, and runoff into rivers, streams, and the ocean. Also called the HYDROLOGIC CYCLE.

WATER HAMMER　　　　　　　WATER HAMMER

The sound like someone hammering on a pipe that occurs when a valve is opened or closed very rapidly. When a valve position is changed quickly, the water pressure in a pipe will increase and decrease back and forth very quickly. This rise and fall in pressures can cause serious damage to the system.

WATER LANCE　　　　　　　WATER LANCE

A pipe on the end of a water hose that is used to hydraulically jet out solids.

WEIR (WEER)　　　　　　　WEIR

(1)　A wall or plate placed in an open channel and used to measure the flow of water. The depth of the flow over the weir can be used to calculate the flow rate, or a chart or conversion table may be used to convert depth to flow. Also see PROPORTIONAL WEIR.

(2)　A wall or obstruction used to control flow (from settling tanks and clarifiers) to ensure a uniform flow rate and avoid short-circuiting.

WEIR (WEER) DIAMETER　　　　　　　WEIR DIAMETER

Many circular clarifiers have a circular weir within the outside edge of the clarifier. All the water leaving the clarifier flows over this weir. The diameter of the weir is the length of a line from one edge of a weir to the opposite edge and passing through the center of the circle formed by the weir.

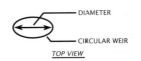

WET OXIDATION　　　　　　　WET OXIDATION

A method of treating or conditioning sludge before the water is removed. Compressed air is blown into the liquid sludge. The air and sludge mixture is fed into a pressure vessel where the organic material is stabilized. The stabilized organic material and inert (inorganic) solids are then separated from the pressure vessel effluent by dewatering in lagoons or by mechanical means.

WET PIT　　　　　　　WET PIT

See WET WELL.

WET WELL　　　　　　　WET WELL

A compartment or tank in which wastewater is collected. The suction pipe of a pump may be connected to the wet well or a submersible pump may be located in the wet well.

X

(NO LISTINGS)

Y

Y, GROWTH RATE Y, GROWTH RATE

An experimentally determined constant to estimate the unit growth rate of bacteria while degrading organic wastes.

Z

ZOOGLEAL (ZOE-uh-glee-ul) FILM ZOOGLEAL FILM

See ZOOGLEAL MAT (FILM).

ZOOGLEAL (ZOE-uh-glee-ul) MASS ZOOGLEAL MASS

Jelly-like masses of bacteria found in both the trickling filter and activated sludge processes. These masses may be formed for or function as the protection against predators and for storage of food supplies. Also see BIOMASS.

ZOOGLEAL (ZOE-uh-glee-al) MAT (FILM) ZOOGLEAL MAT (FILM)

A complex population of organisms that form a slime growth on the sand filter media and break down the organic matter in wastewater. These slimes consist of living organisms feeding on wastes, dead organisms, silt, and other debris. On a properly loaded and operating sand filter, these mats are so thin as to be invisible to the naked eye. Slime growth is a more common term.

SUBJECT INDEX

O

NOTES

NOTES

NOTES

NOTES

NOTES

NOTES

NOTES